Symbols Used in Pedigree Analysis

Male

Female

Mating

Mating between relatives (consanguineous)

I

II

1 2 3

Offspring listed in birth order. Roman numerals symbolize generations; Arabic numbers symbolize birth order within generation.

Monozygotic twins

Dizygotic twins

Offspring of unknown sex

Aborted or stillborn offspring

Deceased offspring

or

Affected individual

or

Propositus (male) or proposita (female); first case in family that was identified.

or

Heterozygotes

X-linked carrier

d.1910 d.1932

Indicates date of death

?

Questionable whether individual had trait

THIRD EDITION

HUMAN HEREDITY

PRINCIPLES AND ISSUES

THIRD EDITION

HUMAN HEREDITY

PRINCIPLES AND ISSUES

Michael R. Cummings
University of Illinois at Chicago
Department of Biological Sciences, Department of Genetics and Institute for the Study of Developmental Disabilities

WEST PUBLISHING COMPANY
St. Paul New York Los Angeles San Francisco

West's Commitment to the Environment

In 1906, West Publishing Company began recycling materials left over from the production of books. This began a tradition of efficient and responsible use of resources. Today, up to 95 percent of our legal books and 70 percent of our college and school texts are printed on recycled, acid-free stock. West also recycles nearly 22 million pounds of scrap paper annually—the equivalent of 181,717 trees. Since the 1980s, West has devised ways to capture and recycle waste inks, solvents, oils, and vapors created in the printing process. We also recycle plastics of all kinds, wood, glass, corrugated cardboard, and batteries, and have eliminated the use of Styrofoam book packaging. We at West are proud of the longevity and the scope of our commitment to the environment.

Production, Prepress, Printing and Binding
by West Publishing Company.

Composition: Graphic World, Inc.
Copyediting: Bill Waller, Chris Thillen
Cover design: Diane Beasley
Cover art: Randy Miyake
Illustrations: Rolin Graphics, Inc.
Indexing: Sandi Schroeder
Interior design: John Edeen

Cover illustration originally appeared in *Touched with Fire.* Reprinted with the permission of The Free Press, a Division of Macmillan, Inc. from *Touched with Fire: Manic Depressive Illness and the Artistic Temperament* by Kay Redfield Jamison. Copyright ©1993 by Kay Redfield Jamison.

COPYRIGHT ©1988, 1991 By WEST PUBLISHING CO.
COPYRIGHT © 1994 By WEST PUBLISHING CO.
610 Opperman Drive
P.O. Box 64526
St. Paul, MN 55164-0526

All rights reserved

Printed in the United States of America

01 00 99 98 97 96 95 94   8 7 6 5 4 3 2 1 0

Library of Congress Cataloging-in-Publication Data

Cummings, Michael R.
Human heredity / Michael R. Cumming.—3rd ed.
p. cm.
Includes bibliographical references and index.
ISBN 0-314-02747-5 (hard)
1. Human genetics. 2. Heredity, Human. I. Title.
QH431.C897 1993
573.2'1—dc20   90-26154
CIP

Figure and Table Credits

CHAPTER 1 Chapter Opener: Visuals Unlimited. Figure 1.1.: Giraudon/Art Resource. Figure 1.2: The Metropolitan Museum of Art, Gift of John D. Rockefeller, 1932. Figure 1.4: Reprinted by permission from p. 115, 118 of *Historical Geology: Evolution of the Earth and Life Through Time,* by Reed Wicander and James S. Monroe, copyright © 1989 by West Publishing Company. All rights reserved. Figure 1.5 (left and right): Wide World Photos. CHAPTER 2 Chapter Opener: Visuals Unlimited. Figure 2.2: © Dr. Monroe Yoder/Peter Arnold, Inc. Figure 2.3: From: Harrison, C. et al. 1983. Cytogen. Cell Genet *35*: 21-27. S. Karger, A. G., Basel. Figure 2.7. (left and middle): Visuals Unlimited. Figure 2.9: Reprinted by permission from p. 2 of *Human Genetics,* Volume 62, copyright 1982. Figure 2.8 Printed with permission of Dr. W. Ted Brown and Karen and Ken Sawyer. Figure 2.13: © Lennart Nilsson. Photo originally appeared in *Behold Man* by Lennart Nilsson (Boston: Little, Brown and Company). CHAPTER 3 Chapter Opener: Yoav Levy/Phototake Figure 3.1: Visuals Unlimited/© Case. Figure 3.2: V. Orel, Medelianum of Moravian Museum. Table 3.4: Reprinted from: Fisher, R. and Yates, F. 6th Ed. 1963. Statistical Tables for Biological, Agricultural, and Medical Research. Longman Group, Essex. CHAPTER 4: Courtesy of Library of Congress. Figure 4.1: From: Woolf, C. and Grant R. 1966 Am J Hum Genet *14*: 391-400. U of Chicago Press. Figure 4.2: Courtesy of Dr. Marilyn Miller, Dept of Ophthalmology, Univ. of Illinois at Chicago. Figure 4.7: Visuals Unlimited. Figure 4.9: From: *Sickle Cell* Scope Publication. Upjohn Co., Kalamazoo, Michigan. Figure 4.10: Reprinted with permission of the Peabody Museum. Figure 4.11: From Menko et al. 1984. Hum. Genet., 1967. 452-454. Figure 4.12: Wide World Photos. Figure 4.14: From Ruffer, M. 1921. *Studies in the Palaeopathology of Ancient Egypt*. U of Chicago Press, Chicago. Figure 4.15: Courtesy of Dr. Ira Rosenthal, Dept. of Pediatrics, Univ. of Illinois at Chicago. Figure 4.22: Photo courtesy of Oncor. Figure 4.24: Wide World Photos CHAPTER 5 Chapter Opener: © Biophoto Assoc./Photo Researchers, Inc. Figure 5.5: Mary Evans Picture Library/Photo Researchers, Inc. Figure 5.7: UPI Bettmann Newsphotos. Figure 5.8: Reproduced with permission from: *Muscle Disorders in Children*, by V. Dubowitz, p. 26, 1978, W.B. Saunders Co. Figure 5.14: From: Zourlas, P. et al. 1965. Clinical histologic and cytogenetic findings in male hermaphroditism. Ostet. Gynec. *25*:768-778. Figure 5.16: From: Imperato-McGinley, J. et al. 1974. Steroid 5-d-reductase deficiency in man. Science *186*:1213-1215. Copyright 1974, Am. Assn. Adv. Sciences. Figure 5.17: From Nature *163*:876 (1949), Figures 2 and 3. Figure 5.18: R. McNerling/Taurus Photos. Figure 5.19: Courtesy of Dr. Gerald Fishman, Dept. of Ophthalmology, Univ. of Illinois at Chicago. CHAPTER 6 Chapter Opener: Visuals

Figure and table credits continue following Index

To those who mean the most,

Lee Ann, Brendan and Shelly, Kerry and Terry

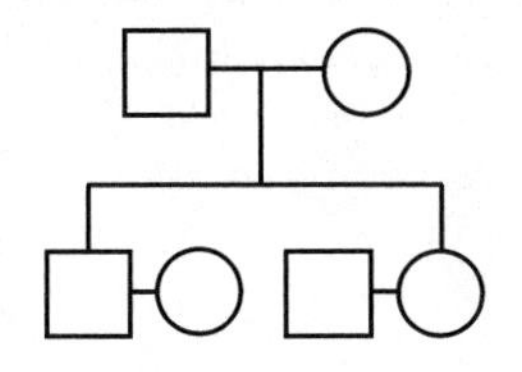

CONTENTS

CHAPTER **11**

MUTATIONS 287

CHAPTER **12**

MUTAGENS, CARCINOGENS, AND TERATOGENS 312

CHAPTER 17

GENES IN POPULATIONS 456

CHAPTER 18

HUMAN DIVERSITY AND EVOLUTION 478

CHAPTER 19

GENETIC SCREENING, AND GENETIC COUNSELING 519

CHAPTER 20

GENETICS, LAW, AND BIOETHICS 541

PREFACE TO THE THIRD EDITION

As with previous editions, the text is written for those who want to learn about human genetics. It is intended for use in one-term courses in human genetics for undergraduate students with little or no previous exposure to college courses in biology or chemistry. Although some chemistry and mathematics are used in the text, they are used in a descriptive context and at an elementary level.

The goals of the third edition are as follows:

- Present the principles of genetics and their application to humans in a clear, concise manner that imparts a working knowledge of genetic concepts.
- Through examples and descriptions of how discoveries are made, help students understand how scientists think and approach problems. Whenever possible, the full names of scientists are used to provide recognition to the men and women whose work has advanced genetics.
- Provide an understanding of the origin and amount of genetic diversity present in the human population and how the forces of selection and evolution have shaped our genotypes and phenotypes.
- Consider the social, ethical, and cultural implications of human genetic principles and how they may apply to individuals and society.

To achieve these goals, clear writing, up-to-date coverage, and flexible organization are emphasized. A conscious effort has been made to present the material in a straightforward and engaging manner, using a minimum of jargon. Previous editions have been used in teaching undergraduate students with a wide range of interests and backgrounds. The comments and suggestions of instructors have been most helpful in clarifying language, presenting concepts, and revising the reading level and use of examples.

The text can be divided into three sections: Chapters 1 through 6 cover the transmission of genetic traits, Chapters 7 through 14 cover gene action and the production of phenotype, and Chapters 15 through 20 cover populations, behavior, evolution, and the social aspects of genetics. Recognizing that most instructors have definite ideas about course organization, the text has been written to accommodate different formats of presentation. After the basics of transmission genetics have been presented, most chapters can be used in any order. Within each chapter, detailed headings let the instructor and the student easily identify central ideas.

Features of the Third Edition

Each week seems to bring new discoveries in genetics, and new techniques and applications of genetic principles are constantly in the news. As a result, it was necessary to thoroughly rewrite and update the entire book.

New Chapter

In response to developments in the field, a new chapter on behavior genetics (Chapter 16) has been added. This chapter presents the genetic models used in the study of behavior, explains the role of animal models for human behavior genetics, and considers both single-gene and multifactorial behavior traits. The chapter ends by assessing the current state of research in human behavior genetics and avenues for future investigation.

New Organization

Chapter 11 (Mutation) has been reorganized, and material on DNA repair has been moved to Chapter 12 (Mutagens, Carcinogens, Teratogens). A new section on the relationship between mutation, the genotype and phenotype, has been added to Chapter 11, reflecting the growth of information about the spectrum of mutations in a single gene that can produce variations in the phenotype. Chapter 18 (Human Diversity and Evolution) has been reorganized for greater emphasis on human origins and the interaction of genetics, anthropology, and archeology in providing insights into the origin and dispersal of the human species.

New Pedagogy

Recognizing that many students have difficulty with genetics problems, a new section on solving genetic problems has been added to Chapter 3 (Patterns of Inheritance). This section takes students through the steps in analyzing and solving genetics problems. In addition, many new problems have been added to Chapter 3 to give students more practice in acquiring this skill.

The art program has been thoroughly revamped to make use of the second-color format in a pedagogically valuable and consistent fashion throughout the text. The color section has been expanded, so that illustrations better presented in color are available to the student. This section is cross-referenced to the appropriate chapter and figure. The color section on the representation of genetic themes in art has been expanded, showing how individuals afflicted with genetic disorders have been portrayed across cultures and time in paintings, sculpture, and other forms of the fine arts.

New Topics

Many new topics have been introduced for the first time, or have received greater emphasis. It is impossible to list them all, but included among these are mitochondrial inheritance; Y-linked genes and pseudoautosomal inheritance; the relationship between chromosomal, gonadal, and phenotypic sex; uniparental disomy; genomic imprinting; chromosomal mapping of phenotypes in monosomy and trisomy; the polymerase chain reaction; the Human Genome Project; gene therapy; trinucleotide repeat mutations (including Huntington disease); anticipation; the relationship between mutation and phenotype in individual genes; molecular basis for some immunodeficiency syndromes; sexual orientation; alcoholism, aggression, and violence; mitochondrial genetics and human evolution; cloning of human embryos; and the origin and dispersal of human populations. Findings reported at the October 1993 meeting of the American Society of Human Genetics have been incorporated at several places in the text, including the chromosomal regions responsible for phenotypes in the cri-du-chat syndrome, measures of genetic variability in the human population using DNA markers, mitochondrial studies on the origin of American Indians, and the RNA binding functions of the gene product encoded at the fragile-X locus.

Special Features

Pedagogical features of the text such as the opening vignettes and Concepts and Controversies boxes have been revised to engage student interest and to provide sidelights to human genetics. Other features such as the chapter outline and marginal glossary have been updated and expanded as study aids for students.

Chapter Outlines

At the beginning of each chapter, an outline provides an overview of the main concepts, secondary ideas, and examples in the order of presentation.

Opening Vignettes

Each chapter begins with a short prologue directly related to the main ideas of the chapter, often drawn from real life. Some of these include accounts of genetic disorders from ancient manuscripts, genetic discoveries, or attempts to use genetics in the service of political ends, such as early attempts at breeding human giants. These have been written to engage the student's interest and demonstrate the relationship between science and its application to everyday life. A significant number of these vignettes are new to this edition, or have been rewritten to make them more useful.

Concepts and Controversies

Within each chapter, boxes are used to provide more information on topics covered in the text, to examine controversies and differences of opinion that often arise when genetic knowledge is transferred into technology and services, and to provide sidelights to genetics as a human endeavor. For this edition, many new boxes have been added, and others have been updated.

Glossary

A glossary in the margin provides students with immediate access to terms as they are introduced in the text. This format allows definitions to be easily identified when studying or preparing for examinations. The same definitions have been gathered into an alphabetical glossary at the back of the text. Because understanding the language of genetics is important to the comprehension of principles, over 350 terms are included.

Summary

Each chapter ends with a summary of the important concepts and main conclusions. For this edition, the summaries have been reorganized into a numbered list to help students master the material.

Questions and Problems

More than 275 end-of-chapter questions and problems are arranged by level of difficulty. Because some quantitative skills are needed in genetics, almost all chapters include some problems that require the student to organize the concepts in the chapter, and to use these concepts in reasoning to a conclusion. These questions also provide the student with a way to assess his or her understanding of the ideas presented in the chapter. Appendix B provides the answers to all end-of-chapter questions and problems.

For Further Reading

Lists of additional readings are presented at the end of each chapter. These include not only review and general articles that are accessible to non-science majors but also the key references for discoveries and papers for advanced readings in the areas covered by the text. In this edition, the readings have been carefully updated just before publication to provide the most up-to-date coverage of the literature.

Ancillary Materials

The unique combination of ancillary materials that accompany this text are designed to aid instructors genetics in preparation of lectures and examinations and to keep up with advances in the rapidly changing field of human genetics.

Newsletter

Because genetics is changing so rapidly, it is difficult to stay current with developments in the field. To keep instructors informed about new and important discoveries in human genetics, a newsletter is published twice yearly. This multipage document is produced with desktop publishing software, and each issue contains about 10 to 12 short, informative articles about recent discoveries in human genetics. Each article is accompanied by bibliographic references to provide access to the relevant literature. The articles are keyed to the chapters and pages where relevant information is presented in the text.

Instructor's Manual

A detailed instructor's manual is available to assist the instructor in preparing lectures and examinations. It contains an outline of each chapter, a list of learning objectives, and additional questions, most of which are presented in a multiple-choice and true-false format. For some chapters, additional material not covered in the text is provided for background or a more detailed consideration of a topic.

Transparency Masters

Over 100 of the most important figures and tables have been reproduced as transparency masters, so that instructors can use them in preparing overhead transparencies.

Acknowledgements

Previous editions of the text have benefitted from the advice and suggestions of many reviewers. I gratefully acknowledge the contributions made by the reviewers of all three editions. In preparing this edition, I was fortunate to have the guidance and counsel of three reviewers who read all or part of the manuscript, and whose advice and collective wisdom have been instrumental in shaping this version of the text: Werner Heim of Colorado College, George Hudock of Indiana University, and H. Eldon Sutton of the University of Texas. Their dedication to the task and scrupulous attention to detail have greatly improved the presentation of concepts, selection of examples, and nuance of language. My gratitude to them goes far beyond these written words.

To all the reviewers who were involved in the preparation of this edition, I extend thanks and gratitude for their efforts and many suggestions to the following people:

Mary Beth Curtis
Tulane University

Patricia A. DeLeon
University of Delaware

Werner G. Heim
Colorado College

George A. Hudock
Indiana University

Beth A. Montelone
Kansas State University

Mary R. Murnik
Ferris State University

Donald J. Nash
Colorado State University

Carol Pou
St. Cloud State University

Elizabeth Savage
University of Alberta

Nancy Shontz
Grand Valley State University

H. Eldon Sutton
University of Texas—Austin

David Weisbrot
William Paterson College

Olivia M. White
University of North Texas

The photographs and art in the text have been contributed by individuals in many parts of the world, all of whom were very kind in complying with requests. Special thanks to Fr. Peter Weigand O.S.B., of St. Anselm's School, Washington, D.C. for providing photographs of Australian aborigines and the staff at Rolin Graphics, whose creative efforts are reflected in the figures.

I also extend special thanks to my colleagues Suzanne McCutcheon, for reviewing and correcting the art and figures, and V. K. Viswanathan, for reviewing the problems. Thanks also to Kanchna Sri for her help in preparing the glossary, and to my graduate students Joyce Ahn, Mondana Faredieh, and Paul Voyda for their patience during the preparation of this edition.

At West Publishing, thanks and gratitude are owed to many people, especially to my editor, Jerry Westby. More than anything it was his insight, encouragement, and persistence that helped bring this text into existence, and his enthusiasm and creative contributions that helped develop this and previous editions. Over the years, his friendship and advice have been the most valuable benefits of this project. Matt Thurber, my production editor, not only worked tirelessly on the usual tasks of galleys, layouts, and proofs, but was enterprising and imaginative in his efforts to acquire illustrations for the color galleries. His cheerful optimism was a constant and steadying influence in the face of what often seemed like insurmountable odds. Thanks also to Dean DeChambeau and Ann Hillstrom. Dean's work analyzing the reviews helped lay out the blueprint for the third edition; Ann's efforts in developing the marketing strategies and materials have helped make the book successful. Thanks to you all.

Michael R. Cummings

Chapter Opening Photographs

Chapter 1	Cave drawings, Algeria, 6,000 B.C.
Chapter 2	Scanning electron micrograph of human chromosome
Chapter 3	Garden peas in pod
Chapter 4	Abraham Lincoln, 16th President of the United States
Chapter 5	Human sex chromosomes
Chapter 6	Child with Down syndrome
Chapter 7	Computer enhanced micrograph of DNA structure
Chapter 8	Bacterial plasmids
Chapter 9	Polyribosomes engaged in translation of mRNA
Chapter 10	Sickled red blood cell
Chapter 11	Child with xeroderma pigmentosum and facial tumors
Chapter 12	Paths of radiation from Chernobyl explosion
Chapter 13	Stadium crowd
Chapter 14	Computer graphic of antibody molecule
Chapter 15	Fredrick the Great: Returning from Maneuvers
Chapter 16	Lord Alfred Tennyson
Chapter 17	Charles Darwin, co-discover of the principle of natural selection
Chapter 18	Body of Bronze age man discovered in receding glacier
Chapter 19	Brain section from Huntington disease patient
Chapter 20	Immigrants arriving at Ellis Island, New York

CHAPTER 1

INTRODUCTION: GENETICS AS A HUMAN ENDEAVOR

CHAPTER OUTLINE

ANCIENT CONCEPTS OF GENETICS

The Domestication of Plants and Animals

Our interest in heredity and the transmission of traits from generation to generation can be traced back more than 10,000 years to preliterate cultures and probably arose earlier. The domestication of plants and animals was a necessary step in the conversion of human societies from hunting and gathering to agriculture and herding. Evidence from animal remains and pictorial representations (Figure 1.1) indicates that dogs, sheep, goats, oxen, camels, and other animals were domesticated some 10,000 to 12,000 years ago. Domestication was undoubtedly a gradual process and may have proceeded from large-scale slaughtering of herds to selective killing of weak or sick members of the herd to the capture of the animals and their maintenance under human control. Once animals were domesticated, heritable traits were recognized and exploited by selective breeding to produce stock with characteristics tailored to human needs. The formation of distinct breeds of an animal is a process that can occur rapidly under domestication. Charles Darwin, in an account of the domestication of the mink, pointed out that a wide array of stocks with different coat colors was produced in only 25 generations.

A large number of cultivated plants were developed at about the same time as domesticated animals (some 10,000 to 12,000 years ago), and the development of cereal crops may have occurred earlier. The existence of two distinct sexes of date palms was known by 5000 B.C. by the Assyrians and Babylonians. The use of artificial fertilization of female flowers with male pollen (Figure 1.2) to control reproduction in the date palm was common in the reign of Hammurabi (2000 B.C.). Records on clay tablets indicate that date pollen was a commercially valuable product at that time. Several hundred varieties of dates were produced by selective breeding, each differing in some property such as size, color, time of maturation, and taste. Many of these strains are still grown in the Middle East, in North Africa, and in California near the Salton Sea.

FIGURE 1.1 Carving from the Old Kingdom (2700–2200 B.C.) tomb of the Egyptian Nefer el Ka-Hay This figure shows the harvesting of papyrus and the herding of cattle, illustrating that the domestication of plants and animals was an ancient practice.

FIGURE 1.2 Artificial pollination of female date palms by a priest wearing a costume. This is depicted in a relief carving done in the reign of the Assyrian king Assurnasirpal II (883–859 B.C.).

We cannot fix the time of origin of domesticated plants and animals exactly or the manner in which these breeds spread from region to region. What is clear is that selective breeding and the development of economically valuable stocks gradually became a conscious human activity. While this process was probably slow in developing, genetic manipulation of animals and plants coevolved with many features of human society and became the base of economic prosperity. For example, ancient poets recount that part of the wealth and reputation of ancient Troy was derived from skills in horse breeding.

Myths and Mutants

Experience with the selection of desirable traits in animals and plants must have led to the recognition of heritable traits in humans. Again, it is unknown when this took place. It is known that accounts of heritable deformities in humans often appear in myths and legends. The connection between defects in animals and traits in humans is present in the myth of the Cyclops.

In *The Odyssey,* a poem written by Homer (around 700 B.C.), the hero Ulysses encounters the Cyclops, a giant with one eye placed in the center of his forehead. This trait must have been recognized as heritable, since the story tells us that there was an entire race of these human-like creatures on an island where Ulysses landed. Interestingly the Cyclops who devoured several of Ulysses' men was also a shepherd who kept flocks of sheep. Among sheep, a relatively common birth defect known as cyclopia results in lambs with one eye placed in the middle of the forehead. In the United States this is the result of pregnant ewes feeding on plants of the genus *Veratum* (the skunk cabbage is a related plant), causing a cyclopic condition in 1% to 25% of the lambs. Plants with similar effects are found in the Eastern Mediterranean region where the ancient Greeks lived. In humans a related disorder, also known as cyclopia, is a heritable trait. Many descriptions of the Cyclops in ancient literature mention that the monster had its nose placed above the eye on the forehead. This is an accurate description of the genetic condition known as cyclopia in humans. From these accounts it seems possible that specific heritable human traits were known in ancient times.

Early recognition that heritable characters are present in humans is evidenced by the fact that many cultures assigned social roles in a hereditary fashion. In the time of Homer, the gift of prophecy was regarded as a heritable trait, and oracles were often members of a single family. Among the Jews, the priesthood was hereditary in the tribe of Levi. In India, the ancient Hindu caste system is based on the assumption that both desirable and undesirable traits are passed from generation to generation and that an individual's worth is determined at birth.

In addition, the religious writings of ancient civilizations show many examples indicating that deformities, physical and mental health, and other characteristics were believed to be inherited. More than 60 birth defects are listed on clay tablets written around 5000 years ago in Babylonia. The sacred scriptures of the Hindu religion provide instructions for choosing a wife, emphasizing that no heritable illness should be present and that the family should show evidence of good character for several preceding generations. The Jewish Talmud contains an accurate description of the inheritance of **hemophilia,** a genetic trait associated with blood clotting defects that is usually expressed only in males. These examples indicate that heritable human traits were not only known but also played a role in shaping social customs and mores.

Hemophilia
A genetic disorder characterized by defective blood clotting.

Pangenesis
A discarded theory of heredity that postulated the existence of pangenes, small particles from all parts of the body that concentrated in the gametes, passing traits from generation to generation and blending the traits of the parents in the offspring.

Epigenesis
The idea that development proceeds by the appearance and growth of new structures.

Preformationism
An idea that an organism develops by the growth of preformed structures already present in the egg.

Homunculus
The preformed, miniature individual thought to be present in the sperm or egg.

THE GREEKS: THE ORIGIN OF WESTERN TRADITION

In Greek civilization much attention was directed at the questions of heredity and evolution in humans. Greek philosophy concerned itself with explaining the origin and organization of the universe and of living systems, including human life. Heredity was regarded as part of the broader field of all biologic systems. Interest in heredity itself was limited to explaining how creatures, including humans, begot offspring that resembled their parents. The search for laws governing reproduction and heredity produced theories that had a profound influence on science that lasted well into the 19th century. We will briefly consider three of these ideas: pangenesis, male and female contributions to offspring, and inheritance of acquired characteristics. **Pangenesis** attempted to explain how characteristics are transmitted from parent to offspring. According to this theory, semen is formed as small particles in every part of the body and travels through the blood to the testicles from where it is transferred during intercourse. Thus according to pangenesis, characteristics gathered from all parts of the body are transmitted to offspring via semen. In this way pangenesis attempts to explain how parental traits such as eye color and hair color are passed on to children. The ancient Greeks believed that females also formed semen in a similar fashion, and this view, although disputed by Aristotle, was influential until the discovery of the mammalian egg in 1827.

In addition to recognizing that semen was involved in the transmission of heritable traits, the Greeks struggled with the question: Which parent contributes more characteristics to the offspring? On this point there was much debate and little agreement. There was some concurrence that both sexes helped to determine the sex of the offspring but disagreement over which sex contributed more to the physical appearance of the offspring. For example, Aeschylus wrote in 458 B.C.:

> The mother of what is called her child is no parent of it, but nurse only of the young life that is sown in her. The parent is the male, and she but a stranger, a friend, who, if fate spares his plant, preserves it till it puts forth.

Most of the Greek philosophers agreed that characteristics acquired through accident or experience were transmitted to the offspring. Aristotle wrote:

> For it has happened that the children of parents who bore scars are also scarred in just the same way and in just the same place. In Chalcedon, for example, a man who had been branded on the arm had a child who showed the same branded letter, though it was not so distinctly marked and had become blurred.

In part it was necessary to introduce the idea of the inheritance of acquired characteristics to explain why children often have traits that are different from those of their parents.

The Romans added little to the basic theories of the Greeks, and medieval scholars adopted the Greek system with little question. In the 17th and 18th centuries, attention was again focused on heredity and development. Two competing theories emerged at this time, epigenesis and preformationism. According to the theory of **epigenesis,** the organs and structures of the adult are not present initially, but arise during the course of development. Opposed to this was the idea of **preformationism,** which held that gametes contained a completely formed miniature adult, known as a **homunculus** (Figure 1.3). Among those adhering to the idea of preformation, ovists held that the egg contained the preformed individual and that the sperm served to stimulate growth. Spermists, on the other hand, believed that the sperm contained the homunculus.

By the end of the 18th century, basic knowledge of gametes and fertilization in plants allowed the use of systematic breeding experiments to produce hybrids and to improve vegetable crops. This emphasis on plants as an experimental system helped set the stage for Gregor Mendel's choice of peas to study heredity in the 1860s.

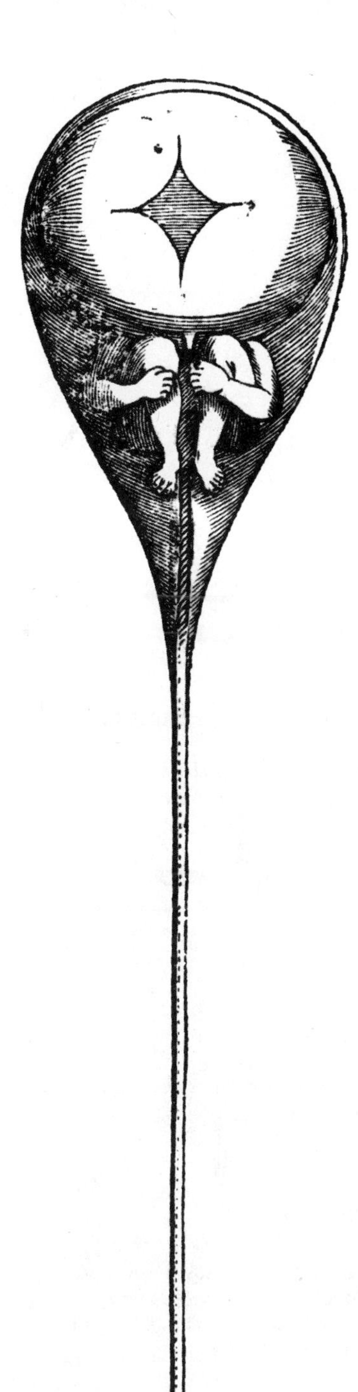

FIGURE 1.3 A preformed human infant or homunculus contained within a sperm. Preformationism, the idea that a gamete contains an intact organism, was first proposed in the late 1600s.

DARWIN AND MENDEL

Up to the middle of the 19th century, the theories formulated by the ancient Greeks to explain heredity and reproduction remained influential. Pangenesis was still regarded as a valid way of explaining the transmission of traits from parent to offspring. The issue of what contributions to the offspring were made by each parent was still unresolved, and the doctrine of the inheritance of acquired characteristics was not only accepted but also had been incorporated into a theory of evolution formulated by Jean-Baptiste Lamarck in the early 1800s (Figure 1.4A).

Although Greek philosophers, among others in the ancient world, speculated that living organisms might have arisen from ancestral forms, serious inquiry into the progressive development of plant and animal species from preexisting species began only in the 18th century. Although discussed, these ideas gained little support, so that in the mid-19th century, the idea that species did not change after their appearance (the *fixity of species*) still predominated.

Against this background, the discovery of the principles of evolution and genetics rank as two of the greatest achievements in the history of natural science. After observing the natural history of a great number of different organisms, Alfred Russel Wallace and Charles Darwin formulated the principle of **natural selection.** They observed that all organisms produce more progeny than can reasonably survive. Individuals among the progeny often contain heritable variations that contribute to their ability for survival and reproduction. Through the process of natural selection, those variations that enhance survival and reproduction gradually spread through a population. Darwin called this process "descent with modification." According to Wallace and Darwin, the spread of favorable variations occurs in response to alterations in environmental conditions. Over long periods, accumulated changes bring about the formation of new species: the process of **evolution** (Figure 1.4B). Those species that do not respond to changes in the environment become extinct. The fossil record indicates that most species that have ever lived are now extinct. As species become extinct, they are replaced by new species that are adapted to survive in the new conditions.

Natural selection
The process by which the organisms possessing heritable variations that make them better adapted to the environment leave more offspring than those organisms not possessing such variations.

Evolution
The appearance of new species of plant and animals.

Adaptation
The process by which an organism adapts to the present environmental conditions. The degree and mechanism by which an organism adapts can be under genetic control.

In the short run, evolution results in **adaptation,** as organisms become more closely matched to the environment in which they live. In the long run, the environment is changing at all times, and organisms must continually adapt to these changes. Environmental changes can be gradual and take place over hundreds or thousands of years or can be abrupt and cataclysmic, as in a volcanic eruption (see Figure 1.5 on page 8). Populations or species that are more diverse have a greater chance of adapting to changes in the environment.

Understanding the forces that bring about variation and diversity are basic to an understanding of evolution. Obviously variations that promote differential survival and reproduction are valuable only if they are passed on from generation to generation. Darwin knew nothing of the process of inheritance but recognized the need for such a mechanism for natural selection to be successful.

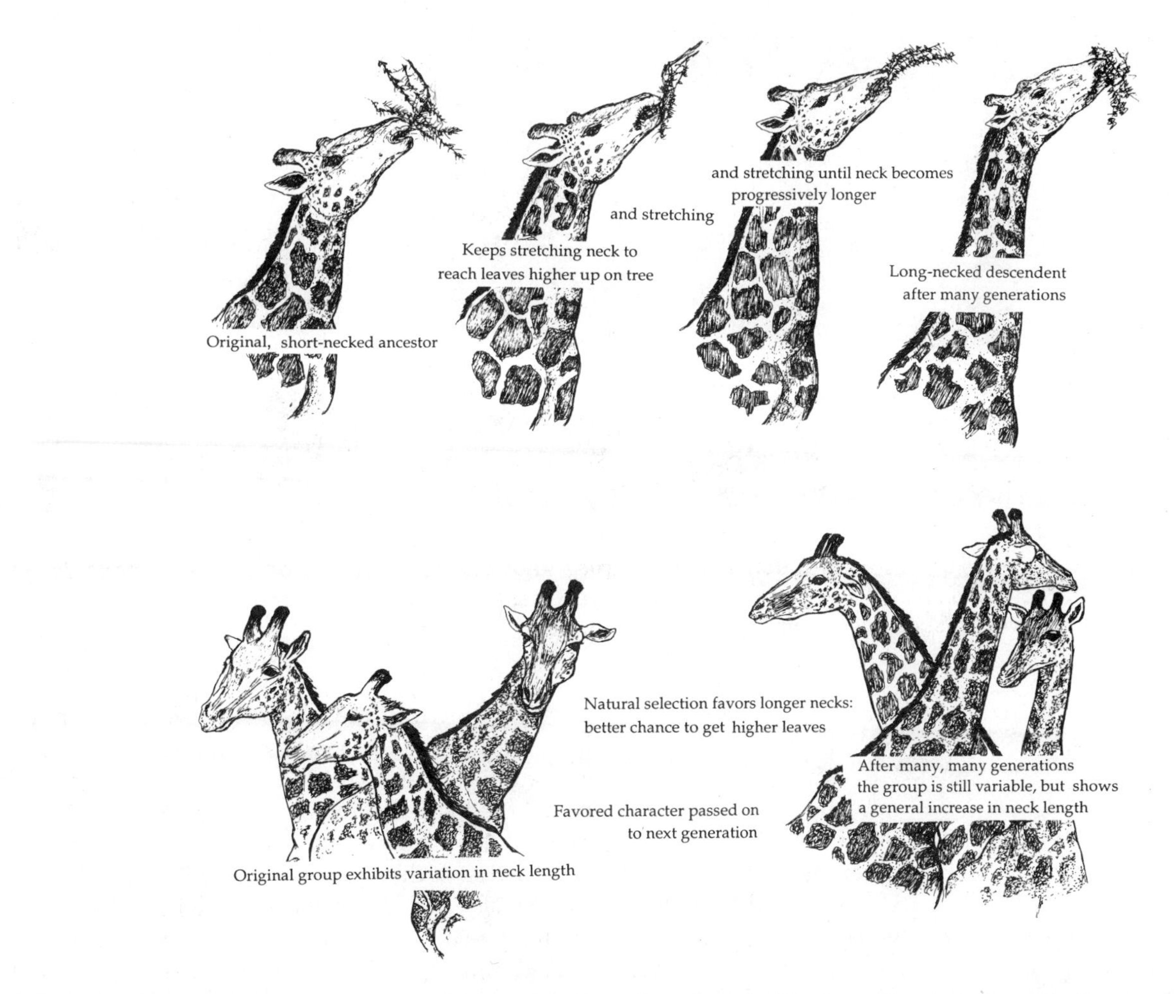

FIGURE 1.4 Contrasting ideas about the mechanism of evolution. (A). According to Lamarck's theory, acquired characteristics can be passed to subsequent generations. Thus, short-necked giraffes stretched their necks to reach higher into trees for food, and according to Lamarck, this acquired trait was passed on to off-spring, who were born with longer necks. (B). According to the Wallace-Darwin theory of natural selection, some giraffes have longer necks as a variant trait. If this trait provides an advantage for feeding, it will be passed on to a greater number of offspring, leading to an increase in the length of giraffe necks over many generations.

In modified form, Darwin embraced the ancient idea of pangenesis as a mechanism of inheritance. He postulated the existence of particles in the blood known as *gemmules* that carried information about hereditary traits to the sex organs. From the sex organs, these gemmules passed physical characteristics from parent to offspring.

In the years immediately following the publication of Darwin's book *The Origin of Species,* the work of an Augustinian monk named Gregor Mendel dramatically changed ideas about inheritance that had gone unchallenged for thousands of years. Working with pea plants, Mendel showed that traits are passed from parent to offspring through the inheritance of factors we now call **genes.** He reasoned that each parent

Genes
The fundamental units of heredity.

FIGURE 1.5 Left: Mt. St. Helen's before eruption. Right: Fir trees knocked down by the force of the volcano's eruption in 1980. Drastic environmental changes caused by natural disasters can result in changes in the genetic structure of populations living in such altered environments.

contributes one factor to the traits shown in the offspring and that pairs of traits separate from each other during the formation of egg and sperm.

Mendel's work restructured our ideas about heredity, correcting many misconceptions. He showed that the male and female contributed equally to the traits of the offspring and that by this mechanism there could be no inheritance of acquired characteristics. In the short space between 1859, when Darwin published *The Origin of Species,* and 1866, when Mendel published his work on the inheritance of traits in pea plants, ideas that had held sway for thousands of years were rejected and replaced with new theories based on specific and fundamental biological mechanisms. The ideas of Darwin and Mendel form the basis of modern biological sciences in general and of evolution and genetics in particular.

Genetics
The scientific study of heredity.

The significance of Mendel's work went unrecognized until 1900. The term **genetics** was coined by William Bateson shortly thereafter to describe the study of inheritance and the expression of inherited traits. Genetics as an organized discipline began in the 20th century and has progressed to the point where we have transferred human genes into animals and are replacing defective human genes with normal genes. Although genetics has progressed rapidly in this century, progress in human genetics has been uneven, and on occasion, human genetics has been badly used for social and political purposes.

THE RISE OF EUGENICS

After Darwin outlined the concept of natural selection and evolution, it became popular to use natural selection to explain many aspects of human society. Francis Galton, a cousin of Charles Darwin, thought that natural selection could be used as a tool to consciously improve the human species. He developed statistical and mathematical tools to study human traits, established the value of twin studies in genetics, and founded **biometrics,** a field that statistically analyzes the variation observed in traits such as height and weight. He also founded **eugenics**, the method of improving the intellectual, economic, and social level of humans by allowing differential reproduction of superior people to prevail over those designated as inferior. Galton studied the inheritance of many traits in families, such as musical ability and leadership, and concluded that such traits were handed down without environmental influence. His findings were summarized in the book *Natural Inheritance,* published in 1889. He proposed that individuals having desirable traits be encouraged to have large families (positive eugenics), while those having traits regarded as undesirable should be discouraged from reproducing (negative eugenics). In this way the evolution of the human species could be controlled and directed by an artificial form of natural selection.

Eugenics
The systematic improvement of humanity by encouraging breeding of those with desirable traits and by discouraging breeding of those with undesirable traits.

Biometrics
The application of statistical methods to problems in biological sciences.

In the United States eugenics took hold at about the time that Mendel's work was rediscovered. While geneticists were confirming Mendelian principles in a variety of plants and animals, disciples of eugenics were upholding the idea of biologic determinism and were attempting to establish that human traits such as criminal behavior, intelligence, slothful living, and drunkenness were completely heritable. This thinking inevitably led to the idea that certain groups of people were biologically inferior and that the reproductive rights of these groups should be terminated by sterilization or institutionalization. The eugenics movement became involved with the passage of legislation at the state and federal level that required sterilization of "socially defective" individuals and restricted immigration from Southern and Eastern Europe and most of Asia to maintain the quality of the human "breeding stock" in the United States.

In Germany, eugenics (known as *Rassenhygiene,* or race hygiene) fused with the new science of genetics (see Concepts & Controversies, page 10) and under the Nazi government gave rise to a campaign that led first to the sterilization and later to the systematic killing of those defined as social defectives. This included the retarded, the physically deformed, and the mentally ill. Later this same rationale was used in an attempt to exterminate entire ethnic groups, such as Gypsies and Jews.

THE EMERGENCE OF HUMAN GENETICS

During and immediately after World War II, research in human genetics focused on the identification of Mendelian traits and the use of mathematical formulas to study genes in different human populations. During this period, human genetics emerged as a separate branch of genetics. Studies on the effects of atomic radiation on residents of Nagasaki and Hiroshima and their offspring by a Japanese-American team of scientists under the leadership of James Neel of the University of Michigan provided a new direction for human genetics. Another group, working under Linus Pauling at Cal Tech, discovered that a genetic condition known as sickle cell anemia

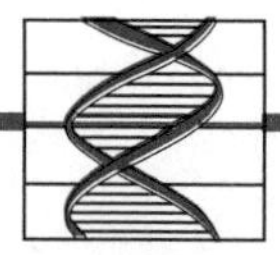

Concepts & Controversies

Genetics, Eugenics, and Nazi Germany

As in the United States and England, eugenics advocates in Germany were concerned with the preservation of "racial" purity. By 1927 the United States had laws that prohibited marriage by "social misfits," provided for compulsory sterilization of criminals and the "genetically unfit," and prohibited immigration of some groups. In Germany the laws of the Weimar government prohibited sterilization, and there were no laws forbidding marriage and restricting immigration of the genetically defective. As a result, several leading eugenicists became allied with the National Socialist Party (Nazis), which advocated forced sterilization and other eugenic measures to preserve the purity of the Aryan "race."

Adolf Hitler and the Nazi party came to power in January 1933. By July of that year a sterilization law was in effect, and before the end of that year it had been amended to broaden its powers. Under the law, those regarded as having lives not worth living, including the feebleminded, epileptics, the deformed, those suffering hereditary forms of blindness or deafness, and alcoholics, were to be sterilized.

By 1938 the policy was extended to include mercy killing *(Gnadentod)* of newborns who were incurably ill with hereditary or birth defects. This program was gradually extended to include children up to 3 or 4 years of age, then adolescents, and finally all institutionalized children, including juvenile delinquents and Jewish children. Killing centers were set up at more than two dozen institutions in Germany, Austria, and Poland to carry out this program. Children were usually killed by poison or starvation.

In 1939 the policy was extended to include mentally retarded and mentally ill adults and genetic defectives. This program, known as T_4, began by killing adults in psychiatric institutions. As increasing numbers were killed, gas chambers were installed at several centers to dispose of the impaired and defective. This practice spread from mental hospitals to include defective individuals at concentration camps and then to whole groups of people in concentration camps, most of whom were Jews, Gypsies, Communists, homosexuals, and political opponents of the government.

This process began with a naive eugenic ideal and, by logical extension, ended with mass murder.

was the result of the synthesis of a defective hemoglobin molecule, giving birth to the field of human biochemical genetics. Human cytogenetics began in 1956 when J. H. Tijo and A. Levan determined that humans carried 46 chromosomes, and in 1959 the chromosomal basis of Down syndrome was identified.

Since then human genetics has made rapid strides. In the span of 35 years, we have learned how to predict the sex of unborn children, to diagnose many genetic disorders prenatally, and to use gene products to prevent the deleterious effects of some genetic diseases. Human embryos can be produced by fusion of sperm and eggs in a laboratory dish and transferred to the womb of a surrogate mother. Embryos can also be frozen for transfer to a womb at a later time. We are now ready to begin correcting genetic defects by inserting normal genes to replace mutant genes, a technique called gene surgery or gene therapy. We can even insert human genes into animals, creating new types of organisms.

THE FUTURE

Although genetic knowledge has been accumulating for thousands of years, almost all of what we know has been discovered in this century. The amount of scientific knowledge is thought to double about every 10 years. If this is the case, then the next few decades will see an explosion of information about genetic mechanisms. More importantly, we are beginning to apply genetic knowledge in ways that will affect us all, and this trend will probably accelerate in the coming years. These applications cover several areas. One of the most visible is the Human Genome Project, an international effort by scientists to map all the genetic information carried by humans at its most elemental level: the 3 billion nucleotide subunits of DNA. This information will generate new knowledge about complex genetic disorders such as heart disease and mental disorders and will provide insight into cellular processes that may help unravel the causes of cancer. The development of biotechnology has produced fundamental changes in the diagnosis of genetic disorders and infectious diseases. It has also led to the development of genetically engineered plants and animals. In the pharmaceutical industry, genetically engineered bacteria are being used to produce human gene products for therapeutic purposes. Human insulin was the first such product, and it is one of over two dozen now available. The application of much of this knowledge raises new ethical questions that we must face and answer in the near future. As a student of human genetics, you have elected to become involved in the search for answers to these important questions.

FOR FURTHER READING

Ahmed, I. ed. 1992. *Biotechnology: A Hope or Threat?* London: St. Martin's Press.

Anderson, W. F., and Dircumakos, E. G. 1981. Genetic engineering in mammalian cells. *Sci. Am.* **245** (July): 106–121.

Bishop, J. E., and Waldholz, M. 1990. *Genome.* New York: Simon and Schuster.

Brackman, A. C. 1980. *A Delicate Arrangement: The Strange Case of Charles Darwin and Alfred Russel Wallace.* New York: Times Books.

Brooks, J. L. 1984. *Just Before the Origin: Alfred Russel Wallace's Theory of Evolution.* New York: Columbia University Press.

Bud, R. 1993. *The Use of Life: A History of Biotechnology.* New York: Cambridge University Press.

Carter, G. S. 1957. *A Hundred Years of Evolution.* London: Sidgwick & Jackson.

Corcos, A. 1984. Reproductive hereditary beliefs of the Hindus, based on their sacred books. *J. Hered.* **75:** 152–154.

Davis, B. D., ed. 1991. *The Genetic Revolution:* Scientific Prospects and Public Perceptions. Baltimore: Johns Hopkins Press.

Dunn, L. C. 1962. Cross currents in the history of human genetics. *Am. J. Hum. Genet.* **14:** 1–13.

Dunn, L. C. 1965. *A Short History of Genetics.* New York: McGraw-Hill.

Hopwood, D. A. 1981. The genetic programming of industrial microorganisms. *Sci. Am.* **245** (Sept.): 91–102.

Kevles, D. J. 1985. *In the Name of Eugenics: Genetics and the Use of Human Heredity.* New York: Knopf.

Kevles, D. J. and Hood, L., eds. 1992. *The Code of Codes: Scientific and Social Issues in the Human Genome Project.* Cambridge, MA: Harvard University Press.

Lifton, R. J. 1986. *The Nazi Doctors: Medical Killing and the Psychology of Genocide.* New York: Knopf.

McKusick, V. J. 1975. The growth and development of human genetics as a clinical discipline. *Am. J. Hum. Genet.* **27:** 261–273.

Müller-Hill, B. 1988. *Murderous Science: Elimination by Scientific Selection of Jews, Gypsies, and others,* Germany, 1933–1945. Oxford: Oxford University Press.

Proctor, R. N. 1988. *Racial Hygiene: Medicine under the Nazis.* Cambridge Mass: Harvard University Press.

Rafter, N. H. 1988. *White Trash.* Boston: Northeastern University Press.

Ryder, M. I. 1987. The evolution of fleece. *Sci. Am.* **256** (January): 112–119.

Stern, C., and Sherwood, E. 1966. *The Origin of Genetics: A Mendel Sourcebook.* San Francisco: Freeman.

Stubbe, H. 1972. *History of Genetics: From Prehistoric Times to the Rediscovery of Mendel's Laws.* Cambridge, MA: MIT Press.

Tijo, H. J., and Levan, A. 1956. The chromosome number of man. *Hereditas* **42:** 1–6.

Torrey, J. G. 1985. The development of plant biotechnology. *Am. Sci.* **73:** 354–363.

Weiss, S. F. 1988. *Race Hygiene and National Efficiency: The Eugenics of William Schallmeyer.* Berkeley: University of California Press.

CHAPTER 2

CHROMOSOMES, MITOSIS, AND MEIOSIS

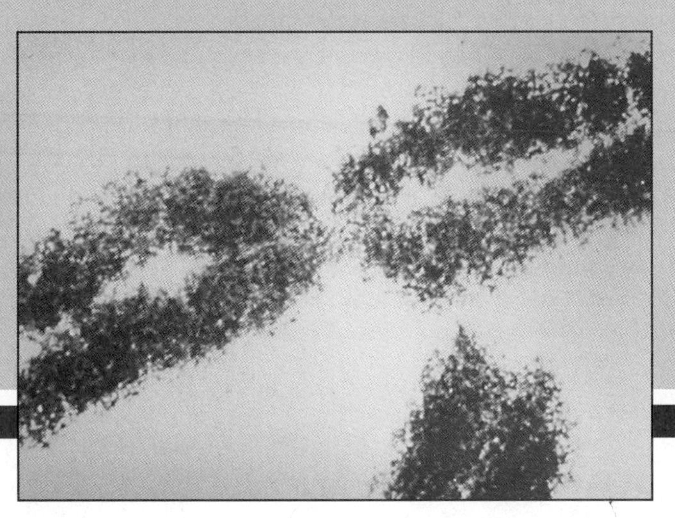

CHAPTER OUTLINE

Emerging from the womb, the newborn human represents the culmination of a series of complex, genetically programmed events that began some 38 weeks earlier at the moment of conception. At birth an infant contains trillions of cells, all derived from the fertilized egg, or zygote, by cell division under control of the genetic information contributed by each parent. Its tissues and organs contain several hundred different cell types, each with a distinctive organization often associated with highly specialized functions.

In spite of the apparent variation in size and shape of the different cell types in the human body, all cells carry the same set of genetic information, and they share a basic architecture. Each is surrounded by a membrane that forms its external boundary, and each possesses a membrane-enclosed nucleus for all or part of its life cycle. In addition, cells possess a variety of internal structures known as organelles that remain similar across the diversity of cell types found in the body. Because the shape, structure, and internal components of each cell are determined by genetic information, a large part of genetics, including human genetics, involves the study of cells. Many genetic diseases are expressed at the cellular level; for example, a mutant gene in sickle cell anemia causes an alteration in the shape of red blood cells, and most of the resulting physical symptoms of this disease are a direct effect of this altered shape.

As cells grow and divide, the genetic information contained within them must be faithfully copied and distributed to their progeny. This is accomplished by the reproduction and distribution of structures within the cell known as chromosomes. The process of chromosome distribution and cell division is known as mitosis. The sperm and egg that fuse to form the zygote are themselves the products of a special form of cell division known as meiosis. Mistakes in either of these processes can have serious genetic consequences, many of which are incompatible with life and may result in spontaneous abortion or miscarriage.

Because cells are the building blocks of the body and because the reproduction and transmission of genetic traits are mediated by cells, we will begin with an outline of their basic structural features. We will also consider the important features of cellular reproduction, with emphasis on the replication and distribution of chromosomes, the carriers of genetic information. Chromosome preparations are important tools in genetic diagnosis, and the methods used in the analysis of chromosome number and structure will be presented. Human reproduction is accomplished through the production of specialized cells known as gametes, and knowledge of how genetic traits are distributed during gamete formation is essential to an understanding of heredity.

THE ORGANIZATION OF CELLS

At the cellular level, all organisms can be classified as one of two types: eukaryotes or prokaryotes. **Eukaryotes** have a **nucleus,** enclosed by a double membrane. **Prokaryotes,** which include bacteria and certain photosynthetic algae, do not contain a nucleus. Viruses are excluded as organisms because they reproduce only inside the cells of organisms and do not have a cellular structure as discussed below. In the following discussion, we will consider some of the aspects of eukaryotic cell structure (Figure 2.1).

Eukaryote
Organism with a nuclear membrane surrounding the genetic material and with other membrane-bound organelles in the cytoplasm.

Nucleus
The membrane-bounded organelle, present in eukaryotic cells, that contains the chromosomes.

Prokaryote
Organism without a nucleus.

Cell Size and Shape

Most of the cells in the human body are too small to be seen without the aid of a microscope. Cell size is usually measured in micrometers (μm). There are 1 million micrometers in a meter, 25,000 per inch, and 1000 in a millimeter.

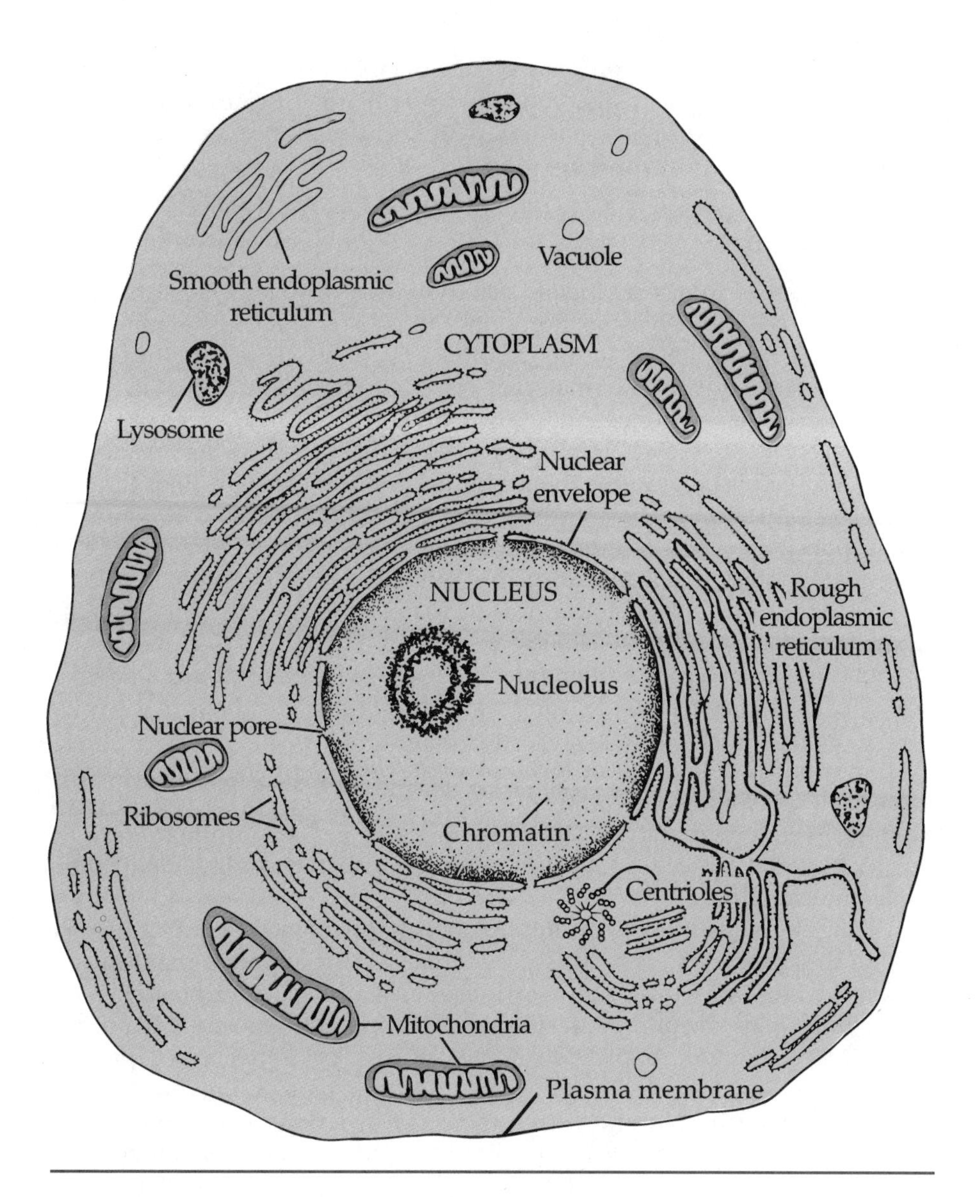

FIGURE 2.1 A composite drawing of a human cell as it would appear under the electron microscope. The shape and distribution of organelles is based on a human liver cell, but not all of the structures shown are actually found in liver cells.

Objects larger than 80 µm in diameter are visible to the unaided eye. Within the human body, cell size varies from small blood cells at around 2 µm to certain nerve cells with processes that may extend up to 3 ft. In the body, most of the cells average between 10 and 30 µm in diameter.

Within the body, cells assume a variety of shapes, and an internal structural network known as the **cytoskeleton** contributes to the shape of a cell. In addition, cells attach to surfaces such as basement membranes or neighboring cells, and these external forces also affect cell shape. While many cells maintain a characteristic shape, others, such as white blood cells, can undergo rapid and extensive changes in conformation. A drawing of a composite human cell is shown in Figure 2.1. The outer plasma membrane

Cytoskeleton
A system of protein microfilaments and microtubules that allows a cell to have a characteristic shape.

Molecule
A structure composed of two or more atoms held together by chemical bonds.

Organelle
A cytoplasmic structure having a specialized function.

encloses the cytoplasm, a complex mixture of **molecules** and supramolecular structures. The cytoplasm also contains a number of specialized subcellular structures known collectively as **organelles.** The most prominent organelle in the cytoplasm is the nucleus.

Cell Structure

Humans, like all living organisms, are composed of cells, which arise from preexisting cells by division. Bordering all cells is the cell membrane, or plasma membrane, about 10 mμ (millimicrons) thick. This structure allows and controls the exchange of materials with the external environment. Gases, water, and some small molecules pass through the membrane easily, while large molecules are moved across by processes that require energy. Molecules in and on the plasma membrane confer a form of molecular identity on all cells. The form and number of these molecules are genetically controlled and are responsible for many important properties of cells, including blood type and compatibility in organ transplants.

Inside the cell, the cytoplasm is a viscous, or semiliquid, material that contains many types of molecules in solution and suspension. Within the cytoplasm, a network of protein microfilaments and microtubules known as the cytoskeleton helps establish and maintain cell shape and serves to anchor the internal structures, such as the organelles.

Mitochondria (singular: mitochondrion)
Membrane-bound organelles present in the cytoplasm of all eukaryotic cells. They are the sites of energy production within cells.

Endoplasmic reticulum (ER)
A system of cytoplasmic membranes arranged into sheets and channels that functions in the synthesis and transport of gene products.

Ribosomes
Cytoplasmic particles composed of two subunits that are the site of protein synthesis.

Nucleolus (plural: nucleoli)
A nuclear region that functions in the synthesis of ribosomes.

Chromatin
The component material of chromosomes, visible as clumps or threads in nuclei examined under a microscope.

Chromosomes
The threadlike structures in the nucleus that carry genetic information.

Organelles known as **mitochondria** are centers of energy transformation. They are somewhat variable in shape but are always enclosed by two double-layered membranes. Mitochondria carry genetic information in the form of circular molecules of DNA. A typical liver cell might contain about 1000 mitochondria.

The **endoplasmic reticulum (ER)** is a network of membranous channels within the cytoplasm. The ER is a dynamic structure predominant in cells that make large amounts of protein; it is often studded with **ribosomes.** The ER is the cellular site of protein synthesis. The internal spaces or channels of the ER are thought to provide a mechanism for transport of proteins throughout the cytoplasm. Ribosomes are the most numerous cellular structures and can be found free in the cytoplasm or attached to the outer surface of the ER or nuclear membrane. Under an electron microscope, ribosomes appear as spherical structures some 15 mμ in diameter but are actually composed of two subunits that function to assemble the proteins necessary for cellular growth and metabolism. The process of protein synthesis is covered in detail in Chapter 9.

The largest and most prominent cellular organelle is the nucleus (Figure 2.2). This structure's double membrane is known as the nuclear envelope. The envelope is studded with pores that allow direct communication between the nucleus and cytoplasm. Within the nucleus, one or more dense regions known as **nucleoli** (singular: **nucleolus**) function in the synthesis of ribosomes. Under an electron microscope, thin strands and clumps of a material known as **chromatin** are seen throughout the nucleus. As the cells prepare to divide, the chromatin condenses into **chromosomes.** The nucleus is of central importance to genetics because it acts to regulate the activity of the cytoplasm. The nucleus can function as a control center because it

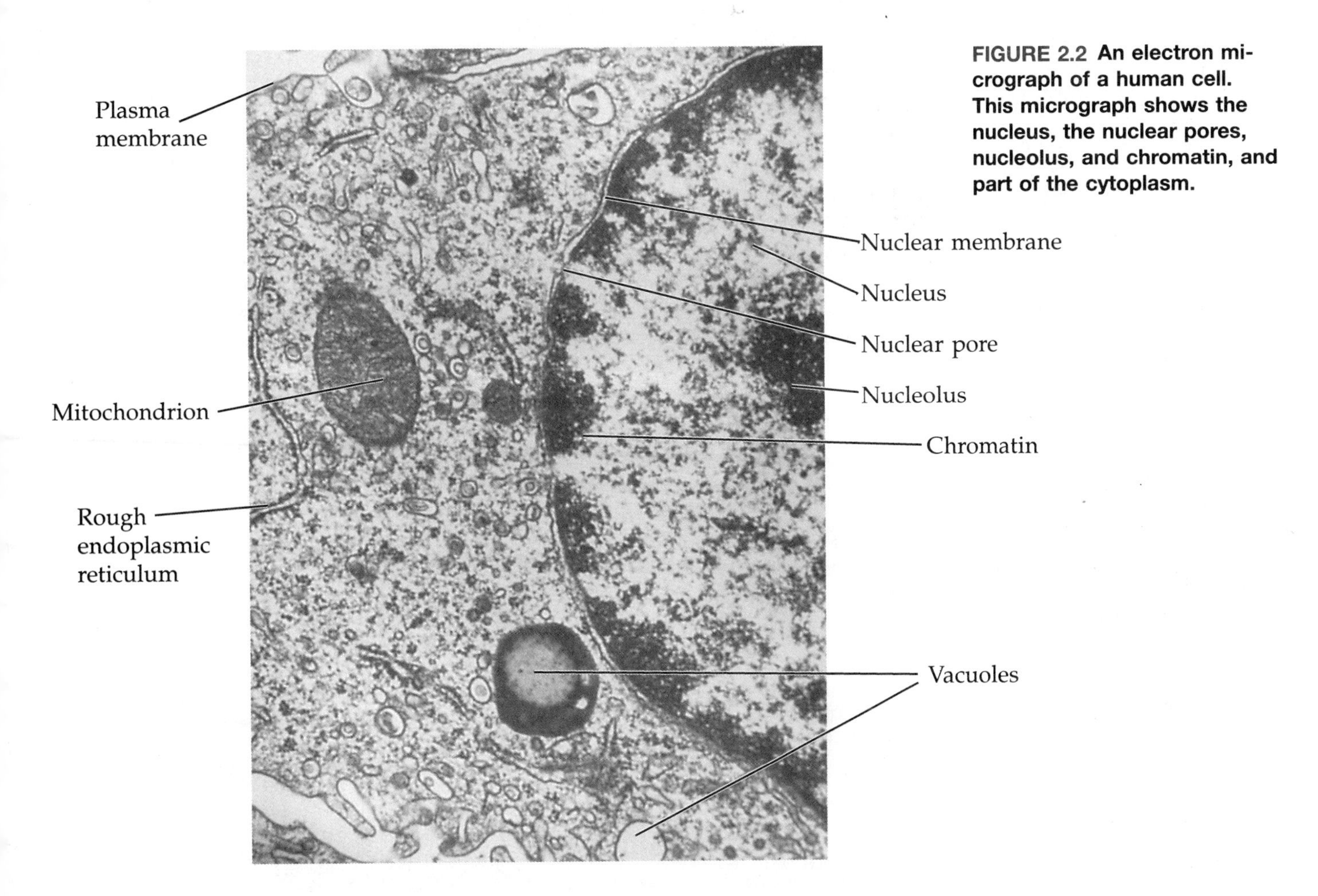

FIGURE 2.2 An electron micrograph of a human cell. This micrograph shows the nucleus, the nuclear pores, nucleolus, and chromatin, and part of the cytoplasm.

contains the instructions that ultimately determine the structure and shape of the cell as well as the range of functions exhibited by the cell. This information, in the form of genes, resides in the chromosomes carried in the nucleus. Because chromosomes carry genetic information, they occupy a central position in human genetics. The correct number of chromosomes in humans was not determined until 1956. In the 1970s, developments in chromosome staining and refined methods of preparing chromosomes for examination led to advances in prenatal diagnosis, genetic screening, and locating or mapping genes to specific chromosome regions.

Identification and Classification of Chromosomes

Chromosomes, though present, are not usually visible in the nuclei of nondividing cells. As a cell prepares to divide, the thin strands of chromatin distributed throughout the nucleus begin to condense into the coiled structures recognizable as chromosomes. As they become visible, certain structural features allow the chromosomes to be distinguished from one another. The number of chromosomes present in the nucleus is characteristic for a given species: the fruit fly *Drosophila melanogaster* has 8, corn plants have 20, and humans contain 46 chromosomes. With rare exceptions, every cell in an individual has the same number of chromosomes. The chromosome number for several species of plants and animals is given in Table 2.1. Note that the number of chromosomes is not related to the evolutionary complexity of the species. Further examination reveals that chromosomes in most eukaryotes occur in pairs. One member of each chromosome pair is derived from the female parent and the other from the male parent. Members of a chromosome pair are known as **homologues.** In many organisms, including humans, there is one pair of **sex chromosomes** (XX in females, and XY in males). The remaining chromosomes also occur in pairs and are known as **autosomes.**

Homologues
Members of a chromosome pair.

Sex chromosome
A chromosome, such as the X and Y in humans that is involved in sex determination.

Autosome
All chromosomes other than the sex chromosomes. In humans, there are 22 pairs of autosomes.

Diploid
The condition in which each chromosome is represented twice, as a member of a homologous pair.

Cells that contain pairs of homologous chromosomes are known as **diploid** cells, and the number of chromosomes carried in such cells is known as the diploid, or $2n$, number of chromosomes. In humans, the diploid number of chromosomes is 46. Certain cells, such as eggs and sperm (gametes), contain only one copy of each chromosome and are referred to as

TABLE 2.1 Chromosome Number in Selected Organisms

Organism	Diploid Number ($2n$)	Haploid Number (n)
Human (*Homo sapiens*)	46	23
Chimpanzee (*Pan troglodytes*)	48	24
Gorilla (*Gorilla gorilla*)	48	24
Dog (*Canis familiaris*)	78	39
Chicken (*Gallus domesticus*)	78	39
Frog (*Rana pipiens*)	26	13
Housefly (*Musca domestica*)	12	6
Onion (*Allium cepa*)	16	8
Corn (*Zea mays*)	20	10
Tobacco (*Nicotiana tabacum*)	48	24
House mouse (*Mus musculus*)	40	20
Fruit fly (*Drosophila melanogaster*)	8	4
Nematode (*Caenorhabditis elegans*)	12	6

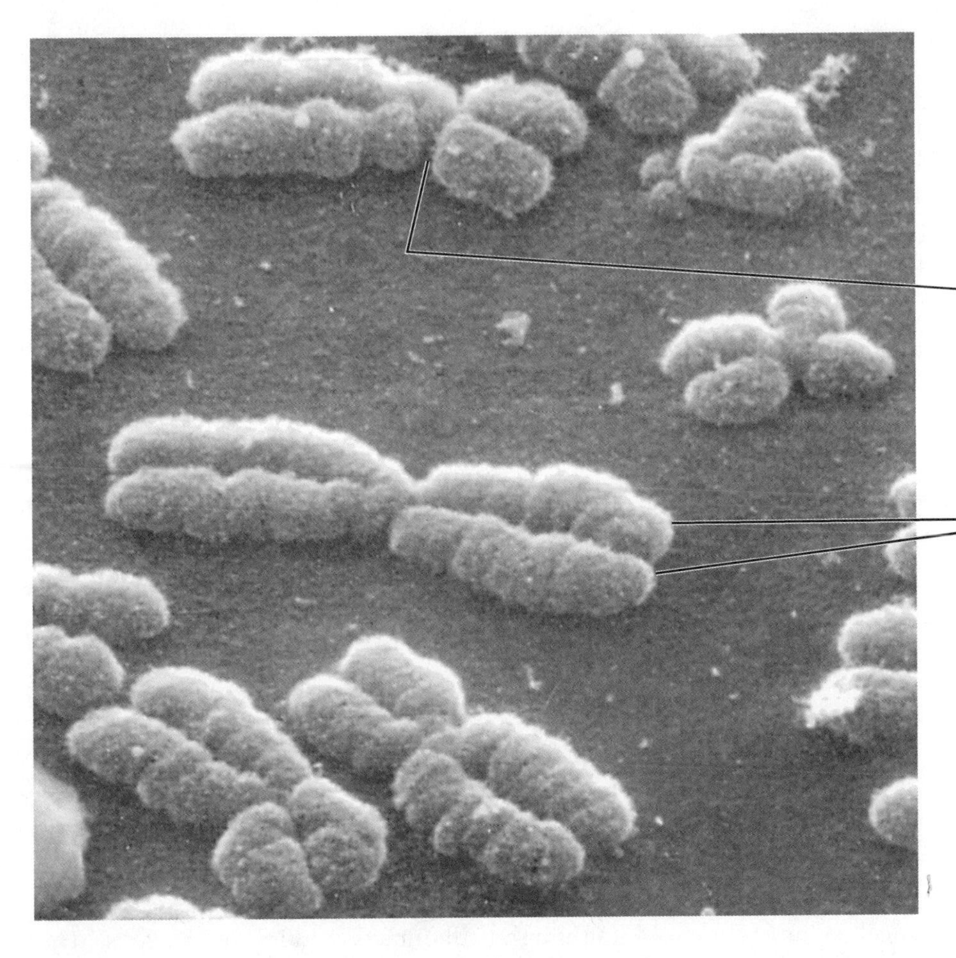

FIGURE 2.3 A scanning electron micrograph of human chromosomes from a dividing cell. The replicated chromosomes appear as double structures, joined at a common centromere.

Haploid
The condition in which each chromosome is represented once, in an unpaired condition.

Zygote
The diploid cell resulting from the union of a male haploid gamete and a female haploid gamete.

Centromere
A region of a chromosome to which fibers attach during cell division. Location of a centromere gives a chromosome its characteristic shape.

Metacentric
A chromosome with a centrally placed centromere.

Submetacentric
A chromosome with a centromere placed closer to one end than the other.

Acrocentric
A chromosome with the centromere placed very close to, but not at, one end.

Telocentric
A chromosome with the centromere located at one end.

haploid cells. The chromosome number in these cells is known as the haploid, or *n*, number of chromosomes. At fertilization the fusion of haploid gametes and their nuclei produces a cell, known as a **zygote,** carrying the diploid number of chromosomes.

Each chromosome contains a specialized region known as the **centromere.** The position of the centromere divides the chromosome into two arms, and its location is characteristic for a given chromosome (Figure 2.3). Chromosomes with centromeres at or near the middle have arms of equal length and are known as **metacentric** chromosomes (Figure 2.4). If the centromere is not centrally located and the arms are unequal in length, the chromosome is known as **submetacentric;** if the centromere is located very close to one end, the chromosome is known as an **acrocentric** chromosome. In **telocentric** chromosomes, the centromere is at one end. No human chromosomes are telocentric. In a later section we will use centromere location and chromosome size and banding patterns, produced by certain stains, to distinguish chromosomes in the human diploid chromosome set.

FIGURE 2.4 Human metaphase chromosomes are distinguished by size, centromere location and banding pattern. The banding of three representative human chromosomes is shown, with centromeres aligned. Chromosome 3 is one of the largest human chromosomes, and because the centromere is centrally located, is designated as metacentric. Chromosome 17 has a submetacentric centromere. Chromosome 21 is the smallest human chromosome and because its centromere is located very close to one end is designated as an acrocentric chromosome.

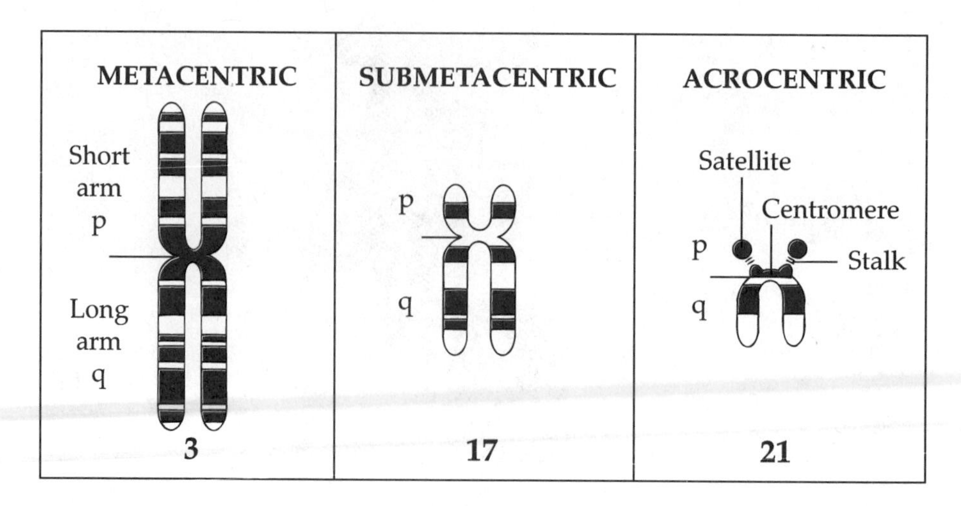

Chromosome Behavior in Mitosis

The Events in Interphase and the Cell Cycle

Interphase
The period of time in the cell cycle between mitotic divisions.

Many cells in the body alternate between states of division and nondivision. The period of nondivision is known as **interphase,** a period that begins immediately following cell division. During interphase the chromosomes are uncoiled and dispersed (although still intact) within the nucleus. The daughter cells that result from cell division are about one half the size of the parental cell. After division these cells enter interphase and initiate a period of growth and synthesis. Interphase is divided into three parts: G1, S, and G2. In G1, many cytoplasmic components, including organelles, membranes, and ribosomes, begin to proliferate. G1 is followed by the S phase, during which a duplicate copy of each chromosome is made. A second period of cellular growth, known as the G2 phase, takes place before the cell enters a new period of division. By the end of G2, the cytoplasmic volume of the daughter cells has reached that of the parental cell. With the chromosomes replicated and cytoplasmic components replenished, the cell then enters a period of division. In this period, known as the mitotic, or M, stage, the cytoplasm and replicated genetic information are distributed to the resulting daughter cells.

Cell cycle
The sequence of events that takes place between successive mitotic divisions.

The events of interphase and mitosis together make up what is known as the **cell cycle** (Figure 2.5). When grown outside the body in cell culture, normal human cells require about 24 hours to complete one cycle. Mitosis takes less than an hour, and therefore most of the time is spent in the G1, S, and G2 stages of interphase. When cells move through the cycle rapidly and continuously in an uncontrolled fashion, this type of growth is associated with cancer (see Concepts & Controversies, page 22).

Mitosis

Cell division represents the second major part of the cell cycle. During this period, two important steps are completed: **mitosis** distributes a complete set of chromosomes to each daughter cell, and **cytokinesis** distributes the cytoplasm to the two daughter cells. The division of the cytoplasm is accomplished by splitting the cell volume into two parts, with each receiving multiple copies of organelles and a sufficient supply of plasma membrane to enclose the new cells. While cytokinesis can be somewhat imprecise and still be operational, the division and distribution of the chromosomes must be accurate and unerring for the cell to function. The chromosomes and the genetic information they contain are precisely replicated during the S stage of interphase, and during mitosis a complete set of diploid chromosomes is distributed to each of the daughter cells. The net result of this replication and distribution is two diploid daughter cells, each of which contains 46 chromosomes that are derived from a single cell with 46 chromosomes. Although the distribution of chromosomes in cell division should be precise, errors in this process do occur, producing cells with an abnormal number or distribution of chromosomes. These mistakes often have serious genetic consequences and will be discussed in detail in Chapter 6.

Mitosis takes place in **somatic cells** and in the **gonial cells** of the ovary and testis. Progeny of the gonial cells undergo meiosis, a form of cell division that will be described below. Although mitosis is a continuous process, for

Mitosis
Form of cell division that produces two cells, each with the same complement of chromosomes as the parent cell.

Cytokinesis
The process of cytoplasmic division that accompanies cell division.

Somatic cells
All cells in the body other than those destined to become eggs or sperm.

Gonial cells
Cells in the ovary or testis that divide by mitosis and give rise to cells destined to undergo meiosis.

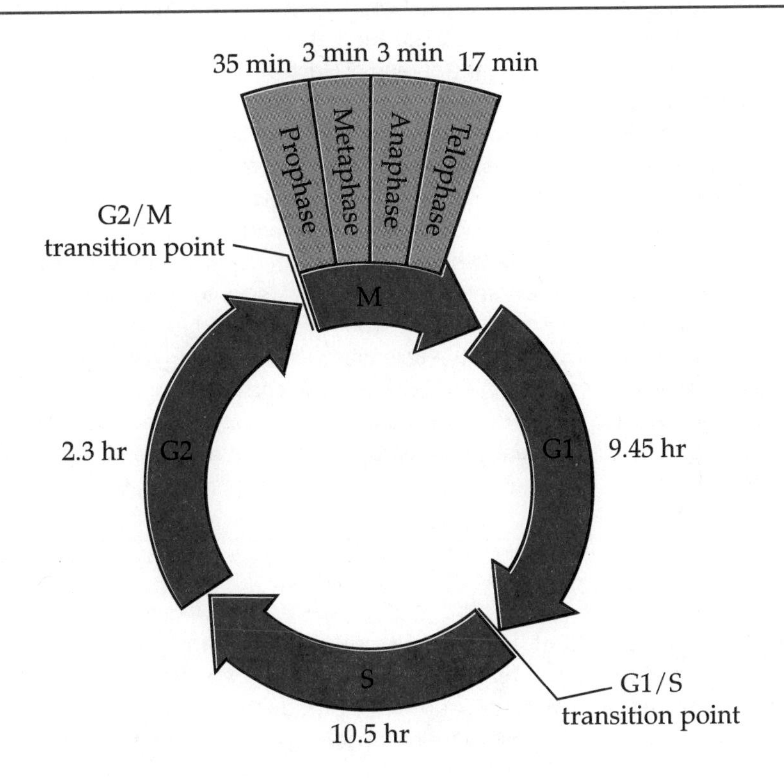

FIGURE 2.5 The cell cycle for a typical eukaryotic cell grown under laboratory conditions. Actual times vary for different cell lines.

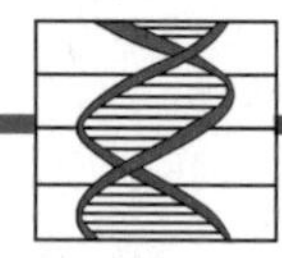

Concepts & Controversies

HeLa Cells: The Ultimate Growth Industry

Many basic advances in human genetics and related fields such as cancer research have come about through the use of cells grown outside the body in what is known as *in vitro* culture (*in vitro* is a Latin phrase that means "in glass"). In the early 1950s many attempts were made to culture human cells using tissues obtained from biopsies or surgical procedures. These early attempts usually met with failure; the cells died after a few days or weeks in culture, mostly without undergoing mitosis. These difficulties continued until February 1951, when a researcher at Johns Hopkins University received a sample of a cervical cancer from a woman named Henrietta Lacks. Following convention the culture was named HeLa by combining the first two letters of the donor's first and last name. This cell line not only grew but prospered under culture conditions and represented one of the earliest cell lines successfully grown outside the body.

Researchers were eager to have human cells available on demand to study the effects of drugs, toxic chemicals, radiation, and viruses on human tissue. For example, it was found that polio virus reproduced well in HeLa cells, providing a breakthrough in the development of a polio vaccine. As improvements in cell culture techniques were made, human cell lines were started from other cancers and normal tissues, including heart, kidney, and liver. By the early 1960s a central collection of cell lines had been established in Washington, DC, and cultured human cells were an important research tool in many areas of biological research.

The first clouds in this happy picture began to gather in 1966 when Stanley Gartler, a geneticist at the University of Washington, discovered that 18 different human cell lines he had analyzed had all been contaminated and taken over by HeLa cells. Over the next 2 years, it was confirmed that 24 of the 34 cell lines in the central repository were actually HeLa cells. Researchers who had spent years studying what they thought were heart or kidney cells had in reality been working with a cervical cancer cell instead. This meant that hundreds or thousands of experiments performed in laboratories around the world were invalid.

As painful as this lesson was, scientists started over, preparing new cell lines and using new and stricter rules to prevent contamination with HeLa cells. Unfortunately this was not the end of the problem. In 1974 Walter Nelson-Rees published a paper demonstrating that five cell lines extensively used in cancer research were in fact HeLa cells. In 1976 another paper announced that 11 additional cell lines, each widely used in research, were also HeLa cells; and in 1981 Nelson-Rees listed 22 more cell lines that were contaminated with HeLa. In all it appears that one third of all cell lines used in cancer research were really HeLa cells. The result was an enormous waste of dollars and resources. It is clear that not enough care was taken to prevent the spread of HeLa cells to other cultured lines. This can be attributed to poor record keeping, mislabeled cultures, and sloppy laboratory techniques. The response of the scientific community has been interesting. Some scientists quickly stepped forward to acknowledge the problem and to retract their conclusions. Others engaged in denials and vituperative arguments, steadfastly refusing to acknowledge the fact that their work was invalid. Henrietta Lacks provided the scientific world with two valuable gifts: a cell line and proof that science is a human endeavor, subject to the frailities of humanity that include not only honesty, candor, and veracity but also ego, fear, and denial.

Prophase
A stage in mitosis during which the chromosomes become visible and split longitudinally except at the centromere.

the sake of discussion it has been divided into four phases: **prophase, metaphase, anaphase,** and **telophase** (Figure 2.6). Throughout the cell cycle, the chromosomes alternately condense and decondense. In an extended configuration, the chromosomal material is dispersed throughout the

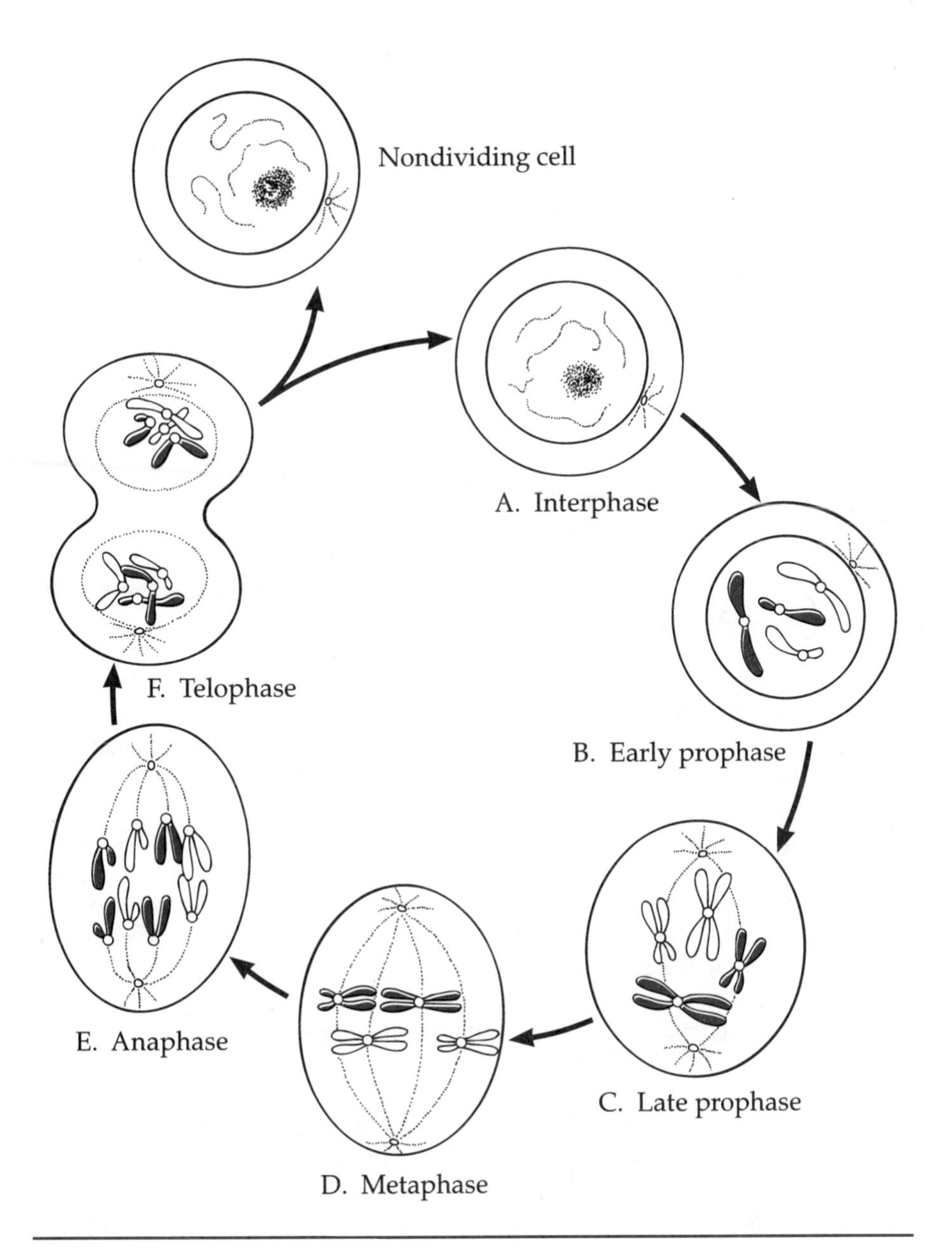

FIGURE 2.6 The stages of mitosis, illustrated with two pairs of homologous chromosomes. In each pair, the maternally donated chromosome is clear, and the paternally derived chromosome is in color.

nucleus and is recognizable as clumps of chromatin. In the G2 stage prior to entry into mitosis, the replicated chromosomes begin to condense, thicken, and shorten. During mitosis the chromosomes coil tightly and are recognizable as distinct structures. As the chromosomes coil, they become thick enough to be recognized under a microscope. When first seen, the chromosomes appear as long, thin, intertwined threads; this marks the beginning of prophase. In the cytoplasm a collection of specialized tubules known as spindle fibers begins to form. As prophase continues, the chromosomes continue to coil and become shorter and thicker; in human

Metaphase
A stage in mitosis during which the chromosomes move and arrange themselves near the middle of the cell.

Anaphase
A stage in mitosis during which the centromeres split and the daughter chromosomes begin to separate.

Telophase
The last stage of mitosis, during which division of the cytoplasm occurs, the chromosomes of the daughter cells disperse, and the nucleus reforms.

Chromatid
One of the strands of a duplicated chromosome, joined by a single centromere to its sister chromatid.

Sister chromatids
Two chromatids joined by a common centromere.

Metaphase plate
The cluster of chromosomes aligned at the equator of the cell during mitosis.

Cell furrow
A constriction of the cell membrane that forms at the point of cytoplasmic cleavage during cell division.

cells, 46 such structures can be seen. At this stage, each chromosome is a double structure, except at the centromere. Each of the structures joined by a centromere is known as a **chromatid.** The two chromatids joined by a single centromere are known as **sister chromatids.** Near the end of prophase, the nucleolus disappears, and the nuclear membrane begins to break down. In the cytoplasm the spindle fibers become fully organized and stretch from centriole to centriole, forming an axis along which mitosis will occur. Figures 2.6B and C illustrate the features of this stage of mitosis.

Metaphase begins when the nuclear membrane has completely disappeared. The two chromatids are still attached at a single centromere, giving chromosomes an X-shaped appearance. The chromosomes become aligned along a plane in midline, or equator, of the cell. This configuration is often referred to as the **metaphase plate,** or **equatorial plate,** and it forms at right angles to the axis formed by the centrioles and the spindle fibers. During metaphase, spindle fibers attach to the centromere of each chromosome. At the completion of metaphase, each chromosome is present on the metaphase plate and has spindle fibers attached to its centromere (Figure 2.6D). At this stage there are 46 centromeres, each attached to two sister chromatids.

Anaphase begins with the longitudinal division of each centromere, producing two chromosomes, one derived from each of the sister chromatids. Spindle fibers attached to the centromeres begin to shorten, pulling each chromosome toward the centriole. The daughter chromosomes begin to separate from each other and move to opposite sides of the cell. During anaphase the chromosome arms trail passively behind the centromeres (Figure 2.6E), making the chromosomes appear V-shaped or J-shaped. By the end of anaphase a complete set of chromosomes is present at each pole of the cell. Although anaphase is the briefest stage of mitosis, it is essential to ensuring that each daughter cell will receive an intact and identical set of 46 chromosomes.

The final stage of mitosis is telophase, characterized by cytokinesis and the formation of nuclei. Cytokinesis begins with the formation of the **cell furrow,** a constriction of the cell membrane that forms at the equator of the cell (Figure 2.7). The constriction gradually tightens and divides the cell in two. During telophase a nuclear membrane forms around each cluster of chromosomes, the spindle fibers disintegrate, and nucleoli reappear (Figure 2.6F). The chromosomes now begin a cycle of decondensation and become dispersed throughout the nucleus. The cycle of chromosome decondensation extends from anaphase through G1, and the cycle of condensation runs from G2 to metaphase.

The major features of mitosis are summarized in Table 2.2.

Significance of Mitosis

Mitosis is an essential process in all multicellular organisms. In humans, 40 to 44 rounds of mitosis accompany the development of the zygote into a full-term infant. Throughout adult life, some cells retain their capacity to divide, while others do not divide after adulthood is reached. Cells in the

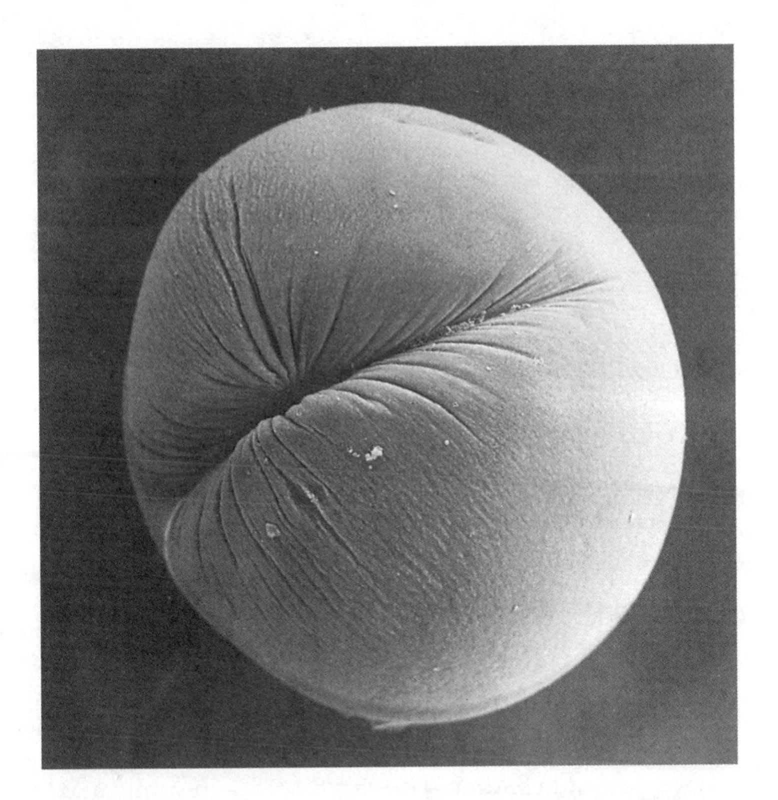

FIGURE 2.7 Cytokinesis in an early frog embryo. The central fold is the cleavage furrow which constricts to divide the cell in two.

TABLE 2.2 Summary of Mitosis

Stage	Characteristics
Interphase	Replication of chromosomes takes place.
Prophase	Chromosomes become visible as threadlike structures; later, as they continue to condense, they are seen to be double, with sister chromatids joined at a single centromere.
Metaphase	Chromosomes become aligned at equator of cell.
Anaphase	Centromeres divide, and chromosomes move toward opposite poles.
Telophase	Chromosomes uncoil, nuclear membrane forms, and cytoplasm divides.

bone marrow move through the cell cycle and divide continually to produce about 2 million red blood cells each second. Cells in the epidermis remain mitotically active and divide to replace dead epidermal cells that are continually sloughed off the surface of the body. The healing of wounds also involves mitosis in the damaged tissues. By contrast, most cells in the nervous system remain permanently in G1 and cannot undergo division. Thus,

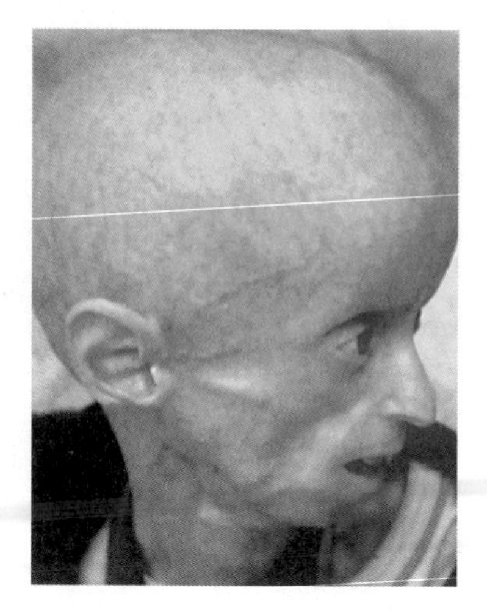

FIGURE 2.8 A 13-year-old female afflicted with progeria.

Hayflick limit
The number of cell divisions that a cultured cell will undergo before dying.

when nerves are damaged or destroyed, they cannot be replaced. For this reason many injuries to the spinal cord result in permanent paralysis. Occasionally, cells escape regulation and grow continuously, forming tumors, including cancer.

The mechanism that determines whether cells are cycling or noncycling operates in the G1 phase of the cell cycle. Although some general features of cell cycle regulation are known, the details remain unknown.

Studies of many organisms have established that each species has a characteristic life span, or longevity. In humans this is estimated to be about 110 to 120 years. The role of mitosis in aging has been demonstrated by studies of cells in culture; this work indicates that cells are programmed to undergo a finite number of cell divisions. Once this number, known as the **Hayflick limit,** is reached, the cells die. Cells cultured from human embryos have a limit of about 50 doublings, a capacity that includes all divisions necessary to produce a human adult and cell replacement during a lifetime. Cells from adults and elderly individuals can divide only about 10 to 30 times in tissue culture before dying. From these studies some scientists have concluded that the cells of many higher organisms, including humans, contain an internal mechanism that determines the maximum number of divisions, which in turn determines the ultimate life span of the individual. The process appears to be under genetic control because several mutations that affect the aging process are known. One of these is a rare heritable condition known as **progeria,** in which affected individuals age rapidly. Seven- or eight-year-old children with this disease physically and mentally resemble individuals of 70 or 80 years of age (Figure 2.8). Affected individuals usually die of old age by the age of 14. **Werner syndrome** is another genetic condition associated with premature aging (Figure 2.9). In

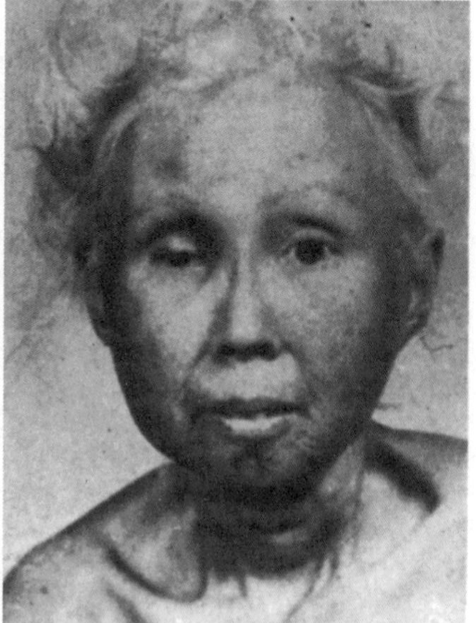

FIGURE 2.9 Photographs of an individual afflicted with Werner syndrome as a teenager (left) and at 48 years of age (right). This individual died at age 57 following a prolonged illness related to aging.

this case, the disease process begins between the ages of 15 and 20 years, and affected individuals die of age-related problems by 45 to 50 years of age.

Progeria
A genetic trait in humans associated with premature aging and early death.

Werner syndrome
A genetic trait in humans that causes aging to accelerate in adolescence, leading to death by about age 50.

THE HUMAN CHROMOSOME SET

Microscopic studies on chromosomes in the last part of the 19th century resolved the question of chromosome number in a wide range of plant and animal species. In other cases, including humans, some difficulties were encountered. Humans have a relatively large number of small, similarly shaped chromosomes, and the cells are difficult to prepare for study. By the early part of this century, most studies concluded that humans had a diploid number of 48 and a haploid number of 24 chromosomes. However, in 1956, J. H. Tijo and A. Levan, studying the chromosomes of aborted embryos, produced definitive evidence that human diploid cells contain 46 chromosomes. Shortly thereafter, other investigators confirmed this and showed that haploid cells in testicular tissues contain 23 chromosomes.

Human chromosomes are most often studied by microscopic examination and photography of cells in mitotic metaphase. For convenience, the chromosomes are cut out from a photographic print and, by convention, are arranged in pairs according to size and centromere location. This construction is known as a **karyotype**. Examination of human karyotypes reveals that one pair of chromosomes is not always homologous. Members of this pair are involved in the process of sex determination and are known as **sex chromosomes**. There are two types of sex chromosomes, X and Y. Females have two homologous X chromosomes, and males have a nonhomologous pair consisting of one X and one Y chromosome. Both males and females have two copies of the remaining 22 pairs of chromosomes known as **autosomes**.

Karyotype
The chromosome complement of a cell line or a person, photographed at metaphase and arranged in a standard sequence.

Sex chromosomes
Chromosomes involved in sex determination. In humans, the X and Y chromosomes are the sex chromosomes.

Autosomes
Chromosomes other than the sex chromosomes. In humans, chromosomes 1-22 are autosomes.

By convention, the chromosomes in the human karyotype are arranged into seven groups, A through G. Each group is defined by chromosome size and centromere location. Without further landmarks it is difficult to distinguish one chromosome from another within each group (Figure 2.9). In the late 1960s and early 1970s, new staining procedures were developed that produce patterns of dark and light bands on each chromosome. Banded chromosomes can be produced by several methods, most commonly by staining with the fluorescent dye quinicrine (to produce Q bands) or by enzyme digestion and staining with Giemsa (G bands). This staining produces consistent band patterns that are different for each chromosome, allowing individual chromosomes to be clearly identified (Figure 2.10). The standardized G-banding pattern for the human chromosome set, known as a **karyogram,** is shown in Figure 2.11. As a result, chromosomes are now identified in karyotypes by number rather than by group letter. Chromosome banding patterns have also been used as the basis for a system of identifying specific chromosome regions on each chromosome (Figure 2.12). In this system the short arm of each chromosome is designated the p arm, and the long arm the q arm. Each arm is subdivided into numbered regions beginning at the centromere. Within each region, the bands are identified by number. Thus any region in the human karyotype can be identified by a

Karyogram
A diagrammatic representation of a karyotype.

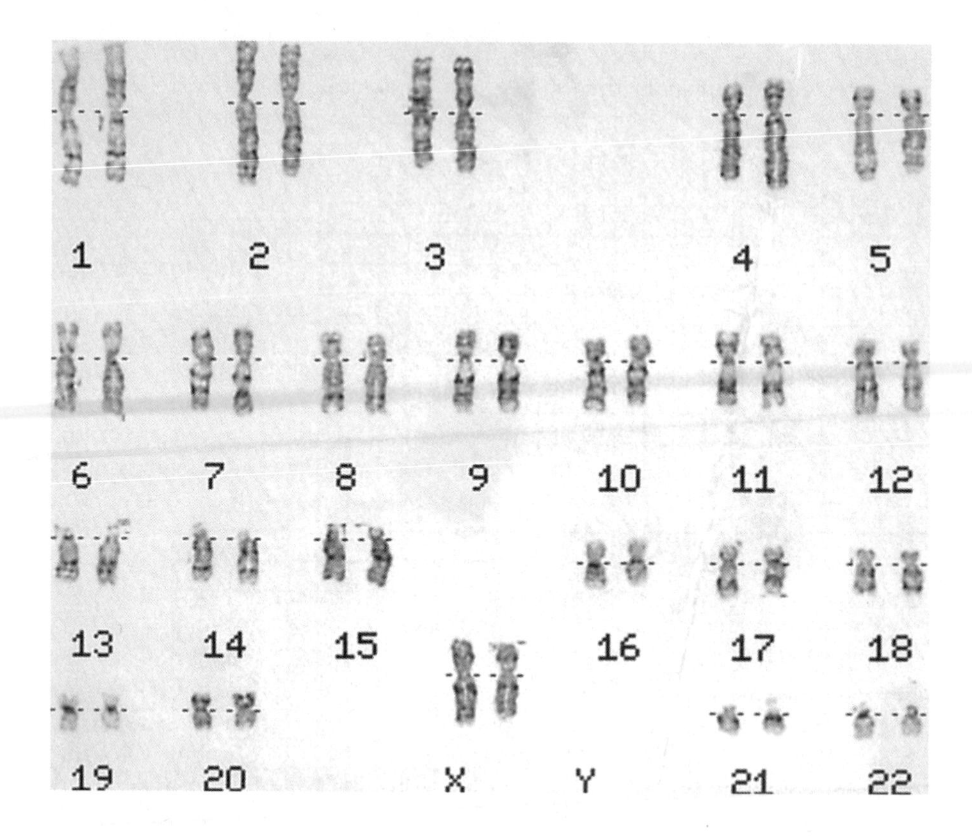

FIGURE 2.10 A karyotype of G-banded human chromosomes. By international convention, chromosomes are arranged in pairs according to size and centromere location (dotted lines). Chromosomes 1–22 are designated as autosomes, and the X and Y chromosome are designated as sex chromosomes. Note that each chromosome has a distinctive banding pattern.

descriptive address such as 1q2.4. The address consists of the chromosome number (1), the arm (q), the region (2), and the band (4) (Figure 2.12).

The development of banding techniques and related methods has provided the clinician and researcher with a powerful tool for chromosome studies (see Concepts & Controversies, page 32). Chromosome studies are usually performed at metaphase, since the chromosomes are maximally condensed at this stage. Preparations can be made from *lymphocytes,* a type of white blood cell that can be induced to divide in the laboratory. Karyotypes can also be prepared from *fibroblasts,* a cell type that can be recovered by skin biopsy. Chromosome analysis has many important applications in human genetics, some of which will be discussed in later chapters. These include prenatal diagnosis, gene mapping, and cancer research.

Chromosome Behavior in Meiosis

Meiosis
The process of cell division during which one cycle of chromosome replication is followed by two successive cell divisions to produce four haploid cells.

Sexual reproduction is a process that is mediated by cell division. All the genetic information we inherit is contained in two cells, the sperm and the egg. These cells are produced by a form of cell division known as **meiosis,** a Greek word meaning to reduce or to diminish. Recall that in mitosis, each daughter cell receives 46 chromosomes. In meiosis, however, members of a chromosome pair are separated from each other to produce haploid cells, or

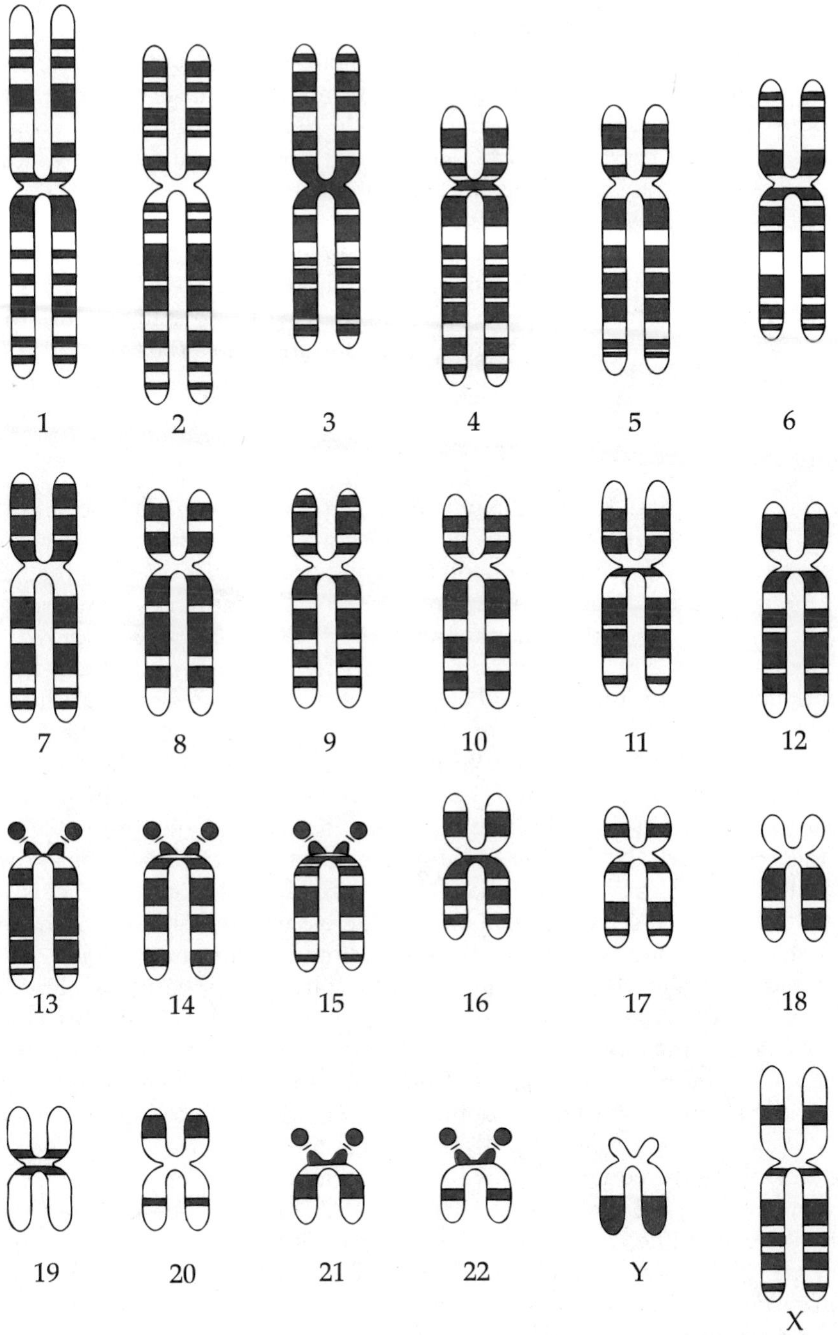

1 2 3 4 5 6
7 8 9 10 11 12
13 14 15 16 17 18
19 20 21 22 Y X

FIGURE 2.11 The banding pattern of human chromosomes.

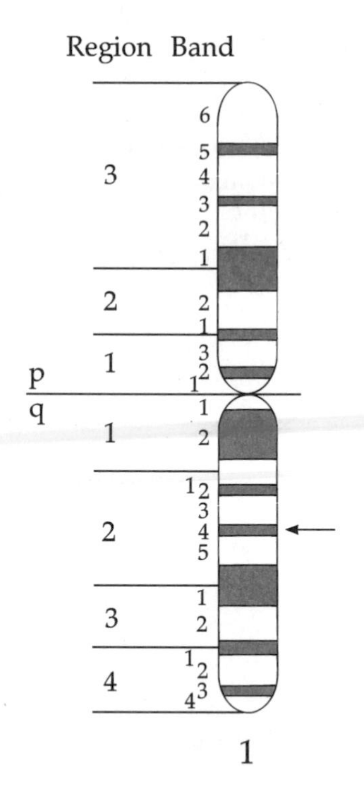

FIGURE 2.12 System of chromosome nomenclature. Each chromosome is identified by size, centromere location and banding pattern. Within a chromosome, the short arm is designated as *p*, the long arm by *q*. Within each arm, regions are designated by a number, and another number is used to designate bands within the region. Band 1q2.4 (marked by an arrow) is called 1q two-four.

gametes, each with 23 chromosomes. Union of the two gametes in fertilization restores the chromosome number to 46 and provides a full complement of genetic information to the zygote (Figure 2.13). Meiosis produces a reshuffling of genetic information, so that offspring are never genetically identical to either parent.

The distribution of chromosomes in meiosis is an exact process; each gamete must contain not just 23 chromosomes but one member of each chromosome pair rather than a random selection of 23 of the 46 chromosomes. How the precise reduction in the chromosome number is accomplished is of central importance in human genetics.

Cells in the testis and ovary that give rise to gametes are diploid and divide by mitosis. Some of the progeny of these germ cells, but no other cells in the body, undergo meiosis. In meiosis, diploid *(2n)* cells undergo one round of chromosome replication followed by two divisions to produce four cells, each with the haploid *(n)* number of chromosomes. The two division events are referred to as meiosis I and II, respectively.

Meiosis I

Meiosis I begins with replication of chromosomes in an interphase cell. As in mitosis, each replicated chromosome has a single centromere. In the first part of prophase I, the chromosomes coil and condense (Figure 2.14). In the next stage of prophase I, members of a chromosome pair (known as homologues) physically associate with each other and lie side by side, a process known as **synapsis** (Figure 2.14). Chromosome pairing usually begins at one or more points along the chromosome and proceeds until the chromosomes are linked together to form **bivalents.** As this process is completed, the sister chromatids of each chromosome become visible (Figure 2.14). The result is a pair of synapsed homologous chromosomes, with each chromosome consisting of two sister chromatids joined by a single centromere. This arrangement, with the four sister chromatids arranged in parallel, is known as a **tetrad.** Near the end of prophase I, the paired homologous chromosomes begin to repel each other but remain lying across each other at cross-shaped or X-shaped regions known as **chiasmata** (singular: chiasma) (Figure 2.14). These structures are the points at which the physical exchange of chromosome material between chromatids within a tetrad can take place. This event, known as **crossing over,** is of great genetic significance and will be discussed later. As the chromosomes separate from each other, the chiasmata appear to move along toward the chromosome ends (Figure 2.14).

In metaphase I (Figure 2.14), the tetrads move to the center of the cell and line up on the equatorial plate. An important feature of meiosis should be noted here. The orientation of each tetrad at metaphase is random. Remember that one chromosome in each pair is paternal, and the other is maternal; there is no fixed pattern of arrangement of these chromosomes with respect to the centrioles (Figure 2.15).

In anaphase I, one half of each tetrad (consisting of one pair of sister chromatids joined at a centromere) moves toward opposite poles of the dividing cell (Figure 2.14), a process known as **disjunction.** This disjunction of chromosomes in the tetrad separates maternal from paternal chromatids

Gamete
A haploid reproductive cell, such as the sperm or egg.

Synapsis
The pairing of homologous chromosomes during prophase I of meiosis.

Bivalent
The structure formed by a pair of synapsed, homologous chromosomes at prophase I of meiosis.

(except at regions where crossing over has occurred), resulting in a reduced number of chromosomes. The random arrangement of each tetrad in metaphase and the separation into dyads is the mechanism that produces **independent assortment,** a principle of genetics we will discuss in Chapter 3. The chromosomes begin to unwind slightly in telophase I, and division of the cytoplasm is complete, producing two cells, each containing 23 dyads. Thus two major events take place during meiosis I: *reduction,* the random separation of maternal and paternal chromatids, and *crossing over,* the physical exchange of segments between homologous chromosomes. In sum, during meiosis I, homologous chromosomes pair in prophase I, often undergo crossing over, and are distributed to daughter cells without division of the centromere.

Tetrad
The four homologous chromatids that are synapsed in the first meiotic prophase.

Chiasmata
The crossing of nonsister chromatid strands seen in the first meiotic prophase. Chiasmata represent the structural evidence for crossing over.

Crossing over
The process of exchanging parts between homologous chromosomes during meiosis; produces new combinations of genetic information.

Disjunction
The separation of chromosomes that occurs in anaphase of cell division.

Independent assortment
The random distribution of members of homologous chromosome pairs during meiosis.

Meiosis II

Although there may be a short interphase between meiosis I and II, there is no duplication of chromosomes, before the cells enter prophase II. In this stage the chromosomes become more condensed (Figure 2.14) and begin moving toward the equator of the cell. At metaphase II the 23 chromosomes, each consisting of a centromere and two sister chromatids, line up on the metaphase plate with spindle fibers attached to centromeres. At the beginning of anaphase II (Figure 2.14) the centromere of each chromosome *divides for the first time,* and the 46 chromatids are converted to independent chromosomes and move to opposite poles of the cell. In telophase II the chromosomes uncoil and become diffuse, the nuclear membrane reforms, and division of the cytoplasm takes place (Figure 2.14). The process of meiosis is now complete. One diploid cell containing 46 chromosomes has undergone one round of chromosome replication and two rounds of division to produce four haploid cells, each containing one copy of each chromosome for a total of 23 chromosomes. The major features of meiosis are summarized in Table 2.3, and Figure 2.16 compares the events of mitosis and meiosis.

Chiasma
The physical overlap of nonsister chromatids seen in first meiotic prophase. These structures represent evidence for crossing over.

Crossing Over

The phenomenon of crossing over is a common feature of meiosis. This process, which is one way to achieve recombination, involves the physical exchange of chromosomal material in prophase I. During prophase I, homologous chromosomes pair with each other, or synapse, and the sister chromatids of each chromosome become visible, forming a tetrad. During this stage, the arms of two nonsister chromatids can overlap, forming a cross-like configuration known as a **chiasma** (plural: chiasmata; Figure 2.17). These sites of overlap are associated with the physical exchange of chromosome segments and the genes they contain. If, for example, a pair of homologous chromosomes carries different forms of a gene (e.g., A and a or B and b), crossing over reshuffles this genetic information, creating new chromosomal combinations.

As a result of the second meiotic division, each of the four chromatids in a tetrad will be distributed to a different haploid cell. In those cells in which a single crossover event has occurred, the result will be two chromosomes

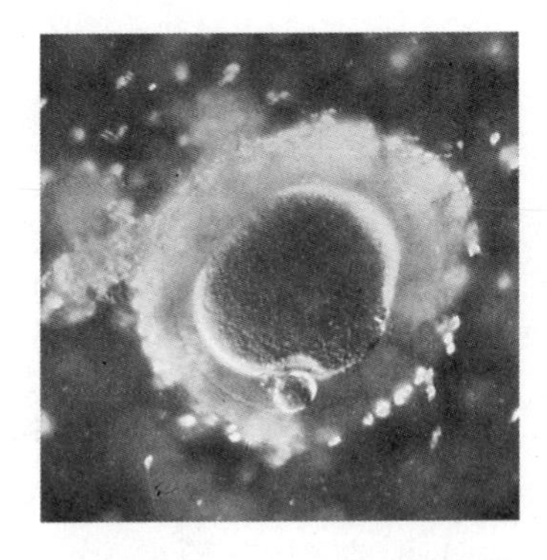

FIGURE 2.13 A photograph of a living human egg immediately after fertilization.

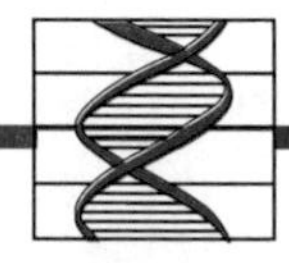

CONCEPTS & CONTROVERSIES

Making a Karyotype

Chromosome analysis can be performed using mitotic cells from a number of sources, including white blood cells (lymphocytes), skin cells (fibroblasts), amniotic fluid cells (amniocytes), and chorionic villus cells. One of the commonest methods begins with collection of a blood sample from a vein in the arm. A few drops of this blood are added to a flask containing a culture medium. Since white blood cells do not normally divide, a mitosis-inducing chemical known as phytohemagglutinin is added to the culture, and the cells are grown for 2 or 3 days at body temperature (37° C) in an incubator. At the end of this time, a drug called colcemid is added to stop dividing cells in metaphase. Over a period of about 2 hours, cells entering mitosis are arrested in metaphase.

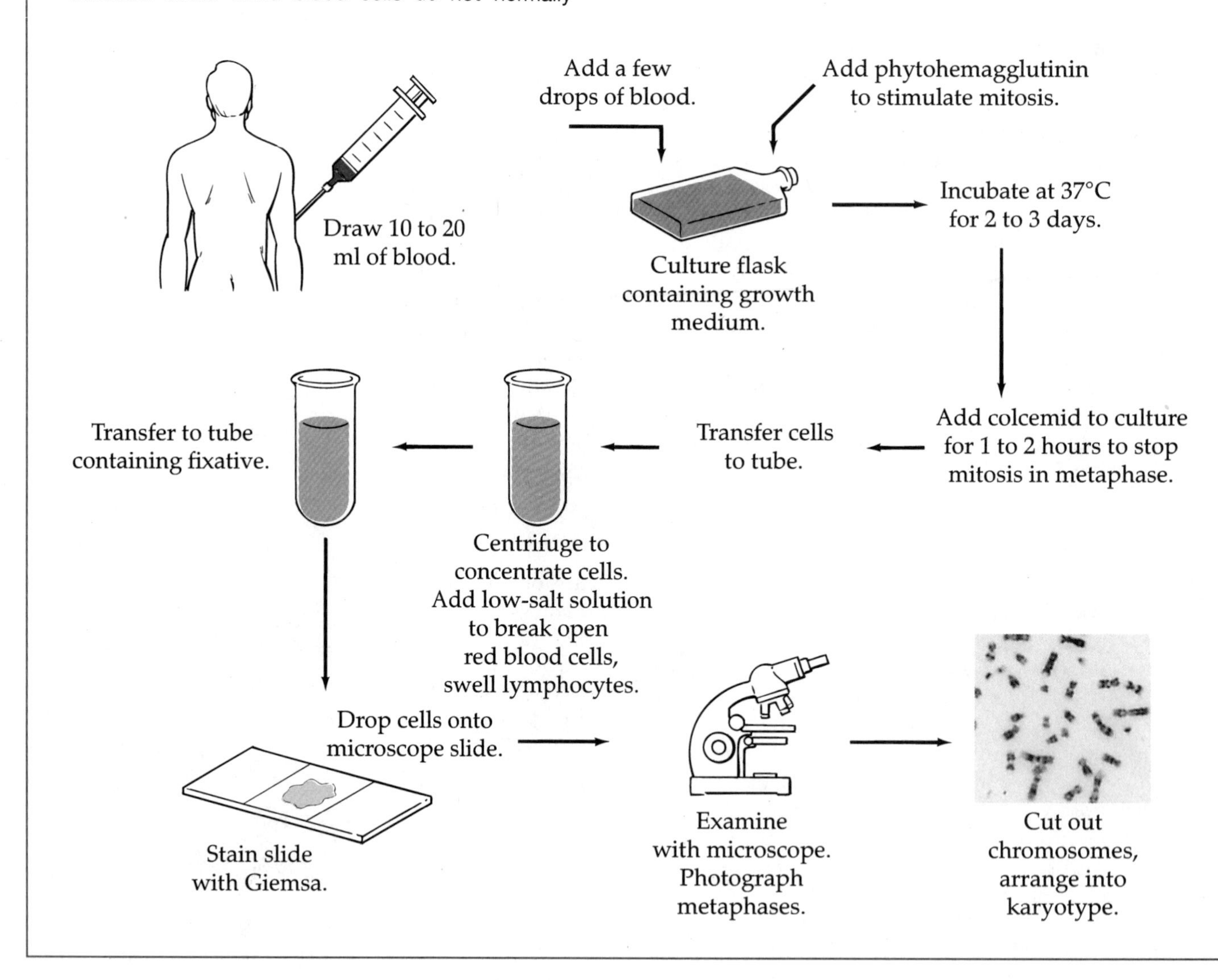

The lymphocytes (white blood cells) are concentrated by centrifugation, and a salt solution is added to break open the nondividing red blood cells and to swell the lymphocytes. This prevents the chromosomes from sticking together and makes the lymphocytes easier to break open. After fixation in a mixture of methanol and acetic acid, the cells are dropped onto a microscope slide. The impact causes the fragile lymphocytes to break open, spreading the metaphase chromosomes. After staining to reveal bands, the preparation is examined with a microscope, and the metaphase array of chromosomes is photographed. A print of this photograph is used to construct a karyotype. Chromosomes are cut from the photograph and are arranged according to size, centromere, location, and banding pattern. The presence of any chromosomal abnormalities will be revealed upon analysis of the karyotype by a trained cytogeneticist.

The process of cutting chromosomes by hand and arranging them into karyotypes is time consuming and tedious. Computer-assisted karyotype preparation is now commercially available. In this system a television camera and a computer are coupled to a microscope. As metaphase spreads are located, the microscopic image is recorded by the television camera, and the image is transmitted to the computer, where it can be analyzed and processed into a karyotype. At the left is a metaphase spread of human chromosomes printed by such a system, and at the right is the computer-derived karyotype. Note that the computer also eliminates any curves in the metaphase chromosomes when processing the image into a karyotype. The use of this equipment shortens the time necessary for karyotype analysis and may lower the cost of this procedure.

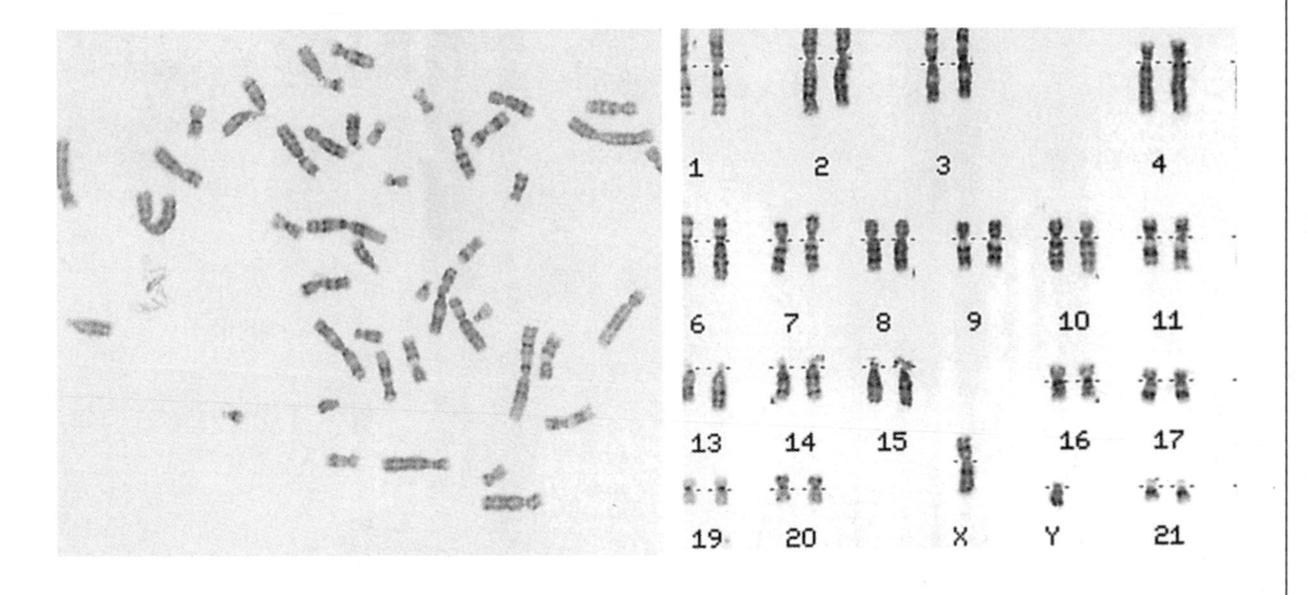

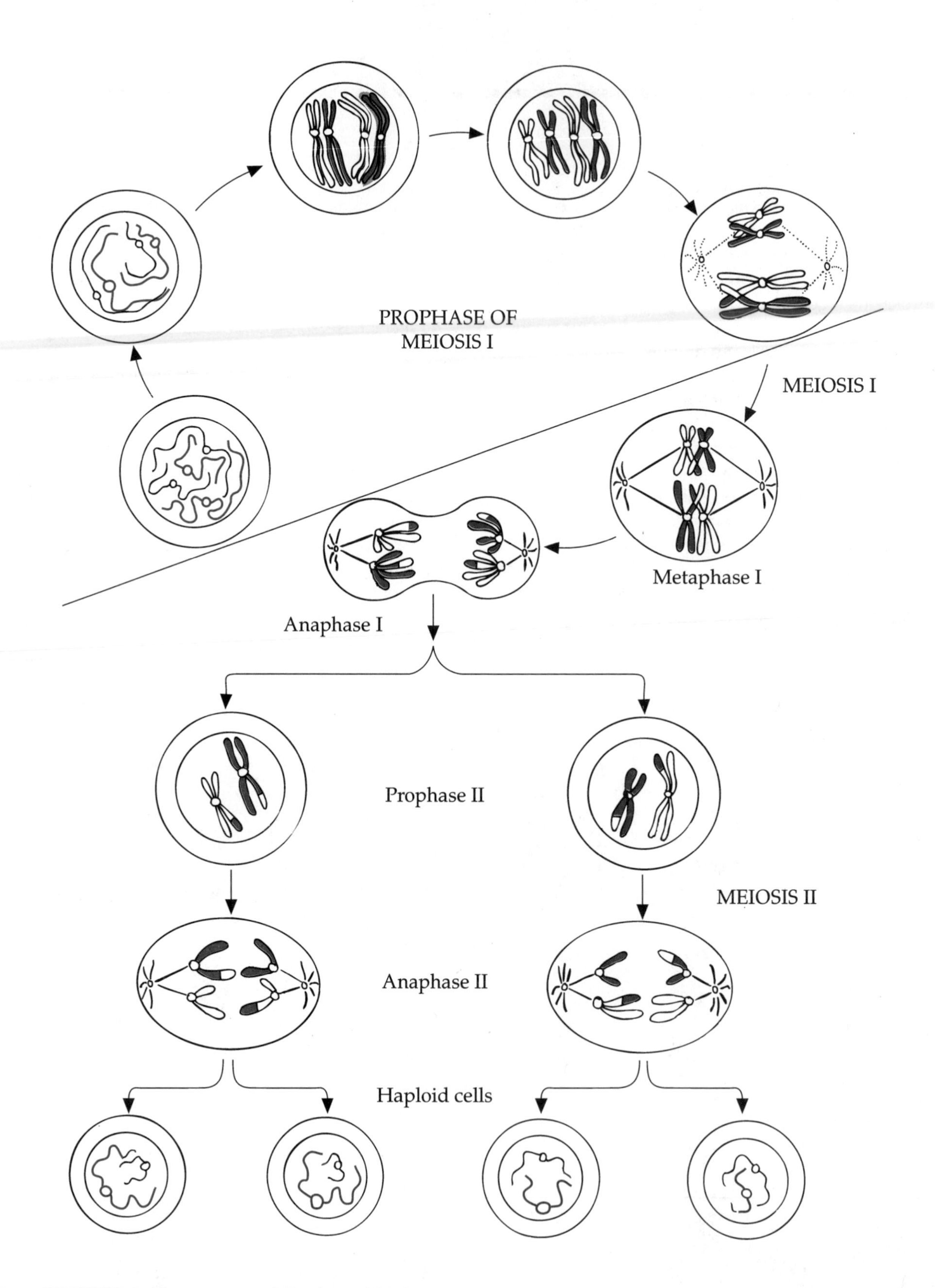
PROPHASE OF
MEIOSIS I
MEIOSIS I
Metaphase I
Anaphase I
Prophase II
MEIOSIS II
Anaphase II
Haploid cells

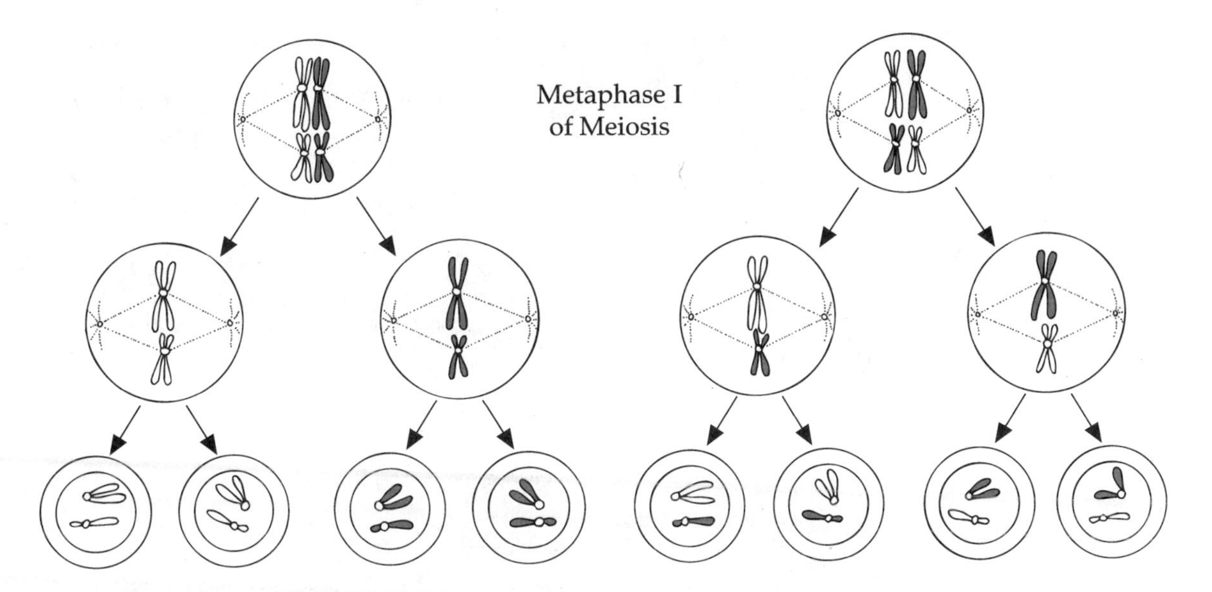

FIGURE 2.15 The alternate arrangement of maternal (*light*) and paternal (*dark*) chromosomes at metaphase I and the resulting arrangements of chromosomes in the gametes.

with new gene combinations and two noncrossover chromosomes that retain the parental gene combinations (Figure 2.17). Multiple crossovers can also occur within a single tetrad, producing even more reshuffling of genes.

Formation of Gametes

Gametogenesis is the formation of mature germ cells, namely ova and spermatozoa. In males the production of gametes, or sperm, by meiosis occurs in the testis and is known as spermatogenesis. Germ cells, or **spermatogonia,** that line the tubules in the testis are mitotically active from puberty until death, and they produce daughter cells called **primary spermatocytes.** The spermatocytes undergo meiosis, and the four haploid cells that result are known as **spermatids.** Each spermatid undergoes a period

Spermatogonia
Mitotically active cells in the gonads of males that give rise to primary spermatocytes.

Primary spermatocytes
Cells in the testis that undergo meiosis.

Spermatids
The four haploid cells produced by meiotic division of a primary spermatocyte.

FIGURE 2.14 (Opposite) The stages of meiosis. Chromosomes become visible as thread-like structures early in prophase I. Homologous chromosomes pair and split longitudinally except at centromeres. While homologous chromosomes are paired, crossing over (a form of recombination) takes place late in prophase I. The chromosomes become distributed at the cell's equator in metaphase I, and in anaphase I, members of pairs separate or disjoin and migrate to opposite ends of the cell. The result is two new cells, but these daughter cells are not genetically equal to each other or to the parental cell, since each contains only one member of each chromosome pair. In addition, crossing over has resulted in the exchange of genetic information between homologous chromosomes, further increasing genetic diversity. In meiosis II, the double-stranded chromosomes align themselves at the equator of the cell, and as in mitosis, the centromeres split and the newly formed chromosomes move apart. The result is four haploid daughter cells, each containing one copy of each chromosome.

TABLE 2.3 Summary of Meiosis

Stage	Characteristics
Interphase I	Chromosome replication takes place.
Prophase I	Chromosomes become visible, homologous chromosomes pair, and sister chromatids become visible; recombination takes place.
Metaphase I	Paired chromosomes align at equator of cell.
Anaphase I	Homologous chromosomes separate; members of each chromosome pair move to opposite poles.
Telophase I	Cytoplasm divides, producing two cells.
Interphase II	Following a brief pause, chromosomes uncoil slightly; this is not a real interphase as such.
Prophase II	Chromosomes re-coil.
Metaphase II	Unpaired chromosomes become aligned at equator of cell.
Anaphase II	Centromeres split; daughter chromosomes pull apart.
Telophase II	Chromosomes uncoil, nuclear membrane reforms, cytoplasm divides, and meiosis is complete.

Sperm
Male haploid gametes produced by morphological transformation of spermatids.

Oogonia
Mitotically active cells that produce primary oocytes.

Primary oocytes
Cells in the ovary that undergo meiosis.

Secondary oocyte
The cell produced by the first meiotic division.

Ootid
The haploid cell produced by meiosis that will become the functional gamete.

Polar body
A cell produced in the first or second division in female meiosis that contains little cytoplasm and will not function as a gamete.

of development into mature **sperm** (Figure 2.18). During this period the nucleus, containing 23 chromosomes, becomes condensed and forms the head of the sperm. All sperm carry 22 autosomes and an X or a Y chromosome. In the cytoplasm a neck and whiplike tail develop, and most of the remaining cytoplasm is lost. In human males, meiosis and sperm production begin at puberty and continue throughout life. The entire process of spermatogenesis takes about 48 days: 16 for meiosis I, 16 for meiosis II, and 16 for the conversion of the spermatid into the mature sperm. The tubules within the testis contain large numbers of spermatocytes, and large numbers of sperm are always in production. A single ejaculate may contain 200 to 300 million sperm, and over a lifetime a male will produce billions of sperm.

In females the production of gametes is known as oogenesis and takes place in the ovary. Germ cells, or **oogonia,** divide by mitosis and form **primary oocytes** that undergo meiosis. However, the cytoplasmic cleavage in meiosis I does not produce cells of equal size (Figure 2.19). One cell, destined to become the oocyte, receives about 95% of the cytoplasm and is known as the **secondary oocyte.** In the second meiotic division, the same disproportionate cleavage results in one cell retaining most of the cytoplasm. The large cell, known as an **ootid,** will become the functional gamete, and the nonfunctional, smaller cells are known as **polar bodies.** Thus in females only one of the four cells produced by meiosis becomes a gamete. All oocytes contain 22 autosomes and an X chromosome.

The timing of meiosis and gamete formation in the human female is different than in the male (Table 2.4). The mitotic divisions of oogonia begin in the ovary of the embryo and are completed at 7 to 8 weeks of gestation. Since no more divisions take place, the female is born with all the primary oocytes she will ever possess. The half million or so primary oocytes enter

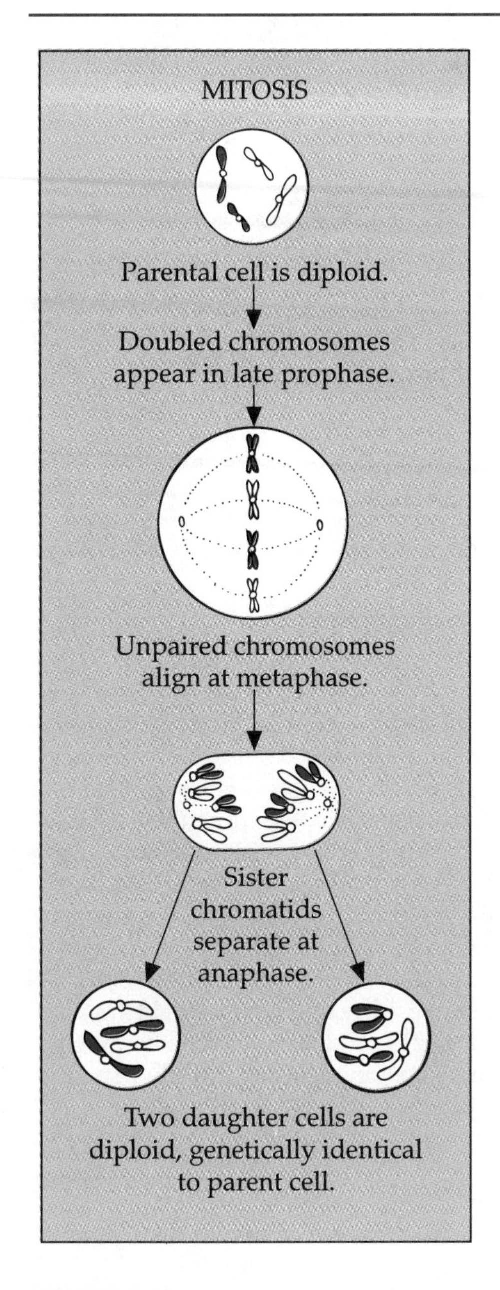

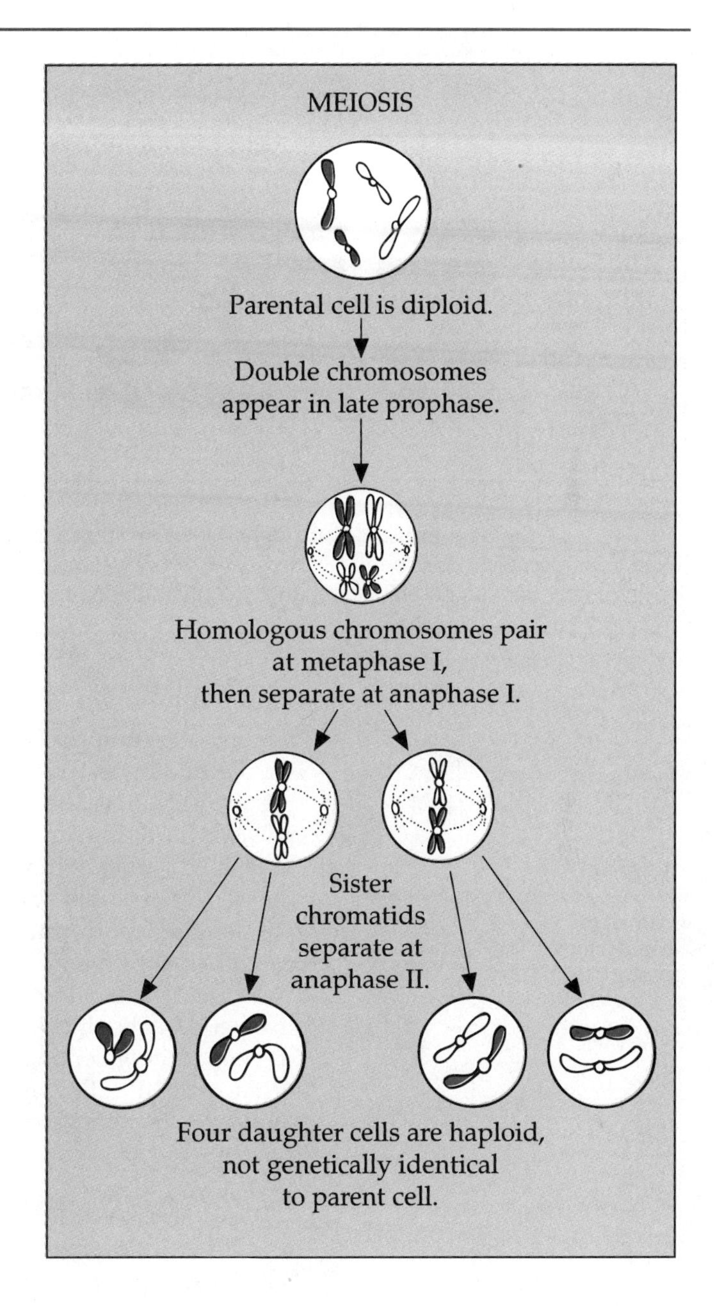

FIGURE 2.16 A comparison of the events in mitosis and meiosis.

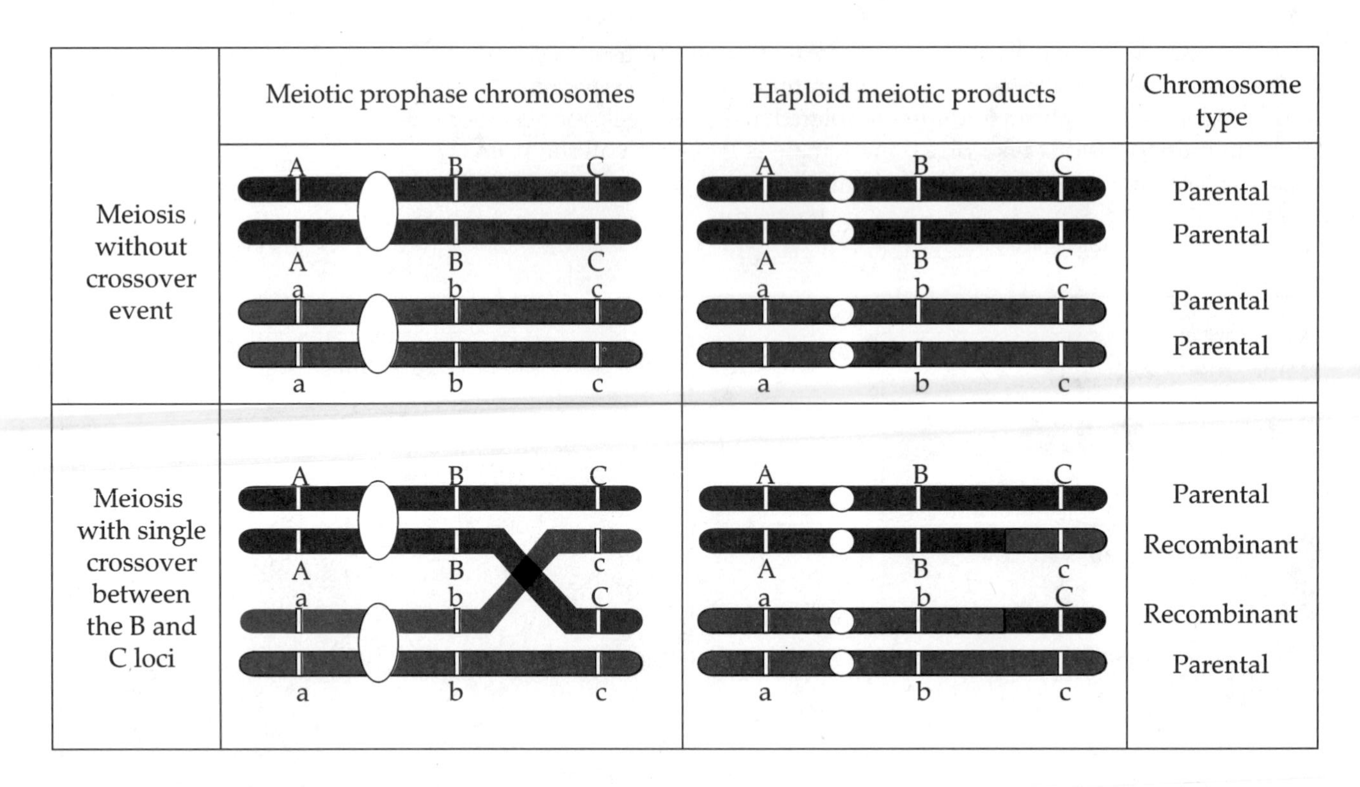

FIGURE 2.17 Meiosis and crossing over. Meiosis without crossing over gives rise to haploid chromosomes with parental gene combinations (upper). A crossover event between nonsister chromatids giving rise to haploid chromosomes with two new genetic combinations and two parental combinations (lower).

Zygote
The diploid cell resulting from the fusion of the haploid male and female gametes.

late meiotic prophase I well before birth and remain in meiosis I until the female undergoes puberty. Under hormonal influence, usually one oocyte per menstrual cycle completes the first meiotic division and is ovulated, or released from the ovary. If the egg is fertilized as it passes down the fallopian tube, it will quickly complete meiosis II, resulting in a diploid **zygote.** If two oocytes are released and fertilized, fraternal twins result. Unfertilized eggs are sloughed off during menstruation. The production of secondary oocytes continues until menopause. Since ovulation occurs once in each menstrual cycle, the female releases about 450 secondary oocytes during the reproductive period of her life. In females meiosis stretches over a range of years, from the initiation of prophase I, while she is still an embryo, to the completion of meiosis II following fertilization. Thus, depending on the span of ovulation, meiosis can take from 12 to 50 years in human females.

Genetic Consequences of Meiosis

It is clear to even the casual observer that children resemble their parents, usually combining some recognizable features from each parent. This variation in the distribution of traits from parent to child and the variation seen between children in the same family is a direct result of meiosis. Recall that genetic information is carried on chromosomes. What meiosis does is produce new combinations of genetic information by the *recombination* of parental chromosomes. The result is that genes carried on different chromosomes recombine by the assortment of chromosomes during meiosis I. How does this happen? As discussed earlier, each chromosome pair consists of one maternally derived (M) and one paternally derived (P) chromosome. With respect to the poles to which the chromosomes will migrate in anaphase I, the alignment of any chromosome pair can be maternal:paternal or paternal:maternal. For any chromosome pair, the chance that a maternal chromosome will migrate to a given pole is 1/2. Since there are 23 such pairs, the chance that all maternal or all paternal chromosomes will go in one direction is $(1/2)^{23}$, or 1/8,388,608. This means

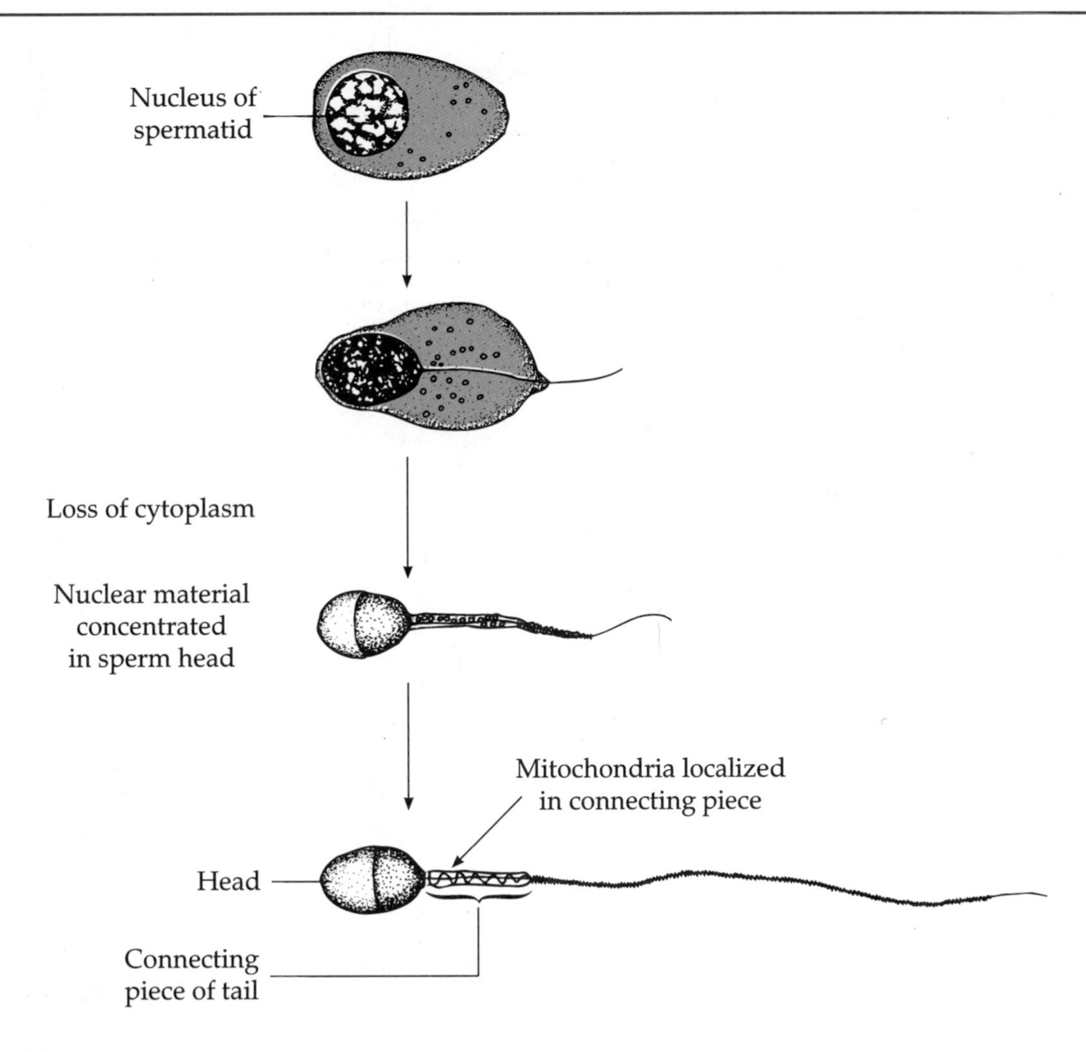

FIGURE 2.18 The process of differentiation associated with the formation of mature sperm.

that the possible number of combinations of maternal and paternal chromosomes that any individual can produce in gametes is 8,388,608. If each parent has this many combinations possible, they could produce more than 7×10^{13} offspring with different combinations of parental chromosomes (see Concepts & Controversies, page 41).

This astronomic number does not take into account the variability generated by the physical exchange of chromosome parts, or crossing over, that takes place during meiosis. Crossing over is a second mechanism through which new combinations of genes are produced by meiosis. Without crossing over, the combination of genes on a particular chromosome would remain coupled together indefinitely. Crossing over allows new and perhaps advantageous combinations of genes to be produced. When the variability generated by crossing over is added to that produced by random chromosome combinations, the number of different genetic combinations that a couple can produce in their offspring has been estimated to be 80^{23}. Obviously the offspring of a couple will represent only a very small fraction of all these possible gamete combinations. For this reason, it is almost

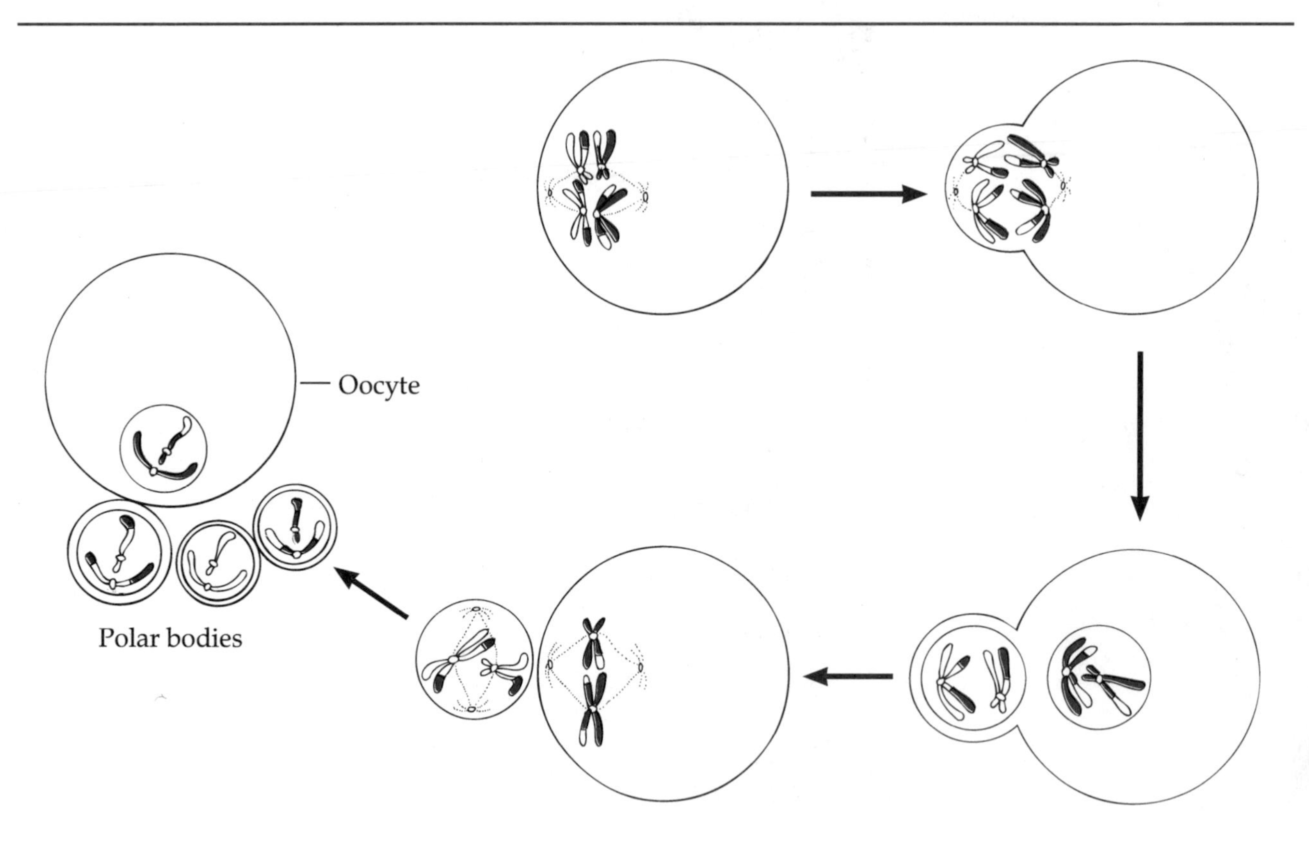

FIGURE 2.19 Unequal cytoplasmic cleavages in female meiosis produce cells of unequal size. As a result, only one of the four meiotic products becomes a gamete. The smaller cells, known as polar bodies, are haploid but do not function as gametes.

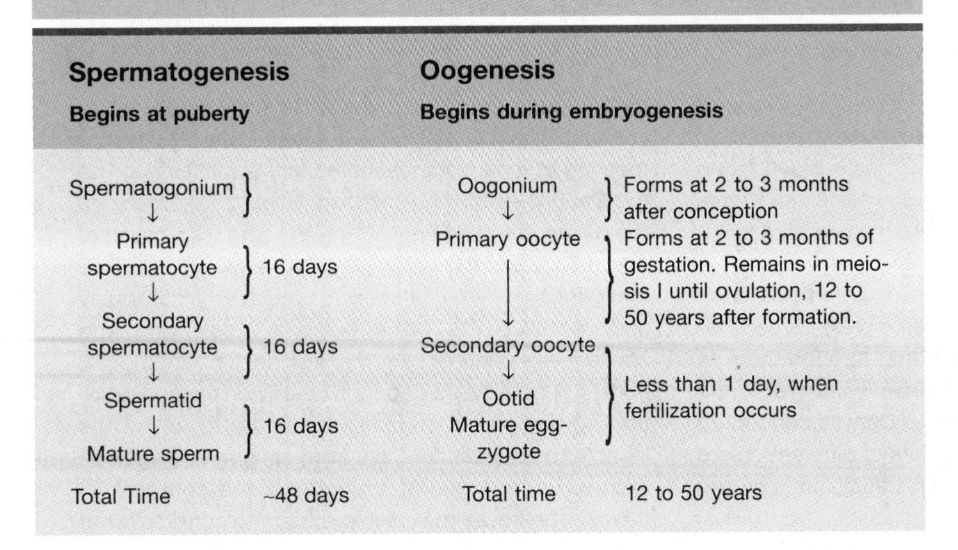

TABLE 2.4 A Comparison of the Duration of Meiosis in Males and Females

Spermatogenesis		Oogenesis	
Begins at puberty		Begins during embryogenesis	
Spermatogonium ↓		Oogonium ↓	Forms at 2 to 3 months after conception
Primary spermatocyte ↓	16 days	Primary oocyte ↓	Forms at 2 to 3 months of gestation. Remains in meiosis I until ovulation, 12 to 50 years after formation.
Secondary spermatocyte ↓	16 days	Secondary oocyte ↓	Less than 1 day, when fertilization occurs
Spermatid ↓ Mature sperm	16 days	Ootid Mature egg-zygote	
Total Time	~48 days	Total time	12 to 50 years

impossible for any two children (aside from identical twins) to be genetically identical.

SUMMARY

1. The cell is the basic unit in the bodies of all organisms, including humans. Since genes control the number, size, shape, and function of cells, the study of cell structure often reveals much about the normal expression of genes.

2. Within the cell the nucleus contains the genetic information that controls the structure and function of the cell. This genetic information is carried in nuclear structures known as chromosomes. The number, size, and shape of chromosomes are species-specific traits. In humans, 46 chromosomes, the $2n$, or diploid, number, are found in most cells, while specialized cells known as gametes contain half of that number, the haploid, or n, number of chromosomes.

3. In humans chromosomes exist in pairs known as homologues. One member of each pair is contributed by the mother and one by the father. Females carry two homologous X chromosomes, while males have an X and a partially nonhomologous Y chromosome. The remaining 22 pairs of chromosomes are known as autosomes.

4. For cells to survive and function, they must contain a complete set of genetic information. This is ensured by replication of each chromosome and distribution of a complete chromosome set in the process of mitosis. Mitosis

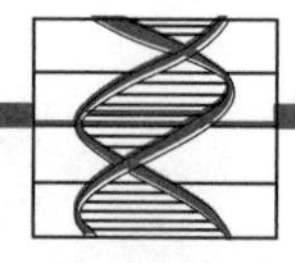

Concepts & Controversies

Of Genes and Genealogy

Our everyday language often preserves fragments of earlier ideas embodied in phrases and expressions. Phrases such as "blood relative," "blue blood," and "full blooded" and the idea that someone like Prince Charles is of "royal blood" are remnants of an ancient idea that blood is involved in the transmission of hereditary traits. According to this idea, particles in the circulatory system (known as gemmules) carried hereditary traits to the sex organs and through them to succeeding generations. As recently as 1871, Sir Francis Galton, a cousin of Charles Darwin, conducted blood transfusions between rabbits of different colors to investigate the effects of such transfusions on the coat color of offspring.

We now know that chromosomes in the gametes are the agents by which hereditary information is transferred from generation to generation, but these phrases remain in our language as a kind of shorthand, encapsulating the idea that traits are passed from generation to generation. The underlying idea, however, is not completely dead. Unfortunately some fervent students of genealogy too often believe that by virtue of descent, someone with a notable ancestor must have many of the same genetic traits possessed by the honorable forebearer.

Interest in genealogy is not new (see figure) and lineage is often an important issue in inheritance or royal succession. Fascination with the study of personal history, or genealogy, is undergoing a resurgence in the United States. This interest includes groups meeting at local libraries, hereditary societies such as the Daughters of the American Revolution, orders for the descendants of English Royalty, and the Church of Jesus Christ of the Latter-day Saints, where theology and genealogy are intimately related. Many individuals find genealogy to be a personally satisfying and challenging pastime. As we will see in Chapter 4, human genetics also depends heavily on the construction of pedigrees, or family histories, to follow the inheritance of specific traits from generation to generation. The process of chromosome assortment in meiosis, however, makes it important to consider what is *not* inherited as well as what is inherited.

Traditional methods of genealogy trace family history through the male line only. This ignores the fact that the female contributes half the genetic information to the offspring in each generation. With respect to genetic information carried in the nucleus, we have all received one half of our chromosomes from our fathers and one half from our mothers. In turn, each of our parents received half of their chromosomes from their parents. Consider a single chromosome carried by a grandparent. The chance that this chromosome was given to one of your parents is 1/2, and the chance that your parent passed it on to you is 1/2. Thus the probability that this chromosome has been passed

represents one part of the cell cycle. During the other part, interphase, a duplicate copy of each chromosome is made. The process of mitosis has been divided into four stages: prophase, metaphase, anaphase, and telophase. Mitosis results in two diploid cells, each with an exact replica of the genetic information contained in the parental cell.

5. Meiosis is a form of cell division that produces haploid cells that contain only the paternal or maternal copy of each chromosome. In meiosis,

from grandparent to grandchild is $1/2 \times 1/2 = 1/4$, or a 25% chance. As ancestors become more distant, the probability that they have contributed a chromosome to you falls off rapidly. There is a better than even chance that at least one of your 16 great-great grandparents have not contributed a single chromosome to you. As the number of ancestors doubles in each preceding generation, one quickly reaches a situation in which the number of ancestors exceeds the number of chromosomes. It is literally impossible, for example, that you have received a chromosome from each of 64 or 128 ancestors, but due to crossing over, it is quite possible that you have received a piece of a chromosome from each of them.

Taking into account that there are many more genes (say 50,000) than chromosomes and that genes can be transferred from one member of a chromosome pair to another by crossing over, similar calculations can be made about the probability of inheriting a gene from an ancestor. If your genealogy reveals an illustrious ancestor more than 500 years ago, such as Charlemagne, Kubla Khan, or a king of Benin, the chance that you have inherited even a single gene from this individual is pretty remote (less than 1 in 1000). All of this is not to discredit genealogy, which is an interesting and important activity, but rather to point out the power of meiosis to reshuffle genes in each generation to produce unique combinations of a diverse heritage.

TO THEIR MOST SACRED MAJESTIES,

GEORGE THE THIRD,

AND

SOPHIA CHARLOTTE,

THIS ATTEMPT TO TRACE
THE ANTIQUITY OF HER MAJESTY'S
ILLUSTRIOUS DESCENT,

IN A FAITHFUL HISTORY OF THE
MEMORABLE ACTIONS OF HER
ROYAL ANCESTORS,

IS MOST HUMBLY INSCRIBED,
BY THEIR MAJESTIES'
MOST OBEDIENT SERVANT,
AND MOST DUTIFUL SUBJECT,

Thomas Nugent.

Cover page of a genealogy prepared for Queen Sophia Charlotte, wife of King George III (1760–1820) of England.

homologous chromosomes synapse to form tetrads. At this time each chromosome consists of two sister chromatids joined by a common centromere. Chromatids can engage in the physical exchange of chromosome segments, an event known as crossing over. In metaphase I, paired, homologous chromosomes align at the equator of the cell; in anaphase I, the homologues are separated and become dyads. Before meiosis II there is no replication of chromosomes, and the unpaired chromosomes align at the metaphase plate. In anaphase II the centromeres divide, moving the daughter chromosomes to opposite poles. The four cells produced in meiosis contain the haploid number (23) of chromosomes.

6. Spermatids, the products of meiosis in males, undergo structural changes to convert them to functional sperm. In female meiosis, division of the cytoplasm is unequal, leading to the formation of one functional gamete and three smaller cells known as polar bodies.

7. Meiosis is a genetically important process. It provides a mechanism for maintaining a constant number of chromosomes from generation to generation. In addition, it generates genetic diversity by reshuffling maternal and paternal chromosomes and by permitting crossing over to occur.

Questions and Problems

1. Assign a function(s) to the following cellular structures:
   - **a.** plasma membrane
   - **b.** mitochondrion
   - **c.** nucleus
   - **d.** ribosome

2. Define the following terms:
   - **a.** chromosome
   - **b.** chromatin
   - **c.** chromatid

3. Originally karyotype analysis relied on chromosome size and centromere placement for discrimination. This resulted in the ambiguous placement of the human chromosomal complement into eight groups. Today each human chromosome is readily identifiable.
   - **a.** What advance led to this improvement in karyotype analysis?
   - **b.** List two ways in which this improvement can be implemented.

4. Karyotype analysis has revealed an interesting phenomenon at 21q13. Describe the location, explaining each term in the address 21q13.

5. A cell that will *not* divide again will most likely be arrested in what phase of the cell life cycle?

6. Identify the stages of mitosis, and describe the important events that occur during each stage.

7. A cell from a human female has just undergone mitosis. For unknown reasons, the centromere of chromosome 7 failed to divide. Describe the chromosomal content of the daughter cells.

8. During which phases of the mitotic cell cycle would the terms *chromosomes* and *chromatid* refer to identical structures?

9. Describe the critical events of mitosis responsible for ensuring that each daughter cell receives a full set of genetic information from the parent cell.

10. It is feasible that evolution could have wrought an alternative mechanism for the generation of germ cells. Consider the meiotic life cycle of a germ cell precursor. If the S phase of this life cycle were "skipped," which meiotic division would no longer be required—meiosis I or meiosis II?

11. We are following the plight of human chromosome 1 during meiosis. At the end of synapsis, how many chromosomes, chromatids, and centromere(s) are present to ensure that chromosome 1 faithfully traverses meiosis?

12. What is the physical structure that has been associated with crossing over?

13. Compare meiotic anaphase I with meiotic anaphase II. Which meiotic anaphase is more similar to mitotic anaphase?

14. Discuss the products of meiosis in human males and females.

15. A human female is conceived on April 1, 1949, and is born on January 1, 1950. Onset of puberty occurs on January 1, 1962. She conceives a child on July 1, 1994. How long did it take for the ovum that was fertilized on July 1, 1994, to complete meiosis (+ / – 5%)?

For Further Reading

Avers, C. J. 1981. *Cell Biology.* 2d ed. New York: Van Nostrand.

Baserga, R., and Kisieleski, W. 1963. Autobiographies of cells. *Sci. Am.* **209** (August):103–110.

Bretscher, M. 1985. Molecules of the cell membrane. *Sci. Am.* **253** (October):100–109.

Cross, P.C. 1993. *Cell and Tissue Ultrastructure: A Functional Perspective.* Englewood Cliffs: Prentice-Hall.

DuPraw, E. 1968. *Cell and Molecular Biology.* New York: Academic.

Fawcett, D. 1966. *The Cell: An Atlas of Fine Structure.* Philadelphia: Saunders.

Karp, G. 1984. *Cell Biology.* 2d ed. New York: McGraw-Hill.

Kessel, R., and Shih, C. 1974. *Scanning Electron Microscopy in Biology: A Student's Atlas of Biological Organization.* New York: Springer-Verlag.

Mazia, D. 1971. The cell cycle. *Sci. Am.* **230** (January):54–64.

McIntosh, J. R., and McDonald, K. L. 1989. The mitotic spindle. *Sci. Am.* **261** (October):48–56.

Padilla, G. M., and McCarty, K. S., eds. 1982. *Genetic Expression in the Cell Cycle.* New York: Academic.

Therman, E. and Susman, M. 1993. *Human chromosomes: Structure, Behavior, Effects.* 3rd ed. New York: Springer-Verlag.

Wagner, R.P., Maguire, M.P. and Stallings, R.L. 1993. *Chromosomes: A synthesis.* New York: Wiley-Liss.

Zimmerman, A. M., and Forer, A., eds. 1982. *Mitosis/Cytokinesis.* New York: Academic.

CHAPTER 3

PATTERNS OF INHERITANCE

CHAPTER OUTLINE

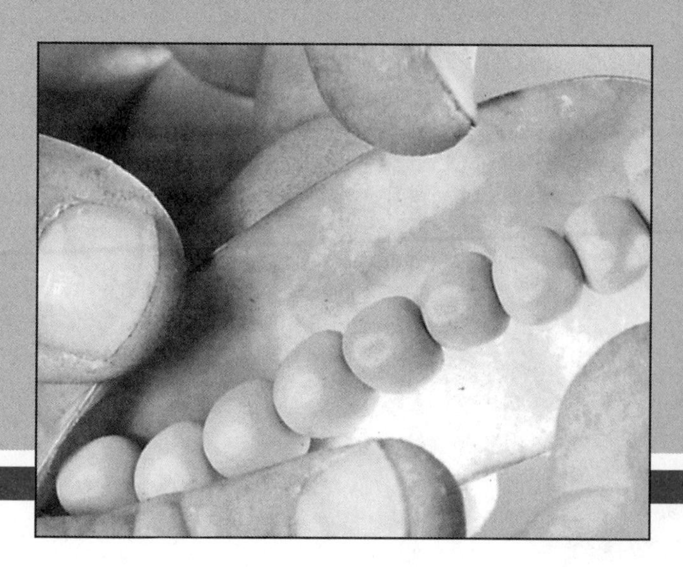

Biologists and, among them, geneticists, study a wide range of organisms, including bacteria, plants, and animals. Often such scientists are regarded as individuals who single-mindedly pursue the inner workings of obscure or exotic organisms. In some cases, in which the organism is associated with a disease, it is easy for everyone to understand why research is both necessary and desirable. No one questions the value of work on the virus that is associated with acquired immune deficiency syndrome (AIDS) in humans. In other cases it is more difficult to understand how something relevant to human genetics can be learned by studying *Escherichia coli,* a bacterium that lives in the nooks and crannies of the human gut, or more to the point of this chapter, by studying the garden pea. The answers to such questions are simple and direct: evolution and experimental design. Although evolution has produced a wide variety of living organisms, they all represent variations on a small number of basic themes that are highly conserved. Examples include the use of a molecule called ATP to capture and to transfer energy, the translation of genetic information into a chain of amino acids that form a functional gene product, and in eukaryotes, the way meiosis works to segregate chromosomes into gametes.

While exceptions to these conserved traits exist, they are rare and serve to point out the similarities shared by all life forms. The similarities make it possible to learn about fundamental processes in humans by studying organisms with simpler levels of organization. For example, many neurobiologists are using a small nematode, *Caenorhabditis elegans*, to study the neural basis of behavior. This organism has been selected for study because its entire nervous system contains only a few hundred cells, compared with the more than 100 billion cells in the human brain. Because of evolutionary conservatism, tracing neural circuits from brain to muscle in this nematode will provide information that may be applied to neural circuits in humans.

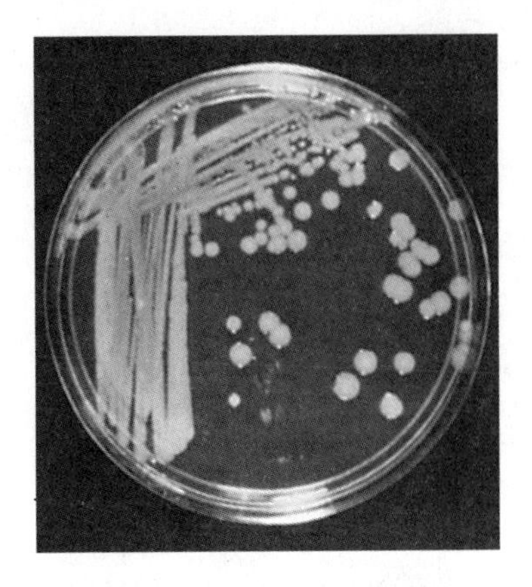

FIGURE 3.1 An agar plate showing colonies of bacterial growth. Each colony contains millions of cells.

The rationale for designing an experiment will often dictate the organism of choice. In scientific investigations, all efforts must be made to reduce or to eliminate error. Many times this means making a large number of independent measurements. If this means that 1 million organisms must be included in a study to get meaningful results, elephants might be a poor choice. On the other hand, 1 million *E. coli* can easily be accommodated on an agar plate the size and shape of a hamburger (Figure 3.1).

The transmission of genetic traits in pea plants, studied by Gregor Mendel in the 19th century, provides the basis for our understanding of how genes are inherited in many organisms, including humans. The inheritance of traits in peas follows a small number of fundamental rules that apply to all eukaryotic organisms that have sexual reproduction. Mendel's work is presented here in a human genetics textbook because it offers a clear demonstration of the rules of inheritance and provides insight into the methodology of scientific investigation.

MENDEL'S EXPERIMENTS

Background

Johann Gregor Mendel was born in 1822 in Hynice, Moravia, a region that is now part of the Czech Republic. He showed great promise as a student, but poverty prevented him from beginning university studies. At the age of 21, he entered the Augustinian monastery at Brno as a means of pursuing his interest in the study of natural history. After completing his monastic studies, Mendel enrolled at the University of Vienna in the fall of 1851. In his first year he took courses in physics, mathematics, chemistry, and the natural sciences.

In botany courses Mendel encountered the new concept that all organisms are composed of cells and that cells are the fundamental unit of all living things. This theory raised several new questions and many controversial issues. One was the question of whether both parents contributed equally to the appearance of the offspring. Because the germ cells in most female plants and animals are so much larger than those of the male, this was a logical and much debated question. Related to this was the question of whether the characters or traits present in the offspring were produced by blending the traits of the parents. If so, in what ratios? Did the female contribute 40%, 50%, 60%, or more of the traits? In 1854 Mendel returned to Brno to teach physics and to begin a series of experiments that were to resolve these questions.

Mendel's success in elucidating the mechanisms of inheritance was not the result of blind luck or accident but rather was the end product of carefully planned experiments. To select an experimental organism, Mendel spent two years studying several species of plants. From this work he concluded that pea plants were well suited to his needs. The plants are easy to grow, take little space, and are inexpensive (Figure 3.2). In addition, the plants can be self-fertilized or cross-fertilized to produce offspring. During this two-year period Mendel studied 34 varieties of garden peas differing in 15 easily distinguished traits. From these he selected for further study varieties that had seven distinctive characters. These varities were all true breeding; that is, self-fertilization gave rise to the same traits in all offspring, generation after generation. The characters selected affected the seeds, pods, flowers, and stems of the pea plants (Table 3.1). Each character studied was represented by two distinct forms, or traits: plant height by tall and short, seed shape by smooth and wrinkled, and so forth.

To avoid errors brought about by small sample sizes, Mendel planned his experiments on a large scale. His subsequent work spanned 10 years and involved 287 genetic crosses producing some 28,000 pea plants. In all his experiments he kept track of each character separately for several generations. He repeated his experiments for each of the traits to confirm his results.

FIGURE 3.2 The monastery garden at Brno, where Mendel carried out his experiments.

TABLE 3.1 Traits Selected for Study by Mendel

Structure Studied	Dominant	Recessive
Seeds	Smooth	Wrinkled
	Yellow	Green
	Gray	White
Pods	Full	Constricted
	Green	Yellow
Flowers	Axial (along stems)	Terminal (top of stems)
Stems	Long	Short

Using his training in mathematics, Mendel analyzed the results of his experiments according to principles of probability.

THE MONOHYBRID CROSSES

To illustrate how Mendel formulated his ideas, we will first examine some of the experiments he performed and the results he obtained. Then we will follow the reasoning that led Mendel to his conclusions and the further experiments that verified his ideas.

In the first set of experiments, Mendel took plants from a true-breeding variety with smooth seeds and crossed them with a variety bearing wrinkled seeds. To do this, he fertilized flowers from one variety using pollen from the other variety (Figure 3.3). The seeds that formed as a result of this pollination were all smooth. This result was true no matter which variety contributed the pollen. The next year Mendel planted the smooth seeds, and when the plants matured, the flowers were self-fertilized. Of these plants, 253 produced a total of 7324 seeds, of which 5474 were smooth and 1850 were wrinkled. He also performed breeding experiments with plants carrying the other six characters and obtained similar results. In these experiments Mendel designated the parents as the P_1 generation and the offspring as the F_1, or first filial, generation. Since the flowers of pea plants have both male and female parts, the F_1 generation was allowed to self-fertilize to produce the F_2 generation. The experiments with seed shape can be summarized as follows:

P_1: smooth × wrinkled
F_1: all smooth
F_2: 5474 smooth 1850 wrinkled

This cross, involving only one character, seed shape, is called a **monohybrid** cross.

Monohybrid cross
A cross between individuals that differ with respect to a single gene pair.

FIGURE 3.3 A diagram of the flower from the pea plant. To cross-fertilize, the anthers must be removed when immature and the flower fertilized with pollen from another plant.

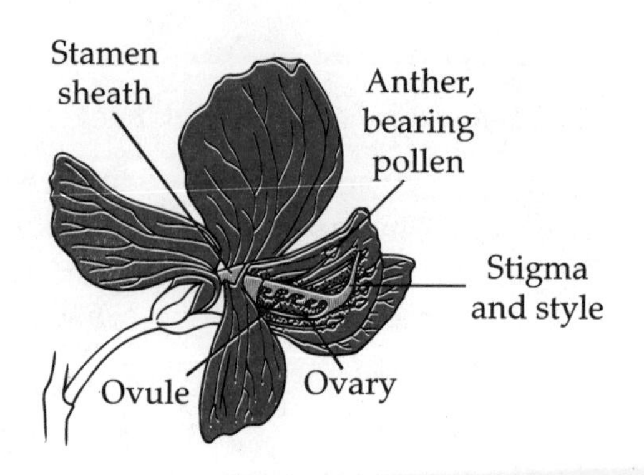

Results and Conclusions

The results of the experiments with all seven characters are similar to those seen in the cross with smooth and wrinkled seeds (Table 3.2) and can be summarized as follows:

1. In all crosses the F_1 generation showed only one of two alternate traits.
2. It did not matter which parental variety was male and which was female; the results were always the same.
3. The character not shown in the F_1 reappeared in the F_2 and was always present in about 25% of the offspring.

These findings were Mendel's first important discovery. The traits he investigated did not blend together in the offspring. Rather they were transmitted in a discrete fashion and remained unchanged from generation to generation. This convinced him that inheritance did not work by blending

TABLE 3.2 Results of Mendel's Monohybrid Crosses

Parental Phenotypes	Characters Seen In F_1	Characters Seen In F_2	F_2 Ratios
Smooth × wrinkled seeds	Smooth	5474 smooth, 1850 wrinkled	2.96:1
Yellow × green seeds	Yellow	6022 yellow, 2001 green	3.01:1
Gray × white seeds	Gray	705 gray, 224 white	3.15:1
Full × constricted pods	Full	882 full, 299 constricted	2.95:1
Green × yellow pods	Green	428 green, 152 yellow	2.82:1
Axial × terminal flowers	Axial	651 axial, 207 terminal	3.14:1
Long × short	Long	787 long, 277 short	2.84:1

the traits of the parents in the offspring. He also observed that the results of his genetic crosses were the same no matter which parent carried which trait, meaning that each parent makes an equal contribution to the genetic makeup of the offspring.

Based on the results of experiments with each of the seven characters, Mendel drew several conclusions:

1. The evidence from the F_1 showed that traits could be hidden, or unexpressed. All the F_1 seeds were smooth, but when these seeds were grown and self-fertilized, they produced some wrinkled offspring. Mendel called this phenomenon dominance: the trait expressed in the F_1 is **dominant.** The trait that is unexpressed in the F_1 but reexpressed in the F_2 is called **recessive.**
2. A comparison of the P_1 smooth plants with the F_1 smooth plants shows that despite identical appearances, their genetic makeup must be different. When P_1 plants are self-fertilized, they give rise only to plants with smooth seeds. However, when F_1 plants are self-fertilized, they give rise to plants with smooth and wrinkled seeds. Consequently it is important to make a distinction between the appearance of an organism and its genetic constitution. The term **phenotype** refers to the observed properties, or appearance, of an organism, and the term **genotype** refers to the genetic makeup of an organism. It is apparent that the P_1 and the F_1 smooth plants have identical phenotypes but must have different genotypes. Mendel postulated that traits are controlled by factors that are passed on to offspring through the gametes.
3. Based on the results of self-fertilization experiments, the F_1 plants must contain factors for both smooth and wrinkled, since both types of progeny are seen in the F_2. The question is, how many factors for seed shape are contained in the F_1 plants? Mendel inferred that both the pollen and the ovule contributed equally to the traits of the offspring, since it did not matter which parent was smooth or wrinkled. In view of this, the simplest assumption (see Concepts & Controversies, page 52) is that each F_1 plant contains two hereditary factors, one for smooth that is expressed and one for wrinkled that remains unexpressed. By extension, each P_1 and F_2 plant must also contain two factors that determine seed shape. Following the reasoning of Mendel, we can use the uppercase letter *S* to represent the dominant factor for smooth and the lowercase *s* to represent the recessive factor for wrinkled to reconstruct the genotypes and phenotypes of the P_1 and F_1, as shown in Figure 3.4.

Dominant
The trait expressed in the F_1 (or heterozygous) condition.

Recessive
The trait unexpressed in the F_1 but reexpressed in some members of the F_2 generation.

Phenotype
The observable properties of an organism.

Genotype
The specific genetic constitution of an organism.

The Principle of Segregation

If factors exist in pairs, some mechanism must exist to prevent these factors from being doubled in each succeeding generation. That is, if each parent has two factors, why doesn't the offspring have four? Mendel reasoned that

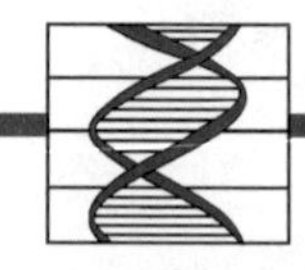

Concepts & Controversies

Ockham's Razor

When Mendel proposed the simplest explanation for the number of factors contained in the F_1 plants in the monohybrid crosses, he was employing a principle of scientific reasoning known as the principle of parsimony, or Ockham's razor.

William of Ockham (also spelled Occam), who lived from about 1300 to 1349, was a Franciscan monk and a scholastic philosopher. He had a strong interest in the study of thought processes and in logical methods. He is the author of the axiom known as Ockham's razor: "Pluralites non est ponenda sine necessitate," or "multiplicity ought not to be posited without necessity." In another translation, this is phrased as "entities must not be multiplied without necessity." In philosophy and theology, this was taken to mean that when constructing an argument, we should not go beyond that which is logically required. While Ockham was not the first to espouse this approach, he used this tool of logic so often and so well in dissecting the arguments of his opponents that it became known as Ockham's razor.

The principle was first used in theology and philosophy and was adapted to the construction of scientific hypotheses in the 15th century. In fact, Galileo used the principle of parsimony to argue that since his model of the solar system was the simplest, it was probably correct. In modern terms the phrase is taken to mean that in proposing a mechanism, use the least number of steps possible. The simplest mechanisms are inevitably the easiest to prove or to disprove by experimental means and the most likely to produce scientific progress.

Mendel concluded that both parents contributed an equal number of hereditary factors for a trait to the offspring. In this case the simplest assumption is that each parent contributed one such factor and that the F_1 plant contained two such factors.

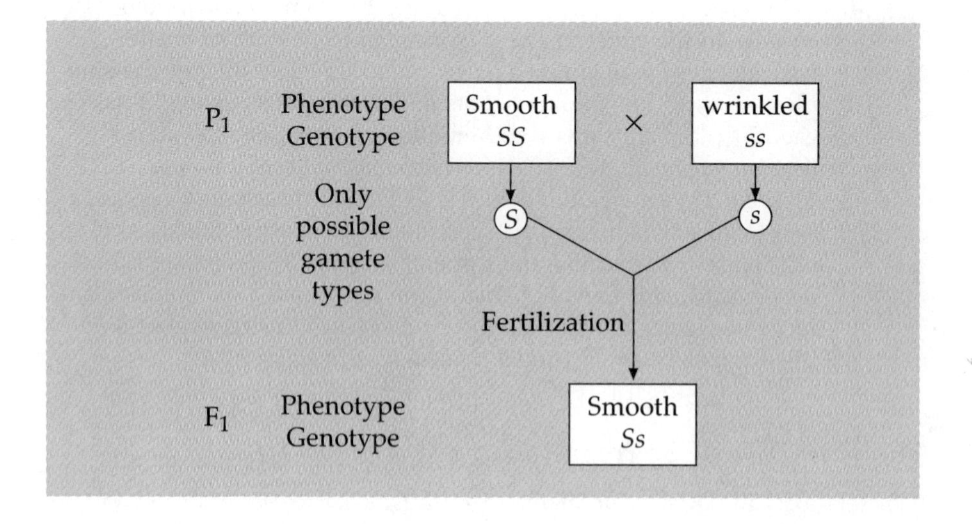

FIGURE 3.4 Distribution of factors for smooth and wrinkled traits in the P_1 and F_1 generations.

members of a pair of factors must separate, or segregate, from each other during gamete formation so that each mature gamete receives one or the other but not both factors. Using *S* to symbolize smooth and *s* to symbolize wrinkled, Mendel's experiments can be represented as in Figure 3.5. The separation of paired factors during gamete formation that results in each gamete receiving one member of a pair is called the principle of **segregation** and is a fundamental law of genetics.

Segregation
The separation of members of a gene pair from each other during gamete formation.

Mendel's experiments with the six other sets of traits can also be explained in this way. His reasoning also makes a further prediction: that two different genotypes should be present in the smooth plants of the F_2 generation. One third of the F_2 smooth plants should carry only smooth factors *(SS)* and give rise only to smooth plants when self-fertilized. The other two thirds of the F_2 smooth plants should carry hereditary factors for both smooth and wrinkled *(Ss)* and give rise to smooth and wrinkled progeny in a 3/4 to 1/4 ratio when self-crossed (Figure 3.6). In addition, the F_2 wrinkled plants should contain only factors for wrinkled and should give rise to only wrinkled progeny if self-fertilized. In fact, Mendel self-fertilized many plants

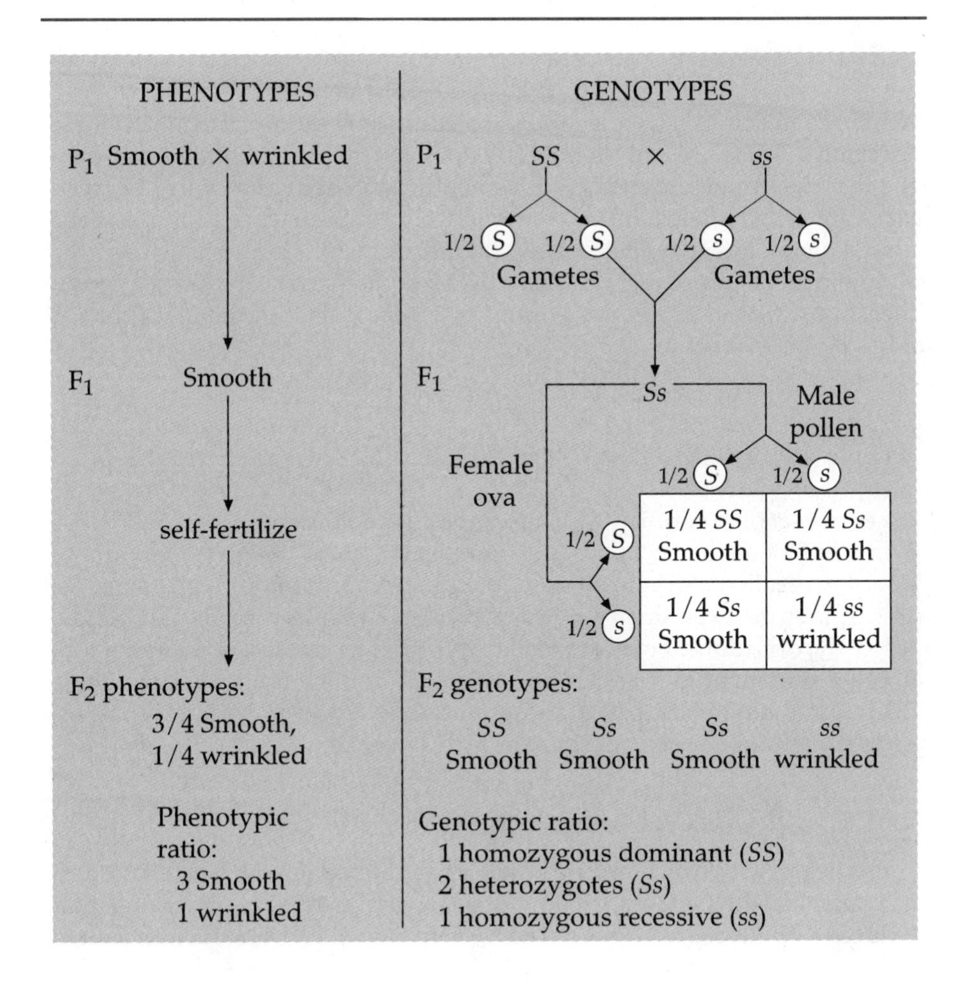

FIGURE 3.5 The segregation of factors into gametes and their random combination in multiple fertilization events. The results produce offspring that explain Mendel's observations on inheritance of seed shape in pea plants.

FIGURE 3.6 Self-crossing the F_2 plants demonstrates the presence of different genotypes among the phenotypically smooth F_2 plants.

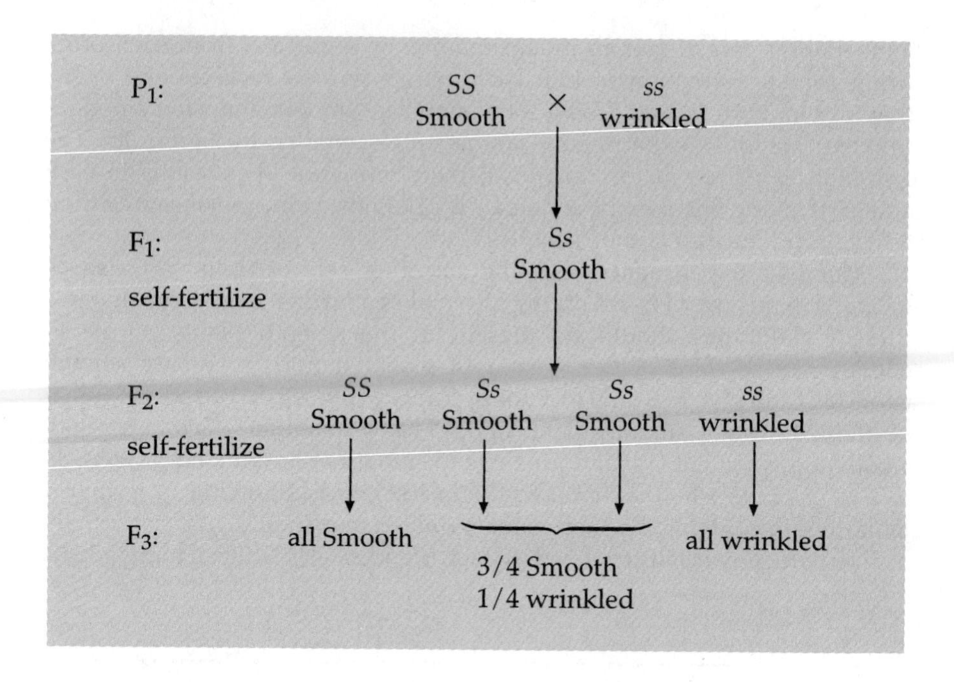

Gene
The fundamental unit of heredity.

Allele
One of the possible alternative forms of a gene, usually distinguished from other alleles by its phenotypic effects.

Homozygous
Having identical alleles for one or more genes.

Heterozygous
Carrying two different alleles for one or more genes.

Dihybrid cross
A cross between individuals who differ with respect to two gene pairs.

from the F_2 generation and several succeeding generations to confirm these predictions. Today we call Mendel's factors **genes.** The alternate forms of a gene that determine different traits (smooth, wrinkled) of a character (seed shape) are called **alleles.** In the example we have been discussing, the gene for seed shape has two alleles, smooth and wrinkled. Individuals carrying two identical alleles of a given gene *(Ss* or *ss)* are said to be **homozygous.** When two different alleles are present in a gene pair *(Ss)*, the individual is said to be **heterozygous.**

The Dihybrid Crosses

In a second set of experiments Mendel investigated the pattern of inheritance for two sets of characters simultaneously. A cross that involves two sets of characters is called a **dihybrid cross.** For the purposes of our discussion, we will consider his experiments with smooth and wrinkled seeds and yellow and green cotyledons. From the results of previous crosses, it is known that smooth is dominant to wrinkled and yellow is dominant to green (see Table 3.2). Smooth is represented by an uppercase *S*, wrinkled by a lowercase *s*, yellow by an uppercase *Y*, and green by a lowercase *y*.

Results and Conclusions

Mendel began with true breeding strains that differed in both seed shape and color: plants with smooth, yellow seeds were crossed with plants that produced wrinkled, green seeds. The F_1 offspring all had smooth and yellow

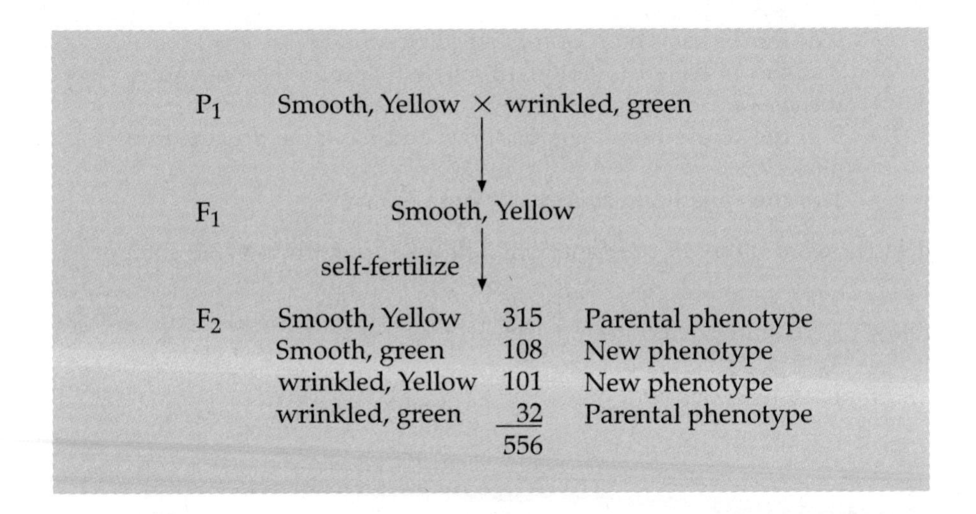

FIGURE 3.7 The phenotypic distribution in a dihybrid cross. Note that all combinations of phenotypes are represented, including the parental combinations and two new combinations.

seeds, confirming that smooth and yellow are dominant traits. He then self-fertilized the F_1 and produced an F_2 consisting of 586 plants that included not only the same phenotypes as the parents but also two new combinations of parental characters (Figure 3.7).

To determine the mode of inheritance of the two genes in a dihybrid cross, Mendel analyzed the results in the F_2 for each trait separately, as if the other trait were not present. If we consider only the seed shape (smooth or wrinkled) and ignore cotyledon color (yellow or green), we expect to obtain 3/4 smooth and 1/4 wrinkled offspring in the F_2. Analyzing the actual results, we find that the total number of smooth seeds is $315 + 108 = 423$ (Figure 3.7). The total number of wrinkled seeds is $101 + 32 = 133$. The proportion of smooth:wrinkled seeds (423:133) is close to 3:1. Similarly, in the F_2 there are $315 + 101 = 416$ yellow seeds and $108 + 32 = 140$ green seeds. These results are also close to a 3:1 distribution.

The Principle of Independent Assortment

Because the self-fertilized smooth, yellow plants of the F_1 generation gave rise to wrinkled, green offspring in the F_2, the F_1 seeds must have contained factors for both wrinkled and green. In other words, the F_1 plants were heterozygous for both seed shape and color. The genotype of the F_1 plants must have been *SsYy*, with the *s* and *y* alleles recessive to *S* and *Y*. To produce the combination of offspring seen in the F_2, the separation of *S* and *s* alleles must have taken place independently of the separation of *Y* and *y* alleles. As a result, the gametes formed in the F_1 plants must have contained all combinations of these alleles in equal proportions: *SY*, Sy, s*Y*, sy. Finally fertilization involving the four types of pollen and ova must have occurred at random, giving 16 possible combinations (Figure 3.8). These are displayed in a box first used by the plant geneticist R. C. Punnett and known as a Punnett square (Figure 3.8). Inspection of the 16 combinations shows that

- 9 of the 16 have both dominant alleles *S* and *Y*.
- 3 of the 16 have the dominant allele *S* and are homozygous for *yy*.
- 3 of the 16 are homozygous for *ss* and have the dominant allele *Y*.
- 1 of the 16 is homozygous for both *ss* and *yy*.

In other words, the 16 combinations fall into four phenotypic categories:

- 9/16 smooth and yellow
- 3/16 smooth and green
- 3/16 wrinkled and yellow
- 1/16 wrinkled and green

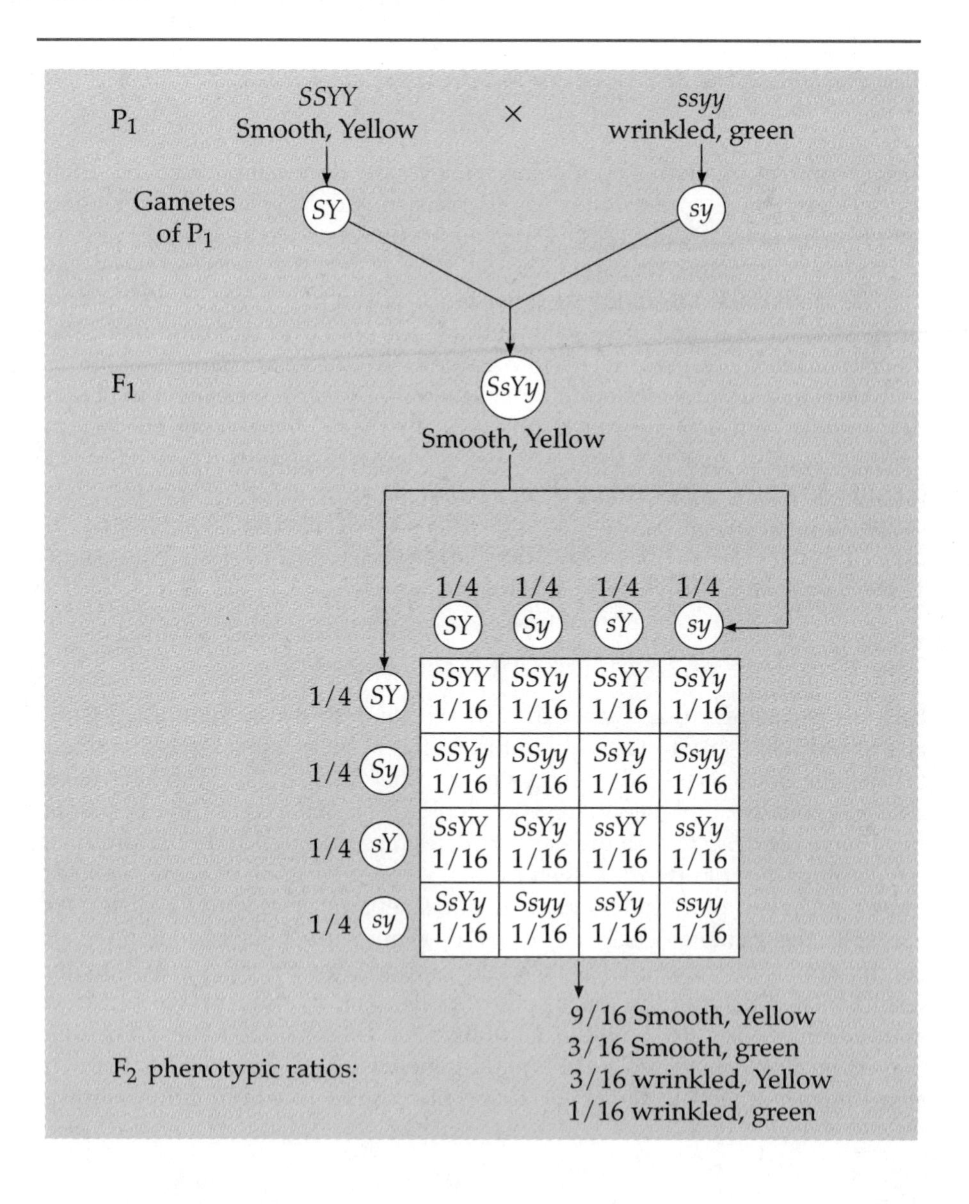

FIGURE 3.8 The independent assortment of factors in the F_1 from a dihybrid cross produces four different kinds of gametes. Random combination of these gametes in fertilization gives rise to 16 different combinations and four phenotypic classes.

Another way of looking at the outcome is to calculate the probability of having any two traits together in a single pea. If the chance that a pea in the F_2 will be smooth is 3/4 and the chance that a pea will be yellow is 3/4, then the chance that any pea will be smooth *and* yellow is $3/4 \times 3/4 = 9/16$. Similarly, if the chance for a yellow pea is 3/4 and that for a wrinkled pea is 1/4, the chance that a pea will be yellow and wrinkled is $3/4 \times 1/4 = 3/16$. By this method the combination of smooth and green is 3/16 (3/4 smooth $\times$ 1/4 green), and that for wrinkled and green is 1/16 (1/4 wrinkled $\times$ 1/4 green).

These phenotypic combinations correspond to the number of phenotypic classes seen in the F_2 and to the proportion of progeny seen in each class, resulting in a 9:3:3:1 ratio (see Concepts & Controversies, page 58–59). What this all means is that during gamete formation, alleles in one gene pair separate independently of the alleles belonging to other gene pairs and that gametes containing all combinations of alleles will be produced in equal numbers. This principle of **independent assortment** is another basic law of genetics.

Independent assortment
The random distribution of genes into gametes during meiosis.

Instead of using a Punnett square to determine the distribution and frequencies of phenotypes and genotypes in the F_2, we can also use what is known as a branch diagram, or forked-line method, which is based on probability. In the dihybrid F_2, the chance that a plant will be smooth is 3/4, and the chance that it will be wrinkled is 1/4. Since each trait is inherited independently, each smooth plant has a 3/4 chance of being yellow and a 1/4 chance of being green. The same is true for each wrinkled plant. Figure 3.9 (p. 60) shows how these probabilities combine to give the phenotypic ratios.

Mendel went on to experiment with the simultaneous inheritance of three or more factors and extended his findings on pea plants to similar studies on corn. In all cases the monohybrid and dihybrid ratios were the same as those obtained with the pea plant.

After 10 years of experimentation, Mendel presented his results at the February and March 1865 meetings of the Natural Science Society in Brno, Czechoslovakia. The text of these lectures was published the following year in the Proceedings of the Society. Although his papers were cited in several

TABLE 3.3 Chi-Square Analysis of Mendel's Data

Seed Shape	Cotyledone Color	Observed Numbers	Expected Numbers (based on a 9:3:3:1 ratio)	Difference *(d)* *(O – E)*
Smooth	Yellow	315	313	+2
Smooth	Green	108	104	+4
Wrinkled	Yellow	101	104	–3
Wrinkled	Green	32	35	–3

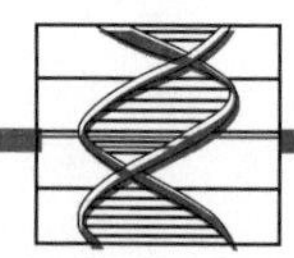

CONCEPTS & CONTROVERSIES

Evaluating Results—The Chi-Square Test

One of Mendel's innovations was the application of mathematics and combinatorial theory to biological research. This allowed him to predict the genotypic and phenotypic ratios in his crosses and to follow the inheritance of several traits simultaneously. If the cross involved two alleles of a gene (e.g., *A* and *a*), the expected outcome was an F_2 phenotypic ratio of 3*A*:1*a* and a genotypic ratio of 1*AA*:2*aA*:1*aa*. What Mendel was unable to analyze mathematically was how well the actual outcome of the cross fulfilled his predictions. He apparently realized this problem and compensated for it by conducting his experiments on a large scale, counting substantial numbers of individuals in each experiment in an attempt to reduce the chance of error.

Shortly after the turn of the century an English scientist named Karl Pearson developed a statistical test to determine whether the observed distribution of individuals in phenotypic categories is as predicted or whether the distribution observed occurs by chance. This simple test is regarded as one of the fundamental advances in statistics and is a valuable tool in genetic research. The method is known as the chi-square (χ^2) test (pronounced "kye square"). In use, this test requires several steps:

1. Record the observed numbers of organisms in each phenotypic class.
2. Calculate the expected values for each phenotypic class based on the predicted ratios.
3. If *O* is the observed number of organisms in a phenotypic class, or category and *E* is the expected number, calculate the difference *d* in each category by subtraction $(O - E) = d$. (Table 3.3)
4. For each phenotypic category, square the difference *d*, and divide by the number expected *(E)* in that phenotypic class.
5. Add all the numbers in step 4 to get the χ^2 value.

If there are no differences between the observed and the expected ratios, the value for χ^2 will be zero. The value of χ^2 will increase with the size of the difference between the observed and the expected classes. The formula can be expressed in the general form

$$\chi^2 = \Sigma\frac{d^2}{E}$$

Using this formula we can do what Mendel could not: analyze his data for the dihybrid cross involving wrinkled and smooth seeds and yellow and green cotyledons that produced a 9:3:3:1 ratio. In the F_2, Mendel counted a total of 556 peas; the number in each phenotypic class is the observed number (Table 3.3). Using the total of 556 peas, we can calculate that the expected number in each class for a 9:3:3:1 ratio would be 313:104:104:35 (9/16 of 556 is 313, 3/16 of 556 is 104, etc). Substituting these numbers into the formula, we obtain

$$\chi^2 = \frac{2^2}{313} + \frac{4^2}{104} + \frac{3^2}{104} + \frac{3^2}{35}$$
$$= 0.371$$

The χ^2 value is very low, confirming that there is very little difference between the number of peas observed and the number expected in each class. In other words, the results are close enough to the expectation that we need not reject them. The question remains, however, how much deviation is permitted from the expected numbers before we will decide that the observations do not fit our expectation that a 9:3:3:1 ratio will be fulfilled. To decide this we must have a way of interpreting the χ^2 value. We need to convert this value into a probability and to ask what is the probability that the calculated χ^2 value is acceptable? In making this calculation we must first establish something called degrees of freedom, *df*, which is one less than the number of phenotypic classes, *n*. In the dihybrid cross we expect four phenotypic classes, so the degrees of freedom are as follows:

$$df = n - 1$$
$$df = 4 - 1$$
$$df = 3$$

Next we can calculate the probability of obtaining the given set of results by consulting a probability chart (Table 3.4). First, find the line corresponding to a *df* value of 3. Look across on this line for the number corresponding to the χ^2 value. The calculated value is 0.37, which is between the columns headed 0.95 and 0.90. This means that we can expect a deviation of this magnitude at least 90% of the time when we do this experiment. In other words, we can be confident that our expectations of a 9:3:3:1 ratio are correct. In general, a *P* value of less than .05 means that the observations do not fit the expected distribution into phenotypic classes and that the expectation needs to be reexamined. The acceptable range of values is indicated by a line in Table 3.4. The use of p = 0.05 as the border for acceptability has been arbitrarily set.

In the case of Mendel's data, there is very little difference between the observed and expected results. Several writers have commented that Mendel's results fit the expectations too closely and that perhaps he adjusted his results to fit a preconceived standard.

In human genetics, the χ^2 method is very valuable and has wide applications. It is used in deciding modes of inheritance (autosomal or sex linked), deciding whether the pattern of inheritance shown by two genes indicates that they are on the same chromosome, and deciding whether marriage patterns have produced genetically divergent groups in a population.

TABLE 3.4 Probability Values for Chi-Square Analysis

	Probabilities								
df	0.95	0.90	0.70	0.50	0.30	0.20	0.10	0.05	0.01
1	.004	.016	.15	.46	1.07	1.64	2.71	3.84	6.64
2	.10	.21	.71	1.39	2.41	3.22	4.61	5.99	9.21
3	.35	.58	1.42	2.37	3.67	4.64	6.25	7.82	11.35
4	.71	1.06	2.20	3.36	4.88	5.99	7.78	9.49	13.28
5	1.15	1.61	3.00	4.35	6.06	7.29	9.24	11.07	15.09
6	1.64	2.20	3.83	5.35	7.23	8.56	10.65	12.59	16.81
7	2.17	2.83	4.67	6.35	8.38	9.80	12.02	14.07	18.48
8	2.73	3.49	5.53	7.34	9.52	11.03	13.36	15.51	20.09
9	3.33	4.17	6.39	8.34	10.66	12.24	14.68	16.92	21.67
10	3.94	4.87	7.27	9.34	11.78	13.44	15.99	18.31	23.21
	← Acceptable →								Unacceptable

From Table IV of Fisher, R., and Yates F. *Statistical Tables for Biological, Agricultural and Medical Research.* 6th ed. 1963. Longman Essex.

FIGURE 3.9 **The phenotypic and genotypic ratios of a dihybrid cross derived by the forked-line method.**

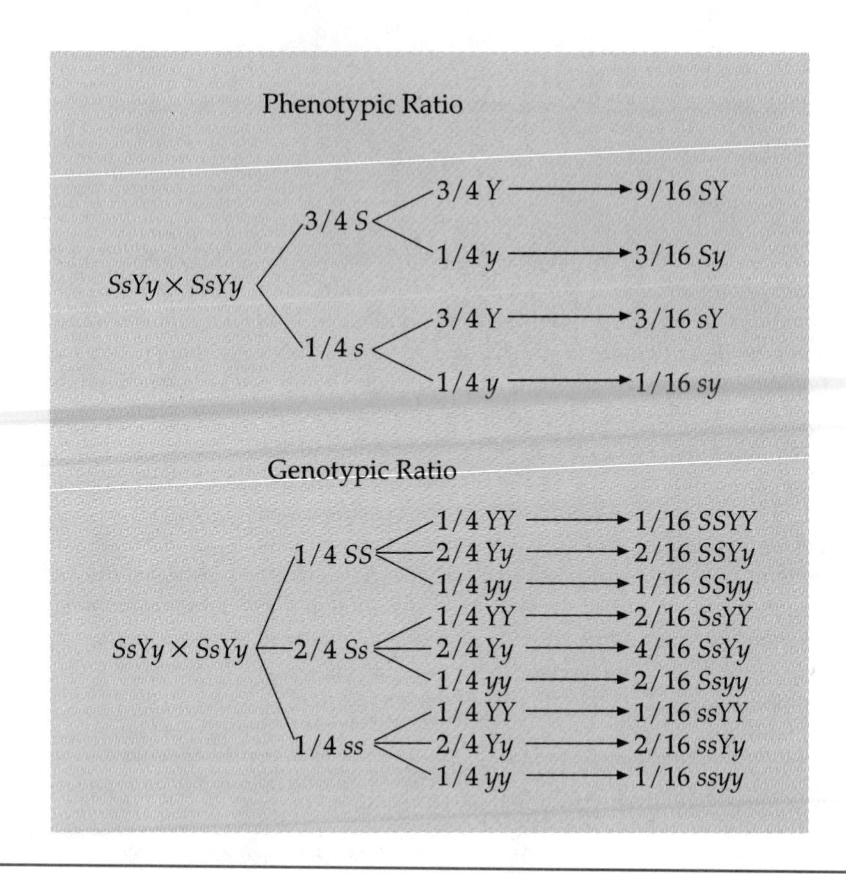

bibliographies, their significance was unappreciated. In 1900 three workers independently confirmed Mendel's discoveries, stimulating a great interest in the study of heredity.

Mendelian Inheritance in Humans

The principles of segregation and independent assortment discovered by Mendel were found to apply to the inheritance of traits in a wide range of organisms, including humans. The first Mendelian trait in humans was identified in 1905, and the discovery of others followed shortly thereafter. To illustrate the segregation of a trait in humans, let us consider the inheritance of a recessive phenotype, oculocutaneous albinism *(a)*, in which melanin pigment is lacking in skin, hair, and eyes (Figure 3.10). The normal allele *(A)* is dominant, and albinism is therefore expressed only in the homozygous recessive condition. In this example, both parents have a pigmented phenotype but are heterozygous for the recessive allele causing albinism. In each parent the dominant and recessive allele separate, or segregate, from each other at the time of gamete formation. Since each parent can produce two different types of gametes (one containing the dominant allele *A* and another carrying the recessive allele *a*), there are four possible combinations

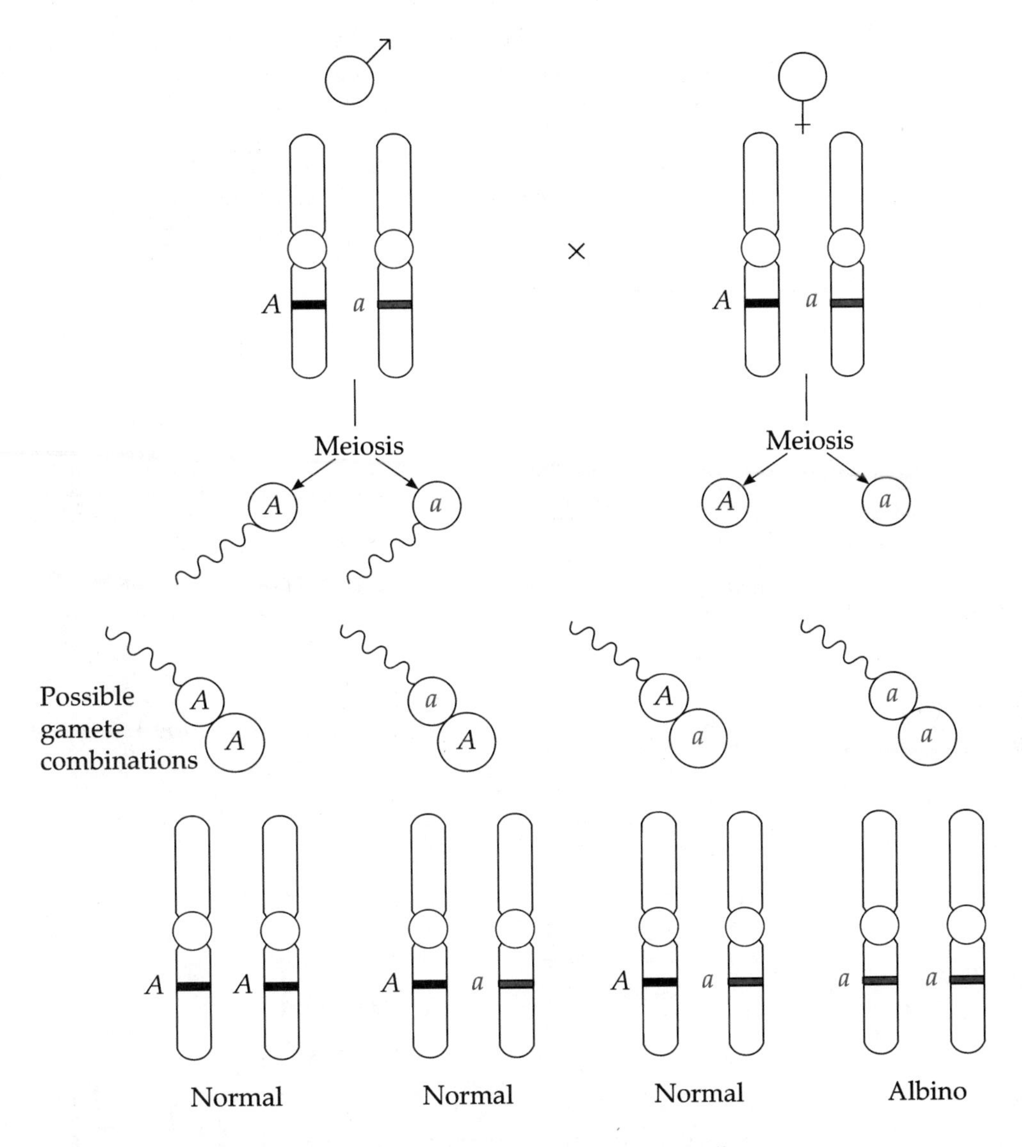

FIGURE 3.10 The segregation of albinism, a recessive trait in humans. In humans, as in pea plants, alleles located on homologous chromosomes separate, or segregate, from each other during gamete formation. (A = normal pigmentation; a = albinism.)

of fertilizations. These four types of fertilization events would result in a phenotypic ratio of 3 pigmented:1 albino, and a genotypic ratio of 1*AA*:2*Aa*:1*aa*. Remember, the expected phenotypic and genotypic ratios are the probabilities of a given outcome. This means that for each child, there is a 3/4, or 75%, chance that the child will have a pigmented phenotype, and there is a 1/4, or 25%, chance that the child will be an albino. Similarly, there is a 2/4, or 50%, chance that each child's genotype will be *Aa;* a 1/4, or 25%, chance it will be *AA;* and a 1/4, or 25%, chance that it will be *aa.*

The simultaneous inheritance of two traits in humans follows the Mendelian principle of independent assortment (Figure 3.11). Let us consider a case in which each parent is heterozygous for albinism *(Aa)* **and** for a

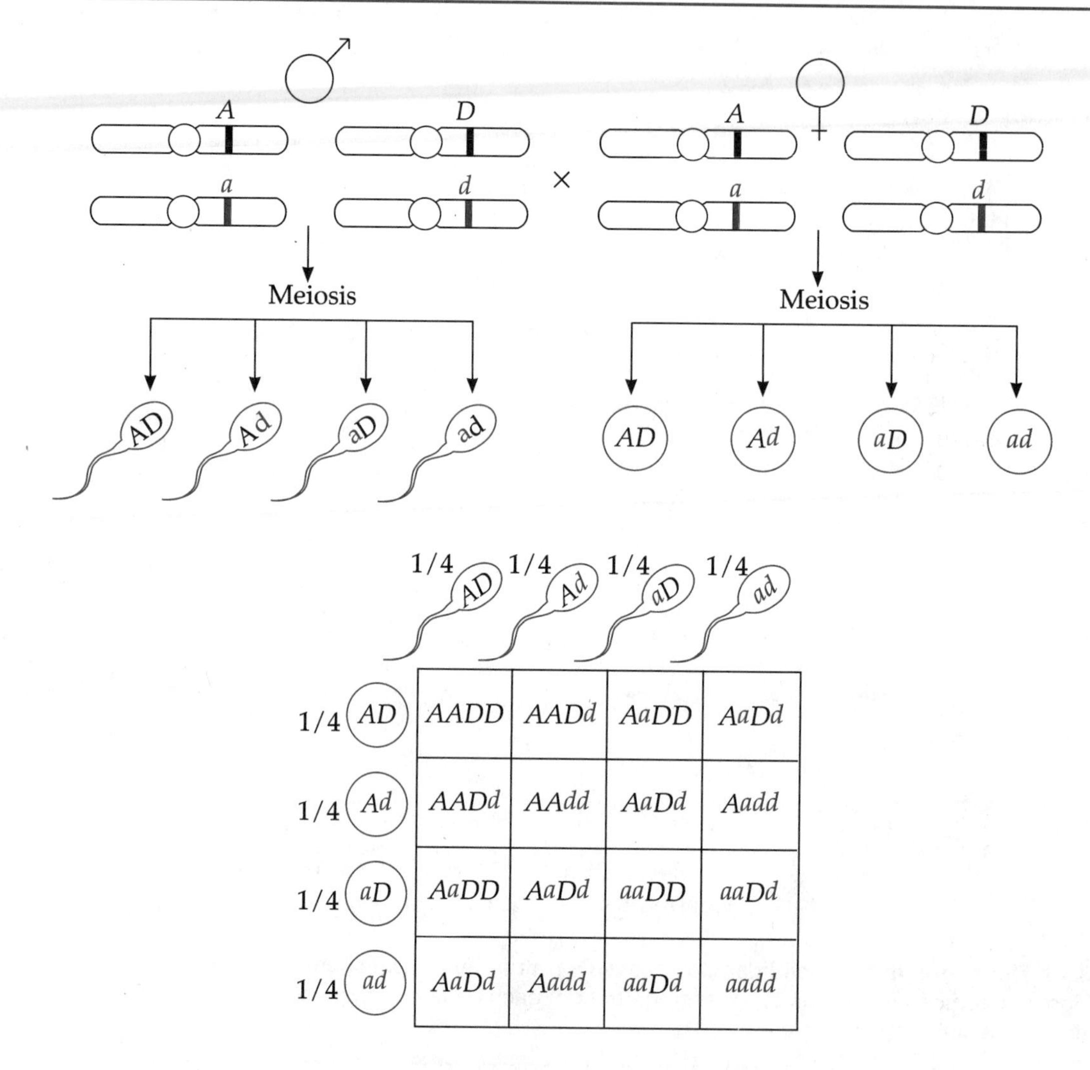

Phenotypic ratio: 9/16 *A_D_*: normal pigment, normal hearing
3/16 *A_dd*: normal pigment, deaf
3/16 *aaD_*: albino, normal hearing
1/16 *aadd*: albino, deaf

FIGURE 3.11 In humans, as in pea plants, members of gene pairs on different chromosomes assort into gametes independently of other gene pairs.

recessive form of hereditary deafness *(Dd)*. As in albinism, the normal allele *(D)* is dominant, and deafness will be expressed only in the homozygous condition *(dd)*. At the time of gamete formation, members of each gene pair assort into gametes independently of all other gene pairs. As a result, each parent will produce equal proportions of four different types of gametes. This results in 16 different combinations of fertilization events and four possible phenotypic classes (Figure 3.11). An examination of the possible genotypes shows that there is a 1/16 chance that a child will express both albinism and deafness. Knowledge of the principles of Mendelian inheritance is used in human genetics to diagnose genetic conditions and to predict the chances of having offspring afflicted with genetic disease.

Work in the early part of this century on humans and other organisms also turned up instances in which the phenotypic patterns of inheritance did not fit into the categories of dominant or recessive phenotypes. We will briefly look at some of these variations in phenotypic expression, and in later chapters we will consider the mechanisms associated with gene action and the production of phenotype.

Variations on Mendelian Ratios

Incomplete Dominance

In the examples we have examined, heterozygotes and dominant homozygotes have had the same phenotype. This type of inheritance is called *complete dominance*. In the case of **incomplete dominance,** the heterozygote has a phenotype that is more or less intermediate to those of either homozygote. One example is the inheritance of flower color in the plant *Mirabilis* (four o'clocks; Figure 3.12). A true breeding variety bearing red flowers crossed to a variety with white flowers will produce all pink flowers in the F_1. It appears from this cross that the parental traits have blended in the offspring. However, if the F_1 plants are self-crossed, they give rise to red, pink, and white offspring in a 1:2:1 ratio. The results of this cross can be explained in terms of Mendelian principles. In this case the phenotypic ratio of 1 red:2 pink:1 white is the same as the genotype ratio of 1*RR*:2*Rr*:1*rr*, and the heterozygote has a phenotype that is different from either homozygote.

Incomplete dominance
Expression of a phenotype that is intermediate between those of the parents.

In humans an example of incomplete dominance is sickle cell anemia, a hemolytic and often fatal heritable condition found primarily in individuals from equatorial Africa or European countries that border on the Mediterranean Sea. Because most black Americans originated in regions of Africa where sickle cell anemia is prevalent, about 8% to 12% of the American black population is heterozygous for this condition. Sickle cell anemia is caused by abnormal hemoglobin molecules that stack together to form aggregates, producing a distortion in the shape of red blood cells (see Figure 4.7). These sickle-shaped blood cells block small blood vessels, causing tissue destruction and often death. While the red blood cells of heterozygotes appear to be normal, they contain a mixture of normal and abnormal hemoglobin. Heterozygotes have no clinical symptoms of sickle cell anemia, and at this level the normal trait is dominant to the sickle cell trait. At the cellular level

FIGURE 3.12 **Flower color in *Mirabilis* illustrates incomplete dominance.**

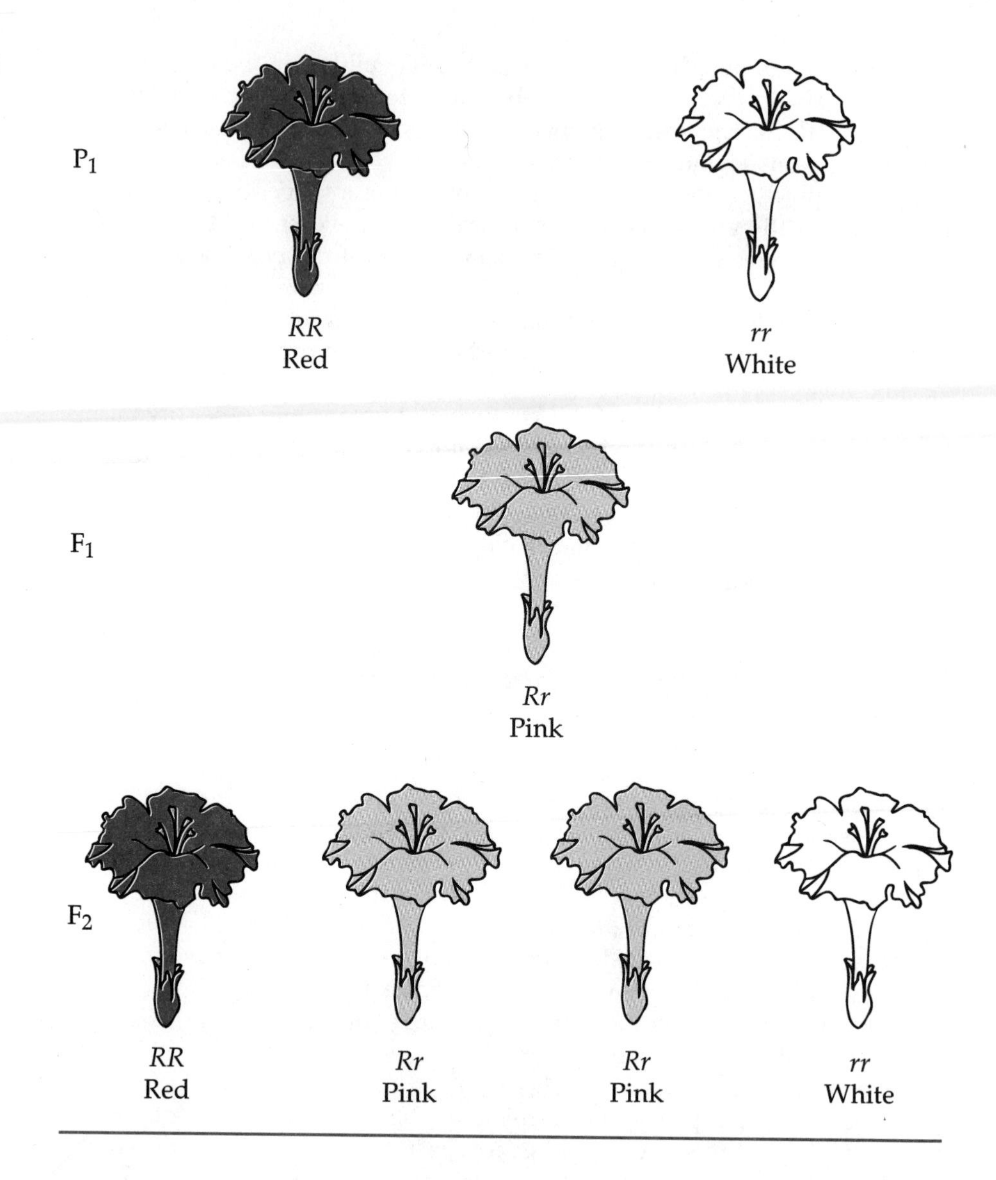

the blood cells of heterozygotes can be distinguished from those of either homozygote by a blood test. If a blood sample is subjected to very low oxygen levels, it causes moderate sickling in the heterozygote, while the recessive homozygote cells show severe sickling and cells from the dominant homozygote show no sickling. Thus at the cellular level the heterozygote phenotype can be distinguished from that of either homozygote and is an example of incomplete dominance.

Codominance

Codominance
Full phenotypic expression of both members of a gene pair in the heterozygous condition.

In a **codominant** situation both alleles present in the heterozygote are expressed fully and equally in the phenotype. The presence of complex polysaccharides on the outer surface of human red blood cell membranes is an example of such a trait. A gene L that controls the presence of a polysaccharide on the surface of red blood cells has two alleles, L^M and L^N. Each allele produces a slightly different form of this molecule on the cell surface. Individuals who are heterozygous ($L^M L^N$) have both the M and N polysaccharide on their red blood cells, while homozygotes have either M or N. Thus neither allele is dominant to the other.

As we will see in a later chapter, dominance refers to a phenotype, and terms such as incomplete dominance and codominance are often ambiguous and depend on the way in which the phenotype is examined. Nonetheless, these examples do not negate or disprove Mendel's discoveries but are simply variants in the way that genes control the production of phenotypes. The principles of allele segregation and independent assortment can still be used to accurately predict the genotypic ratios and the patterns of inheritance of incompletely dominant or codominant alleles. We will consider additional examples of Mendelian inheritance in humans in Chapter 4.

Mendel and Meiosis

In 1900 three scientific papers published independently over a span of three months confirmed Mendel's findings and called attention to the significance of his work. It is generally accepted that 1900 marks the beginning of genetics as a scientific discipline. When Mendel performed his experiments, the behavior of chromosomes in mitosis and meiosis had not been described. By 1900, however, cytology was a well-established field, and the principles of mitosis and meiosis were well known. As the fundamentals of Mendelian inheritance were confirmed to operate in many organisms, it soon became apparent that genes and chromosomes had much in common. Both chromosomes and genes occur in pairs, and in each pair one member is maternally derived and one is paternally derived. In meiosis, members of a chromosome pair separate from each other during gamete formation. In addition, the arrangement of a given chromosome pair at metaphase I in meiosis is independent of the way all other chromosome pairs are arranged. The result is that gametes receive different combinations of maternal and paternal chromosomes. Likewise, members of a gene pair assort themselves into gametes independently of all other gene pairs, producing gametes with all combinations of genes. Finally, fertilization restores the diploid number of chromosomes and two copies of each gene to the zygote.

In 1903 Walter Sutton and Theodore Boveri simultaneously noted the similarities and proposed that chromosomes were in fact the cellular components that physically contained the genes. This **chromosome theory of inheritance** has been confirmed by many experiments in the following decades and is one of the foundations of modern genetics. It is now known that each gene occupies a place, or **locus** (plural: loci), on a chromosome and that each chromosome carries many genes. In humans it is estimated that 50,000 to 100,000 genes are carried on the 23 different chromosomes. Although each gene may have different forms (alleles), any normal individual carries only two such alleles, since he or she has only two copies of each chromosome. Obviously different individuals can have different

Chromosome theory of inheritance
The theory that genes are carried on chromosomes and that the behavior of chromosomes during meiosis is the physical explanation for Mendel's observations on the segregation and independent assortment of genes.

Locus
The position occupied by a gene on a chromosome.

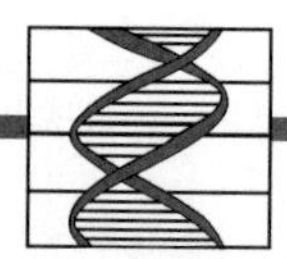

Concepts and Controversies

Solving Genetics Problems

In solving genetics problems, several steps must be followed to ensure success:

1. Analyze each problem carefully to determine what information is provided and what information is requested.
2. Translate the words of the problems into symbols, and assign a meaning to each symbol.
3. Solve the problem using logic and, if necessary, algebra.

One of the basic problems in Mendelian inheritance provides some information about the parental generation (P_1) and asks you to come to some conclusions about the genotypes or phenotypes of the F_1 or F_2 generations. The solution of this kind of problem involves a step-by-step approach:

1. Carefully read the problem, and establish the genotype of each parent.
2. Based on the assigned genotypes, determine what types of gametes can be formed by each parent and in what ratios.
3. Unite these gametes in all combinations. Use a Punnett square if necessary. This will automatically give you all possible genotypes and their ratios for the F_1 generation.
4. If the problem involves an F_2 generation, use all combinations of F_1 individuals as parents for the F_2 generation, and repeat steps 2 and 3 to derive the genotypes and phenotypes of the F_2 generation.

As an example, consider the following problem. The recessive allele *wrinkled (s),* when homozygous, causes peas to have a wrinkled appearance. The dominant allele *Smooth (S)* causes peas to appear smooth when homozygous *(SS)* or heterozygous *(Ss).* What phenotypic ratio would you expect in the F_1 generation in the following cross? One parent is a plant that bears wrinkled seeds. The other is a plant that bears smooth seeds and is the offspring of a true-breeding parent with smooth seeds.

The solution to this problem depends on an understanding of the principles of segregation and the relationship between dominant and recessive phenotypes. The first step is to determine the genotypes of the parental plants. For this, the following information should be considered:

1. Since one parent bears wrinkled seeds and this trait is recessive, this plant must be homozygous for the recessive allele; that is, its genotype is *ss.* The other parental plant bears smooth seeds, and since this plant is the offspring from a true-breeding line of plants with smooth peas, its genotype must be homozygous *SS.*
2. The cross in question is therefore $SS \times ss$. The gametes each parent can make are shown below:

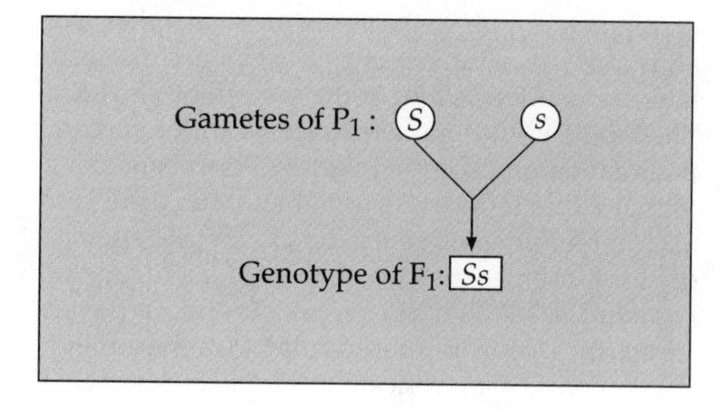

3. Uniting these in fertilization gives the following result:

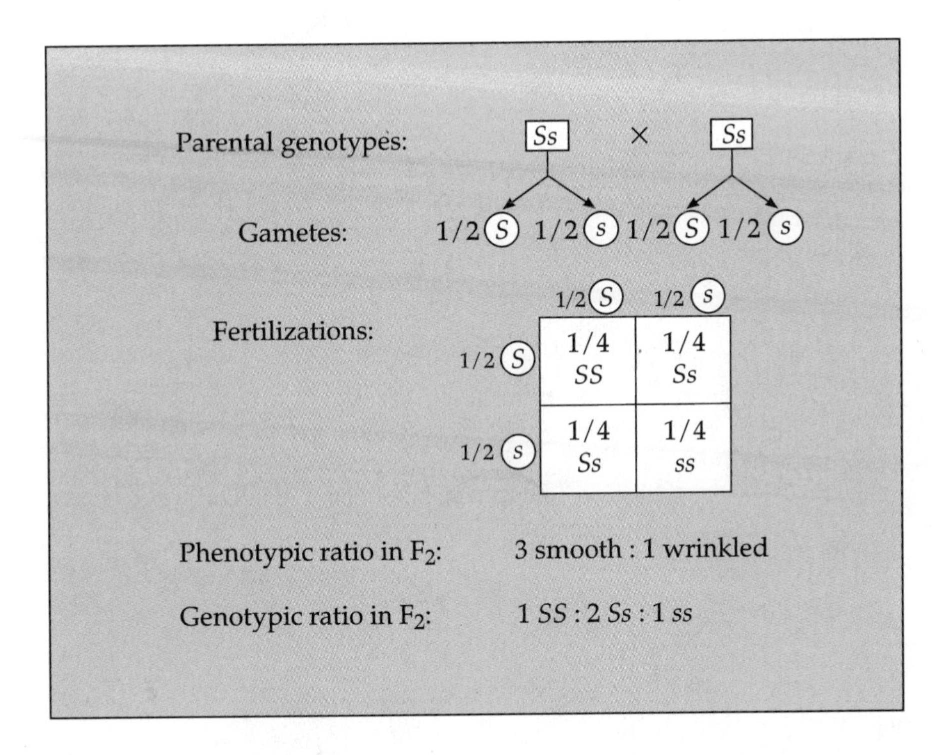

4. In this cross, all the peas produced in the F_1 generation will be phenotypically smooth and genotypically heterozygous *(Ss).* Using these as parents, what are the phenotypes and genotypes of the F_2 generation? To answer this question, first determine what types of gametes and in what ratios each parent can produce. Then combine these gametes in all combinations to produce the F_2 generation. The heterozygous *Ss* parents will each make two kinds of gametes in equal proportions. The gametes and the phenotypic and genotypic combinations in the F_2 generation are shown below:

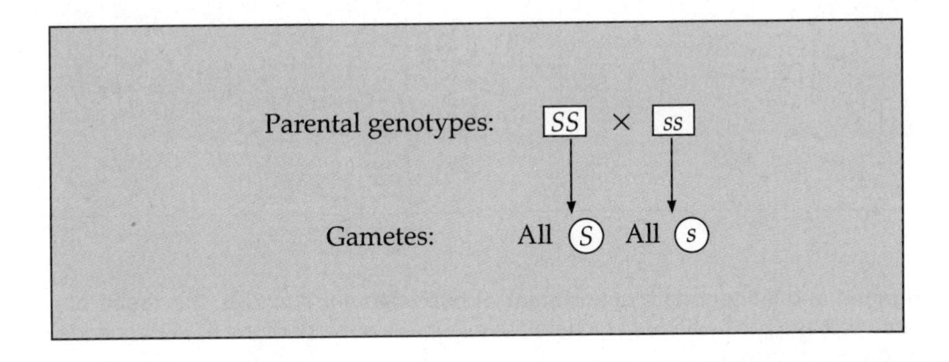

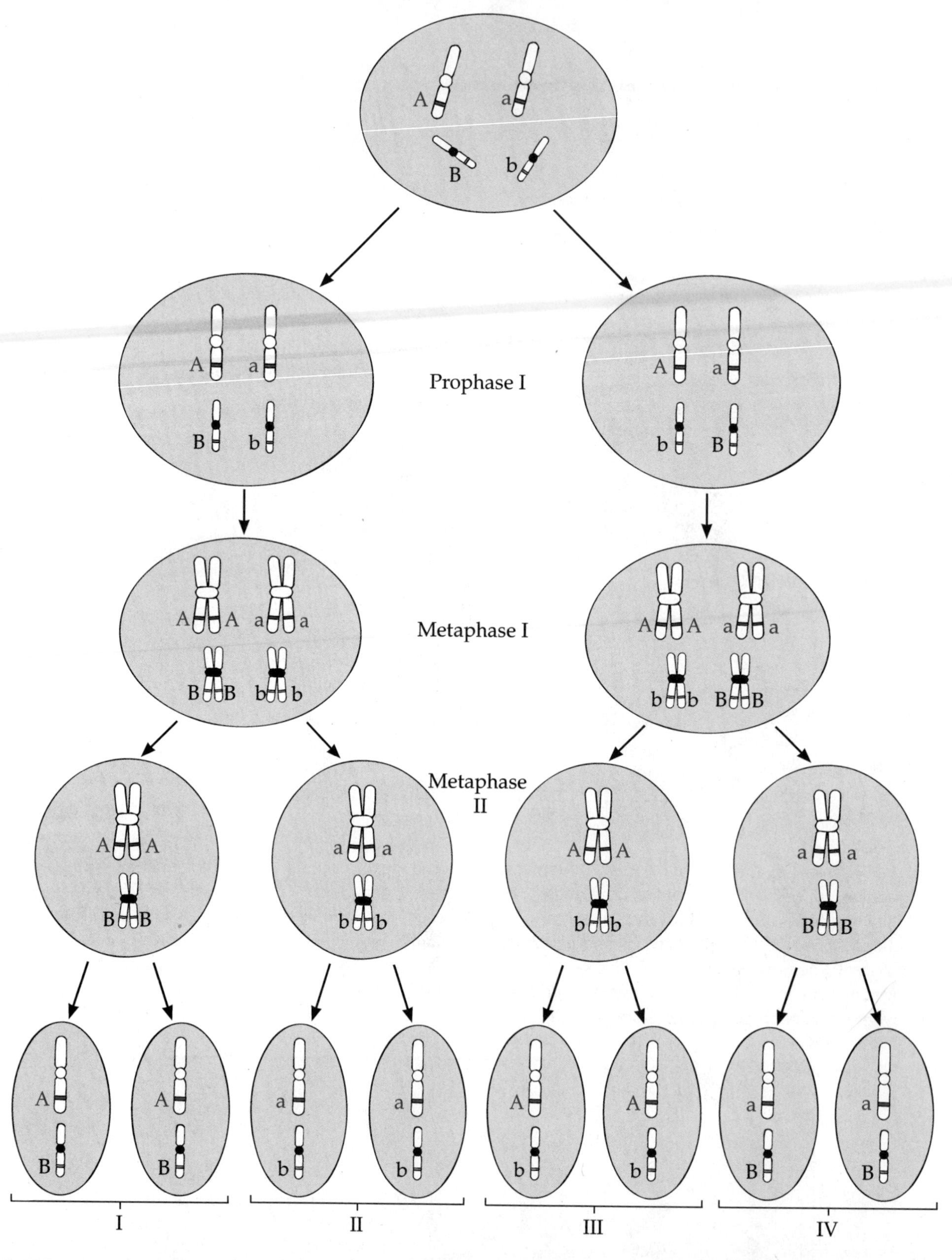

FIGURE 3.13 Segregation and independent assortment of two traits for A and B, the result of chromosome behavior in meiosis. Note that the production of all gamete types in equal proportions (I, IV) depends on the random arrangement of homologues at metaphase I.

combinations of alleles. Figure 3.13 shows the distribution of chromosomes and genes at meiosis.

SUMMARY

1. In the mid-19th century, Gregor Mendel carried out 10 years of experiments on the inheritance of traits in the garden pea, work that provided the foundation for the science of genetics.

2. Mendel postulated that the phenotype is controlled by factors now known as genes. The factors exist in pairs, show dominant/recessive relationships, and undergo separation, or segregation, during the formation of gametes, so that each mature gamete receives one or the other member of a gene pair.

3. In later experiments Mendel discovered that members of a gene pair separate, or segregate, independently of other gene pairs. This independent assortment leads to the formation of all possible combination of gametes.

4. Later work by others showed that Mendelian factors represent alleles, or alternative forms, of a gene and that alleles occupy comparable sites, or loci, on members of a chromosome pair (homologous chromosomes). As members of a chromosome pair separate from each other, the alleles they carry are also separated from each other.

5. The confirmation of Mendel's findings in 1900 coupled with cytologic work on the behavior of chromosomes during meiosis marked the start of genetics as an organized body of knowledge.

6. Although apparent exceptions to the ratios predicted by Mendel were discovered, it is clear that the principles he uncovered can be used to predict the genotypic and phenotypic ratios in these cases and in cases where meiosis is present.

QUESTIONS AND PROBLEMS

1. An organism has the following genotype: *AABb*. What type of gametes can be produced, and in what proportions?

2. Suppose that this organism is mated to one with the genotype *Aabb*. What are the predicted phenotypic ratios and genotypic ratios of the offspring?

3. A woman is heterozygous for two genes. How many different types of gametes can she produce, and in what proportions?

4. An unspecified character controlled by a single gene is examined in pea plants. Only two phenotypic states exist for this trait. One phenotypic state is completely dominant to the other. A heterozygous plant is self-crossed. What proportion of the progeny plants exhibiting the dominant phenotype are homozygous?

5. Two traits are simultaneously examined in a cross of two pure-breeding pea-plant varieties. Pod shape can be either swollen or pinched. Seed color can be either green or yellow. A swollen, green plant is crossed with a pinched, yellow plant, and a resulting F_1 plant is self-crossed. A total of 640 F_2 progeny are phenotypically categorized as follows:

swollen, yellow	360
swollen, green	120
pinched, yellow	120
pinched, green	40

a. What is the phenotypic ratio observed for pod shape? seed color?
b. What is the phenotypic ratio observed for both traits considered together?
c. What is the dominance relationship for pod shape? seed color?
d. Deduce the genotypes of the P_1 and F_1 generations.

6. Sickle cell anemia (SCA) is a human genetic disorder caused by a recessive allele. A couple plan to marry and want to know the probability that they will have an affected child. With your knowledge of Mendelian inheritance, what can you tell them if (a) both are normal, but each has one affected parent and the other parent has no family history of SCA; and (b) the man is affected by the disorder, but the woman has no family history of SCA.

7. Consider the following cross in pea plants, where smooth seed shape is dominant to wrinkled and yellow seed color is dominant to green. A plant with smooth, yellow seeds is crossed to a plant with wrinkled, green seeds. The peas produced by the offspring are all smooth and yellow. What are the genotypes of the parents? What are the genotypes of the offspring?

8. Consider another cross involving the genes for seed color and shape. As before, yellow is dominant to green, and smooth is dominant to wrinkled. A plant with smooth, yellow seeds is crossed to a plant with wrinkled, green seeds. The peas produced by the offspring are as follows: 1/4 are smooth, yellow, 1/4 are smooth, green, 1/4 are wrinkled, yellow, and 1/4 are wrinkled, green. What is the genotype of the smooth, yellow parent? What are the genotypes of the four classes of offspring?

9. Stem length in pea plants is controlled by a single gene. Consider the cross of a true-breeding long-stemmed variety to a true-breeding short-stemmed variety where long stems are completely dominant.
   a. 120 F_1 plants are examined. How many plants are expected to be long stemmed? short stemmed?
   b. Assign genotypes to both P_1 varieties and to all phenotypes listed in (a).
   c. A long-stemmed F_1 plant is self-crossed. Of 300 F_2 plants, how many should be long stemmed? short stemmed?
   d. For the F_2 plants mentioned in (c), what is the expected genotypic ratio?

10. A pea plant that is *AaBb* is self-crossed. A 9:3:3:1 phenotypic ratio is obtained, as expected in a simple dihybrid cross. What is the underlying genotypic ratio? (Assign genotypes to each element of the ratio.)

11. Another character of pea plants amenable to genetic analysis is flower color. Imagine that a true-breeding purple-flowered variety is crossed to a pure line having white flowers. The progeny are exclusively pink flowered. Diagram this cross, including genotypes for all P_1 and F_1 phenotypes. What is the mode of inheritance? Let F = purple and f = white.

12. In determining the mode of inheritance of traits, it is usually informative to self-cross a heterozygote and to examine the progeny phenotypes. Yet most genetics problems define this organism as an F_1, with an apparently superfluous parental generation. Explain why this generation is not "excess baggage."

13. In peas, straight stems (*S*) are dominant to gnarled (*s*), and round seeds (*R*) are dominant to wrinkled (*r*). The following cross (a test cross) is performed: *SsRr* × *ssrr*. Determine the expected progeny phenotypes and what fraction of the progeny should exhibit each phenotype.

14. A strange pea-plant variant is found that has orange flowers. A self-cross of this plant yields the following phenotypes:

red flowers	30
orange flowers	62
yellow flowers	33

What mode of inheritance can you infer for flower color in this pea plant variant?

15. Suppose Mendel had ignored the advice contained in Ockham's razor. Based on a crude estimate of pollen and ovule size, he postulates that one hereditary factor is contributed by the pollen and three hereditary factors are contributed by the ovule. The anthers from a pure variety of pea possessing round seeds are applied to the stigma of a true-breeding variety having wrinkled seeds. Round seeds are completely dominant to wrinkled seeds, and independent assortment is operating in this cross.
   a. 200 F_1 plants are examined. How many plants are expected to bear round seeds? wrinkled?
   b. An F_1 plant bearing round seeds is self-crossed. Of 400 F_2 plants, how many are expected to bear round seeds? wrinkled?

16. In pea plants, long stems are dominant to short stems, purple flowers are dominant to white, and round seeds are dominant to wrinkled. Each trait is determined by a single, different gene. A plant that is heterozygous at all three loci is self-crossed, and 2048 progeny are examined. How many of these plants would you expect to be long stemmed with purple flowers, producing wrinkled seeds?

17. A pea plant exhibits the dominant phenotype for two traits: its seed color is yellow and its pods are swollen. A self-cross produces 178 progeny with the following phenotypes:

yellow, swollen	132
yellow, pinched	46

From this information, can you infer the relevant genotype of the plant that was self-crossed?

18. A plant geneticist is examining the mode of inheritance of flower color in two closely related species of exotic plants. Analysis of one species has resulted in the indentification of two pure-breeding lines: one produces a distinct red flower and the other produces either a very pale yellow color or no color at all, he cannot be sure. A cross of these varieties produces all pink-flowered progeny. The second species exhibits similar pure-breeding varieties; that is, one variety produces red flowers, and the other produces an ambiguous yellow or albino flower. A cross of these two varieties, however, produces orange-flowered progeny exclusively. Analyze the mode of inheritance of flower color in these two plant species.

19. Think about this one carefully. A recessively inherited form of albinism causes affected individuals to lack pigment in their skin, hair and eyes. For this problem we will assume that red hair color in humans is inherited as a recessive trait and that brown hair is dominant. An albino whose parents both have red hair has two children with someone who is normally pigmented and has brown hair. The brown-haired partner has one parent with red hair. The first child is normally pigmented and has brown hair. The second child is albino. What is the hair color (phenotype) of the albino parent? What is the genotype of the albino parent for hair color? What is the genotype of the brown-haired parent with respect to hair color? skin pigmentation? What is the genotype of the first child for hair color and skin pigmentation? What are the possible genotypes of the second child for hair color? What is the phenotype of the second child for hair color? Can you explain this?

20. Discuss the pertinent features of meiosis that provide a physical correlate to Mendel's abstract genetic laws of random segregation and independent assortment.

21. If you are told that being right- or left-handed is heritable and that a right-handed couple is expecting a child, can you conclude that the child will be right-handed?

For Further Reading

Corcos, A. F., and Monaghan, F. 1985. Role of de Vries in the recovery of Mendel's work. I. Was de Vries really an independent discoverer of Mendel? *J. Hered.* **76:** 187–190.

Dahl, H. 1993. Things Mendel never dreamed of. Med. J. Aust. **158:** 247–252.

Dunn, L. C. 1965. *A Short History of Genetics.* New York: McGraw-Hill.

Edwards, A. W. 1986. Are Mendel's results really too close? *Biol. Rev. Cambridge Philos. Soc.* **61:** 295–312.

Finney, D. J. 1980. *Statistics for Biologists.* New York: Chapman & Hall.

Gasking, E. B. 1959. Why was Mendel's work ignored? *J. Hist. Ideas* **20:** 62–84.

George, W. 1975. *Gregor Mendel and Heredity.* London: Priory Press.

Hartl, D. and Orel, V. 1992. What did Gregor Mendel think he discovered? Genetics **131:** 245-253.

Heim, W. G. 1991. What is a recessive allele? *Amer Biol Teacher* **53:** 94–97.

Mather, K. 1965. *Statistical Analysis in Biology.* London: Methuen.

Monaghan, F. V., and Corcos, A. F. 1985. Mendel, the empiricist. *J. Hered.* **76:** 49–54.

Orel, V. 1973. The scientific milieu in Brno during the era of Mendel's research. *J. Hered.* **64:** 314–318.

Orel, V. 1984. *Mendel.* New York: Oxford University Press.

Piegorsch, W.W. 1990. Fisher's contributions to genetics and heredity, with special emphasis on the Gregor Mendel controversy. Biometrics **46:** 915-924.

Pilgrim, I. 1986. A solution to the too-good-to-be-true paradox and Gregor Mendel. *J. Hered.* **77:** 218–220.
Sandler, I., and Sandler, L. 1985. A conceptual ambiguity that contributed to the neglect of Mendel's paper. *Publ. Stn. Zool. Napoli* **7:** 3–70.
Stern, C., and Sherwood, E. 1966. *The Origins of Genetics: A Mendel Sourcebook.* San Francisco: Freeman.
Voipio, P. 1990. When and how did Mendel become convinced of the idea of general, successive evolution? Hereditas **113:** 179-181.
Voller, B. R., ed. 1968. *The Chromosome Theory of Inheritance. Classic Papers in Development and Heredity.* New York: Appleton-Century-Crofts.
Weiling, F. 1991. Historical study: Johann Gregor Mendel 1822-1884. Am. J. Med. Genet. **40:** 1-25.

CHAPTER 4

MENDELIAN INHERITANCE IN HUMANS

CHAPTER OUTLINE

Was Abraham Lincoln, the 16th president of the United States, affected with a genetic disorder? Several writers have speculated that Lincoln had a genetic disease known as Marfan syndrome. The evidence offered to support this idea is based on Lincoln's physical appearance and the report of Marfan syndrome in a distant relative.

Photographs, written descriptions, and medical reports are available to provide ample information about Lincoln's physical appearance. He was 6 ft 4 in tall and thin, weighing between 160 and 180 lb for most of his adult life. He had long arms and legs, with large, narrow hands and feet. Contemporary descriptions of his appearance indicate that he was stoop-shouldered, loose jointed, and walked with a shuffling gait. In addition, he wore eyeglasses to correct a visual problem.

Lincoln's appearance and ocular problems are suggestive of Marfan syndrome, a genetic condition that affects the connective tissue of the body, resulting in a tall, thin individual often affected with a shifted lens in the eye, blood vessel defects, and loose joints.

In 1960 a man diagnosed as having Marfan syndrome was determined to have ancestors in common with Lincoln (the common ancestor was Lincoln's great-great-grandfather). Added to the evidence based on physical appearance, the family history seems to strongly suggest that Lincoln had Marfan syndrome.

Others strongly disagree with this speculation, arguing that the length of Lincoln's extremities and the proportions of his body were well within the normal limits for tall, thin individuals. In addition, although Lincoln had visual problems, an examination of his eyeglasses indicates that he was farsighted, while those with the classical form of Marfan syndrome are nearsighted. Lastly, Lincoln showed no outward signs of problems with major blood vessels such as the aorta.

Did Lincoln really have Marfan syndrome, and should we care? Interest in this issue probably grows from our interest in the lives of historic figures and our fascination with the intimate details of the lives of the famous and the infamous. At the present time there is no solid evidence to suggest that Lincoln had Marfan syndrome, although molecular testing on bone and hair fragments (discussed below) have been proposed. For now we are left only with speculation and inferential reasoning.

The more important question for this chapter is how we can tell when someone is affected with a genetic disorder. Lincoln's family history shows only one documented case of Marfan syndrome in nine generations. Is this enough to decide that he was also affected with this genetic condition?

In this chapter we will show that the principles of inheritance discovered by Mendel in peas also apply to humans. Even though the methods employed in human genetics differ significantly from those used in other organisms, the rules for the inheritance of traits in humans are the same as those for pea plants.

Humans as a Subject for Genetics

Pea plants were selected by Mendel for two primary reasons. First, they can be crossed in all desired combinations; second, each cross is likely to produce large numbers of offspring. In particular, the number of offspring is an important consideration in determining the mode of inheritance. Humans as a subject for genetic analysis satisfy neither of these criteria. In pea plants it is easy to carry out crosses between plants with purple flowers and plants with white flowers and to repeat this cross as often as necessary. For obvious reasons, controlled human matings cannot be made, as in experimental genetics. One cannot ask all albino humans to mate with homozygous normally pigmented individuals and have their progeny interbreed to produce an F_2. In fact, for the most part human geneticists must base their work on matings that have already taken place, whether or not such matings would be the most genetically informative.

Compared with the progeny that can be counted in a single cross with peas, humans produce very few offspring, and these usually represent only a portion of the genetic combinations that exist. If two heterozygous pea plants are crossed ($Aa \times Aa$), about three fourths of the offspring will express the dominant allele *A*, while the recessive allele will be expressed in the remaining one fourth of the progeny. Mendel was able to count hundreds and sometimes thousands of offspring from such a cross and to record progeny in all expected phenotypic classes, establishing a ratio of 3:1. As a parallel, consider two humans, each of whom is phenotypically normal. Suppose this couple have two children, one of whom is a son affected by a genetic disorder. The ratio of phenotypes in this case is 1:1. This makes it difficult to decide whether the trait is carried on an autosome or a sex chromosome and whether it is controlled by a single gene or by two or more genes. One way around this difficulty is to pool results from several families, increasing the number of offspring studied. This, however, introduces biases that must be compensated for in analyzing the results.

These examples should serve to demonstrate that the basic methods of human genetics are observational rather than experimental and require the reconstruction of events that have already taken place rather than the design and execution of experiments to directly test a hypothesis. One of the most basic methods of human genetics is to follow a trait for several generations in a family to infer its mode of inheritance. For this purpose the geneticist constructs family trees, or pedigrees.

Pedigree Analysis

A **pedigree chart** is an orderly presentation of family information in easily readable form. From such a family tree the inheritance of a trait can be followed through several generations, and the genetic status of family members can be determined. Using the principles of Mendelian inheritance, we may then be able to determine whether the trait behaves as a dominant or recessive gene and whether the gene in question is located on an autosome or a sex chromosome. Pedigrees use a set of standardized symbols, many of which are borrowed from genealogy. In constructing a pedigree chart, males are represented by squares (□) and females by circles (○). An individual who exhibits a trait in question is represented by a filled symbol (■ or ●). Heterozygotes, when known, are indicated by half-filled symbols (◨ or ◑). The relationships between individuals in a pedigree is indicated by a series

Pedigree chart
A diagram listing the members and ancestral relationships in a family and used in the study of human heredity.

of lines. A horizontal line between two symbols represents a mating (□—○). Matings between brother and sister or between close relatives are known as consanguineous matings and are symbolized by a double horizontal line. The offspring, listed from left to right in birth order, are connected to each other by a horizontal line (○ ○ □) and to the parents by a vertical line:

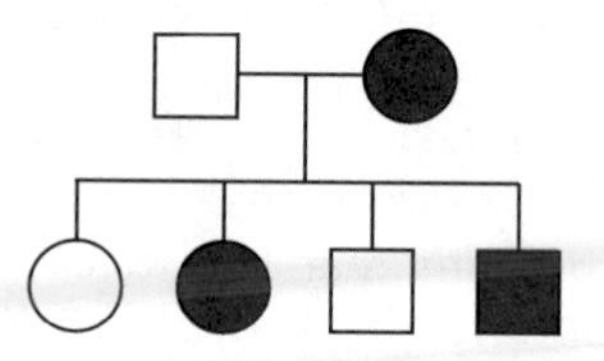

If twins are present as offspring, they are identified either as identical (monozygotic) twins (○○), arising from a single egg, or nonidentical (dizygotic) twins (□○), arising from the fertilization of two eggs.

To identify individuals in a pedigree, a numbering system is employed. Each generation is indicated by a roman numeral, and within a generation, each individual is numbered by birth order using arabic numbers:

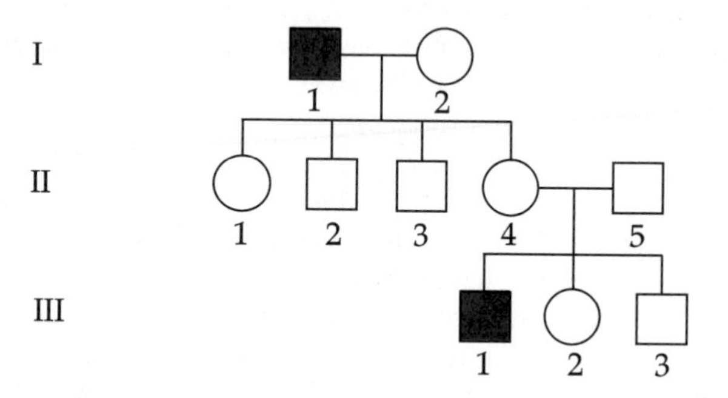

Propositus
An individual affected with a genetic disorder that led to the construction of a pedigree.

Often pedigrees are constructed after a family member afflicted with a genetic trait has been identified. This individual, known as a **propositus** (female: proposita), or **proband,** is indicated on the pedigree by an arrow:

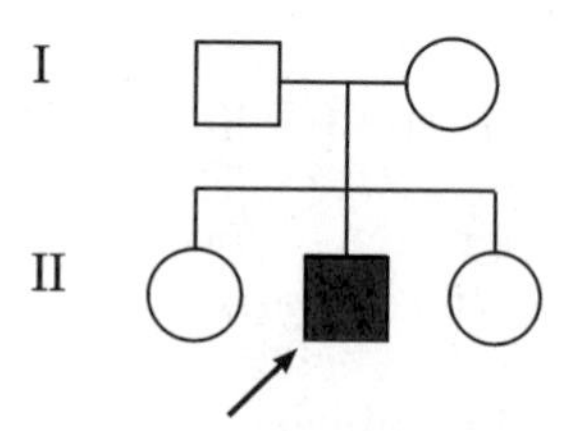

Because pedigree analysis is essentially a reconstruction of a family history, details about earlier generations may be uncertain. If the sex of an offspring is unknown or unimportant, this is indicated by a (◇). In some cases the spouses in a pedigree are omitted if they are not essential to the inheritance of the trait. If there is doubt that a family member possessed the

TABLE 4.1 Symbols Used in Pedigree Analysis

Male

Female

Mating

Mating between relatives (consanguineous)

I

II

1 2 3

Offspring listed in birth order. Roman numberals symbolize generations. Arabic numbers symbolize birth order within generation.

Monozygotic twins

Dizygotic twins

Offspring of unknown sex

Aborted or stillborn offspring

Deceased offspring.

or

Affected individual

or

Propositus (male) or proposita (female); first case in family that was identified.

or

Heterozygotes

X-linked carrier

d.1910 d.1932

Indicates date of death

?

Questionable whether individual has trait

trait in question, this is indicated by a question mark within the symbol. The symbols and terminology used in the construction of pedigrees are summarized in Table 4.1.

Once information on inheritance of a trait over several generations in a family has been collected, a pedigree is constructed. The information in the pedigree is analyzed to determine the mode of inheritance of the trait. The possible modes of inheritance are:

1. autosomal recessive
2. autosomal dominant
3. mitochondrial
4. X-linked dominant
5. X-linked recessive
6. Y-linked

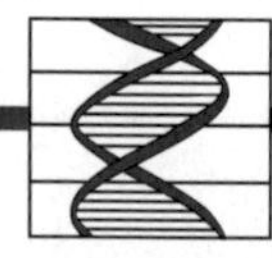

CONCEPTS & CONTROVERSIES

Is Autism a Genetic Disorder?

Although pedigree analysis has been supplemented by techniques of cell and molecular biology, it remains one of the basic methods of human genetics. The information gathered in pedigree construction is used to ascertain whether a trait is genetic and to determine its mode of inheritance. While these steps seem simple and clear-cut, in practice the decisions are often more difficult. To illustrate how pedigree analysis works, let us briefly consider an ongoing question in human genetics: Is autism genetically inherited, and if so, what is its mode of transmission?

Autism is a behavioral condition characterized by abnormal mental and psychologic development. Symptoms include aversion to human contact, language difficulties that may show up as bizzare speech patterns, and repetitive body movements. These characteristics seem to be associated with malfunctions of the central nervous system, and symptoms usually develop before the age of 30 months. The condition affects about 4 of every 10,000 individuals in the general population and in the past was thought to result from severe psychologic trauma. Over the last 10 years, a team of researchers from UCLA and the University of Utah has been conducting a detailed survey of the incidence and inheritance of autism, using almost all families living in the state of Utah as a study group. The survey, compiled in 1989, identified 241 autistic children among the hundreds of thousands screened. In 187 families there was a single autistic child, and in 20 families there were multiple cases. Some pedigrees from the multiple-case families are shown below. Analysis of the survey results indicates that autism is 215 times more frequent among siblings of autistic children than in the general population, for an overall recurrence risk of 8.6%.

The high risk of recurrence in families indicates that at least some forms of autism are genetically controlled. A major problem in this type of study is that the phenotype of autism is difficult to define and may include a number of different diseases that all produce a similar set of symptoms. At the present time there is no biochemical or genetic marker that can be used to confirm that autism is really a single disease.

The pattern of transmission in the 20 families with multiple cases of autism is not easily explained by simple recessive or dominant Mendelian inheritance. Keep in mind that the accuracy of the pedigrees may be influenced by several factors, for example, cases that are too mild to be diagnosed and cases that are so severe that affected individuals die before birth. Inclusion of more families uncovered in other large-scale studies will be necessary to clarify the mode of inheritance. At this juncture the use of pedigree analysis in a population screening has helped make major strides in the identification of autism as a genetic condition rather than an environmentally induced situation. Further work combining refined methods of diagnosis, family studies, and molecular markers will be useful in confirming the mode of inheritance of autism.

SOURCE: Ritvo, E. R., Jorde, L. B., Mason-Brothers, A., Freeman, B. J., Pingree, C., Jones, M., McMahon, W. M., Petersen, B., Jenson, W. R., and Mo, A. 1989. The UCLA–University of Utah epidemiologic survey of autism: Recurrence risk estimates and genetic counseling. *Am. J. Pyschiatry* **146:**1032–1036. Pedigrees are from data presented on p. 1033.

Pedigrees showing the inheritance of autism in two families.

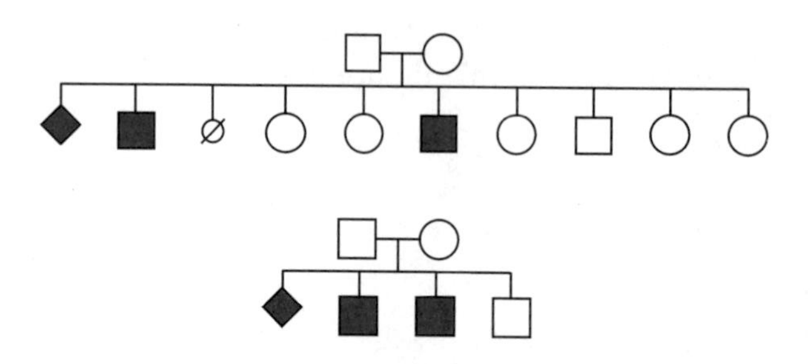

For the analysis, a hypothesis is formed about the mode of inheritance (autosomal recessive, autosomal dominant, etc.), and the information in the pedigree is examined in light of this hypothesis. Those hypotheses that do not fit the observations in the pedigree are discarded. The analysis is completed only when *all* possible modes of inheritance have been considered. If only one hypothesis remains, it is accepted as the mode of inheritance for the trait. If two or more modes of inheritance remain, they are examined to determine which is most likely. This does not mean that alternate hypotheses are ruled out, but selection of the most likely mode of inheritance provides the basis for further work to determine how the trait is inherited.

If a single mode of inheritance can be established, it can be used to predict genetic risk in several situations. These include:

1. pregnancy outcome
2. adult onset disorders
3. recurrence risks in future offspring

In this chapter we will deal with three possible modes of inheritance, autosomal recessive, autosomal dominant and mitochondrial. The others will be considered in Chapter 5.

Autosomal Recessive Traits

Expectations

Two of the most important uses of pedigree analysis are to determine whether a familial trait is genetic and to determine its mode of inheritance. To date, almost 4500 genetic traits have been identified in humans. The chromosomal location of a few hundred of these genes has been determined, and the molecular basis of traits associated with deleterious phenotypes is known in only a small percentage of cases. Based on pedigree analysis, the majority of genetic traits so far identified exhibit a recessive pattern of inheritance. Although human families are small, examination of several generations usually provides enough information to determine if the trait is autosomal and recessive. There are several guidelines for recessive genes in pedigrees, as listed below:

1. For infrequent or rare traits, the pedigree typically involves mating between two unaffected heterozygotes and the production of one or more homozygous offspring.
2. The risk of an affected child from a mating of two heterozygotes is 25%.
3. Two affected (homozygous) individuals usually produce offspring all of whom are affected.
4. Males and females are at equal risk, since the trait is autosomal.
5. In pedigrees involving rare traits, some degree of consanguinity is usually involved.

Table 4.2 lists some autosomal recessive human genetic diseases. A pedigree illustrating a pattern of inheritance typical of autosomal recessive genes is shown in Figure 4.1.

Some genetic traits represent minor variations in phenotype, while others produce more deleterious phenotypes. In a few cases these phenotypes can be life threatening or even fatal. Example of these more severe phenotypes are albinism, cystic fibrosis, and sickle cell anemia.

TABLE 4.2 Some Autosomal Recessive Human Genetic Traits

Albinism	Absence of pigment in skin, eyes, hair
Ataxia telangiectasia	Progressive degeneration of nervous system
Bloom syndrome	Dwarfism, skin rash, increased cancer rate
Cystic fibrosis	Mucous production that blocks ducts of certain glands, lung passages; often fatal by early adulthood
Fanconi anemia	Slow growth, heart defects, high rate of leukemia
Galactosemia	Accumulation of galactose in liver; mental retardation
Phenylketonuria	Excess accumulation of phenylalanine in blood, mental retardation
Sickle cell anemia	Abnormal hemoglobin, blood vessel blockage, early death
Thalassemia	Improper hemoglobin production; symptoms range from mild to fatal
Xeroderma pigmentosum	Lack of DNA repair enzymes, sensitivity to UV light, skin cancer, early death
Tay-Sachs disease	Improper metabolism of gangliosides in nerve cells, early death

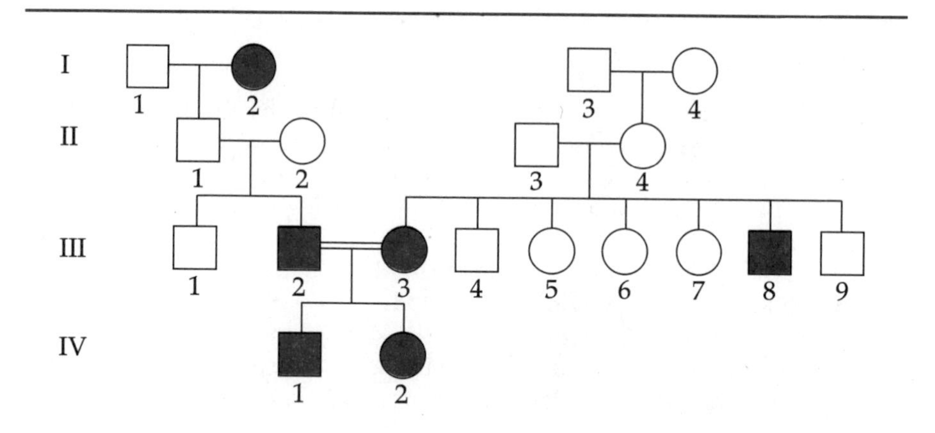

FIGURE 4.1 Pedigree for an autosomal recessive trait. This pedigree fulfills many of our expectations about the inheritance of autosomal recessive traits: most affected individuals have normal parents, about one fourth of the children in large affected families show the trait, both sexes are equally affected, and affected individuals produce only affected children.

FIGURE 4.2 Albino brothers.

Albinism

Albinism
A group of recessive genetic conditions associated with a lack of pigmentation in the skin, hair, and/or eyes.

Albinism is a group of genetic conditions associated with a lack of pigmentation in the skin, hair, and/or eyes. In normal individuals melanin pigment granules are deposited in cells known as melanocytes, found in the skin, hair, and eyes, giving color to these body parts. In albinos the melanocytes are present but contain no pigment because affected individuals are unable to synthesize melanin, a pigment responsible for coloration in the skin, hair and pigmented layer of the retina (Figure 4.2). Two major classes of albinism are known, one affecting only the eye, known as ocular albinism (OA), and the other affecting the eye, skin, and hair, known as ocular-cutaneous albinism (OCA). OCA is characterized by a reduction or absence of pigment in the skin, hair, and eyes. This is accompanied by a reduction in visual acuity and by the presence of involuntary rapid eye movements. The most widely known type of albinism is the autosomal recessive form of OCA. Pedigrees tracing the inheritance of this form of albinism are shown in Figure 4.3 (see Concepts and Controversies, page 83). These pedigrees confirm our expectations about the inheritance of recessive traits in humans. In Figure 4.3A, only albino children are produced by albino parents. In Figure 4.3B, mating between two phenotypically normal individuals produces albino and nonalbino children.

The frequency of albinism varies widely among different human populations. In the U.S. white population, the frequency of OCA is 1 in 37,000; in the U.S. black population the frequency is estimated at 1 in 15,000. Other populations may have higher frequencies: among the San Blas Indians of Panama, the frequency is 1 in 132, and among the Hopi and Navajo Indians in the American Southwest the frequency is 1 in 200. The reasons for these wide variations in allele frequency will be discussed in Chapters 17 and 18.

FIGURE 4.3 Pedigrees for albinism. (A) Showing transmission from affected parents to all offspring. (B) Offspring from heterozygous individuals can be affected or nonaffected.

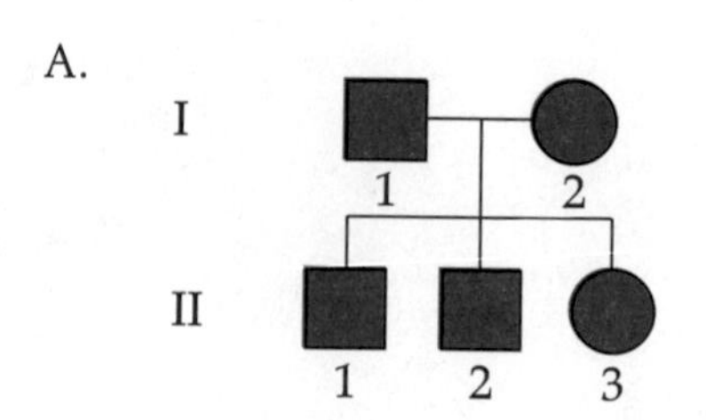

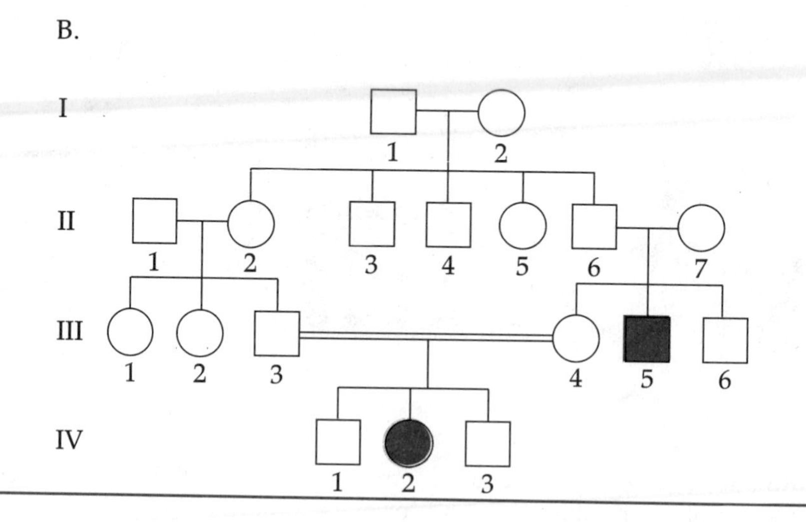

Cystic Fibrosis

Cystic fibrosis
A fatal recessive genetic disorder, common in the U.S. white population, associated with abnormal secretions of the exocrine glands.

Cystic fibrosis (CF) is a disabling and fatal genetic disorder caused by a recessive allele. The main focus of the mutant gene is on exocrine glands, the glands that produce mucus, digestive enzymes, and sweat. Because the sweat glands are defective, they release excessive amounts of salt, and the disease has been diagnosed by analysis of the salt content in sweat, although this is being replaced by molecular diagnosis that directly screens for the mutant gene. According to folklore, midwives would lick the forehead of newborns. If the sweat tasted too salty, they predicted that the child would soon die of lung congestion.

The results of this disease have far-reaching effects on the body because exocrine glands have a number of vital functions. For example, several digestive enzymes are produced in the pancreas and enter the intestine through a duct system. In cystic fibrosis the duct becomes clogged, and food is improperly digested. As a result an affected child often suffers from malnutrition in spite of a normal or even increased appetite and food intake. Eventually the blocked ducts causes the pancreas to form cysts. Over a period of time the gland degenerates and becomes fibrous, giving rise to the name of the disease. A similar pattern of events affects all the exocrine glands of the body. Most patients with cystic fibrosis develop obstructive lung diseases (Figure 4.4). With aggressive therapy that includes respiratory treatments,

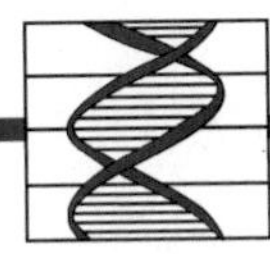

Concepts and Controversies

Was Noah an Albino?

The Biblical character Noah, along with the ark and the animals are among the most recognizable figures in the Book of Genesis. His birth is recorded in a single sentence, and although the story of the ark and flood are recorded later in the book, there is no mention of Noah's physical characteristics. However, other sources contain reference to Noah that are consistent with the idea that Noah may have been one of the first albinos mentioned in recorded history.

In addition to the Bible, the birth of Noah is recorded in other sources. One is the book of Enoch the Prophet, written about 200 BC. This book, quoted several times in the New Testament, was regarded as lost until 1773, when an Ethiopean version of the manuscript was discovered. In describing the birth of Noah, the text relates that his "flesh was white as snow, and red as a rose; the hair of whose head was white like wool, and long, and whose eyes were beautiful." A reconstructed fragment of one of the Dead Sea scrolls also deals with the birth of Noah as an abnormal child born to normal parents.

This fragment of the scroll also provides some insight into the pedigree of Noah's parents and family, as does another reference, the Book of Jubilees. The evidence available from these sources indicates that Noah's father (Lamech) and his mother (Betenos) were probably first cousins. Lamech was the son of Methuselah, and Lamech's wife was a daughter of Methuselah's sister. Recall that some degree of consanguinity is often characteristic of pedigrees involving recessive autosomal conditions like albinism.

If this reading of ancient texts is correct, Noah's albinism is the result of a consanguineous mating, and not only is he one of the earliest albinos on record, but his grandfather Methuselah and Methusaleh's sister are the first recorded heterozygous carriers for a recessive genetic trait.

special diets, and antibiotics, more than 50% of affected individuals reach young adulthood.

Since many affected individuals are sterile, most cases arise as the offspring of phenotypically normal heterozygous parents, and heterozygote detection coupled with genetic counseling can be used effectively to reduce the incidence of this disease. Cystic fibrosis is relatively common in the U.S. white population, with a frequency of 1 in 2000 births. The frequency of heterozygous carriers is about 1 in 22. The disease is rare among U.S. blacks and is estimated at 1 in 100,000 to 1 in 150,000.

After decades of intensive research efforts, the underlying defect in CF was identified in 1989. Classical genetic studies combined with recombinant DNA techniques had previously localized the gene to region q31 of chromosome 7 (Figure 4.5). This region of the chromosome was explored using several methods of genetic mapping, and the CF gene was identified by comparing the genetic organization of a small segment of chromosome 7 in normal and CF individuals.

Genetic and molecular analysis indicates that the product of the CF gene is a protein inserted into the plasma membrane of exocrine gland cells. The protein has a structure similar to other cell membrane proteins, and it has

FIGURE 4.4 Organ systems affected by cystic fibrosis. Sweat glands in affected individuals excrete excessive amounts of sodium and chloride. In the pancreas, thick mucus blocks the passage of digestive enzymes, causing poor digestion and lowered absorption of nutrients. In the lungs and bronchial tubes, thick mucus causes breathing difficulties and bacterial infections. Blockage of reproductive ducts causes infertility.

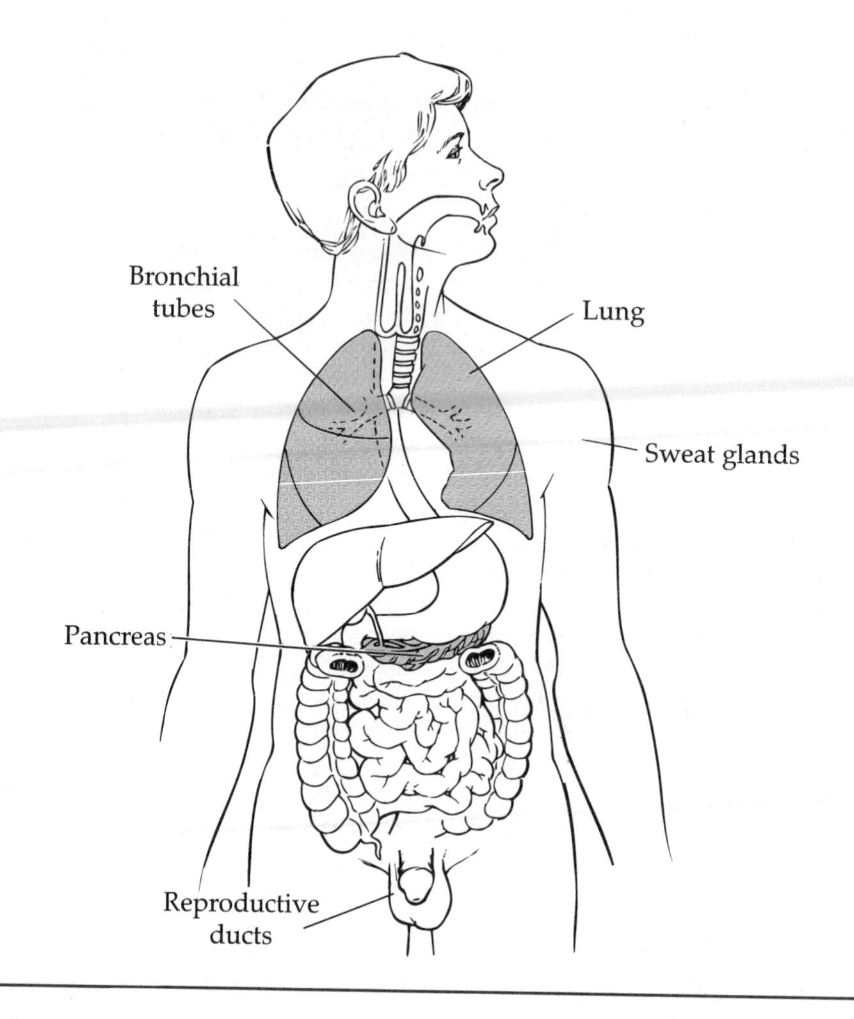

been called the *cystic fibrosis transmembrane conductance regulator,* or CFTR (see Figure 4.6 on page 88). The function of this protein is defective in CF individuals, leading to the development of symptoms characteristic of the disease. It is hoped that further studies of the structure and function of the CF gene and its product will lead to the development of new methods of treatment for this deadly disease, including gene therapy, the replacement of mutant genes with normal genes.

Sickle Cell Anemia

Sickle cell anemia
A recessive genetic disorder, common in the U.S. black population, associated with an abnormal type of hemoglobin, a blood transport protein.

Americans whose ancestors lived in parts of West Africa, around the Mediterranean basin, or in parts of the Indian subcontinent have a high frequency of **sickle cell anemia.** Individuals with this disease have a specific abnormal type of hemoglobin, a protein found in the red blood cells. Normally, this protein functions in the transport of oxygen from the lungs to the tissues of the body, and each red cell contains millions of hemoglobin molecules.

Under conditions of low oxygen tension, the abnormal hemoglobin in sickle cell anemia causes the red blood cells to change shape and become crescent- or sickle-shaped (Figure 4.7). The cells become deformed because abnormal hemoglobin molecules pack together to form rods (Figure 4.8), causing a change in cell shape. The deformed blood cells are fragile and easily broken apart. Since new blood cells are not produced fast enough to replace those that are lost, the oxygen-carrying capacity of the blood is reduced, and anemia results. Victims of sickle cell anemia tire easily and often develop heart failure because of the increased load on the circulatory system. The deformed blood cells also clog small blood vessels and capillaries, and this further reduces oxygen transport, bringing on what is known as a sickling crisis. As oxygen concentration falls, more and more red blood cells become sickled, bringing on intense pain as circulation is impaired. The lack of blood supply in affected areas causes ulcers and sores on the body surface (Figure 4.9). Blockage of the blood vessels in the brain leads to strokes and can result in partial paralysis.

Because of the number of systems in the body affected and the severity of the effects, untreated sickle cell anemia can be lethal. Some affected individuals die in childhood or adolescence, but aggressive medical treatment allows survival into adulthood. As in cystic fibrosis, most affected individuals are children of phenotypically normal, heterozygous parents. In sickle cell anemia, methods for heterozygote detection and prenatal diagnosis are available.

The high incidence of this condition in certain populations is related to the frequency of an infectious disease, malaria. Heterozygotes for sickle cell anemia are partially resistant to malaria. As a result, this mutation is widespread in certain populations. Among U.S. blacks sickle cell anemia occurs with a frequency of 1 in every 500 births, and approximately 1 in every 12 individuals is a heterozygote. The high frequency of this mutation in the black population is a genetic vestige of African origins. The same is true for U.S. residents whose ancestral origins are in lowland regions of Italy, Sicily, Cyprus, Greece, and the Middle East. This mutation in a hemoglobin molecule has a double effect: it causes the serious effects of sickle cell anemia but also has the beneficial effect of conferring resistance to malaria. The molecular basis of this disease is well known and will be discussed in detail in later chapters.

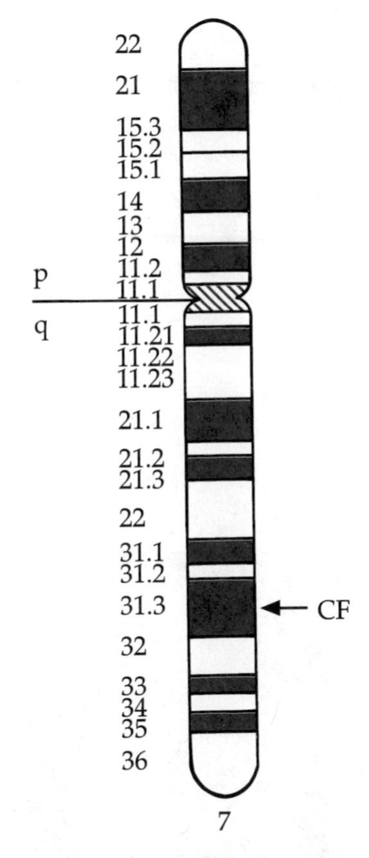

FIGURE 4.5 Diagram of human chromosome 7. The region containing the gene for cystic fibrosis is indicated. The gene maps to 7q31.2-31.3.

Autosomal Dominant Traits

Expectations

In the case of dominant alleles, an individual who is carrying one copy of a dominant allele and is a heterozygote shows expression of a trait. Only in rare cases are dominant genetic disorders present in a homozygous condition. The unaffected individual, on the other hand, carries two recessive alleles. Because human families are small, careful analysis of pedigrees is necessary to determine whether a trait is inherited as a dominant gene. Following the

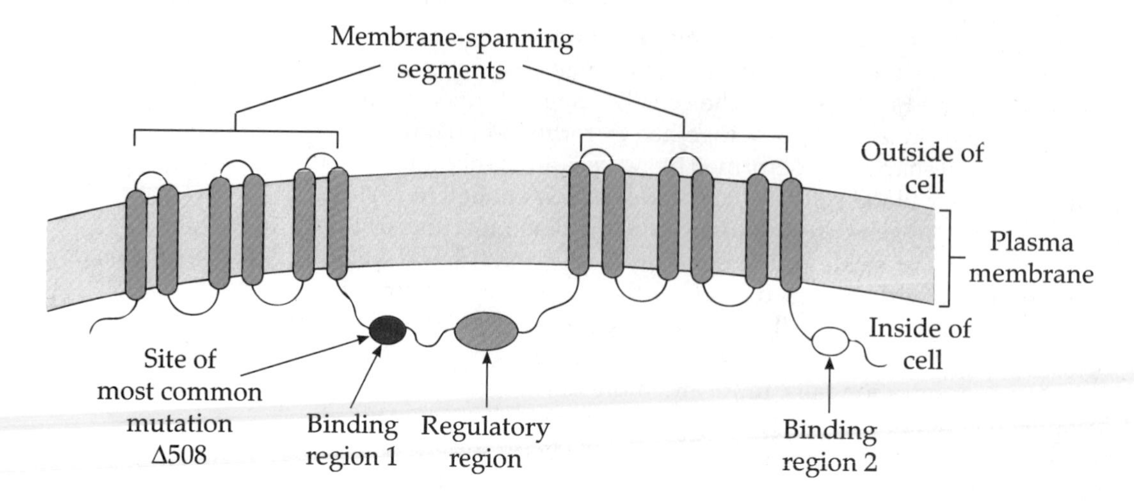

FIGURE 4.6 The CFTR protein contains membrane-spanning segments that function in movement of chloride ions into and out of the cell. The regulatory region regulates the activity of the CFTR molecule in response to intracellular signals. In most cases (about 70%) the protein is defective in shaded region.

principles of Mendelian inheritance, dominant traits in humans should meet several expectations:

1. Every affected individual should have at least one affected parent. Exceptions can occur in cases where the gene has a high mutation rate. Mutation is the sudden appearance of a heritable allele that was not transmitted by the biological parents.
2. An affected person has a 50% chance of transmitting the trait to each child.
3. Males and females should be affected with equal frequency.

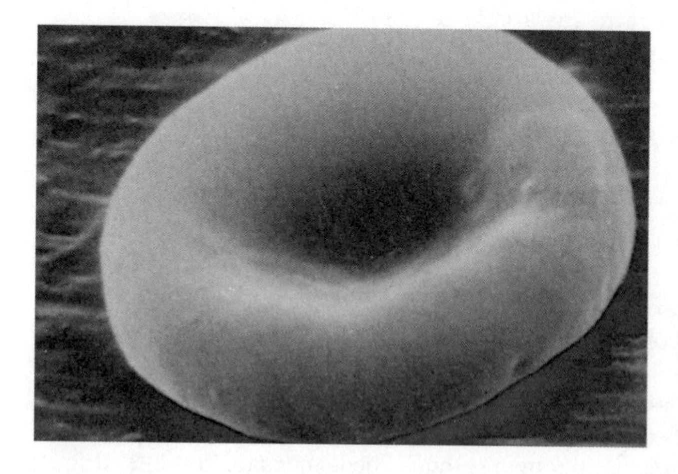

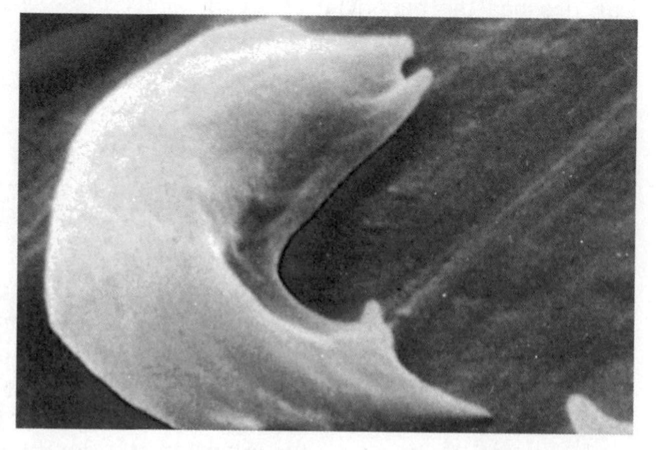

FIGURE 4.7 Normal red blood cells (left). Sickled red blood cells (right).

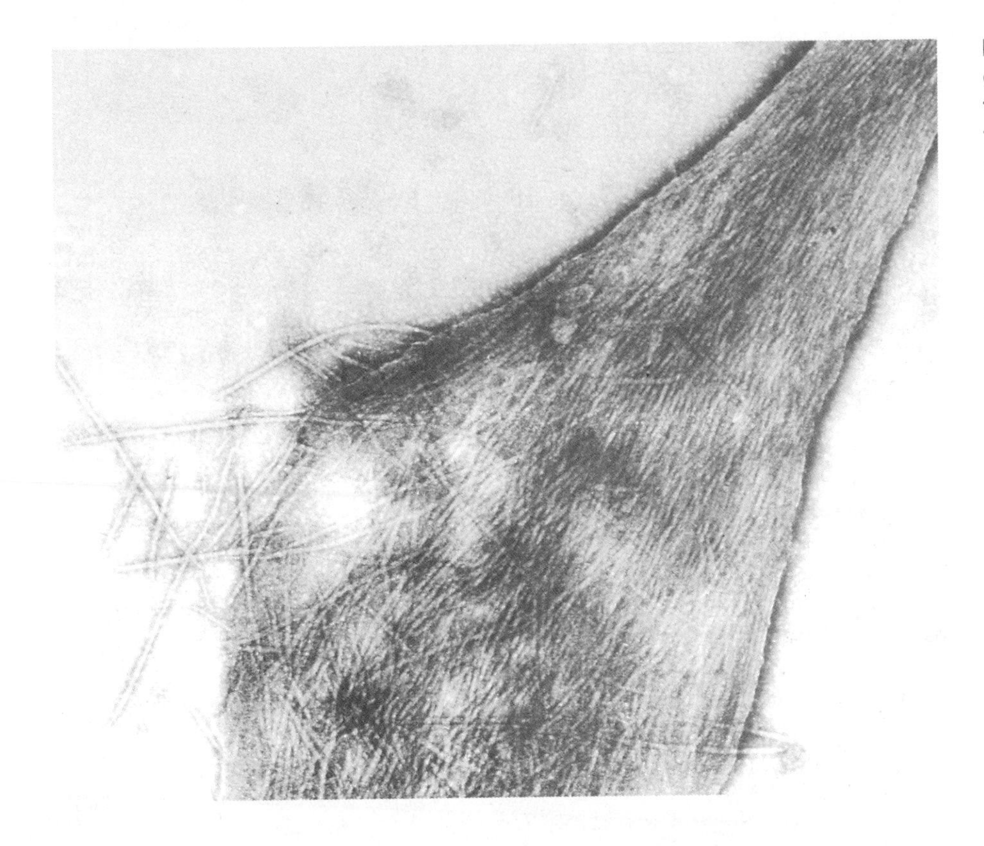

FIGURE 4.8 Sickle cell hemoglobin molecules aggregate to form rod-like structures that coil together to form cable-like structures. The formation of these aggregates in the cytoplasm causes the red blood cell to deform and become elongated or sickle-shaped.

4. Two affected individuals may have unaffected children.
5. In some cases, the homozygous dominant phenotype may be more severe than or different from the heterozygous phenotype.

Table 4.3 lists a number of human genetic disorders caused by autosomal dominant genes. The pedigree in Figure 4.10 is typical of the pattern found in the autosomal dominant condition brachydactyly. This

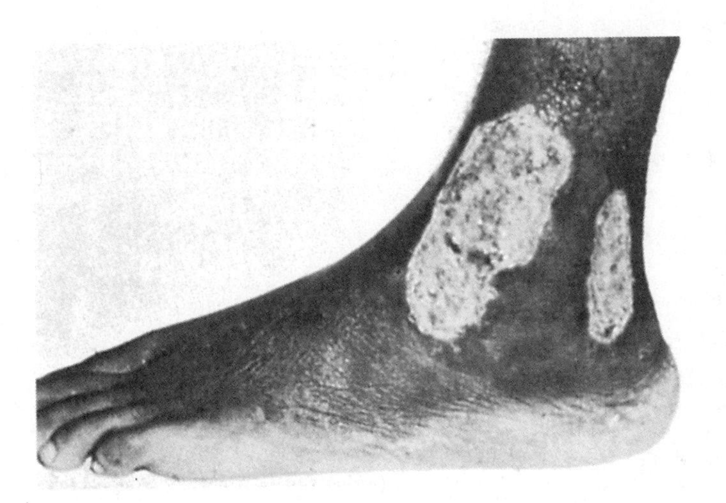

FIGURE 4.9 Ulcers on skin of patient with sickle cell anemia.

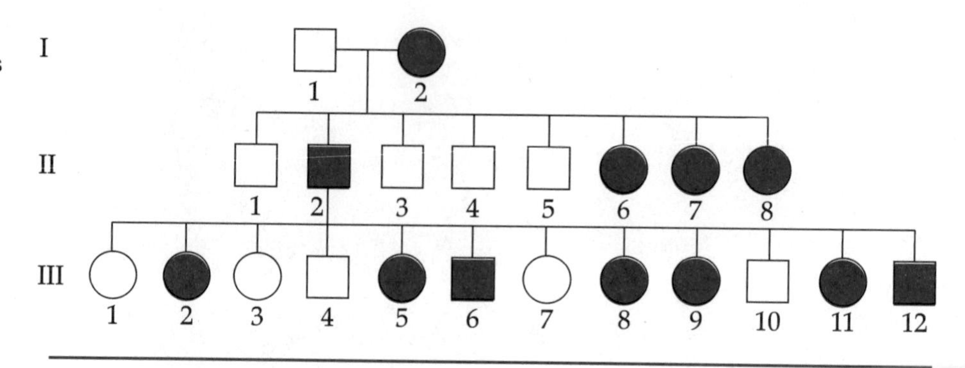

FIGURE 4.10 Pedigree for an autosomal dominant trait. This pedigree is for brachydactyly, a hand defect that was the first Mendelian trait identified in humans.

pedigree was used to confirm the first identification of Mendelian inheritance in humans.

Some genetic traits have phenotypes that are evident as individual variations in body chemistry, indicating that genes exert control over biochemical as well as structural traits. For example, a familial form of **hypercalcemia,** characterized by elevated levels of calcium in the blood, is controlled by a dominant allele. Figure 4.11 represents part of a pedigree collected in a large Dutch family with this condition. The family history shows the pattern expected for an autosomal dominant trait. Often the

Hypercalcemia
A dominant trait tnat causes an elevated level of calcium in the blood.

TABLE 4.3 Some Autosomal Dominant Human Genetic Traits

Achondroplasia	Dwarfism associated with defects in growth regions of long bones
Brachydactyly	Malformed hands with shortened fingers
Campodactyly	Stiff, permanently bent little fingers
Crouzon syndrome	Defective development of midface region, protuding eyes, hook nose
Ehler-Danlos syndrome	Connective tissue disorder, elastic skin, loose joints
Familial hyper-cholesterolemia	Elevated levels of cholesterol; predisposes to plaque formation, cardiac disease; may be most prevalent genetic disease
Familial polycystic kidney disease	Formation of cysts in kidneys; leads to hypertension, kidney failure
Huntington disease	Progresive degeneration of nervous system, dementia, early death
Hypercalcemia	Elevated levels of calcium in blood serum
Marfan syndrome	Connective tissue defect; death by aortic rupture
Nail-patella syndrome	Absence of nails, kneecaps
Porphyria	Inability to metabolize porphyrins, episodes of mental derangement

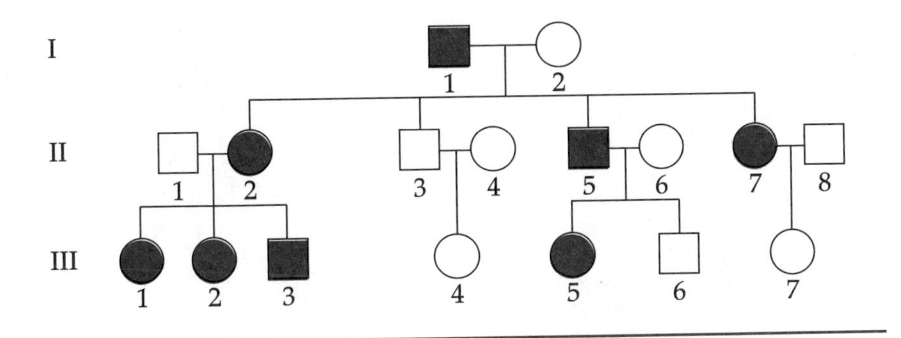

FIGURE 4.11 Pedigree for hypercalcemia in a Dutch family. Pattern is typical for autosomal dominant genes.

phenotype exhibited by a dominant mutation has serious deleterious effects; examples of such mutations are discussed below.

Marfan syndrome
An autosomal dominant genetic disorder that affects the skeletal system, the cardiovascular system, and the eyes.

Marfan Syndrome

Flo Hyman was the star of the U.S. women's volleyball team in the 1984 Olympic Games and helped lead the team to the silver medal. She was widely regarded as the best volleyball player in the world (Figure 4.12). In January 1986, after being taken out of a game in Japan, she slumped to the floor, dead. The autopsy report indicated that she had died from a ruptured aorta caused by **Marfan syndrome.** Some 10 years earlier a basketball player at the University of Maryland, Chris Patton, collapsed and died in a pickup game. The autopsy evaluation again indicated Marfan syndrome as the cause of death. Nationwide it is estimated that 10 to 12 high school and college athletes die each year from the disorder.

This syndrome was first described in 1896 by A. B. Marfan, a French physician. It is an autosomal dominant disorder that affects the skeletal system, the eyes, and the cardiovascular system. The disease is difficult to diagnose and is usually identified only in an autopsy or postmortem examination. Affected individuals tend to be tall and thin (Flo Hyman, for example, was 6 ft 5 in) with long arms and legs. In addition, Marfan patients have long, thin fingers. These characteristics often allow affected individuals to excel in sports like basketball and volleyball, although nearsightedness and defects in the lens of the eye are also common. (Hyman was nearsighted and wore glasses off the court.)

The most dangerous symptoms of Marfan syndrome are those that affect the cardiovascular system, especially the aorta. The aorta arises in the left ventricle of the heart as the main blood-carrying vessel in the body. As it leaves the heart, it arches back and downward, feeding blood via branches to all the major organ systems of the body. Marfan syndrome weakens the connective tissue around the aorta, causing it to enlarge and eventually to split open (Figure 4.13). It is the first few inches of the aorta that are most likely to enlarge and split, and in cases where the enlargement can be detected, it can be repaired by surgery. After her death, Flo's brothers and sisters were examined to detect signs of Marfan syndrome. One of her brothers was found to have an enlargement at the base of the aorta, and it was surgically repaired.

FIGURE 4.12 Flo Hyman, star of the 1984 U.S. women's volleyball team, who died from Marfan syndrome.

FIGURE 4.13 **(A) Normal heart showing the aorta originating in the left ventricle. (B) Enlargement (aneurysm) at base of aorta that ruptures with fatal effect in Marfan syndrome.**

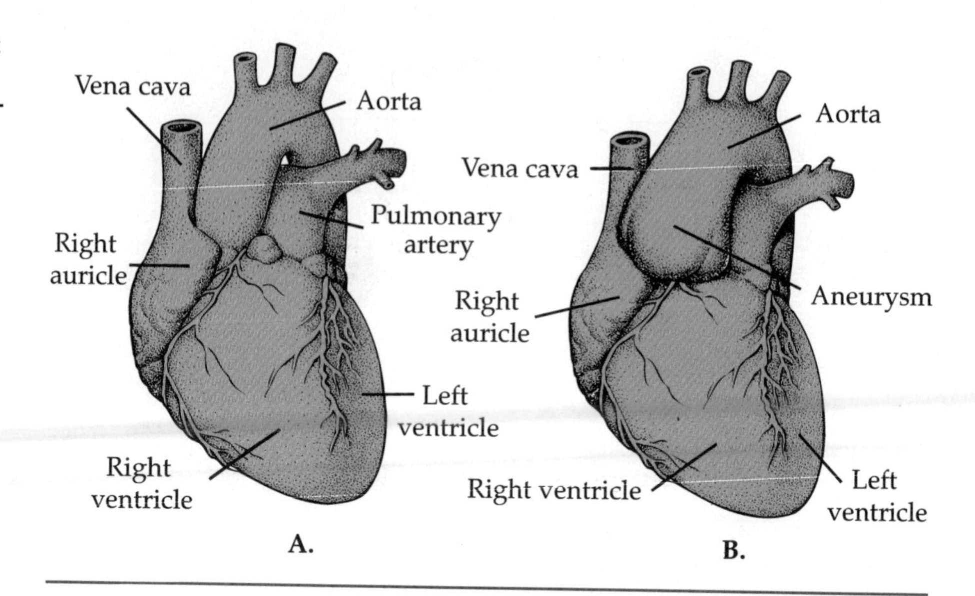

Achondroplasia
An autosomal dominant disorder in which the long bones of the arms and legs fail to develop normally, producing a form of dwarfism.

The defective gene that causes Marfan syndrome was mapped to chromosome 15 in 1989, and in 1991 it was shown to affect fibrillin, a gene product found in the elastic connective tissue of the aorta, the eye, and the sheath covering long bones. The disorder affects males and females with equal frequency and is found in all ethnic groups. About 25% of affected individuals appear in families with no previous history of Marfan syndrome, indicating that this gene undergoes mutation at a high rate. This disorder is present in about 1 in 10,000 individuals. As noted in the chapter introduction, it has been suggested that Abraham Lincoln had Marfan syndrome. Bone fragments and hair from his body are preserved at the National Medical Library in Washington, D.C., and a group of research scientists has proposed that these materials be used in conjunction with molecular techniques to determine whether Lincoln did, in fact, have Marfan syndrome. A committee appointed to review the request has agreed that this material can be tested, but it has recommended that testing be delayed until more is known about the fibrillin gene and the types and distribution of defective alleles in the population.

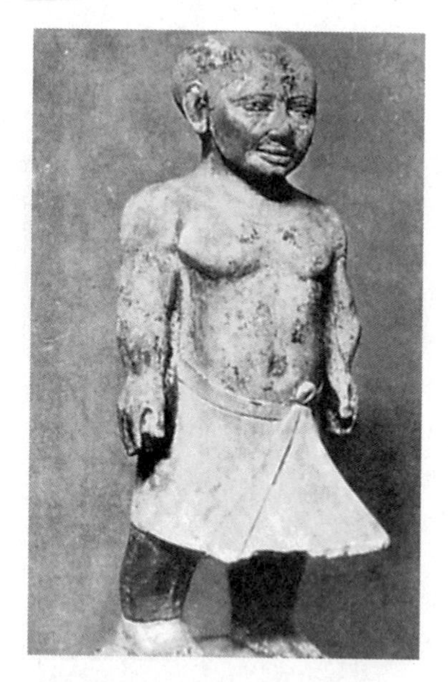

FIGURE 4.14 **Statue of Chnoum-Hotep, an Egyptian from the Vth dynasty afflicted with achondroplasia.**

Achondroplasia

Achondroplasia is an autosomal dominant form of dwarfism that occurs with a frequency of about 1 in 30,000 individuals. In this condition the long bones of the arms and legs fail to develop properly, and the skull may be slightly enlarged. The body trunk is usually of normal size, and individuals with this disorder have a very characteristic appearance. The torso is normal in size, but the arms and legs are greatly reduced. The head is too large in proportion to the body, which shows good muscular development (Figure 4.14). Individuals with this skeletal disorder were known some 5000 years ago. Figure 4.14 shows a statute of Chnoum-Hotep, an Egyptian achondroplastic dwarf who lived in the Vth dynasty (about 2700 BC) and held a high position in the pharoah's court. He was keeper of the wardrobe—in effect, the pharoah's valet. He was buried near Saqqarah, and his tomb was large and well-decorated, another indication of his importance. Figure 4.15 shows a present-day individual with achondroplasia.

Neurofibromatosis

Another dominant trait in humans is **type 1 neurofibromatosis** (NF1). The presence of this autosomal mutant gene produces a highly variable phenotype. Many individuals exhibit spots of abnormal skin pigmentation known as **café-au-lait spots** and the growth of noncancerous tumors in the nervous system. Learning disabilities are also a related phenotypic expression of this mutant. In other cases large tumors produce gross deformities and often compress vital nerves, producing blindness or paralysis. In a few cases death is caused by conversion of the tumors to a cancerous form that quickly invades the nervous system.

Neurofibromatosis
A genetic disorder inherited in a dominant fashion that is associated with tumors of the nervous system.

Café-au-lait spots
Spots of abnormal skin pigmentation that are found on many individuals affected with neurofibromatosis.

Although NF1 is a dominant condition, it often appears in the children of normal parents. NF1 has a frequency of 1 in every 3000 births, and about half of these births are to unaffected parents. This unexpected outcome can be explained by the fact that the NF1 gene has an extremely high rate of mutation, and these cases are produced by new mutational events. There is no effective treatment, except surgery, to remove large tumors.

The gene for NF1 was identified in three stages over a period of just a few years. First, in 1987, genetic analysis indicated that the gene was located somewhere on chromosome 17. In 1989, the location of the gene was narrowed to a small region near the center of the chromosome at 17q11.2 (Figure 4.16). This regional mapping was made possible through the use of molecular markers present on the chromosome that are inherited along with the mutant gene. Finally, in 1990, the gene itself was located within the 17q11.2 region using recombinant DNA techniques designed for gene hunting (these methods will be described in Chapter 8).

Details of the molecular organization of the gene are now known, and the structure of the gene product has been reconstructed. The NF1 gene product is a cytoplasmic protein that acts to transfer signals from cell surface receptors to the cytoplasm. It functions by interacting with other proteins that control cell growth. Information about the function of the gene product in both the normal and mutant conditions may provide the basis for the design of drugs to be used in the treatment of NF. The NF1 gene is very large, occupying a relatively long segment of the chromosome. Because such a gene would provide a target for mutational events, this finding is consistent with the high mutation rate characteristic of NF1 (mutation rates are discussed in Chapter 11).

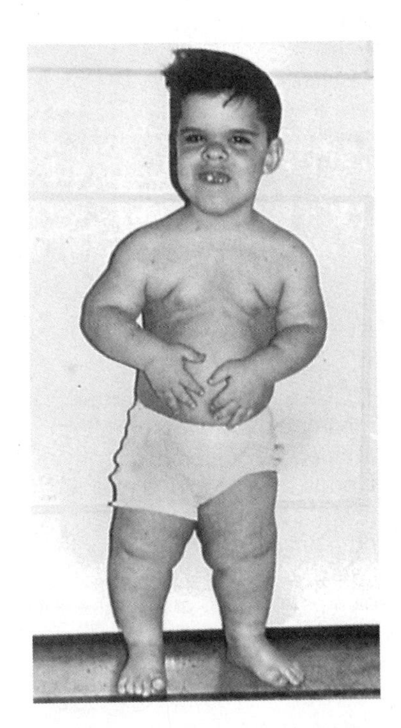

FIGURE 4.15 Present-day individual with achondroplasia.

Since the normal allele of this gene is thought to control important stages of growth and development in the central nervous system, study of this isolated gene should provide information about the molecular events that transform embryonic cells into components of the nervous system. A gene for another, unrelated form of neurofibromatosis has been designated as NF2. In the families studied to date, this form of NF is inherited as a dominant trait associated with tumors of the cranial nerves and deafness. It has been mapped to chromosome 22 using other sets of DNA markers.

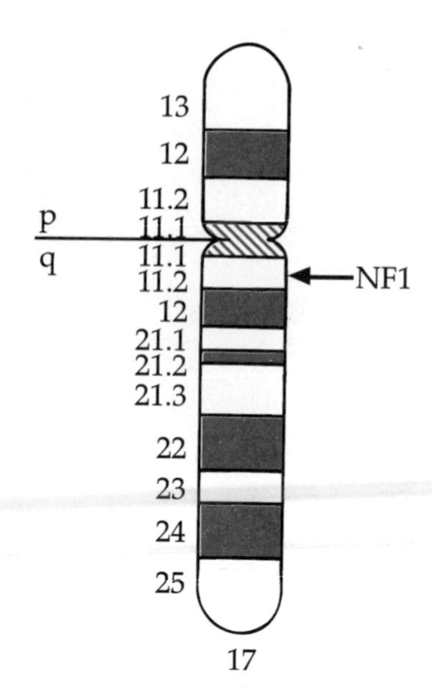

FIGURE 4.16 Diagram of human chromosome 17 showing region that contains the gene for NF. The gene maps to 17q11.2.

MULTIPLE ALLELES

Traits like albinism or achondroplasia provide evidence that the principles of Mendelian genetics apply to humans. However, the incidence of such traits is rare, and not only is it difficult to find such individuals, it is often more difficult to construct meaningful pedigrees for these diseases. Early students of human genetics searched to find traits expressed by everyone, a set of traits with clear-cut classes that could be applied to all racial and ethnic groups. Ideally such traits would be simple to detect and would be unaffected by environment, culture, or diet.

ABO System

ABO blood groups
Three alleles of a gene on human chromosome 9 that specify the presence and/or identity of certain molecules on the surface of red blood cells.

The **ABO blood groups,** discovered by Landsteiner in 1900, provided the first examples of genetic traits that met these standards. Landsteiner observed that mixing blood of different individuals often produced a clumping of cells called agglutination. Like the MN blood groups discussed earlier, the ABO system depends on the presence of certain substances called glycoproteins on the surface of red blood cells. Landsteiner termed these markers isoagglutinins and found they reacted with components of blood serum known as antibodies. Through mixing experiments, Landsteiner identified three isoagglutinogens, A, B, and O, and termed them blood types (Figure 4.17). By 1911 it was known that these blood groups were inherited in a Mendelian fashion, and in 1924, Felix Bernstein showed that the A, B, and O blood types are produced by combinations of three different alleles of the same gene, known as *I* (for isoagglutinogen). Bernstein further demonstrated that the A and B alleles are codominant to each other and that both were dominant to O. Much later it was shown that the A and B alleles control the presence of

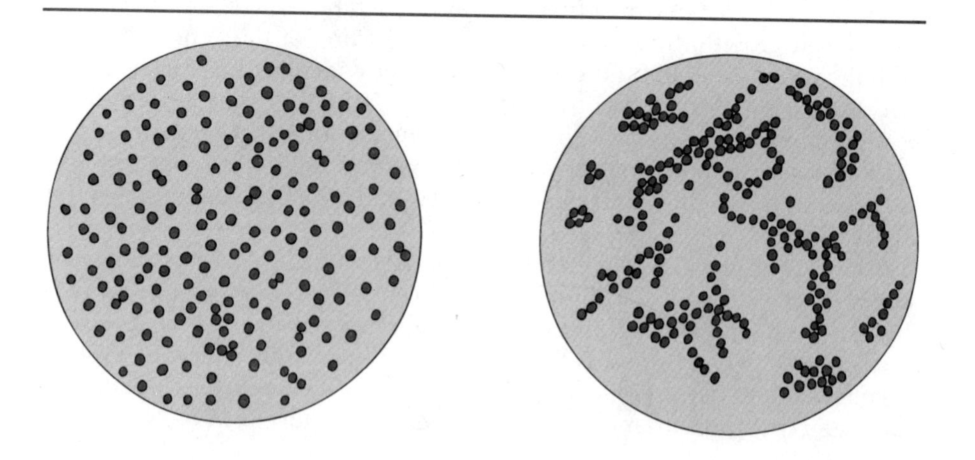

FIGURE 4.17 Blood serum mixing experiments of Landsteiner. When compatible bloods are mixed and examined microscopically *(left)*, there is no reaction. Mixing of incompatible bloods *(right)* leads to clumping. These are of different blood types.

slightly different molecules on the cell surface, while the O allele produces no cell-surface molecules.

The ABO system is an example of a gene with multiple alleles. But since alleles represent different forms of a gene, there is no reason why a given gene cannot have more than two alleles. In fact, when human populations are examined, many genes are found to have multiple alleles. In the case of ABO, the gene *I* has three alleles, I^A, I^B, and I^O. Because individuals can and do carry only two alleles of a gene (one on each homologous chromosome), the *population* rather than the individual serves as the reservoir of multiple alleles. As a result the three alleles of the *I* gene, with I^A and I^B codominant and I^O recessive, when combined two at a time produce six genotypes and four phenotypic classes or blood groups: A, AB, B, and O (Table 4.4). Any given individual has only one of these genotypes and its corresponding phenotype. We will discuss the blood groups more fully in Chapter 14.

TABLE 4.4 ABO Blood Types

Genotypes	Phenotypes
I^AI^A, I^AI^O	Type A
I^BI^B, I^BI^O	Type B
I^AI^B	Type AB
I^OI^O	Type O

Familial Hypercholesterolemia

The autosomal dominant disease familial **hypercholesterolemia** is caused by defective receptors on the cell surface that regulate the uptake and metabolism of cholesterol bound to low-density lipoproteins (LDL). Affected heterozygotes have elevated serum cholesterol levels and usually develop coronary artery disease in the fourth or fifth decade of life. Normally, cholesterol combined with LDL binds to receptors on the cell surface and is taken into the cell. The LDL ends up in the lysosomes, where it is degraded, and the cholesterol is metabolized in several different ways. In familial hypercholesterolemia, this process is disrupted. Two general types of defects are recognized, absent receptors and defective receptors. Several phenotypic classes of defective receptors are known, including a non-functional receptor that is unable to recognize the LDL-cholesterol complex, another with greatly reduced ability to bind the LDL complex, and a third that recognizes and binds the LDL complex but is unable to internalize LDL. Molecular analysis of the low-density lipoprotein receptor gene has revealed more than a dozen alleles at this locus. Each of these defects results in elevated levels of LDL-derived cholesterol in the blood serum and increased deposition of cholesterol plaques in the arterial system, leading to premature coronary heart disease.

Hypercholesterolemia An autosomal dominant genetic disorder associated with the deposition of cholesterol plaques in the arteries of heterozygotes and homozygous dominant individuals; leads to heart disease and early death.

The frequency of heterozygotes for familial hypercholesterolemia in European, Japanese, and U.S. populations is 1 in 500, although the frequency in a region of Quebec has been estimated at 1 in 122. The highest reported frequency to date is 1 in 71 in a community in South Africa. The average frequency of 1 in 500 heterozygotes worldwide makes this disorder the most common dominant genetic condition identified to date, and it is one of the leading causes of heart disease across the world.

Unlike the case with many other dominant conditions, individuals homozygous for familial hypercholesterolemia have been identified. The homozygous phenotype can be distinguished from the heterozygous by the severity of symptoms. Homozygotes have higher serum cholesterol levels than heterozygotes (> 500 mg/dl for homozygotes versus 200 to 400 in heterozygotes), and the other clinical symptoms such as coronary artery disease develop at an accelerated rate, usually by 20 years of age. The average frequency of homozygotes in most populations is estimated to be 1 in a million.

MITOCHONDRIAL INHERITANCE

Mitochondria are cytoplasmic organelles concerned with energy conversion (see Chapter 2). Early in the evolutionary history of life some billions of years ago, ancestors of mitochondria were probably free-living prokaryotes that later formed a symbiotic relationship with primitive eukaryotes. As an evolutionary relic of their free-living ancestry, mitochondria carry DNA molecules that encode genetic information necessary for mitochondrial gene products. Most mitochondria carry several copies of this information, and may have up to 40 DNA molecules, and there are several hundred mitochondria in each cell (red blood cells are an exception; they have none).

The Mendelian patterns of inheritance described earlier in this chapter depends on the transmission of genetic information carried on chromosomes in the nucleus. However, since mitochondria carry genetic information, mutations in these genes can also cause human genetic disorders. Mitochondria are transmitted from generation to generation through the maternal cytoplasm of the egg (sperm lose all cytoplasm during maturation). As a result, mutations in mitochondrial genes are maternally inherited. Both males and females can be affected by such disorders, but only females can transmit these mutant genes from generation to generation (Figure 4.18).

Several genetic disorders are known to be transmitted in this fashion, and all are caused by failures to efficiently convert energy. Thus, the tissues most sensitive to energy supplies are most affected, including the nervous system, skeletal muscle, heart muscle, liver and kidneys. Some of the disorders associated with mitochondria are muscle disorders including myoclonic epilepsy and ragged red-fiber disease (MERRF), maternally inherited myopathy and cardiomyopathy (MMC) and eye disorders including Leber's hereditary optic neuropathy (LHON) and neurologically-associated retinitis pigmentosa (NARP).

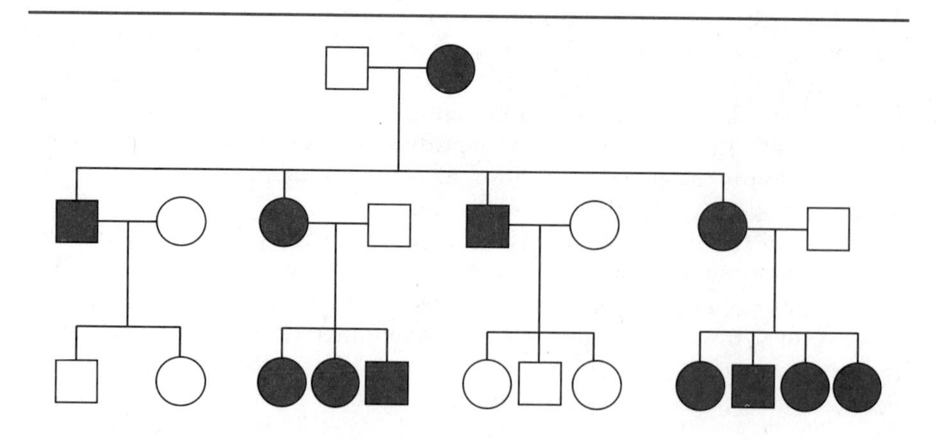

FIGURE 4.18 Pedigree showing maternal transmission of mitochondrial disorder. Both males and females can be affected, but only females can transmit trait to offspring.

Other cases of mitochondrial diseases are sporadic rather than genetic, and mitochondrial mutations acquired during a lifetime are often found in the brains and nervous tissue in autopsies, suggesting that alterations in mitochondria are common and may in fact, be part of the aging process. The mechanisms of mitochondrial mutations will be discussed in Chapter 11.

LINKAGE AND MAPPING

Linkage and Genetic Maps

There are estimated to be 50,000–100,000 genes distributed on the 24 human chromosomes (22 autosomes and the X and the Y chromosome). Genes present on the same chromosome are said to show **linkage.** Because they are on the same chromosome, linked genes tend to be inherited together. To sort out the order and location of genes on each chromosome, geneticists construct maps. A map is a representation of the linear order of genes on a chromosome and the distances between them. In general, there are two kinds of maps, *genetic maps* and *physical maps*. These maps are constructed using different techniques and use different units to measure the distance between genes.

Linkage
A condition in which two or more genes do not show independent assortment. Rather, they tend to be inherited together. Such genes are located on the same chromosome. By measuring the degree of recombination between such genes, the distance between them can be determined.

Genetic map
The arrangement and distance between genes on a chromosome deduced from studies of genetic recombination.

GENETIC MAPS. In a **genetic map,** the genes are arranged in a linear order, and the distance between any two genes is measured by how frequently crossing over takes place between them. In other words, genetic distance is measured by the frequency of crossing over between loci on the same chromosome. The units are expressed as percentage of recombination, with 1 map unit equal to a frequency of 1% recombination. (This unit is also known as a *centimorgan,* or *cM*). It is possible to construct genetic maps because even though two linked genes *tend* to be inherited together, they *do* separate from each other because of crossing over during meiosis. How often this event happens depends on the distance between the genes. The farther apart two genes are on a chromosome, the more likely it is that a crossover event will separate them. Conversely, the closer together two genes are on a chromosome, the less frequently they will be separated by a crossover event. Figure 4.19 shows a pedigree in which linkage is indicated between the gene for the ABO blood group (the isoagglutinogen, or *I* locus) and a condition called nail-patella syndrome. Individuals in which crossing over has occurred are indicated with asterisks. In this pedigree, 4 of the 16 individuals analyzed (25%) show recombination by crossing over. From this pedigree, the distance between the gene for the ABO locus and the gene controlling nail-patella syndrome is 25 map units. For accuracy, either a much larger pedigree or many more small pedigrees need to be examined to determine the extent of crossing over. Actually, when a large series of families is combined in an analysis, the map distance turns out to be about 10 units.

Mapping by observing crossovers is more accurate when the two genes being studied are relatively close together. If two genes are located so far apart on a chromosome that they recombine 50% of the time, the results would be the same as independent assortment. Thus, mapping can directly

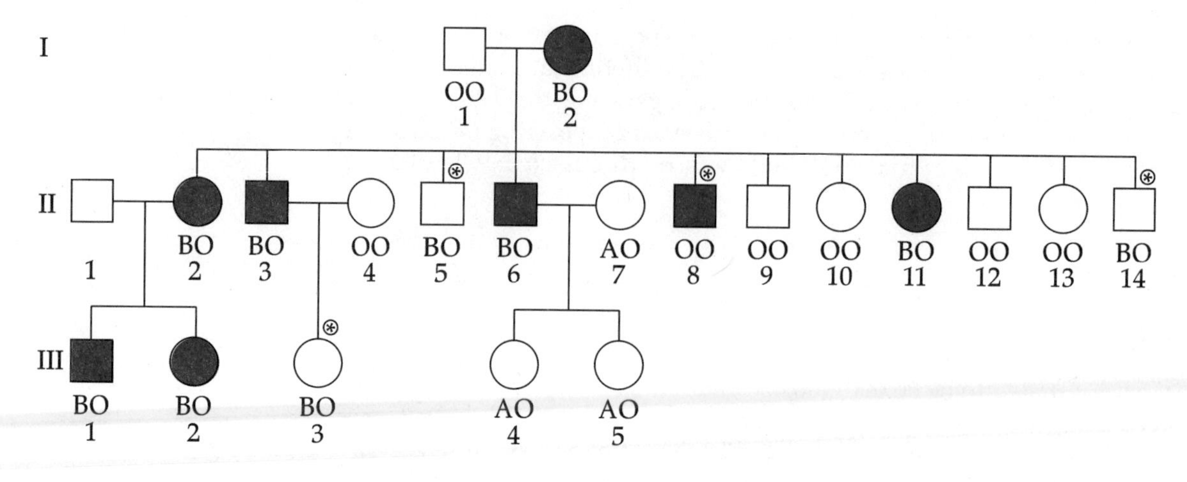

FIGURE 4.19 Pedigree showing linkage between nail-patella syndrome and the ABO locus. Shaded symbols represent those with nail-patella syndrome. Blood types are as shown. Nail-patella syndrome and type B blood tend to be inherited together and are said to show linkage. Individuals marked with an asterisk (II-5, II-8, II-14, and III-3) are recombinant, since they did not inherit both type B blood and nail-patella syndrome.

measure distances of less than 50 map units. Longer distances can be mapped by adding together the distances between intermediate loci.

In many, if not most, human linkage studies, it is difficult to establish linkage and measure genetic distance between genes. Large pedigrees with many offspring in three or more generations in which two genetic disorders are present are rare. In most cases, what are available are families consisting of parents and children, and maybe the grandparents. In these situations, a statistical technique known as the *lod* method is used to measure genetic distances. This measurement is carried out using computer programs with names like LINKMAP, designed to perform linkage analysis. The analysis is performed by taking an observed frequency of recombination between two genes derived from pedigree studies. Then one calculates the probability that the results would have been obtained if the two genes were linked and the probability that the results would have been obtained even if the two genes were *not* linked. The results are expressed as the $\log_{10}$ of the *ratio* of the two probabilities, or lod score (lod stands for the log of the odds).

Using both pedigree analysis and lod scores, genetic maps have been constructed for most of the human chromosomes, although in most cases, there are large distances between genetic markers. A genetic map for a human chromosome is shown in Figure 4.20. Note that the map constructed from meiotic crossovers in males gives the same order as a map constructed using crossovers in females, but the distances between genes in the two maps is very different. The reason is that the frequency of crossovers is different in males and females, and genetic maps, of course, are based on the frequency of crossovers. The reasons for the difference in recombination frequency between males and females are not yet understood.

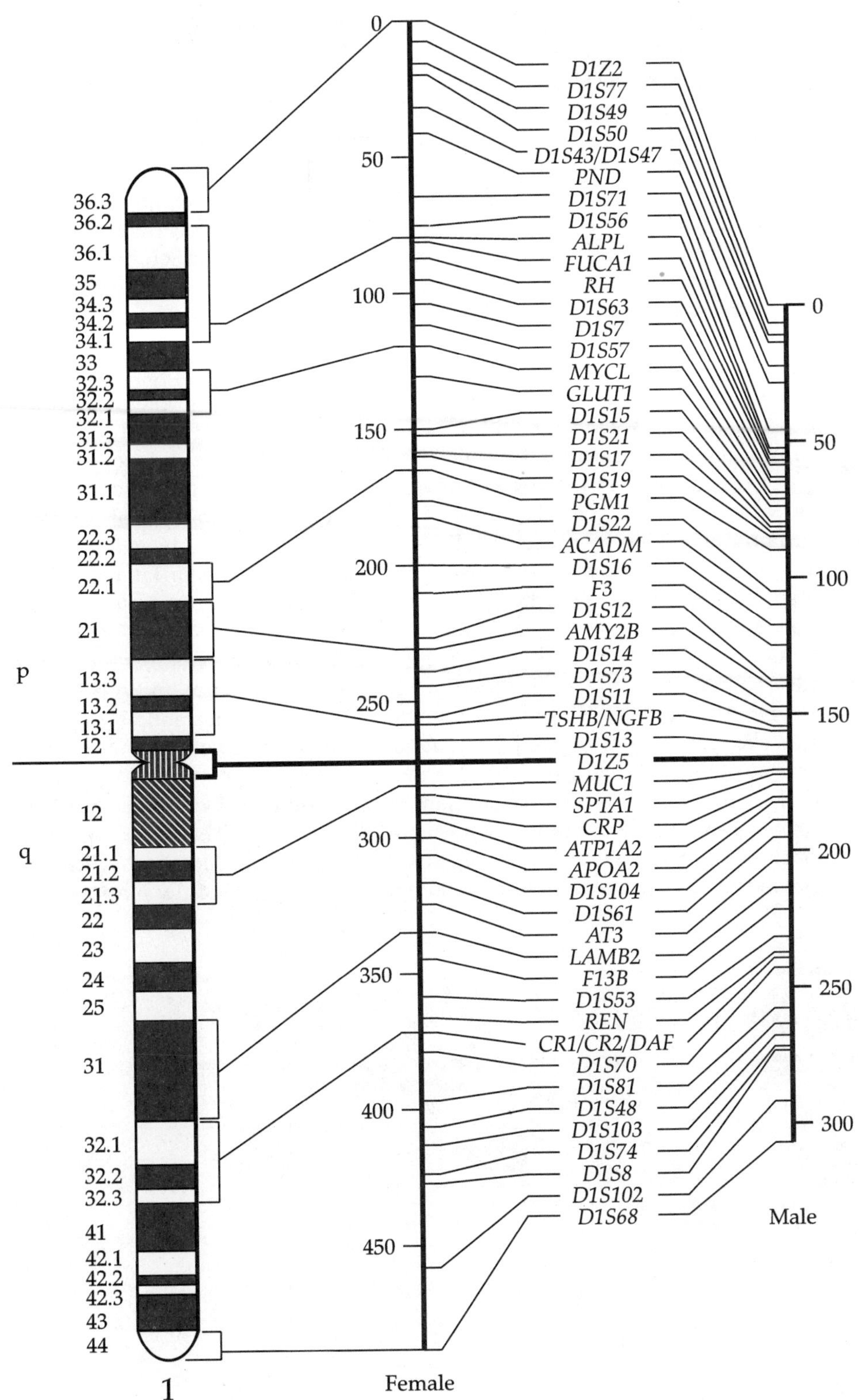

FIGURE 4.20 Genetic map of human chromosome 1. At the left is a drawing of the chromosome. Between the two vertical lines representing the female and male genetic maps are 58 genetic markers (beginning with D1Z2 and ending with D1S68) that have been placed on the map. These markers are not genes, but are molecular variants detected using recombinant DNA technology. Notice that the female genetic map is about 500 cM long, and the male genetic map is just over 300 cM long; this is a result of different frequencies of crossing over in males and females. This map provides a framework for locating genes on the chromosome as part of the Human Genome Project.

PHYSICAL MAPS. A physical map of a chromosome describes the physical location of genes on a chromosome. The construction of physical maps uses several techniques, all of which are different from those used in genetic maps. As a result, physical and genetic maps usually agree on the order of genes on a chromosome but can differ significantly in the distance between genes. Most often, the physical map is more accurate.

Physical mapping proceeds in stages. The first is the assignment of a gene to a particular chromosome. Several techniques are widely used to make such assignments. One such method is called **somatic cell hybridization** (see Concepts and Controversies, page 99). In this method, human and rodent (rat, mouse, or hamster) cells are fused together to produce a hybrid cell that carries a subset of human chromosomes. If a hybrid cell synthesizes a human gene product, it stands to reason that the gene for that product is on one of the human chromosomes present in the hybrid. By comparing the presence or absence of the gene product with the presence or absence of chromosomes, we can deduce which chromosome carries the gene.

Somatic cell hybridization
A method of mapping human genes that uses hybrid cells produced by fusing together cells from two different organisms.

Let us suppose that we have developed five different hybrid cell lines, known as 104, 107, 109, 113, and 117. The presence or absence of human chromosomes in each of these cell lines has been determined by cytogenetic analysis and is represented by a row of pluses and minuses in Table 4.5. Let us suppose further that biochemical analysis has shown that the human form of an enzyme known as β-galactosidase is found in cell lines 104, 107, and 113 but not in cell lines 109 and 117. This is symbolized by a column at the right side of Table 4.5, with P indicating that the enzyme is present and A that the enzyme is absent. Reading down the rows for the presence and absence of each chromosome in the cell lines, note that the pattern of presence and absence of the human form of the enzyme in the cell lines matches the pattern of presence and absence of human chromosome 3. No other human chromosome has this same pattern of distribution. Therefore we can conclude that the gene for β-galactosidase is on chromosome 3. Using somatic cell hybridization, human geneticists have successfully mapped an

TABLE 4.5 Human Chromosome Retained in Hybrid

Hybrid Cell Line	Chromosome Number 1	2	3	4	5	6	7	8	9	10	11	12	13	14	15	16	17	18	19	20	21	22	Enzyme Β-Galactosidase
104	–	+	+	+	+	–	–	–	+	+	–	–	+	+	+	+	–	+	+	–	+	–	P
107	–	+	+	–	–	+	+	–	–	+	–	–	–	+	–	+	–	–	+	–	–	–	P
109	–	+	–	+	–	+	–	+	+	+	–	+	–	+	+	–	+	–	+	+	–	–	A
113	+	–	+	–	–	+	+	–	–	+	+	+	–	+	+	+	+	+	–	–	+	–	P
117	–	–	–	–	–	+	–	+	+	+	+	+	+	–	–	+	–	–	+	–	+	+	A

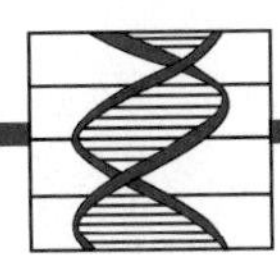

Concepts & Controversies

Somatic Cell Hybridization

In 1960 Georges Barsky mixed two different mouse-cell lines in the same culture flask. From this mixed cell culture a third cell type arose that contained a single nucleus and a chromosome number equal to the sum of the chromosome numbers of the two cell lines placed into the culture. Shortly thereafter other scientists discovered that they could induce cells from different species to fuse by treatment with a virus or a chemical known as polyethylene glycol.

Hybrids formed between human cells and rodent cells have two properties that make them useful in mapping human genes to specific chromosomes: (1) Hybrids lose a few human chromosomes at each mitotic division. The loss of human chromosomes is random and continues until the hybrid has eliminated all but a few human chromosomes. (2) The addition or withholding of chemical supplements to the culture medium can be used to kill all unfused cells and to permit only hybrid cells to grow.

As an example, let us consider the mapping of a specific human gene. Normally both human and hamster cells contain all the enzymes required for purine biosynthesis. One of these enzymes is known as *glycineamide ribonucleotide synthetase (GARS).* A mutant hamster cell line lacks this enzyme and can grow only when purines are added to the cell culture medium. If this mutant hamster cell line is fused with normal human cells and grown in medium without a purine supplement, only hybrid cells that carry the human GARS enzymes will grow (unfused human cells are killed in a separate step). When this fusion was carried out, it was found that the hybrid cells retained a variable number of human chromosomes. Only hybrid cells with chromosome 21 could grow on culture medium without added purines, and all hybrids that carried human chromosome 21 also carried the human form of the enzyme GARS. On the basis of these results, the gene for GARS was assigned to human chromosome 21.

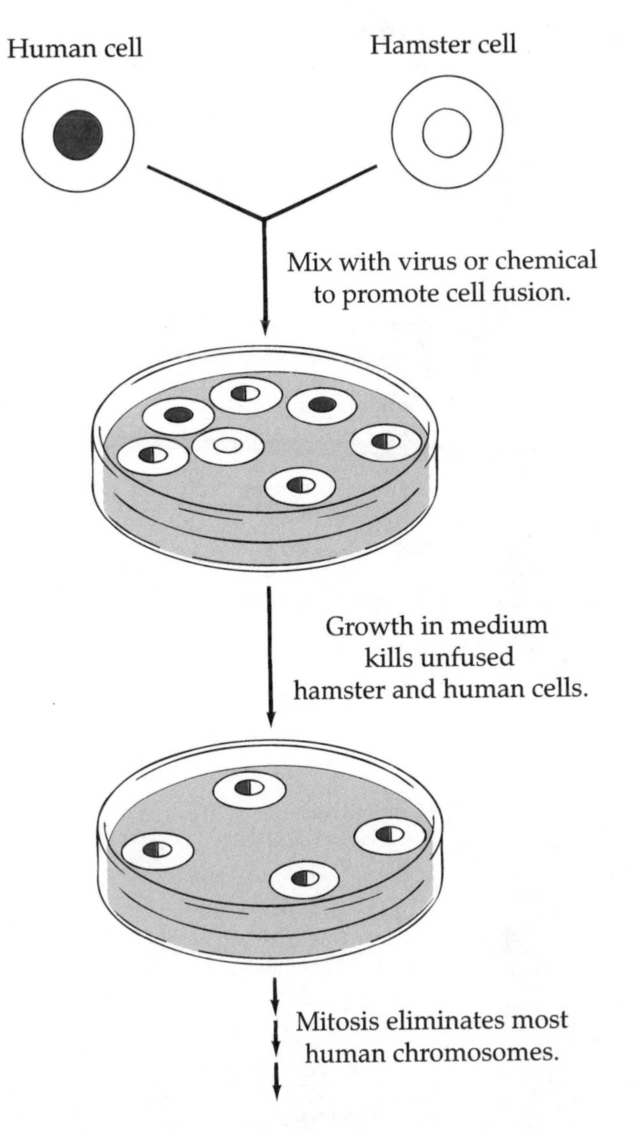

Several hundred human genes have been mapped to a specific chromosome by similar experiments. With improvements in the method it is now possible to transfer single human chromosomes to form a hybrid cell, and a technique known as *DNA-mediated gene transfer* permits the transfer of only a few human genes to a rodent cell. Together these methods can be used to assign a gene to a human chromosome and then to a restricted region of that chromosome.

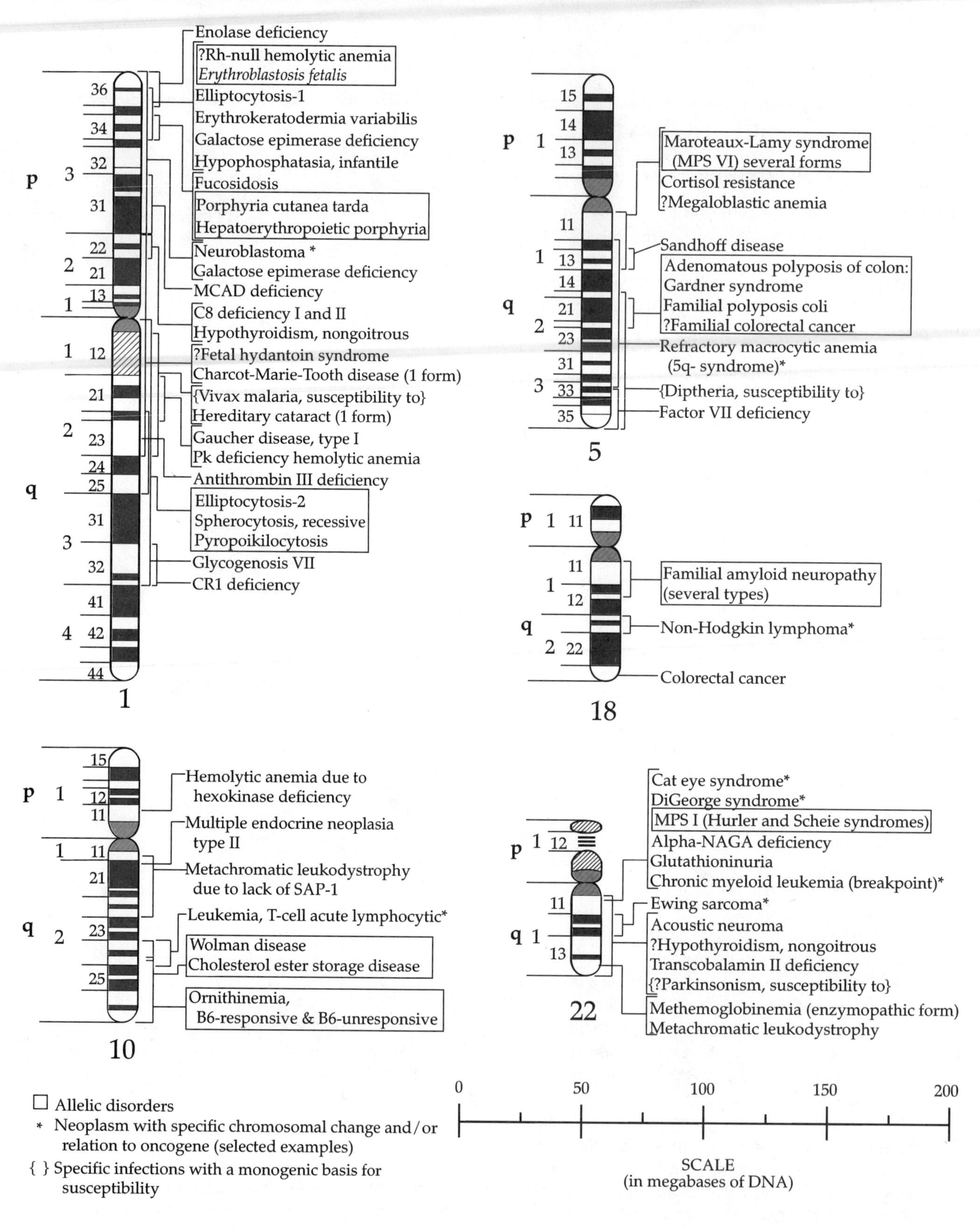

FIGURE 4.21 Maps for five human autosomes, showing the location of genetic disorders mapped to these chromosomes. Similar maps have been constructed for the remaining autosomes and the sex chromosomes.

ever-increasing number of genetic disorders to specific human chromosomes. Some of these are shown in Figure 4.21.

Another technique for assigning genes to chromosomes involves the use of fluorescent or radioactive probes that can bind to specific genes and be visualized in metaphase cells under the microscope. The localization of a gene by this method is known as fluorescent *in situ* hybridization (FISH) (Figure 4.22). In addition to assigning a gene to a specific chromosome, the FISH technique can provide some idea of where on the chromosome the gene is located.

In many cases the gene product associated with a genetic disorder is unknown or cannot be detected in hybrid cells, and in these cases somatic cell hybridization cannot be used to map genes. However, a technique developed through the use of recombinant DNA technology, *restriction fragment length polymorphism* (RFLP) analysis, enables a gene to be mapped even when its product, its function, or both are unknown. This method depends on following the inheritance of RFLP genetic markers and a given genetic disorder in pedigrees to look for linkage in the same way that ABO blood types and nail-patella syndrome were found to be linked. In this case the RFLP molecular markers have been previously assigned to a particular chromosome. By determining which marker is linked to the disease in a pedigree, the gene for the disease can be assigned to a particular chromosome. The location of the genes for cystic fibrosis, neurofibromatosis, Huntington disease, and dozens of other genetic conditions have been mapped in this way. The details of RFLP mapping will be presented in Chapter 8, after more information on molecular genetics has been given.

With the progress brought about by somatic cell hybridization, RFLP analysis, and other techniques, close to 4500 genes and markers have been assigned to human chromosomes, placing geneticists on the threshold of producing a high-resolution map of all human genes. To facilitate this

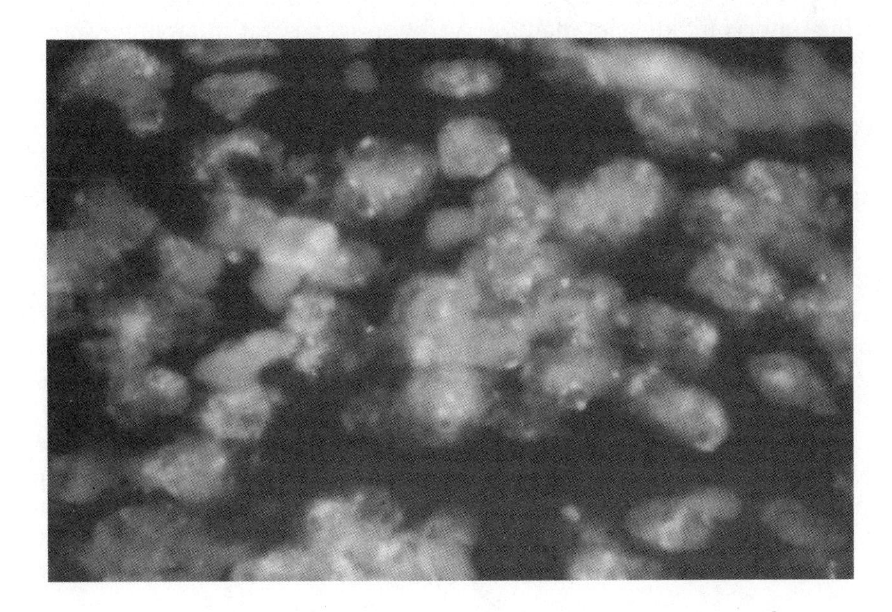

FIGURE 4.22 Localization of gene by *FISH*.

process, an international consortium of government and private agencies has undertaken a large-scale, accelerated project to map all the human genes over the next decade. This endeavor, known as the Human Genome Project, has as its goal the mapping of all gene locations and the determination of the precise order of the DNA subunits, or base pairs, that encode this genetic information. (A genome is the set of genetic information carried by a haploid cell, such as a gamete.) At the molecular level, this means establishing the identity of more than 3 billion base pairs of DNA, roughly the same amount of information contained in a library of 10,000 books. In this country, the Human Genome Project is being conducted jointly by the National Institutes of Health, a federal agency charged with overseeing biomedical research, and by the U.S. Department of Energy (DOE), whose Los Alamos and Lawrence-Berkeley laboratories have led the development of new technology being used in the mapping project.

From the beginning, it has been recognized that information generated by the Human Genome Project will raise social, ethical, and legal questions about the use of genetic information. Responding to these concerns, the National Center for Human Genome Research has established the Ethical, Legal, and Social Implications (ELSI) program to develop safeguards and standards to ensure that genetic information is used in a responsible manner. The social and legal aspects of human genetics will be considered in Chapter 20.

Variations in Gene Expression

Many traits, such as albinism, exhibit a regular and consistent pattern of gene expression. Other traits have variable effects, and in some cases mutant genes may be present but unexpressed. This relatively high level of phenotypic variation cannot be explained by mutation or lack of stability but by developmental interactions. Gene action is often stage- and time-specific during embryonic development. In addition, genes act in an environment with other genes and often form networks or batteries for specific developmental events. This distance between gene and phenotype provides an opportunity for variation within the network to affect gene expression.

Penetrance and Expressivity

The terms *penetrance* and *expressivity* define two different aspects of variation in gene expression. **Penetrance** is the likelihood that a given genotype will produce its characteristic phenotype. It is expressed as a percentage of the individuals with a given genotype who show the expected and characteristic phenotype. If a trait has 100% penetrance, the phenotype is always expressed. If the penetrance is 25%, the expected phenotype is present only 25% of the time. Both genetic and environmental factors can affect penetrance. **Expressivity,** on the other hand, refers to the degree of variation in the phenotypic expression that is present. For example, an inherited disorder such as cystic fibrosis can have severe or mild symptoms. As an example of penetrance, we can consider the trait known as **campodactyly.** This trait, controlled by a

Penetrance
The proportion of individuals with a given genotype who show an expected and characteristic phenotype.

Expressivity
The range of phenotypes resulting from a given genotype.

Campodactyly
A dominant human genetic trait that is expressed as immobile, bent little fingers.

dominant gene, is caused by improper attachment of muscles to bones in the little finger. The result is an immobile, bent little finger. In some individuals both little fingers are involved; in others, only one is affected; and in a small percentage of cases, neither finger is affected. The pedigree in Figure 4.23 illustrates these modes of inheritance. Since the trait is dominant, the expectation is that all heterozygotes and homozygotes would be affected on both hands. The pedigree shows that one individual (III-4) is not affected even though he passed the trait to his offspring.

Penetrance is a phenomenon that can be assessed only in a population of individuals. From the pedigree in Figure 4.23, we can estimate that nine people carry the dominant gene for campodactyly, and penetrance is seen in eight, giving 8/9, or 88%, penetrance. This is only an estimate because II-1, II-2, and III-1 produced no offspring and could also carry the dominant gene with no penetrance. Many more samples from other pedigrees would be necessary to establish a reliable figure for penetrance of this gene.

Expressivity is a term that applies to the degree of expression of a given trait. In the pedigree for campodactyly, some individuals are affected on the left hand, others on the right hand, and in one case both hands are affected. This variable gene expression is not a necessary or intrinsic property of genes but is a result of interactions with other genes and with nongenetic factors in the environment.

Age and Gene Expression

While a large number of genes act prenatally or early in development, the expression of other genes is delayed until later in life. One of the best known examples of such a gene is **Huntington disease** (HD). This disorder, controlled by a dominant gene, has its usual onset between the ages of 30 and 50 years. Affected individuals undergo a progressive degeneration of the nervous system, causing uncontrolled, jerky movements of the head and limbs and mental deterioration. The disease progresses slowly, and death

Huntington disease
A dominant genetic disorder characterized by involuntary movements of the limbs, mental deterioration, and death within 15 years of onset. Symptoms appear between 30 and 50 years of age.

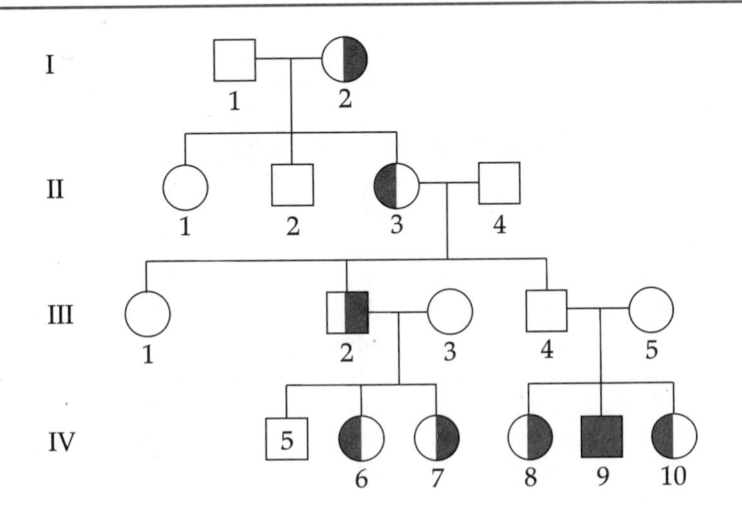

FIGURE 4.23 Pedigree of campodactyly, a dominant autosomal condition that illustrates penetrance and expressivity. In this pedigree, those with two affected hands are indicated with a fully shaded symbol. Those affected only in the left hand are indicated by shading the left half of the symbol. Those affected in the right hand have the symbol shaded on the right half. Unfilled symbols represent unaffected individuals.

FIGURE 4.24 King George III (1738–1820).

Porphyria
A genetic disorder inherited as a dominant trait that leads to intermittent attacks of pain and dementia, with symptoms first appearing in adulthood.

Pleiotropy
The appearance of several apparently unrelated phenotypic effects caused by a single gene.

occurs some 5 to 15 years after onset. This is a particularly insidious disease, since onset usually occurs after the affected person has started a family; his or her children have a 50% chance of developing the disease. Because those who will develop the disease can now be detected before onset of the symptoms, serious questions about such testing are being raised. Some of these issues will be discussed in Chapter 20.

Other genetic disorders also show late onset. **Porphyria,** an autosomal dominant disorder, is caused by an inability to correctly metabolize porphyrin, a component of hemoglobin. As levels of porphyrin increase, some is excreted, producing wine-colored urine. The elevated levels also cause episodes of seizures, intense physical pain, dementia, and psychosis. These symptoms rarely appear before puberty and usually appear in middle age. King George III, the British monarch during the American revolution, suffered from porphyria (Figure 4.24). He suffered his first major attack in 1788 at the age of 50 years. He became delerious and suffered convulsions. His physical condition soon improved, but he remained irrational and confused. Early in 1789 his mental functions spontaneously improved, although his physicians took the credit. Later, after two more episodes, the king was replaced on the throne by his son George IV. He died years later, blind and senile.

Pleiotropy

Many of the genetic diseases we have discussed exhibit multiple effects that are seemingly unrelated. In Marfan syndrome mutation of a single gene produces a weakened aorta, nearsightedness, and malformations of the sternum, or breastbone. Genes that produce several often apparently unrelated effects are called *pleiotropic* genes, and the phenomenon is known as **pleiotropy.** Sickle cell anemia produces a wide range of defects and is also an example of a pleiotropic gene. In sickle cell anemia, as mentioned earlier, the phenotype includes changes in the shape of red blood cells, intense pain, skin ulcers, weakness, and early death. The range of symptoms exhibited in pleiotropy usually arises from a single gene defect but is ultimately the result of developmental and environmental interactions. Pleiotropic effects are the rule rather than the exception in human genetics.

Dominance and Recessiveness

This chapter and earlier chapters have referred to genes as having dominant and recessive alleles and have symbolized them with upper- and lower-case letters (*A* and *a*). This convention is derived from the relationships between the pattern of inheritance of an allele and its phenotypic effect. If the phenotype is manifested only when identical alleles have been inherited from each parent, the trait is recessive. If the trait is present when inherited from only one parent, the allele is said to be dominant. Dominant traits are expressed in the heterozygous condition, and recessive traits are not. While the representation of alleles as dominant or recessive is a convenient way to think about genes, it should be remembered that the terms *dominant* and *recessive* refer to phenotypes and not to the gene itself. As we will see in later chapters, it is often possible to identify heterozygotes for recessive traits by the fact that they produce half as much of a gene product as the homozygous dominant combination. This does not change the recessive nature of the gene, since usually only homozygous recessive individuals are phenotypically affected by the mutant condition.

Recessive traits tend to have more serious phenotypic effects than those controlled by dominant genes. One reason for this is that recessive alleles can remain in the population in the heterozygous condition at a high frequency, affecting only a small number of homozygotes. If, on the other hand, a dominant mutation prevented survival or reproduction, it would quickly be eliminated from the population. As a result, many dominant genetic diseases appear more variable and diffuse in their phenotypic effects. If lethality occurs, it is usually after reproductive age has been reached. In other words, deleterious dominant traits survive because of low penetrance, variable expressivity, and delayed onset.

SUMMARY

1. The inheritance of single gene traits in humans is often referred to as Mendelian inheritance because of the pattern of segregation within families. These traits produce phenotypic ratios similar to those observed by Mendel in the pea plant. While the results of studies in peas and humans may be similar, the methods are somewhat different. Instead of direct experimental crosses, human traits are traced by the construction of pedigrees that follow a trait through several generations of a family.

2. Information in the pedigree is used to determine the mode of inheritance. These modes include: autosomal dominant, autosomal recessive, X-linked dominant, X-linked recessive, and Y-linked.

3. The results of pedigree analysis depend on the distribution of alleles via gametes from parent to child. Since the number of offspring is usually small, large deviations from expected ratios of segregation are often encountered. This effect can be controlled by the examination of pedigrees from a large number of families and pooling the information from many sources to confirm the mode of inheritance for a given trait.

4. Guidelines for pedigree analysis can be used to determine if a trait is inherited in an autosomal recessive fashion. Other guidelines can be used to establish whether the trait is inherited in an autosomal dominant fashion.

5. Several factors can alter an expected pattern of inheritance, including the age of onset, penetrance, expressivity, pleiotropy, and environmental interactions.

6. The distinction between dominant and recessive traits is not always absolute. Although in some genotypes the heterozygote can be identified, the

trait is regarded as recessive because the heterozygote is phenotypically unaffected, while only the homozygous recessive exhibits a distinctive phenotype.

QUESTIONS AND PROBLEMS

1. Define the following pedigree symbols:

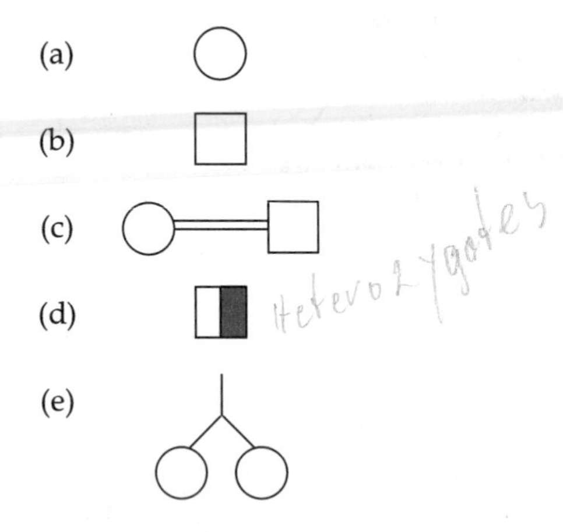

2. Identify the mode of inheritance suggested by the following pedigree:

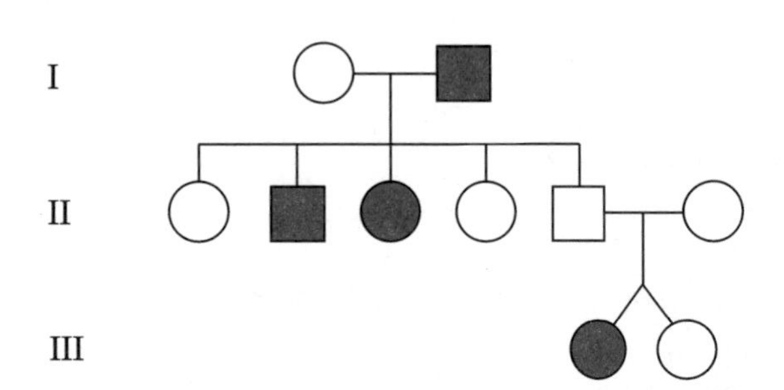

3. What mode of inheritance is suggested by the following pedigree?

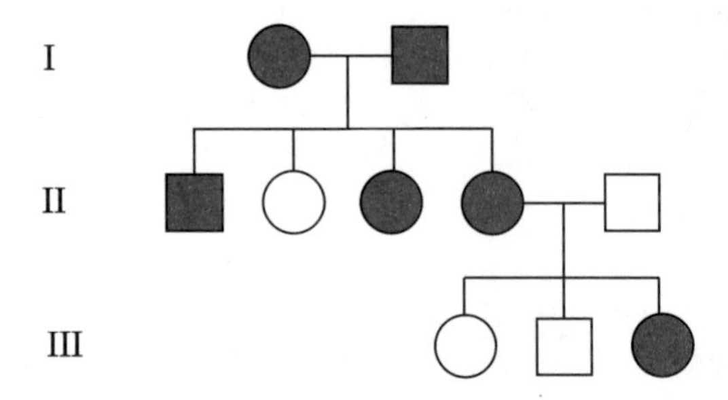

4. Does the indicated individual (III-5) show the trait in question?

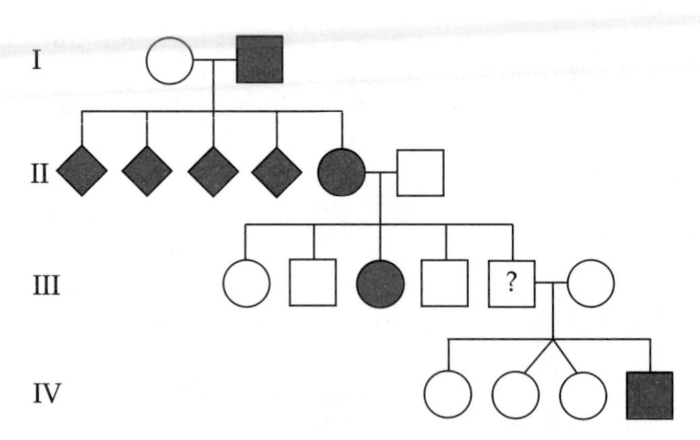

5. Assume that the father of the propositus is homozygous for the trait in question. Explain a mode of inheritance consistent with this pedigree. In particular, explain the phenotype of the propositus.

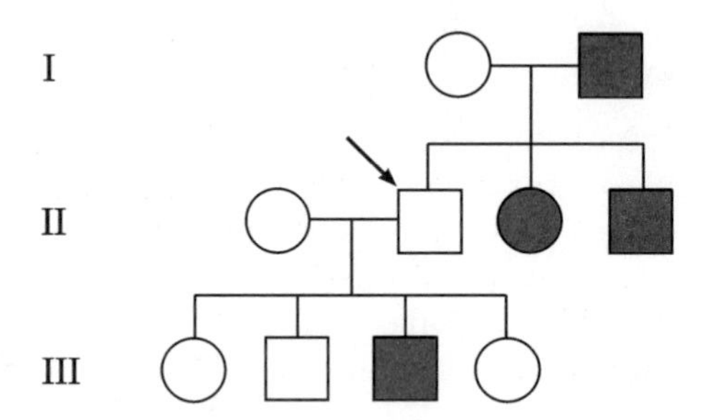

6. Construct a pedigree based on the following information: (1) The proposita exhibits a trait. (2) Neither her husband nor her only sibling (an older brother) exhibits the trait. (3) The proposita has four children by her current husband: the oldest is a boy, followed by a girl, with identical twin girls being the youngest. Only the second oldest child fails to exhibit the trait. (4) The parents of the proposita do not show the trait.

a. What is the mode of inheritance of this trait?
b. Can you deduce the genotype of the husband of the proposita for this trait?

7. Using the following pedigree:
   a. Deduce a compatible mode of inheritance.
   b. Identify the genotype of the individual in question.

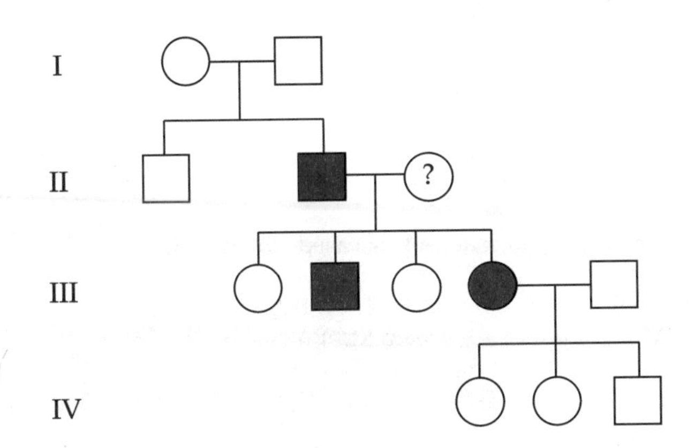

8. From the pedigrees illustrated below, choose the one consistent with the following human diseases/conditions.
   a. ocular-cutaneous albinism
   b. Marfan syndrome
   c. campodactyly

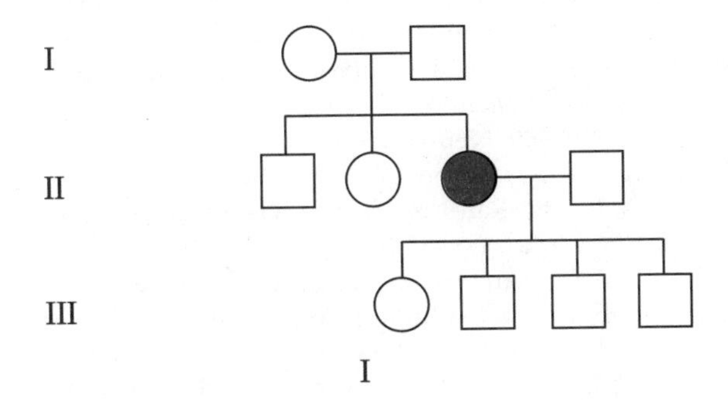

I

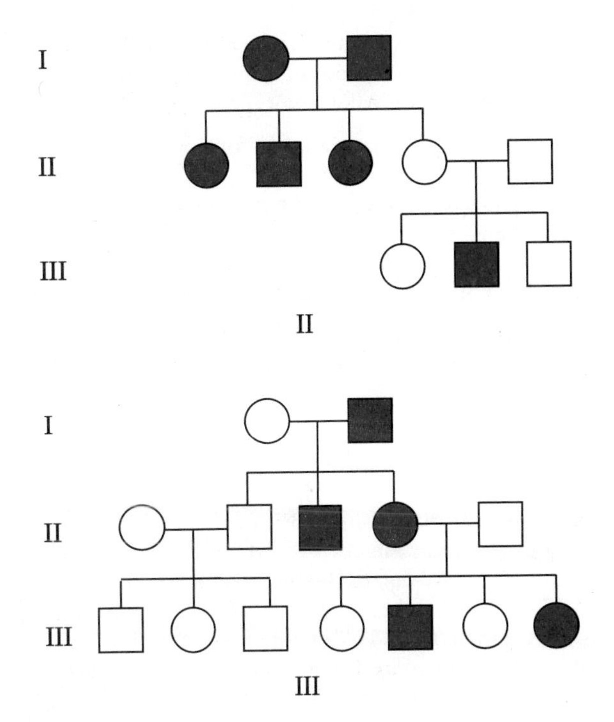

II

III

9. Define penetrance and expressivity.

10. A proposita suffering from an unidentified disease seeks the advice of a genetic counselor prior to starting a family. Based on the following data, the counselor constructs a pedigree encompassing three generations: (1) The maternal grandfather of the proposita suffers from the disease. (2) The mother of the proposita is unaffected and is the youngest of five children, the three oldest being male. (3) The proposita has an affected older sister, but the youngest siblings are unaffected twins (boy and girl). (4) All individuals suffering from the disease have been revealed. Duplicate the counselor's feat.

11. The father of 12 children begins to show symptoms of neurofibromatosis.
    a. What is the probability that Sam, the man's second oldest son (II-2), will suffer from the disease if he lives a normal life span? (Sam's mother and her ancestors do not have the disease.)
    b. Can you infer anything about the presence of the disease in Sam's paternal grandparents?

12. The only daughter of an only son has blood serum possessing antibodies directed against type A and type B blood. Her mother possesses type A antigens. Her paternal grandfather possesses antibodies to type B blood, and he can neither give nor receive blood from his wife due to ABO incompatibility.

However, their son can donate blood to either of them. Construct a pedigree containing this information, and include ABO genotypes.

13. A hypothetical human trait is controlled by a single gene. Four alleles of this gene have been identified: *a*, *b*, *c*, and *d*. Alleles *a*, *b*, and *c* are all codominant; allele *d* is recessive to all other alleles.
    a. How many phenotypes are possible?
    b. How many genotypes are possible?

14. A 54-year-old man begins showing signs of Huntington disease. His daughter is 29 years old and has three children. What is the probability that her youngest child will develop Huntington disease if neither the father's family nor the maternal grandmother has the disease? (Assume a long life for the child.)

15. Contrast a polygenic mode of inheritance with the concept of pleiotropy.

16. How can dominant lethal alleles survive in a population?

For Further Reading

Barnhart, B. J. 1988. The human genome project: a DOE perspective. *Basic Life Sci.* **46**: 161–166.

Cawthon, R. M., Weiss, R., Xu, G., Viskochil, D., Culver, M., Stevens, J., Robertson, M., Dunn, D., Gesteland, R., O'Connell, P., and White, R. 1990. A major segment of the neurofibromatosis type 1 gene: cDNA sequence, genomic structure, and point mutations. *Cell* **62**: 193–201.

Collins, F. S., Ponder, B. A., Seizinger, B. R., and Epstein, C. J. 1989. The von Recklinghausen neurofibromatosis region on chromosome 17: Genetic and physical maps come into focus. *Am. J. Hum. Genet.* **44**: 1–5.

Embury, S. H. 1986. The clinical pathology of sickle cell disease. *Ann. Rev. Med.* **37**:361–376.

Francomano, C. A., Le, P. L., and Pyeritz, R. E. 1988. Molecular genetic studies in achondroplasia. *Basic Life Sci.* **48**: 53–58.

Huntington's Disease Collaborative Research Group. 1993. A novel gene containing a trinucleotide repeat that is expanded and unstable on Huntington's disease chromosomes. Cell **72:** 971-983.

Kinnear, P. E., Jay, B., and Witkop, C. J., Jr. 1985. Albinism. *Surv. Ophthalmol.* **30**:75–101.

Lucky, P. A., and Nordlund, J. J. 1985. The biology of the pigmentary system and its disorders. *Dermatol. Clin.* **3**: 197–216.

Macalpine, I., and Hunter, R. 1969. Porphyria and King George III. *Sci. Am.* **221**(July):38–46.

Peltonen, L. and Kainulainen, K. 1992. Elucidation of the gene defect in Marfan syndrome. Success by two complementary research strategies. FEBS Letters **307:** 116-121.

Prockop, D. J. 1985. Mutations in collagen genes: Consequences for rare and common diseases. *J. Clin. Invest.* **75**: 783–787.

Ramirez, F., Sangiorgi, F. O., and Tsipouras, P. 1986. Human collagens: Biochemical, molecular and genetic features in normal and diseased states. *Horiz. Biochem. Biophys.* **8**: 341–375.

Rommens, J. M., Iannuzzi, M. C., Bat-Sheva, K., Drumm, M. L., Melmer, G., Dean, M., Rozmahel, R., Cole, J. L., Kennedy, D., Hidaka, N., Zsiga, M., Buchwald, M., Riordan, J. R., Tsui, L. C., and Collins, F. S. 1989. Identification of the cystic fibrosis gene: Chromosome walking and jumping. *Science* **245**: 1059–1065.

Rouleau, G. A., Wertelecki, W., Haines, J. L., Hobbs, W., Trofatter, J. A., Seizinger, B., Martuza, R., Superneau, D., Conneally, P. M., and Gusella, J. 1987. Genetic linkage of bilateral acoustic neurofibromatosis to a DNA marker on chromosome 22. *Nature* **329**: 246–248.

Smithies, O. 1993. Animal models of human genetic diseases. Trends Genet. **9:** 112-116.

Stanbury, J. B., Wyngaarden, J. B., and Fredrickson, D. S. 1983. *The Metabolic Basis of Inherited Disease.* 5th ed. New York: McGraw-Hill.

Stern, C. 1973. *Principles of Human Genetics.* 3d ed. San Francisco: Freeman.

Stokes, R. W. 1986. Neurofibromatosis: A review of the literature. *J. Am. Osteopath. Assoc.* **86**: 49–52.

Tsipouras, P., and Ramirez, F. 1987. Genetic disorders of collagen. *J. Med. Genet.* **24**: 2–8.

Wallace, M. R., Marchuk, D. A., Anderson, L. B., Letcher, R., Odeh, H. M., Saulino, A. M., Fountain, J. W., Brereton, A., Nicholson, J., Mitchell, A. L., Brownstein, B. H., and Collins, F. S. 1990. Type 1 neurofibromatosis gene: identification of a large transcript disrupted in three NF1 patients. *Science* **249**: 181–186.

Welsh, M. and Smith, A. 1993. Molecular mechanisms of CFTR chloride channel dysfunction in cystic fibrosis. Cell **73:** 1251-1254.

Yoshida, A. 1982. Biochemical genetics of the human blood group ABO system. *Am. J. Hum. Genet.* **34**: 1–14.

CHAPTER 5

SEX LINKAGE AND SEX DETERMINATION

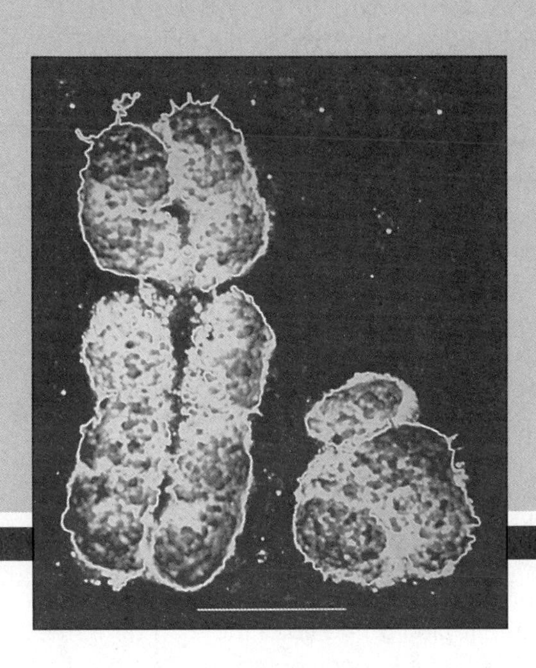

CHAPTER OUTLINE

The Talmud is a compendium of the oral law of the Jewish religion that was assembled by scholars by about A.D. 300. This document, used in teaching, is a mixture of science, law, medicine, and philosophy. In one section of the Talmud it is written:

> If (a woman) circumcised her first son and he died, and she had the second son circumcised and he died, she must not circumcise her third son. It once happened with four sisters from Tzippori that the first

> had her son circumcised and he died, the second sister had her son circumcised and he died, the third sister had her son circumcised and he also died, and the fourth sister came before Rabbi Shimon ben Gamliel and he told her, ''You must not circumcise your son.''

This section and others with similar stories constitute the first known description of hemophilia A, a genetic condition characterized by defective blood clotting. Affected individuals often bleed to death from minor cuts or hemorrhaging caused by a bruise. The gene for this trait is carried on the X chromosome and is said to be sex-linked. According to the Talmud, the sons of such a woman's sisters should not be circumcised either, but the rule does not apply to the sons of her brothers. As we will see later in this chapter, the pattern of inheritance is typical for a sex-linked recessive trait. By the 18th century another X-linked trait, colorblindness, was known to be inherited, and the transmission of this condition could be predicted accurately, although the basis for this pattern of inheritance was unknown.

In previous chapters we examined examples of dominant and recessive traits that are inherited with equal frequency by males and females. For example, there is a 25% chance that the offspring of two heterozygotes for sickle cell anemia will exhibit the trait, and the affected children are just as likely to be male as female. The reason for this is that genes responsible for these traits are located on any of the 22 pairs of homologous chromosomes known as autosomes. Males and females possess the same number and kind of autosomes, and therefore, the ability to transmit these chromosomes via meiosis and random fertilization is equal in both sexes.

As noted earlier, the 23rd pair of chromosomes is not the same in both sexes. Females have a pair of X chromosomes, and males have an X and a Y chromosome. Since the two sexes are not equivalent with respect to this pair of chromosomes, it seems obvious to ask whether they are equivalent with respect to the genes located on these chromosomes. The difference between the X and Y chromosome is emphasized by the fact that the chromosomes are structurally very different.

The X chromosome is a medium-sized, metacentric chromosome with a distinct banding pattern. The Y chromosome is a much smaller, acrocentric chromosome, about 25% as large as the X, although its size is somewhat variable. At meiosis the X and Y chromosomes do not pair, or synapse, along their entire length but only in a limited region at the tip of the short arms. The absence of pairing indicates that there is little genetic homology between the two chromosomes and suggests that the great majority of genes present on the X chromosome are not represented on the Y.

This lack of genetic equivalence between the X and Y chromosomes is responsible for a pattern of transmission known as sex-linked inheritance. For genes on the X chromosome, the pattern is said to be X-linked, and those genes on the Y are said to exhibit Y-linkage. Most examples of sex linkage involve the X chromosome, since few genes are located on the Y chromosome. Because a female carries two copies of the X chromosome, she has two alleles of all X-linked genes and can be heterozygous or homozygous for any of them. A male, on the other hand, has only one copy of the X chromosome, and since the Y chromosome does not carry alleles of most X-linked genes, he cannot be heterozygous or homozygous for most genes on the X chromosome. As a result, males will express all genes on the X chromosome, whether they are dominant or recessive. This condition is of great importance in explaining why males are afflicted with sex-linked recessive genetic disorders more often than females.

In this chapter we will investigate the pattern of inheritance and gene expression for sex-linked traits and then direct our attention to the role of sex chromosomes in determining the phenotypic sex in humans. We will examine the sequence of events in sexual differentiation in which the chromosomally determined sex is translated into the primary and secondary sexual characteristics that distinguish males and females. We will also discuss the related topics of dosage compensation and sex-limited and sex-influenced traits.

X-Linked Inheritance

The empirical observations about **X-linkage** and the inheritance of hemophilia and colorblindness were finally explained in the early part of this century when the X and Y chromosomes were identified and scientists realized that genes were located on chromosomes. The characteristic pattern of sex-linked inheritance derives from the fact that males transmit their X chromosome to all daughters and a Y chromosome to all sons. Females randomly pass on one or the other X chromosome to all daughters and to all sons (Figure 5.1). If a trait is X-linked, a male will pass it to all his daughters, who may be heterozygous or homozygous for the condition. Sons have a 50%

X-Linkage
The pattern of inheritance that results from genes located on the X chromosome.

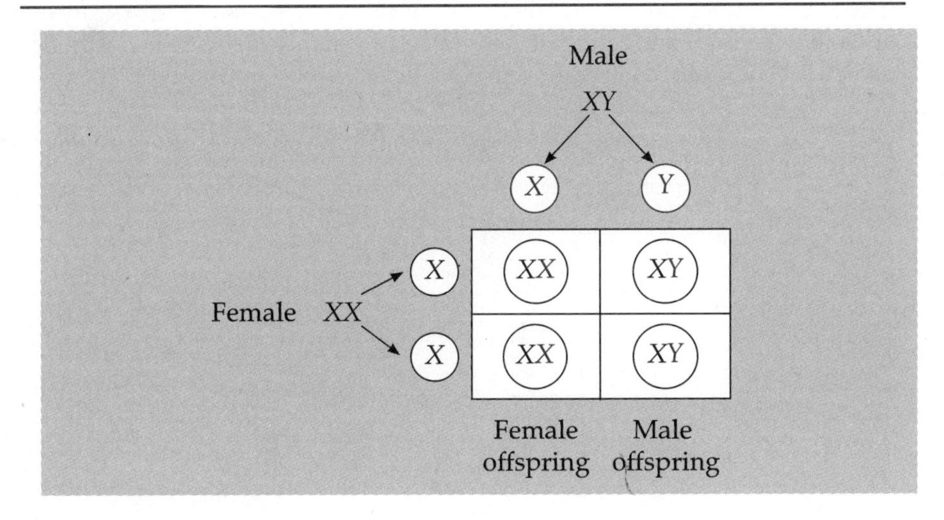

FIGURE 5.1 The segregation of sex chromosomes into gametes. Homogametic females produce gametes containing X chromosomes. Heterogametic males produce two kinds of gametes in equal proportions. Note that female offspring receive an X chromosome from their father and that male offspring receive a Y chromosome from their father. All offspring receive an X chromosome from their mother.

chance of receiving an X-linked recessive trait from their mother. If a trait is Y-linked, it will be passed directly from father to son.

In the following section we will consider examples of sex-linked inheritance and explore the characteristic pedigrees in detail.

X-Linked Dominant Inheritance

Hemizygous
A gene present on the X chromosome that is expressed in males in both the recessive and dominant condition.

Hypophosphatemia
A dominant X-linked condition that produces bone disorders such as rickets.

Females carry two copies of the X chromosome and can be homozygous or heterozygous for X-linked traits. Males carry only one copy of the X chromosome and therefore cannot be homozygous or heterozygous for X-linked traits. The term **hemizygous** is used to define genes present on the X chromosome that are expressed in both the recessive and dominant condition in males. Traits carried by genes on the X chromosome are classified as dominant or recessive based on their pattern of expression in females. Only a small number of dominant traits map to the X-chromosome. One of the best known examples is a phosphate deficiency known as **hypophosphatemia,** which causes a type of rickets, or bowleggedness, that cannot be cured with vitamin D. Dominant X-linked traits have a pattern of transmission that is different from that seen in autosomal or sex-linked recessive genes and have three distinctive characteristics:

1. Affected males produce all affected daughters and no affected sons.
2. A heterozygous affected female will transmit the trait to half her children, with males and females equally affected.
3. On average, twice as many females will be affected than males.

As expected, a homozygous female will transmit the trait to all her offspring. In distinguishing X-linked dominant inheritance from autosomal dominant traits, only the children of affected males are informative. A pedigree for the inheritance of an X-linked dominant trait is shown in Figure 5.2.

X-Linked Recessive Inheritance

COLORBLINDNESS. Until recently colorblindness was paradoxically one of the best known and yet least understood human genetic traits. Although pedigrees involving colorblindness have been known since the 18th century,

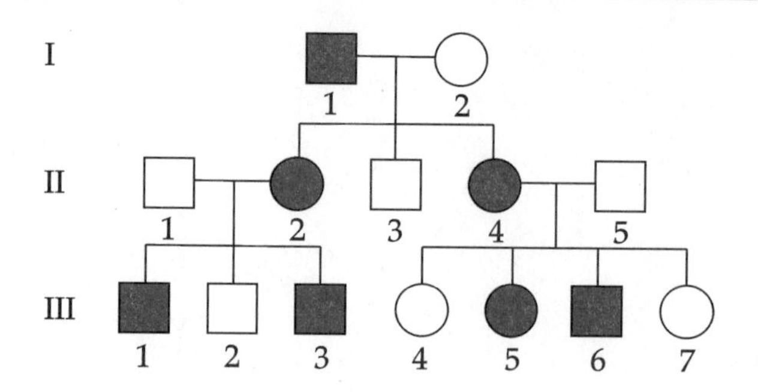

FIGURE 5.2 Pedigree typical for X-linked dominant inheritance.

assignment of the major forms of the condition as X-linked traits was not made until early in this century. Another 75 years were to pass before the molecular basis for these vision defects was resolved. Colorblindness is actually a collection of abnormalities, many of which are extremely rare conditions. The most common forms of colorblindness, known as red-green blindness, affect about 8% of the male population in the United States. (See the second color insert.) Red blindness, or *protanopia*, is characterized by the inability to see red as a distinct color. Green blindness, or *deuteranopia*, is the inability to distinguish green and other colors in the middle of the visual spectrum. Both red blindness and green blindness are inherited as X-linked recessive traits. A rare form of colorblindness, *tritanopia*, or blue blindness, is inherited as an autosomal dominant condition that maps to chromosome 7. All three genes encode the information for members of a class of gene products known as *opsins*. Opsins are present in cone cells of the retinal layer and function by binding to visual pigments in the red-, green-, or blue-cone cells, making the visual pigment/opsin complex sensitive to light of a given wavelength. If, for example, the red-opsin gene product is defective or absent, function of the red cones is impaired, and red colorblindness results. Similarly, defects in the green or blue opsins produce green and blue blindness.

Ironically the clarification of the molecular basis of color vision has revealed that while most individuals carry X chromosomes with a single copy of the red opsin gene, the green opsin genes can be present in one, two, or three copies, increasing the possible numbers of genotypic and phenotypic combinations. Although the situation at the molecular level may be complex, the pedigree of colorblindness phenotypes can be used to demonstrate the results of matings in several combinations (Figure 5.3). These pedigrees demonstrate the patterns of transmission that are observed for sex-linked recessive traits. These patterns can be summarized as follows:

1. Hemizygous males and homozygous females are affected.
2. Phenotypic expression is much more common in males than in females, and in the case of rare alleles, males are almost exclusively affected.
3. Affected males transmit the gene to all daughters but not to any sons.
4. Daughters of affected males will usually be heterozygous and therefore unaffected. Sons of heterozygous females have a 50% chance of receiving the recessive gene.

Some autosomal recessive conditions are expressed only in males. To distinguish X-linked recessive inheritance from these conditions, the phenotype of affected males should be examined. In X-linked recessive traits, all sons of affected males will be unaffected. Table 5.1 lists some sex-linked recessive conditions.

HEMOPHILIA. As mentioned earlier, **hemophilia,** or "bleeder's disease," is characterized by defects in the mechanism of blood clotting. Clot formation is a complex physiologic process, requiring the interaction of many gene products. A defect in any of these steps can cause a delay or lack of clot

Hemophilia
Genetic disorders characterized by defects in the blood clotting mechanism.

formation. In hemophilia A, clotting factor VIII is reduced or missing. Another form of hemophilia, known as hemophilia B, or "Christmas disease," is caused by a defect in clotting factor IX (the disorder is called Christmas disease because it was first described as a genetic disorder in a 5-year-old boy whose name was Christmas). As it happens, the loci for both these genes are on the X chromosome. The patterns of inheritance for X-linked recessive conditions are shown in Figure 5.4. A third form of hemophilia is caused by an autosomal recessive gene. Hemophilia A is the commonest form of X-linked hemophilia and occurs with a frequency of 1 in 10,000 males. Because females have to be homozygous recessive to be hemophiliac, the frequency is much lower, on the order of 1 in 100 million. This low frequency is also due to the fact that a female hemophiliac would have to result from the mating of a heterozygous female with a hemophiliac male (see Figure 5.3). Until recently, few hemophiliac males survived to a reproductive age, lowering the chances for the production of hemophiliac females.

The treatment for hemophilia involves blood transfusions of normal blood carrying the required clotting factors or injection with a freeze-dried

A. Normal male × carrier female

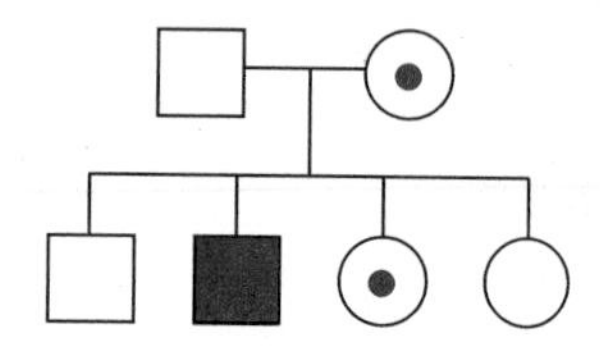

Half of all sons normal.
Half of all sons colorblind.
All daughters normal, but
half are carriers.

B. Colorblind male × normal female

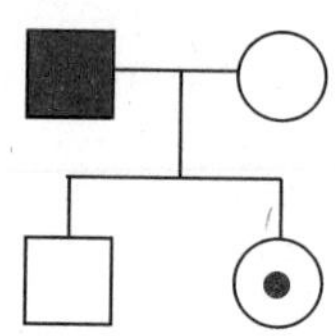

All sons normal.
All daughters normal,
but all are carriers.

C. Colorblind male × carrier female

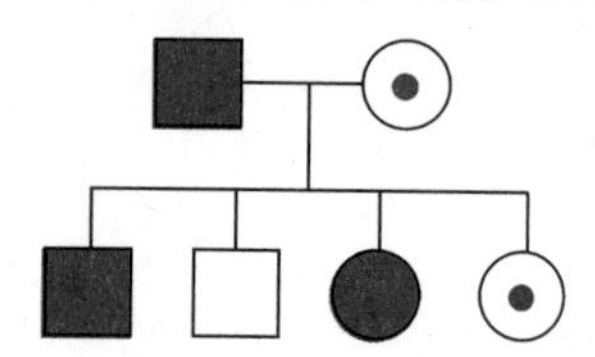

Half of sons colorblind.
Half of sons normal.
Half of daughters colorblind.
Half of daughters carriers.

D. Normal male × colorblind female

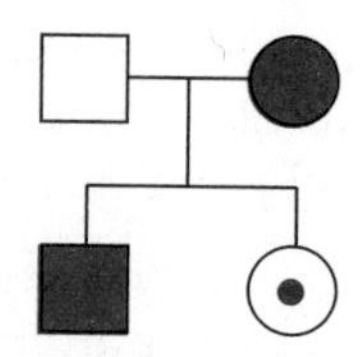

All sons colorblind.
All daughters carriers.

FIGURE 5.3 Four possible modes of inheritance for an X-linked recessive trait, using colorblindness as an example.

TABLE 5.1 Some X-Linked Recessive Genetic Traits

Traits	Phenotype
Adrenoleukodystrophy	Atrophy of adrenal glands, mental deterioration; death 1 to 5 years after onset
Colorblindness	
Deuteranopia	Insensitivity to green light; 60% to 75% of colorblindness
Protanopia	Insensitivity to red light; 25% to 40% of colorblindness
Fabry disease	Metabolic defect caused by lack of enzyme alpha-galactosidase A; progressive cardiac, renal problems, early death
Glucose-6-phosphate dehydrogenase deficiency	Benign condition that can produce severe, even fatal anemia in presence of certain foods, drugs
Hemophilia A	Inability to form blood clots; caused by lack of clotting factor VIII
Hemophilia B	"Christmas disease"; clotting defect cause by lack of factor IX
Icthyosis	Skin disorder causing large, dark scales on extremities, trunk
Lesch-Nyhan syndrome	Metabolic defect caused by lack of enyzme hypoxanthine-guanine phosphoribosyl transferase (HGPRT); causes mental retardation, self-mutilation, early death
Muscular dystrophy	Duchenne-type, progressive; fatal condition accompanied by muscle wasting

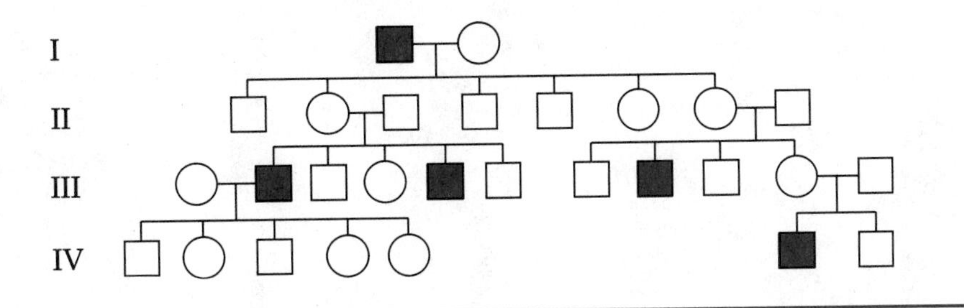

FIGURE 5.4 Pedigree typical for X-linked recessive inheritance.

concentrate prepared from donor serum. Costs for such treatments are high (over $25,000 a year) and are not always effective. The human gene for clotting factor VIII has been isolated and cloned using recombinant DNA techniques, opening the way for commercial production of clotting factor.

This should greatly improve the availability of treatment for hemophilia. In addition, hemophiliacs being treated with clotting factor derived from donated blood are at risk for viral diseases, such as hepatitis and AIDS. The cloned clotting factors are free from contaminating human viruses.

Pedigree analysis indicates that Queen Victoria of England (the granddaughter of King George III) was a carrier for this gene. Since she passed the gene on to several of her children (one affected male, two carrier daughters, and one possible carrier daughter; Figure 5.5), it is likely that the mutation occurred in the X chromosome she received from one of her parents. Although this mutation spread through the royal houses of Europe, the present royal family of England is free of hemophilia, since it is descended from Edward VII, an unaffected son of Victoria (Figure 5.6).

This royal pedigree is of interest for several reasons. First, it is a large and extensive pedigree covering six or more generations, and it clearly shows

FIGURE 5.5 Queen Victoria and some of her descendants. Her daughter Beatrice transmitted hemophilia to the royal family of Spain, and her granddaughter Alix carried the disorder to the royal family of Russia.

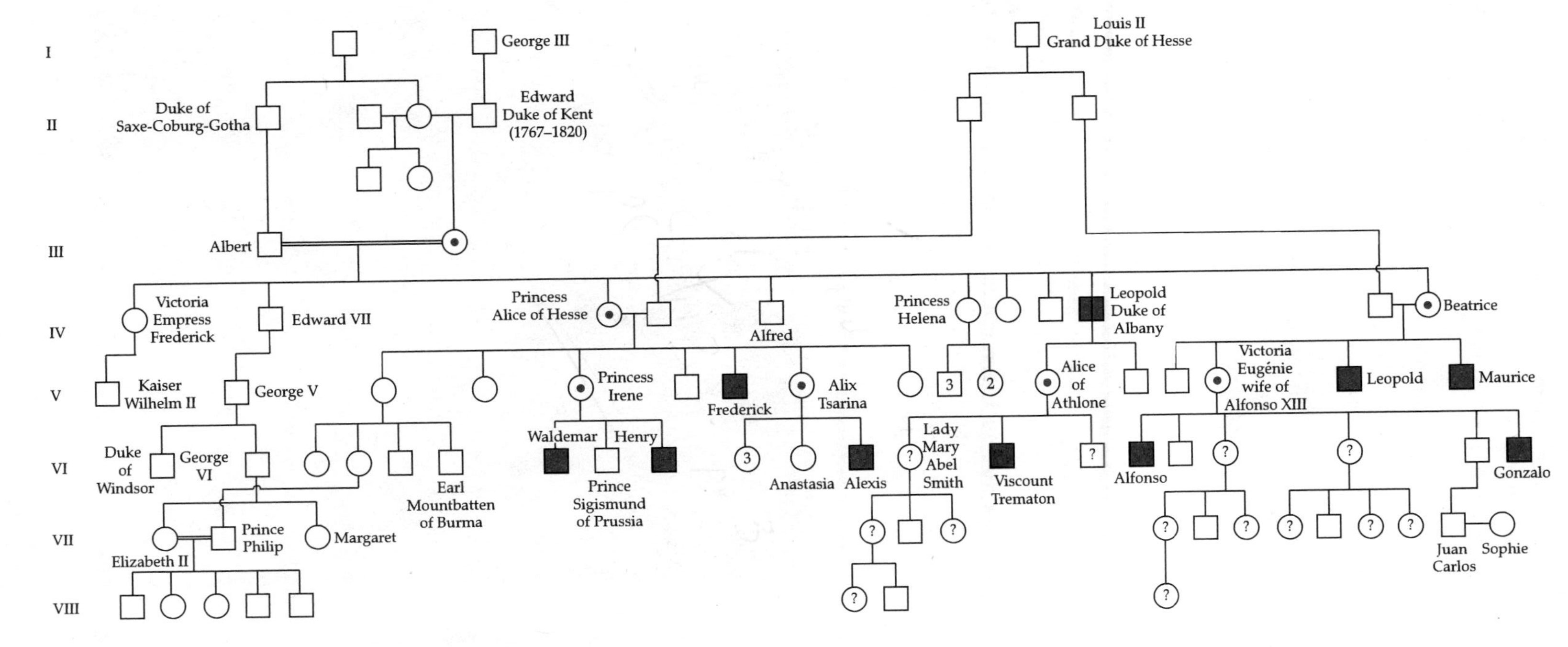

FIGURE 5.6 Pedigree of Queen Victoria showing her descent from George III and the transmission of the sex-linked trait hemophilia.

the pattern of transmission for an X-linked recessive trait through several generations. In contrast, most pedigrees are much smaller, usually consisting of two parents and their offspring, often containing one or two affected sons. Second, in many cases involving recessive traits, it is difficult to pinpoint the origin of the mutation. The pedigree of Queen Victoria is an exception, allowing the source of the mutation to be identified. Third, because the mutation arose in a royal family that intermarried across Europe and Russia, the pedigree is of historical interest.

Perhaps the most important case of hemophilia among Victoria's offspring involved the royal family of Russia. Victoria's granddaughter Alix, a carrier, married Czar Nicholas II of Russia. She gave birth to four daughters and then a son, Alexis (Figure 5.7). Their son, the heir to the throne of the 400-year-old Romanov dynasty, had hemophilia. Frustrated by the failure of the medical community to cure Alexis, the royal couple turned to a series of spiritualists, including the monk Rasputin. While under Rasputin's care, Alexis recovered from several episodes of bleeding, and Rasputin became a powerful adviser to the royal family. Historians have argued that the Czar's preoccupation with Alexis's health and the insidious influence of Rasputin contributed to the revolution that overthrew the throne. Other historians point out that Nicholas II was a weak czar and that revolution was inevitable; but it is interesting to speculate that much of Russian history in the 20th

FIGURE 5.7 The family of Czar Nicholas II. The son, Alexis, who suffered from hemophilia, is seated in front.

century turns on a mutation carried by an English queen at the beginning of the century.

MUSCULAR DYSTROPHY. Although often thought of as a single disease, **muscular dystrophy** is actually a group of diseases that produce similar features, including weakness and progressive wasting of muscle tissue. These diseases occur as both autosomal and X-linked genetic disorders. An X-linked recessive condition known as *Duchenne muscular dystrophy* (DMD) is commonly regarded as the prototype of these diseases. In the United States, DMD affects 1 in 3500 males and usually has an onset between 1 and 6 years of age. Pronounced muscle weakness is one of the first signs of DMD, and affected individuals develop a characteristic set of maneuvers in rising from the prone position (Figure 5.8). The disease progresses rapidly, and by 12 years of age, affected individuals are usually confined to a wheelchair because of muscle degeneration. Death usually occurs by the age of 20 years due to respiratory infection or cardiac failure. A second form of X-linked muscular dystrophy, *Becker muscular dystrophy* (BMD), is distinguished from DMD by later age of onset, milder symptoms, and longer survival.

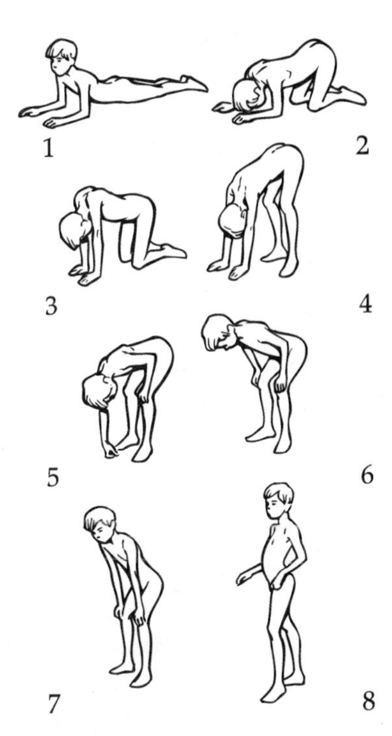

FIGURE 5.8 The characteristic steps (called the Gower sign) by which a child with muscular dystrophy rises from the prone position.

DMD and BMD have been mapped to a locus in band Xp21 in the middle of the short arm of the X chromosome. The gene at this locus encodes a protein that has been named *dystrophin.* Structural studies of this protein suggest that it attaches to the cytoplasmic side of the plasma membrane in muscle cells and works to stabilize the membrane during the mechanical strains of muscle contraction. Presumably in the absence of dystrophin, the plasma membrane of the muscle fiber ruptures during contraction, leading to breakdown and death of muscle tissue.

Analysis of DMD and BMD patients has shown that most individuals with DMD have no detectable amounts of dystrophin present in muscle tissue, while BMD patients synthesize an abnormal, usually shortened dystrophin that is only partially functional. This finding indicates that DMD and BMD represent different allelic forms of the same disease. The gene associated with these two forms of muscular dystrophy has been identified and isolated. This gene is one of the largest yet identified, covering some 2 million base pairs of DNA. Future work on the structure and function of dystrophin will hopefully lead to the development of an effective treatment for muscular dystrophy.

Muscular dystrophy
A group of genetic diseases associated with progressive degeneration of muscles. Two of these, Duchenne and Becker muscular dystrophy, are inherited as X-linked, allelic, recessive traits.

ICHTHYOSIS. Several genetic disorders that affect the skin and cause varying degrees of dryness and scaling are grouped under the term **ichthyosis** (from the Greek *ichthys,* meaning "fish") because the skin of affected individuals often resembles fish scales. One of the most common forms of ichthyosis is X-linked, with a frequency of 1 in 6000 males. While the incidence of the disease is rare in females, a case involving three homozygous sisters was reported in 1980. The underlying defect in X-linked ichthyosis is a deficiency of an enzyme known as steroid sulfatase. The disease usually appears in the first year of life. The X-linked variety of ichthyosis produces symmetric, dark scales on the trunk and the extremities.

Ichthyosis
A group of genetic diseases of the skin that produce scaling and dryness. The common form is inherited as a sex-linked recessive.

In years past people with ichthyosis often worked as circus performers billed as the Alligator Man or Fish Woman.

Y-Linked Inheritance

Y-linked
Genes located on the Y chromosome.

Genes that occur only on the Y chromosome are said to be **Y-linked.** Since only males have Y chromosomes, such traits would appear only in males and be passed directly from father to son. Furthermore, every Y-linked trait should be expressed, since males would be hemizygous for all genes on the Y chromosome. Although the distinctive pattern of inheritance should make such genes easy to identify, there is evidence that only a few Y-linked traits exist. One trait long considered to be an example of Y-linkage is *hairy pinnae,* a condition causing hair growth on the rim of the ear among certain groups in India, but more recent work has cast doubt on this belief.

Are There Any Y-Linked Genes?

The question of Y-linked inheritance can be clarified somewhat by asking two separate but related questions: Are there any genes that can be mapped to the Y chromosome, and do any of these genes show Y-linked inheritance? Recently, a number of genes have been mapped to the Y chromosome. These include a gene that encodes a protein confined to the cell nucleus that shares characteristics with proteins known to regulate gene expression. Another gene mapped to the Y chromosome encodes a gene product found on cell surfaces. Table 5.2 lists some of the genes mapped to the Y chromosome.

So the answer to the first question is yes, there are genes that map to the Y chromosome. The second question relates to genes that are limited to the Y chromosome. To show a Y-linked pattern of inheritance, genes should

TABLE 5.2 Some of the Genes Mapped to Y chromosome

Gene	Product
ANT3 ADP/ATP translocase	enzyme that moves ADP into, ATP out of mitochondria
CSF2RA	cell surface receptor for growth factor
MIC2	cell surface receptor
TDF/SRY	protein involved in early stage of testis differentiation
XE7	protein of unknown function
ZFY	DNA binding protein that may regulate gene expression

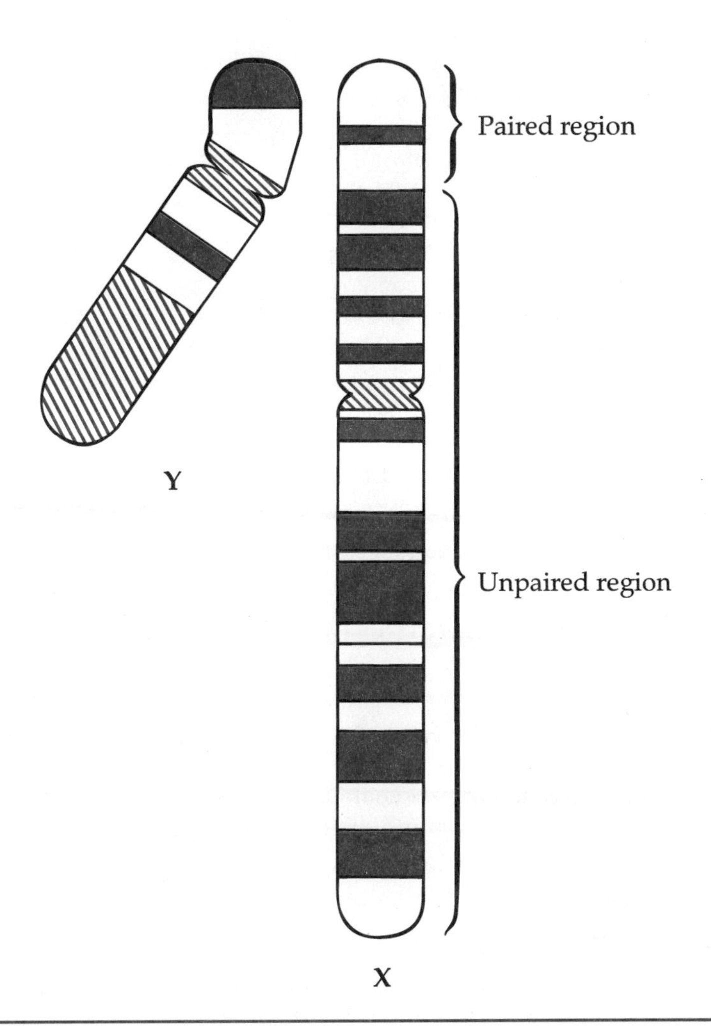

FIGURE 5.9 Diagram of pairing between the short arm of the Y chromosome and the tip of the short arm of the X chromosome.

be found only on the Y chromosome. However, genetic and cytogenetic evidence suggests that some genes may be present on both the X and Y chromosomes. Before the second question can be answered, these genes must be considered.

Pseudoautosomal Inheritance

Recent evidence indicates that the X and Y chromosome may have evolved from a common chromosomal ancestor, and cytogenetic evidence indicates that there is some homologous pairing between the tips of the short arms of the X and Y chromosomes (Figure 5.9). The genetic evidence and the presence of pairing indicate that these chromosome regions may share genes in common. If a gene is located on the short arm of the Y chromosome and has a homologous partner on the short arm of the X chromosome, this gene would show the same pattern of transmission as an autosomal gene. The only way to demonstrate that this gene is located on a sex chromosome would be by genetic mapping studies that show linkage to X and/or Y chromosomal loci. Some of the genes listed in Table 5.2 are found on both the X and Y chromosome. In addition, other DNA sequences of unknown function are present on both the X and Y chromosomes. These genes and DNA sequences have a **pseudoautosomal** pattern of inheritance because even though they are located on the X and Y chromosomes, their pattern of inheritance is the same as autosomal genes.

Pseudoautosomal
The pattern of inheritance for a gene located on both the X and Y chromosomes; in the absence of other information, such genes appear to be inherited in an autosomal fashion.

Since Y-limited genes would be expressed only in males, it is logical to presume that such genes might be involved in male-specific events such as development of the male sexual phenotype. In fact, a gene known as H-Y antigen that is necessary in male development is located on the Y chromosome and shows male-to-male transmission. Overall, therefore, the answer to the second question is also yes. There is at least one Y-linked and Y-limited gene, and other loci-controlling aspects of male differentiation or gametogenesis are probably located there as well. These genes are probably located in the portion of the Y chromosome that is not homologous to the X chromosome.

SEX DETERMINATION IN HUMANS

In humans, as in many other species, there are obvious phenotypic differences between the sexes, known as **sexual dimorphism.** In some organisms the differences are limited to the gonads, while in others, including humans, secondary sex characteristics such as body size, patterns of fat distribution, and amounts and distribution of body hair emphasize the differences between the sexes.

Sexual dimorphism
The presence of morphological traits that characterize males and females.

XY Mechanisms

What determines maleness and femaleness is a complex interaction between genes and the environment. In some cases environmental factors play a major role. For example, in some reptiles such as turtles or crocodiles, sex is determined by the incubation temperature of the eggs. Eggs incubated at higher temperatures produce females; those at lower temperatures produce males. In other reptiles, the opposite is true; higher incubation temperatures produce males, and lower temperatures result in females. In humans, on the other hand, sex determination is primarily associated with the sex chromosomes. As discussed in Chapter 2, females have two X chromosomes (XX) and males have an X and a Y chromosome. This discovery, made in the early part of this century, provided an explanation for sex determination.

Although the XX-XY mechanism of sex determination seems straightforward, it does not provide all the answers to the question of what determines maleness and femaleness. Is a male a male because he has a Y chromosome or because he does not have two X chromosomes? This question was answered about 25 years ago with the discovery that some humans carry an abnormal number of sex chromosomes. Rarely, individuals with only 45 chromosomes (45,X) are born, and these individuals are female and exhibit some characteristic features that we will see in Chapter 6. At about the same time, males carrying two X chromosomes and a Y chromosome were discovered (47,XXY). From the study of these and other individuals with abnormal numbers of sex chromosomes, it is clear that some females may have only one X chromosome and that some males can have more than one X chromosome. Furthermore, anyone with a Y chromosome is almost always male, no matter how many X chromosomes he may have. From studies on the chromosomal status of normal individuals and those carrying abnormal numbers or structural variants of the sex chromosomes, the conclusion is that under normal circumstances, the male phenotype is associated with the presence of a Y chromosome, and the absence of a Y chromosome results in the female phenotype. However, two X chromosomes are required for female development to take place in a normal fashion, and a single X chromosome is required for normal male development.

Sex Ratio

All gametes produced by females contain an X chromosome, while males produce two kinds of gametes in equal amounts, those containing an X chromosome and those containing a Y chromosome. Because the male makes two kinds of gametes, he is referred to as **heterogametic,** and the female is **homogametic,** since she makes only one type of gamete. An egg fertilized by an X-bearing sperm results in an XX zygote that will develop as a female. Fertilization by a Y-bearing sperm will produce an XY, or male, zygote (see Figure 5.1). Clearly, then, it is the male gamete that determines the sex of the offspring.

Because of sex of the offspring is determined by the presence or absence of a Y chromosome and because males produce approximately equal numbers of X- and Y-bearing gametes, males and females should be produced in equal proportions (see Figure 5.1). This proportion, known as the **sex ratio,** changes throughout the life cycle. The sex ratio at conception, known as the primary sex ratio, should be 1:1. While direct determinations are impossible, estimates indicate that more males are conceived than females. The sex ratio at birth, known as the secondary sex ratio, is about 1.05 (105 males for every 100 females). The tertiary sex ratio is the ratio as measured in adults. At 20 to 25 years of age, the ratio is close to 1.00; thereafter, females outnumber males in ever-increasing proportions. The underlying reasons for the higher death rate among males is not known for certain but includes both environmental and genetic factors. Accidents are the leading cause of death among males aged 15 to 35 years, and the expression of deleterious X-linked recessive genes also leads to a higher death rate among males.

Heterogametic
The production of gametes that contain different kinds of sex chromosomes. In humans, males produce gametes that contain X or Y chromosomes.

Homogametic
The production of gametes that contain only one kind of sex chromosome. In humans, all gametes produced by females contain only an X chromosome.

Sex ratio
The relative proportion of males and females belonging to a specific age group in a population.

Sex Differentiation

Chromosomal Sex and Phenotypic Sex

Sex determination by the XX-XY method provides a genetic framework for the developmental events that guide the zygote toward the acquisition of male or female phenotypes. The process of forming male or female reproductive structures depends on several factors, including gene action, interactions within the embryo, interaction with other embryos that may be present in the uterus, and interactions with the maternal environment. As a result of these interactions, the chromosomal sex (XX or XY) of an individual may be different from the phenotypic sex. These differences arise during the course of embryonic and fetal development and can produce a situation in which phenotypic sex is opposite to the chromosomal sex or is intermediate in phenotype or a situation in which characteristics and genitalia of both sexes are present. Thus, the sex of an individual can be defined in terms of chromosomal sex, gonadal sex, and phenotypic sex. In some cases all these definitions are consistent, but in others they are not (see Concepts and Controversies, page 125). To understand these variations and the interactions of genes with the environment, we will consider the events in normal sexual development.

Events in Embryogenesis

The first step in sex differentiation occurs at the moment of fertilization with the formation of a diploid zygote having an XX or XY chromosome constitution. In spite of the fact that the chromosomal or genetic sex of the zygote has been established, the embryo that develops is sexually ambiguous for the first month or so (Figure 5.10). The external genitalia of these early embryos are neither male nor female but indifferent, and internally, both male and female reproductive ducts and associated structures are present. The two internal duct systems are the Müllerian and Wolffian ducts (Figure 5.11). The Müllerian duct system will develop to form the female reproductive ducts, and the Wolffian system gives rise to the male reproductive network of ducts.

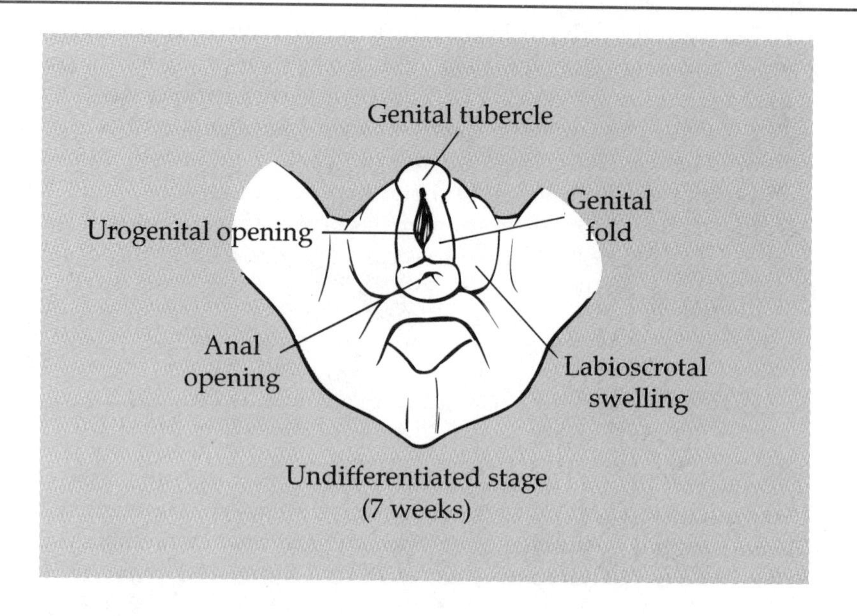

FIGURE 5.10 The undifferentiated genitalia of an early (7-week) human embryo.

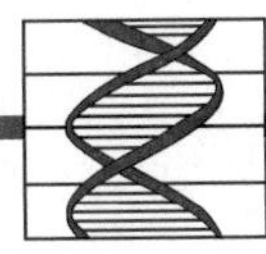

Concepts & Controversies

Sex Testing in International Athletics—Is it Necessary?

Success in amateur athletic competition is often closely tied to financial rewards in professional athletics for the individual, and to national prestige for the nation the athlete represents. Several methods are employed to guard against cheating in competition. One method involves the use of drug testing on urine samples from randomly selected athletes. In the 1960s, concerns about males competing as females to gain an unfair advantage led the International Olympic Committee (IOC) to institute sex testing beginning with the 1968 Olympics. This testing involves sex chromatin or Barr body analysis on epithelial cells recovered by scraping the inner lining of the mouth. Genetic females (XX) should show a single Barr body, and genetic males (XY) should show no Barr bodies. No physical examination is required, and the procedure is non-invasive and safe. If the results call sexual identity into question, a full chromosomal analysis is to follow, and if necessary, a gynecological examination is performed.

In principle, the use of sex chromatin tests sounds straightforward, but in practice, has been notably unsuccessful. The test itself is somewhat unreliable, and leads to both false positives and false negatives. It fails to take into account XY females (with androgen insensitivity), XXY males (Klinefelter syndrome) and XX males. In addition, the test only crudely analyzes chromosomal sex, and does not take into account the anatomical, psychological or sociological aspects of sexual status. Finally, there is no evidence that it has led to the exclusion of any males attempting to compete as females. However, it has caused a substantial number of female athletes to be barred from competition. An analysis of the results of testing on over 6000 female athletes led to the estimate that 1 in 500 female athletes have had to withdraw from competition because of the sex tests.

In response to these criticisms, the two major athletic authorities, the IOC and the International Amateur Athletic Federation (IAAF) have developed different responses. Beginning in 1991, all athletes who compete in IAAF events must have a physical examination conducted under the auspices of their national federation using standardized guidelines established by the IAAF. The examination certifies the physical condition and sexual status of all competitors, and no other tests are required. The IOC has instituted a new sex test, based on the use of recombinant DNA technology to detect the presence of the TDF/SRY gene carried on the Y chromosome. Those who test positive for SRY are classified as males and are ineligible to compete as females. This test was used on a trial basis on 557 athletes at the 1992 Olympic games at Albertville, France. Even this improved technology remains controversial, and at least one geneticist refused to participate in testing at the 1992 Barcelona competition. Critics charge that even with the elegant technology of recombinant DNA, the IOC fails to recognize that in addition to XX females, there are alternative chromosome constitutions that result in a female phenotype. The critics further charge that both the IOC and the IAAF have failed to come to grips with the basic question of why such sex tests are necessary in the first place, and urge that all such tests be abandoned.

At this point, further development is the result of diverging developmental pathways that activate different sets of genes and establish the gonadal sex of the embryo. This process takes place over the next 4 to 6 weeks of development. While it is convenient to think of only two pathways, one leading to males and the other to females, there are many alternate pathways, producing intermediate outcomes in gonadal sex and in sexual phenotypes, some of which we will consider below.

FIGURE 5.11 **The duct system (Wolffian and Müllerian) in an early embryo, and the stages of differentiation in the presence and absence of a Y chromosome.**

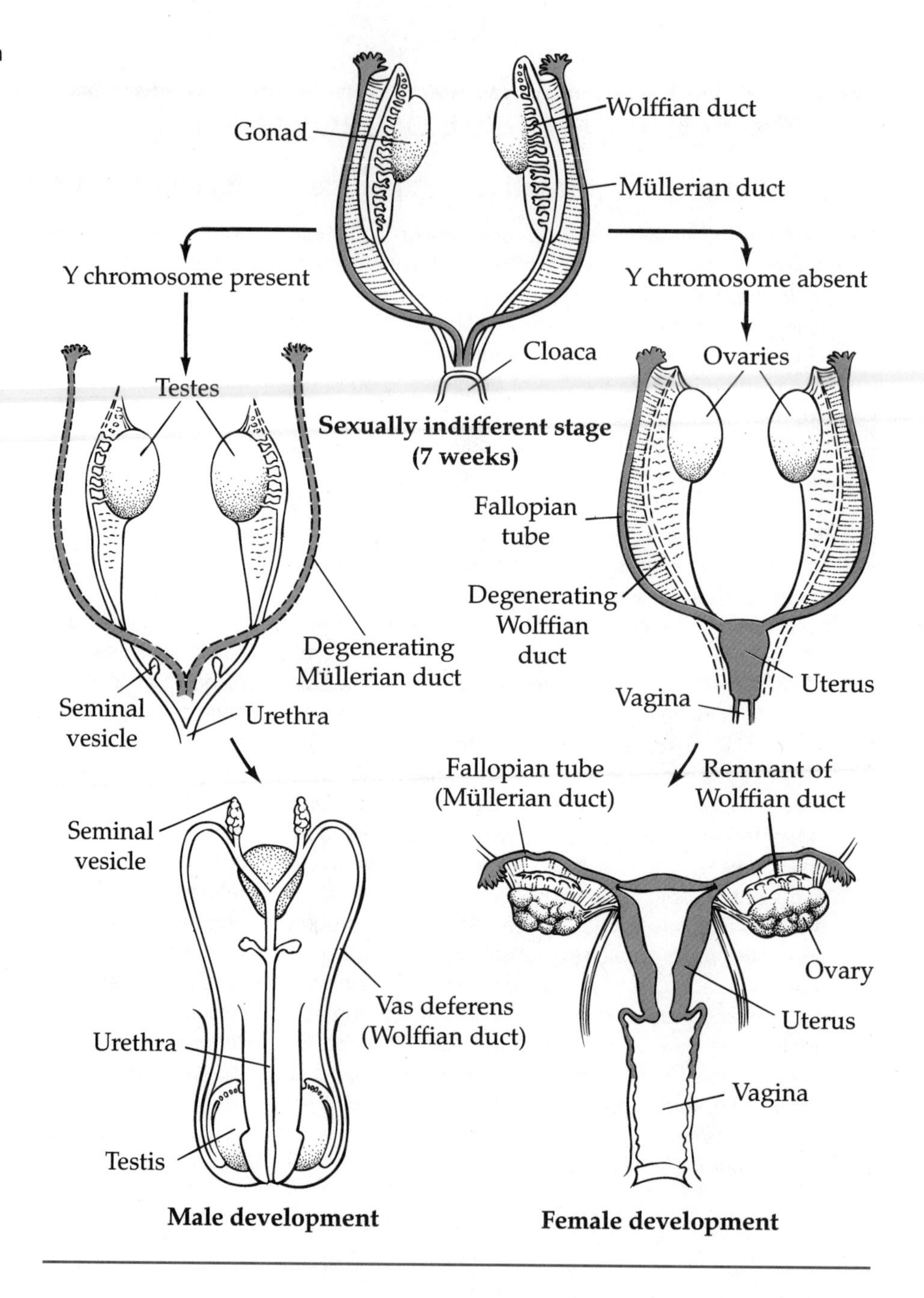

Testis determining factor
A gene located near the end of the short arm of the Y chromosome that plays a major role in causing the undifferentiated gonad to develop into a testis.

In the case of male development, the presence of a Y chromosome causes the indifferent gonad to begin development as a testis. The selection of this developmental pathway is the result of the action of genes on the Y chromosome. These gene products stimulate the growth and differentiation of a subset of cells within the indifferent gonad, causing the development of a testis. A gene known as the **testis-determining factor** (TDF)—also called the *sex-determining region of the Y,* or SRY, gene—mapped to a region on the short arm of the Y chromosome, is thought to play a major role in starting the developmental cascade of gene action that begins testis development. Other genes on the Y chromosome and on autosomes also play important roles at this time.

Once testis development is initiated, cells in the testis secrete two hormones, **testosterone** and the **Müllerian inhibiting hormone (MIH),** and these substances control further sexual development. Testosterone stimulates the development of the male internal duct system (Wolffian ducts), including the epidydimis, seminal vesicles, and vas deferens. MIH inhibits further development of female duct structure and causes degeneration of the Müllerian ducts.

Testosterone
A steroid hormone produced by the testis; the male sex hormone.

Müllerian inhibiting hormone (MIH)
A hormone produced by the developing testis that causes the breakdown of the Müllerian ducts in the embryo.

In the case of female development, the absence of the Y chromosome and the presence of the second X chromosome cause the primitive gonad to develop as an ovary, beginning about the 8th week of gestation. Cells along the outer edge of the gonad divide and push into the interior, forming the ovary. Without the presence of testosterone, the Wolffian duct system degenerates, and in the absence of MIH, the Müllerian duct system develops to form the fallopian tubes, uterus, and parts of the vagina.

After gonadal sex has been established, the third phase of sexual differentiation, the appearance of sexual phenotype, begins. In males, testosterone is metabolized and converted into another hormone, dihydroxytestosterone (DHT), which directs formation of the external genitalia. Under the influence of DHT and testosterone, the genital folds and genital tubercle develop into the penis, and the surrounding labioscrotal swelling form the scrotum.

In females, the genital tubercle develops into the clitoris, the genital folds form the labia minora, and the labioscrotal swellings form the labia majora (Figure 5.12).

In terms of gene action, it is important to note that the development of gonadal sex and sexual phenotype in male and female humans results from different developmental pathways (see Figure 5.13 on page 129). In males, this pathway involves induction by several genes on the Y chromosome, the presence of a single X chromosome, and possibly one or more autosomal genes. In females, this pathway involves the presence of two X chromosomes, the absence of Y chromosome genes, and presumably other autosomal genes. These distinctions indicate that there may be important differences in the way genes in these respective pathways are activated, and they may provide clues in the search for genes that regulate these pathways.

Genetic Control of Sexual Differentiation

As indicated above, developmental pathways that begin with the indifferent gonad present in early gestation often result in a gonadal and/or sexual phenotype that is at variance with the chromosomal sex of XX for females and XY for males. These differences in outcome can result from chromosomal events that exchange segments of the X and Y chromosomes, from mutations that affect the ability of cells to respond to gene products of Y chromosome

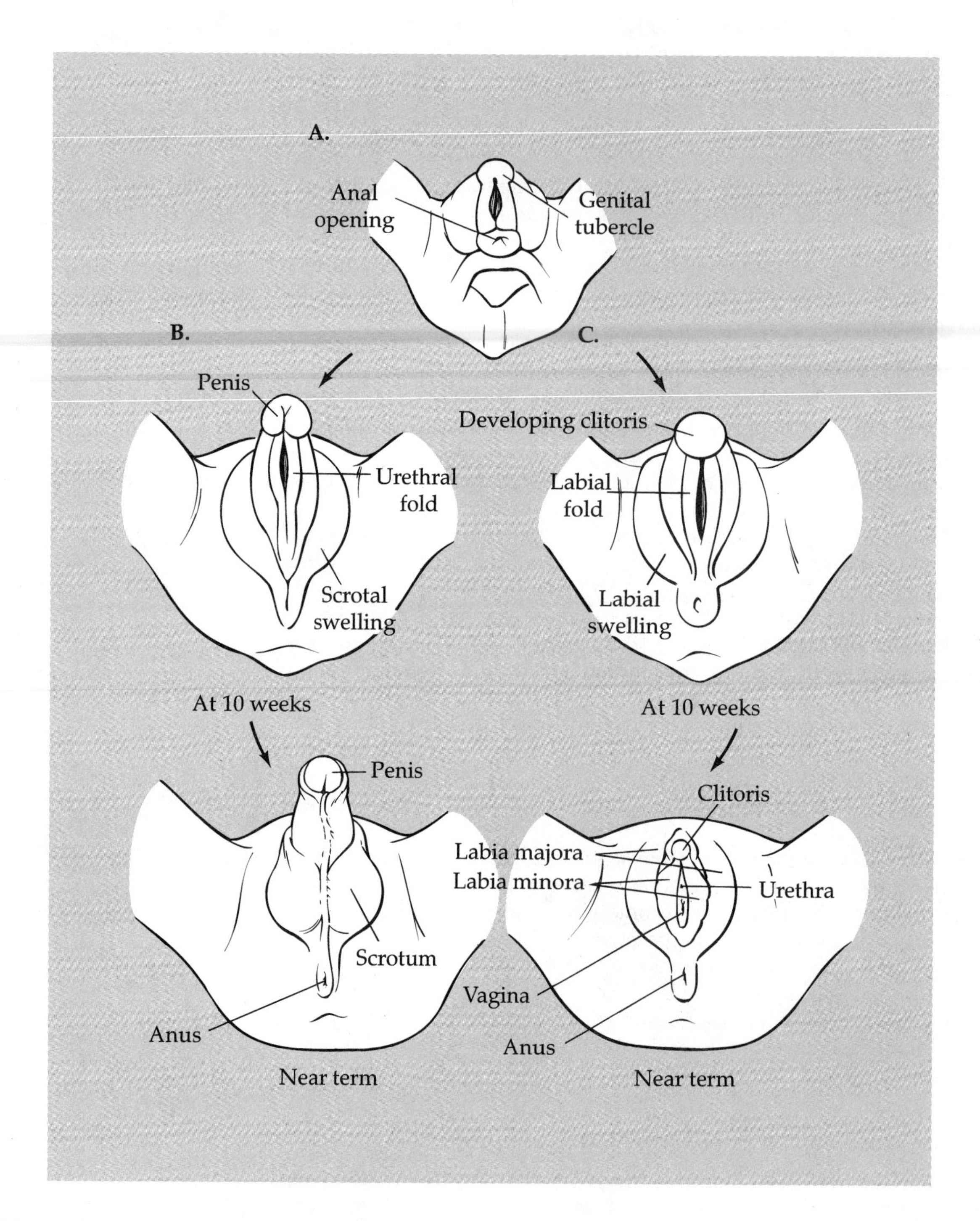

FIGURE 5.12 Steps in the development of phenotypic sex from the undifferentiated stage to the male or female phenotype. Male development takes place in the presence of testosterone and dihydroxytestosterone (DHT). Female development takes place in the absence of these hormones.

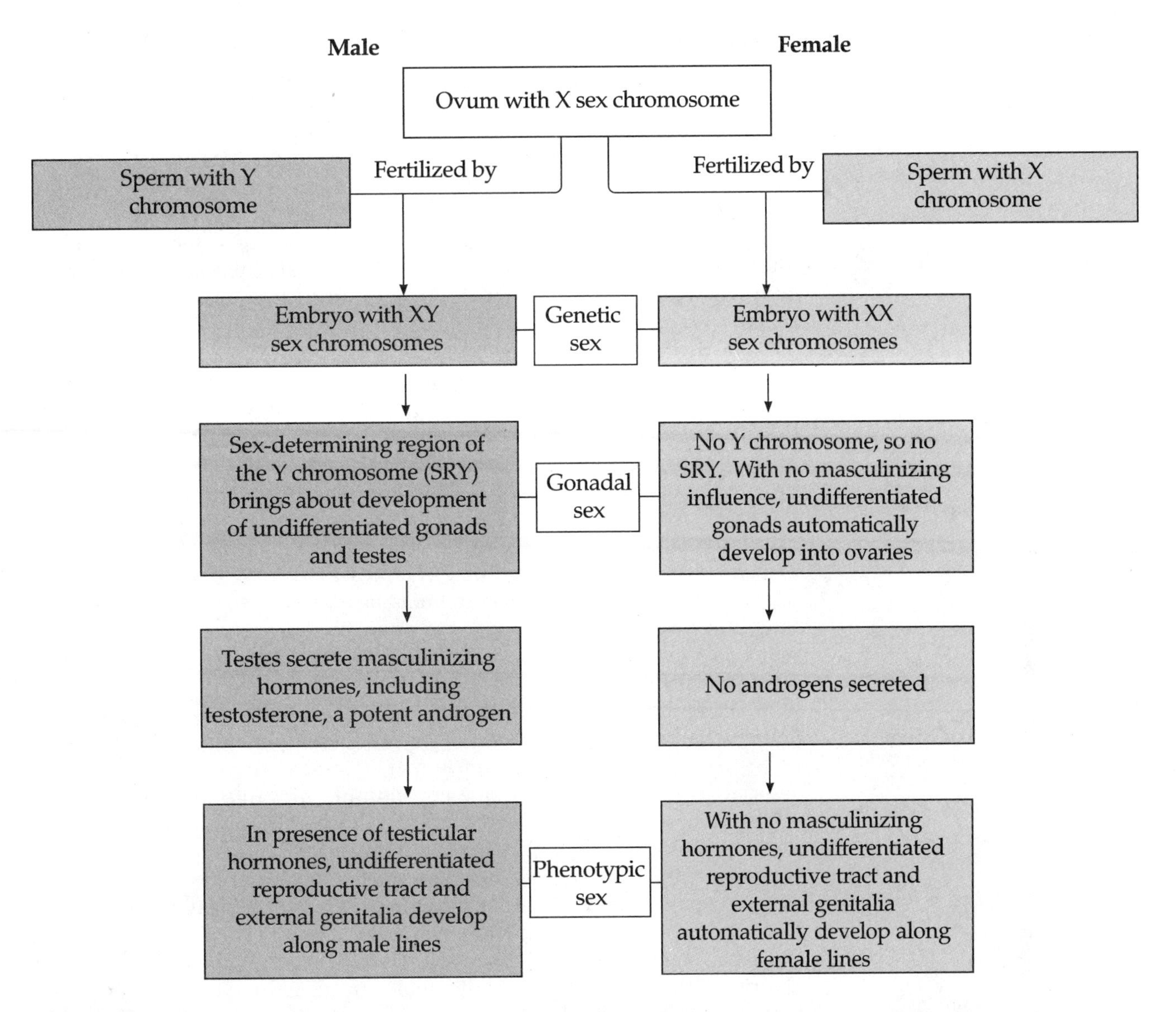

FIGURE 5.13 The major pathways of normal sexual differentiation, and the stages where genetic sex, gonadal sex and phenotypic sex are established.

genes, or from the action of autosomal genes that control events on the X and/or Y chromosome. In addition, interactions between the embryo and maternal hormones in the uterus as well as the presence of other embryos in the uterus can affect the outcome of both the gonadal sex and the sexual phenotype.

Part of our understanding of the mechanisms of sexual development is derived from the study of variations in this process, including single-gene mutations that disrupt sexual development, producing altered sexual phenotypes. We will briefly consider three situations where this a lack of concordance among chromosomal sex, gonadal sex and sexual phenotype.

True hermaphrodites are organisms that possess both ovaries and testes and the associated duct systems. In some species, such as the earthworm, this condition is normal. In humans, a true hermaphrodite is someone with ovarian and testicular tissue present in separate gonads or in a single combined gonad. Cytogenetic examination of several hermaphrodites has shown them to be sex chromosome mosaics, with some cells in the body being XX and others XY or XXY. In other cases, only XY cells were found.

Testicular feminization
An X-linked genetic trait that causes XY individuals to develop into phenotypic females.

Testicular feminization syndrome is an X-linked trait in which chromosomal males develop as females. In this case chromosomal sex (XY) is opposite from phenotypic sex (Figure 5.14). During sexual development in these persons, testis formation is induced normally, and testosterone and MIH production begins as expected. MIH brings about the degeneration of the Müllerian duct system so that no internal female reproductive tract is formed. However, a mutation in this X chromosome gene blocks the ability of cells to respond to testosterone or DHT. As a result, development proceeds as if there is no testosterone or DHT present. The Wolffian duct system degenerates, and the indifferent genitalia develop as female structures. Individuals with this syndrome are often very attractive-looking females with well developed breasts, very little pubic hair, and lack of menstruation (see Concepts & Controversies, page 131).

Pseudohermaphroditism
An autosomal genetic condition that causes XY individuals to develop the phenotypic sex of females.

Pseudohermaphrodites usually have only one type of gonad and ambiguous genitalia. One form of **pseudohermaphroditism** is known to be genetic, caused by an autosomal recessive gene. This condition, associated with XY chromosome constitution, prevents the conversion of testosterone to DHT. In these cases, the Y chromosome causes the development of testes, and the male duct system and internal organs are properly formed from the Wolffian ducts. MIH secretion prevents the development of female internal structures. However, the failure to produce DHT results in genitalia that are essentially female. The scrotum resembles the labia, a blind vaginal pouch is present, and the penis resembles a clitoris. Although chromosomally male, these individuals are raised as females. At puberty, however, masculinization takes place. The testes descend into a developing scrotum, and the phallus develops into a functional penis. The voice deepens, beard growth occurs, and muscle mass increases as in normal males. Biopsy indicates that spermatogenesis is normal. These changes are mediated by the increased levels of testosterone that accompany puberty. This condition is rare, but in a group of small villages in the Dominican Republic (Figure 5.15), over 30 such cases are known. The high incidence of homozygous recessives can be attributed to common ancestry through intermarriage (Figure 5.16). In 12 of the 13 families, a line of descent can be traced to one individual, Altagracio Carrasco (I-3). After several generations in which females changed into males, the local community came to recognize the condition by careful examination of the genitals in infancy. Affected individuals are known locally as *guevedoces*. This mutation and the one that produces testicular

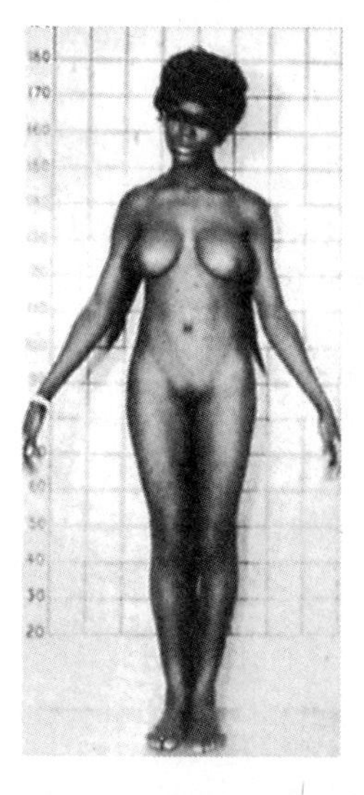

FIGURE 5.14 XY male with testicular feminization.

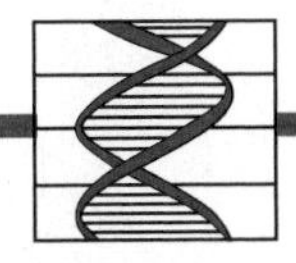

Concepts & Controversies

Joan of Arc—Was It Really John of Arc?

Joan of Arc, the national heroine of France, was born during the Hundred Years War in 1412 in a village in northeastern France. At the age of 13 or 14 years she began to have visions that directed her to help fight the English at Orleans. Following victory she orchestrated the crowning of the new king, Charles VII. Joan was captured by the English during a siege of Paris, and in 1431 she was tried at Rouen for heresy. Although technically her trial was a religious one, conducted by the English-controlled church, it was clearly a political trial. Shortly after being sentenced to life imprisonment, she was declared a relapsed heretic, and on May 30, 1431, she was burned at the stake in the marketplace at Rouen.

In 1455 Pope Callistus III formed a commission to investigate the circumstances of her trial, and a new Trial of Rehabilitation took place over a period of 7 months in 1456. The second trial took testimony from over 100 persons who had known Joan personally.

Extensive documentation from the original trial and the Trial of Rehabilitation exists. The life and career of Joan of Arc has been told, retold, and interpreted in more than 100 plays and literally thousands of books. Although the story is well known, perhaps more remains to be told. For example, from an examination of the original evidence, R. B. Greenblatt has proposed that Joan had the array of physical symptoms associated with testicular feminization. By all accounts, Joan was a healthy female with well-developed breasts. Those living with her in close quarters testified that she never menstruated, and physical examination during her imprisonment indicated that she did not have pubic hair. While such circumstantial evidence is not enough for a diagnosis, it provides the basis for speculation. Undoubtedly it also provides a new stimulus to those medico-genetic detectives who prowl through history attempting to analyze the genetic makeup of the famous, the notorious, and the obscure.

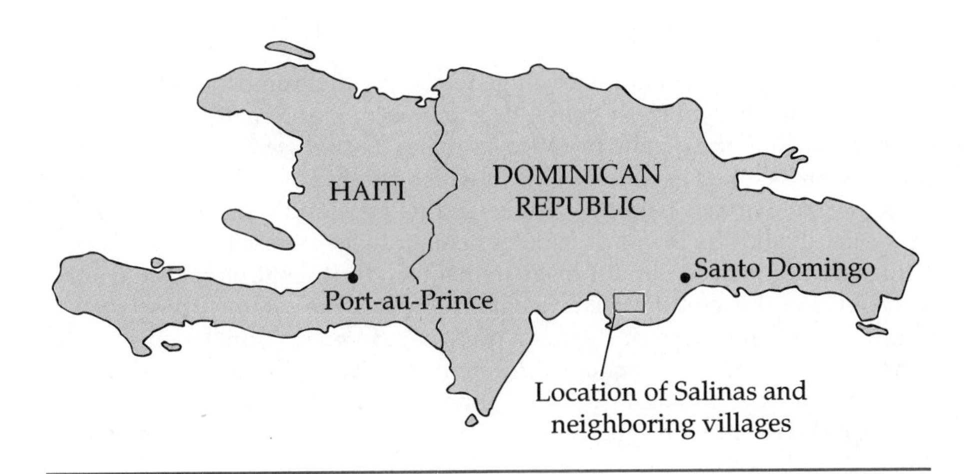

FIGURE 5.15 Salinas, Dominican Republic. This village and others in the area have a high frequency of pseudohermaphroditism.

feminization are evidence for the importance of gene interactions in normal development. Not only is the process of sexual development clearly under genetic control, but the ability to respond to the hormonal environment is critical to normal sexual differentiation.

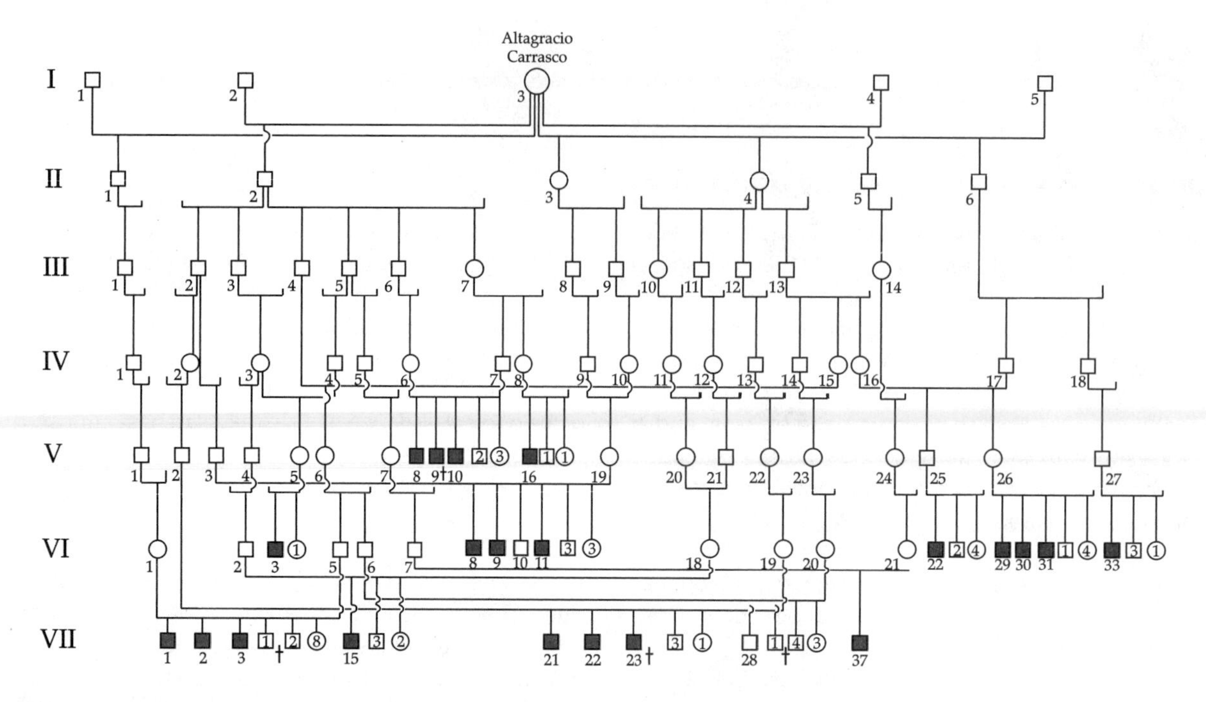

FIGURE 5.16 Pedigree of pseudohermaphroditism among several generations of residents of Salinas, Dominican Republic. Although the condition was first diagnosed in generation V, members of earlier generations were undoubtedly affected.

Dosage Compensation

Genes on the X Chromosome

Since females have two doses of all genes on the X chromosome and males have only one dose of most genes, it would seem that females should have twice as much of these gene products as males. Earlier we considered hemophilia A, an X-linked genetic disease in which clotting factor VIII is deficient. Since normal females have two copies of this gene and normal males have only one, should the blood of females contain twice as much clotting factor VIII as that of males? Careful measurements indicate that females have the same level of this clotting factor as males. In fact, the same is true for all X chromosome genes tested: the level of X-linked gene products is the same in males and females. Somehow, differences in gene dosage are regulated to produce equal amounts of gene products in both sexes. How that is accomplished in humans and how it came to be understood is an interesting story.

Dosage compensation
A mechanism that regulates the expression of sex-linked gene products.

Lyon hypothesis
The proposal that dosage compensation in mammalian females is accomplished by the partial and random inactivation of one of the two X chromosomes.

Barr Bodies and X Inactivation

An explanation of how **dosage compensation** works in humans was put forth by several groups of investigators in 1961 and is generally known as the **Lyon hypothesis,** after Mary Lyon, a British geneticist. Her idea, simply stated, is that dosage compensation is accomplished by inactivating, or turning off, almost all the genes on one of the X chromosomes in females. She based this idea on both genetic and cytologic evidence.

The genetic evidence came from studies on coat color in mice. In female mice heterozygous for X-linked coat-color genes, Lyon observed that coat color was not like that of either homozygote nor was it intermediate to the homozygotes. Instead the coat was composed of patches of the two parental colors in a random arrangement. Males, hemizygous for either gene, never showed such patches and had coats of uniform color. This genetic evidence suggested to Lyon that in heterozygous females, both alleles were active but not in the same cells.

The relevant cytologic observations were made by Murray Barr and his colleagues beginning in 1949. Barr was a physiologist studying nerve function and used nerve cells obtained from cats. He observed that the nuclei of nerve cells from female cats contained a small, dense mass of chromatin located near the nucleolus (Figure 5.17). Nerve-cell nuclei from male cats did not contain this structure, referred to as the Barr body. Lyon suggested that the Barr body is actually an inactivated, condensed X chromosome and that either X chromosome can be inactivated. The Lyon hypothesis can be summarized as follows:

1. In the somatic cells (not the germ cells) of female mammals, one X chromosome is active, and the second X chromosome is randomly inactivated and tightly coiled to form the Barr body.
2. The inactive chromosome can be either paternally derived or maternally derived, and in different cells of the same individuals, different X chromosomes are inactivated.
3. Inactivation takes place early in development. After four to five rounds of mitosis following fertilization, each cell of the embryo randomly inactivates one X chromosome.
4. This inactivation is permanent except in germ cells, and all descendants of a given cell will have the same X chromosome inactivated.

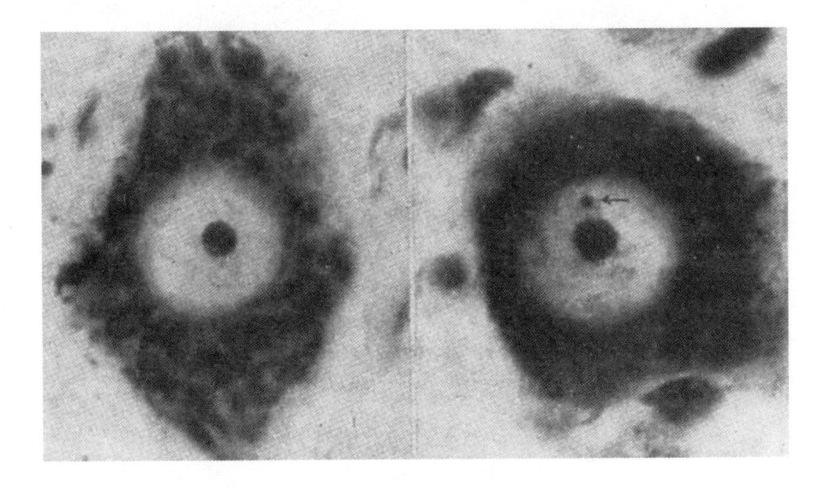

FIGURE 5.17 **(A)** Nucleus from a female cell showing a Barr body. **(B)** Nuclei from male cells show no Barr body.

5. The random inactivation of one X chromosome in females makes males and females equivalent for the activity of X-linked genes.

Females as Mosaics

Mosaic
An individual composed of two or more cell types of different genetic or chromosomal constitution. In this case, both cell lines originate from the same zygote.

The Lyon hypothesis not only explains dosage compensation but postulates that female mammals are actually **mosaics,** constructed of two different cell types, one with the maternal X chromosome active and one with the paternal X chromosome active. This postulate explains the pattern of coat color in the heterozygous mice that Lyon observed. In females heterozygous for X-linked coat-color genes, patches of one color are interspersed with patches of another color. According to the Lyon hypothesis, each patch represents a group of cells descended from a single cell in which the inactivation event occurred.

Another perhaps more familiar example of this mosaicism is the tortoiseshell cat (Figure 5.18 and the second color insert). In cats, a gene on the X chromosome has two alleles, a dominant mutant allele *(O)* that produces an orange/yellow coat color and a recessive normal allele *(o)* that produces a black color. Heterozygous females *(O/o)* have a tortoiseshell coat, with patches of orange/yellow fur mixed with patches of black fur. (White fur in such cats is controlled by a different, autosomal gene.) Tortoiseshell cats

FIGURE 5.18 Tortoiseshell cat showing patches of fur produced by inactivation of one X chromosome. The white fur is a phenotype controlled by a separate gene. (See the second color insert.)

are therefore invariably female, as hemizygous males would be either all orange/yellow or all black. However, in situations where the number of sex chromosomes is abnormal, there are exceptions to this rule.

Mosaicism in human females has been demonstrated in several ways. The first involved the use of two alleles for a gene called G6PD. This gene encodes an enzyme that works to break down sugars. Each allele produces a distinct and separable form of the enzyme. By pedigree analysis, heterozygous females were identified, and skin cells from these females were isolated and grown individually. Each culture, grown from a single cell, showed only one form or the other of the enzyme.

A more direct visualization of mosaicism can be seen in females heterozygous for a rare, X-linked disorder called ocular albinism. In this disorder, male hemizygotes lack retinal pigment. Heterozygous females have an irregular pattern of retinal pigment distribution, made up of patches of pigmented cells mixed with patches of nonpigmented cells (Figure 5.19).

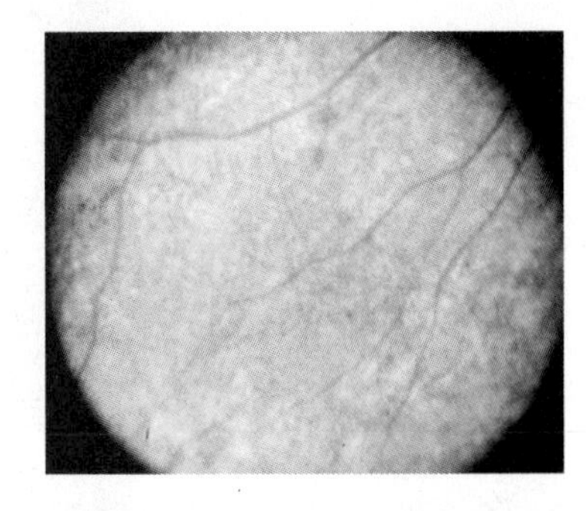

FIGURE 5.19 Pattern of albino and normal tissue in the eye of a human female heterozygous for an X-linked form of albinism. (See the second color insert.)

In humans, random inactivation of the X chromosome occurs very early in embryogenesis, usually at or before the 32-cell stage of development. Given the small number of cells present at the time of inactivation, it is possible that an imbalance in the ratio of inactivated paternal and maternal chromosomes might result, causing females to express X-linked traits for which they are heterozygous. This imbalance has been demonstrated a number of times by observing monozygotic twins, one of whom exhibits an X-linked recessive trait while the other does not (Figure 5.20).

In this family, two female identical twins are heterozygotes for red-green colorblindness through their colorblind father. One of the twins has normal color vision, and the other manifests red-green colorblindness. The colorblind twin has three sons, two who have normal vision and one who is colorblind (see pedigree).

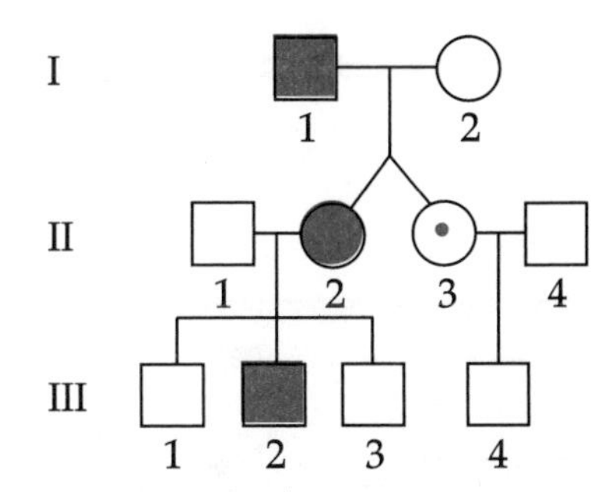

FIGURE 5.20 Pedigree showing monozygotic (MZ) twins discordant for colorblindness (solid symbols). This difference is attributed to skewing of X-chromosome inactivation. In the colorblind twin, almost all the active X chromosomes carry the color-blindness gene, while in the twin with normal vision, almost all the active X chromosomes carry the allele for normal vision.

Analysis of X inactivation was carried out using molecular techniques, allowing the parental origin of active and inactive X chromosomes to be assigned. Testing of skin fibroblasts indicates that in the twin with colorblindness, almost all the active X chromosomes are paternal X chromosomes carrying the colorblindness gene. In the twin with normal vision, the opposite situation is observed; almost all the active X chromosomes are maternal in origin.

The colorblindness in one twin and normal color vision in the other twin are explained by an extremely skewed process of X inactivation associated with twinning, so that one twin was formed from a cluster of cells most of which had a paternal X inactivated, and the other twin originated from cells most of which had the maternal X chromosomes inactivated.

SEX-INFLUENCED AND SEX-LIMITED INHERITANCE

Sex-influenced genes are those that are expressed in both males and females but with frequencies much different than would be predicted by Mendelian ratios. These genes are usually autosomal and serve to illustrate the effect of

Sex-influenced genes
Loci that produce a phenotype that is conditioned by the sex of the individual.

sex on the level of gene expression. Pattern baldness is considered to be an example of sex-influenced inheritance. This trait is expressed more often in males than in females. The gene acts as an autosomal dominant in males and as an autosomal recessive in females. In this case it is thought that the differential expression of pattern baldness in males and females is related to the different levels of testosterone present. In this case the hormonal environment and the genotype interact in determining expression of this gene.

Sex-limited genes
Loci that produce a phenotype that is produced in only one sex.

Sex-limited genes are expressed only in one sex, whether they are inherited in an autosomal or sex-linked pattern. One such gene, an autosomal dominant that controls precocious puberty, is expressed in heterozygous males but not in heterozygous females. Affected males undergo puberty at 4 years of age or earlier. Heterozygous females are unaffected but pass this trait on to half their sons, making it hard to distinguish this trait from a sex-linked gene. Genes dealing with traits such as breast development in females and facial hair in males are other examples of sex-limited genes, as are virtually all other genes dealing with secondary sexual characteristics.

SUMMARY

1. The presence of a pair of chromosomes known as the sex chromosomes in a cell has far-reaching consequences for human genetics. The genetic outcome of sex chromosomes includes sex linkage, which occurs because males are heterogametic and the Y chromosome has few loci in common with the X chromosome. The presence of a heterogametic sex also provides the basis for a 1:1 ratio between the number of males and females.

2. Sex linkage also results in a modification of Mendelian ratios and a characteristic pattern of inheritance. In X-linked inheritance, affected males transmit genes to all daughters, who are usually phenotypically normal but act as carriers and pass the genes to half their sons. Still another consequence of XX females and XY males is that males have only one copy of almost all genes on the X chromosome (as they are hemizygous), and each of these genes is directly expressed, providing the basis of hemizygosity.

3. Although mechanisms of sex determination vary from species to species, the presence of a Y chromosome in humans is normally associated with male sexual development, and the absence of a Y chromosome is associated with female development. Early in development the Y chromosome signals the indifferent gonad to begin development as a testis. Available evidence suggests that one or more genes on the Y chromosome are involved in this process. Further stages in male sexual differentiation, including the development of phenotypic sex, are controlled by hormones secreted by the testis.

4. The genetic evidence for sex linkage allows genes to be assigned to the X chromosome by observing patterns of inheritance. As a result it is easy to assign genes to the X chromosome and to construct a linkage map of the X. Since the Y chromosome lacks equivalent loci present on the X, females seem capable of expressing twice the amount of X-linked gene products. They do not do so because of the mechanism of dosage compensation. Dosage compensation arises by inactivation of one or the other X chromosome in all somatic cells of female mammals. This inactivated chromosome is visible as a sex chromatin or Barr body attached to the inner surface of the nuclear membrane.

5. In cases of sex-influenced and sex-limited inheritance, the sex of the individual affects whether a trait will be expressed and the degree to which the trait will be expressed. This holds true for both autosomal and sex-linked genes. Sex hormones and the developmental history of the individual are thought to modify expression of these genes, giving rise to altered phenotypic ratios.

QUESTIONS AND PROBLEMS

1. The X and Y chromosomes are morphologically different and genetically distinct. However, they do synapse in a limited region at the tip of their short arms, indicating that in this small region they are homologous. If a gene lies in this region, will its mode of transmission be sex-linked or autosomal? Why?

2. A young boy is colorblind. His one brother and five sisters are not. The boy has three maternal uncles and four maternal aunts. None of his uncles' children or grandchildren are colorblind. One of the maternal aunts married a colorblind man. Half of her children, both male and female, are colorblind. The other aunts married men with normal color vision. All of their daughters are normal, but half of their sons are colorblind.
   a. Which of the boy's four grandparents is responsible for transmitting the colorblind trait?
   b. Are any of the boy's aunts or uncles colorblind?
   c. Are either of the boy's parents colorblind?

3. The following diagram is a pedigree of a common trait. Determine whether the mode of inheritance of the gene that is causing the trait is:
   a. autosomal dominant.
   b. autosomal recessive.
   c. X-linked dominant.
   d. X-linked recessive.
   e. Y-linked.

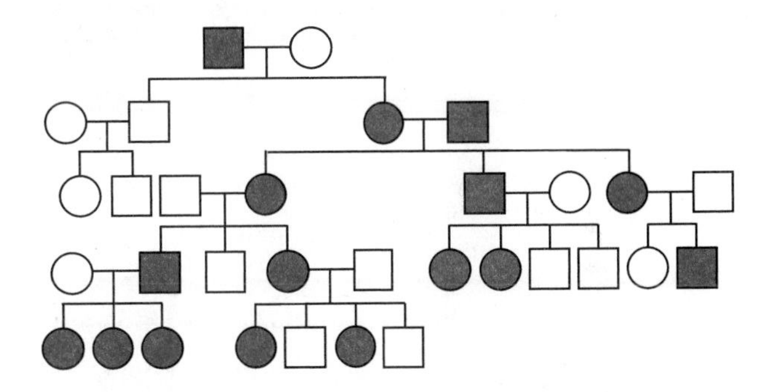

4. The following diagram is a pedigree of a common trait. Determine whether the mode of inheritance of the gene that is causing the trait is:
   a. autosomal dominant.
   b. autosomal recessive.
   c. sex-linked dominant.
   d. sex-linked recessive.
   e. Y-linked.

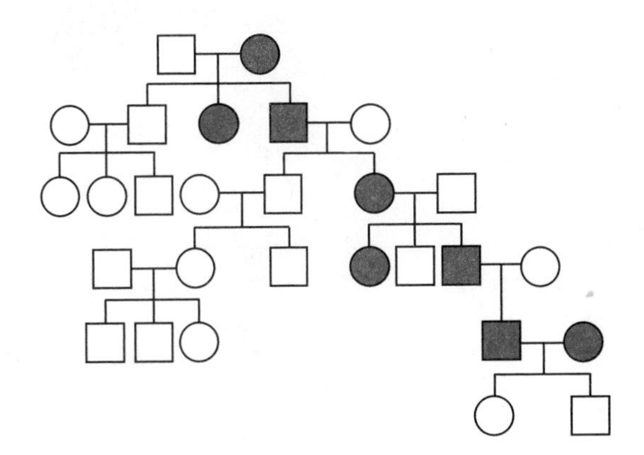

5. The following diagram is a pedigree of a common trait. Determine whether the mode of inheritance of the gene that is causing the trait is:
   a. autosomal dominant.
   b. autosomal recessive.
   c. sex-linked dominant.
   d. sex-linked recessive.
   e. Y-linked.

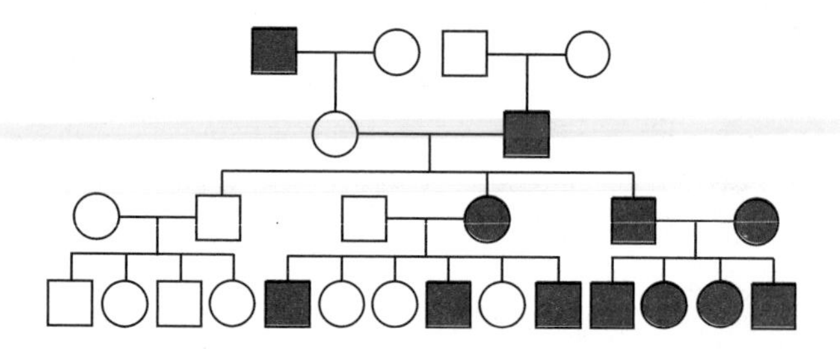

6. The following diagram is a pedigree of a common trait. Determine whether the mode of inheritance of the gene that is causing the trait is:
   a. autosomal dominant.
   b. autosomal recessive.
   c. sex-linked dominant.
   d. sex-linked recessive.
   e. Y-linked.

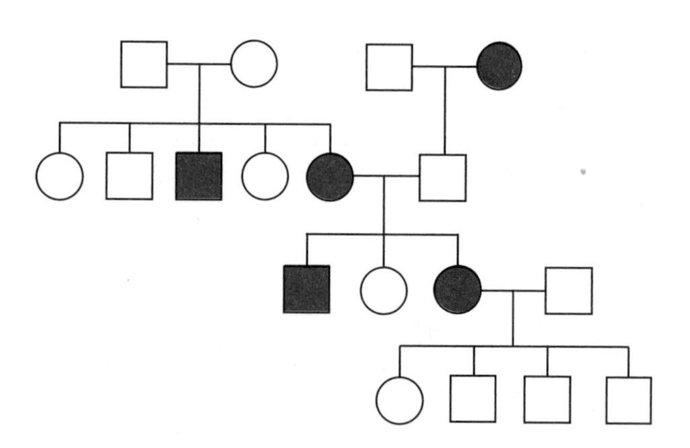

7. Determine whether the following individuals are sterile males, sterile females, fertile males, or fertile females.
   a. XY, homozygous for a recessive mutation in the testosterone gene, which renders the gene nonfunctional
   b. XX, heterozygous for a dominant mutation in the testosterone gene, which causes continuous production of testosterone
   c. XY, heterozygous for a recessive mutation in the MIH gene

8. Individuals with an XXY genotype are sterile males. If one X is inactivated early in embryogenesis, the genotype of the individual effectively becomes XY. Why should not this individual develop as a normal male?

9. It has been shown that hormones interact with DNA to turn certain genes on and off. Use this fact to explain sex-limited and sex-influenced traits.

10. Assume that humanlike creatures exist on Mars. As in the human population on Earth, there are two sexes and even sex-linked genes. The gene for eye color is an example of one such gene. It has two alleles. The purple allele is dominant to the yellow allele. A purple-eyed female alien mates with a purple-eyed male. All of the male offspring are purple-eyed, while half of the female offspring are purple-eyed and half are yellow-eyed. Which is the heterogametic sex?

11. Suppose there was a single gene that determined skin pigmentation and that the gene was located on the X chromosome. Suppose too that this pigmentation gene had two alleles, black *(B)* and white *(W)*. The *B* allele produces a black pigment, which causes skin cells to appear black. The white allele produces no pigment, which causes skin cells to appear white.
    a. If a black man married a white woman, what would their children look like?
    b. If a white man married a black woman, what would their children look like?
    c. If a black-and-white spotted woman married a black man, what would their children look like?

12. Males have only one X chromosome and therefore only one copy of all genes on the X chromosome. Each gene is directly expressed, thus providing the basis of hemizygosity in males. Females have two X chromosomes, but one is always inactivated. Therefore females, like males, have only one functional copy of all the genes on the X chromosome. Again, each gene must be directly expressed. Why then are females not considered hemizygous, and why are they not afflicted with sex-linked recessive diseases as often as males?

13. Use Figure 5.6 to determine if the recessive allele for hemophilia might have been transmitted to

Queen Victoria via the parents of Edward, Duke of Kent, or the parents of the Duke of Saxe-Coburg-Gotha.

14. A colorblind female of blood type A married two different men. One man had type AB blood and was colorblind. The other man had type B blood and normal vision. Three offspring resulted from these marriages. If possible, determine the father of each of the following offspring:
    a. Type A, normal vision, female
    b. Type AB, colorblind, male
    c. Type B, colorblind, female

For Further Reading

Anderson, M. and Kunkel, L. 1992. The molecular and biochemical basis of Duchenne muscular dystrophy. Trends Biochem. Sci. **17**:289-292.

Baron, M., Risch, N., Hamburger, R., Mandel, B., Kushner, S., Newman, M., Drumer, D., and Belmaker, R. 1987. Genetic linkage between X chromosome genetic markers and bipolar affective illness. *Nature* **326:** 289–292.

Brownlee, G. G., 1986. The molecular genetics of hemophilia A and B. *J. Cell Sci. (Suppl.)* **4:** 445–458.

Davies, K. E. 1985. Molecular genetics of the X chromosome. *J. Med. Genet.* **22:** 243–249.

Davies, K. E., Forrest, S., Smith, T., Kenwrick, S., Ball, S., Dorkins, H., and Patterson, M. 1987. Molecular analysis of human muscular dystrophies. *Muscle Nerve* **10:** 191–199.

Goodfellow, P. N., 1986. The case of the missing H-Y antigen. *Trends in Genetics* **2:** 87.

Goodman, R. M. 1979. *Genetic Disorders among the Jewish People.* Baltimore: Johns Hopkins Press.

Greenblatt, R. B. 1981. Case history: Jeanne d'Arc—Syndrome of feminizing testes. *Br. J. Sex Med.* **8:** 54.

Harper, P. S. 1985. The genetics of muscular dystrophies. *Prog. Med. Genet.* **6:** 53–90.

Imperato-McGinley, J., Guerrero, L., Gautier, T., and Peterson, R. G. 1974. Steroid 5-alpha reductase deficiency in man: An inherited form of pseudohermaphroditism. *Science* **186:** 1213–1215.

Kunkel, L. M. 1986. Analysis of deletions in DNA from patients with Becker and Duchenne muscular dystrophy. *Nature* **322:** 73–77.

Lyon, M. F. 1962. Sex chromatin and gene action in the mammalian X-chromosome. *Am. J. Hum. Genet.* **14:**135–148.

Lyon, M. R. 1989. X-chromosome inactivation as a system of gene dosage compensation to regulate gene expression. *Prog. Nucleic Acids Res. Mol. Biol.* **36:** 119–130.

McGuire, J. 1986. The biologic basis of the ichthyoses. *Dermatol. Clin.* **4:** 67–78.

Mendlewicz, J., Simon, P., Sevy, S., Charon, F., Brocas, H., Legros, S., and Vassart, G. 1987. Polymorphic DNA marker on X chromosome and manic depression. *Lancet* **1:** 1230–1232.

Monaco, A. P., 1989. Dystrophin, the protein product of the Duchenne/Becker muscular dystrophy gene. *Trends Biochem. Sci.* **14:** 412–415.

Nathans, J. 1989. The genes for color vision. *Sci. Am.* (February) **260:** 42–49.

Orkin, S. H. 1986. Reverse genetics and human disease. *Cell* **47:**845–850.

Ott, J. 1986. Y-linkage and pseudoautosomal linkage. *Am. J. Hum. Genet.* **38:** 891–897.

Ray, P. N., Belfall, B., Duff, C., Logan, C., et al. 1985. Cloning of the breakpoint of an X;21 translocation associated with Duchenne muscular dystrophy. *Nature* **318:** 672–675.

Sulton, C., Lobaccaro, J., Belon, C., Terraza, A., and Lumbroso, S. 1992. Molecular biology of disorders of sex differentiation. Horm. Res. **38:**105-113.

Tinsley, J., Blake, D., Pearre, M., Knight, A., Kendrick-Jones, J., and Davies, K. 1993. Dystrophin and related proteins. Curr. Opin. Genet. Dev. **3:**484-490.

Whitfield, L., Lovell-Badge, R. and Goodfellow, P. 1993. Rapid sequence evolution of the mammalian sex-determining gene SRY. Nature **364:**713-715.

Williams, M. L. 1986. A new look at the ichthyoses: Disorders of lipid metabolism. *Pediatr. Dermatol.* **3:** 476–486.5.1

CHAPTER 6

CHROMOSOME ABERRATIONS

CHAPTER OUTLINE

In 1953 a French physician named Jérôme Lejeune began working on a condition first described in 1866 by an English physician named John Langdon Down. Individuals affected with this condition have a distinctive physical appearance and are mentally retarded. Lejeune suspected that there was a genetic link to this phenotype.

He began by studying the fingerprints and palm prints of children with the condition, known as Down syndrome, and compared them with the prints of unaffected children. The prints from Down syndrome children showed a high frequency of abnormalities. Because fingerprints and palm prints are laid down very early in development, they serve as a record of events that take place early in embryogenesis. From his studies on the disturbances in the print patterns of affected children, Lejeune became convinced that such significant changes in print patterns were probably not caused by the action of only one or two genes. Instead he reasoned that many genes must be involved; perhaps even an entire chromosome might play a role in Down syndrome.

It was a logical step to examine the chromosomes of Down syndrome children, but Lejeune lacked access to the proper equipment and techniques required. In 1957, in cooperation with a colleague, he began culturing cells from Down syndrome children. Lejeune examined the chromosomes in these cells using a microscope that had been discarded by the bacteriology laboratory in the hospital where he worked. He repaired the instrument by inserting a foil wrapper from a candy bar into the gears so that the image could be focused. His chromosome counts indicated that the cells of Down syndrome individuals contained 47 chromosomes, while those of unaffected individuals contained 46 chromosomes. He and his colleagues published a short paper in 1959 in which they reported that Down syndrome is caused by the presence of an extra chromosome. This chromosome was later identified as chromosome 21.

This remarkable discovery was the first identified human-chromosome abnormality and marked an important turning point in human genetics. The discovery made clear that genetic disorders could be associated with changes in chromosome number, not just mutations of single genes inherited in Mendelian fashion.

Variations in Chromosome Number

At the birth of a child, anxious parents have two questions: is it a boy or a girl, and is the baby normal? The term "normal" in this context means free from all birth defects. The causes of such defects are, of course, both environmental and genetic. Among the genetic causes, we have considered disorders such as sickle cell anemia or Marfan syndrome, caused by the mutation of single genes. Other genetic causes of defects are changes in chromosome number or structural alterations in chromosomes. Such changes may involve entire chromosome sets, individual chromosomes, or alterations within individual chromosomes. Recall that the set of 46 chromosomes present in each somatic cell is referred to as the diploid, or $2n$, number of chromosomes. Similarly the set of 23 chromosomes (constituting the n number) is the haploid set. Together these normal conditions are referred to as the euploid condition.

Variations in the number of haploid sets of chromosomes is known as **polyploidy.** A cell with three sets of chromosomes is triploid, one with four

Polyploidy
A chromosome number that is a multiple of the normal diploid chromosome set.

sets is tetraploid, and so forth. In a later section, we will examine the events that result in polyploidy.

Aneuploidy
A chromosome number that is not an exact multiple of the haploid set.

A change in chromosome number that involves less than an entire diploid set of chromosomes is known as **aneuploidy.** The number of chromosomes present in aneuploidy is not a simple multiple of the haploid set. In humans the most common forms of aneuploidy involve the gain or loss of a single chromosome. The loss of a chromosome is known as monosomy ($2n - 1$), and the addition of a chromosome to the diploid set is known as trisomy ($2n + 1$). Since the discovery of trisomy 21 in 1959 as the first example of aneuploidy in humans, cytogenetic studies have revealed that alterations in chromosome number are fairly common in humans and are a major cause of reproductive failure. It is now estimated that as many as 1 in every 2 conceptions may be aneuploid and that 70% of early embryonic deaths and spontaneous abortions are caused by aneuploidy. About 1 in every 170 live births is at least partially aneuploid, and from 5% to 7% of all deaths in early childhood are related to aneuploidy.

Studies to date indicate that humans have a rate of aneuploidy that is up to 10 times higher than that of other mammals, including other primates. If confirmed, this difference may represent a considerable reproductive disadvantage for our species. Understanding the causes of aneuploidy in humans remains one of the great challenges in human genetics.

POLYPLOIDY

Abnormalities in the number of haploid chromosome sets can arise in several ways: (1) from errors during the process of gamete formation, (2) from events at fertilization, or (3) in the rounds of mitotic division that accompany development of the embryo. Polyploidy can occur through several common mechanisms, including *endoreduplication.* In endoreduplication, cleavage of the cytoplasm fails to follow chromosome replication. This event can occur in either mitosis or meiosis. Recall that mitotic divisions precede meiosis in the ovary and testis. If, during one of these mitotic cycles, the chromosomes replicate and sister chromatids divide but there is no cytoplasmic division, the result is a tetraploid cell containing four copies of each chromosome. If this cell undergoes meiosis in a normal fashion, the result will be gametes that contain the diploid instead of the haploid number of chromosomes. Union between this genetically unbalanced gamete and a normal haploid gamete will produce a triploid zygote (Figure 6.1).

The production of a diploid gamete can also arise through meiotic errors. An error in meiosis I can result in the failure of homologous chromosomes to separate, producing diploid gametes after meiosis II. Alternatively, in meiosis II, if all chromosomes move to the same pole after centromere separation, diploid gametes will also result.

Dispermy
fertilization of haploid egg by two haploid sperm forming a triploid zygote

Another event that can result in polyploidy is **dispermy,** the simultaneous fertilization of a haploid egg by two haploid sperm. The result is a zygote containing three haploid chromosome sets, or triploidy.

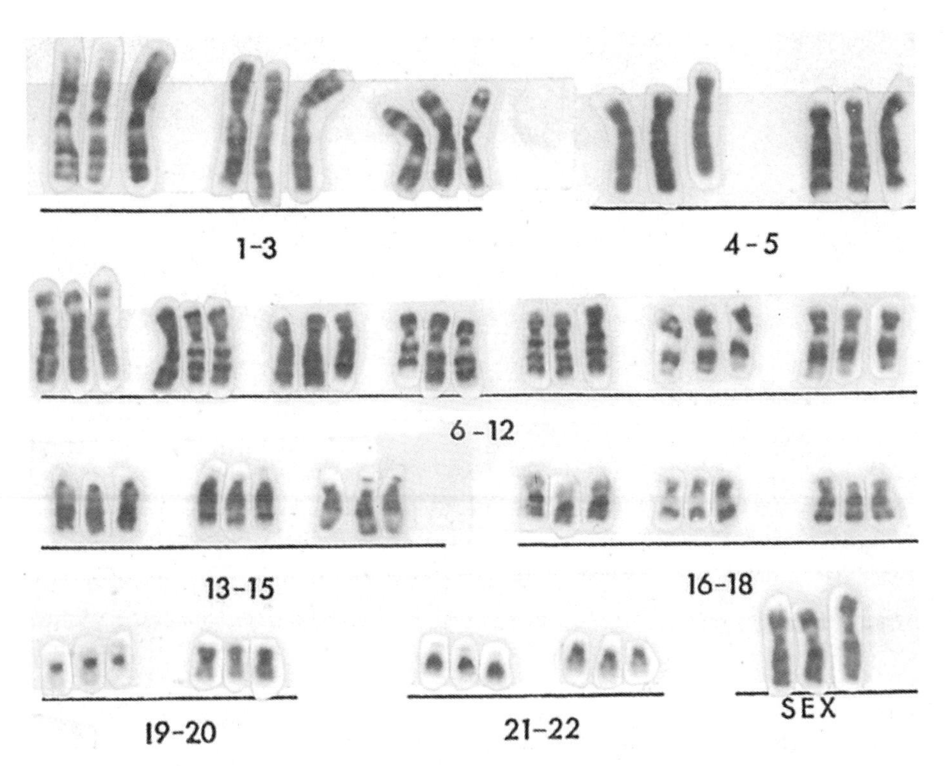

FIGURE 6.1 Karyotype of triploid.

Triploidy

The most common form of polyploidy in humans is **triploidy,** observed in 15% to 18% of spontaneous abortions. Three different types of triploid chromosome sets have been observed: 69,XXY; 69,XXX; and 69,XYY. Approximately 75% of all cases of triploidy have two sets of paternal chromosomes. Failures in male gamete formation do not occur this frequently, and most triploid zygotes probably arise as the result of dispermy. Although biochemical changes that accompany fertilization normally prevent such fertilizations, this system is not fail-safe. However, other mechanisms usually terminate the pregnancy by spontaneous abortion.

Triploidy
A chromosome number that is three times the haploid number, having three copies of all autosomes and three sex chromosomes.

Almost 1% of all known conceptions are triploid, but over 99% of these die before birth, and only 1 in 10,000 live births is triploid. Survival for triploids is limited, and most of those born die within a month. Triploid newborns have multiple abnormalities, including an enlarged head, fusion of fingers and toes (syndactyly), and malformations of the mouth, eyes, and genitals (Figure 6.2). The high rate of embryonic death and failure to survive beyond birth indicates that triploidy is incompatible with life and must be regarded as a lethal condition.

Tetraploidy

Tetraploidy is observed in about 5% of all spontaneous abortions and is reported only rarely in live births. The sex chromosome constitution of all

FIGURE 6.2 **Triploid infant, showing characteristic enlarged head.**

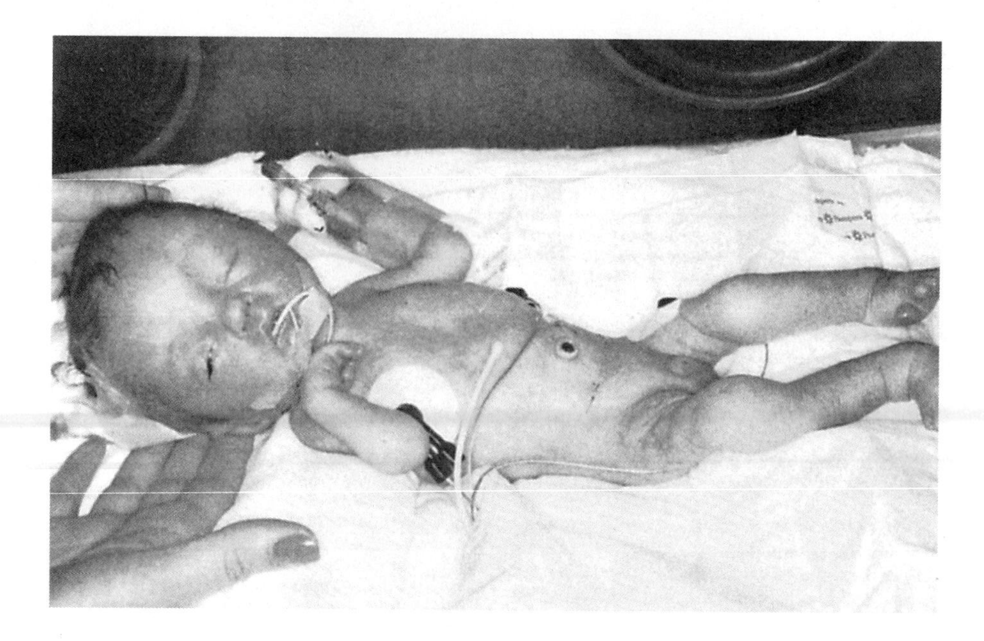

Tetraploidy
A chromosome number that is four times the haploid number, having four copies of all autosomes and four sex chromosomes.

tetraploid embryos is either XXXX or XXYY. **Tetraploidy** can arise in the first mitotic division following fertilization. In this case replication and separation of chromosomes is not followed by cytoplasmic division (cytokinesis). This produces a tetraploid cell by endoreduplication. Subsequent rounds of normal mitosis will result in a tetraploid embryo. If tetraploidy arises sometime after the first mitotic division, two different cell lines will coexist in the embryo, one a normal diploid line and the other a tetraploid cell line. Such mosaic individuals survive somewhat longer than complete polyploids, but the condition is still life threatening.

In summary, polyploidy in humans arises by at least two different mechanisms but is inevitably lethal. It is interesting to note that polyploidy does not involve specific mutations of genes but only changes in the number of gene copies. How this quantitative change in gene number is related to lethality in development is unknown.

ANEUPLOIDY

Nondisjunction
The failure of homologous chromosomes to properly separate during meiosis or mitosis.

As defined above, aneuploidy is associated with the addition or deletion of individual chromosomes from the normal diploid set of 46. Aneuploidy can be caused by several mechanisms, the most important of which is **nondisjunction,** a process in which chromosomes fail to separate properly at anaphase. Although this failure can occur in either meiosis or mitosis, nondisjunction in meiosis is the primary cause of aneuploidy in humans. There are two cell divisions in meiosis, and nondisjunction can occur in either the first or second division, with different genetic consequences (Figure 6.3). Nondisjunction at meiosis I results in gametes that contain both the paternal and maternal members of a chromosome pair and gametes that contain

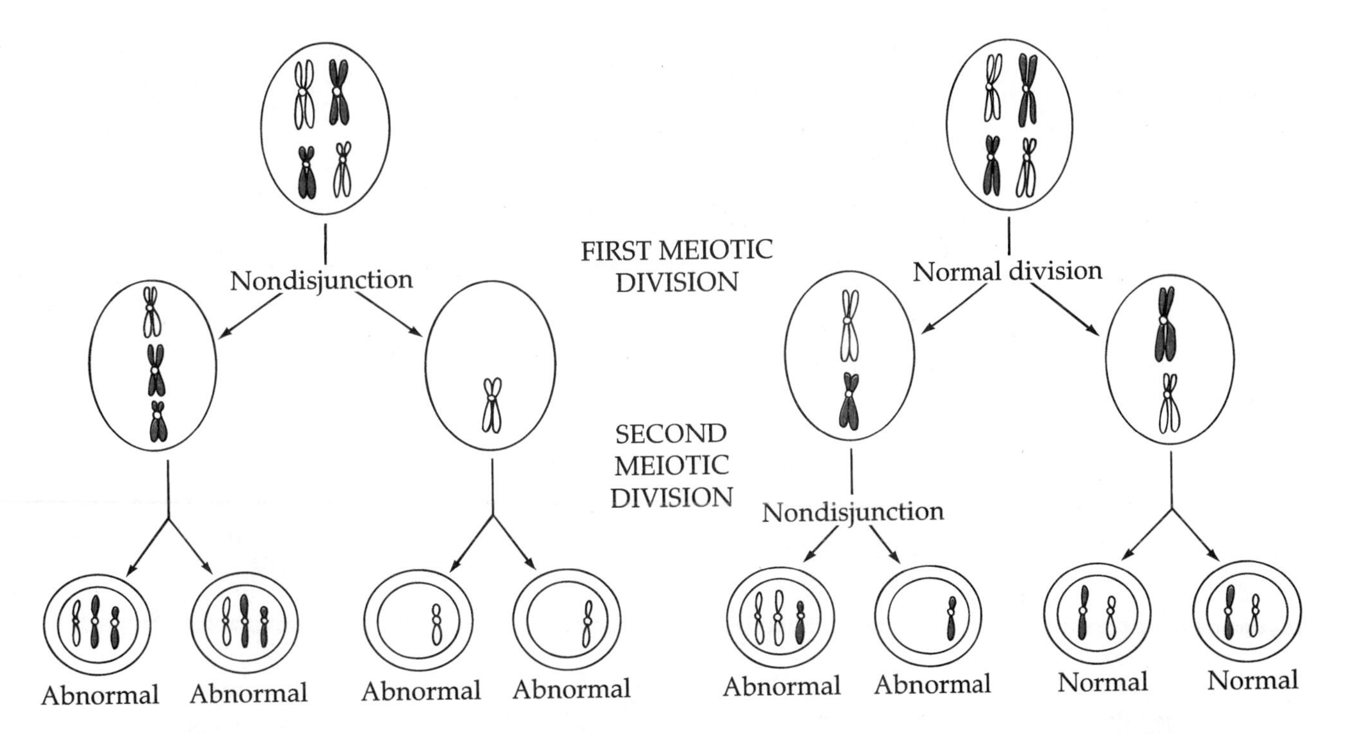

FIGURE 6.3 Nondisjunction at first *(left)* and second meiotic divisions *(right).* If nondisjunction takes place in meiosis I, all gametes are abnormal, containing either both members of a chromosome pair ($n + 1$) or neither member of a chromosome pair ($n - 1$). Nondisjunction at meiosis II produces two normal gametes and two abnormal gametes, one with two copies of a chromosome derived from one homologue and one missing a chromosome.

neither member. If nondisjunction takes place at meiosis II, the result is gametes that contain two copies of the maternal or paternal chromosome and gametes that contain neither a paternal or maternal copy. Gametes lacking one chromosome will produce a monosomic zygote; those containing an extra copy of a chromosome will give rise to a trisomic zygote.

As a group, aneuploid individuals have distinct and characteristic phenotypic features. Those with a given form of aneuploidy tend to resemble each other more closely than their own brothers and sisters. The phenotypic effects of aneuploidy range from minor physical symptoms to devastating and lethal deficiencies in major organ systems. Among survivors, the phenotypic effects often include behavioral deficits and mental retardation. In the following section, we will look at some of the important features of autosomal aneuploid phenotypes. Then we will consider the phenotypic effects of sex chromosome aneuploidy.

Monosomy

Meiotic nondisjunction during gamete formation should result in equal numbers of monosomic and trisomic embryos. However, autosomal monosomies are only rarely observed among spontaneous abortions and live

Monosomy
A condition in which one chromosome of a pair is missing, having one less than the diploid number ($2n - 1$).

births. The likely explanation is that the majority of autosomal monosomic embryos are lost very early, even before pregnancy is recognized.

Trisomy

Trisomy
A condition in which one chromosome is present in three copies while all others are diploid, having one more than the diploid number ($2n + 1$).

Most autosomal trisomies are lethal conditions, but unlike monosomy, the presence of an extra chromosome allows varying degrees of development to occur. Autosomal **trisomy** is found in about 50% of all cases of chromosome abnormalities in fetal death. The findings also indicate that the autosomes are differentially involved in trisomy (Figure 6.4). For example, trisomies for chromosomes 1, 3, 12, and 19 are rarely observed in spontaneous abortions, while trisomy for chromosome 16 accounts for almost one third of all cases. As a group, the acrocentric chromosomes (13–15, 21, and 22) are represented in 40% of all spontaneous abortions. Reasons for this differential involvement include differences in the rate of nondisjunction or in the rate of fetal death before recognition of pregnancy or a combination of factors. Only a few autosomal trisomies result in live birth (trisomy 8, 13, and 18). Trisomy 21 is the only autosomal trisomy that allows survival into adulthood.

Trisomy 13
The presence of an extra copy of chromosome 13 that produces a distinct set of congenital abnormalities resulting in Patau syndrome.

TRISOMY 13 (PATAU SYNDROME) (47, + 13). The condition of **trisomy 13** was discovered in 1960 by cytogenetic analysis of a grossly malformed child. The karyotype indicated the presence of 47 chromosomes, and the extra chromosome was identified as chromosome 13 (47, + 13). Only 1 in

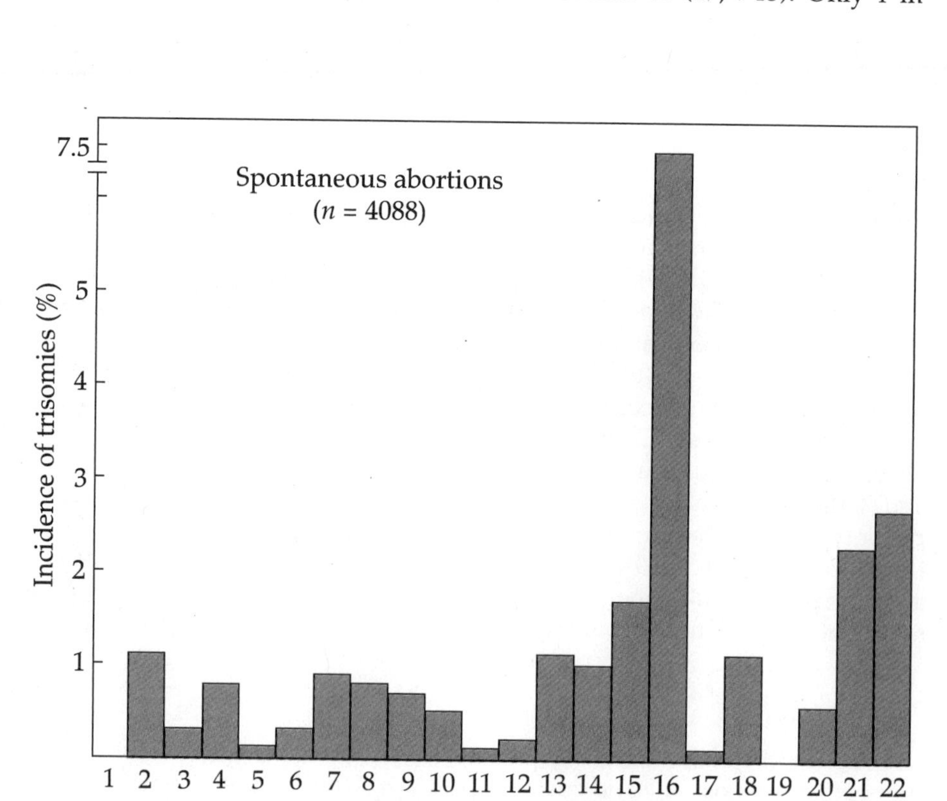

FIGURE 6.4 The involvement of autosomes in trisomy is shown for a study of over 4000 spontaneous abortions.

15,000 live births involves trisomy 13, and the condition is lethal; half of all affected individuals die in the first month, and the mean survival time is 6 months. The phenotype of trisomy 13 involves cleft lip and palate (Figure 6.5), eye defects, polydactyly (extra fingers or toes), and rocker-bottom feet with large protruding heels. Internally there are usually severe malformations of the brain and nervous system, and congenital heart defects. The involvement of so many organ systems indicates that developmental abnormalities begin early in embryogenesis, perhaps as early as the sixth week. Parental age is the only factor known to be related to trisomy 13. The age of parents of children with trisomy 13 is higher (averaging about 32 years) than the average of parents with normal children. The relationship between parental age and aneuploidy will be discussed below.

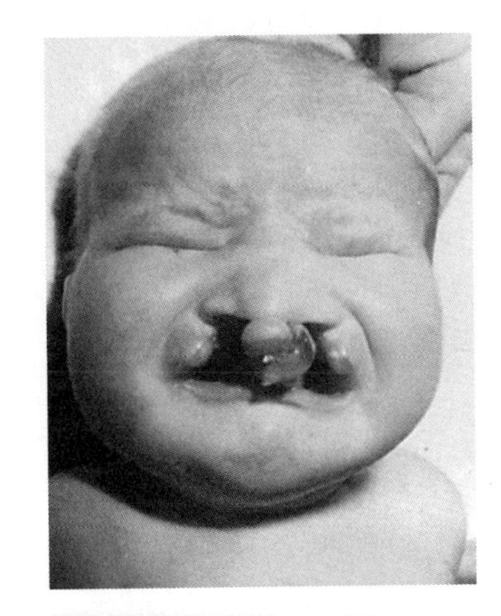

FIGURE 6.5 Infant with trisomy 13, exhibiting cleft lip and palate and small eyes.

TRISOMY 18 (EDWARDS SYNDROME) (47, + 18). In 1960 John Edwards and his colleagues reported the first case of **trisomy 18** (47, + 18). Infants with this condition are small at birth, grow very slowly, and are mentally retarded. For reasons still unknown, 80% of all trisomy 18 births are female. Clenched fists, with the second and fifth finger overlapping the third and fourth fingers, and rocker-bottom feet with protruding heels are also characteristic (Figure 6.6). Heart malformations are almost always present, and death is usually attributed to heart failure or pneumonia. Trisomy 18

Trisomy 18
The presence of an extra copy of chromosome 18 that results in a clinically distinct set of invariably lethal abnormalities resulting in Edwards syndrome.

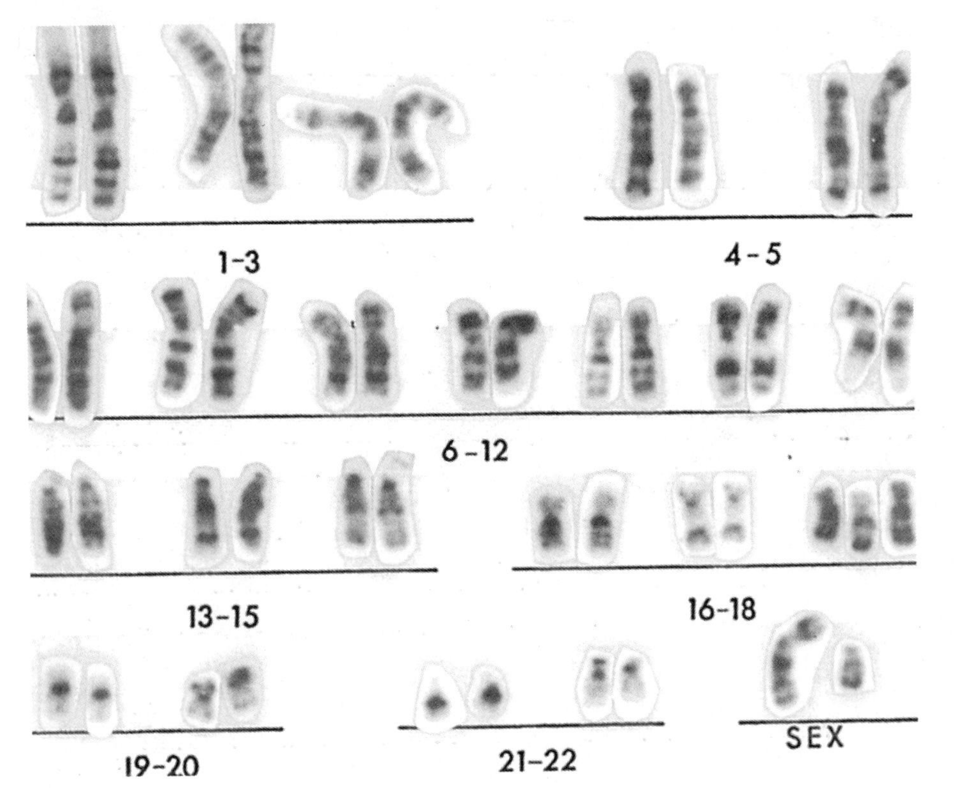

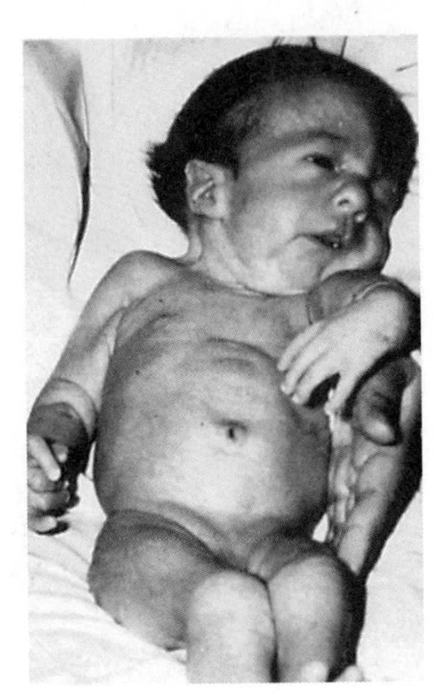

FIGURE 6.6 Infant with trisomy 18 and karyotype. Clenched fists are associated with this syndrome.

occurs with a frequency of 1 in 11,000 live births, and the average survival time is 2 to 4 months. As in trisomy 13, advanced maternal age is a factor predisposing to trisomy 18.

Trisomy 21
Aneuploidy involving the presence of an extra copy of chromosome 21, resulting in Down syndrome.

TRISOMY 21 (DOWN SYNDROME) (47, + 21). The phenotypic features of **trisomy 21** were first described by John Langdon Down in 1866. He called the condition mongolism because of the distinctive fold of skin, known as an epicanthic fold, in the corner of the eye (Figure 6.7). Down subscribed to the belief that the European "race" was superior to all others and that those affected with trisomy 21 represented throwbacks to what he regarded as an earlier, more primitive racial group. Later, to remove the racist implications inherent in the term, Lionel Penrose and others changed the designation to Down syndrome. As described in the chapter opening, the presence of an extra copy of chromosome 21 as the underlying cause of Down syndrome was discovered by Jérôme Lejeune and his colleagues in 1959 and represents the first chromosome abnormality discovered in humans. Trisomy 21 has also been observed in other primate species, notably, the chimpanzee.

Down syndrome is one of the most common chromosome defects in humans and occurs in about 0.5% of all conceptions and in 1 in 900 live births. It is a leading cause of childhood mental retardation and heart defects in the United States. Affected individuals have a wide skull that is flatter than normal at the back (Figure 6.7). The eyelids have an epicanthic fold, similar

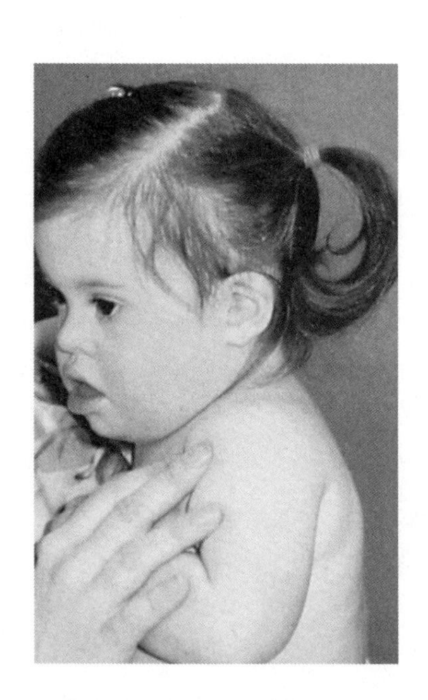

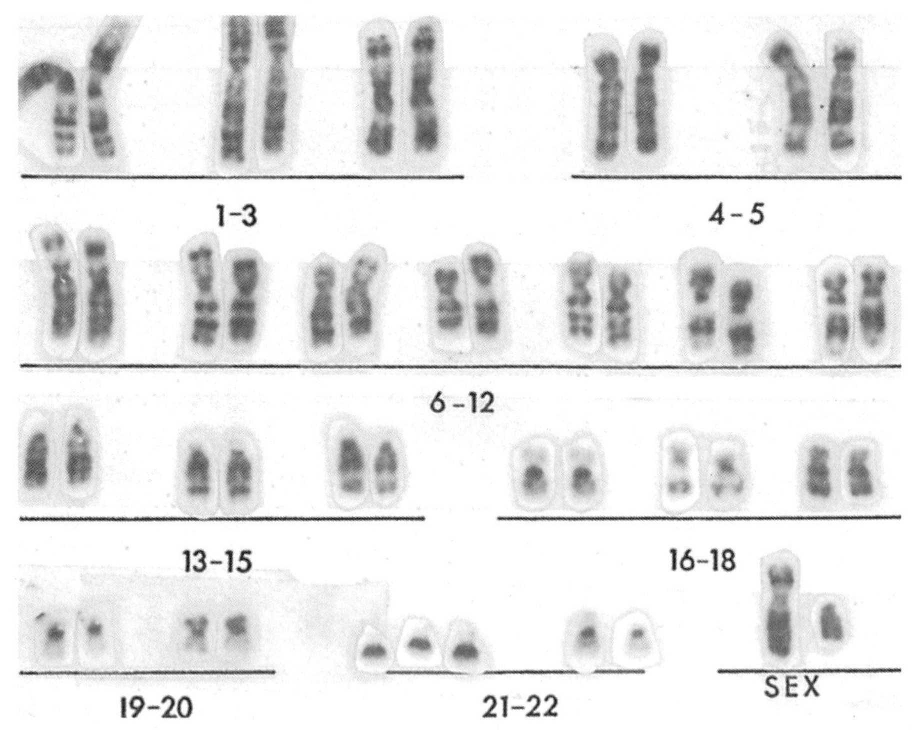

FIGURE 6.7 Child afflicted with trisomy 21 and karyotype.

to those of many Asians, and the iris contains spots, known as Brushfield spots. The tongue may be furrowed and protruding, causing the mouth to remain partially open. Physical growth, behavior, and mental development are retarded, and approximately 40% of all affected individuals have congenital heart defects. In addition, children with Down syndrome are prone to respiratory infections and contract leukemia at a rate 15 times above the normal population. In the last decade improvements in medical care have increased survival rates dramatically, so that many affected individuals survive into adulthood, although few reach the age of 50 years (see Concepts & Controversies, page 150). In spite of these handicaps, many individuals with Down syndrome lead rich, productive lives and can serve as an inspiration to us all.

RISKS FOR AUTOSOMAL TRISOMY. The causes of autosomal trisomies such as Down syndrome are unknown, but a variety of genetic and environmental factors, including radiation, viral infection, hormone levels, and genetic predisposition, have been proposed. To date the only factor clearly related to autosomal aneuploidy is advanced maternal age. In fact, a relationship between maternal age and Down syndrome was well established 25 years before the chromosomal basis for the condition was discovered. The risk of having children with trisomy 21 is low for young mothers but increases rapidly after the age of 35 years. At the age of 20 the incidence of Down syndrome offspring is 0.05%, by age 35 the risk of having a child with Down syndrome has climbed to 0.9%, and at 45 years 3% of all births are trisomy 21 (Figure 6.8). The effect of maternal age on other aneuploidies has been documented, and the relationship between advanced

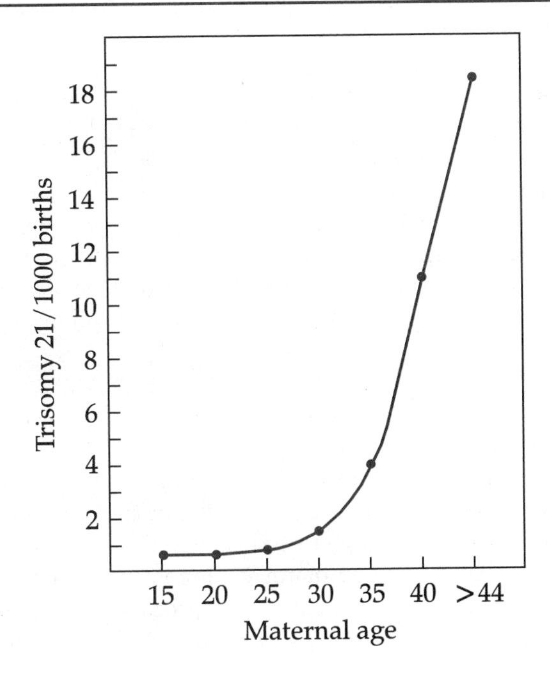

FIGURE 6.8 The relationship between maternal age and the incidence of Down syndrome (trisomy 21). Note that the risk increases greatly after the age of 34 years.

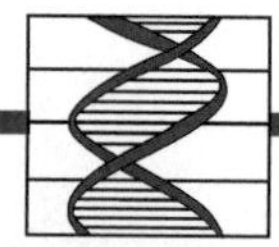

Concepts & Controversies

Alzheimer Disease and Down Syndrome

Alzheimer disease is a degenerative disease of the central nervous system that brings about a progressive mental and physical impairment resulting in death within 5 to 10 years after the onset of symptoms. This condition may affect over 5% of U.S. population over the age of 65 years and is a factor in over 100,000 deaths every year. Alzheimer disease is associated with distinctive structural and chemical changes in the brain and nervous system. Tangles of fibrils develop within the cells of the brain, and a protein known as beta-amyloid accumulates. In addition, there are reduced levels of the chemical messengers associated with the transmission of nerve impulses.

Two groups are known to be at risk for Alzheimer disease: those with a family history of the disease and individuals with Down syndrome. The pathologic changes in the brain and progressive deterioration in behavior that is characteristic of Alzheimer disease occur in many if not all older persons with Down syndrome, leading to the suggestion that a gene or group of genes on chromosome 21 is involved in producing Alzheimer disease. In families in which Alzheimer is present, it is transmitted as an autosomal dominant trait.

Using a series of molecular markers, the gene for the familial form of Alzheimer disease was mapped to chromosome 21 in early 1987. Soon after, it was discovered that the gene encoding the beta-amyloid protein also maps to chromosome 21, and mutations in this gene are responsible for this familial form of Alzheimer disease. Other heritable forms of Alzheimer disease are associated with loci on chromosome 14 and chromosome 19.

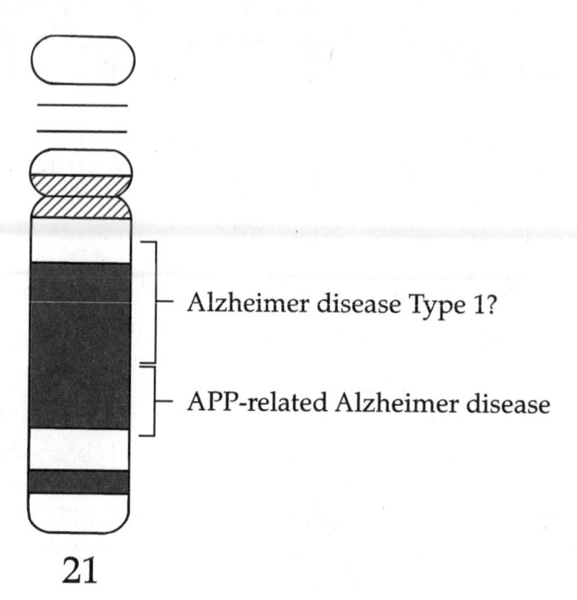

It is not yet known whether Down syndrome individuals develop Alzheimer disease because most of them carry three copies of the genes involved, or whether other factors such as overexpression of the genes are required.

Recently, it has been proposed that some cases of Alzheimer disease may be caused by the slow accumulation of trisomy 21 cells in the nervous system. According to this hypothesis, mitotic nondisjunction causes trisomy 21 cells to reach a threshold that produce the behavioral changes characteristic of Alzheimer disease. Further work on this proposal should shed light on the relationship between Alzheimer disease and Down syndrome.

maternal age and autosomal trisomy is very striking (Figure 6.9). Paternal age has also been proposed as a factor in trisomy, but the evidence is weak, and no clear-cut link has been demonstrated.

The evidence that advanced maternal age is a risk factor for aneuploid offspring comes from studies on the parental origin of nondisjunction events documented by chromosome-banding of the affected child and both parents, and by studies using recombinant DNA technology. Occasionally, some chromosomes exhibit some minor variations that are evident as prominent bands. By examination of banded chromosomes from the trisomic child and the parents, the origin of the nondisjunction can often be determined. For trisomy 21, the nondisjunction event is maternal about 94% of the time and paternal about 6% of the time. In other autosomal trisomies studied, the paternal contribution is about 7%. In all trisomies, the great majority of maternal nondisjunction events occur at meiosis I.

Nondisjunctional events clearly increase as women reach the end of their reproductive period, but the mechanisms controlling this increase remain obscure. One factor may be the duration of female meiosis. In human females all primary oocytes are formed early in development, and all enter the first meiotic prophase well before birth. Meiosis I is not completed until ovulation, so that eggs produced at age 40 have been in meiosis I for over 40 years. During this time metabolic errors or environmental agents may

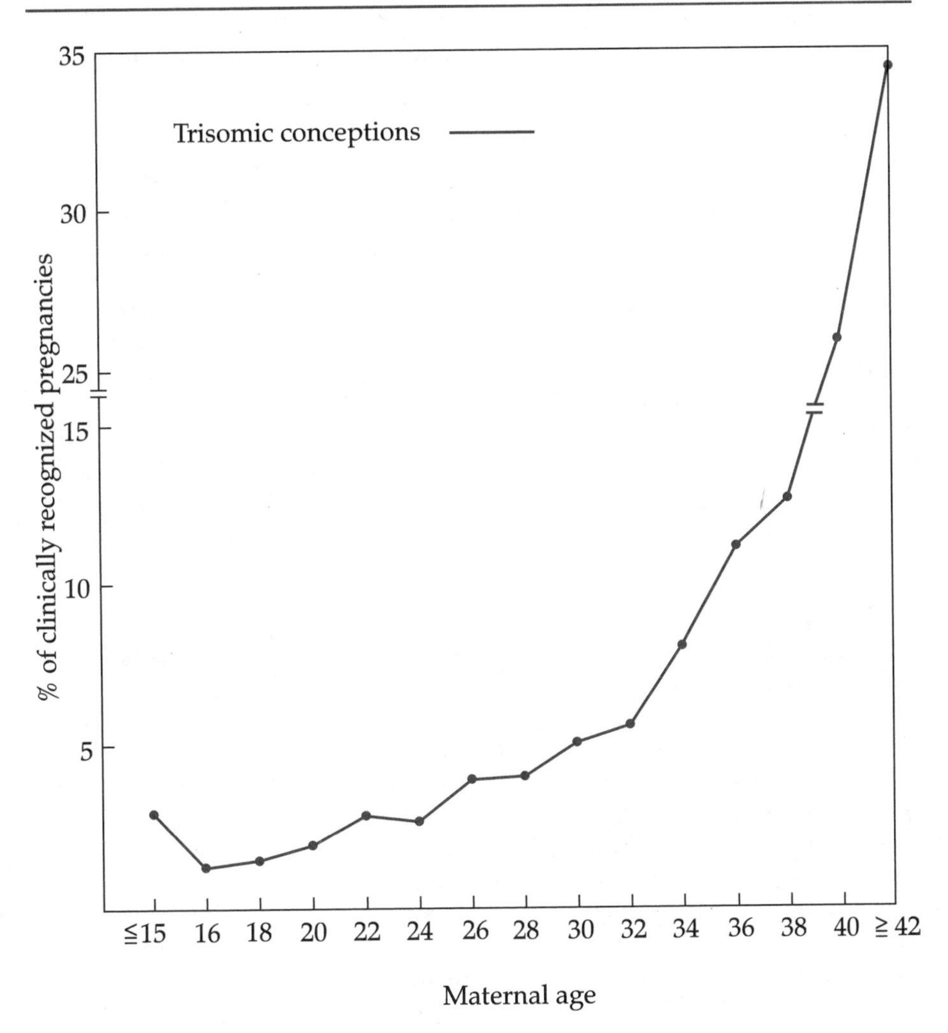

FIGURE 6.9 The relationship between maternal age and the incidence of autosomal trisomy.

damage the cell so that aneuploidy results. However, it is not yet known whether the age of the ovum is directly related to the increased frequency of nondisjunction.

Another factor proposed to account for the increased incidence of aneuploid children with advanced maternal age is related to the interaction between the implanting embryo and the uterine environment. According to this idea, the embryo-uterus interaction normally results in the spontaneous abortion of chromosomally abnormal embryos, a process called maternal selection. As women age, this process may become less effective, allowing more chromosomally abnormal embryos to implant and develop. This second theory to explain the relationship between advanced maternal age and aneuploid offspring is called *relaxed maternal selection*.

Aneuploidy of the Sex Chromosomes

The incidence of aneuploidy involving sex chromosomes in both prenatal deaths and in newborns is higher than for autosomes. Unlike the situation with autosomes, where monosomy is always fatal, monosomy for the X chromosome is a viable condition. Monosomy for the Y chromosome (45, Y), however, is always lethal. The overall incidence of sex chromosome anomalies in live births is 1 in 400 for males and 1 in 650 for females.

Turner syndrome
A monosomy of the X chromosome (45, X) that results in female sterility.

TURNER SYNDROME (45, X). Monosomy for the X chromosome (45, X) was reported as a chromosomal disorder in 1959, but the cytogenetic findings were only the finishing touch to a larger piece of genetic detective work. In 1954, Paul Polani, working on the causes of congenital heart defects, examined three women who suffered from a defect of the aorta usually found only in males. Polani noted that all three women also had phenotypic features originally described by Henry Turner in 1938: short stature, extra folds of skin on the neck, and rudimentary sexual development. Examining cells scraped from the inside of their cheeks, Polani discovered that, like males, these females lacked a sex chromatin, or Barr body. Because of this, he suspected that these **Turner syndrome** individuals might be affected by X-linked traits as frequently as males are affected. Assembling a large group of females lacking ovarian development, including those with Turner syndrome, he tested them for colorblindness, a sex-linked trait. Four of the 25 females tested were colorblind, a result that was much higher than expected in a population of females and similar to that expected in males. In a paper published in 1956, he suggested that Turner syndrome females might have only one X chromosome and in effect, be hemizygous for traits on the X chromosome. After a careful cytogenetic study, Polani and Charles Ford published a paper in 1959 confirming that Turner females are indeed 45, X. The unraveling of the chromosomal basis of Turner syndrome illustrates a basic property of scientific research: work in one area, in this case, congenital heart disease, often leads to significant findings in another field.

Turner syndrome females are short and wide chested with underdeveloped breasts and have rudimentary ovaries (Figure 6.10). At birth a puffiness of the hands and feet is prominent, but this disappears in infancy. As reported by Polani, such individuals also have a narrowing, or coarc-

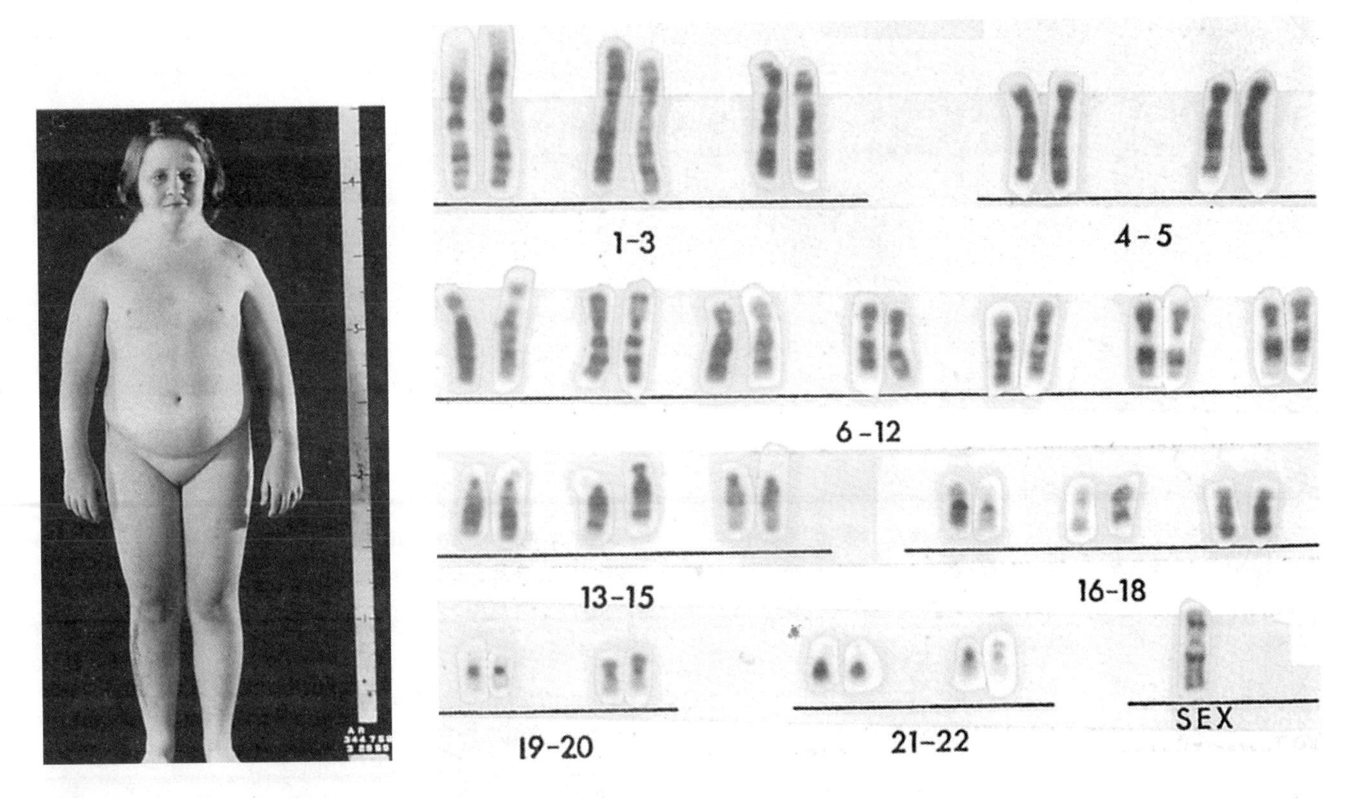

FIGURE 6.10 Individual with Turner syndrome and karyotype. Turner females have short stature, broad chest, and lack of sexual development.

tation, of the aorta. There is no overt mental retardation associated with this syndrome, although evidence suggests that Turner syndrome is associated with reduced skills in interpreting spatial relationships. This chromosome disorder occurs with a frequency of 1 in 10,000 female births, and while affected newborns suffer no life-threatening problems, 95% to 99% of all 45, X embryos die before birth. It is estimated that 1% of all conceptions are 45, X. Another notable feature of this syndrome is that in 75% of all cases, the nondisjunction event originates in the father.

The phenotypic impact of the single X chromosome in Turner syndrome is strikingly illustrated in a case of identical twins, one of whom is 46, XX and the other is 45, X. This situation apparently arose after fertilization and twinning by mitotic nondisjunction. These twins (Figure 6.11) were judged to be identical on the basis of blood types and chromosome-banding studies. They show significant differences in height, sexual development, hearing loss, dental maturity, and performance on tests that measure numerical skills and space perception. While some environmental factors may contribute to these differences, the major role of the second X chromosome in normal female development is apparent. Observations on individuals with Turner syndrome suggest that in females, a second X chromosome is necessary for normal development of the ovary, normal growth patterns, and development of the nervous system. Complete absence of an X chromosome, with or

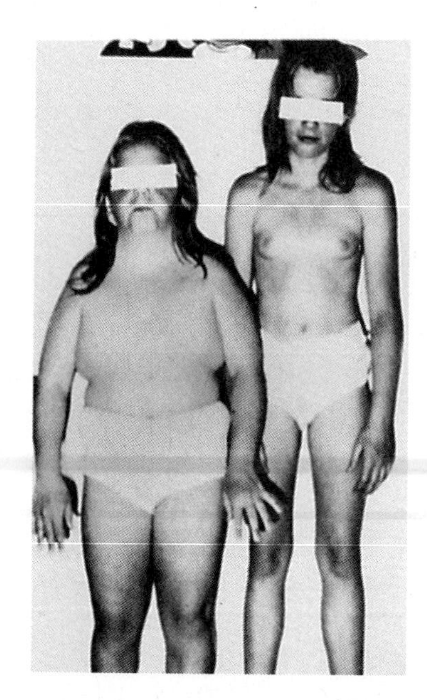

FIGURE 6.11 Monozygotic twins. The twin at left is 45, X Turner syndrome. The twin at right is 46, XX.

without the presence of a Y chromosome, is always lethal, emphasizing the role of the X chromosome as an essential karyotypic component.

KLINEFELTER SYNDROME (47, XXY). The phenotype of Klinefelter syndrome was first described in 1942, and the XXY chromosome constitution was reported in 1959 by Patricia Jacobs and John Strong. The frequency of **Klinefelter syndrome** is approximately 1 in 1000 male births. The phenotypic features of this syndrome do not develop until puberty. Affected individuals are male but show poor sexual development and have very low fertility. Some degree of breast development occurs in about 50% of the cases (Figure 6.12). A degree of subnormal intelligence appears among some affected individuals.

A significant fraction of Klinefelter males are mosaics, with XY and XXY cell lines present in the body. About 60% of the cases are the result of maternal nondisjunction, and advanced maternal age is known to increase the risk of affected offspring. Other forms of Klinefelter include XXYY, XXXY, and XXXXY. The presence of additional X chromosomes in these karyotypes increases the severity of the phenotypic symptoms and brings on clear-cut mental retardation.

XYY karyotype
Aneuploidy of the sex chromosomes involving XYY chromosome constitution.

XYY SYNDROME (47, XYY). In 1965 the results of a cytogenetic survey by Patricia Jacobs of 197 males institutionalized for violent and dangerous antisocial behavior aroused a great deal of interest in the scientific community and the popular press. The findings indicated that nine of these males (about 4.5% of the institutionalized males in the survey) had an **XYY karyotype** (Figure 6.13). These XYY individuals were all above average in height, all suffered personality disorders, and seven of the nine were of subnormal intelligence. Subsequent studies indicated that the frequency of XYY males in the general population is 1 in 1000 male births (about 0.1% of the males in the general population) and that the frequency of XYY individuals in penal and mental institutions is significantly higher than in the population at large. Early investigators associated the tendency to violent criminal behavior with the presence of an extra Y chromosome. In effect, this would mean that some forms of violent behavior were brought about by genetic predisposition. In fact, the XYY karyotype has been used on several occasions as a legal defense (unsuccessfully, so far) in criminal trials. The question is this: is there really a causal relationship between the XYY condition and criminal behavior? There is no strong evidence to support such a link; in fact, the vast majority of XYY males lead socially normal lives. Long-term studies of the relationship between antisocial behavior and the 47, XYY karyotype have been discontinued because of the fear that identifying someone as 47, XYY with potential behavior problems might lead parents to treat those children differently, resulting in the development of behavior problems as a kind of self-fulfilling prophecy. However, until large-scale, long-range studies on this problem are conducted, the question will remain unanswered.

Klinefelter syndrome
Aneuploidy of the sex chromosomes involving an XXY chromosome constitution.

XXX SYNDROME (47, XXX). Approximately 1 in 1000 females are born with three copies of the X chromosome (Figure 6.14). In most cases these females are clinically normal, although there is a slight increase in sterility

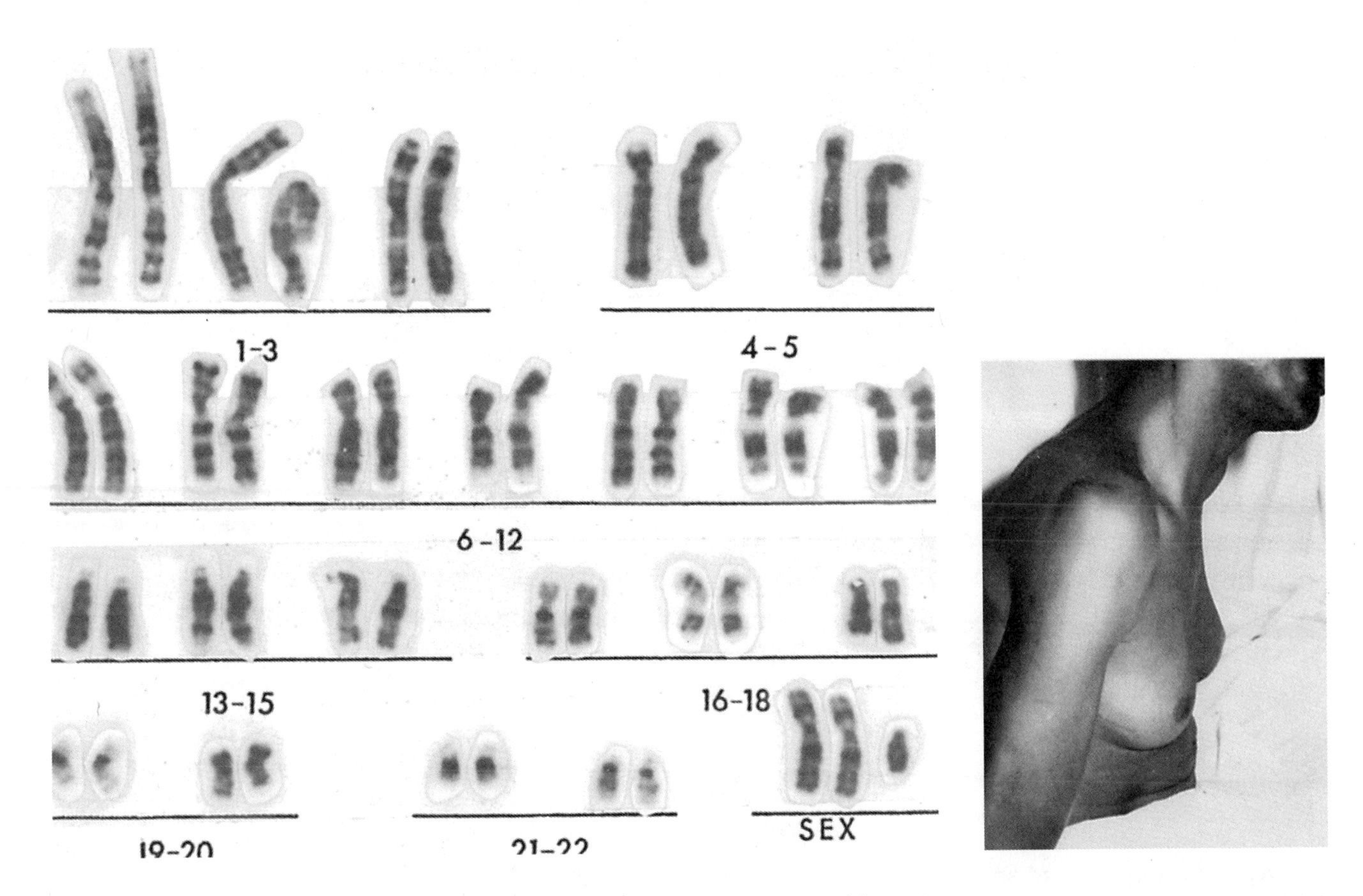

FIGURE 6.12 Klinefelter male showing some breast development, and karyotype.

and mental retardation compared with the population at large. In rare cases, 48, XXXX and 49, XXXXX karyotypes have been reported, and problems of sexual development and mental retardation are severe.

Aneuploidy of the Sex Chromosomes: Some Conclusions

Several conclusions can be drawn from the study of sex chromosome disorders. First, at least one copy of an X chromosome is essential for survival. Embryos without any X chromosomes (44, –XX and 45, OY) are inviable and are not observed in studies of spontaneous abortions. They must be eliminated even before pregnancy is recognized, emphasizing the role of the X chromosome in normal development. The second general conclusion is that addition of extra copies of either sex chromosome produces perturbations of development and causes both physical and mental problems. As the number of sex chromosomes in the karyotype increases, the derangements become more severe, indicating that a balance of sex-chromosome gene dosage and gene products is essential to normal development in both males and females.

FIGURE 6.13 Karyotype of an XYY male.

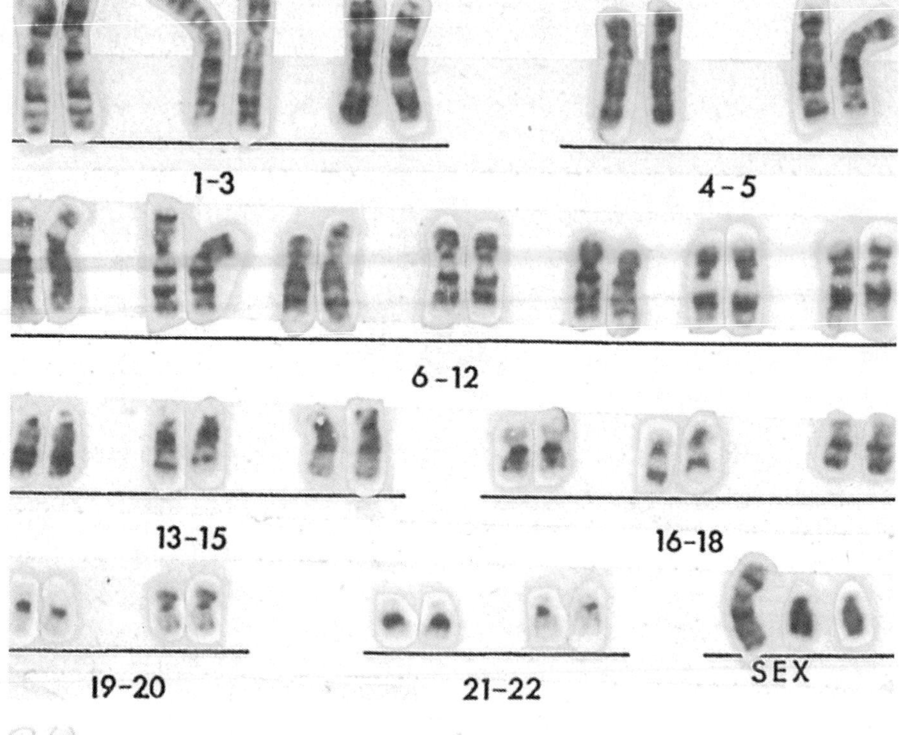

FIGURE 6.14 Karyotype of an XXX female.

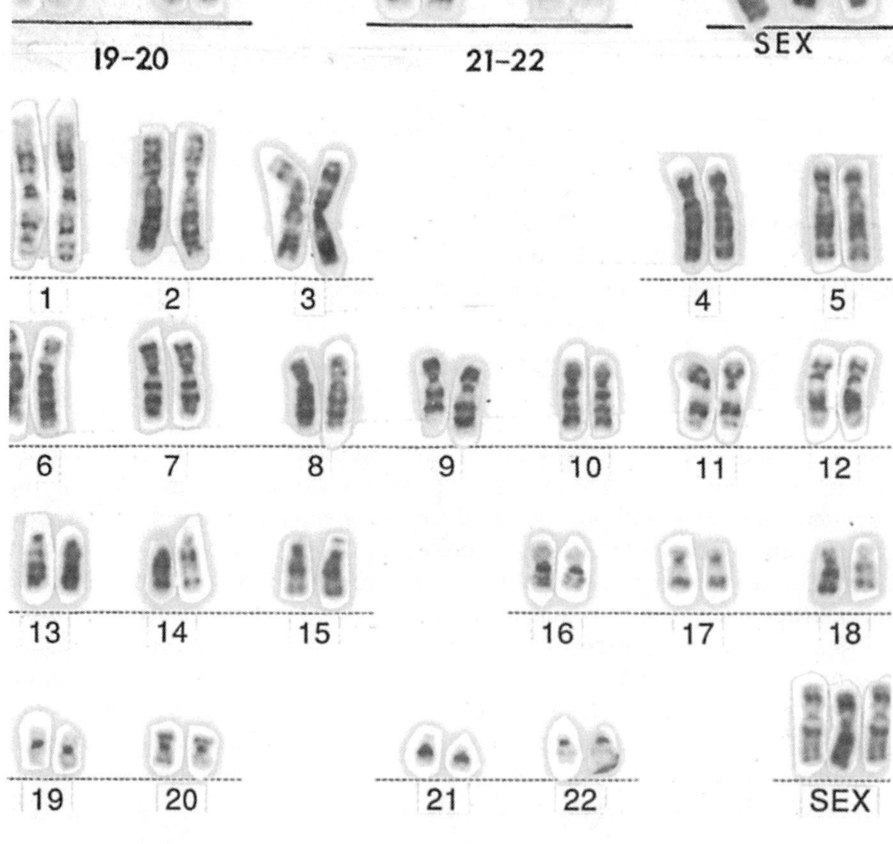

TABLE 6.1 Barr Bodies in Sex Chromosome Anomalies

Karyotype	Number of Barr Bodies
46,XY normal male	0
46,XX normal female	1
45,X Turner syndrome female	0
47,XXY Klinefelter male	1
47,XYY male	0
47,XXX female	2
47,XXXY male	2
48,XXXX female	3
49,XXXXY male	3

Abnormalities in the number of sex chromosomes also provide further cytogenetic evidence for the Lyon hypothesis. Examination of cells from individuals with sex chromosome anomalies show the number of Barr bodies predicted by the Lyon hypothesis. For example, in females with the 45, X Turner syndrome, no Barr bodies are found; in 47, XXY Klinefelter males, one Barr body is found, even though the affected individual is male. Table 6.1 summarizes the association between the number of sex chromosomes and the number of Barr bodies. The general rule, summarized by these cases, is that the number of Barr bodies is one less than the number of X chromosomes.

Variations in Chromosome Structure

Changes in chromosome structure result from the breakage and reunion of chromosome segments. These structural alterations can involve one, two, or more chromosomes. In some cases, rejoining the broken pieces restores the original structure, while in others it creates an array of abnormal chromosomes. Breaks can occur spontaneously through errors in replication or recombination. They can also be produced by environmental agents such as ultraviolet light, radiation, viruses, and chemicals. The alterations produced include **deletions,** or loss of a chromosome segment; **duplications,** which are extra copies of a chromosome segment; **translocations,** which move a segment from one chromosome to another; and **inversions,** or reversal in the order of a chromosome segment (Figure 6.15). Rather than considering how such aberrations are produced, we will concentrate on the phenotypic effects of these structural alterations and on how such changes in chromosome structure can be used to provide information about the location of genes.

Deletion
A chromosomal aberration in which a segment of a chromosome is missing.

Duplication
A chromosomal aberration in which a segment of a chromosome is repeated and therefore is present in more than one copy within the chromosome.

Translocation
A chromosomal aberration in which a chromosome segment is transferred to another, non-homologous chromosome.

Inversion
A chromosomal aberration in which a chromosome segment has been rotated 180° from its usual orientation.

FIGURE 6.15 Structural abnormalities of chromosomes. In a *deletion,* a chromosome segment is lost. In a *duplication,* a chromosome segment is present in more than one copy. In a reciprocal *translocation,* chromosome segments are exchanged. In an *inversion,* a chromosome segment becomes inverted. Inversions that do not involve the centromere (as shown) are called paracentromeric; those that involve the centromere are called pericentromeric.

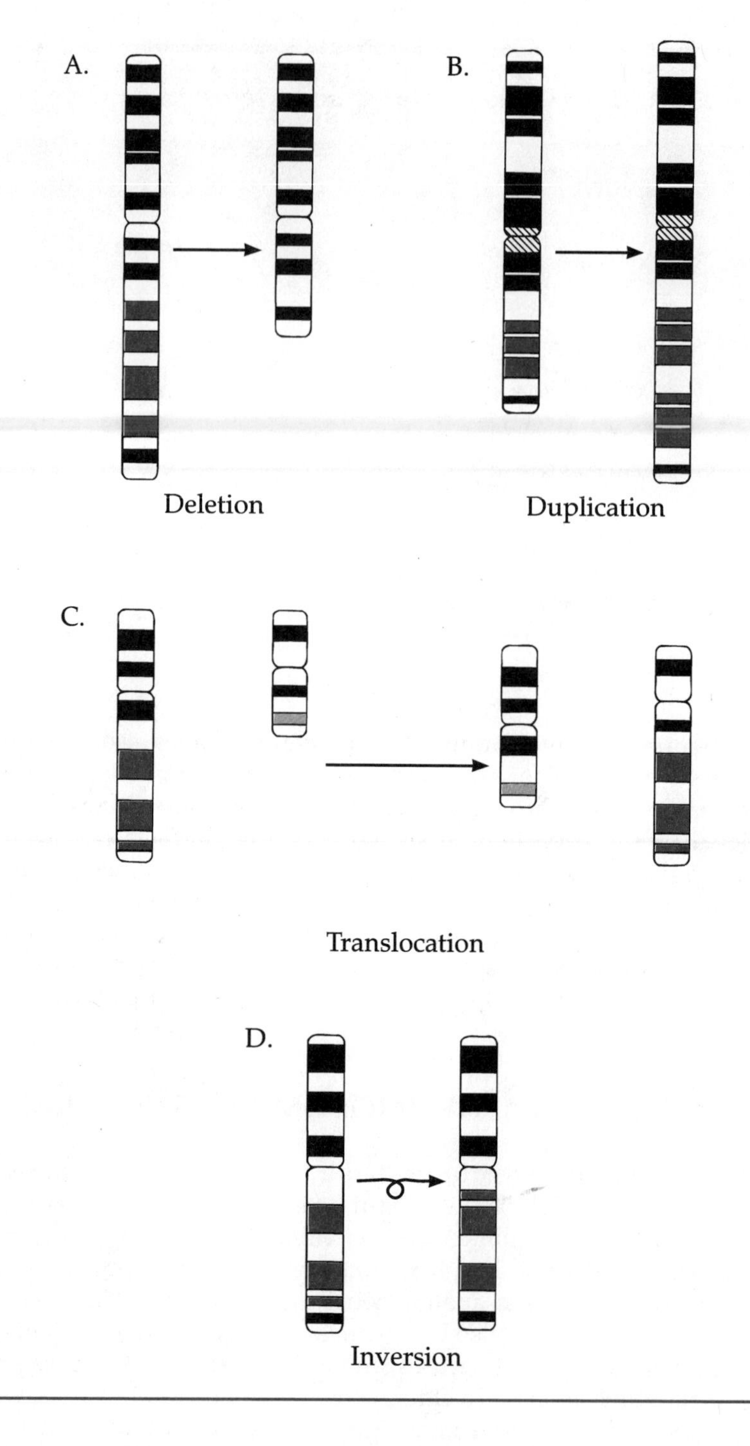

Deletions

Deletion of more than a small amount of chromosomal material will obviously have a detrimental effect on the developing embryo, and deletion of an entire autosome is lethal. Consequently there are only a few viable

TABLE 6.2 Chromosome Deletions

Deletion	Syndrome	Phenotype
4p–	Wolf-Hirschhorn syndrome	Growth retardation, heart malformations, cleft palate; 30% die within 24 months
5p–	Cri-du-chat syndrome	Infants have catlike cry, some facial anomalies, severe mental retardation
11q	Wilms tumor	Kidney tumors, genital and urinary tract abnormalities
13q–	Retinoblastoma	Cancer of eye, increased risk of other cancers
15q–	Prader-Willi syndrome	Infants: weak, slow growth; children adults: obesity, compulsive eating

conditions associated with large-scale deletions. Some of these are listed in Table 6.2.

CRI-DU-CHAT SYNDROME (5p MONOSOMY). An infant carrying a deletion of the short arm of chromosome 5 was first reported in 1963, and the condition has been found to occur in 1 in 100,000 births. A partial deletion on one copy of chromosome 5 makes loci on the normal homologue hemizygous. It is this reduction in gene dosage that is associated with the abnormal phenotype, not the presence of one or more mutant genes. The affected infant is mentally retarded, has defects in facial development, gastrointestinal malformations, and abnormal development of the glottis and larynx (Figure 6.16). As a result of the latter defect, the most characteristic symptom of the disorder is that the cry of affected infants resembles that of a cat, hence the name **cri-du-chat syndrome** (Figure 6.17). This deletion of a chromosome segment affects the motor and mental development of affected individuals but does not seem to be life threatening.

By correlating phenotypes with chromosomal breakpoints in affected individuals, two regions associated with this syndrome have been identified on the short arm of chromosome 5 (Figure 6.16). Loss of chromosome material in the proximal region of 5p15.3 results in abnormal larynx development, while deletions in 5p15.2 are associated with mental retardation and other phenotypic features of this syndrome. This correlation indicates that genes controlling larynx development may be located in 5p15.3, and genes important in the development or function of the nervous system are located in 5p15.2.

Cri-du-chat syndrome
A deletion of the short arm of chromosome 5 associated with an array of congenital malformations, the most characteristic of which is an infant cry that resembles a mewing cat.

PRADER-WILLI SYNDROME. Recently a constellation of physical and mental symptoms known as **Prader-Willi syndrome** has been correlated with deletions in the long arm of chromosome 15. As infants, affected

Prader-Willi syndrome
A deletion of a small segment of the long arm of chromosome 15 that produces a disorder characterized by uncontrolled eating and obesity.

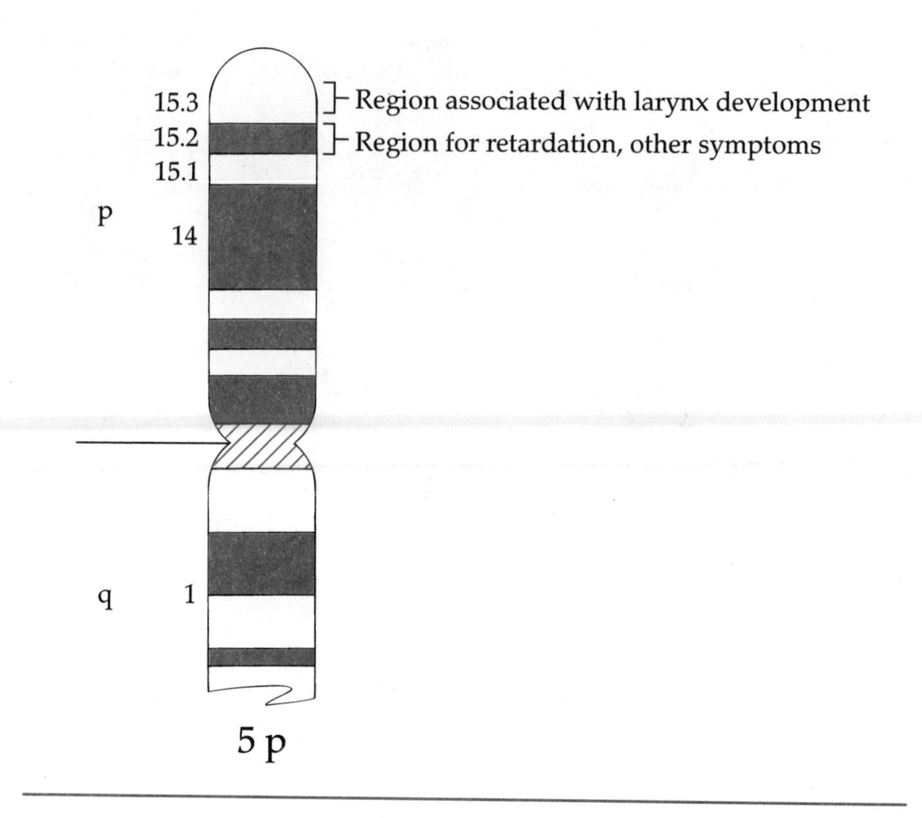

FIGURE 6.16 Cri-du-chat syndrome. Diagram of chromosome 5, showing regions in which deletions produce the cri-du-chat syndrome.

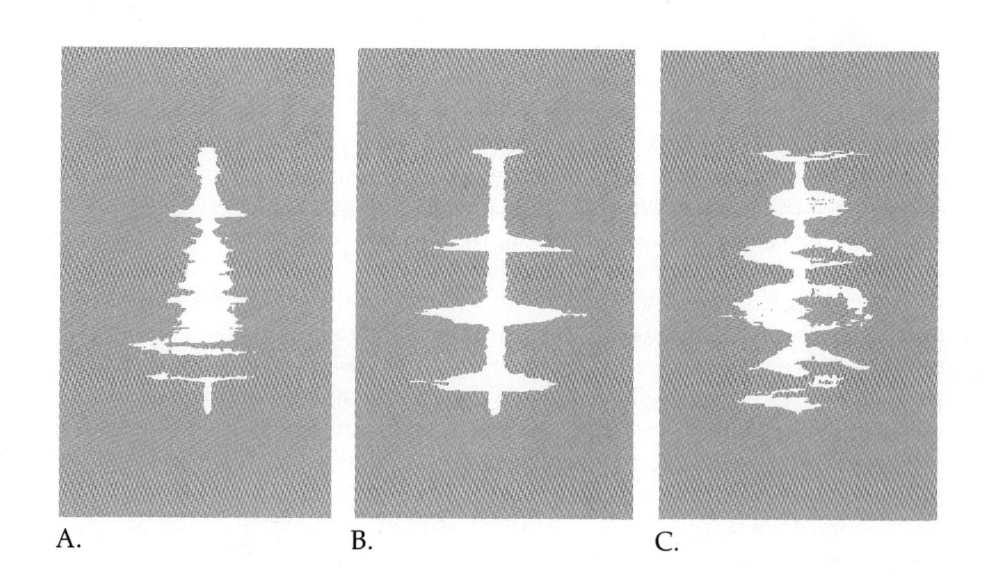

FIGURE 6.17 Sound recording. (A) Normal infant. (B) Cat. (C) Cri-du-chat infant. The cry of the cri-du-chat infant more closely resembles that of a cat than that of a normal infant.

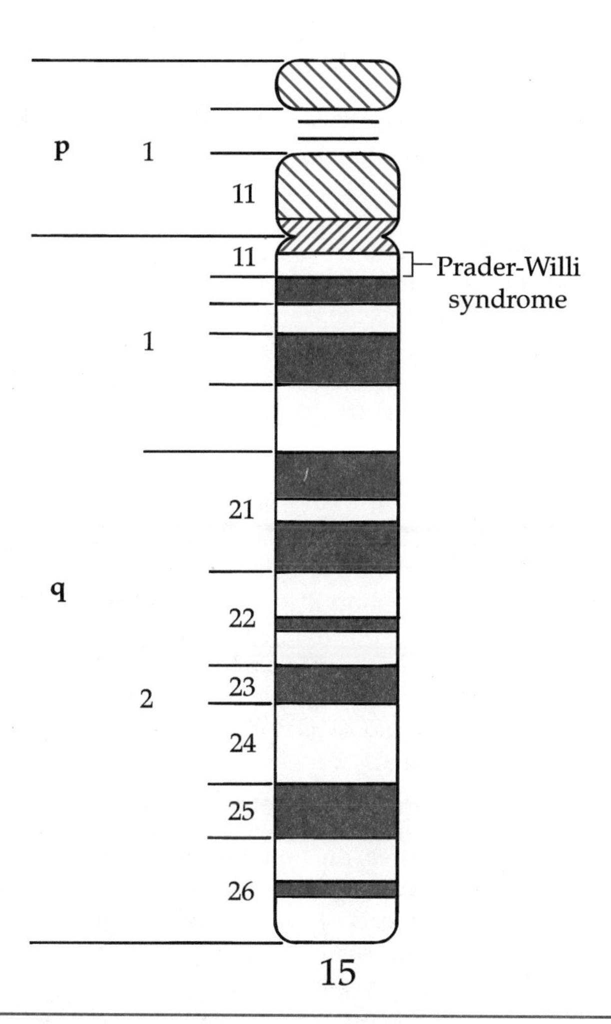

FIGURE 6.18 Diagram of chromosome 15, showing region deleted in Prader-Willi syndrome.

individuals are weak and do not feed well because of a poor sucking reflex. However, by the age of 5 or 6 years, these children develop an uncontrollable compulsion to eat that results in obesity and related health problems such as diabetes. Left untreated, victims will literally eat themselves to death. Other symptoms include poor sexual development in males, behavioral problems, and mental retardation. Careful examination of karyotypes from Prader-Willi patients shows deletions in the long arm of chromosome 15 between bands q11 and q13. The size of the deletion is variable but always includes band 15ql l.2 (Figure 6.18). In about 5% of cases other chromosomal aberrations are found at this site, including duplications and translocations, giving rise to speculation that this region is predisposed to structural instability.

Prader-Willi syndrome is estimated to affect from 1 in 10,000 to 1 in 25,000 people, with males predominating. The cause of the eating disorder is unknown but may be related to disturbances in endocrine function. Treatment includes behavior modification with constant supervision of access to food.

Translocations

Robertsonian translocation Breakage in the short arms of acrocentric chromosomes followed by fusion of the long parts into a single chromosome.

Reciprocal translocation A chromosomal aberration resulting in a positional change of a chromosome segment. This changes the arrangement of genes, but not the number of genes.

Transfer of a chromosome segment to another non-homologous chromosome is known as a translocation. There are two major types of translocations: **reciprocal translocations** and **Robertsonian translocations.** If two non-homologous chromosomes exchange segments, the event is a reciprocal translocation. In such exchanges there is no gain or loss of genetic information, only the rearrangement of gene sequences. In some cases no phenotypic effects are seen, and the translocation is passed through a family for generations (see Concepts & Controversies, page 164). However, during meiosis, cells containing translocations can produce genetically unbalanced gametes containing duplicated or deleted chromosome segments. If these gametes participate in fertilization, the result is embryonic death or abnormal offspring.

About 5% of all cases of Down syndrome involve a Robertsonian translocation, most often between chromosomes 21 and 14. In this type of translocation, there is fusion of the centromeres, and some chromosome material may be lost from the short arms (Figure 6.19). The carrier of such a translocation is phenotypically normal, even though the short arms of both chromosomes may be lost. This carrier parent is actually aneuploid and has only 45 chromosomes. However, because the carrier has two copies of the long arm of chromosome 14 and two copies of the long arm of chromosome 21 (a normal 14, a normal 21, and a translocated 14/21), there is no phenotypic effect. At meiosis the carrier will produce six types of gametes in equal proportions (Figure 6.19). Three of these will result in zygotes that are inviable. Of the remaining three, one will produce a Down syndrome child. Theoretically this means that the chance of producing future children with Down syndrome is one third, or 33%. In practice, however, the observed risk is somewhat lower. It is important to remember that this risk does not increase with maternal age. In addition, there is also a 1 in 3 chance of producing a phenotypically normal translocation carrier, who is at risk of producing children with Down syndrome. For this reason it is important to karyotype a Down syndrome child and his or her parents to determine whether a translocation is involved. This information is essential in counseling parents about future reproductive risks.

Cytogenetic examination of translocation events can also produce information about chromosomal regions that are most critical to the expression of aneuploid phenotypes. For example, the study of a large number of translocations involving chromosome 21 has correlated chromosome regions on the long arm with the symptoms associated with Down syndrome (See Figure 6.20 on page 165). With this information available, efforts are now centered on identifying and characterizing the genes in this chromosome segment to understand how a change in gene dosage produces such serious phenotypic effects.

Translocations are also involved in specific forms of leukemia. The exchange of chromosome segments alters the regulation of a class of genes known as oncogenes whose action is essential to development and maintenance of the malignancy. This topic will be considered in some detail in Chapter 13.

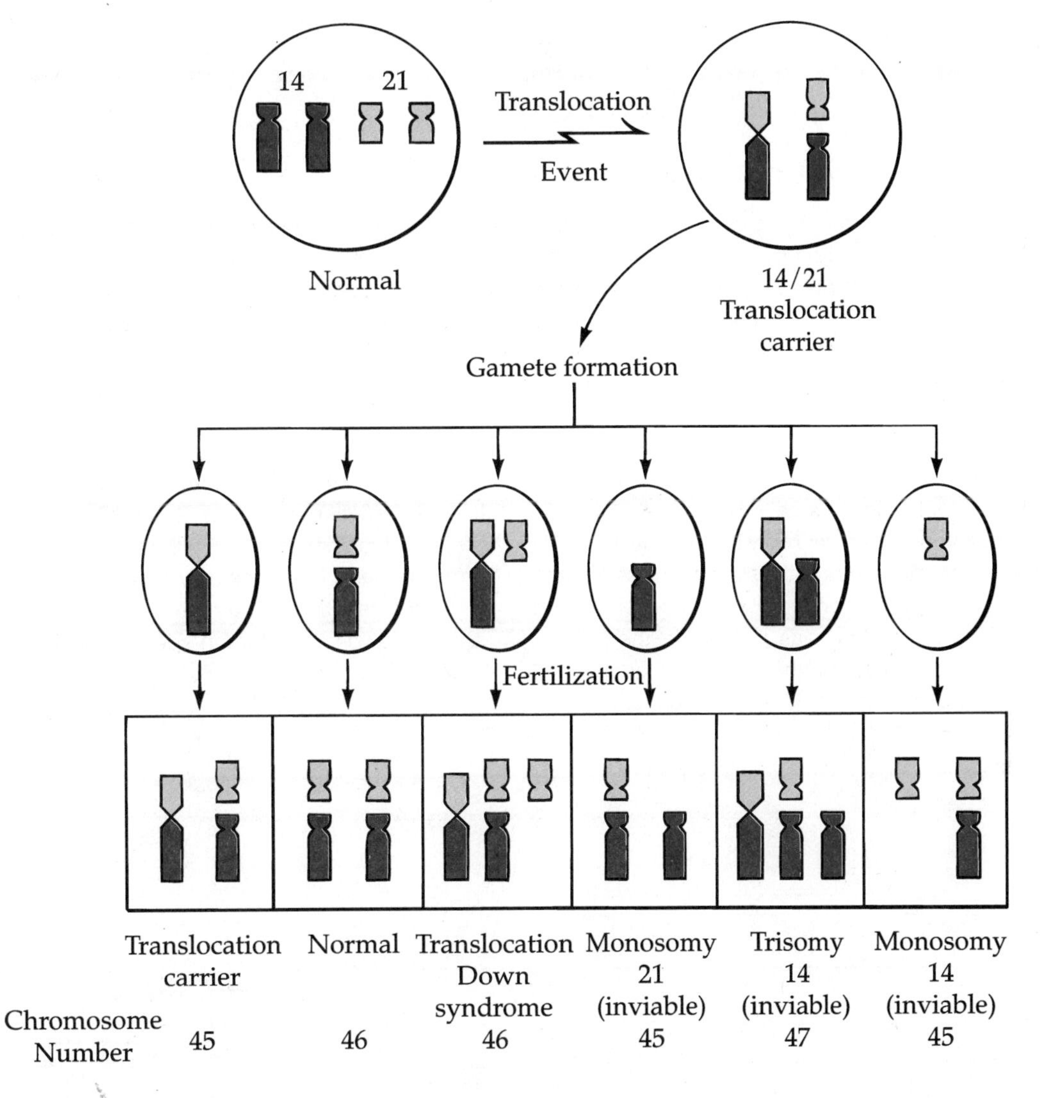

FIGURE 6.19 Meiosis in a 14/21 translocation carrier produces familial Down syndrome.

PRENATAL DIAGNOSIS OF CHROMOSOME ABERRATIONS

The development of prenatal diagnostic techniques represents a major advance in human genetics. It is now possible to precisely diagnose hundreds of chromosomal and biochemical disorders before birth. Prenatal diagnostic techniques now in use include amniocentesis, chorionic villus sampling, ultrasonography, and fetoscopy. Another method that detects and isolates fetal cells in the maternal circulation is now under development and may be used in clinical testing in a few years. In this section, we will examine

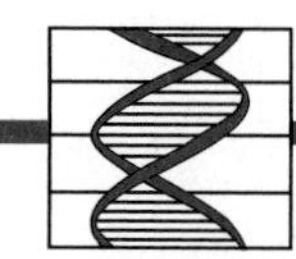

Concepts & Controversies

The Oldest Known Chromosome Aberration

In conducting a cytogenetic analysis to confirm a diagnosis of Down syndrome, Paul Genest and his colleagues observed an abnormal-looking Y chromosome in addition to trisomy 21. The long arm of the Y chromosome contained structures that were identical in appearance to the short arms of the acrocentric chromosomes. In a report in 1967, they hypothesized that this arose as the result of a translocation between an acrocentric chromosome and the Y chromosome. Confirmation for this idea came from the use of specific staining techniques that stain only the short arms of the acrocentric chromosomes and also stain the structures on the long arm of the abnormal Y chromosome. This satellited Y chromosome was present in the phenotypically normal father, brother, paternal uncle, and nephew of the propositus. The research team then screened other relatives of the father and found that the satellited Y chromosome was being passed from father to son as a phenotypically silent chromosome variant. The related family groups surveyed had a single male ancestor in common. This ancestor, Pierre R. (III-3) was born in France in 1635 and emigrated to Quebec in 1665. Because of the structure of the pedigree, he is the only person who could have passed on this chromosome to all the male descendants in the family line. (Filled symbols have a satellited Y; others not tested.) It is impossible to tell whether the satellited Y arose in Pierre R. or was only transmitted through him. Nevertheless, this is the oldest translocation chromosome known in the human species, having originated at least 350 years ago.

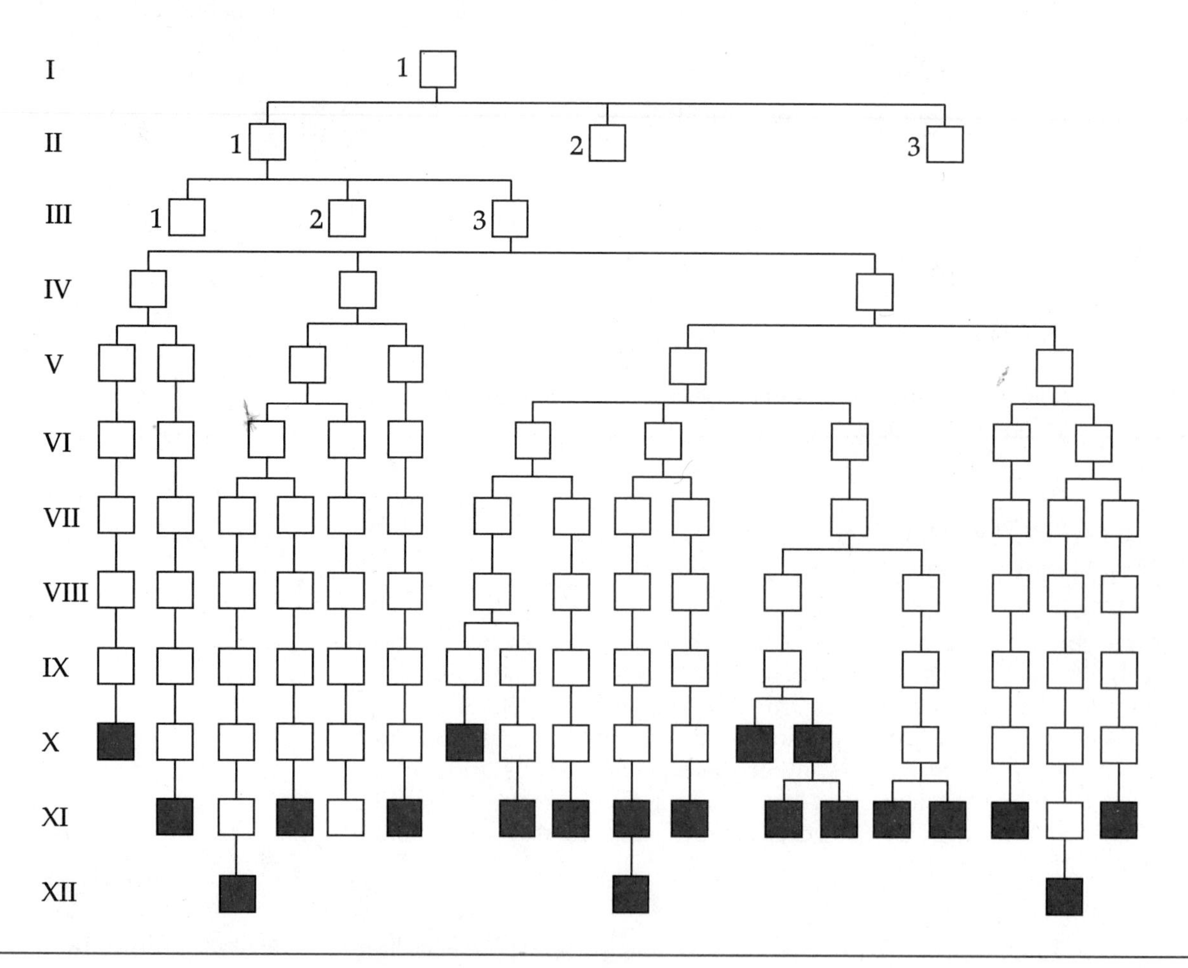

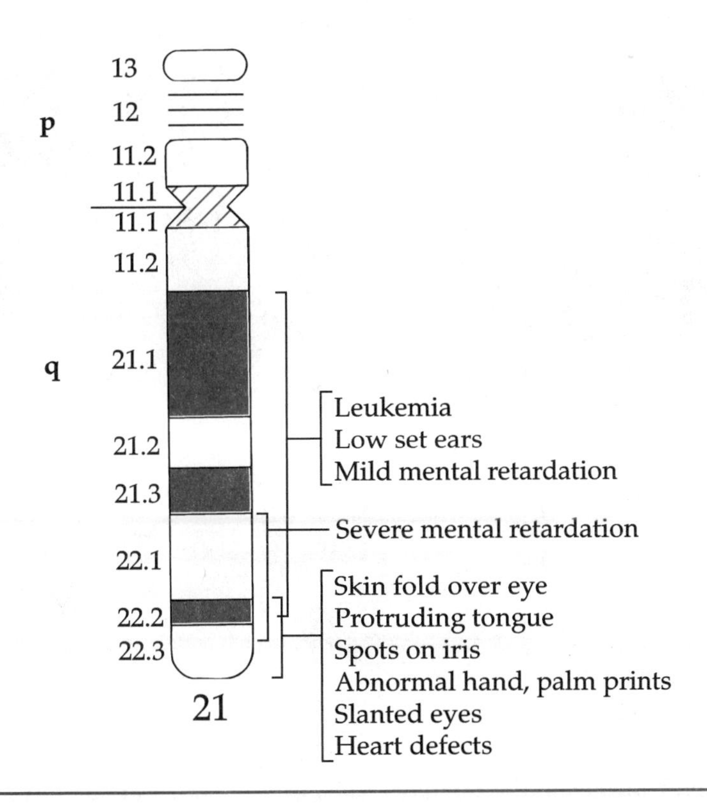

FIGURE 6.20 Diagram of chromosome 21, showing region associated with the phenotype of Down syndrome.

amniocentesis and chorionic villus sampling. Fetal imaging through the use of ultrasonography and fetoscopy will be described in Chapter 19.

Amniocentesis

Until the 1960s amniocentesis was used to diagnose and to follow the progress of disorders such as fetal hemolytic anemia. In the early and mid-1960s, amniocentesis was used for cytogenetic and biochemical analysis using fetal cells and the amniotic fluid to detect genetic disorders. In performing **amniocentesis** the fetus and the placenta are located by ultrasound, and a needle is inserted through the abdominal and uterine wall (avoiding the placenta and fetus) into the amniotic sac (Figure 6.21). Approximately 10 to 30 mL of fluid is withdrawn using a syringe.

Amniocentesis
A method of sampling the fluid surrounding the developing fetus by the insertion of a hollow needle and the withdrawal of suspended fetal cells and fluid; used in the diagnosis of fetal genetic and developmental disorders; usually performed in the 16th week of pregnancy.

The amniotic fluid is mostly composed of fetal urine that contains cells shed from the skin, respiratory tract, and urinary tract of the fetus. The cells can be separated from the fluid by centrifugation (Figure 6.22). The fluid can be tested for biochemical abnormalities that indicate the presence of a genetic defect. More than 100 biochemical disorders can be diagnosed by amniocentesis. The amniotic cells can be assayed for biochemical defects or grown in culture for karyotypic analysis following the procedures outlined in Chapter 2 for metaphase preparations. Once a karyotype has been prepared, it is possible to diagnose the sex of the fetus and the presence of any chromosomal abnormality.

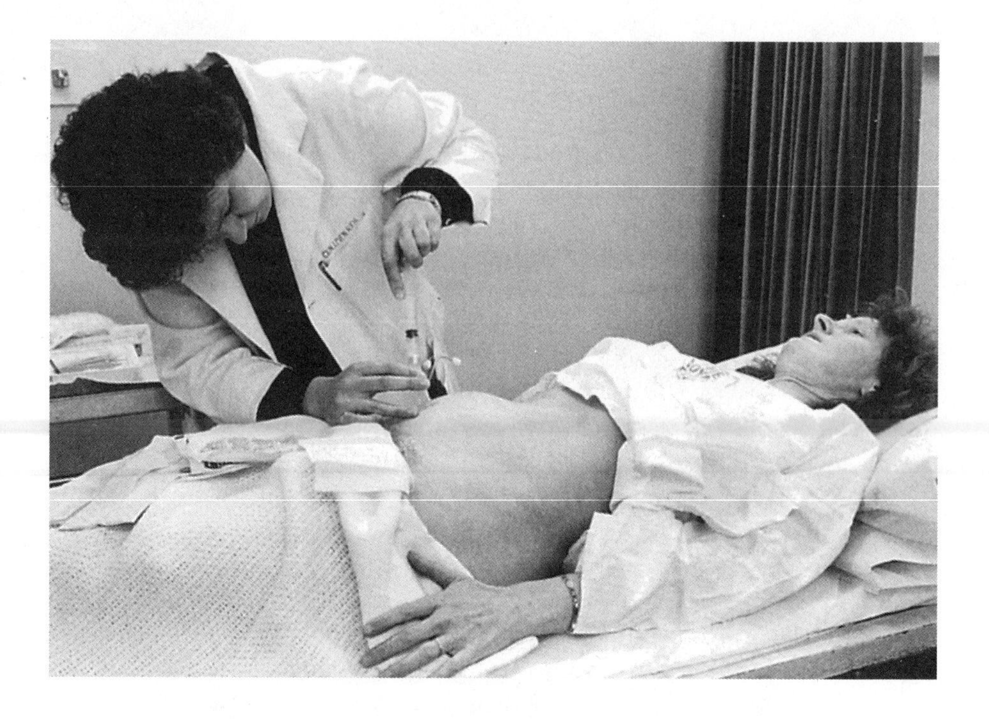

FIGURE 6.21 Amniocentesis being performed.

Several studies in the United States, Britain, and Canada have shown that amniocentesis involves a small but measurable risk to both the fetus and mother. There is a risk of maternal infection and a slight increase (less than 1%) in the probability of a spontaneous abortion. To offset these risks, amniocentesis is normally used only under certain conditions:

1. **Advanced maternal age.** Since the risk for aneuploid offspring increases dramatically after the age of 35 years, this procedure is recommended when the prospective mother is 35 years or older. This risk factor accounts for the majority of all cases in which amniocentesis is used.
2. **Previous aneuploid child.** If a previous child was aneuploid (trisomy is most common), the recurrence risk is about 1% to 2%, and amniocentesis is recommended.
3. **Presence of a balanced chromosomal rearrangement.** If either parent carries a chromosomal translocation or other rearrangement that can cause an unbalanced karyotype in the child, amniocentesis should be considered.
4. **X-linked disorder.** If the mother is a carrier of an X-linked biochemical disorder that cannot be diagnosed prenatally and she is willing to abort if the fetus is male, then amniocentesis is recommended.

Amniocentesis is not usually performed until the 16th week of pregnancy. Before this time, there is very little amniotic fluid, and contamination of the sample with maternal cells is often a complication. Subsequent cytogenetic analysis can require 7 to 10 days, and if abortion is

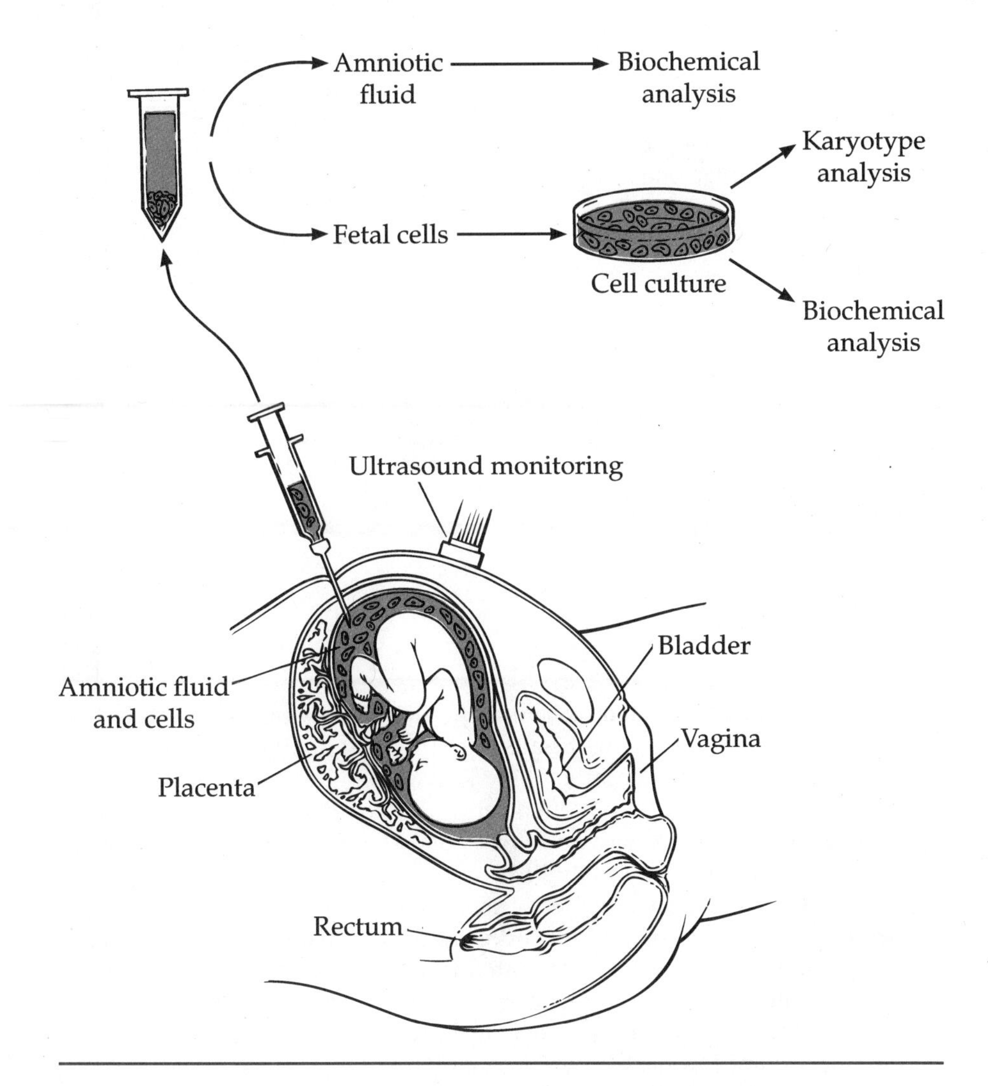

FIGURE 6.22 In amniocentesis, a sample of fluid containing fetal cells is removed. The cells can be separated from the fluid by centrifugation and cultured for karyotypic analysis.

elected, surgical risks become a concern. To allow earlier detection of genetic disorders, an alternative technique known as chorionic villus sampling is available.

Chorionic Villus Sampling

Chorionic villus sampling (CVS) is used to allow the diagnosis of genetic diseases in the developing fetus early in the first trimester of pregnancy. This technique has several advantages over amniocentesis. CVS is usually performed from 8 to 10 weeks of gestation, compared with the 15 to 16 weeks for amniocentesis. Cytogenetic results from CVS are available within a few hours or a few days, and biochemical tests can be performed directly on the sampled fetal tissue rather than after tissue culture, as required following amniocentesis. In use, a small flexible catheter is inserted through the vagina or abdomen into the uterus, guided by ultrasound images. A small sample

Chorionic villus sampling A method of sampling fetal chorionic cells by insertion of a catheter through the vagina or abdominal wall into the uterus. Used in the diagnosis of biochemical and cytogenetic defects in the embryo. Usually performed in the eighth or ninth week of pregnancy.

FIGURE 6.23 In chorionic sampling, a catheter is inserted into the uterus to remove a fetal tissue sample. These cells can be used for karyotypic, biochemical, or molecular analyses.

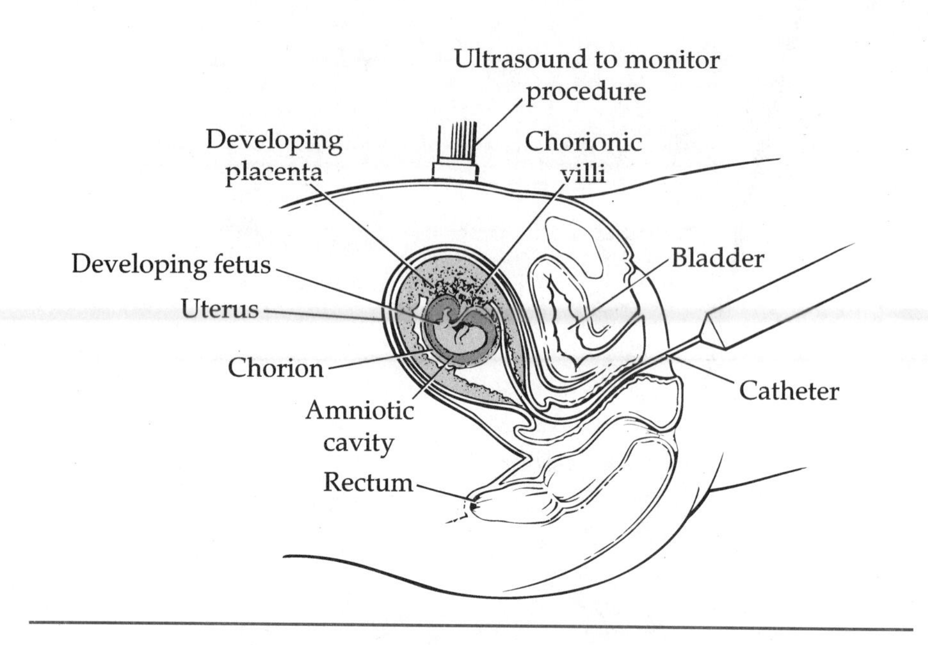

from the chorionic villi, a fetal tissue that will form the placenta, is obtained by suction (Figure 6.23). This tissue is mitotically active and can be used immediately in the preparation of a karyotype. As with amniocentesis, enough material is usually obtained with CVS to allow biochemical testing or the preparation of DNA samples for molecular analysis. The use of DNA analysis for prenatal diagnosis will be discussed in Chapter 8.

CVS is a more specialized technique than amniocentesis and at present is used less often. Although early studies indicated that the procedure posed a higher risk to mother and fetus than amniocentesis, improvements in the instruments and technique have lowered the risk somewhat. CVS offers early diagnosis of genetic diseases, and if termination of pregnancy is elected, maternal risks are lower at 9 to 12 weeks than at 16 weeks.

CONSEQUENCES OF CHROMOSOMAL ANEUPLOIDY

As indicated earlier in the chapter, cytogenetic surveys have shown that chromosomal abnormalities are a major cause of spontaneous abortions (Figure 6.4). Table 6.3 summarizes some of the major chromosomal abnormalities found in miscarriages. These include triploidy, monosomy for the X chromosome (45, X), and trisomy 16. It is interesting to compare the frequency of chromosomal abnormalities found in spontaneous abortions with those seen in live births. Triploidy is found in 17 of every 100 spontaneous abortions but in only about 1 in 10,000 live births; 45, X is found in 18% of chromosomally abnormal abortuses but in only 1 in 7,000 to 10,000 live births.

Surveys of chromosomal abnormalities detected by CVS (performed at about 8 weeks of gestation) and by amniocentesis (performed at about 16 weeks of gestation) show that those detected by CVS are two to five times higher than those detected by amniocentesis, which in turn are about two times higher than those found in newborns. This gradient in the frequency of chromosomal abnormalities over developmental time is evidence that karyotypically abnormal embryos and fetuses are selected against throughout gestation.

Birth defects are another consequence of chromosome abnormalities. The incidences of chromosome aberrations detected in cytogenetic surveys of newborns are shown in Table 6.4. Note that among trisomies, trisomy 16, which is common in spontaneous abortions, is not found among infants, indicating that this condition is completely selected against. Although trisomy 21 is the most common autosomal trisomy in newborns, with a frequency of about 1 in 900 births, data from cytogenetic surveys of spontaneous abortions indicate that about two thirds of such conceptions are lost by miscarriage. Similarly, over 99% of all 45, X conceptions are lost before birth. Overall, while selection against chromosomally abnormal embryos and fetuses is efficient, the data from a number of surveys indicate that there is a significant risk for chromosomal abnormalities and that over 0.5% of all newborns are affected with an abnormal karyotype.

The relationship between the development of cancer in somatic cells of the body and accompanying chromosomal changes is a third consequence of chromosomal abnormalities. An increasing number of malignancies, especially leukemia, are known to be associated with specific chromosomal translocations. In many solid tumors, a wide range of chromosomal abnormalities is present. New evidence suggests that these abnormalities, which include aneuploidy, translocations, and duplications, may arise during a period of genomic instability that precedes or accompanies the

TABLE 6.3 Chromosome Abnormalities in Spontaneous Abortions

Abnormality	Frequency (%)
Trisomy 16	15
Trisomies 13, 18, 21	9
XXX, XXY, XYY	1
Other	27
45, X	18
Triploidy	17
Tetraploidy	6
Other	7

SOURCE: Modified from Hassold, T., 1986 Trends in Genetics *2*: 105–110.

TABLE 6.4 Chromosome Abnormalities in Newborns

Abnormality	Approximate Frequency
45, X	1/7500
XXX	1/1200
XXY	1/1000
XYY	1/1100
Trisomy 13	1/15,000
Trisomy 18	1/11,000
Trisomy 21	1/900
Structural abnormalities	1/400

transition of a normal cell into a malignant one. The mechanisms by which such chromosomal changes can cause or accompany the development of cancer will be discussed in Chapter 13.

FRAGILE SITES

Fragile sites are structural features of chromosomes that become visible under certain conditions of cell culture. They appear as a gap or a break at a characteristic locus and are inherited in a codominant fashion. The fragile sites often produce chromosome fragments, deleted chromosomes, and other alterations in subsequent mitotic divisions. To date 17 heritable fragile sites have been identified in the human genome (Figure 6.24). The molecular nature of fragile sites is unknown but is of great interest, since the sites represent regions that are susceptible to breakage. Almost all studies of fragile sites have been carried out on cells in tissue culture, and at the present time it is unknown whether such sites are expressed in meiotic cells. Most fragile sites do not appear to be associated with any clinical syndrome except for the site on the X chromosome. This site is associated with an X-linked form of mental retardation known as Martin-Bell syndrome, or **fragile-X (fra-X)** syndrome. Males afflicted with this syndrome have long, narrow faces with protruding chins, large ears, enlarged testes, and varying degrees of mental retardation. About 3% to 5% of the males institutionalized for mental retardation have a fragile X chromosome (Figure 6.25). Female carriers show no clear-cut physical symptoms but as a group have a higher rate of mental retardation than normal individuals. The fragile site on the X chromosome is itself the cause of the mental retardation through a newly discovered mechanism of mutation that will be discussed in Chapter 11.

Fragile X
An X chromosome that carries a non-staining gap, or break, at band q27. Associated with mental retardation in males.

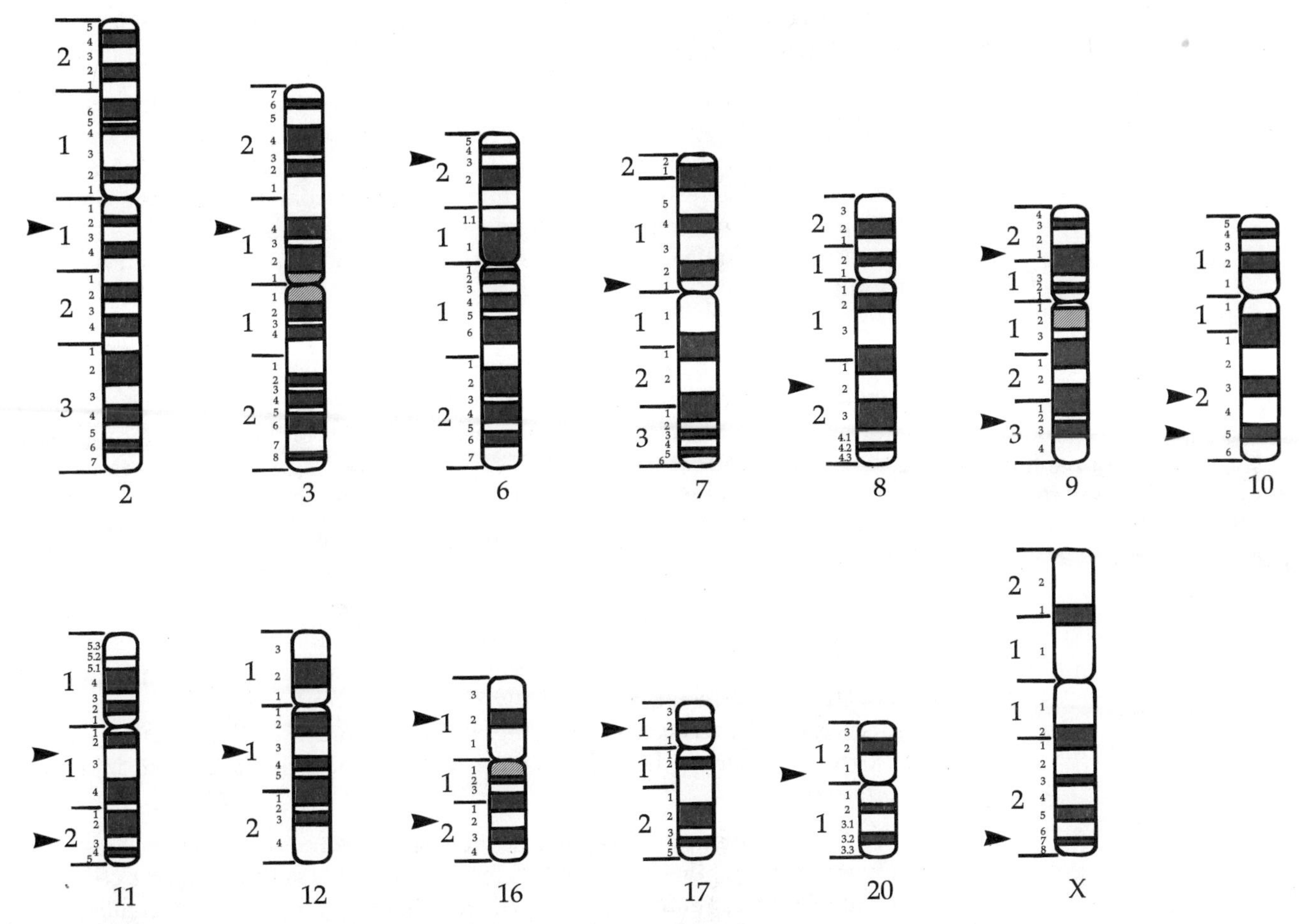

FIGURE 6.24 Location of main fragile sites in the human karyotype.

SUMMARY

1. The study of variations in chromosome structure and number began in 1959 with the discovery that Down syndrome is caused by the presence of an extra copy of chromosome 21. Since then the number of genetic diseases known to be related to chromosome aberrations has steadily increased.

2. The development of chromosome banding and techniques for the resolution of small changes in chromosome structure have contributed greatly to the information that is now available.

3. There are two major types of chromosome changes: a change in chromosome number and a change in chromosome arrangement. Polyploidy, aneuploidy, and changes in chromosome number are major causes of reproductive failure in humans. Polyploidy is rarely seen in live births, but the rate of aneuploidy in humans is reported to be more than 10-fold higher

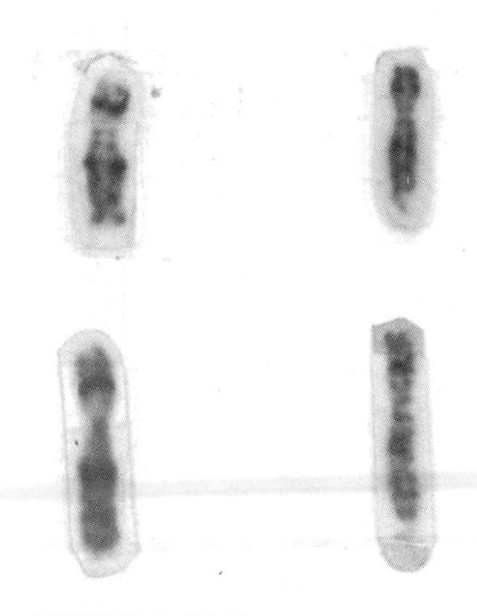

FIGURE 6.25 Normal X chromosomes *(left)* and fragile X chromosomes, showing a characteristic gap at the end of the long arm.

than in other primates and mammals. The reasons for this difference are unknown, but this represents an area of intense scientific interest.

4. The loss of a single chromosome creates a monosomic condition, and the gain of a single chromosome is called a trisomic condition. Autosomal monosomy is eliminated early in development. Autosomal trisomy is selected against less stringently, and cases of partial development and live births of trisomic individuals are observed. Most cases of autosomal trisomy greatly shorten life expectancy, and only trisomy 21 individuals survive into adulthood.

5. Aneuploidy of sex chromosomes shows even more latitude, with monosomy X being a viable condition. Studies of sex chromosome aneuploidies indicate that at least one copy of the X chromosome is required for development. Increasing the number of copies of the X or Y chromosome above the normal range causes progressively greater disturbances in phenotype and behavior, indicating the need for a balance in gene products for normal development.

6. Changes in the arrangement of chromosomes include deletions, duplications, inversions, and translocations. Deletions of chromosome segments are associated with several genetic disorders, including cri-du-chat and Prader-Willi syndromes. Translocations often produce no overt phenotypic effects but can result in genetically imbalanced and aneuploid gametes. We discussed a translocation resulting in Down syndrome that in effect makes Down syndrome a heritable genetic disease, potentially present in 1 in 3 offspring.

7. Fragile sites appear as gaps, or breaks, in chromosome-specific locations. One of these fragile sites, on the X chromosome, is associated with a common form of mental retardation that affects a significant number of males.

QUESTIONS AND PROBLEMS

1. Two hypothetical human conditions have been found to have a genetic basis. The genetic defect responsible for condition 1 is similar in type to the one that causes Marfan syndrome. The defect responsible for condition 2 resembles the one responsible for Edwards syndrome. One of the two conditions results in more severe defects, and death occurs in infancy. The other condition produces a mild phenotypic abnormality and is not lethal. Which condition is most likely lethal and why?

2. Discuss the following sets of terms:
   a. trisomy and triploidy
   b. aneuploidy and euploidy
   c. euploidy and polyploidy

3. A cytology student believes he has identified an individual suffering from monoploidy. The instructor views the cells under the microscope and correctly dismisses the claim.
   a. Why was the claim dismissed?
   b. What type of cells were being viewed?

4. Albinism is caused by an autosomal recessive allele of a single gene. An albino child is born to phenotypically normal parents. However, the paternal grandfather is albino. Exhaustive analysis suggests that neither the mother nor her ancestors carry the allele for albinism. Suggest a mechanism to explain this situation.

5. What is the relative rate of aneuploidy in humans compared with other mammals? Speculate as to what evolutionary role this rate might have.

6. An individual is found to have some tetraploid liver cells but diploid kidney cells. Be specific in explaining how this condition might arise.

7. A spermatogonial cell undergoes mitosis prior to entering the meiotic cell cycle en route to the production of sperm. However, during mitosis the cytoplasm fails to divide, and only one "daughter" cell is produced. A resultant sperm eventually fertilizes a normal ovum. What is the chromosomal complement of the embryo?

8. A teratogen is present at conception. As a result, during the first mitotic division the centromeres fail to divide. The teratogen then loses its potency and has no further effect on the embryo. What will be the chromosomal complement of this embryo?

9. Define chromosomal nondisjunction.

10. Assume a meiotic non-disjunction event is responsible for an individual who is trisomic for chromosome 8. If two of the three copies of chromosome 8 are absolutely identical, when during meiosis did the non-disjunction event take place?

11. What is the genetic basis for:
    a. Edwards syndrome?
    b. Turner syndrome?
    c. Patau syndrome?

12. The majority of non-disjunction events leading to Down syndrome are maternal in origin. Speculate on the possible reasons for females contributing aneuploid gametes more frequently than males.

13. If all the non-disjunction events leading to Turner syndrome were paternal in origin, what trisomic condition might be expected to occur at least as frequently?

14. Define the following terms as they apply to chromosome structure:
    a. deletion
    b. duplication
    c. inversion
    d. translocation

For Further Reading

Borgaonkar, D. S. 1989. *Chromosome Variation in Man: A Catalogue of Chromosomal Variants and Anomalies.* 5th ed. New York: Liss.

Boue, A., Boue, J., and Gropp, A. 1985. Cytogenetics of pregnancy wastage. *Adv. Hum. Genet.* **14:** 1–57.

Croce, C., and Klein, G. 1985. Chromosomal translocations and human cancer. *Sci. Am.* **252** (March): 54–60.

Dellarco, V., Voytek, P., and Hollaender, A. 1985. *Aneuploidy: Etiology and Mechanisms.* New York: Plenum.

Faix, R., Barr, M., Jr., and Waterson, J. 1984. Triploidy: Case report of a live-born male and an ethical dilemma. *Pediatrics* **74:** 296–299.

Friedmann, T. 1971. Prenatal diagnosis of genetic disease. *Sci. Am.* **225** (November): 34–42.

Goldgarber, D., Lerman, M., McBride, O. W., Saffiotti, U., and Gajdusek, D. C. 1987. Characterization and chromosomal location of a DNA encoding brain amyloid of Alzheimer's disease. *Science* **235:** 877–880.

Green, J. E., Dorfmann, A., Jones, S., Bender, S., Patton, L., and Schulman, J. D. 1988. Chorionic villus sampling: Experience with an initial 940 cases. *Obstet. Gynecol.* **71:** 208–212.

Grouchy, J. de. 1984. *Clinical Atlas of Human Chromosomes.* New York: Wiley.

Hassold, T., and Jacobs, P. 1984. Trisomy in man. *Annu. Rev. Genet.* **18:** 69–97.

Khush, G. S. 1973. *Cytongenetics of Aneuploids.* New York: Springer-Verlag.

Knoll, J. H., Nicholls, R. D., Magenis, R. E., Graham, J. M. Jr., Lalande, M., and Latt, S. 1989. Angelman and Prader-Willi syndromes share a common chromosome 15 deletion but differ in parental origin of the deletion. *Am. J. Med. Genet.* **32:** 285–290.

Lafer, C. Z., and Neu, R. L. 1988. A liveborn infant with tetraploidy. *Am. J. Med. Genet.* **31:** 375–378.

Lewis, W. H. (Ed.). 1980. *Polyploidy: Biological Relevance.* New York: Plenum.

Monaco, A. P., and Kunkel, L. 1987. A giant locus for the Duchenne and Becker muscular dystrophy gene. *Trends in Genetics* **3:** 33–37.

Nicholls, R. D., Knoll, J. H., Butler, M. G., Karam, S., and Lalande, M. 1989. Genetic imprinting suggested by maternal heterodisomy in nondeletion Prader-Willi syndrome. *Nature* **342:** 281–285.

Patterson, D. 1987. The causes of Down syndrome. *Sci. Am.* **257** (August): 52–61.

St. George-Hyslop, P., Tanzi, R., Polinsky, R., Haines, J., Nee, L., Watkins, P., Myers, R., Feldman, R., et al. 1987. The genetic defect causing familial Alzheimer's disease maps on chromosome 21. *Science* **235:** 885–890.

Shiono, H., Azumi, J., Fujiwara, M., Yamazaki, H., and Kikuchi, K. 1988. Tetraploidy in a 15-month old girl. *Am. J. Med. Genet.* **29:** 543–547.

Smith, G. F., ed. 1984. *Molecular Structure of the Number 21 Chromosome and Down Syndrome.* New York: New York Academy of Sciences.

Sutherland, G., 1984. The fragile X chromosome. *Int. Rev. Cytol.* **81:** 107–143.

Tanzi, R., Gusella, J., Watkins, P., Bruns, G., St. George-Hyslop, P., Van Keuren, M., Patterson, D., Pagan, S., Kurnit, D., and Neve, R. 1987. Amyloid beta protein gene: cDNA, mRNA distribution, and genetic linkage near the Alzheimer locus. *Science* **235:** 880–884.

Van Broeckhoven, C., Gentre, A., Vandeberghe, A., Horsthemke, B., Backovens, H., Raeymaekers, P., Van Hul, W., et al. 1987. Failure of familial Alzheimer's disease to segregate with the A4-amyloid gene in several European families. *Nature* **329:** 153–155.

Wapner, R. J. and Jackson, L. 1988. Chorionic villus sampling. *Clin. Obstet. Gynecol.* **31:** 328–344.

Warren, S. T., Zhang, F., Licameli, G. R., and Peters, J. F. 1987. The fragile X site in somatic cell hybrids: An approach for molecular cloning of fragile sites. *Science* **237:** 420–423.

Weaver, D. D. 1988. A survey of prenatally diagnosed disorders. *Clin. Obstet . Gynecol.* **31:** 253–269.

Weiss, E., Loevy, H., Saunders, A., Pruzansky, S., and Rosenthal, I. 1982. Monozygotic twins discordant for Ullrich-Turner syndrome. *Am J. Med Genet.* **13:** 389–399.

CHAPTER 7

DNA Structure and Chromosome Organization

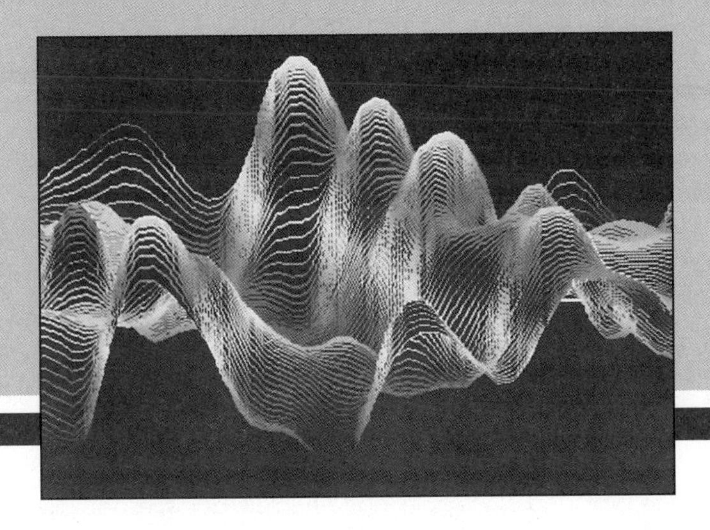

CHAPTER OUTLINE

Early in the 1860s, Frederick Miescher decided to study the chemical composition of human lymph cells (white blood cells) with the hope of understanding something of the nature of cellular mechanisms associated with life. However, such cells were difficult to obtain in the quantities needed for chemical studies. Miescher learned that cells present in the pus of infected wounds were derived from the lymph cells. In the days before antibiotics, wound infections were common, and Miescher visited local hospitals to collect discarded bandages. He scraped the pus from these bandages and developed a method of separating the pus cells from debris and bandage fragments by washing with a salt solution.

In his early experiments Miescher recovered a chemical substance from the nuclei of the pus cells. To analyze this substance further, he decided to first purify the nuclei. As a first step, he broke open the pus cells by treatment with a protein-digesting substance called pepsin. To obtain pepsin, he had to prepare extracts of pig stomachs (a good source of pepsin, which functions in digestion). He then treated the pus cells with the extract of pig stomach for several hours. He noted that a gray sediment collected at the bottom of the flask. Under the microscope this sediment turned out to be pure nuclei. Miescher was therefore the first to isolate and to purify a cellular organelle.

By chemical extraction of purified nuclei, Miescher obtained a substance he called nuclein. Chemical analysis of nuclein revealed that it contained hydrogen, carbon, nitrogen, oxygen, and phosphorus. Miescher showed that nuclein was found in other cell types, including kidney, liver, sperm, and yeast. He regarded it as an important component of most cells. Many years later his nuclein was shown to contain DNA.

At about the same time that Miescher was carrying out his experiments, Mendel outlined the rules for the inheritance of physical traits and developed the notion of what we now call genes. In the 1880s August Weismann and others emphasized the importance of the nucleus in heredity. At the turn of the century Walter Sutton and Theodore Boveri noted that the behavior of genes in inheritance paralleled that of nuclear components (the chromosomes) in meiosis. Later workers confirmed that, in fact, genes are part of chromosomes, and it was generally agreed that the genetic material was to be found in the nucleus. Through all this work, however, the most basic question remained unanswered: What is the nature of the genetic material?

The answer to this question will take us from the level of the gene as the physical unit of heredity to the level of nucleic acid molecules as the chemical components of cells most closely involved with the storage, expression, and transmission of genetic information. The path to this answer was convoluted and long and runs from the experiments of Miescher through 80 years to the experiments of Avery and his colleagues in the 1940s. Over the years, although several lines of evidence provided clues pointing to nucleic acids, especially deoxyribonucleic acid (DNA), as the chemical answer to this problem, most theories were based on proteins as the molecular carriers of genetic information. The general requirements for the genetic material, however, were clear and unambiguous. Any chemical structure proposed as the carrier of genetic information must explain the observed properties of genes: replication, information storage, transfer or expression of the stored information, and mutation. It was not until 1944 that this issue was resolved.

In this chapter we will examine the events that led to the confirmation of DNA as the molecule that carries genetic information, and we will consider the work that led to an understanding of the organization and structure of nucleic acids, in general, and DNA, in particular. We will also explore what is known about how DNA is incorporated into the structure of chromosomes.

CHROMOSOMES AND DNA

From ancient times it has been accepted that genetic information must be physically transmitted from generation to generation. What has evolved in the last 50 years is an understanding of the chemical nature of genetic information and how this information is physically transmitted from parent to offspring. The discovery that genes are on chromosomes focused attention on the nucleus and its chemical components as the carriers of genetic information. Actually, investigations into the biochemistry of the nucleus began about the same time Mendel was formulating the laws of inheritance. Frederick Miescher first separated nuclei from cytoplasm in 1868 and then began to study the chemical composition of the nucleus. He isolated and analyzed a substance he called **nuclein,** which differed from purified proteins in that it contained large amounts of phosphorus and no detectable amounts of sulfur. Later, nucleic acids were recovered from nuclein, and based on the type of sugar contained in the molecule, two variants were identified: **deoxyribonucleic acid (DNA)** and **ribonucleic acid (RNA).**

Nuclein
A mixture of nucleic acids and proteins isolated from nuclei by Miescher.

Deoxyribonucleic acid (DNA)
A molecule consisting of antiparallel strands of polynucleotides that is the primary carrier of genetic information.

Ribonucleic acid (RNA)
A nucleic acid molecule that contains the pyrimidine uracil and the sugar ribose. The several forms of RNA function in gene expression.

Nucleotides
The basic building blocks of DNA and RNA. Each nucleotide consists of a base, a phosphate, and a sugar.

Evidence based on the subcellular distribution of certain dyes and stains revealed that DNA was found mainly in the nucleus, while proteins were found in both the cytoplasm and the nucleus. In spite of this indirect evidence pointing to the involvement of nucleic acids in heredity, most scientists considered proteins to be the logical choice as the genetic material. This idea persisted until about 1950.

Several reasons contributed to this mistaken belief. First, proteins are found in the nucleus and the chromosomes. Second, it was well known that proteins were complex molecules composed of 20 different building blocks called amino acids, while nucleic acids were constituted from only 4 different building blocks called **nucleotides.** In addition, cells contain hundreds or even thousands of different proteins but only two different types of nucleic acids, RNA and DNA. As a result, it was thought that nucleic acids were not complex enough to carry the amount of genetic information necessary for living systems.

The mistaken idea that proteins carried genetic information arose partly because chemical techniques were used to study the composition of the nucleus and cytoplasm. Based on the chemical complexity of proteins and the apparent simplicity of nucleic acids, it seemed logical that a more complex class of molecules would carry genetic information. Unfortunately, these chemical methods could not be used to directly identify which type of cellular molecule can carry genetic information, and no biological technique was available to determine which molecules from the cell are capable of carrying genetic information. This situation changed near the middle of the 20th century, when a team of medical scientists reinvestigated an old observation about the transfer of genetic traits by cellular extracts.

In 1944 Oswald Avery, Colin MacLeod, and MacLyn McCarty published a landmark paper on the chemical nature and biological properties of a substance known as the "transforming factor." This factor, previously described by Fredrick Griffith in the 1920s, was implicated in the transfer of a genetic trait from one strain of bacteria to another. Griffith's work indicated that an unknown substance present in killed bacterial cells was able to cause a specific genetic change in a related bacterial strain.

In a series of experiments that stretched over 10 years, Avery and his colleagues investigated a bacterial strain that causes pneumonia. This *Diplococcus* strain forms smooth, glossy colonies when grown in the laboratory, whereas other strains that do not cause pneumonia grow as colonies with rough-looking edges. The ability to cause pneumonia and grow as smooth colonies is associated with the presence of a thick capsule surrounding the bacterial cell. When an extract of heat-killed smooth bacteria is mixed with living rough cells, a small fraction of the rough cells acquire the ability to form a capsule, grow as smooth colonies, and cause pneumonia. The bacteria that have acquired the ability to form a capsule transmit this trait to all their offspring, indicating that the trait is heritable.

Avery and his colleagues identified DNA as the active component present in the extract from the heat-killed cells. To confirm that DNA was the transforming substance, they treated the preparation with enzymes that destroy protein (protease) and RNA (ribonuclease) before transformation. This treatment removed any residual protein or RNA from the preparation but did not affect the transforming activity. As a final test the preparation was treated with deoxyribonuclease, an enzyme that digests DNA, whereupon transforming activity was abolished.

This work produced two important conclusions. In the bacterium *Diplococcus:*

1. *DNA carries genetic information.* Only DNA is able to transfer heritable information from one strain of *Diplococcus* to another strain.
2. *DNA controls the synthesis of specific products.* Transfer of DNA also results in the transfer of the ability to synthesize a specific gene product (in the form of a capsule).

Further work in 1952 by Alfred Hershey and Martha Chase used viruses to provide important supporting evidence for the role of DNA as the genetic material in the virus T2.

Studies on the genetic properties of DNA in viruses and bacteria did not immediately influence geneticists studying higher organisms. The reasons for this are complex and are partly scientific and partly sociologic. At the experimental level, transformation cannot be performed on eukaryotic organisms, so it was not possible to directly replicate the results of Avery or Hershey with organisms such as *Drosphilia* or mice.

Because of this, acceptance of the idea that DNA is the genetic material in higher organisms was slow and was based mainly on indirect or circumstantial evidence. Since it was generally accepted that the chromosomes contain genetic information, indirect evidence was employed to strengthen the link between DNA and chromosomes. Refined methods of cytochemical staining indicated that not only is DNA largely confined to the nucleus but that it is intimately associated with the chromosomes. Furthermore, DNA is present along the entire length of the chromosomes in a way that corresponds to the distribution of genetic loci. Last, within the cells of a given eukaryotic organism, the concentration of DNA is correlated with the

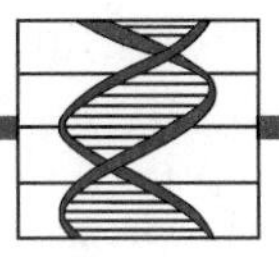

Concepts & Controversies

DNA as a Commercial Product

The ad for the perfume reads: "Where does love originate? Is it in the mind? In the heart? Or in our genes?" A new perfume called DNA has been recently introduced, and is marketed in a helix-shaped bottle. There is no actual DNA in the fragrance, but the molecule is invoked as the genetic material to sell the idea that love emanates from the genes, and that this perfume captures this essence of this idea. Seem strange? Well, how about jewelry that actually contains DNA from your favorite celebrities? The idea behind this line of products is that DNA present, say, in a single hair can be amplified by a process called the polymerase chain reaction. The resulting solution, containing millions of copies of the DNA is then added to small channels drilled into acrylic earrings, pendants, or bracelets. The liquid may be colored to contrast with the acrylic and be more visible. Just as people wear T-shirts carrying representations of Elvis Presley or Albert Einstein, they can now wear jewelry containing DNA from their favorite entertainer, poet, composer, scientist or athlete. For dead heros, the DNA could come from a lock of hair, in fact, a single hair would do. For popular idols that are still with us, a hair or a small blood sample would provide enough DNA for posterity to cherish.

How about music composed from the base sequence of DNA? In this case, the four bases of DNA (adenine, thymine, guanine, cytosine) are translated into musical notes. Long sequences of DNA, retrieved from computer data bases, are converted into notes, transferred to sheet music, and then played by instruments or synthesizers as the music of the genes. Those who admire this form of music say that the DNA sequences present near chromosome centromeres sounds much like the music of Bach or other Baroque composers, but that music from other parts of the genome have a more modern sound.

This fascination with deoxyribonucleic acid may be a little difficult to fathom, but clearly, DNA has captured the public's imagination, and is being used to sell a diverse and ever-increasing array of products. Part of this may derive from the fact that DNA has name recognition. Over the past 40 years, DNA has moved from the scientific literature to articles in popular journals and magazines and even into the comic strips. The term DNA is widely known as and equated with genes, and the essence of life. In a few years, looking back at how DNA captivated attention may seem like another fad, but for now, if you want to sell something, relate it to DNA.

number of chromosomes carried by the cell. In general, somatic diploid (2*n*) cells contain twice the number of chromosomes as haploid (*n*) gametes. Measurements of DNA concentration indicate that somatic cells have twice as much DNA as gametes. With the possible exception of a class of proteins known as histones, no such correlation between chromosome number and molecular concentration is apparent for other cellular components such as proteins, lipids, or carbohydrates. These and other forms of indirect evidence support the idea that DNA is the genetic material of eukaryotic organisms.

Today no one seriously doubts the validity of the circumstantial evidence that DNA is the repository of genetic information in higher organisms. In fact, as awareness of DNA and its role in genetics has grown, DNA is being used to sell products (see Concepts and Controversies, above). Direct and dramatic evidence for the role of DNA as the genetic material has come from the development of **recombinant DNA technology.**

Recombinant DNA technology
Techniques for joining DNA from two or more different organisms to produce hybrid, or recombined, DNA molecules.

In this methodology, DNA segments from another organism such as a human can be spliced into bacterial DNA, and under the proper conditions, this hybrid DNA molecule can direct the synthesis of a human gene product. The synthesis of human proteins in bacteria requires the presence of specific human DNA sequences, providing direct evidence for the role of DNA as the genetic material in higher organisms.

STRUCTURE OF DNA AND RNA

Recognition that DNA plays an important role in genetic processes coincided with efforts to understand the chemical structure of nucleic acids. In the years from the mid-1940s through 1953, several laboratories made significant strides in unraveling the structure of DNA, culminating in the Watson-Crick model for the DNA double helix in 1953. The scientific, intellectual, and personal intrigue that characterized the race to discover the structure of DNA has been documented in a number of books, beginning with *The Double Helix* by James Watson. These personal accounts and histories provide a rare glimpse into the ambitions, jealousies, and rivalries that entangled scientists involved in the dash to a Nobel Prize.

Reviewing Some Basic Chemistry

The structure of DNA in the Watson-Crick model and, in a later chapter, the structure of proteins are described and drawn using chemical terms and symbols. For this reason, a brief review of the terminology and definition of some terms is in order.

All matter is composed of atoms; the different types of atoms are known as elements (of which there are 103). In nature, atoms are rarely found as separate units. More often, they are combined into molecules, which we can define as units of two or more atoms chemically bonded together. Molecules can be represented as formulas that indicate how many of each type of atom are present. The type of atom is indicated by a symbol for the element it represents: H for hydrogen, N for nitrogen, C for carbon, O for oxygen, and so forth. For example, a water molecule, composed of two hydrogen atoms and one oxygen atom, has its chemical formula represented as H_2O:

$$H_2O$$

one oxygen atom
two hydrogen atoms

Many molecules in cells are large and have more complex formulas. A molecule of glucose contains 20 atoms and is written as:

$$C_6H_{12}O_2$$

In molecules, the atomic components are held together by a stable interaction known as a *covalent bond*. In its simplest form, a covalent bond consists of a pair of electrons shared between two atoms. More complex covalent bonds can be formed by sharing two or more electrons. Figure 7.1A shows how such bonds are represented in structural formulas of molecules.

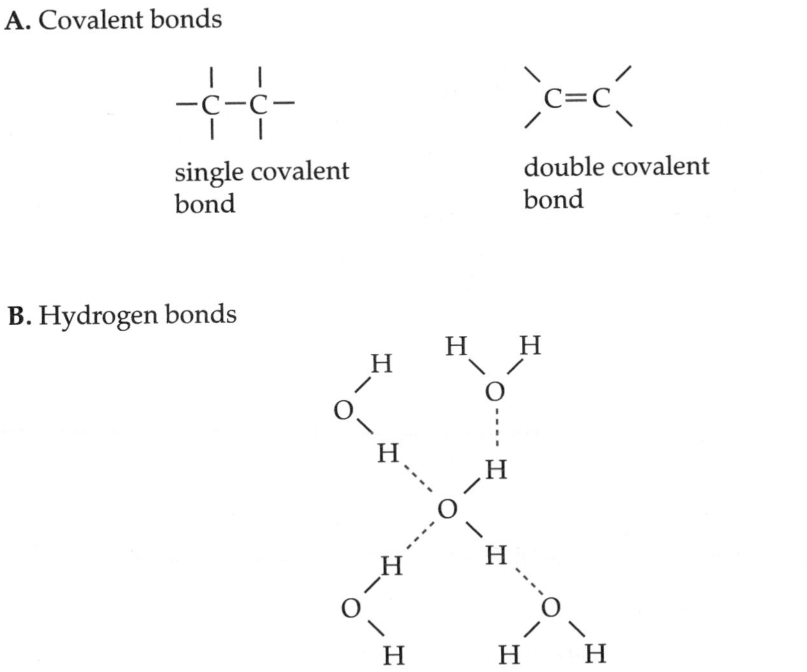

A. Covalent bonds

—C—C—

single covalent bond

C=C

double covalent bond

B. Hydrogen bonds

FIGURE 7.1 Representation of chemical bonds. (A) Covalent bonds are represented as solid lines connecting atoms. Depending on the degree of electron sharing, there can be one covalent bond *(left)* or more than one *(right)*. Once formed, covalent bonds are stable, and are broken only in chemical reactions. (B) Hydrogen bonds are usually represented as dotted lines connecting two or more atoms. As shown, water molecules form hydrogen bonds with adjacent water molecules. These are weak bonds that are easily broken by heat and molecular tumbling and can be reformed with other molecules.

A second type of atomic interaction between molecules involves a weak attraction known as a **hydrogen bond.** In living systems, hydrogen bonds make a substantial contribution to the three-dimensional shape and, therefore, to the functional capacity of biological molecules. Hydrogen bonds are weak interactions between two atoms (one of which is hydrogen), carrying partial but opposite electrical charges. They are usually represented in structural formulas as dotted or dashed lines connecting two atoms (Figure 7.1B).

Hydrogen bond
A weak chemical bonding force between hydrogen and another atom.

When hydrogen forms a covalent bond with certain other atoms such as oxygen or nitrogen, it becomes partly positive. Because of this slight positive charge, the hydrogen is attracted to other atoms bearing a slight negative charge. This weak attraction is called a hydrogen bond. Hydrogen bonds usually form between atoms of different molecules or between atoms at different locations on a large molecule (such as DNA). Although individual hydrogen bonds are weak and easily broken, they function to hold molecules together by sheer force of numbers. As we will see in a following section, hydrogen bonds hold together the two strands in a DNA molecule and they are also responsible for the three-dimensional structure of proteins (Chapter 9).

Nucleotides: The Building Blocks of Nucleic Acids

There are two types of nucleic acids in biological organisms: DNA and RNA. As we have seen, both are made up of subunits known as nucleotides. A nucleotide consists of a **nitrogen-containing base** (either a **purine** or a

Nitrogen-containing base
A purine or pyrmidine that is a component of nucleotides.

Purine
A class of double-ringed organic bases found in nucleic acids.

FIGURE 7.2 The chemical structure of purines. The basic structure of purines is shown first. Below it are the structures for adenine and guanine, with side groups shown in color.

Purine Ring

Adenine

Guanine

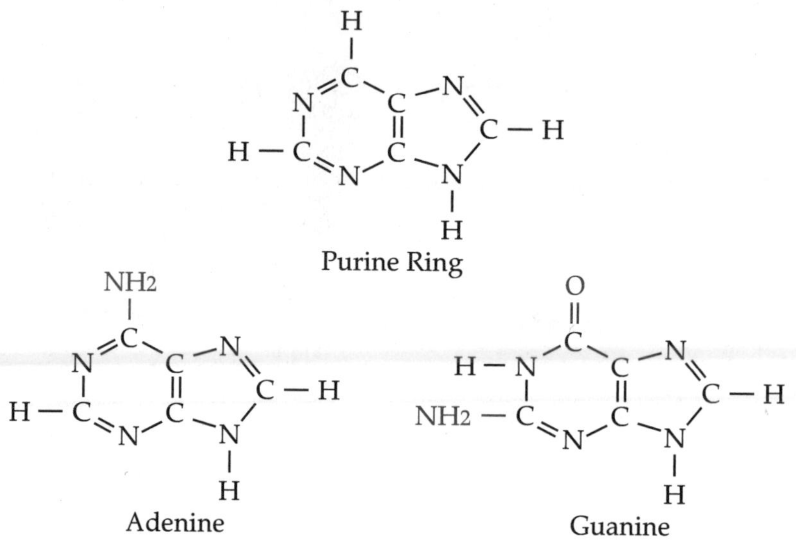

Pyrimidines
A class of single-ringed organic bases found in nucleic acids.

Pentose sugar
A five-carbon sugar molecule found in nucleic acids.

Adenine and Guanine
Purine nitrogenous bases found in nucleic acids.

Cytosine, Thymine, and Uracil
Pyrimidine nitrogenous bases found in nucleic acids.

Deoxyribose and Ribose
Pentose sugars found in nucleic acids. Deoxyribose is found in DNA, ribose in RNA.

pyrimidine), a **pentose sugar** (either ribose or deoxyribose), and a phosphate group. The phosphate groups are strongly acidic and are the source of the designation nucleic *acid.* In the bases, both purines and pyrimidines have the same six-atom ring, but purines have an additional three-atom ring. The purine bases **adenine** (A) and **guanine** (G) are found in both RNA and DNA (Figure 7.2). The pyrimidine bases include **thymine** (T), found in DNA, **uracil** (U), found in RNA, and **cytosine** (C), found in both RNA and DNA (Figure 7.3). Thus, RNA has four bases, (A, G, U, C) and DNA has four bases (A, G, T, C).

FIGURE 7.3 The chemical structure of pyrmidines. The basic structure of pyrimidines is shown first. Below it are the structures for cytosine, thymine, and uracil with side groups shown in color.

Pyrimidine Ring

Cytosine

Thymine

Uracil

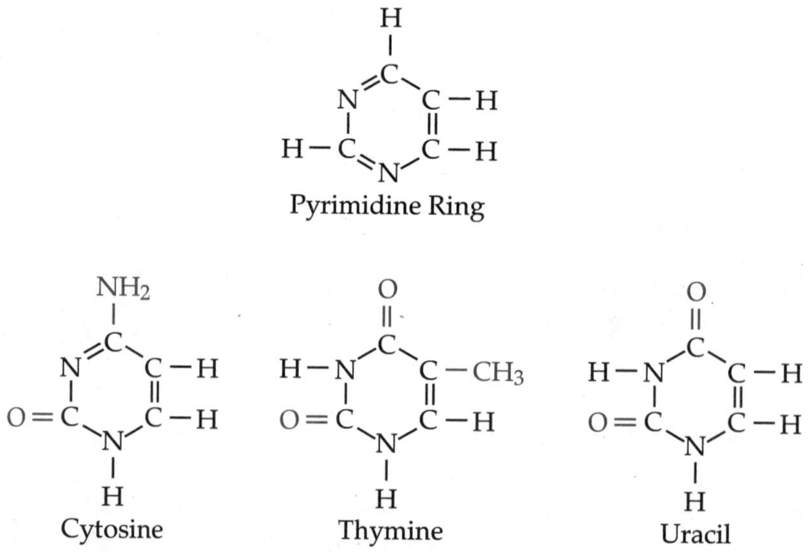

The sugars in nucleic acids have five carbon atoms per molecule. The sugar in RNA is known as **ribose,** and the sugar in DNA is **deoxyribose.** The difference is a single oxygen atom that is present in ribose and absent in deoxyribose (Figure 7.4).

Nucleotides, as mentioned, are molecules composed of a base linked by a covalent bond to a sugar, which in turn is covalently bonded to a phosphate group (Figure 7.5). Nucleotides are named by reference to the base and sugar they contain (Table 7.1).

Two or more nucleotides can be linked by the formation of a covalent bond between the phosphate group of one nucleotide and the sugar of another nucleotide. Chains of nucleotides containing ribose (ribonucleotides) or deoxyribose (deoxyribonucleotides) can be formed in this way (Figure 7.6). Nucleotide chains are directional molecules, with slightly different structures marking each end of the chain. Each chain has one end that carries a phosphate group and is known as the 5′ (pronounced "5 prime") end, and a sugar group at the opposite end, knownas the 3′ ("3 prime") end of the chain (Figure 7.6). These names are the result of a system used by biochemists to number the atoms in the sugar ring (see Figure 7.4). By convention, nucleotide chains are written beginning with the 5′ end, as 5′-CGATATGCGAT-3′, but are usually labeled to indicate polarity.

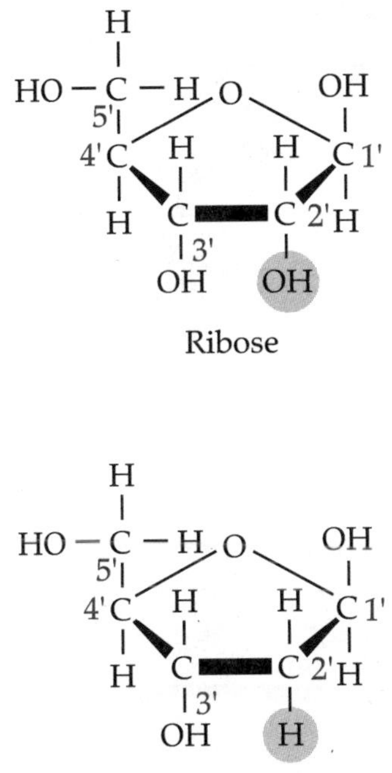

FIGURE 7.4 Chemical structure of ribose and deoxyribose, the pentose sugars in RNA and DNA.

Watson-Crick Model of DNA Structure

Following the work of Avery and his associates, studies on the structure of nucleic acids, especially DNA, gained momentum. In the early 1950s James Watson and Francis Crick began the development of a series of models for the structure of DNA that would incorporate the available information

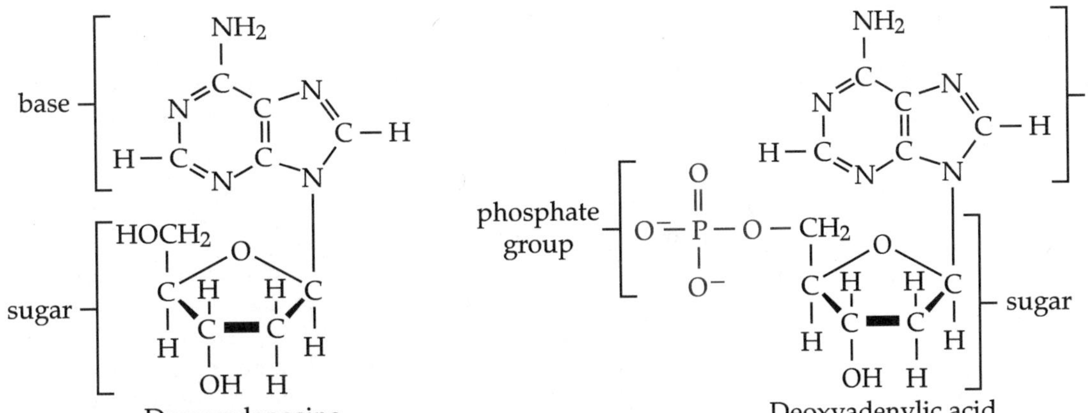

FIGURE 7.5 The structure of the nucleosides and nucleotides of DNA and RNA. Nucleosides are composed of a base and a sugar. Nucleotides are composed of a base, a sugar, and a phosphate group.

TABLE 7.1 Nucleotides of RNA and DNA

RNA
Ribonucleotides
Adenylic acid
Cytidylic acid
Guanylic acid
Uridylic acid

DNA
Deoxyribonucleotides
Deoxyadenylic acid
Deoxycytidylic acid
Deoxyguanylic acid
Deoxythymidylic acid

gathered by physical and chemical methods. Information about the physical structure of DNA was based on the technique of x-ray crystallography. In this process, molecules are crystallized and placed in a beam of x-rays. As the x-rays pass through the crystal, some are deflected as they hit the atoms in the crystal. The pattern of x-rays emerging from the crystal can be recorded on photographic film and analyzed to produce information about the organization and shape of the crystallized molecule.

Between 1950 and 1953, two groups, one in the United States headed by Linus Pauling and the other in England directed by Maurice Wilkins, obtained x-ray crystallographic pictures from highly purified DNA samples. These pictures indicated that the DNA molecule was in the shape of a helix (Figure 7.7). The x-ray films also permitted measurements of the distances between the stacked bases.

Beginning in the late 1940s, Erwin Chargaff and his colleagues extracted DNA from a variety of organisms and using a technique known as chromatography separated and quantitated the amounts of the four bases. The results indicated that DNA from all organisms tested had several common properties, which became known as Chargaff's rule:

1. The total amount of purines (A + G) equals the total amount of pyrimidines (C + T).
2. There is a 1:1 relationship between the amount of adenine and the amount of thymine and a 1:1 relationship between the amount of guanine and the amount of cytosine.

Using the information provided by the physical and chemical studies, Watson and Crick began building wire and metal models of potential DNA structures. After a series of setbacks, they succeeded in producing a model that incorporated all the information described above. Their model has the following features:

1. DNA is composed of two polynucleotide chains running in opposite directions (antiparallel chains) (Figure 7.8).
2. The two polynucleotide chains are coiled around a central axis to form a helix (Figure 7.9). This part of the model fits the x-ray crystallographic results of Rosalind Franklin from the Wilkins laboratory.
3. In each chain, the bases are within the interior of the helix, and the alternating sugar (deoxyribose) and phosphate groups form a "backbone" for the chain on the exterior.
4. The two polynucleotide chains are held together by hydrogen bonds that form between the A and T bases in opposite chains and between the G and C bases in opposite chains. Each set of hydrogen-bonded bases is called a base pair. The pairing of A with T and C with G fits with the results obtained by Chargaff.
5. The base pairing feature of the model results in the two polynucleotide chains of DNA being complementary in base composi-

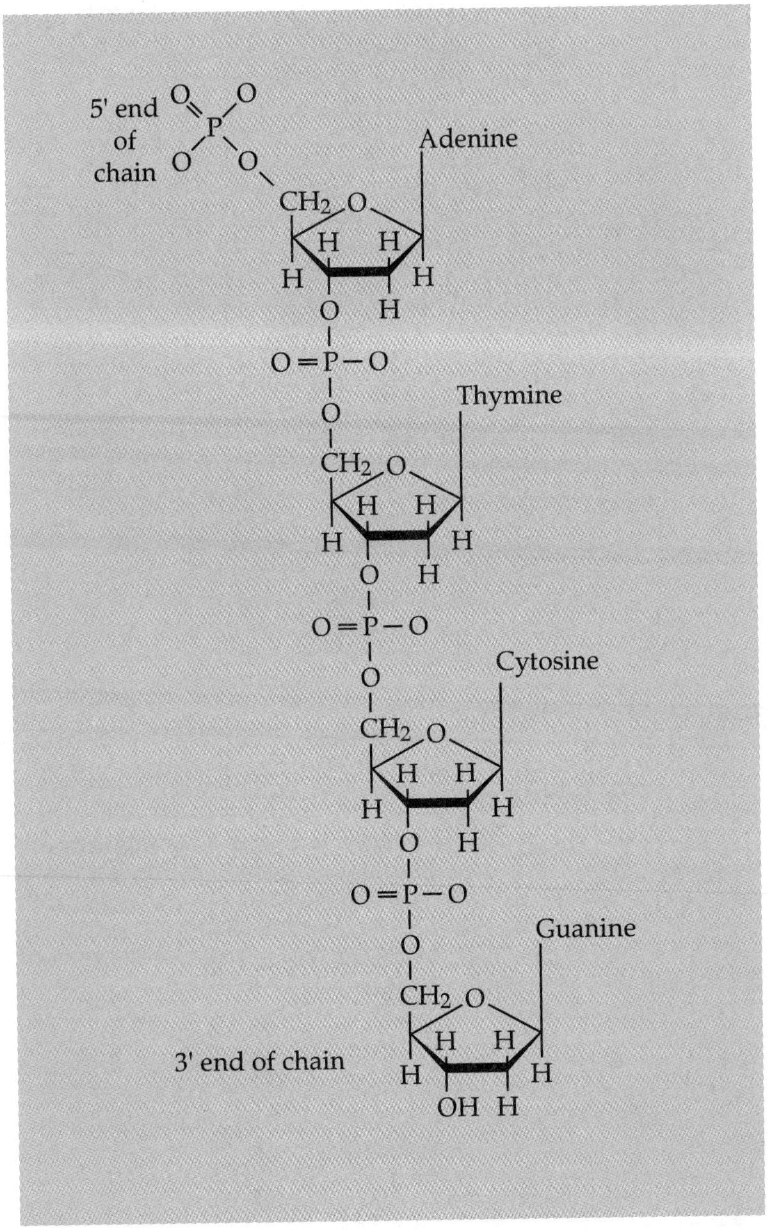

FIGURE 7.6 Linkage of nucleotides by the formation of 5′ to 3′ bonds to form polynucleotide chains.

tion. If one strand has the sequence 5′-TACCAGCG-3′, the opposite strand must be 3′-ATGGTCGC-5′, and the double-stranded structure would be written as:

5′-TACCAGCG-3′
3′-ATGGTCGC-5′

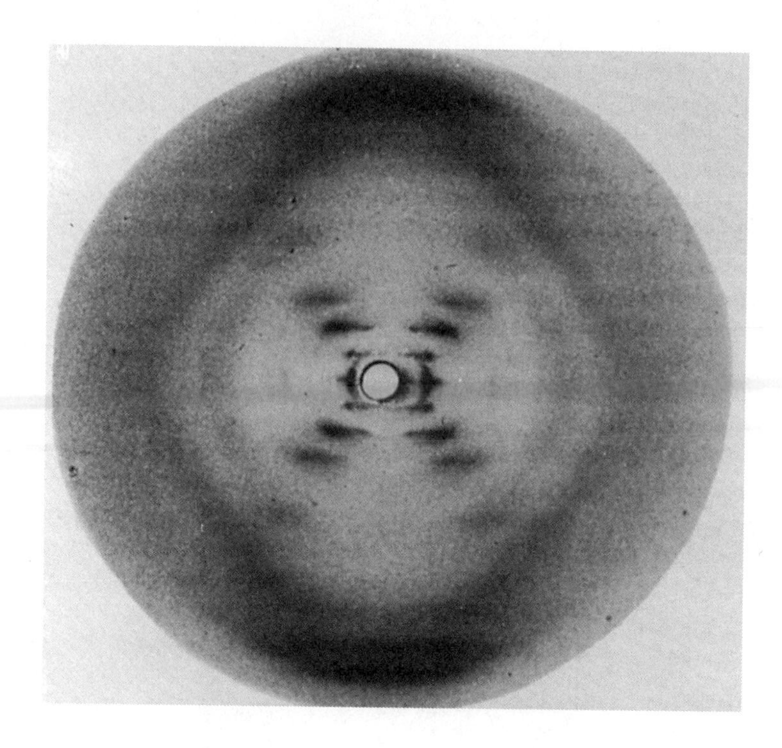

FIGURE 7.7 An x-ray diffraction photograph of crystallized DNA. The central pattern is typical of helical structures, and the dark areas at the top and bottom indicate periodicity of subunits in the molecule.

Three important properties of this model should be considered:

1. In this model, genetic information is stored in the sequence of bases in the DNA. The linear sequence of bases has a high coding capacity; a molecule n bases long has 4^n combinations. That means that a sequence of 10 nucleotides has 4^{10}, or 1,048,576,

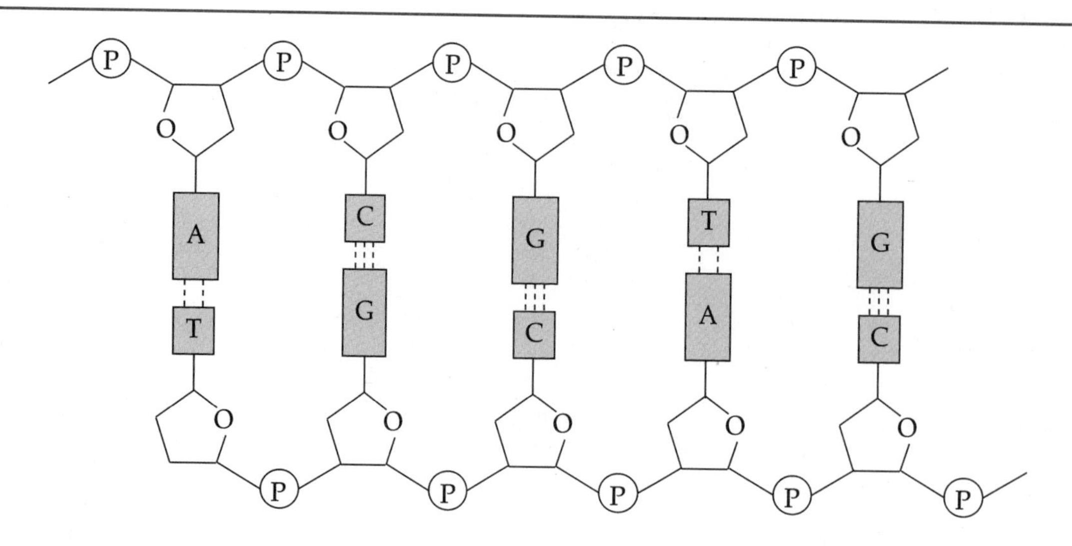

FIGURE 7.8 Two polynucleotide chains in opposite orientation with sugar-phosphate backbones along the outside and the bases on the interior. Bases are hydrogen bonded to those in the opposite strand.

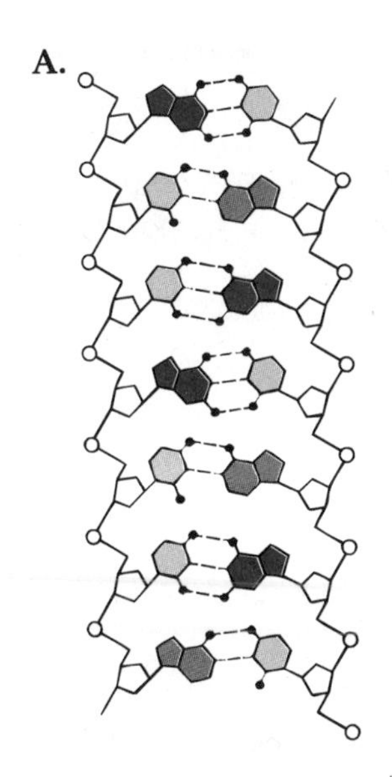

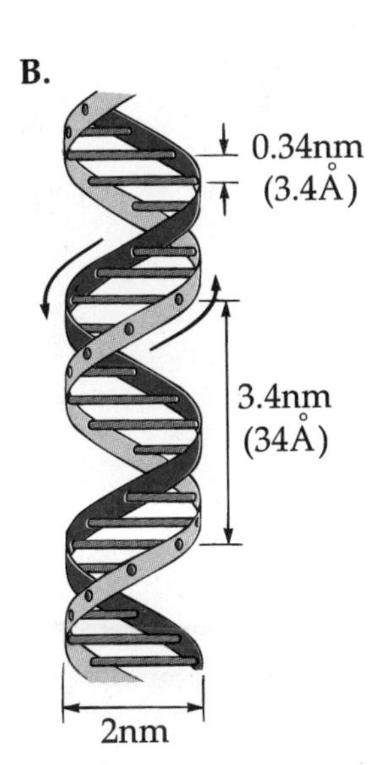

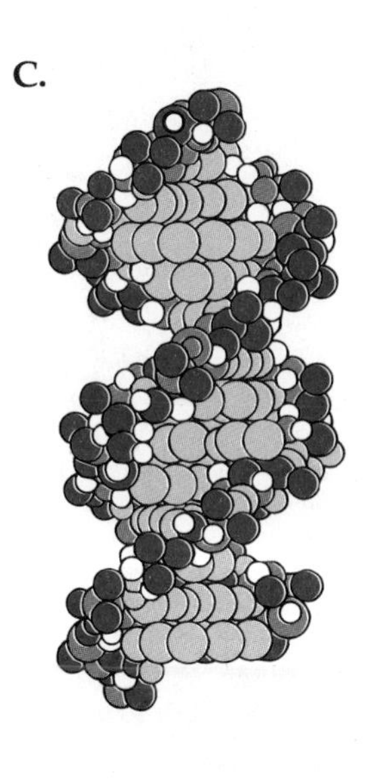

FIGURE 7.9 The double-helix model of DNA proposed by Watson and Crick. (A) The two polynucleotide chains running in opposite directions, with the bases on the interior and the sugar-phosphate groups on the outside. Hydrogen bonding between the bases holds the polynucleotide chains together. (B) The two polynucleotide strands wound around a central axis, forming a helix. In this form of the model, the hydrogen-bonded bases appear as internal rods, and the sugar-phosphate groups as ribbons on the outside. The molecule is 2 nm in diameter, the distance between bases is 0.34 nm, and the distance required for one complete turn of the helix is 3.4 nm. (C) A space-filling model of DNA. The bases on the interior are represented in lighter color, the sugar molecules in dark color, and the phosphate groups in white.

possible combinations. In molecular terms, the complete set of genetic information carried by an organism (its genome) can be expressed as base pairs of DNA (Table 7.2). Genome sizes vary from a few thousand nucleotides (in viruses), encoding only a few genes, to billions of nucleotides, encoding perhaps 50,000 to 100,000 genes (in mammals). In humans, the haploid genome consists of about 3×10^9, or 3 billion, base pairs of DNA, distributed over 23 chromosomes. Chromosome 21, the smallest chromosome in humans, carries about 5×10^7, or 50 million, base pairs of DNA.

2. The model offers a molecular explanation for mutation. Because genetic information can be stored in the linear sequence of bases in DNA, any change in the order or number of bases in a gene can result in a mutation that produces an altered phenotype. This topic will be explored in more detail in Chapters 11 and 12.
3. As Watson and Crick observed, the complementary strands in DNA can be used to explain the molecular basis of the replication of genetic information that takes place prior to each cell division. In such a model, each strand can be used as a template to reconstruct the base sequence in the opposite strand. This topic will be discussed later in this chapter.

TABLE 7-2 Genome Size in Various Organisms

Organism	Species	Genome Size in Nucleotides
Bacterium	*E. coli*	4.2×10^6
Yeast	*S. cerevisiae*	1.3×10^7
Fruitfly	*D. melanogaster*	1.4×10^8
Tobacco Plant	*N. tabacum*	4.8×10^9
Mouse	*M. musculus*	2.7×10^9
Human	*H. sapiens*	3.2×10^9

The Watson-Crick model was described in a brief paper in *Nature* in 1953. Although their model was based on the results of other workers, Watson and Crick correctly incorporated the physical and chemical data into a model that also could be used to explain the properties expected of the genetic material.

At the time there was no evidence to unequivocally support their model, but in subsequent years the model has been confirmed by an enormous amount of experimental work by laboratories worldwide. The 1962 Nobel Prize for Medicine or Physiology was awarded to Watson, Crick, and Wilkins for their work on the structure of DNA. Although much of the x-ray data for the Watson-Crick model was provided by Rosalind Franklin, she did not receive a share of the prize. Although there has been some controversy over this, she could not have shared in the prize, since it is awarded only to living individuals, and Franklin died of cancer in 1958. Her role in the discovery of the structure of DNA is presented in her biography, *Rosalind Franklin and the Discovery of DNA.*

TABLE 7.3 Differences Between DNA and RNA

	DNA	RNA
Sugar	deoxyribose	ribose
Bases	adenine	adenine
	cytosine	cytosine
	guanine	guanine
	thymidine	uracil

Structure of RNA

RNA (ribonucleic acid) is found in both the nucleus and the cytoplasm. While DNA functions as a repository of genetic information, RNA functions in the transfer of genetic information from the nucleus to the cytoplasm. (In a few viruses, RNA also functions to store genetic information.) The nucleotides in RNA differ from those in DNA in two respects. First, the sugar in RNA nucleotides is ribose (deoxyribose in DNA), and the base uracil is used in place of the base thymine (Table 7.3).

Although two complementary strands of RNA can form a double helix similar to the one found in DNA, the RNA present in most cells is single stranded, and a complementary strand is not made. However, RNA molecules can fold back upon themselves and form double-stranded regions. The functions of RNA will be considered in more detail in Chapter 9.

STRUCTURE OF CHROMOSOMES

Although an understanding of DNA structure represents an important development in genetics, it provides no immediate solution to the problem of how DNA is organized in combination with other components such as proteins to form the visible structures known as chromosomes. This problem is significant, since the spatial arrangement of the genetic material plays an important role in controlling the expression of genetic information. In addition, the distribution of DNA into the 46 human chromosomes requires the packing of a little over 2 m (2 million μm) of DNA into a nucleus that measures about 5 μm in diameter. Within this cramped environment, the chromosomes unwind and become dispersed during interphase. In this condition they undergo the process of replication, gene expression, homologous pairing during meiosis, and contraction and coiling to become visible again during prophase. An understanding of chromosome organization is necessary to understand these processes. In contrast, details of the structure and even the nucleotide content of the mitochondrial chromosome are well understood, and they will be considered first.

The Mitochondrial Chromosome

As discussed in Chapter 2, mitochondria are cellular organelles concerned with energy conversion. Mitochondria contain DNA that encodes genetic information, and this mitochondrial chromosome is transmitted predominantly if not exclusively in a maternal fashion. In recent years, a number of genetic disorders associated with mitochondria have been identified (see Chapter 4), and in this pattern of inheritance, all offspring of affected mothers are also affected.

The mitochondrial chromosome has been studied in detail, and the complete DNA sequence is known (Figure 7.10). The structure of the chromosome is simple; it consists of a circular DNA molecule containing 16,569 base pairs and encoding just over 40 genes. The alterations in the mitochondrial genome that lead to genetic disorders will be discussed in Chapter 11. Most mitochondria contain 5 to 10 copies of the mitochondrial chromosome, and because each cell has hundreds of mitochondria, there are thousands of mitochondrial genomes present in each human cell. The DNA of the mitochondrial chromosome is not complexed with proteins and physically resembles the chromosomes of bacteria and other prokaryotes. This similarity reflects the evolutionary history of mitochondria and their transition from free-living prokaryotes to intracellular symbionts.

The Nuclear Chromosomes

While the details of chromosome organization are still elusive, a combination of biochemical and microscopic techniques developed in recent years has provided a great deal of information about the organization and structure of human chromosomes. In humans and other eukaryotic organisms, each chromosome consists of a single DNA molecule associated with protein to

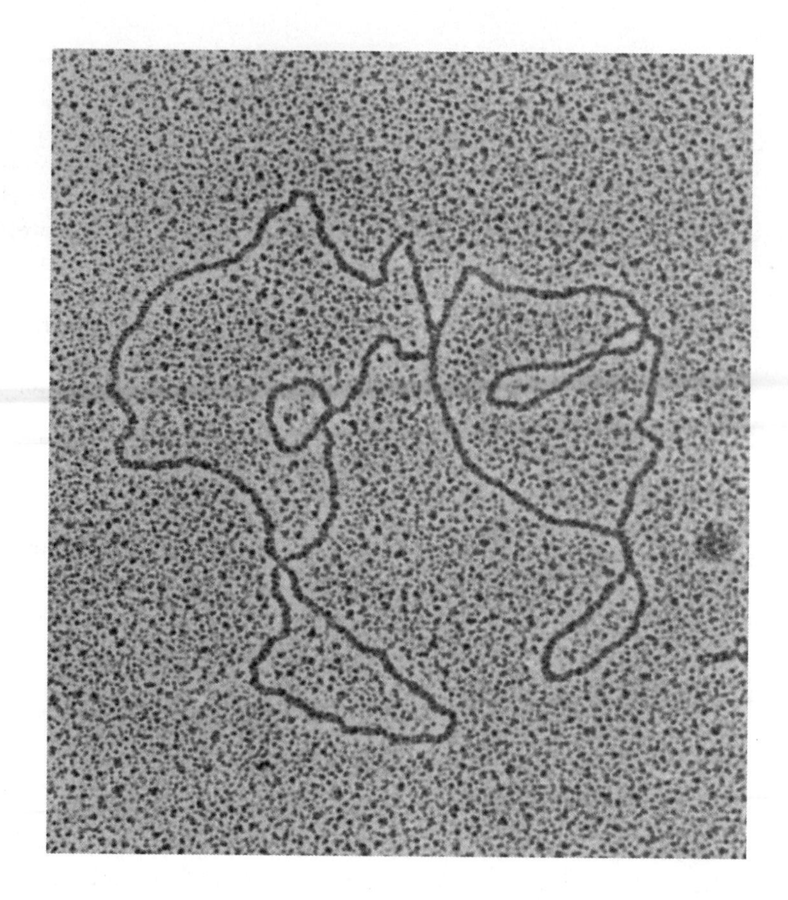

FIGURE 7.10 Mitochondrial genome. The chromosome in mitochondria consists of a circular DNA molecule of 16,596 base pairs. The map positions of several genetic disorders associated with mitochondria are indicated: myoclonic epilepsy and ragged red fiber (MERRF) disease; mitochondrial encephalomyopathy, lactic acidosis, and stroke-like symptoms (MELAS); maternally inherited myopathy and cardiomyopathy (MMC); and neurologically associated retinitis pigmentosa (NARP).

Chromatin
The complex of DNA and proteins that makes up a chromosome.

Histones
Small DNA-binding proteins that function in the coiling of DNA to produce the structure of chromosomes.

Nonhistone proteins
The array of proteins other than histones that are complexed with DNA in chromosomes.

Nucleosomes
A bead-like structure composed of histones wrapped by DNA.

form **chromatin.** The associated proteins are of two types, **histone** and **nonhistone proteins.** The histones play a major role in chromosome structure, while the nonhistones are thought to be involved in gene regulation. Five different types of histones (H1, H2A, H2B, H3, and H4) are complexed with DNA to form the basic building block of chromatin, small spherical bodies known as **nucleosomes** that are connected by thin threads of DNA (Figure 7.11). Each nucleosome consists of about 140 nucleotide pairs of DNA wound around a core composed of eight histone molecules (two each of H2A, H2B, H3, and H4). The DNA threads linking nucleosomes are of variable length, depending on the cell type from which the material is prepared; these threads are associated with histone H1.

Within each nucleosome the winding of DNA around the histones compresses the length of the DNA molecule by a factor of sixfold or sevenfold. However, in the mitotic chromosome the level of compaction is estimated to be 5000- to 10,000-fold. This suggests that there is one or possibly two more levels of organization between that of the nucleosome and the chromosome, which involves folding and/or coiling the DNA molecule. Electron microscopic observations on mitotic chromosomes have partly clarified the subchromosomal levels of organization. Figure 7.12 is a scanning electron micrograph of a mitotic chromosome. At high magnification

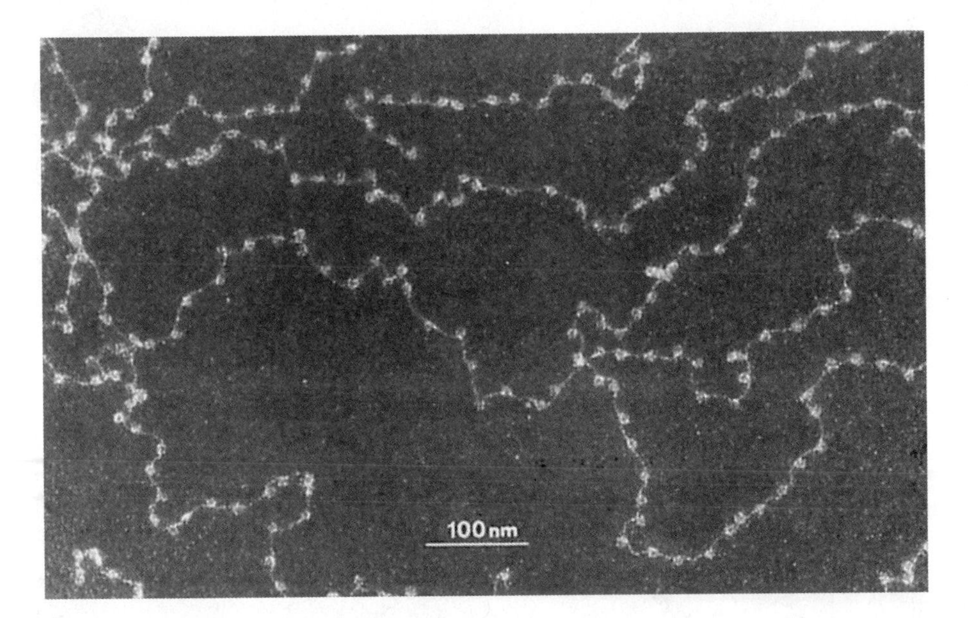

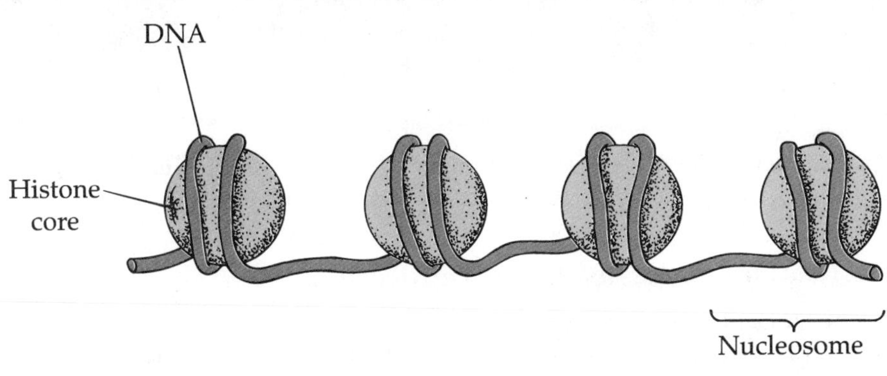

FIGURE 7.11 Nucleosomes. Upper: The association of DNA with histones to form nucleosomes in chromatin derived from chicken blood cells. Lower: A schematic model of the association of DNA with histones to form nucleosomes.

individual fibers with a diameter of 30 nm are observed. These fibers appear to be the result of supercoiling of the chromatin. A model of chromosome organization is shown in Figure 7.13.

DNA Replication

All organisms grow and reproduce through processes that involve cell division. These processes are preceded by replication of DNA. The precise and complete replication of DNA is essential to maintain genetic continuity from cell to cell and generation to generation. Watson and Crick proposed that their model of DNA, with its complementary strands, provided a possible mechanism for replication. If the helix of DNA were unwound, each strand could serve as a **template,** or pattern, for the synthesis of a new, complementary strand. Thus in the two strands of a daughter DNA molecule, one strand would be new, and one would be old (the one used as a template).

Template
The single-stranded DNA that serves to specify the nucleotide sequence of a newly synthesized polynucleotide strand.

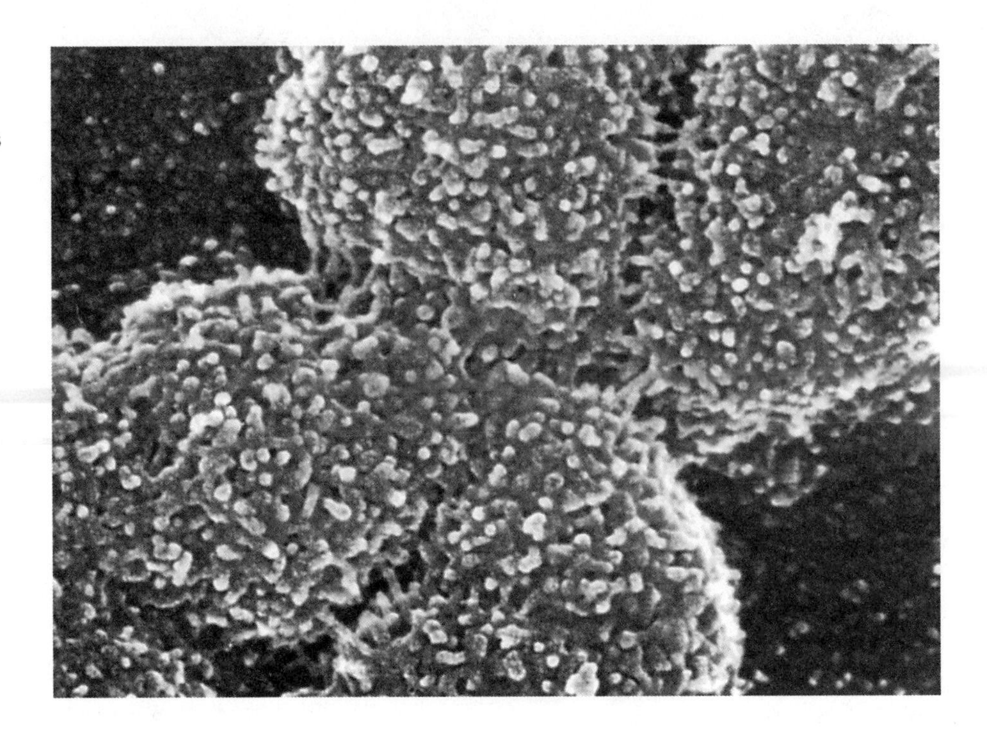

FIGURE 7.12 Scanning electron micrograph of the centromeric region of a human chromosome. Small fibrils can be seen in each sister chromatid.

Semiconservative replication
A model of DNA replication that results in each daughter molecule containing one old strand and one newly synthesized strand. DNA replicates in this fashion.

This model of DNA replication is known as **semiconservative replication,** since one old strand is conserved in each new molecule.

Molecular Aspects of DNA Replication

At the molecular level, DNA replication is a complex process that requires the interaction of several gene products with the DNA double helix. In humans replication begins at sites called origins of replication that are present all along the length of the chromosome. At these sites proteins unwind the double helix by breaking the hydrogen bonds between bases in adjacent strands. This process opens the molecule to the action of a protein known as **DNA polymerase.** After a priming step, DNA polymerase adds nucleotides to a new strand of DNA. The enzyme recognizes the nucleotide on the template strand and adds the correct complementary nucleotide on the newly synthesized strand. However, this enzyme can only add nucleotides to the sugar ends (3′ ends) of other nucleotides. This means that the replicating strand of DNA grows only in a $5' \rightarrow 3'$ direction. Because the polynucleotide chains in DNA run in opposite directions, a continuous strand of new DNA can only be formed from one of the template strands (Figure 7.14). On the other template strand, DNA is synthesized in short, discontinuous segments (Figure 7.14). These short strands are linked together by **DNA ligase** into longer continuous strands.

DNA polymerase
An enzyme that catalyzes the synthesis of DNA using a template DNA strand and nucleotides.

DNA ligase
An enzyme that forms covalent bonds between the 5′ end of one polynucleotide chain and the 3′ end of another chain.

When a continuous strand from one initiation site reaches the phosphate, or 5′, end of another newly synthesized strand, the two ends are linked together into a single strand by DNA ligase. This process of replication on one template strand by continuous elongation and on the other by the

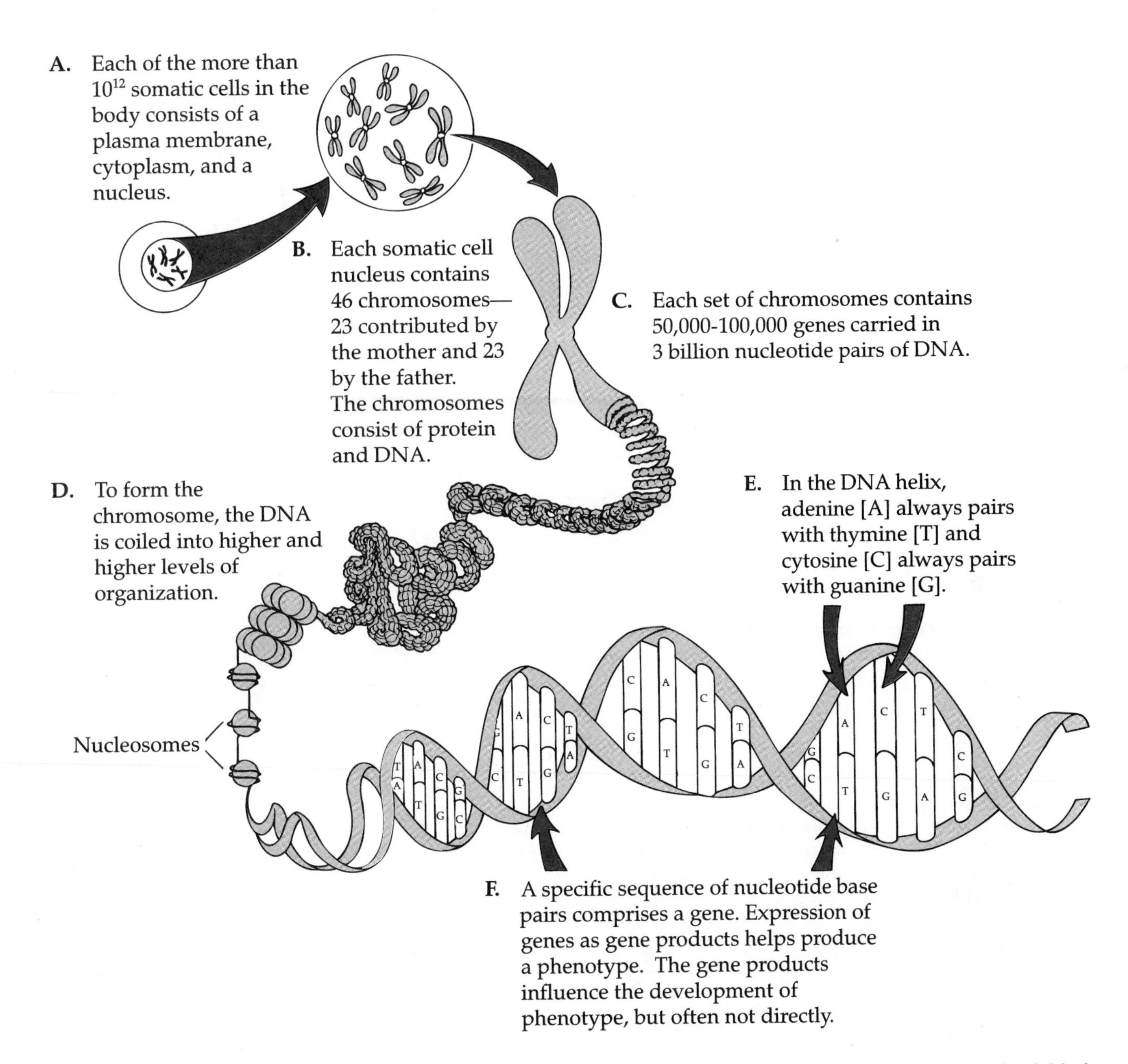

FIGURE 7.13 A model of a human chromosome illustrating the levels of organization from DNA up to the folded fibers that can be seen in Figure 7.12.

formation of discontinuous strands continues until the entire DNA molecule is replicated and the fragments are joined by DNA ligase. The result is two double-stranded DNA molecules with identical base pair sequences. Each molecule has been replicated in a semiconservative fashion and contains one newly synthesized strand and one preexisting strand.

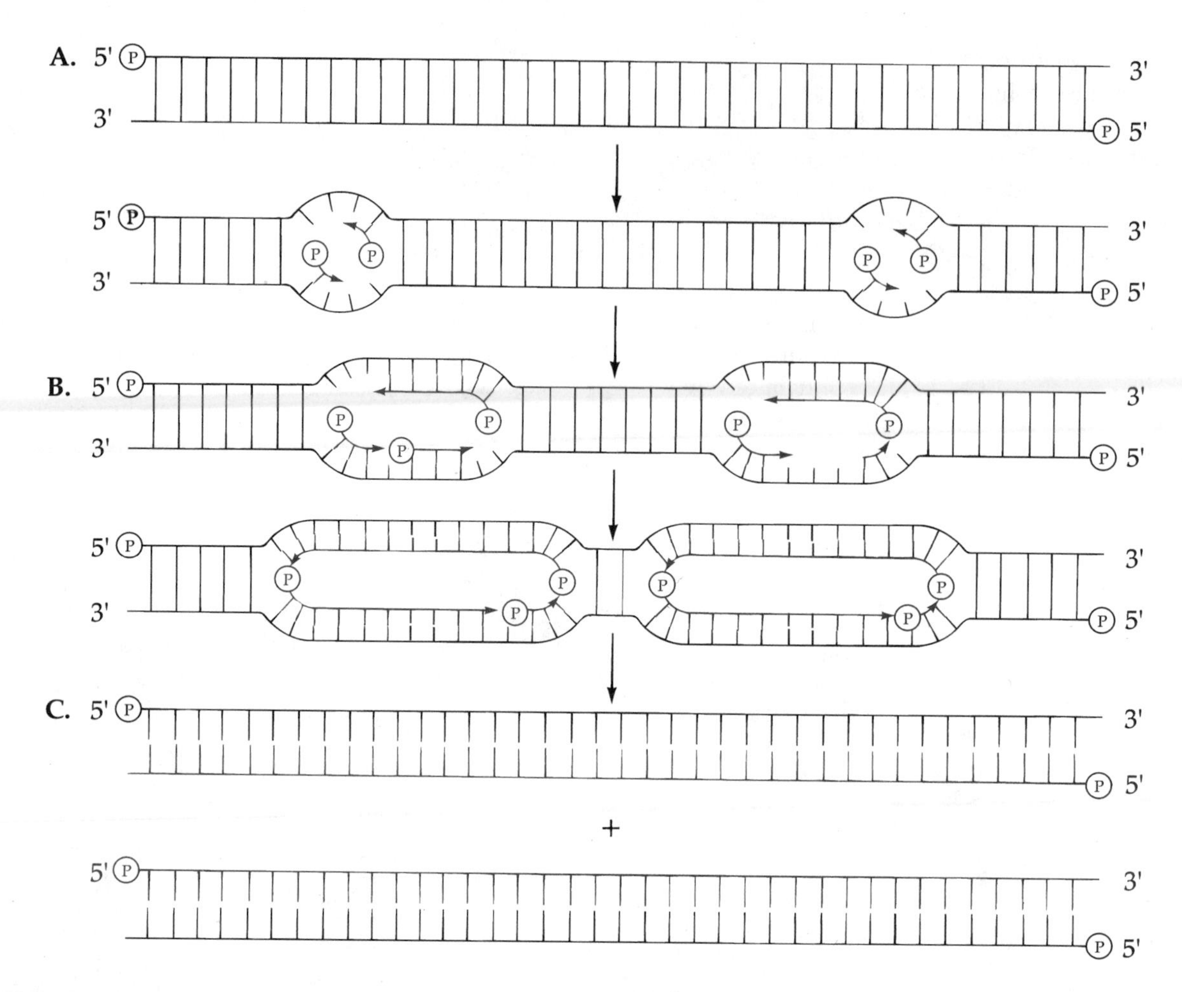

FIGURE 7.14 The steps in DNA replication. (A) Replication is initiated at sites known as origins of replication, where DNA polymerase binds and begins replication after priming. (B) Replication spreads from these sites, with continuous replication on one strand and the synthesis of short segments on the other strand. (C) Enzymes known as DNA ligases join the ends of the replicated strands and the result is two DNA molecules, each composed of a template strand and a new strand.

SUMMARY

1. At the turn of the century scientists identified chromosomes as the cellular components that carried genes. This discovery focused efforts to identify the molecular nature of the gene on the chromosomes and the nucleus. Biochemical analysis of the nucleus began around 1870 when Frederick Miescher first separated nuclei from cytoplasm and described nuclein, a protein/nucleic acid complex now known as chromatin.

2. Although subsequent work described the chemical composition of nucleic acids and their nucleotide components, proteins were regarded as the only molecular component of the cell with the complexity to encode the

genetic information. This changed in 1944 when Avery and his colleagues demonstrated that DNA was the genetic material in bacteria.

3. Recognition of the role of DNA as the molecular basis of heredity stimulated interest in the structure of DNA. In 1953 Watson and Crick constructed a model of DNA structure that incorporated information from the chemical studies of Chargaff and the x-ray crystallographic work of Wilkins and Franklin. They proposed that DNA is composed of two polynucleotide chains oriented in opposite directions and held together by hydrogen bonding to complementary bases in the opposite strand. The two strands are wound around a central axis in a right-handed helix.

4. The mitochondrial chromosome, carrying genes that can cause maternally transmitted disorders, is a circular DNA molecule.

5. Within chromosomes, DNA is coiled around clusters of histones to form structures known as nucleosomes. Supercoiling of nucleosomes may form fibers that extend at right angles to the axis of the chromosome. The structure of chromosomes must be dynamic to allow the uncoiling and recoiling seen in successive phases of the cell cycle, but the details of this transition are still unknown.

6. In DNA replication, strands are copied in a $5' \rightarrow 3'$ direction to produce semiconservatively replicated daughter strands.

QUESTIONS AND PROBLEMS

1. Until 1944, which cellular component was thought to carry genetic information?
   a. carbohydate
   b. nucleic acid
   c. proteinchromatin
   d. lipid
   e. chromatin

2. Summarize the arguments used against the role of nucleic acids as the genetic material.

3. The experiments of Avery and his co-workers led to the conclusion that:
   a. bacterial transformation occurs only in the laboratory.
   b. capsule proteins can attach to uncoated cells.
   c. DNA is the transforming agent and is the genetic material.
   d. transformation is an isolated phenomenon in *E coli.*
   e. DNA must be complexed with protein in bacterial chromosomes.

4. In the experiments of Avery, what was the purpose of treating the transforming extract with enzymes?

5. In analyzing the base composition of a DNA sample, a student loses the information on pyrimidine content. The purine content is A = 27% and G = 23%. Using Chargaff's rule, reconstruct the missing data, and list the base composition of the DNA sample.

6. The basic building blocks of nucleic acids are:
   a. nucleosides.
   b. nucleotides.
   c. ribose sugars.
   d. amino acids.
   e. purine bases.

7. Adenine is a
   a. nucleoside.
   b. purine.
   c. pyrimidine.
   d. nucleotide.
   e. base.

8. Polynucleotide chains have a 5′ and a 3′ end. What groups are found at each of these ends?
   a. 5′ sugars, 3′ phosphates
   b. 3′ sugars, 5′ phosphates
   c. 3′ bases, 5′ sugars

d. 5′ bases, 3′ sugars
e. 5′ phosphates, 3′ bases

9. Summarize the elements of the Watson-Crick model of DNA.

10. The double helix of DNA involves:
    a. two strands coiled in a left-handed helix around a central core.
    b. two polynucleotide strands coiled around each other.
    c. a spiral of two strands.
    d. two strands coiled in a right-handed helix around a central axis.
    e. hydrogen bonding between nucleotides in the same chain.

11. Using Figure 7.6 as a guide, draw a dinucleotide composed of C and A. Next to this draw the complementary dinucleotide in an antiparallel fashion. Connect the dinucleotides with the appropriate hydrogen bonding.

12. Nucleosomes are complexes of:
    a. nonhistone protein and DNA.
    b. RNA and histone.
    c. histones and DNA.
    d. DNA, RNA, and protein.
    e. histone H1 and DNA.

FOR FURTHER READING

Avery, O. T., Macleod, C. M., and McCarty, M. 1944. *J. Exp. Med.* **70:** 137.

Felsenfeld, G. 1985. DNA, *Sci. Am.* **253** (October): 58–67.

Judson, H. F. 1979. *The Eighth Day of Creation: The Makers of the Revolution in Biology.* New York: Simon & Schuster.

Kornberg, A. 1968. The synthesis of DNA. *Sci. Am.* **219** (October): 64–79.

Kornberg, R., and Klug, A. 1981. Nucleosomes. *Sci. Am.* **244** (February): 52–79.

McCarty, M. 1986. *The Transforming Principle: Discovery That Genes are Made of DNA.* New York: Norton.

Olby, R. 1974. *The Path to the Double Helix.* Seattle: University of Washington Press.

Sayre, A. 1975. *Rosalind Franklin and DNA.* New York: Norton.

Watson, J. D. 1968. *The Double Helix.* New York: Atheneum.

Watson, J. D., and Crick, F. H. C. 1953. Molecular structure of nucleic acids: A structure for deoxyribose nucleic acid. *Nature* **171:** 737–738.

Watson, J. D., and Crick, F. H. C. 1953. Genetical implications of the structure of desoxyribonucleic acid. *Nature* **171:** 964–967.

CHAPTER 8

RECOMBINANT DNA AND GENETIC TECHNOLOGY

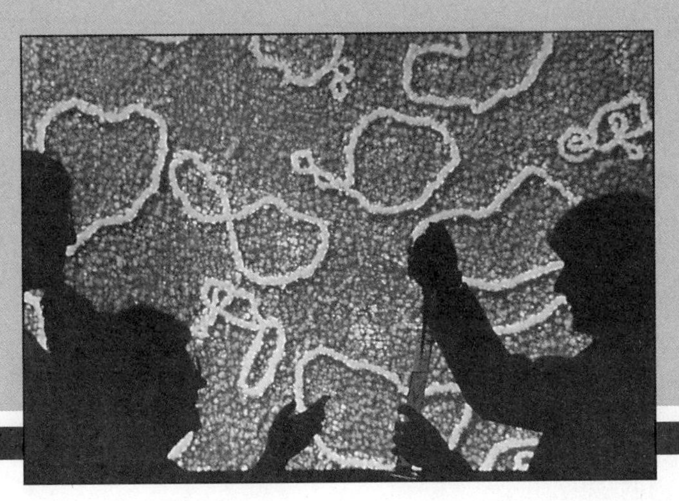

CHAPTER OUTLINE

The story begins simply enough. A boy of Ghanaian descent, born in England and therefore a British citizen, moved to Ghana to live with his father, leaving behind his mother, two sisters, and a brother. Subsequently, he attempted to return to Britain to live with his mother and siblings, and the story gets a little complicated. Immigration authorities suspected that the returning boy was an imposter and was either an unrelated child or possibly the nephew of the boy's mother. Acting on this belief, the authorities denied him residence. The boy's family sought assistance to help establish his identity and to allow him to live in the country of his birth. A series of medical tests was conducted, using genetic markers including ABO and other blood types and more sophisticated techniques such as HLA testing (for details, see Chapter 14). The results indicated that the boy was closely related to the woman he claimed was his mother, but the tests could not tell whether she was the mother or an aunt.

The family's legal counsel then turned to Alec Jeffreys at the University of Leicester to see if the DNA fingerprinting technique developed in Jeffreys' research laboratory could help. To complicate the situation even more, neither the mother's sisters nor the boy's father was available for testing, and the mother was not sure about the boy's paternity. Jeffreys began by taking blood samples from the boy, his reputed siblings, and the woman who claimed to be his mother. DNA was extracted from the white blood cells in each sample and was treated with enzymes that cut the DNA at specific base sequences; the resulting fragments were separated by size. The pattern of these fragments, known as a DNA fingerprint, was analyzed to determine the boy's paternity and maternity. The results show that the boy has the same father as the brother and sisters because they all share paternal DNA fragments (those not contributed by the mother) in common. The most important question was whether the boy and his "mother" were related. Twenty-five fragments found in the woman's DNA fingerprint were also found in the boy's fingerprint, indicating that the two are very closely related and that in all probability the boy is the woman's child. (The chance that they are unrelated was calculated as 2×10^{-15}, or about one in a trillion.) Faced with this evidence, immigration authorities reversed their position and allowed the boy to take up residence with his family.

DNA fingerprinting is but one example of the array of new techniques developed in recent years as a product of the ongoing revolution in genetic technology. This revolution began with the discovery of how to combine DNA from different organisms in a specific and directed way to create recombinant DNA molecules. In this chapter we will examine how this revolution began, how recombinant DNA molecules are constructed, and how this technology is being used in areas of human genetics, biology, medicine, and agriculture.

CLONES

The process of forming identical twins by division of a fertilized egg into two cells, each of which develops into a separate individual, represents a situation in which two genetically identical individuals are generated. Because they are derived from a common ancestor (in this case a fertilized egg or zygote), such twins can be said to be *clones*. **Clones** are molecules, cells, or individuals derived from a single ancestor. Methods for producing cloned organisms are not new; horticultural techniques to produce clones of fruit trees were used in ancient times. Techniques for the production of cloned cells, single-celled organisms, and some animals have all been developed in this century. More recently it has become possible to clone DNA molecules or segments of them.

Clones
Genetically identical organisms, cells, or molecules all derived from a common ancestor.

The development of methods to produce identical copies of DNA molecules in large quantities has had a profound effect on genetic research and has generated applications of this technology across many disciplines, ranging from archaeology to environmental conservation. This chapter reviews the methods by which cloned DNA molecules are prepared and used in genetic research, forensics, medicine, agriculture, and the repair of defective human genes. Before we consider the basic techniques of DNA cloning, we will look briefly at recent advances in the cloning of plants and animals and the possibility that humans themselves can be cloned.

Cloning Plants and Animals

Domesticated plants and animals have been genetically manipulated for thousands of years by selective breeding. Organisms with desirable characteristics were selected and bred together, and the offspring with the best combinations of these characteristics were used for breeding the next generation. While this method seems slow and unreliable, Charles Darwin noted that only a few generations of intense selection were required to produce many varieties of pigeons and minks.

In the 1950s Charles Steward demonstrated that individual carrot cells could be grown in culture and that under the appropriate conditions the cells would form a cell mass known as a callus (Figure 8.1). When transferred to a different medium, calluses could be grown into mature carrots. Using this method on a large scale, one could produce hundreds or thousands of clones from a single carrot. This method, varied somewhat for individual species, has been used to clone plants of several different species. One application of this method has been the development by conventional genetic crosses and selection of a loblolly pine tree that is resistant to disease, that grows rapidly, and that has a high wood content. Cells from this tree were grown individually until they formed a callus and were converted into thousands of copies of the original tree, producing a cloned forest of genetically identical trees that will all mature at the same time, allowing the forest to be harvested for pulpwood on a predictable schedule. This method is just beginning to be used and offers a new approach to timber farming.

The cloning of domesticated animals such as cattle and sheep has moved from the research laboratory to the level of commercial enterprise in the last decade. In this case embryos rather than adults are being cloned. Two methods are available for cloning animal embryos: embryo splitting and nuclear transfer. In embryo splitting, an egg is fertilized in a glass dish (*in vitro*) and allowed to develop into an embryo containing 8-16 cells. The embryo is then divided one or more times by micromanipulation, and the resulting cells form genetically identical embryos that can be implanted into a host uterus for development. This method mimics the way that identical twins or triplets are produced naturally, and can be used to clone any mammalian embryo, including human embryos. The recent splitting of human embryos in the laboratory will be discussed in Chapter 18.

Production of clones by nuclear transfer has the advantage of producing larger numbers of genetically identical embryos. The method uses eggs from which the nucleus has been removed with a fine glass pipette.

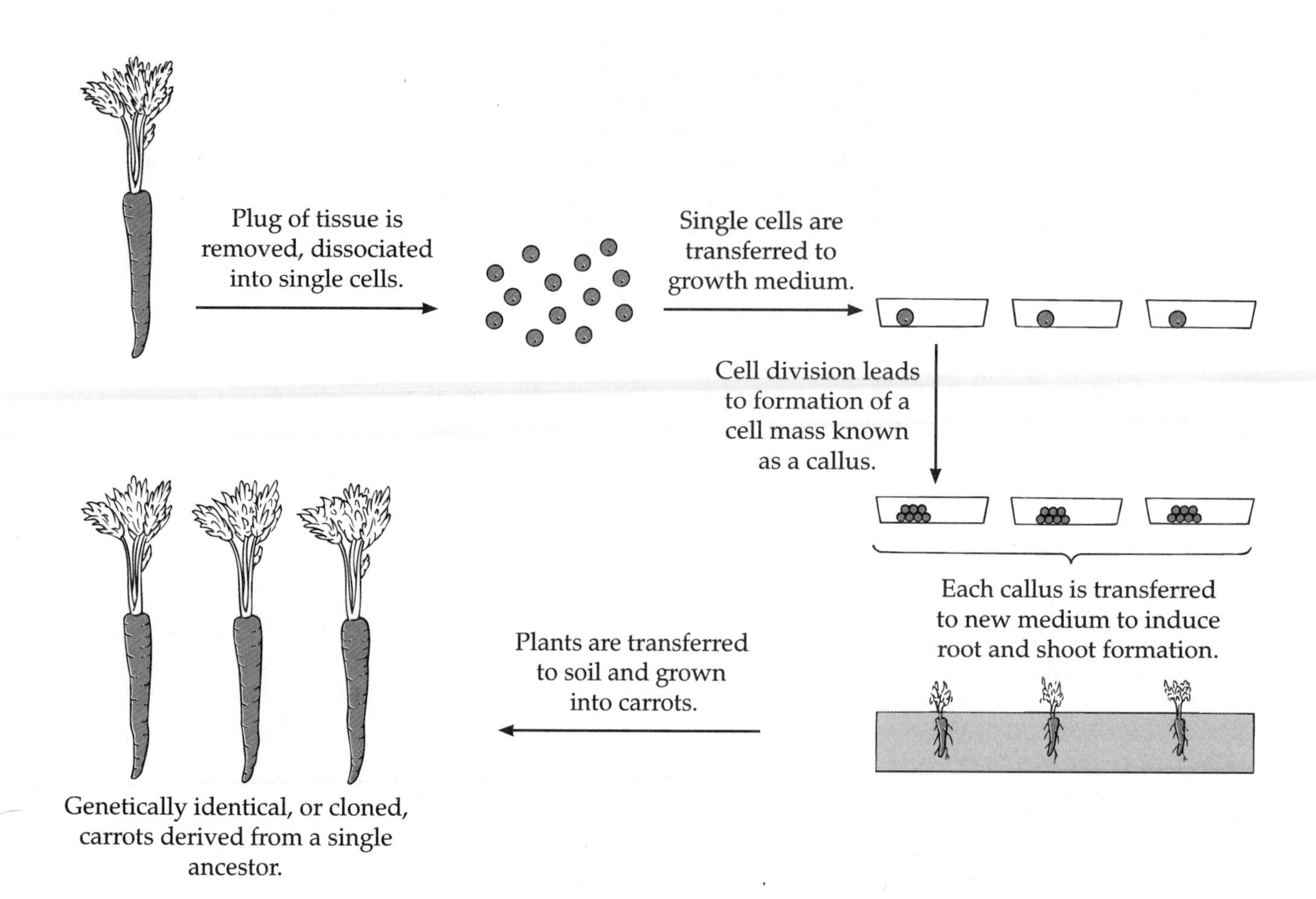

FIGURE 8.1 Carrots cloned by removing a plug of tissue that was dissociated into single cells. The cells were individually plated into dishes containing medium designed to induce cell division. In those dishes where growth occurred, the cells formed a tissue mass known as a callus. Calluses were transferred to a second type of medium designed to promote the development of roots and shoots. Finally, the plants were transferred to soil, producing a number of genetically identical clones, all derived from a single ancestor.

These enucleated eggs are then fused with individual cells taken from a 16-32 cell embryo. In successful experiments the fused cells begin dividing to produce a new embryo and are transplanted into the uterus of a foster mother. If the original embryo contains 16 or 32 cells, then 16 or 32 genetically identical offspring, or clones, can be produced (Figure 8.2). In this way it is possible to produce herds of cloned cattle or sheep with superior wool, milk, or meat production (Figure 8.3).

In 1978 a book entitled *In His Image: The Cloning of a Man,* by David Rorvik, was published. The book tells how a successful industrialist paid to have himself cloned. The author claimed that the story was true, but it quickly became apparent that essential parts of the book were fictional. Nonetheless, the book stirred discussion about the possibility of cloning

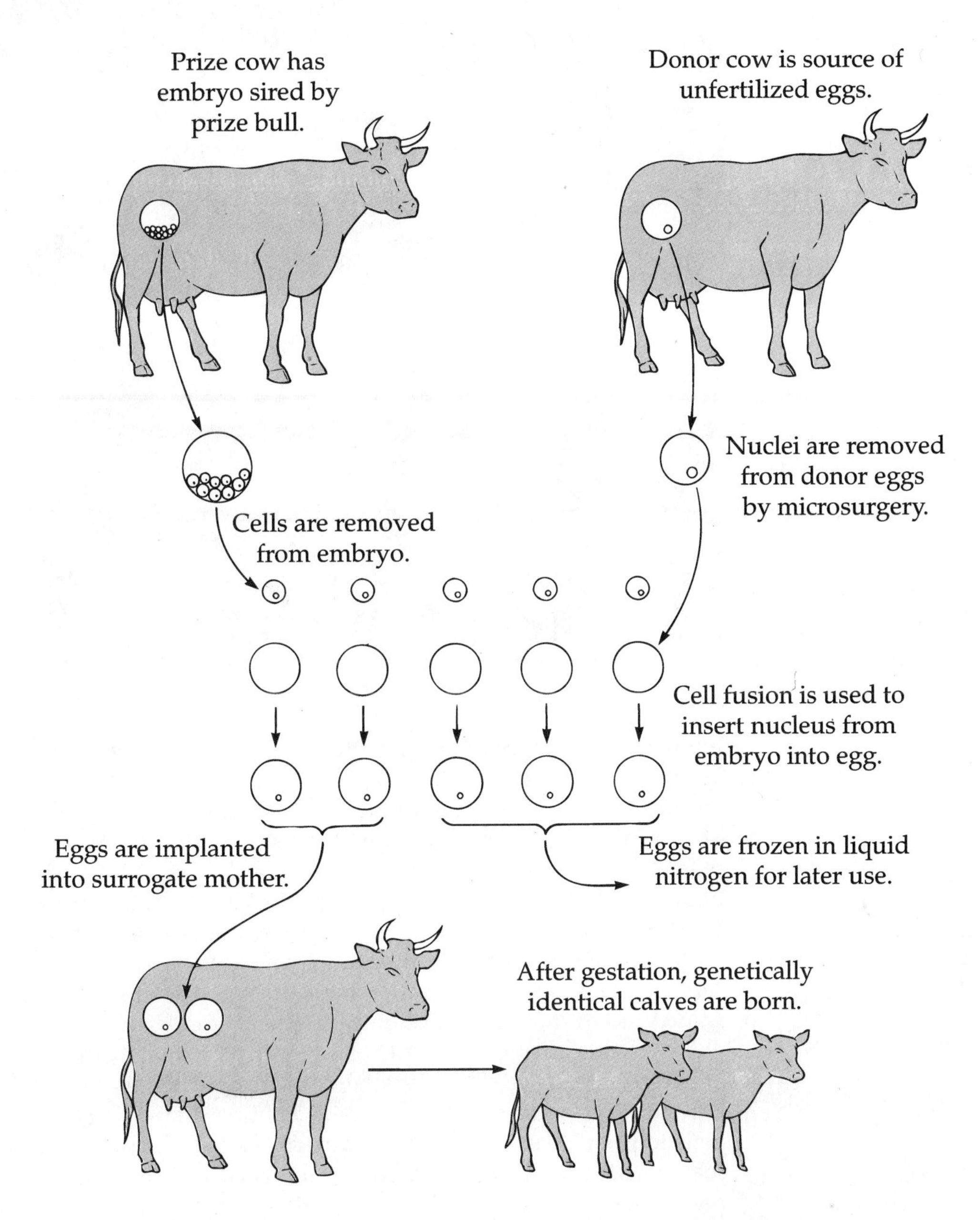

FIGURE 8.2 Several types of animals, ranging from frogs to cattle, have been cloned. First, unfertilized eggs (oocytes) are recovered from donor animals *(upper right),* and nuclei are removed from these cells. Single cells derived from an early stage of embryonic development are fused to each oocyte, transferring an embryonic nucleus into each egg. These eggs, each bearing a genetically identical nucleus, can be frozen in liquid nitrogen for later use, or transplanted into surrogate mothers for development. All offspring of these eggs would be genetically identical copies or clones of the original embryo from which the nuclei were transferred.

FIGURE 8.3 Cloned cattle. These cattle were cloned using blastomere nuclei derived from a single embryo.

humans and the potential legal and ethical problems associated with cloned humans. It is interesting to note that this is not the first recorded account of an attempt to recreate an individual. When Leonardo da Vinci died at the age of 67 years, his half-brother Bartolommeo set out to create a replica of his older brother. For the mother of the new Leonardo, he selected a young peasant girl from the village of Vinci (da Vinci's mother was a peasant girl from this same village), married her, and had a son Pierino. Nurtured by Bartolommeo, Pierino showed great promise as an artist at a young age and served at least two apprenticeships during his adolescence. He quickly gained recognition as an accomplished sculptor. In fact, some of his works were wrongly attributed to Michelangelo. Unfortunately the young man died at the age of 23 years, before he could reach his artistic peak. Although Bartolommeo's attempt at recreating his brother was not based on sound genetic principles, it was extraordinarily successful, providing a reminder that heredity and environment often work closely together to define a phenotype.

The development of methods for cloning plants and animals and gains in *in vitro* fertilization and freezing of gametes and embryos represent significant advances in themselves. When coupled with molecular screening for defective genes, gene transfer techniques, and the prospect of gene surgery, these methods synergistically interact to present us with the possibility of transferring genes between species and even creating new forms of life by combining genetic information from plants and animals in a single organism that can be cloned into thousands of copies. With this background and with these possibilities in mind, we will consider the methodology of genetic engineering and gene cloning.

DNA Cloning and Recombinant DNA Technology

The cloning of DNA molecules or segments of DNA molecules results in the production of a large number of identical DNA molecules, all having a common ancestor. In this sense cloning DNA is similar to cloning whole plants or animals. The methods for cloning DNA molecules, however, are unique to this application and depend on genetic and biochemical discoveries made in the late 1960s and early 1970s. Collectively these techniques are often referred to as *recombinant DNA technology*, or *genetic engineering*. As used below, the term "recombinant DNA" can refer to the artificial association of DNA molecules or parts of DNA molecules that are not found together naturally. Three basic methods are involved in cloning DNA:

1. a method for physically linking DNA segments from different organisms together
2. the use of a self-replicating form of DNA, known as a cloning vector, to be linked with the DNA segment to be cloned
3. The transfer of a vector carrying a DNA segment into a host cell for replication, and a method for the identification of cells containing any desired DNA segment

To provide a basis for understanding how recombinant DNA technology is applied to human genetics, we will briefly review each of these steps in the process of cloning DNA.

DNA-Cutting Enzymes

The development of the array of techniques known as genetic technology is a good illustration of how basic research in one field often has an unexpected impact on unrelated areas. It seems odd that discoveries such as the mapping of genes for cystic fibrosis and muscular dystrophy and the commercial production of human insulin are a direct outgrowth of basic research into how soil bacteria are able to resist infection by viruses, but often, this is how science progresses. The discovery of bacterial enzymes that inactivate invading viruses by cutting the viral DNA into fragments marks the beginning of genetic technology.

In the mid 1970s, Hamilton Smith and Daniel Nathans discovered a series of enzymes known as **restriction enzymes** that attach to DNA molecules and cut both strands of DNA at sites of specific base sequences. This discovery was a key step in the development of recombinant DNA technology. The recognition and cutting site for an enzyme isolated from the bacterium *Escherichia coli*, designated as *Eco*RI, is shown in Figure 8.4. Note that the recognition sequence reads the same on either strand when read in the 5′ to 3′ direction. This recognition sequence is said to be a **palindrome,**

Restriction enzymes
Enzymes that recognize a specific base sequence in a DNA molecule and cleave or nick the DNA at that site.

Palindrome
A word, phrase, or sentence that reads the same in both directions. Applied to a sequence of base pairs in DNA that reads the same in the 5′ to 3′ direction on complementary strands of DNA. Many recognition sites for restriction enzymes are palindromic sequences.

FIGURE 8.4 The palindromic recognition and cutting site in DNA for the restriction endonuclease *Eco*RI.

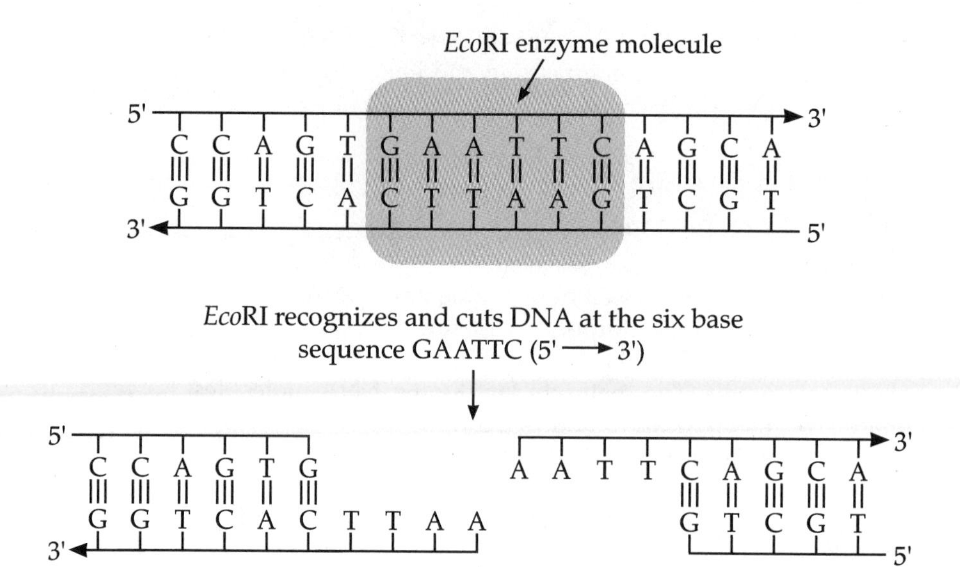

since it reads the same on either strand of the DNA. (Words like "mom," "pop," and "tot" are palindromes, as are phrases such as "tis Ivan on a visit" or "live not on evil.") Because the nucleotide sequence at the cutting site is complementary, the cut ends, or "tails," of the enzymatically treated DNA strands can reassociate with each other or with other DNA molecules with similar tails to produce recombinant DNA molecules. Thus restriction enzymes allow DNA segments from different organisms to be physically joined together. The cutting sites for several restriction enzymes are shown in Table 8.1.

Vectors and Recombinant DNA Construction

Vector
A self-replicating DNA molecule that is used to transfer foreign DNA segments between host cells.

Plasmids
Extrachromosomal DNA molecules found naturally in bacterial cells. Modified plasmids are used as cloning vectors or vehicles.

Vectors are used to carry DNA segments into cells for replication (cloning). Many of the vectors initially used in recombinant DNA work are derived from self-replicating, circular DNA molecules called **plasmids,** found within bacterial cells. One such plasmid, identified as pBR322, is shown in Figure 8.5. The diagram shows the location of sites recognized and cut by restriction enzymes that can be used for the insertion of DNA molecules to be cloned.

If DNA molecules from a plasmid vector and from human cells are cut with *Eco*RI and then placed together in solution, their cut ends can reassociate, or anneal (Figure 8.6). If these cuts are sealed with an enzyme called DNA ligase, recombinant DNA molecules made up of human DNA combined with plasmid DNA will be created. Bacterial cells can be induced to take up the plasmids containing human DNA by treatment with salt solutions. Within the bacterial cell, the recombinant DNA molecule consisting of the plasmid and a segment of human DNA replicates, producing many copies, or *clones*, of itself.

TABLE 8.1 Some Common Restriction Enzymes and Their Cutting Sites

Source	Enzyme	Cleavage Sequence*
Escherichia coli	*Eco*RI	↓ GAATTC CTTAAG ↑
Bacillus amyloliquefaciens	*Bam*HI	↓ GGATCC CCTAGG ↑
Hemophilus influenza	*Hind*III	↓ AAGCTT TTCGAA ↑
Brevibacterium albidum	*Bal*I	↓ TGGCCA ACCGGT ↑
Thermus aquaticus	*Taq*I	↓ TCGA AGCT ↑
Arthobacter luteus	*Alu*I	↓ AGCT TCGA ↑

The steps involved in cloning DNA can be summarized as follows:

1. The DNA to be cloned is isolated and treated with restriction enzymes to produce segments ending in specific sequences.
2. These segments are linked to DNA molecules that serve as vectors, or carriers, producing a recombinant DNA molecule.
3. Vectors commonly used include plasmids, double-stranded DNA molecules that occur naturally and replicate within bacterial cells. Viruses called **bacteriophage,** or **phage** for short, are also used as vectors.

Bacteriophage
A virus that infects bacterial cells. Genetically modified bacteriophage are used as cloning vectors.

FIGURE 8.5 The plasmid-vector pBR322. The location of restriction enzyme sites that can be used in cloning is indicated.

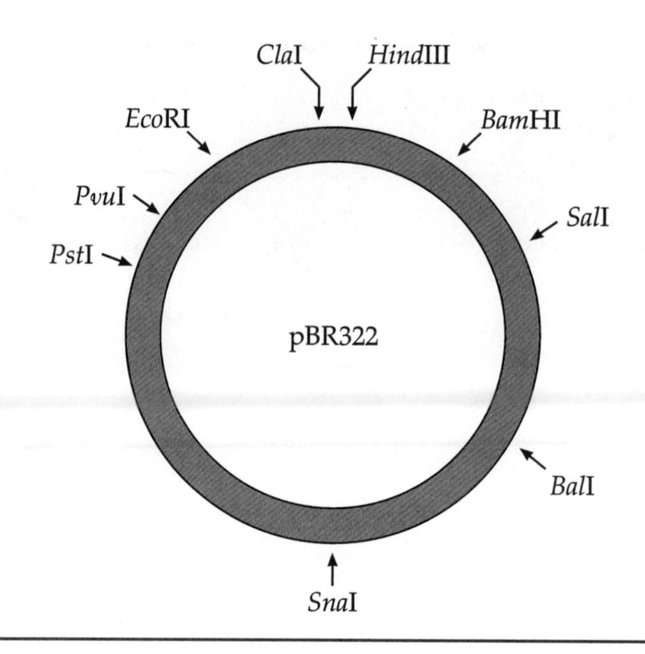

4. Plasmids carrying DNA to be cloned are transferred into bacterial cells, where the recombinant plasmids replicate, producing many copies, or *clones*, of the inserted DNA.

Figure 8.7 presents a summary of these steps.

Genetic Libraries

Library
In recombinant DNA terminology, a collection of clones that contains all the genetic information in an individual. Also known as a gene bank.

Because each cloned fragment of human DNA is relatively small, many separate clones must be created to include a significant portion of the entire human genome. A collection of clones that contains all the DNA sequences in the genome of an individual is known as a **library.** The number of clones required to cover the entire genome depends, of course, on the size of the genome and the number of segments created by the restriction enzyme. A human library composed of cloned fragments 1700 base pairs (1.7 kilobases) in length would require about 8.1 million plasmids to contain the genetic information from one human cell. Viral vectors such as lambda phage can accept much larger DNA inserts and are routinely used in the construction of genetic libraries. A human genetic library can be contained in about 800,000 such viruses that can be stored in a single test tube. Using the techniques of molecular biology, any sequence of interest can be recovered from such a library and used in experimental studies, in clinical application, or for commercial use. Libraries from many organisms are now available, including bacteria, yeasts, crop plants, endangered species, and humans.

Cloned DNA derived from this sequence of steps can be used for a wide range of purposes, including the study of gene structure and organization, gene mapping, gene transfer to produce new plants and animals, and gene therapy to treat genetic diseases. Further refinements of the basic techniques

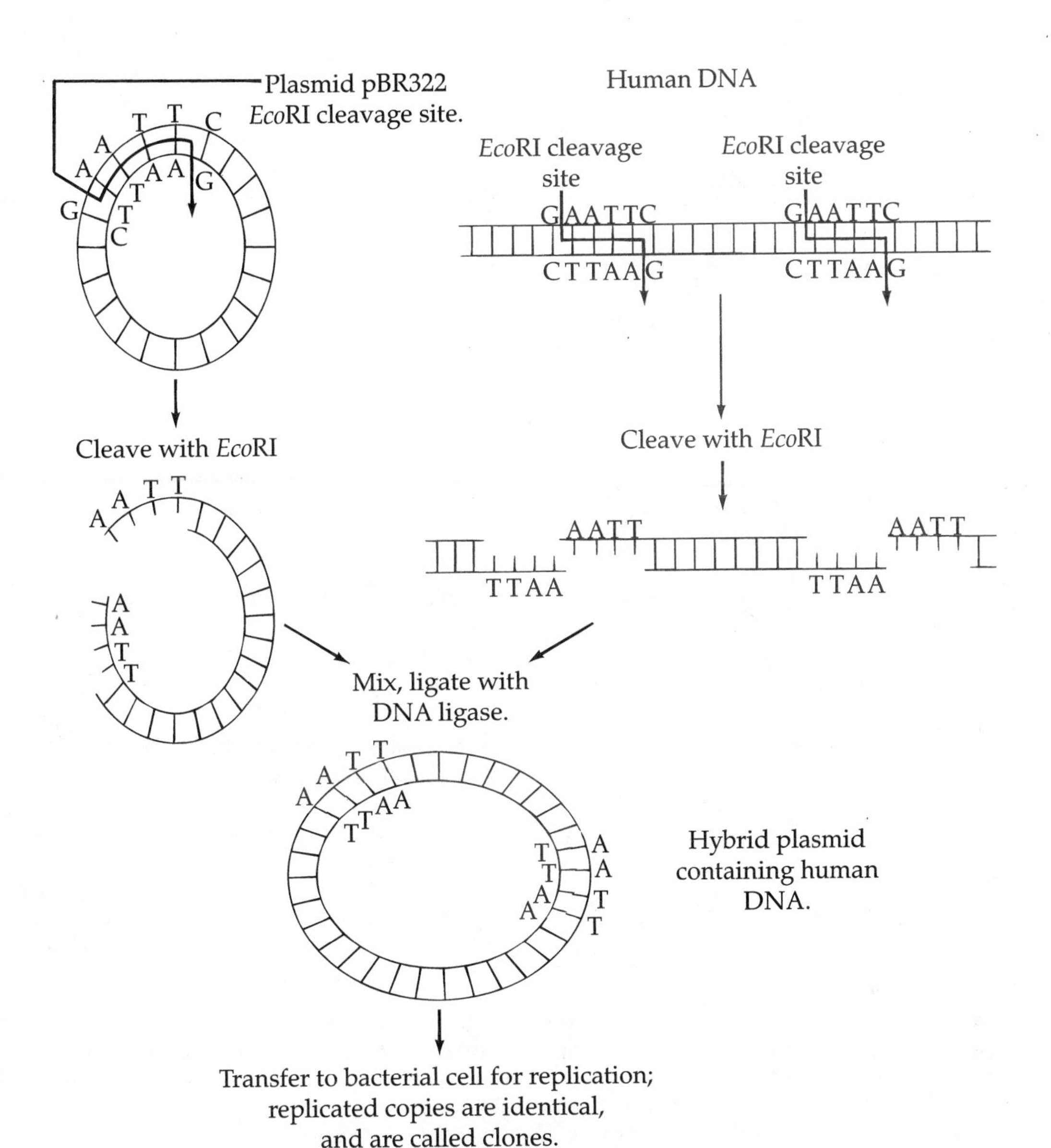

FIGURE 8.6 Construction and cloning of a recombinant plasmid using plasmid DNA and human DNA cut with *Eco*RI and treatment with DNA ligase. The recombinant plasmid can be transferred to a bacterial cell where replicate copies (clones) are made by the process of DNA replication.

of recombinant DNA have allowed the activation of the cloned DNA sequences within bacterial cells to direct the synthesis of gene products to produce commercial quantities of products such as human growth hormone, insulin, vaccines, and even industrial chemicals.

Scientists working in the field of recombinant DNA technology were among the first to recognize that there might be unrecognized dangers in the use and release of recombinant organisms. Accordingly, they called for a

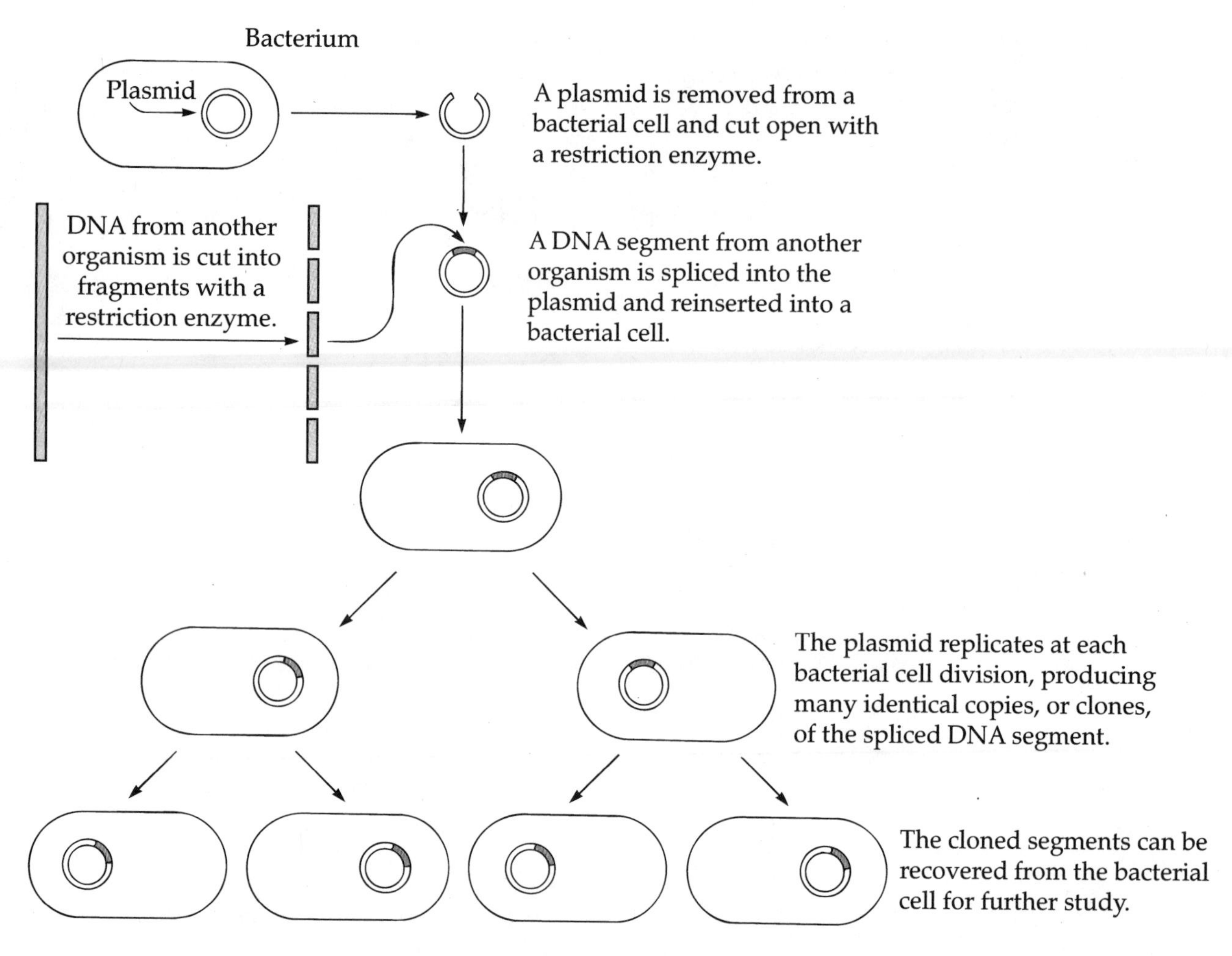

FIGURE 8.7 An overview of the process of cloning a DNA segment. First, plasmid vectors are isolated and cut with a restriction enzyme at a recognition site. Second, DNA from another organism is isolated and cut with the same restriction enzyme, producing a collection of fragments. The DNA fragments are mixed with the plasmid vectors, spliced into the vectors, and reinserted into bacteria cells for growth. The plasmid, carrying a DNA segment from another organism, replicates at each bacterial cell division, producing many copies, or clones, of the DNA segment. These clones can be recovered from the bacterial cells for further use.

moratorium on all such work until issues relating to the safety of genetic engineering could be discussed (see Concepts & Controversies, page 209). After years of discussion and experimentation, there is general agreement that such work poses little if any risk, but the episode demonstrates that scientists are concerned about the risks as well as the benefits of their work.

The Polymerase Chain Reaction

One of the advantages offered by the development of cloning techniques is the ability to produce large amounts of specific DNA sequences for a variety

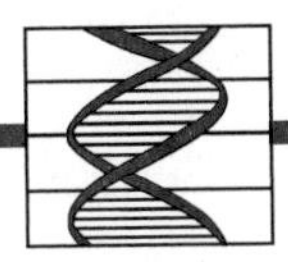

Concepts & Controversies

Asilomar: Scientists Get Involved

The first steps in creating recombinant DNA molecules were taken in 1973 and 1974. Scientists immediately recognized that modifying the genetic information in a bacterium that lived in the human gut *(Escherichia coli)*, could be potentially hazardous. A group of scientists called on the National Academy of Sciences to request that a panel be appointed to study the risks and possible control of recombinant DNA research. A second group published a letter in *Science* and *Nature*, two leading scientific journals, calling for a moratorium on certain kinds of experiments until the potential hazards could be assessed.

Shortly thereafter, an international conference of scientists was held in Asilomar, California, to consider whether recombinant DNA methodology posed any dangers and whether this form of research should be regulated. In 1975 a set of guidelines resulting from this conference was published by the U.S. government, under the direction of the National Institutes of Health (NIH), an agency that regulates the direction of biomedical research in this country. In 1976, the NIH published a stricter set of regulations that prohibited certain kinds of experiments and dictated that other types of experiments were to be conducted only under appropriate conditions of containment that would prevent the release of bacterial cells containing recombinant DNA molecules. The agency also called for research to accurately assess what risks, if any, are posed by the use of such techniques.

In the meantime, legislation was proposed in Congress and at the state and federal levels to regulate or to forbid the use of recombinant DNA technology. The federal legislation was withdrawn after exhaustive sessions of testimony and reports. By 1978 research had shown that the common laboratory strain of *E. coli*, K12, was much safer for use in recombinant DNA research than originally thought. Several investigations showed that K12 could not survive in the human gut and did not survive outside of the laboratory. Further work showed that recombinant molecules had also been produced in nature, without any detectable serious effects.

In 1982 the NIH issued a new set of guidelines that have eliminated most of the constraints on recombinant DNA research. No experiments are currently prohibited, although notification must be sent to the NIH if certain experiments are to be conducted.

It is important to note that the scientists who developed the methods were the first to call attention to the possible dangers of recombinant DNA research, and they did so based only on its *potential* for harm. There were no known cases of the release of unsafe organisms into the environment. Scientists voluntarily shut down some of their own work until the situation could be properly and objectively assessed. Only when independent investigation demonstrated that there was no danger did these experiments resume. This example demonstrates that many scientists care deeply about the consequences of their discoveries and do get involved in issues of social relevance.

of uses. Cloning is not the only way in which a large number of copies of a specific DNA sequence can be produced. Invented in 1986, the polymerase chain reaction (PCR) technique has revolutionized, extended, and in some cases replaced methods of recombinant DNA research.

PCR uses single-stranded DNA as a template for the synthesis of a complementary strand by the enzyme DNA polymerase, in much the same way as DNA replication works in the cell nucleus. (See Chapter 7 for a discussion of DNA replication.) In the PCR reaction, the DNA to be amplified is first heated to break the hydrogen bonds joining the two polynucleotide

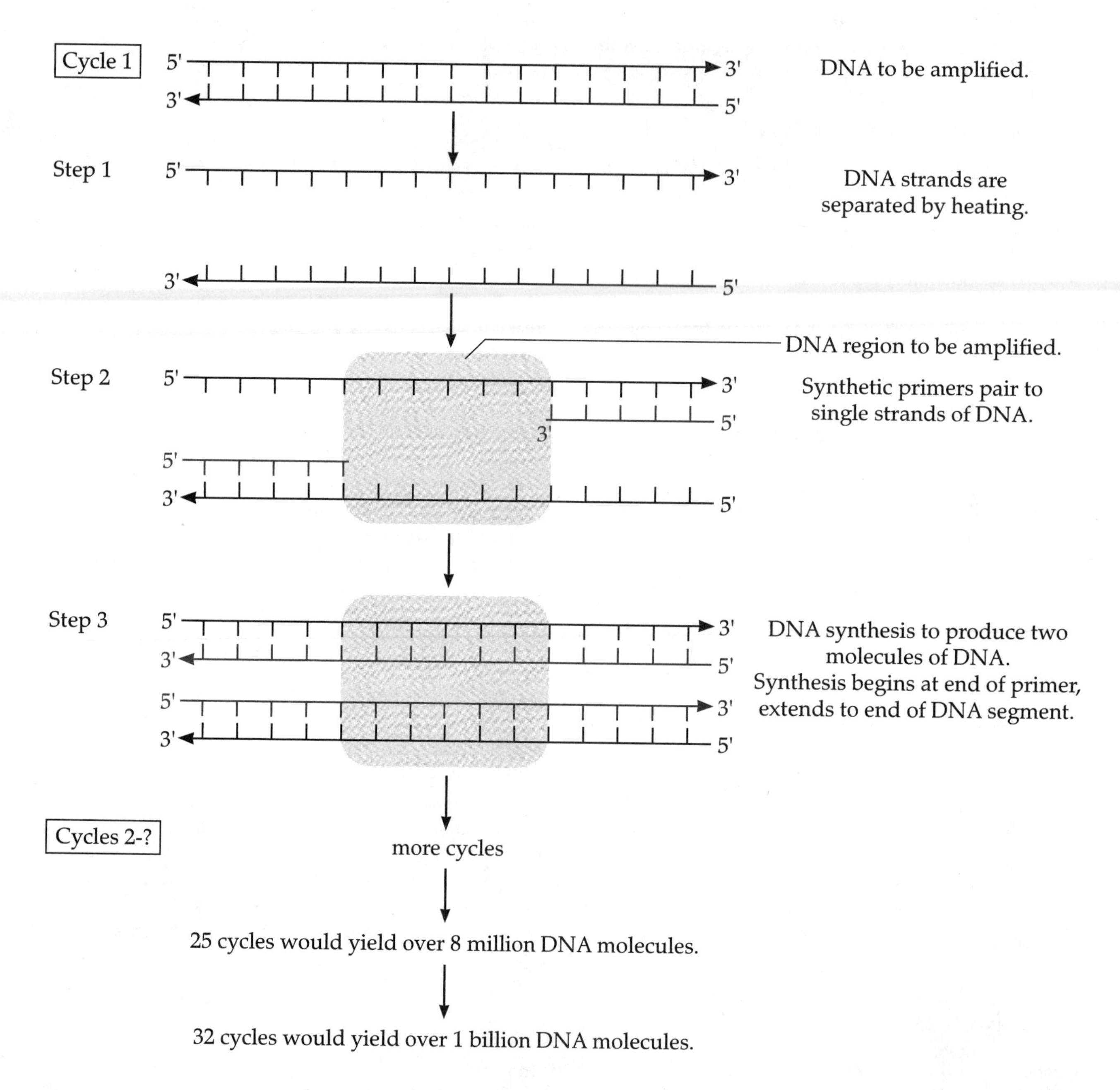

FIGURE 8.8 The process of PCR amplification. In step 1, the DNA to be amplified is heated to separate the molecule into single strands. In step 2, short, single-stranded primers pair with the nucleotides adjacent to the region to be amplified. These primers are made synthetically and are complementary to the nucleotide sequence flanking the region to be amplified. In step 3, enzymes and nucleotides for DNA synthesis are added, and DNA synthesis takes place, beginning at the primers, and using the single-stranded regions as templates. This series of three steps is called a cycle. Cycles are repeated as desired until a sufficient amount of DNA is produced or amplified.

strands, producing single strands (Figure 8.8). Next, short nucleotide sequences that act as primers for DNA replication are added, and these primers anneal to complementary regions on the single-stranded DNAs (Figure 8.8). Primers can be synthesized in the laboratory and are usually 20 to 30 nucleotides long. In the third step, the enzyme DNA polymerase begins at the primers and synthesizes a DNA strand complementary to the primer. This set of three steps is known as a cycle. The entire cycle can be repeated by heating the mixture and separating the double-stranded DNAs into single strands, each of which can serve as a template. The end result is that the amount of DNA present doubles with each cycle. After n cycles , there is a 2^n increase in the amount of double-stranded DNA (Table 8.2).

It is this power to amplify DNA that is one advantage of the PCR technique. The DNA to be amplified does not have to be purified and can be present in minute amounts; even a single DNA molecule can serve as a template. DNA from many sources has been used as starting material for the PCR technique, including dried blood, hides from extinct animals such as the quagga (a zebra-like African animal exterminated by hunting in the late 19th century), single hairs, mummified remains, and fossils.

The oldest DNA used in the PCR reaction to date has been extracted from bees and termites preserved in amber for about 30 million years (Figure 8.9). The DNA amplified from these samples will be used to study the evolution of these insects.

The PCR reaction is also used in clinical diagnosis, forensic applications, and other areas, including conservation. Some of these applications will be discussed in later sections of this chapter.

TABLE 8.2 DNA Sequence Amplification by PCR

Cycle	Number of Copies
0	1
1	2
5	32
10	1024
15	32,768
20	1,048,576
25	33,544,432
30	1,073,741,820

FIGURE 8.9 Stingless bee preserved in amber. Similar specimens were used as source of DNA that was amplified by the PCR method and cloned. The cloned DNA from the extinct bee has been compared to DNA from present bees to study how genes from these insects have changed over time.

Applications of Recombinant DNA Technology

As discussed in Chapter 4, mapping human genes involves the detection of linkage between genes (genetic evidence that two or more genes are on the same chromosome) and the assignment of these groups of linked genes to individual chromosomes. While genes can be directly assigned to the X chromosome by their unique pattern of inheritance, it is more difficult to map genes to individual autosomes. Techniques such as somatic cell hybridization (see Chapter 4) can be used to assign genes to specific chromosomes in cases in which the gene product has been identified or can be detected by an assay. However, mapping many genetic disorders is hampered by the fact that the nature of the mutant gene is unknown, and the only information available is derived from family studies that have established the mode of inheritance as dominant or recessive and from population studies that estimate the frequency of the condition.

Recombinant DNA technology has provided a new method of mapping human genes without the need to know the nature of the mutant gene or the gene product. This method uses restriction enzymes to produce DNA fragments that can serve as genetic markers. These markers can be assigned to specific chromosomes, and genetic disorders can be mapped by showing linkage between these chromosome-specific markers and the disorder.

RFLP Analysis

Recall that restriction enzymes work by recognizing specific base sequences in DNA and by cutting both strands of DNA at these sites. The use of these enzymes in gene mapping depends on the fact that the nucleotide sequence in long stretches of DNA is subject to a small degree of variation. In some cases this variation in base sequence can create or destroy a recognition/cutting site for a restriction enzyme, altering the pattern of cuts made in the DNA. In turn, this gives rise to a detectable variation in DNA fragment length known as a **restriction fragment length polymorphism,** or **RFLP** for short. Such polymorphisms are heritable and are transmitted in a codominant fashion (Figure 8.10). In other words, RFLPs can serve as genetic markers, although the phenotype of an RFLP is not externally visible, such as albinism, or biochemically detectable, such as a blood type. Rather, the phenotype depends on the production of DNA fragments of different size.

Restriction fragment length polymorphism (RFLP)
Variations in the length of DNA fragments generated by a restriction endonuclease. Inherited in a codominant fashion. RFLPs are used as markers for specific chromosomes or genes.

For RFLP analysis, a small blood sample (10 to 20 ml) is withdrawn, and the white blood cells (leukocytes) are separated from the red blood cells (erythrocytes). DNA is extracted from the white cells and treated with a restriction enzyme. As a result, the DNA is cut into hundreds of thousands or millions of fragments called **DNA restriction fragments.** These fragments are placed on an agarose gel and are subjected to an electrical current (a procedure known as electrophoresis). DNA molecules are negatively charged, and as they migrate through the gel toward the positive pole, they are separated by size (smaller fragments migrate faster). The separated

DNA restriction fragment
A segment of a longer DNA molecule produced by the action of a restriction endonuclease.

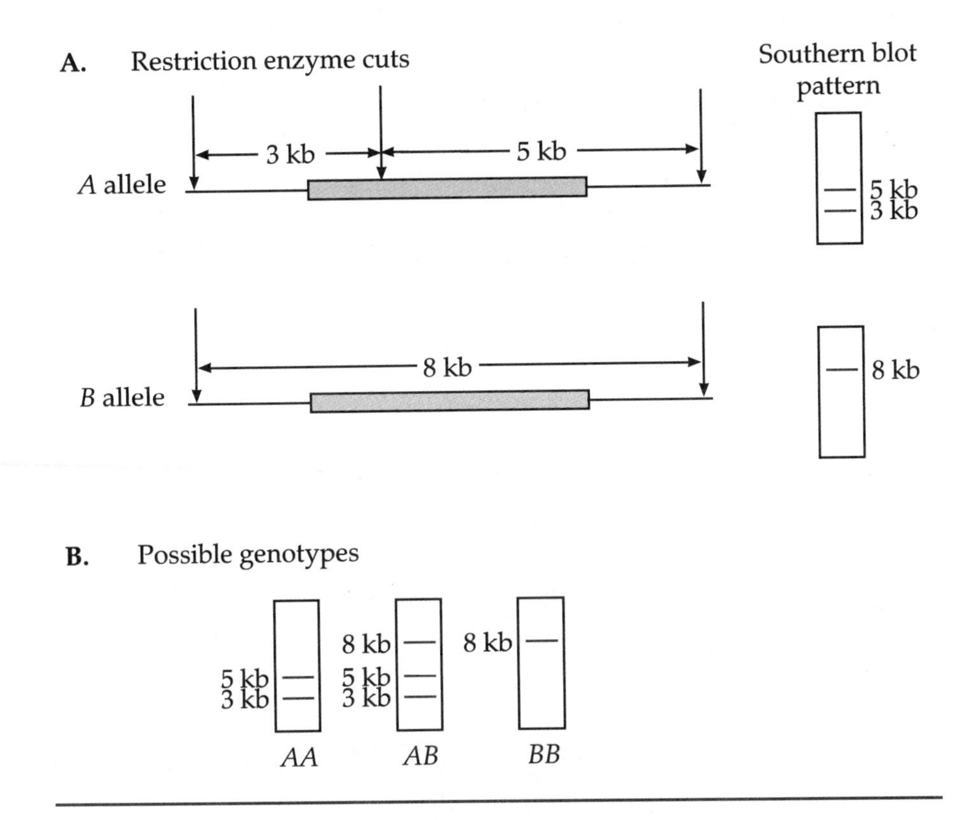

FIGURE 8.10 RFLP analysis. (A) The *A* and *B* alleles represent DNA segments of homologous chromosomes. The thick, shaded areas represent a region that can form double-stranded structures with a cloned DNA probe. Vertical arrows indicate the location of restriction enzyme recognition and cutting sites. In allele *A*, three cutting sites generate two fragments (3 and 5 kilobases). A kilobase (kb) represents 1000 base pairs of a DNA molecule. In allele *B*, two cutting sites are present. Restriction enzyme digestion generates a fragment 8 kb long. (B) Since the alleles are codominant, they can generate three genotypic patterns of fragments on Southern blots.

fragments are then subjected to a procedure known as *Southern blot analysis,* which allows the identification of a subset of restriction fragments.

If a mutational event has eliminated or created a restriction-enzyme cutting site, it can be detected as a change in the length of a DNA fragment, making it larger or smaller (see Figure 8.10). This variation in the cutting pattern is known as an RFLP.

RFLPs can be mapped to specific chromosomes by using somatic cell hybrids that contain different combinations of human chromosomes. DNA from such cells is cut with a restriction enzyme, and the fragment pattern is analyzed to determine whether a certain RFLP cutting site is present or absent. If the pattern of presence and absence of the cutting site matches the pattern of presence and absence of a particular human chromosome, then the RFLP can be assigned to that chromosome in the same way that genes are mapped using somatic cell hybrids (see Table 4.5). Other techniques can also be used to map RFLPs, and several hundred RFLPs have now been identified and mapped. RFLP markers for each human chromosome are available.

Mapping Genetic Disorders

Two things are needed to assign a gene to a particular chromosome: a large, multigenerational family in which a genetic disorder is inherited and a collection of RFLPs (at least one for each human chromosome). First, a pedigree analysis is performed to determine the pattern of inheritance for the

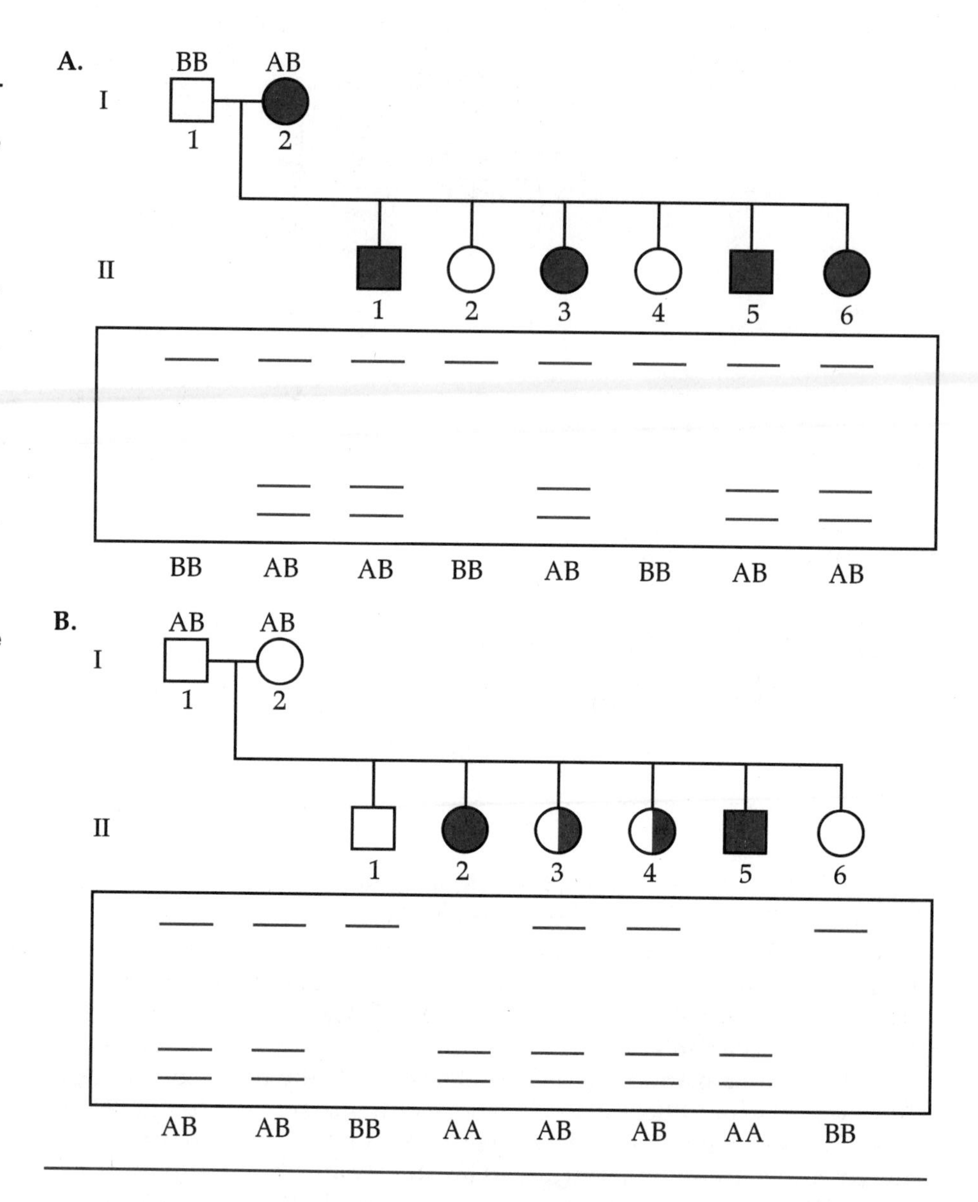

FIGURE 8.11 (A) An example of RFLP pedigree analysis using the A and B alleles described in Figure 8.10 and the inheritance of a dominant trait. A family pedigree is positioned above a gel showing the distribution of the A and B alleles in the family. Affected individuals in the pedigree are indicated by shading. The unaffected father is homozygous BB, and the affected mother is heterozygous AB. Affected children have each inherited a maternal chromosome carrying the A allele (II-1, II-3, II-5, II-6). Children who have inherited the maternal chromosome carrying the B allele (II-2, II-4) are unaffected. Thus the A allele is present on the same copy of the chromosome as the disease locus, and is said to be genetically linked to the disorder. **(B) RFLP analysis involving the A and B alleles and a recessive trait.** In this example neither parent is affected, and two of the children (II-2, II-5) are affected. The parents are each heterozygous AB, and the affected individuals are homozygous AA. Thus, the A allele is linked to the copy of the chromosome carrying the genetic disorder in each of the parents. Since the A allele is linked to the disorder, any of the children carrying the A allele in heterozygous condition (AB) are heterozygotes for the genetic disorder (II-3, II-4). Those children homozygous for the B allele (II-1, II-6) are not carriers of the disorder.

genetic disease and to identify affected individuals (Figure 8.11). Then each family member is tested to determine the pattern of inheritance for chromosome-specific RFLP markers. If the pattern of inheritance for the genetic defect matches the pattern of inheritance for a chromosome-specific RFLP marker, then both the genetic disorder and the RFLP must be on the same chromosome. In this case, the genetic defect can be assigned to the chromosome represented by the RFLP marker (Figure 8.11).

Shortly after the development of RFLP mapping, the technique was used to map the location of genes for Duchenne muscular dystrophy (short arm of the X chromosome), Huntington disease (short arm of chromosome 4), cystic fibrosis (long arm of chromosome 7), neurofibromatosis (chromosome 17), and polycystic kidney disease (chromosome 16). In all these cases the identity of the gene product was unknown when the mapping analysis was undertaken.

It is important to note that RFLP mapping provides more than a method of assigning genes with unknown gene products to particular chromosomes. In these cases the isolation of the gene and the identification of the gene product *begin* with the gene's localization to a particular chromosome. Further analysis with other RFLP markers can be used to identify the precise locus on the chromosome where the gene resides. Once this has been accomplished, the gene itself can be isolated by cloning. Analysis of the base sequence of the cloned gene can be used to establish the boundaries of the gene and to deduce the structure of the gene product. These methods, although time consuming and laborious, have already been used to identify the proteins associated with muscular dystrophy, cystic fibrosis, and neurofibromatosis. Once the proteins are isolated, it is possible to study their mode of action and to use this information to devise therapies to treat these diseases. This method of proceeding from the isolated gene to the gene product to the phenotype is revolutionizing genetic mapping as well as other areas of human genetics, and is known as **positional cloning**. For example, RFLP markers that map close to the loci for genetic disease can be used to detect heterozygotes, or affected fetuses and individuals, before the development of symptoms. The use of presymptomatic diagnosis promises to have a significant impact on clinical genetics. These topics are discussed in more detail in the following sections and in Chapter 19.

Positional cloning
Identifying, mapping, and isolating genes of unknown function by cloning DNA from a region to which the gene has been mapped by RFLP analysis.

The Human Genome Project

The development of recombinant DNA technology and its use in human gene mapping and the identification of genes associated with genetic disorders generated a great deal of interest in producing accurate maps of the human genome. This interest resulted in discussions among scientists and government agencies about initiating a centrally coordinated, large-scale project to determine the location of the 50,000 to 100,000 genes in the human genome and to analyze the structure of these genes at the level of nucleotide sequence. After much consideration, debate, and some controversy, the U.S. Congress launched the Human Genome Project in 1988 as a joint effort between the Department of Energy and the National Institutes of Health (NIH). As mentioned in Chapter 4, the goal of the project is to determine the location of all human genes and analyze the structure of human DNA. In the following years, the project was widened to include the efforts of other nations, and it is now an international effort, coordinated by the Human Genome Organization (HUGO).

The major features of the Genome Project are outlined in Figure 8.12. Although the project is shown as a consecutive series of objectives, work on each stage is progressing in a more or less simultaneous manner in laboratories around the world. The first objective is to develop a high-resolution genetic map of each human chromosome, with genes and markers spaced at an average distance of 1 centimorgan (cM) along the length of the chromosome. (A centimorgan is the distance between two genes that results in crossing over between two genes or markers about 1% of the time.) Genetic

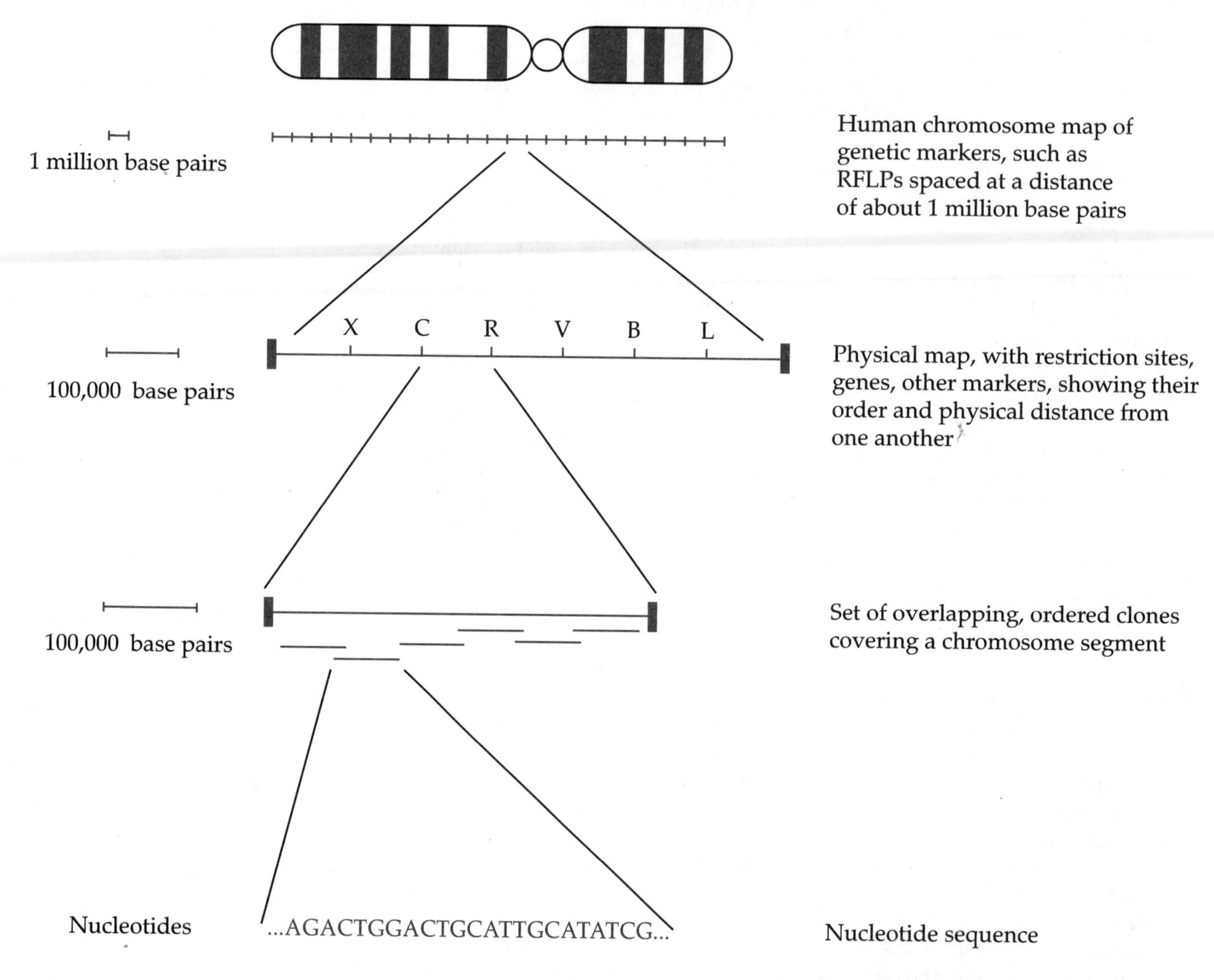

FIGURE 8.12 Major steps in the Human Genome Project. (A) The first goal is the construction of a high-resolution genetic map for each chromosome. In this case, genetic markers are represented by short, horizontal lines along chromosome. (B) From segments of the genetic map, a physical map at a higher resolution will be constructed. Here the thick bars at top and bottom represent genetic markers, and the letters along the vertical line represent restriction enzyme sites that serve as markers. (C) From the physical map, a series of overlapping cloned DNA sequences will be developed (represented by the overlapping vertical lines). (D) Each of these clones will be sequenced, and the nucleotide sequence stored in a computer data bank. When this process has been completed for each interval along the genetic map, and for all the human chromosomes, the nucleotide sequence of the 3 billion nucleotides in the human genome will be known.

maps are constructed by observing how frequently two markers (genes, RFLP sites, etc.) are inherited together and how frequently they are separated by a crossover event. The genetic markers in this map can be identified genes, RFLP sites, and other molecular markers (Figure 8.13). It is estimated that about 3000 markers will be required to produce maps at this level of resolution.

The second phase is the development of a physical map of each chromosome. A physical map shows the actual distance between genetic markers on a chromosome expressed in base pairs of DNA. In humans, the physical distance that corresponds to a centimorgan varies somewhat in different regions of the genome (depending on the frequency of crossovers), but on average, 1 cM is equivalent to about 1 million base pairs of DNA.

Physical maps can be represented in several ways; a cytogenetic map records the location of DNA fragments relative to visible locations or landmarks on the chromosomes. Other physical maps are long-range restriction maps, showing the location of restriction sites at intervals of several million base pairs of DNA. The goal is to produce integrated physical maps for each chromosome with markers spaced at a distance of about 2 million base pairs of DNA (Figure 8.14).

The third stage of the Human Genome Project is the development of a set of overlapping clones that cover the length of each chromosome. The problem is not the generation of such clones; a complete genomic library contains clones covering the entire genome. The difficulty is in identifying clones that overlap with each other and that have continuity over long stretches of DNA. There are still several technical problems to be overcome in establishing collections of overlapping clones. The use of vectors that act as artificial chromosomes when grown in yeast cell hosts (yeast artificial chromosomes, or YACs) allows long segments of human DNA to be cloned, and these can be used as starting material for this collection.

The final stage of the project will be to determine the nucleotide sequence of the entire human genome, using the overlapping clones as a starting point. The development of new technologies for determining DNA sequence will be required to complete this phase of the project, and even with significant advances it is expected that this phase may require most of a decade to complete. It is hoped that the project can be completed by 2005 at a cost of about $3 billion (about $1 per nucleotide).

The implications of this project are far-reaching, and it will have an impact on clinical medicine, genetic counseling, basic research, and treatment for the more than 4000 genetic disorders that affect humans. If a genetic disorder is identified and mapped to a chromosome region, for example, the DNA sequence of that region can be retrieved from computer storage and scanned to identify all genes encoded in this chromosome segment, producing a number of candidate genes. These candidates can be screened by mapping against nearby markers, the gene responsible for the defect can be identified, its amino acid sequence can be determined, and the function of the gene can be deduced, all from information stored in computer data bases.

Information from the project is also expected to have an impact on the identification of genes involved in conditions such as adult-onset diabetes,

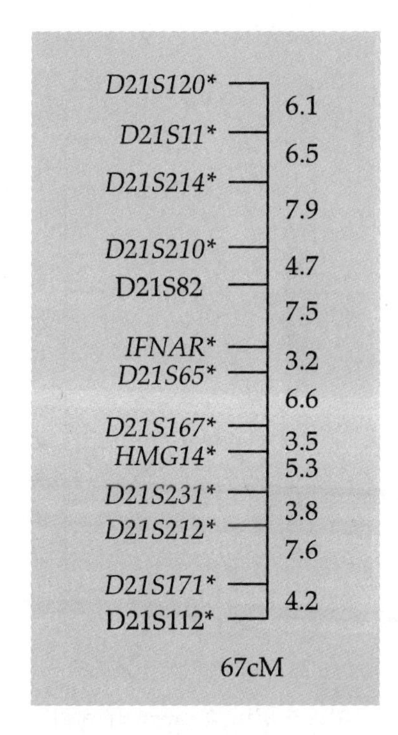

FIGURE 8.13 A genetic map of human chromosome 21. The vertical line represents the chromosome, and the markers are shown at the left. On the right, the distance between markers is shown in centimorgans (cM). The average map for chromosome 21 is 67 cM. Thus, a genetic map shows the order and distance between markers.

some forms of cancer, mental illness, and high blood pressure, where several genes may be involved, and where familial tendencies interact with one or more environmental factors.

The Human Genome Project has raised a number of related legal and ethical issues. For example, how should information about an individual's genetic condition be stored and used? Who should have access to such information? What about the ability to predict the future genetic health of an individual and the gaps between the ability to diagnose and the ability to treat such disorders? Within the Human Genome Project, a program has been established to identify and discuss such issues. This program will consider the implications for individuals and society of mapping and sequencing the human genome, examine the possible outcomes and use of such information, conduct public discussion of the issues, and develop policy options to ensure that the information generated by the project is used for the benefit of individuals and society. Some of these issues will be considered in Chapters 19 and 20.

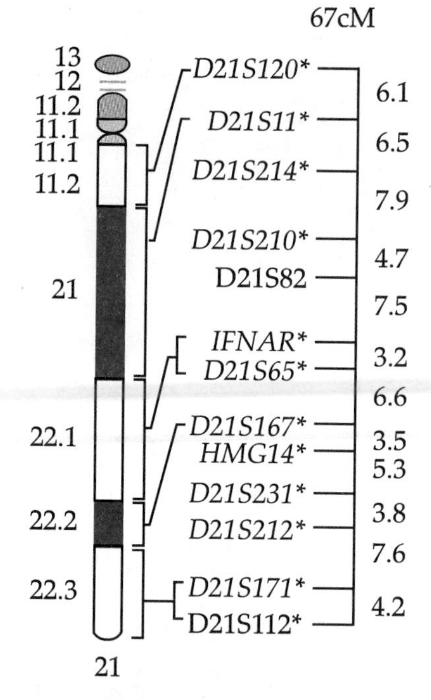

FIGURE 8.14 A physical and genetic map of human chromosome 21. The physical map and map regions are shown to the left, and the genetic map (67 cM in length) is shown to the right. The connecting lines show the correlation between the two maps. For the human genome project, maps with markers spaced at about 2–3 cM will be required.

Diagnosis of Genetic Disorders

As described in Chapter 6, the most widely used methods for prenatal diagnosis of genetic disorders are amniocentesis and chorionic villus sampling. Used in combination with these techniques, recombinant DNA technology can be used to perform direct prenatal genotypic analyses. At present, over 200 genetic disorders can be prenatally diagnosed using recombinant-DNA-based testing.

Sickle cell anemia is an example of a disorder that can be diagnosed prenatally. As we have seen, sickle cell anemia is an autosomal recessive disorder found most often in individuals with ancestral origins in West Africa and the low-lying areas around the Mediterranean Sea. The mutation affects beta globin, one of the protein components of hemoglobin, the oxygen-carrying protein found in red blood cells. Methods of prenatal diagnosis that rely on the detection of the gene product are not effective in this case, because beta globin is normally not produced until a few days after birth. However, the mutation associated with sickle cell anemia changes a single base within the gene and destroys a cutting site for a restriction enzyme, changing the length of restriction fragments (Figure 8.15). This alteration in the restriction fragment pattern allows the direct analysis of genotypes (Figure 8.15).

Prenatal diagnosis can be performed by amniocentesis in the 15th week of development or by chorionic villus sampling in the 8th week of development. Using recombinant DNA technology, a newly developed method allows prenatal diagnosis to be made at 3 to 4 days after fertilization. In this method, called preimplantation genetic testing, human ova are fertilized *in vitro* and allowed to develop in a culture dish for three days. Then, one of the 6 to 8 embryonic cells is removed by microdissection, and the DNA in this single cell is extracted and analyzed to test for a genetic disorder. DNA from the single cell is amplified by the PCR reaction and used to determine the genetic status of the embryo. To date, the method has been used successfully to test for the most common form of cystic fibrosis; embryos not affected with the disorder are transplanted for pregnancy and delivery.

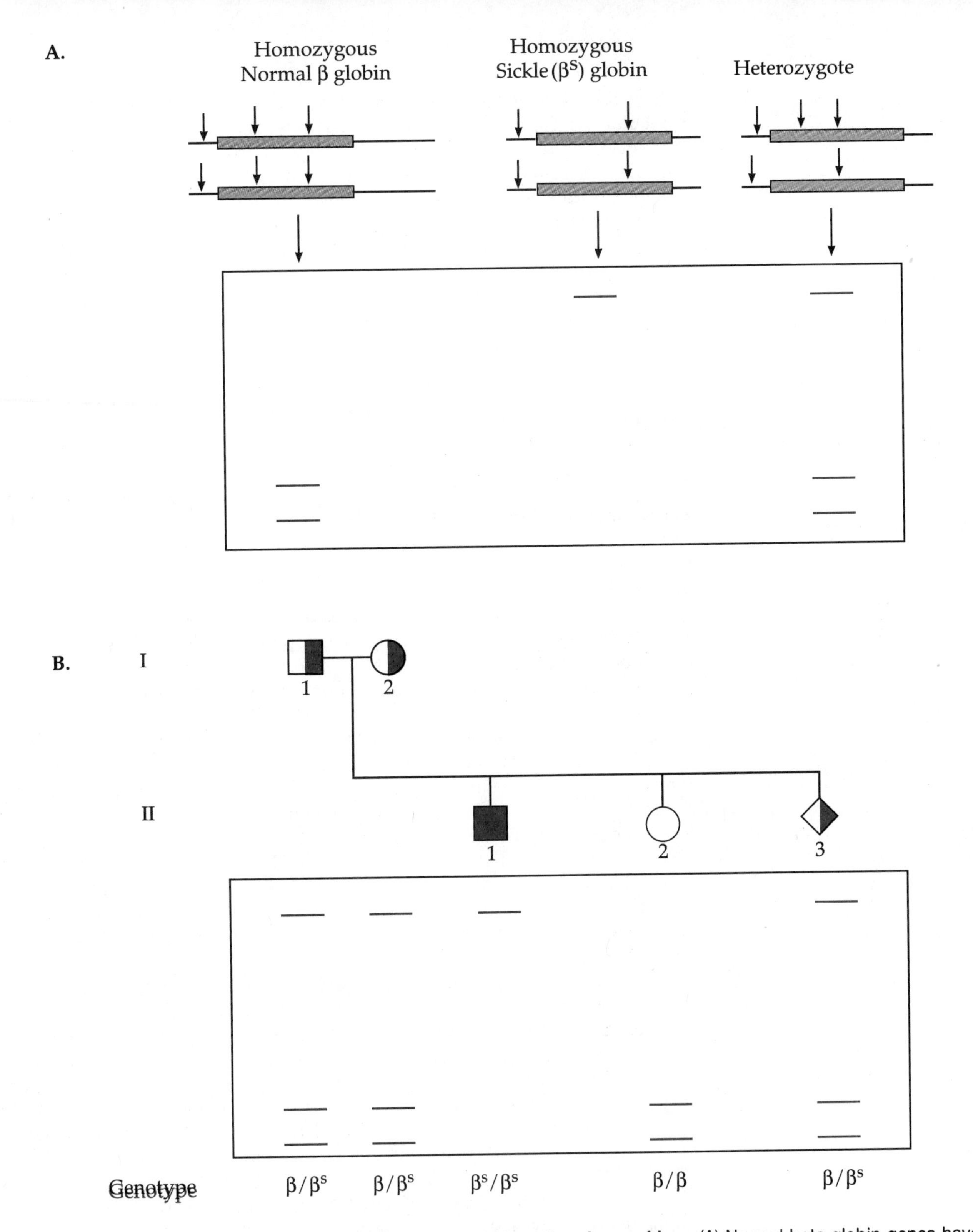

Genotype analysis by restriction fragment length polymorphism. (A) Normal beta globin genes have three restriction enzyme cutting sites (marked by arrows). The mutant (β^5) globin gene contains a single nucleotide change that destroys one cutting site. Fragments from homozygous normal, homozygous sickle cell, and heterozygous individuals can be distinguished from one another by size and number of fragments when separated by gel electrophoresis. (B) In genotypic analysis, DNA is extracted from blood cells or amniotic fluid cells, cut with a restriction enzyme, and the separated fragments of the beta globin genes are detected and analyzed. From the pattern of fragments, the genotypes can be determined. In this family, the parents are both heterozygotes, there is one affected son (II-1), an unaffected homozygous daughter (II-2), and an unaffected, heterozygous fetus (II-3).

Other disorders that can be detected by preimplantation testing include muscular dystrophy, Lesch-Nyhan syndrome, and hemophilia.

Another technique often used in prenatal diagnosis relies on RFLP analysis for genotype prediction. This analysis is used in cases where a genetic disorder has been mapped to a specific region of a chromosome but the gene has not yet been isolated. For several years, cystic fibrosis and Huntington disease were among the disorders diagnosed in this way. Currently, conditions including such as polycystic kidney disease (PCK) is diagnosed in this manner. If RFLP sites map to the same region and are thought to be very close to the gene in question, these markers can be used to predict the genotype of the fetus. Because of possible crossovers between the RFLP and the locus of the disease gene, however, this method is often less certain than a direct analysis of the gene. The uncertainties of genetic screening and its implications will be discussed further in Chapter 19.

DNA Fingerprints

The technique of RFLP analysis was developed in the mid-1970s as a method to study the organization of genes in viruses and yeasts. In the intervening years RFLP analysis has been used not only in mapping human genes but also in many other fields, including forensics and criminal justice, ecology, archaeology, preservation of endangered species, and even dog breeding.

A version of RFLP mapping developed by Alec Jeffreys and his colleagues at the University of Leicester detects variability in short, repeated DNA sequences at 8 to 15 loci scattered on different chromosomes. At these loci, called *variable number tandem repeats* (VNTR), the length of the DNA fragments produced by digestion with restriction enzymes depends on the number of repeats present. At each locus, the number of repeats can vary, ranging from two or three to several hundred. Because there are many different VNTR loci, with possible variation in the number of repeats at each locus, the resulting pattern of bands is unique to each individual (except for identical twins), and Jeffreys coined the term "DNA fingerprint" to describe this pattern of DNA fragments (Figure 8.16). It is this high-resolution version of RFLP analysis that was used in the immigration case described at the beginning of the chapter, and this is the method most commonly used in criminal cases.

For DNA fingerprinting in forensics, DNA is extracted from biologic material gathered at the scene of a crime. This can include blood, semen, hair, and skin tissues. The DNA is cut with restriction enzymes, and the resulting fragments are separated by size on a gel. The pattern of fragments obtained from this DNA is then compared with DNA fingerprints of the victim and suspects in the case. Using the probes developed by Jeffreys, the chances that any two individuals (again with identical twins being the exception) will have the same pattern of fragments can be less than one in a hundred billion (Figure 8.17).

To illustrate the power of DNA fingerprinting, consider a case in England involving a search for the individual who raped and killed two teenage girls. DNA fingerprints from over 4000 men were analyzed and led

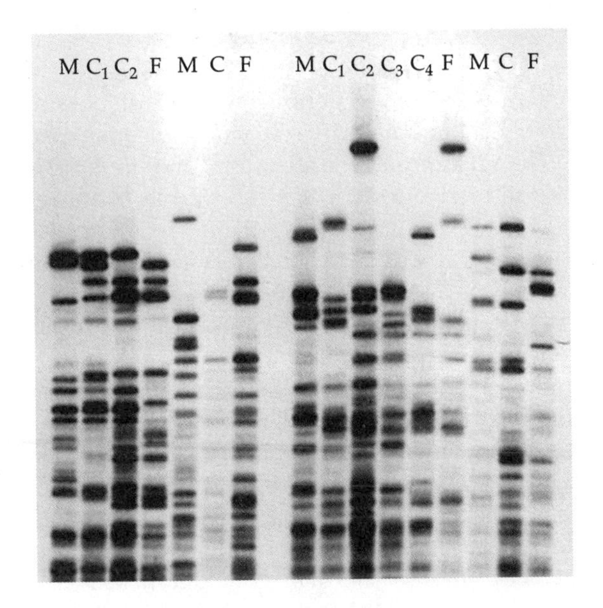

FIGURE 8.16 DNA fingerprints. The bands derived from variations in restriction enzyme cutting sites produce a unique pattern known as a DNA fingerprint. DNA fingerprints for four families are shown with mother (M), children (C), and fathers (F) identified as indicated. Note that band patterns for each individual are unique.

to a confession by the killer. In addition, the DNA fingerprints freed an innocent man jailed for the crimes. (This case was used as the plot for a best-selling novel entitled *The Blooding,* written by Joseph Wambaugh.)

The first person convicted through the use of DNA fingerprinting in the United States was tried in 1987. Since then, DNA analysis has been used in more than 1000 criminal cases. The introduction of DNA fingerprints as evidence has been somewhat controversial, and questions have been raised about the frequencies of VNTR alleles in ethnic populations and subpopulations. (These frequencies are used to calculate the probability of identity.) These issues were addressed in a 1992 report by a committee of the U.S. National Academy of Sciences, but they have not been resolved to the satisfaction of all members of the scientific community. Once the scientific and legal issues are settled, it is likely that this powerful technique will be widely used in the legal system.

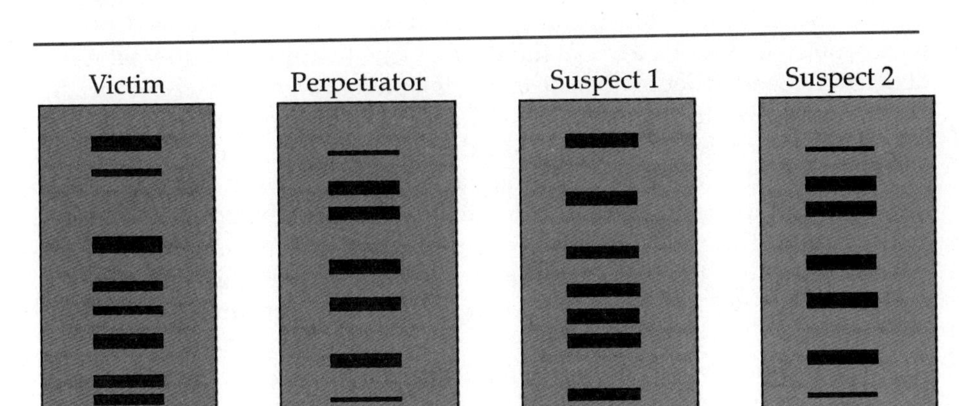

FIGURE 8.17 Diagram of DNA fingerprinting in forensic application. The DNA fingerprint of the victim was obtained from a blood sample and that of the perpetrator from evidence at the crime scene. The DNA fingerprints of two suspects are shown at right. Suspect 2's DNA fingerprint matches the perpetrator's.

Gene Therapy

The availability of cloned human genes has made it possible to consider repairing genetic defects in humans by replacing defective genes with copies of normal genes. The idea of gene surgery, or gene therapy, is not new, but only recently has the technology become available to make this dream a reality.

The use of gene therapy to treat hereditary disorders is an extension of other therapies already in use. In disorders such as PKU and galactosemia, a normal phenotype can be produced by altering the diet of the affected individual. In other disorders, replacing a gene product is an effective treatment. In hemophilia, defective blood clotting is treated by replacing an absent or defective clotting factor with the normal clotting factor. Similarly, diabetes is treated by administering the missing gene product, insulin. The rationale for gene therapy is that instead of treating a heritable disorder with gene products, the defect can be corrected with the gene itself, representing a one-time treatment that cures or greatly relieves the condition.

The type of gene therapy we will consider in this section is called *somatic cell gene therapy,* because the target cells for gene replacement do not include the germ cells or the cells that give rise to germ cells. The benefits of somatic cell gene therapy are limited to the individual being treated and do not extend to future offspring. A brief description of *germ cell gene therapy* is included at the end of this section.

One of the technical problems in gene therapy is an effective method for transferring genes into human cells. A number of methods have been developed, including encapsulation of DNA in lipids to aid transfer across the cell membrane and physical methods such as microinjection or the shooting of DNA-coated microprojectiles into cells. At this stage of gene therapy, the use of viral vectors, especially a class of viruses called retroviruses, is well advanced and is the method of transfer used in the first gene therapy trials.

The first use of gene therapy to cure a genetic disorder involves a recessive condition called adenosine deaminase (ADA) deficiency (Figure 8.18). Individuals lacking this enzyme are born with a disorder of the immune system called severe combined immunodeficiency syndrome (SCID), and they lack a functional immune system. As a result, they die from what would otherwise be trivial infections. Gene therapy trials to treat ADA deficiency began in September 1990; the first patient was a 4-year-old girl. In this treatment, white blood cells called T lymphocytes were isolated from her blood and multiplied by growing in culture. These cells were then infected with a modified virus carrying the human ADA gene, which transferred the gene to the lymphocyte nucleus. After further testing to ensure that the ADA gene was active, the modified cells were infused into the patient's circulatory system. Because of the limited life span of T lymphocytes, this process must be repeated every 3 to 5 months. A second patient began treatment in January 1991, and both have responded well to this form of gene therapy.

Other delivery systems for gene therapy are being developed, and in the near future it may be possible to select from a number of methods to optimize the transfer of genes into target tissues. One such system currently

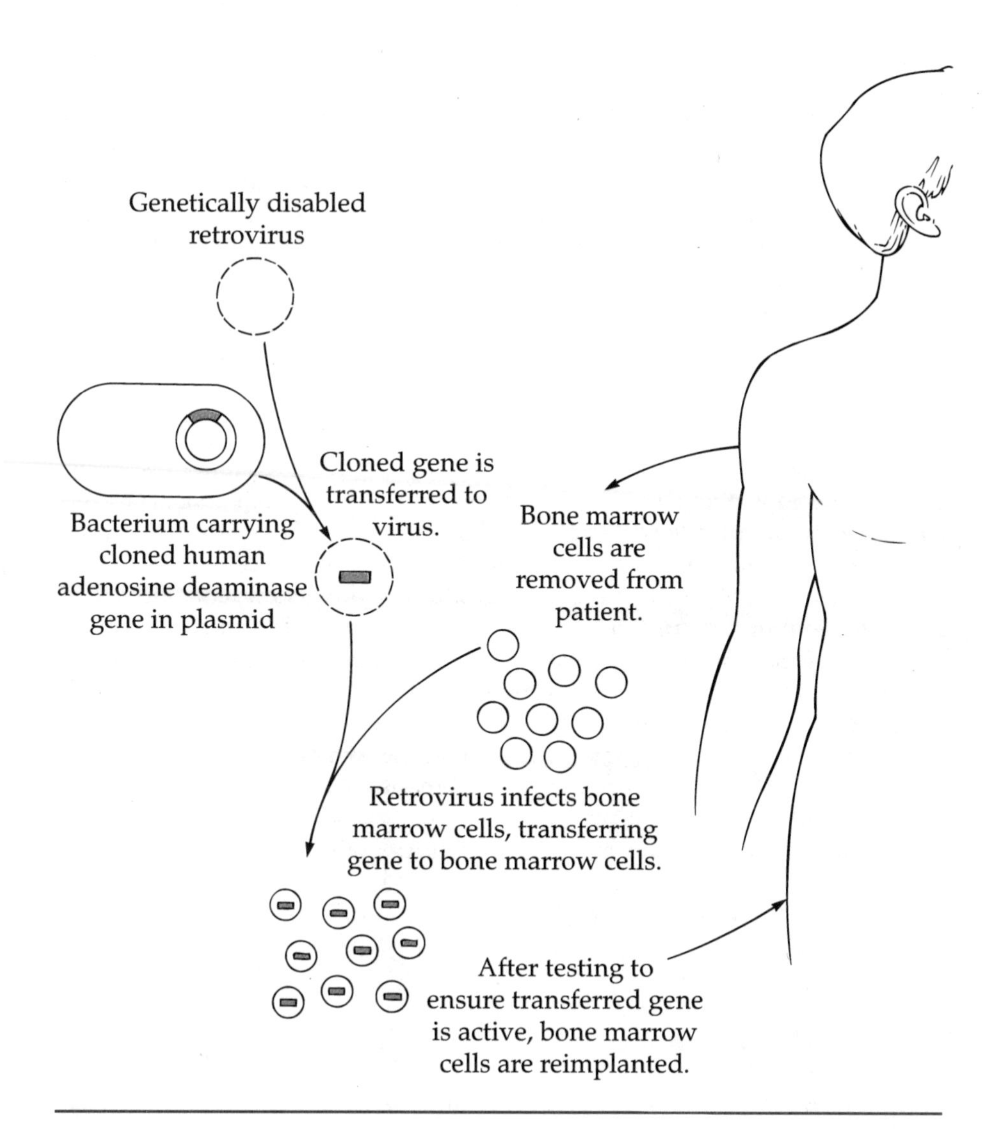

FIGURE 8.18 One approach to gene therapy involves an autologous bone marrow transplant. Bone marrow is removed from an individual affected with a genetic disorder such as severe combined immunodeficiency, caused by an absence of the enzyme adenosine deaminase. The human gene for adenosine deaminase is transferred into a viral vector, which is used to infect the bone marrow cells, transferring the adenosine deaminase gene. Following tests to determine that the gene has been transferred and is active, the bone marrow is reimplanted into the patient.

being tested in clinical trials uses aerosols to deliver genetically altered viruses to lung cells. The heritable form of emphysema is an autosomal recessive condition associated with a deficiency of the enzyme alpha-1-antitrypsin (AAT). Affected individuals develop shortness of breath and a degenerative form of lung disease leading to death between the ages of 35 and 50 years. The adenovirus, which infects lung cells and causes colds, was modified to produce a vector that carries an inserted AAT gene. In gene therapy trials, the virus is transferred to lung cells in an inhalant mist. The modified adenovirus can infect cells but cannot reproduce and cause illness; however, it will carry the AAT gene into the cells, where it will become active, producing AAT. The noninvasive procedure can be repeated as necessary to relieve symptoms.

Another disorder that is a candidate for inhalant therapy is cystic fibrosis (CF). This autosomal recessive condition is associated with lung congestion and excessive mucus production. (See Chapter 4 for a detailed description of cystic fibrosis.) Trials to transfer the cloned CF gene into lung cells by aerosols are now under way.

Although the number of gene therapy trials is growing rapidly, these experimental forms of treatment are not undertaken at random. Proposals for the use of gene therapy receive several levels of review and must be approved by the NIH. At the NIH, the review panels include scientists, lawyers, ethicists, and nonscientists. The recommendations of these committees are sent to the agency's director, who must approve the procedure. At the hospital or medical center where the gene therapy takes place, a local review board must approve the procedure, and it monitors the trials to protect the interests of the patient.

Although somatic cell gene therapy is widely accepted and most of the ethical questions about its use have been resolved, no such consensus has been reached about germ line gene therapy or a form of gene therapy called gene enhancement therapy. In germ line therapy, the target cells are the germ cells or their parental cells, permanently altering the genetic constitution of future generations. At present, such therapy is regarded as unethical, but a number of legal and ethical questions remain unresolved. For example, do parents have a right to have normal children, unaffected by a genetic disorder, if the technology is available?

Similarly, what about gene enhancement, in which gene therapy or the use of gene products produced by recombinant DNA technology might be used not for the treatment of a life-threatening disorder but as a form of self-improvement? For example, should children who are very short receive growth hormone (or a growth hormone gene) to allow them to grow to normal height? What about the administration of such gene products or genes to athletes?

BIOTECHNOLOGY

The biotechnology industry was born in 1976 when a company named Genentech was formed to develop commercial applications of recombinant DNA techniques. Genentech became a public company in 1980 with the sale of stock. By mid-1987, some 350 to 400 biotechnology companies had been formed in the United States, and sales from genetically engineered products reached $150 to $300 million. By 1995 sales from these products are expected to reach $1-2 billion annually.

About 25% of the biotechnology companies are engaged in the development and marketing of human gene products, with over 80 different drugs and chemicals under investigation. Gene splicing is used to manufacture compounds that are naturally produced in the body in minute quantities. Many of these compounds are useful in treating genetic deficiencies or other diseases such as cancer or heart disease but are otherwise unavailable in large quantities. Recombinant DNA technology is also being used to produce large quantities of proteins with other applications (see Concepts and Controversies, page 225).

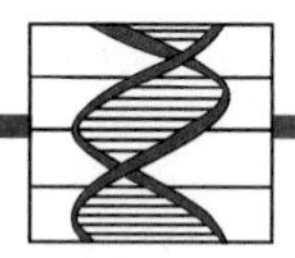

Concepts & Controversies

Sheep Shearing With Mouse Saliva

In sheep-raising regions of the world, including parts of Australia, an annual ritual involves herding the sheep into pens and removing the wool fleece with shears. Because of labor costs, shearing represents about 23% of the cost of producing wool, and skilled shearers are always in demand. In an effort to reduce the cost of producing wool, the Australian Commonwealth Scientific and Industrial Research Organization is working with a pharmaceutical firm on a new approach to harvesting wool.

One of the results of recombinant DNA technology is the ability to produce large amounts of specific proteins, even hormones and growth factors, usually present only in trace amounts. This technology has been applied mainly to produce proteins for the treatment of human diseases, ranging from genetic disorders to heart attacks. As this new application of genetic engineering suggests, ingenious solutions to old problems are also a by-product of this technology.

In the case of sheep shearing, the protein involved is called epidermal growth factor (EGF), originally identified in the saliva of male mice. As the name indicates, this hormone-like substance stimulates the growth of epidermal cells. It does so by binding to the surface of cells and triggering a cascade of intracellular events. The gene for mouse EGF has been cloned and transferred to a host in which the gene is expressed, allowing the recovery of large amount of this protein. When sheep are treated with a dose of EGF, the growth factor interrupts the growth of wool fiber for about 24 hours as hair-follicle cells briefly switch from wool synthesis to cell division. As the wool grows out over the next four to six weeks, the portion of the fiber that was in the hair follicle during treatment contains a weak point. Because of this weak point, the fleece can simply be peeled off the sheep's body without the need for shearing.

Biological wool harvesting, as the process is called, is undergoing trials in Australia, and if it is successful over the next two years, it may go into commercial use. One of the goals of the trials is to adjust the dose and the duration of EGF treatment. The EGF treatment has been so effective that some sheep have had to be fitted with body-sized hair nets to keep their fleece from falling out. So far, shearers harvesting wool by peeling are able to handle more than twice the normal number of sheep in a day, reducing the overall cost of production.

Genes from a variety of organisms have been transferred to plants to make them more resistant to infection and insect pests. In April 1987, field testing began on genetically altered bacteria that prevent frost damage on plants. The development of such products, however, has been fraught with unexpected difficulties (see above Concepts & Controversies). In the following sections commercial applications of biotechnology will be briefly reviewed.

Production of Human Proteins

Insulin is one of a number of gene products used to treat human genetic disorders. Others include growth hormone, blood-clotting factors, and various pituitary hormones. Insulin, used in the treatment of diabetes, has been obtained from animal sources for over 60 years. Other gene products are

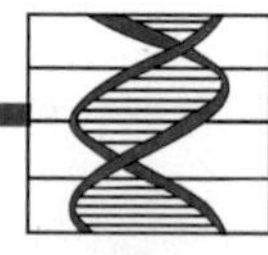

Concepts & Controversies

Recombinant DNA and the Environment

In 1981 Steve Lindow and his colleagues at the University of California Berkeley discovered that frost damage in many crop plants is caused by the presence of bacteria on the leaf surface. The bacteria, belonging to the genus *Pseudomonas,* manufactures a protein that initiates the formation of ice crystals when the temperature falls below 32°F. A mutant strain of *Pseudomonas* isolated in the laboratory was unable to produce this protein, and plants that carried the mutant bacteria on leaf surfaces suffered no frost damage, even when the temperature fell some 10 to 12 degrees below freezing.

In 1982, workers in Lindow's laboratory used recombinant DNA techniques to delete a segment of the gene for the ice-forming protein, permanently inactivating the gene. Because the bacterium with the deleted gene was created by genetic engineering, approval for field testing its ability to prevent frost damage was requested from the National Institutes of Health Recombinant DNA Advisory Committee, and in June 1983, approval was received. However, before field testing could begin, public protests against such testing were held, and a lawsuit was filed to prevent the use of the genetically engineered bacterium outside the laboratory. Opponents to testing expressed concern that release of such bacteria might bring on an unforeseen ecologic disaster.

After 4 years of litigation and negotiation, field trials of Frostban were approved. In April 1987, the Berkeley investigators planted potatoes near Tulelake, California, and sprayed them with the modified bacteria about a month later. The results indicate that there is an 80% improvement in frost resistance and that the bacteria have not spread beyond the area of application. Similar tests on strawberry plants were conducted by Advanced Genetic Systems, a biotechnology company, in fields near Brentwood, California. The results of these tests also demonstrated that these modified bacteria would be effective in reducing the multibillion-dollar cost of frost damage on agricultural crops. This good news is overshadowed, however, by the continuing debate about the safety of releasing genetically engineered organisms into the environment.

One of the main concerns seems to be whether such organisms pose different hazards than organisms modified by other genetic techniques. A panel of the National Academy of Sciences has studied this question and has concluded that such organisms present no special hazards and can be safely used outside the laboratory. Such use is regulated by the U.S. Department of Agriculture and the Environmental Protection Agency, and to date no problems with recombinant organisms have been recorded. However, public resistance to the use of genetically engineered organisms has delayed the use of Frostban, has required lengthy field testing and education programs, and has hampered the development of other altered bacteria as agricultural products by other biotechnology companies.

collected from human sources. Growth hormone can be prepared from pituitary glands dissected from human cadavers, and blood-clotting factors are extracted from donated blood serum. The use of these resources, however, has some serious limitations. Growth hormone prepared from human pituitaries is often contaminated with infectious agents responsible for a fatal degenerative neurologic disease known as Creutzfeldt-Jakob syndrome; and hemophiliacs treated with blood-clotting factors extracted from pooled blood serum are exposed to infection by the human immunodeficiency virus (HIV) associated with AIDS. Recombinant DNA technology makes it possible to obtain large quantities of these therapeutic proteins in pure form at relatively low cost, from a known and controlled source.

Genetic engineering was used to isolate and transfer the human insulin gene into bacterial cells and activate the gene to produce insulin. By using large-scale cultures of such genetically altered bacteria, Eli Lilly and Co. has produced commercial quantities of human insulin since 1982 for use by the more than 2 million Americans afflicted with insulin-dependent diabetes. Using similar techniques, contaminant-free human growth hormone and blood-clotting factors are being produced and used in the treatment of genetic and metabolic diseases, and other human gene products with therapeutic value are rapidly becoming available (Table 8.3).

This first generation of therapeutic molecules produced by recombinant DNA technology has employed bacterial cells as the hosts for transferred human genes. A second generation of vectors and host organisms employs higher plants and animals to produce proteins for medical therapy. As an example of molecular farming, researchers have introduced human genes into sheep so that the proteins encoded by these genes are synthesized only in the mammary glands and are secreted into the milk. In this method human gene products are purified from the milk, with minimum inconvenience to the sheep. The human genes used in these experiments include factor IX, a blood-clotting protein missing in some forms of hemophilia, and alpha-1-antitrypsin (AAT), a deficiency of which is associated with the development of emphysema, a degenerative lung disease.

TABLE 8.3 Pharmaceutical Products Made by Recombinant DNA Technology

Gene Product	Condition Being Treated
Atrial natriuretic factor	Hypertension, heart failure
Epidermal growth factor	Burns, skin grafts
Erythropoieten	Anemia
Factor VIII	Hemophilia
Gamma interferon	Cancer
Hepatitis B vaccine	Hepatitis
Human growth hormone	Dwarfism
Insulin	Diabetes
Interluekin-2	Cancer
Superoxide dismutase	Tissue transplants
Tissue plasminogen factor	Heart attacks

Workers in Europe and in the United States are also developing genetically engineered plants to produce human gene products (Figure 8.19). Greenhouse experiments with enkephalin, a peptide produced in very low quantities by the human nervous system, indicates that human gene products can be produced and stored in plant organs such as seeds and tubers. When the plants are harvested, the proteins can be extracted and purified for marketing. In the case of one such human gene product, it has

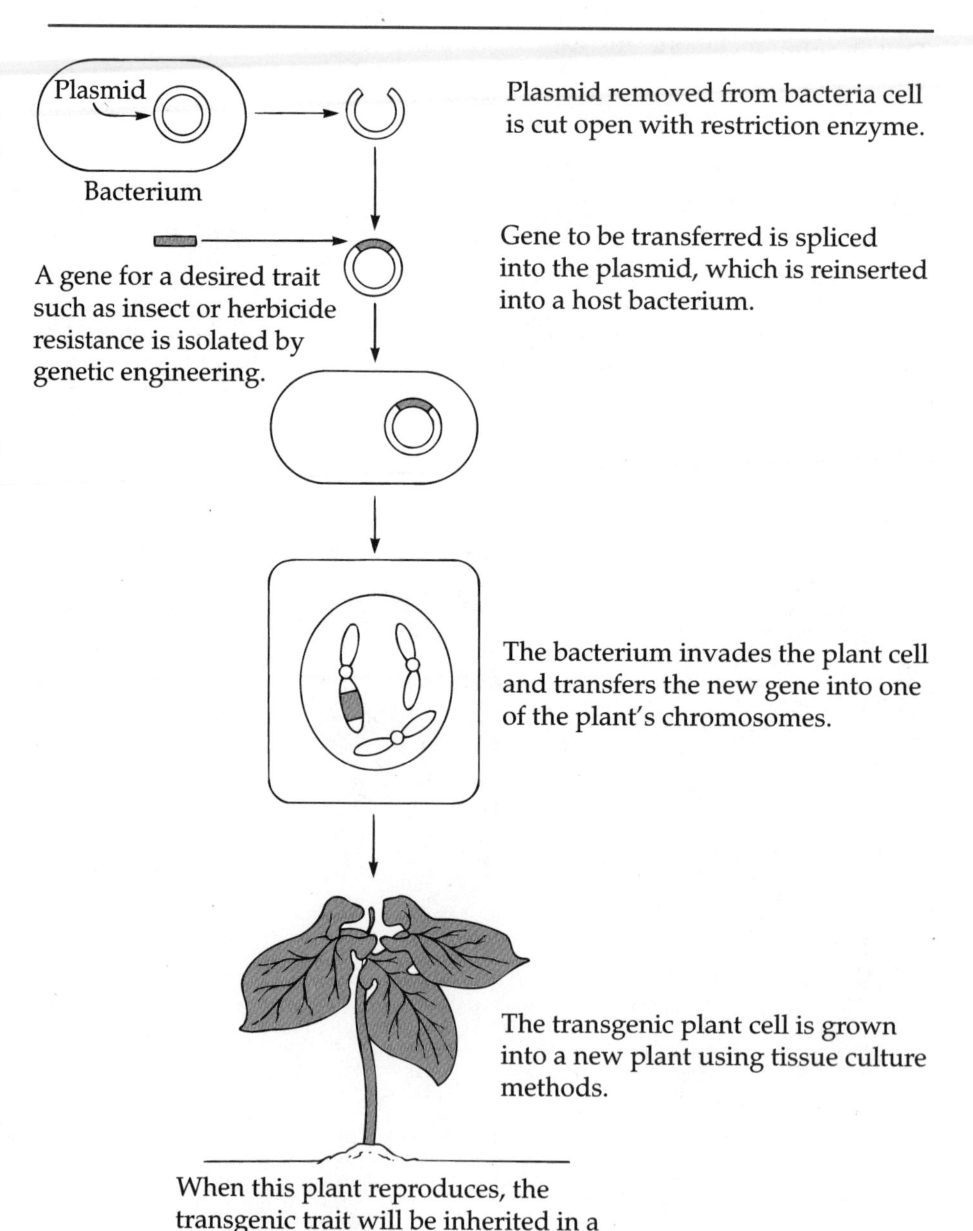

FIGURE 8.19 Directed gene transfer in plants. A plasmid vector useful in plant engineering is isolated and cut with a restriction enzyme. A gene for a desired trait such as insect resistance or herbicide resistance previously identified and isolated by genetic engineering is spliced into the plasmid. This altered plasmid is inserted into a bacterium that invades the plant cell and transfers the new gene into one of the plant's chromosomes. The transgenic plant cell is grown into a callus and then into a new plant using tissue culture methods. When the transgenic plant reproduces, the transferred trait will be passed on in a Medelian fashion.

been estimated that a 1- or 2-acre plot of engineered plants may be enough to supply worldwide demand.

New Plants and Animals

Genetic engineering has been used to transfer genes for herbicide resistance and insect resistance into many crop plants, including tomatoes, potatoes, tobacco, cotton, and alfalfa. The basic scheme involved in this process is outlined in Figure 8.19. At least three federal agencies regulate the development and use of genetically altered food plants, including the Department of Agriculture, the Environmental Protection Agency, and the Food and Drug Administration (FDA). Currently some 45 genetically engineered plants have been approved for field trials, and several more applications are pending. Genetically altered tomato plants have been developed, and tomatoes with improved flavor and ripening characteristics will probably reach the market before 1995.

Tomatoes on the vine become soft because of the action of a gene that becomes activated during the ripening process. Soft, ripe tomatoes do not ship well, so supermarket tomatoes are picked while hard and green, shipped, and treated with ethylene gas to bring on a red color (even though the tomato is not ripe). In the genetically engineered tomato developed by Calgene, the tomato can ripen and develop flavor on the vine, but the softening process is delayed.

To accomplish this, the gene responsible for softening was cloned, and a mirror-image copy of the gene was constructed and inserted into a tomato plant. The activity of the mirror-image gene, in effect, cancels most of the activity of the softening gene, delaying the softening process so the tomatoes can be ripened and then shipped.

Tomatoes represent the first of a number of genetically engineered food plants, including cucumbers and cantaloupes, that will soon be available. The appearance of these foods in stores follows a 1992 ruling by the FDA that genetically engineered foods are to be regulated in the same way as foods produced by conventional genetic crosses.

Efforts are also under way to improve the nutritional composition of plant foods. Neither legumes, such as soybeans, nor grains, such as rice, provide a complete source of essential amino acids. Beans are low in the amino acids methionine and cysteine; grains are low in lysine. Using tobacco as a test system, researchers have successfully transferred genes that increased the methionine content of tobacco plants by 30%. Work is now in progress on a similar transfer to soybeans, and protein-enhanced soybeans, corn, wheat, and alfalfa should be available to consumers within 5 years. These crops would help to prevent nutritional diseases in regions of the world where the population depends on a single crop as a protein source.

In another bit of genetic engineering, human and cow growth hormone genes have been transferred into pigs in an attempt to develop leaner, faster-growing pigs (Figure 8.20). The genes were transferred by injection into newly fertilized eggs that were implanted into a foster mother. Although the **transgenic** (carrying a transferred gene) pigs grow faster on a high-protein diet and are leaner than normal pigs, they also have a number

Transgenic
A cell or organism whose genome has been modified by the introduction of external DNA.

FIGURE 8.20 Transgenic pig. These pigs carry and express human growth hormone genes transferred by microinjection of zygotes.

of problems, such as ulcers, arthritis, and premature death, that are the result of excessive production of growth hormone. Further work on the regulation of hormone production during the first 6 months of growth may eliminate these problems, paving the way for the appearance of *transgenic* pork in the supermarket.

SUMMARY

1. Cloning is the production of identical copies of molecules, cells or organisms from a single ancestor. The development and refinement of methods for cloning higher plants and animals represents a significant advance in genetic technology that will speed up the process of improving crops and the production of domestic animals.

2. These developments have been paralleled by the discovery of methods to clone segments of DNA molecules. This technology is founded on the discovery that a series of enzymes known as restriction endonucleases recognize and cut DNA at specific nucleotide sequences. Linking DNA segments produced by restriction enzyme treatment with vectors such as plasmids or engineered viral chromosomes produces recombinant DNA molecules.

3. Recombinant DNA molecules are transferred into host cells and cloned copies are produced as the host cells grow and divide. A variety of host cells can be used, but the most common is the bacterium, *E. coli.* The cloned DNA molecules can be recovered from the host cells and purified for further use.

4. Cloned DNA molecules are used in a variety of applications, including gene mapping, diagnosis of genetic disorders, forensic applications such as DNA fingerprints, gene therapy and the production of human gene products for therapeutic use.

5. The Human Genome Project uses recombinant DNA technology in an effort to map all of the estimated 50,000 to 100,000 genes carried in human cells, and to determine the nucleotide sequence of the 3 billion base pairs of human DNA. This project is expected to have a great impact on disease diagnosis and treatment.

6. The first clinical trials in gene therapy have begun, and although some technical barriers remain, gene therapy will probably become a common medical treatment within the next decade. Legal and ethical problems associated with the use of gene therapy and the question of gene enhancement must be resolved in parallel with the emergence of this technology.

7. Biotechnology, the commercial application of recombinant DNA methodology, has led to the large scale production of human proteins such as insulin for the treatment of diabetes, and agricultural products such as a tomato with improved flavor and ripening characteristics.

Questions and Problems

1. Restriction enzymes:
   a. recognize specific nucleotide sequences in DNA.
   b. cut both strands of DNA.
   c. often produce single-stranded tails.
   d. do all of the above.
   e. do none of the above

2. A cloned library of an entire genome contains:
   a. the expressed genes in an organism.
   b. all the genes of an organism.
   c. only a representative selection of genes.
   d. a large number of alleles of each gene.
   e. none of the above.

3. The DNA sequence below contains a 6-base palindromic sequence that acts as a recognition and cutting site for a restriction enzyme. What is this sequence? Which enzyme will cut this sequence? Consult Table 8.1.

   CCGAGTAAGCTTAC
   GGCTCATTCGAATG

4. You are given the task of preparing a cloned library from a human tissue culture cell line. What type of vector would you select for this library and why?

5. RFLP sites are useful as genetic markers for linkage and mapping studies as well as disease diagnosis. Keeping in mind that RFLPs behave as genes, what factors are important in selecting RFLPs for use in such studies?

6. In the examples given in this chapter (Figure 8.11), the RFLP is always inherited with (linked to) the genetic disorder. In practice, however, it is sometimes observed that in some progeny the RFLP is not linked to the genetic disorder, reducing its effectiveness in detecting carrier heterozygotes. How do you explain this observation, and what circumstances are likely to affect the frequency with which this nonlinkage occurs?

7. In selecting target cells to receive a transferred gene in gene therapy, what factors must be taken into account?

For Further Reading

Anderson, W. F. 1992. Human gene therapy. Science **256**: 808-813.

Caskey, C. 1993. Presymptomatic diagnosis: a first step toward genetic health care. Science **262**: 48-49.

Collins, F. and Galas, D. 1993. A new five-year plan for the U.S. human genome project. Science **262**: 43-46.

deWachter, M. 1993. Ethical aspects of human germ-line therapy. Bioethics **7**: 166-177.

Friedman, T. 1989. Progress toward human gene therapy. *Science* **244:** 1275–1281.

Gilbert, W., and Villa-Komaroff, L. 1980. Useful proteins from recombinant bacteria. *Sci. Am.* (April) **242:** 74–97.

Harris, J. 1992. *Wonderwoman and Superman: The ethics of human biotechnology.* New York: Oxford University Press.

Hubbard, R. and Wald, E. 1993. *Exploding the Gene Myth.* Boston: Beacon Press.

Kantoff, P. W., Freeman, S. M., and Anderson , W. F. 1988. Prospects for gene therapy for immunodeficiency diseases. *Ann. Rev. Immunol.* **6:** 581–594.

Keveles, D. and Hood, L. 1992. *The Code of Codes: Scientific and Social Issues in the Human Genome Project.* Cambridge MA: Harvard University Press.

Kohn, D. B., Anderson, W. F., and Blaese, R. M. 1989. Gene therapy for genetic diseases. *Cancer Investig.* **7:** 179–192.

Marx, J. L. 1987. Assessing the risks of microbial release. *Science* **237:** 1413–1417.

Morsy, M., Mitani, K., and Clemens, P. 1993. Progress toward human gene therapy. JAMA **270**: 2338-40.

Neufeld, P. J., and Colman, N. 1990. When science takes the witness stand. *Sci. Am.* (May) **262:** 46–53.

Nichols, E. K. 1988. *Human Gene Therapy.* Cambridge, Mass: Harvard University Press.

Olson, M. V. 1993. The human genome project. Proc. Nat. Acad. Sci. **90**: 4338-4344.

Porteus, D. and Alton, E. 1993. Cystic fibrosis: prospects for therapy. Bioessays **15**: 485-486.

Pursel, V. G., Pinkert, C. A., Miller, K. F., Bolt, D. J., Campbell, R. G., Palmiter, R. D., Brinster, R. L., and Hammer, R. E. 1989. Genetic engineering of livestock. *Science* **244:** 1281–1288.

Roberts, L. 1989. Ethical questions haunt new genetic technologies. *Science* **243:** 1134–1135.

Watson, F. 1993. Human gene therapy—progress on all fronts. Trends in Biotechnol. **11**: 114-117.

Weatherall, D. J. 1989. Gene therapy: Getting there slowly. *Br. Med. J.* **298:** 691–693.

White, R., and Lalouel, J.-M. 1988. Chromosome mapping with DNA markers. *Sci. Am.* (February) **258:** 40–47.

CHAPTER 9

GENE ACTION

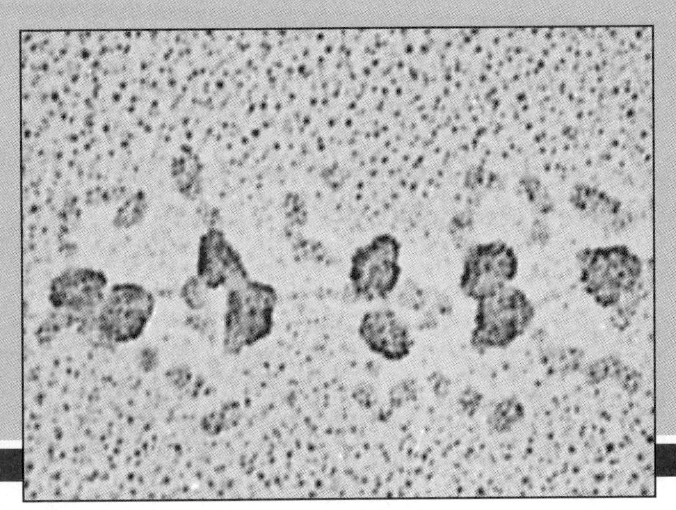

CHAPTER OUTLINE

After a month of feeding on mulberry leaves, the caterpillar loses interest in food, and begins to move its head in a rotating motion. From two glands in the head, a liquid secretion emerges through an opening called a spinneret. On contact with air, this secretion hardens into a thin filament. Using a figure eight movement, the caterpillar weaves this long continuous filament into a cocoon slightly larger than a peanut. If left undisturbed, the caterpillar will develop into an adult moth, called *Bombyx mori*. This continuous filament is a protein called fibroin, and some 5,000 years ago, the Japanese discovered how to use this filament to produce the fabric called silk.

In producing silk, the end of the filament is located and unwound from the cocoon. Each filament may be up to 3000 feet in length. The filaments from several cocoons are wound together to form a silk strand. The strands are reeled into skeins, bundled together to form books of 5–10 pounds, which in turn are packed into bales of about 140 pounds. In ancient times, these bales were shipped from China over a 4,000 mile road to the shores of the Mediterranean Sea and shipped from there to Rome and other destinations in the West. Thus, the establishment of a weaving industry, trade, and part of the culture of East Asia was founded on a single animal protein. This trade continues today, with the United States being the largest consumer of silk.

This example raises several questions relevant to genetics and gene expression. What are proteins? How and where are they made? How are proteins related to genes, and in particular to human genetics? As we will see in this chapter, proteins like silk fibroin, blood clotting factors, and digestive enzymes are gene products, encoded in the nucleotide sequence of DNA. Proteins are the intermediate between genes and phenotype. The phenotypes of a cell, tissue and organism are the result of protein function. When these functions are absent or altered, the result is a mutant phenotype which we describe as a genetic disorder. To discover how a genetic disorder produces an altered phenotype, an understanding of how genes control the synthesis of gene products is necessary. Gene action in turn, begins with an understanding of the structure of DNA.

The Watson-Crick model of DNA structure is the foundation for what is now called molecular biology, or molecular genetics. The model incorporates all the requirements of the genetic material: replication, storage of information, expression of genetic information, and mutability. In the years following this discovery, research has provided details of how DNA functions as the genetic material. In this chapter we will examine the fundamental steps of DNA function in the storage and expression of genetic information. In this and the next chapter, we will be moving up from the level of DNA nucleotide sequence to cellular gene products to the production of phenotype. We have already considered how DNA replicates, and the property of mutability will be explored in Chapter 11.

Genes and Proteins

Proteins play a fundamental role in the relationship between the genetic information stored in DNA (a gene) and the ultimate expression of this information in the phenotype. This was first recognized at the turn of the century when Sir Archibald Garrod proposed a relationship between an inherited disorder and abnormal metabolism. A colleague, William Bateson, concluded that proteins acting as specific catalysts (enzymes) were involved in the production of the phenotype. However, definitive proof for this idea was not available until the 1940s, when George Beadle, Edward Tatum, and their colleagues used the mold *Neurospora crassa* to investigate the relationship between genes and biochemistry. Using *Neurospora,* they were able to show that mutation in a single gene caused loss of activity in a single enzyme. They further demonstrated that enzymes are involved in the production of phenotype. This linkage between mutant genes and mutant enzymes led Beadle and Tatum to formulate the **one gene–one enzyme hypothesis,** which postulates that each gene controls the production and function of one enzyme. Subsequent work on biochemical genetics determined that some proteins are not enzymes and that other proteins are composed of subunits, or polypeptide chains, controlled by different genes. As a result, the one gene–one enzyme hypothesis can be restated as the **one gene–one polypeptide hypothesis.** Thus the scientists of the late 19th century and early 20th century were correct in assuming that proteins are involved with genes. However, as we now know, proteins are the *products* of genes, not the genetic material.

One gene–one enzyme hypothesis
The idea that individual genes control the synthesis and therefore the activity of a single enzyme. This provided the link between the gene and the phenotype.

One gene–one polypeptide hypothesis
A refinement of the one gene–one enzyme hypothesis made necessary by the discovery that some proteins are composed of subunits encoded by different genes.

THE FLOW OF GENETIC INFORMATION

Because proteins are the products of genes, the genetic information encoded in DNA must control the kinds and amounts of proteins present in the cell. Proteins are composed of linear sequences of subunits known as amino acids, joined together by chemical bonds. Since in humans and other higher organisms genes are composed of DNA, a gene can be defined as the linear sequence of nucleotides in DNA that specifies the linear sequence of amino acids in a protein (or polypeptide chain). Another form of nucleic acid, ribonucleic acid, or RNA, serves as an intermediate step between DNA and proteins. There is a directional flow of genetic information from DNA → RNA → protein. In addition, there is the directional flow of genetic information between generations, or **DNA replication.**

DNA replication
The processes of DNA synthesis.

The expression of genetic information stored in the DNA molecule requires several steps. First, the genetic information is copied into a molecule of RNA known as messenger RNA, or mRNA. This step is known **as transcription.** In eukaryotes, including humans, the mRNA moves from the nucleus to the cytoplasm. In the cytoplasm, the information encoded in the base sequence of mRNA is converted into the amino acid sequence of a protein. This step is known as **translation.** The amino acid sequence, in turn, determines the structural and functional characteristics of the protein, which plays a role in phenotype expression. The relationship between DNA, RNA, and protein and the flow of genetic information is summarized in Figure 9.1.

Transcription
Transfer of genetic information from the base sequence of DNA to the base sequence of RNA brought about by RNA synthesis.

Translation
The process of converting the information content in an mRNA molecule into the linear sequence of amino acids in a protein.

Transcription and translation require the interaction of many components, including ribosomes, mRNA, transfer RNA (tRNA), amino acids, enzymes, and energy sources. Amino acids are the subunits of proteins; ribosomes serve as the cytoplasmic workbenches upon which protein synthesis occurs; tRNA molecules are adaptors, recognizing amino acids as well as the nucleotide sequence in mRNA, the gene transcript.

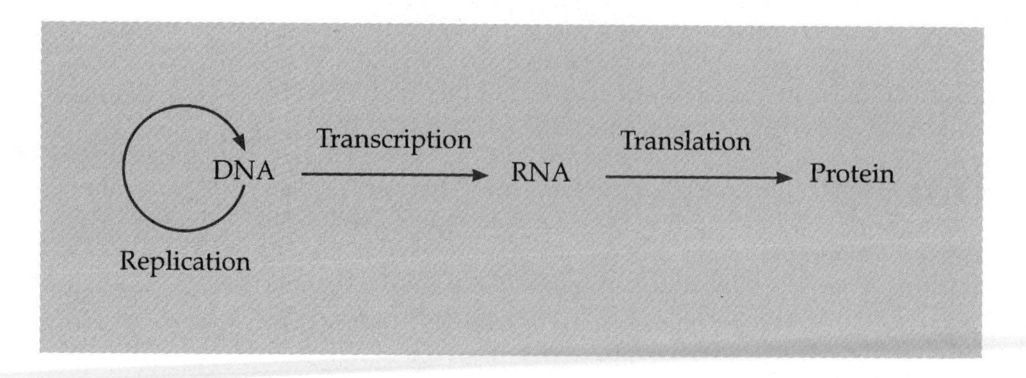

FIGURE 9.1 Information flow from DNA to protein. In this schema, information flows from DNA through RNA to protein via transcription and translation.

Amino group
A chemical group (NH_2) found in amino acids and at one end of a polypeptide chain.

Carboxyl group
A chemical group (COOH) found in amino acids and at one end of a polypeptide chain.

R group
A term used to indicate the position of an unspecified group in a chemical structure.

Peptide bond
A chemical link between the carboxyl group of one amino acid and the amino group of another amino acid.

Polypeptide
A polymer made of amino acids joined together by peptide bonds.

Components of Protein Synthesis

Amino Acids

Proteins are composed of linear chains of amino acids. Twenty different amino acids are used in protein synthesis, and they serve as the building blocks, or subunits, of proteins. The structure of each amino acid includes an **amino group** (NH_2), a **carboxyl group** (COOH), and an **R group** (Figure 9.2). The R groups are chemical side chains that are different for each amino acid and make the amino acids distinct from one another. Some R groups carry a positive charge, some carry a negative charge, and still others are neutral. Table 9.1 lists the 20 amino acids typically found in proteins and the abbreviations commonly used in referring to them. A covalent bond, known as a **peptide bond,** can be formed between the amino group of one amino acid and the carboxyl group of another amino acid (Figure 9.3). Two linked amino acids form a dipeptide, 3 form a tripeptide, and 10 or more form a **polypeptide.** Each polypeptide is a directional molecule with a free amino group at one end known as the **N-terminus** and a free carboxyl group at the other end known as the **C-terminus.**

Amino group
H, H, N, R, C, H, C, O, OH
Carboxyl group

FIGURE 9.2 Diagram of an amino acid showing the amino group, the carboxyl group, and the R group. The R groups are different in each of the 20 amino acids used in protein synthesis.

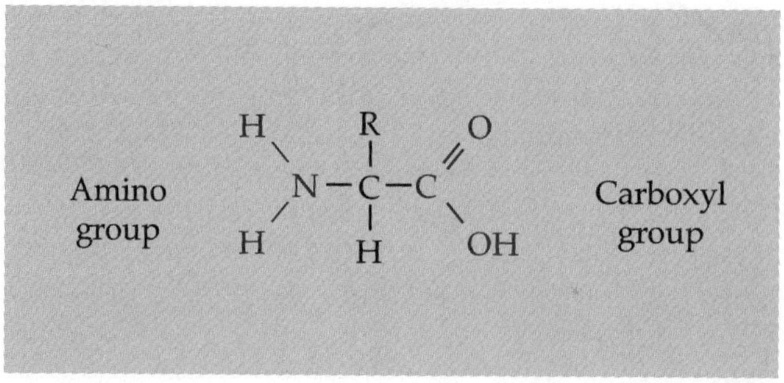

The RNA Components

Information transfer from the nucleotide sequence of DNA to the amino acid sequence of a protein involves RNA. Recall from Chapter 7 that the structure of RNA is different from that of DNA in several important respects. First, the sugar ribose is present instead of deoxyribose, and the base uracil (U) is present in place of thymine (T). In addition, RNA is, for the most part, single stranded. RNA is not self-replicating, and as a result, all forms of RNA are transcribed from DNA. There are three major classes of RNA that play a role in protein synthesis: ribosomal RNA (rRNA), messenger RNA (mRNA), and transfer RNA (tRNA).

RIBOSOMAL RNA (rRNA) AND RIBOSOMES. Over 90% of the RNA in the cell is **ribosomal RNA (rRNA).** Ribosomal RNA is transcribed from DNA and associates with specific proteins to form organelles known as ribosomes that participate in protein synthesis. A typical cell contains thousands of these organelles (Figure 9.4). A ribosome is composed of two subunits, each of which is composed of RNA and protein. These subunits associate to form a functional ribosome.

mRNA. **Messenger RNA (mRNA)** is a single-stranded complementary copy of the DNA sequence in a gene. At any given time, about 5% of the cell's RNA is in the form of mRNA. The sequence of bases in mRNA encodes the information for the sequence of amino acids in the protein product of the gene. Each gene is capable of synthesizing multiple copies of mRNA molecules. After transcription, mRNA precursor molecules are processed into mRNA, which moves to the cytoplasm, binds to ribosomes, and directs the synthesis of a specific protein. The number of different mRNAs that can be synthesized is related to the number of protein-coding genes in the genome. Each message can be reused a number of times before being chemically degraded, so that several copies of a protein can be produced from each mRNA molecule.

tRNA. About 5% of the RNA in the cell is composed of **transfer RNA (tRNA).** Transfer RNA molecules contain 75 to 90 nucleotides and are transcribed from DNA. The structure of tRNA molecules is adapted for two important functions: recognition and binding to a particular amino acid and recognition of a three-base sequence in the mRNA known as a **codon.** This duality ensures that the correct amino acid is inserted at the proper location in the protein being synthesized. The characteristics of the RNA components of protein synthesis are summarized in Table 9.2.

TRANSCRIPTION

Transcription takes place on a DNA strand and results in the production of a complementary, single-stranded RNA molecule. Transcription is accomplished by enzymes known as **RNA polymerases.** In any given gene, RNA polymerase uses only one strand of DNA as a template (the bottom strand

N-terminus
The end of a polypeptide or protein that has a free amino group.

C-terminus
The end of a polypeptide or protein that has a free carboxyl group.

Ribosomal RNA (rRNA)
One of the components of the cellular organelles known as ribosomes.

TABLE 9.1 Amino Acids Commonly Found in Proteins

Amino Acid	Abbreviation
Alanine	ala
Arginine	arg
Asparagine	asn
Aspartic acid	asp
Cysteine	cys
Glutamic acid	glu
Glutamine	gln
Glycine	gly
Histidine	his
Isoleucine	ile
Leucine	leu
Lysine	lys
Methionine	met
Phenylalanine	phe
Proline	pro
Serine	ser
Threonine	thr
Tryptophan	trp
Tyrosine	tyr
Valine	val

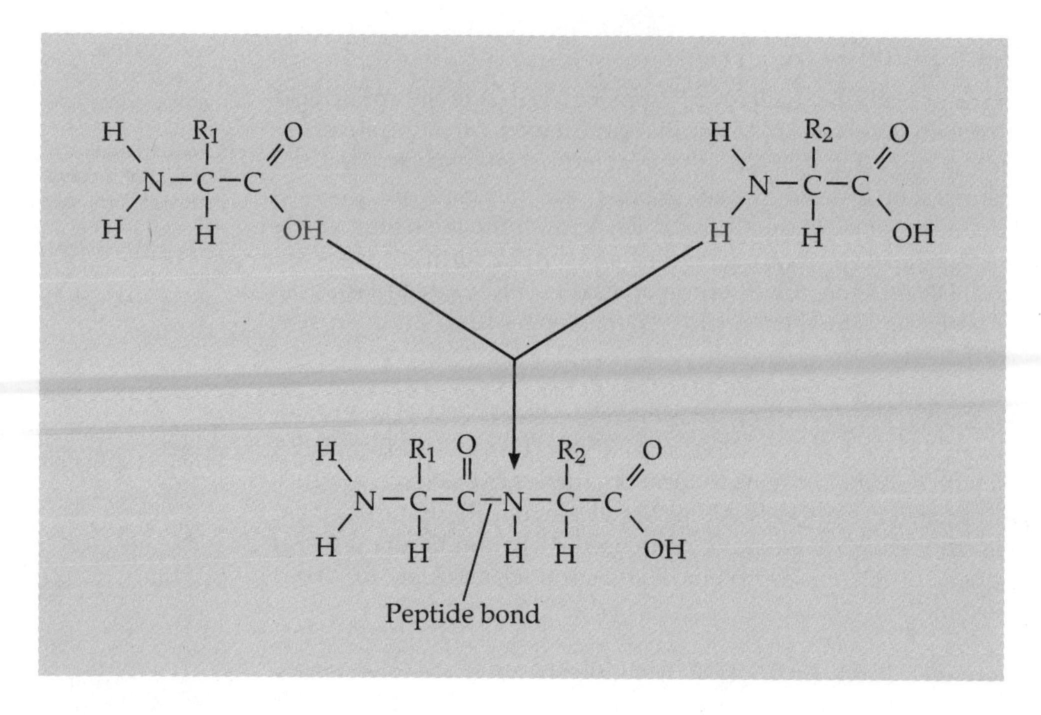

FIGURE 9.3 The chemical change that accompanies the formation of the peptide bond. An equivalent of a water molecule (H_2O) is split off during bond formation. This reaction is catalyzed by an enzyme.

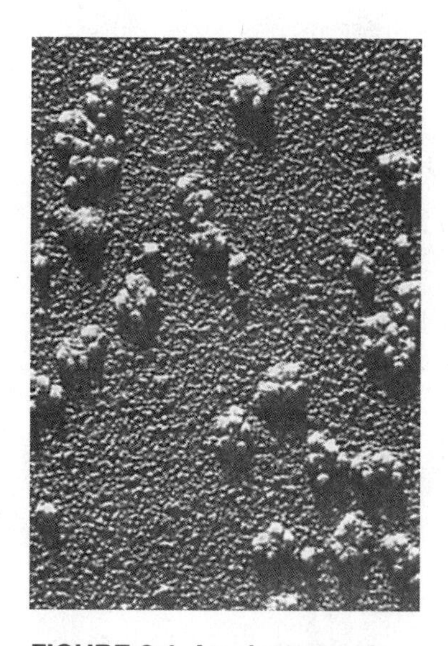

FIGURE 9.4 An electron micrograph showing polyribosomes isolated from a mammalian cell. Groups of 3–7 ribosomes are attached to mRNA to form polysomes.

in Figure 9.5). RNA polymerase uses the nucleotide sequence of a DNA strand to synthesize an RNA molecule that is complementary to the sequence in the template strand. The rules of base pairing in transcription are the same as for DNA replication with one exception: an A on the DNA template specifies a U in the RNA transcript (there is no T in RNA).

Transcription involves four stages: binding, initiation, elongation, and termination (see Figure 9.5 on page 240). Adjacent to the DNA encoding a gene is a region known as the **5′ flanking region,** which contains regulatory sequences. RNA polymerase molecules recognize and bind to specific nucleotides in the flanking region, known as a **promoter** sequence. Once the RNA polymerase identifies and binds to the promoter sequence, the separation of the two strands of DNA and the beginning of RNA synthesis rapidly follow. Initiation of transcription begins at a specific site, on the 3′ DNA template strand, several nucleotides upstream from the first nucleotide triplet that will be translated into an amino acid. Elongation takes place by the addition of nucleotides to the growing RNA strand and the movement of the RNA polymerase along the template strand of DNA (Figure 9.5C).

In humans, transcription occurs at a rate of 30 to 50 nucleotides per second until the transcript reaches another regulatory DNA region known as a *termination* sequence. The termination sequence stops the elongation process and causes the release of both the RNA transcript and the RNA polymerase from the DNA (Figure 9.5D). The length of the finished transcript

TABLE 9.2 RNA Classes

Class	Function	Number of Different Types	Size (Nucleotides)	% of RNA in Cell
Ribosomal RNA (rRNA)	Structural, functional component of ribosomes	3	120 to 4,800	90
Messenger RNA (mRNA)	Carries genetic information from DNA to ribosomes	Many thousands	300 to 10,000	3 to 5
Transfer RNA (tRNA)	Adaptor recognizes nucleotide triplets and amino acids. Transports amino acids to ribosomes	50 to 60	75 to 90	5 to 7

depends on the size of the gene. In humans, most transcripts average about 5000 nucleotides, although lengths up to 200,000 nucleotides have been reported.

In summary, transcription is the transfer of genetic information from the template strand of a DNA molecule to a complementary base sequence in an RNA molecule. Transcription is the first step in the process of gene expression.

In humans, as in other eukaryotes, transcription produces large RNA precursor molecules known as *pre-mRNAs.* These precursors are processed and modified in the nucleus to produce mature mRNAs that are transported to the cytoplasm, where they bind to ribosomes for translation. The organization of human genes and the details of mRNA processing steps will be discussed below.

The Eukaryotic Gene and mRNA Processing

Many protein-coding genes in humans have a complex internal organization and contain **introns,** nucleotide sequences that are transcribed but are not translated into the amino acid sequence of a polypeptide chain. The number of introns in genes varies from zero (histone genes) to 75 (dystrophin, the gene involved in muscular dystrophy) or more. Introns vary in size, ranging from about 100 nucleotides to more than 100,000, and little is known about their function or significance. **Exons** are the nucleotide sequences in a gene that are transcribed, joined to other exons during mRNA processing, and translated into the amino acid sequence of a polypeptide. The internal organization of a typical human gene is shown in Figure 9.6. The combination of exons and introns determines the length of a gene, and often the exons make up only a small fraction of the total length. At this point, it is difficult to say how long the average human gene is, but knowing this would allow geneticists to more accurately estimate the number of genes in the human genome. As we will see in Chapter 11, genes with long introns are particularly prone to mutations.

Messenger RNA (mRNA)
A single-stranded complementary copy of the base sequence in a DNA molecule that constitutes a gene.

Transfer RNA (tRNA)
A small RNA molecule that contains a binding site for a specific type of amino acid and a three-base segment known as an anticodon that recognizes a specific base sequence in messenger RNA.

Codon
A triplet of bases in messenger RNA that encodes the information for the insertion of a specific amino acid in a protein.

RNA polymerase
An enzyme that catalyzes the formation of an RNA polynucleotide chain using a template DNA strand and ribonucleotides.

5′ flanking region
A nucleotide region adjacent to the 5′ end of a gene that contains regulatory sequences.

Promoter
A region of a DNA molecule to which RNA polymerase binds and initiates transcription.

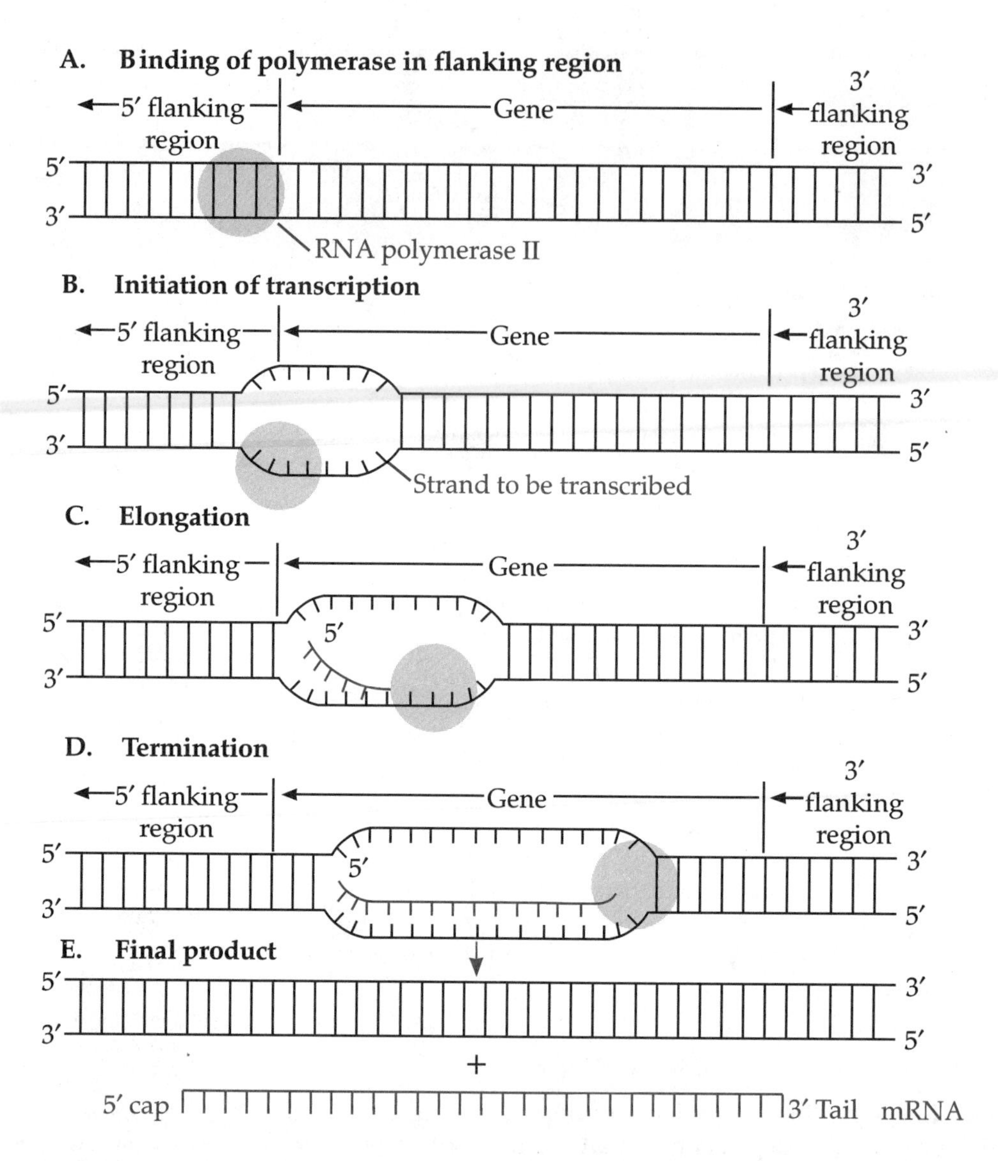

FIGURE 9.5 The process of transcription begins with binding of RNA polymerase to the upstream flanking region of a gene. As initiation begins, the DNA unwinds, and the base sequence of the DNA strand to be copied is transcribed into the complementary base sequence of the single-stranded mRNA. When a termination sequence is encountered, transcription stops, and the polymerase and the mRNA are released from the DNA, which recoils into a double-stranded structure.

Introns
Sequences present in some genes that are transcribed but removed during processing and therefore are not present in mRNA.

The regions on either side of a gene, known as flanking regions, are important in regulating gene expression. The 5′ flanking region, adjacent to the site where transcription begins, contains three control sequences that regulate gene expression. The first, known as a TATA box (named after the sequences it contains), is located some 30 bases upstream from the beginning of the gene. It is thought that the TATA sequence may help correctly position the

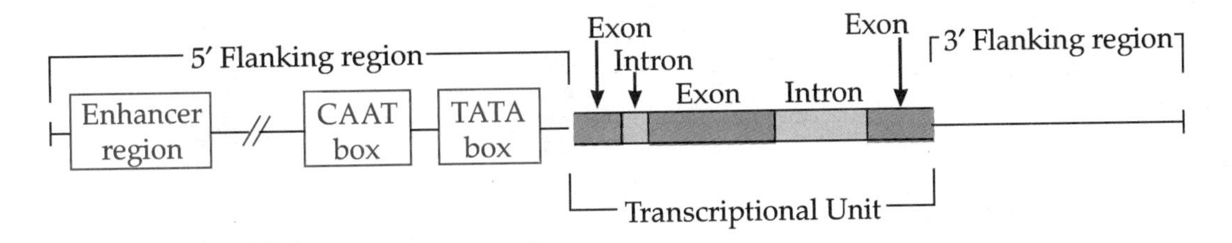

FIGURE 9.6 Organization of a typical eukaryotic gene. The transcription unit consists of all sequences, including noncoding regions, at each end that are transcribed into RNA. Only the shaded portions (exons) will be translated. The 5′ flanking region contains a number of regulatory sequences that help control gene expression.

RNA polymerase to ensure accurate transcription of the gene. A second controlling sequence, called the CAAT box, is about 50 to 100 nucleotides farther away. Still farther away are groups of bases known as enhancers, which are regulatory sequences that stimulate transcription.

Exons
DNA sequences that are transcribed and joined to other exons during mRNA processing and are translated into the amino acid sequence of a protein.

In humans and in other higher organisms, mRNA transcripts found in the nucleus are much larger than those found in the cytoplasm. The precursor mRNA molecules are shortened by removal of introns found between coding regions (exons). The first step in processing the transcript is the addition of nucleotides to both the 5′ and 3′ ends. Modified nucleotides known as a **cap** are added to the 5′ end of the transcript when it is about 30 nucleotides long. After the transcript is released from the DNA, a tail consisting of a string of 100 to 200 A nucleotides (known as a poly-A tail) is added to the 3′ end of the molecule. The cap assists in binding the mRNA to ribosomes during initiation of translation. The role of the poly-A tail is unclear, since some mRNAs lack such modifications.

Cap
A modified base (guanine nucleotide) that is attached to the 5′ end of eukaryotic mRNA molecules.

After addition of the cap and tail, the transcript is then cut at the borders between introns and exons, and the exons are spliced together by an enzyme to form the mature mRNA. While a transcript might be 5000 nucleotides long, the mature mRNA might contain only about 1000 nucleotides or fewer, which means that most of the transcript is discarded (Figure 9.7).

Proper splicing of pre-mRNA is essential for normal gene function. Several human genetic disorders are the result of abnormal mRNA splicing. In β° thalassemia, one or more mutations in introns lead to a low efficiency in splicing, a resulting low level in mRNA, and a corresponding deficiency in beta globin.

THE GENETIC CODE

As outlined above, genetic information is transferred from DNA to a complementary mRNA during transcription. After processing, the mRNA molecule moves to the cytoplasm where the encoded information is translated into the linear series of amino acids in a polypeptide. Francis Crick first proposed this relationship between DNA and RNA, and this concept has been substantiated by a large body of experimental work.

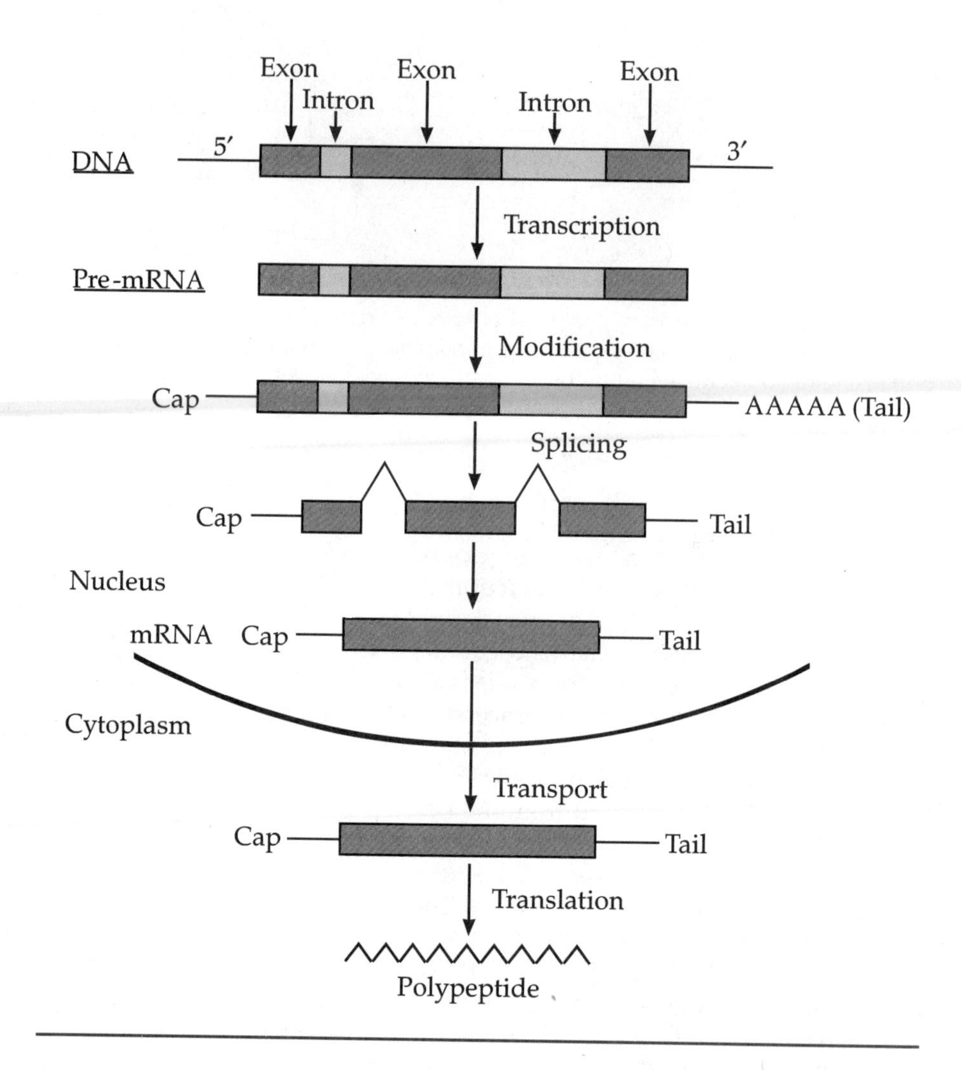

FIGURE 9.7 Steps in the formation of mRNA. One strand of DNA is transcribed into a pre-mRNA. The ends of this molecule are modified and the introns are spliced out. The mature mRNA is then transported to the cytoplasm for translation.

The notion, however, leaves unanswered a basic question: what is the precise nature of the genetic code? DNA contains four bases, A, T, C, and G, while proteins contain 20 different amino acids. How is it that an alphabet of four letters can be arranged to form 20 different words? Obviously, one-letter code words are not sufficient. Two-letter words, AT, CC, TG, and so forth, would form only 16 (4^2) combinations. Using three letters or bases to code for each amino acid would generate 64 (4^3) combinations, 44 more than necessary. In the early 1960s, Crick performed a number of genetic experiments that verified the triplet nature of the code. Using the fact that the genetic dictionary consisted of three-letter code words, scientists began to crack the genetic code in 1961 (see Concepts & Controversies, page 243), and by 1966, all 64 of the possible three-letter code words had been assigned a function. By convention, the gene is described in terms of the base sequence in the mRNA. The three bases in mRNA that specify an amino acid are called *codons*. The bases in DNA complementary to a codon are called *code words*, and the three bases in tRNA that interact with an mRNA codon are called

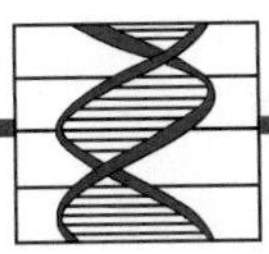

Concepts & Controversies

The Search for the Genetic Code

At the summer 1961 Cold Springs Harbor Symposium, Sidney Brenner reviewed the theory of genetic information transfer as it was understood at the time: DNA contained genetic information in its nucleotide sequence that determined the amino acid sequence of proteins. He said: "Exactly how this specification is accomplished is the main problem of present-day molecular biology, and its solution, the breaking of the genetic code, is the main ambition of many workers in the field."

Neither Sidney Brenner nor anyone else in attendance knew that the breakthrough had already taken place in May of that year in the laboratory of Marshall Nirenberg at the National Institutes of Health outside Washington, DC. Ironically, Nirenberg had applied to attend the symposium but had been turned down.

Early in 1960, working outside the network of the "in group," Nirenberg and a postdoctoral fellow named Johann Heinrich Matthaei began using a cell-free system to study the role of RNA in protein synthesis. Their first successful experiments showed that when RNA from a virus was added to the system, it stimulated protein synthesis, indicating that RNA did have a role in protein synthesis. While the viral work progressed, Matthaei tried using artificial RNA in May 1961 to see if it would also stimulate protein synthesis. These RNAs were polymers of single nucleotides: polyuridylic acid (poly-U), in which all the nucleotides were uridine; poly-A; and poly-AU, a heteropolymer with A and U in random order. When poly-U was added, protein synthesis was stimulated. In subsequent experiments, Matthaei determined that the polypeptide synthesized consisted entirely of phenylalanine. The first genetic code word was identified: poly-U gave rise to polyphenylalanine. Shortly thereafter, they found that poly-C coded for proline.

In December of 1961, Crick and colleagues published an elegant analysis of mutation in a virus establishing that the code consisted of a three-base sequence. Crick wrote: "This [the work of Nirenberg and Matthaei] implies that a sequence of uracils code for phenylalanine, and our work suggests that it is probably a triplet of bases." Several laboratories threw all their efforts into deciphering the code for the other 18 amino acids.

In the summer of 1964, Nirenberg announced that he and another postdoctoral fellow, Philip Leder, had discovered that artificial RNA as short as three bases could stimulate binding of the RNA, the ribosome, and a tRNA-charged amino acid. By knowing the sequence of the RNA and identifying which amino acid bound to the ribosome, the code could be deduced. With this system and one devised by Har Gorbind Khorana of the University of Wisconsin in hand, the code for all 20 amino acids was known by 1966. The standard table shown in Figure 9.11 was devised by Crick to present the code and its amino acids.

In 1968, Nirenberg, Khorana, and Robert Holley (who deduced the structure of tRNA) shared the Nobel Prize for their work on the structure and function of RNA.

anticodons. Because mRNA is translated in the $5' \rightarrow 3'$ direction, codons are always written in the $5' \rightarrow 3'$ direction.

Most amino acids are specified by more than one code word, with serine, leucine, and arginine each having six different codons (Figure 9.8). Three codons (UAA, UAG, and UGA) do not encode any amino acid; instead they function as **stop codons.** Stop codons specify the end of the message and cause the finished polypeptide to be released from the ribosome. The codon AUG, in addition to specifying the amino acid methionine, also functions as

Stop codons
Codons present in mRNA that signal the end of a growing polypeptide chain. UAA, UGA, and UAG function as stop codons.

FIGURE 9.8 The coding dictionary expressed as mRNA codons. The first position is at the 5′ end of each codon. One codon (AUG) for methionine serves as the start codon, while three codons (UAA, UAG, UGA) serve as stop codons.

First Position		Second Position U	Second Position C	Second Position A	Second Position G	Third Position
U		phe	ser	tyr	cys	U
U		phe	ser	tyr	cys	C
U		leu	ser	stop	stop	A
U		leu	ser	stop	trp	G
C		leu	pro	his	arg	U
C		leu	pro	his	arg	C
C		leu	pro	gln	arg	A
C		leu	pro	gln	arg	G
A		ileu	thr	asn	ser	U
A		ileu	thr	asn	ser	C
A		ileu	thr	lys	arg	A
A		met	thr	lys	arg	G
G		val	ala	asp	gly	U
G		val	ala	asp	gly	C
G		val	ala	glu	gly	A
G		val	ala	glu	gly	G

Initiator codon
A codon present in mRNA that signals the location for translation to begin. The codon AUG functions as an initiator codon.

an **initiator codon,** marking the place on mRNA molecules where translation begins.

With few exceptions (mitochondria being one), the genetic code is universal; that is, the code is identical in all organisms, from bacteria to fruit flies to plants to mice to humans. The universality of the genetic code is strong evidence that all living things are closely related and may have a common evolutionary origin.

TRANSLATION

The translation of the base sequence in mRNA into the amino acid sequence of a polypeptide chain requires interaction among many components, each performing a separate, specialized job. For this discussion we will first consider important aspects of the structure of ribosomes and tRNA molecules and will then consider the three steps in translation: initiation, elongation, and termination.

Structure and Function of Ribosomes

Ribosomes are composed of two subunits, each of which contains ribosomal RNA (rRNA) and a number of ribosomal proteins. The large subunit is made up of two rRNA molecules and more than 50 proteins. The small subunit consists of an rRNA molecule and a number of proteins (25 to 30). The subunits are assembled so that the proteins are on the surface, and the rRNAs are mostly in the interior. The small subunit contains a site that binds mRNA

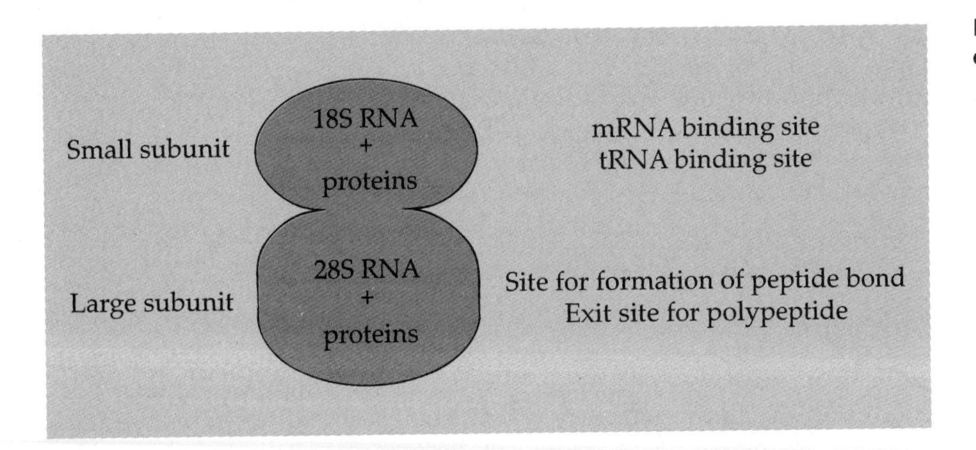

FIGURE 9.9 The components of a eukaryotic ribosome.

and a decoding region that binds a tRNA molecule. The large subunit contains sites where amino acids are joined together and the exit point for the newly synthesized polypeptide chain or protein (Figure 9.9). The process of assembling amino acids into polypeptide takes place on ribosomes.

Structure and Function of tRNA

Although the tRNA molecule is single stranded, the polynucleotide chain folds back upon itself, producing a series of stems and loops (Figure 9.10). The stems contain bases that hydrogen bond to other bases in the chain, forming double-stranded regions in the molecule. The loop at one end of the molecule is the **anticodon loop;** it contains a sequence of three unpaired bases (the anticodon) that are complementary to a three-base sequence in mRNA

Anticodon loop
The region of a tRNA molecule that contains the three base sequence (known as an anticodon) that pairs with a complementary sequence (known as codon) in an mRNA molecule.

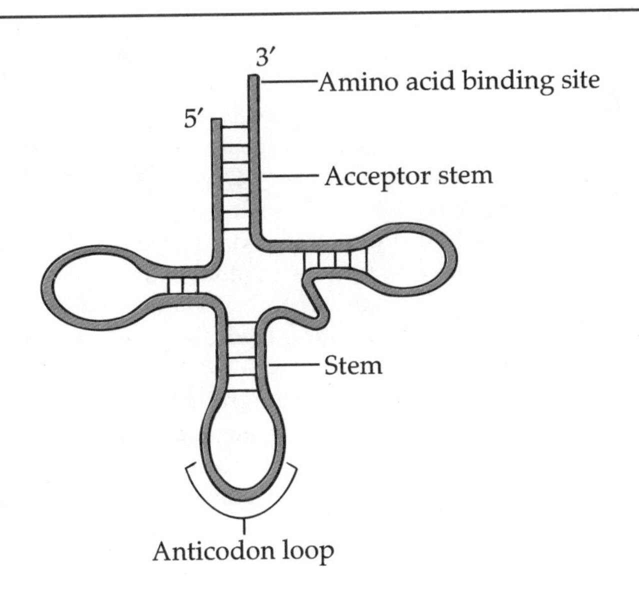

FIGURE 9.10 The two-dimensional structure of a transfer RNA molecule. One end of the molecule binds a specific amino acid, and the other end, containing the anticodon, recognizes the base sequence in an mRNA molecule.

Acceptor stem
The region of a tRNA molecule to which a specific amino acid is attached in an enzymatic reaction.

(the codon). At the other end of the molecule is the **acceptor stem,** where the tRNA becomes covalently linked to a specific amino acid. Each tRNA molecule can bind to only 1 of the 20 different types of amino acids. The binding is controlled by a series of enzymes, each of which recognizes one type of tRNA and one type of amino acid. To participate in protein synthesis, each amino acid must be "charged" by covalently binding to a tRNA molecule.

The Steps in Translation

Protein synthesis is a process that requires the coordinated action of many components. As you proceed through the discussion of this intricate mechanism, remember that these orchestrated workings have a single goal: converting the linear sequence of nucleotides in mRNA into the linear sequence of amino acids in a polypeptide chain. Translation can be divided into three steps: initiation, elongation, and termination.

INITIATION. Initiation is accomplished in two steps (Figure 9.11). In the first, a small ribosomal subunit binds to a transfer RNA molecule carrying methionine. Methionine is a unique amino acid in that it is the first inserted in all polypeptide chains. As part of the first step, the anticodon of methionine tRNA (UAC) binds to the AUG codon at the 5′ end of the mRNA molecule. In the second step, a large ribosomal subunit binds to the small subunit, and the process of initiation is complete.

ELONGATION. The process of elongation begins when a tRNA with its attached amino acid enters the ribosome and binds to the second codon of the mRNA. At this point the two amino acids lie adjacent to each other and an enzyme forms a peptide bond between methionine and the second amino acid (Figure 9.12). The formation of the peptide bond also releases the methionine from its tRNA, and the methionine-tRNA exits from the ribosome. Before further protein synthesis can occur, the ribosome moves precisely three bases along the mRNA to bring the next codon into position for decoding, and the dipeptide (methionine plus the second amino acid) bound by tRNA moves to the now available adjacent site in the ribosome. The third codon, now inside the ribosome, directs the binding of the appropriate tRNA and its amino acid. A peptide bond is formed between the second and third amino acid, and the same sequence of events, including the release of the free tRNA, movement of the ribosome, and shift of the growing polypeptide chain to the first site, is repeated. The ribosome moves down the mRNA in a $5' \rightarrow 3'$ direction. Protein synthesis occurs rapidly, with the addition of 10 to 15 amino acids per second. This means that the beta protein of hemoglobin, consisting of 146 amino acids, is fully synthesized in about 10 seconds.

Once a ribosome has moved away from the 5′ end of the mRNA, another ribosome can associate with the message. Typically then, a single mRNA can be read simultaneously by a number of ribosomes, each of which synthesizes a polypeptide chain. The complex of several ribosomes and an mRNA molecule is known as a **polyribosome,** or **polysome** (Figure 9.13). The

Polyribosome or polysome
The complex formed with an mRNA molecule and several ribosomes.

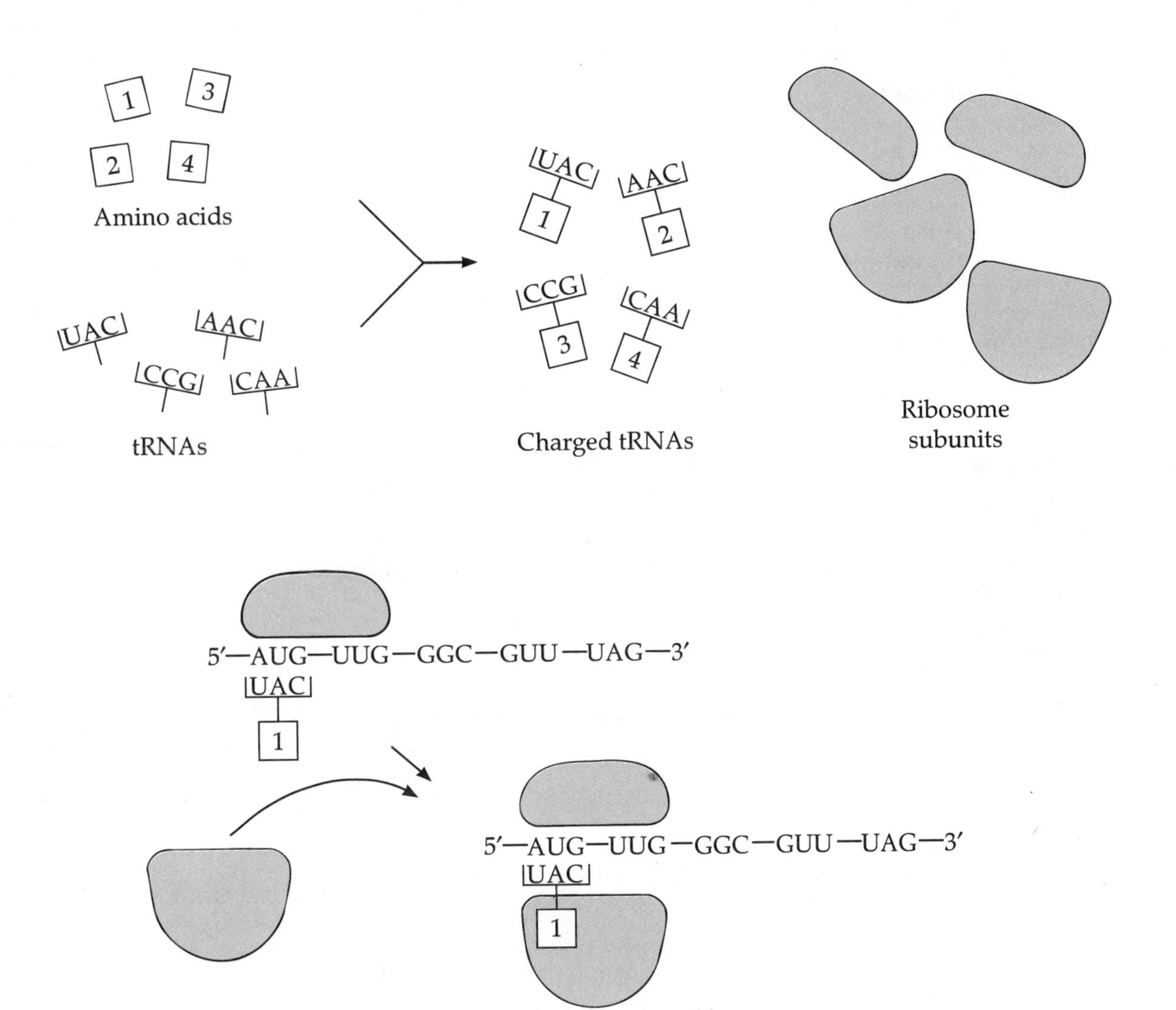

FIGURE 9.11 Steps in the initiation of translation. In this process the small ribosomal subunit, the mRNA, and a charged initiator, tRNA, combine to form the initiation complex. This complex then combines with the large ribosomal subunit, and the process of initiation is completed.

formation of polysomes increases the efficiency of translation and permits the formation of several copies of a polypeptide from a single molecule of mRNA.

TERMINATION. Elongation of the polypeptide chain continues until the ribosome encounters a stop codon (UAA, UAG, or UGA) in the mRNA. There is no tRNA anticodon to any of these, so whenever a stop codon occupies a site on the ribosome, no charged tRNA can bind. The presence of the stop codon also causes a release factor to bind to the ribosome and cleave

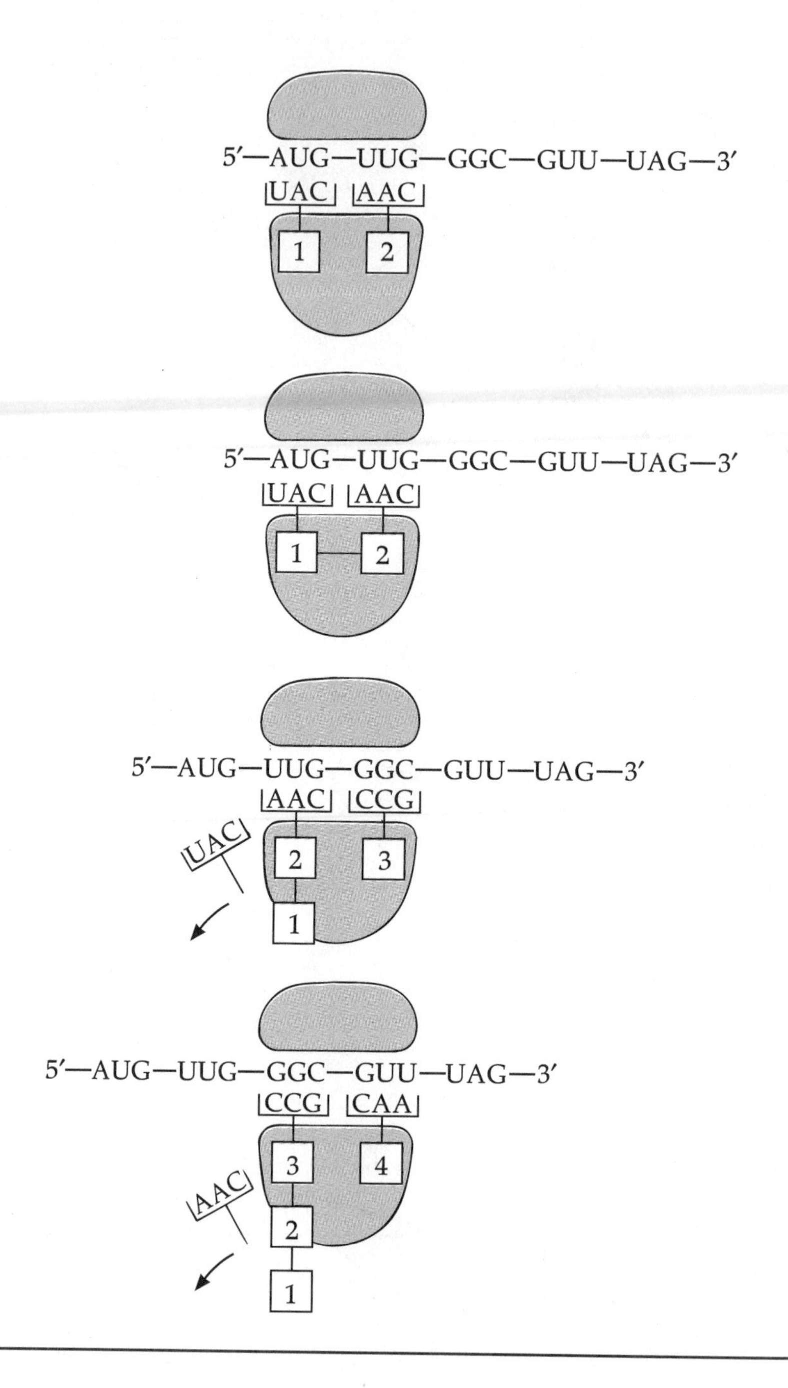

FIGURE 9.12 Steps in the elongation of a polypeptide chain. The second codon directs the binding of the proper charged tRNA. The amino acid attached to this tRNA is then linked to the adjacent amino acid by the formation of a peptide bond. The tRNA in the first site loses its attachment to the amino acid and is released. The nascent polypeptide, its attached tRNA, and the mRNA translocate to the first site, and the process is repeated.

the polypeptide from the tRNA, releasing the polypeptide product, the mRNA, and the final tRNA. Finally, the ribosome dissociates, and the two subunits separate (Figure 9.14). The role of antibiotics in dissecting the process of protein synthesis is discussed in the Concepts & Controversies on page 250.

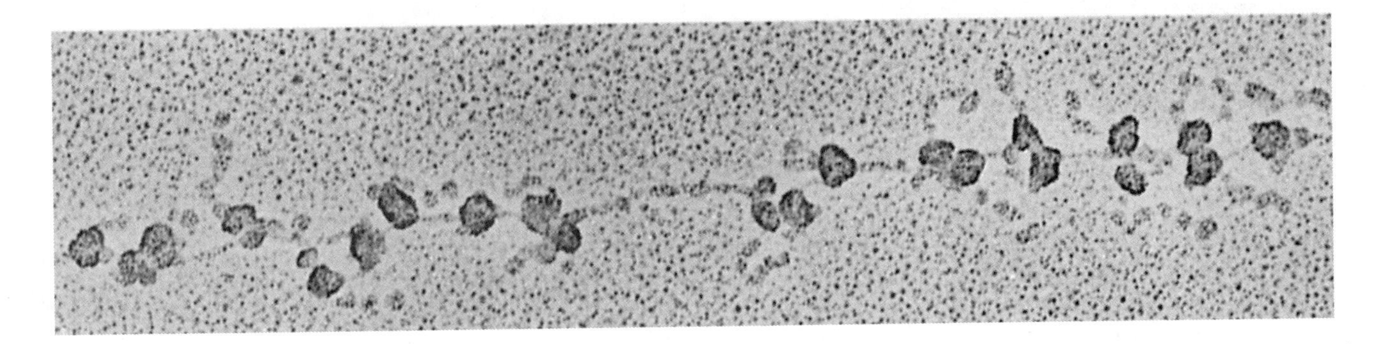

FIGURE 9.13 **An electron micrograph of ribosomes engaged in translation.**

The Polypeptide Product

The linear chain of amino acids assembled on the ribosome is known as a polypeptide. As this structure is released from the ribosome, it assumes a three-dimensional shape. During this process the polypeptide may be chemically modified or interact with other polypeptide chains. When the polypeptide is functional, it can rightly be called a protein.

Four levels of protein structure can be identified. The linear sequence of amino acids in the backbone of the protein is the **primary structure** (Figure 9.15). The combination of amino acid sequences that can be present in the primary structure of a protein is truly astronomic. Since each position can be occupied by any of 20 amino acids, the number of different combinations is

Primary structure
The amino acid sequence in a polypeptide chain.

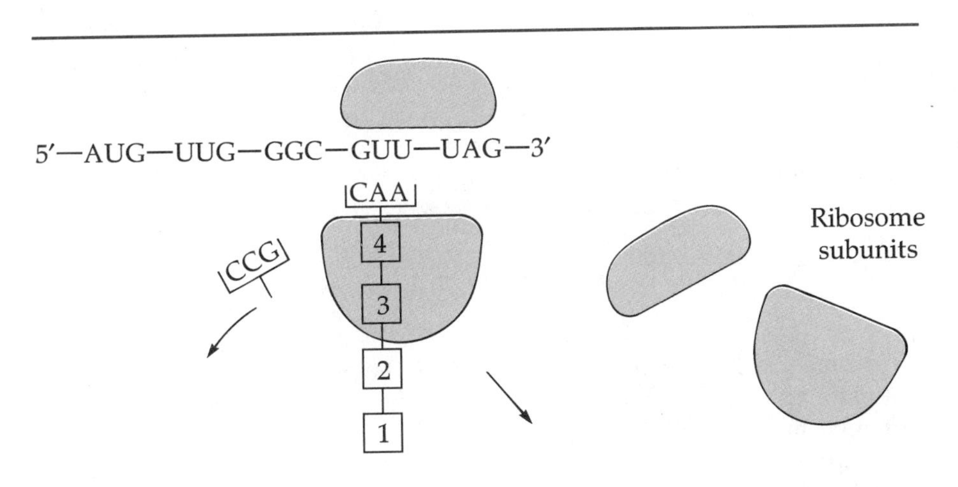

FIGURE 9.14 **Steps in the termination of protein synthesis.** As the termination codon reaches the ribosome, no charged tRNA can bind. The presence of the termination codon causes the complex to fall apart, releasing the polypeptide product, the mRNA, and the ribosomal subunits. Although only a tripeptide is shown, most polypeptides consist of several hundred amino acids.

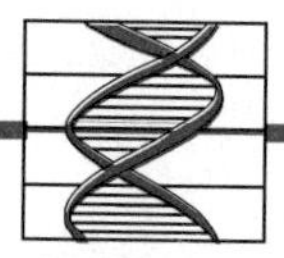

Concepts & Controversies

Antibiotics and Protein Synthesis

Antibiotics are chemicals produced by microorganisms that provide them with a defense mechanism. The most effective antibiotics work by interfering with essential biochemical or reproductive processes. Thus many antibiotics block or disrupt one or more stages in protein synthesis. Some of these are listed below.

Tetracyclines are a family of chemically similar antibiotics used to treat a range of gram-positive and gram-negative bacterial infections. Tetracyclines are antibiotics that interfere with the process of initiation. Specifically, tetracycline interacts with the small ribosomal subunit to prevent the binding of the tRNA anticodon in the first step in initiation. As it turns out, eukaryotic ribosomes as well as bacterial ribosomes are sensitive to the action of tetracycline. Since tetracycline cannot pass through the plasma membrane of eukaryotic cells but does penetrate bacterial cells, it is useful in treating bacterial infections.

Streptomycin is an antibiotic that is used clinically to treat serious bacterial infections. Functionally it binds to the small ribosomal subunit but does not prevent initiation or elongation. When streptomycin binds to the ribosomal subunit, it alters the interaction between codons in the mRNA and anticodons in tRNA so that incorrect amino acids are incorporated into the growing polypeptide chain. In addition, streptomycin causes the ribosome to fall off the mRNA at random, preventing the synthesis of complete protein molecules.

Puromycin is an antibiotic that is not used clinically but has played an important role in elucidating the mechanism of protein synthesis. The structure of puromycin closely resembles that of the tRNA-amino acid complex (a "charged" tRNA). As a result it is able to enter the ribosome and become incorporated into the growing polypeptide chain. However, no further amino acid incorporation takes place, and polypeptide synthesis is prematurely terminated. Subsequently the shortened polypeptide with an attached puromycin falls off the ribosome.

Chloramphenicol was the first broad-spectrum antibiotic introduced for clinical use. Eukaryotic cells are resistant to its actions, and it was widely used to treat bacterial infections. However, its use has since been limited to external applications and serious infections because of its potentially lethal side effects on bone marrow. Chloramphenicol binds to the large ribosomal subunit in prokaryotes and specifically inhibits the enzymatic reaction that leads to the formation of the peptide bond. This results in the termination of protein synthesis.

Thus almost every stage of protein synthesis during initiation and elongation can be inhibited by one antibiotic or another, and work on the design of new antibiotics often requires detailed knowledge of cellular events such as protein synthesis.

20^n, where n is the number of amino acids in the protein. For a peptide chain 5 amino acids long, 20^5, or 3,200,000, different molecules, each with a different amino acid sequence, can be produced. Since the average polypeptide is 300 amino acids in length, then 20^{300} different combinations are possible.

The sequence of amino acids, determined by the order of nucleotides in DNA, specifies the other levels of structure in the protein. The next two levels of structure are determined to a great extent by the side group (R group) of each amino acid. The R groups interact with each other, by forming hydrogen bonds, or covalent bonds, causing the polypeptide chain to fold

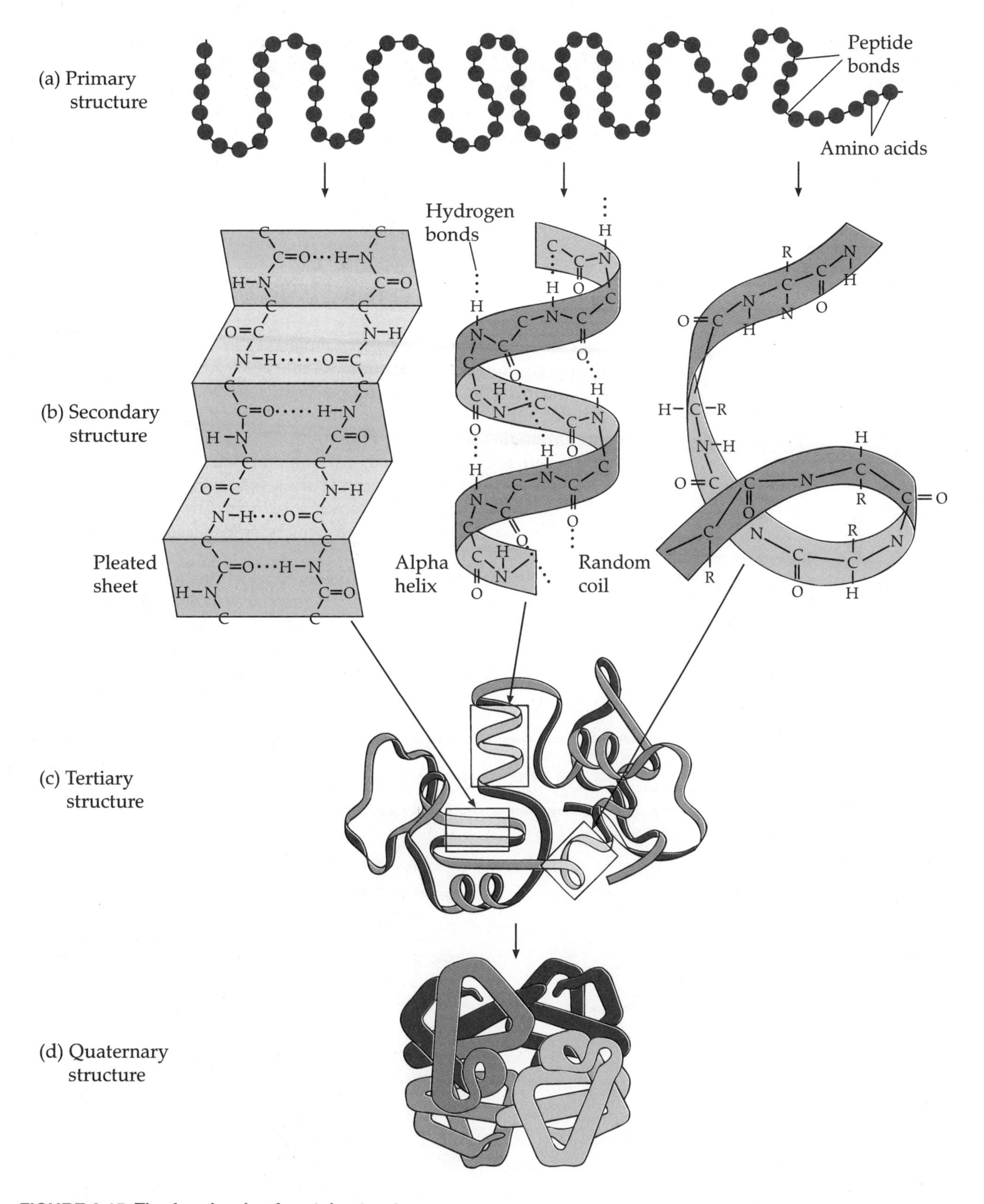

FIGURE 9.15 The four levels of protein structure.

Secondary structure
The pleated or helical structure in a protein molecule that is brought about by the formation of bonds between amino acids.

Tertiary structure
The three-dimensional structure of a protein molecule brought about by folding on itself.

Quaternary structure
The structure formed by the interaction of two or more polypeptide chains in a protein.

itself into a three-dimensional configuration. **Secondary structure** refers to a repeating configuration, such as helical shapes or pleated sheets, that arise by interaction of amino acid R groups. Most proteins exhibit a mixture of the helical and pleated secondary structure (Figure 9.15).

Further folding of the protein chain produces a **tertiary** level of structure (akin to folding a spring back on itself). The **quaternary** level of structure is characteristic of functional proteins that are composed of more than one polypeptide chain. Hemoglobin is a well-studied example of a protein with four levels of structure (Figure 9.15).

FUNCTIONS OF PROTEINS

Proteins are the most abundant class of molecules in the cell and participate in a wide range of distinct functions (Table 9.3). Actin and myosin, for example, are contractile proteins found in muscle cells. Hemoglobin is a transport protein that shuttles oxygen to cells. The immune system depends on protein antibodies to identify and to remove foreign invaders such as bacteria from our bodies. Connective tissue and hair are rich in structural proteins such as collagen and keratin. Histones are structural proteins that complex with DNA to form nucleosomes, a substructure found in chromosomes. Many hormones, such as insulin (which regulates sugar metabolism), are proteins.

The largest group of proteins, however, functions as enzymes. These molecules act as catalysts in biochemical reactions, increasing the rate and lowering the energy required to carry out a biochemical reaction. Most enzymes function to catalyze a single biochemical reaction. This capacity is the result of the three-dimensional shape, or conformation, of the protein.

TABLE 9.3 Functional Diversity of Proteins

Type	Example	Function
Enzyme	DNA polymerase	DNA replication and repair
Hormone	Insulin	Uptake of glucose into cells
Receptor	Estrogen receptor	Binds estrogen in cytoplasm, moves into nucleus
Structure	Collagen	Component of cartilage, tendons, bone and other tissues
Contraction	Myosin	Muscle contraction
Immunity	Gamma globulin	Antibody that binds and inactivates foreign substance
Transport	Hemoglobin	Carries oxygen to cells, tissues

The configuration of the enzyme generates the **active site,** which binds the molecules (known as substrates) that are to undergo a reaction (Figure 9.16). The correlation between enzymes and genetic disorders will be explored in more detail in Chapter 10.

Active site
The portion of a protein that is required for enzymatic function.

It should be emphasized that the function of enzymes and of all proteins is ultimately dependent on the amino acid sequence or primary structure of the polypeptide chain. Changes in the amino acid sequence can result in a new arrangement of R groups that can drastically alter the shape and ultimately the function of proteins. We will discuss the phenotypic consequences of these changes in Chapter 11.

SUMMARY

1. The idea that genetic disorders result from biochemical alterations was proposed by Garrod in the first decade of this century.

2. Evidence that gene action is mediated by proteins that control biochemical reactions was provided by Beadle and his colleagues. In a variety of organisms, they demonstrated that mutations result in the loss of enzymatic activity, producing a mutant phenotype. Beadle proposed that genes function by controlling the synthesis of proteins and that protein function is responsible for the production of the phenotype.

3. Crick later summarized the molecular relationship between DNA and proteins: DNA makes RNA, which in turn makes protein.

4. The processes of transcription and translation require the interaction of many components, including ribosomes, mRNA, tRNA, amino acids, enzymes, and energy sources. Ribosomes serve as the workbenches upon which protein synthesis occurs; tRNA molecules are adaptors, recognizing amino acids as well as the nucleotide sequence in mRNA, the gene transcript.

5. In transcription, one of the DNA strands serves as a template for the synthesis of a complementary strand of RNA. The information transferred to RNA is encoded in triplet sequences of nucleotides. Of the 64 possible triplet codons, 61 code for amino acids, and 3 serve as stop codons.

6. Translation requires the interaction of charged tRNA molecules, ribosomes, mRNA, and energy sources. Within the structure of the ribosome, the anticodons of the charged tRNA molecules bind to complementary codons in the mRNA. The ribosome moves along the mRNA, producing a growing polypeptide chain. At termination this polypeptide is released from the ribosome and undergoes a conformational change to produce a functional protein.

7. Four levels of protein structure are recognized, three of which are a result of the primary sequence of amino acids in the backbone of the protein chain. Although proteins perform a wide range of tasks, enzyme activity is one of the primary tasks. Enzymes function by lowering the energy of activation required in biochemical reactions. The products of these biochemical reactions are inevitably involved in the production of phenotype.

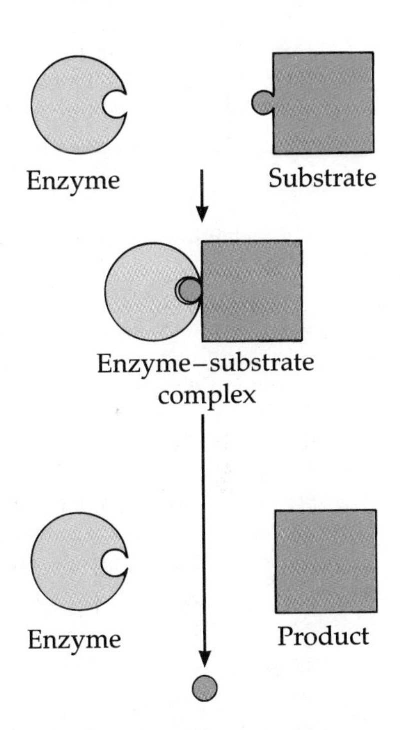

FIGURE 9.16 Diagrammatic representation of the active site of an enzyme and the interaction with the substrate. The formation of the enzyme-substrate complex lowers the energy required to carry out the conversion of the substrate to the product.

QUESTIONS AND PROBLEMS

1. How do mutations in DNA alter proteins?

2. The transcript and protein made by several mutant genes were examined. The results are given below. Determine where in the gene the mutation lies the 5′ flanking region, exon, intron, cap and tail, or ribosome binding site.
   a. normal length transcript, normal length nonfunctional protein
   b. normal length transcript, no protein made
   c. normal length transcript, normal length mRNA, short nonfunctional protein
   d. normal length transcript, longer mRNA, longer nonfunctional protein
   e. transcript never made
   f. transcript made and rapidly degraded

3. The following segment of DNA codes for a protein. The uppercase letters represent exons. The lowercase letters represent introns. The lower strand is the template strand. Draw the primary transcript and the mRNA resulting from this DNA:

GCTAAATGGCAaaattgccggatgacGCACATTGACTCGGaatcgaGGTCAGATGC
CGATTTACCGTtttaacggcctactgCGTGTAACTGAGCCttagctCCAGTCTACG

4. The following is a portion of a protein:

   met-trp-tyr-arg-gly-pro-thr-

   Various mutant forms of this protein have been recovered. Using the normal and mutant sequences, determine the DNA and mRNA sequences that code for this portion of the protein and explain each of the mutations.
   a. met-trp-
   b. met-cys-ile-val-val-leu-gln-
   c. met-trp-tyr-arg-ser-pro-thr-
   d. met-trp-tyr-arg-gly-ala-val-ile-ser-pro-thr-

5. Write the mRNA and polypeptide encoded by the following DNA. The upper strand is the sense strand:

   TACGAACAGTGTCGGTTACCCATT
   ATGCTTGTCACAGCCAATGGGTAA

6. The list below contains a DNA molecule and the mRNA and protein it encodes. Fill in the missing nucleotides and amino acids. Which is the template strand?

```
DNA      AAA      A  T        GA
DNA            G       CTC G   GAA
mRNA     UU   A    U  U         A
Protein    arg          leu
```

7. Some mutations are suppressed by a second mutation. Explain how this might work, assuming that the second mutation occurs in the DNA that encodes a tRNA.

8. The following is the structure of glycine. Draw a tripeptide composed exclusively of glycine. Label the N-terminus and C-terminus. Draw a box around the peptide bonds.

$$H_2N-\underset{\displaystyle H}{\overset{\displaystyle H}{\underset{|}{\overset{|}{C}}}}-C\begin{matrix}\nearrow O \\ \searrow OH\end{matrix}$$

9. Mutant ribosomes were isolated from different cells. Describe the most likely defect in each of the following ribosomes:
   a. unable to bind mRNA
   b. unable to join amino acids together
   c. unable to bind tRNA

10. Write the anticodon(s) for the following amino acids:
    a. met
    b. trp
    c. ser
    d. leu

11. Each of the following functions either in transcription or in translation. List each item in the proper category, transcription or translation: RNA polymerase, ribosomes, nucleotides, tRNA, pre-mRNA, amino acyl synthetase, DNA, A site, TATA box, anticodon, enhancer, amino acids.

12. Enzyme X normally interacts with substrate A and water to produce compound B.

a. What would happen to this reaction in the presence of another substance that resembles substrate A and was able to interact with enzyme X?
b. What if a mutation in enzyme A changed the shape of the active site?

13. Can a mutation change a protein's tertiary structure without changing its primary structure? Can a mutation change a protein's primary structure without affecting its secondary structure?

14. Explain the role of proteins in the relationship between DNA and phenotype.

For Further Reading

Alberts, B. A., Bray, D., Lewis, J., Raff, M., Roberts, K., and Watson, J. 1983. *Molecular Biology of the Cell.* New York: Garland Press.

Baralle, F. E. 1983. The functional significance of leader and trailer sequences in eucaryotic mRNAs. *Int. Rev. Cytol.* **85**: 71–106.

Beadle, G. W., and Tatum, E. L. 1941. Genetic control of biochemical reactions. *Neurospora. Proc. Natl. Acad. Sci. USA* **27**: 499–506.

Bielka, H. (Ed.). 1983. *The Eukaryotic Ribosome.* New York: Springer-Verlag.

Chambon, P. 1981. Split genes. *Sci. Am.* **244** (May): 60–71.

Crick, F. H. C. 1962. The genetic code. *Sci. Am.* **207** (October): 66–77.

Danchin, A., and Slonimski, P. 1985. Split genes. *Endeavour* **9**: 18–27.

Darnell, J. 1985. RNA. *Sci. Am.* **253** (October): 68–87.

Doolittle, R. F. 1985. Proteins. *Sci. Am.* **253** (October): 88–99.

Garrod, A. 1902. The incidence of alkaptonuria: A study in chemical individuality. *Lancet* **2**: 1666–1670.

Hamkalo, B. 1985. Visualizing transcription in chromosomes. *Trends Genet.* **1**: 255–260.

Koshland, D. E. 1973. Protein shape and control. *Sci. Am.* **229** (October): 52–64.

Noller, H. F. 1985. Structure of ribosomal RNA. *Ann. Rev. Biochem.* **53**: 119–162.

Rossman, M. G., and Argos, P. 1981. Protein folding. *Ann. Rev. Biochem.* **50**: 497–532.

CHAPTER 10

BIOCHEMICAL GENETICS

CHAPTER OUTLINE

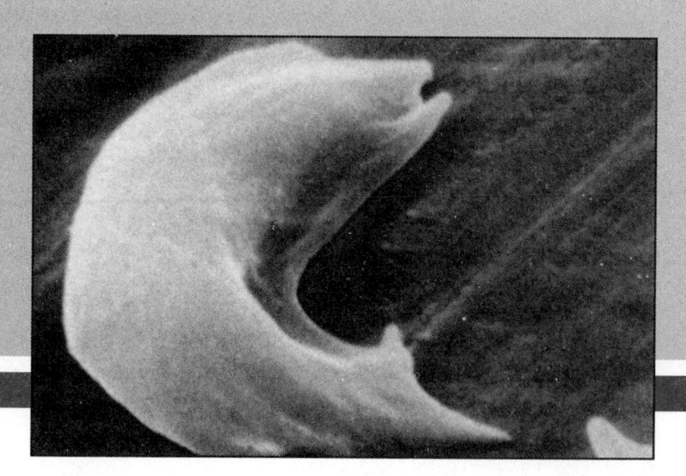

A young Norwegian mother had two children who were both mentally retarded. The first child, a girl, did not learn to walk until nearly two years of age, and spoke only a few words. Most peculiar was the fact that the child had a musty odor, which could not be eliminated by bathing. Her younger brother, was similarly slow to develop, and never learned to walk or talk. He had the same musty odor as his sister. To explain why both her children were retarded and had a musty odor, she went from doctor to doctor, but to no avail. Finally in the spring of 1934, the persistent mother took the two children now aged four and seven years, to Dr. Asbjorn Fölling, a biochemist and a physician.

Because the urine from these children had a musty odor, he first tested the urine for signs of infection, but there was none. He did find something in the urine that reacted with ferric chloride, producing a green color. Beginning with 20 liters of urine collected from the children, he began a series of biochemical extractions to isolate and identify the unknown substance. After five weeks, he was able to purify the compound, and then spent another six weeks identifying the substance as phenylpyruvic acid. To confirm that the compound in the urine was indeed phenylpyruvic acid, he synthesized and purified phenylpyruvic acid from organic chemicals, and showed that both compounds had the same physical and chemical properties.

With his knowledge of chemistry, he hypothesized that the phenylpyruvic acid was produced by a disorder that affected the breakdown of the amino acid phenylalanine. He further postulated that this biochemical abnormality might cause the phenotype of retardation. To confirm this, he examined the urine of several hundred patients in nursing homes and schools for the retarded and found phenylpyruvic acid in the urine of eight retarded individuals, but never in the urine of normal individuals. Within six months after the tenacious mother approached him, Dr. Fölling had prepared a manuscript for publication describing a condition that is a prototype for metabolic genetic disorders. This condition, called phenylketonuria, helped establish the relationship between biochemistry and the phenotype.

As we discussed in the last chapter, the relationship between genes and phenotype is complex and may be composed of many steps. The discovery of biochemical mutations in fungi by George Beadle and Edward Tatum in 1941 was an important stage in understanding how gene expression results in a phenotype. Their ideas can be summarized in the following statements:

- Biochemical processes in all organisms are under genetic control.
- These biochemical processes can be resolved into a series of individual, interconnected biochemical reactions.
- Each of these individual reactions is regulated by a single gene that controls the production, function, and specificity of a single enzyme.
- Mutation or alteration of a single gene changes the ability of the cell to carry out a single biochemical reaction.

This one gene–one enzyme hypothesis was modified by later work with hemoglobins and can be restated as the one gene–one polypeptide hypothesis.

In Chapter 9 we considered how DNA encodes information for the chemical structure of proteins; in this chapter we will consider the relationship between proteins and the phenotype, using examples that emphasize the diverse role proteins play in living systems.

The Role of Proteins

Chemists began studying enzymes in the early part of the 19th century. In 1811, K. S. Kirchhoff described a substance he extracted from wheat with the capacity to convert starch into sugar. This substance was isolated and purified some 20 years later by A. Payen and J. F. Persoz. They named this substance diastase (from the Greek word *diastasis,* meaning "separation"), since it separated sugar from starch. These workers also linked diastase to the synthesis and storage of starch in plant metabolism and pointed out that diastase could be used for the industrial production of sugars. In 1873 W. Kuhne introduced the term "enzyme" to describe substances that carried out fermentation. However, the notion that life processes such as metabolism or respiration could be carried out by organic molecules such as proteins was hotly debated in the latter part of the 19th century. On one hand, the vitalists contended that living systems were more than the sum of their parts and were infused with an immaterial force or spirit that made them alive. The embryologist Hans Driesch called this force entelechy. In the 20th century the French philosopher Henri Bergson called this force an *élan vital.* Mechanists and reductionists, on the other hand, argued that the properties and performances of living organisms could be explained by the action of lower levels of organization, such as cells and molecules. As late as 1900, the chemist R. Willstätter (a vitalist) contended that although enzymes were bound to proteins, they could not themselves be proteins. According to vitalists, a mere organic molecule could not have a property attributed to a living system. The decisive step to resolve this argument came in 1926 when James Sumner crystallized the enzyme urease. Chemical analysis of this purified enzyme showed it to be a protein. By 1931, more than 30 different enzymes had been purified, and all were proteins. From this point, rapid progress was made in demonstrating that cellular metabolic events are associated with chains of biochemical reactions known as metabolic pathways.

Proteins are the most numerous and multi-functional class of macromolecules in the cell. They are essential to all cellular structures and biological processes carried out in every cell type. Proteins are components of membrane systems and the internal skeleton of cells. They form the glue that holds cells and tissues together, carry out biochemical reactions, destroy invading microorganisms, and act as hormones, receptors, and transport molecules. Even the replication of genetic information is carried out through the action of proteins. The importance of proteins to living systems is reflected in their name. The term "protein," coined in the 19th century, is from the Greek word *proteios,* translated as "being of first importance."

The functional diversity of proteins is matched only by their structural diversity. Each protein is composed of a specified number of amino acids arranged in a specific order. As outlined in the last chapter, the primary structure of a protein is ultimately responsible for its three-dimensional shape and its functional specificity. Heritable errors in the information encoding the amino acid sequence can have consequences that range from insignificant to lethal. In Chapter 9 we considered how DNA encodes information for the chemical structure of proteins; in this chapter we will consider the relationship between proteins and phenotype, using examples that emphasize the diverse roles proteins play in living systems.

Metabolic Pathways and Genetic Disorders

Since most proteins function as enzymes, we will first consider genetic disorders caused by defects in enzyme activity. Enzymes are biological catalysts that carry out specific types of biochemical reactions. Enzymes function to convert molecular **substrates** into **products** through a biochemical reaction (Figure 10.1). Enzymatic reactions do not occur at random or in isolation; they are organized into one or more series of reactions known as *biochemical pathways* (Figure 10.2). The sum of all biochemical reactions in the cell is referred to as **metabolism,** and biochemical pathways are often referred to as metabolic pathways. In a metabolic pathway, the product of one reaction serves as the substrate for the next reaction in the pathway. Failure to carry out a single biochemical reaction in a pathway will result in a shutdown of all following reactions in the pathway beyond that point. It also

Substrate
The specific chemical compound that is acted upon by an enzyme.

Product
The specific chemical compound that is the result of enzyme action. In biochemical pathways, a compound can serve as the product of one reaction and the substrate for the next reaction.

Metabolism
The sum of all biochemical reactions by which cells convert and utilize energy

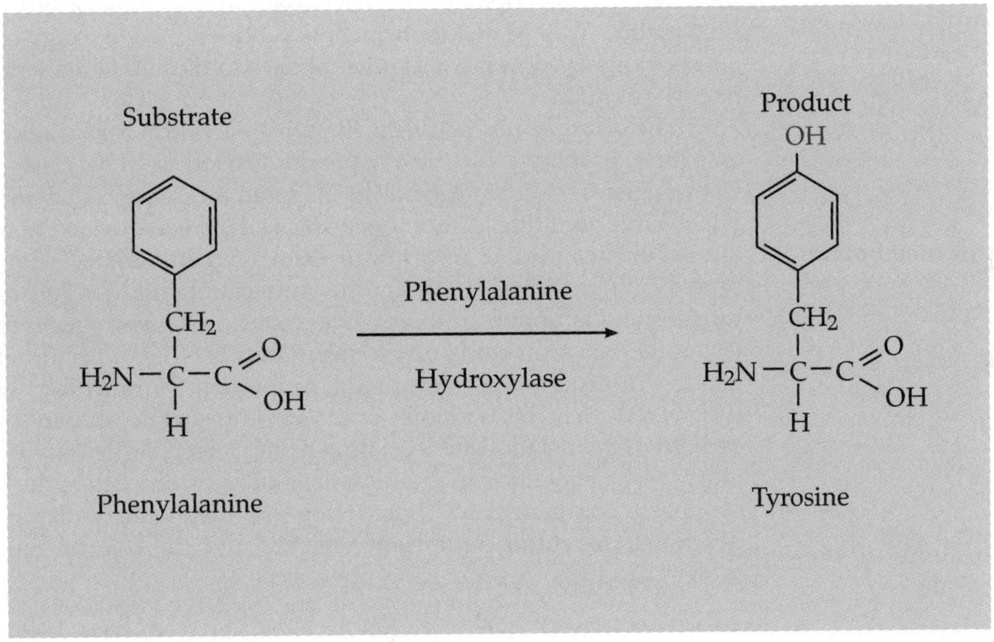

FIGURE 10.1 Each step in a metabolic pathway involves a single biochemical reaction in which a precursor is converted to a product. In this reaction, the amino acid phenylalanine acts as a substrate for the enzyme phenylalanine hydroxylase and is converted to the product, the amino acid tyrosine.

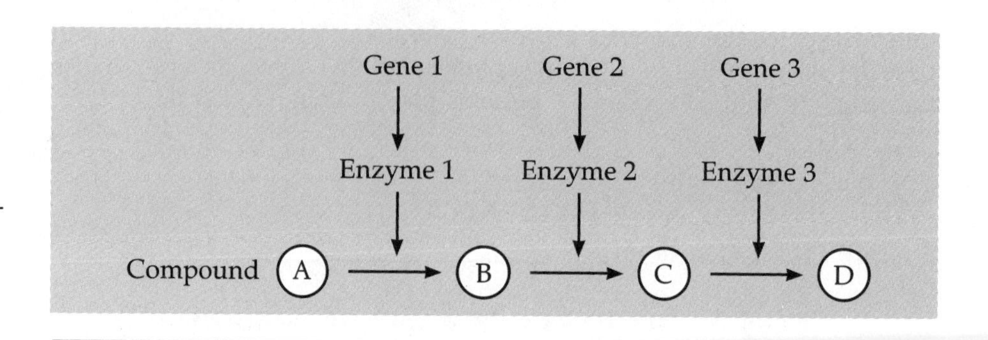

FIGURE 10.2 Representation of a sequence of reactions in a biochemical pathway. In this pathway, compound A is converted to B, B is converted to C, and C is converted to D. Each step is controlled by a single enzyme that catalyzes the reaction. Each enzyme is the product of a single gene.

results in the accumulation of substrates and products in the pathway before that point.

Because biochemical reactions are organized into pathways, a mutation that causes loss of activity in a single enzyme can have phenotypic effects in a number of ways. First, the accumulation of one or more precursors in the pathway may be detrimental. The accumulated precursors may also cause overuse of alternate, minor pathways, resulting in the accumulation of toxic metabolic products. Second, the lack of a metabolic product may be harmful, or the blocked reaction may prevent a subsequent reaction in the pathway that produces a necessary product. Keep in mind, however, that not all metabolic blocks necessarily lead to an abnormal phenotype. In some cases mutation produces a phenotypically neutral variation that becomes widely distributed in the population. Such variations are known as **polymorphisms.** Enzymatic errors of metabolism thus produce a wide range of phenotypic effects, ranging from inconsequential ones to those that are lethal prenatally or early in infancy.

Polymorphism
The occurrence of two or more genotypes in a population in frequencies that cannot be accounted for by mutation.

The relationship between human genetic disorders and metabolism was first proposed by Sir Archibald Garrod in 1901 (see Concepts & Controversies, p. 261). Garrod investigated a disease called alkaptonuria in which large quantities of homogentisic acid are excreted in the urine. Garrod termed this condition an **"inborn error of metabolism."** His work represented a pioneering study in the application of Mendelian genetics to humans and in understanding the relationship between genes and biochemical reactions. Unfortunately a direct link between specific enzymes and genetic disorders was not made in humans until 1952, when it was discovered that individuals with an autosomal recessive disorder of carbohydrate metabolism have little or no detectable activity of the enzyme glucose-6-phosphatase in their livers or kidneys. Since then the biochemistry of dozens of genetic defects involving metabolic pathways has been described, providing important insights into the role of genes and their protein products in generating phenotypes.

Inborn error of metabolism
The concept advanced by Archibald Garrod that many genetic traits are the result of alterations in biochemical pathways.

Amino Acid Metabolism

Most eukaryotes, including humans, can synthesize some of the 20 amino acids found in proteins. The amino acids that cannot be synthesized by the body and must be included in the diet are called **essential amino acids.** For

Essential amino acids
Amino acids that cannot be synthesized in the body and must be supplied in the diet.

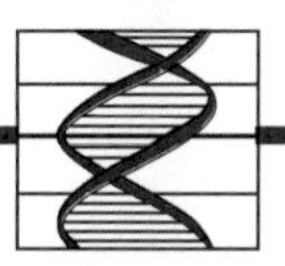

Concepts & Controversies

Garrod and Metabolic Disease

Sir Archibald Garrod was a distinguished physician, professor at Oxford, and physician to the royal family. Garrod was widely trained: he was skilled in biology and biochemistry as well as in clinical medicine. From his own investigations, he showed that alkaptonuria is characterized by the excretion of large amounts of a substance known as homogentisic acid, which causes the urine to turn black upon exposure to air.

Garrod proposed that the presence of homogentisic acid was caused by a metabolic block, which prevented further metabolism. He speculated that the metabolic block was caused by an enzyme deficiency. His investigations showed that 60% of affected individuals were the result of first-cousin marriages, but that the parents were unaffected. This led him to propose that this condition was caused by a recessive Mendelian trait. His work on alkaptonuria and other metabolic diseases was published as a book, *Inborn Errors of Metabolism*, that is now regarded as a milestone in genetics.

Why was Garrod's work not immediately appreciated? Perhaps because his work spanned three disciplines that were almost totally insulated from one another: genetics, biochemistry, and medicine. Biochemists had no interest in heredity, physicians regarded the conditions he studied as too rare to be important, and geneticists cared little about either metabolism or medicine.

humans, there are nine essential amino acids: histidine, isoleucine, leucine, lysine, methionine, phenylalanine, threonine, tryptophan, and valine.

One of these essential amino acids, phenylalanine, serves as the beginning point for a network of metabolic pathways, several of which are associated with human genetic disorders (Figure 10.3). A block in the conversion of phenylalanine to tyrosine in the first step of the pathway results in the disease known as **phenylketonuria,** or **PKU.** This disease has an incidence of about 1 in 12,000 births, and is most often associated with a deficiency of the enzyme phenylalanine hydroxylase.

Phenylketonuria (PKU) An autosomal recessive disorder of amino acid metabolism that results in mental retardation if untreated.

Affected individuals exhibit a range of neurological problems, including enhanced reflexes, convulsive seizures, and mental retardation. Because the skin pigment melanin is also a product of the blocked metabolic pathway (Figure 10.3), PKU victims usually have lighter hair and skin color than their siblings or other family members.

The phenotypic effects of this metabolic block are not caused by failure to convert phenylalanine to tyrosine, but by the accumulation of high levels of phenylalanine and other compounds in the pathway leading from phenylalanine to phenylpyruvic acid (Figure 10.3). As phenylalanine accumulates, it blocks the uptake of several amino acids by the developing brain. These include tyrosine, leucine, isoleucine, valine, tryptophan, histidine, and methionine. This imbalance in amino acid levels in the brain causes brain damage, mental retardation, and the associated neurological symptoms.

Children with PKU are unaffected before birth because excess phenylalanine that accumulates during intrauterine development is metabolized by

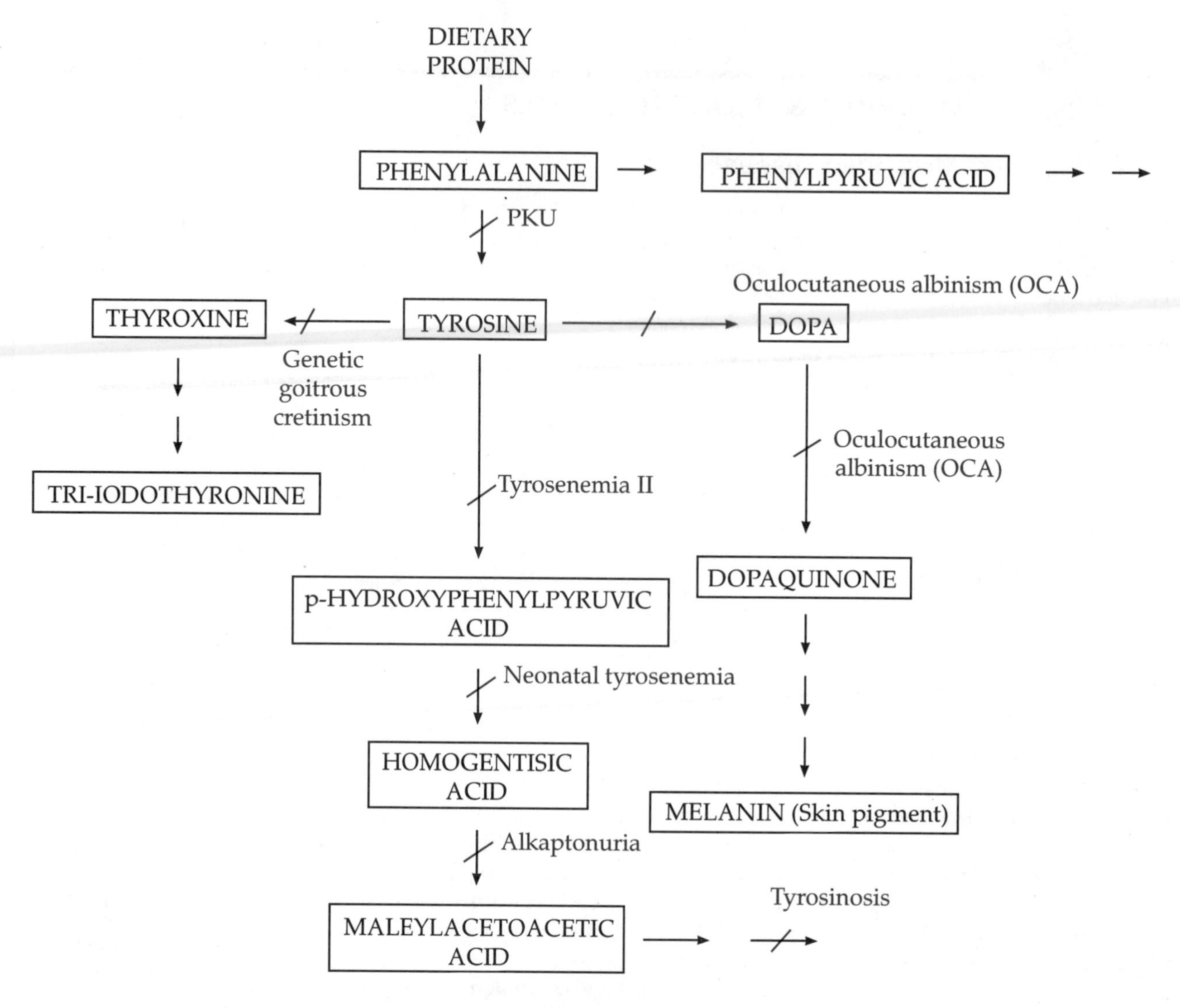

FIGURE 10.3 The metabolic pathways beginning with the essential amino acid phenylalanine. Metabolic blocks in this pathway caused by mutations lead to genetic disorders such as PKU, tyrosinosis, albinism, and alkaptonuria.

maternal enzymes. Such infants suffer neurologic damage and become retarded only after birth when fed on a normal diet containing phenylalanine. Restricting phenylalanine intake prevents mental retardation and damage to the rest of the nervous system and allows individuals to develop in a relatively normal fashion. This treatment, started in the 1960s, is widely used and has been successful in reducing the effects of this disease.

However, managing PKU by controlling dietary intake is both difficult and expensive (see Concepts & Controversies, page 263). A major problem is that phenylalanine is an amino acid present in many protein sources, and it is impossible to eliminate all protein from the diet. In the PKU diet, a

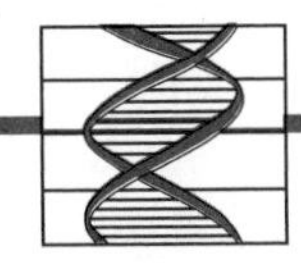

CONCEPTS & CONTROVERSIES

Dietary Management of Metabolic Disease

In several metabolic diseases, modifications of the diet are essential to prevent serious adverse effects such as mental retardation and death. Dietary treatment serves to replace essential metabolites or to prevent the buildup of intermediates. Unfortunately, not all metabolic diseases can be treated in this way. The major disorders for which dietary control is available are phenylketonuria, galactosemia, tyrosenemia, homocystinuria, and maple syrup urine disease.

The dietary products for each disease are, of course, different, but they are usually available in two forms, one for infants and one for older children that contains somewhat higher levels of proteins and other nutrients. For amino acid disorders such as phenylketonuria, products prepared from enzymatically digested protein preparations or mixtures of amino acids are used. These products also contain fats, usually in the form of corn oils, and carbohydrates from sugar, corn starch, or corn syrup. Vitamin and mineral supplements are also added. In one of the commercial products for PKU, casein, a protein extracted from milk is digested with enzymes into amino acids. This mixture is poured over an activated charcoal column, which removes phenylalanine, tyrosine, and tryptophan. These last two amino acids are added to the amino acid mixture along with a source of fat, carbohydrates, vitamins and minerals. The powder is then used at each meal to provide a source of amino acids. A typical menu for a school-age child is shown in the next column.

Previously, the restrictive diet was followed for 4 to 6 years. The rationale was that development of the nervous system is completed by this age and that elevated levels of phenylalanine have only a minimal impact. The decision was also partly financial; the diet costs more than $5000 a year. Many clinicians now prefer to continue the diet until adolescence. This decision is based on research indicating that early termination of therapy can be deleterious, leading to reduced intellectual development and abnormal changes in electroencephalogram patterns. It seems certain that the age at which the diet should be discontinued will now undergo careful reappraisal.

Breakfast

- ⅔ cup ready-to-eat rice cereal
- ½ banana
- 6 oz formula

Lunch

- ½ can vegetable soup
- 3 crackers
- 1 cup fruit cocktail
- 4 oz formula

Dinner

- 2 cups low-protein noodles
- ½ cup meatless spaghetti sauce
- 1 cup salad (lettuce)
- French dressing
- 4 oz formula

Snack

- ½ cup popcorn
- 1 tablespoon margarine

synthetic mixture of amino acids (with very low levels of phenylalanine) is used as a protein substitute. The challenge is to maintain a level of phenylalanine in the blood that is high enough to permit normal development of the nervous system and yet low enough to prevent mental retardation. Neurological development is thought to be completed by the age of 6 years, at which time the restrictive diet can be withdrawn (but see Concepts & Controversies above).

All states now require screening of newborns for PKU, so the number of untreated cases is very low. Screening and treatment allows PKU homozygotes to lead essentially normal lives. However, as PKU children have matured and reached reproductive age, an unforeseen problem has developed. All children of PKU homozygous females have turned out to be mentally retarded. Apparently the high levels of phenylalanine in the maternal circulation cross the placenta and damage the nervous system of the developing fetus in a way that is independent of the child's genotype. It is now recommended that PKU females return to a low phenylalanine diet before conception and maintain it throughout pregnancy. We will examine this problem in more detail in Chapter 18.

Several other genetic diseases are associated with metabolic blocks in the pathways leading from phenylalanine. The pathway from tyrosine leads to the production of the thyroid hormones thyroxine and triiodothyronine. A block in this pathway causes the recessive autosomal disease **genetic goitrous cretinism** (Figure 10.3). In this disorder the newborn child is unaffected because maternal thyroid hormones are able to cross the placenta and promote normal growth. However, in the weeks following birth, physical development is slow, mental retardation occurs, and the thyroid gland greatly enlarges. This condition is caused by the failure to synthesize a metabolic end product (a hormone) and not by the accumulation of a metabolic intermediate as in PKU. This form of cretinism can be treated by administration of the metabolic end product, thyroid hormone, obtained from animal sources.

Genetic goitrous cretinism
A hereditary disorder in which the failure to synthesize a needed hormone produces physical and mental abnormalities.

In this same network of pathways, the failure to convert homogentisic acid to maleylacetoacetic acid is the cause of the recessive condition **alkaptonuria.** This was the disorder first investigated by Garrod at the turn of the century. He postulated that it was under genetic control; this idea was confirmed in 1958 when the enzyme responsible for the conversion, homogentisic acid oxidase, was identified. Because of the metabolic block, excess homogentisic acid accumulates in the body and is excreted in the urine. The excess acid is converted to a dark pigment in cartilage areas that are exposed to light. As a result there is often discoloration of the ears, tip of the nose, palate, and whites of the eyes. Deposition in cartilage also produces a form of arthritis in later life.

Alkaptonuria
A relatively benign autosomal recessive genetic disorder associated with the excretion of high levels of homogentisic acid.

The most common form of **albinism** is an autosomal recessive condition known as *oculocutaneous albinism* (OCA), associated with a lack of pigment (melanin) synthesis in the skin, hair, and eyes. Garrod suggested in 1908 that albinism might be a genetic disorder. It is now known that OCA results from a lack of activity of the enzyme tyrosinase (Figure 10.3). Tyrosinase is a copper-containing enzyme that catalyzes the first two steps in melanin biosynthesis: the oxidation of tyrosine to DOPA and the conversion of DOPA to dopaquinone, a precursor to melanin. As a result, individuals with OCA lack melanin pigmentation and are phenotypically albino. In this condition and the others we have discussed, a mutation in DNA results in an absent or defective gene product that, in turn, results in an abnormal phenotype.

Albinism
In humans, a complex of genetic disorders associated with the inability to synthesize the pigment melanin. The most common form is inherited as an autosomal recessive trait.

Genetic Disorders in Society, History, and Art

It is difficult to speculate when the existence of heritable traits in humans became recognized. As described in Chapter 1, heritable disorders such as cyclopia often appear in myths and legends from a diverse array of cultures. In many societies, social roles were assigned in a hereditary fashion, from prophets and priests to kings and queens. This recognition of what were regarded as heritable traits helped shape the development of culture and social customs.

In ancient civilizations, the birth of malformed individuals was regarded as a sign of war or famine rather than the result of a malformation or genetic disorder. For example, more that 60 types of birth defects are recorded on clay tablets from ancient Babylonia, along with the dire meaning of such births. In later societies, ranging from that of the Romans to 18th-Century Europe, some malformed individuals (such as dwarfs) were regarded as curiosities rather than figures of impending doom, and were highly prized by royalty as courtiers and entertainers.

Whether from a sense of curiosity, fear, or perhaps the urge to record the many variations of the human form, both famous and anonymous individuals afflicted with genetic disorders have been portrayed in painting, sculpture, and other forms of the visual arts. What is striking about these portrayals is that in spite of the differences in time or culture, they are often very detailed and highly accurate. In fact, across time, culture, and artistic medium, affected individuals often resemble each other more closely than they do their siblings or peers. In many cases, the portrayals allow the condition to be clearly diagnosed at a distance of several thousand years (see Figure 4.14, for a case of achondroplastic dwarfism in an Egyptian who lived almost 5000 years ago).

The following section presents a series of artistic representations of individuals afflicted with genetic disorders. This section reflects the long-standing link that exists between science and the arts in many cultures, not as a gallery of freaks or monsters, but to remind us that being human encompasses a wide range of conditions.

A more thorough discussion of genetic disorders in art can be found in the book *Genetics and Malformations in Art* by J. Kunze and I. Nippert, published by Grosse Verlag, Berlin, in 1986.

■ **Tutankhamun,** artist unknown (circa 1350 B.C.). The Pharaoh Tutankhamun showing minor breast development and sagging abdominal wall, features often found in Klinefelter syndrome.

■ **The Riddle of Nijmegan,** Abraham van Wessel (1658). In this painting, which hangs in the town hall of Nijmegan, The Netherlands, a genetic riddle is posed.

The woman portrayed cradling an old man says: Listen to my declaration. The two in red are my father's brothers. The two in green are my mother's brothers. The two in white are my children and I, mother, have from these six the father as my husband, without consanguinity relations forbidding this.

The two sons in red say: We would hate it would not be known that our niece was given to our father, because she is not our father's niece what nobody would easily guess.

The two in green say: it is strange to see in this picture that he is our natural father and married our niece but this does not regret us.

The two in white say: the old man is father of all of us. The lady is mother of both of us. But say how can it be that our brothers are our mother's uncles.

The riddle, which can be solved by constructing a pedigree, is a good exercise in genetic problem-solving.

Twaer ons leet soo waer achter bleven
Onse lucht en waer onsen vader gegeven
Want sy en is niet onses vaders lucht
Twelck niemant en sal geraden licht
Het is wonder te mercken in deser figuren
Want hy is onsen vader inder naturen
Ende heeft onse lucht getoont
Nochtans ons des niet en beroont

■ **Eugenia Martonez Vellago,** Juan Carreno de Miranda (1614-1685). A child portrayed as Bacchus exhibiting short stature, short extremities, obesity, and facial features associated with Prader-Willi syndrome.

■ **Mother with deformed child,** Goya (1746-1828). The child on the mother's lap has a lack of limb development characteristic of Robert syndrome, an autosomal recessive disorder associated with functional defects of chromatin in the pericentromeric regions.

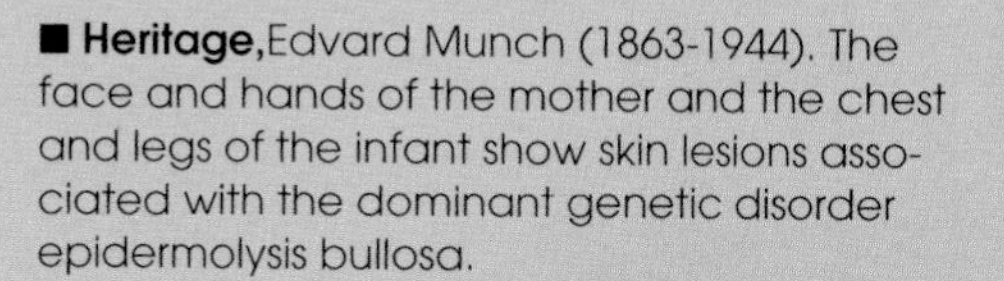

■ **Heritage,**Edvard Munch (1863-1944). The face and hands of the mother and the chest and legs of the infant show skin lesions associated with the dominant genetic disorder epidermolysis bullosa.

Member of the Gonzales Family, artist unknown (painted between 1579-81). this individual shows the phenotype associated with the autosomal recessive disorder langinous hypertrichosis, which causes the development of excess facial hair.

Isabella Eugenia with dwarf, artist unknown. A dwarf with features of STH deficiency.

At the Moulin Rouge, Henri Toulouse-Lautrec (1864-1901). Toulouse-Lautrec, pictured in the center background, standing next to the tall gentleman in the top hat, exhibited many of the symptoms associated with the recessive disorder, pynkodystosis, including brittle bones and failure of the fontanel (soft spot on a baby's skull) to close.

Marriage

■ **Portrait of Count Boruwalski,** artist unknown (1759). The body proportions of the Count are consistent with those seen in pituitary dwarfism I (primordial dwarfism), where the proportion of body parts to one another are normal, but the individual is a dwarf. This trait is inherited as an autosomal recessive, but some cases are sporadic. The disorder is caused by a deficiency of growth hormone, with all other endocrine functions being normal.

■ **Marriage à la Mode,** by William Hogarth (circa 1750). This painting shows the negotiations involved in a marriage contract in 18th century England. The importance of family history in such negotiations is emphasized by the scroll at the right, depicting William the Conqueror, with a family tree emerging from his navel.

■ **Joan of Arc,** Antoine Dufor (painted about 1505). From the physical descriptions provided at the trials of Joan of Arc, including her lack of pubic hair and amenorrhea, it has been speculated that she had the sex-linked recessive disorder testicular feminization.

■ **AmongThose Left,** Ivan Albright, Le Lorraine (1928-29). The blacksmith portrayed here has many of the features of Noonan syndrome, an autosomal dominant trait associated with short stature, shield-like chest, low-set ears, and webbed neck. This diagnosis, made from a portrait, was confirmed by examination of a great-grandson of the blacksmith who also had Noonan syndrome, illustrating that portraits are often accurate representations of genetic disorders.

Carbohydrate Metabolism

Carbohydrates are organic molecules that include sugars, starches, glycogens, and celluloses. The simplest carbohydrates are sugars known as monosaccharides (Figure 10.4). Those with five carbon atoms are known as pentoses, and we have already considered the role of pentoses in nucleotide structure. The sugars with six carbon atoms are known as hexoses and include glucose, galactose, and fructose. These sugars are important as

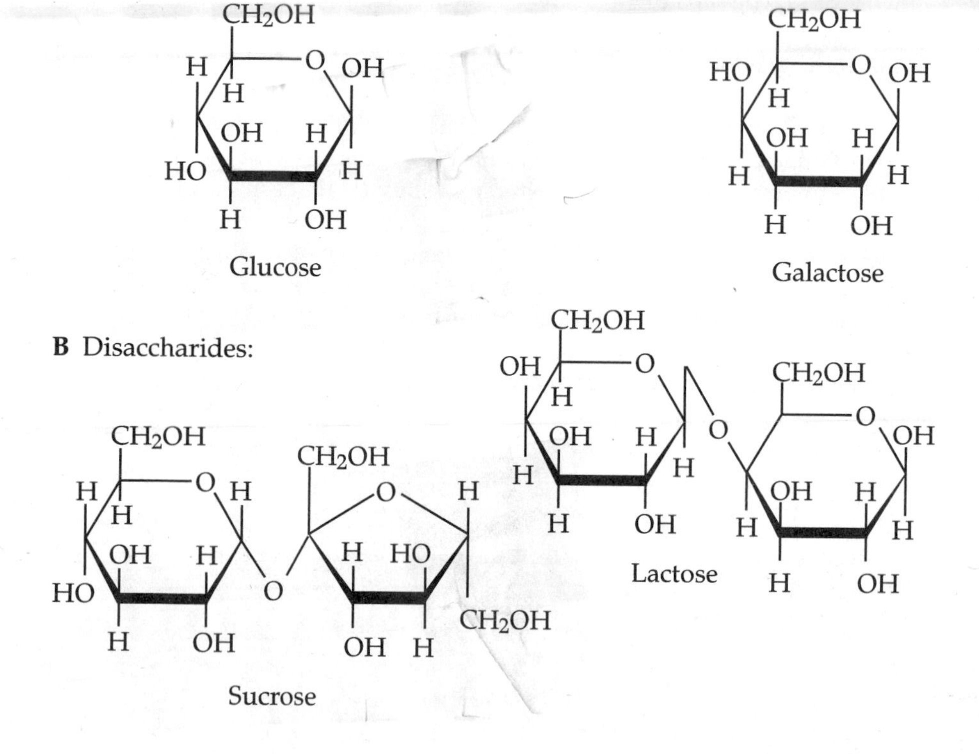

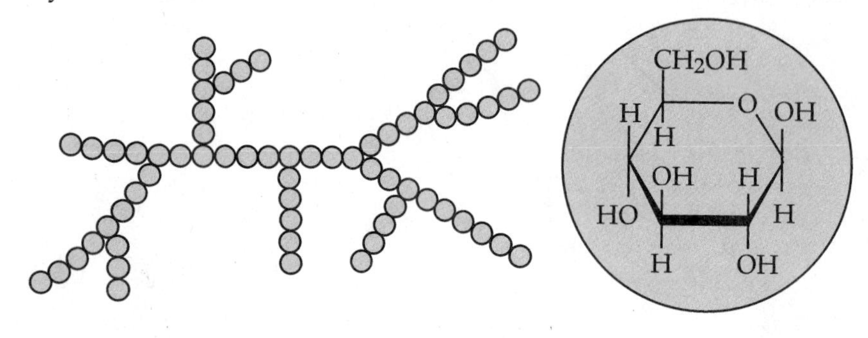

FIGURE 10.4 Some common carbohydrates. (A) Chemical structure of the monosaccharides glucose and galactose. Each contains the same number of carbon, hydrogen, and oxygen atoms, arranged in a different fashion. (B) Structure of the disaccharides sucrose and lactose. (C) Portion of the structure of the polysaccharide glycogen. Each circle represents a glucose molecule *(right).*

metabolic energy sources. A combination of two monosaccharides produces a disaccharide (Figure 10.4). Some common disaccharides include maltose (two glucose units, used in brewing beer), sucrose (a glucose and fructose unit, the sugar you buy at the store), and lactose (a glucose and a galactose unit, found in milk). Larger combinations of sugars form polysaccharides, including glycogen, starch, and cellulose (Figure 10.4). In animals, including humans, the principal storage form of carbohydrate is glycogen, a molecule composed of long chains of glucose units.

As mentioned earlier, the first human genetic disorder linked with an enzyme deficiency is a defect in splitting sugar molecules from glycogen (glycogen storage disease type I). Many different enzymes are required for the reactions that convert other sugars into glucose, for those that store glucose as glycogen, and for the reactions that release glucose from glycogen. Metabolic blocks in any of these reactions can have serious consequences. Some of the genetic disorders associated with the metabolism of glycogen are listed in Table 10.1.

Galactosemia
A heritable trait associated with the inability to metabolize the sugar galactose. If it is left untreated, high levels of galactose-1-phosphate accumulate, causing cataracts and mental retardation.

Several genetic disorders are associated with the metabolism of simple sugars. **Galactosemia** is an autosomal recessive disease that results from the inability to metabolize galactose, a monosaccharide that is part of lactose, the sugar in human milk (Figure 10.5). It occurs with a frequency of 1 in 57,000 births and is caused by lack of the enzyme galactose 1-phosphate uridyl transferase. In the absence of this enzyme, the metabolic intermediate galactose 1-phosphate accumulates to toxic levels in the body. Homozygous recessive individuals are unaffected at birth but develop symptoms a few days later. They begin with gastrointestinal disturbances, dehydration, and loss of appetite; later symptoms include jaundice, cataract formation, and mental retardation. In severe cases the condition is progressive and fatal, with death occurring within a few months; but mild cases may remain

TABLE 10.1 Some Inherited Diseases of Glycogen Metabolism

Type	Disease	Metabolic Defect	Inheritance	Phenotype
I	Glycogen storage disease—Von-Gierke disease	Glucose-6-phosphatase deficiency	Autosomal recessive	Severe enlargement of liver, often recognized in second or third decade of life; may cause death due to renal disease
II	Pompe disease	Lysosomal glucosidase deficiency	Autosomal recessive	Accumulation of membrane-bound glycogen deposits. First lysosomal disease known. Childhood form leads to early death.
III	Forbes disease, Cori disease	Amylo 1, 6 glucosidase deficiency	Autosomal recessive	Accumulation of glycogen in muscle, liver. Mild enlargement of liver, some kidney problems.
IV	Amylopectinosis, Andersen disease	Amylo 1, 4 transglucosidase deficiency	Autosomal recessive	Cirrhosis of liver, eventual liver failure, death.

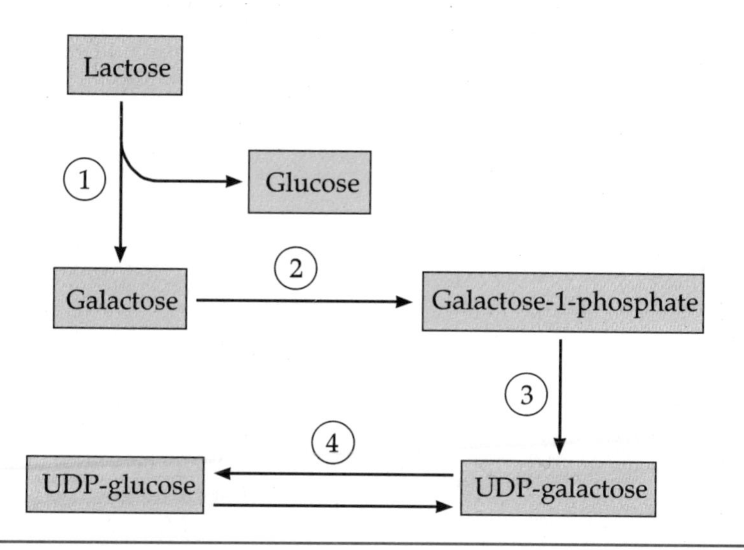

FIGURE 10.5 Metabolism of lactose, the main sugar of milk. In this pathway lactose is hydrolyzed by lactase (enzyme 1) into glucose and galactose. The second step, using galactose as a substrate, is catalyzed by galactokinase (enzyme 2) and produces galactose-1-phosphate. This, in turn, is converted into UDP-galactose by galactose-1-phosphate uridyl transferase (enzyme 3). Genetic loss of this enzyme causes a buildup of galactose-1-phosphate and results in the phenotype of galactosemia.

undiagnosed for many years. A galactose-free diet and the use of galactose- and lactose-free milk substitutes and foods leads to a reversal of all symptoms. But unless treatment is started within a few days of birth, mental retardation cannot be prevented.

Galactosemia is also an example of a multiple allele system. In addition to the normal allele, *G*, and the recessive allele, *g*, a third allele, known as G^D (the Duarte allele, named after Duarte, California, the city in which it was discovered), has been found. Homozygous G^D/G^D individuals have only half the normal enzyme activity but show none of the symptoms of the disease. The existence of three alleles produces six possible genotypic combinations, with enzyme activities ranging from 100% to 0% (Table 10.2). This disease can be detected in newborns, and mandatory screening programs in some states test all newborns for galactosemia.

TABLE 10.2 Multiple Alleles of Galactosemia

Genotype	Enzyme Activity	Phenotype
G^+/G^+	100%	Normal
G^+/G^D	75%	Normal
G^D/G^D	50%	Normal
G^+/g	50%	Normal
G^D/g	25%	Borderline
g/g	0%	Galactosemia

Unlike the dietary restrictions for PKU, which can be lifted after a few years, it is recommended that galactosemics remain on a galactose-free diet throughout life because the buildup of galactose 1-phosphate that occurs on a normal diet is highly toxic and cataracts are a result. For those individuals homozygous for the G^D allele, a galactose-free diet is advised for at least 2 years.

Two other disorders of carbohydrate metabolism, fructosuria and pentosuria, are noteworthy because they demonstrate that not all metabolic blocks have devastating effects. Although these genetic disorders are caused by the lack of a specific enzyme and the accumulation of metabolic intermediates, they do not produce clinically significant symptoms. The phenotypes in these diseases are characterized by high levels of sugars in the blood and urine. **Fructosuria** is an autosomal recessive condition with a frequency of about 1 in 130,000 and is caused by a lack of the enzyme fructokinase in the liver, kidney, and intestine. **Pentosuria** is caused by the failure to convert the 5-carbon sugar xylulose to xylitol by the lack of the enzyme xylitol dehydrogenase. This autosomal recessive condition is found almost exclusively in Jews of East European (Ashkenazi) descent. The incidence is estimated to be 1 in 2500 among Jews in the United States and 1 in 5000 among Israeli Jews.

Fructosuria
An autosomal recessive condition associated with the inability to metabolize the sugar fructose, which accumulates in the blood and urine.

Pentosuria
A relatively benign genetic disorder of sugar metabolism characterized by the accumulation of xylulose in the blood and urine.

Nucleic Acid Metabolism

Inherited disorders of nucleic acid metabolism fall into two general categories; those concerned with the synthesis and/or reuse of purines and pyrimidines and those that are caused by a defect in DNA repair mechanisms (Table 10.3). In this section we will consider only one of the former: **Lesch-Nyhan syndrome.** Defects in nucleic acid metabolism that have their primary effects on the immune system will be considered in Chapter 14, and those disorders affecting DNA repair will be examined in Chapter 11.

Lesch-Nyhan syndrome
An X-linked recessive condition associated with a defect in purine metabolism that causes an overproduction of uric acid.

In 1964, M. Lesch and W. L. Nyhan described an inherited disorder characterized by renal failure, spastic movements, mental retardation, high levels of uric acid, and a strong tendency for self-mutilation. They postulated that this condition was caused by an inborn error in purine metabolism, and in 1967, J. E. Seegmiller and his colleagues demonstrated that the disease was caused by a lack of the enzyme hypoxanthine-guanine phosphoribosyl transferase (HGPRT) (Figure 10.6). Pedigree analysis has established that the disease is an X-linked recessive that affects 1 in 10,000 males. Hemizygous males are unaffected at birth, but delays in motor development appear at about 3 months. Later, spastic movements become apparent, but the most striking feature of this disease is the compulsive self-mutilation that usually appears between 2 and 4 years of age. Unless restrained, affected individuals bite off pieces of their fingers, lips, and cheeks. Affected individuals have an excessive production of purines and uric acid. The drug allopurinol is often used to lower uric acid levels and to reduce kidney damage; but no treatment is currently available to control the neurological symptoms, and the disease is progressive and fatal. The human gene for HGPRT has been cloned, and gene replacement therapy has been proposed for this condition.

TABLE 10.3 Some Inherited Diseases of Nucleic Acid Metabolism

Disease	Metabolic Defect	Inheritance	Phenotype
Lesch-Nyhan syndrome	Hypoxanthine-guanine phosphoribosyltransferase	X-linked recessive	Mental retardation, motor impairment, self-mutilation, kidney failure, early death
Hereditary xanthinuria	Xanthine oxidase	Autosomal recessive	Urinary purine is xanthine rather than uric acid; xanthine crystals in muscles; painful, debilitating disease
Severe combined immunodeficiency	Adenosine deaminase	Autosomal recessive	Reduced or absent immunoglobulins, defective immune response, reduced growth, recurrent infections, early death
Defective T-cell immunity	Purine nucleoside phosphorylase	Autosomal recessive	Normal level of immunoglobulins, severe lack of lymphocytes, recurring infections of respiratory system
Orotic aciduria	Oritidine-5′-phosphate decarboxylase and/or orotate phosphoribosyl transferase	Autosomal recessive	Lethargy; failure to grow and develop, moderate behavioral developmental delays, some mental retardation, excretion of high levels of orotic acid
Xeroderma pigmentosum	Defective DNA repair after UV damage	Autosomal recessive	Exposure to sunlight causes skin lesions, pigment deposition, and malignancies.
Ataxia telangiectasia	Defective DNA repair after gamma (x-ray) irradiation	Autosomal recessive	Skin lesions all over body, sensitivity to x-rays, some mental retardation, increased risk of cancer

Lipid Metabolism

The biochemistry of lipids is a large and complex area of human metabolism, and many hereditary diseases associated with the synthesis or breakdown of lipids have been reported. Several of these are listed in Table 10.4. In this section we will concentrate on two of these disease, **Tay-Sachs disease** and **Sandhoff disease,** because they illustrate the sometimes complex relationship between a polypeptide gene product and a functional protein.

A fatal disease of early childhood associated with progressive mental deterioration and blindness was first described by the British opthalmologist W. Tay in 1818 and again by the American neurologist B. Sachs in 1896. Affected infants are normal at birth, but by the age of 6 months they become listless and weak and show difficulty in feeding. Sudden noises provoke an exaggerated startle reaction, with rapid extension of both arms. There is a progressive loss of motor function, with delays in learning to sit and stand; few Tay-Sachs children ever walk. In most patients a characteristic cherry-red

Tay-Sachs disease
A fatal condition, controlled by an autosomal recessive gene, that causes deposition of unmetabolized intermediates in the nervous system.

Sandhoff disease
A genetic disorder inherited as an autosomal recessive associated with a defect in production of hexosaminidase B, causing symptoms similar to those of Tay-Sachs disease.

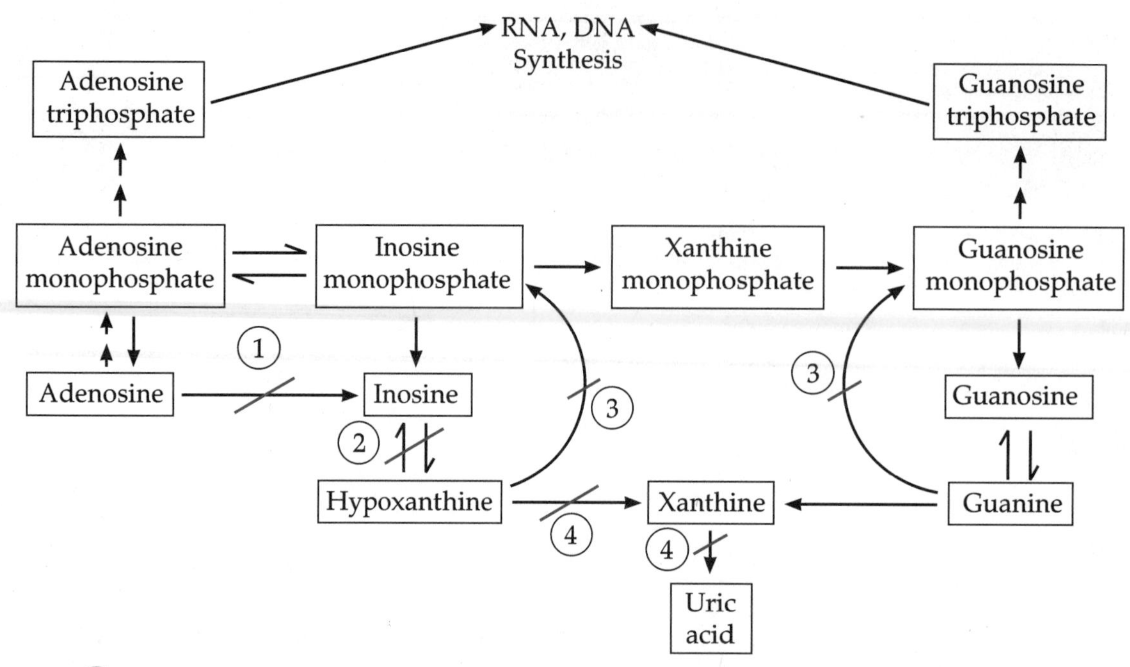

FIGURE 10.6 Biosynthetic pathways involved in purine metabolism. The metabolic blocks for several diseases involving nucleic acids are indicated.

spot develops on the retina of the eye. Around 12 months of age there is a rapid decrease in mental and motor functions and onset of blindness, deafness, rigidity, and brain enlargement. Death usually occurs around 3 years of age from pneumonia or a lung infection.

Tay-Sachs disease is inherited as an autosomal recessive and in most populations has a frequency of less than 1 in 100,000, with a heterozygote frequency of around 1 in 300. However, in some populations the frequency of the disease is much higher. Among Jews of East European descent (Ashkenazi), as many as 1 in 30 individuals are carriers. There can be several reasons for large differences in gene frequency among populations, and these will be explored in Chapter 17.

Ultrastructural studies of the brain and nervous system of Tay-Sachs patients show degeneration of cell structures and the accumulation of membranous whorls within cells. Biochemical analysis of this stored material shows high concentrations of a lipid called GM2-ganglioside. Other workers

TABLE 10.4 Some Inherited Diseases of Lipid Metabolism

Disease	Metabolic Defect	Inheritance	Phenotype
Niemann-Pick disease	Sphingomyelinase	Autosomal recessive	Enlarged liver, progressive loss of motor and intellectual functions, increased levels of sphingomyelin, cholesterol in brain, liver; death by 1 to 4 years
Farber disease	Ceramidase	Autosomal recessive	Swelling of joints, impaired growth, subcutaneous nodules; death by 1 to 2 years
Fabry disease	Alpha galactosidase	X-linked recessive	Onset in late childhood or early adolescence; skin lesions, episodes of intense pain in extremities, progressive renal failure; death before middle age
GM1-gangliosidosis	Beta galactosidase	Autosomal recessive	Progressive loss of motor function is mental retardation, convulsions; death by 2 years
GM2-gangliosidosis (Tay-Sachs disease)	Hexosaminidase A	Autosomal recessive	Normal at birth, progressive loss of motor functions, development of mental retardation, cherry-red spot on retina, convulsions, rigidity; death by 3 years
Sandhoff disease	Hexosaminidase A, B	Autosomal recessive	Same as Tay-Sachs
Wolman disease	Acid esterase	Autosomal recessive	Persistent vomiting, diarrhea, enlarged liver, elevated levels of triglycerides in liver, spleen; death by 1 year

showed that Tay-Sachs patients lacked an enzyme known as hexosaminidase. Two forms of this enzyme, known as hexosaminidase A and B were discovered. Tay-Sachs patients are lacking hexosaminidase A activity but have normal levels of hexosaminidase B.

In 1968 K. Sandhoff described a patient with the same symptoms as Tay-Sachs disease, but this individual was totally lacking *both* the A and B forms of hexosaminidase. Subsequent work showed that this disease is also inherited in an autosomal recessive fashion. This discovery stimulated a great deal of research into the structure and function of the hexosaminodases and the nature of the molecular defect in both Tay-Sachs and Sandhoff disease. This effort has employed the techniques of biochemistry, immunology, and somatic cell hybridization. While the story has turned out to be more complex than anticipated, several general conclusions can be drawn.

Hexosaminidase A is composed of two different subunits, alpha and beta. Hexosaminidase B is composed only of beta units. Another form of hexosaminidase, known as hex S, is composed only of alpha units. One way to demonstrate the relationships between these forms of the enzyme is through dissociation-reassociation experiments (Figure 10.7). Purified hex A

FIGURE 10.7 Dissociation/reassociation experiments with hexosaminidase A. The random reassociation of alpha and beta subunits results in the formation of all three forms of hexosaminidase.

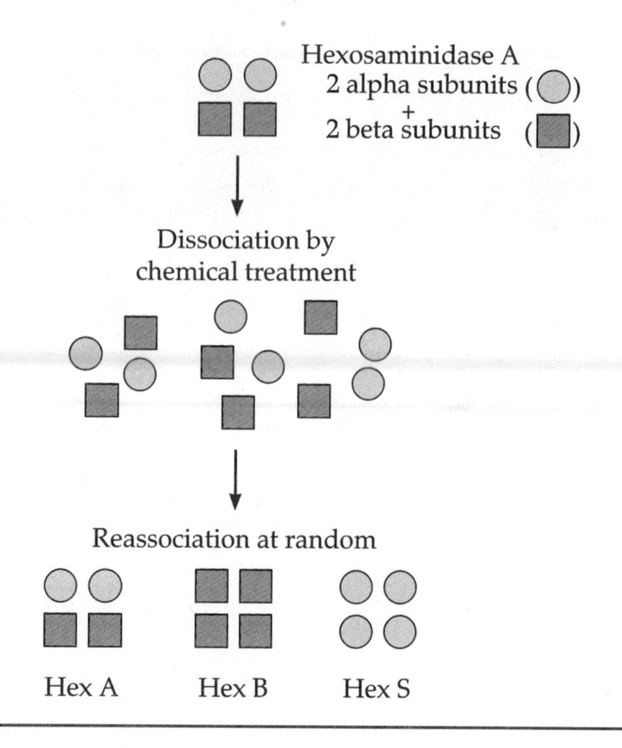

can be separated into its subunits by chemical treatment. If these subunits are permitted to reassociate at random, all three forms, hex A, hex B, and hex S, will result. If a mixture of hex B and hex S is dissociated and reassociated, hex A, hex B, and hex S are formed.

Although several models are possible, most of the evidence indicates that each form of hexosaminidase is composed of four subunits: hex A with $2\alpha + 2\beta$ subunits, hex B with 4β, and hex S with 4α subunits. The gene coding for the α subunit is located on chromosome 15, and the structural gene for the β subunit is located on chromosome 5. Thus hex A is an example of a functional protein composed of polypeptides from two different genes, organized at the quaternary level of protein structure.

Receptor Proteins

Although many, and perhaps most, proteins function as enzymes, proteins also act in other roles, including signal receptor and transducer. These functions usually take place in the plasma membrane of the cell, and mutations in receptor function can have drastic consequences. For example, in testicular feminization (discussed in Chapter 5), a defect in the ability to detect and bind the hormone testosterone leads to a complete change in sexual phenotype, causing a genotypic male to develop into a phenotypic female.

Familial hypercholesterolemia
A dominant autosomal genetic condition associated with a defect in cellular receptors that function in cholesterol metabolism. Affected individuals are susceptible to heart disease and early death.

Another genetic disorder associated with a defect in a cellular receptor is **familial hypercholesterolemia,** an autosomal dominant condition with multiple alleles discussed in Chapter 4. Ingested cholesterol is packaged into particles called low-density lipoproteins (LDL). LDLs are removed from the circulatory system by receptors that project from the cell surface. The major site of uptake and metabolism of cholesterol is the liver. The LDLs are taken into the cell, and the cholesterol is used within the cell for a variety of synthetic reactions, including membrane synthesis. The receptor is recycled and returned to the surface to bind more LDL (Figure 10.8). If the receptors are absent or defective, LDL builds up in the blood and is deposited on the artery walls as atherosclerotic plaque, causing heart disease. Familial hypercholesterolemia is inherited as an autosomal dominant. Heterozygotes have half the usual number of receptors and twice the normal level of LDL; they begin to have heart attacks in their early 30s. About 1 in 500 individuals is a heterozygote. Homozygotes (about 1 in a million people) have two faulty genes for receptor synthesis and have no functional receptors. These individuals have LDL levels about six times normal, with heart attacks beginning as early as 2 years of age. Heart disease is unavoidable by the age of 20 years, with death in most cases by the age of 30 years. Other genetic disorders associated with receptors are listed in Table 10.5.

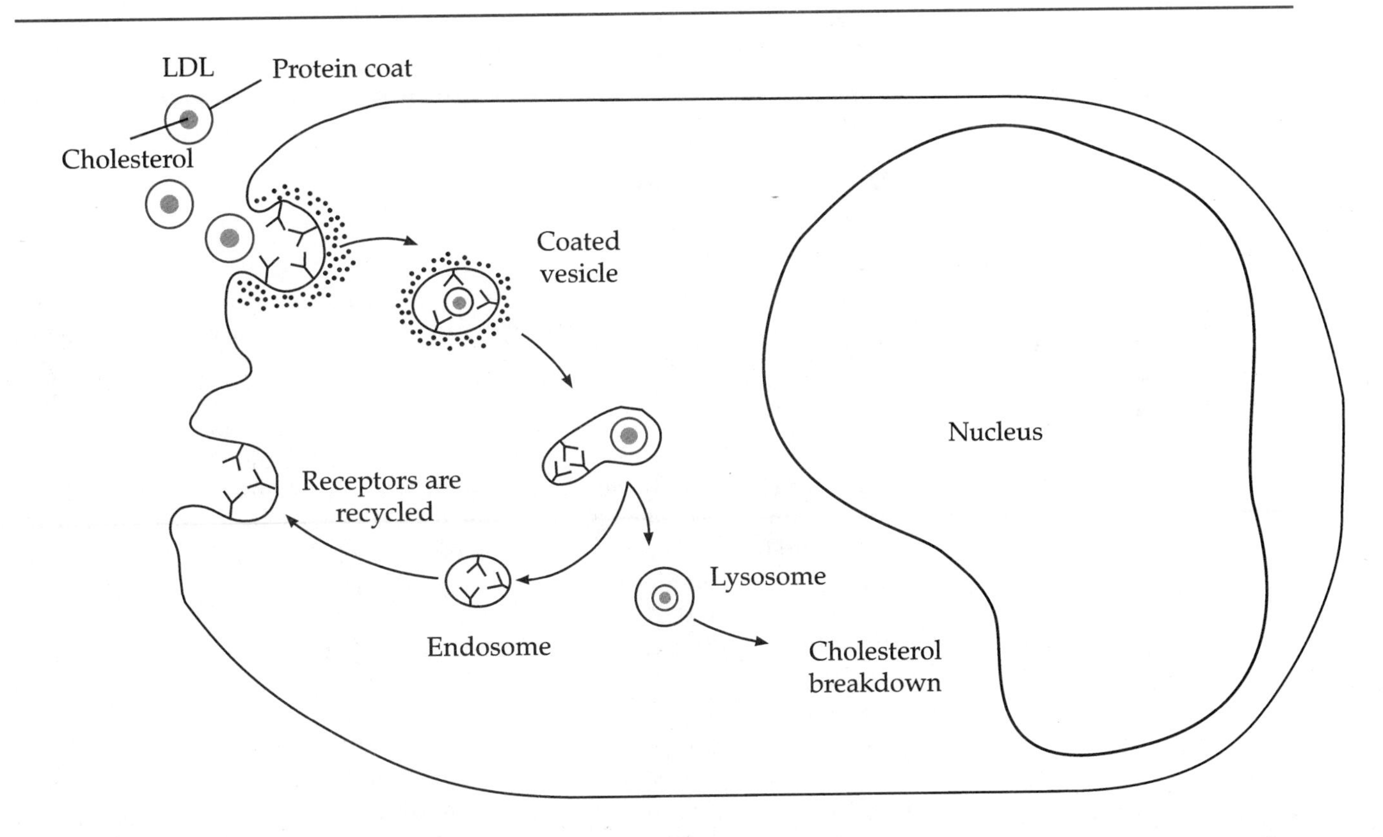

FIGURE 10.8 Recycling of LDL receptors. Low-density lipoproteins containing cholesterol bind to receptors on cell surface and become internalized as coated vesicles. The coated vesicle forms lysosomes containing the LDL and endosomes, containing the receptors. Cholesterol is metabolized in the lysosomes, and the endosomes fuse with the plasma membrane, returning the receptors to the cell surface.

TABLE 10.5 Some Heritable Traits Associated with Defective Receptors

Disease	Defective/Absent Receptor	Inheritance	Phenotype
Familial hypercholesterolemia	Low-density lipoprotein (LDL)	Autosomal dominant	Elevated levels of cholesterol in blood, atherosclerosis, heart attacks; early death
Pseudohypoparathyroidism	Parathormone (PTH)	X-linked dominant	Short stature, obesity, round face, mental retardation
Diabetes insipidus	Vasopressin	X-linked recessive	Failure to concentrate urine; high flow rate of dilute urine, severe thirst, dehydration; can produce mental retardation in infants unless diagnosed early
Testicular feminization	Testosterone	X-linked recessive	Transformation of genotypic male into phenotypic female; malignancies often develop in intra-abdominal testes

TRANSPORT PROTEINS: THE GLOBIN MODEL

Hemoglobin, an iron-containing protein molecule found in red blood cells, is involved in the transport of oxygen from the lungs to the cells of the body. The genetics of hemoglobin occupy a central position in human genetics. The study of hemoglobin led to an understanding of the molecular relationship between genes, proteins, and human disease. In addition, the study of hemoglobin in sickle cell anemia provided the first evidence that mutation results in a change in the amino acid sequence of proteins. The organization of the globin gene clusters has also illustrated the mechanism of evolution at the molecular level and the regulation of gene activity during development. Heritable defects in globin structure or synthesis are well understood at the molecular level and are truly "molecular diseases," as Linus Pauling has called them. In this section we will consider the structure of the hemoglobin molecule, the organization of the globin loci, and some disorders related to globin structure and synthesis.

Hemoglobin is a protein composed of four polypeptide chains, and each chain is associated with a heme group. Heme is an organic molecule that contains an iron-binding site to which oxygen can attach (Figure 10.9). Although there are several different kinds of hemoglobin, the heme group is the same in all cases.

Pseudogenes
A nonfunctional gene that is closely related (by DNA sequence) to a functional gene present elsewhere in the genome.

Adult hemoglobin (designated Hb A) is composed of two different polypeptide chains: 2 α chains and 2 β chains (Figure 10.10). The α and β

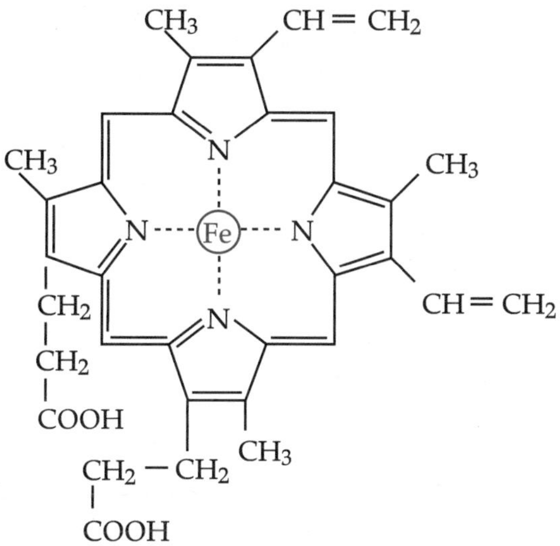

FIGURE 10.9 Structure of the heme group. This planar molecule is inserted into the folds of every globin molecule. The complexed iron at the center is important in binding oxygen and carbon dioxide for transport.

chains are encoded in separate genes on different chromosomes. Adult red blood cells also contain another form of hemoglobin known as Hb A2, composed of 2 alpha chains and 2 delta chains. The delta chains are closely related to the beta chains. Hb A2 constitutes 2% to 3% of the total hemoglobin in a red blood cell. Normally there are between 4.2 and 5.9×10^{12} red blood cells in each liter of blood, and each cell is replaced every 10 to 20 days, so hemoglobin production is an important metabolic function.

Hemoglobinopathies
Disorders of hemoglobin synthesis and function.

Thalassemias
A group of heritable hemoglobin disorders associated with an imbalance in the production of alpha and beta globins.

The α genes are located on the short arm of chromosome 16 (Figure 10.11). In the α cluster there are two copies of the α gene (designated $alpha_1$ and $alpha_2$) and three other genes: the embryonic zeta gene and two pseudogenes, pseudozeta and pseudoalpha-1. **Pseudogenes** are nonfunctional genes that have sequence homology to other genes but that contain mutations that prevent their expression. The beta gene cluster is located on the short arm of chromosome 11 and contains several family members (Figure 10.12). In order, these genes are the embryonic epsilon gene, ^{G}gamma and ^{A}gamma, a beta pseudogene, the delta gene, and the beta gene.

The alpha and beta globin genes have a similar internal organization and contain three exons and two introns (Figure 10.13). The alpha and beta gene products are similar in size and amino acid composition. The alpha globin polypeptide is 141 amino acids long, and the beta globin molecule is 146 amino acids in length. The amino acid sequences of the two polypeptides are very similar (see Figure 10.14 on page 277). Because of the similarities in amino acid sequence, the alpha and beta genes fold into similar configurations, each cradling the heme group within internal folds of the polypeptide chain.

The heritable disorders of hemoglobin fall into two categories: the **hemoglobinopathies,** or hemoglobin variants, which involve quantitative and qualitative changes in the amino acid sequence of globin polypeptides, and the **thalassemias,** characterized by imbalances in globin synthesis.

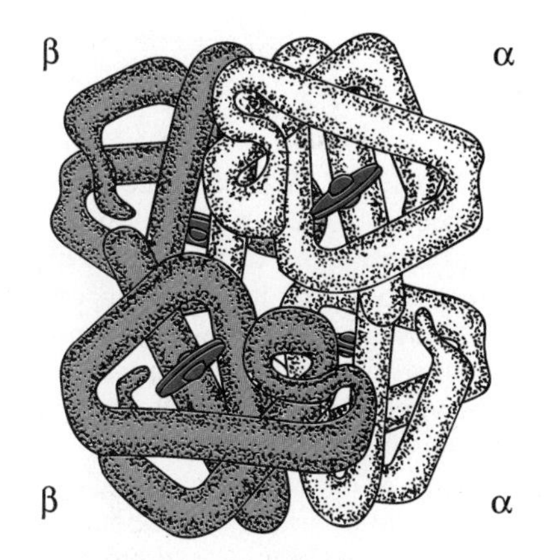

FIGURE 10.10 The quaternary structure of an intact hemoglobin molecule. Two alpha and two beta polypeptide chains interact to form the functional molecule.

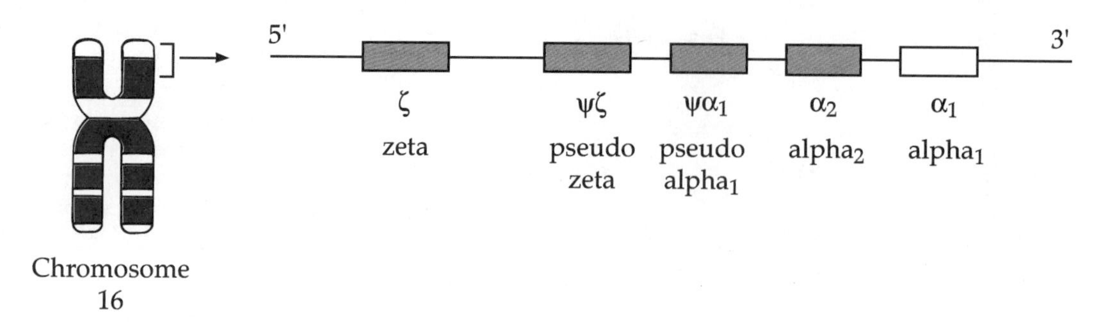

FIGURE 10.11 Chromosomal location and molecular organization of the alpha globin gene cluster on chromosome 16.

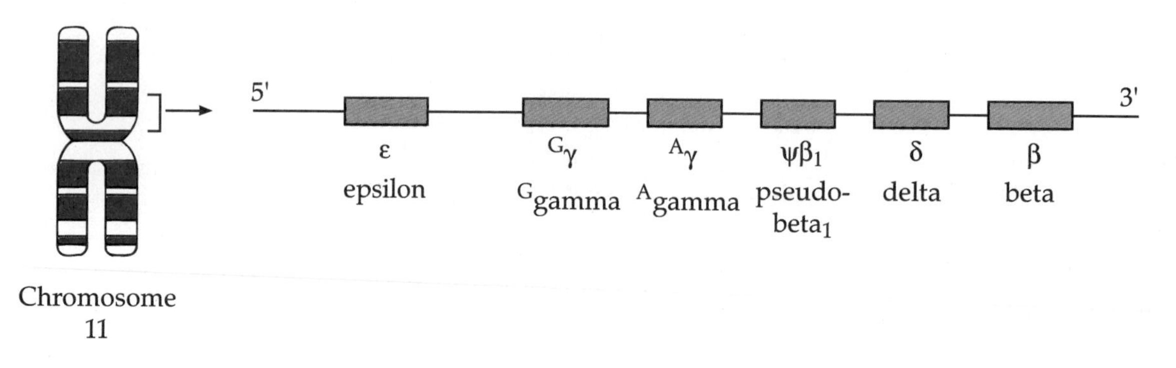

FIGURE 10.12 Chromosome location and molecular organization of the beta globin gene cluster on chromosome 11.

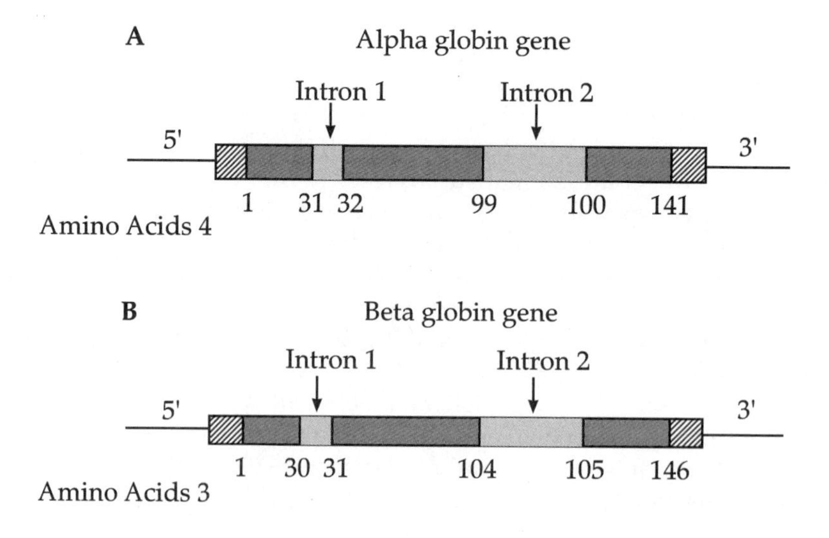

FIGURE 10.13 (A) Detailed structure of the alpha globin gene, showing organization of the exons and introns. The numbers along the bottom represent the amino acids in the polypeptide product. (B) Detailed structure of the beta globin gene. Note that the exons and introns in the beta gene are different in size from those in the alpha gene.

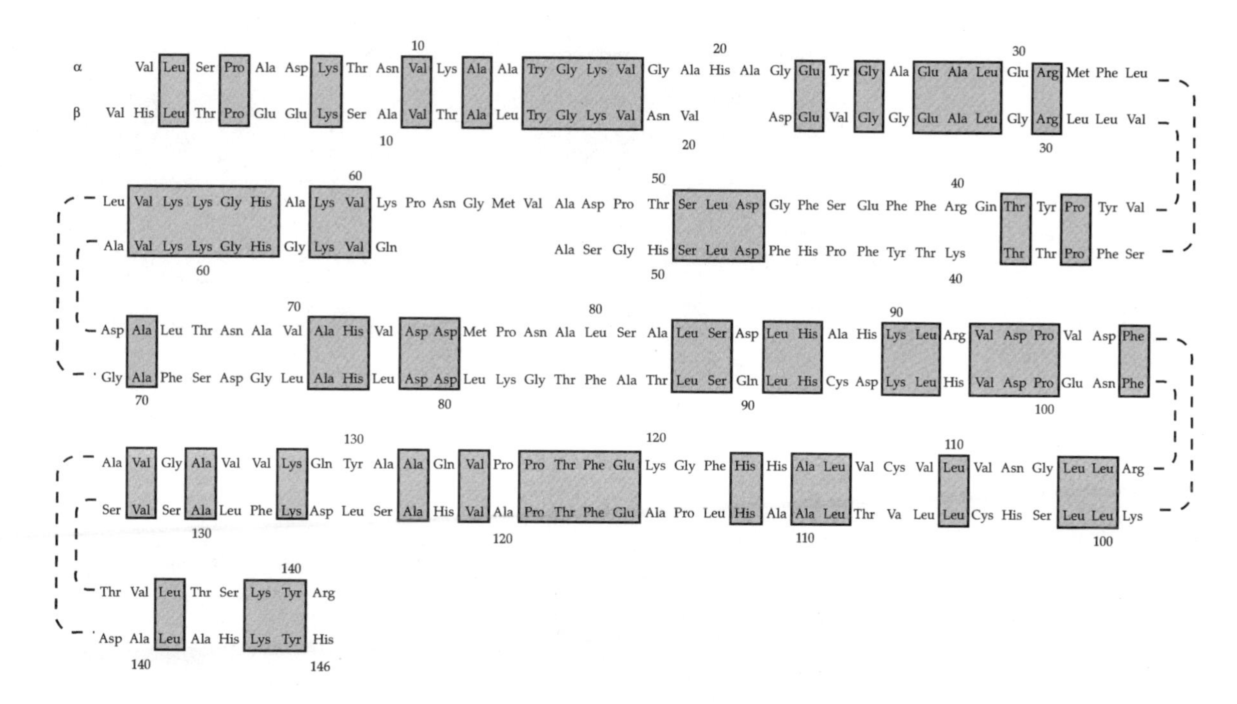

FIGURE 10.14 Amino acid sequence of the alpha *(top)* and beta *(bottom)* globin genes. Shaded boxes indicate regions of homology. Gaps at positions 54–58 in the alpha globin gene are the result of alignment to produce maximum homology.

Hemoglobin Variants

Well over 300 variants of hemoglobin have been described, each originating as the result of genetic mutation. Over 90% of the known variants are caused by the substitution of one amino acid for another in the globin chain, and over 60% of these are found in beta globin.

The simplest type of variant involves the substitution of one amino acid for another. The vast majority of these variants have been described in only one family or in small populations. However, this category also includes the Hb S and Hb C variants, which occur with high frequencies in certain populations and are associated with severe (Hb S) or mild (Hb C) phenotypic symptoms. Some of these variants are listed in Table 10.6. Note that many variants have no detectable phenotypic effect.

Hb S is an abnormal hemoglobin that becomes insoluble in the deoxygenated state. The insoluble molecules polymerize into long tubular structures that distort the membrane of the red blood cell (Figure 10.15). The result is the characteristic sickle-shaped cell associated with the autosomal recessive disease sickle cell anemia (SCA). The deformed blood cells break easily, producing anemia; the sickled cells also clog blood vessels, producing tissue damage (see micrographs of sickled cells in Chapter 4). Chemical differences between Hb A and Hb S were first demonstrated by Linus Pauling and his colleagues in 1949. They observed that these two forms of

TABLE 10.6 Beta Globin Chain Variants with Single Amino Acid Substitutions

Hemoglobin	Amino Acid Position	Amino Acid	Phenotype
A_1	6	glu	Normal
S	6	val	Sickle cell anemia
C	6	lys	Hemoglobin C disease
A_1	7	glu	Normal
Siriraj	7	lys	Normal
San Jose	7	gly	Normal
A_1	58	tyr	Normal
Hb Boston	58	his	Reduced O_2 affinity
A_1	145	cys	Normal
Bethesda	145	his	Increased O_2 affinity
Fort Gordon	145	asp	Increased O_2 affinity

hemoglobin differed in their rate of migration in an electrical field. Since the R groups on many amino acids are electrically charged, they concluded that the amino acid structure of the two molecules must be different. This difference was confirmed by Vernon Ingram, who demonstrated that the

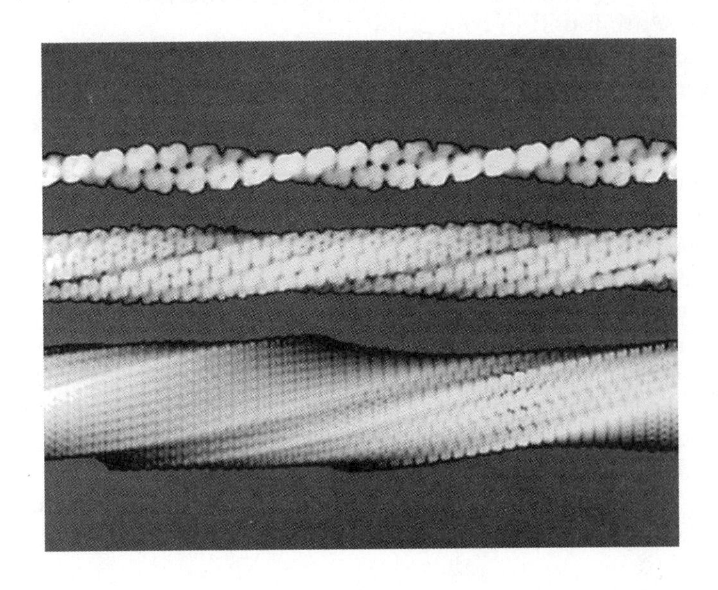

FIGURE 10.15 A computer-generated image of sickle cell hemoglobin stacked into rodlike structures. *Top:* a pair of interlocking fibers composed of hemoglobin. *Middle:* seven pairs of fibers forming the structure that distorts cell shape in sickle cell anemia. *Bottom:* Larger structure composed of many fibers of stacked hemoglobin molecules.

difference between Hb A and Hb S was a single amino acid substitution at position 6 in the beta chain. This alteration in a single amino acid is the molecular basis of sickle cell anemia. All the symptoms of the disease and its inevitably fatal outcome, if untreated, derive from this alteration of 1 amino acid out of the 146 found in beta globin.

It is interesting to compare the effects of another amino acid substitution at position 6 in the beta globin chain. Hb C is slightly insoluble in the deoxygenated state. It forms intracellular crystals that make the erythrocyte membrane more rigid and lead to a reduced life span for the affected red blood cells. This produces a mild form of anemia, with accompanying enlargement of the spleen, but there is almost never a need for clinical treatment for this condition. The sequence of the first seven amino acids in Hb A, Hb S, and Hb C are shown below:

	1	2	3	4	5	6	7
Hb A:	val	his	leu	thr	pro	glu	glu
Hb S:	val	his	leu	thr	pro	val	glu
Hb C:	val	his	leu	thr	pro	lys	glu

Not all amino acid substitutions in the globin genes are associated with a genetic disorder. The known substitutions at position 7 (Table 10.6) in the beta globin chain have no detectable effects, although these variations are inherited in the same autosomal recessive fashion as sickle cell anemia.

Thalassemias

The **thalassemias** are a group of inherited disorders in which the synthesis of globin polypeptide chains is reduced or totally lacking. This results in abnormal combinations of globin chains (e.g., a tetramer of four alpha chains) that can have serious and even fatal consequences. Thalassemias are common in several parts of the world, especially the Mediterranean region and Southeast Asia, where 20% to 30% of the population can be affected. The name "thalassemia" is derived from the Greek word *thalassa,* for "sea," emphasizing the fact that this condition was first described in people living around the Mediterranean Sea.

Two types of thalassemia have been described: alpha thalassemia, in which the synthesis of alpha globin is reduced or absent, and beta thalassemia, in which the synthesis of beta chains is affected (Table 10.7). Both types can have more than one cause, and although inherited as autosomal recessives, both types have some effects in the heterozygous condition.

Knowledge that each copy of chromosome 16 carries two functional copies of the alpha globin gene is necessary to understand **alpha thalassemia** (Figure 10.16). One form, known as alpha thalassemia-1, is caused by a deletion of both copies of the alpha gene from one chromosome. In the heterozygous condition, the synthesis of alpha chains is reduced to half the usual level. The homozygote carries a complete deletion of the alpha genes. In both cases non-alpha chains are present in excess and form tetramers of four identical chains (β_4, γ_4). These abnormal hemoglobins cause a mild anemia in the heterozygotes. The homozygous form of alpha thalassemia-1 is fatal, causing intrauterine death, usually at an advanced stage of gestation.

Alpha thalassemia
Genetic disorder associated with an imbalance in the ratio of alpha and beta globin caused by reduced or absent synthesis of alpha globin.

TABLE 10.7 Summary of Thalassemias

Type of Thalassemia	Nature of Defect
α-Thalassemia-1	Deletion of two alpha globin genes/haploid genome
α-Thalassemia-2	Deletion of one alpha globin gene/haploid genome
δ-B-Thalassemia	Deletion of beta and delta genes/haploid genome
Nondeletion α-Thalassemia	Absent, reduced, or inactive alpha globin mRNA
β^0-Thalassemia	Absent, reduced, or inactive beta globin mRNA. No beta globin produced.
β^+-Thalassemia	Absent, reduced, or inactive beta globin mRNA. Reduced beta globin production.

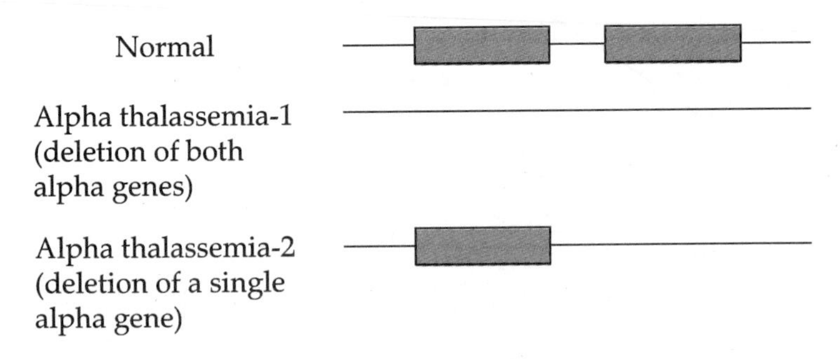

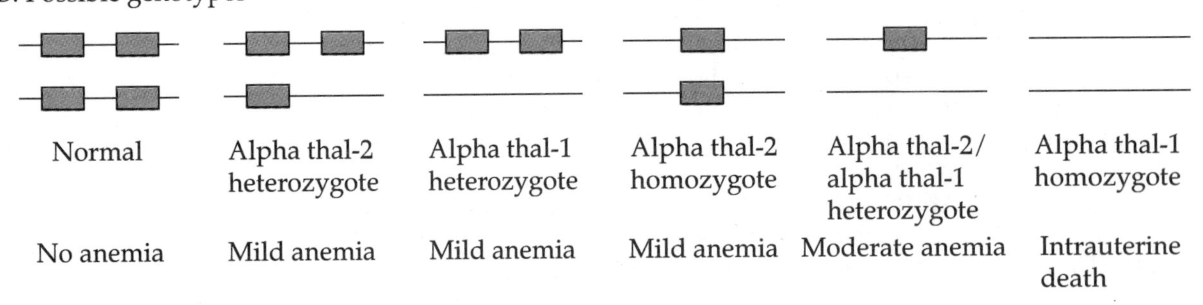

FIGURE 10.16 Mutations that result in alpha thalassemia. (A) Normal configuration of alpha globin genes on chromosome 16. In alpha thalassemia-1, both copies are deleted. In alpha thalassemia-2, one copy is deleted. (B) The possible genotypes involving the normal and mutant alleles and the resulting phenotypes.

Alpha thalassemia-2 is the result of a deletion of one copy of the alpha globin gene. The heterozygote (carrying three copies of the alpha gene) has no detectable symptoms, and the homozygote with two copies of the alpha gene (one on each chromosome) has a mild anemia (Figure 10.16).

The basis of **beta thalassemia** is more subtle and usually does not involve deletion of the structural gene. In some forms of beta thalassemia the underlying defect lies in the conversion of pre-mRNA into a mature mRNA molecule. In β^0 thalassemia, for example, a mutation in one or more of the introns interferes with normal mRNA splicing events, resulting in very low levels of functional mRNA and, in turn, low levels of beta globin. This disease has played an important role in our understanding the details of mRNA processing and translation.

Beta thalassemia
Genetic disorder associated with an imbalance in the ratio of alpha and beta globin caused by reduced or absent synthesis of beta globin.

Structural Proteins: Inherited Disorders of Connective Tissue

Inherited disorders of connective tissue include *osteogenesis imperfecta,* which affects collagen, Marfan syndrome (described in Chapter 4), which affects fibrillin, and *Ehlers-Danlos* syndrome, characterized by hyperflexible skin and joints; the molecular basis of this disorder is still unknown.

Collagen, the protein affected in osteogenesis imperfecta, is one of the most abundant proteins in the body and is an important structural component of connective tissue. It is found in bone, cartilage, tendons, skin, and the arteries of the circulatory system. There are at least 12 types of collagen, and collagen genes are distributed on several different chromosomes. The basic structure common to all types of collagen is a triple helix of three polypeptide chains (Figure 10.17). The three polypeptide chains are first assembled into a precursor molecule called *procollagen,* which is cleaved at each end to yield collagen.

Osteogenesis imperfecta (OI) is a group of inherited connective disorders characterized by collagen defects. Type I OI, associated with a defect in type I collagen, is the most common form of this disorder, and it is inherited as an autosomal dominant trait. Affected individuals have extremely brittle bones that fracture easily, may develop deafness caused by abnormalities of the bones in the middle ear. In addition, the whites of the eyes often appear blue.

The triple helix of type I collagen consists of a 2:1 ratio of alpha 1 to alpha 2 collagen. In one form of type I OI, one copy of the alpha 1 gene is inactive, resulting in a 50% decrease in the amount of type I collagen produced (Figure 10.17). With 50% less collagen, the bones are weaker, are unable to withstand mechanical stress, and fracture easily. The whites of the eyes appear blue because the sclera (a thin fibrous tissue that covers the whites of the eye) is thinner than normal and somewhat transparent, allowing the veins underneath to show through.

Transcriptional Regulators

A class of proteins that is emerging as an important contributor to genetic disorders includes those that regulate gene expression. These *transcription*

FIGURE 10.17 Defects in osteogenesis imperfecta (OI). Type II OI is associated with severe bone deformations and neonatal death. This form of OI is caused by mutations in the type I collagen gene.

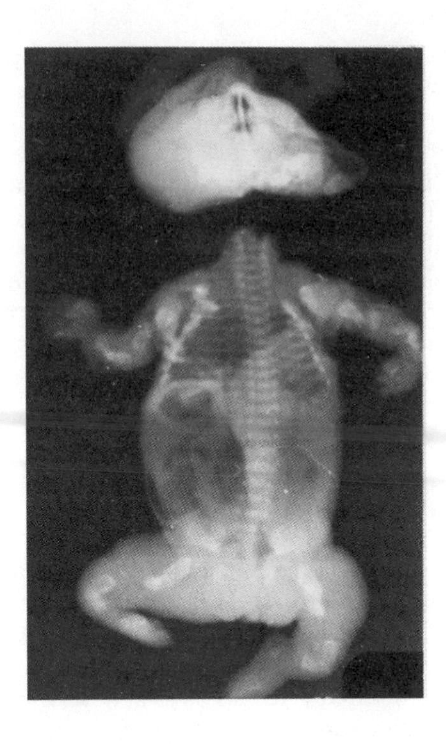

factors are proteins that bind to DNA and control the activity of specific genes or classes of genes. More than 10 families of these DNA-binding proteins have been described so far. Each family has a distinct molecular configuration that permits binding to DNA. The largest such family is the *zinc-finger* proteins, with tandemly repeated, folded loops of amino acids (extended like fingers), each stabilized by zinc ions (Figure 10.18). Over 200 zinc-finger genes have been identified, and all are thought to be involved in the regulation of transcription and the control of cellular growth and differentiation.

As more disorders are characterized at the molecular level, mutant zinc-finger genes are emerging as important contributors to human genetic disorders. Two such disorders will be briefly described.

Xeroderma pigmentosum (XP) is a group of autosomal recessive disorders characterized by sensitivity to sunlight and the development of ultraviolet-light-induced skin cancers (see Chapter 11). Mutations in at least seven different genes (designated as XP-A through XP-G) can cause the symptoms of XP, and all are associated with the inability to repair damage to DNA caused by ultraviolet light. The gene for XP-A has been cloned, and from its nucleotide sequence it has been determined that the gene product is a zinc-finger protein. Its mechanism of action is not yet known, but the protein may bind to DNA to directly repair damage to DNA, or it may activate genes involved in DNA repair.

An autosomal dominant disorder known as *Greig cephalospondyly syndrome* (GCPS) affects limb and facial development, causing fusion of fingers and toes, as well as abnormalities of the face and skull. The gene,

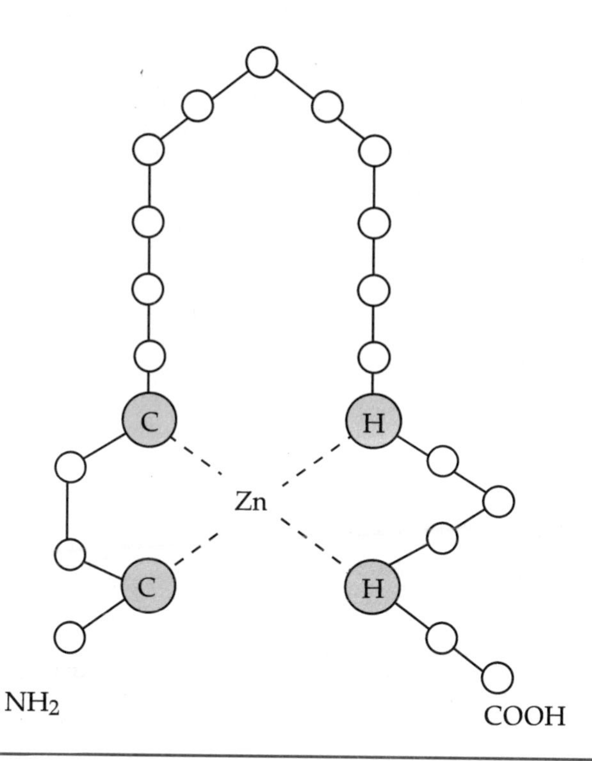

FIGURE 10.18 Structure of a zinc-finger protein. Cysteine (C) and histidine (H) residues hold a zinc ion while the rest of the amino acids extend in a finger-like projection. Zinc-finger proteins bind to DNA and activate transcription.

mapped to chromosome 7p13, encodes a zinc-finger protein known as GLI-3. The pattern of GLI-3 expression and the timing of its expression are consistent with the role of this protein as a regulator of limb and facial formation during prenatal development.

A heritable disorder that affects kidney formation, Wilms tumor (WT-1), is also caused by a mutant zinc-finger gene that maps to chromosome 11p13; it is described in Chapter 13.

The role of zinc-finger proteins in programming gene transcription and the discovery that mutations in this group of genes can cause developmental defects indicate that investigation of DNA-binding proteins may be important in understanding mutations that exert their phenotypic effect during prenatal development or during infancy.

Summary

1. In the early part of this century, Sir Archibald Garrod's studies on the human metabolic diseases cystinuria, albinism, and alkaptonuria provided the first hints that gene products control biochemical reactions. These diseases and many others are the result of mutations that cause metabolic blocks in biochemical pathways. In alkaptonuria, Garrod argued, the normal metabolic reaction is blocked by the lack of a needed enzyme. He speculated that the inability to carry out this reaction is the result of a recessive Mendelian gene.

2. Unfortunately, Garrod's insights were appreciated only in retrospect. In the late 1930s, George Beadle, along with his colleagues Boris Ephrussi and Edward Tatum, began to study the genetic control of biochemical reactions in flies and fungi. Their work made it clear that single mutations led to the loss of activity in a single enzyme, an idea known as the one gene-one enzyme hypothesis. Subsequent work refined this into the one gene-one polypeptide hypothesis of today. Upon receiving the Nobel prize in 1958, Beadle said that he and Tatum had only "rediscovered what Garrod had seen so clearly."

3. Later investigations clearly identified enzyme defects in a large number of human metabolic diseases, and the role of mutations in nonenzymatic proteins became clear. In 1949, James Neel identified sickle cell anemia as a recessive disease, and Linus Pauling began to study the physical properties of hemoglobin, leading to Vernon Ingram's discovery that the molecular basis of sickle cell anemia is a single change in the amino acid sequence of a polypeptide chain. Logically this might involve changes in the nucleotide sequence of the genetic material, and this idea has been confirmed several times over.

4. Defects in receptor proteins, transport proteins, structural proteins, and other nonenzymatic proteins are able to cause phenotypic effects in the heterozygous state and show an incompletely dominant or dominant pattern of inheritance.

5. The examples of metabolic diseases outlined in this chapter serve to reinforce the notion of the biological uniqueness of the individual. The constellation of genes present within each individual is the result of the random combination of parental genes as well as the sum of changes brought about by recombination and mutation. This genetic combination confers a distinctive phenotype upon each of us. Garrod referred to this metabolic uniqueness as chemical individuality. Understanding the molecular basis for this individuality remains one of the great challenges of human biochemical genetics.

Questions and Problems

1. Many individuals with metabolic diseases are normal at birth but show symptoms shortly thereafter. Why?
2. List the ways in which a metabolic block can have phenotypic effects.
3. Phenylketonuria and alkaptonuria are both autosomal recessive diseases. If a person with PKU marries a person with AKU, what will the phenotype of their children be?
4. Knowing that individuals who are homozygous for the G^D allele show no symptoms of galactosemia, is it surprising that galactosemia is a recessive disease? Why?
5. Use Figure 10.6 to answer this question. Severe immunodeficiency can be inherited as an autosomal or sex-linked recessive. A woman is a carrier for Lesch-Nyhan syndrome and the sex-linked form of severe combined immunodeficiency disease. She marries a normal man.
   a. List the phenotypes and phenotypic ratios of their children, assuming that the mutant alleles for both these diseases are on the same X chromosome.

b. What will the phenotypes of the children be if the mutant alleles are not on the same X chromosome?

Questions 6–8 refer to the following hypothetic pathway in which substance A is broken down to substance C by enzymes 1 and 2. Substance B is the intermediate produced in this pathway:

$$A \xrightarrow{\text{enzyme 1}} B \xrightarrow{\text{enzyme 2}} C$$

6. a. If an individual is homozygous for a null mutation in the gene that codes for enzyme 1, what will be the result?
   b. If an individual is homozygous for a null mutation in enzyme 2, what will be the result?

7. a. If the first individual in question 6 married the second individual, would their children be able to catabolize substance A into substance C?
   b. Suppose each of the aforementioned individuals were heterozygous for an autosomal dominant mutation. List the phenotypes of their children with respect to compounds A, B, and C. (Would the compound be in excess, not present, etc.?)

8. An individual is heterozygous for a recessive mutation in enzyme 1 and heterozygous for a recessive mutation in enzyme 2. This individual marries an individual of the same genotype. List the possible genotypes of their children. For every genotype, determine the activity of enzyme 1 and 2, assuming that the mutant alleles have 0% activity and the normal alleles have 50% activity. For every genotype, determine if compound C will be made. If compound C is not made, list the compound that will be in excess.

9. Explain why there are variant responses to drugs and why they act as heritable traits.

10. a. Do all the mutations that give rise to the hemoglobinopathies lie in exons?
    b. Would you expect the mutations that give rise to beta thalassemia to lie in the exons of the beta globin gene? Why?

Questions 11 to 15 refer to a hypothetic metabolic disease in which protein E is not produced. Lack of protein E causes mental retardation in humans. Protein E's function is not known, but it is found in all cells of the body. Skin cells were taken from eight individuals who cannot produce protein E and were grown in culture. The defect in each of the individuals is due to a single recessive mutation. Each individual is homozygous for her or his mutation. The cells from one individual were grown with the cells from another individual in all possible combinations of two. After a few weeks of growth the mixed cultures were assayed for the presence of protein E. The results are given in the following table. A plus means that the two cell types produced protein E when grown together (but not separately), while a minus means that the two cell types still could not produce protein E:

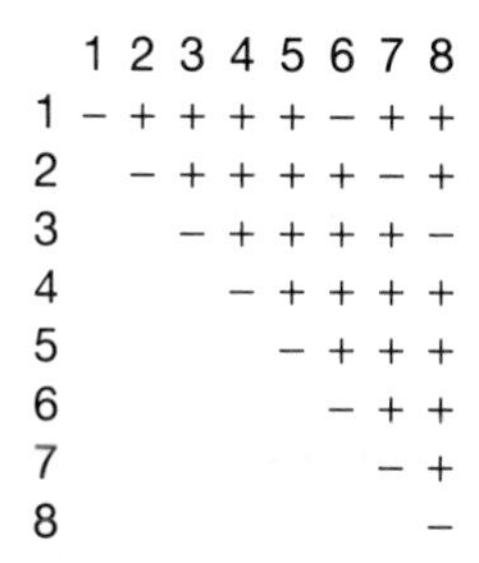

	1	2	3	4	5	6	7	8
1	−	+	+	+	+	−	+	+
2		−	+	+	+	+	−	+
3			−	+	+	+	+	−
4				−	+	+	+	+
5					−	+	+	+
6						−	+	+
7							−	+
8								−

11. a. Which individuals seem to have the same defect in protein E production?
    b. If individual 2 married individual 3 would their children be able to make protein E?
    c. If individual 1 married individual 6 would their children be able to make protein E?

12. Assuming that these individuals represent all possible mutants in the synthesis of protein E, how many steps are there in the pathway to protein E production?

Compounds A, B, C, and D are known to be intermediates in the pathway for production of protein E. To determine where the block in protein E production occurred in each individual, the various intermediates were given to each individual's cells in culture. After a few weeks of growth with the intermediate, the cells were assayed for the production of protein E. The results for each individual's cells are given in the following table. A plus means that protein E was produced after the cells were given the intermediate listed at the top of the column. A minus means that the cells still could not produce protein E even after being exposed to the intermediate at the top of the column:

	Compounds				
Cells	A	B	C	D	E
1	−	−	+	+	+
2	−	+	+	+	+
3	−	−	−	+	+
4	−	−	−	−	+
5	+	+	+	+	+
6	−	−	+	+	+
7	−	+	+	+	+
8	−	−	−	+	+

13. Draw the pathway leading to the production of protein E.

14. Denote the point in the pathway in which each individual is blocked.

15. a. If an individual who is homozygous for the mutation found in individual 2 and heterozygous for the mutation found in individual 4 marries an individual who is homozygous for the mutation found in individual 4 and heterozygous for the mutation found in individual 2, what will be the phenotype of their children?
    b. List the intermediate that would build up each of the types of children who could not produce protein E.

For Further Reading

Antonarakis, S., Kazazian, H., and Orkin, S. 1985. DNA polymorphisms and molecular pathology of the human globin gene clusters. *Hum. Genet.* **69:** 1–14.

Beadle, G. W. 1945, Biochemical genetics. *Chem. Rev.* **37:** 351.

Bearn, A. G., and Miller, E. D. 1979. Archibald Garrod and the development of the concept of inborn errors of metabolism. *Bull. Hist. Med.* **53:** 317–324.

Benson, P. F., and Fenson, A. H. 1985. *Genetic Biochemical Disorders.* New York: Oxford University Press.

Childs, B. 1970. Sir Archibald Garrod's conception of chemical individuality: A modern appreciation. *N. Engl. J. Med.* **282:** 4–5.

Eisensmith, R., Okano, Y., Dasovitch, M., Wang, T., Güttler, F., Lou, H., Guldberg, P., Lichter-Konecki, U., Konecki, D., et al, 1992. Multiple origins for phenylketonuna in Europe. Am. J. Hum. Genet. **51**: 1355-65.

Galjaard, H. 1980. *Genetic Metabolic Diseases: Early Diagnosis and Prenatal Analysis.* New York: Elsevier/North-Holland.

Garrod, A. E. 1902. The incidence of alkaptonuria: A study in chemical individuality. *Lancet* **2:** 1616–1620.

Hobbs, H., Brown, M. and Goldstein, J. 1992. Molecular genetics of the LDL receptor gene in familial hypercholesterolemia. Human Mutat. **1**: 445-466.

Ingram, V. M. 1957. Gene mutations in human hemoglobin: The chemical differences between normal and sickle cell hemoglobin. *Nature* **180:** 326–328.

King, R. A., and Olds, D. P. 1985. Hairbulb tyrosinase activity in oculocutaneous albinism: Suggestions for pathway control and block location. *Am. J. Med. Genet.* **20:** 49–55.

Kivirkko, K. 1993. Collagens and their abnormalities in a wide spectrum of diseases. Ann. Med. **25**: 113-126.

McKusick, V. 1992. *Mendelian Inheritance in Man: catalogs of autosomal dominant, autosomal recessive and x-linked phenotypes.* 10th ed., Baltimore: Johns Hopkins Press.

Neel, J. V. 1949. The inheritance of sickle cell anemia. *Science* **110:** 64–66.

Pauling, L., Itoh, H., Singer, S. J., and Wells, I. C. 1949. Sickle cell anemia: A molecular disease. *Science* **110:** 543–548.

Scriver, C. R., Beaudet, A. L., Sly, W. S. and Valle, D. 1989. *The Metabolic Basis of Inherited Disease.* 6th ed. New York: McGraw-Hill.

Scriver, C. R., and Clow, C. L. 1980. Phenylketonuria and other phenylalanine hydroxylation mutants in man. *Ann. Rev. Genet.* **14:** 179–202.

Sculley, D., Dawson, P., Emmerson, B. and Gordon, R. 1992. A review of the molecular basis of hypoxanthine-guanine phosphoribosyltransferase (HPRT) deficiency. Hum. Genet. **90**: 195-207.

Spritz, R. A., Strunk, K. M., Giebel, L. B., and King, R. A. 1990. Detection of mutations in the tyrosinase gene in a patient with Type LA oculocutaneous albinism. *New Engl. J. Med.* **322:** 1724–1728.

Vella, F., 1980. Human hemoglobins and molecular disease. *Biochem. Educ.* **8:** 41–53.

Weatherall, D.J. 1993. Molecular medicine: towards the millenium. Trends Genet. **9**: 102.

CHAPTER 11

MUTATION

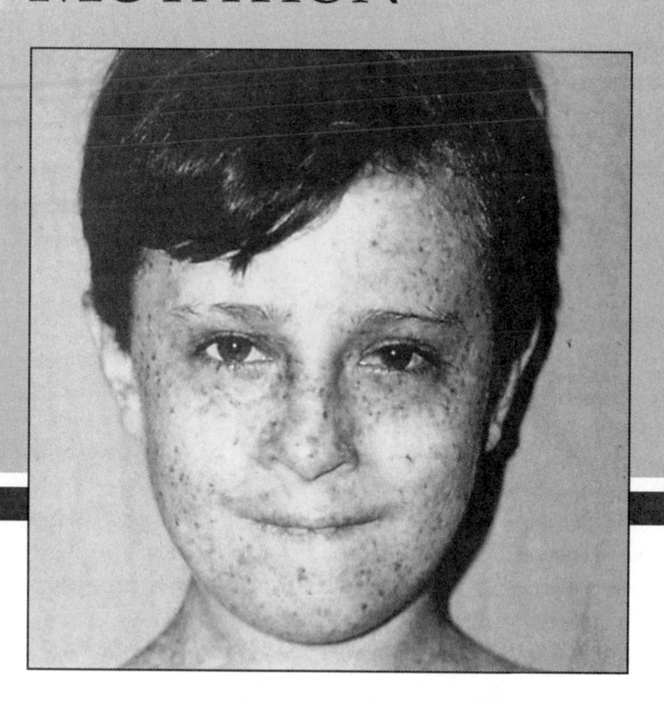

CHAPTER OUTLINE

Hearing a description of how red blood cells change shape in sickle cell anemia, Linus Pauling recalled that

> the idea occurred to me that sickle cell anemia was a molecular disease, involving an abnormality of the hemoglobin molecule determined by a mutated gene.

Early in 1949, Linus Pauling and his student Harvey Itano began a series of experiments to determine whether there was a difference between normal hemoglobin and sickle cell hemoglobin. They obtained blood samples from persons with sickle cell anemia and from unaffected individuals and prepared concentrated hemoglobin solutions from these samples. The solution of concentrated hemoglobin was placed in a tube fitted with an electrode at each end, and an electrical current was passed through the tube. Hemoglobin from individuals with sickle cell anemia migrated toward the cathode, indicating that it has a positive electrical charge. Samples of normal hemoglobin migrated toward the anode, indicating that it has a net negative electrical charge. In the same year, James Neel, working with sickle cell patients in the Detroit area, had demonstrated that sickle cell anemia was caused by an autosomal recessive trait, establishing the genetic link to this condition. In their paper, Pauling and his colleagues built on Neel's finding, and concluded that the disease (and the heterozygous condition known as sickle cell trait) derived from a mutant gene involved in the synthesis of the hemoglobin molecule. This discovery provided the foundation for human biochemical genetics, and later played a key role in understanding the nature of mutation.

After establishing the structure of DNA, Francis Crick was anxious to establish that mutant genes produced proteins with amino acid sequences that differed from the protein produced by the normal allele. He persuaded Vernon Ingram to look for such differences in proteins. After some unsuccessful attempts, Ingram settled on hemoglobin, because it was highly probable that the difference in mobility between normal and sickle cell hemoglobin was caused by differences in amino acids. Beginning with concentrated hemoglobin preparations, Ingram first cut the protein into pieces using the enzyme trypsin. He separated the thirty resulting fragments, and noted that normal hemoglobin and sickle cell hemoglobin differed only in one fragment, a peptide about ten amino acids in length. Ingram then began to study the amino acid sequence in this fragment, and in 1957 reported a difference of only a single amino acid (glutamine in normal hemoglobin and valine in sickle cell hemoglobin) between the two proteins. This finding confirmed the relationship between a mutant gene and a mutant gene product, but raised a more basic question: What is the nature of mutation?

From a genetic point of view, mutation can be defined as any heritable change. These changes are the source of the genetic variation in humans and other organisms. The results of mutations can be classified in a number of ways. Mutations that produce dominant alleles are expressed in the heterozygous condition; mutations to recessive alleles are expressed only when homozygous. The effects of mutation can also be ranked in categories such as the severity of phenotype or age of onset. For our purposes, two general categories of mutations can be distinguished: chromosomal aberrations and gene, or point, mutations. We considered chromosomal changes in Chapter 6. In this chapter we will discuss only those changes that occur within a single gene, specifically, alterations in the sequence or number of nucleotides in DNA. We will first consider how such mutations are detected and then investigate at what rate these mutations take place. Finally, we will examine how mutation works at the molecular level.

Detecting Mutations

Armed with a definition of mutation, the question is, how do we know that a mutation has taken place? In haploid organisms such as bacteria, experiments can be designed so that only new mutants will form colonies on an agar plate. In humans, the appearance of a dominant mutation in a family can be detected in a fairly straightforward manner. However, mutation of a dominant allele to a recessive one can be detected only in the homozygous condition, posing a problem for human geneticists. As we will see in a later chapter, a recessive mutant allele can be carried in the heterozygous condition at a high frequency in a population. Because the human genome cannot be manipulated in genetic crosses to detect such heterozygotes, the methods of detecting whether a recessive mutation has occurred are indirect and inferential.

If an affected individual appears in an otherwise unaffected family, the first problem is to determine whether the trait is the result of mutation or nongenetic factors. For example, if a mother is exposed to the virus causing rubella (a form of measles) early in pregnancy, the fetus may develop a range of physical and neurologic symptoms that superficially resemble those seen in some metabolic genetic disorders. This condition, however, is not the result of mutation but rather the effect of the virus on the developing fetus. In general, the detection of mutations depends on pedigree analysis and the study of births in a family line.

If an allele is dominant, is fully penetrant, and appears in a family with no history of this condition over several generations, we can presume that a mutation has taken place. An example of such a trait is presented in the pedigree in Figure 11.1. In this case a severe blistering of the feet appeared in one out of six children, although the parents were unaffected. The trait was

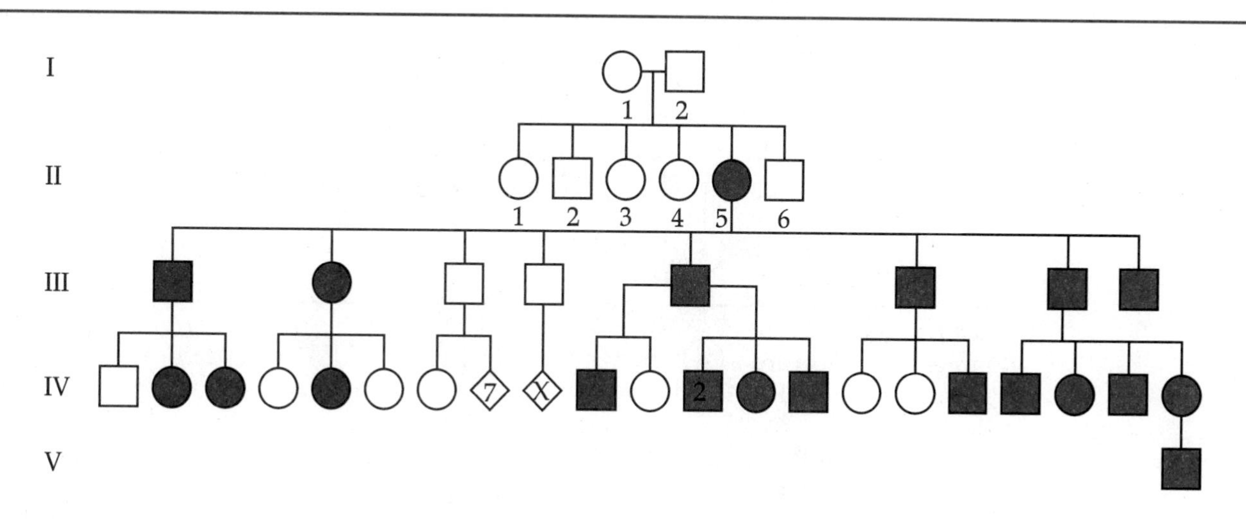

FIGURE 11.1 Pedigree of foot blistering, showing appearance of a dominant trait (II-5), and its transmission through several generations. For clarity, spouses are not shown in this pedigree after generation I.

transmitted by the affected female to six of her eight children and was passed to the succeeding generation in the manner expected for an autosomal dominant condition. A reasonable explanation for this pedigree is that a mutation to a dominant allele occurred and that II-5 was heterozygous for this dominant allele. On the other hand, there are a number of uncertainties that can affect this conclusion. For example, if the father of this child was not the husband in the pedigree but an affected male, then it would only seem that a mutational event had taken place.

If mutation results in an allele that is recessive and sex linked, it often can be detected by examination of males in the family line. However, it is difficult to determine whether a heterozygous female who transmits a trait to her son is the source of the mutation or is only passing on a mutation that arose in an ancestor. The X-linked form of hemophilia that spread through the royal families of Western Europe and Russia seems to have originated with Queen Victoria (Figure 11.2; see also Figure 5.6 for a more extensive pedigree). An examination of the pedigree shows that none of her male ancestors (her grandfather was King George III) had hemophilia. However, one of her sons was affected, and at least two of her daughters were carriers. Since Victoria transmitted the trait to a number of her children, it is reasonable to assume that she was a heterozygous carrier. Her father was not affected, and there is nothing in her mother's pedigree to indicate that she was a carrier. It is likely, therefore, that Victoria received a newly mutated allele from one of her parents. We can only speculate as to which parent.

If an autosomal recessive trait appears suddenly in a family, it is usually difficult or impossible to trace the trait through a pedigree to identify the person or even the generation in which the mutation first occurred, since only homozygotes are affected. In contrast to other conditions, autosomal recessive heterozygotes can remain undetected for generations.

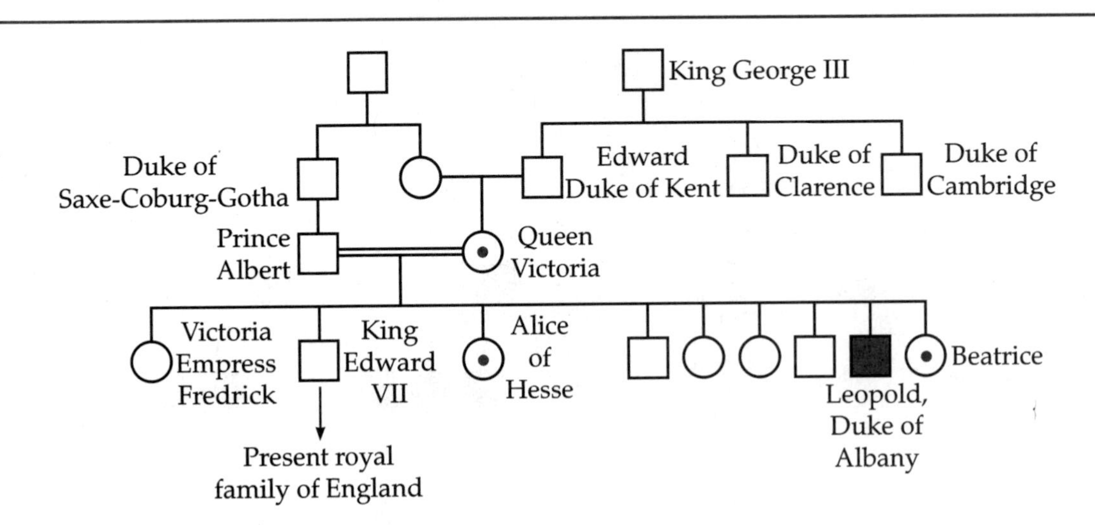

FIGURE 11.2 Pedigree of Queen Victoria showing her ancestors and children. Since she passed the gene for hemophilia to three of her children, she was probably a heterozygote rather than the source of the mutation.

Measuring Mutation Rates

Studies of human pedigrees such as those discussed above indicate that mutation does take place in the human genome. The available evidence suggests that it is a rare event, but is it possible to measure the rate of mutation? Knowing the underlying rate of mutation would allow geneticists to monitor the rate over time to determine whether it is increasing, decreasing, or remaining the same. In organisms such as bacteria, fungi, *Drosophila,* and mice, direct experimental measurements have provided reliable information about the rate of mutation at a number of loci. For humans, the estimation of mutation rates is more difficult, and the information that is available has been gathered for only a small number of genes.

Mutation rates can be expressed as the number of mutated alleles per locus per generation. One way of measuring this rate is to consider the number of new mutant alleles per given number of gametes. Suppose that for a certain gene, 4 out of 100,000 births show mutation from a recessive to a dominant allele. Since each zygote producing these births carries two copies of the gene, we have sampled 200,000 copies of the gene. Because the mutations are dominant, the 4 births represent 4 mutated genes (we are assuming that the newborns are heterozygotes carrying only one mutant allele). In this case, then, the mutation rate is 4/200,000, or 2/100,000, and in scientific notation would be written as 2×10^{-5}/per allele/per generation.

Mutation rate
The number of events producing mutated alleles per locus/per generation.

If the locus were X-linked and if 100,000 male births were examined and 4 mutants were discovered, this would represent a sampling of 100,000 copies of the gene (since the males have only one copy of the X chromosome). Excluding contributions from female carriers, the mutation rate in this case would be 4/100,000, or 4×10^{-5}/per allele/per generation.

Direct Measurement of Mutation

For dominant alleles that fulfill certain conditions, it is possible to measure the rate of mutation directly. To ensure accuracy in the measurement, the trait selected must:

- never be produced by recessive alleles
- always be fully expressed and completely penetrant so that mutant individuals can be identified
- have clearly established paternity
- never be produced by nongenetic agents such as drugs or infection
- be produced by dominant mutation of only one locus

Although it is sometimes difficult to determine when these conditions fully apply, one dominant allele, achondroplasia, fulfills most of these requirements. This is a dominant form of dwarfism that produces short arms and legs and an enlarged skull (see Chapter 4). A definitive diagnosis by x-ray examination can be done shortly after birth. Since 1941 several surveys have used this gene to estimate the mutation rate in humans. A recent survey covers more than 240,000 births worldwide. In this survey 7 achondroplastic births to unaffected parents were observed out of 242,257 births recorded. Thus the mutation rate for achondroplasia has been calculated at 1.4×10^{-5}.

Although the mutation rate for achondroplasia can be measured directly, it is not clear whether this is a typical rate of mutation for human genes. Perhaps we can observe the rate of mutation in this gene only because it has an inherently high rate of mutation. In addition, not all cases of mutation may be reported. Therefore it is important to measure mutation rates in a number of different genes before making any general statements. As it turns out, two other dominant mutations have widely different rates of mutations. Neurofibromatosis, an autosomal dominant condition, is characterized by pigmentation spots and tumors of the skin and nervous system (described in Chapter 4). About 1 in 3000 births are affected individuals. Many of these births occur in families with no history of neurofibromatosis, indicating that this locus has a high mutation rate. In fact, the mutation rate in this disease has been calculated to be as high as 1 in 10,000 (1×10^{-4}), one of the highest rates so far discovered in humans. For Huntington disease, the dominant condition that causes progressive degeneration of the nervous system (see Chapter 4), the mutation rate has been calculated as 1×10^{-6}, a rate some 100-fold lower than that for neurofibromatosis and 10-fold lower than that reported for achondroplasia. Table 11.1 gives some estimates of mutation rates in human genes. These mutation rates average out to about 1×10^{-5}. Note that the genes listed in the table are all inherited as autosomal dominant or X-linked traits. It is almost impossible to measure directly the mutation rates in autosomal recessive alleles by inspection of phenotypes,

TABLE 11.1 Human Mutation Rates

Trait	Mutants/Million Gametes	Mutation Rate
Achondroplasia	10	1×10^{-5}
Aniridia	2.6	2.6×10^{-6}
Retinoblastoma	6	6×10^{-6}
Osteogenesis imperfecta	10	1×10^{-5}
Neurofibromatosis	50–100	$0.5–1 \times 10^{-4}$
Polycystic kidney disease	60–120	$6–12 \times 10^{-4}$
Marfan syndrome	4–6	$4–6 \times 10^{-6}$
Von Hippel–Landau syndrome	<1	1.8×10^{-7}
Duchenne muscular dystrophy	50–100	$0.5–1 \times 10^{-4}$

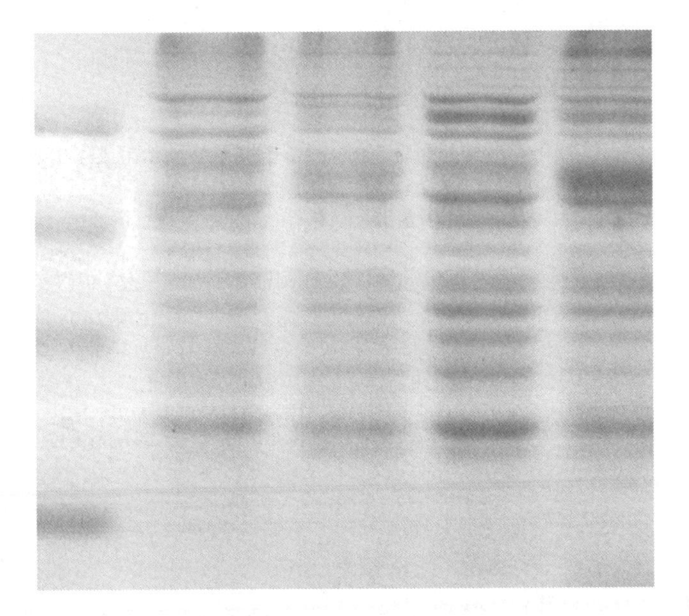

FIGURE 11.3 Electrophoresis of proteins. Proteins placed in an electrical field on a gel separate from each other based on their net electrical charge and size. Each band on the gel represents a different type of protein.

but in recent years molecular techniques have been used to detect the appearance of new mutations, including those in recessive alleles.

One method detects altered gene products as an indicator that a mutational event has occurred. Proteins migrate through a gel in an electric field at a rate that is proportional to the size of the molecule and the net electrical charge on the protein. This technique, known as *electrophoresis,* uses gels made of a variety of materials, including starch, agarose, or polyacrylamide, each with different pore characteristics. The net charge on a protein is a function of amino acid composition; some amino acids are positively charged, others are negatively charged, and still others are neutral. If a mutation results in a change in the amino acid sequence of a protein and alters the electrical charge on the molecule, the proteins produced by the normal and mutant alleles can be separated by electrophoresis (Figure 11.3).

This method has been used to determine whether exposure to radiation near the center of atomic bomb detonations in Hiroshima and Nagasaki produced an increase in mutation rates. This study will be discussed in more detail in Chapter 12.

Another molecular technique used to detect mutations is the polymerase chain reaction (PCR), described in Chapter 8. This method depends on having some knowledge of the molecular organization of the gene and of the nucleotide sequence of the gene or adjacent regions. In spite of these stringent requirements, the PCR has been used to detect independently arising mutations in a number of genes, such as HPRT deficiency, responsible for the Lesch-Nyhan syndrome. However, since mutations in only a relatively small number of genes have been studied by these recombinant DNA techniques, not enough information is available to provide a reliable estimate of the mutation rate in recessive genes.

Indirect Measurement of Mutation

Several indirect methods for measuring mutation rates have been developed. They rely on statistical procedures derived from the principles of population genetics and will not be discussed in detail. The rationale for the indirect methods depends on the observation that in many mutant traits, the affected individuals leave few if any offspring. Since the gene is not passed on to the offspring, over a period of time the trait should be eliminated from the population. However, since the genes for these traits are not decreasing in the population, mutation must be replacing those abnormal alleles lost by reproductive failure. Statistical measurements can be used to calculate the rate of mutation that is necessary to maintain a constant gene frequency.

Average mutation rates in humans can also be estimated by comparison with the mutation rate in mammals used in laboratory research. The mouse, *Mus musculus,* has a genome about the same size as the human genome and presumably carries a similar number of genes. Observations on spontaneous mutations in laboratory mice and the experimental production of mutations indicate that the rate of mutation in the mouse is about 1.1×10^{-5}. The indirect and comparative methods support the observation that the average mutation rate in humans is about 1×10^{-5}.

Still, many geneticists feel that the genes in which the mutation rate can be measured might be those that have an inherently high rate of mutation. Other factors such as the size of a gene may also influence mutation rates. Larger genes may mutate more frequently than smaller genes because they contain more DNA. To reduce any potential bias, most human geneticists prefer to use a more conservative estimate of the mutation rate in humans and generally use the number 1×10^{-6} as the average mutation rate.

Gametic Mutations

With some rough estimates of the mutation rate in hand, it is possible to ask how many new mutations each person may produce. If we assume a mutation rate of 1×10^{-6} per locus and also assume that there are 50,000 (5×10^4) loci in the human haploid gene set, then the expected number of mutations produced by each individual per gamete is:

$$(1 \times 10^{-6}) \times (5 \times 10^4) = 5 \times 10^{-2} = 0.05$$

Since each gamete contains a haploid gene set, on average, about 5%, or 1 in 20, gametes carry a new mutation. We can also rephrase the question to ask what the chances are of a child receiving a mutation produced by either parent. Because each individual is the product of two gametes, we must multiply the mutation frequency by a factor of 2:

$$0.05 \times 2 = 0.1$$

This means that there is a 1 in 10, or 10%, chance that any diploid zygote carries a new mutation. To put this into perspective, approximately 3,800,000 children are born each year in the United States. Of this number, about 380,000 carry a new mutation that is not present in either parent.

This may seem like an alarmingly high number of mutations. However, the impact of this mutational rate is diminished by several factors. First, most mutations are recessive, meaning that new mutations are not immediately expressed. Second, as we will learn when we consider the genetics of populations, most recessive traits are carried in the heterozygous condition rather than the homozygous state. Third, many mutations are deleterious, and homozygotes are less likely to live to reproductive age or to leave many offspring. For example, homozygous Tay-Sachs individuals die as young children and cannot pass the trait to the next generation. In this case the only way to produce homozygous individuals is by the mating of heterozygotes. Fourth, some mutations have little or no phenotypic impact and are regarded as neutral. (See the discussion of hemoglobin variants below.)

In making these calculations, we must remember that we do not know with certainty what the actual mutation rate is, nor do we really know how many genes are in the human genome. Because the numbers we have used are subject to large errors, our calculations may be too high or too low. For the moment, however, they represent a fair approximation. As measured in a relatively small number of genes, the mutation rate in humans varies over a hundred- or thousand-fold range per gene/per generation (see Table 11.1). Three factors affect mutation rates in genes and contribute to the observed variation:

1. *Size of the gene.* Larger genes present larger targets for mutational events. Neurofibromatosis (NF-1) has a high mutation rate and is an extremely large gene. The NF-1 protein contains over 2000 amino acids, but including the noncoding regions, the gene extends over 300,000 base pairs of DNA. The gene for Duchenne and Becker muscular dystrophy is the largest gene identified to date in humans, and contains over 2 million base pairs. Both these genes have high rates of mutation.
2. *Nucleotide sequence.* In some genes, short nucleotide repeats are present in the DNA. In the gene for fragile-X syndrome (Chapter 6), a CGG repeat within the first coding region (exon) is present in 6 to 50 copies in unaffected individuals. Symptoms begin to appear in those with more than 52 copies and become more severe as the number of CGG repeats increases. The presence of these repeats may predispose a gene to mutation at a higher rate.
3. *Spontaneous chemical changes.* Among the bases present in DNA, cytosine is especially susceptible to chemical changes that can lead to changes in DNA. These and other chemical changes will be discussed below and in the next chapter. In the case of cytosine, genes rich in G-C base pairs are more likely to undergo spontaneous chemical changes than those rich in A-T pairs. All of these factors draw attention to the fact that mutation involves changes in DNA. In the following section we will consider the nature and consequences of some of these changes.

Mutation at the Molecular Level: DNA as a Target

At the molecular level, mutations can involve substitutions, insertions, or deletions of one or more bases in a DNA molecule. Those mutations that involve an alteration in the sequence but not the number of nucleotides in a gene are called **nucleotide substitutions.** Generally, such substitutions involve one or a small number of nucleotides. A second type of mutation causes the *insertion* or *deletion* of one or more bases. Since codons are composed of three bases, changing the number of bases alters the sequence of all subsequent codons and results in large-scale changes in the amino acid sequence of the protein product of such mutated genes. We will begin by examining the simpler form of mutation, the substitution of one nucleotide for another, and will then consider mutations involving the addition or deletion of bases.

Nucleotide substitutions
Mutations that involve substitutions, insertions or deletions of one or more nucleotides in a DNA molecule.

Nucleotide Substitutions

Several hundred variants of the alpha and beta proteins of human hemoglobin that contain single amino-acid substitutions are known. These provide many well-studied examples of the effects of nucleotide substitutions on protein structure and function. Nucleotide substitutions in coding regions can have a number of outcomes, some of which are described in the following section. In this discussion, keep in mind that the term "codon" refers to the sequence of three nucleotides in mRNA that codes for an amino acid.

Missense mutations are single nucleotide changes that cause the substitution of one amino acid for another in a protein. This substitution may or may not affect the function of the gene product and may or may not have phenotypic consequences. In the gene coding for beta globin (Figure 11.4), a single nucleotide substitution in codon 6 from GAG (glu) $\rightarrow$ GUG (val) results in sickle cell anemia, an autosomal recessive disorder marked by the production of abnormally shaped red blood cells and a potentially lethal phenotype. (See Chapters 4 and 10 for detailed descriptions of this condition.) Another nucleotide substitution in the same codon from GAG (glu) $\rightarrow$ AAG (lys) results in a condition known as HbC (Hemoglobin C), associated with a mild set of clinical symptoms. In a third beta globin variant called Hb Makassar, the codon at the sixth position is changed from GAG (glu) $\rightarrow$ GCG (ala), a substitution that causes no clinical symptoms and is regarded as a harmless hemoglobin variant.

Missense mutation
A mutation which cause the substitution of one amino acid for another in a protein.

In these examples, the resulting polypeptides differ only in the amino acid at position 6: HbA has glutamic acid (glu), HbS has valine (val), HbC has lysine (lys), and Hb Makassar has alanine (ala). The sequence of the other 145 amino acids in the polypeptide is unchanged. In these cases, single nucleotide changes in the sixth codon of the beta globin gene result in phenotypes that range from harmless (Hb Makassar), to the mild clinical symptoms of HbC, to the serious and potentially life-threatening consequences of HbS and sickle cell anemia.

Hb A	1	2	3	4	5	6	7	8
DNA	CAC	GTG	GAC	TGA	GGA	CTC	CTC	TTC
mRNA	GUG	CAC	CUG	ACU	CCU	GAG	GAG	AAG
Amino acids	val	his	leu	thr	pro	glu	glu	lys

Hb C	1	2	3	4	5	6	7	8
DNA	CAC	GTG	GAC	TGA	GGA	TTC	CTC	TTC
mRNA	GUG	CAC	CUG	ACU	CCU	AAG	GAG	AAG
Amino acids	val	his	leu	thr	pro	lys	glu	lys

Hb S	1	2	3	4	5	6	7	8
DNA	CAC	GTG	GAC	TGA	GGA	CAC	CTC	TTC
mRNA	GUG	CAC	CUG	ACU	CCU	GUG	GAG	AAG
Amino acids	val	his	leu	thr	pro	val	glu	lys

FIGURE 11.4 DNA code word, mRNA codon, and amino acid sequence of Hb A, Hb C, and Hb S. Single base substitutions in codon 6 produce the two variant forms of beta globin.

Nonsense mutations cause a change from 1 of the 61 codons that specify an amino acid to one of the three termination codons (see Figure 9.6). This usually leads to premature termination of translation and the formation of shortened polypeptide chains. In the beta globin variant McKees Rock, the last two amino acids are missing, and the protein is only 143 amino acids long. The change in codon 144 UAU (tyr) → UAA (termination) results in a beta chain that is shorter by two amino acids. This change has little or no effect on the function of the beta globin molecule as a carrier of oxygen. However, some nucleotide substitutions can produce more drastic changes in polypeptide length. In one form of beta thalassemia (thalassemia results from an imbalance in the production of alpha and beta globin polypeptides; see Chapter 10 for a description of this disorder), a change at codon 39 from CAA (gln) → UAA (termination) produces a nonfunctional shortened polypeptide only 38 amino acids long. In individuals homozygous for this mutation, no beta globin is produced, resulting in a condition known as beta-zero-thalassemia.

Nonsense mutation
A mutation that changes an amino acid-specifying codon to one of the three termination codons.

Sense mutations produce longer-than-normal proteins by changing a termination codon into one that codes for amino acids. Several hemoglobin variants with longer-than-normal globin molecules are shown in Table 11.2. In each case, the extended polypeptide chain can be explained by a single nucleotide substitution in the normal termination codon. In Hb Constant Springs—1, for example, the alpha chain termination codon UAA at position 142 is changed to CAA (gln), and as a result, 31 additional amino acids are added to the alpha chain before another termination codon is encountered in the mRNA.

Sense mutation
A mutation that changes a termination codon into one that codes for an amino acid. Such mutations produce elongated proteins.

Nucleotide substitutions can also affect other aspects of gene expression, including *transcription, splicing and processing* of pre-mRNA, and even *translation* of the mRNA into a polypeptide. Substitutions in the region adjacent to the beginning of a gene (the 5′-flanking region) can increase or

TABLE 11.2 Alpha-Globins with Extended Chains Produced by Nucleotide Substitutions

Hb	Abnormal Chains
Constant Springs—1	gln (142) + 30 amino acids
Icaria	lys (142) + 30 amino acids
Seal Rock	glu (142) + 30 amino acids
Koya Dora	ser (142) + 30 amino acids

decrease the rate of transcription or even abolish transcription of the neighboring gene.

As described in Chapter 9, correct processing of pre-mRNA molecules in the nucleus requires precise alignment of coding regions (exons) and noncoding regions (introns) at splice junctions. Nucleotide substitutions at intron-exon boundaries can interfere with normal splicing, resulting in an mRNA that retains an intron or is missing an exon. When this aberrant mRNA is translated, a nonfunctional polypeptide is usually produced. Abnormal splicing of mRNA is involved in some cases of beta thalassemia, and in homozygotes it can result in a complete lack of beta globin production.

Nucleotide substitutions *within* an exon or an intron (instead of at the boundaries) can create new splice sites, resulting in altered polypeptides. Hb E, a fairly common variant of beta globin, is caused by a nucleotide substitution within an exon (at codon 26) that creates a new splice site at codon 26. The result is formation of a nonfunctional polypeptide.

Recall that translation begins at the 5′ end of the mRNA with an AUG (met). In humans and some other eukaryotes, there is another sequence just upstream of the AUG codon that controls the rate at which translation occurs. In one case, nucleotide substitutions in this region reduce the rate of translation of mRNA for alpha globin, resulting in alpha thalassemia.

The array of phenotypes produced by single nucleotide substitutions in the globin genes illustrates the direct relationship between the gene and its phenotype. The examples above demonstrate that the phenotypic outcome of a single nucleotide substitution depends on its position with respect to the gene. Some substitutions result in a nonfunctional gene product, or even the absence of a gene product with severe phenotypic consequences, while other substitutions have no noticeable phenotypic effects. The mechanisms by which nucleotide substitutions produce their effects include amino acid substitution, creation or elimination of termination codons, changes that affect processing of messenger RNA, and alterations in the rate of transcription or translation. It is important to remember that not all base substitutions necessarily result in a change in the gene product. For example, if a mutation changes a codon from CCU → CCC, there will be no change in the amino acid sequence of the gene product, since both codons specify the amino acid proline. Such changes are real mutations, however, since they represent heritable changes in DNA. In this case the genetic code, with its built-in redundancy, buffers the gene product against the effect of mutation. It has been estimated that a substantial percentage of nucleotide substitutions cause no amino acid changes, while other forms of mutation may be more devastating to the phenotype.

Deletions and Insertions

The category of mutations that includes deletions and insertions is somewhat diverse, ranging from mutations that involve the insertion or deletion of a single nucleotide to ones that involve the deletion or duplication of an entire gene. As more genes are analyzed at the molecular level, deletions and insertions are emerging as a major cause of genetic disorders, accounting for 5% to 10% of all known mutations. Deletions or insertions of one to three bases produce what are known as **frameshifts.** These are the result of the insertion or deletion of nucleotides within the coding sequence of a gene. Since codons consist of groups of three bases, adding or subtracting a base from a codon changes the sense of all subsequent codons. This results in a change in the amino acid sequence of the protein encoded by the mutated gene. Suppose that a codon series read as the following sentence:

Frameshift mutation
Mutational events in which one to three bases are added to or removed from DNA, causing a shift in the codon reading frame.

THE FAT CAT ATE HIS HAT

An insertion in the second codon destroys the sense of the remaining message:

insertion

THE FAA TCA TAT EHI SHA T

In a similar fashion, a deletion in the second codon can also generate an altered message:

THE FTC ATA TEH ISH AT

deletion

In mutations characterized by base substitutions, usually only one amino acid in the protein is altered. In frameshift mutations, however, the addition or deletion of a single base can cause large-scale changes in the amino acid composition of the polypeptide chain, and usually leads to a nonfunctional gene product. Table 11.3 lists a number of hemoglobin variants with extended chains that result from frameshift mutations. In these examples, the frameshifts occur near the end of the gene and have a minimum impact on the function of the gene product. In hemoglobin α, the mRNA codons for the last few amino acids are as follows:

Position number	138	139	140	141	TER
mRNA codon	UCC	AAA	UAC	CGU	UAA
Amino acid	ser	lys	tyr	arg	

TABLE 11.3 Globins with Extended Chains Produced by Frameshift Mutation

Name	Type	Abnormal Chains
Hb Wayne 1	α	Normal to residue 138 + 8 amino acids
Hbv Wayne 1	β	Normal to residue 146 + 11 amino acids
Hb Cranston	β	Normal to residue 144 + 11 amino acids

In the α extended chain variant, hemoglobin Wayne, a deletion in the last base of codon 139 results in the production of a frameshift:

Position number	138	139	140	141	142
mRNA codon	UCC	AAU	ACC	GUU	AAG
Amino acid	ser	asn	thr	val	lys

The deletion of a single base causes a shift in the codon reading frame so that the normal termination codon UAA adjacent to codon 141 is split up into two separate codons, causing new amino acids to be added until another stop codon (generated by the deletion) is reached. The result is an alpha chain variant with 146 amino acid residues instead of 141. The other extended chain variants in Table 11.3 are also the result of frameshift mutations that destroy the normal termination codon and result in an extended polypeptide chain.

Larger deletions and insertions that do not involve frameshifts are characteristic of a number of genetic disorders. For example, the autosomal recessive disorder cystic fibrosis is caused by mutations in the gene that encodes a plasma membrane protein called (cystic fibrosis transmembrane regulator) CFTR. About 70% of all mutations in the CFTR gene involve a deletion of a codon for phenylalanine at position 508. CFTR proteins with this deleted amino acid do not insert into the plasma membrane; as a result, chloride ion migration is disrupted, and the obstructive accumulation of mucus in lungs and gland ducts results.

Trinucleotide Repeats

Trinucleotide repeats
A form of mutation associated with the expansion in copy number of a nucleotide triplet in or near a gene.

Trinucleotide repeats are a recently discovered class of insertion mutations associated with a number of genetic disorders, including fragile X syndrome, myotonic dystrophy and Huntington disease. The discovery of repeats in the fragile X gene (FMR1) helps explain many unique aspects of the inheritance and the phenotypic expression found in this disorder.

In fragile X syndrome, the mutation is expressed cytologically as a break or gap in the X chromosome at q27.3, and expressed phenotypically as a set of clinical symptoms that include mental retardation. Mothers of affected males are heterozygous carriers. Although such females pass the fragile X chromosome to 50% of their offspring, in some cases the phenotype has a low degree of penetrance in males. Males who inherit the mutant allele but have a normal phenotype are called *transmitter males*. Carrier mothers of transmitter males are normal, and have a low risk of having fragile X children. Daughters of transmitter males, on the other hand, have a higher risk of having affected children, leading to the type of pedigree shown in Figure 11.5.

The apparent paradox in the pattern of fragile X inheritance has been partially explained by the discovery of a length variation in the FMR1 gene. There is a repeated trinucleotide CGG sequence in the first coding region of the FMR1 gene. Normal individuals have from 6 to 52 copies of this sequence in the FMR1 gene, while individuals with more than 230 copies of this sequence have the symptoms associated with the fragile X syndrome. Those with an intermediate number of copies (ranging from 60-200) are unaffected carriers. Those FMR1 alleles that carry an intermediate number of copies are called *premutation* alleles; carriers of these alleles are themselves unaffected, but their children and grandchildren are at high risk for being affected. When intermediate alleles are transmitted by males, the number of CGG repeats is more likely to remain constant or even decrease; when transmitted by females, the number of CGG repeats is more likely to increase, abolishing expression of the FMR1 gene (Figure 11.6), and resulting in fragile X syndrome.

Early evidence indicates that the FMR1 gene product may be a RNA binding protein, whose absence alters the normal expression of a cascade of genes, resulting in the wide range of phenotypic symptoms of fragile X

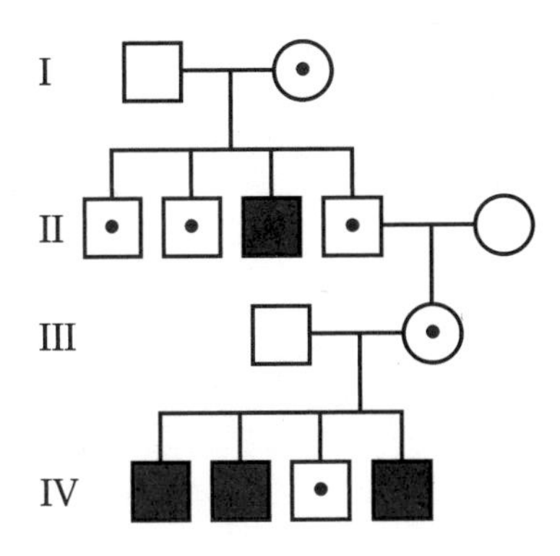

FIGURE 11.5 A pedigree illustrating inheritance of fragile X syndrome. Mothers (I-1) of normal transmitting males (II-4) are phenotypically normal and tend to have few offspring with fragile X syndrome. The daughters (III-2) of transmitting males, however, are at high risk of having affected children. Expansion of CCG repeats in premutation alleles is more likely by inheritance from transmitter males, conferring risk on the offspring in generation IV.

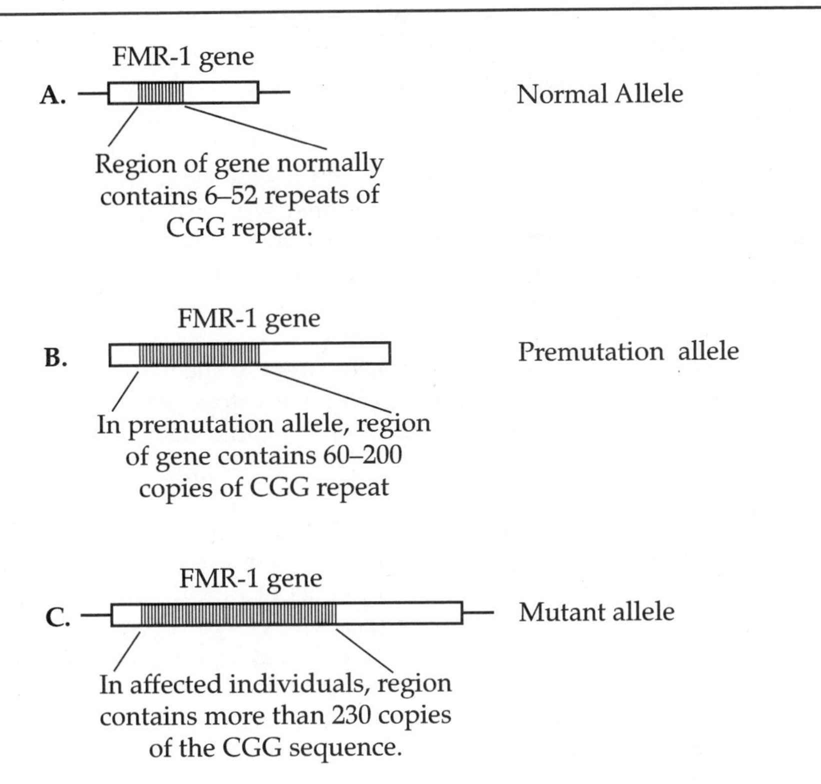

FIGURE 11.6 Expansion of FMR-1 gene. (A) The gene for fragile-X syndrome (FMR-1) normally contains 6–50 copies of a CGG repeat within the first exon. (B) In individuals affected by this disorder, this region expands to include 52–300 copies of the repeat. How expansion occurs and how it brings about the altered phenotype are still unknown.

TABLE 11.4 Mutations with Expanded Trinucleotide Repeats

Gene	Triplet that Expands	Normal Copy Number	Copy Number in Disease
Fragile-X syndrome	CGG	6-50	52-300
Huntington disease	CAG	11-34	42 to more than 100
Myotonic dystrophy	CTG	5-27	50 to more than 2,000
Spinal bulbar atrophy	CAG	17-26	40-52
Spinocerebellar ataxia	CAG	25-36	43-81

syndrome. As the trait is transmitted through generations, there is a successive increase in the size of the trinucleotide repeat (Figure 11.6), accompanied by an earlier age of onset and increase in severity of the symptoms.

Other disorders caused by an increase in trinucleotide repeats are listed in Table 11.4. In myotonic dystrophy (DM), the mutation is caused by expansion of a CTG repeat adjacent to a structural gene. As with fragile X syndrome, there is a progressive earlier onset of the disorder in succeeding generations, and a correlation with the size of the expanded repeat, the age of onset and the severity of symptoms. In contrast to the situation in fragile X syndrome, expansion of repeats in myotonic dystrophy and the onset of clinical symptoms is more likely with male transmission than with female transmission.

The other three disorders (Huntington disease, spinobulbar muscular atrophy and spinocerebellar ataxia) associated with expansion of trinucleotide repeats show several similarities. First, all are progressive neurodegenerative disorders inherited as autosomal dominant traits. Second, all have expanded CAG repeats, and third, all show a degree of correlation between the size of the repeat and the age of onset.

Anticipation
The occurrence of a genetic disorder at earlier ages and with increasing severity in successive generations.

The appearance of symptoms at earlier ages in succeeding generations is called **anticipation**, and was first noted for myotonic dystrophy early in this century. Although carefully documented by clinicians, the phenomenon of anticipation was discounted by geneticists because genes were regarded as highly stable entities, with only occasional mutations. The discovery of staged expansions of nucleotide repeats as the molecular basis of anticipation indicates that in these disorders, initial changes to a repeat sequence in or near a gene (generating intermediate numbers of repeats and premutation alleles) increases the chances that further changes in repeat number will occur, creating alleles with full mutations. This finding means that the concept of mutation must now be modified to incorporate the existence of unstable genome regions that undergo these dynamic changes. In other words, when some regions of the genome (those containing trinucleotide repeats) undergo an expansion in number, this event enhances the chance that further expansions, with corresponding changes in phenotype will occur. This also means that other disorders, especially those associated with non-classical segregation patterns that have been traditionally explained by terms such as incomplete penetrance or variable expression may need to be re-examined to establish a possible relationship between expansion of trinucleotide repeats and the phenotype.

Mutations, Genotypes, and Phenotypes

One of the first genetic disorders to be analyzed at the molecular level was sickle cell anemia. As outlined above, this disorder is caused by a single nucleotide substitution in codon 6 of the beta globin gene. This nucleotide substitution changes the amino acid at position 6 from glutamic acid to valine, producing a distinctive set of clinical symptoms. All affected individuals and all heterozygotes have the same nucleotide substitution, leading to the conclusion that specific genetic disorders were caused by specific mutations.

As it turns out, sickle cell anemia is probably an exception, rather than the rule. Molecular analysis of other genes reveals that more often, a spectrum of mutations is possible within a single gene, often accompanied by variations in the phenotype (Figure 11.7). As the molecular basis of mutations in a large number of genes becomes known, it is common to find that a genetic disorder is caused by many different mutational events in a given gene. For example, beta thalassemia is a genetic disorder in which the synthesis of the beta chain of hemoglobin is reduced or absent, causing an imbalance in hemoglobin production. Clearly, any mutational event that interferes with the production or function of the beta globin polypeptide results in the phenotype characteristic of beta thalassemia (Figure 11.6). Molecular analysis of the beta gene in those affected with beta thalassemia has revealed a number of different types of mutations, including nucleotide substitutions and alterations in DNA that affect control of mRNA processing and control of transcription and translation.

Mutation and Phenotypic Outcome

The situation where a genetic disorder is caused by a range of mutational events affecting a single gene is a form of *genetic heterogeneity*. In

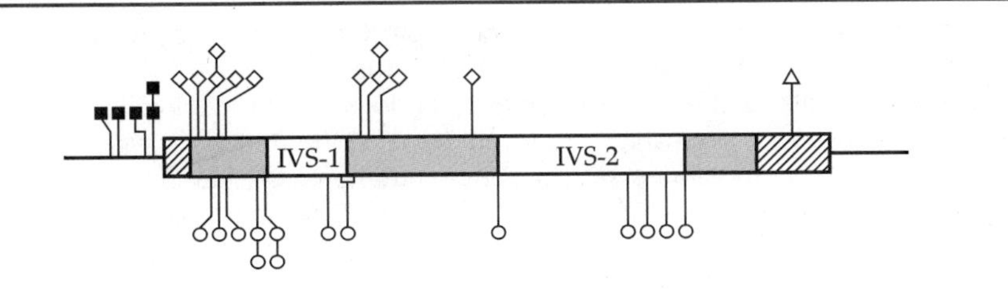

FIGURE 11.7 Diagram of the gene for beta globin showing the type and location of 30 mutations, each of which results in the phenotype of beta thalassemia. The hatched areas at each end of the gene represents regions cleaved during pre-mRNA processing; the shaded areas represent exons and the introns are indicated by IVS-1 and IVS-2. The symbols used for the types of mutations are: (■) mutations deficient in transcription, (◇) frameshift and nucleotide substitutions, (○) pre-mRNA processing mutations, and (△) RNA cleavage mutations.

fact, as more becomes known about mutations at the molecular level, it seems generally true that most human genetic disorders are heterogeneous. Cystic fibrosis provides a clear example of the variable nature of mutations in a single gene. From the analysis of over 30,000 mutant cystic fibrosis (CFTR) genes, more than 300 different mutations have been described. These include single amino acid substitutions (missense mutations), premature termination of translation (nonsense mutations), deletion of single amino acids, larger deletions involving one or more exons, frameshift mutations, and splice-site mutations. These mutations are distributed in all regions of the gene (Figure 11.8), strengthening the idea that any mutational event that interferes with expression of a gene will produce an abnormal phenotype.

Like many other genetic disorders, cystic fibrosis can cause individuals to exhibit a wide variety of clinical symptoms, leading to the conclusion that the type and location of mutations in a gene may ultimately determine the phenotype. The correlation between genotype and phenotype has been investigated for a number of specific mutations in the CFTR gene. In some mutations, such as the 508 deletion (present in 70% of all cases of CF), the CFTR protein is produced but does not associate with the plasma membrane. As a result, regulation of chloride ion transport is absent, and clinical

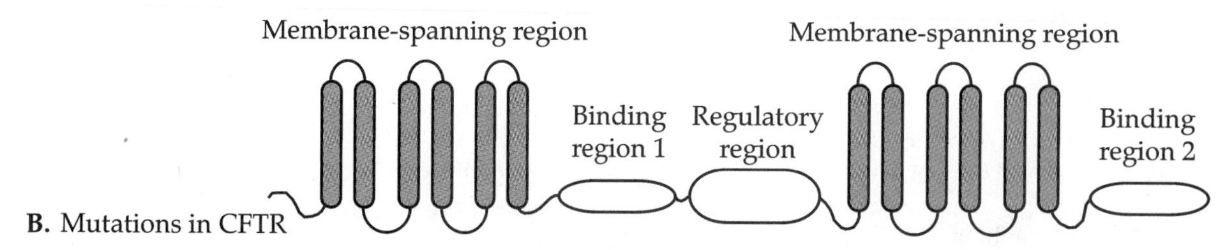

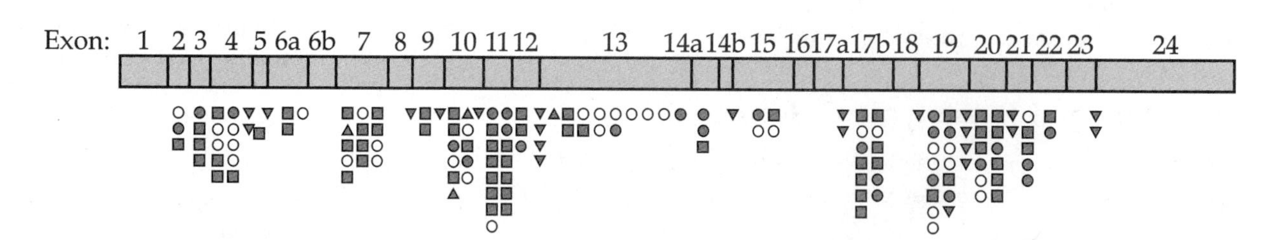

FIGURE 11.8 Mutations in the cystic fibrosis gene. (A) The cystic fibrosis gene product (CFTR) regulates the passage of chloride ions into the cell. The CFTR protein contains two regions that span the plasma membrane of the cell, two binding regions that respond to intracellular signals, and a regulatory region (R domain) that controls activity of the protein. (B) The exons of the CFTR gene are shown, below which are listed some of the mutations that have been reported for this gene, including deletions (▲), missense mutations (■), nonsense mutations (●), frameshift mutations (O), and splicing mutations (▼). Any of these mutations result in the clinical symptoms of cystic fibrosis.

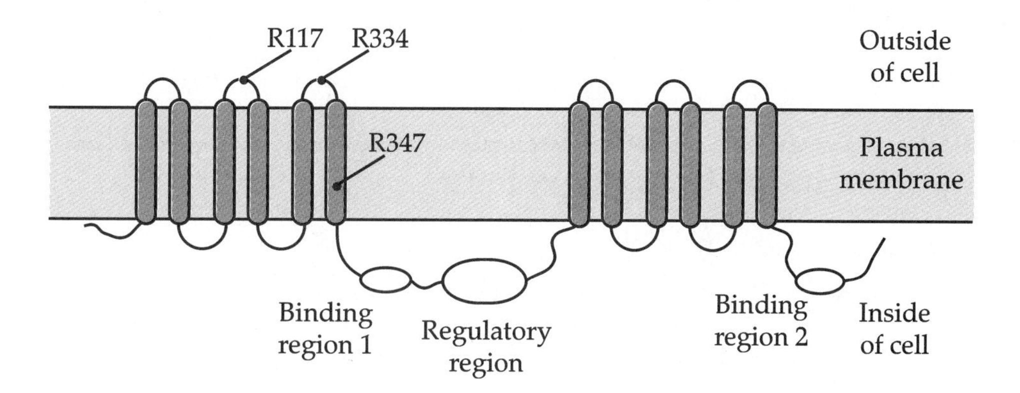

FIGURE 11.9 Mutations in the cystic fibrosis gene product that confer partial function and produce mild symptoms. These mutations are less common, and affect only about 2% of all CF individuals. In each case, the three mutations (R117, R334, and R347) are missense mutations. These mutations allow between 5–30% of normal activity for the gene product and mild symptoms. The most common mutation, a deletion in binding region 1 (present in 70% of all cases) is completely nonfunctional and is associated with severe symptoms.

symptoms are severe. In other mutations (Figure 11.9), the CFTR protein is produced and inserted into the membrane but is not fully functional. These mutations are associated with a milder form of cystic fibrosis. Thus, the genotype (in the form of the molecular nature of the mutation) determines the degree of function of the gene product, which in turn determines the phenotype.

A similar relationship between the genotype and phenotype is also apparent from an analysis of Duchenne muscular dystrophy (DMD) and Becker muscular dystrophy (BMD). Both disorders result from mutations in a gene on the X chromosome that encodes a protein called dystrophin. In DMD, dystrophin is usually absent, while in BMD, an abnormal, only partially functional form of dystrophin is present. The clinical phenotypes of DMD and BMD are also different. The onset of DMD usually occurs before the age of 6 years; symptoms are severe and progress rapidly, with death by age 20. In BMD, onset is delayed until the second or third decade of life, symptoms are milder, and affected individuals frequently survive to an advanced age.

At the molecular level, both forms of the disease are associated with deletions, and there is no apparent relationship between the extent of the deletion and the severity of the symptoms. In DMD, however, deletions usually alter the reading frame of the mRNA (frameshift mutation), resulting in the absence of functional dystrophin. In BMD, deletions do not alter the reading frame, allowing production of a partially functional protein, and milder symptoms. These disorders illustrate the relationship between genotype and phenotype, and they also demonstrate that an understanding of the molecular basis of a mutation can be used to predict the phenotypic outcome of the mutational event (for an exception, see Concepts and Controversies, p. 306).

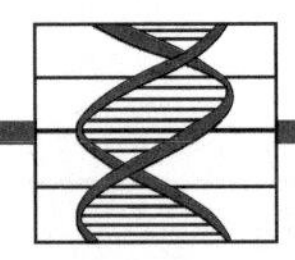

Concepts & Controversies

One Gene, Two Genetic Disorders

Consider two seemingly unrelated fatal genetic disorders. One is a rare condition called fatal familial insomnia; it affects both males and females, usually in the fifth decade of life. Symptoms begin with long bouts of insomnia and muscle twitches and progress rapidly to hallucinations, fevers, stupor, coma, and death. The second condition is Creutzfeld-Jacob syndrome; the genetic form of this disorder affects individuals in adulthood and is a neurodegenerative disease associated with loss of memory, dementia, seizures, and death some 3 to 12 months after the onset of symptoms.

Geneticists working on these two disorders were led to the same genetic locus on the short arm of chromosome 20, where a protein called prion-related protein is encoded. Not only were the researchers led to the same gene, they discovered that each disorder was caused by an identical mutation of the *same* codon. In this gene, codon 178 normally codes for the amino acid aspartic acid, but in both fatal familial insomnia and Creutzfeld-Jacob syndrome, a mutation causes codon 178 to encode the amino acid asparagine (a GAC $\rightarrow$ AAC nucleotide substitution).

Puzzled by this result, workers began to focus their attention on other regions of the gene. By comparing the amino acid sequence of the prion-related protein in individuals affected with fatal familial insomnia to that of the protein from those affected with Creutzfeld-Jacob syndrome, it was discovered that the two disorders differed at codon 129. Codon 129 is a polymorphic site in this protein, encoding methionine in some alleles and valine in others. By itself, this polymorphism is harmless. In combination with a mutation in codon 178, however, codon 129 determines which fatal disease affected carriers will develop. If methionine is specified by codon 129 and there is a mutation at codon 178, fatal familial insomnia results. If valine is encoded at 129, mutation at 178 results in Creutzfeld-Jacob syndrome.

It is not yet clear how the interaction of the polymorphic amino acid at position 129 interacts with the amino acid inserted by mutation at codon 178 to produce one or the other fatal disorder, but it is speculated that protein folding may be altered, causing the different outcomes. Both disorders are associated with brain lesions, but the distribution of lesions is different. It is not yet clear whether the type of lesions is the same, but if so, the mutation at codon 178 may specify the lesion, and the polymorphism at codon 129 may specify the pattern of lesions.

At least one other pair of genetic disorders is caused by mutations within the same gene, although not at the same codon. These disorders also map to chromosome 20, to a locus that encodes a protein called GNAS1. This protein is normally present in the cytoplasm and receives chemical signals from hormones and transmits this signal to the cytoplasm and nucleus of the cell. A mutation that causes a deficiency of GNAS1 causes a hormone-related disorder called pseudohypoparathyroidism. Another mutation that produces an excess of GNAS1 results in a second hormone-related disorder called McCune-Albright syndrome.

These examples serve to reinforce the idea of a fundamental relationship between a gene, a gene product, and a phenotype. As exemplified by the polymorphism in prion-related protein, some regions of a protein are less critical than others and can undergo amino acid substitutions with no phenotypic consequences, whereas other regions are critical for proper function and cannot be changed without devastating effects. In the second set of disorders, it is clear that the amount of a gene product is often a critical factor in determining the phenotype.

Genomic Imprinting

Humans, like other diploid organisms, carry two copies of each gene, one received from the mother, the other from the father. Normally there is no difference in the expression of the two copies. But in certain genetic disorders, recent evidence suggests that expression of alleles may differ, depending on whether they are inherited from the mother or the father. This differential expression is called **genomic imprinting** and is caused by the modification of a localized chromosomal region.

Genomic imprinting
Phenomenon in which the expression of a gene depends on whether it is inherited from the mother or the father. Also known as genetic or parental imprinting.

The first evidence for imprinting in mammals came from experimentally transplanting haploid germ cell nuclei in mice to produce zygotes containing two female or two male haploid genomes. Experimental mouse zygotes with a genome of male origin develop abnormal embryonic structures, but have normal placentas, whereas zygotes with a female-derived genome develop normal embryonic structures and abnormal placentas. In the mouse, both conditions are lethal, leading to the conclusion that both a maternal and a paternal genome are required for normal development and that different sets of genes are inactivated during the formation of eggs and sperm.

In humans, triploid embryos that have two paternal genomes and one maternal genome are composed mostly of placental tissue; the opposite condition (two maternal genomes and one paternal genome) results in embryos with only a small, rudimentary placenta.

Over the past few years, it has been discovered that genomic imprinting plays a role in specific human disorders. One of the clearest examples involves Prader-Willi syndrome (PWS) and Angelman syndrome (AS). Most cases of PWS, an autosomal recessive disorder characterized by obesity, uncontrolled appetite and mental retardation, is associated with a small deletion in the long arm of chromosome 15 (see Figure 6.17). In about 40% of cases however, no deletion can be detected. Using molecular markers, these non-deletion cases are found to have inherited two maternal copies of chromosome 15. This condition, where both copies of a given chromosome are inherited from a single parent, is called **uniparental disomy.** In this situation, the clinical symptoms of Prader-Willi syndrome are not caused by a mutation or deletion of a chromosome region, but by the presence of two maternal copies of chromosome 15. Interestingly, in cases of PWS that do involve a deletion, the deleted region is always from the paternal copy of the chromosome, leaving the individual with only a single, maternal copy of the region.

Uniparental disomy
A condition where both copies of a chromosome are inherited from a single parent.

Angelman syndrome is characterized by severe mental retardation; uncontrollable, jerky puppet-like movements; and seizures of laughter. Cytogenetic analysis of affected individuals reveals that about 50% of affected individuals have a small deletion in the q11q13 region of the long arm of chromosome 15, the same region that is deleted in PWS. Studies using molecular markers indicate that the deletion is carried in the maternal copy of chromosome 15, leaving the affected individual with only a paternal copy of the region. Many non-deletion cases of AS are associated with paternal disomy of chromosome 15. In these cases, it appears that the disorder is caused by the presence of two paternal copies of the chromosome rather than mutation or deletion of a gene.

These findings indicate that both a paternal and maternal copy of this region of chromosome 15 are required for normal development, that absence of a paternal copy results in PWS, and that absence of a maternal copy results in AS.

Genomic imprinting is involved in the expression of other genetic disorders, including retinoblastoma (chromosome 11) and Wilms tumor (chromosome 13). These disorders are discussed in Chapter 13. In these two disorders, loss of both copies of a gene allows tumor formation to occur. In a small number of cases, loss of only one copy of the gene allows tumor formation to occur. When only one copy is lost and tumor formation results, it is always the maternal copy of the gene that is missing. Somehow, the loss of the maternal copy of the normal gene is more critical to the development of a tumor than the loss of a paternal copy, providing evidence for the role of genomic imprinting in tumor development.

Imprinting does not affect all regions of the genome, but appears restricted to certain segments of human chromosomes 4p, 8q, 17p, 18p, 18q and 22q. Imprinting should not be considered as a mutation or permanent change in a gene or a chromosome region; what is affected is the *expression* of a gene, not the gene itself. Thus, imprinting does not violate the Mendelian principles of segregation or independent assortment. Imprinting is not permanent; remember that a copy of a given chromosome received by a female from her father will be transmitted as a maternal chromosome in the next generation. Thus in each generation, previous imprinting must be erased, and a new pattern of imprinting imposed, defining the newly imprinted region as either paternal or maternal for transmission to the next generation. Imprinting events are thought to take place during gamete formation and are transmitted to all tissues of the offspring. Although imprinting is not strictly a mutational event, it does involve differential chemical modification of DNA in paternally and maternally derived chromosomes. The precise mechanism of imprinting remains unknown, but the process may play an important role in gene expression and may be part of the molecular explanation of the phenomenon of incomplete penetrance.

SUMMARY

1. The study of mutations is an essential part of genetics. Mutation is the ultimate source of genetic variation and provides the geneticist with a means of studying genetic phenomena. Without mutations it would be difficult to determine whether a trait is under genetic control at all and impossible to determine its mode of inheritance.

2. Mutations can be classified in a variety of ways using criteria such as morphology, biochemistry, and degrees of lethality. Somatic mutations may drastically affect an individual, but such mutations are not heritable. Only mutations in the germ cells or gametes are passed on to the next generation and are of genetic significance.

3. The detection of mutations in humans is more difficult than in experimental organisms such as bacteria and fruit flies. As a class, dominant mutations are the easiest to detect because they are expressed in the heterozygous condition. Accurate pedigree information can often identify the individual in whom the mutation arose.

4. It is more difficult to determine the origin of sex-linked recessive mutations, but an examination of the male progeny is often informative. If the mutation in question is an autosomal recessive, it is almost impossible to identify the original mutant individual.

5. Studies of mutation rates in a variety of dominant and sex-linked recessive traits indicate that mutations in the human genome are rare events, occurring about once in every 1 million copies of a gene. The cumulative effect of this rate generates between 6 and 7 million mutant genes in each generation. The impact of this mutational load is diminished by several factors, including the redundant nature of the genetic code, the recessive nature of most mutations, and the lowered reproductive rate or early death associated with many genetic diseases.

6. Studies on the molecular basis of mutation have provided evidence for a direct link between gene, protein, and phenotype. Mutations can arise spontaneously, as the result of an error in DNA replication, or as the result of atomic shifts in nucleotide bases. Environmental agents, including chemicals and radiation, also cause mutations. Frameshift mutations cause a change in the reading frame of codons, often resulting in dramatic alterations in the structure and function of polypeptide products.

7. Genomic imprinting alters expression of genes depending on whether they are inherited maternally or paternally. Imprinting has been implicated in a number of disorders, including Prader-Willi and Angelman syndromes. Not all regions of the genome are affected, and only segments of chromosomes 4, 8, 17, 18, and 22 are known to be imprinted. Genes are not permanently altered by imprinting, and must be re-imprinted in each generation during gamete formation.

QUESTIONS AND PROBLEMS

1. Define the following terms:
  a. frameshift
  b. mutation rate
  c. somatic mutation

2. Familial retinoblastoma, a rare dominant autosomal defect, arose in a a large family that had no prior history of the disease. Consider the following genealogy (the colored symbols represent individuals):

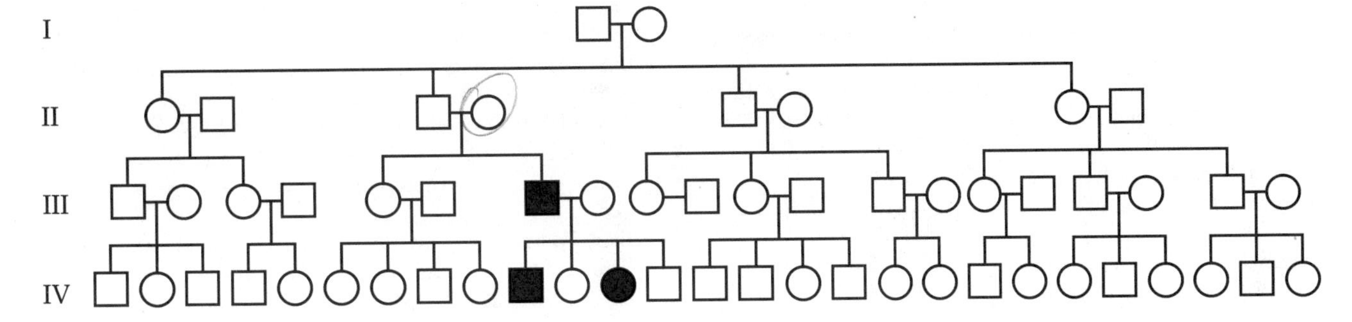

a. Circle the individual(s) in which the mutation most likely occurred.
b. Is this individual affected by the mutation? Justify your answer.
c. Assuming that the mutant allele is fully penetrant, what is the average percentage of an affected individual's offspring that will be affected?

3. Achondroplasia is a rare dominant autosomal defect resulting in dwarfism. The unaffected brother of an individual with achondroplasia is seeking counsel on the likelihood of his being a carrier of the mutant allele. What is the probability that the unaffected client is carrying the achondroplasia allele?

4. Tay-Sachs disease is a recessive autosomal disease. Since affected individuals do not often survive to reproductive age, why has Tay-Sachs disease persisted in human population?

5. Below is an idealized pedigree for the inheritance of a genetic disorder that shows paternal imprinting. An imprintable allele is inherited as a Mendelian trait, but expression is determined by the sex of the transmitting parent. In paternal imprinting, there will be no expression of the trait when it is transmitted by the father. In other words, there is a phenotypic effect only when the trait is transmitted by the mother. In this situation, there will be a number of carriers that do not express the trait. In the idealized world represented by this pedigree, there are equal numbers of males and females expressing the trait in each generation and equal numbers of non-expressing male and female carriers in each generation. Redraw the pedigree as it would appear for the same trait showing maternal imprinting.

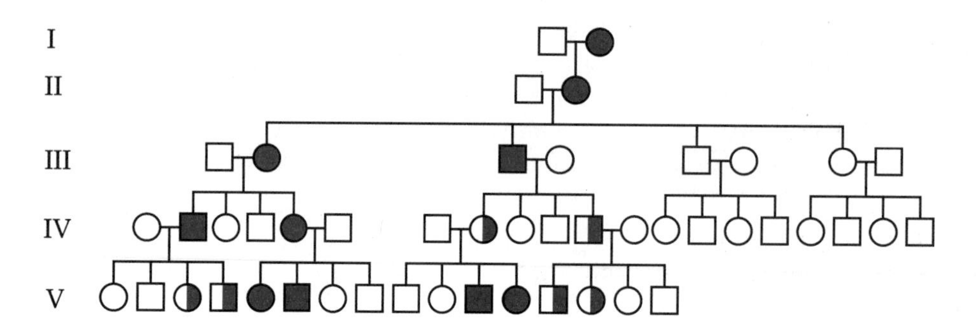

For Further Reading

Arlett, C. F. 1986. Human DNA repair defects. *J. Inherit. Metab. Dis.* **9** (Suppl.): 69–84.

Brunner, H., Brüggenwirth, H., Nillesen, W., Jansen, G., Hamel, B., Hoppe, R., deDie, C., Höwler, C., vanDost, B., Wieringa, B., Ropers, H., and Smeets, H. 1993. Influence of sex of the transmitting parent as well as of parental allele size on the CTG expansion in myotonic dystrophy (DM). Am. J. Hum. Genet. **53:** 1016–1023.

Cleaver, J. E. 1989. DNA repair in man. *Birth Defects* **25:** 61–82.

Cohen, M. M., and Levy, H. P. 1989. Chromosome instability syndromes. *Adv. Hum. Genet.* **18:** 43–149, 365–371.

Deering, R. A. 1962. Ultraviolet radiation and nucleic acids. *Sci. Am.* (December) **207:** 135–144.

Engel, E. 1993. Uniparental disomy revisited: the first twelve years. Am. J. Med. Genet. **46:** 670–674.

Hanawalt, P. C., and Haynes, R. H. 1967. The repair of DNA. *Sci. Am.* **216** (February): 36–43.

Harper, P., Harley, H., Reardon, W. and Shaw, D. 1992. Anticipation in myotonic dystrophy: new light on an old problem. Am. J. Hum. Genet. **51:** 10–16.

Huntington's Disease Collaborative Research Group. 1993. A novel gene containing a trinucleotide repeat that is expanded and unstable on Huntington's Disease chromosomes. Cell **72:** 971–983.

Lawn, R. M., Efstradiatis, A., O'Connell, C., and Maniatis, T. 1980. The nucleotide sequence of human beta globin gene. *Cell* **21:** 647–651.

Liebhaber, S. A., Goossens, M., Poon, R., and Kan, Y. W. 1980. The primary structure of the alpha globin gene cloned from normal human DNA. *Proc. Natl. Acad. Sci. USA* **73:** 7054–7056.

Lindahl, T. 1982. DNA repair enzymes. *Ann. Rev. Biochem.* **51**: 61–88.

Müller, H. and Scott, R. 1992. Hereditary conditions in which the loss of heterozygosity may be important. Mutat. Res. **284:** 15–24.

Nicholls, R. 1993. Genomic imprinting and uniparental disomy in Angelman and Prader-Willi syndromes: a review. Am. J. Med. Genet. **46:** 16–25.

Orkin, S. H., and Kazazian, H. H. 1984. The mutation and polymorphism of the beta globin gene and its surrounding DNA. *Ann. Rev. Genet.* **18**: 131.

Orr, H., Chung, M., Banfi, S., Kwiatkowski, T., Servadio, A., Beaudet, A., McCall, A., Duvick, A., Rnaum, L and Zoghbi, H. 1993. Expansion of an unstable trinucleotide repeat in spinocerebellar ataxia type 1. Nature Genetics **4:** 221–226.

Richards, R. and Sutherland, G. 1992. Fragile X syndrome: the molecular picture comes into focus. Trends Genet. **8:** 249–255.

Schwartz, E., and Surrey, S. 1986. Molecular biologic diagnosis of the hemoglobinopathies. *Hosp. Pract.* **9:** 163–178.

Tsui, L-C. 1992. The Spectrum of cyctic fibrosis mutations. *Trends Genet.* **8**: 392–398.

CHAPTER 12

MUTAGENS, CARCINOGENS, AND TERATOGENS

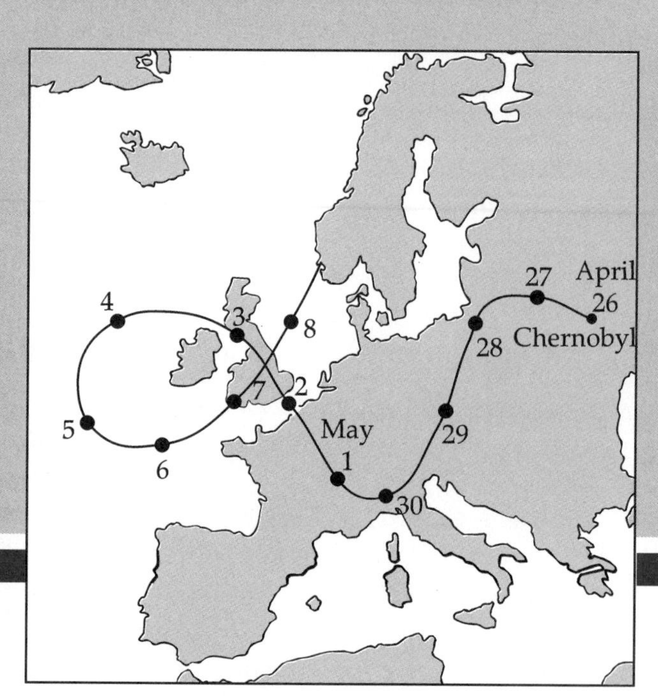

At 1:23 A.M. local time on April 26, 1986, a series of explosions blew the roof off the No. 4 reactor at the Chernobyl nuclear power station outside the city of Kiev in Ukraine. The village of Pripyat, 10 kilometers from the reactor site, was evacuated on the 27th, and 10 days later 135,000 people were removed from a 30-kilometer zone around the reactor. Within 6 weeks, 17 deaths had been reported in the former Soviet Union, all related to radiation exposure.

The cloud of radioactivity released by the reactor moved across much of Western Europe in the next 2 weeks (see map). In the first week of May, high levels of radioactivity were detected in drinking water in Scotland. This was followed by elevated levels of radioactivity in vegetables, cow's milk, and livestock. In the United States increased levels of radioactivity were first found on May 15. The increases were low, on the order of 0.02% of the recommended limit for such contamination. In addition, low levels of radioactivity were detected in the air at 12 testing stations throughout the country. As a result of the accident at Chernobyl, almost everyone in the Northern Hemisphere will be exposed to increased levels of radiation.

This incident, which received worldwide attention, raises several questions. What are the effects of radioactivity on health? Should we be concerned about barely detectable increases in the amount of radioactivity to which we are exposed? Finally, what are the genetic effects, if any, of radioactivity?

Radiation causes mutations, which we previously defined as changes in the base sequence of DNA. Mutations in DNA can take place spontaneously during DNA replication, and can also be induced by a variety of external agents, such as heat, radiation, or chemicals. These agents are called mutagens. When mutations are the result of the action of a mutagen such as ultraviolet light or a chemical, the mutation is classified as induced rather than spontaneous. Mutagens can act in several ways. They may bind directly with the bases in DNA, damage the molecule by causing breaks, or interfere with repair systems. It is important to remember that spontaneous and induced mutations are random events. In both cases the nature of the mutation produced is a matter of chance. Mutagens act only to increase the *rate* at which a mutation takes place, not to direct which genes are mutated.

In this chapter we will be concerned with the biologic and genetic effects of mutagens on humans. We will consider two major classes of mutagens: radiation and chemicals. In both cases we will look at the nature of the mutagens and what is known of their action. We will also briefly examine physical or chemical agents that cause abnormalities during human development. These agents are known as teratogens. The changes they cause are not heritable, although the outcome often resembles the phenotype of genetic diseases.

RADIATION

In physical terms, **radiation** is a process by which energy travels through space. There are two main forms of this energy: electromagnetic and corpuscular. Electromagnetic radiation is best described as waves of electrical or magnetic energy, while corpuscular radiation is composed of atomic and subatomic particles that move at high speeds and cause damage when they collide with other particles, including biological molecules.

Radiation
The process by which electromagnetic energy travels through space or a medium such as air.

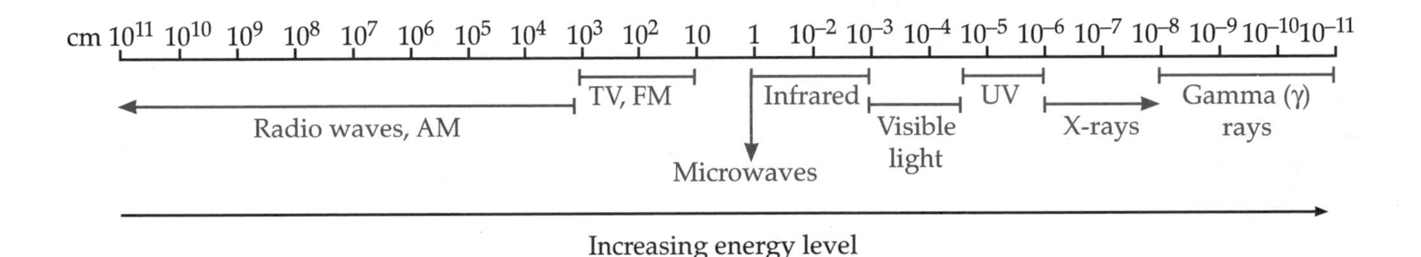

FIGURE 12.1 The electromagnetic spectrum arranged by wavelength and its relationship to energy levels.

Ionizing radiation
Electromagnetic or corpuscular radiation that generates ions as it passes through matter.

Both forms of radiation are known as **ionizing radiation** because they produce chemically and physically reactive ions when they interact with biological molecules. Ions are electrically charged atoms or molecules produced by an imbalance of atomic particles. However, not all mutagenic forms of radiation produce ions. Ultraviolet light is a powerful mutagen of lower energy than ionizing radiation.

Figure 12.1 shows the electromagnetic spectrum. In this spectrum, the energy level is inversely proportional to the wavelength; that is, long wavelengths have low energy, and short wavelengths have high energy. At the long end of the spectrum are low-energy radio waves with wavelengths measured in thousands of kilometers. At shorter wavelengths the radiation increases in energy level. Microwaves and radar have wavelengths of about 1 cm. Near the middle of the spectrum is the portion that humans perceive as visible light. Just beyond the range of visible light, the energy levels in the range of ultraviolet light are high enough to become mutagenic (see Concepts & Controversies, page 315). Beyond the ultraviolet region, radiation has enough energy to promote ion formation and to cause direct damage to DNA molecules. At very short wavelengths, x-rays have high penetrating power and can easily damage DNA in germ cells and gametes.

Corpuscular, or particulate, forms of radiation are composed of particles such as atomic nuclei or electrons moving at high speeds. For our purposes, we will consider only two forms, α particles and β particles. Their

TABLE 12.1 Properties of Ionizing Radiation

Type	Relative Penetration	Relative Ionization	Range in Tissue
α particle (2 neutrons + 2 protons)	1	10,000	Micrometers
β particle (electrons)	100	100	Micrometers–millimeters

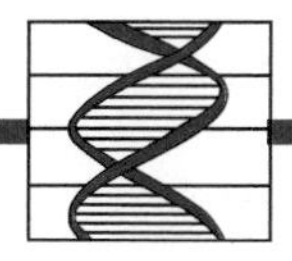

Concepts & Controversies

The Hole in the Sky

In the mid-1970s research linked the destruction of the ozone layer in the atmosphere to chlorofluorocarbons (CFCs), a class of chemicals used in air conditioning, aerosol sprays, insulation, and a range of other products. The ozone layer functions as a shield that prevents most of the sun's ultraviolet radiation from reaching the earth's surface. These gaps appear in the Antarctic spring, when increasing sunlight provides the energy for atmospheric chlorine to break apart ozone molecules. The ozone-depleted region forms during mid to late September and disperses later in the spring, beginning in late October or early November. Similar but less dramatic changes occur over much of the earth's atmosphere in a seasonal cycle. In response to this threat, the United States and Canada and several other countries banned the use of CFCs in aerosol sprays in the late 1970s.

In a draft document dealing with depletion of the ozone layer, the U.S. Environmental Protection Agency has estimated that the increase in ultraviolet radiation resulting from a thinning of the ozone layer will produce as many as 40 million additional cases of skin cancer and 800,000 additional cancer deaths in the United States over the next 88 years. This works out to more than 450,000 cases of skin cancer and 9,000 additional cancer deaths each year. The EPA also reported that increased ultraviolet radiation will cause damage to agricultural crops and wildlife.

The Montreal Conference of 1987 was the first to deal with the problem of ozone depletion. Fifty-six nations agreed to gradually lower the production and use of CFCs, with a 50% reduction in the use of these compounds by 1999. Since then, the gap in the ozone layer over the Antarctic has expanded much faster than predicted, and other chemicals, including carbon tetrachloride and methyl chloroform, have been recognized as threats to the ozone layer. This prompted the United Nations to sponsor the London Conference of June 1990 to speed the elimination of CFCs and related compounds. The representatives of almost 100 nations who attended the conference agreed to completely phase out use and production of CFCs by the year 2000 and to eliminate the production and use of all other depleting chemicals by the year 2020. Even so, it is expected that it will take at least a century before the ozone layer is repaired.

This agreement is the first signed by both developed and developing nations in response to a global problem. It is expected to be used as a model for other environmental problems, such as the greenhouse effect and the destruction of tropical rain forests.

In spite of this agreement, halting production and use of ozone-depleting chemicals has not yet stopped loss of the ozone layer. In early October 1993, ozone concentrations over Antarctica reached an all-time low, breaking the record set the previous year. In addition, the area of ozone-depleted atmosphere grew by 15% to cover 23 million square kilometers (an area equal in size to North America). Some of this loss may be due to debris pumped into the stratosphere by the eruption of Mt. Pinatubo, a volcano in the Philippines, but the rate and extent of ozone loss over the last few years remains a serious problem.

precise meaning and atomic composition are less important than their effects, which we will examine later. Recall that atoms are composed of a positively charged center known as the nucleus and an outer shell of negatively charged electrons. Alpha particles are positively charged and consist of two protons and two neutrons. Beta particles are negatively charged electrons, displaced from orbit around atomic nuclei. The energy levels and relative ability of ionizing radiation (electromagnetic and corpuscular) to penetrate biological material are shown in Table 12.1.

Measurement of Radiation

Dosimetry
The process of measuring radiation. For biologic work, the usual units are the roentgen, rad, rem, gray, and sievert.

Roentgen
The amount of ionizing radiation that produces 2.083×10^9 ion pairs in 1 cc of air.

rad
The radiation absorbed dose; a measurement of the radiation absorbed, as opposed to the amount produced; an amount equal to 100 ergs of energy absorbed per gram of irradiated tissue.

rem
The roentgen equivalent man; the amount of ionizing radiation that has the same biological effect as one rad of x-rays.

A variety of units is used to measure radiation; these measurements are collectively called **dosimetry.** The first unit used to measure radiation was the **roentgen** (R), named after Wilhelm Roentgen, who discovered x-rays. It is a measure of the number of ions produced when x-rays pass through a volume of air.

For studies on humans and other biologic systems, it is important to measure the amount of radiation absorbed; this unit was originally called the **rad** (radiation absorbed dose). The absorbed dose is considered important in predicting the outcome in an irradiated system. Note that the rad measures the amount of radiation absorbed but not the amount of biologic damage produced by the radiation. Table 12.1 shows that the same amount of different forms of radiation (same absorbed dose) can produce vastly different levels of damage (as measured in ionization). For example, 1 rad of α particles produces much greater biologic and genetic damage than 1 rad of γ rays. Therefore, another unit of radiation known as the **rem** was used to measure radiation in terms of biologic damage. One rem of radiation, whether from electromagnetic sources or particulate sources, produces the same level of biologic damage, even though this may involve vastly different levels of radiation. The typical range of biologic damage was usually measured in millirems. Each millirem is 1/1000 of a rem. These units and their definitions are summarized in Table 12.2. Recently, a new international standard (IS) has been adopted that replaces the rad, rem, and other units. In this system, the gray (Gy) replaces the rad, and the sievert (Sv) replaces the rem. The conversion for these units is as follows:

$$1 \text{ Gy} = 100 \text{ rad}$$
$$1 \text{ Sv} = 100 \text{ rem}$$

TABLE 12.2 Units of Radioactivity

Parameter Measured	Unit	Definition of Unit
Amount of radioactivity	Disintegration rate in curies (ci) or in becquerels (Bq)	1 Ci = 3.7×10^{10} disintegrations per second 1 Bq = 2.7×10^{-11} Ci
Exposure	Roentgen (R)	1 R = 2.58×10^{-4} coulombs per kilogram
Absorbed dose	Rad	1 rad = 0.01 joule per kilogram; 1 gray (Gy) = 100 rad
Biological damage	Rem	Rad × quality factor (Q), which is a measure of biological effect; 1 sievert (Sv) = 100 rem

One other factor that is important in measuring the effects of radiation is the time over which the dose of radiation is received. The **dose rate** is a measure of the radiation dose/unit time. As a general rule, large doses delivered over a short time are of concern because of their immediate biologic effects, while low doses are of concern for their long-term biologic and genetic effects. Time, however, does not diminish the effects of radiation; the effects of radiation are cumulative.

Dose rate
The amount of ionizing radiation delivered to a specific tissue area or body per unit time.

SOURCES OF RADIATION

Exposure to radiation is unavoidable. All forms of life are exposed continuously to radiation. Electromagnetic and corpuscular forms of radiation have been present since the birth of the universe. These natural sources of radiation are referred to as background radiation. We are also exposed to manufactured sources of radiation from medical testing, nuclear testing, nuclear power, and consumer goods.

Background

There are two possible sources of **background radiation:** (1) cosmic rays and high-energy particles from our sun and from outer space; (2) radioactive elements in the earth, air, water, and soil and radioactive elements in our own bodies and in our food.

Background radiation
Ionizing radiation in the environment that contributes to radiation exposure.

The relative effect of cosmic rays varies with geographic location and altitude. The atmosphere acts to absorb part of the energy of cosmic rays by collisions with atoms of atmospheric gas, so individuals at lower altitudes receive lower doses (Table 12.3). If you live in Denver (5000 feet), you receive twice as much of this radiation as people living at sea level in Miami. Averaged over the general population, the dose from cosmic radiation is between 25 and 75 mrem per year.

TABLE 12.3 Cosmic Ray Dose at Various Altitudes

Elevation (ft)	Approximate Dose (mrem/yr)
0–500	26–27
500–1000	27–28
1000–2000	28–31
2000–4000	31–39
4000–6000	39–52
6000–8000	52–74

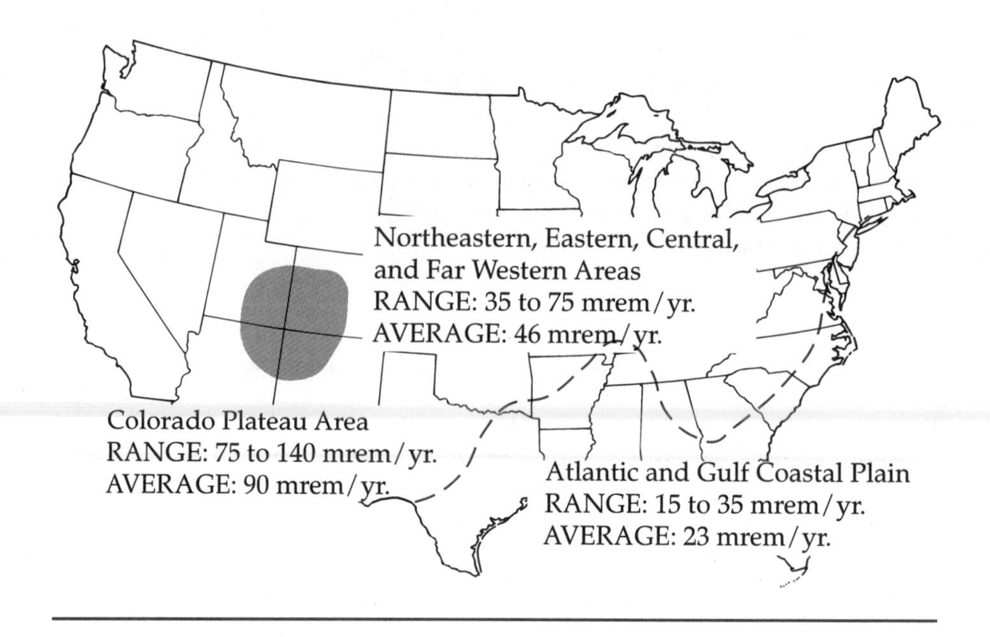

FIGURE 12.2 Distribution of terrestrial radiation in the United States. The doses shown are averages.

In the soil, radioactive elements such as uranium and thorium contribute to the background dose of radiation. As expected, the amount of exposure varies with geographic location (Figure 12.2). In the United States, the dose from terrestrial radiation averages about 50 mrem per year but varies from 20 to 90 mrem per year. In the Colorado Plateau region (Figure 12.2), the soil is rich in uranium, and residents of this area may have exposures of greater than 100 mrem per year from the soil. In some areas of the world, the figure is much higher than this. In a district in Kerala, India, the soil is rich in thorium, and the dose received by residents ranges from 200 to 2000 mrem per year.

Building materials made from terrestrial products are also radioactive and contribute to background exposure. Wood structures tend to give the lowest levels of radiation, with brick, concrete, and granite giving the highest. Buildings also act as closed containers that trap levels of radiation inside, especially radon gas, a radioactive decay product derived from uranium. Radon occurs naturally in the environment but seeps into houses and buildings from the ground beneath the foundation and is of growing concern as an environmental hazard in the United States (Figure 12.3). Research to date has been sparse, but as many as 10% of all houses in the United States (about 4 million structures) could have unacceptably high levels of radon. Because radioactivity occurs naturally in the environment, the food we ingest and the air we inhale contains radioactivity that accumulates in our body (Table 12.4).

Medical Testing

Aside from natural sources, the greatest exposure to radiation comes from medical diagnosis and treatment. It is estimated that 65% of the U.S. population is exposed each year to medical or dental x-rays. The

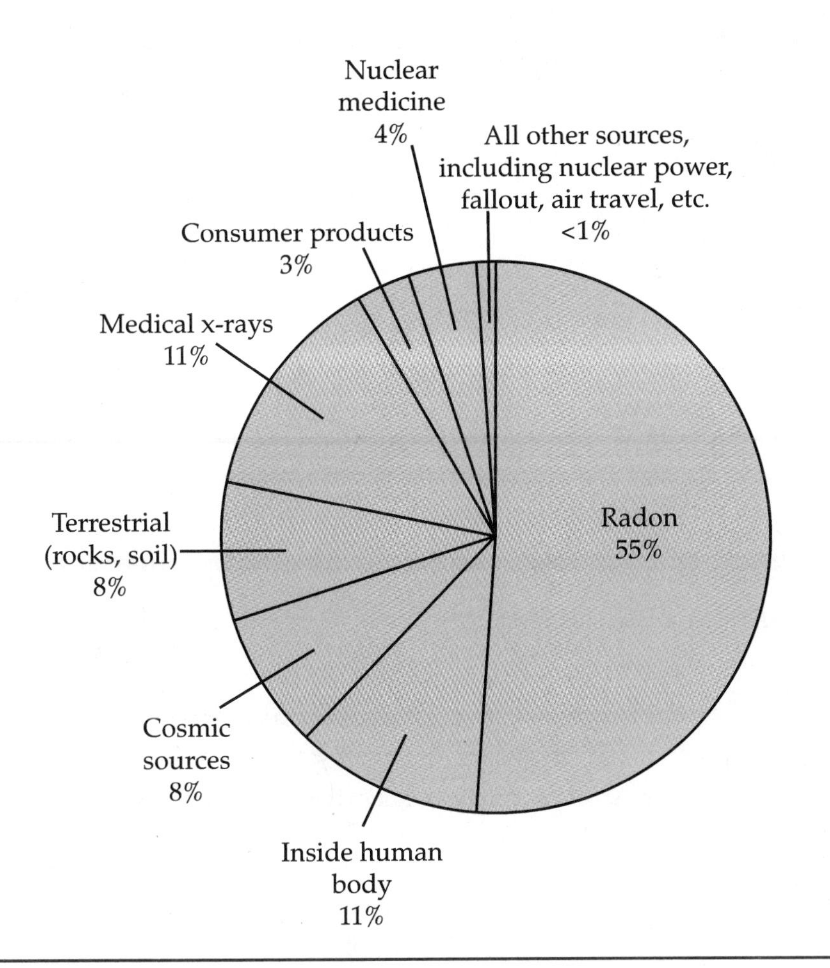

FIGURE 12.3 Radiation exposure to the U.S. population. Until recently it has been estimated that individuals residing in the United States are exposed to 150 to 200 mrem of radiation per year, with the average dose taken to be 170 mrem. The National Council on Radiation Protection and Measurements has raised this estimate to 360 mrem. This change is the result of new information about the presence and levels of radon gas in the home and workplace. According to the new estimates, 82% of the total exposure of 360 mrem comes from natural background sources. The remaining 18% is derived from man-made radiation. Radon accounts for 55% of the total exposure (about 200 mrem) and is the major source of radiation exposure to the U.S. population.

TABLE 12.4 Sources and Amounts of Radiation Accumulated by the Body, in mrem/yr

Radiation Source	Soft Tissue	Bone Marrow
^{3}H	~0.001	~0.001
^{14}C	0.7	0.7
^{40}K	19	15
^{87}Rb	0.3	0.6

TABLE 12.5 Radiation Dose to Adult Bone Marrow in Medical and Dental X-Ray Procedures

Procedure	Dose Per Exam (mrad)
Skull	78
Cervical spine	52
Chest	10
Ribs	143
Upper GI series	535
Lumbar spine	347
Gall bladder	168
Small bowel series	422
Barium enema	875
Abdomen (kidney, uterus, bladder)	147
Dental	21

dose delivered in various medical and dental procedures is shown in Table 12.5. The annual per capita dose from these sources is about 38 mrem.

TABLE 12.6 Projections of Whole-Body Exposure to the U.S. Population from Nuclear-Weapons-Testing Fallout

Year	Dose (mrem)
1963	13
1969	4.0
1980	4.4
1990	4.6
2000	4.9

Nuclear Testing and Nuclear Power

From the 1940s through the 1960s, several nations, including the United States, conducted atmospheric testing of nuclear weapons, depositing large quantities of radioactive materials into the environment as fallout. Although the testing of nuclear weapons is now done underground (except by China and France), debris from earlier testing will remain a source of exposure to the U.S. population for some years to come. Because China and France still conduct atmospheric testing of nuclear weapons, levels of exposure to radiation may rise slowly in the future. The current and projected exposure to the U.S. population is shown in Table 12.6.

There are almost 100 nuclear power plants in the United States. In addition, there are about 150 other nuclear reactors used for research and development. Another 200 or so reactors are used by the U.S. military as propulsion units for ships and submarines. The current rate of exposure from nuclear power plants is much less than 1 mrem per year for persons living within 10 miles of such facilities and is barely detectable for those living outside this limit.

TABLE 12.7 Radiation Dose from Selected Consumer Products, in mrem/yr

Product	Dose to Person Using Product	Average Dose to Population
Luminous wristwatches	1–3	0.2
Luminous clocks	9	0.5
Television sets	0.6	0.3
Construction materials	7	3.5

Miscellaneous Sources of Exposure

A variety of consumer products use or produce radiation and can be a factor in exposure, including television sets, smoke detectors, luminous dial watches, and airport luggage inspection systems. Some of these exposures are listed in Table 12.7.

Because cosmic radiation exposure increases as a function of altitude, airplane travel can contribute to overall exposure. For those who fly coast-to-coast in the United States, the average exposure to radiation is 1.0 mrem per trip.

Average Annual Exposure to Radiation

A summation of radiation exposure received by the U.S. population is shown in Figure 12.3. About 80% of the total exposure of 360 mrem comes from natural background sources. Radon is now known to account for 55% of the total exposure (about 200 mrem) and is the major source of radiation exposure to the U.S. population. The remaining 20% is derived from man-made radiation. This exposure includes medical procedures, consumer products, and all other sources.

Biologic Effects of Radiation

The biologic effects of radiation involve several levels of organization, including molecules, organelles, and cells. Since the cell is about 80% water, ionizing radiation often generates **free radicals,** in the form of ionized hydrogen, or hydroxyl radicals (OH), derived from water. These react chemically to form other radicals or hydrogen peroxide (H_2O_2). Such molecules are highly reactive and can destroy the structure of proteins and DNA (Figure 12.4). In turn, the damage produced by radiation leads to malfunctions in metabolic processes and cell death. These properties of

Free radical
An unstable and highly reactive molecule resulting from the interaction of ionizing radiation with water.

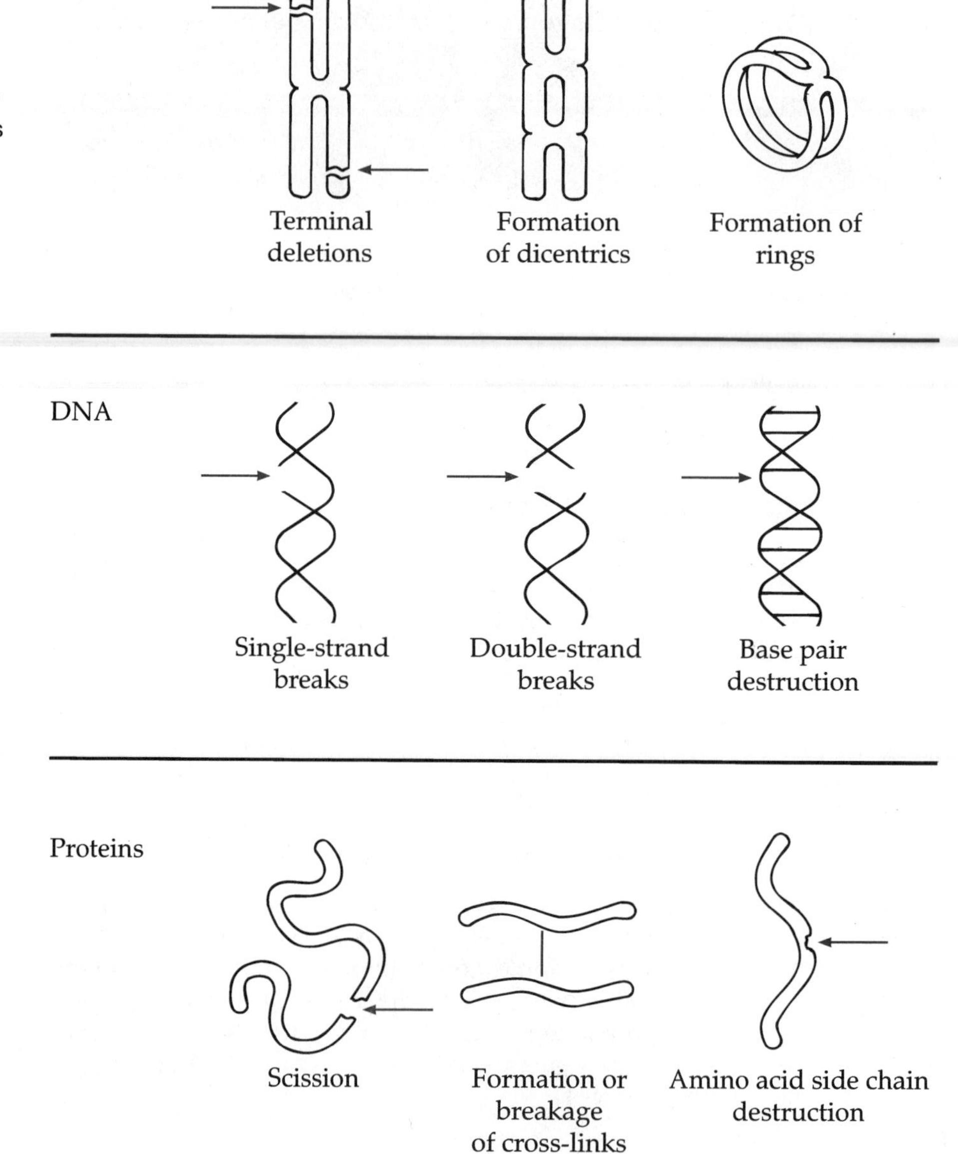

FIGURE 12.4 Effect of radiation on chromosomes, DNA, and proteins. Ionizing radiation can act directly or indirectly through free radicals created by interaction with water molecules in the cell.

radiation can also be used to kill insects, fungi, and bacteria in food (see Concepts & Controversies, page 323).

Large doses of radiation (above 400 rem) act through the destruction of dividing cells. Whole-body exposure to high doses kills mitotically active cells in the bone marrow that supply red and white blood cells to the circulation and also kills the actively dividing cells lining the gastrointestinal tract. Whether the outcome is death or recovery depends in large part on the ability of surviving cells to replace those destroyed by irradiation. A whole-body dose of radiation equivalent to about 450 rem (4.5 Sv) will kill

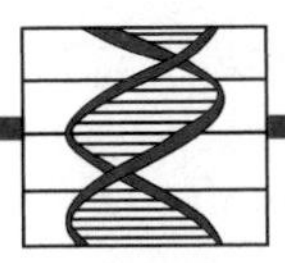

Concepts & Controversies

Irradiated Food

Over the past 40 years, research has demonstrated that treatment of food with radiation can be used for food preservation and decontamination. Irradiation of food can be used to prevent sprouting of root crops such as potatoes, to extend the shelf life of many foods, including fruits and vegetables, to destroy harmful microorganisms in meat and fish, and to decontaminate food ingredients such as spices.

In the process of irradiation, food is placed on a conveyor and is moved to a sealed, heavily shielded chamber where it is exposed to radiation from a radioactive source, x-rays or an electron beam, by remote control. The food itself does not come into contact with the radioactive source and is not itself made radioactive. Relatively low doses up to 100 kilorad (1 kilorad equals 1000 rad) are used to inhibit sprouting of potatoes, killing parasites in pork and insects in fruits and vegetables. Doses of 100 to 1000 kilorad are used to retard spoilage in meat, poultry and fish, and higher doses (1000 to 10,000 kilorad) are used to sterilize foods. Worldwide, about 55 commercial installations conduct food irradiation. The volume of food irradiated per year varies from country to country, ranging from a few tons of spices to hundreds of thousands of tons of grains.

Irradiated food has been routinely used by NASA to feed astronauts while in space, and irradiated foods are sold in more than 20 countries around the world. In the United States, in a series of decisions beginning in 1964, the Food and Drug Administration (FDA) has approved irradiation as a treatment for spices, herbs, fruits and vegetables including white potatoes, and pork and chicken. All irradiated food sold in the United States must be labeled as such with an identifying logo (shown here). To date, however, public concerns about radiation have prevented widespread acceptance of irradiated foods in U.S. markets. Advocates of irradiated food point out that this method can lower or eliminate the use of many chemical preservatives now in use, lower food costs by reducing spoilage, and reduce the incidence of food-borne illnesses such as salmonella poisoning. Those opposed to food irradiation argue that irradiation produces chemical changes in food and that the safety of these radiolytic products is in question, although such chemical changes also occur in foods that are prepared in other ways. Opponents also point out the possibility that irradiation might produce radiation-resistant or mutant forms of microorganisms.

Is irradiated food safe? It is clear that food treated with ionizing radiation is not made radioactive and that irradiation is a technology with great promise, but more research is necessary to resolve the question of whether irradiation affects food quality. Given the public response to date, consumer education will be required before irradiation of food becomes widely adopted in the United States.

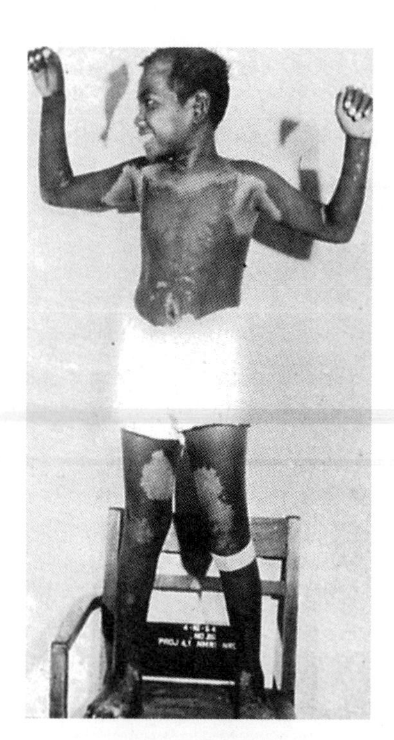

FIGURE 12.5 A child suffering from exposure to radiation released during U.S. atomic testing in the South Pacific in 1954. A wind shift carried the radiation cloud to the populated Marshall Islands, 500 miles from the test site at Bikini atoll.

TABLE 12.8 Biological Effects of Radiation

Dose (rem)	Symptoms	Two-Day White Blood Count	Prognosis
0–25	None	Normal	Good
100–200	Nausea, vomiting	20–50% reduction	Shoud recover
200–400	Nausea, vomiting	50–75% reduction	Most recover; rest die
400–600	Nausea, vomiting, diarrhea, bleeding	90% reduction	50% or more will die

50% of the individuals receiving this dose. This dose is known as the **LD_{50}** (lethal dose to 50% of those exposed; Table 12.8). Lower doses, between 100 and 200 rem, produce nausea, vomiting, burns to exposed skin (Figure 12.5), and a 50% reduction in the white blood cell count. This last effect is associated with a reduced immunity against infection and is a factor in survival.

LD_{50}
The radiation dose that will kill half the members of a population within a specific time.

Radiation can also have longer term biologic effects through somatic mutations, specifically, the induction of cancer. As a result of the Chernobyl nuclear power plant accident (Figure 12.6), the U.S. Nuclear Regulatory Commission has estimated that around 14,000 additional cancer deaths can be anticipated in Western Europe over the next 70 years. No one disputes that the exposure to radiation will cause more cancer deaths, although there is some disagreement over the number of deaths anticipated.

Genetic Effects of Radiation

As a mutagen, radiation increases the frequency of chromosome aberrations and other mutational events (see Figure 12.4). At the molecular level, DNA damage can occur by several mechanisms, including strand breaks and base deletions. At the chromosomal level, radiation may produce breaks that result in deletions, translocations, and chromosome fragments. Radiation can also induce aneuploidy through chromosome loss.

In assessing the genetic risk associated with ionizing radiation, a number of factors need to be considered. First, what constitutes a genetically significant dose of radiation? To determine what constitutes a genetically significant dose, one method measures the level of exposure required to obtain a 100% increase in the observed rate of mutation. This level is called the **doubling dose.**

Doubling dose
The dose of ionizing radiation that doubles the spontaneous mutation rate in the organism or species being studied.

In the doubling dose method, the amount of radiation necessary to double the spontaneous mutation rate is determined. While it is not feasible to conduct such experiments in humans, studies on genetic disorders in children born to survivors of the U.S. atomic bombings of Hiroshima and Nagasaki are available (Figure 12.7). In this work, several indicators of

FIGURE 12.6 Aerial photograph of the Chernobyl nuclear reactor after the accident in April 1986.

genetic effects were used, including stillbirths, major congenital defects, deaths among live-born infants, malignancies, frequencies of chromosomal aberrations and rearrangements, frequency of mutations that alter protein charge and/or function, and sex ratios among exposed mothers. From

FIGURE 12.7 The destruction caused by the atomic bomb dropped by the United States on Hiroshima, Japan, in 1945.

calculations of the amount of radiation reaching the gonads of the parents, it is possible to estimate that the genetic doubling dose for acute radiation exposure ranges from 170 to 220 rem (1.7 to 2.2 Sv).

In most cases, however, humans are exposed to low-level background radiation with occasional bursts of medical exposure. While it is admittedly difficult to extrapolate from the acute dose of radiation delivered over a short time in the atomic bombings to calculate the human doubling dose to chronic radiation (low dose delivered over a long time), it is estimated that the chronic doubling dose is between 340 and 450 rem (3.4 and 4.5 Sv). The estimates of the human doubling dose derived from the studies of the atomic bomb victims is significantly lower than those calculated by extrapolation from experiments in animals such as mice, suggesting that humans may be more resistant to the effects of radiation than previously thought.

Given an estimate of the doubling dose for chronic radiation of 395 rem (halfway between 340 and 450 rem), let us ask how close to the doubling dose is the average amount of radiation received by an adult by the mid-reproductive age of 30 years:

$$\text{Annual dose} = 360 \text{ mrem} \times 30 \text{ years} = 10{,}800 \text{ mrem total exposure}$$

If the total exposure is 10,800 mrem and the doubling dose is 395,000 mrem, the average person would receive:

$$\frac{10{,}800}{395{,}000} = 0.02734 = 2.7\%$$

In other words, over 30 years (about a generation), the average U.S. resident receives about 2.7% as much radiation from all sources as is needed to double the mutation rate in a generation. Of this amount, only about 0.54% is received from radiation generated by human activity.

Chemical Mutagens

Everyone is exposed to a large number of chemicals every day. Most often, chemical exposure is thought of only in terms of manufactured chemicals, but everything we eat is composed of chemicals. To resist infections by microorganisms or fungi and attack by insects, plants synthesize a wide range of defensive chemicals, many of which are mutagenic to humans. As a result, our diet is the major source of exposure to chemicals, even without including preservatives or additives. In addition to naturally-occurring mutagens found in our diet, many manufactured chemicals act as mutagenic agents.

The field of chemical mutagenesis began in 1942 when Charlotte Auerbach and J. M. Robson discovered that nitrogen mustard, a component of military poison gas, caused mutations. By the end of World War II, 30 to 40 such mutagenic compounds were known. The problem of identifying and restricting exposure to mutagenic industrial chemicals is enormous. There are presently over 6 million chemical compounds known, and almost 500,000 are used in manufacturing processes. More than a thousand new compounds

Base analogue
A purine or pyrimidine that differs in chemical structure from those normally found in nucleic acids and that is incorporated into nucleic acids in place of the normal base.

Promutagen
A nonmutagenic compound that is a metabolic precursor to a mutagen.

are formulated each year for use in research or manufacturing, and we know too little about the mutagenic effects of most of them. To coordinate efforts in identifying mutagenic chemicals, the International Commission for Protection against Environmental Mutagens and Carcinogens was founded in 1977. This organization is dedicated to the development of standards of evaluation and of regulations for the use and distribution of mutagenic chemicals.

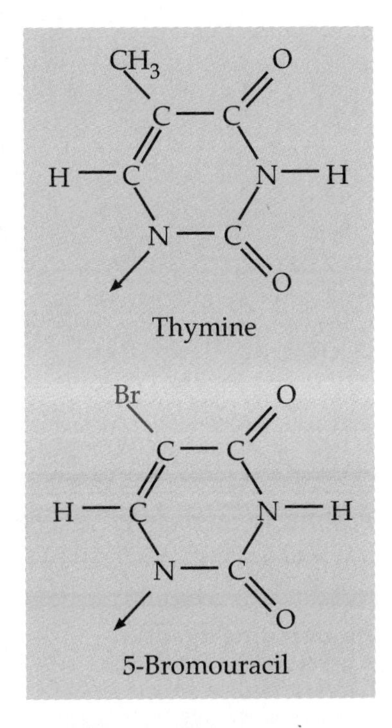

FIGURE 12.8 Structural similarity between thymine and 5-bromouracil. The bromine (Br) group is about the same size as the CH_3 group on the thymine.

Action of Chemical Mutagens

Unlike radiation, which penetrates directly from the environment to damage DNA in somatic cells and germ cells, chemicals follow a more indirect route. As a result, methods of exposure and metabolic interactions often play important roles in determining the activity of chemicals as mutagens. The principal methods of exposure to chemicals are ingestion, inhalation, or absorption through the skin. Once inside the body, chemicals or their metabolic products must be able to penetrate the nucleus and cause changes in DNA.

Chemicals act as mutagens in several ways, and they are often classified by the type of damage they cause to DNA. **Base analogues** are mutagenic chemicals that substitute for bases during nucleotide synthesis and DNA synthesis. Figure 12.8 compares the structure of a base analogue, 5-bromouracil, with that of thymine. If 5-bromouracil is incorporated into DNA in place of thymine (T), a 5-Br-A base pair is generated. In a subsequent round of replication, 5-Br may change its structure slightly and pair with guanine (G), producing a 5-Br-G base pair. After another round of replication, the net result is an A-T to G-C transition (Figure 12.9). Among base analogues found in food, caffeine is probably the most common, although because caffeine is rapidly excreted, it does not pose much of a threat as a mutagen.

Other chemical mutagens cause mutational events by altering the configuration of bases within a DNA molecule, causing them to resemble other bases. For example, the deamination of cytosine (C) converts it to uracil (Figure 12.10). What initially was a G-C pair is converted to a G-U pair. When present in DNA, uracil has the pairing properties of thymine (T). Thus after subsequent rounds of replication, the original G-C base pair is converted into an A-T pair (Figure 12.11). Experimentally, nitrous acid is used to induce deamination of cytosine into uracil. Although nitrous acid is not found in the environment outside certain manufacturing sites, nitrates and nitrites used in the preservation of meat, fish, and cheese are converted into nitrous acid during digestion. On their own, nitrates and nitrites are not considered mutagenic, but they can be activated or converted into mutagens by metabolic reactions or changing chemical conditions. Such compounds are known as **promutagens.** The mutagenic risk from nitrates and nitrites in food does not seem high but has proven difficult to assess with any degree of accuracy.

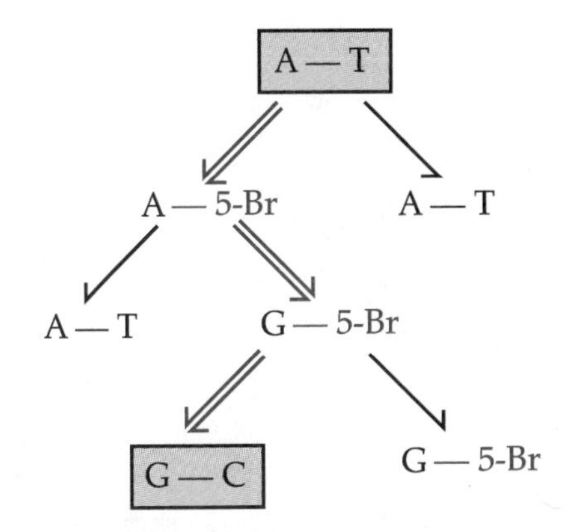

FIGURE 12.9 Rounds of replication showing incorporation of 5-Br and the transition it produces in subsequent rounds of replications. Boxes mark the beginning (A-T) and resulting (G-C) base pairs.

Mutagens that bind directly to DNA can result in frameshift mutations rather than base substitutions. Chemicals that insert themselves into DNA can distort the structure of the double helix, resulting in addition or deletion

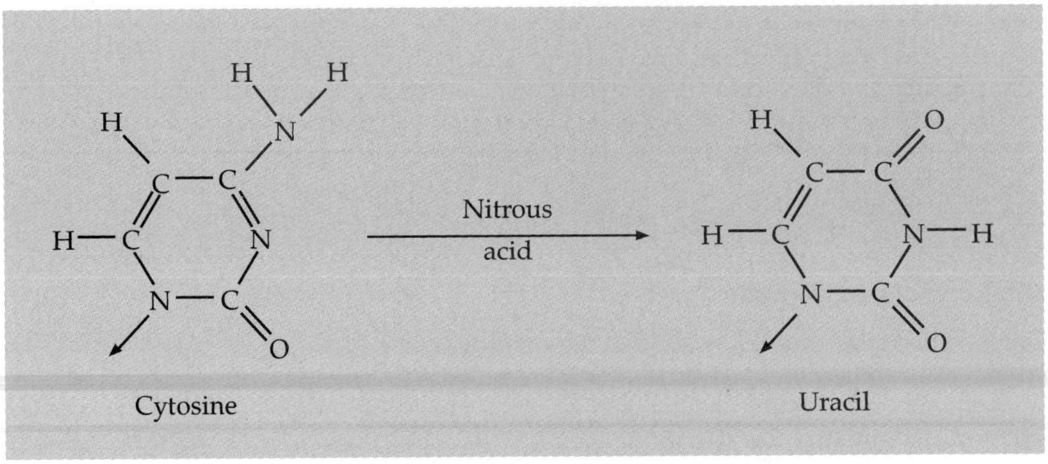

FIGURE 12.10 The process of deamination of cytosine to produce uracil.

Intercalating agent
A molecule that inserts itself between the base pairs in a DNA molecule. Some intercalating agents disrupt the alignment and pairing of bases in DNA, resulting in mistakes during replication.

of bases during DNA replication. The structure of two such **intercalating agents,** acridine orange and proflavin, are shown in Figure 12.12. Although the exact mechanism is unknown, intercalating agents apparently wedge themselves into a DNA molecule between base pairs, prying apart the ordered stacking of bases in the molecule. When replication takes place in this disturbed region, deletion or insertion of incorrect bases or both can take place. The result is the production of frameshift mutations.

In addition to specific alterations in DNA structure such as base substitutions and frameshifts, a wide range of chemicals can cause other less specific types of damage to DNA. These include DNA strand breaks (single, double, or both) and crosslinking within or between strands. Chemicals such as peroxides and free radicals can induce breaks or crosslinking of DNA strands.

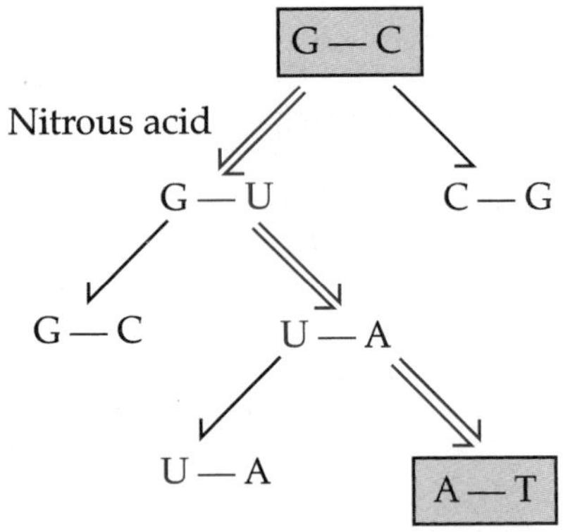

FIGURE 12.11 Rounds of replication following the deamination of cytosine to uracil, showing transitions from a G-C base pair to an A-T pair.

Detecting Mutagens

Several systems have been devised to screen chemicals to detect those that are mutagens. Experimental organisms or cultured cells are exposed to the suspected mutagen and assayed for the appearance of genetic effects. These systems have the advantage of being relatively inexpensive and are not time-consuming. Table 12.9 lists some of the organisms and experimental systems used in screening mutagens. The advantages of microorganisms in evaluating mutagens will be discussed below.

Ames Test

One of the most widely used mutagenic screening systems was devised by Bruce Ames and his colleagues. It is used to rapidly screen chemicals for their mutagenic capacity. This procedure uses four tester strains of the bacterium

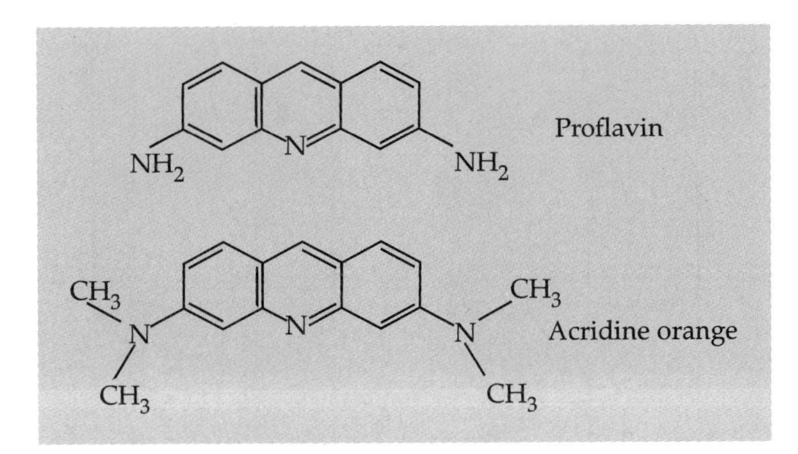

FIGURE 12.12 Molecular structure of intercalating agents that produce frameshift mutations.

Salmonella typhimurium that were genetically constructed for use in mutagenesis screening. One strain is used to detect base pair substitutions, and the other three detect frameshift mutations.

All four strains require the amino acid histidine for growth (they are called *his*$^-$), and the assay detects the mutation of the *his*$^-$ gene to *his*$^+$. In use, the cells are exposed to a mutagen and then plated on a medium lacking histidine. If the mutagen has caused the *his* defect to be corrected, the bacteria can then grow on histidine-free medium. The number of colonies detected on the plate indicates the number of mutations that have occurred and is a measure of the degree of mutagenesis (Figure 12.13).

In some cases, metabolism can produce mutagenic chemicals from inactive precursors or promutagens. Detection of promutagens in the **Ames test** is achieved by first mixing the chemical with rat liver extracts. The enzymes present in the extract generate metabolic products to which the tester strains are exposed.

Ames test
A bioassay developed by Bruce Ames and his colleagues for identifying mutagenic compounds.

TABLE 12.9 Mutagenic Screening Systems

	Damage Detected		
Organism	**Chromosome Aberrations**	**Nondisjunction**	**Mutation**
Escherichia coli			+
Salmonella typhimurium			+
Vicia faba (bean plant)	+	+	
Drosophila	+	+	+
Mammalian cell cultures	+	+	+

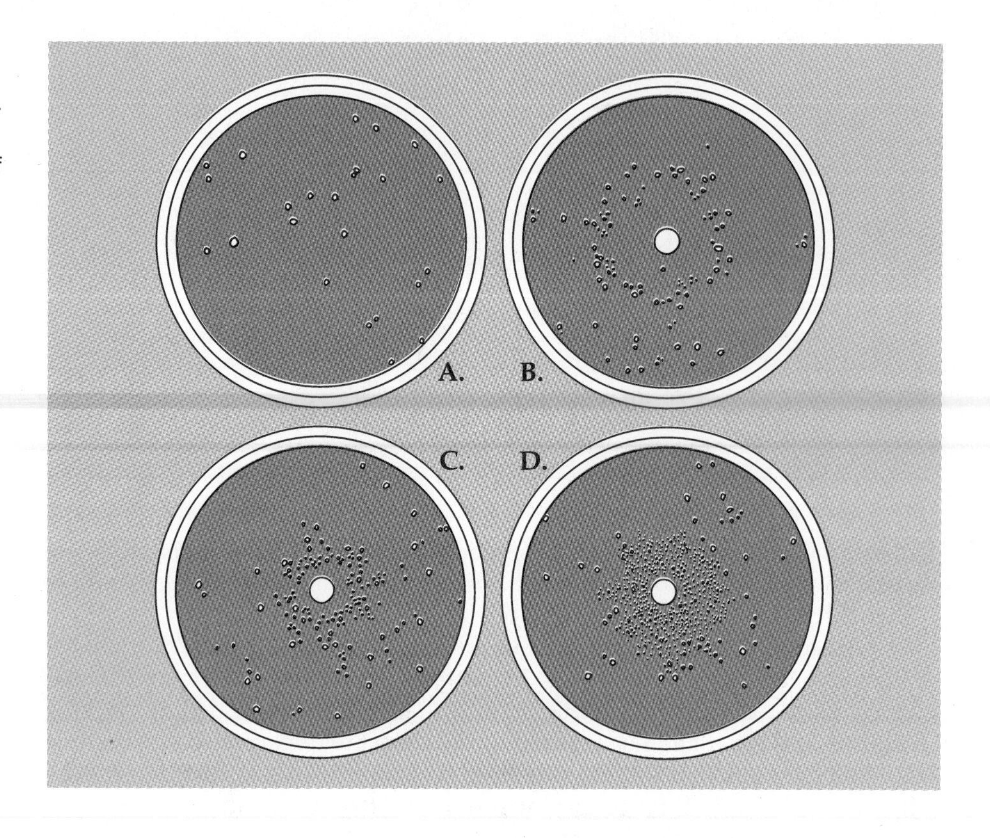

FIGURE 12.13 The Ames test. Dish A contains no mutagens. The colonies present are the result of spontaneous mutation. The paper disks in the center of dishes B, C, and D contain chemicals being tested for mutagenicity. The number of colonies (over and above those counted in dish A) indicate the degree of mutagenicity associated with each chemical.

As an example of how the Ames test works, let us consider a fire retardant chemical called Tris, used until the mid-1970s to treat children's pajamas and other clothing. When tested for mutagenicity in the Ames test, the kind of results shown in Figure 12.14 were obtained, indicating that Tris is a powerful mutagen. Further tests indicated that Tris and its metabolites appeared in the urine of children who wore treated clothing, indicating that it was absorbed through the skin. Based on these findings, the use of Tris is no longer allowed.

It is widely believed that cancer is the result of a mutagenic event, so it is not surprising that many chemicals identified as mutagens in the Ames test are also powerful carcinogens. Cancer can be regarded as the end product of one or more somatic mutation events. For example, if a mutagenic event knocks out a repair system in a somatic cell, the accumulation of future mutations may lead to a malignant transformation. We will consider the genetic and environmental aspects of cancer in the next chapter.

DNA Repair Mechanisms

Fortunately, not every mutation that occurs results in a permanent alteration of the genome. A number of genetically controlled systems repair damage to DNA. This repair function was first observed by cytologists, who noted that

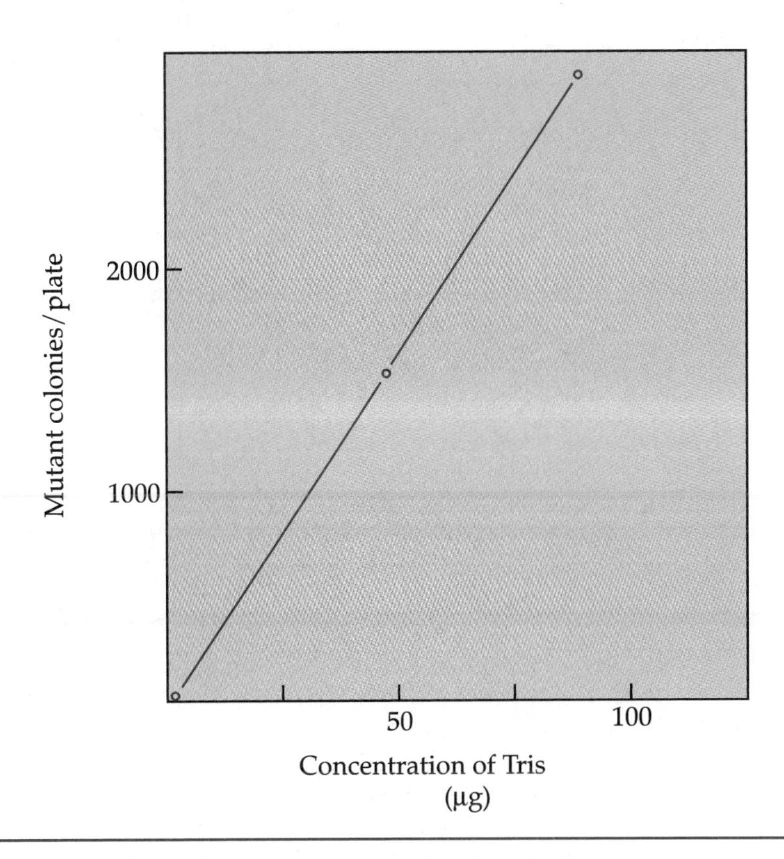

FIGURE 12.14 Mutagenicity of Tris flame retardant in the Ames test. As the concentration of Tris increases, the number of mutant colonies dramatically rises.

chromosome breaks, whether generated spontaneously or by exposure to chemicals or radiation, often rejoin with no apparent detrimental effects. At the molecular level, damage to DNA can be classified into two types: single nucleotide changes that alter the nucleotide sequence but that do not change the physical structure of the DNA molecule, and structural alterations or distortions that can interfere with replication or transcription. Such changes include pairing between bases on the same strand or the introduction of nicks on one or the other strand.

One type of repair system can correct errors made during DNA replication. Other repair systems recognize damage or structural modifications to the DNA molecule. These systems fall into several general categories including **excision repair** and **direct repair.** Exposure of DNA to ultraviolet light (from sunlight, tanning lamps, or other ultraviolet lamps) causes adjacent thymine molecules in the same DNA strand to pair with each other, forming **thymine dimers** (Figure 12.15). These dimers cause a regional distortion of the DNA molecule and can interfere with normal replication processes. Their continued presence in DNA can cause the insertion of incorrect bases during replication. This problem can be corrected by two different repair mechanisms. One is excision repair, which cuts the dimers and some adjacent nucleotides out of the affected DNA strand. The gap created by this excision is filled in by DNA synthesis using the opposite strand as a template (Figure 12.16A). Finally, an enzyme

Excision repair
Mechanism of DNA repair that cuts thymine dimers and some adjacent nucleotides out of a DNA strand and repairs the gap using the other strand as a template.

Direct repair
Mechanism of DNA repair involving reversal of damage. Light-induced breakage of improper bonds in thymine dimers is an example.

Thymine dimer
A molecular lesion in which chemical bonds form between a pair of adjacent thymine bases in a DNA molecule.

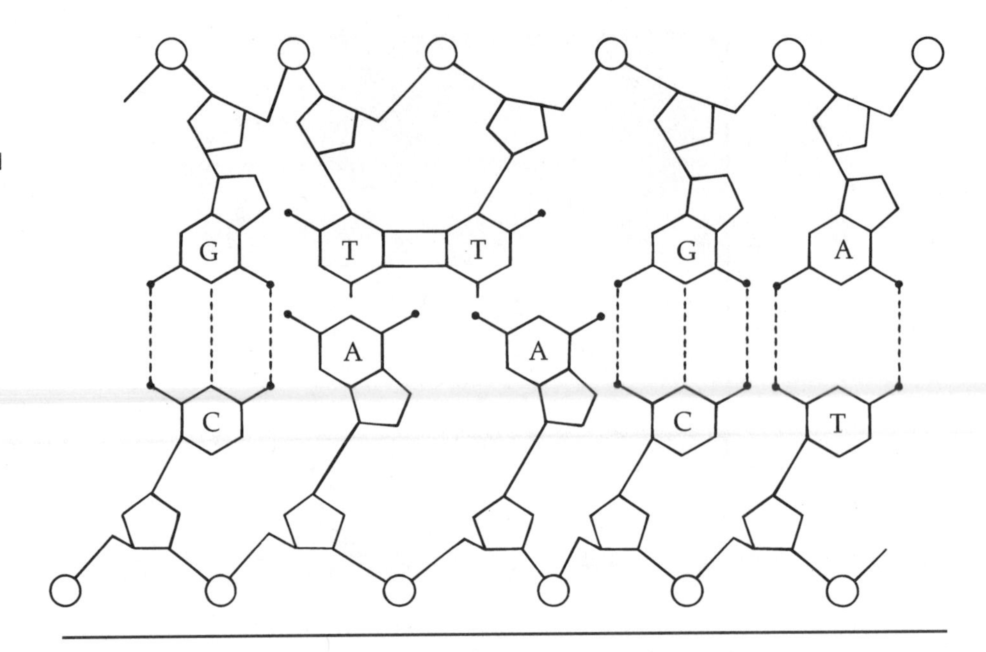

FIGURE 12.15 Thymine dimers in DNA. These structures, produced by ultraviolet light, cause a distortion in the molecular structure of DNA and are highly mutagenic unless removed.

known as *DNA ligase* seals gaps or nicks and leaves a continuous nucleotide strand.

Photoreactivation enzyme
An enzyme that removes thymine dimers induced by ultraviolet light from DNA molecules.

A second form of thymine dimer repair (direct repair) involves an enzyme known as the **photoreactivation enzyme.** This enzyme binds to the thymine dimer and cleaves the bond between the thymine molecules, reversing the effect of the ultraviolet light. Although the enzyme can bind to thymine dimers in the dark, it requires a photon of light to cleave the dimer (Figure 12.16B).

In humans, the existence of DNA repair systems and genetic disorders involving DNA repair have been inferred from the clinical symptoms associated with these disorders.

Genetic Disorders and DNA Repair

Xeroderma pigmentosum
A class of genetic disorders characterized by sensitivity to sunlight, skin pigmentation, and skin cancer. Individuals with this disorder are unable to properly repair damage to DNA caused by ultraviolet light.

XERODERMA PIGMENTOSUM. Because the mechanism of DNA repair is under genetic control, it too is subject to mutation. One such disease, **xeroderma pigmentosum** (XP), has provided a great deal of information on the mechanisms of DNA repair in humans. This disease is an autosomal recessive disorder with a frequency of 1 in 250,000. Affected individuals are extremely sensitive to sunlight (which contains ultraviolet radiation). Even short exposure causes dry, flaking skin, the appearance of heavy pigmentation, and disfiguring malignant tumors (Figure 12.17). This condition is severe, and early death from cancer is the usual outcome. The basis of this condition has been studied using cultured skin cells from affected individuals. The cultured cells from XP patients are deficient in excision repair of thymine dimers. In other words, when the ultraviolet radiation in sunlight produces thymine dimers in skin cell DNA, XP patients are unable to repair this damage, which then leads to the development of malignant tumors.

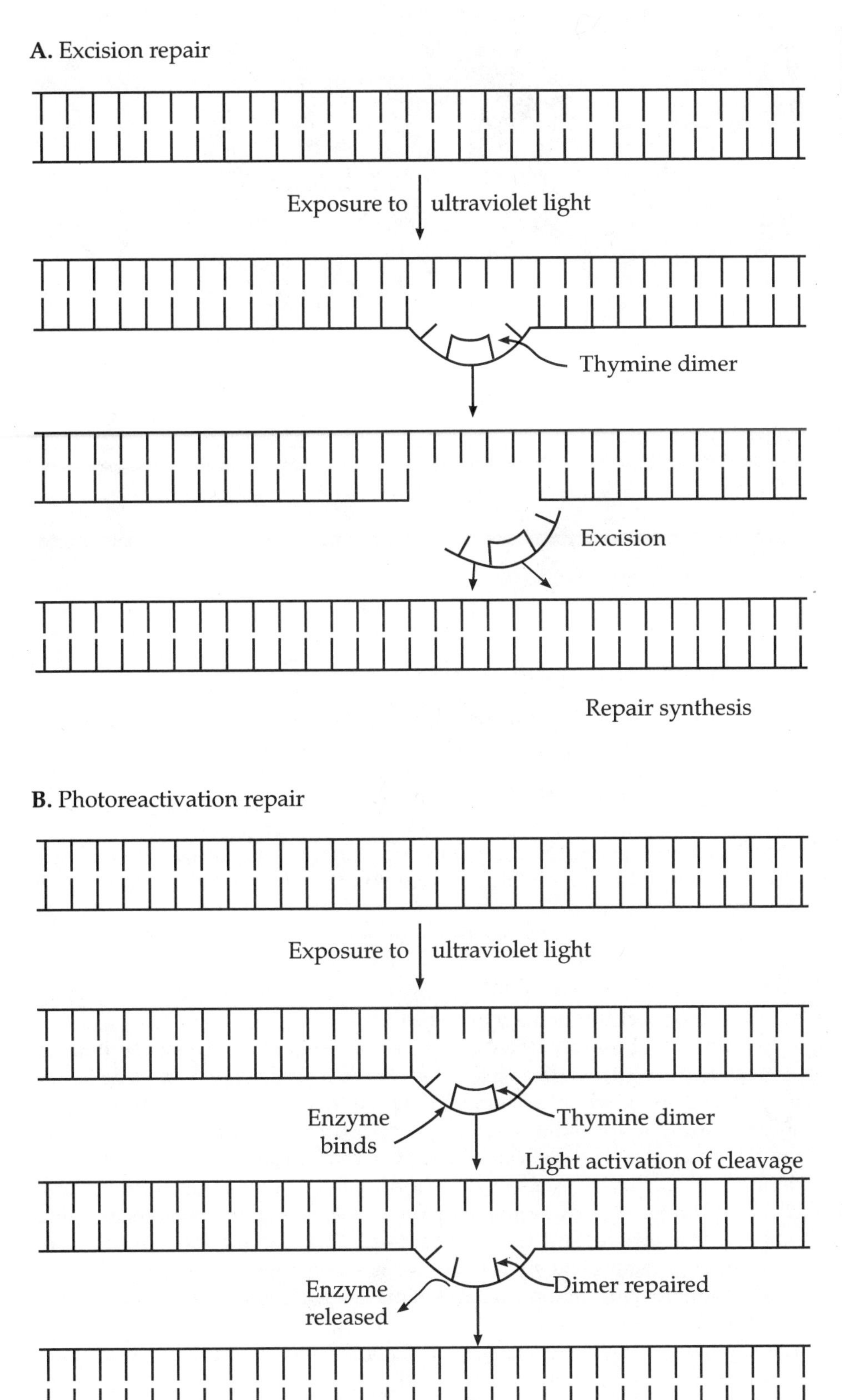

FIGURE 12.16 Repair of thymine dimers. (A) Excision repair of thymine dimers. (B) Photore-activation repair of thymine dimers.

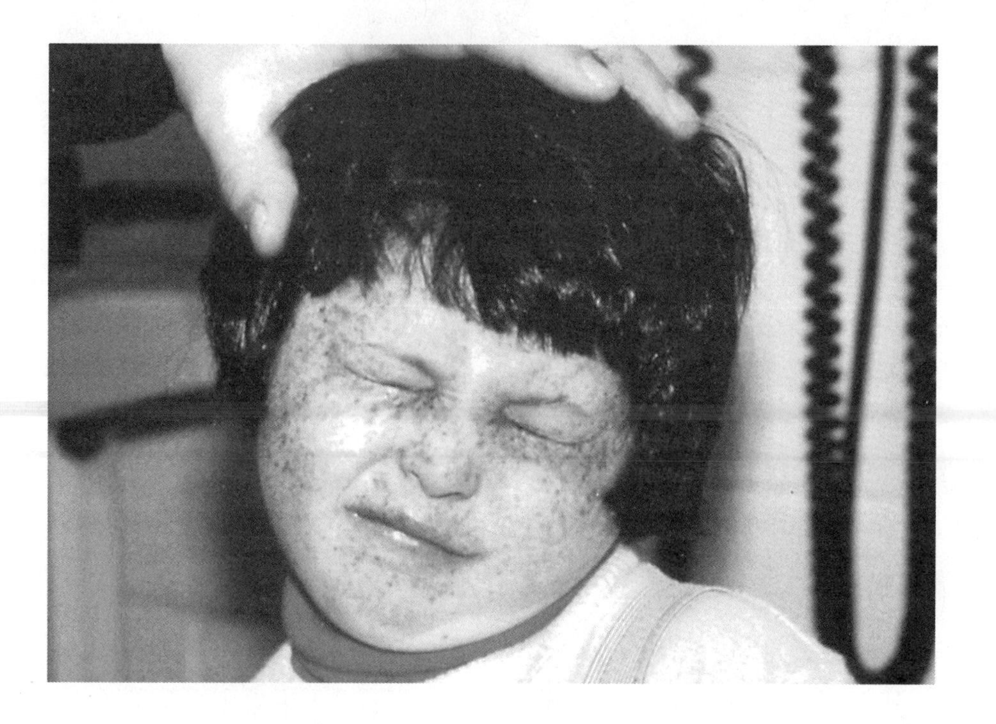

FIGURE 12.17 An individual afflicted with xeroderma pigmentosum.

Analyses of the excision repair defects in a large number of XP patients indicates that as many as seven different genes may be involved in the production of XP. It is not yet clear whether the products of these XP genes are directly involved in excision repair or are accessory proteins.

The complexity of XP indicates that DNA repair in humans is a complex process involving a number of different genes. A mutation in any of these genes leads to the clinical symptoms, skin tumors, and death. This disease also provides a strong correlation between mutation and the development of cancer.

This work also establishes that individuals unaffected by XP probably experience a high rate of DNA damage when exposed to sunlight. However, under normal circumstances this exposure activates one or more DNA repair systems to correct the damage (see below).

Even in those unaffected by XP or other DNA repair defects, overexposure to ultraviolet radiation in the form of sunlight or tanning lamps may overload the repair system so that skin tumors result. Currently over 400,000 new cases of skin cancer are reported in the United States each year, the vast majority of which are caused by overexposure to sunlight. The thinning of the ozone layer of the atmosphere by the release of industrial chemicals may be partly responsible for an increase in the incidence of skin cancer (see Concepts and Controversies, p. 335). The most deadly form of skin cancer is malignant melanoma, which originates in melanocytes, the cells in the epidermis that produce skin pigment. The incidence of melanoma has doubled since 1980, and it is now the most common form of cancer affecting women between the ages of 25 and 29 and the ninth most common

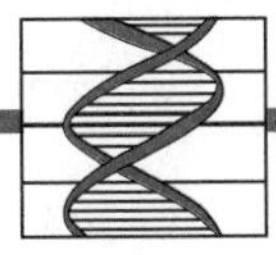

CONCEPTS & CONTROVERSIES

Using DNA to Monitor Ozone Depletion

Depletion of the ozone layer that protects the earth's surface from ultraviolet radiation was first detected over the Antarctic region, but thinning of the ozone layer over much of the earth's surface has now been confirmed. A reliable, inexpensive device to monitor and quantify the increased ultraviolet radiation resulting from ozone depletion would be useful in predicting the amount of damage caused to plants, animals, and ecosystems and in clarifying the relationship between ozone depletion and skin cancer. Ideally, such a system could be set up and operated under conditions where conventional electronic equipment cannot be used.

Researchers in Florida have developed a DNA dosimeter to be used in measuring the amount of ultraviolet light reaching the earth's surface. To construct the device, DNA is extracted from human cells grown in the laboratory. The purified DNA is suspended in water and sealed in small quartz tubes, about 0.5 in. in diameter and about 4 in. long. The tube is completely filled, and the ends are sealed. Quartz is used because it allows ultraviolet light to pass through into the interior, whereas conventional glass screens out most if not all ultraviolet light. Each device costs less than a dollar to make.

Exposure of DNA to ultraviolet light causes the formation of thymidine dimers, so the dosimeter is exposed to sunlight for a period of time, and the DNA is analyzed to determine the extent of thymidine dimer formation. This provides a direct measurement of the amount of DNA damage caused by the ultraviolet light received during the period of exposure. In preliminary tests carried out in the Bahamas, ocean-borne dosimeters were found to be more sensitive than conventional instrumentation, and they detected higher levels of DNA damage than those predicted by other means. This may allow DNA dosimeters to predict the worst-case scenario with respect to the amount of ultraviolet radiation, since the DNA in the dosimeters is not protected by cellular structures and no DNA repair systems are present in the tube. In a new, 3-year study, the dosimeters are being evaluated as part of a personal monitoring system. If all goes well, in the future you may put on your ultraviolet DNA dosimeter along with your sunglasses before venturing out into the noonday sun.

form of cancer in the United States. Almost all cases can be prevented by avoiding excessive exposure to the sun and by using sunscreens that block absorption of ultraviolet light.

RELATED DISEASES. **Cockayne syndrome** is another genetic disorder characterized by sensitivity to sunlight. This is an extremely rare autosomal recessive disease that first appears between 6-12 months of age. It is associated with mental retardation, dwarfism, neurologic degeneration, and early death. In Cockayne syndrome there is no increased incidence of cancer as there is with XP. However, this may be the result of the shortened life span of Cockayne patients. Studies of cultured cells from Cockayne individuals indicates that mutations in more than one gene can produce the symptoms of this syndrome, and that at least two and possibly as many as five different genes may be involved. The most common form, Type B, is associated with a defect in excision repair. The gene for this form of Cockayne syndrome

Cockayne syndrome
A rare autosomal disorder characterized by sensitivity to light, mental retardation, and early death.

encodes a protein that unwinds the DNA molecule (DNA helicase). This protein is not required for cell survival, but is involved in the repair of transcribed DNA sequences (genes).

Ataxia telangiectasia
A heritable disorder associated with sensitivity to x-rays and the inability to repair DNA damage caused by ionizing radiation.

Ataxia telangiectasia (AT) is a rare autosomal recessive condition with a frequency estimated at 1 in 40,000 to 1 in 100,000. This disease was first identified as a chromosome breakage syndrome. In addition to chromosome breaks, AT is associated with hypersensitivity to ionizing radiation such as x-rays. AT cells are defective in their ability to repair damage to DNA caused by x-rays (Figure 12.18). AT patients are cancer prone, again supporting a relationship between mutation and malignant growth. However, the use of x-rays to treat cancer in AT individuals can be fatal because of their sensitivity to ionizing radiation. There is genetic evidence for the existence of four genes that can produce the phenotype associated with AT. Loci for three of these genes have been found on chromosome 11, with one of the forms, AT-D mapped to 11q22-23. Although the genes have not yet been identified or isolated, flanking markers are available, and it may soon be possible to offer genetic counseling to affected families.

Fanconi anemia
A rare autosomal disorder characterized by a reduction in circulating blood cells, and chromosome aberrations. Affected individuals are sensitive to x-rays and other ionizing radiation. DNA repair is defective.

Bloom syndrome
An autosomal recessive condition characterized by dwarfism, decreased immunity and sensitivity to sunlight. Chromosome fragility and translocations indicate defective DNA repair.

Also linked to defects in DNA repair are two other genetic disorders: **Fanconi anemia** and **Bloom syndrome.** These autosomal recessive conditions share some phenotypic features in common. In both there is an increased frequency of chromosome breaks and an increased risk of cancer. Patients with Fanconi anemia are especially sensitive to ionizing radiation, such as x-rays, and to chemicals that induce cross-linking of bases in DNA. The nature of the repair defect in these two syndromes is still unresolved, but the results obtained so far suggest that the repair mechanisms associated with chromosome breaks may be different from those involved in thymine dimer repair. Recent work indicates that Bloom syndrome may involve DNA ligase, an enzyme involved in DNA replication and repair.

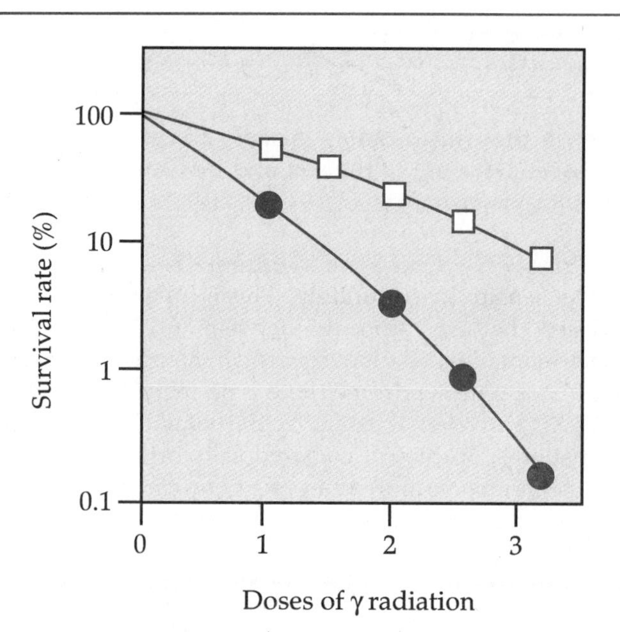

FIGURE 12.18 Survival of normal cells (□) and AT cells (●) when exposed to ionizing radiation such as x-rays.

TABLE 12.10 Rates of DNA Damage in a Mammalian Cell

Damage	Events/hr
Depurination	580
Depyrimidation	29
Deamination of cytosine	8
Single-stranded breaks	2300
Single-stranded breaks after depurination	580
Methylation of guanine	130
Pyrimidine (thymine) dimers in skin (noon Texas sun)	5×10^4
Single-stranded breaks from background ionizing radiation	10^{-4}

Assessing the Risk of DNA Damage

In this chapter we have spent many pages examining the effects of radiation and chemicals on the base sequence and physical structure of DNA. Yet in the previous chapter we concluded that mutation is a rare event, occurring only a few times in a million gene copies. Even an increase in the mutation rate caused by radiation or chemicals does not produce massive changes in the number of mutations generated.

Several explanations for this discrepancy between assaults on DNA and the observed mutation rate are possible. One possibility is that there is only a remote chance that radiation or mutagens actually have an opportunity to attack DNA and to generate a mutagenic event. In this case the details of mutagenic events we have examined would be of interest mostly to the scientists who study these relatively esoteric events, and there would seem to be little danger from exposure to mutagens. A second possibility is that DNA repair systems are extremely active and efficiently prevent the accumulation of mutagenic damage. If this is the case, it would be prudent to know how mutagens work, so that they can be avoided. Table 12.10 gives the estimated rate of damage to DNA in a typical mammalian cell at 37° C (body temperature). The rate of damage accumulation from background levels of radiation and chemicals is so high that within the normal life span of the cell, all of its DNA would be completely destroyed. Fortunately, humans are endowed with a number of highly efficient repair systems, as shown in Table 12.11. However, because the rate of background damage is so high, it might be fairly easy to overload the repair systems. One type of damage, the formation of thymine dimers, is thought to be a major cause of cell death, mutation, and transformation. XP patients have an 80% reduction

TABLE 12.11 Maximum DNA Repair Rates in a Human Cell

Damage	Repairs/hr
Single-stranded breaks	2×10^5
Pyrimidine dimers	5×10^4
Guanine methylation	10^4–10^5

in capacity to repair such dimers and a 10,000-fold greater risk of having cancer. This indicates that if the thymine dimer repair system is overloaded, mutagenic events can occur with a high frequency.

TERATOGENS

Teratogen
Any physical or chemical agent that brings about an increase in congenital malformations.

Teratogens are agents that produce abnormalities during the course of embryonic or fetal development. Unlike mutagens, they do not produce *heritable* changes but instead produce nongenetic birth defects. As we will see below, many mutagens can also act as teratogens, as can some genetic disorders such as phenylketonuria (PKU). In 1960 only four or five agents were known to be teratogens. The discovery that a tranquilizer, thalidomide, caused limb defects in unborn children helped focus attention on this field. Today, 30 to 40 teratogenic agents are known, with another 10 to 12 compounds strongly suspected.

Unfortunately, little or nothing is known about the mechanisms by which most of these agents produce fetal damage. Table 12.12 lists some known and suspected teratogenic agents that produce defects in developing human embryos and fetuses. The first group, ionizing radiation, has already been discussed in detail. Obviously, pregnant women should avoid all

TABLE 12.12 Human Teratogens

Known	Possible
Radiation	Cigarette smoking
Fallout	High levels of vitamin A
X-rays	Lithium
	Zinc deficiency
Infectious Agents	
Cytomegalovirus	
Herpesvirus II	
Rubella virus	
Toxoplasma gondii (spread in cat feces)	
Maternal Metabolic Problems	
Phenylketonuria (PKU)	
Diabetes	
Virilizing tumors	
Drugs and Chemicals	
Alcohol	
Aminopterins	
Chlorobiphenyls	
Coumarin anticoagulants	
Diethlystilbestrol	
Tetracyclines	
Thalidomide	

unnecessary x-rays, and no dose should be delivered to the abdomen of any woman of childbearing age unless she is known not to be pregnant. At present, the use of diagnostic ultrasound is not considered to be teratogenic.

Among infectious agents, viruses, such as the measles virus, and herpesvirus II, which is associated with genital herpes, can cause severe brain damage and mental retardation in a developing fetus. Evidence to date suggests that this action of herpes occurs only when the mother becomes infected with herpes *during* pregnancy. The damage caused by herpes does not appear when the mother has been infected before pregnancy and has a recurring attack during pregnancy. Herpesvirus I (associated with cold sores) does not appear to be teratogenic. Maternal metabolic abnormalities such as PKU can also be teratogenic.

As we saw in Chapter 10, dietary treatment has allowed PKU homozygotes to develop normally. Many of these individuals are now becoming pregnant, and because the special diet has been discontinued, high levels of serum phenylalanine cross the placenta and cause severe brain damage to the developing fetus. The use of a phenylalanine-restricted diet during pregnancy has not been entirely successful in correcting this problem, and the National Institute of Child Health and Human Development has instituted a multiyear study to develop a proper program of treatment for pregnant PKU homozygous women. Until the results of this study are available, it is currently recommended that women with PKU resume a phenylalanine-free diet several months before pregnancy.

Fetal Alcohol Syndrome

Fetal alcohol syndrome
A constellation of birth defects caused by maternal drinking during pregnancy.

Among teratogenic drugs and chemicals, prenatal exposure to alcohol is by far the most serious and the most widespread. Alcohol consumption during pregnancy results in spontaneous abortion, growth retardation, facial abnormalities (Figure 12.19), and mental retardation. This collection of defects is known as **fetal alcohol syndrome** (FAS). In milder forms, the condition is known as fetal alcohol effects. The incidence of these preventable birth defects is difficult to estimate but is calculated to be 1.9 per thousand births for FAS and 3.5 per thousand for fetal alcohol effects. While this incidence may seem low, the economic and social consequences are very significant. In the United States the cost for surgical repair of all facial abnormalities and treatment of all sensory and learning problems and mental retardation is more than $320 million per year, part of which goes to victims of fetal alcohol syndrome. For example, about 11% of the total budget for treatment of institutionalized mentally retarded individuals is for those affected by FAS.

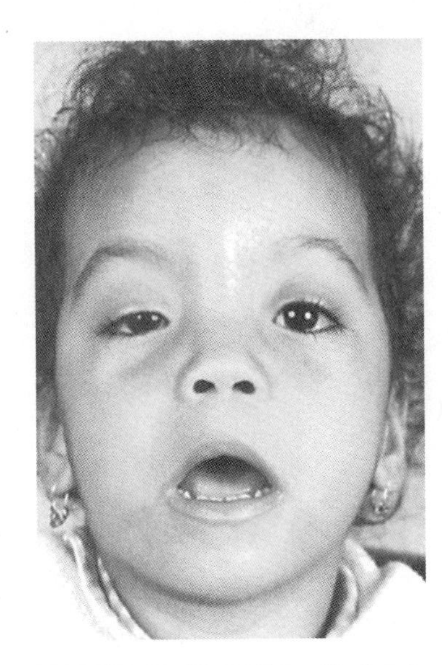

FIGURE 12.19 Child with fetal alcohol syndrome.

Research indicates that the teratogenic effects of alcohol can occur at any time during pregnancy, particularly late in the first trimester. Even in the third trimester, alcohol can cause severe growth retardation. One ounce of absolute alcohol (the amount contained in two mixed drinks) consumed per day in the third trimester reduces the birth weight of the fetus by about 160 grams. This means that the equivalent of two mixed drinks in one day at any time in the last 3 months of pregnancy will reduce the birth weight by 5%. Each date of similar alcohol consumption will further reduce the birth weight another 5% from the normal weight. Low birth weight is associated with high rates of neonatal death. Newborns below 50% of normal birth weight have a 45% mortality rate.

Alcohol works to produce its effects on the fetus by causing the constriction of the blood vessels of the placenta and umbilical cord, reducing the fetal oxygen supply. Measurements on placentas recovered from normal births indicate that umbilical blood vessels constrict in the presence of alcohol concentrations as low as 0.05%, about the amount of alcohol contained in 1 to 1.5 drinks. The oxygen deprivation may also play a role in the behavioral defects and mental retardation associated with FAS. In other words, fairly low levels of blood alcohol can have a serious effect on the development of a fetus, and pregnant women should avoid all alcohol during pregnancy.

While the actions of alcohol as a teratogen are now well known, it is hoped that future work will resolve the degree of risk involved with chemicals and substances that are suspected teratogens and will identify new agents among the thousands of chemicals in present use. More importantly, it is necessary to investigate the genetic basis of susceptibility to teratogenic agents, especially drugs and chemicals.

SUMMARY

1. Mutations can arise spontaneously through several mechanisms, including errors in DNA replication, or they can be induced by a variety of environmental agents, including radiation and chemicals.

2. Ultraviolet light and ionizing radiation are powerful mutagenic agents. Ultraviolet light works by causing the formation of thymine dimers in DNA. More energetic forms of radiation, such as x-rays and α and β particles, penetrate tissue and can cause mutagenic effects by damage to DNA or by splitting of water. Thus, radiation can be mutagenic directly by interaction with DNA or indirectly by generating ions and free radicals that damage DNA.

3. The radiation dose received by the human population includes exposure from cosmic, terrestrial, and manufactured sources of radiation. Experimental systems and observations on humans exposed to radiation have made it possible to estimate the effects of ionizing radiation on the human mutation rate. Although these calculations are subject to many errors, they offer an approximation of the mutational rate induced by current levels of exposure.

4. Mutations can also be induced by a wide range of chemicals. Only a small fraction of the millions of chemicals in our food or in use have been evaluated for their mutagenic effects. Chemical mutagens work by changing the purine or pyrimidine structure of DNA. This results in nucleotide substitutions or frameshift mutations.

5. DNA repair mechanisms offer a defense against the damage caused by mutagens. These repair mechanisms can correct errors that arise during replication or after exposure to environmental agents such as ultraviolet light.

6. A number of genetic disorders are caused by defects in DNA repair systems, indicating that the process of DNA repair is complex and is probably under the control of many genes.

7. Mutagenic agents can also act as teratogens and disrupt developmental events, producing birth defects. The molecular action of most teratogens is not well understood.

QUESTIONS AND PROBLEMS

1. Replication involves a period of time during which the DNA is particularly susceptible to the introduction of mutations. If nucleotides can be incorporated into DNA at a rate of 20 nucleotides/second and the human genome contains 3 billion nucleotides, how long will replication take? How might this time be reduced?

2. In what direction does the proofreading function remove improperly incorporated nucleotides?

3. Distinguish between spontaneous and induced mutations.

4. Discuss two mechanisms that will lead to induced mutations.

5. A bacterial geneticist was attempting to isolate a bacterial strain containing a mutation in gene *X*. It took him 3 years to wade through cells possessing a variety of spontaneous mutations in different genes before finding one that had a mutation in gene *X*. Needing strains possessing different mutations in gene *X* for comparative studies, he decided to induce mutations to shorten the time this mutant hunt would take. Assess the rationale employed by this scientist.

6. To illustrate the different ways by which radiant energy can product mutations, discuss the mechanisms by which x-rays and ultraviolet light interact with DNA.

7. A radioactive atom emits an α particle. After emission, has the atomic number (not atomic mass) changed? Explain.

8. Contrast the following radiation units and identify which unit is most useful in assessing the potential for genetic damage: roentgen, rad, and rem.

9. You are given the choice of receiving 0.1 sievert of either α particles or γ rays. Indicate your preference, if any, and defend it.

10. Name two sources of background radiation.

11. An individual received 30 rems of radiation in 1 hour from a leak at a nearby nuclear power facility. An x-ray technician also received 30 rems of radiation from leaks, but over a period of 10 years. Which individual, if either, should be more concerned?

12. Which cells of the body are preferentially killed by large doses of radiation, and how is this used to advantage in cancer therapy?

13. Identify a chemical mutagen that induces
    - **a.** A-T to G-C substitutions
    - **b.** G-C to A-T substitutions

14. What is a frameshift mutation, and how can intercalating agents effect these alterations?

15. **a.** A known quantity of a suspected chemical mutagen is added to the appropriate strain of *Salmonella typhimurium* and an Ames test is conducted. Equal numbers of cells are plated on medium containing histidine and medium lacking histidine. A total of 10,000 colonies arise on the plate supplemented with histidine, and 200 are found on the plate lacking histidine.
    b. The experiment is repeated exactly with a second suspected mutagen. The results show 10,000 colonies on the plate supplemented with histidine and 100 colonies on the plate lacking histidine. Which compound, A or B, is the stronger mutagen, if either? Explain.

For Further Reading

Ames, B. 1974. Identifying environmental chemicals causing mutation and cancer. *Science* **204:** 587–593.

Ames, B., McCann, J., and Yamasaki, E. 1975. Method for detecting carcinogens and mutagens with the *Salmonella*/mammalian microsome mutagenicity test. *Mutation Res.* **31:** 347–364.

Auerbach, C., and Kilbey, B. 1971. Mutations in eukaryotes. *Ann. Rev. Genet.* **5:** 163–218.

Barnes, D.E., Lindahl, T. and Sedgwick, B. 1993. DNA repair. Curr. opin. Cell Biol **5**: 424-433.

Castellani, A. (Ed.). 1985. *Epidemiology and Quantitation of Environmental Risk in Humans for Radiation and Other Agents.* New York: Plenum.

Committee for the Compilation of Materials on Damage Caused by the Atomic Bombs in Hiroshima and Nagasaki. 1981. *Hiroshima and Nagasaki.* New York: Basic Books.

Committee on the Biological Effects of Ionizing Radiation. 1980. *The Effects on Populations of Exposure to Low Levels of Ionizing Radiation.* Washington, D.C.: National Academy Press.

Denniston, C. 1982. Low level radiation and genetic risk estimations in man. *Ann. Rev. Genet.* **16:** 329–356.

Deveret, R. 1979. Bacterial tests for potential carcinogens. *Sci. Am.* (August) **241:** 40–49.

Drake, J., Glickman, B., and Ripley, L. 1983. Updating the theory of mutation. *Am. Sci.* **71:** 621–630.

Friedberg, E. 1992. Xeroderma pigmentosum, Cockayne's syndrome, helicases, and DNA repair: what's the relationship? Cell **71**: 887-889.

Gofman, J. W., and O'Connor, E. 1985. *X-rays: Health Effects of Common Exams.* San Francisco: Sierra Club Books.

Haseltine, W. 1983. Ultraviolet light repair and mutagenesis revisited. *Cell* **33:** 13–17.

Hoeijmakers, J.H. 1993. Nucleotide excision repair. II: From yeast to mammals. Trends Genet. **9**: 211-217.

Lindahl, T. 1982. DNA repair enzymes. *Ann. Rev. Biochem.* **51**: 61–88.

McCann, J., Choi, E., Yamasaki, E., and Ames, B. 1975. Detection of carcinogens as mutagens in the *Salmonella*/microsome test: Assay of 300 chemicals. *Proc. Natl. Acad. Sci. USA* **72:** 5135–5139.

Moss, T., and Sills, D. 1981. The Three Mile Island nuclear accident: Lessons and implications. *Ann. N.Y. Acad. Sci.* **365:** Entire issue.

National Council on Radiation Protection and Measurements. 1987. *Ionizing Radiation Exposure of the Population of the United States: Recommendations of the National Council on Radiation Protection and Measurements.* Bethesda, Md.: NCRP Report #93.

National Council on Radiation Protection and Measurements. 1989. *Exposure of the U.S. Population from Occupational Radiation: Recommendations of the National Council on Radiation Protection and Measurements.* Bethesda, Md.: NCRP Report #101.

Neel, J. V., Schull, W. J., Awa, A. A., Satoh, C., Kato, H., Otake, M., and Yoshimoto, Y. 1990. The children of parents exposed to atom bombs: Estimates of the genetic doubling dose of radiation for humans. *Am. J. Hum. Genet.* **46:** 1053–1072.

Obe, G. (Ed.). 1984. *Mutations in Man.* New York: Springer-Verlag.

Sankaranarayanan, K. 1982. *Genetic Effects of Ionizing Radiation in Multicellular Eukaryotes and the Assessment of Genetic Radiation Hazard in Man.* Amsterdam: Elsevier Biomedical.

Sankaranarayanan, K. 1993. Ionizing radiation, genetic risk estimation and molecular biology: impact and inferences. Trends Genet. **9**: 79-84.

Schull, W., Otake, M., and Neel, J. 1981. Genetic effects of the atomic bombs: A reappraisal. *Science* **213:** 1220–1227.

Shepard, T. 1983. *Catalog of Teratogenic Agents.* 4th ed. Baltimore: Johns Hopkins University Press.

Singer, B., and Kusmierek, J. 1982. Chemical mutagenesis. *Ann. Rev. Biochem.* **51:** 665–694.

Sugimura, T., Kando, S., and Takebe, H. (Eds.). 1982. *Environmental Mutagens and Carcinogens.* New York: Liss.

Venitt, S., and Parry, J. M. (Eds.). 1984. *Mutagenicity Testing: A Practical Approach.* Oxford, England: IRI Press.

Wagner, H. N., and Ketchum, L. E. 1989. *Living with Radiation: The Risk, the Promise.* Baltimore: Johns Hopkins University Press.

Weeda, G., Hoeijmakers, J., and Bootsma, D. 1993. Genes controlling nucleotide excision repair in eukaryotic cells. *Bioessays* **15:** 249–258.

CHAPTER 13

GENES AND CANCER

CHAPTER OUTLINE

As you sit in a classroom, a crowded stadium, or a concert hall, look around. According to the American Cancer Society, about one in three of the people you see around you will develop cancer at some point in their life, and about one in four will die from cancer. Each year about 500,000 individuals die of cancer, a rate of about 1 death per minute, and over 1 million new cases of cancer are diagnosed annually in the United States (Table 13.1). Currently over 10 million individuals are receiving medical treatment for cancer in U.S. hospitals and medical centers.

TABLE 13.1 Estimated New Cases of Cancer in the United States, 1993

Site	Number of New Cases
Skin	>700,000
Lung	170,000
Colon-rectum	152,000
Breast (female)	182,000
Prostate	165,000
Urinary	79,500
Uterus	44,500
Pancreas	27,700
Ovary	22,000

Cancer is a complex group of diseases that affect many different cells and tissues in the body. It is characterized by the uncontrolled growth and division of cells and by the ability of these cells to spread, or metastasize, to other sites within the body. Unchecked, the growth and metastasis result in death, making cancer a devastating and feared disease. Improvements in medical care have reduced deaths from infectious disease and have led to increases in life span, but these benefits have also helped make cancer a major cause of illness and death in our society. Because the risk of many cancers is age-related and because more Americans are living longer, they are at greater risk of developing cancer.

Although our society seems preoccupied with cancer, the idea that it is a disease of modern civilization is not accurate. Ancient Egyptian and Indian documents indicate that cancer was recognized as a life-threatening disease over 3500 years ago. The Greek physician Galen (A.D. 131–201) provided clear descriptions of cancers and referred to them as *karkinos* or *karkinomas,* terms that translate into the Latin word *cancer.*

Although significant advances have been made in the last decade, the underlying mechanisms responsible for the origin of cancer are still unknown. For centuries cancer was considered to be a contagious disease, and those with cancer were regarded as contaminated. Such individuals were shunned and thought to be untouchable. Scientific evidence gathered over the last hundred years has disproved this idea about cancer, but even today this belief is deeply ingrained in our culture. In addition, cancer patients have been fired or been denied employment or access to health insurance because of the high cost of medical care. As a result, federal legislation has been proposed that would prevent many forms of discrimination against cancer victims.

The link between cancer and mutation was forged early in this century by Theodore Boveri, who proposed that normal cells mutate into malignant ones because of changes in chromosome constitution. Four lines of evidence have been developed to support the idea that cancer has a genetic origin. (1) More than 50 forms of cancer are known to be inherited to one degree or another. (2) As outlined in Chapter 12, the Ames test has shown that most carcinogens are also mutagens. (3) Work with cancer-associated viruses has revealed the presence of a class of mutant genes, known as oncogenes, that promote and maintain the growth of a tumor. (4) As Boveri proposed, discrete chromosomal abnormalities are found in particular forms of cancer, especially leukemia. While a genetic event may give rise to cancer, *this event most often arises in somatic tissue and not germ line cells, and therefore the cancer itself is not passed on to offspring* although an increased risk of cancer can be inherited. The environment and behavior can also play a significant role in the genesis of cancer. In this chapter we will examine what is known about the relationship between genetics and cancer, with emphasis on genetic diseases that are associated with high rates of cancer. We will also consider tumor-suppressor genes and oncogenes and their role in causing and supporting malignant transformation in cells. We will examine the relationship between leukemia and chromosomal aberrations. Finally, we will analyze the interaction between cancer and the environment to explain how a multitude of factors may initiate the multistep process required for the development of malignant growth.

TUMORS

Before we consider cancer, some distinction should be made between tumors and cancer. Tumors are abnormal growths of tissue; most of the time, tumors arise in somatic tissues. *Benign* tumors are self-contained, noncancerous growths that do not spread to other tissues (are not metastatic) and are not invasive. Benign tumors usually cause problems by increasing in size until they interfere with the function of neighboring organs.

Cancer can be defined as a malignant tumor. Such tumors have several distinguishing characteristics. First, cancers are usually *clonal* in origin; that is, they are descendants from a single cell (usually in somatic tissue). Second, cancers develop in a step-wise fashion over time, through a series of genetic alterations that result in more aggressive growth with each mutation. Third, cancers are invasive and become *metastatic;* that is, cells can detach from the primary tumor and move to secondary sites where new malignant tumors are formed. The property of metastasis is conferred as a result of mutational changes in the cell.

Hereditary Forms of Cancer

The existence of families with high rates of specific cancers has been known since early in the 19th century. Families with high frequencies of cancer such as breast cancer or colon cancer have been reported many times since then, but in most cases no clear-cut pattern of inheritance has been identified. How is it, then, that some families have a much higher than average rate of cancer? Many explanations have been offered, including multiple gene inheritance, environmental agents, or chance alone (Table 13.2).

Recent advances in cancer research have provided some clues about the relationship between mutant genes and the cellular events that lead to tumor formation. As a result of these advances, cancer is now viewed as a disease that develops in stages, consisting of a small number of individual mutational events that can be separated by long intervals. While the exact number of steps is unknown and probably varies for different cancers, experimental evidence suggests that as few as two mutational events may be sufficient to cause a cell to become cancerous (Figure 13.1). In the majority of cancer cases, mutational events accumulate randomly over a period of years, correlating with the age-related incidence of many cancers.

In those forms of cancer that show a heritable predisposition, the first mutation is present in the germ cells and is transmitted genetically (Table 13.3) (Concepts and Controversies, p. 348). The second or subsequent mutations are acquired by somatic cells through spontaneous replication errors or exposure to environmental agents that cause genetic damage, resulting in cancer. Accordingly, what is inherited in these cases is not cancer itself but a predisposition to cancer that requires at least one additional mutational event to trigger the disease. For some forms of cancer, individuals who carry the primary mutation have a 100,000-fold increase in risk of developing cancer. On the other hand, not all individuals who inherit the first mutation will develop cancer. If the second mutational event does not occur, then no tumor will develop. In the following discussion, two classes of genes will be considered: **tumor suppressor genes,** which normally function to suppress cell division, and **proto-oncogenes,** which normally act to promote cell division. Mutations that alter the function, amount, or timing of action of these two classes of genes are associated with the development of cancer.

Tumor suppressor gene
A gene that normally functions to suppress cell division.

Proto-oncogene
A gene that normally functions to control cell division and may become a cancer gene (oncogene) by mutation.

TABLE 13.2 Factors in Cancer

Genetic	Nongenetic
Single gene mutations	Environmental carcinogens
Chromosome aberrations	Abnormal hormone levels
	Diet
	Chance

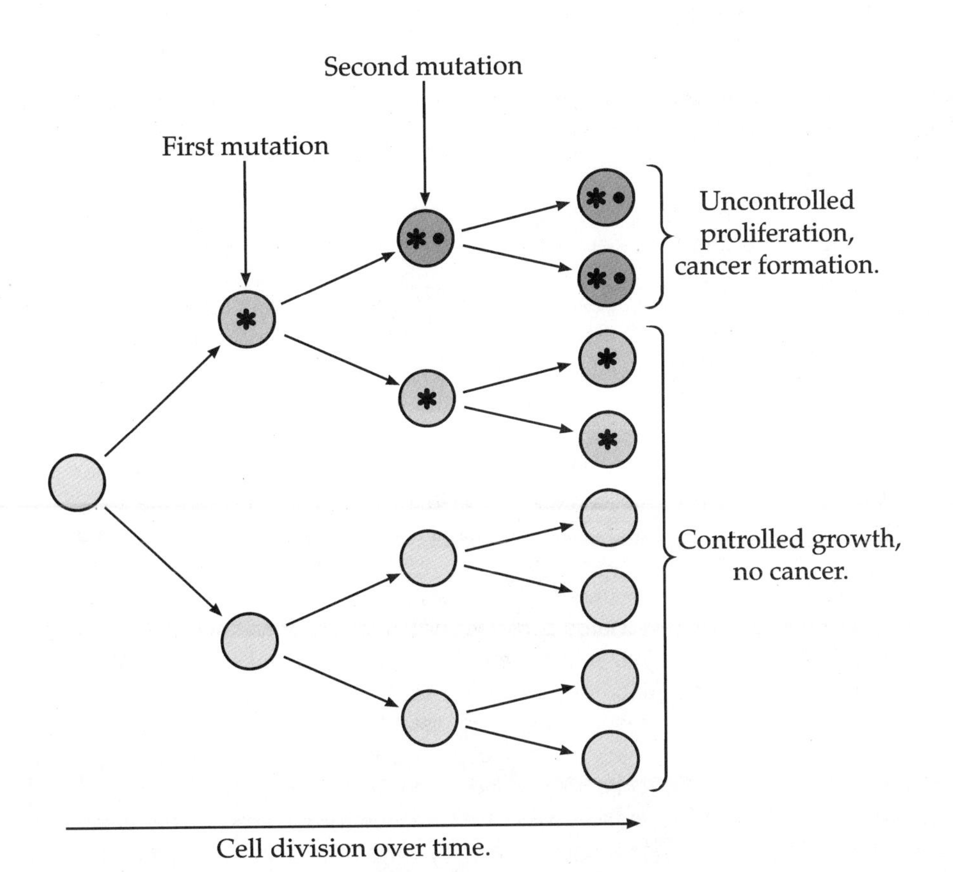

FIGURE 13.1 During repeated cycles of mitoses, cells can acquire mutations. Two independently arising mutations can lead to uncontrolled cell proliferation and cancer.

TABLE 13.3 Genetic Disorders that Predispose to Tumor Formation

Disorder	Chromosome
Early-onset familial breast cancer	17q
Familial adenomatous polyposis	5q
Gorlin syndrome	9q
Hereditary non-polyposis colorectal cancer	2p
Li-Fraumeni syndrome	17p
Multiple endocrine neoplasia type 1	11q
Multiple endocrine neoplasia type 2	10
Neurofibromatous type 1	17q
Neurofibromatous type 2	22q
Retinoblastoma	13q
Von Hippel-Lindau disease	3p
Wilms tumor	11p

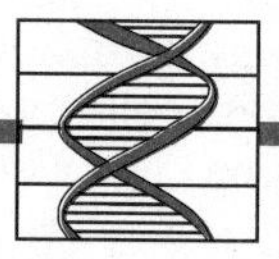

Concepts & Controversies

The Genetic Link to Breast Cancer

For women in the United States, the odds of getting breast cancer are 1 in 8. It is the most common form of cancer in women; 46,000 women die and 182,000 new cases are diagnosed each year. Although epidemiological factors may be involved in breast cancer, geneticists have struggled to answer the question: Is there a genetic predisposition to breast cancer? After over 20 years of work, the answer is yes. Although it is involved in only about 5% of all cases, a gene located on the long arm of chromosome 17 at q21.2 is responsible for susceptibility to a form of breast cancer that has an age of onset in the third and fourth decades of life. About 1 in 200 females inherit this gene, and of these, 80% to 90% will develop breast cancer. The gene, named BRCA1, has not yet been identified or isolated, but its location has been narrowed to a region containing about 2 million to 3 million base pairs (enough DNA to encode over 100 genes).

The search for this gene began with the analysis of epidemiological studies in the 1970s by Mary Claire King and her colleagues to identify families with a history of breast cancer. About 15% of the 1500 families surveyed had multiple cases of breast cancer; these results gave rise to a model that predicted that among the 1500 families, about 5% of the breast cancers were related to an autosomal dominant gene. However, the model also predicted that among the 15% with multiple cases, fully two thirds of these were random cases not related to a genetic cause. The problem in screening for a gene was to distinguish multiple-case families with a genetic cause from the other multiple-case families where the disease was simply random. Since this was impossible, King began testing as many families as possible.

The search began in the 1970s using protein markers to establish linkage between breast cancer and the marker by testing family members from multiple-case families. By the mid-1980s, as recombinant DNA methods became available, King and her colleagues switched to DNA markers and the PCR technique for screening. Finally in 1990, the 183rd marker to be used, an RFLP marker from chromosome 17 called D17S74, was tested on family members from 23 large pedigrees with a history of breast cancer. This marker showed clear linkage with breast cancer. Other laboratories quickly confirmed linkage using the same marker and also established linkage between the marker and ovarian cancer. The investigators formed a consortium, and using DNA from members of 214 families with either breast or ovarian cancer, narrowed the region to a segment on the long arm of chromosome 17. Current work is focused on identifying and isolating the gene and deducing the amino acid sequence of the gene product.

It is hoped that understanding the nature and action of this susceptibility gene will shed light on how breast cancer develops in the more common, noninherited form of the disease. The markers identified near the gene are already being used in genetic screening and counseling to identify the 1 in 200 carrying the mutant form of BRCA1 who are at high risk for breast cancer.

Tumor Suppressor Genes and Cancer

As we have seen, the events of interphase and mitosis make up the cell cycle. (Review the components of the cell cycle in Chapter 2.) Passage of cells through the cycle is regulated at two points: the transition between G2 and M (the G2/M transition) and at a point in G1 just before cells enter S known as the G1/S transition (see Figure 2.5). Tumor suppressor genes act at one or both these points to inhibit cell division. These genes and/or their gene products must be absent or inactive for cell division to take place. If tumor suppressor genes become deleted or inactivated by mutation, control over cell division is lost, and the cell can proliferate in an unchecked fashion.

The following examples will describe how mutations in tumor suppressor genes are involved in the development of cancer. A genetic model of cancer based on a series of mutations that results in colon cancer will be presented, integrating what is known about tumor suppressor genes and control of the cell cycle.

Retinoblastoma

Retinoblastoma (RB) is a cancer of the eye, affecting the light-sensitive retinal cells. It occurs with a frequency of 1 in 14,000 to 1 in 20,000 births, and although it may be present in infancy, it is most often diagnosed between the ages of 1 and 3 years. Two forms of retinoblastoma can be distinguished. The first is a hereditary susceptibility (accounting for 40% of all cases) that is inherited as an autosomal dominant trait. In families in which the susceptibility is inherited, offspring have a 50% chance of receiving the mutant RB gene, and 90% of these individuals will develop retinoblastoma, usually in both eyes. In addition, those carrying the mutation are at high risk of developing other cancers, especially osteosarcoma and fibrosarcoma. The second form of retinoblastoma, accounting for 60% of all cases, is sporadic, and affected individuals usually develop tumors only in one eye and are not at high risk for other cancers.

Retinoblastoma
A malignant tumor of the eye that arises in the retinal cells, usually occurring in children. Associated with a deletion on the long arm of chromosome 13.

Using this information about the distribution of the disease, Alfred Knudson and his colleagues proposed a two-step model for the development of retinoblastoma. According to this model, retinoblastoma develops when two mutant copies of the RB gene are present in a single retinal cell. If the first mutation is inherited, all cells of the body, including the retinal cells, would carry the mutant RB gene. A second mutational event in the normal RB gene in any retinal cell would result in retinoblastoma (Figure 13.2). Since all retinal cells carry a mutant gene, this form of the disease is more likely to involve both eyes and to occur at an earlier age than the sporadic form. On the other hand, in the sporadic form of the disease, two independent mutations must occur in a single retinal cell for a tumor to develop. Because the chance that mutations of both RB genes will take place in the same cell is low, sporadic cases are more likely to involve only one eye; and because of the time required to acquire two separate mutations, this form of retinoblastoma will arise later in life.

Surveys of patients with retinoblastoma have confirmed these predictions. Most cases that involve only one eye occur at a later age than bilateral cases, and pedigree analysis indicates that most cases involving two eyes are inherited, while the vast majority of unilateral cases are sporadic and noninherited.

The RB gene is located on chromosome 13 at 13q14 (Figure 13.3), and it encodes a protein (pRB) 928 amino acids in length that is confined to the nucleus. The protein is found not only in cells of the retina but also in almost all other cell types and tissues of the body, including actively dividing and

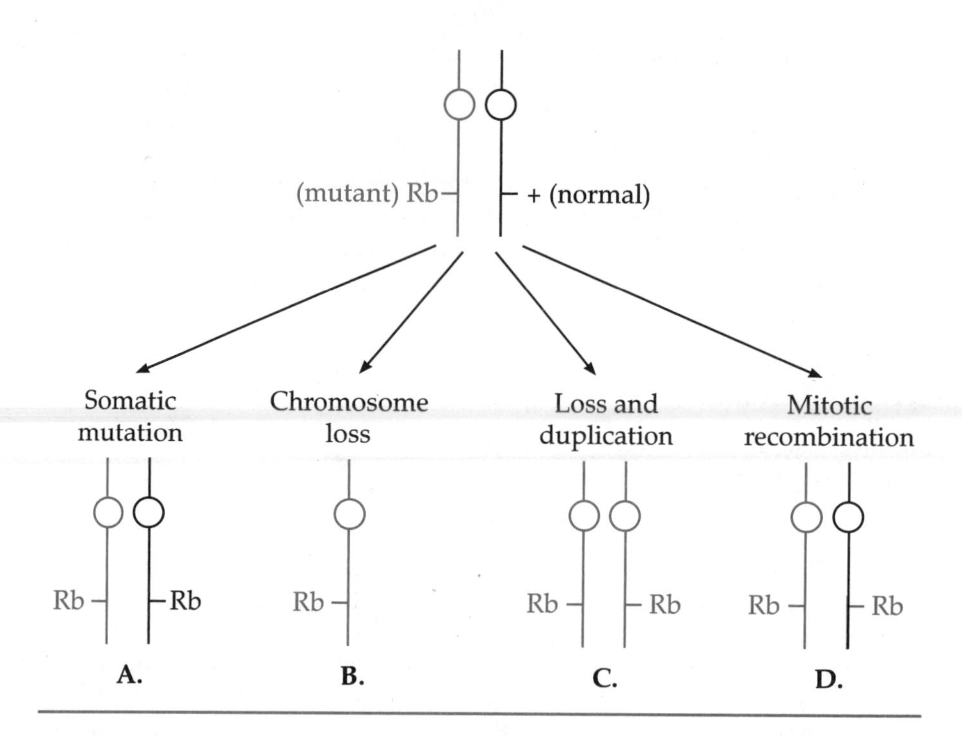

FIGURE 13.2 Some of the ways in which a mutational event can bring about the second mutation necessary for the development of retinoblastoma. (A) A somatic point mutation may convert the normal (+) allele to a mutant (rb) allele. (B) The chromosome carrying the normal allele might be lost. (C) Alternately, the chromosome carrying the normal allele might be lost, and the chromosome carrying the mutant allele might be duplicated. (D) A recombinant event in a mitotic cell followed by chromosome segregation can produce a homozygous condition.

resting cells. The retinoblastoma protein is present at all stages of the cell cycle, and its activity is regulated in synchrony with the cell cycle. The protein acts as a molecular switch, controlling progression through the cell cycle. During G1, if the protein is active, it acts to suppress passage from G1 into S, halting cell division. Conversely, if pRB becomes inactive, it allows cells to pass through the S phase, through G2, and on to mitosis. If both copies of the RB gene are deleted or become mutated in a retinal cell, the absent or defective pRB cannot regulate cell division, and the cell begins to divide in an uncontrolled fashion, forming a tumor.

What happens if both copies of the RB gene become mutated in a cell type other than retinal cells? In the case of one type of bone cell, the result is a cancer known as osteosarcoma. (Recall that those carrying one inherited mutant RB gene are predisposed to osteosarcoma.) In fact, cultured osteosarcoma cells have been used to provide evidence for the role of the RB gene in cancer. Analysis using recombinant DNA techniques indicates that osteosarcoma cells carry two mutant copies of the RB gene and produce no retinoblastoma protein. If a cloned copy of a normal RB gene is transferred into the osteosarcoma cells, pRB is produced, and cell division stops.

In other experiments, when osteosarcoma cells are injected into a cancer-prone strain of mice, tumors are produced. However, no tumors are produced if the osteosarcoma cells are first modified by transfer of a copy of the normal RB gene and are producing pRB. The results of these experiments reinforce the idea that the retinoblastoma protein plays a central role in the regulation of the cell cycle.

Wilms Tumor

Wilms tumor (WT) is a kidney cancer with primary expression in children between 1 and 4 years of age; it occurs with a frequency of 1 in 10,000 births and is found in two forms, a noninherited sporadic form and as an autosomal dominant genetic disease. The WT gene maps to band p13 of chromosome 11 (Figure 13.3). In inherited cases, therefore, a gene in a specific region of chromosome 11 is mutated or inactivated, requiring only one additional mutation to initiate tumor formation. The second mutational event could occur by any of several mechanisms, including chromosome loss, translocation, or deletion, or by mutations within the gene itself. Using a combination of recombinant DNA techniques and cytogenetics, several laboratories have studied the nature of the second mutational event. The results show that cytologically visible chromosomal changes account for just over half of the second mutations. The remaining fraction, around 45% of the cases, have either microdeletions or point mutations. The somewhat unexpected results indicate that mutational events in somatic cells that result in WT can be caused by large-scale chromosomal changes about as often as single-base changes. Further work is necessary to determine whether a similar ratio of chromosomal aberrations and point mutations is characteristic of other inherited predispositions to cancer.

Wilms tumor
A malignant tumor of the kidney.

Molecular mapping techniques have mapped the WT gene to the 11p13 region of chromosome 11. This gene is expressed only in fetal kidney during nephron formation, the tissue in which the Wilms tumor arises. The polypeptide encoded by this gene has structural features identical to those

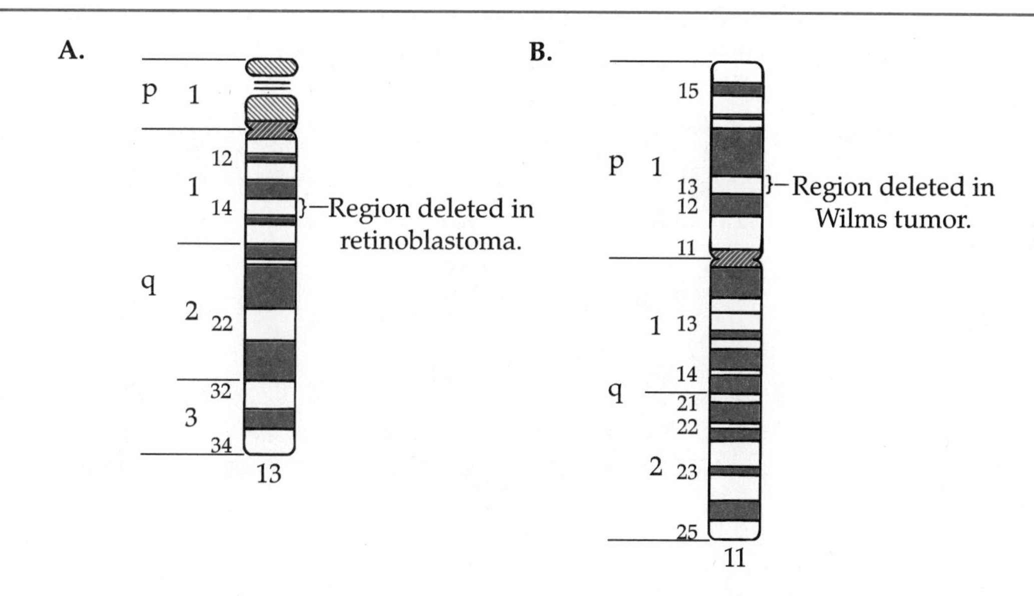

FIGURE 13.3 Chromosomal regions and cancer. (A) A diagram of chromosome 13 showing the region associated with deletions in retinoblastoma. (B) A diagram of chromosome 11 showing the region associated with deletions in Wilms tumor.

found in proteins that regulate transcriptional activity in the nucleus. The structure of the polypeptide, its pattern of expression, and the timing of expression make it attractive to speculate that the Wilms tumor locus encodes a gene that normally regulates transcription of one or more specific genes in the fetal kidney over a defined developmental time period. It seems likely that the normal WT gene product functions to switch off cell proliferation or switch on cell differentiation that results in the formation of kidney structures. How mutation or inactivation of the Wilms tumor gene results in the formation of a malignant growth is not yet known, but it seems important to identify and characterize the gene products that control the tissue-specific expression of this gene as well as the gene or genes regulated by this locus.

Colon Cancer

Familial adenomatous polyposis (FAP)
A dominant condition associated with the development of growths known as polyps in the colon. These polyps often develop into malignant growths, causing cancer of the colon and/or rectum.

Cancer of the colon and rectum is one of the most common forms of cancer in the United States (Table 13.4). There are two major forms of genetic predisposition to colon cancer: **familial adenomatous polyposis (FAP)** and hereditary nonpolyposis colon cancer. The first accounts for only about 1% of all cases of colon cancer, but has been useful in deriving a genetic model for colon cancer described below. The second predisposition, hereditary nonpolyposis colon cancer, accounts for about 15% of all cases, and is associated with a form of genetic instability, described in the next section. To clarify the role of inheritance in colorectal cancer, Randall Burt and his colleagues at the University of Utah studied a large pedigree with more than 5000 members covering six generations. This family contains clusters of siblings and relatives with colon cancer. Like many other families, this one shows no definite pattern of inheritance for the cancer. However, as part of this study, about 200 family members were examined for the presence of

TABLE 13.4 Colon and Rectal Cancer in the United States

Estimated new cases, 1993	
Colon	109,000
Rectum	43,000
Total	152,000
Mortality (estimated deaths, 1993)	
Colon	50,000
Rectum	7,000
Total	57,000
5-Year survival rate (early detection)	
Colon	91%
Rectum	85%

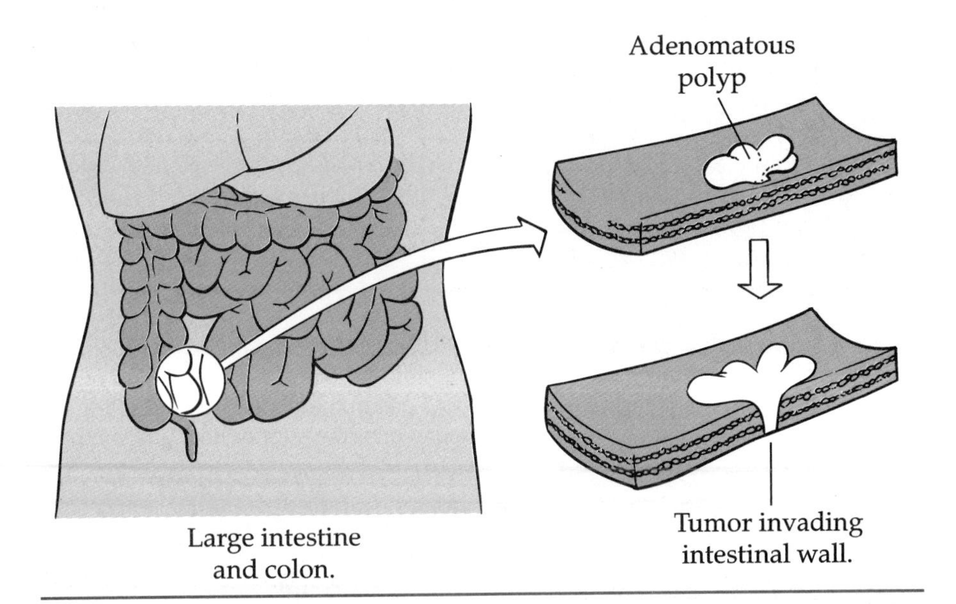

FIGURE 13.4 Adenomatous polyps, which usually precede or accompany colon cancer.

intestinal growths known as adenomatous polyps (Figure 13.4). In FAP individuals, these benign tumors, also known as **adenomas** or **polyps**, usually precede or accompany colon cancer and are regarded as precursors of the malignant condition. When the presence of intestinal growths and colon cancer are considered together as a single phenotype, an autosomal dominant pattern of inheritance is clear. The results also show that the allele for the occurrence of intestinal growths and cancer may have a relatively high frequency in the general population (3/1000).

Adenoma
A tumor or growth found in glandular tissues.

Polyp
A growth attached to the substrate by a small stalk. Commonly found in the nose, rectum, and uterus.

RFLP analysis was used to localize the gene for FAP to chromosome 5. It is thought that the FAP gene may regulate cell growth, so that in the normal homozygote no growth of adenomas occurs. In the heterozygote, cell growth in the colon is not completely regulated, causing production of polyps, resulting in FAP syndrome. This result suggests that the normal FAP allele is a tumor suppressor gene, and that mutations in several other genes cause the transition from adenoma to tumor. Genetic analysis has determined that most early adenomas show only a single genetic change, while most colonic carcinomas carry four or five mutations, with middle and late stage adenomas showing an intermediate number of mutations (Figure 13.5). In general, deletions of chromosome 17p and 18q usually occur at later stages of tumor development, while deletions of 5p and mutations in 12p take place in early stage adenomas. For example, the loss of a segment of 17p is seen in more than 75% of colorectal carcinomas, but such loss is seen only infrequently in early adenomas.

These findings have been incorporated into a genetic model of colon cancer (Figure 13.5). According to this model, the development of colorectal cancer is the result of a series of mutations that accumulate over time in a colon cell. Each mutation confers a slight growth advantage on the cell, first allowing it to proliferate and form an adenoma, then enlarging the adenoma through a series of stages, eventually allowing a cell to escape from cell cycle

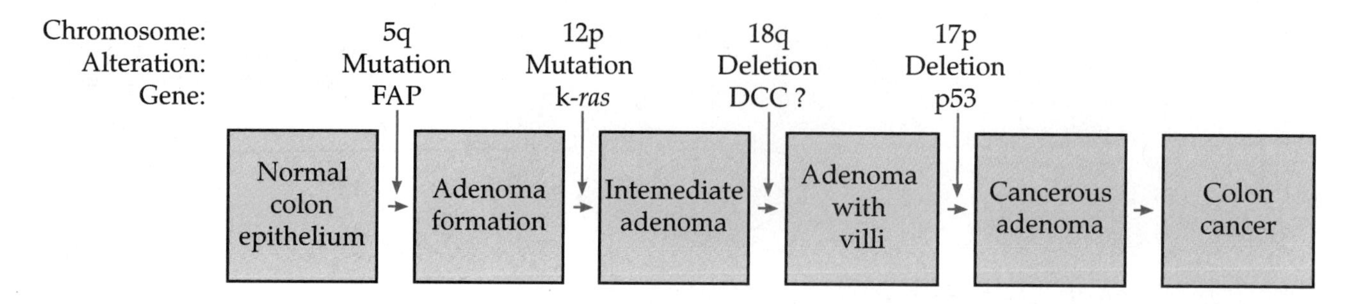

FIGURE 13.5 Genetic model for colon cancer. In this multiple-step model, the first step is the loss or inactivation of both FAP alleles on chromosome 5 (in familial cases, one mutant FAP allele is inherited), leading to the formation of benign adenomas. Subsequent mutations in genes on chromosomes 12, 18, and 17 cause the transformation of the benign adenoma into a cancerous adenoma, resulting in colon cancer. In these mutations, the sum of changes is more important than the order.

control to form a malignant cell that proliferates into a tumor. In later stages, additional mutations allow the tumor cell to become metastatic and break away to form a tumor at a remote site. It is important to note that although these mutational events often occur in a specific sequence, it is the sum of the accumulated changes that is critical to the development of malignant tumors from adenomas.

Molecular analysis using recombinant DNA techniques has identified an increasing number of cases in which tumor development involves a minimal number of mutational events at specific chromosomal sites, often on different chromosomes (Table 13.5).

Genetic Instability and Cancer

Most cancers are associated with multiple mutations that accumulate over time. If one of these mutations is inherited, fewer mutations are required to cause cancer, resulting in a genetic predisposition to cancer. One of the

TABLE 13.5 Number of Mutations Associated with Specific Forms of Cancer

Cancer	Chromosomal Sites of Mutations	Minimal Number of Mutations Required
Retinoblastoma	13q14	2
Wilms tumor	11p13	2
Colon cancer	5q, 12p, 17p, 18q	4 to 5
Small-cell lung cancer	3p, 11p, 13q, 17p	10 to 15

unresolved questions about the development of colon cancer and other cancers is how and why the multiple mutations required for tumor formation accumulate. Recently, another gene associated with colon cancer has been mapped to chromosome 2. This gene, called familial colon cancer (FCC), is unique because a mutational event in this gene apparently sets off a cascade of other mutations throughout the genome in groups of short, repetitive DNA sequences called **microsatellites**. Analysis of colon cancers carrying a mutant form of FCC show mutations at thousands of microsatellites throughout the genome. It is thought that this gene may promote colon cancer by making parts of the genome unstable and prone to mutation, raising the chances of mutation in the genes in the pathway to colon cancer. Estimates indicate that as many as 1 in 200 individuals may carry a mutant FCC gene, making it one of the most common causes of a genetic disease.

Microsatellites
Short nucleotide sequences (like CA or CG) repeated dozens or hundreds of times at a single site, and present at thousands of sites in the genome.

The FCC gene has not yet been identified or isolated, and its mechanism of action remains unknown. However, markers near the gene are available, making it possible to identify those who have inherited the mutant allele and are at high risk for colon cancer. Eventually, using this or other markers, it may be possible to do widespread population screening to identify those at risk, and eliminate some or all of the 55,000 annual deaths from colon cancer.

Genetic instability has also been associated with mutations in the p53 gene (one of the genes in the pathway to colon cancer). This form of genetic instability is called **amplification**, which is an increase in gene copy number. In cell lines carrying one or two copies of the normal p53 gene, amplification is undetectable, and there is no escape of cell cycle control. Loss or mutation of both p53 alleles results in widespread amplification and loss of cell cycle control. Taken together, these studies and those on the FCC gene suggest that normal cells contain genes that when mutated, cause widespread genomic instability, increasing the probability that mutations leading to tumor cell formation will occur. The concept of genetic control over genome instability is an important development in understanding the genetics of cancer associated with both tumor suppressor genes and oncogenes.

Amplification
Process by which genes are disproportionately replicated in the genome, greatly increasing their copy number relative to other genes.

Oncogenes and Cancer

Viruses

The role of viruses in cancer began with the work of Peyton Rous on a malignant tumor found in chickens known as a **sarcoma.** In 1911 Rous found that cell-free extracts from these tumors would induce tumor formation when injected into healthy chickens. The agent in the extract was later identified as a virus, known as the Rous sarcoma virus (RSV) (Figure 13.6). At the time, Rous's work was criticized by many of his colleagues, who claimed that his extract was not cell-free and that he was simply transferring cancer cells that grew in the injected chickens to form tumors. As a result, Rous gradually abandoned the project. Several decades later, RSV was identified as the cause of the tumors and became one of the most widely studied animal tumor viruses. In belated recognition of his pioneering work on viral tumors, Rous was awarded the Nobel Prize in 1966 (at the age of 85).

Sarcoma
A cancer of connective tissue. One type of sarcoma in chickens is associated with the retrovirus known as the Rous sarcoma virus.

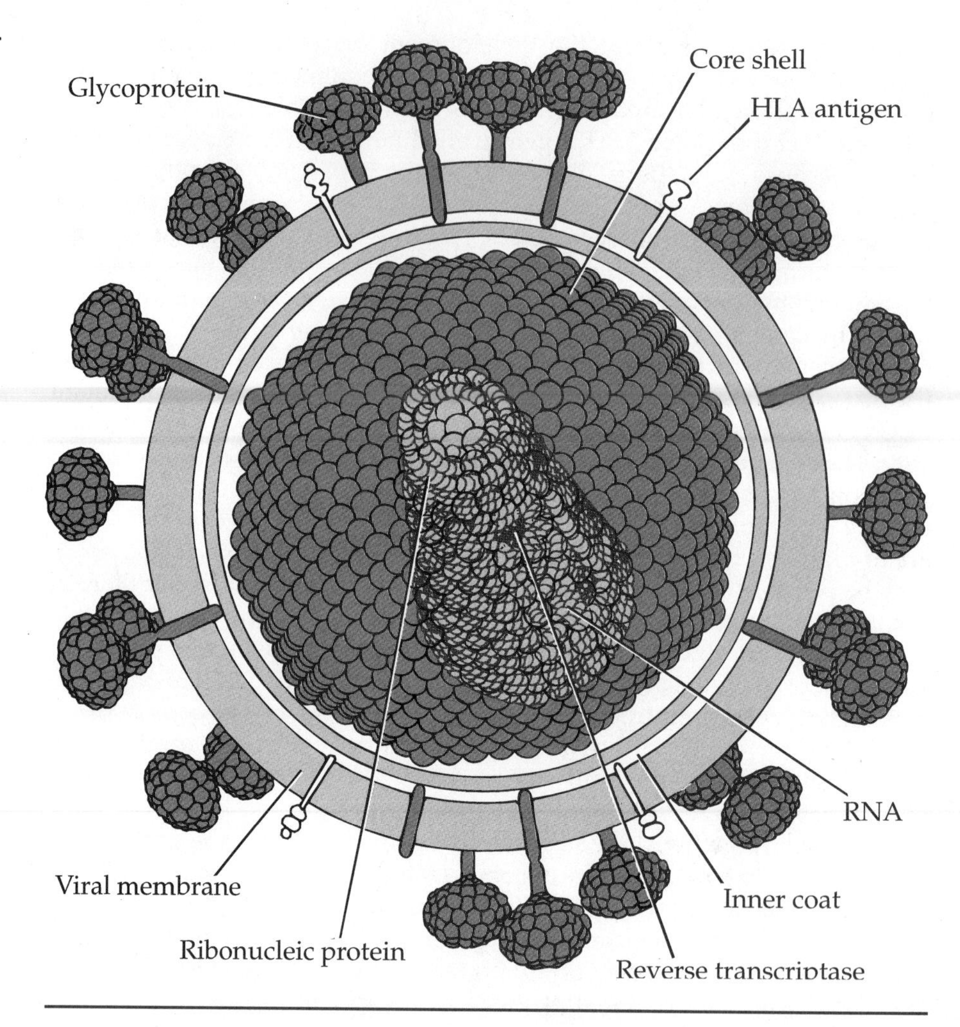

FIGURE 13.6 A drawing representing the structure of the Rous sarcoma virus, a representative retrovirus.

A virus is a submicroscopic particle that invades living cells and reproduces, using its own mRNA and the protein synthesizing machinery of the host cell. Viruses are composed of an external protein coat, or capsid that encloses a viral genome (chromosome) composed of nucleic acid. In addition, many viruses that invade animal cells are surrounded by a meshwork envelope of protein, lipid, and carbohydrate (Figure 13.6). The nucleic acid of a virus can be either DNA or RNA. The genetic information of most viruses consists of a small number of genes, ranging from 3 or 4 to around 50. These genes encode proteins that function in the replication of viral nucleic acid, convert the cell to the production of viral components, or are coat proteins. Several hundred different viruses have been identified and are responsible for an array of human diseases, including polio, measles, chickenpox, genital herpes, AIDS, and even the common cold.

Viruses can be grouped into two general categories, DNA viruses or RNA viruses, depending on the nature of the genetic material. Cancer-causing viruses have now been found in both groups (Table 13.6). The DNA

TABLE 13.6 Viruses and Cancer

DNA Cancer Viruses	RNA Cancer Viruses
SV40	Rous sarcoma virus
Polyoma	Mouse mammary tumor virus
Adenovirus	

tumor viruses include SV40 (simian virus 40), first identified in monkey tumors, and polyoma, a virus that produces tumors in several species of animals, including mice, hamsters, and rats. Some RNA viruses are known as **retroviruses** because the single-stranded RNA viral genome must return to the form of a double-stranded DNA molecule before replication proceeds. Tumor-causing retroviruses have been found in the chicken, mouse, rat, hamster, and other species. The Rous sarcoma virus is an example of a tumor-producing retrovirus. The results of work with RSV and similar viruses have been important in understanding the origins of human cancer.

Retrovirus
Viruses that use RNA as a genetic material. During their life cycle, the RNA is transcribed into DNA. The name retrovirus symbolizes this backward order of transcription.

Oncogenes

The ability of the Rous sarcoma virus to cause tumor formation in chickens is due to the presence of a single gene (Figure 13.7). Since this gene is associated with the ability of the virus to induce tumors, it has been called an **oncogene.** The discovery of the oncogene in RSV was a central and important event in cancer research because it indicated that cancer could be caused by changes in a small number of genes. Over the years other types of RNA tumor viruses have been discovered, and many forms of animal tumors, including mouse leukemia, cat leukemia, and mouse breast cancer, were shown to be caused by retroviruses. Like RSV, some of these RNA tumor viruses carry a single gene that causes cancer. These genes are of several different types, but all have the common property of being oncogenic. Around 50 different oncogenes have been identified to date. Some of these are listed in Table 13.7. Oncogenes are named for the virus in which they were discovered. In Rous sarcoma virus, the gene is known as v-*src;* the oncogene in avian erythroblastosis virus is known as v-*erb,* and so forth.

Oncogene
A gene that induces or continues uncontrolled cell proliferation.

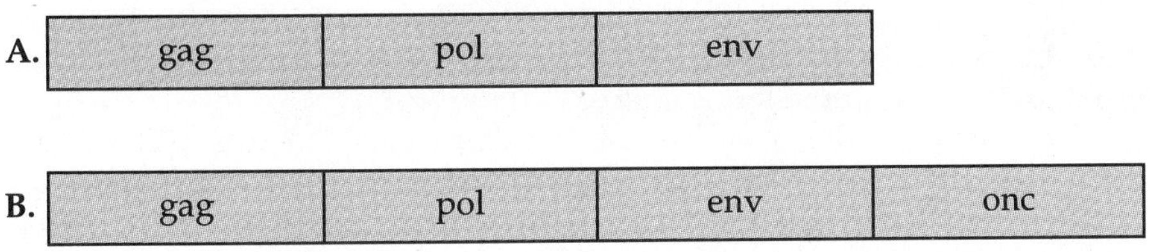

A.	gag	pol	env	
B.	gag	pol	env	onc

FIGURE 13.7 (A) Genetic organization of a retrovirus that is not capable of causing cancer growth. The three genes carried by this virus enable the virus to replicate itself, but not to produce tumors. (B) Chromosome and gene organization of a tumor-promoting retrovirus, carrying an *onc* gene. The onc gene is derived from a gene of the host cell, and confers upon the virus the ability to produce uncontrolled cell growth and tumor formation upon infection of another host cell.

TABLE 13.7 Retroviral Oncogenes and Human Proto-Oncogenes

Viral Oncogene	Associated Tumor	Human Proto-oncogene	Human Chromosome
v-*src*	Sarcoma	c-*src*	20
v-*fps*/v-*fes*	Sarcoma	c-*fps*/c-*fes*	15
v-*yes*	Sarcoma	c-*yes*	?
v-*ros*	Sarcoma	?	?
v-*ski*	?	?	?
v-*myc*	Carcinoma, sarcoma, myelo-cytoma	c-*myc*	8
v-*erb*-A	?	c-*erb*-A	17
v-*erb*-B	Erythroleukemia, sarcoma	c-*erb*-B	7
v-*myb*	Myeloblastic leukemia	c-*myb*	6
v-*rel*	Lymphatic leukemia	?	?
v-*mos*	Sarcoma	c-*mos*	8
v-*abl*	B-cell lymphoma	c-*abl*	9
v-*fos*	Sarcoma	c-*fos*	14
v-*raf*	?	c-*raf*-1	3
v-Ha-*ras*/v-*bas*	Sarcoma, erythroleukemia	c-Ha-*ras*-1	11
		c-Ha-*ras*-2	X
v-Ki-*ras*	Sarcoma, erythroleukemia	c-Ki-*ras*-1	6
		c-Ki-*ras*-2	12
v-*fms*	Sarcoma	c-*fms*	5
v-*sis*	Sarcoma	c-*sis*	22

Since not all retroviruses carry oncogenes, where did such genes originate? Are they viral in origin, or might they represent genes captured by the virus from host animal cells? Research has clearly demonstrated that oncogenes carried by retroviruses are derived from the host genome during viral infection. These normal sequences are called proto-oncogenes, or cellular oncogenes (c-*onc*). Proto-oncogenes are normal genes, present in all cells, that have the potential to cause cancer if they are mutated or if their usual pattern of expression is altered.

But what is the usual function of such genes? For the most part, proto-oncogenes are associated with cell growth, cell division, and cell differentiation and their gene products are present throughout the cell (Table

■ Computer graphics model of antibody molecule (Chapter 14).

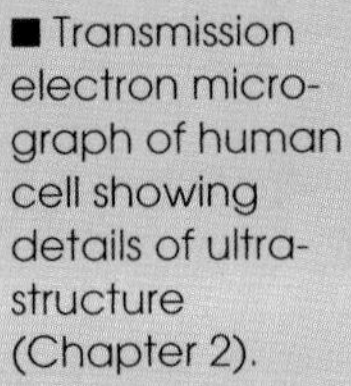

■ Transmission electron micrograph of human cell showing details of ultrastructure (Chapter 2).

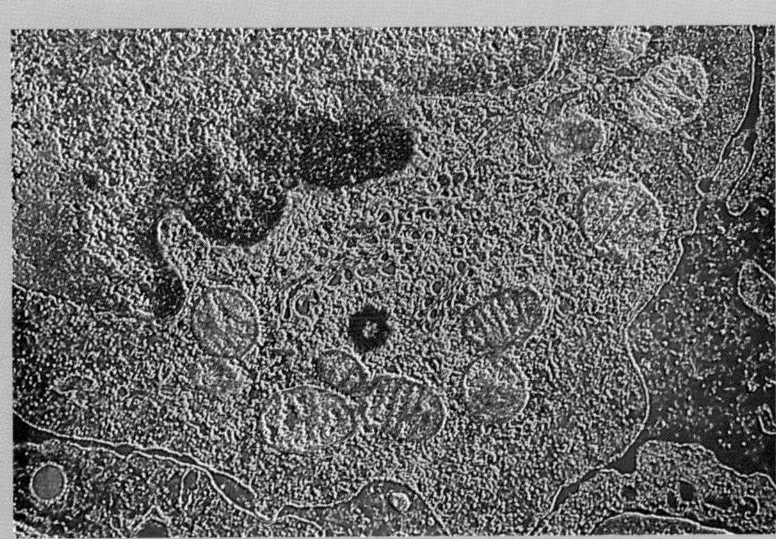

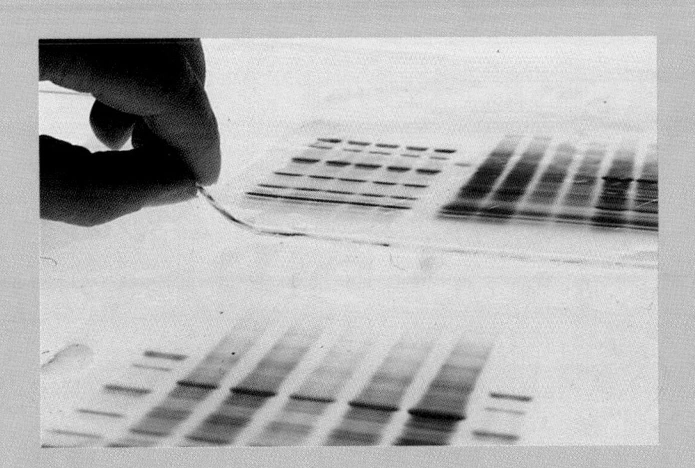

■ Gel showing separation of DNA fragments generated by restriction enzyme digestion (Chapter 8).

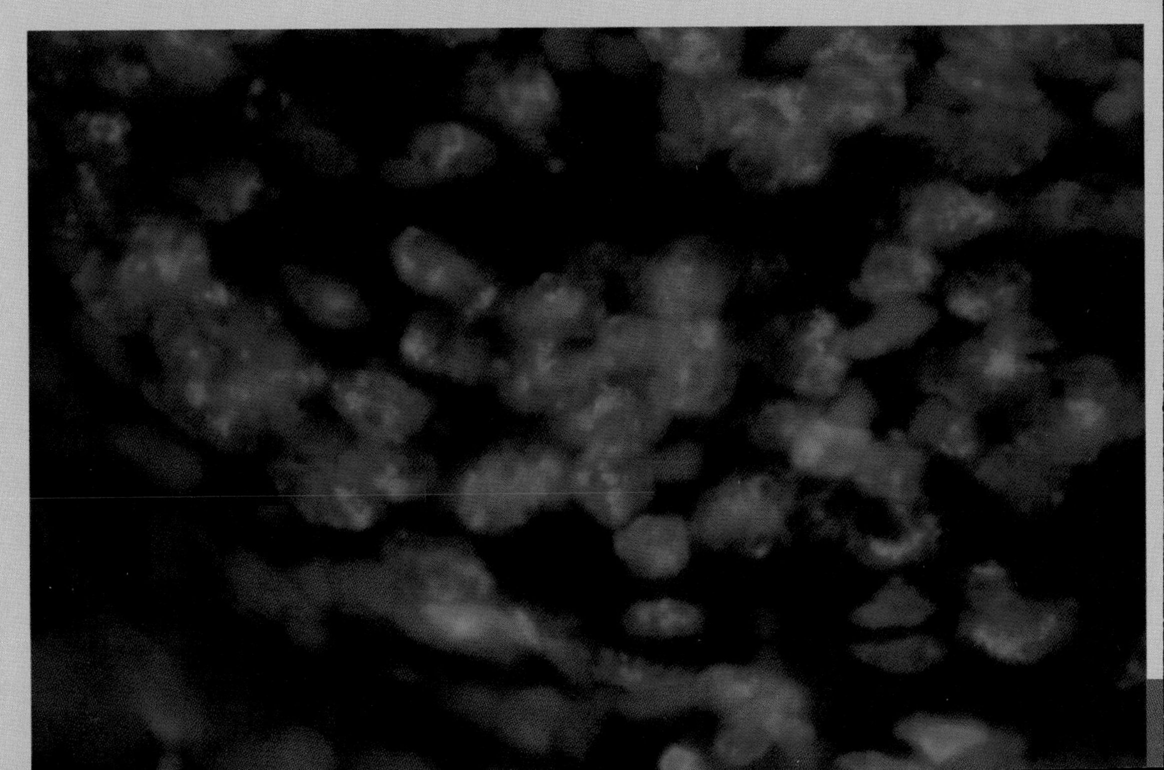

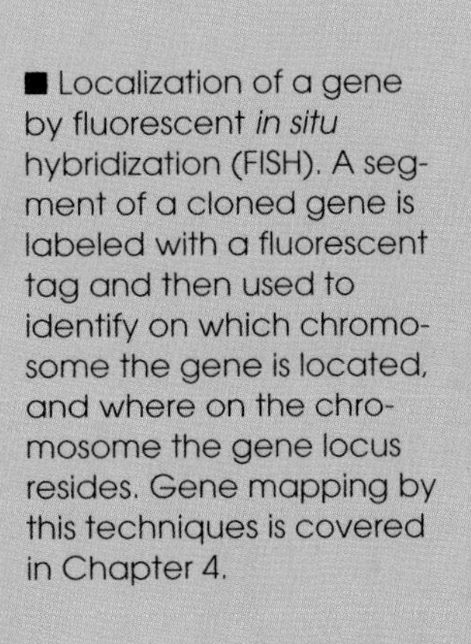

■ Localization of a gene by fluorescent *in situ* hybridization (FISH). A segment of a cloned gene is labeled with a fluorescent tag and then used to identify on which chromosome the gene is located, and where on the chromosome the gene locus resides. Gene mapping by this techniques is covered in Chapter 4.

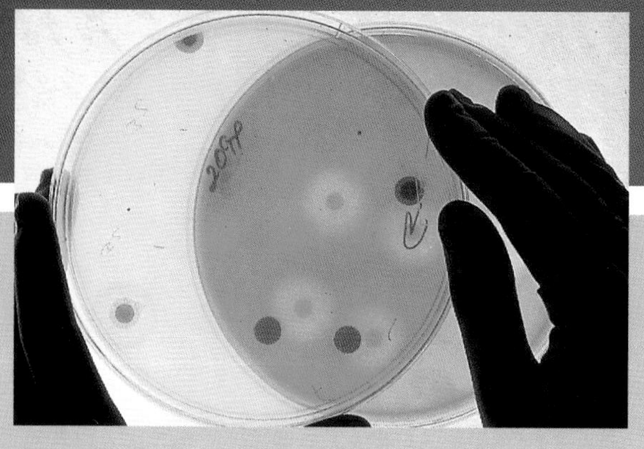

■ Colonies of bacteria on a petri plate. Each colony is descended from a single cell, and therefore, each colony is considered to be a clone. Clones are described in Chapter 8.

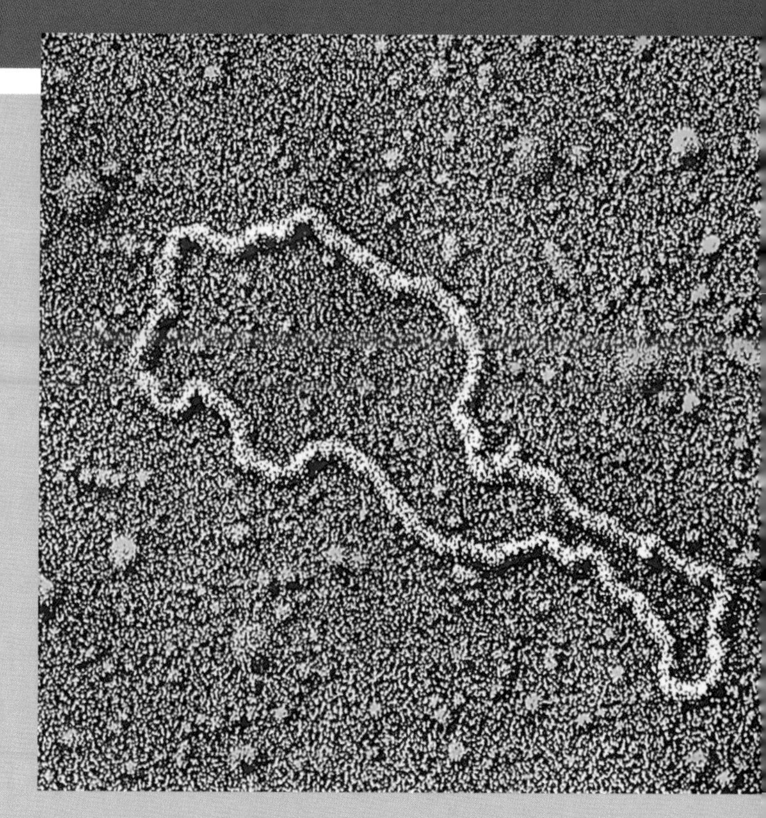

■ Plasmid isolated from a bacterial cell. Such plasmids are used as vectors in DNA cloning (Chapter 8).

■ Cloned plants derived from single cells removed from a common ancestor (Chapter 8).

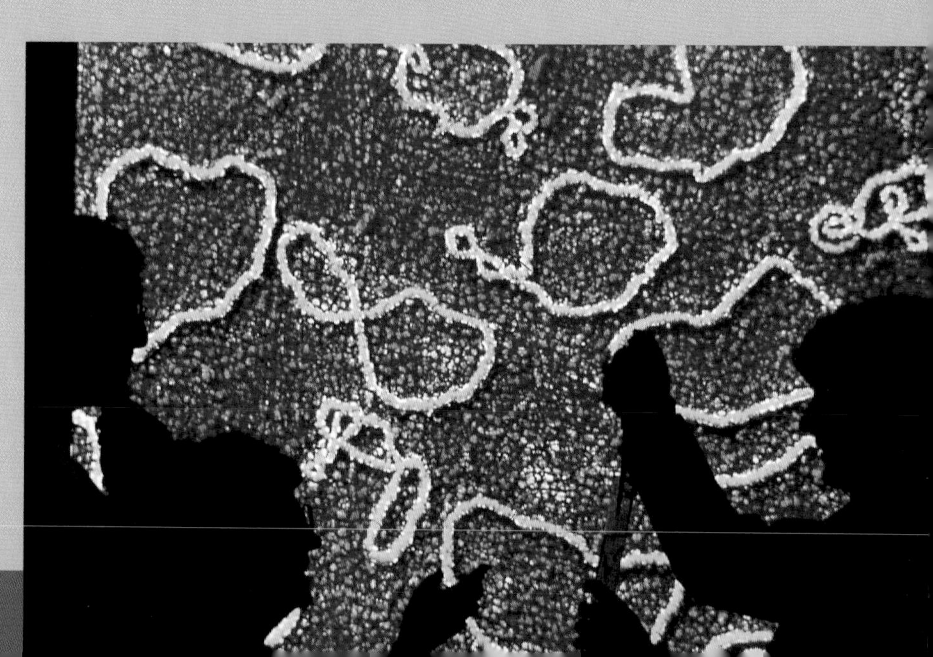

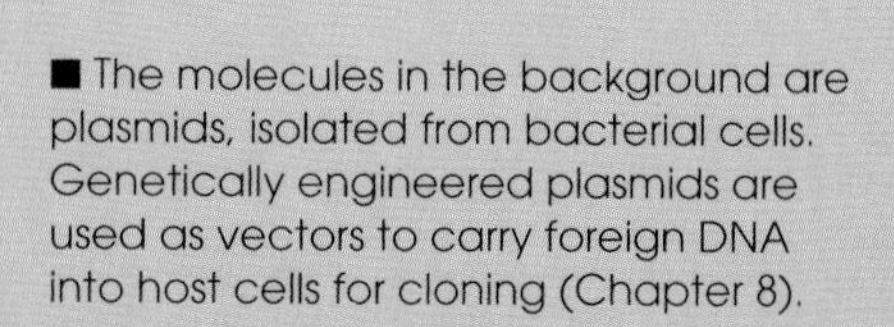

■ The molecules in the background are plasmids, isolated from bacterial cells. Genetically engineered plasmids are used as vectors to carry foreign DNA into host cells for cloning (Chapter 8).

■ Hands and face of nine-week fetus as seen by fetoscopy (Chapter 19).

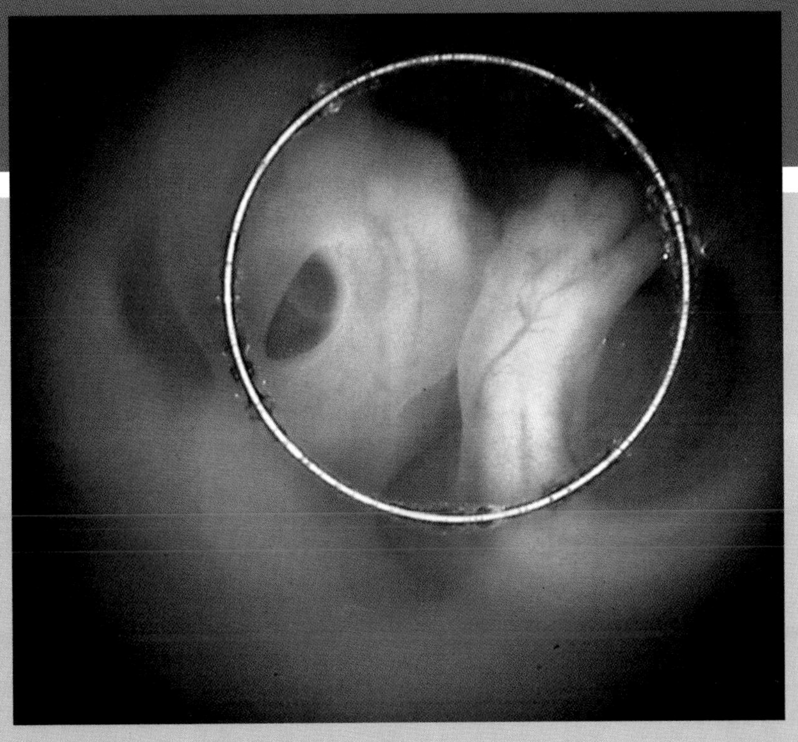

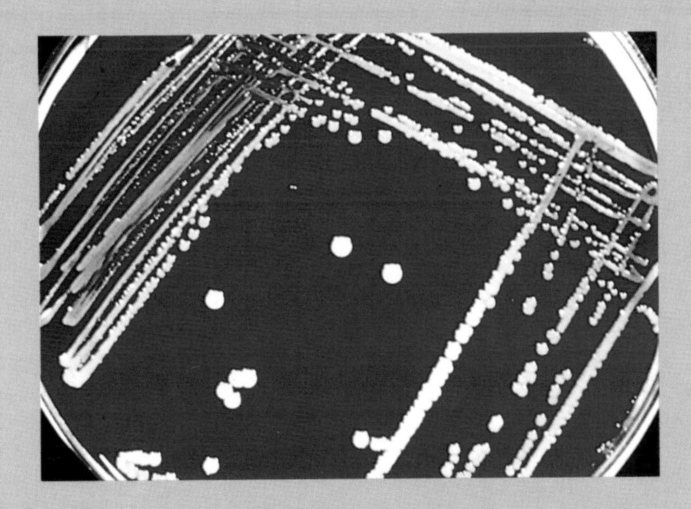

■ Bacterial Colonies on a petri plate. Each colony is a clone, derived from a single ancestor (Chapter 8).

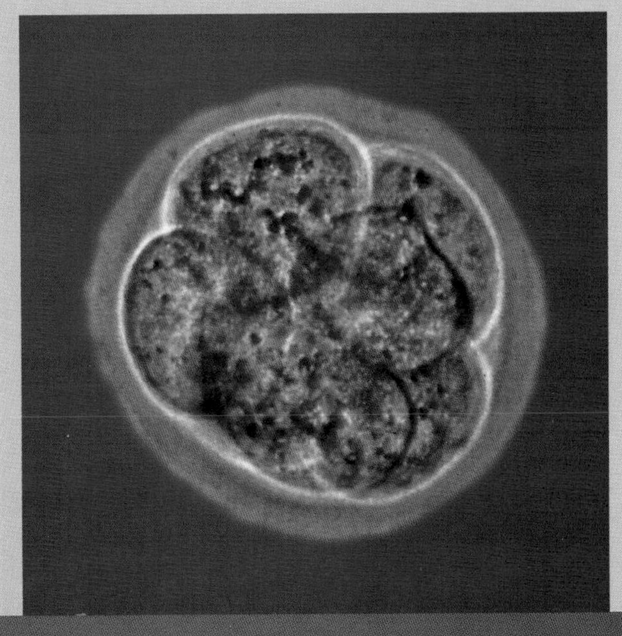

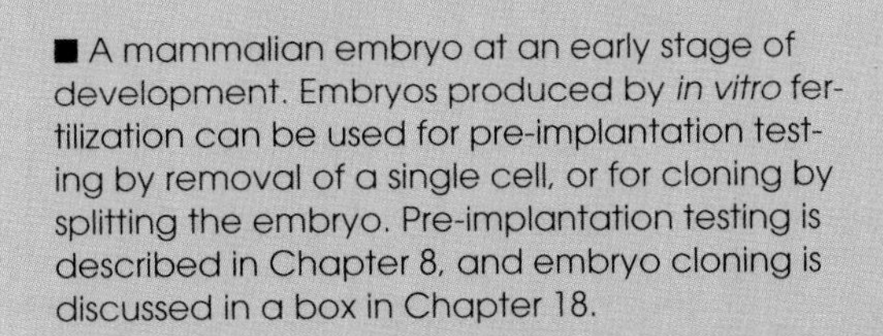

■ A mammalian embryo at an early stage of development. Embryos produced by *in vitro* fertilization can be used for pre-implantation testing by removal of a single cell, or for cloning by splitting the embryo. Pre-implantation testing is described in Chapter 8, and embryo cloning is discussed in a box in Chapter 18.

Genetics: A Photo Gallery

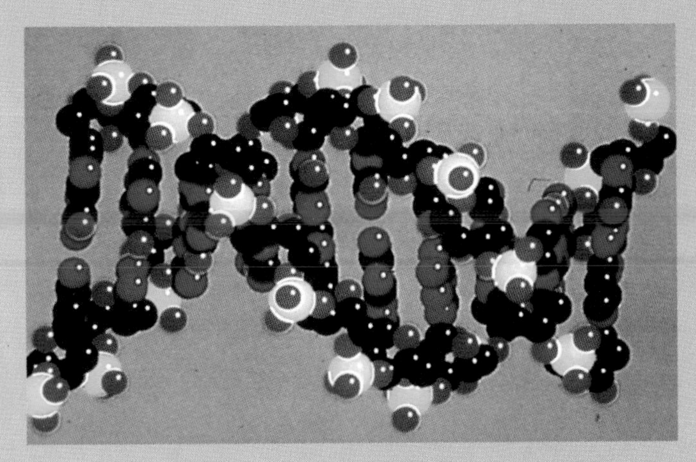

■ Computer-generated model of DNA (Chapter 7).

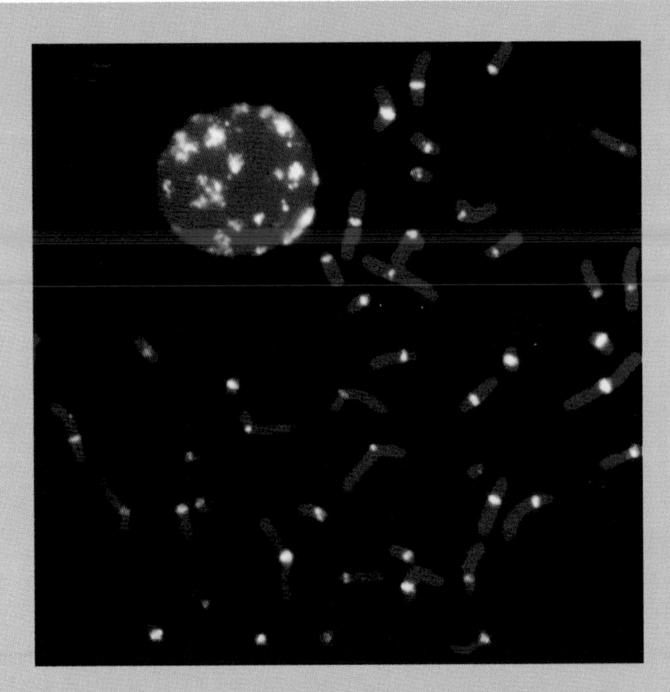

■ Interphase nucleus and metaphase chromosomes stained to show the centromeres in yellow and the chromosomes in red (Chapter 2).

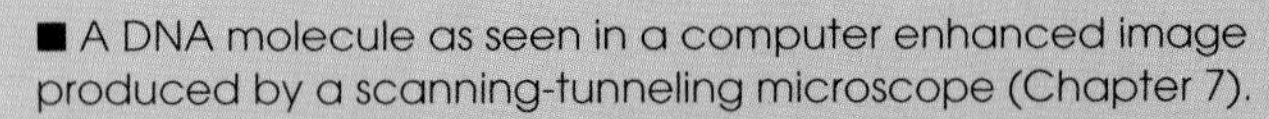

■ A DNA molecule as seen in a computer enhanced image produced by a scanning-tunneling microscope (Chapter 7).

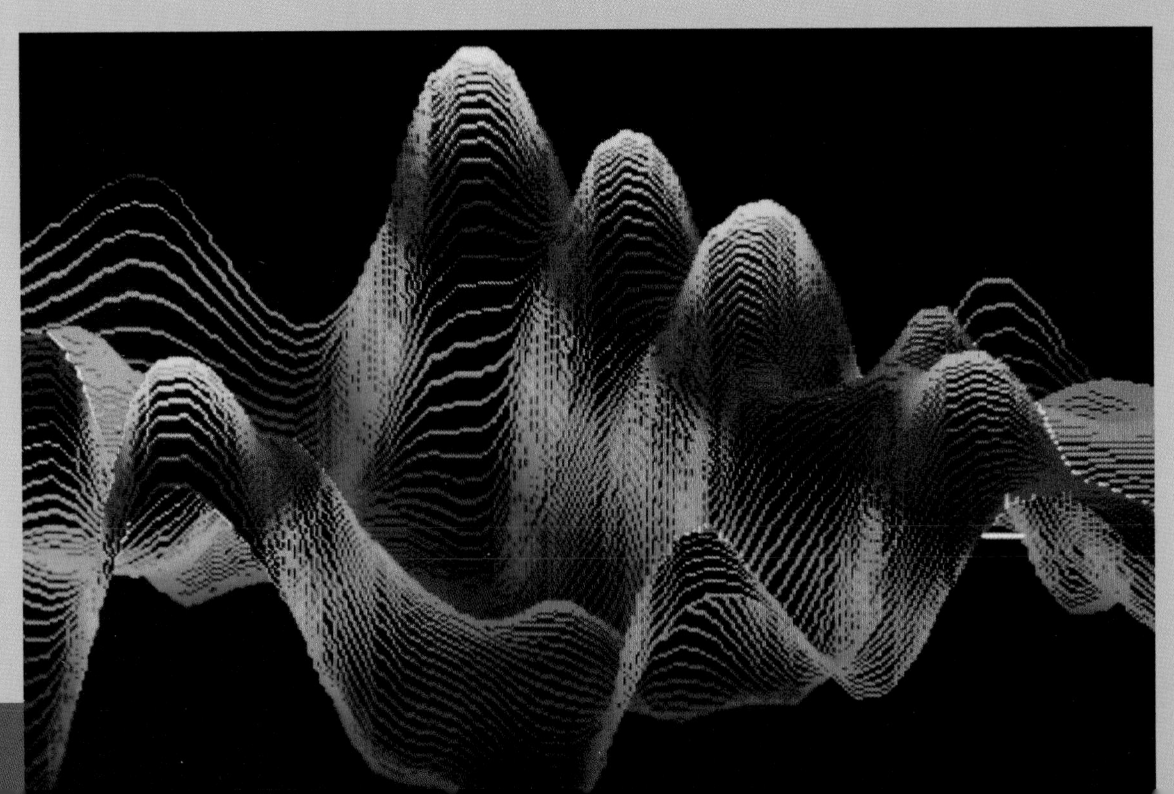

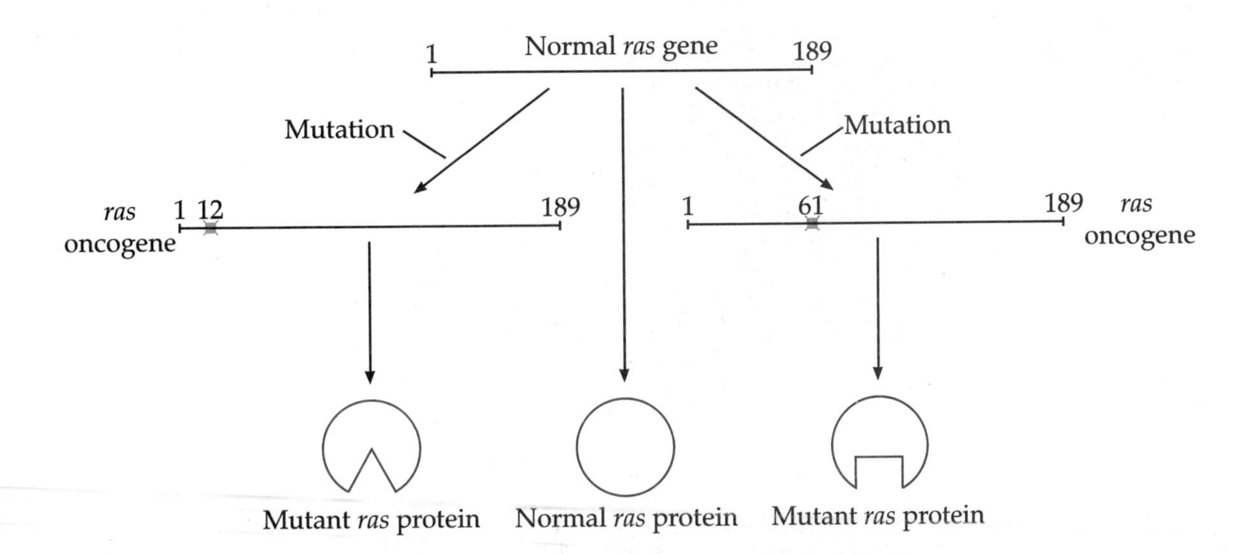

FIGURE 13.8 The *ras* proto-oncogene is a normal component of the genome and makes a gene product necessary in regulating cell growth. Mutations at positions 12 or 61 cause the formation of an oncogene that produces altered gene products, supporting the growth of cancer cells.

only in the tumor cells and *not* in the normal tissue of the patient. Somewhat surprisingly, the amino acid substitution in all 12 mutant genes occurred at either amino acid 12 or amino acid 61 (Figure 13.8). Work using x-ray analysis of protein crystals has shown that substitution of the glycine at position 12 with any other amino acid disrupts the structure of the protein and prevents its folding into the proper configuration. The altered protein is not able to function in signal transfer across the cell membrane, and the cell escapes from growth control, becoming cancerous.

In sum, proto-oncogenes normally function to promote cell division. Mutations that cause an increase in the number of copies of the gene (gene amplification) or the amount of gene product or that extend the time over which the gene acts can cause a cell to escape cell cycle regulation and begin uncontrolled growth and division. The result will be a clone of cells that may undergo additional genetic alterations, resulting in the formation of a malignant tumor.

Knowledge about the molecular organization of oncogenes and their products can be used to develop new methods for the diagnosis and treatment of cancer. If altered oncogene products are released into the blood, monoclonal antibodies against these proteins could be used to detect cancer at a very early stage. Similar tests have already been developed for a gene product released into the bloodstream by breast cancer cells. New strategies for treatment may also be derived from knowledge about oncogenes themselves. At the clinical level, if molecular analysis reveals the presence of multiple copies of oncogenes in cases of lung cancer, breast cancer, or cervical cancer, aggressive treatment and therapy are indicated. Laboratory studies on cultured cells indicate that if an active oncogene is switched off, the cell

13.8). For example, the DNA sequence of the *sis* oncogene closely matches that of a proto-oncogene that encodes a protein known as platelet-derived growth factor (PDGF). Similarly, the DNA sequence of the oncogene *erb*-B is related to that of a gene that encodes a cellular receptor for another growth factor. Given that the majority of proto-oncogenes may act to regulate cell growth and division, it is not difficult to imagine that oncogenes represent alterations in the structure or regulation of such genes and lead to uncontrolled proliferation and cancer.

The version of the oncogenes carried by retroviruses are called *v-onc* genes. Retroviruses that carry a *v-onc* gene are able to infect and transform a host cell into a malignant tumor cell. Although oncogenes were discovered in viruses, only a few rare human forms of cancer are caused by virally transmitted oncogenes. In most cases, the conversion from proto-oncogene to oncogene takes place inside a somatic cell, without intervention by a virus. What then, is the difference between a proto-oncogene in a normal cell and an oncogene in a cancer cell? Many differences are possible, including mutations that produce an altered gene product and those that cause underproduction or overproduction of the normal gene product. In fact, all these types of alterations have been found in human oncogenes or in their adjacent regulatory regions. For example, the normal *ras* gene encodes information for a protein that is 189 amino acids in length and is associated with the plasma membrane of the cell. It functions in the reception and transfer of growth-inhibiting signals across the cell membrane.

Analysis of 12 different *ras* oncogenes isolated from human tumors reveals that in each case a single-base change differentiates the mutant oncogene from the proto-oncogene found in normal cells. In all 12 tumors, a point mutation altered a single base in a codon, leading to a single amino acid substitution in the gene product. The presence of this altered gene product, in turn, initiated the transformation of a normal cell into a cancerous cell. Furthermore, the mutant gene and the mutant oncoprotein are found

TABLE 13.8 Cellular Localization of Proto-oncogene (*c-onc* and *v-onc* Gene Products

Gene	Location of *c-onc* Proteins	Location of *v-onc* Proteins
abl	nucleus	cytoplasm
erb B	plasma membrane	plasma membrane and Golgi
fps	cytoplasm	cytoplasm and membranes
myc	nucleus	nucleus
ras	membranes	membranes
src	membranes	membranes

■ Family with pigmented and albino members (Chapter 4).

■ Australian aborigine displaying tawny hair, a genetically transmitted trait (Chapter 18).

■ Calico cat, with fur patches resulting from inactivation of an X chromosome (Chapter 5).

■ Cave drawings. This Cave drawing from Algeria was made about 6,000 B.C., and illustrates the relationship between humans and animals, about the time many species were being domesticated. The role of human activity in animal genetics is discussed in Chapter 1.

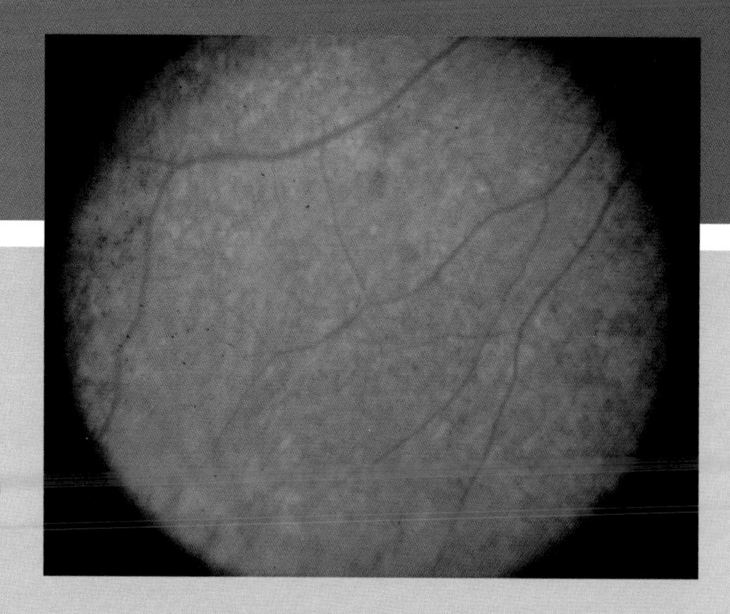

■ Retina of a female heterozygous for an X-linked form of albinism, showing patches of pigmented, nonpigmented cells (Chapter 5).

■ Individuals with color blindness traits perceive colors differently. Someone with normal color vision would see the scene on the left. Someone with red-green color blindness would see the scene on the right (Chapter 5).

■ Bee preserved in amber. Insects preserved in amber and other fossilized material have been used as sources to recover DNA from extinct organisms. Bees of this age, some 35-40 million years old, are among the oldest organisms from which DNA has been extracted. The use of recombinant DNA techniques recovering this DNA is discussed in Chapter 8.

■ The well-preserved body of a man was discovered in September, 1991 at the edge of a receding glacier in the Italian Alps. Several techniques established that the body is approximately 5,300 years old. Artifacts and DNA from the body will be valuable in providing information about the origin and dispersal of early human populations in Western Europe. The origin and dispersal of human populations is described in Chapter 18.

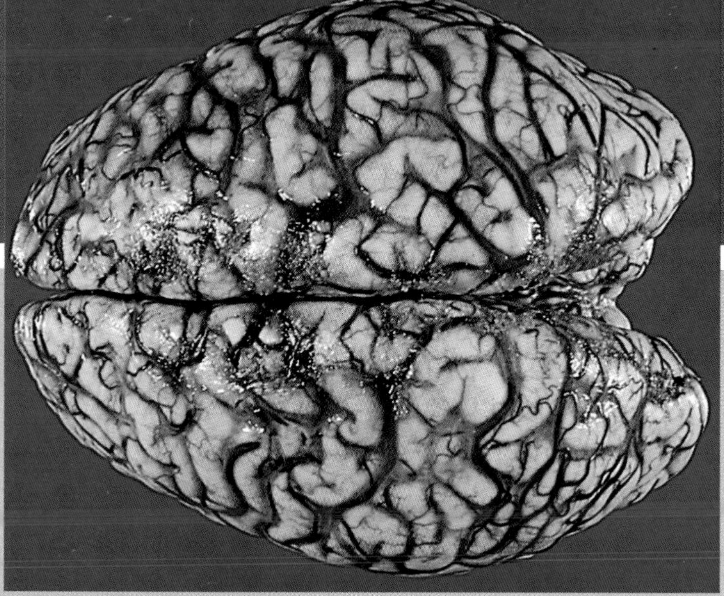

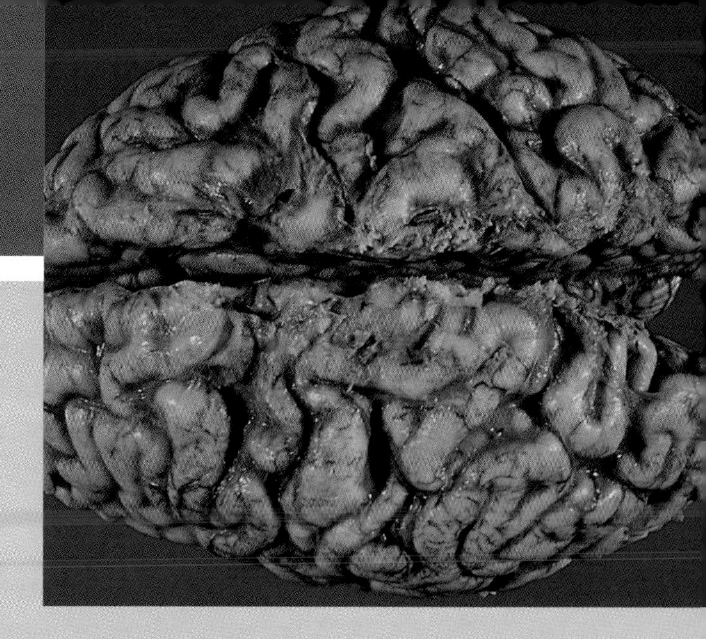

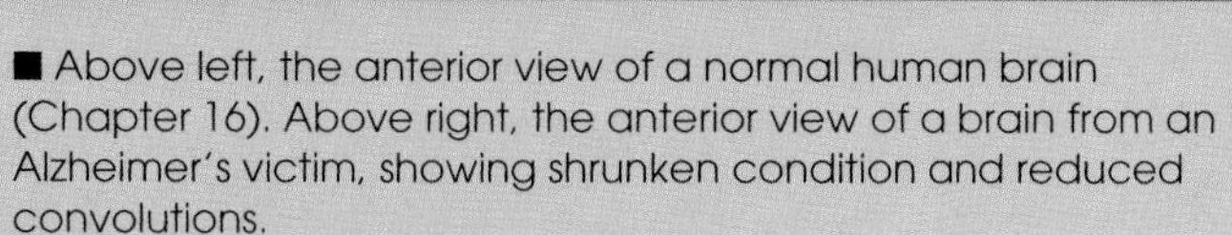

■ Above left, the anterior view of a normal human brain (Chapter 16). Above right, the anterior view of a brain from an Alzheimer's victim, showing shrunken condition and reduced convolutions.

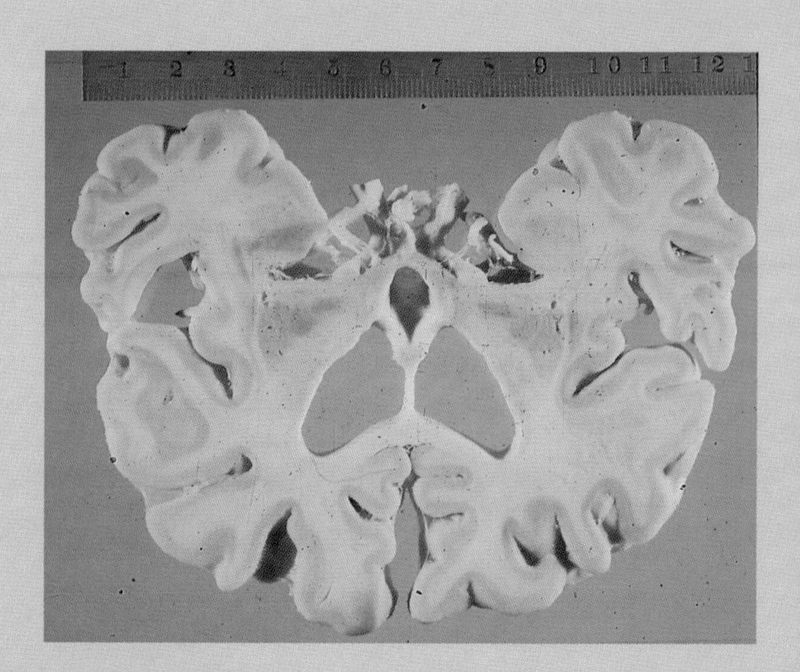

■ Brain lesions in Huntington disease. In the brains of individuals afflicted with Huntington disease, many cells show shape changes. Among regions affected are those involved in motor control. Huntington disease is discussed in Chapter 16.

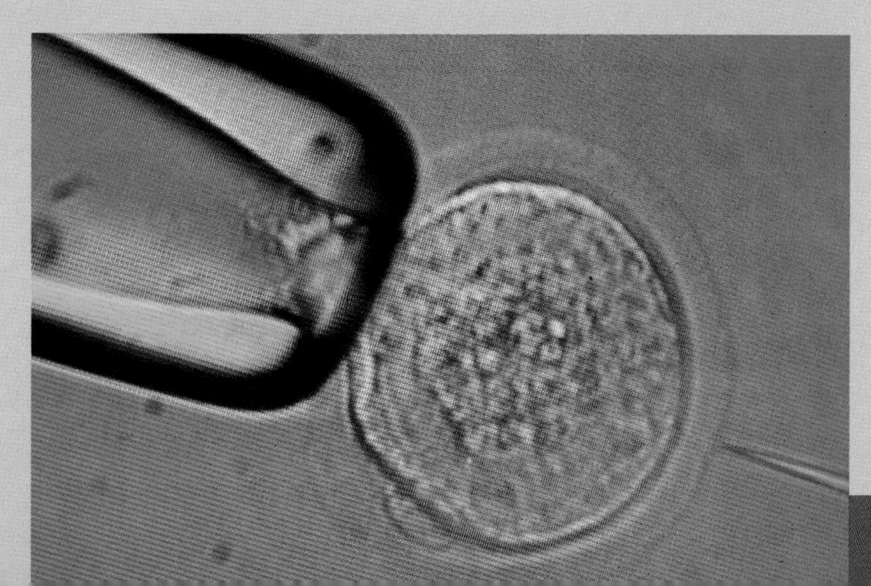

■ Injection of DNA. Cloned DNA can be transferred to human embryos by microinjection (Chapter 8).

Chromosome Instability Syndromes

Individuals with the autosomal recessive condition known as xeroderma pigmentosum have an enzymatic defect that prevents the repair of damaged DNA. Details of this system were presented in Chapter 12. Such individuals are extremely sensitive to ultraviolet light and are prone to develop skin cancer, especially in areas of the body that have been exposed to sunlight.

Several other genetic disorders are associated with chromosomal instability and cancer. Among these, Bloom syndrome and Fanconi anemia are both autosomal recessive conditions that result in a large number of chromosome breaks, gaps, rearrangements, and translocations. Children with Bloom syndrome are characterized by low birth weight, short stature (dwarfism), and a variety of malignancies, including leukemia and intestinal cancer. Fanconi anemia is associated with a marked reduction in circulating blood cells, abnormal pigmentation of the skin, and greatly increased risks of leukemia.

A fourth autosomal recessive disorder, ataxia telangiectasia (AT), also causes chromosomal breaks; but in contrast to the preceding diseases, the breaks in this syndrome consistently involve chromosome 14 and, to a lesser extent, chromosome 7. Patients with ataxia telangiectasia have increased rates of leukemia, skin cancer, and cancers of the lymphatic system such as Hodgkin disease (Table 13.9).

It is tempting to speculate that the common thread in all these chromosome instability syndromes is the inheritance of a germ line mutation that predisposes to cancer by reducing the ability to repair damage to DNA, and that subsequent mutagenic events, such as exposure to sunlight in xeroderma pigmentosum or exposure to x-rays in AT bring about the development of cancer. However, until more is known about the molecular basis of these diseases and the nature of the genes involved, the relationship to the multistep theory of cancer will remain attractive but uncertain.

TABLE 13.9 Human Genetic Disorders Associated with Chromosome Instability and Cancer Susceptibility

Disorder	Inheritance	Chromosome Damage	Cancer Susceptibility	Hypersensitivity
Ataxia telangiectasia	Autosomal recessive	Translocations on 7, 14	Lymphoid, others	X-rays
Bloom syndrome	Autosomal recessive	Breaks, translocations	Lymphoid, others	Sunlight
Fanconi anemia	Autosomal recessive	Breaks, translocations	Leukemia	X-rays
Xeroderma pigmentosum	Autosomal recessive	Breaks	Skin	Sunlight

becomes nonmalignant. Further work on the mechanism of gene switch-off may result in development of a new class of anticancer drugs directed against the regulatory regions of specific genes. Alternatively, since the protein products of oncogenes have been characterized, it may be possible to block or to reduce the action of such proteins, causing the cancer cell to become quiescent.

CHROMOSOMES AND CANCER

Down Syndrome

As outlined in Chapter 6, Down syndrome is caused by the presence of an extra copy of chromosome 21. This quantitative change in genetic information is not associated with any single metabolic defect that is diagnostic for trisomy 21 or with any known gene mutation. In addition to defects in cardiac structure and in the immune system, children with Down syndrome are 18 to 20 times more likely to develop **leukemia** than children in the general population. Down syndrome children contract the same types of leukemia and in the same proportions as children in the general population. However, the survival curve for the two groups is dramatically different. As shown in Figure 13.9, there is a 30% survival for non–Down syndrome individuals with leukemia after 10 years, while none of the Down syndrome children with leukemia survived beyond 5 years. Thus the presence of an extra copy of chromosome 21 not only increases the risk of a particular form of cancer but also exerts a decisive influence on the outcome of this disease.

Leukemia
A form of cancer associated with uncontrolled growth of leukocytes (white blood cells) or their precursors.

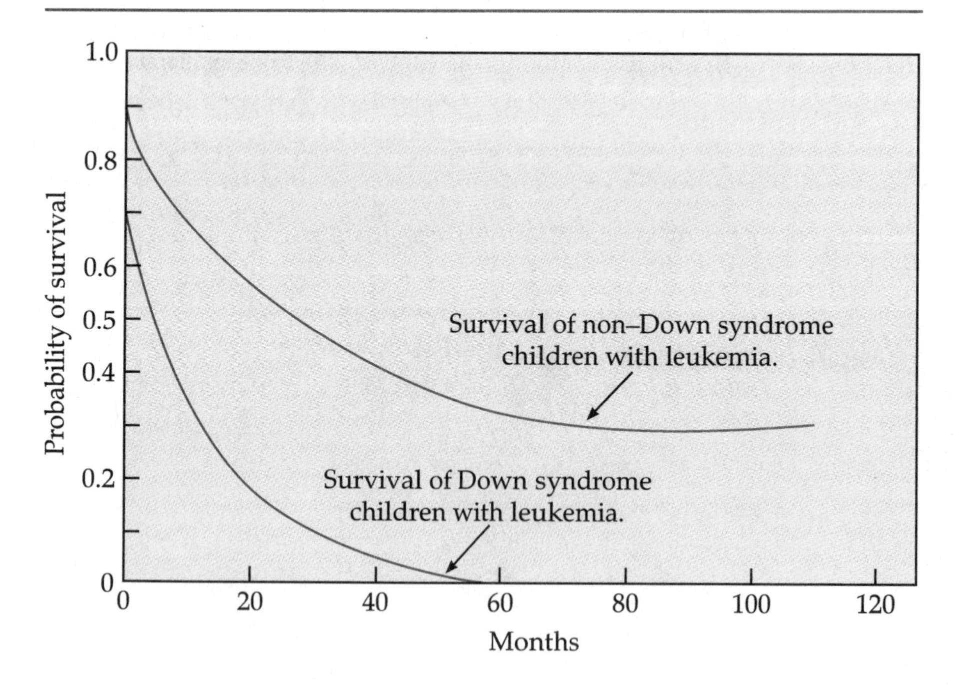

FIGURE 13.9 Survival of children with Down syndrome (lower line) and normal children with leukemia. Not only does Down syndrome predispose to leukemia, but it also affects the outcome of therapy.

Chromosome Aberrations and Leukemia

The relationship between chromosomes and cancer was first proposed by Boveri in 1917. Since then chromosome aberrations have been found in the cells of many tumor types, including those experimentally induced by radiation or chemical carcinogens. For many years these aberrations were regarded as a result of the malignant condition. One such aberration found in patients with chronic myeloid leukemia was originally thought to involve a deletion of part of chromosome 22 and was named the **Philadelphia chromosome,** after the city in which it was discovered. Somewhat later it was determined that the Philadelphia chromosome was actually a translocation (an exchange of chromosome parts) involving chromosomes 9 and 22 (Figure 13.10). More importantly, this defect was consistently and specifically associated with this form of leukemia. This discovery by Janet Rowley of the University of Chicago was the first example of a chromosome translocation accompanying a human disease.

Subsequent work demonstrated that other cancers were also associated with specific translocations, including acute myeloblastic leukemia, Burkitt lymphoma, and multiple myeloma (Table 13.10). Specific chromosomal abnormalities may be associated with solid tumors as well, but since blood cells and their precursors are easy to obtain and to culture, most progress has been made using these tissues. Most often the chromosomal defects are deletions of a specific band or a reciprocal exchange of chromosome parts

Philadelphia chromosome An abnormal chromosome produced by an exchange of portions of the long arms of chromosome 9 and 22.

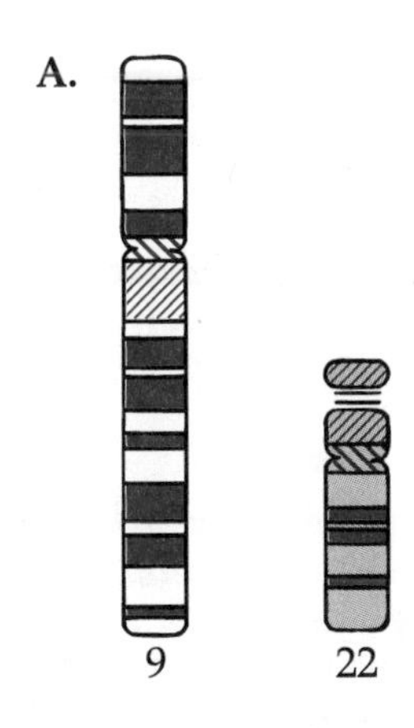

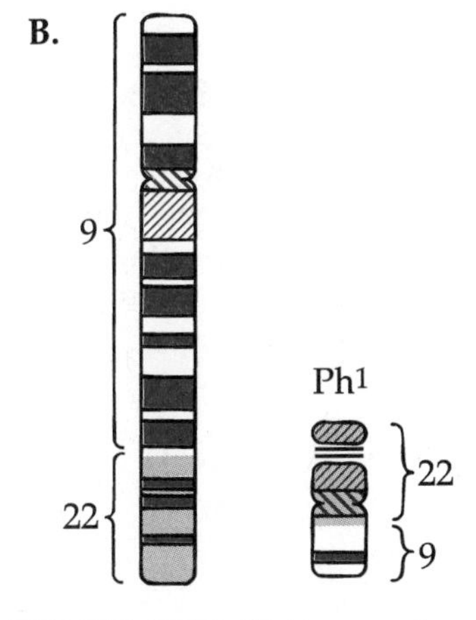

FIGURE 13.10 (A) Human chromosomes 9 and 22. (B) The reciprocal translocation between chromosome 9 and 22 produces the Philadelphia chromosome (Ph[1]) associated with chronic myelogenous leukemia.

TABLE 13.10 Chromosome Translocation Associated with Human Cancers

Chromosome Translocation	Cancer
t(9;22)	Chronic myelogenous leukemia (Philadelphia chromosome)
t(15;17)	Acute promyelocytic leukemia
t(11;19)	Acute monocytic leukemia, acute myelomonocytic leukemia
t(1;9)	Pre–B-cell leukemia
t(8;14), t(8;22), t(2;8)	Burkitt lymphoma, acute lymphocytic leukemia of the B-cell type
t(8;21)	Acute myelogenous leukemia, acute myeloblastic leukemia
t(11;14)(q13;q32)	Chronic lymphocytic leukemia, diffuse lymphoma, multiple myeloma
t(14;18)	Follicular lymphoma
t(4;11)	Acute lymphocytic leukemia
t(11;14)(p13;q13)	Acute lymphocytic leukemia

between different chromosomes. The consistent finding that certain forms of cancer are associated with specific chromosome abnormalities suggests that these aberrations are in some way causally related to the development of the malignancy. There is strong evidence that chromosome rearrangements are not by-products of the malignant condition but instead represent an important step in the development of this form of cancer. The genetic and molecular basis for this role is now becoming clear as the field of cancer cytogenetics merges with the molecular biology of oncogenes. As oncogenes became identified in human cells using the techniques of molecular biology, the chromosomal locations of the normal proto-oncogenes were systematically mapped by cytogeneticists. It soon became clear that many of these genes were located at or very near the break points of chromosomal translocations involved with specific cancers. Examples of these break points and oncogene loci are shown in Figure 13.11. It is logical to suspect that such breaks might alter the expression of proto-oncogenes and commit a cell to oncogenesis.

Chronic myelogenous leukemia (CML) is a genetic disorder of lymphoid cells associated with the fusion of two genes present at the breakpoints of the 9; 22 translocation characteristic of CML. The Abelson oncogene c-*abl* maps at the breakpoint on chromosome 9, and the *bcr* gene

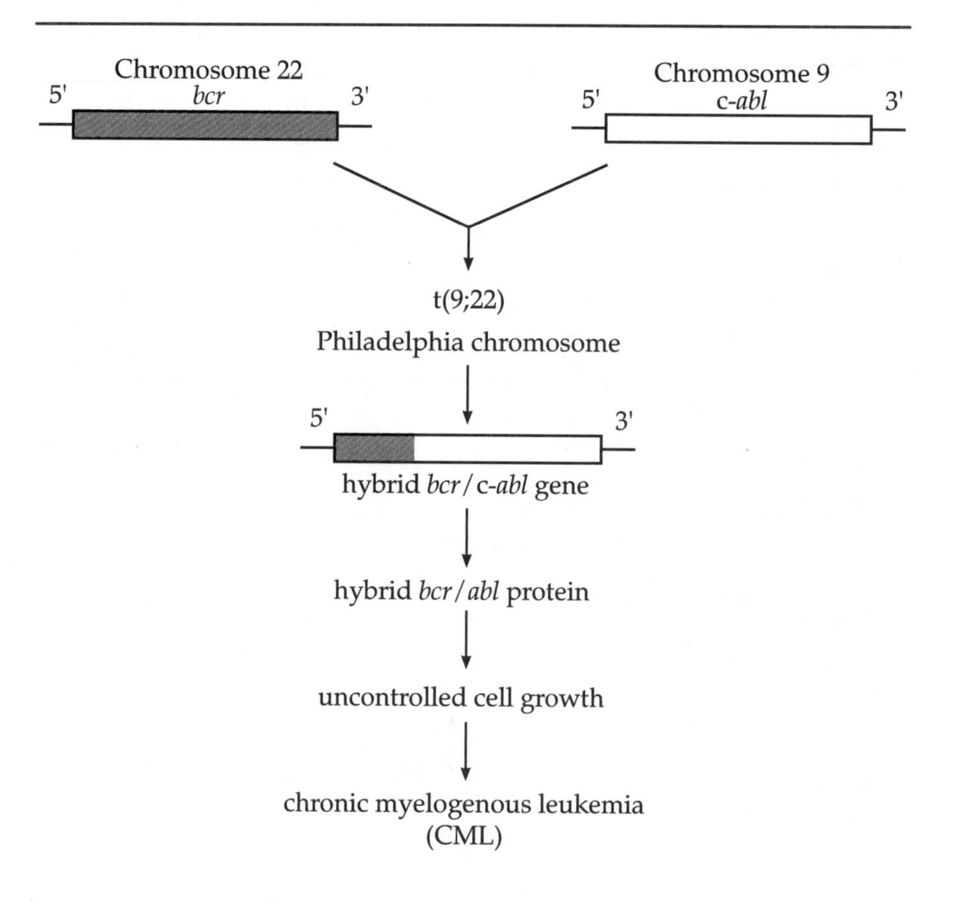

FIGURE 13.11 Gene fusion associated with the 9; 22 translocation in somatic lymphoid cells. In this event, the *bcr* gene on chromosome 22 fuses with the c-*abl* gene on chromosome 9. The resulting hybrid gene is transcriptionally active, and produces a fusion protein that acts to stimulate cell division in lymphoid cells. The overproduction of lymphoid cells results in chronic myelogenous leukemia.

maps at the breakpoint on chromosome 22. The translocation produces a hybrid gene that has *bcr* sequences at the beginning of the gene, and most, but not all, of the normal c-*abl* sequences at the end of the gene (Figure 13.11). This hybrid gene is transcribed and the message is translated to produce a fusion protein in which the *bcr*-encoded amino acids activate the amino acids in the region encoded by the c-*abl* gene. The hybrid protein switches on cell division in the lymphoid cells, resulting in CML.

Cancer and the Environment

Epidemiology

Research into the relationship between the environment and cancer has been conducted for over 50 years, but it is only in recent years, with the development of more sophisticated data gathering and analysis, that this work has provided solid evidence for the relationship between the environment and cancer. **Epidemiology** is the study of factors that control the presence or absence of a disease. It is an indirect and inferential science and provides correlations between factors and the existence of a disease such as cancer. These correlations provide working hypotheses that must be confirmed in laboratory experiments on animal models and then in carefully controlled clinical trials with humans (see Concepts & Controversies, page 367).

Epidemiology
The study of the factors that control the presence, absence, or frequency of a disease.

Typically, epidemiology begins with a large-scale study on the incidence of cancer across a number of populations. If statistically significant differences are found, further studies seek to identify the factors that systematically correlate with these differences. The results of many such studies illustrate that there are widespread geographic variations in cancer cases and mortality that are presumably correlated with environmental factors (Table 13.11). Many cases of cancer in the United States are thought to be related to our physical surroundings, personal behavior, or both. The degree of environmental contribution is a matter of some dispute, but estimates indicate that at least 50% of all cancer can be attributed to the environment.

Occupational Hazards

The relationship between occupation and cancer was first noted in 1775 by the English physician Percival Potts. Potts noted that London chimney sweeps had a high rate of scrotal cancer, caused by exposure to soot and coal tars. In addition to coal tars, more than 65,000 chemicals are in commercial and industrial use in the United States, and fewer than 1% of them have been adequately tested for their ability to cause cancer. Occupational exposure to some chemicals is known to cause cancer, but because of the long lag between the time of exposure and the onset of cancer, identification of those at risk is often difficult. A study by the federal government has estimated that occupational exposure to only a few materials currently in use may account

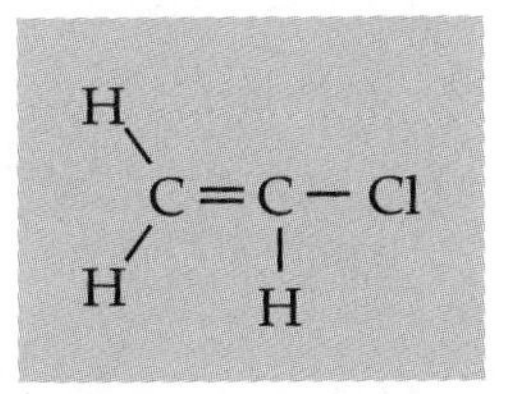

FIGURE 13.12 Chemical structure of vinyl chloride. This molecule can form long polymeric chains to produce polyvinyl chloride, a plastic widely used in manufacturing.

TABLE 13.11 Age-Adjusted Cancer Death Rates per 100,000 Population

Country	All Sites Male	All Sites Female	Country	All Sites Male	All Sites Female
United States	216.9(18)*	137.7(19)	Ireland	219.0(17)	158.8(4)
Argentina	215.9(19)	139.3(17)	Israel	174.7(32)	145.9(13)
Australia	212.4(21)	125.4(29)	Italy	228.5(15)	126.0(27)
Austria	247.3(9)	147.6(12)	Japan	190.0(27)	109.3(33)
Canada	214.9(20)	136.3(20)	Luxembourg	269.0(4)	148.3(11)
Chile	191.3(26)	148.5(10)	Netherlands	266.2(5)	144.8(14)
Costa Rica	178.6(31)	138.3(18)	New Zealand	172.0(35)	154.4(7)
Cuba	172.1(34)	119.1(31)	Nicaragua	22.9(48)	35.9(46)
Denmark	233.3(13)	172.3(1)	Norway	193.8(25)	130.4(23)
Dominican Republic	53.7(45)	48.1(45)	Poland	214.2(22)	126.2(26)
Egypt	39.0(47)	18.8(48)	Portugal	180.1(29)	108.0(34)
England and Wales	248.7(8)	156.9(5)	Puerto Rico	157.3(37)	95.1(39)
Finland	243.9(10)	113.9(32)	Romania	161.7(36)	106.0(35)
France	255.7(6)	120.8(30)	Scotland	275.0(1)	172.0(2)
Germany, FR	242.0(11)	152.0(8)	Singapore	249.6(7)	130.4(23)
Greece	188.3(28)	103.6(36)	Sweden	198.0(23)	141.2(15)
Guatemala	68.6(43)	78.8(42)	Switzerland	230.3(14)	134.2(22)
Hong Kong	235.2(12)	125.5(28)	Thailand	46.2(46)	30.4(47)
Hungary	269.3(3)	162.6(3)	Uruguay	271.8(2)	156.2(6)
Iceland	150.7(39)	140.7(16)(23)	Venezuela	135.0(40)	128.9(25)

*Figures in parentheses are order of rank.

for 18% to 38% of all cancer cases in the next few decades. Among these materials is vinyl chloride.

Vinyl chloride (Figure 13.12) is a gas first discovered in 1837 and now used in the manufacture of many plastic items. Over 8 billion pounds is produced each year. Vinyl chloride is used to make polyvinyl chloride (PVC), which in turn is used to manufacture products as diverse as floor tile, bottles, food wrap, and insulation. In laboratory experiments, vinyl chloride was shown to cause cancer in rats at doses as low as 50 parts per million (ppm) and in particular to cause a rare form of liver cancer known as angiosarcoma. When these experiments were conducted in 1970, the permissible exposure

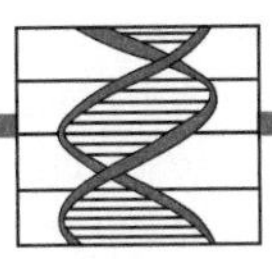

Concepts & Controversies

Epidemiology, Asbestos, and Cancer

Asbestos is a fibrous mineral known since ancient times. The emperor Charlemagne is reported to have had a tablecloth made of asbestos that was thrown into a fire after meals, emerging from the flames unharmed, to the amazement of his guests. In more modern times, asbestos has been widely used in manufactured products and is present in our homes, schools, and automobiles. It has been used in brake linings, ceiling tiles, wallboard, textiles, ironing board covers, and even kitchen gloves. Almost everyone in the United States has been exposed to asbestos in one form or another. To determine the possible role of asbestos in cancer, an epidemiological study compared the cause of death in a group of asbestos workers with the causes of death in an age- and sex-matched group of an equal number of individuals selected from the general population. Some data from this study are shown in the table. A total of 444 deaths were observed among the sample of asbestos workers, while only 301 occurred in the group drawn from the general population. In determining whether a disease like cancer is linked to asbestos, the number of cancer deaths observed in asbestos workers is divided by the number of cancer deaths observed in the general population. If the number of cancer deaths in the two groups is the same, the ratio will be about 1.0. If the number of cancer deaths among asbestos workers is greater than the population at large, then the ratio will be greater than 1.0. The results in the table show a 386% increase in cancer deaths among asbestos workers compared with general population, and a 762% increase in deaths from lung cancer.

This circumstantial evidence was then used as the basis of laboratory experiments using animals such as rats and mice. The animals were exposed to various levels of asbestos fibers and monitored for the development of cancer. Another group of animals served as a control group and was not exposed to asbestos. Cancers in this group provide a baseline measurement of cancer rates among the experimental animals. In the case of asbestos, the correlation between asbestos and cancer was upheld in animal experiments, leading the government to set standards for maximum permissible exposure to asbestos.

to vinyl chloride workers was 500 ppm (10 times higher than the levels found to cause cancer in animals). Surveys of workers in U.S. vinyl chloride plants showed that several workers were diagnosed as having angiosarcoma, a form of cancer that is very rare in the general population. Finally in 1974, permissible exposure levels were reduced to 50 ppm and in 1975 to 1 ppm. The Food and Drug Administration also banned the use of PVC in beverage containers. Although workers in vinyl chloride plants receive the highest levels of exposure, finished plastics always contain entrapped vinyl chloride gas in amounts that can be significant, and the effects of this exposure on the population at large has yet to be assessed.

Cancer and Society

The American Cancer Society estimates that 85% of the lung cancer cases among men and 75% of the cases among women are related to smoking. Smoking produces cancers of the oral cavity, larynx, esophagus, and lungs and accounts for 30% of all cancer deaths. Most of these cancers have very low survival rates. Lung cancer, for example, has a 5-year survival rate of

10%, a number that has not changed significantly in 20 years. Cancer risks associated with tobacco are not limited to smoking; the use of snuff or chewing tobacco carries a 50-fold increased risk of oral cancer. These risks are established through epidemiological studies (see Concepts & Controversies, page 367).

Aside from melanoma, there are about 400,000 cases of skin cancer reported in the United States every year, almost all of which are related to ultraviolet exposure from sunlight or tanning lamps. The incidence of skin cancer is increasing rapidly in the population, presumably as a result of an increase in outdoor recreation. Skin color is a trait controlled by several genes and is associated with a continuous variation in phenotype. The genetics of such traits will be discussed in a later chapter. In humans, traits controlled by several genes typically produce a wide range of phenotypes, and skin color is no exception. Epidemiologic studies have shown that lightly pigmented individuals are at much higher risk for skin cancer than heavily pigmented individuals. This observation supports the idea that genetic characteristics can affect the susceptibility of individuals or subpopulations to environmental agents that cause a specific form of cancer.

Over the past decade, epidemiologic and laboratory studies have increasingly focused on the role of diet and nutrition in the development of cancer. Many investigators believe that 30% to 40% of all cancers are related to diet. The association between breast cancer and environmental factors has been intensively investigated for at least 20 years. Unfortunately, the results in this case emphasize the shortcomings of epidemiology as a way of establishing cause-and-effect relationships. Epidemiological studies indicate that women in North America and Europe have a five-fold higher risk for breast cancer than women in Asia and less developed areas of the world. Among the correlations, the relationship between dietary fat and breast cancer seemed strongest (Figure 13.13). This correlation was supported by the observation that if women from a lower risk area such as Japan or Mexico move to the United States, familial breast cancer rates increase within a generation or two, indicating the role of an environmental factor. However, more recent epidemiological evidence has cast doubt on the hypothesis that dietary fat is a risk factor for breast cancer. The Nurses Health Study, started in 1976, including more than 120,000 women, has followed the intake of dietary fat compared with medical histories and has found no correlation between fat intake and breast cancer. Similar results were obtained in a smaller study conducted by investigators at the National Cancer Institute.

Because many other dietary factors, including total calories consumed, protein intake, and trace elements, might be correlated with breast cancer, the dispute about diet and breast cancer is ongoing. In the meantime, the National Academy of Sciences has issued cautionary diet guidelines. These include the reduction of fat intake to 20% or less of total calories, with increases in the intake of high-fiber foods such as whole grains and vegetables rich in vitamins A and C.

In spite of the dispute over the relationship between dietary fat and breast cancer, most research indicates that behavior patterns contribute heavily to the burden of cancer cases and represent about 40% of all

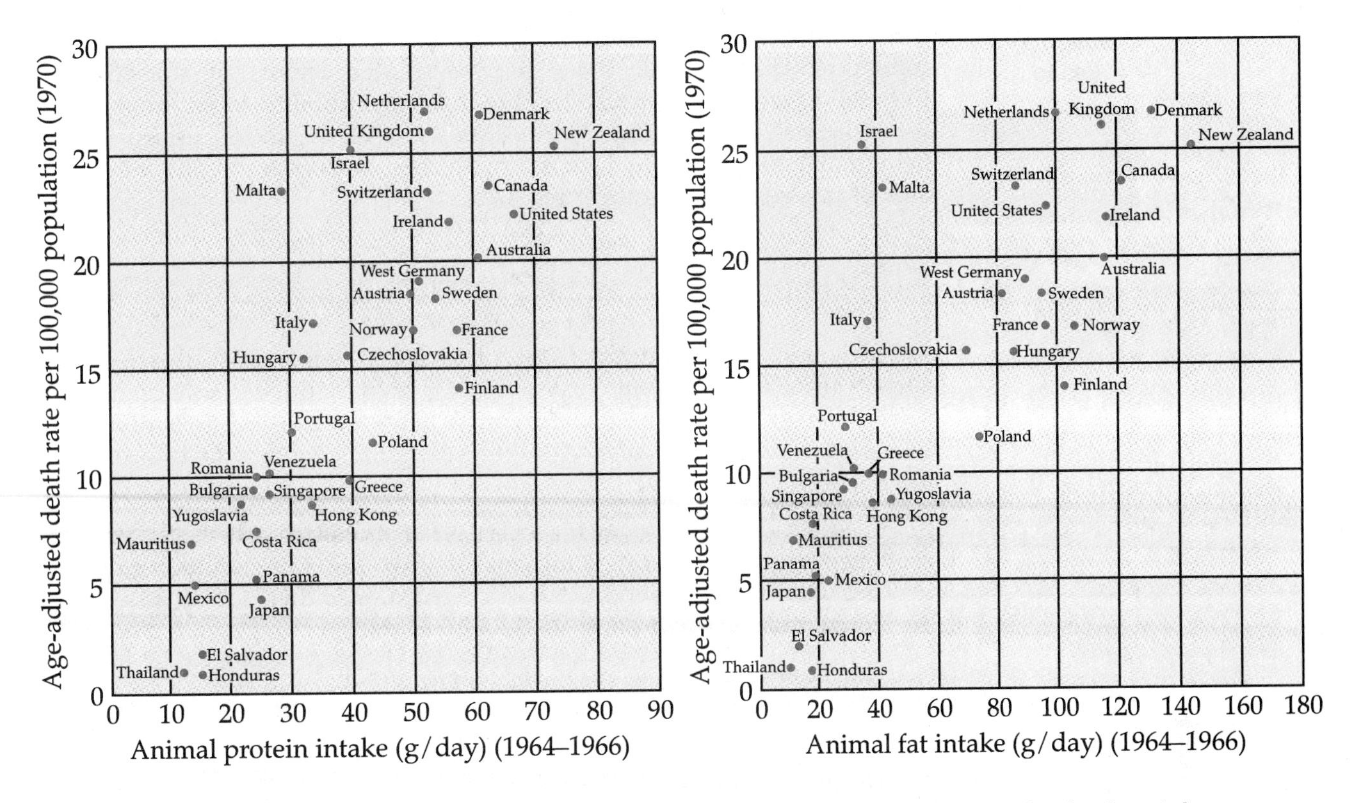

FIGURE 13.13 The relationship between breast cancer and the intake of animal fat and animal protein.

cancer deaths, nearly all of which could be eliminated by alterations in lifestyle.

Genetic Models for Cancer Susceptibility

Genetic factors may play several roles in determining individual susceptibility to environmental carcinogens. Disorders mentioned earlier in this chapter, including xeroderma pigmentosum (XP) and Fanconi anemia, represent examples of inherited conditions that exhibit hypersensitivity to environmental agents that promote cancer. Xeroderma patients develop skin cancer upon exposure to levels of ultraviolet light that have little or no effect on non-XP individuals, and Fanconi patients are hypersensitive to x-ray-induced cancers.

Evidence for the involvement of specific enzymes in cancer susceptibility come from studies on smokers and lung cancer. In one study, smokers who smoked few cigarettes and yet developed lung cancer were compared with heavy smokers with no signs of lung cancer. Those patients who developed cancer had significantly elevated metabolism in the debrisoquine pathway, while heavy smokers without cancer had much lower activity. Debrisoquine-4-hydroxylation is a metabolic process that converts carcinogens in tobacco smoke into highly active chemicals that promote tumor formation. Thus, genetic traits that control the activity or efficiency of metabolic pathways can affect individual susceptibility to environmental cancer agents. The identification and study of such metabolic pathways may help to diagnose individuals at risk for certain cancers and represent a new area of investigation in cancer research.

SUMMARY

1. Although some cancers show hereditary tendencies, there is no single cancer that shows a consistent pattern of Mendelian inheritance. In several cases, such as retinoblastoma, a gene for cancer susceptibility is inherited, with at least one additional mutation required to bring about malignancy.

2. These diseases serve as the basis for a model in which cancer is a multistep process, requiring a minimum of two separate mutational events to occur within a cell to produce cancer. According to this model, cancer cells are clonal descendants from a single mutant cell.

3. The study of two classes of genes, tumor suppressor genes and oncogenes, has established the relationship between cancer, the regulation of cell growth and division, and the cell cycle.

4. The discovery of tumor suppressor genes that normally act to inhibit cell division has provided insight into the regulation of the cell cycle. These gene products act at control points in the cell cycle at G1/S or G2/M, and deletion or inactivation of these products causes cells to continuously divide.

5. The discovery of an oncogene in the Rous sarcoma virus led to the identification of almost 75 oncogenes and their normal cellular counterparts, proto-oncogenes. Proto-oncogenes may serve to regulate cell growth and are converted to oncogenes by alterations in activity or by mutations that produce a defective gene product. Human tumors express at least two oncogenes and occasionally up to seven such genes. Since all human tumors show alterations in oncogenes, it is possible that a limited number of molecular pathways lead to the development of cancer.

6. Other human disorders, including Down syndrome, are associated with high rates of cancer, and this predisposition may result from the presence of an initial mutation or genetic imbalance that moves cells closer to a cancerous state.

7. It is now apparent that many cancers are environmentally induced. Occupational exposure to minerals and chemicals poses a cancer risk to workers in a number of industries, and the widespread dissemination of these materials poses an undefined but potentially large risk to the general population. Social behavior is responsible for about 40% of all cancer cases in the United States, the majority of which are preventable.

8. Evidence is accumulating that individuals have different susceptibilities to environmental agents that cause cancer and that these differences have a genetic basis.

QUESTIONS AND PROBLEMS

1. It is often the case that a predisposition to certain forms of cancer is inherited. Examples are familial retinoblastoma and colon cancer. What does it mean to have inherited an increased probability to acquire a certain form of cancer? What subsequent event(s) must occur?
2. Theodore Boveri predicted that malignancies would often be associated with chromosomal mutation. What lines of evidence substantiate this prediction?
3. Viral oncogenes are often transduced copies of normal cellular genes—they become oncogenic when their gene product or its regulation is subtly altered. What is the role of cellular proto-oncogenes, and how is this role consistent with their implication in oncogenesis?
4. What oncogene has been implicated in Burkitt lymphoma? What mutational event is typically associated with this tumor?
5. Outline the rationale behind gene transfer experiments. What oncogenes are likely to be identified in such a study?
6. What are some factors that epidemiologists have associated with a relatively high risk of developing cancer?
7. What is the probability that a phenotypically normal couple will have a child with retinoblastoma if one of the parents has a sibling with bilateral familial retinoblastoma?
8. The risk of developing many types of cancer increases as a function of age. Why might this be so?
9. Discuss the relevance of epidemiologic and experimental evidence in recent governmental decisions to regulate exposure to asbestos in the environment.

FOR FURTHER READING

Benedict, W. F., Xu, H. J., Hu, S. X., and Takahashi, R. 1990. Role of the retinoblastoma gene in the initiation and progression of human cancer. *J. Clin. Invest.* **85:** 988–993.

Birrer, M., and Minna, J. 1989. Genetic changes in the pathogenesis of lung cancer. *Ann. Rev. Med.* **40:** 305–317.

Bishop, J. M. 1982. Oncogenes. *Sci. Am.* (March) **246:** 80–92.

Bodmer, W. F., Bailey, C. J., Bodmer, J., Bussey, H., Ellis, A., Gorman, P., Lucibello, F., Murday, V., Rider, S., Scambler, P., Sheer, D., Solomon, E., and Spurr, N. 1987. Localization of the gene for familial adenomatous polyposis on chromosome 5. *Nature* **328:** 614–616.

Burt, R., Bishop, D. T., Cannon, L. A., Dowdle, M. A., Lee, R. G., and Skolnick, M. H. 1985. Dominant inheritance of adenomatous polyps and colorectal cancer. *N. Engl. J. Med.* **312:** 1540–1544.

Cairns, J. 1985. The treatment of diseases and the war against cancer. *Sci. Am.* (November) **253:** 51–59.

Call, K. M., Glaser, T., Ito, C. Y., Buckler, A. J., Pelletier, J., Haber, D. A., Rose, E. A., Kral, A., Yeger, H., Lewis, W. H., Jones, C., and Housman, D. E. 1990. Isolation and characterization of a zinc finger polypeptide gene at the human chromosome 11 Wilms tumor locus. *Cell* **61:** 509–520.

Cancer Facts and Figures. 1993. New York: American Cancer Society.

Committee on Diet, Nutrition and Cancer, Assembly of Life Sciences, National Research Council. 1982. *Diet, Nutrition, and Cancer.* Washington, D.C.: National Academy Press.

Corbett, T. H. 1977. *Cancer and Chemicals.* Chicago: Nelson-Hall.

Croce, C., and Klein, G. 1985. Chromosome translocations and human cancer. *Sci. Am.* (March) **252:** 54–60.

Digweed, M. 1993. Human genetic instability syndromes: single gene defects with increased risk of cancer. Toxicol. Letters **67**: 259-281.

Fearon, E. R., and Vogelstein, B. 1990. A genetic model for colorectal tumorigenesis. *Cell* **61:** 759–767.

Gallie, B. L., Squire, J. A., Goddard, A., Dunn, J. M., Canton, M., Hinton, D., Zhu, X. P., and Phillips, R. A. 1990. Mechanism of oncogenesis in retinoblastoma. *Lab. Invest.* **62:** 394–408.

Hamm, R. D. 1990. Occupational cancer in the oncogene era. *Br. J. Indust. Med.* **47:** 217–220.

Land, H., Parada, L. F., and Weinberg, R. A. 1983. Cellular oncogenes and multistep carcinogenesis. *Science* **222:** 771–778.

Lee, W-H., Bookstein, R., and Lee, E. 1988. Studies on the human retinoblastoma susceptibility gene. *J. Cellular Biochem.* **38:** 213–227.

Maitland, N., Brown, K., Poirier, V., Shaw, A., and Williams, J. 1989. Molecular and cellular biology of Wilms tumor. *Antican. Res.* **9:** 1417–1426.

Murphee, A. L., and Benedict, W. F. 1984. Retinoblastoma: Clues to human oncogenesis. *Science* **223:** 1028–1033.

Roth, J. S. 1985. *All about Cancer.* Philadelphia: Strickland.

Rous, P. 1911. Transmission of a malignant new growth by means of a cell-free filtrate. *J. Am. Med. Assoc.* **56:** 1981.

Simone, C. C. 1983. *Cancer and Nutrition.* New York: McGraw-Hill.

Solomon, E., Voss, R., Hall, V., Bodmer, W. F., Jars, J. R., Jeffreys, A. J. , Lucibello, F., Patel, I., and Rider, S. 1987. Chromosome 5 allele loss in human colorectal carcinomas. *Nature* **328:** 616–619.

Sparkes, R. S. 1985. The genetics of retinoblastoma. *Biochim. Biophys. Acta.* **780:** 95–118.

Sugimura, T. 1990. Cancer prevention: underlying principles and practical proposals. *Basic Life Sci.* **52:** 225–232.

Vogelstein, B., Fearson, E. R., Kern, S., Hamilton, S., Preisinger, A., Nakamura, Y., and White, R. 1989. Allelotypes of colorectal carcinomas. *Science* **244:** 207–211.

Weinberg, R. A. 1983. Molecular basis of cancer. *Sci. Am.* (November) **249:** 126–142.

Weinberg, R. A. 1985. The action of oncogenes in the cytoplasm and nucleus. *Science* **203:** 770–776.

Wynder, E. L., and Rose, D. P. 1984. *Diet and Breast Cancer. Hospital Practice* (November) **4:** 73–88.

Yunis, J. J. 1983. The chromosomal basis of human neoplasia. *Science* **221:** 227–236.

CHAPTER 14

IMMUNOGENETICS

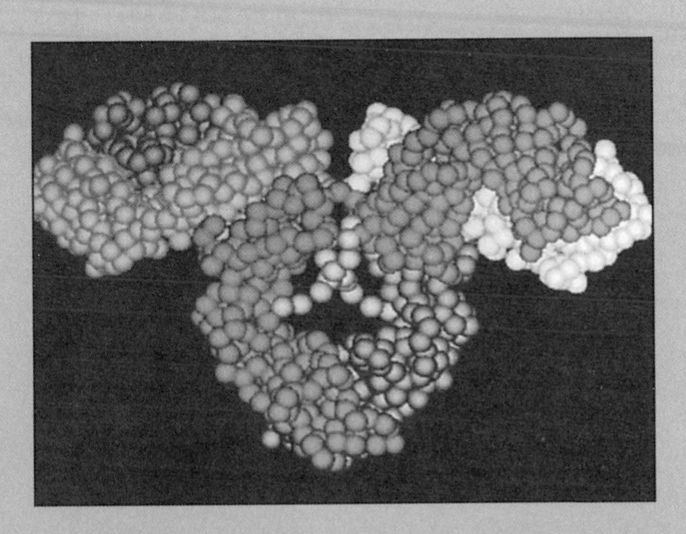

CHAPTER OUTLINE

Approximately 2500 years ago, a mysterious plague swept through Athens, killing thousands of residents. The Greek historian Thucydides wrote a careful account of the epidemic, noting that only those who had recovered from the disease could safely care for the sick and dying. As he noted: ''For no one was ever attacked a second time, or with a fatal result.'' In other words, those who recovered had become immune to the disease. Physicians in ancient China observed that people who contracted smallpox were resistant to further attacks of this disease. Partially successful attempts were made in the eighth century A.D. to transfer this resistance to uninfected individuals by injecting them with fluid obtained from those with the disease.

The first successful and safe method of transferring resistance to a disease was developed about a thousand years later in the late 18th century by Edward Jenner, an English physician. He observed that people who developed cowpox, a disease that resembled smallpox, became resistant to infection with smallpox. He tested this by inoculating a boy with pus from a lesion of a cowpox patient and some weeks later followed this with an inoculation from a patient with smallpox. The boy did not develop smallpox, and the method, called vaccination (from *vaccus,* the Latin word for "cow"), became an effective tool in controlling this disease. In fact, vaccination is such an effective means of controlling this disease that in 1980 the World Health Organization declared the world free from smallpox because of an intensive, worldwide vaccination program. Since the virus causing smallpox cannot propagate outside the body of an infected individual, the disease has disappeared. Today only laboratory workers conducting experiments on this virus need to be vaccinated.

The basic principles underlying vaccination were discovered by Louis Pasteur in the 19th century. In studying bacteria that cause a form of cholera in chickens, Pasteur found that aged cultures lost their ability to cause disease but retained the ability to induce immunity when used in vaccinations. The bacteria in the aged cultures had become attenuated, or less viable. Using the principle of attenuation, he went on to develop a vaccine for rabies.

The principle of all vaccines is fundamentally the same as that discovered by Pasteur. An attenuated or inactivated infectious agent is injected, conferring immunity to the disease caused by the agent. Some refinements of this method showed that *toxins* secreted by disease-causing bacteria could also be used to develop a vaccine.

Today, vaccines are widely used to prevent a wide range of diseases, including diphtheria, smallpox, whooping cough, tetanus, and measles. However, the study of the body's system of disease resistance, known as immunology, is a young and developing science. Many of the basic mechanisms in the body's response to infectious disease are as yet unknown. New diseases still arise to remind us of our limited knowledge; some are conquered, while others remain frustratingly remote. For example, in the early 1970s, women began to develop a condition known as toxic shock syndrome. By 1980 several thousand cases and over 25 deaths from this condition had been reported. After much effort, it was determined that the use of certain brands of highly absorbent tampons led to vaginal infections by the bacterium *Staphylococcus aureus.* In this case the bacteria secreted a toxin, enterotoxin F, that produced the symptoms of fever, rash, diarrhea, low blood pressure, shock, and, in some cases, death. Other new diseases, however, remain somewhat a mystery. In 1981 an infectious disease known as acquired immunodeficiency syndrome, or AIDS, appeared in the United States. Affected individuals exhibit rare forms of cancer, pneumonia, and other infections. There is, however, one common factor in all cases of AIDS: a complete and irreversible breakdown in the immune system. This breakdown allows the development of a number of infections, one or more of which become fatal.

As an introduction to immunology and immunogenetics, we will first review the components of the immune system and the cells and tissues involved in the immune response. Then we will examine four areas of immunology that are of genetic significance: transplants, blood groups, genetic diseases of the immune system, and autoimmune diseases.

Phagocyte
A cell with the ability to surround and engulf viruses and microorganisms.

T cells
White blood cells that originate in bone marrow and undergo maturation in the thymus gland.

B cells
White blood cells that originate in bone marrow and mature in the bone marrow.

Antibodies
A class of proteins produced by B cells that couple or bind specifically to the class of proteins that stimulate the immune response.

Antigen
A foreign molecule, virus, or cell that stimulates the production of antibodies.

Components of the Immune System

The concept of an immune system is deceptively simple: recognize self from nonself, and destroy nonself. In practice, however, the immune system is a complex network of intercommunicating cells and specialized molecules whose inner workings have proven difficult to unravel. What is somewhat ironic is that the immune system is constructed from a small number of different cell types. One of these, the **phagocyte,** has the ability to surround, engulf, and destroy organisms and cellular debris in the blood stream (Figure 14.1). Invading bacteria, viruses, and fungi are eliminated by these cells. A second cell type, **T cells,** are a specialized form of white blood cells involved in many aspects the immune response (Figure 14.2). Another class of white blood cell known as **B cells** produce **antibodies,** which are specialized protein molecules. Antibodies recognize and bind to foreign molecules and cells, marking them for destruction and removal from the body. These foreign molecules and cells are collectively called **antigens.**

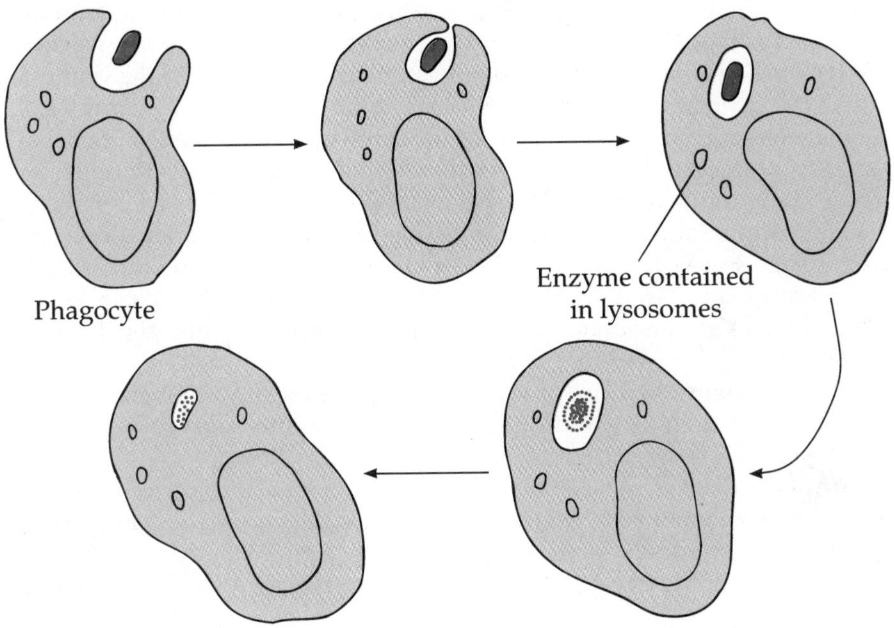

Phagocyte

Enzyme contained in lysosomes

FIGURE 14.1 In phagocytosis, the phagocyte makes contact with its target and, by means of cytoplasmic extensions, surrounds and engulfs it. In a second step, the phagocyte secretes enzymes to destroy its prey. These enzymes are contained in lysosomes that fuse with the encapsulated foreign material and subsequently digest it.

FIGURE 14.2 An electron micrograph of a T lymphocyte. They are small, spherical cells dominated by a large nucleus. The membrane is studded with specialized receptors that detect a specific antigen.

Cellular immunity
Immune responses carried out by cells rather than by antibodies in the blood serum.

Humoral immunity
Immune responses that are mediated by antibody (immunoglobulin) molecules.

The cells of the immune system are components of a two-part system of self-defense: **cellular immunity** and **humoral immunity.** Cellular immunity involves the phagocytes, a type of T cell, and cytotoxic, or killer, cells mobilized in direct response to invading pathogens. Humoral immunity involves the production and circulation of antibodies in the blood and lymph systems by B cells (humor is a medieval term for body fluids) and the recognition of antigens by T cells.

Phagocytes, T cells, and B cells

Phagocytes are produced in the bone marrow and are found not only in the blood but in all other parts of the body as well. Phagocytes, and in particular one type called a **macrophage,** are an essential component of the immune system. Macrophages arise from stem cells in the bone marrow (Figure 14.3). Their action represents a basic part of the cellular immune system, but they also play a larger and crucial role in initiating the action of every other cell type in the immune response. When a macrophage encounters a foreign cell, it wraps around and encloses the invading cell in its membrane. Then the macrophage secretes enzymes to break down and destroy the foreign cell. The macrophage also signals T cells that foreign cells or substances are present. This signal activates the T cells and brings them into the immune response.

Macrophages
Large, white blood cells that are phagocytic and involved in mounting an immune response.

T cells (Figure 14.3) are produced in the bone marrow. Immature T cells migrate to the thymus gland, where they mature into subtypes of T cells: *helper* cells, *suppressor* cells, and *killer* cells.

We will discuss two classes of T cells: T4 helper/inducer cells and T8 cytotoxic/suppressor cells. Mature T4 cells leave the thymus, enter the blood stream and the lymph system, and circulate through the body as part of the immune system. Mature T8 cells migrate to the lymph nodes and also take up residence in the spleen.

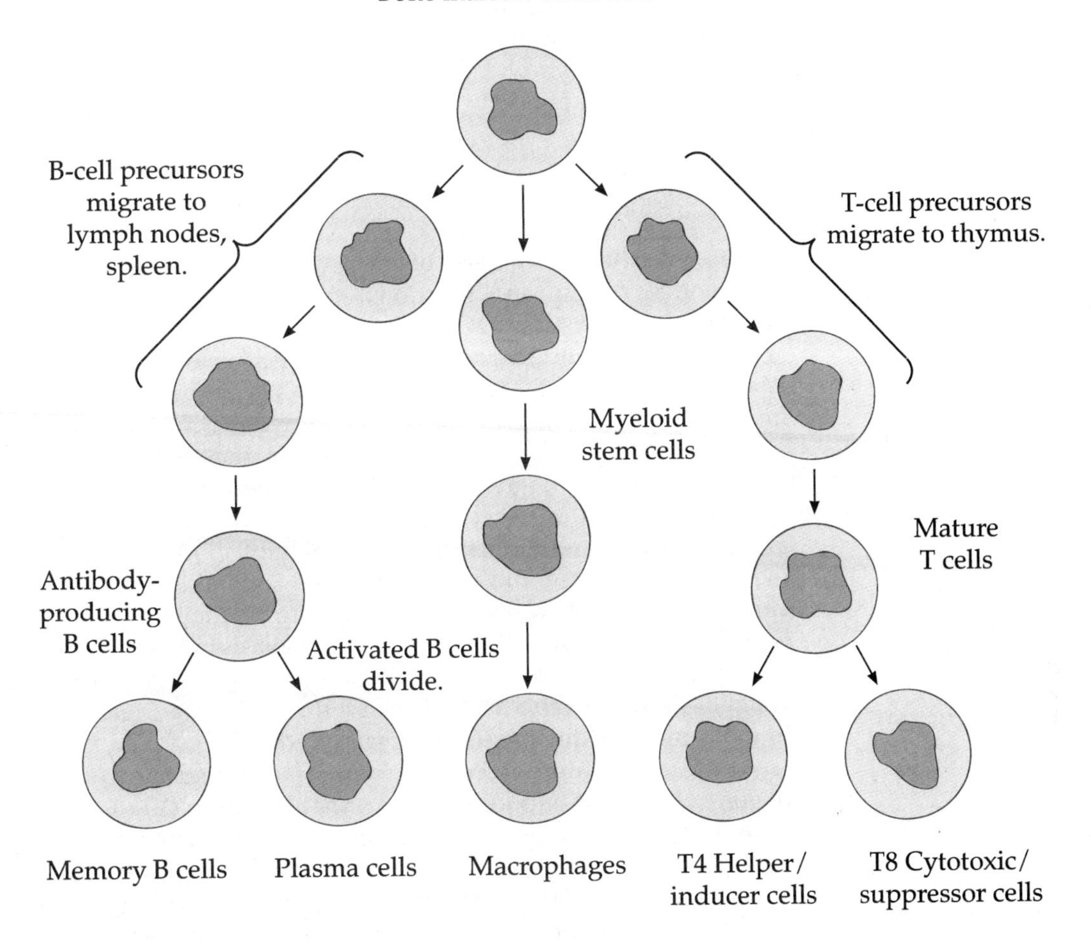

FIGURE 14.3 The origin, differentiation, and migration of cell types in the immune system.

Another cell type, known as **natural killer cells,** is also involved in the immune system. These cells circulate in the body and recognize a wide range of foreign cells but primarily identify and destroy cancer cells and cells infected with viruses. They are called natural killers because they do not have to be activated by helper T cells to respond to the presence of malignant cells.

All T cells have one thing in common: the ability to recognize the identity of a specific invading organism or substance. This recognition is achieved by means of cell-surface molecules known as **antigen receptors.** The receptors present on different T cells are genetically determined to recognize one and only one type of antigen. Because there are more than a million substances that are potential antigens, this means that there is an equal number of T cell subtypes, each genetically programmed to recognize a specific antigen. The macrophages, cytotoxic T cells, and natural killer cells are the major components of the cellular immune system.

Natural killer cells
White blood cells that originate in bone marrow and are able to kill invading microorganisms without activation by cells of the immune system.

Antigen receptor
Cell surface receptor on cells of the immune system that allow recognition of antigens.

B cells are white blood cells that act as the principal component of the humoral immune system. B cells are produced in the bone marrow. From here they migrate to the spleen and to the lymph nodes throughout the body. Like the T lymphocytes, B cells are also antigen-specific. B cells are genetically programmed to produce antibodies when triggered by helper T cells.

Antibodies

Antibodies belong to a class of molecules known as the *immunoglobulins*. Based on structural and functional differences, there are five classes of immunoglobulins (Ig): IgG, IgA, IgM, IgD, and IgE (Table 14.1). IgG antibodies are associated with immunological memory (discussed below). About 80% of antibodies in the circulatory system belong to the IgG class of immunoglobulins. IgA antibodies are secreted across plasma membranes and play a role in the immune response in the respiratory and digestive tracts. IgM antibodies are associated with the early stages of the humoral response and are the first class of antibodies secreted by B cells. The function of the IgD class of antibodies is unknown. IgE antibodies are associated with the allergic response. Individuals afflicted with allergies have a hypersensitive immune system that overproduces IgE antibodies.

The IgG molecule, like all Ig classes, consists of four polypeptide chains: two **H chains** and two **L chains** (Figure 14.4). The differences in molecular weights shown in Table 14.1 reflect the different organizational states of the Ig molecules and their associated carbohydrates.

H chains contain about 440 amino acids and are composed of two regions: a **constant region (C)** and a **variable region (V).** The first 100 amino acids at the N-terminus of the protein have a different amino acid sequence in different classes of H chains and make up the variable region. The remaining amino acids are invariable within each class of H chains and make up the constant region. The H chains are encoded by genes on the long arm of chromosome 14.

H chain
The larger polypeptide in antibody molecules; specifies the class to which the immunoglobulin belongs.

L chain
The smaller polypeptide in antibody molecules.

Constant region (C)
The region of an H or L chain closer to the C-terminus that is invariant within a given class of antibody.

Variable region (V)
A region at the N-terminus of an H or L chain that contains different amino acid sequences, even from one immunoglobulin molecule to another.

TABLE 14.1 Properties of Human Immunoglobulins

Immunoglobulin	Mol. Wt.	Serum Concentration (mg/100 ml)
IgG	150,000	700-1700
IgA	160,000	150-400
IgM	950,000	50-190
IgD	180,000	0.3-30
IgE	200,000	<0.1

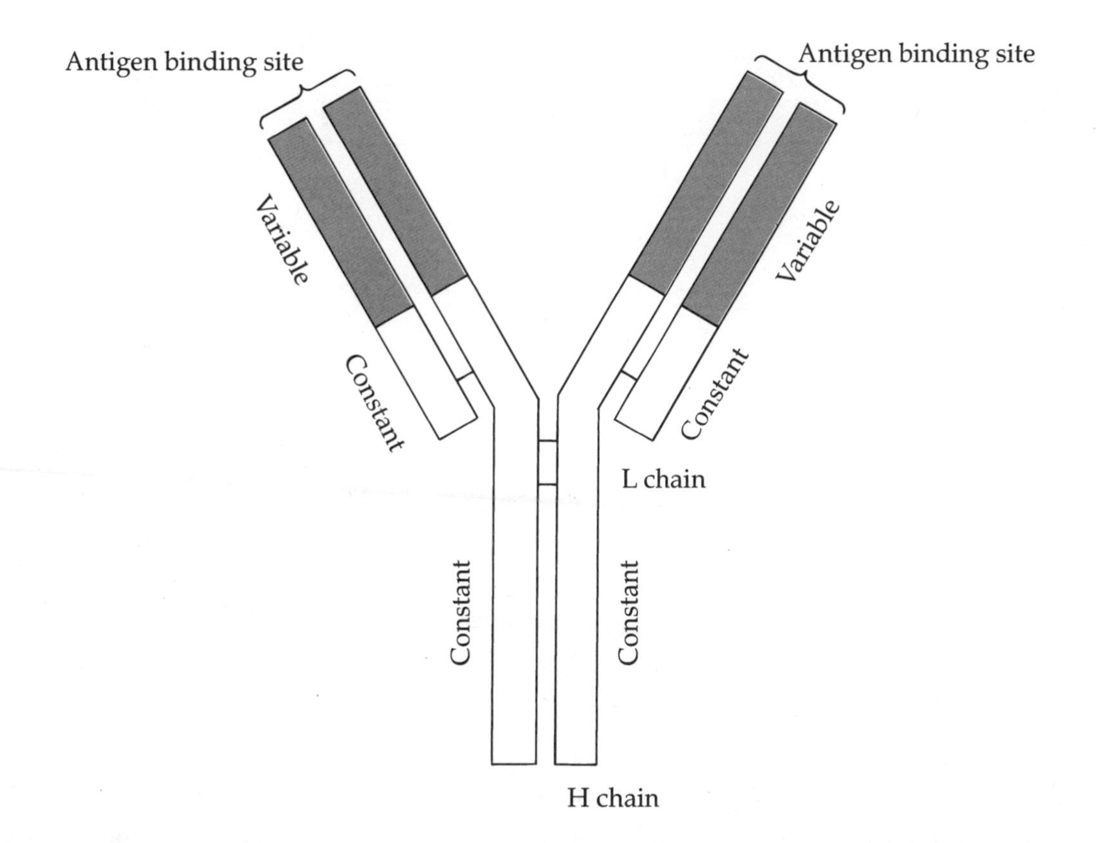

FIGURE 14.4 Structure of an antibody from the IgG fraction of blood serum. The variable segments at the ends of the L and H chains form the antigen binding site.

In the light chains, the variable region consists of the first 110 amino acids, with the remaining amino acids making up the constant region. There are two classes of light chains: one encoded by the *kappa* genes located on chromosome 2, and those encoded by the *lambda* genes, on chromosome 22.

In an IgG antibody molecule, two light chains and two heavy chains are joined by chemical bonds formed by amino acid R groups within each chain (Figure 14.4). In the Y-shaped molecule, the variable regions of the H and L chains at the end of each arm form the **antigen combining site,** which binds to a specific antigen.

Antigen combining site
The site at which the antibody will combine with, or bind to, the antigen; it is composed of the variable regions of the H chain and the L chain.

Antibody Formation

The process of antibody formation remains one of the least understood genetic processes, and this area of research is at the forefront of modern biology. The basic question is how antibody diversity is encoded within the genome. Since there are literally billions of combinations of H and L chains that produce antibodies, it seems unlikely that each combination is separately encoded in the genome. Several theories have been put forward to explain how antibody diversity can be genetically encoded.

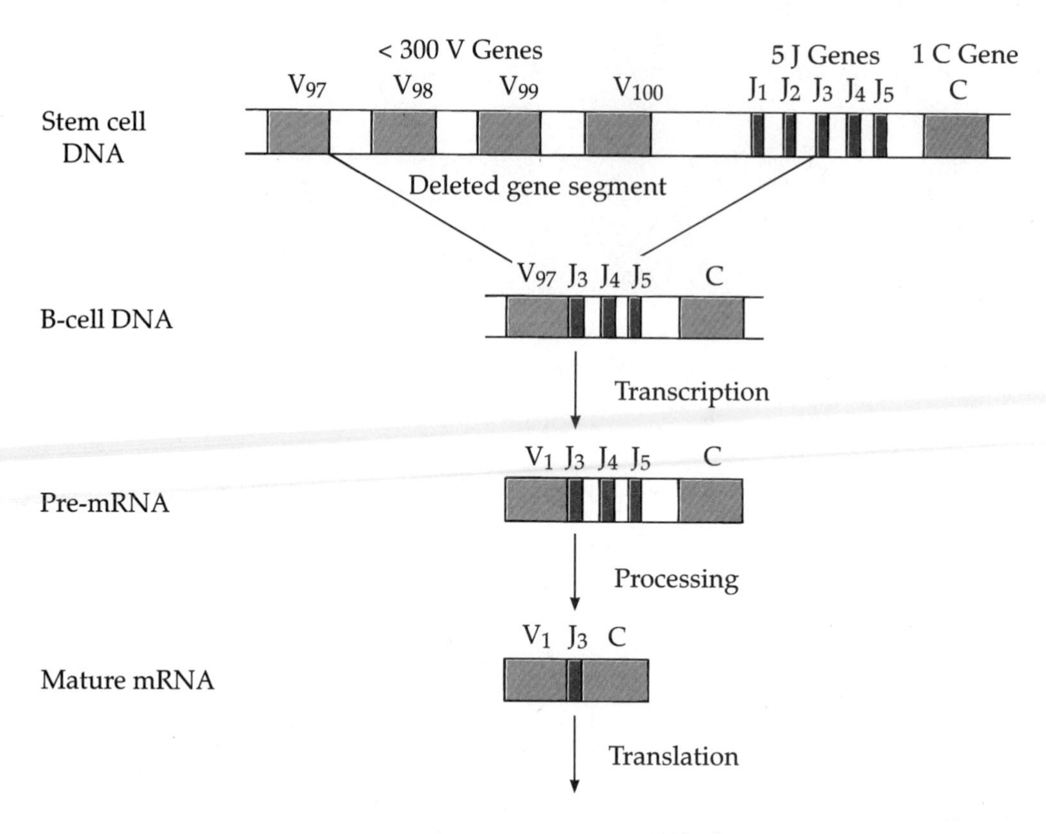

FIGURE 14.5 In germ cells, kappa light chain genes contain several hundred V regions (only four are shown here), five J regions, and one C region. As B cells mature, randomly selected V and J regions are fused to a C region by a process involving DNA cutting and splicing.

The available evidence supports one of these theories, called the *recombination theory,* which involves DNA rearrangements. These rearrangements take place during B-cell maturation, before antibody genes are transcribed and antibody production begins. As an example, let us consider the light chain kappa genes. These genes contain three regions: The V and J regions encode the variable region of the L chain, and the C region encodes the constant region (Figure 14.5). In the germ cells and B-cell precursors, each kappa L gene is composed of 70 to 300 V segments, about 5 J segments, and 1 C segment. These segments are located some distance apart on the chromosome.

As the B cell matures, one of the several hundred V genes is randomly selected and joined to one of the J segments and the adjacent C segment by a recombination event. This fused gene now encodes a specific kappa L chain. In a mature B cell, this gene is transcribed and translated to form an L chain that becomes part of an antibody molecule. The genes for the lambda L chains and the H chains are also composed of segments that undergo recombination during B cell maturation. Because these recombination events take place at random in each B cell during maturation, the result is the production of many different variant antibody chains. Each B cell, however, has been programmed by the recombination events involving the L genes and the H genes to produce only one of the many possible variants. Thus, antibody diversity is the result of genetic events that shuffle a number of basic components into a large number of combinations. In addition to the recombination events described, other events that take place during B-cell maturation expand antibody diversity, allowing production of billions of possible antibodies from several hundred basic gene segments in three gene classes.

THE IMMUNE RESPONSE

Humoral Immune Response

Now that we have looked at the cellular and molecular components of the immune system, let us consider how these elements interact during an immune response (Figure 14.6). Suppose that a wandering macrophage makes contact with an infecting virus. The macrophage engulfs and destroys some copies of the virus. In the process some of the antigens from the virus are displayed on the surface of the macrophage. These surface antigens are grouped and presented to a helper T cell that recognizes the macrophage by the presence of identifying surface molecules and the viral antigen. During the process of recognition the antigen receptor on the T cell binds to viral antigens on the surface of the macrophage. This recognition process causes the helper cell to become activated. The activated T helper cell identifies and activates the class of B lymphocytes that encode antibodies against the viral antigen.

The activated B cell divides a number of times. Some of the progeny become plasma cells that synthesize and secrete the antibody specific for the antigen recognized by the T helper cell. These plasma cells live for only a few days but during their life span synthesize and secrete large quantities of antibody. Other progeny of the activated B cell become memory cells. These cells enter and remain in the circulatory system and will provide a faster response to the antigen, should it ever appear again.

The antibody secreted by the plasma cells binds to the antigen on the surface of the infecting virus, marking it for rapid destruction by phagocytes. This process, known as the **immune response,** is monitored by T8 suppressor cells. When the immune response is no longer needed, the suppressor cells turn off the action of T4 helper cells and prevent them from triggering the activity of B cells in the lymph nodes. By secreting certain chemicals, the suppressor cells can also turn off antibody production in B cells and can stop the entire immune response. As the immune response slows under the direction of the T8 suppressor cells, most of the T and B cells specific to this antigen begin to die. However, a large number of antigen-specific T and B cells remain in the circulatory system and lymph nodes. These cells are known as **memory** cells, since they "remember" the antigen that was encountered. If the same organism reinvades the body and presents the original antigen to the immune system, there will be a substantial number of T cells and B cells already present to remove the antigen. This will result in

Immune response
The activation of the immune system caused by the presence of a foreign substance, or antigen.

Immunological memory
The capacity of the immune response to mount a rapid and vigorous response to a second contact with an antigen.

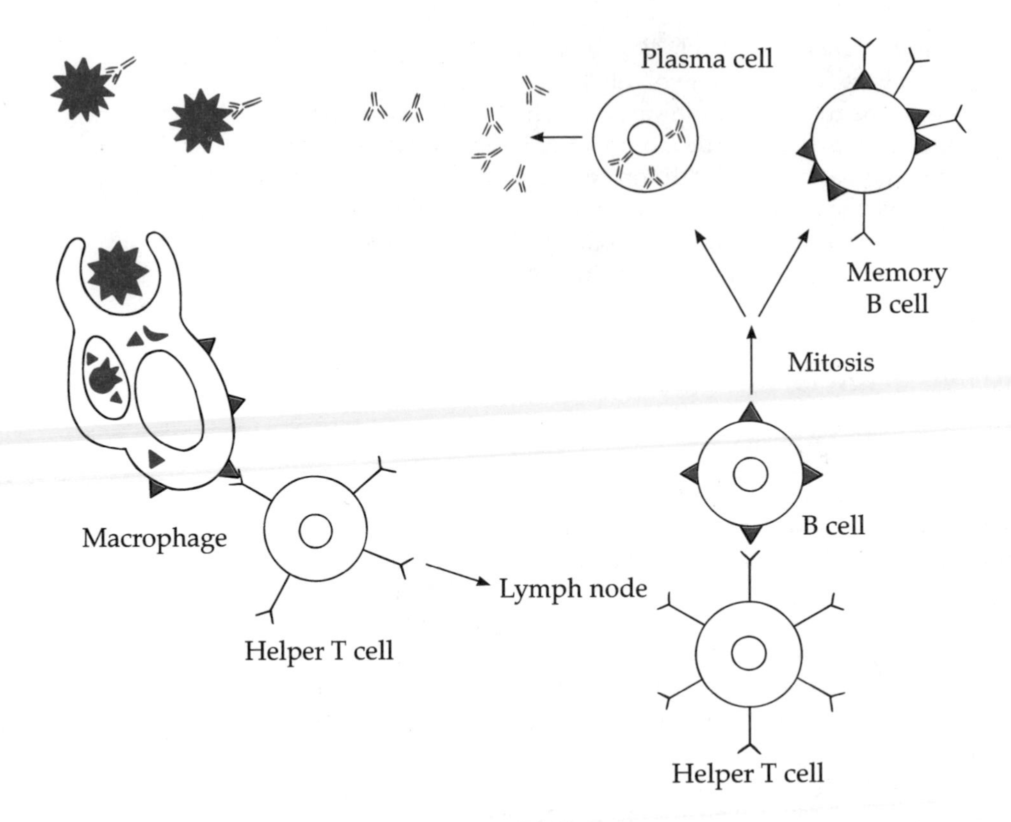

FIGURE 14.6 An overview of the humoral immune response. A macrophage contacts an infecting virus and, in the process of phagocytosis, displays some of the viral antigens on its cell surface. This converts the macrophage to an antigen-presenting cell that is recognized by a T4 helper cell. The helper cell becomes activated by this contact and moves to a lymph node, where it finds and activates a B cell that can produce an antibody to this antigen. The activated B cell divides, giving rise to short-lived plasma cells that produce large quantities of antibody before dying. Other progeny become B memory cells, ready to respond in case of a second infection by the same virus. The secreted antibodies bind to the virus, marking it for rapid destruction by phagocytic cells.

a rapid and massive immune response to the pathogen. Under these circumstances there usually will be no infection. The development of memory cells in response to another infection provides an explanation for the observation recorded by Thucydides 2500 years ago.

Cellular Immune Response

We have concentrated on an example of humoral immunity in which macrophages make contact with the pathogen. The immune response can also be initiated by T4 helper/inducer cells. In the cellular immune response, body cells infected with a virus often display viral antigens on their surface. The T4 cells can detect the presence of viral antigens by means of their cell-surface antigen receptors. Once activated, the inducer cells trigger the maturation of immature T cells to combat the infection. T8 cytotoxic cells activated by T helper cells can identify infected cells by the presence of viral antigens on their surface. These cytotoxic T cells attach to the infected cell and initiate a series of reactions that penetrate the membrane, releasing the cytoplasmic contents and killing the cell. Any viruses released from the cells are scavenged by phagocytes. The central role of the T4 helper/inducer cells in the immune response is summarized in Figure 14.7.

The cellular components of the immune system communicate through a series of chemical signals. As a group, these communication molecules are known as **lymphokines.** When T cells make contact with an invading organism, they release a lymphokine known as **migration inhibitory factor,** which signals macrophages to stay in the vicinity to assist in the upcoming immune response. The macrophages then release **interleukin 1,** which causes T cells at the infection site to release **interleukin 2.** Interleukin 2 stimulates more T cells to assist in the immune response. In the lymph nodes, B cells are stimulated to produce antibodies by lymphokines released by the T4 helper cells. The lymphokines interleukin 1 and interleukin 2 help bring the immune response to peak activity. The lymphokines secreted by suppressor cells act to turn down and to switch off the immune response.

Lymphokines
Glycoprotein molecules that are used as chemical signals to communicate among cells in the immune system.

Migration inhibition factor
A lymphokine that inhibits the migration of macrophages.

Interleukin 1
A lymphokine secreted by macrophage cells that triggers lymphokine (interleukin 2) release by T cells at the site of an infection.

Interleukin 2
A lymphokine released by T cells at the site of an infection that stimulates other T cells to participate in an immune response.

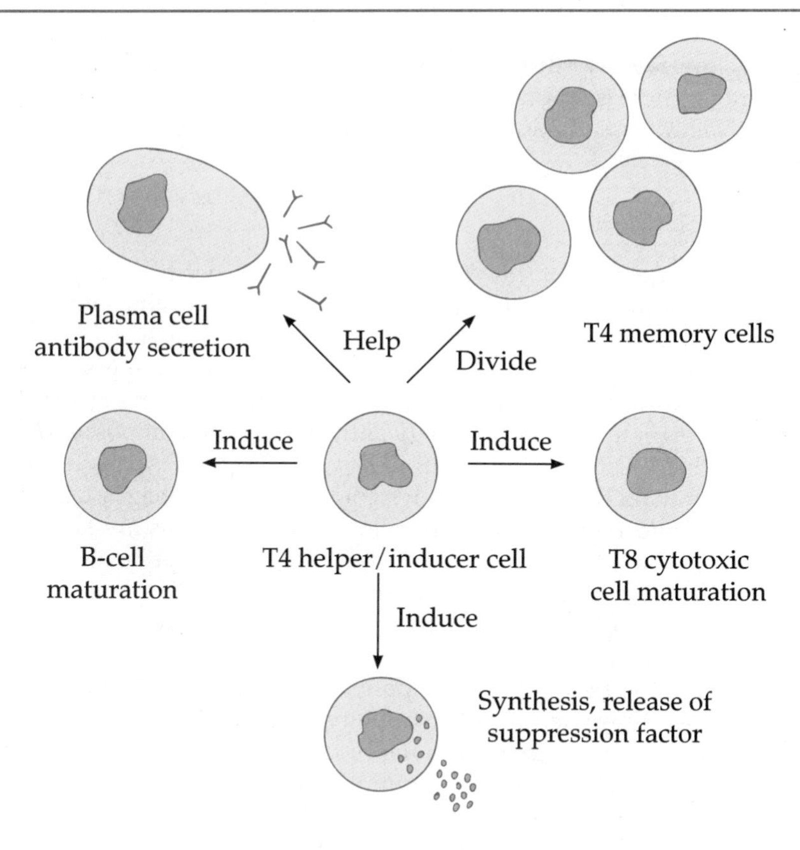

FIGURE 14.7 The central role of T4 helper/inducer cells in the immune response. This role includes the stimulation of the B cells. Activated B cells divide, and some differentiate into plasma cells that actively secrete antibody, aided by the T4 cells. Once stimulated, some T4 cells divide and produce a group of memory cells that "remember" the antigen they encountered, maintaining a state of readiness. The activated T4 cells also stimulate the maturation of cytotoxic T8 cells. As the infection ends, the T4 cell stimulates the T8 suppressor cells to release a suppressor factor that slows and stops the immune response.

Transplantation Genetics

Documents from ancient Egypt indicate that organ and tissue transplants were attempted 3500 years ago. However, in those attempts the transplants survived only a short time before rejection. Gradually it became clear that for a transplant to be successful, a certain compatibility must exist. This compatibility can be demonstrated by comparing grafts between identical twins and more unrelated individuals. Grafts or transplants between identical twins are almost always successful, while those between more unrelated individuals are almost always unsuccessful.

We now know that the basis for success or failure in transplants is the presence or absence of an immune response. A graft of skin from brother to sister will be rejected in about 3 weeks. If a second graft is attempted, it will be rejected in a few days, indicating the role of the immune system in the process of rejection. The grafted tissue acts as an antigen to the recipient's immune system, which mobilizes cytotoxic T cells to cause graft rejection. The graft also stimulates the development of immunologic memory, so the second graft is rejected in a shorter time.

As knowledge of the immune system has improved, methods to circumvent the process of rejection have been developed. Chemical and radiologic suppression of the immune response has permitted the transplantation of vital organs such as kidney, heart, lung, and liver. However, the relatively low rate of survival in some of these transplants serves as a reminder that there is still much to be learned about the role of the immune system in the transplantation (see Concepts & Controversies, page 385).

The HLA System

HLA complex
A cluster of genes located on chromosome 6 that are concerned with the acceptance or rejection of tissue in organ grafts and transplants.

In humans, the genetics of transplantation involve a specific gene set known as the **HLA complex,** located on the short arm of chromosome 6. This complex contains four major genes, designated HLA-A, HLA-B, HLA-C, and HLA-D (Figure 14.8). The HLA-A, -B, and -C genes encode glycoproteins found on the surface of all nucleated cells and are known as Class I antigens. The HLA-D region encodes Class II antigens and is composed of three subregions: HLA-DR (D-related), HLA-DQ, and HLA-DP. Class II antigens are present only on the surface of B lymphocytes, macrophages, activated T lymphocytes, and other specialized cells. Class II antigens are structurally different from Class I antigens and consist of two polypeptide chains rather than glycoproteins. Class III genes located between HLA-B and HLA-D regions encode genes of the complement system, which plays an important role in cell-mediated immunity.

The HLA complex is the most polymorphic genetic system known in humans. Diversity in the histocompatibility system is accomplished by the presence of a large number of alleles at each locus rather than by recombination events involving a number of gene segments as in the antibody genes. There are 23 recognzied alleles of HLA-A, 47 alleles of HLA-B, 8 in the C locus, 14 in DR, 3 in DQ, and 6 in DP. Each allele encodes a distinct antigen, designated by a letter that stands for the locus and a

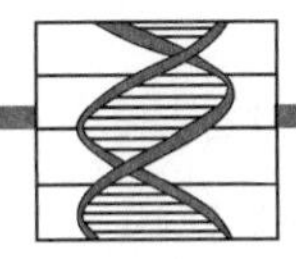

CONCEPTS & CONTROVERSIES

Animal-Human Transplants

On October 26, 1984, physicians at Loma Linda University Medical Center in California transplanted the heart from a 7-month-old baboon into a 3-week-old human infant known as Baby Fae. The infant was born with a heart defect that is inevitably fatal, and at the time of the transplant she was critically ill. Although it was possible that a human heart could have been found for the transplant, a xenograft, a transplant between different species, was recommended. Several factors entered into this decision, including the fact that the surgeon had extensive experience in performing xenografts in animals, and there were no reports of successful human heart transplants in newborns. The transplant functioned for 20 days until an immune response mounted by the infant caused heart failure. This transplant had raised many controversies among physicians and the public at large. Among these issues are the suitability of using animals as a source of organs for human transplantation, the choice of a baboon instead of a primate more closely related to humans, and the question of whether immunological typing of animals is valid for human transplantation.

There is no question that there is a great need for organs for transplantation. There are now some 80,000 individuals on kidney dialysis in the United States, many of whom are waiting for donor organs for transplantation. Only about 7000 such kidneys are transplanted each year. The wait is so long that dialysis patients are beginning to voluntarily withdraw from dialysis, choosing death rather than waiting for a kidney they feel will never come. Recent studies indicate that some 12,000 adults would benefit from heart transplants if a sufficient donor supply were available. In addition, there are some 25,000 children like Baby Fae born every year with congenital heart disease, many of whom would benefit from transplants.

The present system of supplying organs from accident victims and cadavers cannot meet present demands. To remedy this, workers have studied xenografts in animals for the last 20 years. The procedure is still highly experimental, and the question arises whether it should be used on humans. If further human trials are conducted, should they involve terminally ill patients, as has been done for artificial hearts? In addition, several technical difficulties present themselves. The ability of drugs like cyclosporine to prevent rejection of xenografts is still not understood.

Tissue matching between primates and humans is also an area of concern. Studies have shown that baboon cells cannot be accurately tested for histocompatibility antigens using reagents designed for human tissue, making it difficult to identify a suitable animal donor. While there is a close relationship between apes and humans, the baboon is more distantly related than other animals such as the chimpanzee. Although chimpanzees may be more suitable donors, they are becoming scarce as their natural habitats are destroyed. This raises the related question of whether animals should be killed to provide organs for human transplants or whether efforts should first be directed at generating a sufficient increase in the supply of human organs.

Because of the uncertainty surrounding animal-human transplants, continued research in this area is needed, under guidelines that outline clearly stated goals.

number that signifies the allele. For example, A1 is allele number 1 at the A locus, B27 is the allele number 27 at the B locus, and so forth.

The HLA genes are inherited in a Mendelian codominant fashion. Each parent contributes one chromosome number 6 containing its HLA genes to the offspring. Because the assorted HLA alleles are located close together on

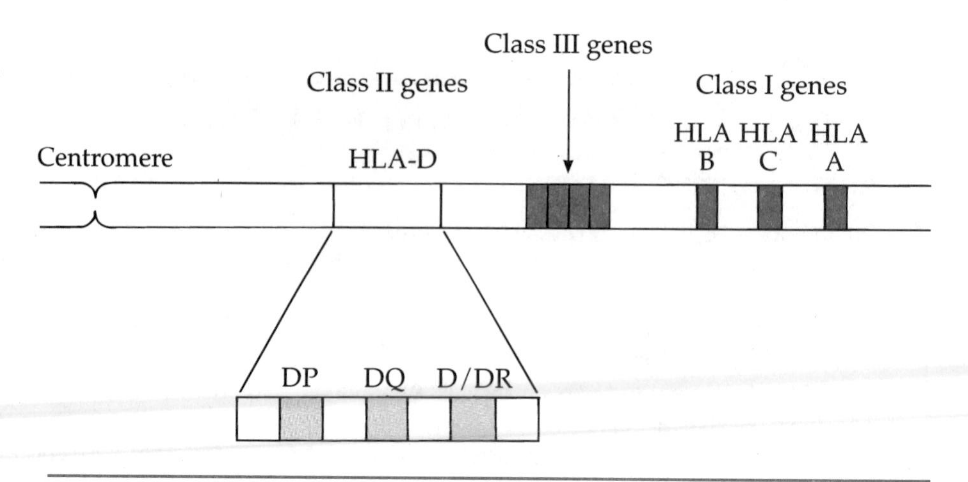

FIGURE 14.8 The HLA complex on chromosome 6, showing the arrangement of the four major genes and flanking genes.

Haplotype
A set of closely linked genes that tend to be inherited together, such as the HLA complex.

a single chromosome, there is rarely any recombination between them, and they tend to be inherited together. The group of HLA alleles on a single chromosome is known as a **haplotype.** Since humans are diploid, we all possess two haplotypes (Figure 14.9). Because of the number of alleles for each locus, there are literally millions of combinations of HLA alleles. In spite of the many random combinations that are possible, the HLA antigens are nonrandomly distributed in the human population. For example, allele B27 is found in 4% of American blacks but in 8% of American whites. B27 has a very low frequency in the Japanese but very high frequencies in some American Indian tribes.

The transplantation of organs and tissues in humans depends to a large extent upon the degree of matching HLA types. HLA typing is the currently accepted method of matching a suitable donor with a recipient in an organ transplant. Among unrelated donors and recipients, the chance for an HLA match is about 1 in 200,000. Figure 14.10 demonstrates the importance of HLA matching in kidney transplants. Where HLA types were closely matched between donor and recipient, over 90% of the kidneys were still functioning after 4 years. When HLA haplotypes were not matched, fewer than 50% of the grafts were still functioning after 48 months. HLA antigens are the major genes involved in distinguishing self from nonself. HLA Class I antigens are involved in antigen recognition for killer T cells, and HLA Class II antigens must be presented to the helper T cell along with the foreign antigen to activate the helper cell.

Graft versus Host Disease

In most transplants the immune system of the recipient reacts against the HLA antigens on the cells of the donor. If the donor cells are not properly matched, they are rejected. If, however, the immune system of the recipient is nonfunctional because of a genetic disorder (as in severe combined immunodeficiency) or administration of radiation and chemicals (as in the treatment of leukemia), a bone marrow transplant can restore an immune

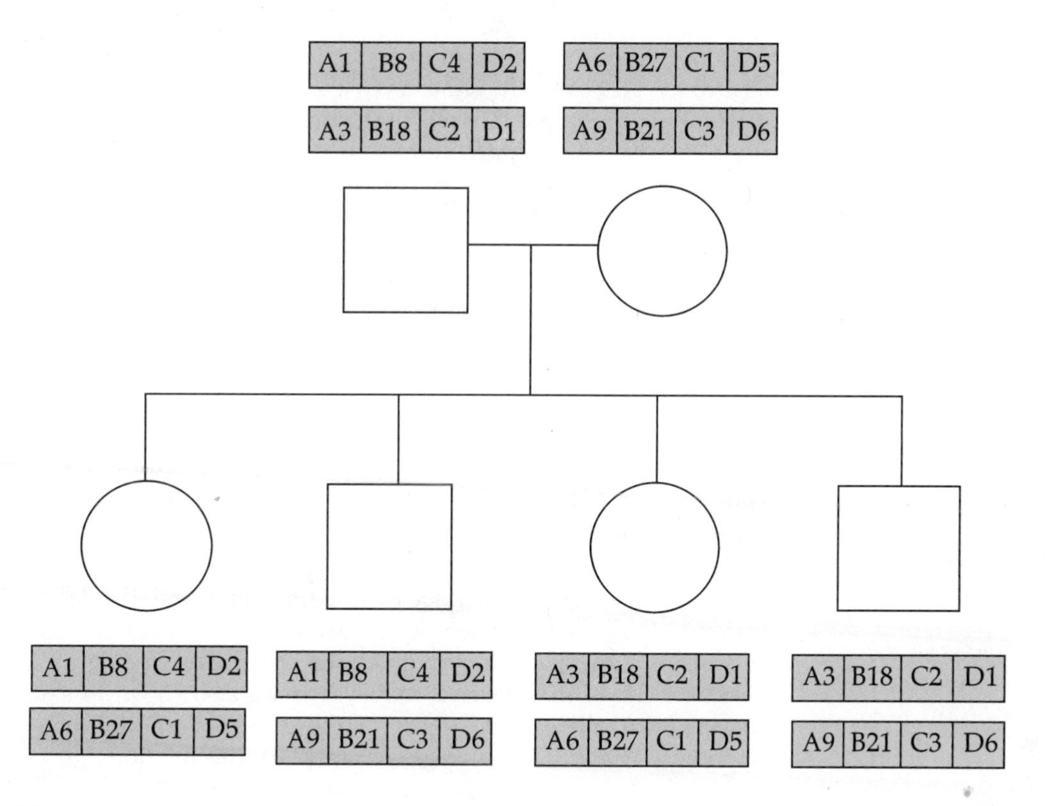

FIGURE 14.9 The transmission of HLA haplotypes. Each haplotype contains four major loci, each encoding a different antigen.

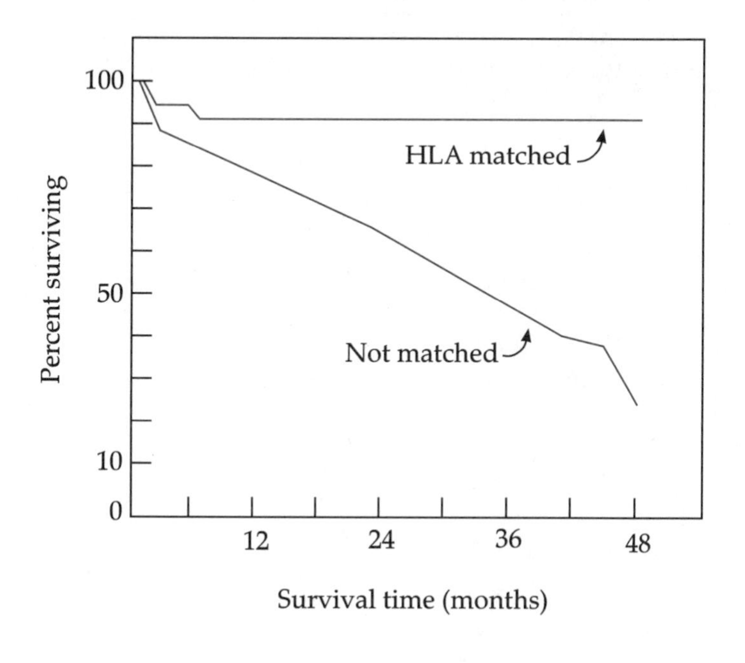

FIGURE 14.10 The outcome of kidney transplants with *(upper)* and without *(lower)* HLA matching.

response. Even in this form of transplantation, however, it is important to ensure an HLA match. Otherwise, the immunologically competent donor cells can react against the cells of the host, sometimes with fatal results. This reaction, known as **graft versus host disease (GVHD),** is a case of the transplant rejecting the host. For many years this reaction prevented the successful transplantation of bone marrow to treat genetic disease of the immune system and other conditions such as leukemia and aplastic anemia. There are now several methods that allow suppression of the immune response of the graft, resulting in the use of bone marrow grafts as the treatment of choice in some conditions.

Graft versus host disease (GVHD)
A situation that arises when a transplant, such as bone marrow, containing immunocompetent cells mounts an immune response against the host body, which is unable to protect itself.

Disease Associations

In studying the distribution of HLA alleles in the population, a relationship between certain HLA alleles and specific diseases has been discovered. For example, more than 90% of the individuals with the connective tissue disease **ankylosing spondylitis** carry the HLA-B27 allele. This disease is a chronic inflammatory condition associated with joints and the spine, leading to fusion of the joints between vertebrae. The available evidence suggests that B27 class I antigens are involved in promoting the joint inflammation characteristic of this disease. Other diseases also show an association with certain HLA alleles, and some of these are listed in Table 14.2. Several hypotheses have been proposed to explain the relationship between HLA

Ankylosing spondylitis
An autoimmune disease that produces an arthritic condition of the spine; associated with HLA allele B27.

TABLE 14.2 HLA and Disease

Disease	HLA Allele	Frequency in Patients %	Frequency in General Population %	Relative Risk Factor
Ankylosing spondylitis	B27	>90	8	>100
Congenital adrenal hyperplasia	B47	9	0.6	15
Goodpasture syndrome	DR2	88	32	16
Juvenile rheumatoid arthritis	DR5	50	16	5
Multiple sclerosis	DR2	59	26	4
Pernicious anemia	DR5	25	6	5
Psoriasis	B17	38	8	6
Reiter's syndrome	B27	75	8	50
Rheumatoid arthritis	DR4	70	28	6
Systemic lupus erythematosus (SLE)	DR3	50	25	3

alleles and disease, but so far none has proven to be valid. It appears that most of these disorders result from a combination of HLA and non-HLA effects, with environmental activation, possibly by infection. For example, ankylosing spondylitis may be triggered in B27 patients by an infection, and the bacterium *Klebsiella* has been implicated in the onset of this disease.

Blood Groups

Human blood groups were first investigated in the early part of this century and quickly became an important part of human genetics because they provided a set of polymorphic markers for genetic studies. Later, blood groups became important as markers for transfusions and transplantations. Over 20 different blood groups are known in humans, most of which are useful as genetic markers, and only a few of which are of clinical importance. In earlier chapters we briefly discussed blood types, including the MN group and the ABO group. Our focus then was genetic polymorphisms and multiple alleles. In this section we will examine the ABO and Rh groups and their relation to the immune system.

ABO Blood Types

The ABO blood groups were discovered in 1900 by Karl Landsteiner, a Viennese physician and Nobel Prize winner. He distinguished four separate blood types on the basis of an agglutination reaction. He mixed serum (the liquid portion of blood) from one individual with red cells (erythrocytes) from another individual. If the serum and cells belong to different blood groups, the cells often clump together, or **agglutinate.** This reaction can be observed in a drop of serum with the naked eye. By conducting agglutination tests on a large number of individuals, four phenotypic classes, or blood types, can be identified: A, B, O, and AB. Those with type A have antigen A on the surface of their red blood cells, group B individuals have antigen B, type AB have both A and B antigens, and type O individuals have neither A nor B. Mendelian inheritance of these blood groups was demonstrated in 1911, and in 1924, Felix Bernstein showed that the genetic factors for A, B, and O were not separate genes but alleles of a single gene, designated as isoagglutinin, now known to be located on chromosome 9. The alleles are symbolized as I^A, I^B, *and* I^O *but are often written in a shorthand form omitting the gene symbol I.* In this allelic series, A and B are codominant, and O is recessive to both A and B. Unlike the antibodies produced by B cells in response to foreign antigens, those of the ABO system are secreted continuously. The genotypes, phenotypes, and antigens present are shown in Table 14.3.

Agglutinate
To clump together, as red blood cells do in the presence of a specific antibody.

There is a reciprocal relationship between the antigens on the surface of the blood cells and the antibodies present in the serum of the same individual. If a person has A antigen on erythrocytes, the serum contains antibodies against B, and vice versa. This relationship is summarized in Table 14.4. The reason for the development of antibodies and the inverse

TABLE 14.3 ABO Blood Types			
Blood Group Phenotype	**% in Population**	**Genotype**	**Antigens on Cells**
A	42	*AA, AO*	A
B	10	*BB, BO*	B
AB	4	*AB*	A, B
O	45	*OO*	None

relationship is not clear but is an important consideration in blood transfusions and in tissue and organ transplants. In theory, the transfer of blood is safe under the following conditions:

1. The serum of the recipient does not contain antibodies for the blood cell antigens of the donor.
2. Whether the donor serum contains antibodies for the recipient's blood is irrelevant. The donor's antibodies are so diluted in the blood of the recipient that they have no effect.

Thus, AB individuals have no serum antibodies and can receive blood of any type, and type O individuals have no cell antigens and can give blood to anyone. These relationships are summarized in Table 14.5. In practice, however, cross-matching of donor and recipient is done by typing both the serum and the cells. Unless there is an emergency, blood selected for transfusion is of the same ABO type as the recipient.

TABLE 14.4 ABO Blood Type Antibodies	
Blood Type	**Antibodies in Serum**
A	Antibodies against B
B	Antibodies against A
AB	No antibodies against A or B
O	Antibodies against A and B

TABLE 14.5 Transfusions in ABO System				
Blood Type of Recipient	**Donor Blood Type (Red Cells)**			
	A	**B**	**AB**	**O**
A	+*	–	–	+
B	–	+	–	+
AB	+	+	+	+
O	–	–	–	+

*A plus indicates that a transfusion is acceptable; a minus indicates that it is not acceptable.

Rh Blood Types

The Rh blood group, discovered by Levine in 1939, is also important in transfusions and plays a dominant role in a condition known as **hemolytic disease of the newborn (HDN).** The blood group was named Rh because it was discovered in *rh*esus monkeys. The Rh blood group is genetically complex; for the sake of discussion, we will consider it as a simple two-allele system: Rh^+, dominant to the second allele, Rh^-. Rh^+ individuals make an antigen that is present on the cell surface. Rh^- individuals do not make this antigen. Rh alleles are nonrandomly distributed in human populations. Among white Americans, 85% are Rh^+, and 15% are Rh^-. Among American Indians and Asians, almost 100% of the population is Rh^+.

Hemolytic disease of the newborn (HDN)
A condition that results from Rh incompatibility and is characterized by jaundice, anemia, and an enlarged liver and spleen. Also known as erythroblastosis fetalis.

HDN begins during fetal development and results from an immunologic incompatibility between mother and fetus when the mother is Rh^- and the fetus is Rh^+. If some fetal blood enters the maternal circulation, the Rh^+ antigen present on the surface of the fetal cells stimulates the formation of antibodies by the mother's immune system. Usually this happens during the birth process, so the first pregnancy causes antibody production in the mother, but the first offspring escapes HDN. If a subsequent pregnancy involves an Rh^+ fetus, antibodies present in the Rh^- mother cross the placenta and destroy the red blood cells of the fetus. This causes anemia, jaundice, cerebral damage, mental retardation, and death.

About 10% of all pregnancies in the United States involve HDN, but because of other factors, only about 1% of the cases result in anemia. Before the development of an effective treatment, HDN was one of the 10 leading causes of death to newborns. Now Rh^- mothers are given a commercial Rh antibody preparation called Rhogam immediately after the birth of their first Rh^+ child. The antibodies administered to the mother combine with any fetal Rh^+ cells that are present and destroy them before the maternal immune system can make any antibodies. This treatment must be repeated in all subsequent births involving Rh^+ children.

Because the Rh^- mother can also be sensitized by a miscarriage or an abortion, Rhogam is often administered before the birth of the first child. It is important to note that Rh^- females can also be sensitized to Rh antigen by blood transfusions. If an Rh^- female receives a transfusion of Rh^+ blood before her first pregnancy, she will produce antibodies against the Rh antigen. If her first pregnancy involves an Rh^+ child, the antibodies will cross the placenta and produce HDN in her first pregnancy.

Genetic Diseases of the Immune System

Several genetic diseases are now known to represent defects in one or more parts of the immune system. These diseases represent deficiencies in the production and/or function of specific cell types or in other elements of the immune system. These diseases provide a valuable opportunity to study the function of the immune system in the absence of one of its components. However, not all immunodeficiency syndromes are the result of mutations;

many are produced by external factors, such as infectious organisms, malnutrition, burns, and drug therapies. We will consider some examples of genetic disorders in the immune system, including one in which B cells are absent, one in which T cells are absent or defective, and one in which both T and B cells are absent.

X-linked agammaglobulinemia A rare, sex-linked, recessive trait characterized by the total absence of immunoglobulins and B cells.

X-Linked Agammaglobulinemia (XLA)

The first recognized immunodeficiency disease was described in 1952 by Ogden Bruton, a physician at Walter Reed Army Hospital. He examined a young boy who had suffered at least 20 serious infections in the preceding 5 years. Blood tests indicated that this child had no circulating antibodies. As other patients were discovered, they all had similar characteristics: Affected individuals are usually boys, who are highly susceptible to bacterial infections. Either the B cells were completely absent or immature B cells were unable to mature and produce antibodies. Without functional B cells, there can be no circulating antibodies produced. On the other hand, there are nearly normal levels of T cells. In other words, humoral immunity is absent, but cellular immunity is normal. This X-linked disease usually appears 5 to 6 months after birth, when maternal antibodies transferred to the fetus during pregnancy disappear, and when the infant's B-cell population normally begins to produce antibodies. Patients with this syndrome are highly susceptible to pneumonia and streptococcal infections and pass from one life-threatening infection to another.

Individuals with XLA lack mature B cells, but they have normal populations of B-cell precursors, called pre–B cells, indicating that the defect is in the transition from pre–B cell to B cell. Using recombinant DNA techniques, the XLA gene was mapped to Xq21.3-Xq22 and was identified as an enzyme belonging to a class known as protein kinases. In the cytoplasm, protein kinases help transmit chemical signals that originate outside the cell. It is known that B-cell development depends on signals from other cell types, and it appears that the gene product that is defective in XLA plays a critical role in B-cell maturation. Understanding the role of this kinase in B-cell development may allow future use of somatic gene therapy to treat this disorder.

T-cell immunodeficiency A disease characterized by a complete lack of T-cell immunity but normal B-cell immunity; may be inherited as an autosomal recessive trait.

T-Cell Immunodeficiency Syndromes

A condition associated with a deficiency in T cells that is probably inherited as an autosomal recessive disease has been described in a small number of cases. In affected individuals there is a gradual decline in the number and activity of lymphocytes beginning at the age of 3 months due to a lack of T cells. All patients tested have a deficiency in nucleoside phosphorylase, an enzyme that is involved in purine metabolism. It has been suggested that this metabolic defect plays a role in the onset of this condition, although the mechanism remains obscure. Individuals with T-cell deficiency suffer recurring viral and fungal infections.

A nongenetic form of T-cell deficiency is also known. This condition, known as DiGeorge's syndrome, is caused by a developmental defect that results in the absence of the thymus gland. Because immature T cells migrate from bone marrow to the thymus gland where they mature, affected individuals have no T cells and are lacking all traces of cellular immunity.

A genetic defect in only one type of T cell is also known. In an autosomal recessive condition known as Chédiak-Higashi syndrome, affected individuals are susceptible to recurring infections and a high incidence of cancer. Normal populations of T and B cells are present, but natural killer cells are completely absent. These cells recognize and kill virus-infected and malignant cells. In the absence of this protection, affected individuals develop cancer.

Severe Combined Immunodeficiency Syndrome

Severe combined immunodeficiency syndrome (SCID) is a genetically heterogenous condition with both autosomal and X-linked forms. In this condition, there is an absence of both humoral and cell-mediated immunity; the affected individual has no functional immune system. About half of all cases in the United States are due to the X-linked form (XSCID), 35% are autosomal recessive with an unknown cause, and the remaining 15% are inherited as an autosomal recessive disorder caused by a deficiency of the enzyme adenosine deaminase (ADA).

Severe combined immunodeficiency syndrome (SCID) A disease characterized by a complete lack of ability to mount an immune response; inherited as an X-linked recessive and in another form as an autosomal recessive trait.

In XSCID, no T cells are present; B cells are present but do not produce antibodies. Affected individuals are subject to recurring viral, bacterial, and fungal infections beginning at about age 3 to 5 months. Death is usually caused by pneumonia, but measles and other viruses and fungal infections have also caused deaths. Approximately 200 children are born with this disorder each year (out of approximately 3 million births in the United States), and if treatment is unsuccessful, most die by the age of 7 months.

The longest known survivor with this condition was a patient named David, the "boy in a bubble," who died at the age of 12 years (Figure 14.11). He was the second son afflicted with SCID born in his family and was isolated in a sterile plastic bubble for all but the last 15 days of his life. It was hoped that during isolation an immune system would develop spontaneously and that a natural immunity would be acquired. When this did not occur, a bone marrow transplant from his older sister was attempted. Unfortunately, David died of complications following the transplant. Marrow transplants, however, have been used successfully to treat others with this disease.

In the autosomal recessive form of SCID associated with ADA deficiency, there are no T cells, and if B cells are present, they do not produce antibodies. As with XSCID, there are recurring infections, and death usually occurs by the age of 7 months. ADA-deficient SCID was selected to be the first genetic disorder treated by somatic gene therapy (described in Chapter 8). More recently, an infant who was prenatally diagnosed with this condition underwent a form of somatic gene therapy in an attempt to effect a permanent cure. After birth, stem cells (precursors to the T and B cells) were recovered from blood drawn from the placenta. Using a viral vector, a normal copy of the ADA gene was transferred to the stem cells, and after the cells

FIGURE 14.11 **David, the "boy in a bubble," longest known survivor of SCID.**

I were grown in culture, they were transfused into the infant. It is hoped that the stem cells will migrate from the circulatory system and take up residence in the bone marrow, where they can produce normal T and B cells.

Acquired Immunodeficiency Syndrome (AIDS)

The Centers for Disease Control, a U.S. Government agency located in Atlanta, Georgia, issue a weekly newsletter known as the *Morbidity and Mortality Weekly Report* that is a monitor of the nation's health. In the issue for June 5, 1981, a new listing appeared foreshadowing what has become a deadly cycle of infectious disease. This listing monitors the appearance, spread, and progress of a disease known as **Acquired Immunodeficiency Syndrome,** or **AIDS.**

Acquired Immunodeficiency Syndrome (AIDS)
An infectious disease characterized by the loss of T4 helper lymphocytes, causing an inability to mount an immune response.

The symptoms of AIDS include recurring fevers, weight loss, multiple viral and fungal infections, and rare forms of pneumonia and cancer. The symptoms are similar to those seen in genetic immunodeficiency syndromes, especially the deficits in cell-mediated immunity. In fact, AIDS is accompanied by the depletion of lymphocytes, especially the T4 helper cells.

AIDS is caused by infection with a retrovirus, first identified by French scientists and called **Human Immunodeficiency Virus (HIV)** (Figure 14.12). HIV selectively infects the T4 helper/inducer cells of the immune system. Like other retroviruses, HIV has RNA as its genetic material. After infection, the RNA is copied into DNA by a viral enzyme, reverse transcriptase. The newly copied DNA moves into the nucleus and can integrate itself into human chromosomes. In this stage the virus can remain inactive for several years. When the virus is activated, the viral genes are transcribed by the host T4 cell, viral proteins are synthesized, and new viruses are assembled. The replicated viruses bud off from the cell membrane, causing the host T4 cell to collapse and die. HIV-infected cells that are not producing virus can be

Human Immunodeficiency Virus (HIV)
A retrovirus that infects T4 lymphocytes, precipitating a loss in the ability to mount an immune response.

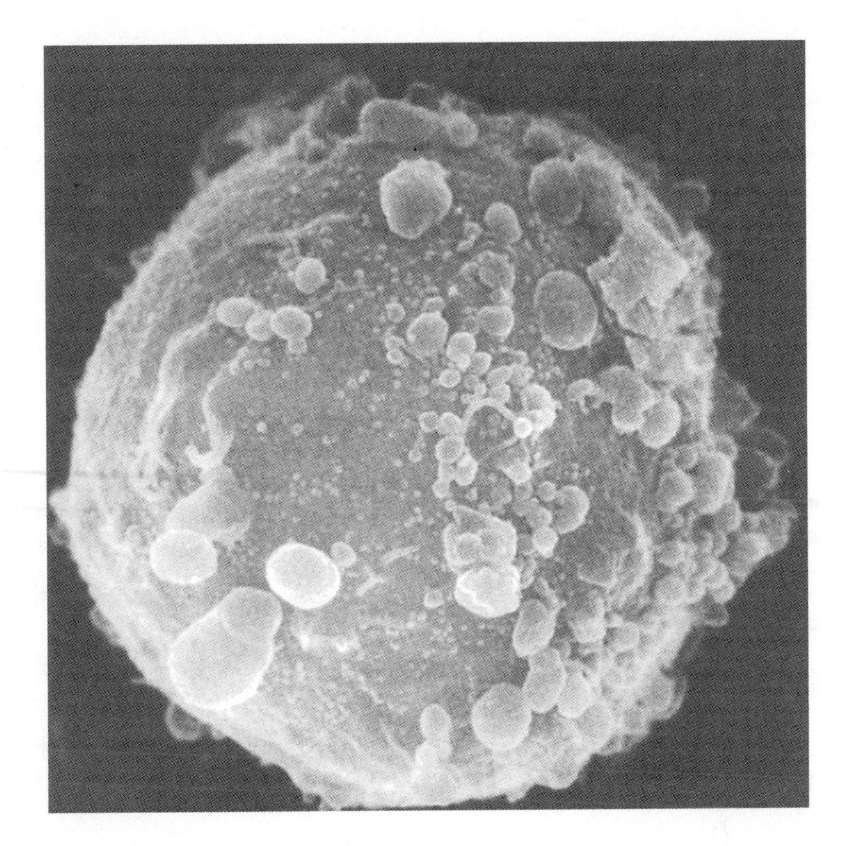

FIGURE 14.12 Electron micrograph of HIV, the virus associated with AIDS, in the process of invading a T4 lymphocyte.

killed by the immune system if they display viral proteins on their surfaces. In addition, infected T cells fuse to form multicellular clusters that further deplete the T-cell population.

In the immune system there are normally two to three times as many T4 helper cells as there are T8 suppressor cells. In AIDS the situation is reversed because the supply of T4 cells is depleted. In fact, in advanced cases, the level of T4 cells can be undetectably low. This may be partly caused by masking of the biochemical identity of the T4 cells by HIV. Because the T4 cells act as the "on" switch for the immune system (see Figure 14.7), infected individuals cannot mount an immune response. The T4 cell count and functional capacity of the remaining cells is an accurate measure of the disease's progress. When the T4 cell count falls below 400 cells/mm^3 of blood, (normal is 800 to 1000), the immune system begins to fail. This stage is usually accompanied by fungal or viral infections of the mucous membranes and skin. In the last stages of the disease, the T4 cell count falls below 200. This collapse of the immune system leaves the body vulnerable to the development of fatal infections, such as pneumocystic pneumonia caused by a protozoan, *Pneumocystis carinii*, and the development of cancers such as Karposi's sarcoma, an otherwise rare malignancy of blood vessels and internal organs.

Initially it was thought that 30% to 40% of HIV-infected individuals would develop AIDS within 5 years of infection. More recent work indicates that almost all infected individuals will develop AIDS and die prematurely.

Infected individuals form a pool from which the virus is transmitted to others by sexual contact (including artificial insemination), transfusion of blood products, or by sharing unsterilized needles used for drug injection. The virus also can cross the placenta from an infected mother to the fetus or be transmitted by breast feeding.

HIV-infected individuals can be detected because after infection the immune system produces a small amount of antibody against the viral protein coat. This amount is ineffective against the infection, but it can be identified by a blood test. Tests on stored blood serum gathered from all parts of the world indicate that HIV has only recently infected the human population. Serum gathered in Africa in the 1950s provides the oldest samples to show the presence of antibody to the AIDS virus. HIV appears to have spread through Central Africa in the early 1970s and appeared in Haiti in the late 1970s. From here it may have spread to the United States. The origin of the virus is still unknown. Currently it is thought that a related retrovirus known as•SIV, found in African green monkeys, was transmitted to humans through biting, and after a series of mutations through intermediate types, became HIV. A large research effort is directed at the development of a vaccine for AIDS. Progress has been difficult because HIV exhibits a great deal of genetic variability that results in significant differences in viral proteins that form the outer shell of the virus. This means that a vaccine against one strain of the virus would be ineffective against other strains. Unless a way is found to circumvent this genetic variability, development of an effective vaccine will be difficult.

AUTOIMMUNE DISEASES

Immune tolerance
A condition of nonreactivity toward molecules or cells of the body that might be expected to induce an immune response.

Normally a state of truce known as **immune tolerance** exists between the body and the immune system. Tolerance toward the body's own proteins develops soon after birth, and in this process the immune system learns to distinguish self from nonself. The mechanism by which this learning takes place is not clear, but once this distinction has been made, the immune system does not react against the cells and tissues of the body. When immune tolerance breaks down, the result is an autoimmune response. The immune system literally turns against the body, and cytotoxic T cells begin to proliferate and to destroy cells in vital organs. This process can be the result of an injury, an infection, or a genetic predisposition.

A connective tissue disease known as systemic lupus erythematosus (SLE) is caused by a malfunctioning immune system. In this condition B cells proliferate and manufacture large quantities of antibodies, including those directed against host antigens. These antigens target cells in the kidneys, heart, and nervous system for destruction. The result is severe damage to the affected organs. The underlying mechanisms in SLE is unknown but may involve a disturbance in the balance between B cells that produce antibodies and the suppressor T cells that control the activity of the B cells. There is a genetic component to SLE associated with the HLA system. HLA alleles DR2 and DR3 are found in about 50% of SLE patients but only in about 25% of the normal population.

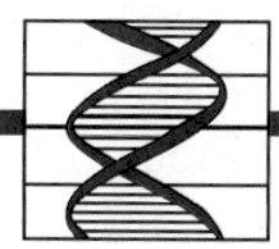

Concepts & Controversies

Gene Therapy for Diabetes by Transplantation

The use of insulin to treat diabetes, developed in the 1920s, is an early example of gene therapy by product replacement. This form of therapy has been successful in controlling diabetes but is unable to stop the gradual onset of kidney failure, blindness, heart disease, and early death. While other methods of gene therapy now under development rely on the use of cloned copies of normal genes, a new treatment for diabetes has taken another approach: transplantation of cells in the pancreas that make insulin.

Insulin is produced in clusters of cells known as the islets of Langerhans, embedded in the pancreas. About 1% of the pancreas is composed of islet cells. Over a period of several years, Dr. Kevin Lafferty at the University of Colorado at Denver and Dr. Paul Lacy at Washington University in St. Louis have developed methods for isolating intact and functional islet cells from the pancreas. Experiments in animals have shown that transplantation of islet cells is able to reverse experimentally induced diabetes.

Clinical trials have started at both institutions, transplanting islet cells into type I diabetes. Currently, islet cells are being transplanted into diabetics who have received a kidney transplant and are taking drugs such as cyclosporine to help prevent rejection of the kidney. The antirejection drugs also help to prevent rejection of the islet cells. If these trials are successful, it is hoped that islet cells can be injected through the abdominal wall into the spleen or abdominal fat layers to bypass the need for major surgery. Coupled with HLA testing to identify those predisposed to type I diabetes, this method of transplantation offers hope that the scourge of diabetes can be brought under control.

In the last few years it has become apparent that one form of diabetes is an autoimmune disease. Insulin-dependent, or type I, diabetes is the most serious form of the disease and usually starts in childhood. Some 2 million Americans have type I diabetes and must take daily insulin injections. Insulin is a hormone that regulates the metabolism of sugars and their conversion to energy. When insulin is absent, high sugar levels build up in the blood, causing the initial symptoms of thirst and excessive urine production. As the disease progresses, serious consequences result, including weight loss, kidney failure, blindness, limb amputation, and heart disease. Many diabetics die within the first 10 years, and about half die within 40 years of onset of the disease. Type I diabetes develops when the immune system attacks and kills the cells in the pancreas that produce insulin. This destructive process takes place over a period of years and results in a partial or complete loss of the ability to produce insulin (see Concepts & Controversies, above). HLA alleles DR-3 and DR-4 are associated with a high risk of this form of diabetes. Those with HLA DR-3 or DR-4 have a 15-fold increased risk of diabetes. If both DR-3 and DR-4 are present, the risk is increased by 30-fold. Other autoimmune diseases are listed in Table 14.6.

Type II diabetes, or adult-onset diabetes, is a milder form of the disease that is not caused by an autoimmune response. It affects 10 million Americans and is normally controlled by oral medications or diet. It is estimated that an additional 10 million to 12 million Americans are affected by type II diabetes but are undiagnosed.

Research into the workings of the immune system is still in its infancy. Recent breakthroughs include the development of monoclonal antibodies and the identification of lymphokines as the molecular means of communication within the immune system.

TABLE 14.6 Some Autoimmune Diseases

Some Autoimmune Diseases
Addison disease
Autoimmune hemolytic anemia
Diabetes mellitus—insulin-dependent
Graves' disease
Membranous glomerulonephritis
Multiple sclerosis
Myasthenia gravis
Polymyositis
Rheumatoid arthritis
Scleroderma
Sjögren's syndrome
Systemic lupus erythematosus

SUMMARY

1. Immunity to disease has been known since ancient times, but it is only in the last decade that strides have been made in understanding the body's ingenious and complex system of self-defense. The immune system is somewhat unusual in that it is not identified with any major organ system. It is composed of a small number of cell types located throughout the body. Each cell type performs a number of functions and communicates through chemical messages known as lymphokines with other cells in the immune system and with the rest of the body.

2. The critical event in the immune response is the distinction between self and nonself. Assisting in this task are recognition molecules known as antibodies. These proteins are present in literally millions of different forms that recognize and bind to cells and molecules identified as nonself.

3. Much is now known about the molecular organization of antibody genes. The diverse antibody types are produced by shuffling and reshuffling combinations of gene segments.

4. The genetics of organ and tissue transplantation are controlled by a gene complex located on chromosome 6 known as the HLA complex. The four major genes in this group have such a large number of alleles that the combination of HLA alleles carried by an individual constitutes a form of genetic signature. Matching of HLA alleles is critical in successful organ transplants. Certain HLA alleles are associated with specific diseases and can be used for diagnosis and to predict what disease one might contract.

5. ABO antigens interact with the immune system to produce four separate blood types. Transfusions are possible only when there is compatibility between donor and recipient. Antigen incompatibility in the Rh system may produce hemolytic disease of the newborn, in which the fetus is regarded as an invader by the maternal immune system.

6. Genetic control of the immune system is illustrated by the discovery of a number of traits that affect its development and function. These heritable diseases can affect a single cell type, a single subsystem, or the entire immune system.

7. In addition to genetic immune diseases, a contagious and highly fatal form of immunodeficiency known as AIDS is becoming widespread. AIDS is associated with infection by a retrovirus, HIV.

8. Autoimmune diseases represent a case of mistaken identity, where self is confused with nonself. The initiation of autoimmune disease may be triggered by an infection but results in destruction of cells and tissues of the body.

Questions and Problems

1. Define attenuation, and discuss its importance in combating immunological diseases.
2. Identify the components of cellular immunity, and define their roles in the immune response.
3. What type of cell can signal T cells to assist in an immune response, and what chemical messenger does it use?
4. Antibodies have long been a subject for study because of the staggering variety of antigens to which they are capable of responding. What other element of the immune system is equally adept at antigen recognition?
5. How many polypeptide chains are present in a fully functional antibody molecule belonging to the IgG class? the IgM class?
6. Describe the genetic basis of antibody diversity.
7. Identify three chemical messengers involved in the immune response, and discuss their roles.
8. In the human HLA system there are 23 HLA-A alleles, 47 for HLA-B, 8 for HLA-C, 14 for HLA-DR, 3 for HLA-DQ, and 6 for HLA-DP. How many different human HLA genotypes are possible?
9. What mode of inheritance has been observed for the HLA system in humans?
10. The following data were presented to a court during a paternity suit: (1) the infant is a universal donor for blood transfusions; (2) the mother bears antibodies against the B antigen only; (3) the alleged father is a universal recipient in blood transfusions. Can you identify the ABO genotypes of the three individuals? Can the court draw any conclusions?
11. An individual has an immunodeficiency that prevents helper T cells from recognizing the surface antigens presented by macrophages. As a result, the helper T cells are not activated, and they, in turn, fail to activate the appropriate B cells. At this point, is it certain that the viral infection will continue unchecked?
12. A couple have the following HLA genotypes:

    Male A12, B36, C7, DR14, DQ2, DP5/A9, B27, C3, DR14, DQ3, DP1
    Female A8, B15, C2, DR5, DQ2, DP3/A2, B20, C3, DR8, DQ1, DP3

    What is the probability that an offspring will suffer from ankylosing spondylitis if:
    a. The child is male and has HLA-A9?
    b. The child is female and has HLA-A9?
    c. The child is male and has HLA-DP1?
    d. The child is male and has HLA-C7?
13. Graft versus host disease has no counterpart in blood transfusion reactions. Explain why.
14. Define lymphokines.
15. Assume the Rh character is controlled by a single gene having alleles that show complete dominance relationships at the phenotypic level. An Rh^{+} father and an Rh^{-} mother have 8 boys and 8 girls, all Rh^{+}.
    a. What are the Rh genotypes of the parents?
    b. Should they have been concerned about hemolytic disease of the newborn?

For Further Reading

Arnett, F. 1986. HLA genes and predisposition to rheumatic diseases. *Hosp. Pract.* **20:** 89–100.

Atkinson, K. 1990. Chronic graft-versus-host disease. *Bone Marrow Transplant.* **5:** 69–82.

Bosma, M. J. 1989. The SCID mutation: Occurrence and effect. *Curr. Topics Microbiol. Immunol.* **152:** 3–9.

Britton, C. 1993. HIV infection. Neurol. Clin. **11:** 605–624.

Buckley, R. 1992. Immunodeficiency diseases. JAMA **268:** 2797–2806.

Burrows, P. and Cooper, M. 1993. B-cell development in man. Curr. Opin. Immunol. **5:** 201–206.

Buse, J. and Eisenbarth, G. 1985. Autoimmune endocrine disease. *Vitam. Horm.* **42:** 253–315.

Cooper, D. N., and Clayton, J. F. 1988. DNA polymorphisms and the study of disease associations. *Hum. Genet.* **78:** 229–312.

Hood, L., Dronenberg, M., and Hunkapiller, T. 1985. T cell antigen receptors and the immunoglobulin supergene family. *Cell* **40:** 25–229.

Hood, L., Weissman, I., and Wood, W. 1978. *Immunology.* Menlo Park, CA: Benjamin-Cummings.

Hors, J., and Gony, J. 1986. HLA and disease. *Adv. Nephrol.* **15:** 329–351.

Iseki, M. and Heiner, D. 1993. Immunodeficiency disorders. Pediatr. Rev. **14:** 226–236.

Jerne, N. 1973. The immune system. *Sci. Am.* **229** (July): 52–60.

Jerne, N. 1985. The generative grammar of the immune system. *Science* **229:** 1057–1059.

Kindt, T., and Capra, J. 1984. *The Antibody Enigma.* New York: Plenum.

Kotzin, B., Leung, D., Kappler, J., and Marrack, P. 1993. Superantigens and their potential role in human disease. Adv. Immunol. **54:** 99–166.

Leder, P. 1982. The genetics of antibody diversity. *Sci. Am.* **246** (May): 102–115.

McDivitt, H. 1986. The molecular basis of autoimmunity. *Clin. Res.* **34:** 163–175.

Puck, J. 1993. X-linked immunodeficiencies. Adv. Hum. Genet. **21:** 107–144.

Tonegawa, S. 1985. The molecules of the immune system. *Sci. Am.* **253** (October): 123–131.

Vetri, D., Vorechovsky, I., Sideras, P., Holland, J., Davies, A., Flinter, F., Hammarstrom, L., Kinnon, C., *et al.*, 1993. The gene involved in X-linked agammaglobulinaemia is a member of the *src* family of protein-tyrosine kinases. Nature **361:** 226–233.

Vladutiu, A. 1993. The severe combined immunodeficient (SCID) mouse as a model for the study of autoimmune diseases. Clin. Exp. Immunol. **93:** 1–8.

Yamamoto, F., Clausen, H., White, T., Marken, J., and Hakomori, S. 1990. Molecular genetic basis of the histo-blood group ABO system. *Nature* **345:** 229–233.

CHAPTER 15

POLYGENIC INHERITANCE

CHAPTER OUTLINE

In 1713, a king ascended the throne of Prussia to begin one of the largest military buildups of the 18th century. In the space of 20 years, King Frederick William I (a cousin of King George II of England), ruler of fewer than 2 million citizens, enlarged his army from around 38,000 men to just under 100,000 troops. Compare this with the neighboring kingdom of Austria, with 20 million inhabitants and an army of just under 100,000 men, and you will understand why Frederick William was regarded as a military monomaniac. The crowning glory of this military machine was the King's personal guard, known as the Potsdam Guards, or Potsdam Grenadiers. This unit was composed of the tallest men obtainable. Frederick William was obsessed with having giants in this guard, and his recruiters used bribery, kidnapping, and smuggling to fill the ranks of this unit. It is said that members of the guard could lock arms while marching on either side of the King's carriage. Many members were close to

7 ft tall. The height of his guards may seem more impressive when you realize that the average height of the population was probably around 5 ft 5 in.

Frederick William was also rather miserly, and because this recruiting was costing him millions of thalers, he decided it would be more economical to simply breed giants to serve in his elite unit. To accomplish this, he ordered that every tall man in the kingdom was to marry a tall, robust woman, with the expectation that the offspring would all be giants. Unfortunately this idea was a frustrating failure. Not only was it slow, but most of the children were shorter than their parents. While continuing this breeding program, Frederick William reverted to kidnapping and bounties, and he also let it be known that the best way for foreign governments to gain his favor was to send giants to be members of his Guard. This experiment in human breeding ended when Frederick William's son, Frederick the Great, disbanded the Potsdam Guards, allowing the foreign members to return to their homes.

Polygenes and Multifactorial Traits

What exactly went wrong with Frederick William's experiment in human genetics? The idea sounds good, and we have the advantage of knowing that when Mendel intercrossed true-breeding tall pea plants, the offspring were all tall. Even when heterozygous tall pea plants are crossed, three-fourths of the offspring are tall.

The problem is that height in pea plants is controlled by a single gene pair, while in humans, height is determined by several gene pairs and is said to be **polygenic.** The tall and short phenotypes in pea plants are easily classified into two distinct categories, with no overlap, and are examples of **discontinuous variation.** In measuring height in humans, it is difficult to classify individuals into one or two categories. Instead, height in humans is an example of a phenotype that shows **continuous variation.** Unlike Mendel's pea plants, people are not either 18 in or 84 in tall; in other words, the phenotypic classes are not clear-cut. Traits showing continuous variation that are controlled by two or more different genes are **polygenic traits.**

Understanding the distinction between discontinuous and continuous traits was an important step in genetics and is based on acceptance of the idea that there is interaction among genes and between genes and the environment. All genes can interact with the environment, and they differ only in the degree of this interaction. **Multifactorial traits** are those that involve two or more genes and strong interaction with the environment. In this chapter we will examine traits controlled by genes at two or more loci. We will also consider how nongenetic factors such as the environment affect gene expression. The degree of environmental or genetic effects on a trait can be estimated by measuring *heritability.* We will consider this concept and the use of twins as a means of measuring the heritability of a trait. In the last part of the chapter we will examine a number of human polygenic traits that have been the subject of political and social controversy.

Polygenic
A phenotype that is dependent upon the action of a number of genes.

Discontinuous variation
Phenotypes that fall into two or more distinct, nonoverlapping classes.

Continuous variation
A distribution of phenotypic characters that is distributed from one extreme to another in an overlapping, or continuous, fashion.

Polygenic trait
A phenotype resulting from the action of two or more genes.

Multifactorial traits
Traits that result from the interaction of one or more environmental factors and two or more genes.

Continuous Variation

Mendel was not the only scientist in the late 19th century experimenting with the inheritance of traits. In one series of experiments, Josef Kölreuter crossed tall and dwarf tobacco plants. The F_1 plants were all intermediate in height to the parents. When self-crossed, the F_1 produced an F_2 that contained plants of many different heights. Some of the F_2 were as tall or short as the parents, but most of the F_2 were intermediate in height when compared with the parents (Figure 15.1). Mendel's results with pea plants, however, produced evidence of discontinuous variation (Figure 15.1).

Shortly after the turn of the century, traits in a variety of organisms were found to show continuous variation in phenotype. In each of these cases, the offspring had a phenotype that seemed to be a blend of the parental traits.

In the first decade of this century, geneticists debated whether the phenomenon of continuous variation could be reconciled with the inherit-

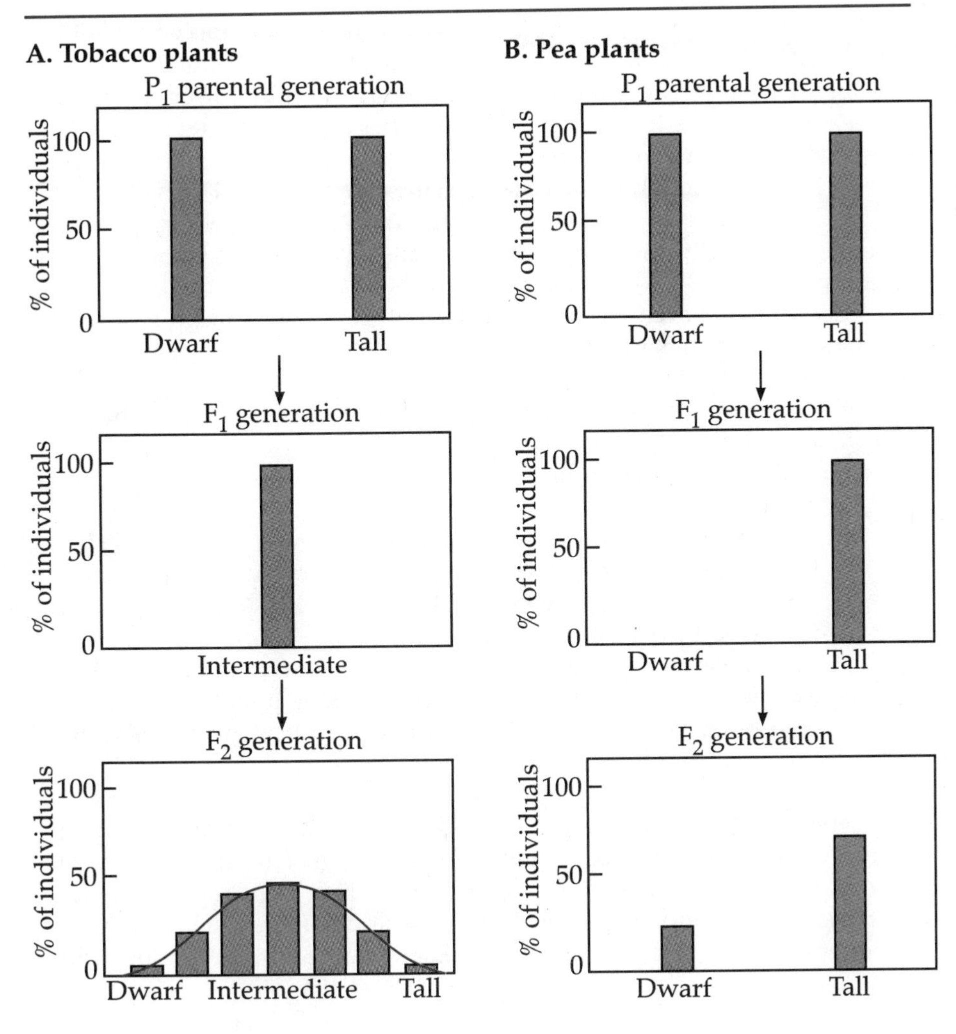

FIGURE 15.1 Comparison of continuous and discontinuous traits. (A) Histograms showing the percentage of plants of different height in crosses between tall and short tobacco plants carried to the F_2 generation. (B) Histograms showing the percentage of plants of different height in crosses between tall and short pea plants carried to the F_2 generation.

ance of Mendelian factors or whether this apparent blending of traits signaled the existence of another mechanism of inheritance. This argument is of importance to human genetics because many human traits and many genetic disorders show continuous variation. In the years immediately after the rediscovery of Mendel's work only a few single-gene markers were known in humans, and interest centered on discovering whether "social" traits such as alcoholism, feeblemindedness, and criminal behavior were inherited. Since these traits were not inherited in the ratios seen for traits in pea plants, many geneticists discounted the importance of Mendelian inheritance in humans. In fact, the biomathematician Karl Pearson is reported to have said that "there is no truth in Mendelism at all." Unfortunately, the work of Mendel and Garrod had little or no effect on human genetics in the first 3 decades of this century because interest in the social aspects of genetics predominated.

Between 1910 and 1930, the controversy over continuous variation was resolved. Experimental work with corn and tobacco plants conclusively demonstrated that continuous variation could be explained by Mendelian factors. These findings revealed that traits determined by a number of alleles, each of which makes a small contribution to the phenotype, will exhibit a continuous distribution of phenotypes in the F_2 generation. In this distribution, known as a normal curve, a small number of organisms are represented at the extreme range of phenotypes (*e.g.*, very short or very tall). Most organisms in the population exhibit phenotypes between the extremes; their distribution follows what statisticians call a "normal curve" (Figure 15.2). This pattern of inheritance, known as polygenic or quantitative inheritance, is additive, since each allele adds an incremental amount to the phenotype.

Continuous variation produced by polygenic inheritance has several distinguishing characteristics:

- Traits are usually quantified by measurement rather than by counting.
- Traits may be controlled by the effect of multiple alleles of one or a few major genes.
- Traits may be controlled by two or more genes, each of which contributes in an additive way to the phenotype. The effect of individual additive alleles may be small, with some alleles making no contribution.
- The phenotypes of polygenic and multifactorial traits vary in expression. This variation is produced by gene interaction and by environmental factors and is best analyzed in populations rather than in individuals.

Polygenic inheritance is an important concept in human genetics. Traits such as height, weight, skin color, and intelligence may be under polygenic control. In addition, congenital malformations such as neural tube defects, cleft palate, and club foot as well as genetic disorders such as diabetes, hypertension, and behavioral disorders are polygenic, or multifactorial, traits.

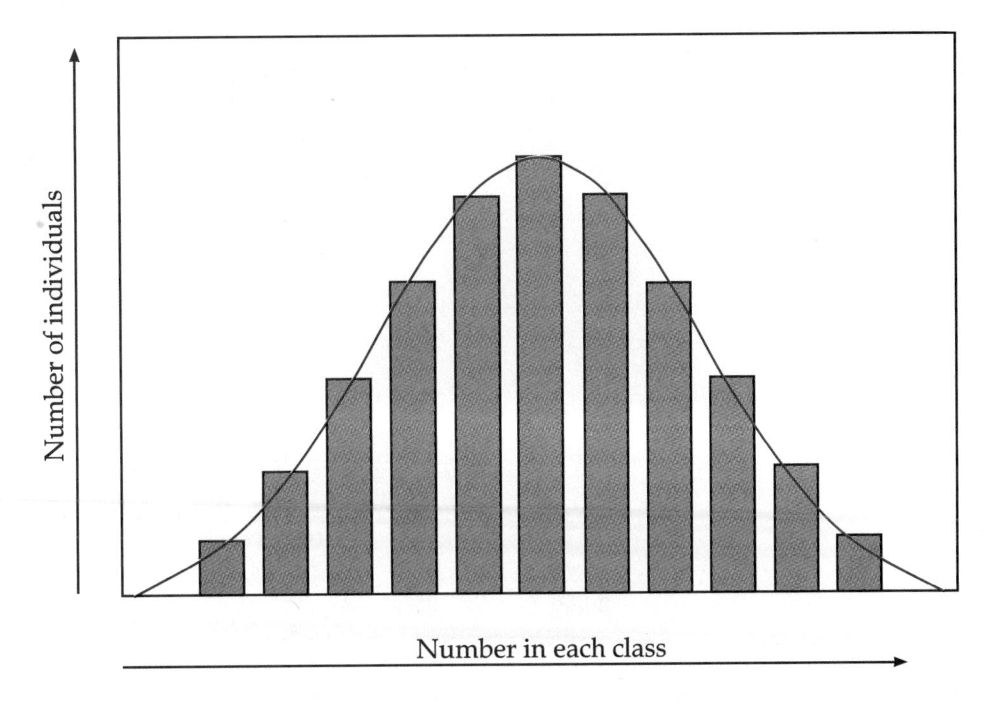

FIGURE 15.2 A normal curve is characterized by few individuals at the extremes, with clustering around the average or mean value of the distribution.

The normal distribution of phenotypes and F_2 ratios in traits involving two, three, and four genes is shown in Figure 15.3. As the number of genes controlling the trait increases and phenotypic variation due to the environment increases, the results generate a continuous distribution. As more loci become involved in a trait, the number of phenotypic classes increases (Figure 15.3). As the number of classes increases, there is less phenotypic difference between each class. This means that there is a greater chance for environmental factors to override the small difference between classes. For example, exposure to sunlight can alter skin color and obscure phenotypic differences.

Normally, the results of phenotypic measurements in polygenic inheritance are expressed as a frequency diagram. Figure 15.4 on page 407 shows a frequency distribution for height in humans.

Regression to the Mean

One of the distinguishing characteristics of polygenic traits is that most of the offspring of extreme phenotypes (tall × tall, for example) show a phenotype intermediate to each of the extremes. This phenomenon is known as **regression to the mean.**

Regression to the mean
In a polygenic system, the tendency of offspring of parents with extreme differences in phenotype to exhibit a phenotype that is the average of the two parental phenotypes.

As an example, let us consider King Frederick William I's attempt to breed giants for his elite guard unit. For the sake of simplicity, we will assume

FIGURE 15.3 The distribution of F_2 phenotypes obtained by crossing F_1 heterozygotes in polygenic systems with two to four gene pairs, each with two alleles. As the number of phenotypic classes increases, the distribution of phenotypes approaches a normal curve.

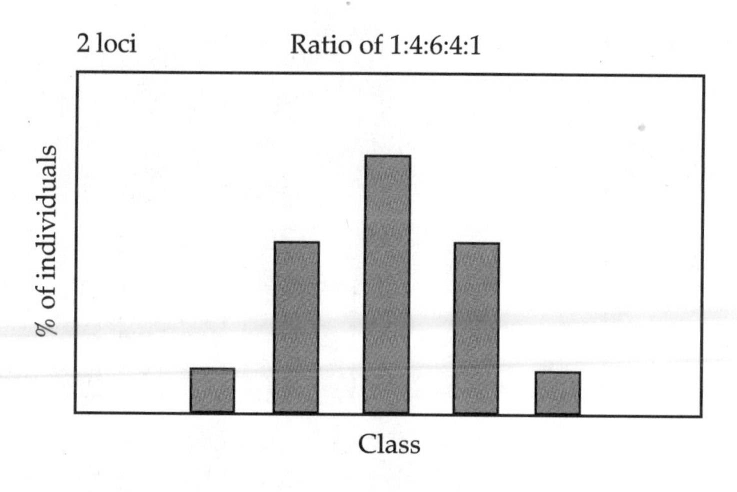

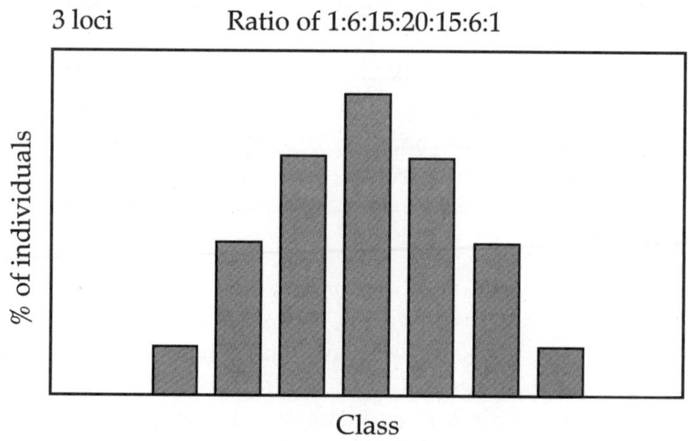

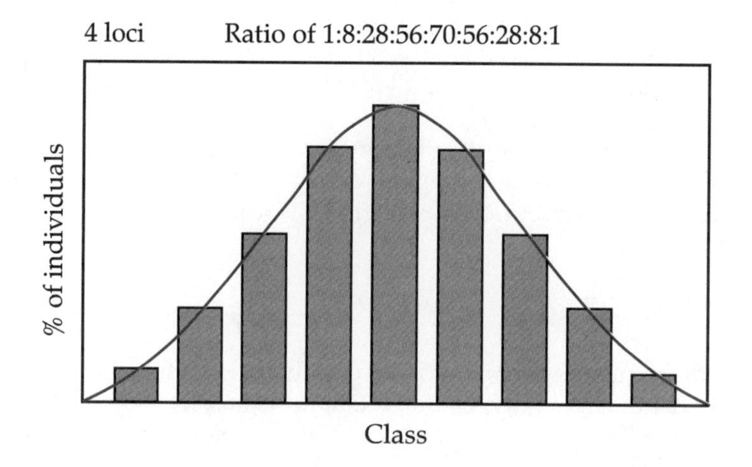

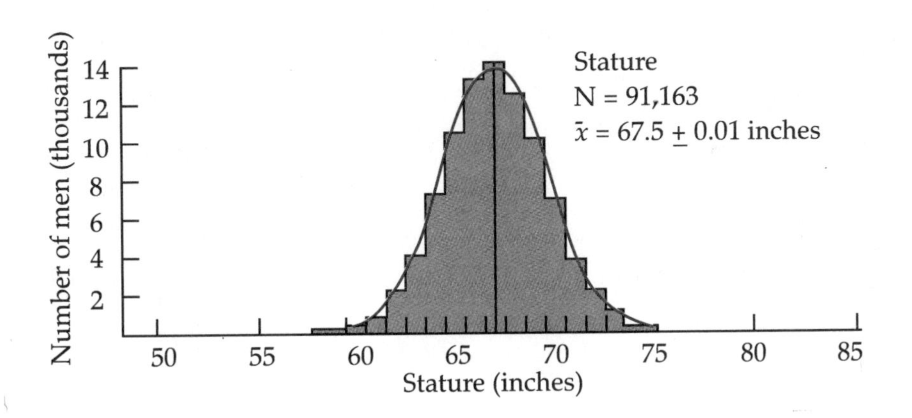

FIGURE 15.4 Stature in English males called for military service in 1939. The mean is indicated by a solid vertical line. A bell-shaped distribution, calculated from the data, is shown as a dotted line. (From: Harrison, G. A., Weiner, J. S., Tanner, J. M., and Barnicot, N. A. *Human Biology.* Oxford: The Clarendon Press, 1964.)

that complete dominance is in effect. (In reality, incomplete dominance is probably involved in controlling height in humans.) Let us further assume that his breeding program used individuals at least 6 ft in height. In this *simplified* example, let us assume that the dominant alleles A, B, and C in the homozygous dominant *or* heterozygous condition each add 4 in above a base height of 6 ft. The recessive alleles *a, b,* and *c* add nothing above the base height. Thus, an individual with the genotype *aabbcc* is 6 ft tall (representing the short extreme), and an individual with the genotype *AaBbCc* or *AABBCC* is 7 ft tall (representing the tall extreme). Suppose that a 7-ft member of the guard carrying the genotype *AaBbCc* is mated to a 6-ft 8-in woman with the genotype *AaBbcc*. The outcome of this cross is diagrammed in Figure 15.5. In this case, there are eight different paternal gametes and four different maternal gametes possible. The result is 32 fertilization combinations (8 × 4) producing four phenotypic classes. The phenotypes and ratio of offspring are diagrammed in Figure 15.6. Approximately 72% of the offspring will be 6 ft 8 in or shorter, and the chance of having a 7-ft-tall child is only about 28%. In the population we are considering, one extreme is represented by the minimum height of 6 ft, and the other extreme is the giant at 7 ft. The height of most of offspring tends toward the average or mean height (6 ft 6 in) between the two extremes. Because *A, B,* and *C* are dominant, the distribution is skewed toward the dominant extreme, but nonetheless the average height of the offspring will tend toward the mean of 6 ft 6 in. In succeeding generations, further regression to the mean will occur. As it turns out, many of the members of Frederick William's guard unit were tall because of endocrine malfunctions and were not genotypically equipped to produce tall offspring under any circumstances.

Regression to the mean is brought about not only by dominance and additive effects but also by gene interaction and environmental effects. If these other factors were included in our crosses, and if the genotypes included individuals of average height (say, 5 ft 4 in), the regression toward the mean would be even more pronounced. Genetically the way to produce the desired offspring would be to inbreed the giants for many generations (say 200 years or so) to produce homozygous *AABBCC* individuals to be used as parents for giants.

FIGURE 15.5 (A) Gametes produced by a 7-ft male parent and a 6-ft 8-in female parent. (B) Punnett square showing the 32 genotypic combinations and four phenotypic classes that result. The genotypes resulting in the tallest phenotypes are shaded.

A.

P_1 Parents:	7-ft male AaBbCc	x	6-ft 8-in female AaBbcc
Types of gametes:	ABC, ABc, AbC, Abc, aBC, aBc, abC, abc		ABc, Abc, aBc, abc

B.

♂ AaBbCc \ ♀ AaBbcc	ABc	Abc	aBc	abc
ABC	AABBCc	AABbCc	AaBBCc	AaBbCc
ABc	AABBcc	AABbcc	AaBBcc	AaBbcc
AbC	AABbCc	AAbbCc	AaBbCc	AabbCc
Abc	AABbcc	AAbbcc	AaBbcc	Aabbcc
aBC	AaBBCc	AaBbCc	aaBBCc	aaBbCc
aBc	AaBBcc	AaBbcc	aaBBcc	aaBbcc
abC	AaBbCc	AabbCc	aaBbCc	aabbCc
abc	AaBbcc	Aabbcc	aaBbcc	aabbcc

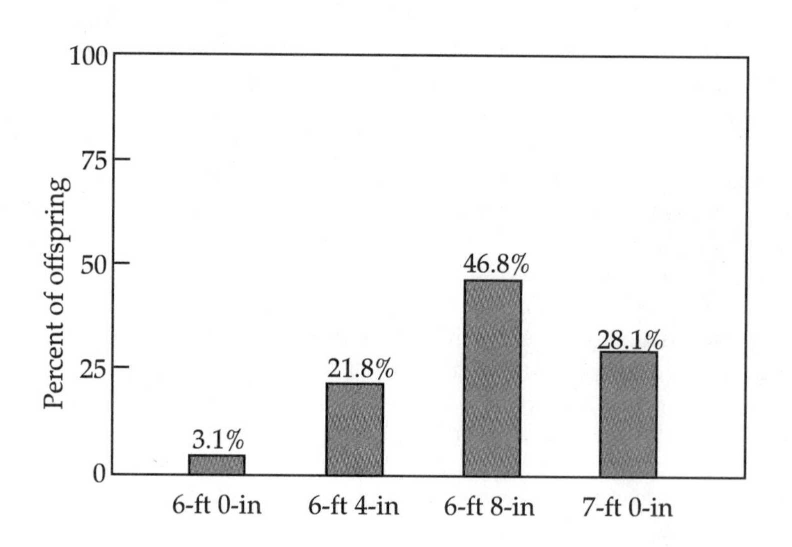

FIGURE 15.6 Frequency distribution of height classes from crosses in Figure 15.5. Height of the offspring shows regression to the mean.

Genes and the Environment

We have looked at the interaction between genes and the environment in previous chapters but always in the context of single genes. The term "expressivity" is used to describe variations in gene expression, which may be due in part to interaction with the environment.

In considering the interaction of genes and the environment with respect to polygenes, it is useful to review some essential concepts. The genotype represents the genetic constitution of an individual; it is fixed at the moment of fertilization and, barring mutation, is unchanging. The phenotype is the sum of the observable characteristics of an organism. The phenotype is variable and undergoes continuous change throughout the life of the organism. The environment of a gene is composed of genetic factors, including all other genes in the genotype, their effects and interactions, and all nongenetic factors, whether physical or social, that have the capacity to interact with the genotype. At any given moment in its life, an organism represents the sum of all the interactions between genes and environment that have occurred since fertilization.

In assessing the interaction between the genotype and the environment, as in all science, you have to ask the right question (see Concepts & Controversies p. 410). Suppose the question is posed as "How much of a given trait in an individual is caused by heredity, and how much by environment?" Remember that the phenotype is a product of interaction between the genotype and the environment. Because each individual has a unique genotype and has been exposed to a unique set of environment conditions, it is impossible to quantitatively evaluate the phenotype's genetic and environmental components. Thus, for a given individual, the question as posed cannot be answered.

Threshold Effects

Although the degree of interaction between a genotype and the environment can be difficult to estimate, family studies indicate that such interactions do occur. In some multifactorial traits, there does not seem to be a continuous distribution of phenotypes; individuals are either affected or not. Congenital birth defects such as club foot or cleft palate are examples of traits that are distributed in a discontinuous fashion but are, in fact, multifactorial. A model has been used to explain the expression of such traits. In this model, liability for a genetic disorder is distributed in a normal curve among individuals (Figure 15.7). This liability is caused by a number of genes, each contributing to the liability in an additive fashion. Those individuals with a liability above a threshold will develop the genetic disorder if exposed to certain environmental conditions (Figure 15.7). In other words, environmental conditions are most likely to have the greatest impact on genetically predisposed individuals.

The threshold model is useful in explaining the occurrence of certain disorders and congenital malformations. Evidence for such a threshold in any given disorder is indirect and comes mainly from family studies. The

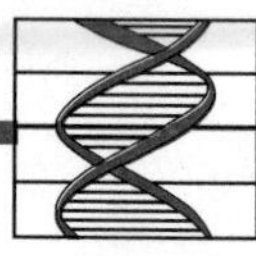

Concepts & Controversies

Baseball Genes and Baseball Environment

Some traits are certainly familial; the degree to which they may be genetically determined is often a matter of question. In the latter half of the 19th century, Sir Francis Galton began tracing traits through families to determine whether there was a clear-cut pattern of inheritance. As a modern example of the type of trait he studied, consider the case of the Boone family and baseball talent. Ray Boone played 13 years in the major leagues and is now a scout for the Boston Red Sox. His son Bob Boone has played 18 years in the majors and is now the manager for the Tacoma Tigers, the AAA team for the Oakland Athletics. Bob's son Bret was drafted by the Seattle Mariners. Bret's brother Aaron is currently playing college ball at USC.

It would seem that major league baseball talent runs in the Boone family. To assess the degree of interaction between the genotype (say, the ability to play baseball) and environmental factors (such as growing up in a family of major league ball players), geneticists would examine the phenotypes and the amount of phenotypic variation (from high baseball ability to no baseball ability) present in a population rather than just the individuals in one family. The results of such a study would be interpreted in terms of the concept of heritability, which provides a description of the genetic influences present among a population living in a particular set of environmental conditions. It is important to remember that heritability does not imply a fixed genetic contribution to a trait in the way that we can determine the genetic contribution to cystic fibrosis or sickle cell anemia but, rather, describes how much of the phenotypic variation in baseball ability is caused by genetic factors.

Boones by the Numbers

BRET BOONE

Born: 4-6-69. **Home:** Villa Park, Calif. **Ht.:** 5-10. **Wt.:** 170. **Bats/Throws:** Right.
2B-SS, Southern California 1988-1990

MAJOR LEAGUES	AVG	AB	R	H	2B	3B	HR	RBI	BB	SO	SB
Two years	.205	180	20	37	6	0	6	2	82	78	17

BOB BOONE

Born: 11-19-47. **Home:** Villa Park, Calif. **Ht.:** 6-2. **Wt.:** 205. **Bats/Throws:** Right.
3B-P, Stanford 1966-69; Drafted by Philadelphia Phillies, 6th round, June 1969
C, Philadelphia 1972-81, California 1982-88, Kansas City 1989-1990

MAJOR LEAGUES	AVG	AB	R	H	2B	3B	HR	RBI	BB	SO	SB
18 years	.254	7245	679	1838	303	26	105	826	663	608	38

RAY BOONE

Born: 7-27-23. **Home:** El Cajon, Calif. **Ht.:** 6-0. **Wt.:** 172. **Batted/Threw:** Right.
SS, Cleveland 1948-53; 3B, Detroit 1954-58; 1B, Milwaukee, 1960.

MAJOR LEAGUES	AVG	AB	R	H	2B	3B	HR	RBI	BB	SO	SB
13 years	.275	4587	645	1260	162	46	151	737	608	463	21

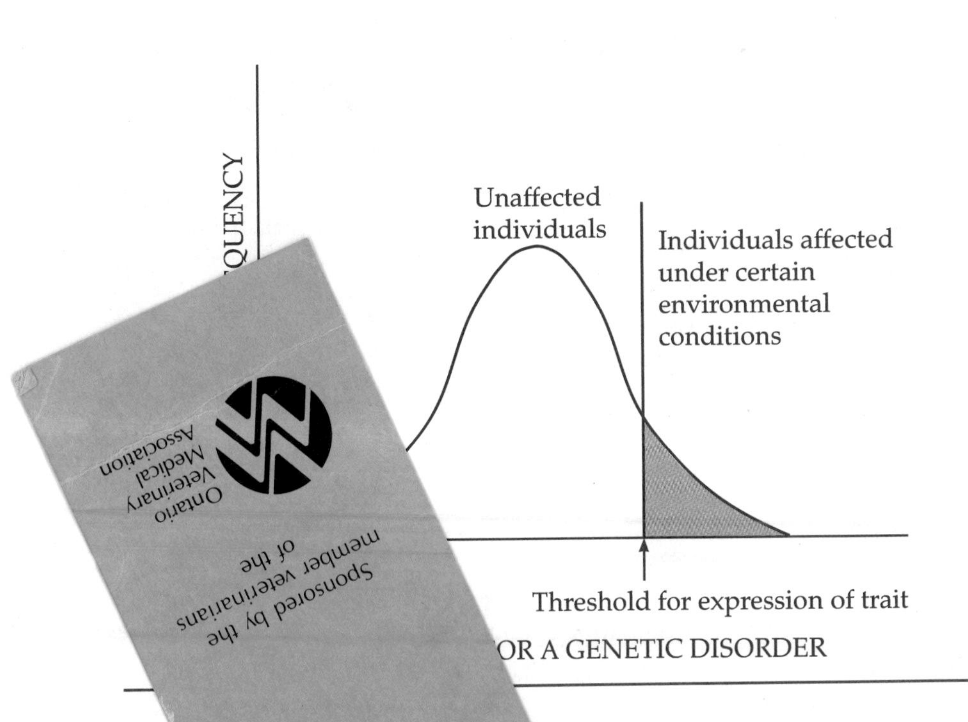

FIGURE 15.7 Threshold effect. A simple model to explain the discontinuous distribution of multifactorial traits. In this model, liability for a multifactorial genetic disorder is distributed among individuals in a normal curve. This liability is caused by a number of genes, each acting in an additive fashion. Those individuals with a liability above a certain threshold will be affected if exposed to certain environmental conditions.

freque ... es of affected individuals is compared with th ... he general population. In a family, first-deg ... s), have one half of their genes in common, ... one fourth, third-degree relatives have one e ... n, and so forth. As the degree of relatedness ... y that individuals will share the same combin ... ci. According to the threshold model, the ri ... o decrease as the degree of relatedness decr ... patterns for some congenital malformations is ... a relationship of declining risk as degree of r ...

The multifact ... only indirect evidence for the effect of genotyp ... interaction between the genotype and the env ... ul, however, in genetic counseling, where it i ... risks in families with certain congenital malfo ... sorders.

Interactions of Genoty

To assess accurately the degre ... notype and the environment, we must exami ... notype that is exhibited by a population of in ... at individual members of the population. This ... d from two sources: (1) the presence of differen ... population and (2) the presence of different en ... enotypes have been expressed. Assessing the r ... ction of

TABLE 15.1 Familial Risks for Multifactorial Threshold Traits

Multifactorial Trait	Risk Relative to General Population MZ Twins	First Degree Relatives	Second Degree Relatives	Third Degree Relatives
Club foot	300x	25x	5x	2x
Cleft lip	400x	40x	7x	3x
Congenital hip dislocation (females only)	200x	25x	3x	2x
Congenital pyloric stenosis (males only)	80x	10x	5x	1.5

phenotypic variability in a population is embodied in the concept known as heritability.

HERITABILITY

Genetic variance
The phenotypic variance of a trait in a population that is attributed to genotypic differences.

The phenotypic variance of a trait in a population that is attributed to differences in the environment.

The variation in phenotype that is the result of different genotypes is known as **genetic variance.** Any variation that occurs between individuals of the same genotype in a population is known as **environmental variance.** The heritability of a trait, symbolized by *H*, is that proportion of total variance caused by genetic differences. Heritability is always a variable, and it is not possible to obtain an absolute heritability value for any given trait. The value obtained depends on the population being measured and the amount of environmental variability present at the time of measurement. Remember that heritability is a population phenomenon and applies to groups, not to individuals. In general, if the heritability is 100% ($H = 1.0$), the environment has no effect, and all variation seen in the population is genetic. If the heritability is zero ($H = 0.0$), all variation is due to the environment.

It is difficult to find multifactorial traits for which the degree of heritability can be accurately assessed, because of interactions between genes and the environment. The analysis of fingerprint patterns is one trait that has been used to assess the heritability of a human trait. Fingerprint patterns were first studied systematically in the 19th century when the physiologist Purkinje classified prints into nine basic patterns, similar to the system used today. Later, Sir Francis Galton recognized that fingerprint patterns are inherited and remain the same throughout life, making them valuable as genetic markers.

Fingerprint patterns are laid down in the first 3 months of embryonic development and are the products of genetic and environmental factors. This means that everyone, including identical twins, has a unique set of fingerprints. Even though identical twins share the same set of genes and occupy the same uterus, they are each subject to slightly different environments. These different environmental factors are responsible for creating different fingerprint patterns.

Fingerprints are composed of skin ridges, called dermal ridges. During embryonic development, these ridges form sets of loops, whorls, and arches. Similar patterns are formed from the ridges on the palms, toes, and soles. Analysis of these patterns is known as **dermatoglyphics** (literally: writing on the skin). The lines in the palms, known as flexion creases (the heart, life, and head lines of palmistry) are formed at the same time as the dermal ridges (Figure 15.8).

Dermatoglyphics
The study of the skin ridges on the fingers, palms, toes, and soles.

Today fingerprints are classified as loops, whorls, and arches (Figure 15.9) and by 10-digit ridge counts. Ridge counts are useful in the study of phenotypic variance and heritability. They are easily and objectively measured and once established are not subject to social and environmental factors.

Heritability in humans is calculated from observations made among relatives. We know the fraction of genes that related individuals have in common: one half between parents and children, one fourth between grandparents and children, and so forth. These relationships were expressed by Galton as **correlation coefficients.** In studying genetic relatedness, the degree of relatedness can be expressed as correlation coefficients. The half-set of genes received by a child from its parent corresponds to a correlation coefficient of 0.5. The genetic relatedness of identical twins is 100%, and this

Correlation coefficient
A measure of the degree to which variables vary together.

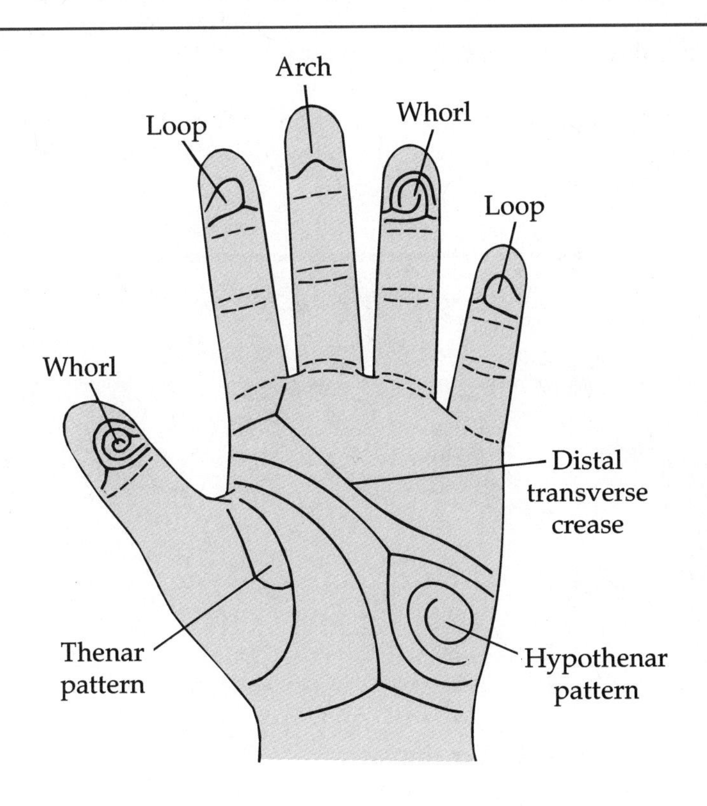

FIGURE 15.8 The dermatoglyphic pattern of the palm and fingers.

FIGURE 15.9 The three basic types of fingerprints: (A) Arch. (B) Loop. (C) Whorl. In (B) and (C), the number of ridges crossing the superimposed line provides the ridge count.

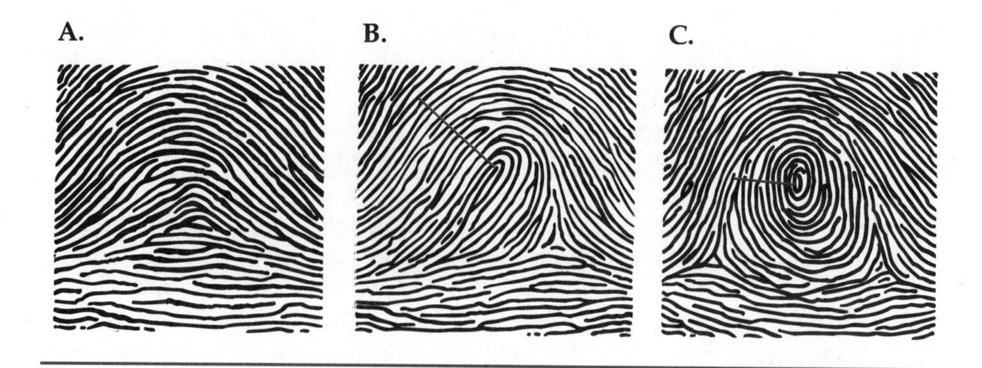

is expressed as a correlation coefficient of 1.0. Unless a mother and father are related by descent, they should be genetically unrelated, and the correlation coefficient for this relationship is 0.0.

Using correlation coefficients, one study of total ridge counts (TRC) sampled 825 British males and 825 British females. The results (Table 15.2) show the relationships among those examined, the observed correlation coefficients, and the theoretic (expected) coefficients. The results provide no information about the number or nature of the genes involved, but the almost total agreement between the observed and theoretic values indicate that TRC is almost totally under genetic control, with environmental factors playing only a minor role.

Statistical analysis of the information on ridge counts between parents and offspring can be used to estimate the heritability of ridge counts. These estimates indicate a heritability of about 0.95. A heritability value of 0.95 means that most of the phenotypic variation in fingerprint ridges is genetically transmitted from parent to offspring. The small amount of nongenetic variation helps explain why identical twins have different fingerprint patterns.

TABLE 15.2 Correlations and Heritability for Fingerprint Ridge Counts

Relationship	Expected Correlation	Observed Correlation	Heritability
Mother-child	0.50	0.48	0.96
Father-child	0.50	0.49	0.98
Siblings	0.50	0.50	1.0
Dizygotic twins	0.50	0.49	0.98
Monozygotic twins	1.00	0.95	0.95

From: Holt, S. B. 1961. *Br. Med. Bull.* **17**:247–250.

Comparing the degree of known genetic similarity to the degree of observed phenotypic variability provides an estimate of heritability. However, this method has one main problem: The closer the genetic relationship, the more likely it is that the relatives share a common environment. In other words, parents and children may be similar because they have one half of their genes in common but may also be similar because they share a similar environment. To bypass this problem, human geneticists seek out situations in which genetic and environmental influences are clearly separated. For example, monozygotic twins are genetically identical. If such twins are separated at birth and raised in different environments, genetic variability is constant, and the environment is varied. Comparison of traits in unrelated adopted children with those of natural children in the same family represents a situation with a constant environment and maximum genetic variability. The study of twins and adopted children is therefore an important tool in measuring heritability in humans.

Twin Studies

Galton, the founder of eugenics, a movement to improve humans through selective breeding, pointed out in 1875 the value of studying twins to obtain information about the effects of environment on heredity. Before examining the results of such studies, we need to look briefly at the biology of twinning. There are two types of twins, **monozygotic (MZ)** (identical) and **dizygotic (DZ)** (fraternal). Monozygotic twins originate from a single fertilization event: a single egg fertilized by a single sperm. Monozygotic twins arise as the result of a division into two separate embryos at an early stage of development. Additional splitting is also possible (see Concepts & Controversies, page 416). This separation may take place at the time when the zygote undergoes its first mitotic division or at any time up to the first 2 weeks of development. Conjoined twins (so-called Siamese twins) are thought to arise from an incomplete division of the embryo. Because they arise from a single fertilization event and in the absence of nondisjunction, MZ twins are of the same sex and carry the same genetic markers such as blood types. Dizygotic twins originate from two fertilization events: two eggs, ovulated in the same menstrual cycle, are each fertilized by a sperm. DZ twins are no more related than other pairs of siblings, with half their genes in common.

Monozygotic (MZ) twins
Twins derived from a single fertilization event involving one egg and one sperm; such twins are genetically identical.

Dizygotic (DZ) twins
Twins derived from two separate and nearly simultaneous fertilization events, each involving one egg and one sperm. Such twins share, on average, 50% of their genes.

For heritability studies it is essential to identify a pair of twins as MZ or DZ. In the early years of this century, twins were considered identical if they were of the same sex. When this was shown to be unreliable, the condition of the fetal membranes was used as a diagnostic test. However, this method must be used with caution. The developing fetus is surrounded by two membranes: an inner **amnion** and an outer **chorion,** which is attached to the placenta. Only when there is a single placenta, a single chorion, and separate amnions can the twins be considered monozygotic.

Amnion
A fluid-filled sac within which human development occurs.

Chorion
A membrane outside and surrounding the amnion, from which projections known as villi extend to the uterine wall, forming the placenta.

An accurate diagnosis of twins as MZ or DZ is made only by extensive tests. Comparison of many traits with absolute correlation between individuals can be used to identify twins as MZ. Divergence for one or more traits means that the twins are DZ. Among the characters used are blood groups,

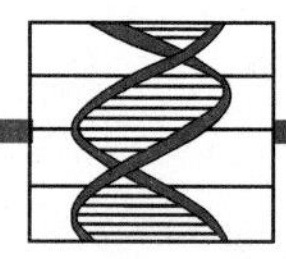

CONCEPTS & CONTROVERSIES

Twins, Quintuplets, and Armadillos

Because of the way in which they are formed, monozygotic twins are genetically identical and closely resemble each other. The process of embryo splitting that gives rise to MZ twins can be considered a form of human asexual reproduction. In fact, another mammal, the nine-banded armadillo produces litters of genetically identical, same-sexed offspring. In the case of the armadillo, a single fertilized egg splits into two, and often the daughter embryos divide, resulting in litters of two to six genetically identical offspring.

In humans, multiple births occur, but only rarely. About 1 in 7500 births are triplets, and 1 in 658,000 births are quadruplets. In many cases, both twinning and multiple fertilizations occur in multiple births. For example, in triplets, two of them may have developed from embryo splitting, and the third from an independent fertilization event. Hormones are often used as fertility drugs. They work by inducing multiple ovulations in a menstrual cycle, and multiple births, up to septuplets, have been reported after use.

Embryo splitting can also result in human multiple births. The Dionne quintuplets, born in May 1934, are the first recorded case in which all five members of a set of quintuplets survived. From studies of blood tests and physical similarities, it appears that these same-sexed quints arose as the result of a single fertilization event and several embryo cleavages according to the scheme presented below:

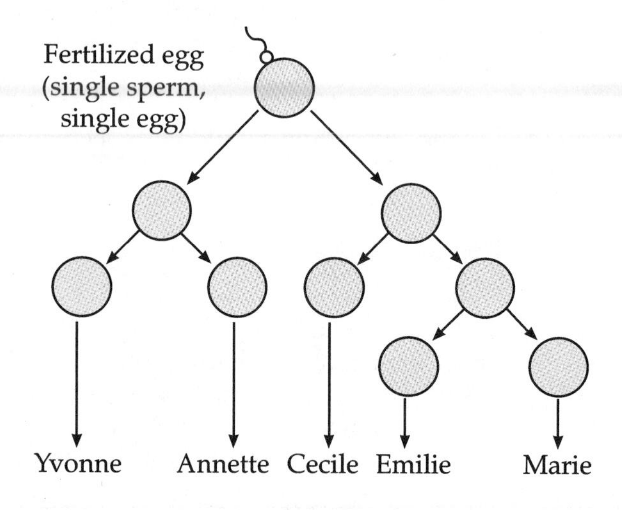

From this it would seem that MZ twins, the Dionne quintuplets, and armadillos have something in common: they are all the result of embryo splitting. Thus, embryo splitting as a reproductive mechanism has been conserved, although at a reduced level, in the evolution from the primitive marsupial mammals to the placental mammals like humans.

HLA haplotypes, sex, eye color, hair color, fingerprints, palm and sole prints, DNA fingerprinting, and RFLP analysis.

A simple method for evaluating phenotypic differences between twins is the use of traits that can be scored as present or absent rather than measured quantitatively. For such traits twins are *concordant* if both exhibit the trait and *discordant* if only one twin has the trait. As noted, MZ twins have 100% of their genes in common, while DZ twins, on average, have 50% in common. For any genetically determined trait the correlation in MZ twins should be higher than that in DZ twins. If the trait is completely heritable, the concordance should be 1.0 in MZ twins and close to 0.5 in DZ twins. In evaluating the results of twin studies, it is the degree of difference between **concordance** in MZ twins versus DZ twins that is important. The greater the

Concordance
Agreement between traits exhibited by both twins.

TABLE 15.3 Concordance Values in Monozygotic (MZ) and Dizygotic (DZ) Twins

Trait	Concordance Values (%)	
	MZ	DZ
Blood types	100	66
Eye color	99	28
Mental retardation	97	37
Hair color	89	22
Down syndrome	89	7
Handedness (left or right)	79	77
Epilepsy	72	15
Diabetes	65	18
Tuberculosis	56	22
Cleft lip	42	5

difference, the greater is the resultant heritability. Table 15.3 lists concordance values in twins for a variety of traits. Examination of the table shows that the concordance value for cleft lip in MZ twins is higher than that for DZ twins (42% versus 5%). Although this difference suggests a genetic component to this trait, the value is so far below 100% that environmental factors (perhaps teratogens) are obviously important in the majority of cases. In all cases, concordance values must be interpreted cautiously.

Concordance values can be converted to heritability values through a number of statistical formulas. Heritability values derived in this manner are given in the last column of Tables 15.2 and 15.5. Remember that heritability is a relative value, valid only for the population measured and only under the environmental conditions in effect at the time of measurement. Heritability determinations made within one group cannot be compared with heritability measurements for the same trait in another group, since the two groups differ in genotypes and environmental variables in unknown ways.

Twins and Obesity

Studies and heritability estimates of twins and adopted children are necessarily indirect methods to study traits that are multifactorial (polygenic with an environmental contribution). These studies are usually based on correlations rather than direct demonstration of cause and effect and as such are subject to a number of uncertainties. To illustrate this we will consider the question of whether heredity plays a significant role in obesity.

Obesity is a trait that can be said to "run" in families. Obesity is also a national health problem. A federal study has estimated that 26% of the adult U.S. population between the ages of 20 and 75 years is overweight. These individuals are at higher risk for diseases such as high blood pressure, elevated levels of cholesterol in the blood, coronary artery disease, and diabetes.

In a recent study the heritability of obesity was assessed in 1974 pairs of MZ twins and in 2097 pairs of DZ twins who were born between 1917 and 1927 and who served in the armed forces. Weights and heights measured at induction into the armed forces were compared with similar measurements taken 25 years later. Obesity was measured by body mass index (BMI; weight in kilograms divided by height in meters squared), and by reference to tables listing ideal weight-height relationships. Table 15.4 shows the concordance for five levels of obesity. At induction, concordance for MZ twins was much higher than that for DZ twins at all five levels of obesity. Twenty-five years later, the concordance levels for MZ twins were still much higher than those for DZ twins. Using appropriate statistical methods, the concordance values can be converted to heritability. The values obtained are listed in Table 15.5. The result shows high heritability values for obesity, suggesting that it is under strong genetic control.

However, let us consider some potential problems with this type of study that can affect the conclusions. First, the study included only men who had passed a preinduction physical that sorted out those with marked obesity, skewing the population studied to less obese individuals. This would tend to underestimate the contribution of heredity to obesity. Second, the study did not include women or children or men excluded from military service for other causes, limiting the ability to generalize the conclusions.

TABLE 15.4 Concordance Values for Obesity in Twins

% Overweight	% Concordant at Military Induction		% Concordant 25 Years Later	
	MZ	DZ	MZ	DZ
15	61	31	68	49
20	57	27	60	40
25	46	24	54	26
30	51	19	47	16
40	44	0	36	6

From: Stunkard, A. J., Foch, T. T., and Hrubec, Z. 1986. *JAMA* **256:**51–54.

TABLE 15.5 Heritability Estimates for Obesity in Twins (from Several Studies)

Condition	Heritability
Obesity in children	0.77–0.88
Obesity in adults (weight at age 45)	0.64
Obesity in adults (body mass index at age 20)	0.80
Obesity in adults (weight at induction into armed forces)	0.77
Obesity in twins reared together or apart	
Men	0.70
Women	0.66

Last, the study did not attempt to directly study the role of environmental factors such as diet in obesity. Still, the conclusions indicate that obesity is under strong genetic control.

Perhaps the most effective way of separating the effects of genes and environment and of controlling external factors is the study of identical and fraternal twins reared under different environmental conditions. In a study of body-mass index in identical and fraternal twins reared together or apart, heritability values for body-mass index were calculated as 0.70 for men and 0.66 for women (Table 15.5), indicating that genetic factors make a substantial contribution to obesity.

Another method of assessing the role of genetic factors and the environment in obesity is the study of adopted unrelated children, who bring maximum genotypic differences into a family that is subjected to a relatively constant environment. Studies comparing obesity in adopted children with obesity in the biologic and adopted parents confirms the role of genetic factors as important influences on obesity and assigns a minor role to the family environment. For example, a study completed in 1989 compared obesity in 3580 adoptees and their biologic siblings, who were reared separately. The results of this study indicate that there is a strong relationship between obesity in adoptees and their full siblings, even though they were raised in different environments. This reinforces the conclusion mentioned earlier that heredity plays an important role in obesity. The analysis also indicates that polygenic inheritance can account for body mass, ranging from thin to obese. The results are consistent with other studies that show the existence of a major recessive gene for obesity that accounts for 35% of the variation in body mass, with other genes contributing in a polygenic manner an additional 42% of the variation. The remaining 23% of the variation cannot be explained by genetic factors.

Human Polygenic Traits

Skin Color

The theory of polygenic inheritance in humans was first tested by Charles and Gertrude Davenport, leading figures in the American eugenics movement. For their study, conducted between 1910 and 1914, they collected information on skin color in black-white marriages in Bermuda and in the Caribbean. To measure skin color, they used a top with a disk composed of colored sectors of various sizes (Figure 15.10). The colors used were black, white, red, and yellow. When the top was spun, the colored sectors blended together and produced a color that could be matched to a given skin color by changing the size of the black- and white-colored sectors. Based on the size of the black sector, individuals were assigned to one of five categories,

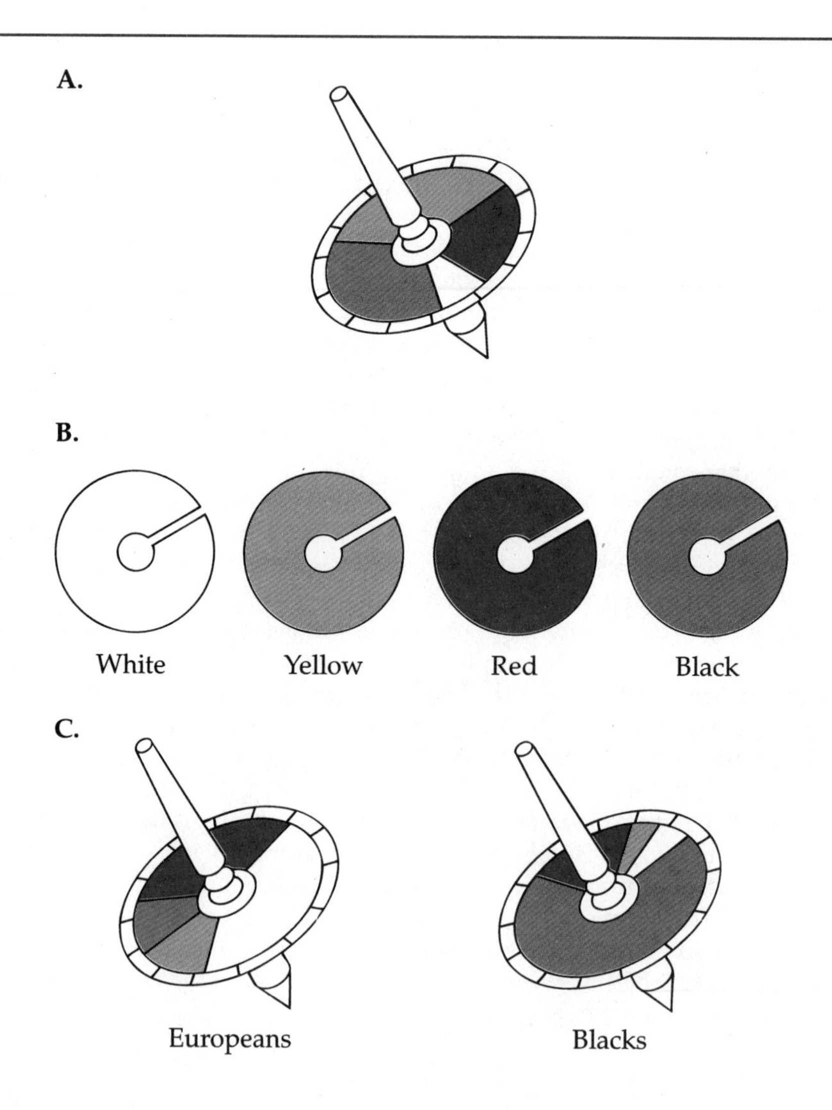

FIGURE 15.10 The top used by the Davenports to measure human skin color. (A) Top with disks in place. (B) The disks used to measure skin color. (C) Arrangement of disks to show skin color of Europeans and North American blacks. When the top is spun, the colors blend to produce a color used to match that of the skin.

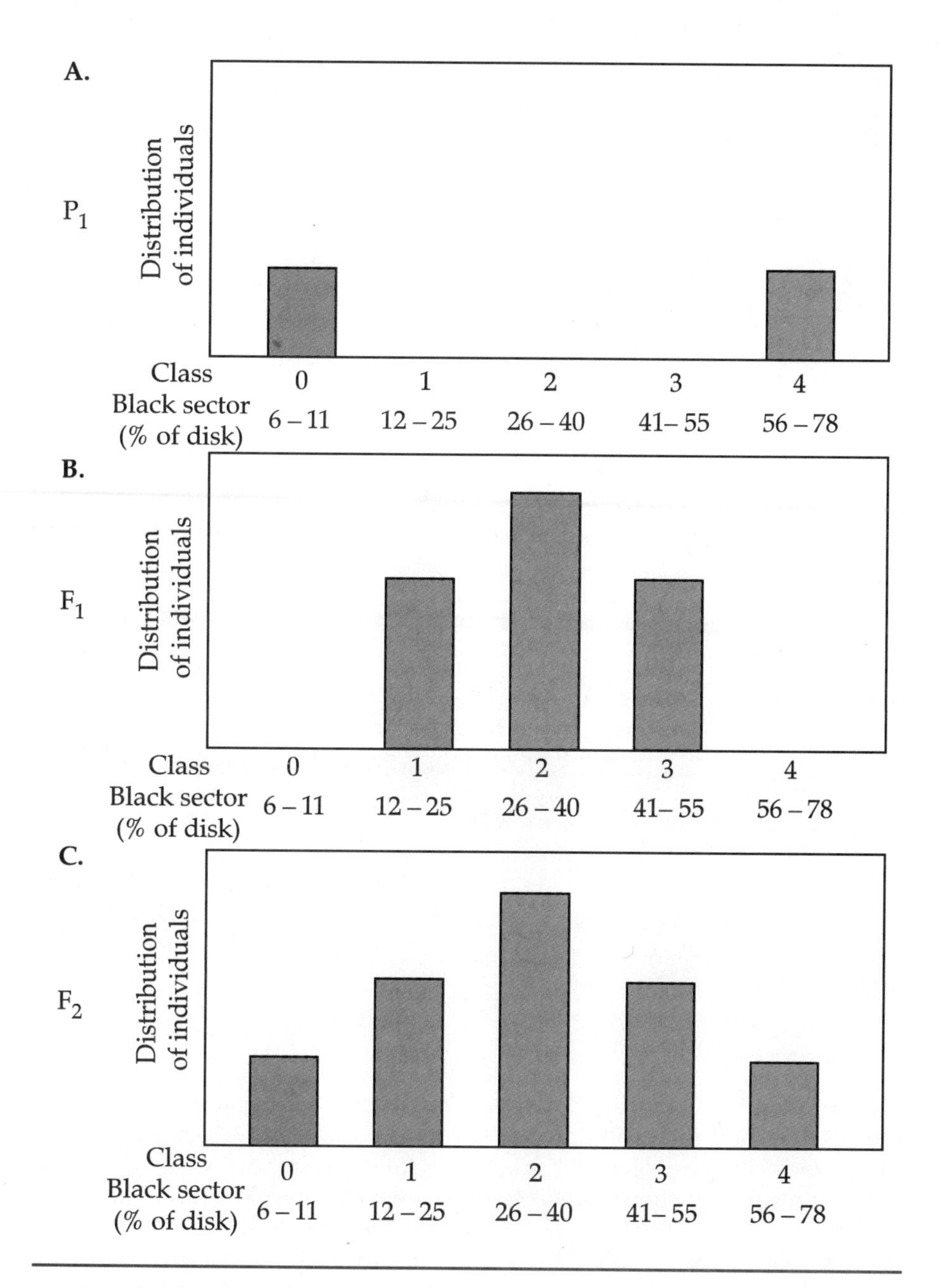

FIGURE 15.11 Frequency diagrams of skin color testing results. (A) Parents: white male and black female. (B) Color values of seven children from parents in (A). (C) Skin studies on 32 children of parents in (B). Color values cluster near the mean value.

0 to 4. The results illustrate several properties of polygenic traits. For example, the offspring (F_1) of such marriages have skin color values intermediate between those of their parents. In the F_2 generation, a small number of children were as white as one grandparent, a small number were as black as the other grandparent, and most were distributed between these extremes (Figure 15.11). Because individuals in the F_2 could be grouped into five classes, the Davenports hypothesized that skin color was controlled by two gene pairs. Each class represented a genotype that resulted from the

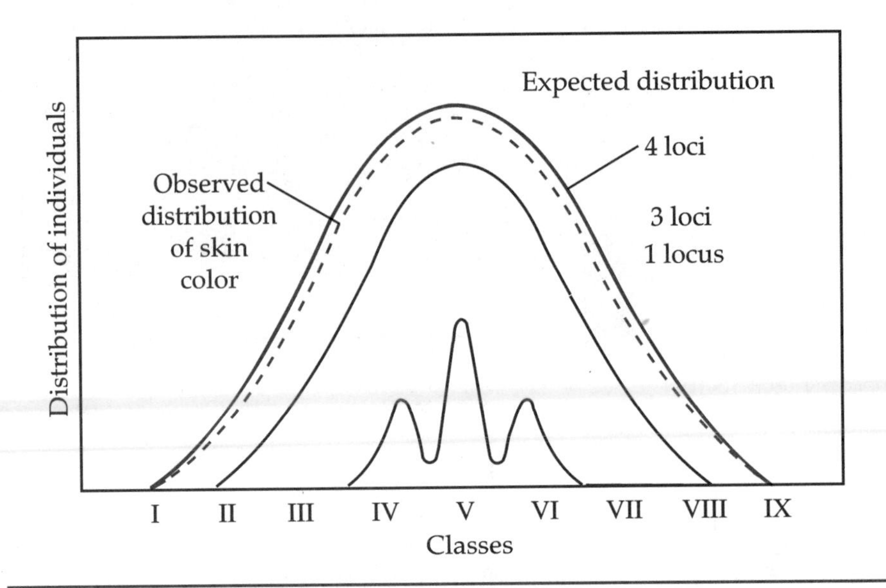

FIGURE 15.12 The distribution of expected phenotypes in human skin color controlled by one, three, or four loci. The observed values, measured with a melanometer, correlate strongly with the results predicted in a four-locus system.

segregation and assortment of two gene pairs. Suppose these genes are A and B respectively. Class 0 with the lightest skin represents the genotype aabb, Class 1 has the genotype Aabb or aaBb, and so forth, up to Class 5, which is homozygous dominant (AABB) and has the darkest skin.

Subsequent work using a more sophisticated instrument that measures the reflection of light from the skin surface (a melanometer) has shown that skin color is actually controlled by more gene pairs, and the current estimate is that between three and six gene pairs are involved. The data are most consistent with a model involving four genes, each with two alleles and no dominance (Figure 15.12).

Intelligence

The idea that intelligence is a distinct entity that can be quantified began in the late 18th and early 19th centuries with the development of phrenology. In the view of phrenologists (which we know to be incorrect), the brain contains a series of compartments, each of which controls a single function, such as musical ability, courage, or intelligence. According to phrenology, because each area of the brain has an assigned mental function, the intellectual or musical capacities of an individual can be determined by an examination of the shape and size of a particular region of the skull.

In the latter part of the 19th century, many scientists related intelligence to overall brain size. Craniometry, the measurement of brain size, then became the dominant means of assessing intelligence. In this method the cranium of a skull was filled with small lead shot (about the size of a BB). Then the shot was emptied into a graduated cylinder, and the cranial capacity was measured in cubic centimeters. Alternatively, brains were removed after death and weighed. Larger or heavier brains or both were thought to denote higher intelligence. This method also proved to be incorrect.

At the turn of the 20th century, the French psychologist Alfred Binet began his studies on intelligence by measuring cranial capacity. He eventually abandoned this approach as unreliable and turned to psychologic rather than physical methods for measuring intelligence. He began by trying to identify children whose poor performance in the classroom indicated a need for special education. He developed a series of simple tasks that he felt were related to basic mental processes such as comprehension, direction (sorting), and correction. The tests were designed to avoid learned skills such as reading and were a mixture of tasks requiring differing abilities. Eventually he assigned an age level to each task, and each child would begin by performing the simplest tasks and progressing in sequence until the tasks became too difficult. The age assigned for the last task successfully performed became the child's mental age, and the intellectual age was calculated by subtracting the mental age from the chronologic age. Binet's test became the basis of the Stanford-Binet intelligence tests in use today. Later another psychologist, Wilhelm Stern, divided mental age by chronologic age, and the number was known as the **intelligence quotient,** or IQ. If a child of 7 years (chronologic age) was able to successfully perform tasks for a 7-year-old but not tasks for an 8-year-old, a mental age of 7 would be assigned. To determine the IQ for this child, divide mental age by chronologic age:

Intelligence quotient (IQ)
A score derived from standardized tests that is calculated by dividing the individual's mental age (determined by the test) by his or her chronologic age, and multiplying the quotient by 100.

$$\frac{\text{mental age} = 7}{\text{chronologic age} = 7} = 1.0 \times 100 = 100 = \text{IQ}$$

Multiply the quotient by 100 to eliminate the decimal point, and we obtain an IQ of 100 as the average for any given age.

The substitution of psychologic for physical methods in measuring intelligence did not change the basic idea that intelligence is an entity with a biologic basis that can be quantified. In fact, the use of IQ tests has strengthened the assumption that IQ measures a fundamental, genetically determined physiologic or biochemical property of the brain related to intelligence. The question to be resolved is whether psychologic methods such as IQ tests can measure intelligence any more accurately than the discredited physical methods of phrenology or craniometry.

To determine whether IQ tests or any other method accurately measure intelligence, we must first define intelligence in a way that it can be objectively measured, as can height, weight, or fingerprint ridge counts. Properties such as abstract reasoning, mathematical skills, verbal expression, ability to diagnose and to solve problems, and creativity are often cited as important components of intelligence. Unfortunately, there is no evidence that any of these properties are measured by an IQ test, and there is no objective way to quantitate such components of intelligence.

In fact, the underlying assumption that IQ tests measure a fundamental property of the brain related to intelligence has never been shown to be true. There may be dozens of distinct neural processes involved in mental functions, many of which may contribute to intelligence. In spite of the rapid advances in neurobiology, little or no information is available about the biologic basis of any of these processes. Without knowledge of the mechanisms involved, it is difficult to devise quantitative measurements about the outcome of these mechanisms.

The values obtained in IQ measurement may, however, have significant heritable components. Much of the evidence for heritability of IQ scores comes from the use of twin studies.

Some of the major twin studies used in measuring the heritability of IQ have been disputed. Cyril Burt, a British psychologist, published a series of papers from 1909 to 1972 in which he argued that intelligence is primarily innate and inherited. Using twin studies he concluded that the average correlation for intelligence in twin pairs is 0.77. Beginning in 1976, other workers uncovered evidence that Burt fabricated a great deal of his data and even invented the existence of colleagues who were listed as co-authors on some of his papers. Unfortunately, Burt's fabrications have somewhat confused the issue of genetic contributions to intelligence.

IQ and Race

The idea that intelligence is biologically determined and the misuse or misunderstanding of the limits of heritability estimates have led to the conclusion that differences in IQ among different racial and ethnic groups are genetically determined. On standardized IQ tests, blacks score an average of 15 points lower than the average white score of 100, while those of Asian ancestry score an average of 7 points above 100. These differences are consistent across different tests, and the scores themselves do not seem to be a serious issue. The controversy is over what causes these differences. Are such differences genetic in origin, or do they reflect environmental differences, or are both factors at work? If both, to what degree does inheritance contribute to the differences?

Heritability studies have been used to support the argument that intelligence is mainly innate and inherited, citing heritability values of 0.8 for intelligence. In some cases, however, the reasoning used to support this argument misuses the concept of heritability. Recall that a measured heritability of 0.8 (for example) means that 80% of the phenotypic variation being measured is due to genetic differences *within that population*. Differences observed between two populations in heritability for a trait simply cannot be compared because heritability measures only variation within a population at the time of measurement and cannot be used to estimate genetic variation between populations. In other words, we cannot use heritability differences among groups to conclude that there are genetic differences among those groups.

Many environmental factors also affect IQ scores, including age, birth order, nutrition, and culture. In support of the role of environment in IQ, one study, known as the Texas Adoption Project, has examined IQ in 300 families that received adoptive children at birth. The IQ of 469 adopted children was compared with the IQs of the biologic mother and the adoptive mother. The results show a stronger correlation between the IQ of the child and the IQ of the biologic mother than between the child's IQ and that of the adoptive mother. Although this conclusion seems to support the role of heredity in IQ, the correlations were so small that fully 90% of the IQ variability was unaccounted for. If the contribution of the biologic and adoptive fathers is assumed to equal that of the mothers, 80% of the variability is unexplained. This variability may be caused by environmental factors beyond the adoptive parents and include siblings, friends, school, and other factors.

It is quite evident that both genetic and environmental factors make important contributions to intelligence. Clearly, the relative amount each contributes cannot be measured accurately at this time. Several points about this debate should be kept in mind. First, it is clear that IQ test scores cannot be equated with intelligence. Second, IQ scores are not fixed and can be changed significantly with training in problem solving and, in fact, change somewhat throughout the life of an individual. Variation in IQ scores is quite wide, and measurements in one racial or ethnic group overlap those of other groups.

The problem in discussing the differences in IQ scores arises when the quantitative differences in scores are converted into qualitative judgments used to rank groups as superior or inferior. Genetics, like all sciences, progresses by the formulation of hypotheses that can be rigorously and objectively tested. When intelligence can be objectively defined and measured, then genetic methods can be used to quantitatively approach the subject.

SUMMARY

1. In this chapter we have considered genes that affect traits in an additive or quantitative manner. The pattern of inheritance that controls metric characters (those that can be measured in a quantitative fashion) is called polygenic inheritance because two or more genes are usually involved.

2. The distribution of polygenic traits through the population follows a bell-shaped, or normal, curve. Parents whose phenotypes are near the extremes of this curve usually have children whose phenotypes are less extreme than the parents' and are closer to the mean of the population. This phenomenon, known as regression to the mean, is characteristic of systems in which the phenotype is produced by the additive action of many genes.

3. Variations in the expression of polygenic traits are often due to the action of environmental factors. Because of this, organisms that share a common genotype will show a range of phenotypic variation.

4. The impact of environment on genotype can cause genetically susceptible individuals to exhibit a trait in a discontinuous fashion even though there is an underlying continuous distribution of genotypes for the trait.

5. The degree of phenotypic variation produced by the genotype in a given population can be estimated by calculating the heritability of a trait. Heritability is estimated by observing the amount of variation among relatives having a known fraction of genes in common. MZ twins have 100% of their genes in common and when raised in separate environments provide an estimate of the degree of environmental influence on gene expression.

6. Heritability is a variable, validly calculated only for the population under study and the environmental condition in effect at the time of the study. It provides an estimate of the degree of genetic variance within a population, and heritability values cannot be compared among populations because of differences in genotypes and environmental factors.

7. In twin studies, the degree of concordance for a trait is compared in MZ and DZ twins reared together or apart. MZ twins result from the splitting of an embryo produced by single fertilization event, while DZ twins are the products of multiple fertilization events. While twin studies can be useful in determining whether a trait is inherited, they cannot provide any information on the mode of inheritance or the number of genes involved.

8. Many human traits are controlled by polygenes, including skin color, intelligence, and aspects of behavior. The genetics of these traits has often been misused and misrepresented in the service of ideologic, or political ends.

9. The inheritance of skin color in humans is best explained by a four-gene additive model without dominance. These genes control the synthesis and deposition of melanin in epidermal cells. Intelligence is most probably a polygenic trait that is strongly influenced by the environment. Our understanding of the genetics of intelligence is hampered by the prevailing assumption that it is a single entity that can be expressed as a single number and by the lack of a demonstrated means of measuring this quality.

QUESTIONS AND PROBLEMS

1. The text outlines some of the problems Frederick William I encountered in his attempt to breed tall Potsdam Guards.
   a. Why were the results he obtained so different from those obtained by Mendel with short and tall pea plants?
   b. Why were most of the children shorter than their tall parents?
   c. What role might the environment have played in causing Frederick William problems, especially in a time when nutrition varied greatly from town to town and from family to family?
   d. Do you think his experiment would have worked better if he had ordered brother-sister marriages within tall families instead of just choosing the tallest individuals from throughout the country?

2. As it turned out, one of the tallest Potsdam Guards had an unquenchable attraction to short women. During his tenure as guard he had numerous clandestine affairs. In each case children resulted. Subsequently, some of the children, who had no way of knowing that they were related, married and had children of their own. Assume that height is determined by two pairs of genes. The genotype of the 7-ft-tall Potsdam Guard was *A′A′B′B′*, and the genotype of all his 5-ft clandestine lovers was *AABB*, where an *A′* or *B′* allele adds 6 in to the base height of 5 ft conferred by the *AABB* genotype.
   a. What were the genotypes and phenotypes of all the F_1 children?
   b. Diagram the cross between the F_1 offspring, and give all possible genotypes and phenotypes of the F_2 progeny.

3. Describe why there is a fundamental difference between the expression of a trait that is determined by polygenes and the expression of a trait determined monogenetically.

4. Define genetic variance.

5. Define environmental variance.

6. How is heritability related to genetic and environmental variance?

7. Why are relatives used in the calculation of heritability?
8. What main problem is encountered when relatives are used in the calculation of heritability?
9. Why are monozygotic twins, reared apart, so useful in the calculation of heritability?
10. What is the importance of the comparison of traits between adopted and natural children in the determination of heritability?
11. If monozygotic twins show complete concordance for a trait whether they are reared together or apart, what does this suggest about the heritability of the trait?
12. If there is no genetic variation within a population for a given trait, what is the heritability for the trait in the population?

For Further Reading

Aldhous, P. 1992. The promise and pitfalls of molecular genetics. Science **257**: 164-165.

Balmor, M. G. 1970. *The Biology of Twinning in Man*. Oxford, England: Clarendon.

Benirschke, K. 1972. Origin and clinical significance of twinning. *Clin. Obstet. Gynecol.* **15:** 220–235.

Bouchard, C., Tremblay, A., Despres, J.-P., Nadeau, A., Lupien, P. J., Theriault, G., Dussault, J., Moorjani, S., Pinault, S., and Fournier, G. 1990. The response to long-term overfeeding in identical twins. *New Engl. J. Med.* **322:** 1477–1481.

Bouchard, C., Përusse, L. 1993. Genetics of obesity. Ann. Rev. Nutrition **13**: 337-354.

Feldman, M. W., and Lewontin, R. 1975. The heritability hangup. *Science* **190:** 1163–1166.

Harrison, G. A., and Owens, J. J. T. 1964. Studies on the inheritance of human skin color. *Ann. Hum. Genet.* **28:** 27–37.

Mackintosh, N. J. 1986. The biology of intelligence? *Br. J. Psychol.* **77:** 1–18.

Moll, P., Burns, T. and Laver, R. 1991. The genetic and environmental sources of body mass index variability: The Muscatine ponderosity family study. Am. J. Hum. Genet. **49**: 1243-1255.

Sorenson, T. I., Price, R. A., Stunkard, A. J., and Schulsinger, F. 1989. Genetics of obesity in adult adoptees and their biological siblings. *Br. Med. J.* **298:** 87–90.

Stern, C. 1970. Model estimates for the number of gene pairs involved in pigmentation variability in Negro-Americans. *Hum Hered.* **20:** 165–168.

Stunkard, A. J., Foch, T., and Hrubec, Z. 1986. A twin study of human obesity. *JAMA* **256:** 51–54.

Stunkard, A. J., Harrus, J. R., Pedersen, N. L., and McClearn, G. E. 1990. The body-mass index of twins who have been reared apart. *New Engl. J. Med.* **322:** 1483–1487.

CHAPTER 16

GENETICS OF BEHAVIOR

CHAPTER OUTLINE

Sometime in the late 1850s, an eight-year-old boy and his father, a physician, drove along a road through the woods on eastern Long Island. They met two women walking along the road, and this chance encounter had a profound effect on the young boy. Years later, he would recall that meeting:

> I recall it as vividly as though it had occurred but yesterday. It made a most enduring impression upon my boyish mind, an impression

> every detail of which I recall today, an impression which was the very first impulse to my choosing chorea as my virgin contribution to medical lore. Driving with my father through a wooded road leading from East Hampton to Amagansett, we suddenly came upon two women, mother and daughter, both tall, thin, almost cadaverous, both bowing, twisting, grimacing. I stared in wonderment, almost in fear. What could it mean? My father paused to speak to them and we passed on. Then my Gamaliel-like instruction began; my medical education had its inception. From this point on, my interest in the disease has never wholly ceased.

The young boy, George Huntington, went on to study medicine at Columbia University. In 1872, a year after graduation, he published an account of this disorder, which became known as Huntington's chorea, or as it is called today, Huntington disease. In the paper, his summary of the condition is a model of brevity and clarity:

> There are three marked peculiarities in this disease: 1. Its hereditary nature, 2. A tendency to insanity and suicide, 3. Its manifesting itself as a grave disease only in adult life.

Huntington accurately described the pattern of inheritance of the disorder in a way that is consistent with an autosomal dominant trait. Unfortunately, having described the condition, Huntington turned his attention in other directions, and little more was accomplished in the next 88 years until recombinant DNA techniques were used to map the chromosome locus to the short arm of chromosome 4.

In several ways, conditions such as Huntington disease (HD) seem to be a model for genetic disorders that affect behavior. The pattern of autosomal dominant inheritance is well-defined and clear-cut. The gene was one of the first to be mapped to a chromosomal locus by RFLP mapping, demonstrating the power of molecular techniques in the analysis of human behavior. Recently, the gene was isolated by positional cloning, one of the newer genetic techniques generated by recombinant DNA technology. Lastly, the molecular basis of mutation in the HD gene represents a new class of mutations that to date are known to affect only the nervous system. Thus it would seem that following the methods used in the study of Huntington disease, researchers could identify, map, and isolate many genes affecting human behavior.

Unfortunately, searching for single genes (like the HD gene) that play a role in behavioral variations and disorders may have only limited success. One of the most difficult problems in human behavioral genetics is defining the phenotype. While Huntington disease has a defined phenotype and progression, many behavioral traits do not. In addition, the phenotype of some conditions, such as schizophrenia (well defined as a medical condition), may be genetically heterogeneous and actually include several genetic disorders, each with a similar phenotype. Finally, many behavioral traits are multifactorial, and the phenotype is determined by several genes and by environmental interaction, with no single gene having a major effect.

To understand the issues in behavior genetics and how decisions about phenotypic definitions, genetic models of behavior, and the roles of methods influence both the speed and the outcome of this research, we begin this chapter by discussing some of the models and methods of human behavior genetics. From there, we briefly consider animal models, where single-gene effects on behavior have been well-documented. Next, we discuss single genes that affect human behavior through their effect on the nervous system, and then consider more complex traits, and those that have the greatest social impact. Finally, we summarize the current state of human behavior genetics and the ethical, legal, and social imolications of this research.

MODELS, METHODS, AND PHENOTYPES

The genetic control of behavior in humans has been clearly demonstrated. However, observation and pedigree analysis indicate that not all behaviors are inherited as simple Mendelian traits, demonstrating the need for genetic models that can explain observed patterns of inheritance. To a large extent, the model of inheritance proposed for a trait determines the methods used to follow its pattern of inheritance, and the techniques that can be pursued in mapping and isolating the gene or genes responsible for the trait's characteristic phenotype. This is true of all traits, behavioral and otherwise. In the case of behavior, however, many traits—especially those with social impact—have complex phenotypes and are not inherited as single-gene traits, making it necessary to first consider how such traits might be inherited.

Genetic Models of Inheritance and Behavior

Several models for genetic effects on behavior have been proposed. The question of models for a behavioral trait boils down to one of how many genes control a given trait (Table 16.1). The simplest model is a single gene, dominant or recessive, that affects a well-defined behavior. In fact, several human behavioral traits—including Huntington disease, Lesch-Nyhan syndrome, fragile-X syndrome, and others—can be described by such a model. Some multiple-gene models are also possible. The simplest of these is a polygenic additive model in which two or more genes contribute equally in an additive fashion to the phenotype. This model has been proposed (along with others) to explain schizophrenia (the inheritance of additive polygenic traits was considered in Chapter 15). In a variation of this model, one or more genes might have a major effect, with other genes making smaller contributions to the phenotype. Still another multigene model involves the interaction of alleles at different loci to produce a new phenotype. This form of gene interaction, known as **epistasis,** has been well-documented in experimental genetics, although it has not yet been invoked to explain a human behavior trait.

Epistasis
A form of gene interaction in which one gene affects the expression of another

TABLE 16.1 Models for Genetic Analysis of Behavior

Model	Description
Single gene	
Polygenic trait	Additive model, with two or more genes
	One or more major genes with other genes contributing to phenotype
Multiple gene	Interaction of alleles at different loci generates a unique phenotype.

In each of these models, the environment can make significant contributions, and the study of behavior must take this into account (see Concepts & Controversies, page 432). In some cases, this means developing methods that combine different approaches in order to study the genetic basis of behavior and assess the role of the environment in the development of the phenotype.

Methods of Studying Behavior Genetics

For the most part, the methods in behavior genetics follow the classical pattern for human genetics. If the model proposed involves a single gene, pedigree analysis and linkage and segregation studies including the use of RFLP markers and other methods of recombinant DNA technology are the most appropriate methods. However, because many behavior traits may be polygenic, twin studies are a prominent feature of human behavior genetics. Concordance and heritability values based on twin studies have been used to conclude that there is a genetic component to mental illnesses such as manic depression and schizophrenia, as well as behavior traits such as sexual preference and alcoholism. The results of such studies should be interpreted with caution, since they are subject to the limitations inherent in interpreting heritability (see Chapter 15) and are often conducted on small sample sizes, where minor variations can have a disproportionately large effect on the outcome.

To overcome these problems, behavior researchers are attempting to extend and adapt twin studies to improve this method as a genetic tool. One innovation is the study of the offspring of twins. This second-generation research, using both concordant and discordant traits, can provide clues about the maternal effects on a phenotype and confirm the existence of genes predisposing to a certain behavior. Twin studies are also being coupled to recombinant DNA techniques to search for behavior genes, and this combination may prove to be a powerful method for identifying such genes.

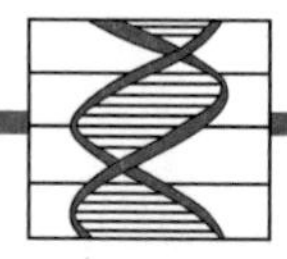

CONCEPTS & CONTROVERSIES

Behavior Genetics: The Problem with Model Selection

It is clear that many behavior traits exhibit a familial pattern of inheritance. This observation, along with twin studies and adoption studies, clearly indicates that there is a genetic component to many behavioral disorders such as manic depression and schizophrenia. In most cases, a simple pattern of Mendelian segregation cannot be demonstrated. Researchers are then faced with selecting a possible model of inheritance. Using this presumed mode of inheritance, further choices are made as to the methods to be used in the study. A common strategy is to find a family in which several members are affected, and in which the behavior appears to be inherited as a recessive or an incompletely penetrant dominant trait. One or more markers are then used in linkage analysis to further study the trait, and to identify the chromosome carrying the mutant gene. However, if an incorrect assumption is made about the mode of inheritance for the trait, then further work may produce errors. Reports of loci on the X chromosome and on chromosome 11 for manic depression were established in family studies, but were not supported by further work on other families, and in some cases additional work excluded the chromosomes previously identified as sites of a locus for manic depression.

To demonstrate some of the methodological pitfalls associated with behavior genetics and model selection, one recent family study examined the behavior of attending medical school to determine if it fits a familial pattern consistent with a genetic model. In this study of 249 first- and second-year medical students, the frequency of attending medical school by first-degree relatives was 13%, compared to 0.22% in the general population. Thus, the overall risk for first-degree relatives (over the age of 18) for medical school attendance was 61 times higher than the general population. Although this result indicates that the trait is strongly familial, the pattern of inheritance is not clear. To resolve this, the researchers used standard statistical methods for complex segregation analysis. The results of their analysis provide strong evidence for familial transmission of the trait, and reject the model of no transmission. While several models were supported by the analysis, including a polygenic model with a major gene effect, the most parsimonious model (see box on Ockham's razor in Chapter 3) was that of recessive inheritance. Using a further set of tests for the analysis of Mendelian transmission, the researchers were able to demonstrate that the recessive model of inheritance was just at the borderline for acceptance or rejection. With these results, it could be argued that other larger studies would confirm that attendance at medical school is controlled by a recessive gene.

While it is true that genetic factors may partly influence the decision to enter medical school, it is unlikely that this decision is controlled by a single recessive gene, as supported by the family studies and the segregation analysis. The purpose of this study was not to cast doubt on the methods used in behavior genetics, but to point out the problems inherent in accepting simple explanations for the transmission of complex traits that depend on both genetic and environmental factors, and to illustrate the need for precise definition of the phenotypes being studied.

Phenotypes: What Is Behavior?

As we mentioned earlier, a basic problem that limits progress in human behavior genetics is an accurate definition of the trait under consideration. The definition must be precise enough to distinguish the behavior from other, similar behaviors and from the behavior of the control group. For some mental illnesses, clinical definitions are well-established by guidelines such as the Diagnostic and Statistical Manual of Mental Disorders of the American

Psychiatric Association, and include specific criteria. For other behaviors, the phenotypes are often poorly defined, and may be unrelated to the underlying biochemical and molecular basis of the behavior. For example, alcoholism can be defined as the development of characteristic deviant behaviors associated with excessive consumption of alcohol. Is this definition explicit enough to be useful as a phenotype in genetic analysis? Is there too much room for interpretation of what is deviant behavior, or of what is excessive consumption? As we will see, whether the behavioral phenotype is narrowly or broadly defined can affect the outcome of the genetic analysis, and even the model of inheritance for the trait.

Animal Models: The Search for Behavior Genes

Several approaches have been used in experimental genetics to study behavior. In one method, two closely related species or two strains of the same species are studied to detect variant behavior phenotypes. Genetic crosses are used to establish whether the behavior is inherited, and if so, to determine the pattern of inheritance. In another approach, individuals exhibiting variant behavior are isolated from a population and interbred to establish a strain with a distinct behavioral pattern. As mentioned earlier, genetic crosses can be used to establish the pattern of inheritance and the number of genes controlling the phenotype. More recently, the effects of single genes on behavior have been studied. In some cases, these studies have led to the isolation and cloning of genes that affect behavior. In the following sections, we describe some examples of behavior genetic studies on experimental organisms.

Open-Field Behavior

In the mid-1930s, the emotional and exploratory behaviors of mice were tested by studying open-field behavior. When mice are introduced into a brightly lit environment, some actively explore the environment while other mice are apprehensive, do not move about, and have elevated rates of urination and defecation. This behavior pattern is under genetic control, since strains exhibiting both types of behavior have been established.

To test the genetic components of this behavior, studies beginning in the 1960s used an enclosed, illuminated box with the floor marked into squares (Figure 16.1). Exploration is tested by counting the movements in different squares, and emotion quantified by counting the number of defecations. The BALB/cJ strain is homozygous for a recessive albino allele, shows low exploratory behavior, and is highly emotional. The C57BL/6j strain has normal pigmentation, is active in exploration, and shows low emotional behavior.

If these two strains are crossed, and the offspring interbred each generation, each generation beyond the F_1 will contain both albino and normally pigmented mice. When tested for open-field behavior, pigmented

FIGURE 16.1 Open-field behavior in mice.

mice behaved like the pigmented C57 parental line, showing active exploration and low emotional behavior; the albino mice behaved like the BALB parental line, with low exploratory activity and high emotional behavior, indicating that the albino gene affects behavior as well as pigmentation.

Tests for heritability indicate that the albino gene accounts for 12% of the genetic variance in exploratory activity, and 26% of the variance in emotional activity. These results indicate that open-field behavior is a polygenic trait, with additive contributions by the controlling genes.

Learning in *Drosophila*

The fruit fly, *Drosophila*, offers several advantages for the study of behavior, including the ability for short-term learning (Table 16.2). To learn, flies are presented with olfactory cues, one of which is accompanied by an electric shock. Flies learn to avoid the odor associated with the shock. Mutant screens have identified a number of mutant genes that affect this learning ability, including the mutants *dunce*, *turnip*, and *rutabaga*.

These single-gene mutants demonstrate the connection between learning and an intracellular signal molecule called cyclic AMP (cAMP). In the cell, cAMP activates a cascade of biochemical reactions that controls gene transcription, and in cells of the nervous system, the responses associated with learning. Cyclic AMP is produced by the enzyme adenyl cyclase (Figure 16.2); the *rutabaga* mutation contains a missense mutation that destroys adenyl cyclase, halting synthesis of cAMP. The *turnip* mutation destroys the activity of a protein that activates adenyl cyclase, and the *dunce* mutation affects the pathway through which adenyl cyclase is recycled. The clustering of these mutations in related biochemical pathways provides strong evidence

TABLE 16.2 Some Behavior Mutants of *Drosophila*

Category	Mutation	Phenotype
Learning	*dunce*	Cannot learn conditioned response.
	turnip	Impaired in learning conditioned response.
	rutabaga	Impaired in several types of learning and memory.
Sexual Behavior	*fruitless*	Males court each other.
	savoir-faire	Males unsuccessful in courtship.
	coitus-interruptus	Males stop copulation prematurely.
Motor Behavior	*flightless*	Lacks coordination for flying.
	sluggish	Moves slowly.
	wings up	Wings are held perpendicular to body.

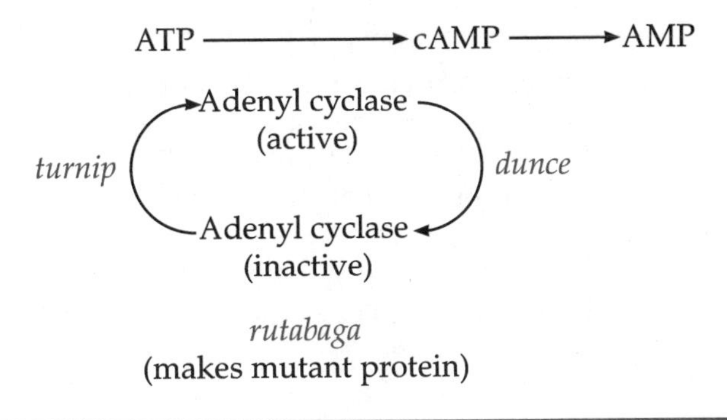

FIGURE 16.2 Metabolic pathway of cyclic AMP (cAMP). The enzyme adenyl cyclase exists in an active form and an inactive form. The *turnip* mutation interferes with the process of activating the adenyl cyclase; the *dunce* mutation prevents the adenyl cyclase from being recycled to the inactive form. The *rutabaga* mutation is a mutation of the structural gene for adenyl cyclase, and encodes a defective protein with no enzymatic activity.

for the involvement of cAMP in learning. Experiments in other organisms support this finding, and imply that some aspects of learning and memory in humans may be controlled by cyclic nucleotides.

Single-Gene Effects on Human Behavior

In Chapter 10, we discussed the role of genes in metabolism. As described in that chapter, mutations that disrupt metabolic pathways or interfere with the synthesis of required gene products can influence the function of cells, and in turn, produce an altered phenotype. If the affected cells are part of the nervous system, then alterations in behavior may be part of the phenotype.

In fact some genetic disorders do affect cells in the nervous system, and in turn, affect behavior. In PKU, for example, brain cells are damaged by excess levels of phenylalanine, which prevent the uptake of other amino acids, causing mental retardation and other behavioral deficits.

In this section we discuss several single-gene defects that have specific effects on the development, structure, and/or function of the nervous system; and as a consequence, affect behavior. Following this we discuss more complex interactions between the genotype and behavior, where the number and action of genes is less well-described, and where effects on the nervous system may be more subtle.

Charcot-Marie-Tooth Disease

Charcot-Marie-Tooth disease
A heritable form of progressive muscle weakness and atrophy. One form, CMT-I, can be produced by mutation at any of three loci.

This disorder of the peripheral nervous system is characterized by the onset of weakness and atrophy of lower leg muscles and may progress to involve arm muscles (teeth are not involved; H. H. Tooth described the condition in 1886). Loss of sensation in the feet and hands may also occur. Affected nerve fibers lose their myelin sheath, and the speed of nerve impulse conduction is greatly reduced. One form of this disorder (CMT-I) is inherited as an autosomal dominant disorder; mutations at any of three loci are associated with this form. One of these has been mapped to chromosome 17, one to chromosome 1, and a third locus is on an as yet unidentified autosome. Individuals with Type I CMT begin to show loss of control over foot movement during middle childhood, and develop muscle wasting and loss of sensation in the lower legs. Muscle atrophy and loss of sensation in hands appears later. Walking becomes difficult, and leg braces are often needed.

The form of CMT-I that maps to chromosome 17 encodes a membrane-associated protein, expressed in cells that form the myelin nerve sheath in the peripheral nervous system. Altered expression of this protein impairs peripheral nerve function, which in turn leads to muscle atrophy. The gene products encoded by the other two loci associated with CMT-I have not been identified.

Friedreich Ataxia

Friedreich ataxia
An autosomal recessive disorder associated with progressive degeneration of the brain, spinal cord, and some peripheral nerves. Onset usually occurs in childhood and results in early death.

This trait, inherited as an autosomal recessive condition, is a progressive neurodegenerative disorder that causes slow degeneration of parts of the brain, including the cerebellum and brain stem. The spinal cord and some peripheral nerves also undergo degeneration. Affected individuals begin to show signs of unsteady gait and uncoordinated limb movements sometime between the ages of 5 and 15 years. Structural degeneration of the brain and spinal cord follow, and death usually occurs in the third decade of life. The frequency of the disorder is about 1 in 22,000, although some populations have a higher frequency because of founder effects.

The gene for Friedreich ataxia (FRDA) maps to the long arm of chromosome 9, in the region 9q12–q13 (Figure 16.3). Recently, a gene from this region has been identified as a candidate for the FRDA locus. This gene encodes a protein expressed in the brain, but not in nonneuronal tissues such

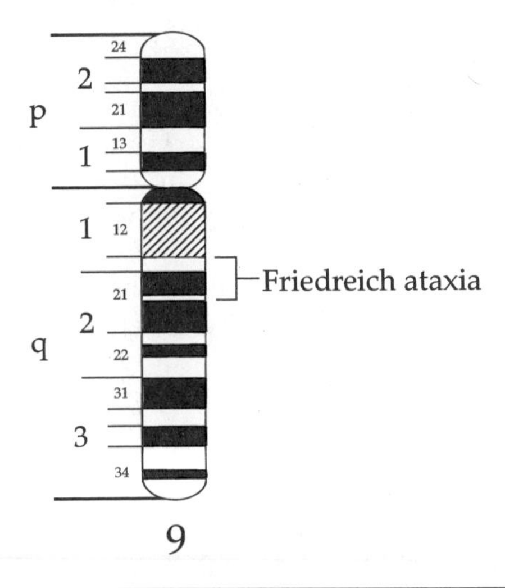

FIGURE 16.3 Locus for Friedreich ataxia on human chromosome 9.

as liver. The amino acid of the protein has been deduced from the order of nucleotides in the mRNA. Searches of computer data bases indicate that this protein is not similar to any known proteins, but does contain a membrane-spanning segment of amino acids, meaning that it is probably membrane-associated and may be involved in the reception or transduction of signals into and/or out of the cell. More work is needed to identify the function of the protein, and to establish the presence of mutant forms of the gene in those affected by FRDA.

Menkes Kinky-Hair Disease

Menkes kinky-hair disease An X-linked recessive disorder of copper metabolism causing neurological impairment, progressive failure of the nervous system, and death in early childhood.

This X-linked recessive disorder, first described in 1962, is a disorder of copper metabolism. Copper is an element essential to life, and is normally present in the body bound to certain proteins that participate in metabolic reactions and those that form connective tissue. Menkes kinky-hair disease (MNK) is characterized by failure to grow immediately after birth and by the appearance of severe neurological impairment within a month or two. No treatment is available, and death occurs from failure of the nervous system between 1 and 10 years of age, with most deaths occurring between the ages of 6 months and 3 years.

The hair of affected individuals is stubby, twisted, or kinky, and of varying diameter along the length of the hair shaft. This abnormality is caused by the defective formation of molecular links in keratin, the protein of hair. The formation of these links during hair growth requires normal serum levels of copper, which is affected in this disorder.

The gene for MNK has been mapped to Xq12–q13, and encodes a copper-binding membrane protein involved in the transport of copper across cell membranes. In many cases, the mutation involves nucleotide deletions, inactivating the gene. This abnormal transport of copper causes motor disorders, severe mental retardation, and degeneration of brain tissues.

Huntington disease
An autosomal dominant disorder associated with progressive neural degeneration and dementia. Adult onset is followed by death 10–15 years after symptoms appear.

Huntington Disease

This autosomal dominant disorder is usually first expressed in mid-adult life as involuntary muscular movements and jerky motions of the arms, legs, and torso. As the condition progresses, there are personality changes, agitated behavior, and dementia. Most affected individuals die within 10 to 15 years after onset of symptoms.

Brain autopsies of affected individuals show damage to several brain regions, including those involved with motor activity. In many cases, the cells in affected regions are altered in shape or destroyed (Figure 16.4).

Although the gene has been isolated, the gene product has not yet been identified. Brains from affected individuals exhibit metabolic abnormalities, and accumulate excessive amounts of a neurotoxic chemical, quinolonic acid. The metabolite is derived from the amino acid tryptophan, and is produced by the enzyme 3-hydroxyanthranilate oxygenase, which is also elevated in the brains of affected individuals. In Huntington disease, it appears likely that abnormal metabolism leads to destruction of brain cells, which in turn affects behavior.

Aggressive Behavior and Brain Metabolism

Recently, preliminary studies have established a link between a mutant gene causing abnormal brain metabolism and forms of aggressive behavior. This finding grew out of a study of one form of X-linked mental retardation associated with behavioral abnormalities, and represents a condition with a direct link between a single gene defect and a phenotype with aggressive and/or violent behavior. A multi-generational affected family showed 14 males with a mild form of mental retardation (Figure 16.5). In eight males where detailed information was available, all showed mild mental retardation or were borderline mentally retarded. All eight showed a characteristic

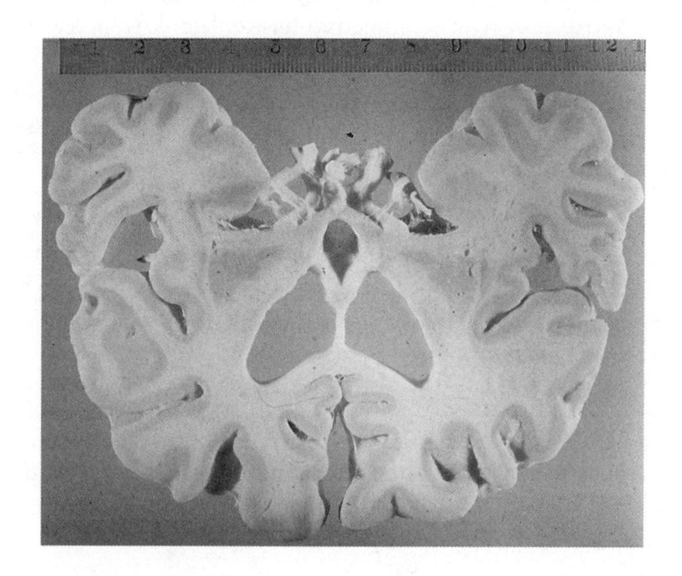

FIGURE 16.4 Damage to brain in Huntington Disease.

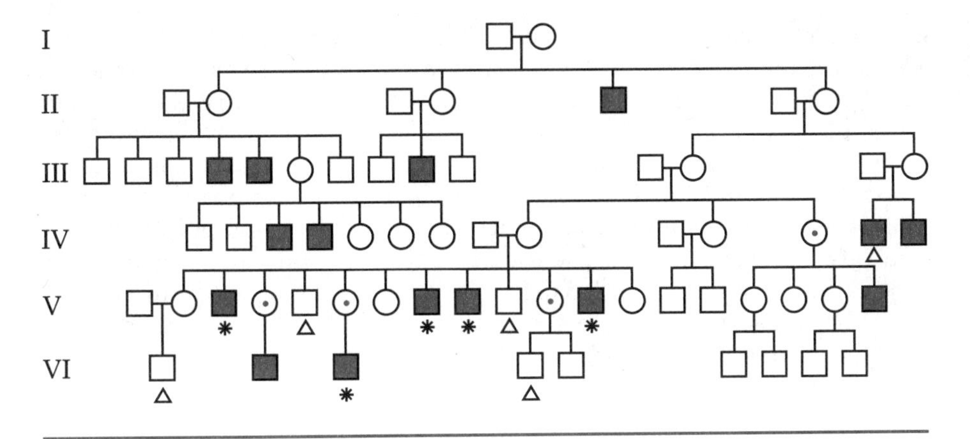

FIGURE 16.5 Segregation of mental retardation, aggressive behavior, and mutation in the monoamine oxidase type A gene (MAOA). Affected males are indicated by the solid symbols. Males marked with an asterisk are hemizygous for the mutant form of MAOA, males marked with a triangle are known to carry the normal allele, and females marked with a dot in the symbol have been tested and shown to be heterozygous carriers.

pattern of aggressive and often violent behavior triggered by anger, fear or frustration. The behavior responses varied widely in levels of violence and time, but included acts of attempted rape, arson, stabbings, and exhibitionism.

Using three separate molecular marker systems, the locus for this form of behavior was mapped to the short arm of the X chromosome in the region of Xp11.23–11.4. One of the structural genes in this region is an enzyme for metabolism of a class of chemicals called neurotransmitters that transmit signals within the brain and nervous system. Urinalysis of the eight affected individuals indicated abnormal levels of several metabolites associated with enzyme monoamine oxidase type A (MAOA). Based on this result, it was postulated that the eight affected males studied may carry a mutation for this enzyme, and that lack of MAOA activity is associated with their behavior pattern.

A subsequent study of the MAOA structural gene in five of the eight affected individuals shows that they all carry a mutation which changes a codon for glutamine into a termination codon. This mutation destroys activity of the MAOA enzyme. In addition, this nucleotide substitution has been found in two obligatory female heterozygotes, and is not present in 12 unaffected males in this pedigree. Lack of MAOA activity in affected males alters metabolism of certain neurotransmitters as reflected in altered urinary concentration of metabolites. Other studies in both humans and animals have concluded that altered metabolism of these same neurotransmitters is associated with aggressive and impulsive behavior.

Further work is needed to determine whether mutations of MAOA are associated with altered behavior in other pedigrees and in animal model systems. In addition, the interaction of MAOA with external factors such as diet, drugs and environmental stress remain to be established. The identification of a specific metabolic defect associated with this behavior pattern suggests that biochemical or pharmacological treatment for this disorder may be possible.

While the use of recombinant DNA markers in linkage studies as a way of identifying single genes that affect behavior has been successful in the case of Huntington disease and the other disorders just described, in other cases, it has led to erroneous results. In a 1987 linkage study using DNA markers, a gene for manic depression was mapped to a region of chromosome 11. Later, individuals from the study group who did not carry the markers developed manic depression, indicating the lack of linkage between markers on chromosome 11 and the gene or genes for manic depression. Similarly, a report on the linkage between DNA markers on chromosome 5 and schizophrenia was found to be coincidental, or at best to apply only to a small, isolated population. These early failures to find single genes that control these disorders have led to the reevaluation of single-gene models for many behavior traits, and to the development of alternative models, as described in the next section.

THE GENETICS OF MOOD DISORDERS AND SCHIZOPHRENIA

Mood disorder
A group of behavior disorders associated with manic and/or depressive syndromes

Mood
A sustained emotion that influences perception of the world

Affect
Pertaining to emotion or feelings

Unipolar disorder
An emotional disorder characterized by prolonged periods of deep depression

Bipolar disorder
An emotional disorder characterized by mood swings that vary between manic activity and depression

Mood disorders, also known as affective disorders, are psychological conditions in which there are profound disturbances of emotions. **Moods** are defined as sustained emotions; **affects** are short-term expressions of emotion. These disorders are characterized by periods of prolonged depression **(unipolar disorder)** or by cycles of depression that alternate with periods of elation **(bipolar disorder).**

Schizophrenia is a collection of mental disorders characterized by psychotic symptoms, delusions, thought disorders, and hallucinations, often called the schizoid spectrum. Schizoid individuals suffer from disordered thinking, inappropriate emotional responses, and social deterioration.

Mood disorders and schizophrenia are complex, often difficult to diagnose, have genetic components, and are widespread conditions. Heredity is regarded as a predisposing factor in both types of conditions, but the mode or modes of inheritance are unclear, and the role of social and environmental factors unknown. Nonetheless, genetic components of these conditions are emerging, and despite recent setbacks in identifying single genes, progress is being made in forming genetic models of these disorders.

Mood Disorders: Unipolar and Bipolar Illnesses

The lifetime risk for a clinically identifiable mood disorder is 8% to 9%. Depression (unipolar illness), the most common of these disorders, accounts for about 10% of all patients seen in nonpsychiatric clinical settings. Depression is more common in females (about a 2:1 ratio), begins in the fourth or fifth decade of life, and is often a protracted or recurring condition. Depression has several characteristics, including weight loss, insomnia, poor concentration, irritability and anxiety, and lack of interest in surrounding events.

About 1% of the U.S. population suffers from bipolar illness. The age of onset is during adolescence or the second and third decades of life, and males and females are at equal risk for this condition. Manic activity is characterized by hyperactivity, acceleration of thought processes, low attention span, creativity, and feelings of elation or power.

The link between mood disorders and genetics is derived from twin studies, with concordance of 57% in MZ twins and 14% in DZ twins. In addition, the rates of mood disorders in adopted individuals and their biological parents support the role of genetic factors. The evidence for genetic factors is stronger for bipolar disorders than for unipolar disorders.

Several attempts have been made to map genes for bipolar illnesses. Studies have presented evidence for an X-linked form of bipolar disorder, and an autosomal locus on chromosome 11 was also proposed as a locus for bipolar mood disorder. However, in both cases, extension of the analysis to include new individuals—and the appearance of symptoms in individuals not carrying the RFLP markers—invalidated the results, and the studies were retracted.

The failures to find linkage between genetic markers and manic depression does not undermine the role of genes in bipolar illness, but means that new strategies of linkage analysis may be required to identify the gene or genes involved. In fact, a new strategy, called an association study, is being used in a worldwide effort to screen for genes controlling bipolar illness. Association studies use DNA markers, but follow the inheritance of the marker and the disorder (bipolar illness in this case) in unrelated individuals affected with the disorder rather than following the trait in large families. The idea is to identify portions of the genome that are more common in affected individuals than in those without the trait. Once these regions have been identified, they will be closely studied to search for genes responsible for bipolar illness.

In addition to its elusive genetic nature, bipolar illness remains a fascinating behavioral disturbance because of its close association with creativity (see Concepts and Controversies, p. 443). Many great artists, authors, and poets have been afflicted with manic depressive illness (Figure 16.6). Studies on the nature of creativity have shown that the thought patterns of the creative mind parallel those of the manic stage of bipolar illness. In her book, *Touched with Fire,* Kay Jamison explores the relationship among genetics, neuroscience, and the lives and temperaments of creative individuals including Byron, Van Gogh, Poe, and Virginia Woolf.

Schizophrenia

Schizophrenia
A behavioral disorder characterized by disordered thought processes and withdrawal from reality. Genetic and enviornmental factors are involved in this disease.

Schizophrenia is a relatively common disorder, affecting about 1% of the population. The disorder usually appears in late adolescence or early adult life. Because of its prevalence, it has been estimated that half of all hospitalized mentally ill and mentally retarded individuals are schizophrenic, and that up to 25% of all hospital beds are occupied by affected individuals.

Schizophrenia is a disorder of the thought processes rather than of mood. Diagnosis is often difficult, and there is notable disagreement on the

FIGURE 16.6 Portrait of Virginia Woolf. This author and poet was affected by manic depressive illness. Like others similarly affected, she often commented on the relationship between creativity and her illness.

definition of schizophrenia because it has no single distinguishing feature and causes no characteristic brain pathology. Some features of the disorder include

- Psychotic symptoms, including delusions of persecution
- Disorders of thought; loss of the ability to use logic in reasoning
- Perceptual disorders, including auditory hallucinations (hearing voices)
- Behavioral changes, ranging from mannerisms of gait and movement to violent attacks on others
- Withdrawal from reality and inability to participate in normal activities

Several models have been proposed for schizophrenia, and in general fall into two groups: models in which biologic factors (including genetics) play a major role and environmental factors are secondary; and conversely, models in which environmental factors are primary and biologic factors are secondary. Some evidence points to metabolic differences in the brains of schizo-

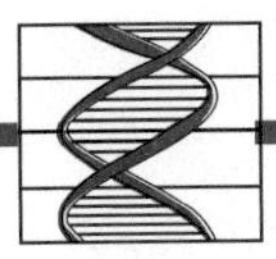

Concepts and Controversies

The Link Between Madness and Genius

As long ago as the 4th century BC, Aristotle observed that talented philosophers poets and artists tend to have mental problems. Since then, this idea has become part of our popular wisdom and is summarized by the statement that "There is a thin line between genius and madness." In recent years, a substantial amount of evidence has accumulated indicating that distinguished artists, composers, authors and poets suffer from mood disorders, particularly depression and manic depression at a rate 10 to 30 times more frequently than the general population. From these studies, it now appears that there may be a link between bipolar disorder and creativity. These studies have been documented in several recent books, such as *Touched With Fire*, by Kay Jamison, *The Price of Greatness* by Arnold Ludwig, and *The Broken Brain* by Nancy Andreasen.

The pedigree on the cover of this text traces the history of mental disturbances in the Tennyson family. Through seven generations, each generation had members suffering from mental instability ranging from unstable moods through bipolar disorders to insanity. The poet Alfred Lord Tennyson, himself affected by life-long bouts of depression, was surrounded by family members affected with mental instability. This list includes two of his brothers, his sister, his father, two uncles, an aunt, and his grandfather. Along with the family temperament went a consuming passion and recognized genius for poetry and verse. Although today Alfred is the best known of the family poets, his brothers Charles and Fredrick published volumes of poetry, and each won prestigious awards for translating ancient Greek poems and classics. Alfred's siblings and his aunt and uncles also wrote verse and one of Alfred's nieces became a poet and playwright.

As neurobiologists seek to explain the link between creative impulses and bipolar disorder, several characteristic features of manic depression are emerging. Brain imaging studies have shown that different regions of the brain are affected during bouts of depression and during manic stages. In the brains of manic depressive individuals, there are characteristic and distinctive patterns of metabolism in the prefrontal cortex, the part of the brain associated with intellect. In addition, there are blood flow differences between affected and unaffected individuals. Taken together, these and other studies indicate that neural connections in the brains of manic depressives may have a different pattern than unaffected individuals, and that the transition from depression to mania back to depression may stimulate mental activity and creativity.

From a gentic standpoint, family studies and twin studies strongly argue that a predisposition for manic depression is inherited, although no single gene controlling this trait has yet been identified. The search for the genetic basis of mood disorders is continuing, employing different models of control. Perhaps in the near future, the relationship between artistic creativity and genetics can be explained by gene action and/or interaction in the brain and nervous system.

phrenics compared to those of normal individuals (Figure 16.7), but it is unclear whether these differences are genetic. Overall, however, the best evidence supports the role of genetics as a primary factor in schizophrenia, with environmental factors needed for full expression. The influence of genotype on schizophrenia can be seen by examining risk factors for relatives of schizophrenics (see Figure 16.8 on page 445). Overall, relatives of affected individuals have a 15% chance of developing the disorder (as opposed to 1% among unrelated individuals). Using a narrow definition

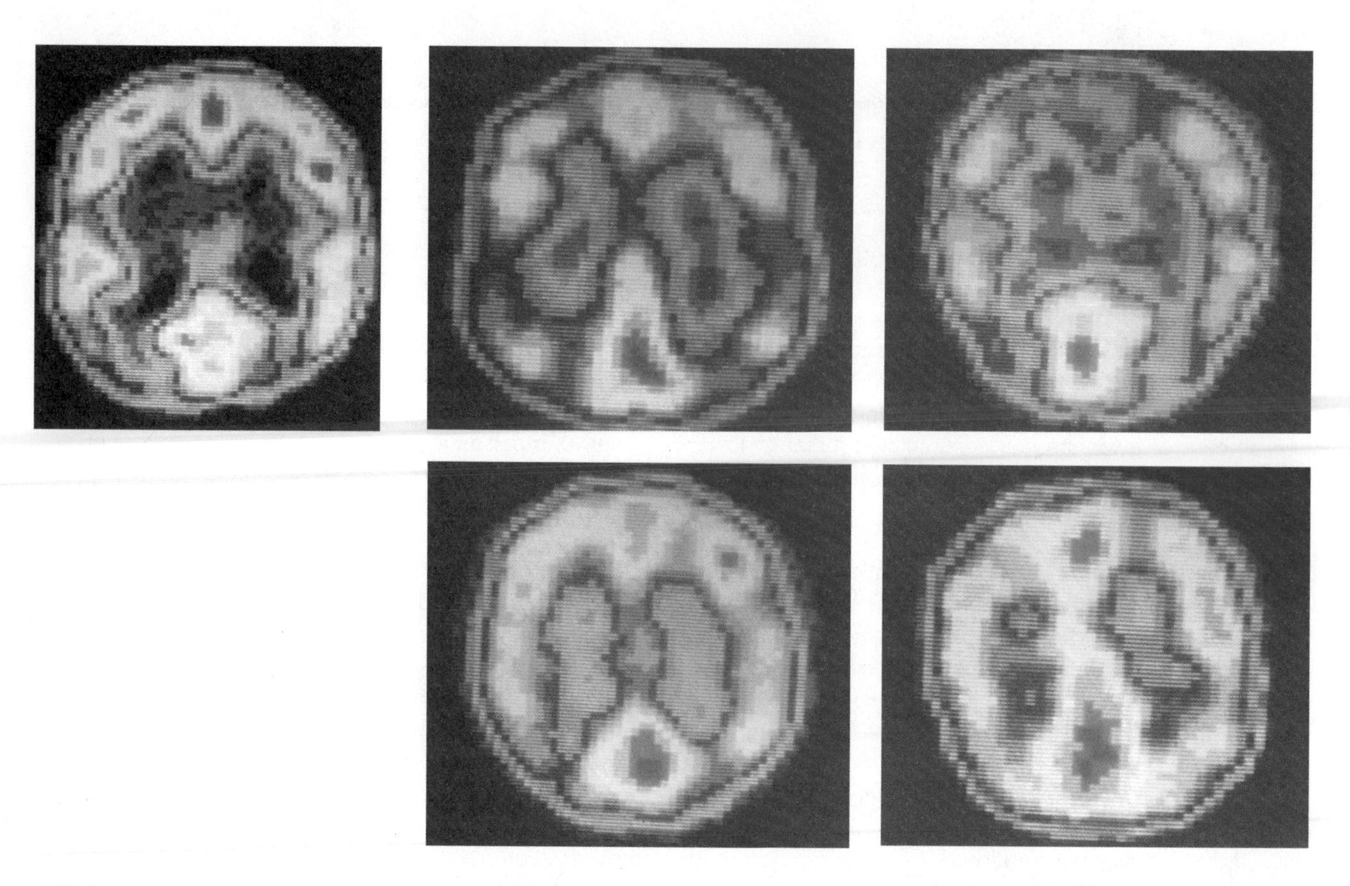

FIGURE 16.7 Brain metabolism in a set of MZ quadruplets, all of whom suffer from varying degrees of schizophrenia. These scans of glucose utilization, visualized by positron emission tomography (PET) show low metabolism rates in the frontal lobe (top of each scan) compared to non-schizophrenics. The frontal lobe is the location of human cognitive ability.

of schizophrenia, the concordance value for MZ twins is 55% versus 10% for DZ twins. MZ twins raised apart show the same level of concordance as MZ twins raised together. If a broader definition is used, combining schizophrenia and borderline or schizoid personalities, the concordance for MZ twins approaches 100%, and the risk for sibs, parents, and offspring of schizophrenics is about 45%.

The mode of inheritance of schizophrenia is unknown. Pedigree and linkage studies have suggested loci on the X chromosome, and a number of autosomes—including chromosomes 5, 11, and 19—have been identified as sites of genes contributing to this condition. Although all the linkage studies have been contested and contradicted by other studies, several tentative conclusions can be drawn about the genetics of schizophrenia. Genetic models indicate that a polygenic model with a single major gene making most of the contribution is consistent with family studies as well as with other observations, including monozygotic concordance and incidence of the disorder.

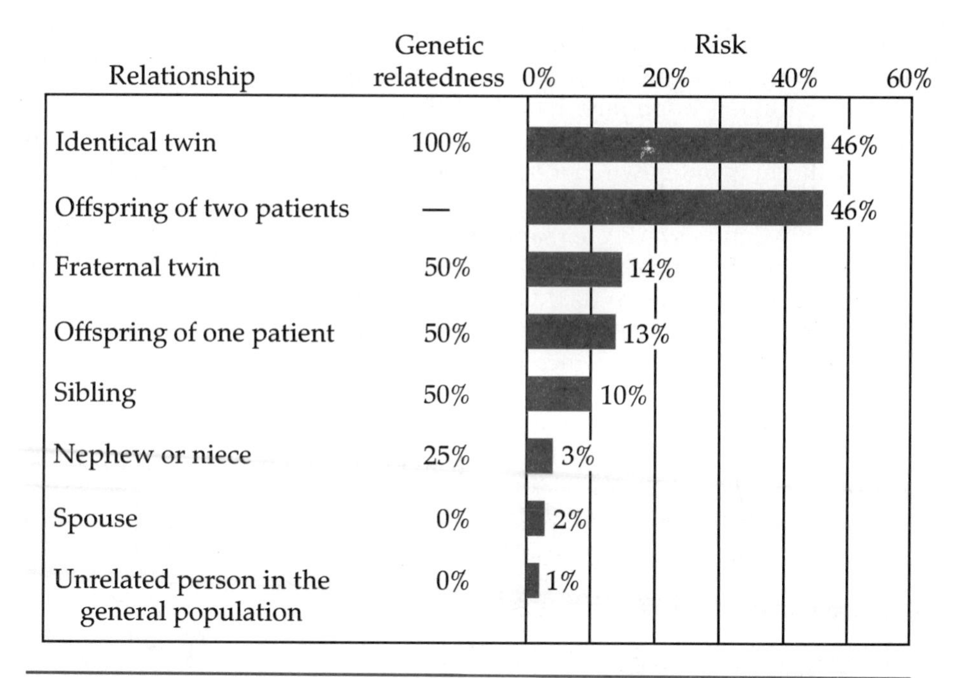

FIGURE 16.8 Lifetime risk of schizophrenia for varying degrees of relationship. The observed risks are more compatible with a multifactorial mode of transmission than a single-gene or polygenic model with a major gene effect.

Schizophrenia is probably genetically heterogenous, and several loci may be involved, even if there is a major gene effect. Studies with chromosomal translocations and DNA markers identified a locus on chromosome 5 that may be responsible for one type of schizophrenia, and linkage studies with DNA markers show an association between pseudoautosomal RFLP markers (see Chapter 5 to review pseudoautosomal inheritance) and schizophrenia, indicating that a locus on the sex chromosomes may play a role. For schizophrenia, familial linkage studies appear to be more valuable than association studies. Some eighteen research centers in Europe are currently screening for major genes controlling bipolar illness and schizophrenia using about 150 DNA markers from all regions of the genome. The recognition and definition of subphenotypes (also known as endophenotypes) under the term schizophrenia will also be needed to genetically analyze this puzzling condition.

Other Multifactorial Behavior Traits

Tourette Syndrome

Tourette syndrome (GTS) as a familial disorder was first described by Gilles de la Tourette in 1885. This condition is characterized by both motor and behavioral disorders. About 10% of affected individuals have a family history of the condition; males are affected more frequently than females (3:1), and onset is usually between 2 and 14 years of age. The disorder is characterized

Tourette syndrome A behavioral disorder characterized by motor and vocal tics and inappropriate language. Genetic components are suggested by family studies showing increased risk for relatives of affected individuals.

by episodes of motor and vocal tics that can progress to more complex behaviors involving a series of grunts and barking noises. The vocal tics include outbursts of profane and vulgar language and parrot-like repetition of words spoken by others. Because of variable expression, the incidence of the condition is unknown, but it has been suggested that the disorder may be very common.

Family studies indicate that biological relatives of affected individuals are at significantly greater risk for GTS than relatives of unaffected controls. Linkage studies on over 1000 GTS families have been performed, and a recent review of these studies proposed that a model of autosomal dominant inheritance with incomplete penetrance and variable expression is most compatible with the linkage results. A worldwide collaborative effort using over 600 DNA markers to map a major gene associated with GTS has excluded over 85% of the genome as a site for the GTS gene. If a major gene for GTS exists, it should be identified in the next few years. However, as more of the genome is excluded in this search, it is possible that the genetic model of autosomal dominant inheritance will need to be reexamined.

Alzheimer Disease

Alzheimer disease A heterogenous condition associated with the development of brain lesions, personality changes, and degeneration of intellect. Genetic forms are associated with loci on chromosomes 14, 19, and 21.

The behavioral symptoms of Alzheimer disease (AD) begin with loss of memory and a progressive dementia that involves disturbances of speech, motor activity, and recognition. There is an ongoing degeneration of personality and intellect, and eventually, affected individuals are unable to care for themselves.

Brain lesions (Figure 16.9) accompany these behavioral changes, and were first described by Alois Alzheimer in 1906. These brain lesions are

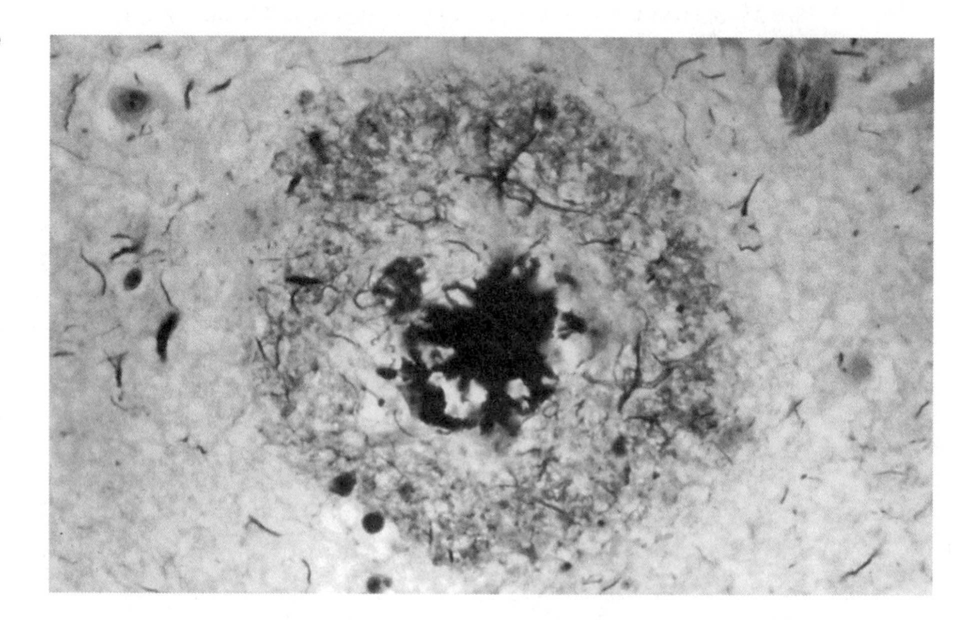

FIGURE 16.9 Lesion or plaque in brain of individual with Alzheimer syndrome. The deposit of protein is surrounded by a ring of degenerating nerve cells.

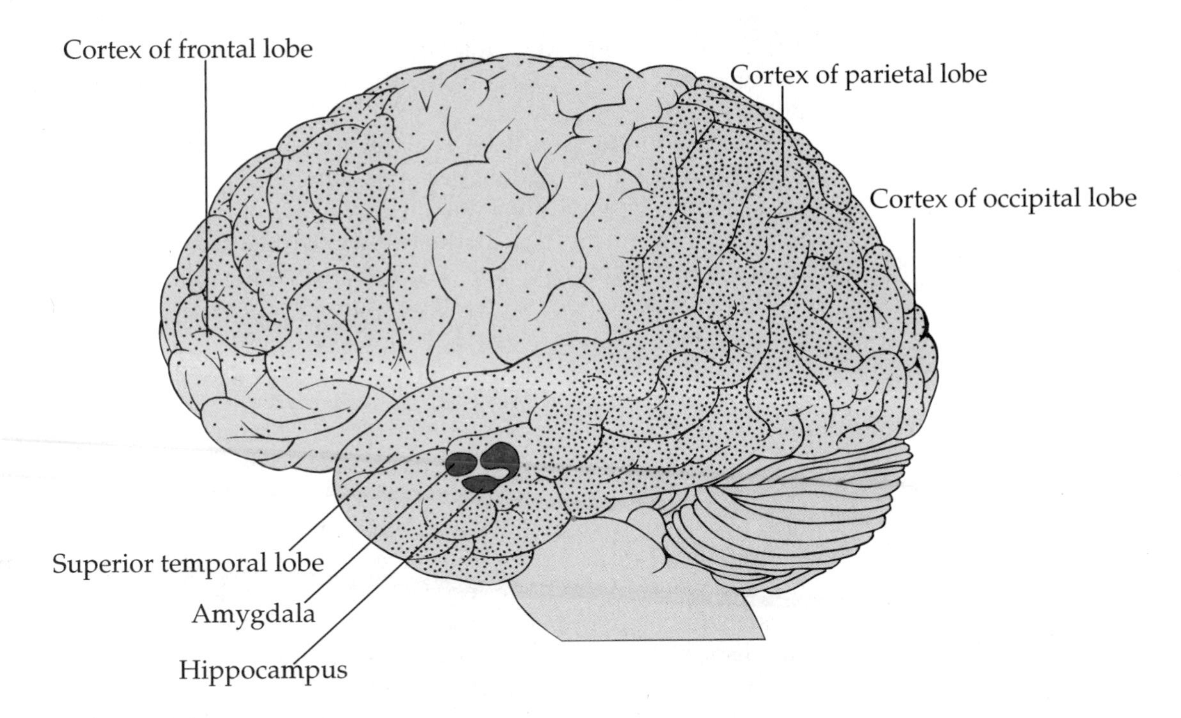

FIGURE 16.10 Location of brain lesions in Alzheimer disease. Plaques are most heavily localized in the amygdala and hippocampus. These brain regions are part of the limbic system, a region of the brain involved in controlling memory.

formed from a protein fragment called amyloid beta-protein, which accumulates outside cells in aggregates known as senile plaques. The proteins and cells that become embedded in plaques cause the degeneration and death of nearby neurons, affecting selected regions of the brain (Figure 16.10). Formation of senile plaques is not specific to AD; almost everyone who lives to the age of eighty will have such lesions. The difference between normal aging of the brain and Alzheimer disease appears to be the number of such plaques (greatly increased in AD) and the time of accumulation (decades earlier in AD).

It is estimated that Alzheimer disease affects 10% of the U.S. population over the age of 65 and 50% of those over the age of 80, and that the cost of treatment and care for those affected is over $80 billion.

The genetics of Alzheimer disease is somewhat complex; less than 50% of all cases can be directly traced to genetic causes, indicating that the environment may play a significant role in this disorder. We will first examine the genetic evidence and then discuss some of the proposed environmental factors associated with this progressive and degenerative disorder. Studies of families with members affected with Alzheimer disease indicate that the familial incidence is 43%, with the age of onset ranging from 25 years to 85 years of age.

The gene encoding the beta protein (known as APP) is located on the long arm of chromosome 21, and mutations of this gene are responsible for an early-onset form of AD. Mutations in APP are inherited as an autosomal dominant trait, and this form of the disorder is called Alzheimer-disease 1 (AD 1).

Other individuals with early-onset Alzheimer disease and those with late-onset disease have mutant genes on other chromosomes, or a combination of genetic and environmental factors. A second form of early-onset AD, perhaps accounting for the majority of all early-onset cases, has been mapped to the long arm of chromosome 14. One form of late-onset Alzheimer disease (AD 2) has been mapped to the long arm of chromosome 19, in the region 19q12–q13.2.

It is not clear whether all cases of AD have a genetic explanation, with other genes yet to be discovered, or whether environmental factors are important in the development of this behavioral disorder. In an analysis of 232 families in which AD is segregating, it was concluded that components of AD are multifactorial, indicating an important role for other genetic and nongenetic factors. Among the nongenetic factors that may play a role in the development of AD, two have received the most attention. Aluminum is found deposited in the brain lesions associated with AD; thus, it has been suggested that aluminum intake and differences in the uptake, transport, and metabolism of this metal may play a role in the development of AD.

Prion
An infectious protein that is not a virus, but is the cause of several disorders, including Creutzfeld-Jakob syndrome

One of the proteins present in the brain lesions of AD is similar to that found in infectious proteins called **prions**. Prions are known to cause neurodegenerative diseases in domesticated animals, and are associated with several human neurological diseases including kuru and Creutzfeld-Jacob disease. It has been proposed that exposure to prions and prion infection may be a factor in the development of AD.

It has already been established that AD is genetically heterogenous, and that mutations at any of several loci can produce the AD phenotype. In addition, the fact that many cases of AD cannot be traced to a genetic source may indicate that there is more than one cause for AD, and the role of nongenetic influences and their mechanisms should be considered to define all the risk factors for this debilitating condition.

Alcoholism

As a behavioral disorder, the excessive consumption of alcohol has two important components. First, consumption of large amounts of alcohol can cause cell and tissue damage to the nervous system and other tissues of the body, and result in altered behavior, including hallucinations, delirium, and loss of recent memory. These effects are secondary to the behavior that results in a decline in the ability to function in social settings, the workplace, and the home.

It is estimated that 75% of the adult U.S. population consumes alcohol, and that of these, about 10% will be classified as alcoholics, with a male:female ratio of about 4:1. From the genetic standpoint, alcoholism is most likely a genetically influenced, multifactorial (genetic and environmental) disorder. The role of genetic factors in alcoholism is indicated by a number of findings:

- In mice, selection experiments indicate that alcohol preference can be selected for; some strains of mice will select 75% alcohol over water, while others will shun all alcohol.
- There is a 25% to 50% risk of alcoholism in sons and brothers of alcoholic men.
- There is a 55% concordance rate for alcoholism in MZ twins, with a 28% rate in same-sex DZ twins.
- Sons adopted by alcoholic men show a rate of alcoholism more like that of their biological father.

The biochemical pathways involved in the metabolism of alcohol are known, but mutations in genes controlling these pathways are unlikely to play a major role in alcoholism. The nature of the genetic influence on alcoholism and its site or sites of action are unknown; segregation analysis in families with alcoholic members has produced evidence against the Mendelian inheritance of a single major gene, and for multifactorial inheritance. Other researchers have advanced a single-gene model and reported an association between an allele of a gene encoding a neurotransmitter receptor protein (called D2) and alcoholism. This evidence is based on the finding that in brain tissue, the A1 allele of the D2 gene was found in 69% of the samples from severe alcoholics, but only 20% of the samples from nonalcoholics, implying that an impaired function of the A1 allele is involved in alcoholism.

Subsequent linkage and association studies on the A1 allele have produced conflicting results. In some studies, there was no statistical difference between the incidence of the A1 allele between alcoholics and nonalcoholics; in other studies there was a difference, although of borderline statistical significance. Taken together, however, the available studies have failed to show a link between the A1 allele and alcoholism. They have also failed to show any relationship between abnormal neurotransmitter metabolism or receptor function and alcoholic behavior.

The search for genetic factors in alcoholism illustrates the problem of selecting the proper genetic model for analysis of behavioral traits. Although segregation and linkage studies indicate there is no major gene for alcoholism, association studies have provided a weak correlation between the D2 gene and alcoholism. However, this single-gene model has produced no evidence for a cause-and-effect relationship between the gene and behavior. If a multifactorial model involving a number of genes, each with a small additive effect, is invoked, the problem becomes more complicated. How do you prove or disprove that a given gene contributes, say, 10% to the behavioral phenotype? At present, the only method would involve studying thousands of individuals to find such effects.

Sexual Orientation

In sexual behavior, most humans are heterosexual and show a preference for the opposite sex, but a fraction of the population is homosexual and shows a preference for sexual activity with members of the same sex. These behavioral variations in sexual activity have been recorded since ancient times, but biologic models for these behaviors have been proposed only recently.

Among the biologic models, the role of genetics in sexual orientation has been investigated by twin studies and adoption studies. In one recent twin study that employed 56 MZ twins, 54 DZ twins, and 57 genetically unrelated adopted brothers, concordance for homosexuality was 52% for MZ twins, 22% for DZ twins, and 11% for the unrelated adopted sibs. Using these values to calculate heritability assuming different numbers of genes involved, overall heritability was calculated to range from 31% to 74%. In another study investigating homosexual behavior in women employing 115 twin pairs and 32 genetically unrelated adopted sisters, heritability calculated from concordance values was found to range from 27% to 76%. These results indicate a strong genetic component to homosexual behavior, but have been challenged on the grounds that the results can be affected by the phrasing of the interview questions and by the methods used to recruit participants. However, the average heritability estimates from these studies parallels those from the Minnesota Twin Project studying MZ twins separated at birth and reared apart. Further twin studies are needed to determine whether the heritability values are accurate. If confirmed, the studies to date indicate that homosexual behavior is a multifactorial trait involving several genes as well as unidentified environmental components.

Recent work using RFLP markers and homosexual behavior has indicated linkage between one subtype of homosexuality and markers on the long arm of the X chromosome (Figure 16.11). In this study, a two-step

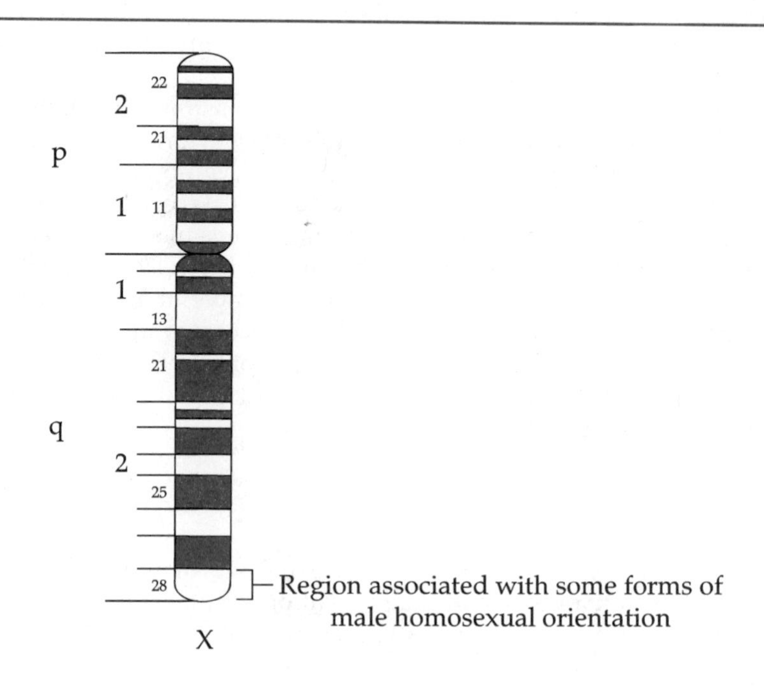

FIGURE 16.11 Region of the X chromosome found by linkage analysis to be associated with at least one form of male homosexual orientation.

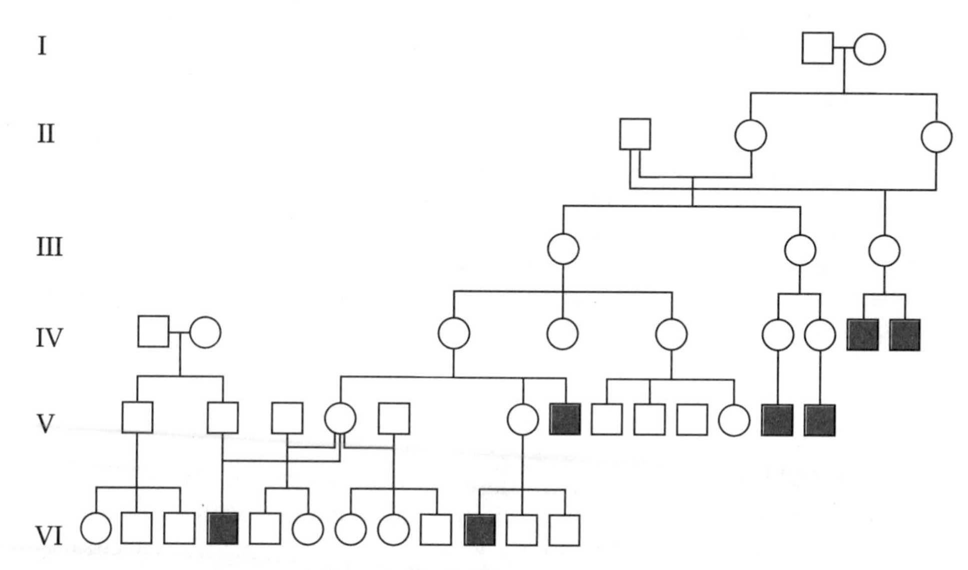

FIGURE 16.12 Pedigree showing apparent maternal transmission of male homosexuality. Affected males (■), unaffected males (□).

approach was employed. First, family histories were collected from 114 homosexual males. From 76 randomly selected individuals, pedigree analysis was performed, using interviews with male relatives to ascertain sexual preference. The results indicated that maternal uncles and sons of maternal aunts had a higher rate of homosexual orientation than paternally related individuals. This indicated the possibility of maternal inheritance. A further pedigree analysis was conducted using 38 families in which there were two homosexual brothers, with the idea that this might show a stronger trend for maternally controlled inheritance. The results from this analysis do show a stronger trend toward maternal inheritance, and an absence of paternal transmission (Figure 16.12).

Using information from the pedigree analysis, the second part of the study employed DNA markers to determine whether an X-linked locus or loci might be involved in the transmission of homosexual behavior in these families. A total of 22 DNA markers from the X chromosome were tested in the 38 families with two homosexual brothers, and linkage was detected with markers from the distal region of the long arm, in the Xq28 region (Figure 16.11). This region is large enough to contain several hundred genes, and more work will be needed to confirm the linkage relationship and to search the region for a locus affecting sexual orientation.

Two important factors related to this study need to be mentioned. First, seven sets of homosexual brothers did not coinherit all the markers in the Xq28 region, indicating that genetic heterogeneity or nongenetic factors may play a significant role in this behavioral variation. Second, if confirmed, the study cannot estimate what fraction of homosexual behavior is related to the Xq28 region, nor whether lesbian sexual behavior is influenced by this region. In spite of its preliminary nature, this work represents one of the best indications that specific genetic factors influence at least one subtype of male sexual behavior.

Summing Up: The Current Status of Human Behavior Genetics

In reviewing the current status of human behavior genetics, several elements are apparent. Almost all studies of complex human behavior have provided only indirect and correlative evidence for the role of specific genes. Segregation studies and heritability estimates indicate that these complex traits most likely involve multiple genes, either with a major gene effect or with several genes contributing in an additive way to the phenotype. Searches for single-gene effects have to date proven unsuccessful, and initial reports of single genes controlling bipolar illness, schizophrenia, and alcoholism have been retracted, unconfirmed, or at best limited to small isolates.

The multifactorial nature of these complex traits means that methodologies for the investigation of genes with small, incremental effects will have to be refined, and new combinations of the classical methods of twin studies and adoption studies with molecular methods will have to be devised. By their nature, twin studies and adoption studies involve small numbers of individuals. For example, less than 300 pairs of MZ twins raised apart have been identified worldwide. Where traits involve multiple genes, replication of results to confirm gene assignments can require detailed examination of thousands of individuals in hundreds of families. This process is necessarily slow and labor-intensive.

Perhaps newer approaches such as association studies in combination with new and quantitative ways of defining phenotypes can be used to provide a meaningful way of dissecting the genetic components of a behavioral phenotype.

Even the limited evidence currently available indicates that the environment plays a significant role in the behavioral phenotype. As confirmation of the role of genes in behavior becomes available, investigations on the role of environmental factors cannot be neglected. The history of human behavior genetics in the eugenics movement of the early part of this century provides a lesson in the consequences of overemphasizing the role of genetics in behavior. Attempts to provide single-gene explanations for behavioral "traits" inhibited the growth of human genetics as a discipline.

The increasing evidence for the involvement of specific genes in controlling human behavior traits has implications outside the laboratory for society at large. As discussed in Chapter 8, the Human Genome Project has raised questions about how genetic information will be disseminated and used, who will have access to this information, and under what conditions. For genes affecting behavior, these same concerns need to be addressed. If genes for alcoholism or homosexuality can be identified, should this information be used to predict an individual's future behavior patterns? Will this information be used to discriminate in employment or insurance?

Many behavioral phenotypes such as Huntington or Alzheimer disease are clearly regarded as abnormal. Few would argue against the development of treatments for intervention and perhaps prevention of these conditions. When do behavior phenotypes move from being abnormal to being a variant? If there is a connection between bipolar illness and creativity, to what extent should this condition be treated? If genes that influence sexual orientation are identified, will this behavior be regarded as a variant, or as a condition that should be treated and/or prevented?

While research can provide information on the biological factors that play a role in determining human behavior, it cannot provide answers to questions of social policy. Those answers have to be formulated using information from research to formulate social policy and laws.

SUMMARY

1. Many forms of behavior represent complex phenotypes with multifactorial inheritance. Single genes that affect behavior do so as a consequence of their effect on the development, structure, and function of the nervous system.

2. The methods used to study inheritance of behavior encompass both classical methods of linkage and pedigree analysis, newer methods of recombinant DNA analysis, and new combinations of techniques such as twin studies combined with molecular methods. Refined definitions of behavior phenotypes are also being used in the genetic analysis of behavior.

3. Results from work on experimental animals indicate that behavior is under genetic control, and have provided estimates of heritability. The molecular basis of single-gene effects in some forms of behavior has been identified, and provide useful models to study gene action and behavior.

4. Several single-gene effects on human behavior are known; most of these affect the development, structure, or function of the nervous system and consequently affect behavior.

5. Bipolar illness and schizophrenia are common behavior disorders, each affecting about 1% of the population. General models of single-gene inheritance for these disorders have not been supported by extensive studies of affected families, and more complex forms of multifactorial inheritance seem likely.

6. Other multifactorial traits that affect behavior include Tourette syndrome, Alzheimer disease, alcoholism, and sexual orientation.

7. Twin studies in combination with the use of molecular markers have identified a region of the X chromosome that may affect one form of homosexual behavior. If this study can be confirmed, and the gene or genes identified, this combination of methods may be useful as a model for future work on identifying genes that affect behavior.

Questions and Problems

1. The two main affective disorders are manic depression and schizophrenia. What are the essential differences and similarities between these disorders?

2. In a long-term study of over 100 pairs of MZ and DZ twins separated shortly after birth and reared apart, one of the conclusions was that "general intelligence or IQ is strongly affected by genetic factors." The study concluded that about 70% of the variation in IQ is due to genetic variability (review the concept of heritability in Chapter 15). Discuss this conclusion, and include in your answer the relationship between IQ and intelligence, and to what extent these conclusions can be generalized. In evaluating their conclusion, what would you like to know about the twins?

3. One of the models for behavioral traits in humans involves a form of interaction known as epistasis. In a simplified example involving two genes, the expression of one gene affects the expression of the other. How might this interaction work, and what patterns of inheritance might be shown?

4. Perfect pitch is the ability to name a note when it is sounded. In a study of this behavior, perfect pitch was found to predominate in females (24 out of 35 in one group). In one group of 7 families, 2 individuals had perfect pitch. In 2 of these families, the affected individuals included a parent and a child. In another group of 3 families, 3 or more members (up to 5) had perfect pitch, and in all 3 families, 2 generations were involved. Given this information, what if any conclusions can you draw as to whether this behavioral trait might be genetic? How would you test your conclusion? What further evidence would be needed to confirm your conclusion?

5. Opposite to perfect pitch is tune deafness, the inability to identify musical notes. In one study, a bimodal distribution in populations was found, with frequent segregation in families and sib pairs. The author of the study concluded that the trait might be dominant. In a family study, segregation analysis suggested an autosomal dominant inheritance of tune deafness with imperfect penetrance. One of the pedigrees is presented below. On the basis of the results, do you agree with this conclusion? Could perfect pitch and tune deafness be alleles of a gene for musical ability?

6. If you discover a single-gene mutation that leads to an altered behavioral phenotype, what would you suspect about the nature of the mutant gene and the focus of its action?

7. What are the advantages of using *Drosophila* for the study of behavior genetics? Can this organism serve as a model for human behavior genetics? Why or why not?

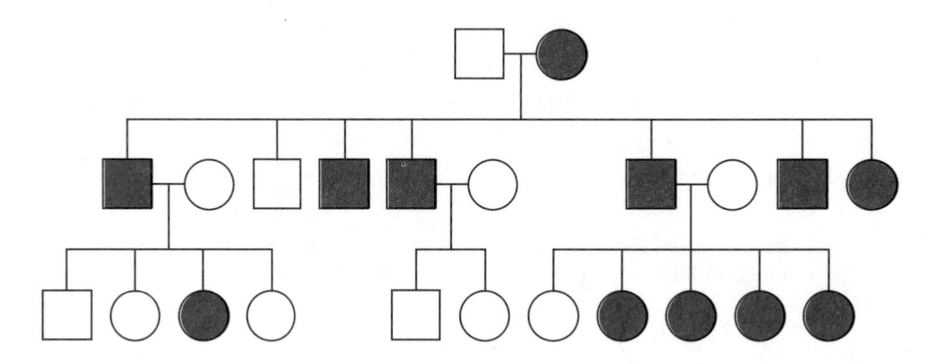

For Further Reading

Aldhous, P. 1992. The promise and pitfalls of molecular genetics. *Science* **257:** 164–165.

Aston, C. and Hill, S. 1990. Segregation analysis of alcoholism in families ascertained through a pair of male alcoholics. Am. J. Hum. Genet. **46:** 879–887.

Blum, K., Noble, E., Sheridan, P., Montgomery, A., Ritchie, T., Jagadeeswaran, P., Nogami, H., Briggs, A., and Cohn, J. 1990. Allelic association of human dopamine D(2) receptor gene in alcoholism. *J.A.M.A.* **263:** 2055–2060.

Bolos, A., Dean, M., Lucas-Derse, S., Ramsburg, M., Brown, G., and Goldman, D. 1990. Population and pedigree studies reveal a lack of association between the dopamine D(2) receptor gene and alcoholism. *J.A.M.A.* **264:** 3156–3160.

Brunner, H., Nelen, M., vanZandvoort, P., Abeling, N., van Gennip, A., Walters, E., Kulper, M., Ropers, H. and van Dost, B. 1993. X-linked borderline mental retardation with prominent behavioral disturbance: phenotype, genetic localization, and evidence for disturbed monoamine metabolism. Am. J. Hum. Genet. **52:** 1032–1039.

Brunner, H., Nelen, M., Breakfield, X., Ropers, H. and van Oost, B. 1993. Abnormal behavior associated with a point mutation in the structural gene for monoamine oxidase A. Science **262:** 578–580.

Comings, D., Comings, B., Devor, E., and Cloninger, C. 1984. Detection of major gene for Gilles de la Tourette syndrome. *Am. J. Hum. Genet.* **36:** 586–600.

Coryell, W., Endicott, J., Keller, M., Andreasen, N., Grove, W., Hirschfield, R., and Scheftner, W. 1989. Bipolar affective disorder and high achievement: a familial association. *Am. J. Psychiatry* **146:** 983–988.

Devore, E., and Cloninger, C. 1989. The genetics of alcoholism. *Ann. Rev. Genet.* **23:** 19–36.

Duclos, F., Boschert, U., Sirugo, G., Mandel, J-L., Hen, R., and Koenig, M. 1993. Gene in the region of Friedreich ataxia locus encodes a putative transmembrane protein expressed in the nervous system. *Proc. Nat. Acad. Sci.* **90:** 109–113.

Haines, J. 1991. The genetics of Alzheimer disease—a teasing problem. *Am. J. Hum. Genet.* **48:** 1021–1025.

Hamer, D., Hu, S., Magnuson, V., Hu, N., and Pattatucci, A. 1993. A linkage between DNA markers on the X chromosome and male sexual orientation. *Science* **261:** 321–327.

Jamison, Kay. 1993. *Touched With Fire: Manic Depressive Illness and the Artistic Temperament.* New York: Free Press.

Kendler, K., Heath, A., Neale, M., Kessler, R., and Eaves, L. 1992. A population-based twin study of alcoholism in women. *J.A.M.A.* **268:** 1877–1882.

Lawrence, S., Keats, B., and Morton, N. 1992. The AD1 locus in familial Alzheimer disease. *Ann. Hum. Genet.* **56:** 295–301.

Levay, S. 1991. A difference in hypothalamic structure between heterosexual and homosexual men. *Science* **253:** 1034–1037.

Patel, P., Roa, B., Welcher, A., Schoener-Scott, R., Trask, B., Pentao, L., Snipes, G., Garcia, C., Francke, U., Shooter, E., Lupski, J., and Suter, U. 1992. The gene for the peripheral myelin protein PMP-22 is a candidate for Charcot-Marie-Tooth disease type 1A. *Nature Genet.* **1:** 159–165.

Pericak-Vance, M., Bebout, J., Gaskell, P., Yamaoka, L., Hung, W-Y., Alberts, M., Walker, A., Bartlett, R., Haynes, C., Welsh, K., Earl, N., Heyman, A., Clark, C., and Roses, A. 1991. Linkage studies in familial Alzheimer disease: evidence for chromosome 19 linkage. *Am. J. Hum. Genet.* **48:** 1034–1050.

Powledge, T. 1993. The genetic fabric of human behavior. *Bioscience* **43:** 362–367.

Powledge, T. 1993. The inheritance of behavior in twins. *BioScience* **43:** 420–424.

Schellenberg, G., Paijami, H., Wigsman, E., Orr, H., Goddard, K., Anderson, L. Nemens, E., White, J., Alonso, M., et al. 1993. Chromosome 14 and late-onset familial Alzheimer disease (FAD). Am. J. Hum. Genet. **53:** 619–628.

Tumer, Z., Tommerup, N., Tonneson, T., Kreuder, J., Craig, I., and Horn, N. 1992. Mapping of the Menkes locus to Xq13.3 distal to the X-inactivation center by an intrachromosomal insertion of the segment Xq13.3–q21.2. *Hum. Genet.* **88:** 668–672.

van de Wetering, B., and Heutink, P. 1993. The genetics of the Gilles de la Tourette syndrome: a review. *J. Lab. Clin. Med.* **121:** 638–645.

Vulpe, C., Levinson, B., Whitney, S., Packman, S., and Gitschier, J. 1993. Isolation of a candidate gene for Menkes disease and evidence that it encodes a copper-transporting ATPase. *Nature Genet.* **3:** 7–13.

CHAPTER 17

GENES IN POPULATIONS

CHAPTER OUTLINE

Arguments in which a father and son take opposite sides are neither new nor even unusual. All offspring occasionally disagree with their parents. It is rare, however, when such arguments result in ideas that are regarded as controversial some 175 years later. One such disagreement, between Thomas Malthus and his father, arose in the years following the French Revolution. This event, like the American Revolution that preceded it, was hailed by many (including the elder Malthus) as the beginning of a new era for humanity. According to this view, throwing off the repressions of monarchy presented an opportunity for unlimited progress in social, political, and economic areas. His son Thomas argued that to the contrary, even if all social impediments were removed, there were immutable natural constraints that would limit progress. These limits would always result in the continuation of poverty and misery as part of the

human condition. His essay was first published in 1798 as *The Essay on the Principle of Population as It Affects the Future Improvement of Society.* Later versions of this work were expanded and published as several editions of a book and also appeared in the 1824 supplement to the *Encyclopedia Britannica.*

In his essay, Malthus noted that populations grow in a geometric fashion, with the human population doubling in size every 25 years or so (Figure 17.1). Resources such as living space and food supply are more limited and grow slowly, if at all. This means that population size will rapidly outstrip the ability of the environment to support a continual increase in the birthrate. When this point is reached, constraints such as war, disease, and starvation begin to limit population growth by increasing the death rate. The use of voluntary constraints on population growth such as delayed marriage, celibacy, and birth control can help limit population growth and bring about a reduction in human suffering and an improvement in living conditions. However, the existence of sexual passion and human nature cause most people to ignore such voluntary constraints. According to the younger Malthus, the result is a continuation of unrelieved poverty and marginal living conditions, even in the most prosperous of nations.

In the 19th century, the writings of Malthus were used to argue that social reform and welfare were useless, since poverty was the result of natural law and not social inequities. In the 20th century, the debate over Malthus continues. Beginning in the 1960s, population growth was again regarded as a global threat that would override technologic progress, leading to a lowered standard of living and an increase in poverty and social problems. Movements calling for zero population growth and universal policies of birth control arose in the United States and other Western countries. More recently, China

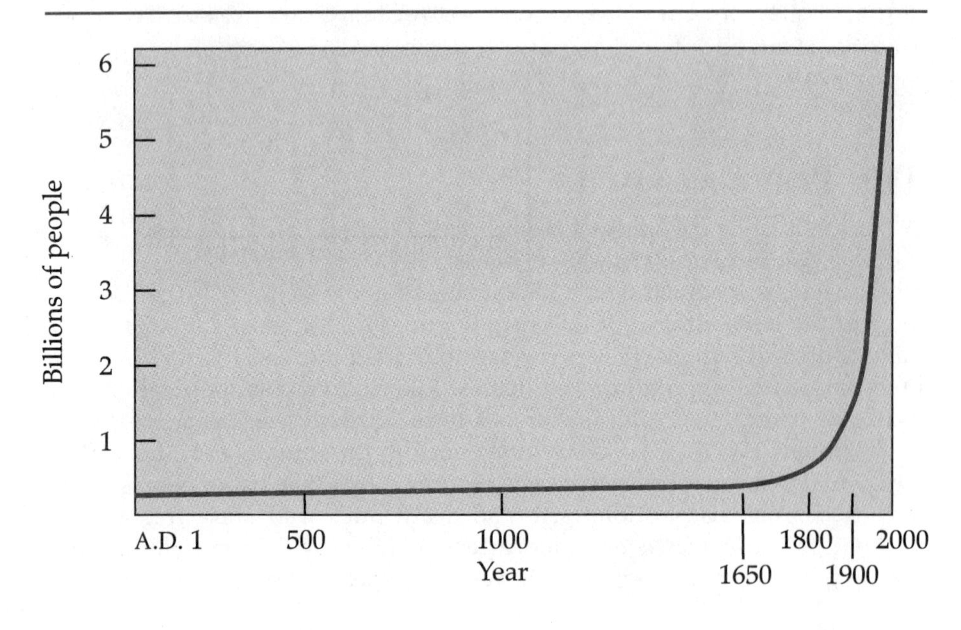

FIGURE 17.1 Growth of the world population. At the beginning of the Christian era (A.D. 1), the world population was estimated to be 200 million. It took 1800 years to reach 1 billion. The population reached 2 billion by 1930 and 4 billion by 1975. Sometime in July 1987 the estimated population of the planet passed 5 billion inhabitants. By the year 2000 the population is expected to increase to 6 billion.

instituted a policy of 1 couple: 1 child. Many developing nations regard such policies as attempts to limit their potential for growth and economic expansion and, in the extreme, as policies designed to result in extinction for smaller, poorer countries.

It is clear that Malthus correctly foresaw the implications and dangers of uncontrolled population growth, and he was one of the first to deal with the dynamics of populations. The relationship between a population and its environment postulated by Malthus was noted by Charles Darwin and Alfred Russel Wallace and incorporated into their idea of natural selection as the mechanism of evolutionary change. In considering the effects of limited resources on populations, Darwin and Wallace concentrated their attention on those that lived rather than those that perished. Both observed that individuals carrying advantageous hereditary variations were most likely to survive and to leave the greatest number of offspring. Neither Darwin nor Wallace had any knowledge of how this variation was generated, but they recognized its importance. After Mendel's work was widely recognized, it became obvious that genes were the basis of this variation and that the genetic organization of populations was an essential part of the mechanism of evolution. Today the study of populations is closely tied to the study of evolution. Populations are regarded as a collection of allele frequencies and evolution as the result of changes in allele frequencies.

In this chapter we will consider the population as a unit and its genetic organization, the methods of measuring allele frequencies, and the ways in which the genetic structure of the population directly affects the incidence of human genetic disease. We will begin with a definition of populations and their subunits and the concept of populations as reservoirs of genetic diversity. We will then consider how allele frequencies can be measured in populations by means of the Hardy-Weinberg law and the application of this law to answer practical questions about the frequency of recessive and sex-linked disorders and the frequency of heterozygotes in a population.

The Population as a Genetic Reservoir

Humans are distributed over a wide range of geographic areas that include most of the land surfaces of the earth (Figure 17.2). Because humans live in many different regions, the species is subdivided into locally interbreeding units known as **populations,** or **demes.** Like individuals, populations are dynamic: They have a life history of birth, growth, and response to the environment, and they can reach senescence and eventually may die. Unlike individuals, populations can also be described by other parameters such as density, spatial distribution, birth and death rates, and allele frequencies. Populations obviously contain more genetic diversity than individuals. For example, no single individual can have blood types A, B, and O. Only a group

Population
A local group of organisms belonging to a single species, sharing a common gene pool; also called a deme.

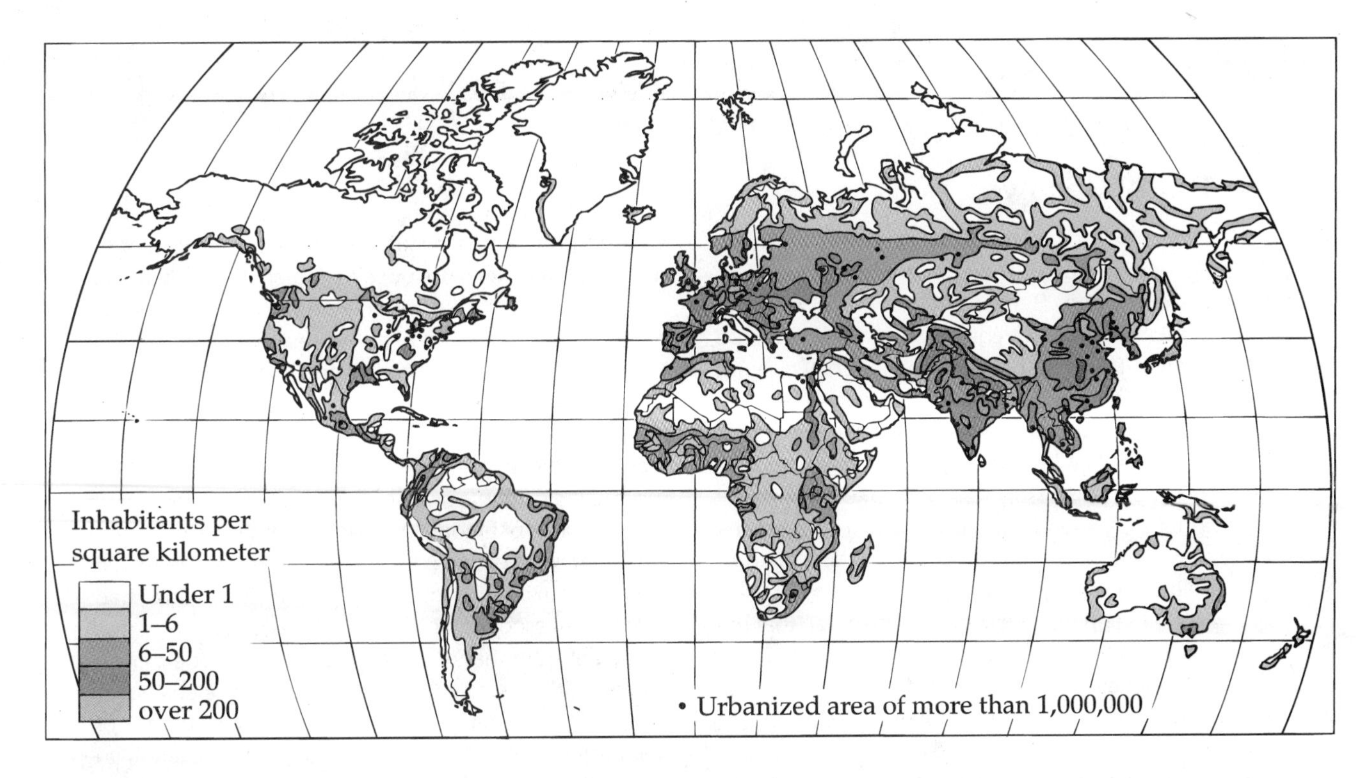

FIGURE 17.2 A map of human population density.

of individuals has the genetic capacity to carry all three blood types. The set of genetic information carried by a population is known as the **gene pool.** For a given gene such as the ABO locus, the pool includes all the A, B, and O alleles present in the population. Zygotes produced by one generation represent samples withdrawn from the gene pool to form the next generation. The gene pool of this new generation is descended from the parental pool but for a variety of reasons, including chance, may have different distributions of allele frequencies than those in the parental pool. Through alterations in allele frequency, a population can change and evolve while maintaining its existence over a period of time. The long-term effect of changes in the genetic structure of a population is evolutionary change.

Gene pool
The set of genetic information carried by the members of a sexually reproducing population.

MEASURING ALLELE FREQUENCIES

Populations are dynamic in structure; they may undergo drastic changes in size through expansion, or they may decline and have only a few or no survivors. Birthrates, disease, migration, and climate are among the factors that influence the size of a population. As these factors change, the genetic structure of a population can also change (see Concepts & Controversies, page 460). The problem is how to measure the genetic structure of a population and how to tell when allele frequencies undergo change. One way

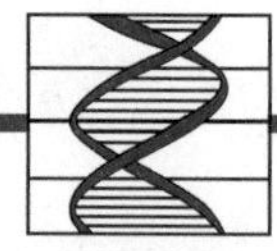

Concepts & Controversies

The Thrifty Genotype

The Pima Indians of the American Southwest have one of the highest rates of adult diabetes of any population in the United States. Between 42% and 66% of the adult population over the age of 35 years is affected with diabetes. This disorder, accompanied by obesity, developed in the Pima only in this century and became recognized as a serious health threat only after 1950. Diabetes is usually followed by blindness, kidney failure, and heart disease.

How is it that an inherited condition with so many deleterious effects can be present in a population at such a high frequency? Some 30 years ago, James Neel speculated on this question. He observed that the incidence of diabetes is very low in primitive hunter-gatherer societies, such as the Pima were before the 20th century. He also noted that females prone to diabetes become sexually mature at an earlier age and that this might provide a reproductive advantage for diabetics. In addition, such females give birth to larger-than-average babies (increased birth weight is linked to increased survival).

From these observations Neel postulated that in the feast-or-famine diet of hunter-gatherers, the diabetic represents a "thrifty" genotype that is more efficient in the use of food. In diabetics, more insulin is released after every meal to metabolize as much of the available glucose as possible. According to Neel, as the age of the individual increases, these repeated releases of insulin produce a response in the form of an antagonist that stops insulin action. But since such overloading with carbohydrates occurs so infrequently in hunter-gatherers, this does not pose a problem to the diabetic.

We now know that there is no such antagonist as envisioned by Neel, but as he reviewed the idea in 1982, he proposed that "although incorrect in detail, it may have been correct in principle." The notion of an antagonist can be replaced by several hypotheses, including the suggestion that adult diabetics produce more insulin than nondiabetics. This quick response prevents loss of blood glucose through the kidneys and was highly effective in hunter-gatherer societies. It offered maximum use of available food and had reproductive advantages. With the conversion of the Pima diet from that of a hunter-gatherer to one high in refined carbohydrates, the release of extra insulin over and over again causes a reduction in the number of cell-surface insulin receptors. This, in turn, leads to diabetes.

Thus, the diabetic may represent a genotype that was well adapted to the environmental conditions (hunter-gatherer) that prevailed for hundreds of thousands of years. Now, in the short span of less than 100 years, environmental conditions have changed dramatically, and this genotype is at a distinct disadvantage where carbohydrate-rich foods are freely available.

It is an interesting hypothesis and would be an example of natural selection in action. As we learn more about the causes of adult-onset diabetes, this evolutionary idea bears watching.

to do this is to search for the appearance of new phenotypes. As we will see later, most often this is an inefficient method. If the trait is recessive, the new allele has probably spread to many members of the population in a heterozygous form before it appears as a new phenotype in a homozygous recessive individual.

Allele frequency The percentage of all alleles of a given gene that are represented by a specific allele.

The most direct way of monitoring the genetic structure of a population is the measurement of allele frequencies. In our discussion, the term **allele frequency** will mean the frequency with which alleles of a given genetic locus are present in the population. As several following examples will show, allele frequencies are not the same as genotype frequencies.

Allele frequencies cannot always be determined directly because we most often observe phenotypes rather than genotypes. However, if we consider codominant alleles, phenotypes are equivalent to genotypes, and we can determine allele frequencies in a direct manner. Earlier we examined the MN blood group in humans. In this case the gene *L* on chromosome 4 has two alleles, L^M and L^N, that control the M and N blood types, respectively. Each allele controls the synthesis and presence of an antigen on the surface of red blood cells, independent of the other allele. Thus individuals may be type M (L^ML^M), type N (L^NL^N), or type MN (L^ML^N). The genotypes, blood types, and immunologic cross reactions of the MN blood groups are shown in Table 17.1.

In a codominant system such as the MN blood group, allele frequency can be determined by simply counting the number of each allele present in a given phenotype. For example, in a population of 100 individuals, suppose that blood typing shows there are 54 MM homozygotes, 26 MN heterozygotes, and 20 NN homozygotes. The 54 MM individuals represent 108 M alleles (54 individuals, each carrying 2 L^M alleles). The 26 MN heterozygotes represent an additional 26 L^M alleles, for a total of 134 L^M alleles ($108 + 26 = 134$) out of a total of 200 alleles (100 individuals, each with 2 alleles $= 200$). The frequency of the M allele is $134/200 = 0.67$, or 67%. The frequency of the N allele can be calculated in the same way by counting 40 L^N alleles in the homozygotes (20 individuals, each with 2 L^N alleles) and an additional 26 L^N alleles in the heterozygotes, for a total of $66/200 = 0.33$, or 33%. Table 17.2 summarizes this method of calculating gene frequencies in codominant populations, and Table 17.3 lists the frequency of L^M and L^N alleles in several human populations.

The preceding example used a codominant gene system for the measurement of allele frequencies. However, most human genes exhibit dominant or recessive phenotypes. If one allele is recessive, then the heterozygote and the homozygous dominant individuals have identical phenotypes. In this situation, it is not possible to determine allele frequencies by counting alleles, since we cannot determine how many heterozygotes are

TABLE 17.1 MN Blood Groups

Genotype	Blood Type	Antigens Present	Antibody Reactions
L^ML^M	M	M	Anti-M
L^ML^N	MN	M, N	Anti-M, Anti-N
L^NL^N	N	N	Anti-N

TABLE 17.2 Determining Allele Frequencies for Codominant Genes by Counting Alleles

Genotype	MM	MN	NN	Total
Number of individuals:	54	26	20	100
Number of L^M alleles:	108	26	0	134
Number of L^N alleles	0	26	40	66
Total	108	52	40	200

Frequency of L^M in population: 134/200 = 0.67 = 67%

Frequency of L^N in population: 66/200 = 0.33 = 33%

TABLE 17.3 Frequencies of L^M and L^N Alleles in Various Populations

Population	Genotype Frequency (%)			Allele Frequency	
	MM	MN	NN	L^M	L^N
U.S. Indians	60.00	35.12	4.88	0.776	0.224
U.S. blacks	28.42	49.64	21.94	0.532	0.468
U.S. whites	29.16	49.38	21.26	0.540	0.460
Eskimos (Greenland)	83.48	15.64	0.88	0.913	0.087

present in the population. However, a mathematical formula can be used to determine allele frequencies when one or more alleles is recessive and a number of conditions (described below) are met. This method was developed independently by Godfrey Hardy and Wilhelm Weinberg in 1908 and is known today as the **Hardy-Weinberg law.**

Hardy-Weinberg law
The statement that allele frequencies and genotype frequencies will remain constant from generation to generation when the population meets certain assumptions.

THE HARDY-WEINBERG LAW

In the years immediately after Mendel's work became widely known, there was a great deal of debate about whether Mendelism applied to humans. One of the first Mendelian traits identified in humans was a dominant mutation known as brachydactyly (see Chapter 4). It was proposed that such a dominant gene arising in a population would in time automatically produce a 3:1 phenotypic ratio, with most of the population showing brachydactyly. Since this dominant trait did not exhibit a 3:1 ratio in the population, there was reason to doubt the validity of Mendelism as it applied to humans. Both Hardy and Weinberg recognized that this reasoning was false because it failed to distinguish between the *mode of inheritance* (in this case a dominant trait with a 3:1 ratio) and the *frequency* of the dominant and recessive alleles in the population. The mathematical model developed independently by Hardy and Weinberg provides a means for estimating the frequency of alleles in a population and describes how alleles combine to form genotypes. This model is based on a number of assumptions, including the following:

- The population is large.
- There is no selection for or against genotypes.
- Mating within the population is random.
- Other factors such as mutation and migration are absent or rare events and can be ignored.

These assumptions make the Hardy-Weinberg method less exact than counting alleles directly as in a codominant system. Since the assumptions of the Hardy-Weinberg method only rarely exist in natural situations allele frequencies determined in this way are regarded as estimates.

Let us illustrate how the model works by considering a population carrying an autosomal gene with two alleles, A and a. The frequency of the dominant allele A in gametes is represented by p, and the frequency of the recessive allele a in gametes is represented by q. Since there are no other alleles, $p + q = 1.0$ (or 100%). A Punnett square can be used to represent the genotypes produced by the random combination of gametes carrying these alleles:

Eggs \ Sperm	$A(p)$	$a(q)$
$A(p)$	AA (p^2)	Aa (pq)
$a(q)$	Aa (pq)	aa (q^2)

In the combination of gametes that produces the next generation the chance that both the egg and sperm will carry the A allele is $p \times p = p^2$. The chance that the gametes will carry unlike alleles is $(p \times q) + (p \times q) = 2pq$. The chance that a homozygous recessive combination of alleles might result is $q \times q = q^2$. It is important to note that while the value p^2 is the chance that both gametes carried an A allele, p^2 is also a measure of the frequency of the homozygous AA genotype. Similarly, $2pq$ is a measure of heterozygote frequency, and q^2 is the frequency of homozygous recessive (aa) individuals.

In other words, the distribution of genotypes in the next generation can be expressed as

$$p^2 + 2pq + q^2 = 1$$

where 1 means 100% of the genotypes present. This formula represents the Hardy-Weinberg law, which states that both allele frequencies and genotype frequencies will remain constant from generation to generation in a large, interbreeding population where mating is at random and there is no selection, migration, or mutation. The formula can be used to calculate the frequency of the *A* and the *a* allele or the frequency of the various genotypes in a population. To show how the model works, let us begin with a population for which we already know the frequency of the alleles. Suppose we have a large, randomly mating population in which the frequency of an autosomal dominant allele *A* is 60% and the frequency of the recessive allele *a* is 40%. This means that $p = 0.6$ and $q = 0.4$. Since *A* and *a* are the only two alleles, the sum of $p + q$ equals 100% of the alleles:

$$p\ (0.6) + q\ (0.4) = 1.$$

In the gametes of the population that will produce the next generation, 60% of the gametes carry the dominant allele *A*, and 40% carry the recessive allele *a*. The distribution of genotypes in the next generation is shown in the following Punnett square:

Eggs	Sperm: *A* ($p = 0.6$)	Sperm: *a*($q = 0.4$)
A ($p = 0.6$)	*AA* ($p^2 = 0.36$)	*Aa* ($pq = 0.24$)
a ($q = 0.4$)	*Aa* ($pq = 0.24$)	*aa* ($q^2 = 0.16$)

In the new generation, 36% ($p^2 = 0.6 \times 0.6$) of the offspring will have a homozygous dominant genotype *AA*, 48% ($2pq = 2[0.6 \times 0.4]$) will be heterozygous *Aa*, and 16% ($q^2 = 0.4 \times 0.4$) will have a homozygous recessive genotype, *aa*. We can also use the Hardy-Weinberg equation to calculate the frequency of the *A* and *a* alleles in the new generation. The frequency of *A* is

$$p^2 + \frac{1}{2}(2pq)$$

$$0.36 + \frac{1}{2}(0.48)$$

$$0.36 + 0.24 = 0.60 = 60\%$$

For the recessive allele *a*, the frequency is

$$q^2 + \frac{1}{2}(2pq)$$

$$0.16 + \frac{1}{2}(0.48)$$

$$0.16 + 0.24 = 0.40 = 40\%$$

Because $p + q = 1$, we could have calculated the value for a by subtraction:

$$p + q = 1$$
$$q = 1 - p$$
$$q = 1 - 0.60$$
$$q = 0.40 = 40\%$$

Meaning of Genetic Equilibrium

Note that the frequency of *A* and *a* in the new generation is the same as in the parental generation. Populations in which the allele frequency of a given gene remains constant from generation to generation are in **genetic equilibrium** for that gene. This does not mean that the population is in a state of equilibrium for *all* alleles. On the contrary, if forces such as mutation, selection, or migration are operating on other genotypes, then the other alleles will change in frequency from one generation to the next.

Genetic equilibrium
The situation when the frequency of alleles for a given gene remains constant from generation to generation.

The presence of a Hardy-Weinberg equilibrium also illustrates why dominant alleles of genes do not increase in frequency as new generations are produced. In the case of brachydactyly, if conditions for the Hardy-Weinberg law are met, a genetic equilibrium will be established. Because allele and genotype frequencies determine phenotype frequencies, brachydactyly will not increase in the population and reach a 3:1 frequency, but instead will be maintained at an equilibrium frequency from generation to generation.

In addition, genetic equilibrium also helps to maintain genetic variability in the population. In the example above, at equilibrium we can count on 60% of the alleles for the gene in question being dominant and 40% being recessive in generation after generation. The presence and maintenance of genetic variability is important to the process of evolution.

USE OF THE HARDY-WEINBERG LAW IN HUMAN GENETICS

The Hardy-Weinberg law is one of the foundations of population genetics and has many applications in human genetics and human evolution. We will consider only a few of its uses, primarily those that apply to the measurement of allele frequencies and genotype frequencies.

Autosomal Codominant Alleles

As outlined earlier, in the case of autosomal codominant traits, the allele frequency and genotype frequency can be determined directly from the phenotype, since each genotype gives rise to a distinctive phenotype, as in the MN blood groups. For this mode of inheritance, allele and genotype frequencies can be determined by counting individuals in the population. Further, there is no need to make assumptions about Hardy-Weinberg conditions.

Autosomal Dominant and Recessive Alleles

In the case of autosomal dominant and recessive traits, the homozygous dominant and heterozygous genotypes cannot be distinguished on the basis of phenotype. For this mode of inheritance, if random mating is assumed, the allele frequencies can be estimated from the frequency of the homozygous recessive individuals in the population. For example, cystic fibrosis is an autosomal recessive trait, and homozygous recessive individuals are easily identified in the population by their distinctive phenotype. Suppose that in a population, it is determined that 1 in 2500 individuals are affected with cystic fibrosis. These individuals have the recessive genotype *aa*, and the frequency of this genotype in the population is given by q^2 in the Hardy-Weinberg equation. Thus, the frequency of the *a* allele in the population is given as the square root of q^2:

$$q^2 = 1/2500 = 0.0004$$
$$q = 0.0004$$
$$q = 0.02 = 1/50$$

Once we know the frequency of the *a* allele is 0.02 (2%), we can calculate the frequency of the normal, dominant allele A:

$$p + q = 1$$
$$p = 1 - q$$
$$p = 1 - 0.02$$
$$p = 0.98 = 98\%$$

Therefore in this population the distribution of the alleles for cystic fibrosis is 98% *A* and 2% *a*.

X-Linked Traits

In estimating the allele frequency for the autosomal recessive trait cystic fibrosis, one of our underlying assumptions was that the frequency of either the dominant *A* allele or the recessive *a* allele is the same in both sperm and eggs. The situation with respect to X-linked traits is somewhat different. Because human females carry two X chromosomes, they carry two copies of all genes on the X. Males, on the other hand, have only one X chromosome and are hemizygous for all loci on the X chromosome. This means that genes on the X chromosome are not distributed equally in the population: females carry two thirds of the total number, and males carry one third. However, because males are hemizygous for all traits on the X chromosome, the allele frequency for X-linked traits in the population is simply the frequency of males with the mutant phenotype. For example, in the United States, about 8% of males are colorblind. Therefore, the frequency of the allele for colorblindness in the population is 0.08.

Because females have two copies of the X chromosome, genotypic frequencies for X-linked traits in females can be calculated using the Hardy-Weinberg equation. If colorblindness in males occurs with a frequency of 8% ($q = 0.08$), then in females we would expect colorblindness (homozy-

TABLE 17.4 Frequency of X-Linked Traits in Males and Females

Males with Trait	Females with Trait
1/10	1/100
1/100	1/10,000
1/1000	1/1,000,000
1/10,000	1/100,000,000

gous recessive) to occur with a frequency of q^2, or 0.0064 (0.64%). Therefore in a population of 10,000 males, 800 would be colorblind, but in a population of 10,000 females, only 64 would be colorblind. This example emphasizes again the fact that males are at much higher risk than females for deleterious traits carried on the X chromosome. Table 17.4 lists comparative values for the frequency of X-linked traits in males and females.

Multiple Alleles

Until now we have discussed the determination of frequencies in gene systems having only two alleles. For genes such as the ABO blood system, however, three alleles of the isoagglutinin locus *(I)* are present in the population. In this system, A and B are codominant, and both are dominant to O. This system has six possible genotypic combinations:

AA, AO, BB, BO, AB, OO

Homozygous AA individuals and AO individuals are phenotypically identical, as are BB and BO individuals. This results in four phenotypic combinations, known as types A, B, AB, and O.

The Hardy-Weinberg law can be used to calculate both the allele frequencies and genotype frequencies for this three-allele system by the addition of another term to the equation. For the three blood group alleles:

$$p(\mathrm{A}) + q(\mathrm{B}) + r(\mathrm{O}) = 1$$

In other words, the sum of the frequency of the A, B, and O alleles accounts for 100% of the alleles for this gene present in the population. The genotypic frequencies will be given by the equation:

$$(p + q + r)^2$$

The allele frequency values for A, B, and O can be estimated from the distribution of phenotypes in a population, if random mating is assumed.

If we know or have estimated the frequency of the A, B, and O alleles for a given population, we can then calculate the genotypic and phenotypic

TABLE 17.5 Genotype and Phenotype Frequencies in Multiple-Allele Systems

Genotype	Genotypic Frequency	Phenotype	Phenotypic Frequency
AA	$p^2 = (0.38)^2 = 0.14$	A	0.53
AO	$2pr = 2(0.38 \times 0.51)$		
	$= 0.39$		
BB	$q^2 = (0.11)^2 = 0.01$	B	0.12
BO	$2qr = 2(0.11 \times 0.51)$		
	$= 0.11$		
AB	$2pq = 2(0.38 \times 0.11)$	AB	0.08
OO	$= 0.082$		
	$r^2 = (0.51)^2 = 0.26$	O	0.26

frequencies for all allelic combinations. The genotypic combinations can be calculated by an expansion of the equation:

$$p^2\ (AA) + 2pq\ (AB) + 2pr\ (AO) + q^2\ (BB) + 2qr\ (BO) + r^2\ (OO) = 1$$

Table 17.5 lists the frequencies for A, B, and O in different populations in the world. Figure 17.3a shows the geographic distribution of ABO alleles. By using the equations shown above and the values in Table 17.5, we can calculate the genotypic and phenotypic frequencies for the populations shown in Figure 17.3b and 17.3c.

Estimating Heterozygote Frequency

An important application of the Hardy-Weinberg law is the estimation of heterozygote frequency in a population. In the genetic structure of human populations, the majority of deleterious recessive genes are carried in the heterozygous condition. For recessive traits, to calculate the frequency of heterozygous genotypes, we usually begin by counting the number of homozygous recessive individuals in the population. For example, earlier in the chapter we calculated the allele frequency in cystic fibrosis, an autosomal recessive trait with a frequency of 1 in 2500 among white Americans. (The disease is much rarer among American blacks and Asians). Homozygous individuals can be distinguished from the rest of the population by clinical symptoms that indicate defects in the function of exocrine glands. (To review the symptoms, see Chapter 4.) The frequency of the homozygous recessive

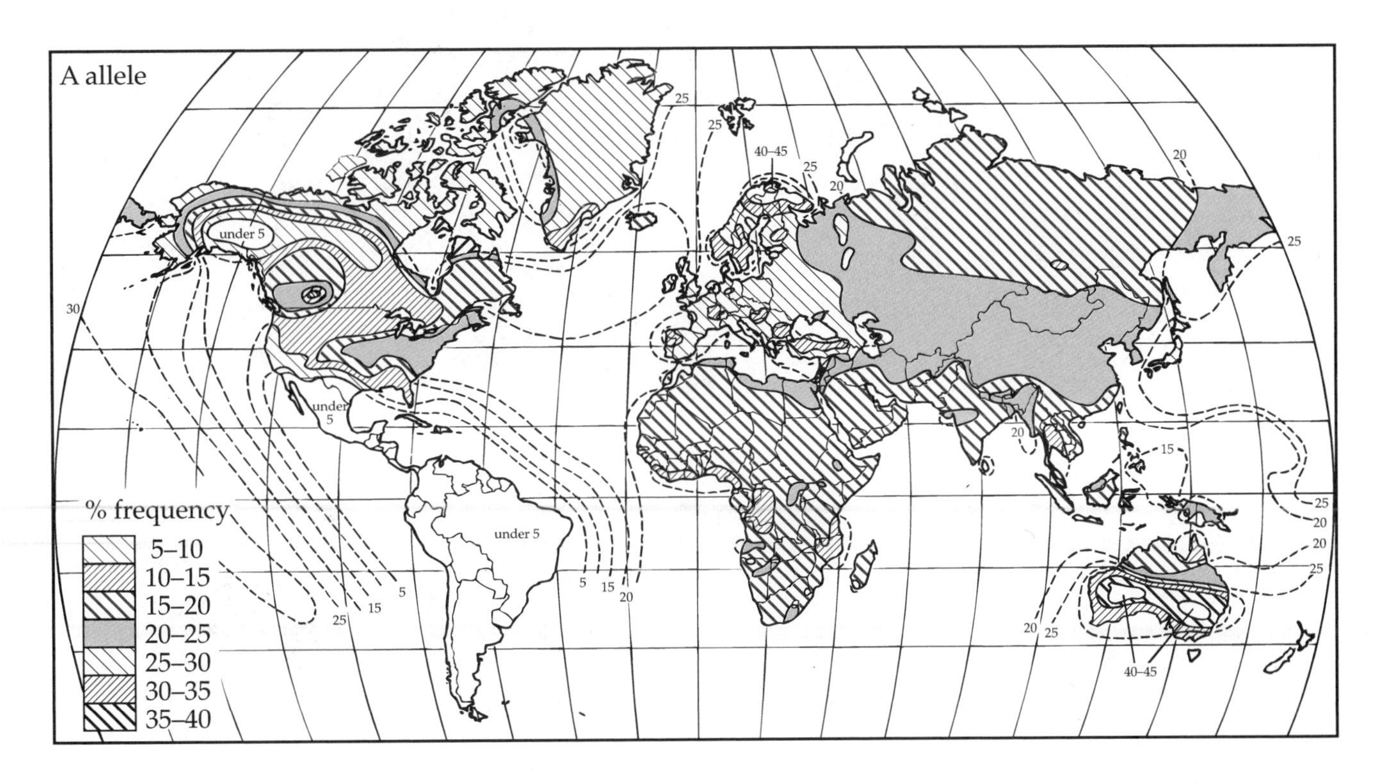

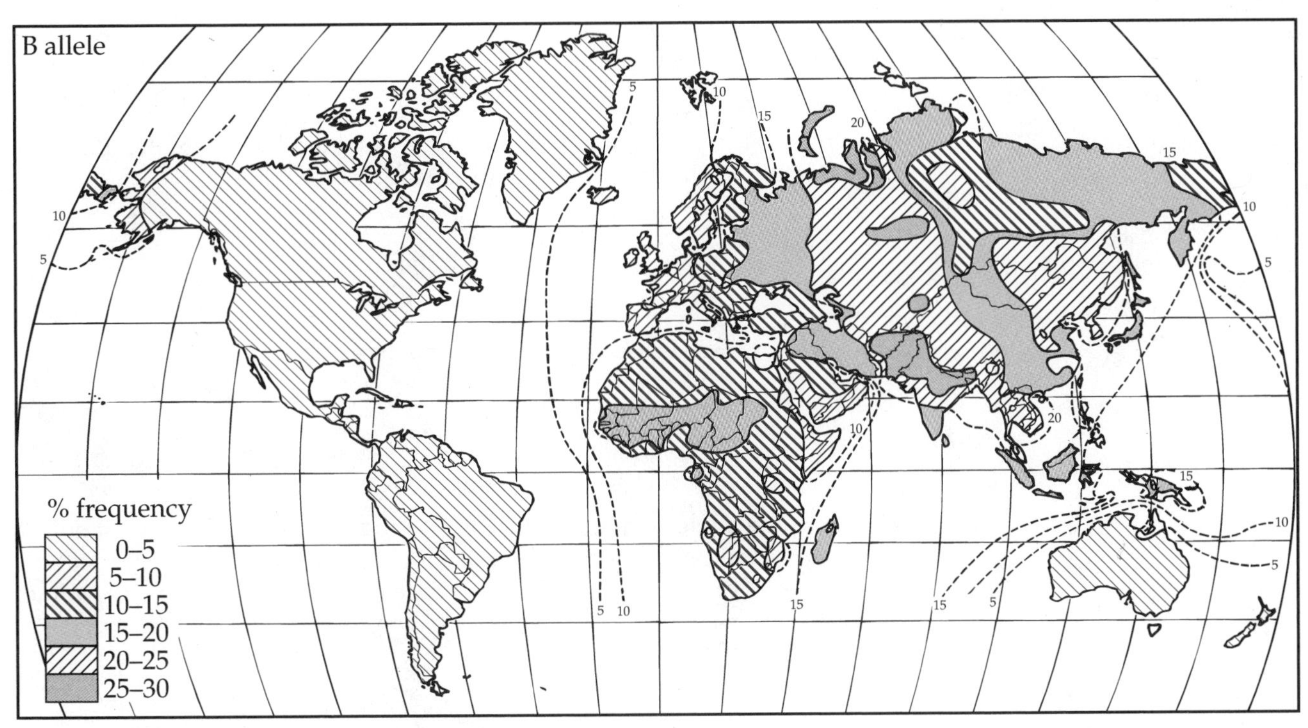

FIGURE 17.3 The geographic distribution of the ABO alleles in the indigenous world population. (A) Distribution of the A allele. (B) Distribution of the B allele. **(continued)**

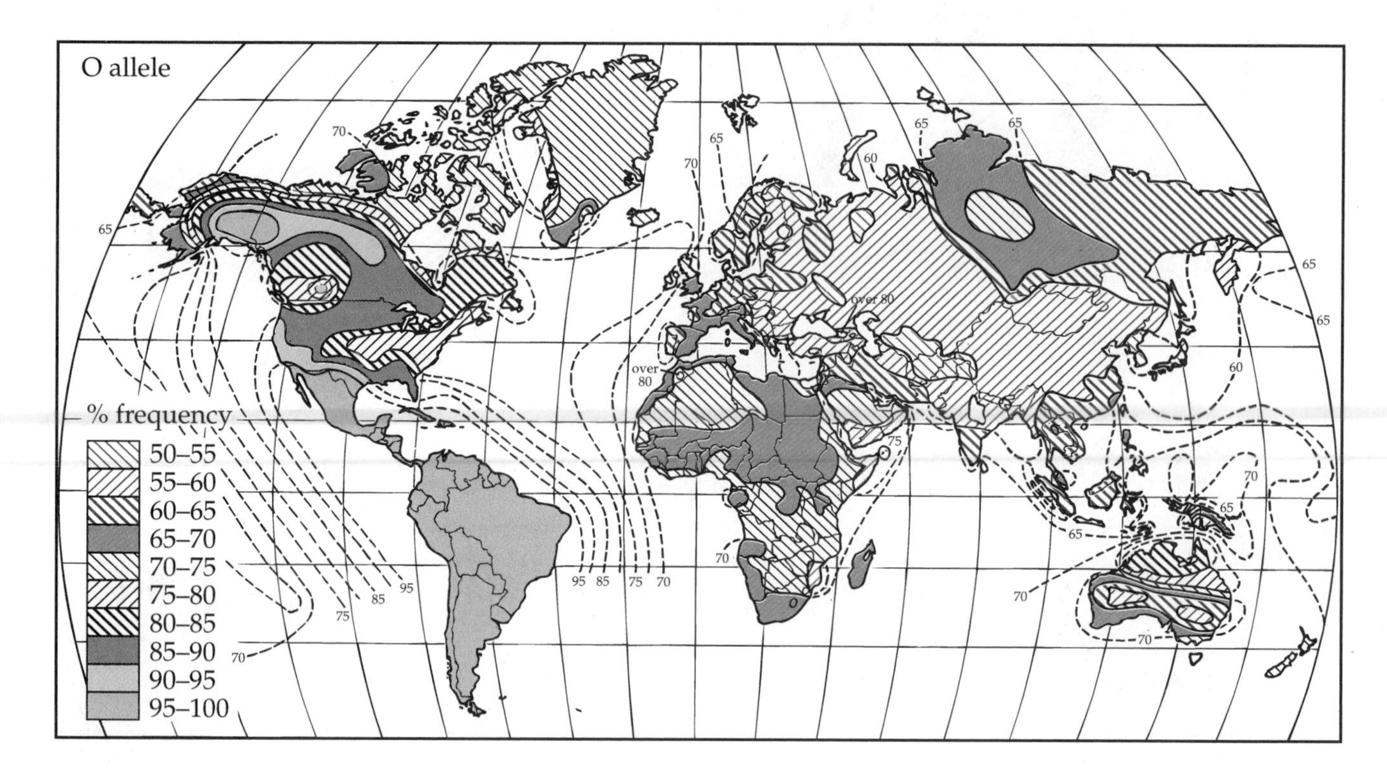

FIGURE 17.3 cont'd (C) Distribution of the O allele.

genotype is 1 in 3500, or 0.0004, and is represented by q^2. The frequency of the recessive allele q in the population can be calculated as

$$q = \sqrt{q^2}$$
$$q = \sqrt{0.0004}$$
$$q = 0.02 = 2\%$$

Since $p + q = 1$, we can calculate the frequency of p:

$$p = 1 - q$$
$$p = 1 - 0.02$$
$$p = 0.98 = 98\%$$

Knowing the allele frequency, we can use the Hardy-Weinberg equation to calculate genotype frequencies. Recall that in the Hardy-Weinberg equation, the frequency of the heterozygous genotype is given by $2pq$. Using the values we have calculated for p and q, we can determine the frequency of heterozygotes:

$$2pq = 2(0.98 \times 0.02)$$
$$2pq = 2(0.0196)$$
$$\textit{Heterozygote frequency} = 0.039 = 3.9\%$$

This means that about 4%, or about 1 in 25, white Americans carry the gene for cystic fibrosis.

Similarly, sickle cell anemia is an autosomal recessive trait that affects 1 in 500 black Americans. Using the Hardy-Weinberg equation, we can calculate that 8.5% of black Americans, or 1 in every 12, are heterozygous carriers for this trait. Table 17.6 lists the frequencies of heterozygous carriers for recessive traits with frequencies ranging from 1 in 10 to 1 in 10,000,000. Table 17.7 lists the heterozygote frequencies for some common human autosomal recessive traits.

Many people are surprised to learn that heterozygotes for recessive traits are so common in the population. In most cases this is because they assume that if a trait like albinism occurs in 1 in 10,000 individuals, the number of heterozygotes must also be rather low. In fact, 1 in 50 individuals is a heterozygote for albinism (Table 17.7), and there are about 200 times as many heterozygotes as there are homozygotes.

What are the chances that two heterozygotes will marry and produce an albino child? We can calculate how this occurs as follows: The chance that two heterozygotes will marry is $1/50 \times 1/50 = 1/2500$. Since they are heterozygotes, the chance that they will produce a homozygous recessive offspring is $1/4$. The chance that they will marry and produce an albino child is therefore $1/2500 \times 1/4 = 1/10{,}000$. In other words, for the trait to be present in 1 in every 10,000 individuals, 1 in 50 *must* be heterozygous carriers of the recessive allele.

Once the frequency of either allele is known, we can calculate the frequency of the homozygous genotypes as well as the heterozygotes. The

TABLE 17.6 Heterozygote Frequencies for Recessive Traits

Frequency of Homozygous Recessives (q^2)	Frequency of Heterozygotes Individuals ($2pq$)
1/100	1/5.5
1/500	1/12
1/1000	1/16
1/2500	1/25
1/5000	1/36
1/10,000	1/50
1/20,000	1/71
1/100,000	1/158
1/1,000,000	1/500
1/10,000,000	1/1582

TABLE 17.7 Heterozygote Frequency for Some Recessive Traits in the United States

Trait	Heterozygote Frequency
Cystic fibrosis	1/22 whites, much lower in blacks, Asians
Sickle cell anemia	1/12 blacks, much lower in most whites and in Asians
Tay-Sachs disease	1/30 among descendants of Eastern European Jews, 1/350 among others of European descent
Phenylketonuria	1/55 among whites, much lower in blacks and those of Asian descent
Albinism	1/10,000 in Northern Ireland; 1/67,800 in British Columbia

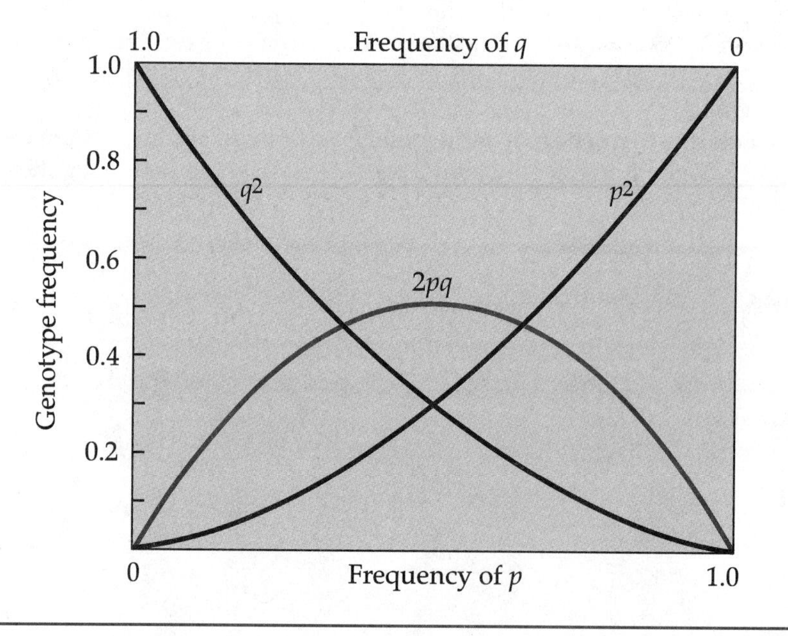

FIGURE 17.4 The relationship between genotype frequency and gene frequencies in the Hardy-Weinberg equation.

frequency of the genotypes depends on the allele frequency. Figure 17.4 shows the relationship between gene frequency and genotype frequency. As the frequencies of p and q move away from zero, the percentage of heterozygotes in the population increases rapidly. This is a further illustration of the fact that in traits such as albinism, the majority of those carrying the recessive allele are heterozygotes. For most deleterious conditions, the presence of a large number of heterozygotes helps reduce the impact of genetic disease on the population.

Anthropology and Population Structure

In many situations, anthropologists use properties such as language, dress, social customs, taboos, and food sources to determine the relative distances between cultures and to determine the ancestral sources of contemporary population groups. The analysis of allele frequencies and genotypes has proven to be a valuable tool in these studies, and the application of population genetics to anthropology has given rise to a discipline known as **anthropological genetics.**

Anthropological genetics The union of population genetics and anthropology to study the effects of culture on gene frequencies.

The relationship between linguistic or cultural differences and biologic differentiation is illustrated by the population structure on Bougainville Island in the Solomon Islands (Figure 17.5). This island, roughly 130 miles long and 40 miles at its maximum width, is home to 17 major languages (Figure 17.6). Most of these languages are spoken over a range of only about 10 miles, and many are subdivided into distinct dialects or even sublanguages occupying much smaller ranges. The local populations on this island traditionally raise food in gardens and have little social or political contact with neighbors outside their immediate area. This linguistic and other anthropologic evidence suggests that local groups are quite isolated from one another. In the absence of strong selection or other forces, these populations should also be genetically distinct from one another. To examine this possibility, blood-type phenotypes were determined in 18 villages represent-

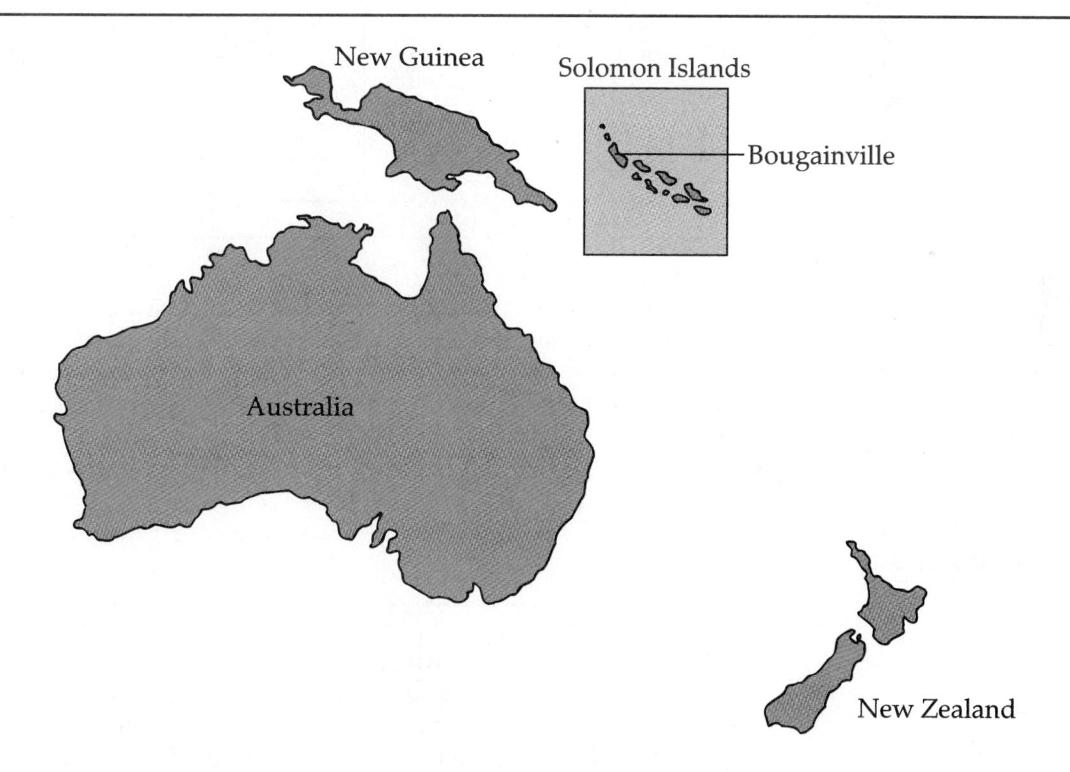

FIGURE 17.5 Map of the southwest Pacific showing the Solomon Islands and the location of Bougainville.

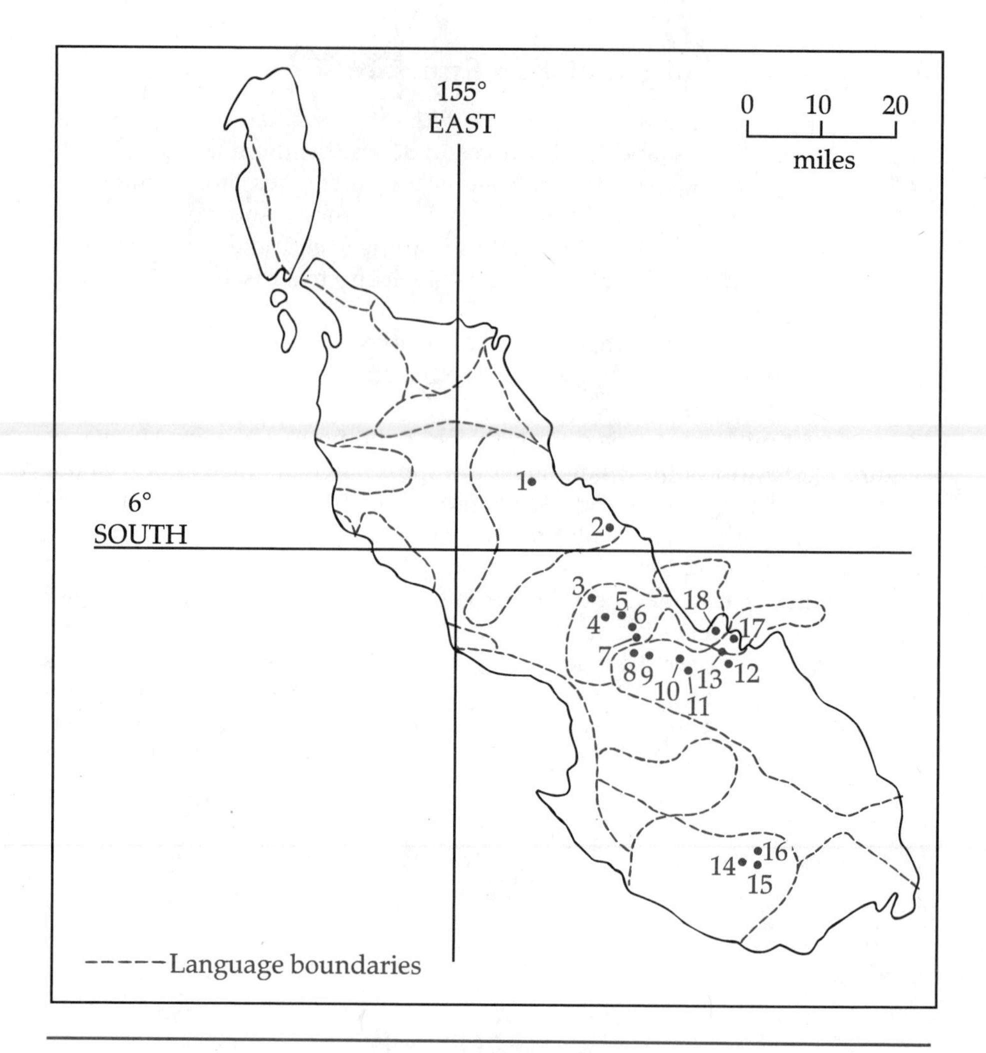

FIGURE 17.6 Language divisions on Bougainville. Villages included in the genetic study are numbered 1 through 18. Blood type data for these groups are shown in Table 17.8.

ing eight of these language groups. Some of these data are presented in Table 17.8. There is a wide range in the observed frequencies of blood types in the 18 villages. As we will see in Chapter 18, if there were interbreeding between or among these populations, it would equalize blood-type frequencies within a generation or two. The divergent ABO phenotypes indicate that the populations are genetically isolated, supporting the idea that they are also culturally isolated.

SUMMARY

1. The work of Charles Darwin and Alfred Russel Wallace on evolution and natural selection was influenced by Malthus's essay on population dynamics. Both Darwin and Wallace recognized that the struggle for existence favored those carrying variations that helped them adapt to the existing conditions, allowing them to leave more offspring.

TABLE 17.8 Observed ABO Phenotypes in 18 Villages on Bougainville Island

Group	ABO Phenotype (%)			
	A	O	B	AB
1	63.7	16.3	10.3	9.4
2	47.9	52.0	0.00	0.00
3	41.6	58.3	0.00	0.00
4	41.5	58.4	0.00	0.00
5	71.2	28.7	0.00	0.00
6	56.4	43.5	0.00	0.00
7	41.9	58.06	0.00	0.00
8	45.6	53.5	0.78	0.00
9	40.6	59.3	0.00	0.00
10	41.2	57.7	0.004	0.004
11	43.5	56.4	0.00	0.00
12	53.9	40.8	0.043	0.008
13	40.0	41.6	15.0	0.033
14	22.4	56.2	15.7	5.61
15	51.8	35.0	10.5	2.6
16	38.4	50.0	7.69	3.8
17	23.0	47.7	24.7	4.5
18	47.5	26.4	15.8	10.1

2. In the early decades of this century, genes were recognized as the agents causing such variations, giving rise to the field of population genetics. After the mathematic and theoretic basis of this field was established, experimentalists began to study gene frequencies in populations rather than in the offspring of a single mating. This work has produced the basis for our understanding of genetic evolution.

3. The Hardy-Weinberg law provides a means of measuring gene frequencies within populations and determining whether the population is in equilibrium.

4. Measurements with the Hardy-Weinberg equation assume that the population is large and randomly interbreeding and that factors such as mutation, migration, and selection are absent. The presence of equilibrium

in a population explains why dominant traits do not replace their recessive alleles.

5. The Hardy-Weinberg law can also be used to measure the frequency of heterozygotes in a population and to establish when gene frequencies are shifting in the population. The conditions leading to changing gene frequencies in a population are those that produce evolutionary change.

Questions and Problems

1. Draw a graph showing the difference between a population that grows at a geometric rate and one that is only permitted to grow at a constant arithmetic rate. Be sure to label each axis.
2. What are some of the natural constraints referred to by Malthus that keep the size of human populations in check?
3. Define the following terms: population, gene pool, gene frequency, and genotype frequency.
4. **a.** Explain the connection between changes in population gene frequencies and evolution, and relate this to the observations made by Darwin and Wallace concerning natural selection.
   b. Do you think populations can evolve without changes in gene frequencies?
5. Design an experiment to determine if a population is evolving.
6. Suppose you are monitoring the allelic and genotypic frequencies of the MN blood group locus in a small human population. You find that for 1-year-old children the genotypic frequencies are MM = 0.25, MN = 0.5, and NN = 0.25, while the genotypic frequencies for adults are MM = 0.3, MN = 0.4, and NN = 0.3.
   a. Compute the M and N allelic frequencies for the 1-year-olds and adults.
   b. Are the gene frequencies in equilibrium in this population?
   c. Are the genotypic frequencies in equilibrium?
7. Drawing on your newly acquired understanding of the Hardy-Weinberg equilibrium law, point out why the following statement is erroneous: "Since most of the people in Sweden have blond hair and blue eyes, the genes for blond hair and blue eyes must be dominant in that population."
8. In a population where the females have the allelic frequencies $A = 0.35$ and $a = 0.65$ and the male frequencies are$A = 0.1$ and $a = 0.9$, how many generations will it take to reach Hardy-Weinberg equilibrium for both the allelic and genotypic frequencies? Assume random mating, and show the allelic and genotypic frequencies for each generation.
9. Explain why a population carries more genetic diversity than an individual.
10. If a trait determined by an autosomal recessive allele occurs at a frequency of 0.25 in a population, what are the allelic frequencies? Assume Hardy-Weinberg equilibrium, and use "*A*" and "*a*" to symbolize the dominant wild-type and recessive alleles, respectively.
11. Five percent of the males of a population express a sex-linked recessive trait.
    a. What are the frequencies of the dominant and recessive alleles?
    b. What are the genotypic frequencies for the males and females in the population?
12. In a given population, the frequencies of the 4 phenotypic classes of the ABO blood groups are found to be A = 0.33, B = 0.33, AB = 0.18, and O = 0.16. What is the frequency of the O allele?
13. In Table 17.3 which pairs of populations appear to be most closely related in terms of their allelic frequencies at the MN locus? What scenario would you propose to account for these relationships?
14. For each population in Table 17.3 use the given allelic frequencies to determine if the genotypic frequencies are in general agreement with the Hardy-Weinberg law.
15. Using Table 17.6, determine the frequencies of p and q that result in the greatest proportion of heterozygotes in a population.

For Further Reading

Barrantes, R., Smouse, P. E., Mohrenweiser, H. W., Gershowitz, H., Azofeifa, J., Arias, T. D., and Neel, J. V. 1990. Microevolution in lower Central America: Genetic characterization of the Chibcha-speaking groups of Costa Rica and Panama, and a consensus taxonomy based on genetic and linguistic affinity. *Am. J. Hum. Genet.* **46:** 63–84.

Bodmer, W. F., and Cavelli-Sforza, L. L. 1976. *Genetics, Evolution and Man.* San Francisco: Freeman.

Chakraborty, R., Smouse, P. E., and Neel, J. V. 1988. Population amalgamation and genetic variation: Observations on artificially agglomerated tribal populations of Central and South America. *Am. J. Hum. Genet.* **43:** 709–725.

Constans, J. 1988. DNA and protein polymorphism: Application to anthropology and human genetics. *Anthropol. Anz.* **46:** 97–117.

Deevey, E. S., Jr. 1960. The human population. *Sci. Am.* (September) **203:** 194–205.

Feldman, M. W., and Christiansen, F. B. 1985. *Population Genetics.* Palo Alto, CA: Blackwell Scientific.

Friedlaender, J. S. 1975. *Patterns of Human Variation.* Cambridge, Mass.: Harvard University Press.

Mettler, L. E., Gregg, T. G. and Schaffer, H. E. 1988. *Population Genetics and Evolution.* 2nd ed. Englewood Cliffs, N.J.: Prentice-Hall.

Neel, J. V. 1978. The population structure of an Amerindian tribe, the Yanomama. *Ann. Rev. Genet.* **12:** 365–413.

Relethford, J. H. 1985. Isolation by distance, linguistic similarity and the genetic structure on Bougainville Island. *Am. J. Physiol. Anthrop.* **66:** 317–326.

Romeo, G., Devoto, M., and Galietta, L. J. 1989. Why is the cystic fibrosis gene so frequent? *Hum. Genet.* **84:** 1–5.

Spiess, E. B. 1989. *Genes in Populations.* 2nd Ed. New York: Wiley.

Woo, S. L. 1989. Molecular basis and population genetics of phenylketonuria. *Biochem.* **28:** 1–7.

CHAPTER 18

HUMAN DIVERSITY AND EVOLUTION

CHAPTER OUTLINE

On Saturday, June 30, 1860, 6 months after the publication of the *The Origin of Species* by Charles Darwin, the British Association sponsored a meeting to hear a talk by John W. Draper, an American who was to speak on ''intellectual developments of Europe considered with reference to the views of Mr. Darwin.'' The talk was destined to become a footnote in history. The clergymen who opposed Darwin's challenge to orthodoxy intended to use the meeting to launch a frontal attack on what they termed ''the monkey theory.'' Darwin's book implied that humans were of animal origin, and the bishop of Oxford, Samuel Wilberforce, was prepared to rebut this idea with all the means at his disposal. It was the clash of ideas between Bishop Wilberforce and an English biologist, Thomas Huxley, that was to make this meeting memorable.

The meeting was moved to a larger room to accommodate the hundreds, many of them clerics, who wanted to hear Darwin's ideas debunked. Draper spoke for about an hour, but it is doubtful whether many paid him much attention. After Draper's talk, Bishop Wilberforce arose to address the crowd. By all accounts, he was a gifted and dynamic speaker. He set about attacking Darwin's arguments in derisive and disparaging tones. The audience responded favorably, cheering him on. As his remarks came to a thundering conclusion, he turned to Huxley and asked whether it was through his grandfather or grandmother that he (Huxley) claimed descent from a monkey. Huxley rose and in a quiet voice gave a brief summary of Darwin's observations about the diversity of life and the theory of descent with modification that had grown from these observations. He pointed out that it was the best synthesis of ideas about species that had been advanced to date. He closed by turning to the bishop and remarking that he would not be ashamed to have a monkey for an ancestor but would be ''ashamed to be connected with a man who used great gifts to obscure the truth.'' The audience, realizing that the bishop had said very little of substance, responded with applause, which made it clear that Huxley had bested the Bishop.

This debate marks the opening of the battle over Darwin and his ideas. It helped focus the attention of the general public on the ideas of natural selection and evolution that were to become such powerful forces in science and society in Victorian England and in America.

ORIGINS OF DIVERSITY

The ABO blood system was discovered by Landsteiner and his colleagues early in this century. Blood types represent the first genetic system that did not depend on screening large populations to identify a small number of affected individuals. Blood types are carried by everyone and can be analyzed on a microscope slide from a few drops of blood. Shortly after this discovery, it was suggested that the ABO blood system could be used in the analysis of allele frequencies in different racial groups. By 1919 the first reports on populational differences in ABO alleles were published (Figure 18.1). Because of the importance of this information in blood transfusions, a large quantity of information about ABO allele frequencies quickly became available, and maps of the allelic distributions were constructed (Figure 18.2). The ABO system is an example of a genetic **polymorphism,** because there are two or more distinct alleles of the gene present in the population. If an allele is present in a frequency greater than 1%, that locus is regarded as polymorphic.

Polymorphism
The occurrence of two or more genotypes in a population in frequencies that cannot be accounted for by mutation alone.

The study of genetic polymorphisms provides information about the amount of genetic variability that is present in a population. In this chapter we will examine the forces that generate genetic diversity and those that

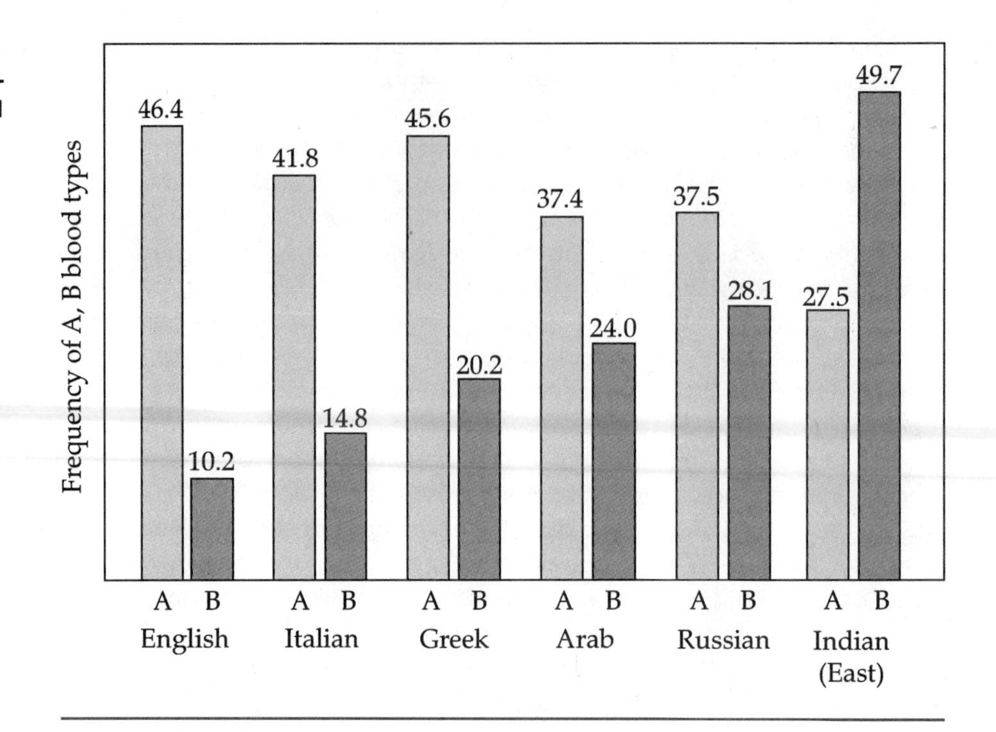

FIGURE 18.1 The distribution of A, B, and O alleles for various populations as measured in 1919. (The value for O is obtained by adding the values for A and B, then subtracting from 100.)

cause it to increase or decrease in a population. We will also consider examples of cultural practices that bring about changes in allele frequencies and the forces of natural selection that bring about evolutionary change.

Mutation

In each generation the gene pool is reshuffled to produce the genotypes of the offspring. In most cases the individuals present in a population at any given moment represent only a small fraction of all the genotypic combinations that can be constructed from the reservoir of genetic information present in the gene pool. Under these circumstances genetic variability is produced by recombination and Mendelian assortment alone. In other words, while a large number of new combinations of alleles can be produced, these processes do not produce any *new* alleles. Mutation is the ultimate source of all new alleles and is the origin of all genetic variability. It is important to stress that mutation is entirely a random process and can affect any genetic locus, although some genes appear to be more vulnerable to mutation than others.

The mechanisms and rates of mutation in humans were presented in Chapter 11. Although the effect of mutation on allele frequencies and genetic variability can be complex, let us consider a situation in which mutation is occurring at a single locus in a single direction ($a \rightarrow A$). To determine the effect of mutation on changing allele frequencies, we must first be able to measure the rate at which mutations take place. Using dominant genes, we previously calculated that the mutation rate in humans is about 1×10^{-5}.

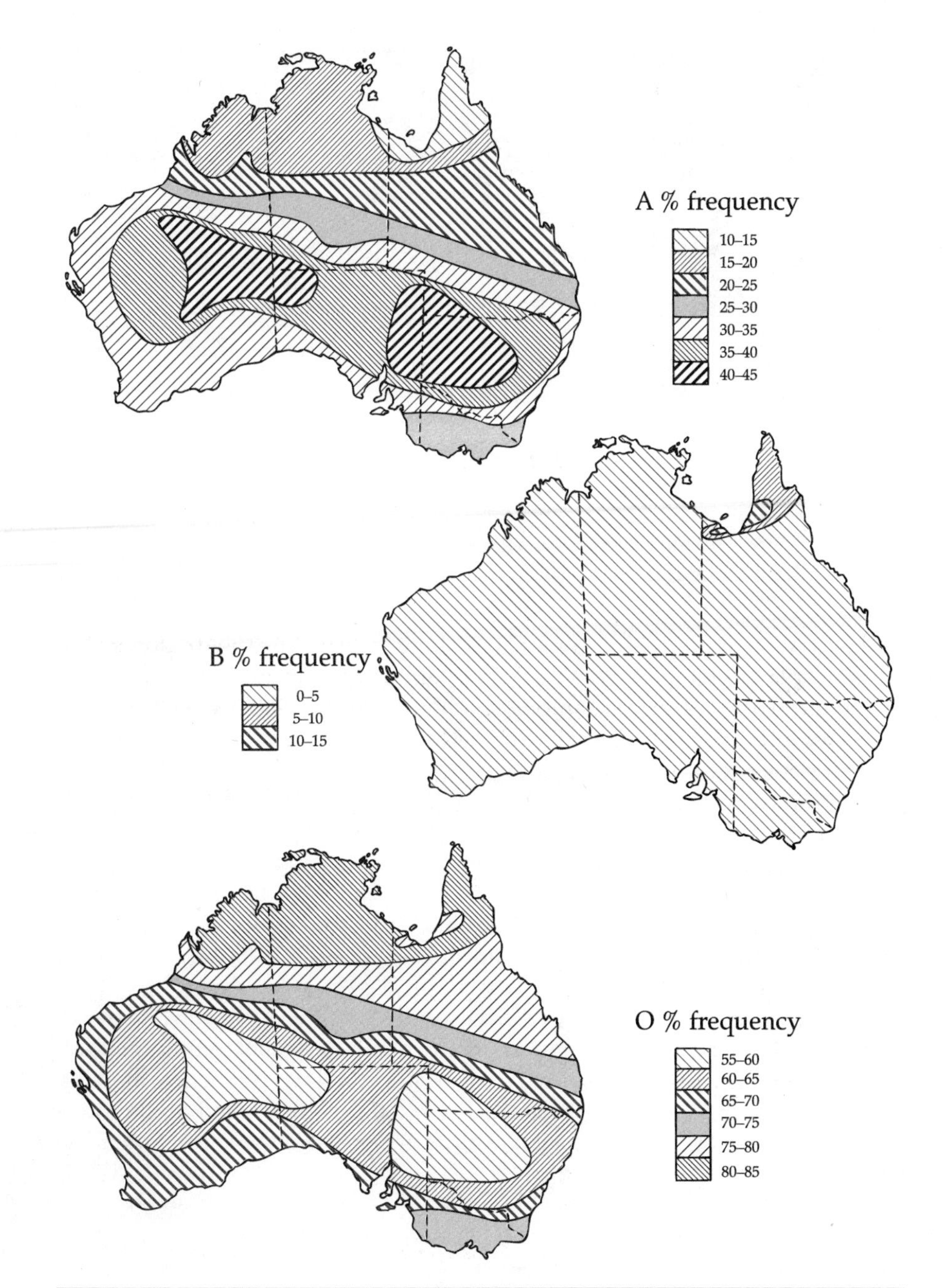

FIGURE 18.2 A reconstruction of the geographic distribution for the ABO alleles in the prehistoric populations of Australia.

Once the rate of mutation is known, we can use the Hardy-Weinberg law to calculate the change in allele frequency that occurs in each generation. The dominant condition known as achondroplasia, which produces dwarfism, has been known in Egyptian populations for over 2500 years, or about 125 generations, allowing 20 years per generation. If copies of the gene for achondroplasia were introduced into the gene pool only by mutation in each

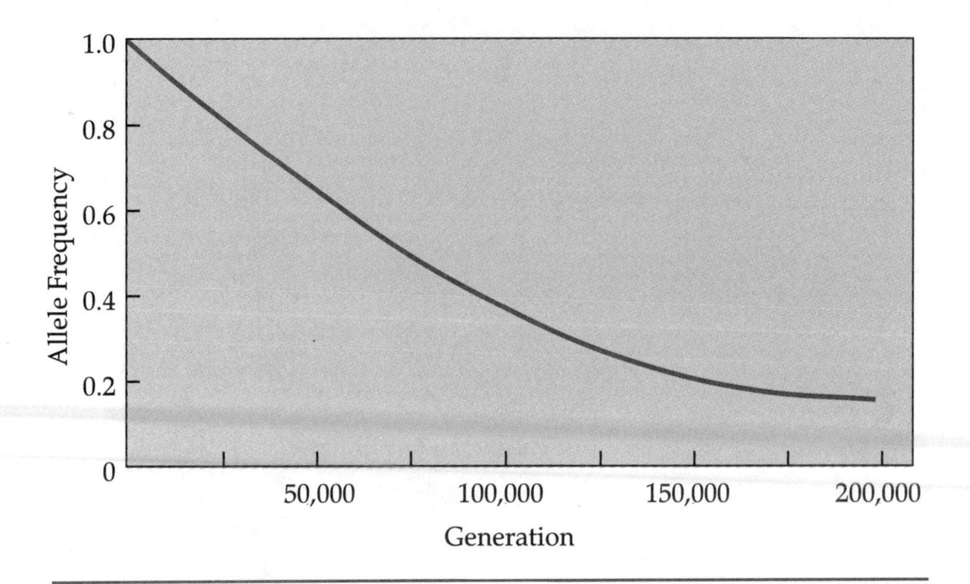

FIGURE 18.3 The rate of replacement of a recessive allele *a* by a dominant allele *A* by mutation alone. Even though the initial rate of replacement is relatively high, it would take about 70,000 generations, or 1.4 million years, to drive the frequency of the recessive allele from 1.0 to 0.5.

generation beginning 2500 years ago, how much has the frequency of achondroplasia changed over this time period? Should we expect a higher frequency of achondroplasia among residents of an ancient city such as Cairo than a recently established city like Houston? Let us assume that initially only homozygous recessive individuals of normal stature were present in the population (with the genotype *dd*), and mutation has added new mutant *(D)* alleles to each generation at the rate of 1×10^{-5}. Figure 18.3 shows the theoretic rate of change in allele frequency over time induced by this rate of mutation. To change the frequency of the normal allele *(d)* from 1.0 to 0.5 with a mutation rate of 1×10^{-5} will require about 70,000 generations, or 1.4 million years. Thus the frequency of achondroplasia need not be any higher in Houston than in Cairo.

Even if exposure to radioactivity or chemical mutagens were to increase, the overall effect of mutations on allele frequency would be very small. Mutation is the only source of genetic variability, but acting alone, it has a minimal effect on allele frequencies. If, on the other hand, mutation is accompanied by other factors such as drift or selection, the impact on allele frequencies may be much greater.

Drift

In an earlier chapter we considered the problem that human families often have small numbers of offspring, making it difficult to establish that genetic traits follow a 3:1 or 9:3:3:1 ratio. In human genetics, the large sample sizes needed to verify genetic ratios are obtained by pooling information from many family studies. In human population genetics, it is necessary to sample a large number of unrelated individuals to establish genotype frequencies and to apply the principles of the Hardy-Weinberg law. Occasionally, however, a population is formed from a small group of individuals, known as founders. The allelic forms of genes carried by these founders, whether they are advantageous or detrimental, will become established in the new population. These events take place simply by chance and are known as **founder effects.**

Founder effects
Allele frequencies established by chance in a population that is started by a small number of individuals (perhaps only a fertilized female).

Table 18.1 illustrates this situation in the extreme case in which a population is started by two heterozygous individuals who produce only two offspring. The first column of the table lists the possible genotypic combinations of the two offspring. In the combinations listed in the first two rows, either the *A* allele or the *a* allele is eliminated in a single generation. In the middle two rows, the combinations preserve the parental allele frequencies. In the last two rows, the combinations alter the frequency of *A* and *a* dramatically from that in the parental generation. The last column of the table lists the probability that such genotypic combinations in the offspring will be produced. In all, the table indicates that most of the possible genotypic combinations in the offspring will result in large-scale changes in allele frequencies in the subsequent generation.

These random changes in allele frequency that occur from generation to generation are examples of **genetic drift.** In addition to founder effects, genetic drift can occur in small, stable populations and by temporary but drastic reductions in population size. This effect, called a **population bottleneck** is often caused by natural disasters. In extreme cases, drift can lead to the elimination of one allele from all members of the population. Small interbreeding groups on isolated islands often provide examples of genetic drift. The population history for one such isolated island, Tristan da Cuhna (Figure 18.4), is known in detail. Located in the southern Atlantic Ocean at latitude 37° south, longitude 12° west, it is one of the most remote locations on the planetary surface. It is 2900 km from Capetown and 3200 km from Rio de Janeiro. The island was occupied by the British in 1816 to prevent the rescue of Napoleon, who was in exile on the island of St. Helena, 2400

Genetic drift
The random fluctuations of allele frequencies from generation to generation that take place in small populations.

Population bottleneck
Fluctuation in size that occurs when a large population is drastically reduced in size and then expands again; often results in an altered gene pool as a result of genetic drift.

TABLE 18.1 Combinations and Frequency of Offspring Produced by Heterozygous Parents *Aa* × *Aa*

Genotypic Combinations of Offspring	Allele Frequency A	Allele Frequency a	Probability of Genotypic Combination
AA and *AA*	1.0	0.0	1/16
aa and *aa*	0.0	1.0	1/16
AA and *aa*	0.5	0.5	2/16
Aa and *Aa*	0.5	0.5	4/16
AA and *Aa*	0.75	0.25	4/16
Aa and *aa*	0.25	0.75	4/16

FIGURE 18.4 Location of Tristan da Cuhna, first discovered in 1506 by a Portuguese admiral, and St. Helena, site of Napoleon's last exile.

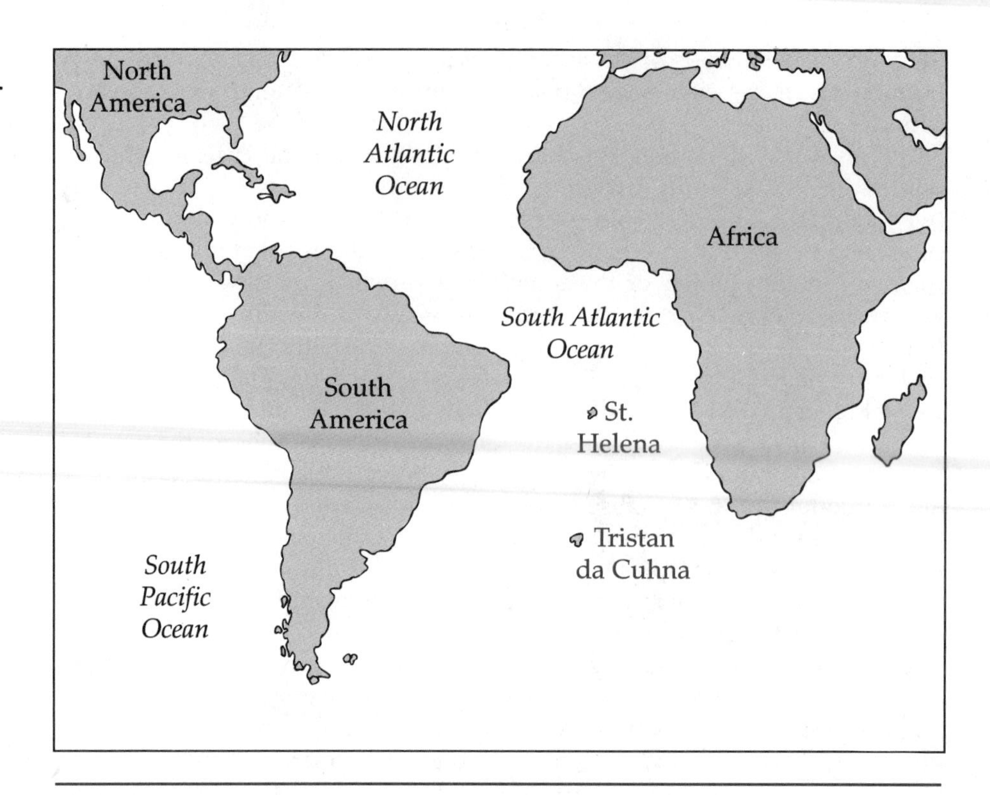

Clinodactyly
An autosomal dominant trait that produces a bent finger.

TABLE 18.2 Homozygous Markers among Tristan Residents
Transferrins
Phosphoglucomutase
6-phosphogluconate dehydrogenase
Adenylate kinase
Hemoglobin A variants
Carbonic anhydrase (2 forms)
Isocitrate dehydrogenase
Glutathione peroxide
Peptidase A, B, C, D

SOURCE: Data from Jenkins, T., Beighton, P., and Steinberg, A. G. 1985. *Ann. Hum. Biol.* **12:** 363–371, Table 2.

kilometers to the north. After Napoleon's death, as the British garrison departed, Corporal William Glass, his wife, and two daughters received permission to remain. Others joined Glass at intervals, and the development of the isolated and highly inbred population that formed here can be traced with great accuracy.

In 1961 a volcanic eruption forced the removal of all 294 residents to England. While in England, the residents were tested for many genetic polymorphic traits to determine the effects of isolation and inbreeding on the structure of the island's gene pool. As might be expected, one of the effects of inbreeding is an increase in homozygosity for recessive traits. Table 18.2 lists genetic markers for which all the islanders tested are homozygous. In 1963 almost all the residents returned to the island where they and their descendants remain.

The present population of about 300 individuals was derived from a small number of original settlers, and only seven family names are represented on the island. Because of the founder effect, traits carried by the early settlers are often found in a large fraction of the present population. On Tristan, a deformity of the fifth finger, known as **clinodactyly,** is present at a high frequency. This autosomal dominant trait is especially prominent in members of the Glass family. Recall that the first permanent residents of the island were, in fact, William Glass, his wife, and two daughters. The high frequency of this gene in the present island population can be explained by its presence in one of the original colonists. Marie Williams, who arrived on the island in 1827, probably carried an allele of the enzyme glucose-6-

phosphate dehydrogenase (G6PD) that is now widely distributed among members of the present population.

These examples illustrate how genetic drift can be responsible for changing allele frequencies in populations that are isolated, inbred, and stable for long periods of time. Most human populations, however, do not live on remote islands and are not subject to prolonged isolation and inbreeding. Yet there are many widespread differences in the distribution and frequency of alleles among populations, indicating that other factors must be at work. As we will see in the following section, the most powerful of these factors is **selection.**

Selection
The forces causing differential reproduction, thereby bringing about changes in the frequencies of alleles and genotypes in populations.

Selection

In populations that follow the Hardy-Weinberg law, allele frequencies do not change from generation to generation. This stability, or equilibrium, is brought about by several factors, among them that all genotypes in the population are equally viable and fertile. (See Chapter 17 for a list of the assumptions inherent in the Hardy-Weinberg model.) In formulating their thoughts on evolution, Darwin and Wallace recognized that not all members of a given population can be equally viable or fertile, given the competition for limited resources such as food and mates. By virtue of the wide range of different genotypes present in the population, some individuals will be better adapted to the environment than others. These better-adapted individuals have increased chances of leaving more offspring than those with other genotypes. The ability of a given genotype to survive and to reproduce is known as its **fitness.** By definition, the fitter genotypes are those most able to survive and reproduce. As a result, they make a larger contribution to the gene pool of the next generation than other, less fit genotypes.

Fitness
A measure of the relative survival and reproductive success of a given individual or genotype.

In time this differential reproduction of better adapted, or fitter, individuals leads to changes in allele frequencies within the population. The process of differential reproduction of fitter genotypes is known as **natural selection.** Darwin recognized that selection is the primary mechanism that leads to evolutionary divergence and the formation of new species. Note that the basis of selection is differential reproduction of the fittest genotypes, not only the survival of the fittest.

Natural selection
The differential reproduction shown by some members of a population that is the result of differences in fitness.

In applying the principle of selection to human populations, we must look for local factors that can bring about a change in allele frequencies. If such forces are present in one location and not another, populations subject to these forces will eventually show differences in the frequency of one or more alleles.

One of the best documented examples of the interaction between selection and genotype frequency is the relationship between the sickle cell allele and malaria. Sickle cell anemia is an autosomal recessive condition caused by a single amino acid substitution in beta globin, one of the protein components of hemoglobin. Individuals with this condition are subject to a wide range of clinical problems (for a review of the symptoms, see Chapter 4). In spite of the fact that there is a high childhood death rate for untreated homozygotes, the allele is present in very high frequencies in certain populations. In some West African countries, 20% of the population may be heterozygous for this trait, and in regions along rivers, such as the Gambia, almost 40% of the population is heterozygous. In the absence of other factors, it is difficult to understand why this fatal disease has not been eliminated from the population.

In 1949 the geneticist J. B. S. Haldane proposed that heterozygotes for another form of anemia, beta thalassemia, might be protected from malarial infection. His suggestion was based on the observation that the geographic distribution of malaria and thalassemia were very similar. Malaria is a disease caused by infection by a protozoan parasite known as *Plasmodium falciparum* and the disease is transmitted to humans by infected mosquitoes. Victims of malaria are weakened by the disease and suffer recurring episodes of illness throughout life. Individuals with malaria are more likely to become victims of other diseases, often with fatal results. Thus malaria victims have a reduced fitness. To residents of some countries, malaria may seem like an exotic and rare disease, but more than 2 million people die from malaria each year, and more than 300 million individuals worldwide are infected with malaria. Because of population growth in affected regions and the rise of drug-resistant strains of *Plasmodium*, the incidence of malaria is actually increasing rather than decreasing.

The geographic distributions of malaria and sickle cell are shown in Figure 18.5. This correlation suggests that sickle cell anemia may confer resistance to malaria, and experiments on human volunteers have confirmed that this is the case. In heterozygotes and in homozygotes, the presence of sickle cell hemoglobin (Hb S) in red blood cells causes a physical alteration in the membrane of erythrocytes and makes the cells resistant to infection by the malarial parasite. As a result, the heterozygote genotype is fitter than the homozygous normal genotype because of its resistance to malarial infection. Even though the homozygous sickle cell individual is also resistant to malaria, the heterozygote is fitter than the sickle cell homozygote because the heterozygote has none of the clinical symptoms of sickle cell disease. Selection therefore favors the survival and differential reproduction of heterozygotes. Because heterozygotes for this autosomal recessive trait outnumber homozygotes, only a relatively small selective advantage for the heterozygote is needed to overcome the selective disadvantage of the homozygous condition. In the case of sickle cell anemia, it has been estimated that a 20% increase in heterozygote fitness is enough to balance an 85% decrease in fitness for sickle cell homozygotes. As a result of this increased fitness, heterozygotes make a larger contribution to the gene pool of the offspring than the other genotypes, and the Hb S allele is spread through the population and maintained at a high frequency. In contrast, if the selective force (malaria) is eliminated by mosquito control or by migration, the frequency of the allele should decline.

The effects of selection on allele frequency can also be seen by analysis of subpopulations geographically separated from one another but all derived from one ancestral population. Various emigrations and disper-

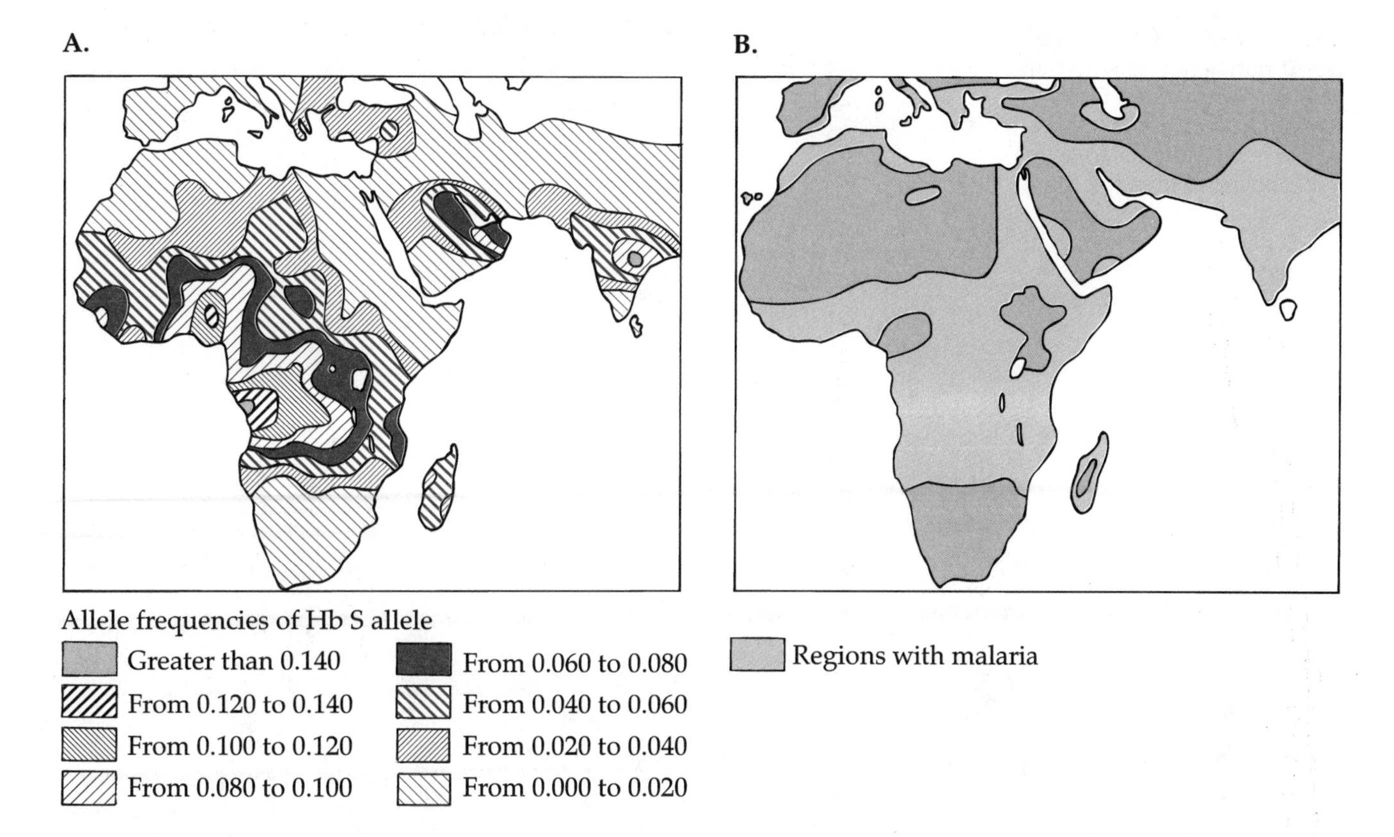

FIGURE 18.5 (A) The distribution of sickle cell anemia in the Old World. (B) The distribution of malaria in the Old World.

sions of the Jewish people from ancient Israel have occurred over the last 2500 years. These populations took up residence in areas ranging across Europe, North Africa, the Middle East, and Asia. The frequency of the X-linked condition, glucose-6-phosphate dehydrogenase (G6PD) deficiency in these populations is very different (Figure 18.6). Note that the frequency ranges from almost zero in populations from Central Europe to around 70% in Kurdish Jews. There are two likely explanations for such differences in allele frequency: drift and selection. The geographic distribution of G6PD deficiency is similar to that for malaria. This distribution and the fact that these dramatic differences in allele frequency arose in the short span of 100 to 125 generations indicate a role for selection. In this case the selective force once again is malaria. Malarial parasites reproduce well only in cells that contain the enzyme G6PD, offering protection against malaria to homozygous females and hemizygous males who are G6PD deficient. If we assume that the ancestral population had a low frequency of this allele, selection has caused a dramatic change in frequency in a relatively short time.

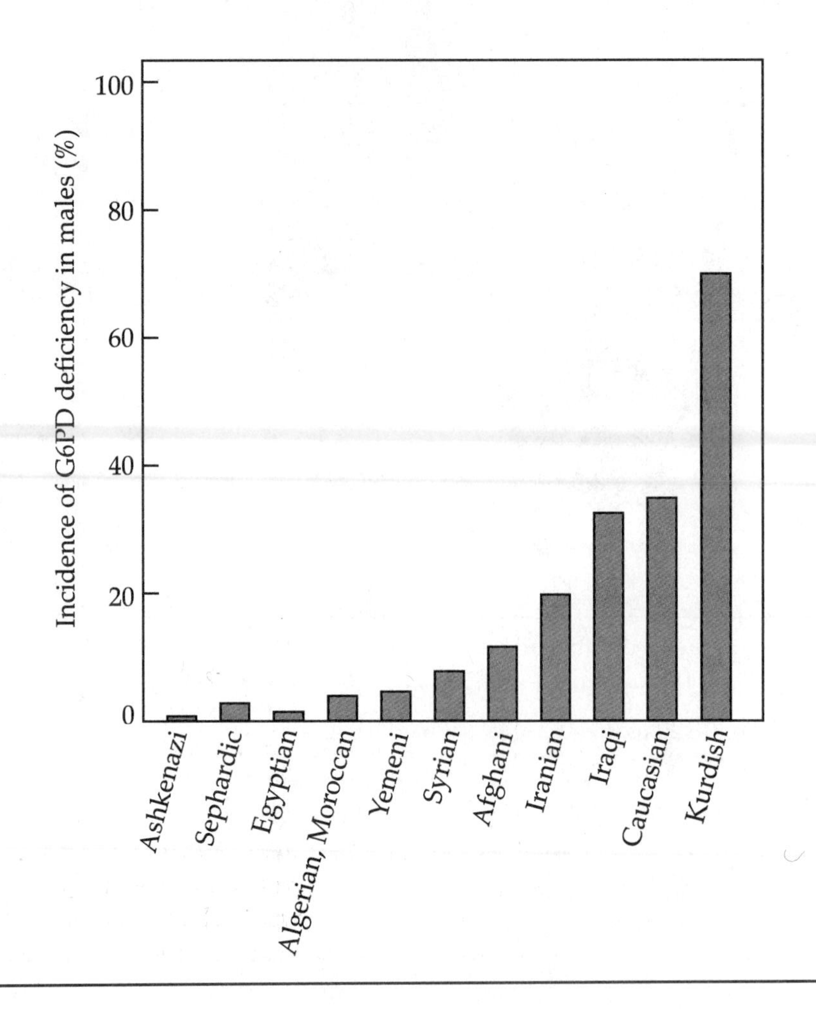

FIGURE 18.6 Distribution of G6PD deficiency in various Jewish populations. Since intermarriage with native populations is rare, the differences in frequency are attributed to the action of natural selection. G6PD deficiency confers increased resistance to malaria (see text for detailed explanation).

CULTURE, SOCIETY, GENES, AND DIVERSITY

In addition to the selective forces such as disease and the forces of nature, human activities can also influence the frequency and distribution of genes and their alleles. Human activities such as discovery and invention that improve adaptation are collectively known as culture. Beginning in the 16th century, the New World was rapidly populated by Europeans seeking better social, economic, and religious opportunities. These large-scale migrations created new combinations of genotypes.

Social customs and rules are powerful forces that dictate and shape human mating patterns. In most societies incestuous matings are forbidden. At the other end of the spectrum, social pressure often limits mate selection to those with some common bonds, such as language, religion, geography, and economic status. In effect, these social constraints prevent random mating, and this behavior can rapidly change genotype frequencies. In this section we will consider how two human activities, migration and mate selection, bring about such changes.

Migration

Migration is the method by which the human species has distributed and redistributed itself across the surface of the planet. The original populations of North and South America arrived via a land bridge in what is now the Bering Strait as small groups of hunter-gatherers and families. Over a period of centuries, the population spread from the Arctic Circle to the Antarctic Circle via further migration. By the 16th century, improvements in transportation (a product of culture) made it possible for humans to migrate to any portion of the globe in less than a lifetime, and Europeans began to populate the New World in increasing numbers. By the 20th century, millions of migrants were arriving annually, increasing the amount of gene flow. Gene pools that were originally geographically distant became mixed through random mating. Thus, migration is a force that can alter the genetic structure of populations.

As a force in evolution, migration has two primary effects. First, it acts in a manner opposite to genetic drift to break down genetic differences between populations. Drift is usually a significant factor in areas of low population density or among isolated populations. It tends to make populations genetically different from one another. Migration, on the other hand, reverses this trend, increases the size of the gene pool, and promotes gene flow among population groups. In another sense, migration is similar to mutation in that it can bring about the introduction of new alleles into a population. Because migration rates are usually much higher than rates of mutation, migration is more effective in changing allele frequencies than is mutation.

Migration can also have a dramatic impact on allele frequency when it is coupled with other factors such as genetic drift. The island of Mauritius is situated in the Indian Ocean east of Madagascar (Figure 18.7). Around 1800, August de Bourbon, a nobleman, became dissatisfied with conditions in postrevolutionary France and migrated to the island with his wife. One of his grandsons, Pierre de Bourbon, developed Huntington disease and died at the age of 51 shortly after the turn of the present century. Records and diaries indicate that earlier family members were probably also affected with this condition. At the present time, 6 members of the de Bourbon family are afflicted with this dominantly inherited fatal condition, and 19 others are known to have died from this disease. Pedigree analysis has identified another 25 family members who are at risk of this disease. Because there are only 13,000 Europeans on the island, the rate of Huntington disease in this subpopulation is one of the highest in the world. In this case you might say that the French Revolution brought about a rapid and dramatic change in allele frequency on Mauritius.

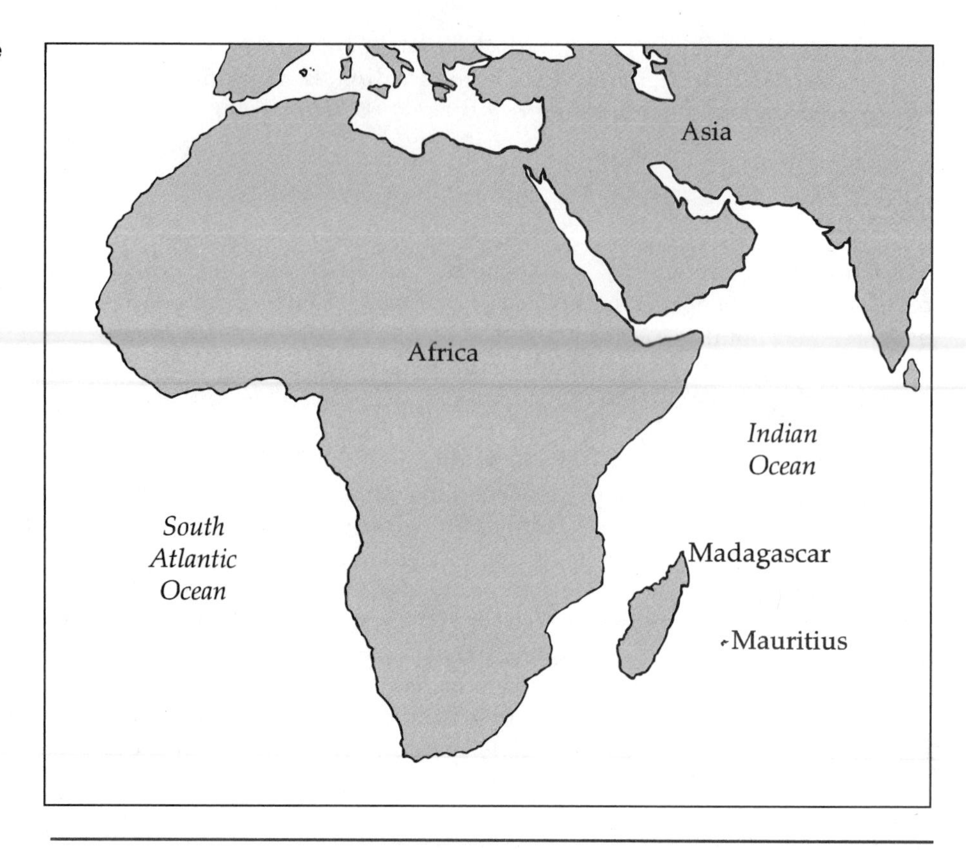

FIGURE 18.7 On Mauritius the frequency of Huntington disease is among the highest in the world.

Mate Selection

Assortative mating
Reproduction in which mate selection is not at random but instead is based on physical, cultural, or religious grounds.

Inbreeding
Production of offspring by related parents.

Incest
Sexual relations between parents and children or between brothers and sisters.

Consanguineous matings
Matings between two individuals who share a common ancestor in the preceding two or three generations.

In a population at genetic equilibrium, one of the assumptions of the Hardy-Weinberg law is that mating is at random. This ideal condition rarely occurs in human populations, partly as a result of geographic proximity and partly as a result of social structure and cultural limitations. One form of nonrandom mating is called **assortative mating.** Among humans, cultural factors such as common language, physical characteristics, economic status, and religion are often important in mate selection. To the extent that these factors are partially genetically controlled, mating is nonrandom and can influence allele frequency.

Inbreeding is another form of nonrandom mating and involves mating between related individuals sharing common genes. In humans the most extreme form of inbreeding is **incest,** mating between parents and children or between siblings. In almost all societies, strictures against incest are common. It seems plausible to suggest that such cultural prohibitions arose from the observation that such matings resulted in increased numbers of abnormal offspring. If this hypothesis is correct, incest taboos represent a cultural force that affects allele frequency in human populations.

However, in some societies, incest or mating with close relatives was common among royalty. In the Ptolemic dynasties of Egypt, such **consanguineous matings** were frequent (Figure 18.8). The genetic effect of

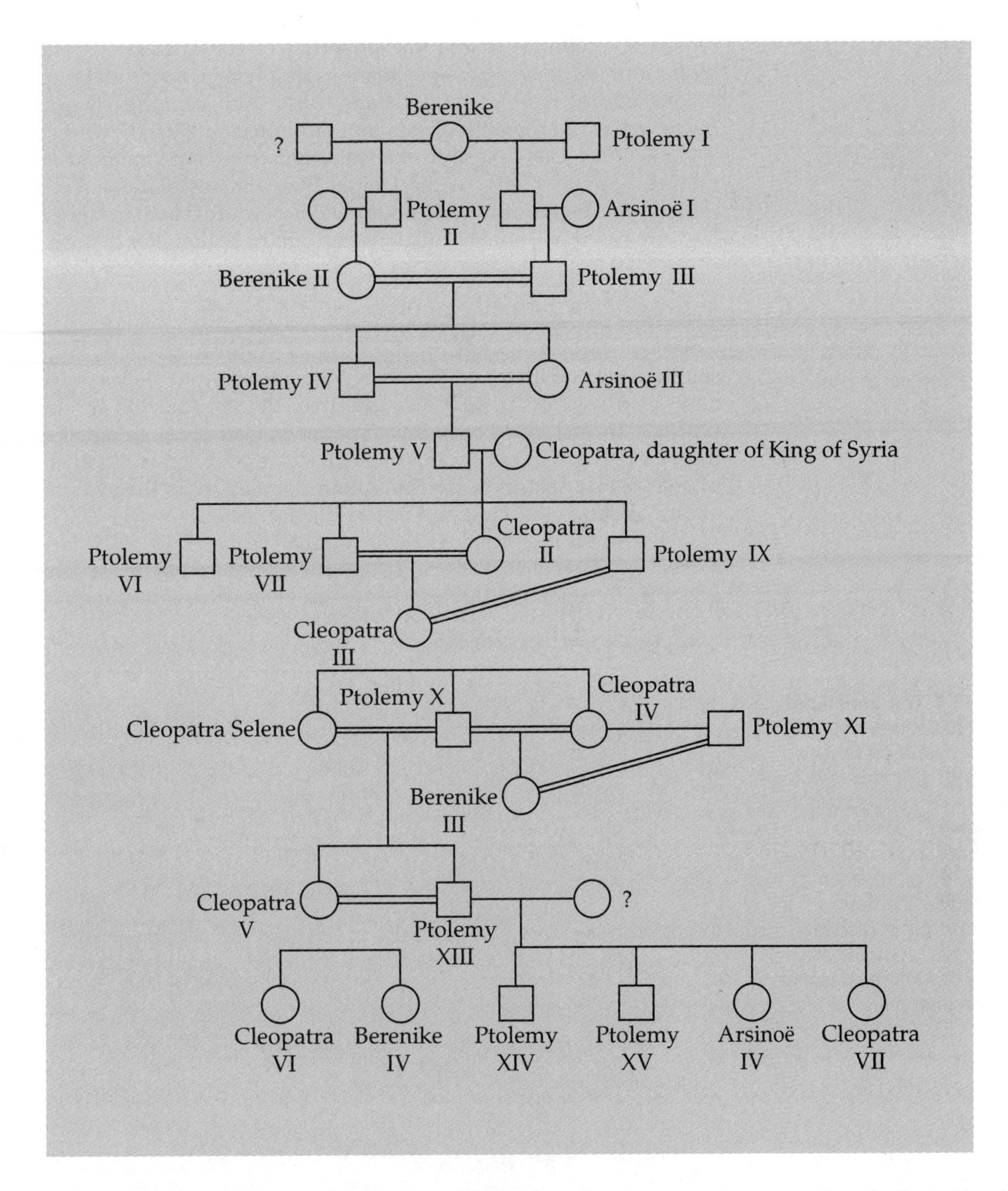

FIGURE 18.8 Pedigree of the Ptolemaic dynasty of Egypt. Incest and consanguineous matings (indicated by double lines) were common. Cleopatra VII, the great Cleopatra, is most famous for her affairs with Julius Caesar and Marc Antony. She had one son by Caesar and three children by Antony.

consanguinity is an increase in homozygosity and a decrease in heterozygosity (Figure 18.9). Table 18.3 shows the frequency of some autosomal recessive conditions among the offspring of consanguineous matings. In modern times, consanguineous matings involving cousins were common in many cultures during the 19th and early 20th centuries (Figure 18.10). However, as the mobility of the population has increased, the incidence of such matings has decreased, contributing to a more random system of matings (Figure 18.11). In the United States, many states have laws that regulate the degree of consanguinity that is permitted in marriages, but these laws are based on social or religious customs or both rather than on genetics.

Culture as a Selective Force

Collective human activity, in the form of cultural practices, can act as a selective force to alter the frequency of genotypes and alleles (see Concepts and Controversies, p. 495). For example, the introduction of agricultural practices in Africa has been linked to the spread of malaria and a parallel increase in the frequency of the allele for sickle cell anemia. Another example of culture as a selective force in human populations is the evolution of the ability of adults to digest and utilize dietary lactose.

Lactose is the principal sugar in milk (human milk is 7% lactose) and serves as a ready energy source. In spite of its importance in nutrition, lactose is not easily absorbed through the wall of the small intestine. In the small

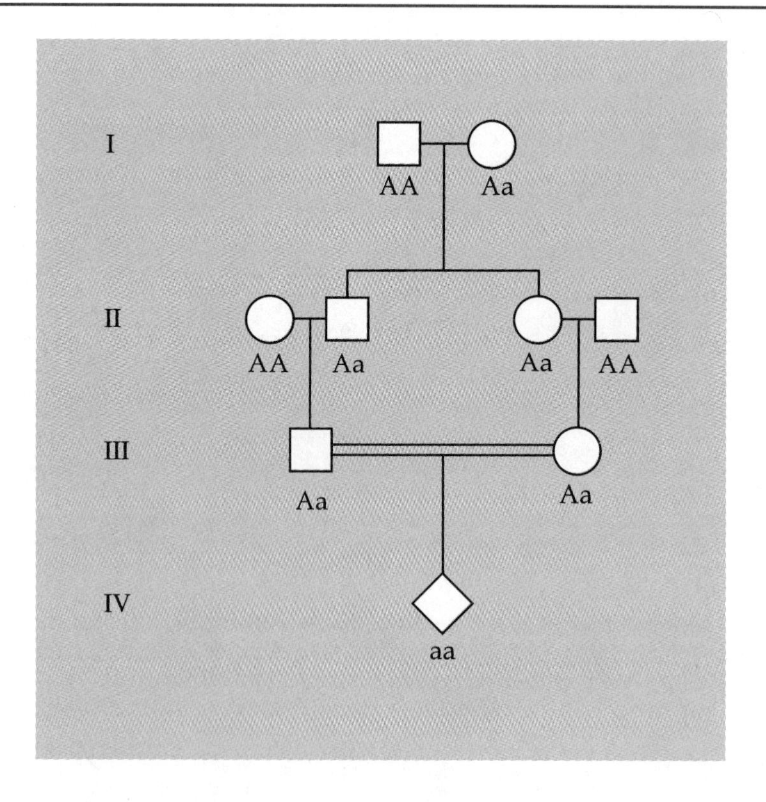

FIGURE 18.9 The genetic effect of consanguineous matings as shown by a first-cousin mating. In this hypothetical case, a recessive allele *a* with deleterious effects in homozygotes is passed through four generations. Inbreeding (symbolized by the double horizontal line) in generation III combines this recessive allele into a homozygote by descent from a common ancestor.

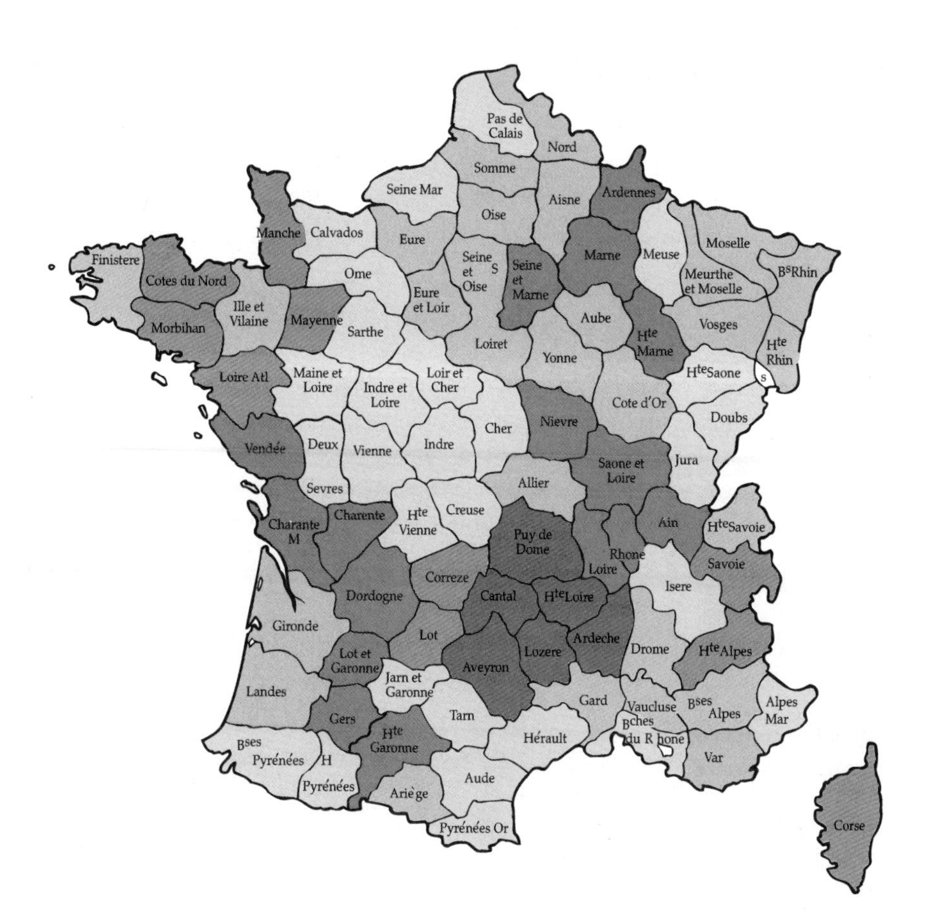

FIGURE 18.10 Relative levels of consanguinity in France between 1926 and 1930. Darker shades represent increasing levels of consanguinity.

TABLE 18.3 Percentage of First-Cousin Parents in Children with Recessive Genetic Diseases

Disease	% of First-Cousin Parents
Albinism	10
PKU	10
Xeroderma pigmentosum	26
Alcaptonuria	33
Icthyosis congenita	40
Microcephaly	54

SOURCE: Reed, S. 1980. *Counseling in Medical Genetics.* 3rd ed. New York: Alan Liss, p. 77, Table 10-1.

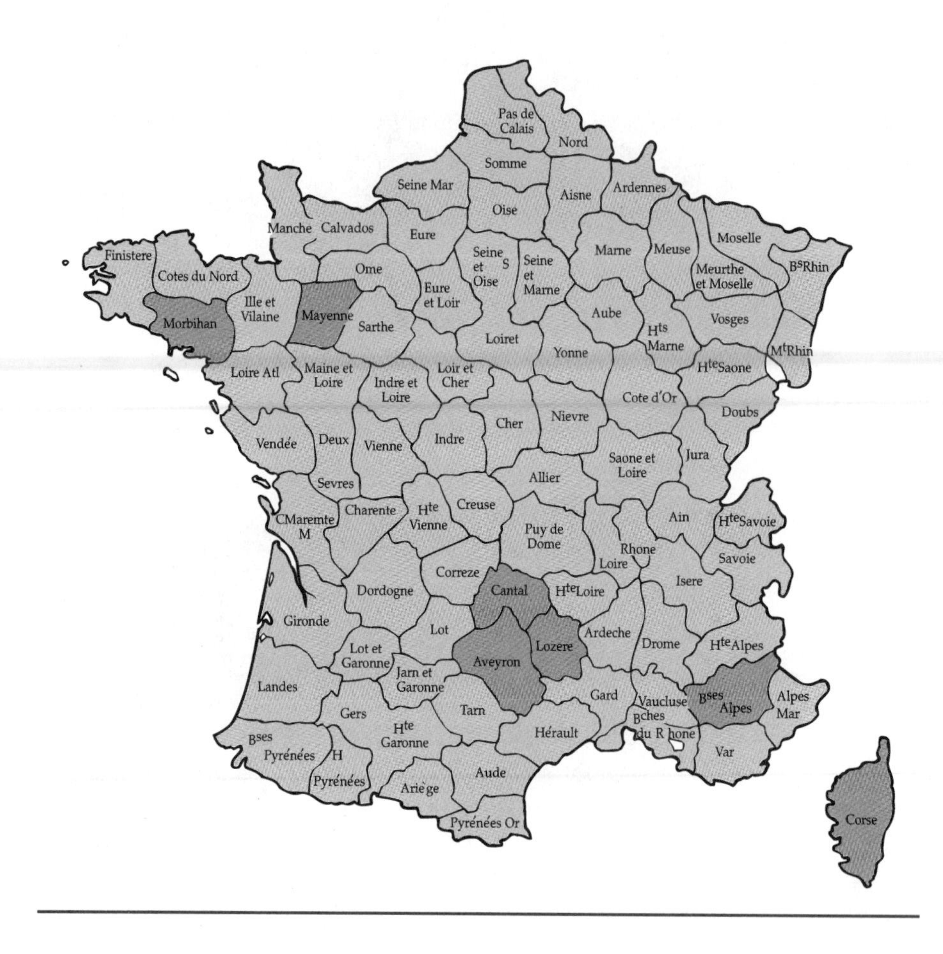

FIGURE 18.11 Relative levels of consanguinity in France between 1956 and 1958. Darker shades represent increasing levels of consanguinity.

intestine of infants, however, the enzyme lactase is secreted and converts lactose into its components, glucose and galactose, sugars that are easily absorbed by the intestine. In most humans the production of lactase declines at about the time of weaning, and as children grow into adolescents and adults, they are unable to absorb lactose—they are termed lactose malabsorbers (LM). Many such adults exhibit lactose intolerance, a reaction by the body to the presence of lactose, characterized by flatulence, cramps, diarrhea, and nausea. In these individuals, undigested lactose passes from the small intestine into the colon, where it exerts an osmotic effect to draw water from the surrounding tissues. In the colon, lactose is digested by coliform bacteria, producing gas and the resulting diarrhea.

There are however, some human populations in which lactase continues to be produced throughout adulthood, and these individuals are classed as lactose absorbers (LA). Population surveys of lactose absorption in populations ranging from hunter-gatherers to industrialized urban areas show that the frequency of lactose absorption varies from 0.0 to 100% (Table 18.4). Genetic evidence indicates that adult lactose absorption (and the adult production of lactase) is inherited as an autosomal dominant trait.

Several hypotheses have been advanced to explain the wide-ranging differences in lactose absorption in various populations. One of these relies

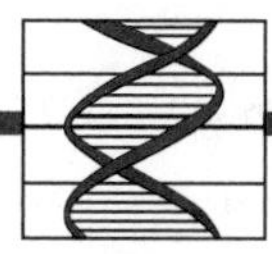

CONCEPTS & CONTROVERSIES

Genetic Diversity and Embryo Cloning

Human activity in the form of culture has been changing gene frequencies for several thousand years. The introduction of agriculture into Africa has been linked to the spread of malaria and the selection of heterozygotes for sickle cell anemia. Other cultural practices such as dairy herding have also been associated with changing gene frequencies. Recently, a new advance in technology offers the possibility of altering gene frequencies in offspring by cloning human embryos. Embryo splitting to form twins or multiple embryos is a natural process, but it is now possible to split embryos in the laboratory, following *in vitro* fertilization. This development is an application of methods already in use to produce multiple embryos for livestock and othr agricultural animals, and is not really a technological breakthrough.

What are the genetic consequences of cloning human embryos? Within a single family, if multiple children are produced from a single embryo, it would lead to a reduction in the amount of genetic diversity present in members of a pedigree, much the same as consanguineous matings reduce heterozygosity. However, unless embryo cloning is practiced on a large scale and over a long time period, there is little likelihood that it will have genetic consequences for the species.

If this development is not really a technological breakthrough, and has little chance of bringing about large scale genetic changes, why has the addition of this technique to the stockpile of alternate reproductive technologies created such a stir? After all, identical twins, triplets and even quadruplets are not unknown, and in some cases, not uncommon.

The answer is unclear, but it may be that embryo cloning will serve to bring forward the issue of what constitutes ethical use of alternate reproductive technologies and re-invigorate the process of forming public policy on these issues. Once again, this development demonstrates that policy and law lag far behind the pace of genetic technology, and that the gap may be growing wider. As outlined in the next two chapters, embryo cloning is only one of many issues that need serious discussion and resolution.

on cultural practices to explain the spread of the LA allele. According to this hypothesis, human populations originally resembled other land mammals and were lactose intolerant as adults. As the practice of dairy herding developed in some groups, those carrying the LA allele had the selective advantage of being able to derive nutrition from milk. This improved their chances of survival and success in leaving offspring. As a result, the cultural practice of maintaining dairy herds provided the selective factor that conferred an increased advantage on the LA genotype.

Surveys of populations provide a strong correlation between a cultural history of dairying and the frequency of lactase persistence. For example, the Tuareg of North Africa are nomadic herders who have been in the Central Sahara for 2000 to 3000 years (Figure 18.12). Other food is often unavailable, and adult consumption of several liters of milk per day is common. In this population, there is a high frequency of the LA allele. In fact, in all populations with high LA frequencies (60% to 100%), there is a history of dairying often traced back for over 1000 years. In contrast, among populations such as those in the tropical forest belt of Africa, where sleeping

TABLE 18.4 Lactose Absorption in Various Populations

Population	Percent Lactose Absorbers (LA)
Eskimos—Greenland	15
!Kung—Africa	2.5
Tuareg—Africa	85
Bantu—Africa	0
Arabs—Saudi Arabia	86
Sephardic Jews—North Africa	38
Danes—Europe	98
Czechs—Europe	100
U.S.—Asians	3
Aborigines	15
U.S.—Blacks	25

FIGURE 18.12 The Tuareg people of North Africa.

sickness, carried by the tsetse fly inhibits herding, the frequency of lactose absorption is low (0% to 20%). Taken together, the evidence from genetics, anthropology, and geography supports the idea that the variation in frequency of lactose absorption found in present human populations is derived from cultural practices acting as a selective force on allele frequencies. The presence of this mode of selection makes humans unique among land mammals in having populations with a high frequency of adult lactose absorbers.

The Frequency of Human Genetic Disorders

Many genetic disorders are disabling or fatal, so why are they so common? In other words, what keeps natural selection from eliminating the deleterious alleles responsible for these diseases? In analyzing the frequency and population distribution of human genetic disorders, it is clear that there is no single answer. One conclusion, drawn from the Hardy-Weinberg law, is that rare lethal or deleterious recessive alleles survive because the vast majority of them are carried in the heterozygous condition and thus are hidden in the gene pool. Other factors, however, can cause the differential distribution of alleles in human populations, and several of these will be discussed below.

In some cases, such as X-linked Duchenne muscular dystrophy (DMD), almost all affected individuals die without reproducing, and the mutant gene should be eliminated from the population. However, because the mutation rate for DMD is high (perhaps as high as 1×10^{-4}), mutation replaces the DMD alleles carried by affected individuals who die without reproducing. Thus, the frequency of the DMD allele in a population represents a balance between those alleles introduced by mutation and those removed by the death of affected individuals. Mutation rates can also be used to explain the maintenance of the autosomal dominant disease achondroplastic dwarfism at high frequencies in the population.

In analyzing the origin and distribution of Huntington disease (HD) on the island of Mauritius, we have already seen how the founder effect is able to raise the frequency of a deleterious gene in an isolated or restricted population. However, this phenomenon cannot explain how detrimental alleles are able to spread through large, historically well-established populations at a high frequency. Explanation of these conditions usually involves a heterozygote advantage. We have already examined the classic case of heterozygote advantage in the relationship between malaria and sickle cell anemia. Untreated sickle cell homozygotes have a higher childhood death rate; the more numerous, surviving heterozygotes are resistant to malarial infection. There are similar advantages for carriers of thalassemias and glucose-6-phosphate dehydrogenase (G6PD) deficiency in areas where malaria is endemic.

For some other genetic diseases, if there is a selective advantage for heterozygotes, it is less obvious or perhaps no longer observable. Tay-Sachs is an autosomal recessive disease that is fatal in early childhood. Although it is rare in most populations, there is a 10-fold increase in Tay-Sachs disease in some populations of Ashkenazi Jews, and the frequency of heterozygotes can be as high as 11%. There is some indirect evidence to suggest that

Tay-Sachs heterozygotes are more resistant to tuberculosis, a disease endemic to cities and towns, where most of the European Jews lived. As in sickle cell anemia, the death of homozygous Tay-Sachs individuals is the genetic price paid by the population that allows the higher fitness and survival of the more numerous heterozygotes.

The high incidence of cystic fibrosis (CF) has been a more difficult case to explain. In European and European-derived populations, cystic fibrosis affects about 1 in 2000 births, and the frequency of heterozygotes is about 1 in 22. Until recently, this autosomal recessive disease has been lethal in early adulthood, and almost all cases have been the result of matings between heterozygotes. Several hypotheses have been advanced to explain the frequency of this disease, including a high mutation rate, higher fertility in heterozygotes, and genetic drift. None of these hypotheses has made a convincing case to explain the high frequency of this deleterious gene. In fact, recent evidence indicates that there is not a high mutation rate at this locus, that there is no higher fertility in heterozygotes, and that the probability of genetic drift is very low. At the physiologic level, CF impairs chloride ion transport in secretory cells, and heterozygotes have a reduced level of chloride transport. In many parts of the world, today as in the past, bacterial diarrhea contributes significantly to infant mortality. Because certain toxins produced in bacterial diarrhea cause an oversecretion of chloride ions, it has been postulated that infant CF heterozygotes are more likely to survive such illnesses that kill by electrolyte depletion and dehydration. Now that the CF gene has been cloned, it will be possible to examine this hypothesis in more detail.

Thus it appears that many different factors contribute to the frequency of genetic diseases in human populations and that each deleterious gene must be analyzed individually. In some cases the frequency of genes that occupy long stretches of DNA or are at hypermutable sites is maintained by mutation. In other cases migration and founder effects can greatly increase the frequency of deleterious alleles. From an evolutionary perspective, natural selection favors certain heterozygote carriers of fatal genetic disorders, while affected homozygotes bear the burden associated with conferring advantages on other genotypes.

The Question of Race

Genetic and Non-genetic Definitions

Race
A genotypically distinct subgroup of a species.

In the 19th century biologists and anthropologists used the term **"race"** to describe groups of individuals within a species that could be distinguished in some phenotypic way from other groups within that species. For example, two-spotted beetles and four-spotted beetles found in the same population might be classified as separate races, even when some two-spotted and four-spotted bettles might be siblings. With the development of population genetics as a separate discipline in the 1930s and 1940s, a genetic concept of race began to replace the morphologic or phenotypic concept. Races were then regarded as local populations with significant differences in the average frequency of a variety of genes. These groups were known as **geographic races.** From our earlier discussions it should seem obvious that over time, two isolated populations that were originally genetically identical can become genetically different from each other. Ultimately, however, the classification of populations into geographic races is subjective and depends on the degree of genetic difference required to satisfy the individual making the determination.

Geographic races
Races or subspecies separated from one another by geographic barriers.

The concept of geographic race has not really clarified the difficulty of adequately defining what we mean by the term "race" as it applies to humans. It is clear that there are major physical differences among humans. Residents of the Kalahari Desert are rarely mistaken for close relatives of Aleutian seal hunters. In spite of these apparent and obvious physical differences, is there any need for or value in classifying humans into racial groups? More to the point, is there a genetically useful and legitimate use for the concept of human races? To the biologist studying nonhuman organisms, races represent groups that have become genetically separated to the extent that they cannot be regarded only as populations but have not yet separated enough to be regarded as distinct species. If we wish to apply this criterion to humans, we need to examine the degree of genetic differences that exist among populations.

Are There Human Races?

In a sense it is easy and perhaps logical to assume that the observable and distinct physical differences that exist between groups of humans are evidence for significant genetic differences between these groups. Over the last 30 years, a large amount of information has become available on the degree of genetic variation present in the human genome and how this variation is distributed among population groups.

Using techniques of immunology and protein electrophoresis, about 200 different gene products have been studied in some detail. To date, almost 75% of all the gene products examined by these physical and chemical techniques are invariant. Such proteins are called **monomorphic** because they occur in only one form. The remaining 25% exist in two or more forms, each encoded by a separate allele (Figure 18.13). These *polymorphic* proteins are present in different frequencies in diverse population groups. It is these polymorphic loci that account for the genetic differences between individuals and populations. A polymorphic locus we have already considered in other contexts is the ABO system, consisting of three alleles. As shown in Table 18.5, there are significant differences in the frequency of these alleles in different populations.

Monomorphic
Showing only one form.

Extension of this kind of analysis to other polymorphic genes has clearly demonstrated that almost all genetic diversity is accounted for by variation within populations rather than by differences between populations. Using biochemical criteria, only about 7% of all variation in the human species can be explained by genetic differences between what are classified as racial groups. About 8% of the variation is accounted for by differences between populations within racial groups, and the remaining 85% is accounted for by variation between individuals within those racial groups.

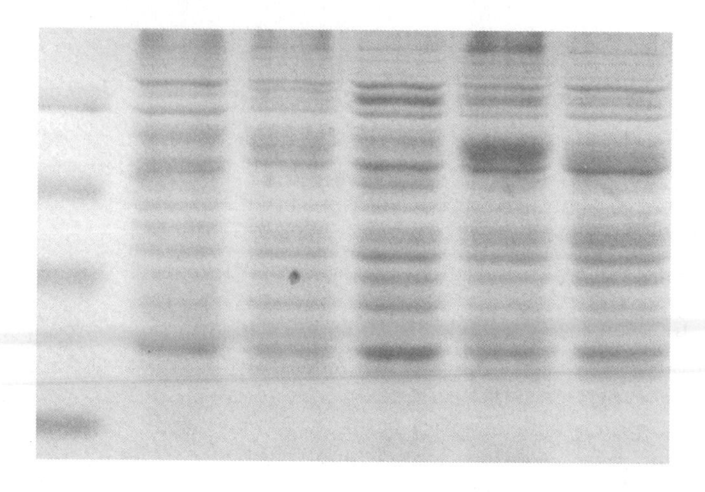

FIGURE 18.13 Protein polymorphisms as revealed by gel electrophoresis. Each band represents one polymorphic form. Both homozygotes and heterozygotes are shown.

More recently, genetic variation within and between populations has been studied using DNA markers in five major human groups. This group included 59 different populations from around the world (22 Caucasian, 17 African, 8 Asian, 8 Hispanic and 4 American Indian) with more than 11,900 individuals tested. This analysis indicates that 98.5% of the observed variation lies between individuals within populations, about 0.8% of the observed variation is between racial groups, and about 0.6% of the variation is between populations in the same racial group. These results indicate that individual variations in DNA profiles overwhelm any interpopulation differences, no matter how the populations are ethnically or racially classified.

In some ways these results run counter to our own everyday experience. From observation it appears that there is a great range of phenotypic variation among humans that would justify the classification of humans into racial groups. Looking at the genetic evidence, it is possible that the genes analyzed to date do not include those that are responsible for the phenotypic variation we observe, such as skin color, body form, and facial appearance. The other possibility is that such genes are included in the small amount of genetic variation we can measure between groups. Looking at the evidence from DNA studies, it would be difficult to classify individuals into racial groups based on variations in DNA polymorphisms, leaving the question of human races unanswerable at this level of analysis.

Diversity and Human Evolution

In this chapter and in other parts of the book we have come across many human traits that are polymorphic. The evidence indicates a great deal of genetic variation is present in the human genome. All this variation has been introduced by the process of mutation. To be detectable as a polymorphism in a substantial fraction of the population, the mutant allele must somehow spread through the population, since the rate of mutation is too low to account for the observed levels of gene frequencies. Natural selection and

TABLE 18.5 Frequency of ABO Alleles in Various Populations

Population	Frequency A(p)	B(q)	O(r)
Armenians	36.0	10.4	53.6
Basques	25.5	—	74.5
Eskimos	35.5	4.6	59.9
Belgians	27.0	5.9	67.1
Danes	29.4	7.7	62.9
Greeks	22.9	8.2	68.9
Poles	25.9	14.0	60.1
Russians (Urals)	29.5	19.5	51.0
Russians (Siberia)	13.0	25.1	61.9
Russians (Tadzhikistan)	21.1	37.1	41.7
Sinhalese (Sri Lanka)	14.0	15.2	70.8
Indians (Assum)	19.2	11.1	69.7
Indians (Madras)	16.5	20.5	63.0
Chinese (Hong Kong)	19.1	19.1	61.8
Japanese	26.2	18.3	55.5
Nigerians (Ibo)	13.2	9.5	77.3
Nigerians (Yoruba)	13.8	14.6	71.6
Upper Voltans (Burkina Faso)	14.8	18.2	67.0
Kenyans	17.2	14.0	68.8

drift are the primary mechanisms by which alleles spread through local population groups.

One of the ways to monitor the spread of alleles through a population is to study polymorphisms that exhibit a geographical gradient. This distribution may result from migration patterns, selection, or a combination of factors. An example of such a distribution is the gradient of blond or tawny hair among Australian aborigines (Figure 18.14). The midpoint of this gradient is located in the west-central desert of Australia (Figure 18.15), and the trait has spread to much of the west coast. The trait is inherited in a codominant fashion.

It is difficult to imagine what selective forces might act directly on a hair-color allele to enable this allele to spread through much of the aborigine population of western Australia. It is possible that tawny hair is only a

FIGURE 18.14 An Australian aborigine near Kolomburu in Western Australia, exhibiting a polymorphism of hair color.

phenotypic by-product of a gene that controls a biochemical trait with more direct selective significance. The aborigines live in a harsh desert environment, and perhaps this allele contributes to an increase in fitness or interacts with other genes to increase fitness, or perhaps those with tawny hair are more often selected as mates.

It seems remarkable that there is so much genetic variability among individual humans and yet so little genetic difference among geographically separate populations. In fact, as mobility increases, it seems apparent that there will be a significant decrease in the genetic differences among populations.

Gene Flow Among Populations

There has been a long-standing interest in both anthropology and human genetics in estimating gene flow among divergent populations as a means of studying and reconstructing the origin and history of hybrid populations formed when European and non-European populations come into contact. The best documented and now classic example is the gene flow into the American black population from Europeans, but other populations have been studied as well.

Most of the black population in the United States originated in West Africa, and the majority of the white population arrived from Europe. In African populations, the frequency of the Duffy blood group allele Fy^o is close to 100%, while in Europeans this allele has a frequency close to zero. Europeans are almost all Fy^a or Fy^b, and these alleles have very low frequencies in native African populations. By measuring the frequency of the

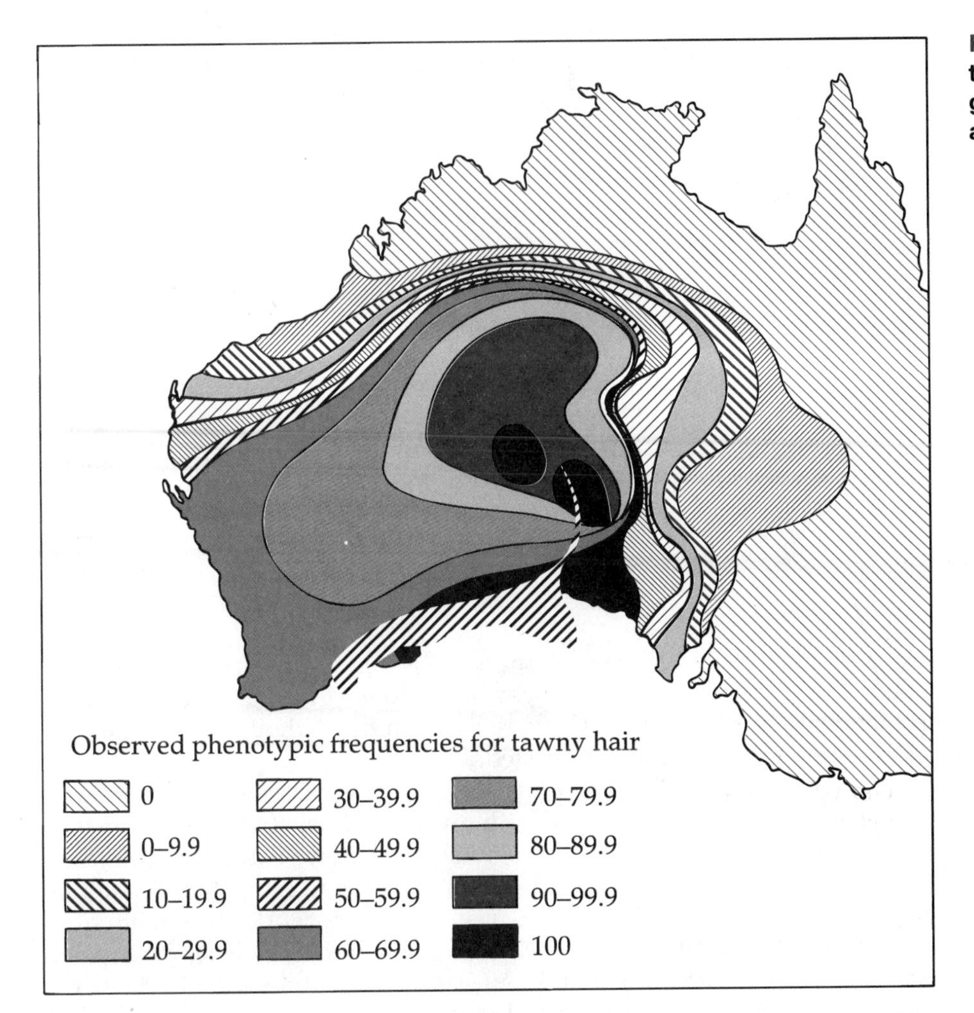

FIGURE 18.15 Clinal distribution of the tawny hair color gene among Australian aborigines.

Fy^a and Fy^b alleles in the U.S. black population, we can obtain an estimate of how much genetic mixing has occurred over the last 250 years. Figure 18.16 shows the frequency of Fy^a among black populations in West Africa and in several locations in the United States. Using this as an average gene, we can calculate that in some Northern cities about 20% of the genes in the black population are derived from Europeans.

The frequency of PKU (phenylketonuria) in U.S. blacks has been estimated to be 1 in 50,000, about one third that of the U.S. white population. Using haplotype analysis, the origin of the PKU mutation and the phenylalanine hydroxylase (PAH) gene in U.S. blacks has been studied. Results suggest that about 20% of the PAH genes in U.S. blacks originated from a Caucasian population, while the rest are likely to be of West African origin.

A more comprehensive study using 15 polymorphic loci instead of a single gene estimated the proportion of European genes in U.S. blacks from the Pittsburgh region. Fifty-two segregating alleles at these loci were studied,

FIGURE 18.16 **Distribution of the Fy^a allele among African and American black populations.**

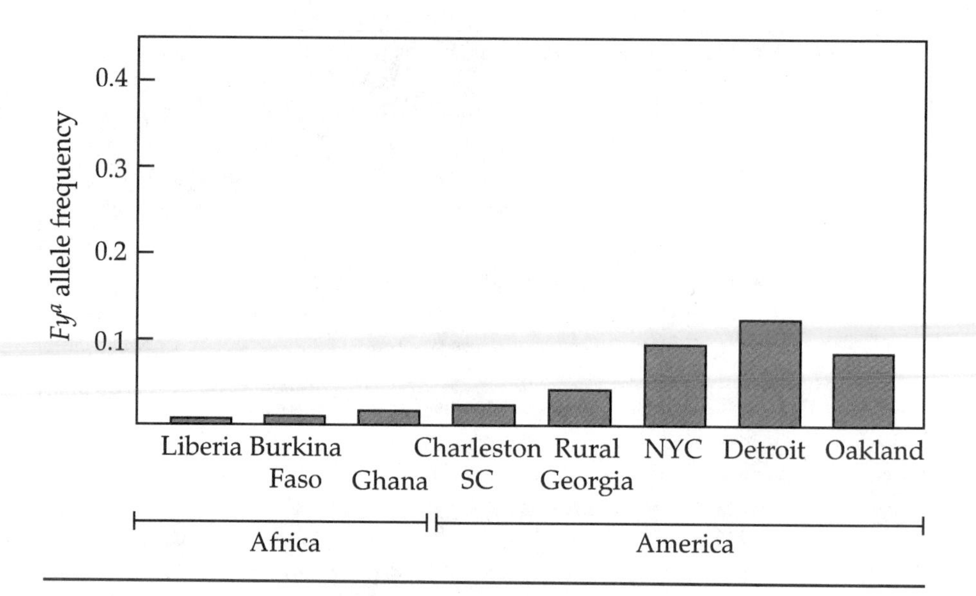

indicating that the proportion of European genes in this black population is about 25%. Because this estimate used 18 unique alleles of African origin, it is thought to be more precise than previous studies, although the population studied was geographically restricted.

Another approach to the study of gene flow between Caucasian and U.S. black populations employed both nuclear markers (ABO, MN, Rh, etc.) and mitochondrial markers to examine gene flow. Since mitochondrial markers are maternally inherited, they can be used to estimate the maternal contribution to the population under study. In this work, the results from nuclear alleles indicate that about 25% of the genes in the black population studied (60 U.S. blacks) originate from Caucasians. Studies of mitochondrial markers indicate that the contribution from Caucasian females is equal to that from Caucasian males.

However, not all contacts between populations leads to a reduction in genetic differences. Using a combination of genetics and historical demography, researchers have investigated the origin and extent of European-American admixture in the Gila River American Indian community of central Arizona. Allele frequencies for eight loci were studied in the Gila Bend populations of the Pima and Pagago tribes and analyzed for the presence of Caucasian alleles. The demographic survey examined records and pedigrees of 9,616 individuals in the Gila River Indian community to reconstruct the degree of non-Indian admixture.

The results from the genetic study indicate a total European-American admixture of 0.054 (5.4%), while the demographic data indicate an admixture of 0.059. The first contact between Europeans and the Gila River community occurred in 1694. Using 20 years as the estimate for one generation, about 15 generations of contact have occurred, with the gene flow being about 0.4% year, indicating that the Gila River American Indian community has retained almost 95% of its native gene pool, even after close contact with other gene pools for 300 years.

Primate Evolution and Human Origins

Population genetics can provide information about the genetic mechanisms associated with speciation and evolution, but the process itself is difficult to observe, mainly because of the time scale over which it operates. Natural selection acts to bring about the formation of new species by acting to adapt the genotypes present in the population to the environment. According to one idea about the mechanism of species formation, the process begins when a population becomes divided into small, reproductively isolated breeding groups. Over a long period these isolates are exposed to the forces of selection, and the best adapted genotypes contribute differentially to the gene pools of subsequent generations, bringing about a gradual change in gene frequencies. Eventually this process can result in race formation. If the period of isolation continues, the races may continue to genetically diverge from one another, resulting in the formation of new species.

Primate evolution from early Miocene Epoch, some 20 to 25 million years ago, to the appearance of modern man some 100,000 years ago has been reconstructed based mainly on the fossil record. Primates ancestral to both humans and apes are found as fossils from the Miocene (Figure 18.17). These primates, known as **hominoids,** originated in Africa between 4 and 8 million

Hominoid
Members of the primate superfamily *Hominoidea*, including the gibbons, great apes, and humans.

Era	Period	Epoch	Millions of years bp	Groups present
Cenozoic	Quartenary	Holocene		*H. sapiens*
			0.01	
		Pleistocene		*H. erectus*
			2.0	
	Tertiary	Pliocene		Hominids
			5.3	
		Miocene		Hominoids
			25	
		Oligocene		
			37	
		Eocene		
			58	
		Paleocene		Primates
			66	

FIGURE 18.17 A geologic time scale showing a summary of primate evolution. The divergence of the hominoids probably began in the late Oligocene. The fossil records of hominoids currently extend back to around 4 million years before the present (bp), but the ramapiths may have appeared almost 7 million years ago.

years ago and underwent a series of rapid evolutionary steps, giving rise to a diverse array of species. This diversity, coupled with the fragmentary fossil record, makes it difficult to draw conclusions about phylogenetic relationships among the early hominoids and to identify the ancestral line leading to modern hominoids.

Hominids
A member of the family Hominidae, which includes bipedal primates such as *Homo sapiens.*

Up until recently, it was generally accepted that a group of hominoids known as ramapiths gave rise to the **hominids,** a line of primates leading to humans. The ramapiths lived on the ground and in trees in an environment that was a combination of woods and open grasslands. Hominids can be distinguished from ramapiths by several characteristics: changes in jaw structure and teeth, probably accompanied by dietary changes; large brain size; and upright walking (bipedal locomotion). Now, however, several lines of evidence ranging from molecular biology to palentology indicate that hominid line did not branch off from the ramapiths, but instead arose from an unknown ancestor.

The earliest known hominids are a group called the australopithecines, known from the fossil record of about 4 million years ago. The most primitive species of the australopithecines identified to date is *Australopithecus afarensis,* which shares many characteristics with ancestral hominoids. This species, found in East Africa, has a combination of ape-like and human-like characteristics. Although members of this species had bipedal locomotion like modern humans, the body proportions were ape-like, with short legs and relatively long arms. Perhaps the most famous specimen of *Australopithecus* is a 40% complete skeleton known as Lucy, recovered in the 1970s. Although there are significant gaps in the fossil record, there is general agreement that *Australopithecus afarensis* is the ancestral stock from which all other known hominid species are derived.

By about 2 million years ago, the australopithecines were not alone in East and South Africa. The fossil record indicates that by then, there were at least 3 and perhaps 6 or more species of hominids, occupying different or overlapping habitats. The hominid species of this time period belonged to one of two groups: the australopithecines were characterized by relatively small brains and large heavy cheeks and teeth, and a second group characterized by relatively large brains and smaller facial structure and teeth. The relatively large-brained species are grouped in the genus *Homo.* The earliest recognized member of this genus is *Homo habilis,* first appears in fossils from about 2 million years ago, and were taller and had a much larger brain than other hominids. They were also expert toolmakers, and in the opinion of many palentologists and anthropologists, represent the line leading to our species, through another hominid, *Homo erectus* (Figure 18.18).

The Transition to Modern Humans

As noted above, the fossil record provides an incomplete picture of human evolution, and there are significant gaps with no fossil evidence at all. As DNA technology has emerged, it has provided the means to explore the evolutionary history of individual genes across time, and molecular genetics is now being used to study the evolutionary history of the human species. By comparing the nucleotide sequence of a gene from humans with that of

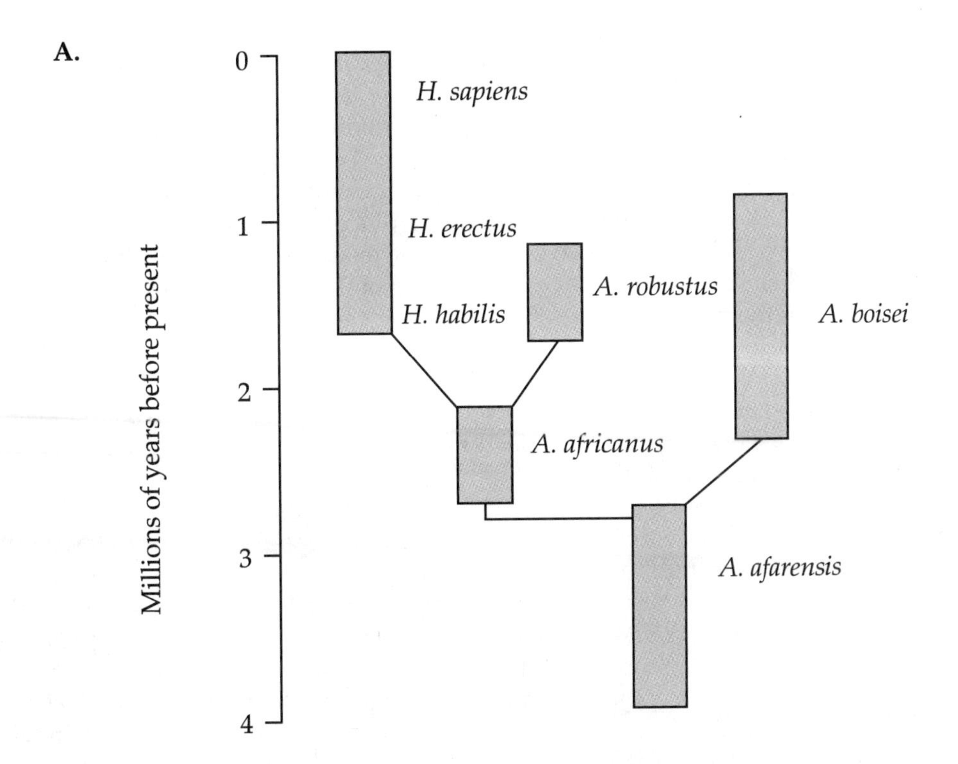

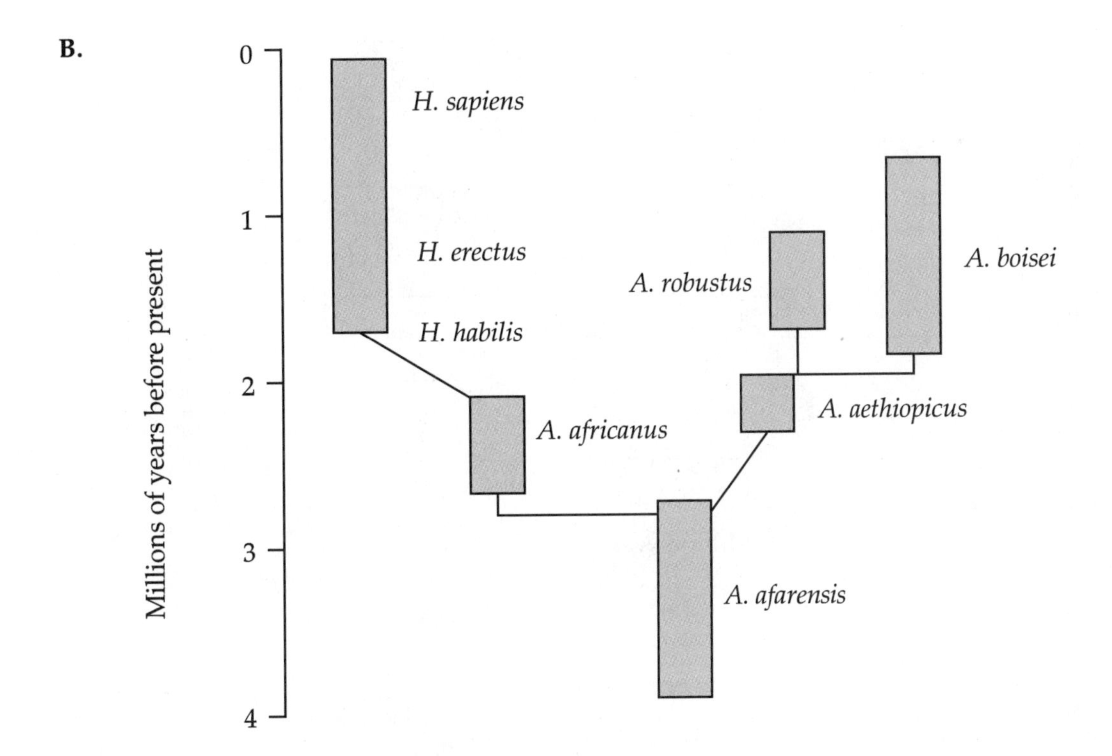

FIGURE 18.18 Two of the many phylogenies that have been proposed to explain hominid evolution and the origin of the human species. In both, *A. afarensis* is proposed as common ancestor, with *A. africanus* as intermediate to the appearance of the genus *Homo*.

other animals, plants, and even fossils, it is possible to reconstruct the evolutionary history of a particular gene and to estimate how long ago related species diverged from a common ancestor. Some genes have evolved and diverged extensively, while others, although dispersed to many different species, have remained relatively unchanged over millions of years.

We will briefly survey the results of molecular studies as they apply to two aspects of human evolution: the relationship between humans, apes, and their ancestors and the early stages of divergence following the appearance of the human species. By comparing the relatedness of genomic DNA from various hominoids, an evolutionary tree can be constructed (Figure 18.19). According to this scheme, the line leading to the gorillas branched off somewhere between 7 and 11 million years ago. This was followed by the split between the lines leading to chimpanzees and humans that occurred between 5.5 and 7.7 million years ago.

Most of the controversy surrounding the origin of the human species is caused by the incomplete nature of the fossil evidence following the split between the human line and the chimpanzee line. Comparisons of DNA relatedness are unable to provide any additional evidence in this area because there are no living hominids other than those in the present species of humans. Studies of nucleotide substitutions in specific genes has allowed an estimate of the minimum population size that diverged from *Homo erectus* to form *H. sapiens*. Studies of polymorphisms at the HLA locus have shown that a band of at least 10,000 individuals probably formed the core population over the period of 10,000 or more generations it took for *H. sapiens* to separate

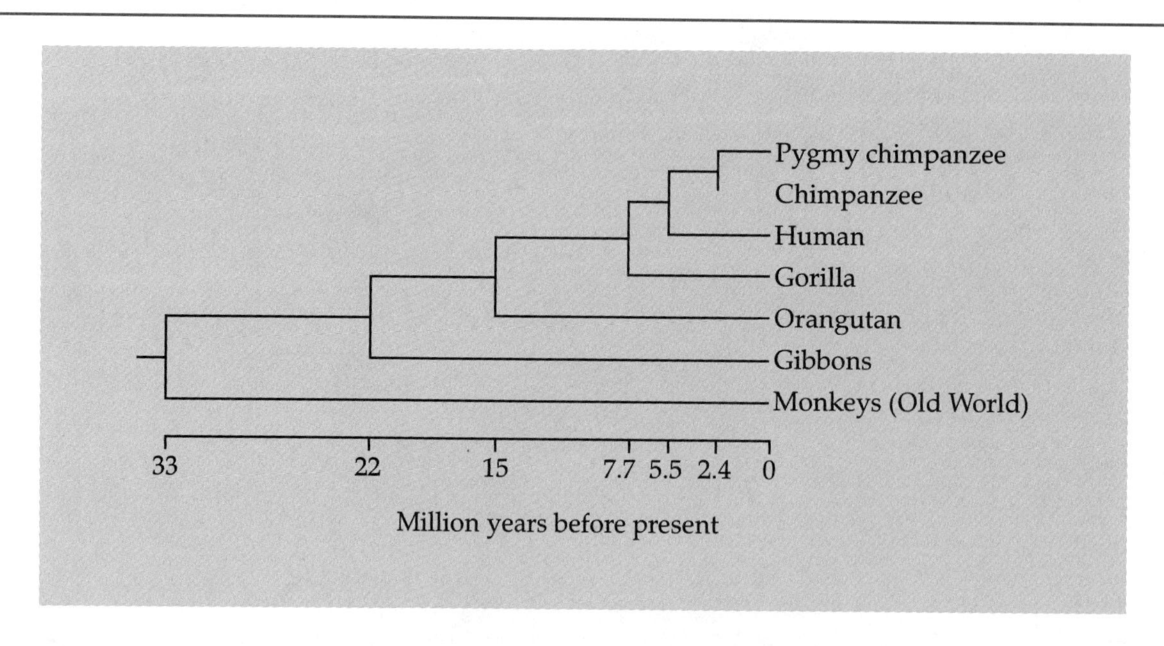

FIGURE 18.19 Evolutionary relationships and approximate time of divergence for hominoids based on DNA relationships. According to this scheme, humans and chimpanzees are more closely related to each other than either are to the gorilla. The human line diverged from that leading to chimpanzees some 5.5 to 7.7 million years ago.

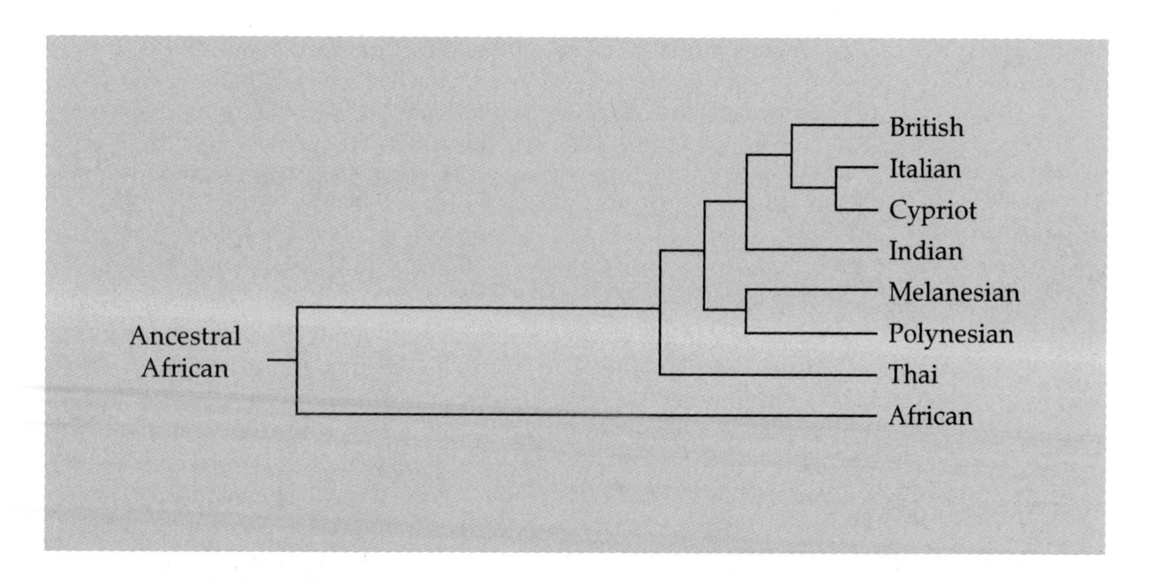

FIGURE 18.20 Divergence in African and non-African populations based on DNA polymorphisms. This diagram shows that an ancestral African population gave rise to present-day Africans as well as to a non-African line that diverged into present-day European and Asian populations. As data from more genes and more populations are examined, this diagram will probably be refined, but its basic plan will probably not be altered.

from *H. erectus*. Other studies of DNA polymorphisms at the beta globin locus have provided confirmation for the idea that *H. sapiens* arose in Africa and diverged to form non-African populations about 50,000 to 100,000 years ago (Figure 18.20).

The combined evidence from the fossil record and molecular studies has reinforced the main lines of evidence in the attempt to reconstruct the origins of our species. Many details remain to be elucidated, and because the record is so incomplete, there is often controversy among those who interpret the fossil evidence and among those who conduct molecular studies as well as arguments between the two groups. As molecular techniques become refined and applied to the problem of human origins, it seems likely that the record of evolution written in our genes will provide the key to understanding our origins.

Between *H. habilis* and our species there is abundant evidence for *H. erectus*. Most of the remains of this species have come from geologic layers in the mid-Pleistocene age, some 700,000 years ago, although there is evidence that this line may have originated as much as 2 million years ago. This species was widely distributed, ranging across North Africa and Eastern and Southeastern Asia. The fossils known popularly as Java man and Peking man belong to this species.

The development of *H. erectus* is characterized by physical changes already mentioned (large brain capacity, height), as well as evidence of cultural development, including sophisticated toolmaking and the use of fire.

H. sapiens evolved from *H. erectus* perhaps 300,000 to 500,000 years ago in a form intermediate between that of *H. erectus* and *H. sapiens*. Anatomically modern humans *(Homo sapiens)* first appeared some 200,000 years ago. Their appearance was marked by a great change in toolmaking and culture. A wide range of stone tools were produced, including engravers for marking stone and bone. Burial sites and evidence recovered from caves indicates that religion in the form of animal worship and provisions for the dead were common.

Recent Human Evolution

Many questions about the transition to modern humans remain unanswered. What exactly is the relationship between *H. erectus* and *H. sapiens?* Was there a rapid transition from *H. erectus* to *H. sapiens?* Was this transition brought about by changes in the environment or by the development of technology or by both? Is there a role for punctuated evolution in the emergence of modern humans? While there has been much speculation on these questions, they remain unanswered for the moment. But the application of molecular techniques provides another approach to these questions.

From the evidence provided by fossils and artifacts, there is agreement that groups belonging to *H. erectus* moved out of Africa about 1 million years ago and spread through Europe and Asia. What currently divides paleoanthropologists is the question of where *H. sapiens* originated. There are currently two opposing ideas about the origin of modern humans. One, based on the use of RFLP analysis in mitochondrial DNA, argues that after *H. erectus* moved out of Africa, populations that remained behind continued to evolve and gave rise to *H. sapiens* about 200,000 years ago. From here, modern humans spread to all parts of the world, replacing the lineages descended from *H. erectus,* including *H. sapiens Neandertal.*

This conclusion is based on studies that have identified 182 mitochondrial types (based on their combination of RFLP alleles) gathered from individuals worldwide. The underlying assumption in this study is that base changes in mitochondrial DNA accumulate at a constant rate, providing a "molecular clock" that can be calibrated by studying the fossil record. If this assumption is correct, the degree of relatedness between two individuals can be measured by the number of differences between their mitochondrial DNAs. These differences are used to construct a diagram known as a phyletic tree, indicating relationships (Figure 18.21). The tree leads back to a single ancestral mitochondrial lineage, originating in Africa. Calculations using the rate at which nucleotide substitutions generate new RLFP markers indicate that this tree originated about 200,000 years ago, from a population that could have consisted of as few as 10,000 individuals.

The second hypothesis about the origin of *H. sapiens* postulates that after *H. erectus* spread from Africa over Europe and Asia, modern humans arose in multiple regions as part of an interbreeding network of lineages descended from the original colonizing populations of *H. erectus,* not from a single origin in Africa. Support for this idea is derived from a combination of genetic and fossil evidence showing a gradual transition from archaic to modern humans. This evidence indicates that this transition took place in multiple sites outside of Africa.

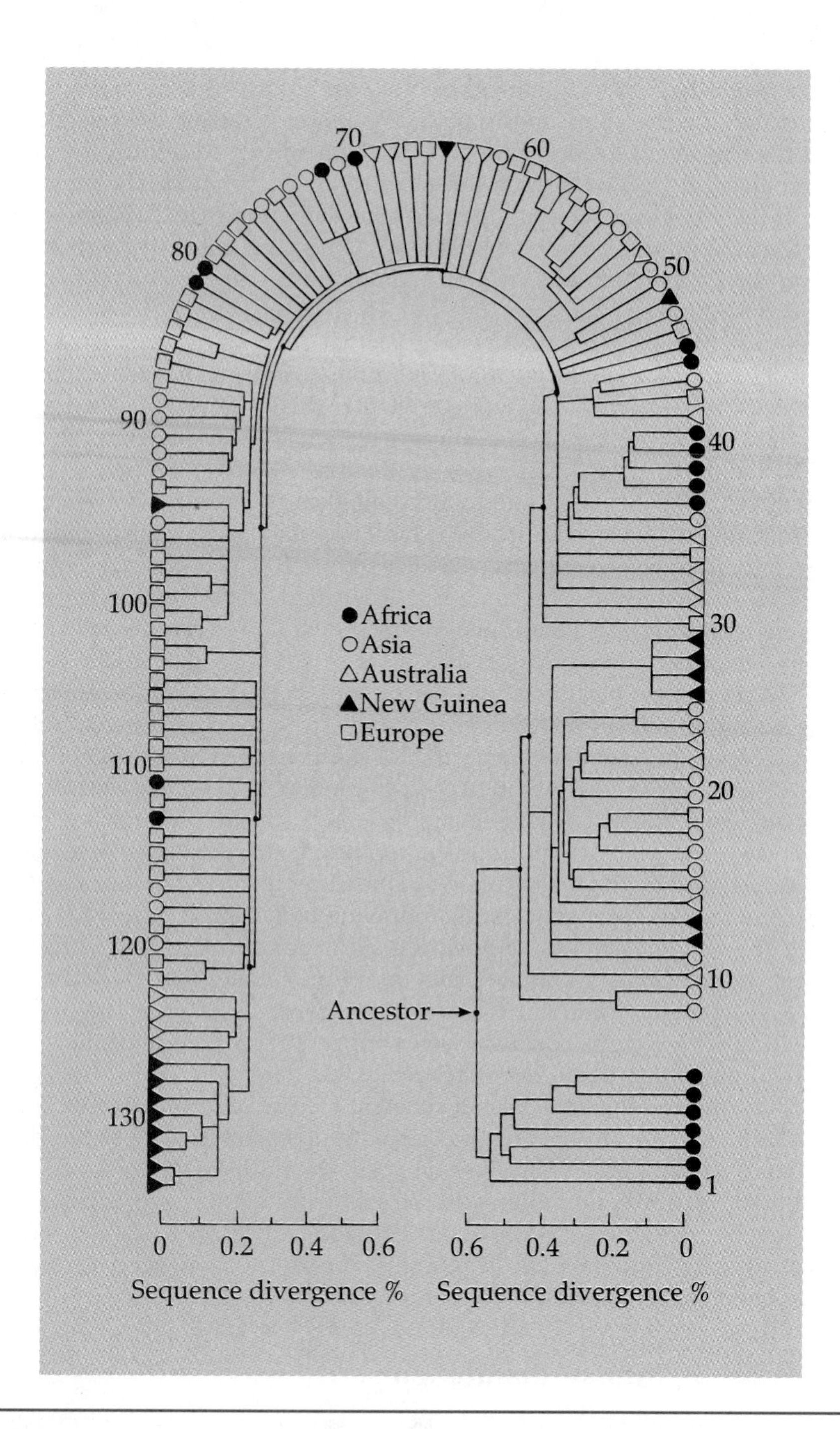

FIGURE 18.21 Phyletic tree of mitochondrial DNA differences constructed from 134 individuals. As with nuclear genes, this diagram shows that an ancestral population gave rise to present-day Africans as well as to a non-African line that was the origin of all other populations studied.

The two opposing ideas are hotly debated, and there has been an unfortunate tendency to resort to personal invective. While these alternate explanations for the appearance of modern humans show the insights that can be gained through analysis of the available data, they also point up the limitations of the genetic and other techniques being used. For example, further work has shown that other, very different phyletic trees can be constructed for the mitochondrial RFLP markers, raising serious doubts about the validity of a single mitochondrial lineage. In addition, other work has recently called into question the accuracy of the timing of the molecular clock. If the clock is inaccurate, perhaps the mitochondrial divergence took place after the migration of *H. erectus* from Africa, and before the appearance of *H. sapiens*, and is actually evidence that confirms the multi-center origin of modern humans.

Another problem in human evolution that genetics can be used to study is how populations of the genus *Homo* underwent geographic expansion from Africa to all parts of the world. Whether this expansion occurred once (with *H. erectus* about 1 million years ago) or twice (*H. erectus* followed by *H. sapiens* about 200,000 years ago), diffusion requires growth of local populations, probably associated with biological or cultural innovations or both, followed by migrations from the local population. Because populations were relatively small during this period (10,000 to 100,000 years ago), random genetic drift was a powerful force in establishing allele frequencies, and it probably laid down the genetic geography seen today as differences in allele frequencies across the world (see below).

The most recent transition, starting about 10,000 years ago and known as the Neolithic transition, involves a change from hunter-gatherer cultures to those based on agriculture and animal breeding. Evidence suggests that this transition began in a region of the Middle East and spread slowly in all directions from there. Of genetic significance is the question whether the diffusion of farming was cultural (the technique spread, but people stayed in place), or demic (the technique was spread by farmers who moved from the site of origin through existing populations). Demic diffusion would reduce the genetic drift in populations and blur the genetic distinctions between populations. To answer this question, genetic maps, plotting the degree of genetic variability, have been constructed for the world's population. These maps correlate well with geographic maps of the spread of agriculture, constructed from archaeological data.

Thus, migrations have been a constant feature of human evolution for over a million years, and in conjunction with other forces such as mutation, drift, and natural selection, have shaped the human genome and the differential distribution of alleles. In the following section, we will consider how genetics, in tandem with other techniques, has been used to follow demic diffusion and how the present-day pattern of allele distributions represents evolutionary relics of past migrations.

Dispersal of Human Populations

Genetics, anthropology, and archaeology are now being used in combination to reconstruct the origins and ancestry of human populations, to answer questions about how and when groups of humans became dispersed across the globe, and to explain the origins of allelic differences between populations. In particular, the development of recombinant DNA technology has provided information about the molecular basis for population differences in allele frequencies, providing a window into the evolution of human populations.

The origins, migrations, and interrelationships of the American Indians have been a subject of much controversy. It is generally agreed that the Americas were populated by migration from Asia between 12,000 and 75,000 years ago, during the last period of glaciation, but there has been little agreement about the number of migrations and their timing. Using linguistic and anthropological criteria four groups of American Indians have been identified: (1) Eskimos and Aleuts (Eskaleuts), (2) Nadene-speaking North American Indians, (3) other North American Indians, and (4) South American Indians. In some classifications the last two are often grouped together as the Amerinds. Genetic studies addressing these issues have analyzed both mitochondrial and nuclear genomes.

Analyses of base sequence variations in mitochondria indicate that all American Indian groups share a common mitochondrial DNA (mtDNA) lineage with populations in East Asia, the presumed origin of migrants who populated the Americas. The mtDNA studies reveal the presence of four mtDNA lineages among American Indians. Among these, the Nadenes represent a separate group and are of recent origin, having arrived in North America 8500 to 12,000 years ago. The mtDNA results also show that the present Amerinds (groups 3 and 4, above) are derived from at least two population groups founded at different times; the oldest of these crossed from Asia between 21,000 and 42,000 years ago.

If alleles of the immunoglobulin heavy chain are used to study populations of American Indians and contemporary Asian populations, four separate migrations from Asia are indicated (Figure 18.22). Based on this allelic study and other biological evidence, including anatomy, the Eskaleuts and Nadenes are thought to represent two distinct populations that separately entered North America from Siberia 8000 to 12,000 years ago. The South American Indians are more genetically distant from the other populations and are probably descendants of the oldest migrants, who entered the Americas more than 20,000 years ago.

The results from two types of genetic studies showing waves of migration provide strong support for previously controversial classifications based on linguistics and other anthropological standards. Until recently, the oldest accepted archaeological evidence based on skeletal remains and artifacts dated from 13,000 or 14,000 years ago. The most recent archaeological evidence and the genetic evidence taken together strengthen the conclusion that humans were present in the Americas between 40,000 and 50,000 years ago. Accordingly, these three techniques are interacting in a synergistic fashion to provide new information about the origins and migrations of the native populations of the Americas and are transforming our understanding of human evolution in the New World.

In Europe, other studies have uncovered genetic relics of previous patterns of human migration. A large-scale study of 19 loci and 63 alleles carried out at over 3000 locations in Europe has resulted in the identification of 33 boundaries that represent sharp genetic changes (Figure 18.23). Across a boundary, there is an abrupt shift in genotype combinations and allele

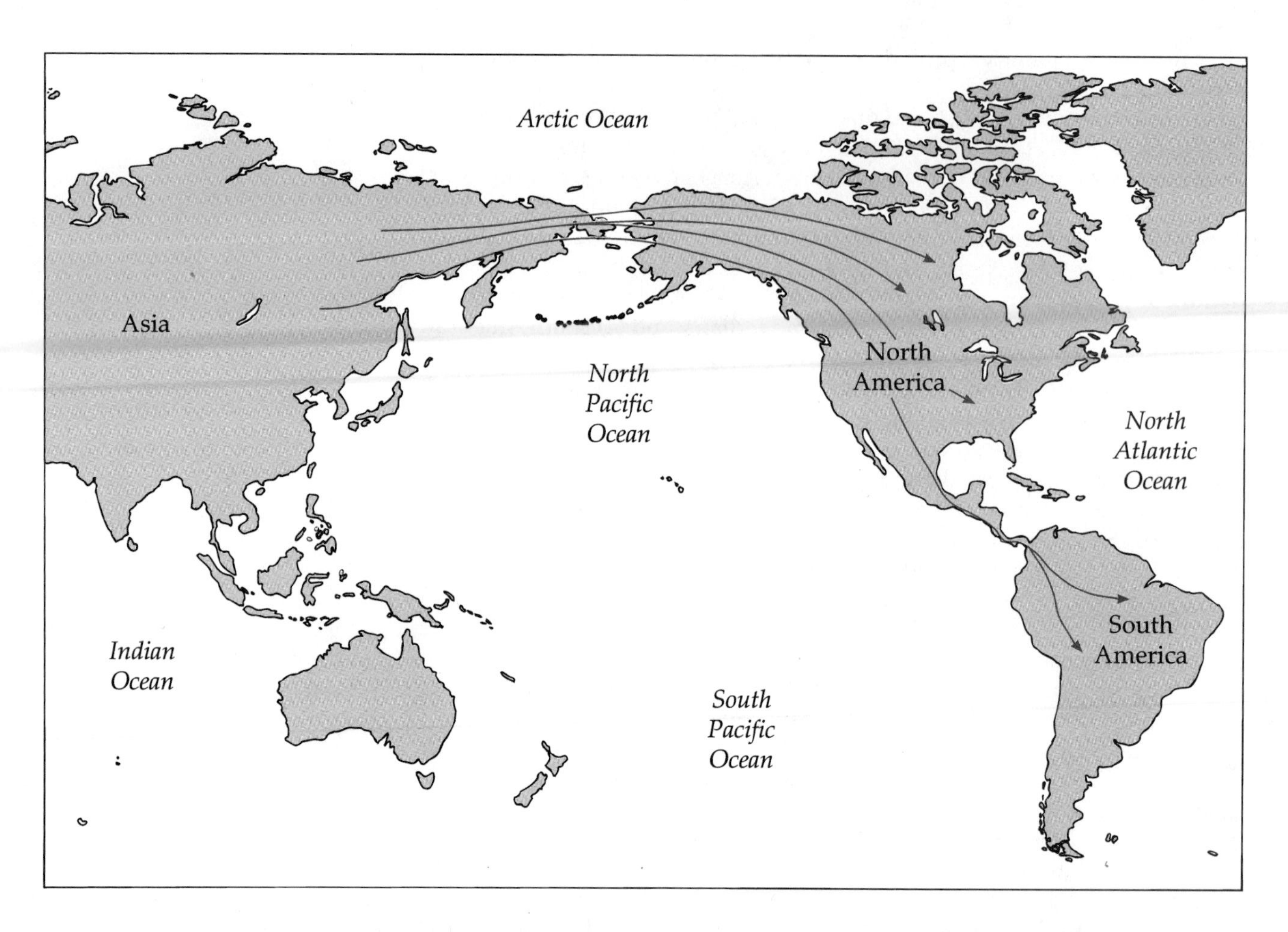

FIGURE 18.22 The waves of migration from Asia to North America. At the time of migration, a land bridge connected the two continental land masses. Migrations could have occurred between 12,000 and 75,000 years ago.

frequencies, suggesting that little mixing of the populations has occurred. Of the 33 boundaries, 22 are physical boundaries (mountains or ocean) over which there is little genetic exchange. However, the other 11 genetic boundaries are not associated with physical barriers but represent linguistic barriers separating populations. (Some of the 22 physical boundaries also have linguistic barriers.) The results of this study suggest that the genetic structure of the populations in Europe is not caused by adaptation to local environmental conditions but reflects the diverse origins of populations that came into contact through migrations. The results further indicate that language is an effective barrier to gene flow and acts as a selective force to establish and maintain genetic differences between populations. In spite of the fact that European populations have been in contact for hundreds or thousands of years, there has been little breakdown of allelic differences, and with the persistence of language barriers, there is little likelihood that these boundaries will disappear in the near future.

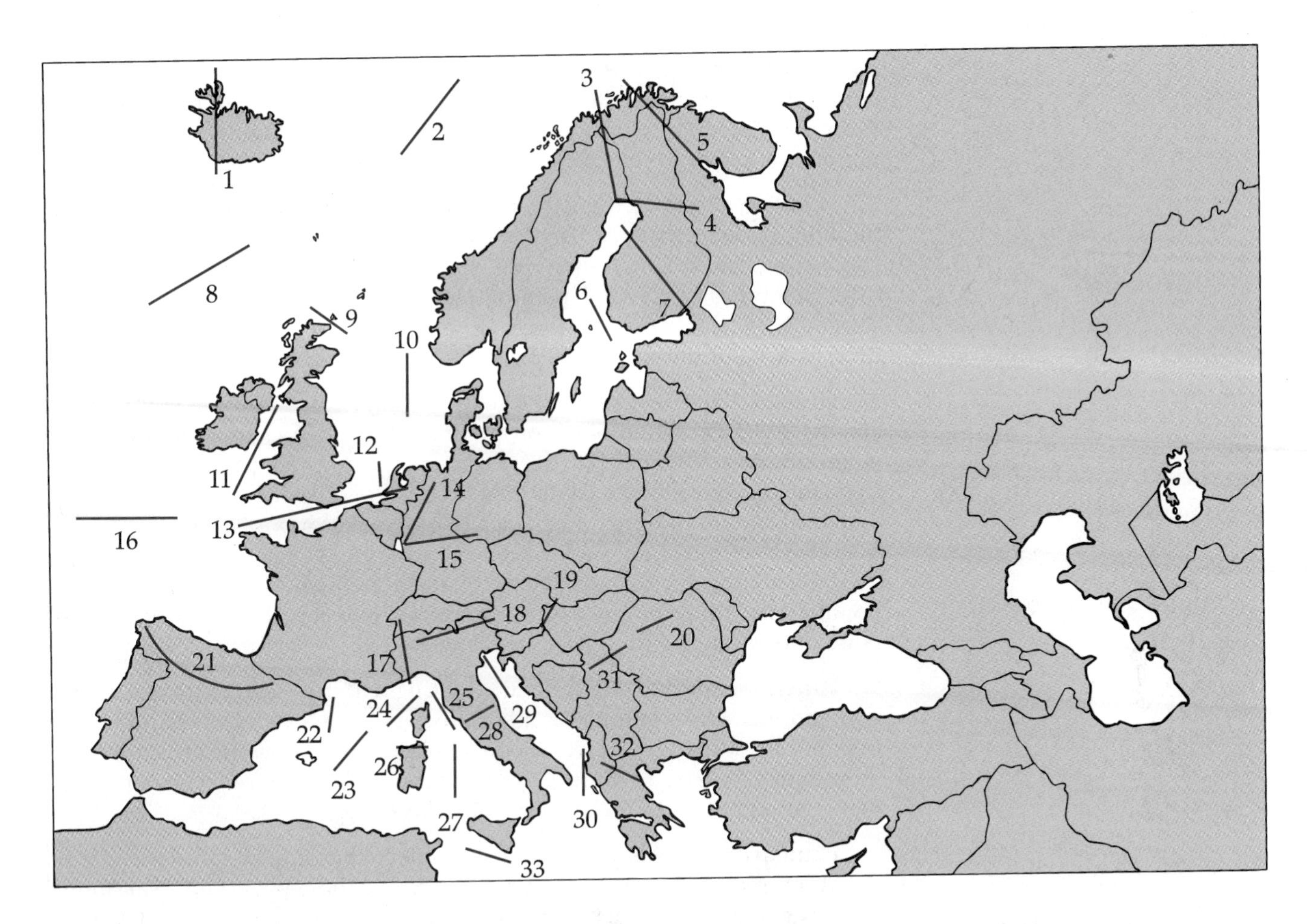

FIGURE 18.23 Genetic/linguistic barriers in Europe. Genetic boundaries mark regions of abrupt shifts in allele frequencies. Of the 33 genetic barriers identified, 31 correspond to linguistic boundaries. Two genetic boundaries (through Iceland and Greece) do not correspond to modern linguistic boundaries. The correlation between language and genetics suggests that language may play a major role in establishing and maintaining genetic boundaries.

SUMMARY

1. Studies of protein polymorphisms reveal that the human gene pool contains a large amount of genetic diversity. Around 25% of all loci examined are polymorphic, and it has been estimated that the average diploid genotype is 10% heterozygous. Because all forms of polymorphisms are not revealed by the methods used to study proteins, these estimates are somewhat conservative, and it is possible that more than 60% of all loci exhibit polymorphisms.

2. All genetic variants originate by mutation, but mutation is an insignificant force in bringing about changes in allele frequency. Other forces, including drift and selection, act on the genetic variation present in the gene pool and are primarily responsible for changing the frequency of alleles in the population. Drift is a random process that acts in small, isolated populations to change allele frequency from generation to generation. Examples include island populations or those separated from general population by socio-religious practices.

3. Selection acts to increase the reproductive success of fitter genotypes. As these individuals make a disproportionate contribution to the gene pool of succeeding generations, genotypes change. The differential reproduction of fitter genotypes is known as natural selection. Darwin and Wallace identified selection as the primary force in evolution, leading to evolutionary divergence and the formation of new species.

4. Humans have developed a strategy of adaptation known as culture. This activity has a profound effect on the interaction of the gene pool with the environment and can also act as a force that brings about changes in allele frequencies. While both plant and animal species employ strategies of dispersal and migration, human technology makes it possible for humans to migrate to any place on earth within a short time. The net effect of this short migratory time is the reduction of allele frequency differences among populations. In some instances, migration into small, relatively isolated populations can rapidly generate changes in allele frequencies.

5. Patterns of culture also influence mating patterns, tending to make them nonrandom. This process of assortative mating can also change allele frequencies, although slowly. Food choices, coupled with selection over thousands of years, are thought to have influenced the frequency of a gene that enables adults to use milk as a food source.

6. Human diversity is perhaps most visible in the wide range of phenotypic variations seen in the human population. Using these phenotypic variations, many schemes have been proposed to classify humans into different types or racial groups. The underlying assumption in recent attempts to classify humans into racial groups is that large differences seen in phenotypes must be a reflection of large differences in genotypes and allele frequencies. Detailed examinations of genetic polymorphisms at a wide range of loci show that this assumption is not correct and that there is more genetic difference between individuals of the same group than there is between groups.

7. The fossil record from the Miocene Epoch can be used to trace the evolution of hominoids and hominids, primates that gave rise to the present human species. The incomplete nature of the fossil record makes it difficult to construct a phylogeny, but techniques of molecular biology are now being used to clarify the origin of humans.

8. A combination of linguistics, archaeology, anthropology and genetics is being used to reconstruct the dispersal of human populations across the globe. The evidence available suggests that North and South America were populated by four waves of migrations sometime in the last 12,000 to 75,000 years ago. In Europe, the alignment of genetic and linguistic barriers suggests that culture, in the form of language can be a force in establishing and maintaining genetic differences between populations.

Questions and Problems

1. Distinguish between mutations and polymorphisms. How are polymorphisms used in the study of genetic variation?
2. Why is it that mutation, acting alone, has little effect on gene frequency?
3. What is the relationship between founder effects and genetic drift?
4. Darwin's theory of natural selection has been popularly summarized as "survival of the fittest." Is this an accurate description of natural selection? Why or why not?
5. What are examples of nonrandom mating, and how does this behavior affect genetic equilibrium?
6. How has the concept of race evolved over the last 100 years?
7. In your opinion, is it genetically useful to classify humans into racial groups? Why? Is it useful to classify humans into racial groups for other purposes? What are they and why?
8. How do you think the development of culture affected the process of human evolution? Is our present culture affecting selection? Can you give specific examples?

For Further Reading

Ayala, F. 1984. Molecular polymorphism: How much is there, and why is there so much? *Dev. Genet.* **4:** 379–391.

Bahn, P. 1993. 50,000 year old Americans of Pedra Furada. Nature **362**: 114–115.

Barbujani G. and Sokol, R. 1990. Zones of sharp genetic change in Europe are also linguistic boundaries. Proc. Nat. Acad. Sci. **87**: 1816–1819.

Bodmer, W. F., and Cavelli-Sforza, L. L. 1976. *Genetics, Evolution and Man.* San Francisco: Freeman.

Cavalli-Sforza, L., Menzonni, P., and Piazza, A. 1993. Demic expansion and human evolution. Science **259**: 639–646.

Durham, W. 1991. *Coevolution: Genes, Culture and Human Diversity.* Stanford: Stanford University Press.

Gould, S. J. 1982. Darwinism and the expansion of evolutionary theory. *Science* **216:** 380–387.

Gyllensten, U. B., and Erlich, H. A. 1989. Ancient roots for polymorphism at the HLA-DQ alpha locus in primates. *Proc. Nat. Acad. Sci.* **86:** 9986–9990.

Kennedy, K. 1976. *Human Variation in Space and Time.* Dubuque, Iowa: Brown.

Lewontin, R. 1974. *The Genetic Basis of Evolutionary Change.* New York: Columbia University Press.

Little, B. B., and Malina, R. M. 1989. Genetic drift and natural selection in an isolated Zapotec-speaking community in the valley of Oaxaca, Southern Mexico. *Hum. Hered.* **39:** 99–106.

Mayr, E. 1963. *Animal Species and Evolution.* Cambridge, Mass.: Harvard University Press.

Molnar, S. 1983. *Human Variation.* Englewood Cliffs, N.J.: Prentice-Hall.

Relethford, J. H. 1988. Heterogeneity of long-distance migration in studies of genetic structure. *Ann. Hum. Biol.* **15:** 55–63.

Roberts, D. F. 1988. Migration and genetic change. Raymond Pearl lecture 1987. *Hum. Biol.* **60:** 521–539.

Romero, G., Devoto, M., and Galietta, L. J. 1989. Why is the cystic fibrosis gene so frequent? *Hum. Genet.* **84:** 1–5.

Schanfield, M. 1992. Immunoglobulin allotypes (GM and KM) indicate multiple founding populations of native Americans: evidence of at least four migrations to the New World. Human Biol. **64**: 381–402.

Semino, O., Torrino, A., Scozzari, R., Brega, A., De Benedictis, G., and Santachiara-Benerecetti, A. S. 1989. Mitochondrial DNA polymorphisms in Italy: III. Population data from Sicily: A possible quantitation of maternal African ancestry. *Ann. Hum. Genet.* **53:** 193–202.

Shields, G., Schmiechen, A., Frazier, B., Redd, A., Voevoda, M., Reed, J., and Ward, R. 1993. mtDNA sequences suggest a recent evolutionary divergence for Beringian and northern North American populations. Am. J. Hum. Genet. **53**: 549–562.

Smith, J. M. (Ed.). 1982. *Evolution Now: A Century after Darwin.* San Francisco: Freeman.

Stanton, W. 1960. *The Leopard's Spots.* Chicago: University of Chicago Press.

Torroni, A., Schurr, T., Cabell, M., Brown, M., Neel, J., Larsen, M., Smith, D., Vullo, C., and Wallace, D. 1993. Asian affinities and continental radiation of the four founding native American mtDNAs. Am. J. Hum. Genet. **53**: 563–590.

Torroni, A., Sukernik, R., Schurr, T., Starikovskaya, Y., Cabell, M., Crawford, M., Comuzzie, A., and Wallace, D. 1993. mtDNA variation of aboriginal Siberians reveals distinct genetic affinities with native Americans. Am. J. Hum. Genet. **53**: 591–608.

Towne, B., and Hulse, F. S. 1990. Generational changes in skin color variation among Habbani Yemeni Jews. *Hum. Biol.* **62:** 85–100.

Wallace, D. and Torroni, A. 1992. American Indian prehistory as written in the mitochondrial DNA: a review. Human Biol. **64**: 403–416.

Yunis, J. J., and Prakash, O. 1982. The origin of man: A chromosomal pictoral legacy. *Science* **215:** 1525–1530.

CHAPTER 19

GENETIC SCREENING, AND GENETIC COUNSELING

CHAPTER OUTLINE

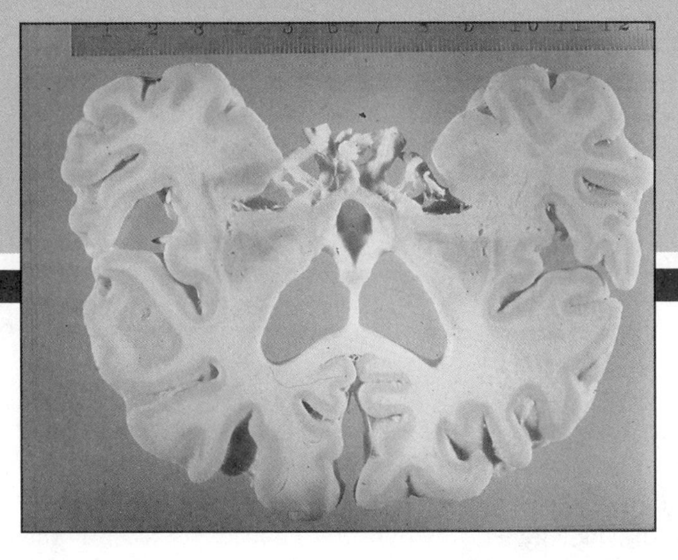

Huntington disease, as we have seen, is an autosomal dominant condition associated with the gradual loss of motor coordination, degenerative personality changes, and progressive dementia. The disease develops slowly and leads to death within 15 years after its onset. Most affected individuals develop symptoms of the disease around the age of 40, beginning with involuntary limb tremors and behavior changes. Through a combination of pedigree analysis and molecular biology, the gene controlling the disease has been mapped to chromosome 4. Genetic markers in and near the gene have been identified, and individuals at risk for the disease can be identified by means of an RFLP marker test using a small blood sample.

Suppose that your uncle, age 49, has recently been diagnosed as having Huntington disease. His physician calls you to inform you of the diagnosis and indicates that you have a 25% chance of developing the disease. You are 25 years old, married, and have one child. Should you undergo testing to determine whether you are at risk? Can you live with the outcome if the results indicate that you will develop the disease? Do you want your children to be informed of your condition? Do you want them to be tested? Do you want them to know if they will develop the disease? Should you have any more children? Suppose that you decide not to be tested but your child is tested and is found to have the Huntington disease gene. You are the only parent who could have transmitted the mutant allele. You now have information about your genotype that you may not have wanted. Does this mean you can prevent your child from being tested?

This scenario and the questions it raises are not hypothetical. Although Huntington disease affects only a small fraction of the population (3 or 4 out of 100,000), it is now possible to identify those who will develop this devastating disease (with 95% accuracy). Helping individuals to understand the implications of genetic screening and to make informed decisions about actions to be taken is called genetic counseling.

In this chapter we will survey the field of genetic screening, exploring the rationale, methods, and economics of this discipline and the potential for its use and misuse. We will also consider the role of the genetic counselor in providing information and education to those who undergo genetic screening. With the growth in genetic technology, these fields will have a great impact on our own lives and personal decisions and on those of our family members and friends.

GENETIC SCREENING

Genetic screening can be defined as the systematic and organized search for individuals of certain genotypes. Traditionally the term genetic screening has meant the detection of persons who have or may carry a genetic disease or who are at risk of producing a genetically defective child. More recently the term has been expanded to include the search for those who may have a genetic susceptibility to environmental agents. Genetic screening is conducted for a variety of reasons. Prenatal screening is often used as the basis for selective abortion; newborn screening is conducted for diagnosis and treatment of a range of metabolic disorders. Adult screening for carrier status is of use in genetic counseling and family planning. Adult screening is also used for occupational screening to detect individuals who may have an inherited susceptibility to materials or conditions in the workplace.

Population screening to detect those exposed to an infectious disease such as tuberculosis is a well-established and effective means of controlling such diseases. New testing programs for autoimmune deficiency syndrome (AIDS) are being implemented, and more inclusive testing programs for AIDS are being proposed. Unlike public health screening, genetic screening has several aspects that are unique and need to be recognized. First, identification of an individual with or at risk for a genetic disorder often leads to the discovery of other affected or at-risk individuals within the same family. Second, screening often identifies individuals who will develop genetic disorders later in adult life. When the genetic defect diagnosed is a traumatic and fatal one, such as Huntington disease, this knowledge often has serious personal and social effects. Third, the results of genetic screening often have a direct impact on the offspring of the screened individual.

To consider genetic screening, we first need to determine what types of screening are possible and under what circumstances screening is carried out.

Prenatal Screening

The purpose of prenatal screening is the detection of genetic disease and birth defects in the unborn child. Over 200 single-gene disorders can be diagnosed in the prenatal condition (Table 19.1). In most cases the conditions are rare, and genetic testing is done when there is a family history of the disease. For some conditions, such as Tay-Sachs disease or sickle cell anemia, biochemical tests can be conducted on the parents to determine if either is a carrier. If the tests for both parents are positive, the fetus has a 25% chance of being

TABLE 19.1 Some Metabolic Diseases and Birth Defects That Can Be Diagnosed by Prenatal Testing

Acatalasemia	Mannosidosis
Adrenogenital syndrome	Maple syrup urine disease
Chédiak-Higashi syndrome	Marfan syndrome
Citrullinemia	Muscular dystrophy, X-linked
Cystathioninuria	Niemann-Pick disease
Cystic fibrosis	Oroticaciduria
Fabry disease	Progeria
Fucosidosis	Sandhoff disease
Galactosemia	Spina bifida
Gaucher disease	Tay-Sachs disease
G6PD deficiency	Thalassemia
Homocystinuria	Werner syndrome
I-cell disease	Xeroderma pigmentosum
Lesch-Nyhan syndrome	

affected. In such cases prenatal testing can determine whether the fetus is a recessive homozygote afflicted with the disease. Similarly, if the mother is known to be a carrier for certain deleterious, X-linked genetic disorders, testing is indicated.

For other genetic conditions, such as Down syndrome, caused by the presence of an extra copy of chromosome 21, direct examination of the chromosome constitution of the fetus is the most direct way to determine whether the fetus is affected. In this case testing is not carried out because of a familial history of genetic disease or detection of heterozygotes in the parents but usually because of advanced maternal age. Since the risk of Down syndrome increases dramatically with maternal age (see Chapter 6), cytogenetic testing is recommended for all pregnant females who are older than 35 years of age.

In addition to genetic diseases, some birth defects associated with abnormal embryonic development can be diagnosed prenatally. Among these is a class of defects associated with the failure to correctly form the neural tube, a structure that arises in the first 2 months of development. One such defect, spina bifida, is a condition in which the spinal column is open or partially open (Figure 19.1). In cases of neural tube defects, the condition can be diagnosed accurately by testing the amniotic fluid for elevated levels of α-fetoprotein. In about 80% of cases, α-fetoprotein levels in the maternal blood serum are also elevated. A blood test using maternal blood can be used to identify those mothers for whom further tests, such as amniocentesis, are recommended.

Several methods are used in obtaining samples or images used in prenatal testing. These include amniocentesis, chorionic villus biopsy, ultrasonography, and fetoscopy. In addition, a variety of techniques is used to analyze the samples, including cytogenetics, biochemistry, and recombinant DNA technology. Because recombinant DNA technology is able to analyze the genome directly, it is the most specific and sensitive method

FIGURE 19.1 The end of the neural tube, known as a meningocele, projecting from the base of the spine in an infant with spina bifida.

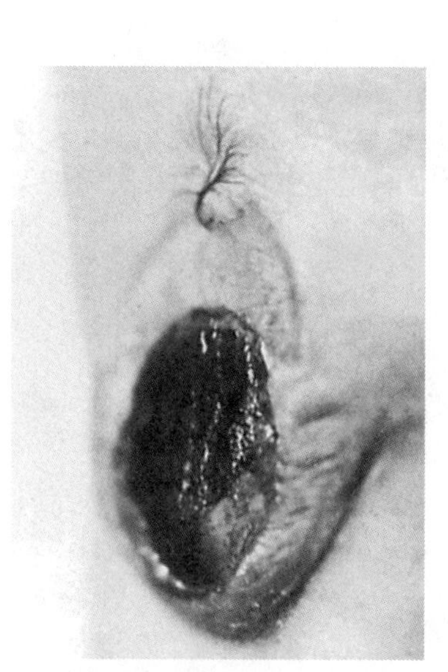

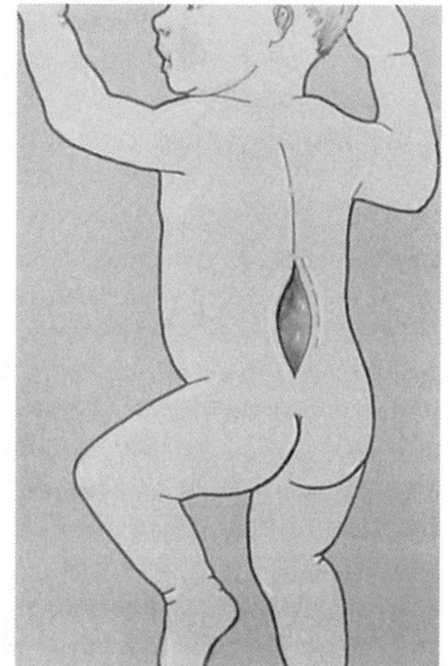

currently available. The accuracy, sensitivity, and ease with which recombinant DNA technology can be used to assemble a profile of the genetic diseases and susceptibilities carried by an individual have raised a number of legal and ethical issues that have yet to be resolved.

We have already considered two methods of obtaining samples for genetic screening (amniocentesis and chorionic villus biopsy; see Chapter 6) and have also discussed the methods used in genetic analysis (Chapters 6 and 8). Here we will examine ultrasonography and fetoscopy, two alternate methods of obtaining samples, and will then consider the consequences and problems that arise from the use of such tests.

Ultrasonography is a technique based on sonar detection methods developed and refined during World War II. For prenatal diagnosis, a transducer that emits pulses of energy is placed on the abdominal surface over the enlarged uterus (Figure 19.2). As the ultrasound strikes the surface of the fetus, some sonic waves are reflected from the fetus to the transducer. These reflected waves are electronically converted to images displayed on a screen (Figure 19.3).

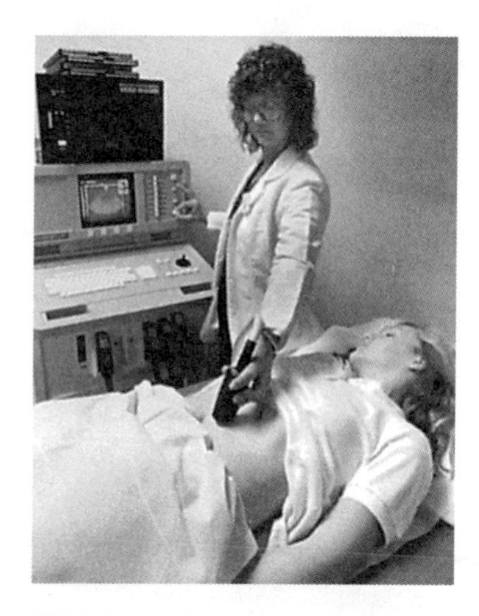

FIGURE 19.2 A pregnant woman undergoing ultrasonography.

Ultrasound can be used to diagnose multiple pregnancy, determine fetal sex, and identify neural tube defects and skeletal disorders, limb malformations, other central nervous system defects, and congenital heart defects. It is also used to guide the needle used in amniocentesis and in carrying out fetal transfusions. Since its introduction into obstetrics in the 1950s, the use of ultrasonography has become widespread because it eliminates the need for fetal exposure to ionizing radiation (in the form of x-rays). At the present time one third to one half of all developing fetuses in

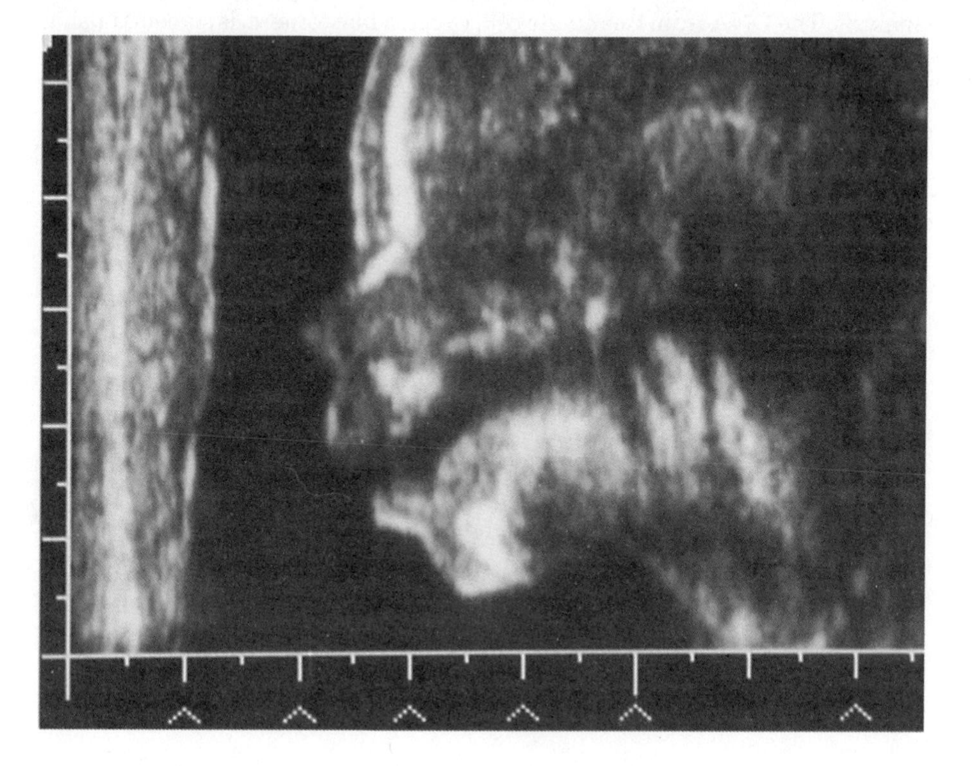

FIGURE 19.3 Fetus visualized by ultrasound.

the United States (at least 1 million fetuses) are examined by ultrasonic waves. In spite of this widespread use, little information is available about risks to the developing fetus from exposure. The National Institutes of Health have called for detailed studies to determine what risks are associated with ultrasonography. To date no studies have produced conclusive evidence regarding the safety of ultrasonic waves. In the absence of evidence for safety, ultrasonography should be used with caution.

Fetoscopy is the direct visualization of the fetus by means of a fiber-optic device known as an endoscope. In this procedure a hollow needle is inserted into the amniotic cavity, and the device is introduced through the needle. The image is transmitted to a video screen and can be viewed and recorded from the screen. (See the photo in the color insert, which shows the face and hand of a fetus in the 9th week of pregnancy, obtained by fetoscopy.) The technique is most useful in cases in which the condition to be diagnosed cannot be detected by cytogenetic or biochemical means. Fetoscopy can also be used to obtain samples of fetal blood, allowing diagnosis of some genetic diseases such as hemophilia and certain forms of thalassemia. Fetoscopy poses a danger to the fetus, however, and there is a 2% to 5% chance of spontaneous abortion with the use of this method.

A new method of prenatal genetic screening, combining microsurgery and recombinant DNA technology, is being developed and has already been used to perform genetic testing on human preimplantation embryos at the 6- to 10-cell stage of development. In this procedure embryos derived from *in vitro* fertilization are incubated until they reach a multicellular stage of development. Then, using a micromanipulator, a hole is made in the area surrounding the embryo (zona pellucida), and a single cell is removed for analysis. The DNA from this single cell, called a blastomere, is screened using the polymerase chain reaction (PCR) to detect the presence of mutant genes such as those for muscular dystrophy or hemophilia. Operated embryos continue to develop *in vitro* and after uterine transfer develop to full term.

Genetic testing is being pushed back further to the stage of the developing gamete in a method known as preconception genetic testing. This method is designed for situations in which one or both parents are carriers for a genetic disorder. In this procedure eggs that have completed the first meiotic division are recovered from the ovary by laparoscopy (the insertion of a fiber-optic tube through the abdominal wall). Recall that at the first meiotic division there is an unequal cytoplasmic division (Chapter 3) and that the cell receiving a small amount of cytoplasm is called the polar body. The polar body is isolated by microsurgery, and the DNA is analyzed to determine whether the defective gene carried by the mother is present in the polar body. If it is, the egg does not carry the defective gene, and the oocyte can be used for *in vitro* fertilization. On the other hand, if the polar body does not carry the defective gene, it must be in the egg, which is then discarded. This technique was first used to screen oocytes from a couple heterozygous for alpha$_1$-antitrypsin deficiency (AAT) before *in vitro* fertilization, and it is being used to test for other disorders as well.

In spite of the many genetic disorders and birth defects that can be detected with prenatal testing, the technique has some limitations. These include measurable risks to the mother and fetus, including infection, hemorrhage, fetal injury, and spontaneous abortion. Conventional strategies for the use of prenatal testing will not always detect the majority of certain defects. In Down syndrome, for example, amniocentesis is recommended for all mothers over the age of 35 years. (In reality, only a small percentage of pregnancies to women over 35 are screened by amniocentesis.) However, some 65% of all Down syndrome births occur to mothers under the age of 35. The differential distribution of Down syndrome births reflects discrepancies in the number of pregnancies to women under and over the age of 35. Younger mothers may have 65% of the Down syndrome children, but they also have 93% of all births. Older women have about 7% of all children but 20% of the Down syndrome births, emphasizing once again the relationship between maternal age and increased risk of Down syndrome.

In the case of neural tube defects, some 90% of all affected infants are born to parents with no family history of such conditions. Thus, testing couples who have had an affected child will have little effect on the overall rate of prenatal detection for this birth defect. On the other hand, screening of all pregnant females is not cost-effective or possible, given the limited number of prenatal screening clinics.

Newborn and Carrier Screening

The autosomal recessive condition phenylketonuria (PKU) was the first genetic disease to be screened in newborns. Newborn screening for PKU is now mandatory in all states and in 20 foreign countries. Over 100 million children have been screened for this disease, and more than 10,000 affected individuals have been identified and treated. Although PKU is a relatively rare disorder (1 in 12,000), it is severe, imposes a large personal and financial burden, and can be easily screened using an inexpensive test. In addition, the discovery that the effects of the disease can be controlled by dietary treatment was instrumental in establishing mandatory screening programs. PKU is the model disease in newborn screening programs, and most of the other diseases screened in newborns are metabolic deficiencies. Some of the other metabolic diseases included in newborn screening programs are listed in Table 19.2.

Carrier screening is the identification of phenotypically normal individuals who are heterozygous for an autosomal recessive or X-linked recessive disease. Screening for carriers of two genetic conditions have been carried out on a large scale: Tay-Sachs disease and sickle cell anemia. The development of screening programs for these two diseases has been made possible by three factors. First, the diseases occur mainly in defined populations. Tay-Sachs carriers are found most frequently among Jews of East European origin, and sickle cell carriers are commonest in U.S. blacks of West African origin. Second, carrier detection for these disorders is inexpensive and rapid. Third, the existence of prenatal testing gives couples at risk the option of having only unaffected children.

Tay-Sachs disease is an autosomal recessive condition with an incidence in the general population of 1 in 360,000. In Ashkenazi Jews, the rate is almost 100 times higher (1 in 4800 births). In the 1970s, carrier screening programs were undertaken to identify heterozygotes in the United States and other

TABLE 19.2 Genetic Diseases Commonly Screened in Newborns

Galactosemia
Phenylketonuria
α_1-antitrypsin deficiency
Homocystinuria
Tyrosinosis
Sickle cell anemia

countries. In the first 10 years of screening, over 300,000 individuals were tested. Of these, 268 couples were identified in which both members were carriers and had not yet had an affected child. The programs were coupled with counseling sessions that provided education about the risks of having an affected child, the availability of prenatal screening, and reproductive options. None of these screening programs is mandatory, although some states have laws that require couples to be informed that screening for TSD is available. In 1970 there were 50 to 100 Tay-Sachs births annually in the United States. Because of screening programs, there are now fewer than 10 such births each year.

Sickle cell anemia is an autosomal recessive condition that differentially affects black Americans and whites whose family origins are in the lowlands of the Mediterranean Sea, including Sicily, Italy, Greece, Lebanon, and Israel.

Several blood tests were developed in the 1960s that can differentiate individuals who are carrier heterozygotes and affected homozygotes. In 1971 Connecticut instituted a program of screening black schoolchildren in grades 7–12, with parental consent, for **sickle cell trait,** a term used to designated heterozygous carriers. In 1972 federal funding was used to establish the National Sickle Cell Anemia Control Act, part of which was designed to establish carrier screening programs. As a result of this federal legislation, screening programs to detect carriers of the sickle cell trait were set up nationwide. Some of the programs were compulsory, requiring black children to be screened before attending school; others required screening before obtaining a marriage license. Professional football players were screened, as were cadets at the Air Force Academy, in Colorado Springs, where heterozygotes were excluded from enrollment. The assumption was that heterozygotes might undergo sickling of red blood cells at high altitudes under reduced oxygen concentrations. This policy was reversed under threat of lawsuit in 1981. Other individuals, testing as positive heterozygotes, were reportedly turned down for insurance and employment, even though carriers do not have any inherent health problems.

Sickle cell trait
The symptoms shown by those heterozygous for sickle cell anemia.

Some of the programs were criticized for laxity in confidentiality of records and the failure to provide counseling to those identified as heterozygotes. In the late 1970s, many of the sickle cell screening programs were cut back or reorganized, and currently only 10 states offer sickle cell screening.

Against the background of an uneven record of previous carrier screening programs, a debate is emerging about universal screening to detect carriers for cystic fibrosis (CF). Cystic fibrosis (see Chapter 4 for a detailed discussion) is an autosomal recessive condition that occurs in all ethnic groups, although at different rates (Table 19.3). About 30,000 individuals in the United States are affected by CF, and 8 million others are heterozygous carriers.

The questions surrounding CF screening are legal, ethical, and economic. Part of the reason for the debate is that CF screening is one of the first recombinant DNA–based tests to become available, and decisions about its implementation and availability will set precedents for the scores of DNA-based tests that will be available in the next few years. With respect to CF screening, it appears that the question is not if there should be screening, but when and how such screening should be implemented. In 1992, approx-

TABLE 19.3 Cystic Fibrosis Among Live Births in the U.S.

Population	Incidence at Birth
Caucasian	1 in 2,500
Hispanic	1 in 9,600
African American	1 in 18,000
Asian American	1 in 90,000

SOURCE: Office of Technology Assessment 1992

imately 63,000 individuals were screened to determine their genetic status for CF. This represented a seven-fold increase over 1991 but was still far short of the 6 million pregnancies per year for which CF screening can be performed.

The concerns surrounding CF screening itself, and CF screening as a model for other forms of genetic testing, resolve into a number of issues. These include:

1. Standards of care. Should screening be offered to everyone, or only those with a family history?
2. Confidentiality and discrimination. Who will have access to test results? Will identification as a carrier result in exclusion from health insurance coverage?
3. Quality and reliability of tests. How accurate are the tests? How often do false positive or false negative results occur?
4. Cost effectiveness. What proportion of the population must participate in screening for it to be cost effective? How can participation be encouraged?

Still to be resolved are questions about who is qualified to do genetic testing and screening, how the costs of screening will be recovered, and how the public can be educated about the procedures and evaluation of risk. Some of these issues will be considered in a later section of this chapter.

Occupational Screening

Two classes of genetic tests are used for genetic screening in the workplace: tests to screen workers for genetically determined susceptibility to specific environmental agents that cause disease, and tests to monitor the amount of genetic damage actually produced in susceptible workers. We will consider only the first class of tests. Genetically determined sensitivity to certain environmental agents is well documented; in other cases the relationship between genetic factors and environmental agents is uncertain or is based on inadequate information.

Approximately 50 different genetic traits have been related to susceptibility to environmental agents. Some of these are listed in Table 19.4. In

TABLE 19.4 Some Genetic Factors Affecting Susceptibility to Environmental Agents

Trait	Agents
G6PD deficiency	Oxidants such as ozone, nitrogen dioxide
Sickle cell trait	Carbon monoxide, cyanide
Thalassemias	Lead, benzene
Erythrocyte porphyria	Lead, drugs including sulfanilomide, barbiturates
Gout	Lead
Sulfite oxidase deficiency	Sulfite, bisulfite, sulfur dioxide
Wilson disease	Copper, vanadium
Pseudocholinesterase variants	Carbamate insecticides, muscle relaxants
Cystinuria	Heavy metals

many of these cases, the traits are rare, testing for carriers is difficult, or the relationship to a specific environmental agent is based on inadequate data. To justify use in occupational screening, the traits must be present in a sufficient fraction of the workforce, and the trait should be associated with a clear risk to carriers. Most often the trait must be present in at least 1% of the workforce, and exposure should result in serious illness or risk of death for inclusion in a program of screening. Table 19.5 contains a list of traits that fulfill many of these requirements.

The current status of genetic testing in the workplace is difficult to evaluate. In 1973 a report identified five genetic conditions as suitable for occupational screening: serum α_1-antitrypsin deficiency, glucose-6-phosphate dehydrogenase deficiency, carbon disulfide sensitivity, hypersensitivity to organic isocyantes, and sickle cell trait (heterozygotes). Recommendations for these as candidates for testing were made because these conditions met "the prerequisites for industrial applications of bettering job assignment, improving coverage of industrial air limits, and hence reducing risk to worker health." Unfortunately, the recommendations for screening were not entirely based on a firm scientific foundation. In spite of its shortcomings, this report was influential in establishing many industrial screening programs.

In 1983 the federal Office of Technology Assessment (OTA) conducted a thorough review of genetic screening in the workplace. The report, entitled *The Role of Genetic Testing in the Prevention of Occupational Disease,* indicates that there are clear-cut and well-known relationships between some genetic traits and agents present in the workplace. For example, G6PD deficiency hemizygotes or homozygotes may develop severe anemia when exposed to chemical oxidizing agents. Most of these relationships are based on *in vitro*

TABLE 19.5 Genetic Traits Associated with Workplace Hazards

Trait	Environmental Agent	Status of Interaction
Glucose-6-phosphate dehydrogenase (G6PD) deficiency	Primaquine, fava beans	Definite
Methemoglobin reductase deficiency	Nitrates, acetanilide, amines, sulfanomides	Definite
N-Acetyl transferase deficiency	Isoniazid hydrochloride, dapsone, hydralazine	Definite
PTC nontaster	Thiouria, related compounds	Possible
Slow alcohol metabolism	Ethanol toxicity	Possible
Sickle cell trait (heterozygotes)	Low oxygen concentration	Inconclusive

exposure of cells from G6PD-deficient individuals to these chemicals. While many of these same chemicals can be encountered in industrial settings, the report calls for studies that directly assess whether exposure to these chemicals in the workplace actually poses any hazards to G6PD-deficient individuals. The report concludes that in the absence of direct evidence of harm to workers actually in contact with such agents, there is no justification for occupational screening programs of any kind. Critics of this report argue that sufficient information exists from laboratory studies on cells and tissues to show direct harm to certain genotypes upon exposure and that little or no research is being done on workers in the workplace. In effect, they charge, workers are being used as guinea pigs, and testing will start only after workers with susceptible genotypes have been seriously affected.

In spite of this controversy, the OTA report found that 17 of the nation's largest firms had instituted genetic screening programs and that 59 others planned to initiate occupational screening tests in the following 5 years. These tests will presumably be used to exclude some individuals from employment and to determine job assignments for others. Clearly, the issue of occupational screening will be of increasing importance in the coming years. Unfortunately, the rush to install screening programs for AIDS and illegal drugs may set a precedent that will lead directly to the use of genetic screening in the workplace before the scientific and legal basis for such testing has been thoroughly considered.

Reproductive Screening

Artificial insemination by a donor (AID) is often used in cases in which male fertility is low or, more rarely, to avoid genetic risks to the offspring by the

TABLE 19.6 Some Traits Uncovered in Screening 676 Potential Sperm Donors

Trait	Number
Polyposis coli (predisposes to cancer)	3
Ankylosing spondylitis	2
Epilepsy	3
Manic-depressive psychosis	2
Dominant renal disease (not defined)	1
Severe hip dislocation	3

From: Selva, J., Leonard, C., Albert, M., Auger, J., and David, G 1986. Genetic screening for artificial insemination by donor (AID). *Clin. Genet.* **29**:389–396

transmission of a genetic defect. Some 172,000 women are artificially inseminated each year, resulting in 65,000 births. Of these, 30,000 births are the result of insemination with anonymously donated sperm obtained through physicians or sperm banks. In all cases, care must be taken to prevent genetic defects from being transmitted through the donor sperm. Since everyone carries some deleterious mutations, genetic defects in the donors cannot be eliminated. The problem is to ensure that no genetic disorders are transmitted by the donor sperm, on the one hand, and not to eliminate too many donors, on the other hand. In one study of over 600 potential semen donors, 6% were excluded as donors based on a detailed screening procedure. Of these, 2.6% were excluded for cytogenetic reasons and 3.4% for genetic reasons. The chromosomal abnormalities detected included breaks, translocations, partial aneuploidy, and the presence of fragile sites. Table 19.6 lists some of the genetic traits uncovered in family histories of the prospective donors. They include single gene traits and polygenic or familial traits such as epilepsy.

In another study, the OTA surveyed 15 sperm banks and 367 physicians to determine whether sperm donors had been screened for infectious diseases, genetic defects, or both. According to the report, published in 1987, 14 sperm banks tested all donors for presence of the human immunodeficiency virus (HIV), while the other tested only men from high-risk groups. Twelve of the sperm banks screened for transmissible diseases; 13 screened for genetic diseases. Interestingly, only 44% of the physicians tested for HIV, and fewer than 30% tested for transmissible diseases such as syphilis or hepatitis. Moreover, only 48% screened for any genetic defects.

The most disturbing aspect of the report indicates that physician screening of sperm donors for genetic diseases is unreliable. Twenty-five percent of the physicians said they would accept sperm from a healthy donor with a family history of Huntington disease. Huntington disease is an autosomal dominant disorder that does not appear until the individual is over 40 years of age, while most sperm donors are under 30 years of age. On the other hand, 49% of physicians would reject a healthy sperm donor with a family history of hemophilia. Recall that hemophilia is an X-linked trait expressed from birth. Healthy males do not carry the trait and are incapable of passing it on to their offspring. The report has sparked calls for the Food and Drug Administration (FDA) to require physicians and sperm banks to screen sperm samples for HIV and to use fresh sperm only when the donor is known to the recipient. In the meantime, the results from these surveys indicate some degree of genetic risk associated with artificial insemination.

Cost-Benefit Analysis in Genetic Screening

There are a number of genetic disorders, like PKU, for which corrective treatment is available but which must be diagnosed early (even before symptoms appear) for treatment to be effective. Other genetic disorders, like Tay-Sachs disease or Down syndrome, cannot be treated but can be diagnosed by amniocentesis early enough for elective abortion. In instituting genetic screening programs, the question of cost must be given some consideration. While the family of a PKU child might think that newborn screening programs should be continued at any cost, public administrators usually consider whether money expended in such programs benefits the state in terms of dollars and cents.

The actual calculation of costs and benefits is complex and includes not only the direct cost of the program but also the contributions made to the state by a treated PKU patient and the increase in the well-being of a patient and the family. To perform a simpler analysis, let us presume that the cost of analysis for PKU is $1.50 per test, including overhead and salaries. Assuming complete coverage of births in the United States, 4 million births per year will be screened at a total cost of $6 million. If the incidence of PKU is 1 in 12,500, about 320 PKU births will occur in the total of 4 million. Dietary treatment will cost about $5000 per year and will be maintained for an average of 10 years. Thus the cost of treatment for the 320 PKU births that occur annually is 320 × $5000 × 10, or $16 million. Since the total cost must include both testing and treatment, the cost for the PKU program is $22 million ($6 million for testing and $16 million for treatment). From the point of view of the state, the alternative to screening and treatment is placement of the untreated, retarded individual in a state institution. Placing the annual cost of maintaining severely retarded individuals at a *very* conservative $40,000 and the life expectancy of institutionalized PKU patients of 30 years, the cost to the state of housing the 320 affected individuals produced in a year is 320 × $40,000 × 30, or $384 million. Clearly, the screening and treatment program is 17 times less expensive than institutionalizing untreated, afflicted individuals. This rather detached method of computing costs and benefits does not include the human costs to patient and family and other intangibles that derive from having treated children function as normal members of society.

Quebec established a Network of Genetic Medicine in 1969 to screen, treat (where possible), and provide follow-up services for a variety of genetic diseases. The goal of the network is to offer all pregnant women in the province prenatal and newborn screening and effective treatment where possible.

The network was subject to a cost-benefit analysis for the years 1969–1985. This analysis incorporated the actual costs of the operation and attempted to quantify many of the tangible, major benefits. Table 19.7 summarizes the information for 1980 to 1985 and makes projections through the year 2000. Clearly the social and economic benefits of the program far outweigh the costs incurred, even when benefits have been conservatively estimated.

The Impact of Genetic Screening and Counseling

The development of genetic screening and counseling programs has provided many benefits to individuals and society at large. But it has also created a number of associated problems and has raised serious questions about whether screening should be mandatory, who is to have access to the results of screening, and whether individuals identified as carriers of genetic

TABLE 19.7 Cost-Benefit Analysis of Medical Network in Quebec

Year	Cost (Thousands of Dollars)	Benefits	Net Difference
1980	33,685	36,182	2,497
1981	36,020	39,929	3,909
1982	38,222	43,481	5,259
1983	40,301	46,483	6,542
1984	42,262	50,007	7,745
1985	44,111	52,981	8,870
1990	51,885	64,944	13,059
2000	61,962	78,125	16,163

From: Dagenais, D., Courville, L., and Dagenais, M. 1985. A cost-benefit analysis for the Quebec Network of Genetic Medicine. *Soc. Sci. Med.* **20:**601–607

defects are socially stigmatized. In this section, we will briefly examine several aspects of these problems.

Personal Consequences

The information that one is a carrier of a genetic disease often has a devastating psychologic effect. Many identified carriers suffer a loss of self-image and regard themselves as worthless. This feeling is often reinforced by the feelings of family members toward carriers. For example, in some parts of rural Greece, marriages are arranged by parents and relatives. In one village in which screening for sickle cell was conducted, carriers were regarded as unsuitable marriage partners for anyone, not just other carriers.

To counter these effects, screening programs must be coupled with effective counseling programs for carriers, their families, and the general public. The education process must stress that carriers are not at risk for the disease, nor should they be prevented from marrying other carriers. Options for matings between heterozygotes should be carefully distinguished, including adoption, artificial insemination, and prenatal diagnosis coupled with selective abortion.

The effect of genetic testing on childbearing decisions has been documented in a number of studies. In one such study, couples at risk for having children afflicted with a severe form of beta thalassemia were counseled about the availability of prenatal diagnosis. Before such services were available, couples known to be at risk (through birth of an affected child or carrier screening) had stopped having children altogether, and almost all pregnancies that occurred were reported as accidental. Of these pregnancies, 70% were terminated for fear of having a child with beta thalassemia. The availability of prenatal diagnosis brought about a significant change in childbearing decisions in such couples. In fact, reproductive patterns returned to almost normal levels, and less than 30% of all pregnancies were terminated because of thalassemia. Other surveys have reported similar findings, emphasizing the impact of genetic testing on individual lives.

Predictive genetic testing for autosomal dominant fatal genetic disorders that first appear in middle age (Huntington disease, polycystic kidney disease) has been evaluated to determine its psychological and social impact. Recently, the Canadian Collaborative Study of Predictive Testing reported on the psychological consequences of predictive testing for Huntington disease (HD). This form differs from other genetic testing in that it requires other family members to be tested to produce informative results. Consequently, a request for testing has impact on the other members and forces them to consider whether they wish to be tested. In the study reported by the Canadian group, 200 individuals with an affected parent were followed after HD testing. They were separated into three groups: those with increased risk, those with no change in risk (mostly from uninformative test results), and those identified as having a decreased risk. The results suggest that testing has positive benefits for many participants. Clearly, those in the low-risk group showed an increase in well-being and psychological health. However, those in the increased risk group did not show a negative response to their condition. In fact, they reported less depression and an increased sense of well-being 12 months after testing. It appears that knowledge of status, whether for increased or decreased risk, is of psychological value, while those with uncertain status remain susceptible to depression and have a lowered sense of well-being. Further tests and follow-up will be required to determine whether positive psychological effects are a hallmark of predictive genetic testing.

Social Consequences

At the broader level we can compare the response of the Jewish community to screening for Tay-Sachs disease with some of the responses in the black community to sickle cell screening. In Tay-Sachs disease screening, the program was voluntary and welcomed by the community, which was involved in its planning and implementation. An adequate educational and counseling program accompanied the screening. In contrast, screening for sickle cell disease (homozygous recessive individuals afflicted with the disease) and sickle cell trait (heterozygous individuals unaffected by the disease) was largely mandated by law, and prominent members of the black community were not involved in the planning and initiation of screening. The origin of the program from outside the community coupled with sporadic problems (lack of confidentiality and inadequate education and counseling) generated suspicion and resentment about the screening program.

Many of these problems could have been avoided by better planning and implementation of this large-scale screening program. Voluntary rather than mandatory participation coupled with adequate education and counseling and community involvement would undoubtedly have eased many fears and suspicions. Perhaps it would be better to offer testing and counseling to those who request it rather than screening large groups to identify and to label individuals as "carriers." Others argue that only mandatory screening programs can be effective. If, say, only 10% of those at risk take advantage of screening programs, the program is ineffective, and the cost-benefit ratio would not justify the existence of the program.

The lessons from earlier attempts at sickle cell screening are particularly important in light of new discoveries about sickle cell disease. A recent study has revealed that children with sickle cell disease who are under 3 years of age have poor resistance to bacterial infections, particularly those caused by *Streptococcus.* Children with sickle cell anemia have a 15% chance of dying from a bacterial infection during the first 3 years of life. The study also demonstrated that doses of the antibiotic penicillin were highly effective in preventing illness and death.

These findings have prompted a National Institutes of Health (NIH) panel to recommend that sickle cell screening be made available to all newborns, whether or not they are members of high-risk ethnic groups. This recommendation has been made because it would avoid errors in classifying individuals as members of certain ethnic groups and because screening is easy and inexpensive. Sickle cell screening can be done using a small blood sample and materials costing $0.22 per test. According to the panel, the expenditure of approximately $880,000 per year (4 million births × $0.22 per test) would produce a 15% reduction in the mortality rate among small children with sickle cell anemia. Recently, a panel convened by the U.S. Public Health Service has recommended that a program to screen all newborns in the U.S. for sickle cell anemia be implemented. If this recommendation is accepted, newborn screening will begin in the next few years.

It remains to be seen whether sickle cell screening programs will be adopted in any of the 40 states that do not currently test for this condition. This recommendation by the NIH panel also raises the question of whether such screening should be mandatory or voluntary and whether the states should screen all newborns or only those in high-risk groups. These issues are certain to be debated once again by community groups and state legislatures in the near future.

Legal Problems and Consequences

Genetic screening has raised a number of legal issues, many of which have not yet been resolved. We will explore some of these issues in more detail in the last chapter. Here we will consider several questions about genetic screening to illustrate that genetic methodology and practice are several steps ahead of legislation, legal decisions, and social consensus.

If a child is born with a genetic defect that can be diagnosed prenatally, can the physician be held responsible for not informing the parents that prenatal screening for this defect is available? If an insurance company pays for a genetic screening test, does it have the right to know the results of the test? Can health or life insurance companies require genetic testing as a condition for obtaining insurance? Should individuals who test positive for Huntington disease or other genetic disorders be denied health or life insurance?

The Occupational Safety and Health Administration has categorized 24 chemicals that may be associated with reproductive hazards and excludes all fertile women from jobs that involve exposure to these chemicals. Is this protection or a form of sexual discrimination? Does the fact that these rules may apply to 20 million jobs change your answer?

These questions illustrate that many problems involving the development and use of genetic screening need to be resolved. These issues involve science as well as sociology, law, and ethics. In decisions about genetic screening, the rights of individuals must be considered and balanced against the rights of employers and society. Health policy is an area in which all citizens need to be educated and informed. As the constellation of genetic screening tests grows, these problems must be met and solved.

GENETIC COUNSELING

Genetic counseling is a clinical service that deals with the risk or occurrence of genetic disorders. This process involves the assessment of risk or recurrence of genetic disorders, the developmental history of affected children, and treatments that might be available for the trait and communication of recurrence or risks to future offspring. The roles of the counselor are analysis of family history, risk assessment, and education for those seeking assistance, testing, or advice. Final decisions are always left to those being counseled.

How Does Counseling Work?

In most cases couples are referred for genetic counseling after the birth of an affected child. In this situation the parents usually want to know the risk to any subsequent children. Others who seek counseling are concerned about familial conditions that might affect their children or are planning marriage and wish to know what risks might arise in the genotypes of their children.

The process of counseling begins with the construction of a pedigree derived from a family history taken in an interview. Cytogenetic or biochemical testing of the couple may also be done at this time. Using the results of genetic tests and an analysis of the pedigree, the counselor establishes whether the trait is genetically determined. If this is the case, the next step is risk assessment. In estimating risk, it should be remembered that about 2% of all children are born with a serious genetic problem. This level is the background risk for all couples and is also called random risk. Conditions that can be regarded as high risks include dominant conditions (50% risk if one parent is heterozygous), simple autosomal recessives (25% when both parents are heterozygotes), and certain chromosomal translocations. Often conditions are difficult to assess because they involve polygenic traits or conditions with high mutation rates (like neurofibromatosis). Some risk factors are summarized in Table 19.8.

TABLE 19.8 Some Risk Factors in Families with One Affected Child

Trait	Risk of More Affected Children
Autosomal recessive	25%
Autosomal dominant	50%
Rare, sex-linked recessive	0% females; 50% males
Chromosome abnormality	<1 to 100%
Genetic anomaly; not a simple mode of inheritance	Generally <10%
Nongenetic malformation	2%

Once the process of risk assessment has been completed, the counselor must communicate this information to those being counseled. While this sounds somewhat straightforward, several barriers commonly arise in this process. Many individuals have an incomplete knowledge of elementary probability and think that a 1 in 15 risk means that they can safely have 14 additional unaffected children after the birth of an affected infant. Others will interpret a risk of 1 in 1000 as significant, believing that since there is a chance for recurrence, it will necessarily happen.

Emotional strife can also inhibit effective understanding of recurrence risks. Many parents of affected children are confused, angry, and filled with guilt. Some fathers are unable to accept the fact that they are carriers of a deleterious trait and will deny paternity.

Often those being counseled are unfamiliar with the basic concepts of biology and do not understand how genes, proteins, or cell-surface antigens are related to their defective child. The counselor must strive to overcome these barriers so that the parents are able to make an informed decision about future reproductive choices. The counselor can also provide information about reproductive alternatives such as adoption, artificial insemination, *in vitro* fertilization, egg donation, and surrogate motherhood (see Concepts & Controversies, page 537).

As more and more genetic defects can be detected by heterozygote and prenatal screening and as these techniques become more available, the role of the genetic counselor will become more important. The services of genetic counselors are a relatively new feature of the health care system. Typically, counselors are graduates of a 2-year master's degree program and are trained in course work, clinical work, and laboratory methods. They also receive training in ethical, social, and legal issues related to genetic disorders. Certification for counselors is offered by the American Board of Genetic Counseling. Most counselors work at university medical centers or at large

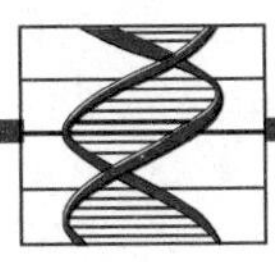

Concepts & Controversies

The Business of Making Babies

While relatively new methods are revolutionizing the fields of genetic screening and genetic counseling, technology already available is rapidly making human conception a part of private enterprise. One in six U.S. couples, totaling over 3 million couples of childbearing age, are classified as infertile. The process of *in vitro* fertilization (IVF) began in the United States in 1981 at the Medical College of Virginia at Norfolk. Since then over 150 additional clinics using this and related techniques have opened. Many of these clinics are associated with university medical centers, but others are operated as freestanding businesses. Some are companies that have sold stock to raise start-up money or cover operating costs. It is estimated that it requires $500,000 to $1 million to start a clinic and that 50 to 60 fertilization attempts per month are necessary for the company to be profitable. Each attempt at IVF costs between $4000 and $5000, and multiple attempts (four to six) are usually required for success.

In vitro fertilization is a technique in which egg maturation is induced with drugs and the mature eggs are recovered from the ovary, fertilized in a laboratory dish (recall that *in vitro* means, literally, in glass), and implanted into the women's uterus. Eggs are usually recovered by laparoscopy. A small incision is made in the abdomen, and a fiber-optic device is used to obtain the mature eggs. Alternatively an ultrasonically guided needle is inserted into the vagina and moved up to the ovaries to remove the eggs. This procedure is fast (about 15 minutes) and requires no surgery. If extra eggs are recovered, they are fertilized, and the resulting embryos are stored frozen in liquid nitrogen. This eliminates the need to retrieve eggs every month for fertilization. If fertilization is successful, the extra embryos can remain in storage for implantation at a later time or can be donated to another couple.

Several companies, like IVF Australia, are already open in several U.S. locations, and it is possible that licensing agreements could result in a small number of companies dominating the field through franchising arrangements. Some investment analysts and those involved in the clinics predict that the IVF business will grow from a $500-million-dollar business to a $6-billion annual business in 5 years. As the technology for genetic screening and genetic counseling expands the range and accuracy of these techniques, a genetic testing and counseling industry might arise.

hospitals in metropolitan areas. As genetic testing and screening methods proliferate, it is hoped that genetic counseling services will become available to an increasing number of those who request genetic testing.

Summary

1. Genetic screening is the search for individuals of a particular genotype. In prenatal and newborn screening, several considerations are of importance. Treatable diseases are favorable for screening even when they are rare in the population. These include PKU, galactosemia, and maple syrup urine disease.

2. Prenatal screening can also detect chromosome abnormalities, such as Down syndrome, and birth defects, such as spina bifida, that may have a genetic component. One of the considerations in prenatal screening is the identification of the risk group to be screened. The frequency of Down syndrome increases rapidly as maternal age increases over 35 years, and it is easy to see that screening should be made available to all pregnant females over the age of 35. Yet most Down syndrome births occur to younger mothers because they have many more children than older mothers. Should screening be made available to all mothers for Down syndrome? Given the limited resources available now for such screening, is this cost effective? Several techniques are used in prenatal diagnosis, and each carries a risk to both the mother and the fetus. Less invasive methods such as the detection of fetal cells in the maternal circulation are under development, making prenatal screening for some diseases safer and more economical.

3. Carrier screening is the search for heterozygotes who may be at risk of producing a defective child. An increasing number of autosomal recessive diseases can be screened by molecular probes, including sickle cell anemia, Huntington disease, cystic fibrosis, and Duchenne muscular dystrophy. Large-scale carrier screening has been conducted for two autosomal recessive diseases that affect discrete population segments: Tay-Sachs disease and sickle cell anemia. The programs have been technically successful but were accepted somewhat differently in the affected segments of the population. A study of their implementation and the community reaction to them will be valuable in the design and planning of other carrier screening programs.

4. Occupational screening is used to detect individuals who are genetically susceptible to agents in the workplace that can cause the development of disease. While a number of agents that can cause adverse reactions have been identified and a number of diseases can be screened, no large-scale studies have been conducted to establish that such chemicals are harmful to sensitive individuals in the workplace. Because occupational screening can be used to exclude individuals from employment, its implementation should be restricted to those genetic conditions in which a danger has been clearly demonstrated.

5. The rapid development of methodology for genetic screening has generated a number of problems, including whether screening should be voluntary or mandatory and whether the results of screening tests can be used to deny services such as insurance of health care coverage. These issues will undoubtedly be the subject of much debate and a fair amount of legislation in the next few years.

6. Genetic counseling is a service that undertakes the accurate assessment of a family history to determine the risk of genetic disease in subsequent children. In most cases this is done after the birth of a child affected with a genetic disorder, but in other cases counseling is entirely retrospective. Decisions about whether to have additional children or to undergo abortion or even to marry are always left to those being counseled.

Questions and Problems

1. List the types of genetic screening covered in this chapter, and briefly summarize the unique characteristics of each type.
2. The measurement of α-fetoprotein levels is used to diagnose neural tube defects. For every 1000 such tests, approximately 50 positive cases will be detected. However, up to 20 (40%) of these cases may be false positives. In a false positive, the α-fetoprotein level is elevated, but the child has no neural tube defect. Your patient has undergone testing of the maternal blood for α-fetoprotein, and the results are positive. She wants to abort a defective child but not a normal one. What are your recommendations?
3. Would you support a tax increase in your state to institute a genetic program similar to the Quebec Network of Genetic Medicine? State your answer from an economic and social viewpoint.
4. The reaction to screening for Tay-Sachs disease and sickle cell anemia offers an interesting contrast in the institution and administration of genetic screening programs. Cystic fibrosis is an autosomal disease that mainly affects the white population, and 1 in 20 whites are heterozygotes. Now that the gene has been mapped to chromosome 7, assume that RFLP markers are available to diagnose heterozygotes. Should a genetic screening program for cystic fibrosis be instituted? Should this be funded by the federal government? Should the program be voluntary or mandatory, and why?
5. As a genetic counselor, you are visited by a couple who wish to have children. There is a history of a deleterious, recessive trait in males in the woman's family but not in the man's family. The couple are convinced that because his family shows no history of this genetic disease, they are at no risk of having affected children. What steps would you take to assess this situation and educate this couple?
6. A couple have had a child born with neurofibromatosis. They come to your genetic counseling office for help. After taking an extensive family history, you determine that there is no history of this disease on either side of the family. The couple want to have another child and want to be advised about risks of another child with neurofibromatosis. What advice do you give them?

For Further Reading

Agency for Health Care Policy and Research. Sickle Cell Disease Panel. Sickle cell disease: comprehensive screening and management in newborns and infants. Rockville, MD, Public Health Service, Department of Health and Human Services, April, 1993.

Billings, P. and Beckwith, J. 1992. Genetic testing in the workplace: a view from the USA. *Trends in Genet.* **8:** 198–202.

Calabrese, E. J. 1986. Ecogenetics: Historical foundation and current status. *J. Occup. Med.* **28:** 1096–1102.

Dagenais, D., Courville, L., and Dagenais, M. 1985. A cost-benefit analysis for the Quebec Network of Genetic Medicine. *Soc. Sci. Med.* **20:** 601–607.

Draper, E. 1991. *Risky Business: Genetic testing and exclusionary practices in the hazardous workplace.* New York: Cambridge University Press.

Emery, A. E. H., and Pullen, I. 1984. *Psychological Aspects of Genetic Counseling.* New York: Academic.

Fuhrmann, W., and Vogel, F. 1983. *Genetic Counseling.* 3rd ed. New York: Springer-Verlag.

Gibbs, R. A., and Caskey, C. T. 1989. The application of recombinant DNA technology for genetic probing in epidemiology. *Ann. Rev. Pub. Health* **10:** 27–48.

Hodgson, S. V., and Bobrow, M. 1989. Carrier detection and prenatal diagnosis in Duchenne and Becker muscular dystrophy. *Br. Med. Bull.* **45:** 719–744.

Jinks, D. C., Minter, M., Tarver, D. A., Vanderford, M., Hejtmancik, J. F., and McCabe, E. R. B. 1989. Molecular genetic diagnosis of sickle cell disease using dried blood specimens on blotters used for newborn screening. *Hum. Genet.* **81:** 363–366.

Johnson, A., and Godmilow, L. 1988. Genetic amniocentesis at 14 weeks or less. *Clin. Obstet. Gynecol.* **31:** 345–352.

Kolata, G. 1986. Genetic screening raises questions for employers and insurers. *Science* **232:** 317–319.

Modell, B., and Kuliev, A. 1993. A scientific basis for cost-benefit analysis of genetics services. *Trends in Genet.* **9:** 46–52.

Murray, R. F. 1986. Tests of so-called genetic susceptibility. *J. Occup. Med.* **28:** 1103–1107.

Reed, S. 1980. *Counseling in Medical Genetics*. 3rd ed. New York: Liss.

Rowley, P. 1984. Genetic screening: Marvel or menace? *Science* **225:** 138–144.

Sandovnick, A., and Baird, P. 1985. Reproductive counseling for sclerosis patients. *Am. J. Med. Genet.* **20:** 349–354.

Selva, J., Leonard, C., Albert, M., Auger, J., and David, G. 1986. Genetic screening for artificial insemination by donor (AID). *Clin. Genet.* **29:** 389–396.

Sommer, S. S., Cassady, J. D., Sobell, J. L., and Bottema, C. D. 1989. A novel method for detecting point mutations or polymorphisms and its application to population screening for carriers of phenylketonuria. *Mayo Clin. Proc.* **64:** 1361–1372.

U.S. Congress, Office of Technology Assessment, *Cystic Fibrosis and DNA Tests: Implications of Carrier Screening*, OTA-BA-532. Washington, D.C.: U.S. Government Printing Office, August, 1992.

Uzych, L. 1986. Genetic testing and exclusionary practices in the workplace. *J. Public Health Policy* Spring 1986: 37–57.

Wapner, R. J., and Jackson, L. 1988. Chorionic villus sampling. *Clin. Obstet. Gynecol.* **31:** 328–344.

Williams, C., Weber, L., Williamson, R., and Hjelm, M. 1988. Guthrie spots for DNA-based carrier testing in cystic fibrosis. *Lancet* ii: 693.

CHAPTER 20

GENETICS, LAW, AND BIOETHICS

CHAPTER OUTLINE

"Give me your tired, your poor, your huddled masses longing to breathe free.'' This line from the poem by Emma Lazarus is engraved on the Statue of Liberty standing on Liberty Island in Upper New York Bay, the port of entry for millions of immigrants in the early part of this century. The statue and its engraved lines symbolize America as the land of freedom and opportunity open to all. The last line reads: ''Send these, the homeless, tempest-tost to me. I lift my lamp beside the golden door.'' After 1921, however, the golden door began closing, and in 1924, it was slammed shut to immigrants from many parts of Europe. (Immigration from Asia had been cut off earlier.)

After 1924, those who were lucky enough to enter were asked to read and to sign a declaration that included the statement below:

> That I have had the following excludable classes explained to me, and that, except as hereinafter noted I am not a member of any one of the following classes of individuals excluded from admission to the United States under the immigration laws: (1) idiots; (2) imbeciles; (3) feebleminded; (4) epileptics; (5) insane persons; (6) persons having had previous attacks of insanity; (7) persons with constitutional psychopathic inferiority; (8) persons with chronic alcoholism; (9) paupers; (10) professional beggars; (11) vagrants; (12) persons afflicted with tuberculosis; (13) persons afflicted with a loathsome or dangerous contagious disease; (14) persons convicted of, or who admit committing, a crime involving moral turpitude; (15) polygamists; (16) anarchists; (17) persons who believe in or advocate the overthrow by force or violence of the Government of the United States or the assassination of public officials, or the unlawful destruction of property, or who have ever held or advocated such views; (18) persons inadmissible under the provisions of section 3 of the act of February 5, 1917; (19) persons inadmissible under the provisions of the act of October 16, 1918, as amended; (20) prostitutes; (21) procurers; (22) contract laborers; (23) persons likely to become public charges.

It is not clear how many individuals voluntarily declared themselves to be feebleminded, idiots, or imbeciles, but the story of how those items came to be on an immigration declaration is one that involves the interaction of genetics, eugenics, and politics over a decade or more of American history.

It is difficult to describe the relationship between genetics, the law, and social policy as it now exists without a discussion of the eugenics movement in the United States. In the first decades of this century, genetic principles were used as a guide to draft a wide range of social legislation at both the state and federal levels. Many of the biologically based laws drafted during the reform-minded Progressive Era continue to have strong impact on patterns of immigration and reproductive rights. Knowledge of the social climate in which these laws were developed and the consequences that followed from their adoption will help prepare us to consider the coming onslaught of legislation that is being proposed to regulate new genetic technology as it relates to reproduction, gene surgery, and the commercial patenting of life forms. As in the Progressive Era, these laws will personally affect each of us.

The Beginning: Eugenics and the Law

History

Darwin's *The Origin of Species*, published in 1859, contained few references to any aspect of human evolution. However, the concept of natural selection promulgated by Darwin had a great impact on the study of human society. Natural selection provided a framework within which all social progress could be explained as the result of selection of the fit and elimination of the unfit. Sir Francis Galton (Figure 20.1), a cousin of Darwin, was convinced that natural selection was at work in human evolution. To confirm this, he studied the pedigrees of successful individuals, including artists, composers, diplomats, and military officers. He found that a great many of their ancestors and offspring alike were also talented and successful. Completely discounting the role of the environment, Galton concluded that talent or genius was solely an inherited trait. As he phrased it, nature was more important than nurture. The idea that all human traits are genetically determined is known as **hereditarianism.** Having arrived at this conclusion, Galton then reasoned that because human talents and abilities were hereditary, encouraging the differential reproduction of the gifted and discouraging procreation among the inferior members of society would enhance the talents of future generations. The idea of differential reproduction among different classes did not originate with Galton. Some 2500 years ago, Greek philosophers speculated on the use of genetics for social planning, and Plato's dialogues explore the same idea. The use of controlled reproduction to try to improve the human species is known as **eugenics,** a term first used by Galton in 1883. Encouraging the reproduction of gifted and talented individuals is an example of positive eugenics; discouraging or preventing reproduction of the unfit is known as negative eugenics.

Hereditarianism
The idea that human traits are determined solely by the genotype, ignoring the contribution of the environment.

Eugenics
The attempt to improve the human species by selective breeding.

Elaborating on the role of natural selection in human society, writers following Galton's lead argued that although civilization was a product of natural selection, it had progressed to the point that it was inhibiting further evolutionary improvements in mankind. In a book entitled *The Survival of the Unfittest*, C. W. Armstrong summarizes this view as follows: "By public and private charity, we encourage the weak, the defective, the shiftless and the unsuccessful not only to continue in their present condition, but to breed freely." He continues: "In other words, we deliberately hand over the advantages Nature intended for the successful to those whose lives, from one course or another, are a failure; enabling them to live and reproduce with greater ease than the healthy, the intelligent and the strong." He concludes: "No surer way could possibly be devised for bringing about the survival of the unfittest than taking from the fit and giving to the unfit." To reinstate the effects of natural selection in our society, selective breeding of humans would be required, and steps would have to be taken to discourage or to prevent reproduction among the unfit.

FIGURE 20.1 Sir Francis Galton, cousin of Charles Darwin and founder of the eugenics movement.

Galton's views, circulated in a series of magazine articles, resulted in a book, *Natural Inheritance*, first published in 1889. In collecting information for this book, Galton offered prizes for the most complete family histories that included such traits as artistic ability, temperament, and reputation. His views were influential and became widely accepted in many circles of society. The eugenics movement, based as it was on Darwinian principles, spread rapidly. A National Eugenics Laboratory was established in London, and the Eugenics Education Society was formed. Membership included many prominent members of the clergy, physicians, and scientists. The society published the *Eugenics Review,* a journal to educate the public about the benefits of eugenics. The confirmation of Mendel's work on inheritance in 1900 was, to an extent, the final piece of the puzzle. Mendelian inheritance provided a scientific basis for eugenics. The principles of Mendelian inheritance were viewed as a simple set of rules by which humans could produce a superior form of the species in the foreseeable future.

In America, eugenics and genetics were first united in the American Breeder's Association, founded in 1903 by biology professors and stock breeders to encourage Mendelian research. One of the committees set up by the association was the Eugenics Committee, which was to "investigate and report on heredity in the human race" and to "emphasize the value of superior blood and the menace of inferior blood."

Charles Davenport, chairman of the Eugenics Committee, was instrumental in setting up the Eugenics Record Office at Cold Springs Harbor, New York, which later served as the center of the eugenics movement in America. The office trained eugenic field workers, who in turn went to prisons, asylums, and reformatories to construct pedigrees of the inmates. This information was sent back to the office, where it formed the basis of a repository of information on human heredity. Unfortunately, Davenport and many others in the eugenics movement thought that most human traits were controlled by single genes, just as Mendel had reported for traits such as seed color and height in pea plants. As a result, human traits such as insanity, alcoholism, morality, tuberculosis, religious preference, love of the sea, and quickness of temper were all considered to be inherited as single Mendelian traits, most of which were recessive (Figure 20.2). The emphasis in all these studies was on the gene; little thought was given to the effects of environment.

The attitude that all aspects of human life were rooted in heredity led many experts in public health and social reform to conclude that those who were poor, criminal, mentally ill, retarded, or simply unsuccessful all carried genes that predetermined their conditions. This thinking confirmed the

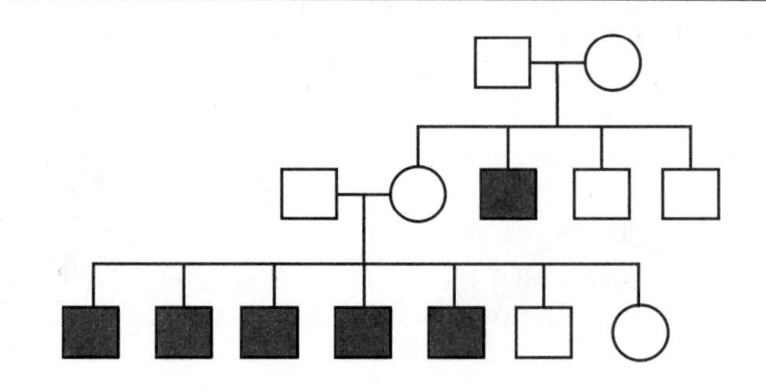

FIGURE 20.2 Pedigree showing inheritance of nomadism, or love of wandering. The wandering instinct was regarded as a fundamental human trait. The pedigree indicates that nomadism is a sex-linked recessive trait.

feelings of the entrenched upper and middle classes that the poor and the immigrants flooding the United States in the first decade of this century were not only socially but also biologically inferior. Madison Grant, a leading eugenicist, expresses this view in his book *The Passing of the Great Race:*

> These new immigrants were no longer exclusively members of the Nordic race as were the earlier ones who came of their own impulse to improve their social conditions. The transportation lines advertised America as a land flowing with milk and honey and the European governments took the opportunity to unload upon careless, wealthy and hospitable America the sweepings of their jails and asylums. The result was that the new immigration, while it still included many strong elements from the north of Europe, contained a large and increasing number of the weak, the broken and the mentally crippled of all races drawn from the lowest stratum of the Mediterranean basin and the Balkans, together with hordes of the wretched, submerged populations of the Polish Ghettos. Our jails, insane asylums and almshouses are filled with this human flotsam and the whole tone of American life, social, moral and political has been lowered and vulgarized by them.

To Grant, native Americans were those with family origins in Northwest Europe, those of what he called Nordic blood.

Having accepted that most human social problems were rooted in genetics, it was natural to think that eugenics would provide a means of eliminating social problems along with the socially unfit by raising the biologic quality of future generations through selective breeding (Figure 20.3). To allow this to occur without competition from the prolifically

FIGURE 20.3 In the early part of this century, eugenics exhibits were a common feature at county fairs, state fairs, and similar events. Such exhibits provided education about genetics and eugenics and often included contests to find the eugenically perfect family.

breeding lower classes, inferior groups had to be identified, labeled, and controlled. This could be best accomplished by regulating immigration and by controlling the reproduction of those defectives already in the United States. Thus the eugenics movement, from about 1905 through 1933, became a powerful and influential force in the development of laws and legal decisions formulating social policy. This influence reached a peak in the passage of federal laws restricting immigration from Southern and Eastern Europe and the implementation of forced sterilization as a eugenic policy.

Marriage Laws

The first law in the United States regulating marriage for the purposes of preventing reproduction among those regarded as defective was passed in Connecticut in 1896. This law forbade marriage or extramarital relations between a man and a woman either of whom was epileptic, imbecilic, or feebleminded, when the woman was under 45 years of age. By 1905, five other states had similar laws. Typically, marriage by epileptics, criminals, alcoholics, the insane, and the feebleminded was made a violation of state law. Many of these laws were promoted and supported by local groups such as physicians, church organizations, and charities. By the time the eugenics movement was fading away in the mid-1930s, 41 (out of 48) states had laws preventing marriage by several groups defined as socially or genetically defective or both, usually including the feebleminded and insane, alcoholics, and epileptics. Few of these laws were rigorously enforced, but they served to provide publicity about eugenics and to educate the public about the perceived dangers of unregulated reproduction.

Sterilization

In the United States, eugenics became closely identified with laws to sterilize those labeled as socially defective. Within the eugenics movement there was a split over the use of sterilization as a eugenic measure. Many opposed sterilization on moral grounds, but expert testimony by others in the eugenics movement was instrumental in passage of sterilization laws in many states. A committee of the American Breeder's Association, known as the Committee to Study and Report on the Best Practical Method of Cutting Off the Defective Germplasm in the American Population, concluded that 10% of the American population was socially inadequate. To counteract the prolific breeding of these genetically defective individuals, the committee proposed that they should be segregated from the rest of the American gene pool by confinement in institutions or by sterilization. In the opinion of the committee, this would remove most of the undesirable genes from the gene pool in two generations.

The first sterilization law was passed in 1907 in Indiana, where the sterilization of criminal males by vasectomy was pioneered. In the next 10 years similar laws were passed in 15 other states. Under these laws, men were to be sterilized by vasectomy and women by tubal ligation. Most of these laws were punitive, providing for sterilization of those labeled as genetic defectives as well as those convicted of certain crimes. In Missouri, for example, one proposed law called for sterilization of those convicted of rape, highway robbery, stealing chickens, and automobile theft.

These early laws were criticized for not allowing due process and protection of constitutional rights. Through the 1920s and 1930s, revisions in sterilization laws removed most of these objections. However, in spite of these reforms, there was widespread feeling that in matters of genetic defects, the rights of the state far outweighed the rights of the individual. A prominent New York physician, W. J. Robinson, expressed these views in his book *Eugenics, Marriage and Birth Control: Practical Eugenics,* published in 1922. He writes: "There are certain conditions that cannot be influenced by the environment, and it is in these conditions that the state has a right to step in and prevent propagation and the pollution and corruption of the race." On the rights of those judged to be genetically defective: "No casuistry, no sophistry can offer any argument against the sterilization of such defectives. It is the acme of stupidity, in my opinion to talk in such cases of individual liberty, of the rights of the individual. Such individuals have no rights. They have no right in the first instance to be born, but having been born, they have no right to propagate their kind."

The *Buck vs. Bell* decision (see Concepts & Controversies, page 548) by the U.S. Supreme Court in 1927 established the legal principle that it is within the power of the state to use forced sterilization for eugenic reasons. In rendering the decision of the Court, Justice Oliver Wendell Holmes emphasized the right of the state to employ sterilization:

> We have seen more than once that the public welfare may call upon the best citizens for their lives. It would be strange if it could not call upon those who already sap the strength of the state for these lesser sacrifices, often not felt to be such by those concerned, in order to prevent our being swamped with incompetence. It is better for all the world, if instead of waiting to execute degenerate offspring for crime, or to let them starve for their imbecility, society can prevent those who are manifestly unfit from continuing their kind. The principle that sustains compulsory vaccination is broad enough to cover cutting the fallopian tubes.

What such reasoning fails to take into account is that sterilization is an ineffective eugenic measure. In the case of genetic disorders exhibited only in homozygous recessive individuals, sterilization will have almost no effect on the frequency of the disease. This is because most copies of the gene are carried by heterozygotes (see Figure 17.5) who are phenotypically normal. For example, if a recessive disorder is present in the population at a frequency of 1 in 40,000 and homozygotes are prevented from reproducing for 100 generations (about 2500 years), the frequency of the disorder in the population at the end of that time would be 1 in 41,660. In the case of many dominant conditions, such as Huntington disease, the onset of the disorder does not occur until middle age, well after the onset of reproductive age. Sterilization laws are still on the books in about 20 states, and although they are sporadically enforced, they remain as testimony to the effectiveness of the eugenics movement.

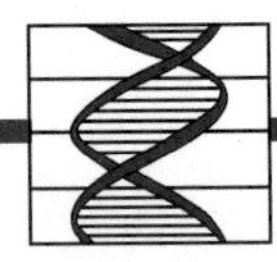

Concepts & Controversies

The *Buck vs. Bell* Decision

The U.S. Supreme Court decision of 1927 in the case of Carrie Bell confirmed the right of the state to enforce mandatory sterilization laws. Since these laws are still on the books in over a dozen states and since sterilizations have been carried out in the recent past, the issues raised in this case are of concern.

Carrie Bell was born to a mother judged to be feebleminded and was adopted by a family in Charlottesville, Virginia, at the age of 4 years. She attended school through the sixth grade and then took a job cleaning house. After becoming pregnant, she was committed, in January 1924, to the State Colony for Epileptics and the Feebleminded, where she gave birth. In the fall of 1924, Carrie was selected to be the first individual sterilized under a new Virginia law. At the trial in the Circuit Court of Amherst County, Harry Laughlin, a well-known eugenicist, concluded that Carrie's mother had a mental age of about 7 years, that Carrie had a mental age of 9 years, and that her baby was probably feebleminded. As to the genetics of the situation, her ancestry could not be traced with any accuracy because she belonged to "the shiftless, ignorant and worthless class of anti-social whites of the South." Nonetheless, he concluded that her mental state and immoral behavior were inherited. Another witness testified that feeblemindedness was a simple Mendelian recessive trait. The court ordered her sterilized, and the Virginia Supreme Court upheld the order. The U.S. Supreme Court also upheld the order, with Justice Oliver Wendell Holmes delivering the majority opinion (only one justice dissented).

In considering the case, it has been argued that the decision was made on the basis of the available scientific knowledge and that even if such a mental condition is inherited in a polygenic fashion, preventing reproduction of all such individuals would reduce the incidence of this trait in subsequent generations. Critics of this decision have pointed out that almost 30 years after the dissemination of Mendelian principles, the criteria necessary for simple recessive inheritance were well known and only loosely applied in this case.

Immigration

The racism that became associated with the eugenics movement was most evident in attempts to restrict immigration. The rationale for immigration quotas was based on the argument that heredity alone was responsible for all aspects of the human personality and human behavior and that environmental effects were negligible. Thus immigrant groups with prolific reproductive patterns and high frequency of deleterious mutations were likely to lower the level of American life. Since these traits could not be changed by the environment, the only means of protecting the American way of life was to exclude such immigrants from this country. This thinking was not based on phenotypes such as skin color and instead equated race with ethnic origins and social and economic classes. By virtue of natural selection, the elite classes, with origins in Northwest Europe, were regarded as the most biologically fit. This view derived from the assumption that social standing and economic wealth was derived solely from heredity and not the environment. The high levels of unemployment, poverty, and crime among recent immigrants was taken as evidence of genetic inferiority. To maintain this status quo, it was thought that the less fit and inferior should not be allowed to mix with the American race because it would inevitably lead to the degeneration of the race. Madison Grant and others defined the American race as those derived from English, Scotch, Dutch, and German ancestors. (Although the Irish are from this region, they were excluded because they were Catholic.)

In response to an outcry from upper- and middle-class constituents, Congress established a joint committee in 1907 to consider the effects of immigration on the quality of American life and society. The report of the committee, issued in 1911, confirmed the impression that new immigrants were less biologically fit and led to the imposition of a literacy test for all new immigrants.

After World War I, new sentiment arose for restrictive immigration laws, and new evidence was available to support the need for such legislation. IQ tests administered to all who entered the Army in World War I showed that recent immigrants, mainly from Southern and Eastern Europe, had IQ scores significantly lower than those of older immigrant groups. Even though the data showed that IQ scores increased in a linear fashion with years of residence in the United States and an increasing ability to use the English language, the scores were interpreted to mean that recent immigrants were of inferior racial and genetic quality.

In addition, surveys of those confined to institutions for the insane and feebleminded were reported in a manner designed to show that a disproportionate number were from Eastern and Southern Europe. Expert testimony by leading authorities on eugenics indicated that many recent immigrants also carried recessive deleterious traits that would become evident in their homozygous recessive children and contaminate the American gene pool. Harry H. Laughlin, a prominent eugenicist who was appointed as the expert eugenics agent of the U.S. House of Representatives Committee on Immigration and Naturalization, summarized this view in testimony before the committee in 1922: "Making all logical allowances for environmental conditions, which may be unfavorable to the immigrants, the recent immigrants as a whole, present a higher percentage of inborn socially inadequate qualities than do the older immigrants."

Based on what were presented as established scientific principles of biology and genetics, Congress enacted the Immigration Act of 1924, which was signed by President Calvin Coolidge. The law required that immigration from European countries be restricted to a number equal to 2% of the U.S. residents counted in the census of 1890 who had been born in that country. Since large-scale immigration from Southern and Eastern Europe began later than 1890, the law in effect closed the door to America for millions and marked the end of the idea that America was a land of opportunity open to all. In spite of the racist overtones to this aspect of the eugenics movement, not all in the eugenics movement agreed with the validity of the arguments presented at the immigration hearings.

Forty-one years later, in 1965, the Cellar Act, sponsored by Rep. Emanuel Cellar of New York, reformed the immigration laws. Currently 20,000 immigrants per year from one country are allowed entry to the United States, with immigrants admitted in the order in which they applied.

With this background in mind, let us now consider some contemporary problems relating to genetics and the law.

NEW GENETIC TECHNOLOGY AND THE LAW

Over the last 10 years, developments such as genetic screening, *in vitro* fertilization, and embryo transfer have been rapidly transferred from the research laboratory to medical practice and private industry. In many cases these developments have replaced or challenged traditional community standards with unconventional methods of reproduction or selective abortion of fetuses carrying genetic disorders. A subsequent surge in legal decisions and legislation has arisen to restrict and to regulate the use of this technology and to provide legal safeguards and guidelines.

Law and Genetic Disorders

Beginning in the 1970s, federal laws dealing with genetics were enacted. The first was the National Sickle Cell Anemia Control Act of 1972. The intent of the law was to detect cases of sickle cell anemia, but the test also detected heterozygous carriers. In the same year, the National Cooley's Anemia Control Act provided screening for cases of beta thalassemia.

These laws represented a significant departure for federal genetic programs, moving from research into direct support for genetic services. The laws were structured to avoid mandatory or coercive screening, and they attempted to reduce discrimination against carriers (although critics charged that the laws were ineffective in this regard). In 1976, Congress reorganized, broadened, and renamed the legislation, calling it the National Sickle Cell Anemia, Cooley's Anemia, Tay-Sachs and Genetic Diseases Act. (The name was later shortened to the National Genetic Diseases Act.)

This law embodied several elements important to screening programs now under consideration, including confidentiality of records and voluntary participation. It also stressed that genetic counseling should be available for all screening participants. The consolidated law was expanded to include funding for basic and applied research, training, counseling, and education programs.

In 1981, federal support for genetic screening and genetic services was dramatically reduced, and the National Genetic Diseases Act was replaced by the Maternal and Child Health Block Grant, which left decisions about fund allocation and genetic services to individual states. In addition, the block grant program halted the use of federal funds for genetic services, training, and public education.

A presidential body appointed in 1983 (the President's Commission for the Study of Ethical Problems in Medicine and Biomedical and Behavioral Research) examined federally supported genetic services and the ethical, social, and legal implications of genetic screening. The commission expressed concern about the lack of federal funding for genetic services and identified many of the issues currently being raised about cystic fibrosis screening. Unfortunately, little progress has been made over the ensuing decade in resolving the concerns raised by advances in genetic research.

The most recent legislation dealing with genetics, as we have seen, is the Human Genome Project, a 10- to 15-year initiative (estimated to cost $3 billion) to map and sequence all the genes in the human genome. One of the assumptions of this program is that an understanding of genetic disorders at the molecular level will lead to advances in diagnosis and therapy for the more than 4000 identified genetic conditions. The Human Genome Project also includes funds to examine the ethical, legal, and social issues related to advances in genetics generated by the program.

Several states have enacted laws that specifically deal with genetic disorders. These include Arizona, California, Illinois, Iowa, Kentucky, Louisiana, Maryland, Missouri, New Jersey, New York, Oregon, Virginia, and Wisconsin. In general, these laws deal with genetic testing, genetic screening, counseling, and discrimination in employment against individuals with genetic disorders. The California Hereditary Disorders Act of 1990 is one of these laws. It deals with access to genetic services, confidentiality of genetic information, discrimination against affected individuals and carriers, the voluntary nature of screening programs, the reproductive rights of those at risk of passing a genetic disorder to their offspring, and professional and public education programs about genetics. Although it has been widely proposed as a model for other states, similar laws have been enacted in only a small number of states.

As outlined above, the relationship between genetic technology and social standards as embodied in the law is undergoing rapid change, and many separate issues are involved. Some specific issues we have not yet considered include wrongful birth, wrongful life, elective abortion, embryo freezing, cloning and transfer, organ selling, and protection of newborns. In the following sections we will examine a cross section of these issues as they relate to genetics and the law.

Wrongful Birth and Wrongful Life

The development of prenatal genetic testing has made it possible for couples at risk for having a child with a genetic disease to know beforehand whether a pregnancy involves an affected fetus. This technology has brought with it an increased responsibility on the part of physicians to inform couples who may be at risk that prenatal testing is available. Many lawsuits have been brought against physicians who did not provide such information or did not properly perform prenatal tests.

In 1980 one such suit was brought in California by parents of a Tay-Sachs child. Both parents were descended from East European Jews and knew they were at risk for having a child affected by Tay-Sachs disease. They underwent tests for this disease and, based on the results of this test, were informed that they were not at risk for having such a child. The couple subsequently conceived and gave birth to a child who soon developed the sluggish behavior, convulsions, and blindness associated with Tay-Sachs disease. In the court case the infant (through her parents) sued the laboratory that had performed the test, claiming that she had a right to be born healthy or not to be born at all. In effect, the suit charged that hers was a wrongful life. The court agreed and found that the child (again through her parents) was entitled to damages. The legal concept of wrongful life has been upheld in only three states (California, New Jersey and Washington) courts and denied in others. At least eight states prohibit malpractice suits based on a claim of wrongful life.

A related legal concept, wrongful birth, is a term used to denote legal action on the part of the parents themselves for the birth of a defective child that results from the failure to diagnose a genetic defect in the fetus or another member of the family in a timely and accurate manner. These claims can result in special damages awarded for medical expenses and other costs associated with an affected child. Again, not all courts have recognized such actions, and at least three states, Minnesota, Utah, and South Dakota, have enacted laws that limit lawsuits for wrongful birth.

The Sale of Human Tissue

Although the voluntary donation of blood is widely accepted and practiced in the United States, it is also legal to buy and to sell blood and blood products such as plasma and clotting factors. In fact, the annual worldwide sales of blood plasma alone are estimated to be in the range of $2 billion. In the United States it is also legal to sell human sperm. At least one sperm bank is a public, for-profit corporation with its stock traded on the American Stock Exchange. However, the overall situation involving the sale of human tissues and organs is somewhat confused by the fact that while it is legal to sell some tissues such as blood and sperm, it is illegal to sell other tissues and organs such as bone marrow, corneas, hearts, kidneys, and livers. The 1984 National Organ Transplant Act prohibits the buying and selling of certain organs and provides for fines of up to $50,000 or 5 years in prison or both for violators convicted under this law. Under present law organ donors can specify the recipient but cannot be compensated for the donation.

However, the law of supply and demand for organs, the increasing use of live organ donors rather than cadavers, and the development of a new generation of drugs that suppress transplant rejection by the recipient's immune system have led to calls for the reexamination of this issue and for the institution of a regulated, commercial trade in human organs, possibly using the blood plasma business as a model for the development of this endeavor.

In other areas, recent court decisions are in the process of establishing the rights of donors in the legal business of selling human cell lines and products derived from them. While undergoing treatment for leukemia at UCLA Medical Center, a man named John Moore had his enlarged spleen removed in 1976. Using cells from this material, researchers at UCLA isolated a virus that appears to be associated with a rare form of leukemia. This tissue also gave rise to a permanently dividing cell line derived from Moore's T cells. This cell line produces a lymphokine (see Chapter 14 for a discussion of lymphokines) called GM-CRF that stimulates the growth of two other types of blood cells, granulocytes and macrophages. Activated granulocytes and macrophages are thought to be important in identifying and killing cancer cells. This cell line, called Mo (from the first two letters of Moore's name), and its lymphokine were patented by UCLA in 1984.

Arrangements were made with Genetics Institute of Cambridge, Massachusetts, and with Sandoz Pharmaceutical Corporation of East Hanover, New Jersey, to develop the cell line and to produce GM-CRF, with an estimated commercial value of $3 billion. Moore filed suit against both UCLA and Sandoz, claiming ownership of the cells taken from his spleen and asking to share in the profits derived from the sale of the cells or the products derived from them. In 1988 the California Court of Appeals ruled in favor of Moore, and the defendants appealed the case. In July 1990 the California Supreme Court ruled that patients do not have property rights over tissues or cells removed from their bodies that are used to develop drugs or medicines. This ruling applies only in California but is the first in the nation to establish a legal precedent over the commercial development and use of human tissue.

Although the National Organ Transplant Act of 1984 prevents the sale of organs such as a kidney, lung, or heart, current laws permit the sale of human tissues and cells but do not define whether donors have property rights to their cells and do not establish ownership of such materials. Appearing before a science and technology subcommittee of the U.S. House of Representatives, many biomedical scientists testified that patients sign consent forms that give hospitals the right to dispose of surgically removed organs and tissues and argued further that the cells are altered during research and cannot be claimed by the patient. Others argue that a patient should not be viewed as a product and that patients whose tissues are used as sources of commercial products are entitled to a share of the profits derived from the sale of his or her tissues. This case is not an isolated incident. Of 939 patents applied for by U.S. medical institutions over a 5-year period, 211 (22%) included human tissues. The decision in the Moore case will have a significant effect on these patent applications and the associated biotechnology industry and, in a broader sense, on the use and disposition of fetal tissues, frozen embryos, and the process of donating tissues and cells.

Patenting Life Forms

In a landmark 1980 ruling, the U.S. Supreme Court decided that Dr. Ananada Chakrabarty, then an employee of the General Electric Co., could patent a bacterium that digests crude oil. In its decision the Court said that the intent of Congress in establishing patent law was that patents should cover anything under the sun made by human hand. In the intervening years the Patent Office has issued over 200 patents for genetically engineered organisms, many of which are bacteria.

In spring of 1987 the Patent Office issued a related decision that has broad legal and ethical implications. This ruling allows the Patent Office to issue patents covering nonhuman, multicellular organisms, including animals. The ruling does not cover animals produced by natural breeding, only those produced by genetic engineering. For example, human growth hormone genes have been introduced into pig embryos in the hope that such

FIGURE 20.4 Chimeric hybrid produced by fusion of a sheep embryo with that of a goat.

pigs would grow faster. Some of the progeny have incorporated the human gene into their genome and have passed it on to their progeny. Such genetically altered pigs could be patented. In another project, a goat embryo was fused with a sheep embryo to produce a chimera called a "geep" (Figure 20.4). Other techniques, such as embryo splitting, are also covered by this ruling.

The use of techniques such as gene or organismal cloning, embryo fusion, and recombinant DNA methodology makes it possible to mix genes from plants, animals, humans, and bacteria and to transfer these to animal embryos to create new organisms designed for a specific purpose. The Patent Office ruling recognizes the existence of such technology and will allow such custom-designed animals to be patented. While humans who have undergone gene surgery or gene transfer are excluded from this ruling, it opens the way to patent the method of treating human genetic disorders by gene transfer.

In April 1988 the Patent Office issued a patent to Harvard University for a genetically engineered mouse carrying an oncogene. The patent covers not only the mouse but its offspring as well, since the spliced oncogene is a heritable trait. The presence of the oncogene in the cells of the mouse makes it susceptible to many forms of cancer, and these altered mice will be used to test carcinogenic effects of chemicals present in the workplace or the environment. The mice can also be used to test compounds that may confer resistance to the development of cancer.

By extension, the Patent Office ruling allows genes to be patented. In fact, application for a patent covering the cystic fibrosis gene (CFTR) has been filed. The rush to patent the results of genetic research has not been limited to entrepreneurs or corporations. The National Institutes of Health, one of the government agencies sponsoring the Human Genome Project, has attempted to patent all identified DNA sequences that are capable of being transcribed (potential genes) as they are discovered.

Decisions by the Patent Office have served to increase the controversy over legal and ethical issues in genetic engineering. Critics of the office fear it will allow large biotechnology companies to become the eventual owners of all livestock genomes, leading to a further reduction in the genetic diversity of the plants and animals that provide our food supply. Supporters feel that the actions of the Patent Office will provide new incentives to invest in U.S. agriculture and to help revolutionize the food industry.

DNA Markers and Genetic Testing

Genetic testing using DNA markers allows the diagnosis of adult-onset heritable disorders many years before symptoms appear. The ability to test healthy individuals to determine whether they are at risk of developing a serious or possibly fatal genetic disorder such as polycystic kidney disease, familial hypercholesteremia, or Huntington disease has raised a number of ethical and legal issues related to the use of presymptomatic genetic testing.

The International Huntington Association and the World Federation of Neurology have issued a joint set of ethical guidelines for use in predictive testing for Huntington disease, and the Canadian Collaborative Study of

Predictive Testing for Huntington Disease has examined many of the legal and ethical dilemmas that are an outgrowth of such predictive testing. Because such testing can have serious social and psychological repercussions for individuals and families, we will briefly consider some of the issues that need to be addressed before predictive testing for a range of late-onset disorders becomes widely available.

Two of the major issues associated with predictive genetic testing are informed consent and confidentiality. In keeping with the standards established for medical procedures, individuals inquiring about predictive testing should receive relevant, up-to-date information concerning all aspects of testing, consequences, and alternatives in order to make an informed, voluntary decision about the procedure. In many cases, however, predictive testing cannot be offered to an individual who may be at risk unless other family members or relatives are also tested. Legal issues with respect to this unique aspect of predictive testing have not been resolved. For example, can relatives or family members be forced to undergo testing in order to provide information about the genetic status of another person? Can a child be compelled by a parent to undergo testing?

Because predictive testing can involve relatives, disclosure of test results about at-risk individuals can have a serious impact on others in the family. To consider an extreme example, suppose one member of a pair of MZ twins requests testing for a late onset disease, but the other does not. Obviously, the results of this test will apply to both individuals. Can the clinic or agency administering the test require that both twins give consent before one twin is tested? As an extension, how far does confidentiality associated with test results extend? Should potential employers or insurance companies have access to test results without the consent of the tested individual? Can companies request that predictive testing be done without the knowledge of the employee using blood samples taken as part of routine physical examinations? Answers to these and other questions need to be developed by studying the impact of predictive testing on individuals and families and by developing guidelines for such testing to ensure that it is used in an appropriate manner.

As discussed in Chapter 8, forensic applications of DNA markers have been used in the United States since 1987. Since then, evidence provided by the technique has resulted in more than a dozen convictions and has been used in more than 1000 criminal cases nationwide. To take full advantage of this method, some states, including California, are starting archives of DNA fingerprints collected from convicted sex offenders. However, as the courts have gained more experience with the use of DNA typing as evidence, some controversy has developed over the use of the technique.

The ability to determine that two samples produce the same band pattern and that the pattern is rare enough to identify a single individual are central to the use of DNA fingerprinting. Unlike DNA typing used for diagnostic purposes, forensic applications must unambiguously determine an allele pattern from among millions of possible patterns, rather than just the two parental combinations. Because of technical problems inherent in the process of separating DNA fragments, the required degree of resolution is often difficult to obtain. In addition, a particular DNA fingerprint should be rare enough to absolutely identify a given individual. The frequency of the allele combinations in a fingerprint depend on the frequency with which the individual RFLP alleles are present in the population. At present, data on the population distributions of various RFLP alleles may not be reliable enough to calculate accurately the frequency of a particular DNA fingerprint. In addition, some populations may have very different frequencies for some alleles, resulting in significant changes in the fingerprint frequency for that population.

These flaws in the use of DNA typing have become apparent only after the method has achieved widespread use. Several agencies, including the National Academy of Sciences, have proposed standards for the forensic use of DNA fingerprints. These standards may include certification for those performing the test, licensing DNA typing laboratories, and standardizing the methodology. Installing such safeguards should ensure that genetic technology will serve the best interests of all individuals in the criminal justice system.

The Future: Genetics, Law, and Bioethics

Many of the benefits of genetic technology, especially genetic testing, raise serious issues that will affect all of us. Using genetic and cytogenetic markers, prenatal testing can identify fetuses with Down syndrome, cystic fibrosis, Tay-Sachs disease, and literally hundreds of other genetic disorders. If the therapeutic abortion of fetuses with these conditions can be justified, what about an individual who can lead an active and productive life but who will develop Huntington disease at age 50?

Genetic engineering has made it possible to produce large quantities of human growth hormone that is used to treat various forms of dwarfism. Now that the supply of this hormone is no longer a problem, should it be used only to treat dwarfism, or should treatment be extended to children who are shorter than average? What about the administration of hormones to produce very tall individuals who may excel at some sports such as volleyball or basketball? These are real issues that must be confronted and solved in a way that protects the interests of the children being treated yet takes advantage of the advances in biotechnology.

The ethical dilemmas involved with genetic technology can be illustrated by considering two examples that deal with the interaction of genetics and reproduction, sex selection and organ donation.

Although the sex of a fetus can be detected by amniocentesis, ultrasound, or chorionic villus sampling, methods of prefertilization sex selection are available. These methods depend on the separation of X-bearing from Y-bearing sperm followed by artificial insemination. Originally developed in university-associated research laboratories, these methods include separation in a centrifugal field (X sperm are heavier than Y sperm), separation in an electrical field, or allowing sperm to migrate over a column of albumin. Private companies now provide these services and advertise that the method is 75% to 80% effective in preselecting the sex of a child.

Many societies have strong traditions of sex preference, with males preferred 4:1 in some cases. In the United States a survey of college students indicated that 44% would like to make use of sex-selection techniques if they were widely available. From the available information, it appears that the general use of sex selection would result in an imbalance in the sex ratio, with an excess of males. Although it is often argued that the use of such methods will eventually balance the sex ratio, others claim that an excess of males might pose social risks that include increases in violence and crime.

In view of the uncertain outcome that will accompany the use of sex selection, do prospective parents have the right to use this method without regard to its social consequences? Or does the state have the right to control the use of sex selection to maintain a socially acceptable balance of males and females? India has recently enacted a law that outlaws prenatal sex determination for the purpose of aborting the socially less desired female fetuses. Can the state use sex selection as a form of population control, levying taxes on parents who have too many children of one sex, in effect making this reproductive choice part of its social policy?

Organ Donors

Leukemia is a disease that strikes approximately 27,000 Americans every year. The American Cancer Society estimates that just over 2000 of these cases will involve children and teenagers. While chemotherapy is the most effective treatment for many forms of leukemia, bone marrow transplantation is often recommended. Recently the case of a teenager with leukemia has served to focus attention on the difficulties surrounding advances in transplant technology. In this case a bone marrow transplant was the only hope for the girl's survival, but no suitable donors were available. Because siblings have a one in four chance of providing the appropriate HLA tissue type (see Chapter 14), the parents of the patient decided to conceive a child with the hope that there would be a tissue match. Because the mother was in her early 40s, there was a significant risk of an aneuploid child. The infant was born in April 1990, and her tissue matched that of her sister. At the age of 14 months, the infant served as a marrow donor. The transplant was successful, and her older sister now has a 95 percent chance of long term survival. This is not an isolated case; at least thirty other cases of children being conceived to serve as marrow donors for siblings are known.

Since the infant donors are below the age of reason, do they have rights to the use of their body tissues? Are the parents able to act in the best interests of the infant donor in making the decision for a transplant? What if the decision is to use a kidney from the infant instead of a replaceable tissue such as bone marrow? Does the state have any right to intervene on behalf of the donor or the recipient?

These issues illustrate the problems that we must meet and solve in the near future. To do this effectively, we must first have an informed public, knowledgeable about the basic facts of genetics. Second, these informed individuals must be active participants in the discussion and formulation of new policies and laws related to genetics.

EPILOGUE

In medieval times, kings ruled by divine right, and the lot of the populace was thought to be a manifestation of God's will. As monarchies were replaced by other forms of government, Malthusian doctrine replaced divine will, and the downtrodden and defective were simply the product of the natural and inevitable mathematical relationship between the logarithmic growth of populations and the slower growth of resources. As the errors of Malthusian thinking became evident in the 19th century, hereditarianists came to the fore, convinced that certain people were predestined to be failures and defectives, dominated, as it were, by their genotypes. In this country the interface of genetics, law, and society early in this century produced policies of enforced sterilization, restrictive immigration, and institutionalization of those found to be genetic defectives. As the century closes, we will again have to grapple with difficult choices about the applications, legality, and morality of genetic knowledge and technology. The difficulty of our situation can be summarized in the words of an eminent geneticist, the late Theodosius Dobzhansky: "If we enable the weak and the deformed to live and promulgate their kind, we face the prospect of a genetic twilight. But if we let them die or suffer when we can help them, we face the certainty of a moral twilight."

FOR FURTHER READING

Armstrong, C. W. 1927. *The Survival of the Unfittest.* London: Daniel.

Bajema, C. J. (Ed.). 1976. *Benchmark Papers in Genetics.* Vol. 5, *Eugenics Then and Now.* Stroudsburg, Penn.: Dowden, Hutchinson & Ross.

Cowan, R. S. 1985. *Sir Francis Galton and the Study of Heredity in the Nineteenth Century.* New York: Garland.

Cravens, H. 1978. *The Triumph of Evolution: American Scientists and the Heredity-Environment Controversy 1900–1941.* Philadelphia: Johns Hopkins University Press.

Grant, M. 1921. *The Passing of the Great Race.* 4th ed., rev. New York: Scribner.

Haller, M. H. 1984. *Eugenics: Hereditarian Attitudes in American Thought.* New Brunswick, N.J.: Rutgers University Press.

Kevles, D. J. 1985. *In the Name of Eugenics: Genetics and the Use of Human Heredity.* New York: Knopf.

Ludmerer, K. M. 1972. *Genetics and American Society: A Historical Appraisal.* Baltimore: Johns Hopkins University Press.

Pace, N. 1982. *The Excess Male.* Norfolk/Virginia Beach, Va.: Donning.

Robinson, W. J. 1922. *Eugenics, Marriage and Birth Control: Practical Eugenics.* New York: Critic and Guide Society.

Warren, M. A. 1985. *Gendercide: The Implications of Sex Selection.* Ottowa: Rowsman, Allanheld.

APPENDIX A

PROBABILITY

Mendel's use of mathematics to analyze the results of his experiments is frequently overlooked as an important contribution to biology, when in fact, his application of mathematical reasoning to the analysis of data helped transform an observational and descriptive science into a quantitative and experimental one. At the time Mendel carried out his experiments, statistics and statistical methods were not highly developed. What Mendel did in analyzing the results of his crosses was to convert the numbers of individuals with particular genotypes or phenotypes into ratios. From these ratios, he was able to deduce the mechanisms of inheritance.

As we now know, the ratios of Mendel are the result of random segregation and assortment of genes into gametes during meiosis and their union at fertilization in random combinations. This randomness provides an element of chance in the outcome, that prevents us from making exact predictions. In counting pea seeds in the F_2, we may expect ¾ of the seeds to be yellow, but we cannot be absolutely certain that the first seed in an unopened pod will be yellow. The rules of probability can however, help us guess how often such an event will take place.

Most people have an innate sense of probability that seems part of common sense. For example, almost everyone would agree with the idea that a January snowfall is more probable in Minneapolis than in Miami. Other aspects of probability also seem obvious. When a coin is flipped, the probable outcomes are heads or tails. In the birth of a child, we expect the outcome to be a boy or a girl.

Unfortunately, the use of intuition alone in matters of probability is not always reliable. For example, what would you say is the probability that in a crowd of 20 people, 2 individuals share the same birth date? Considering that there are 365 days in the year (excluding leap year), intuition may say that it is not very likely. In fact, in a group of 20, there is almost an even chance that 2 people will share the same birthday. The probability of a shared birthday for groups of various sizes is as follows: for a group of 23, the probability is 51%; for a group of 30, it is 71%; for 40, it is 89%; and for 50, there is a 97% chance that two people will share the same birthday. We will not explore the mathematical reasoning behind this probability, but it is based on the fact that if one person can have any of the 365 days for his or her birthday, the second person can have any of the remaining 364 days, the third person may have any of the remaining 363 days, and so on.

From the example above, it should be clear that in order to be useful in genetics and in science, this intuitive sense of probability needs to be expressed in more quantitative terms. The use of a quantitative approach to probability allows us to assign a numerical value to the probability that a given event will occur, and prevents us from leaping to conclusions about the possible outcome of genetic crosses. In quantifying probability, let us begin at the limits. For example, if an event is certain to occur, it has a probability of 1; if the event is certain not to occur, then the probability is 0. In genetics as in most other areas, we usually deal with events that are a mixture of degrees of certainty. While it is certain that we will all die (a probability of 1), when we die is less certain and therefore must be assigned a probability somewhere between 0 and 1. Insurance companies spend a great deal of time and effort in attempting to determine such probabilities, although we may prefer not to think about them.

In general terms, we can express the probability (p) of an event as the proportion of times that such an event occurs (r) out of the number of times that the event can occur (n):

$$p = r/n$$

In other words, if an event occurs r times in n trials, the probability that the event will take place is r/n. This probability is somewhere between the limits of 0 and 1. If we toss a coin, it may land with heads up or tails up. The probability that it will land with heads up is:

$$p = r/n = 1/2$$

Likewise, the probability of a child being a boy or a girl is 1/2. Other events have different probabilities. In a pair of dice, each die has six faces. When a die is thrown, the probability of any of the faces being up is:

$$p = r/n = 1/6$$

In a deck of 52 cards, the probability of drawing any given card (the ace of spades for example) is:

$$p = r/n = 1/52$$

In roulette, the wheel contains the numbers 1-36 plus 0 and 00. The probability of the ball landing on any number is therefore:

$$p = r/n = 1/38$$

In a monohybrid cross, the probability that an offspring of the self-fertilized F_1 pea plant will have a dominant phenotype is:

$$p = 3/4$$

In considering probability, we must not only consider the probability of one type of outcome, but also the probability of other outcomes. If an event has a probability of p, the probability of an alternative outcome is $q = 1 - p$. In other words, the sum of the probability of p and q equals 1. In the examples above, the probability of drawing an ace of spades is 1/52; the probability of drawing another card is 51/52, which when added to 1/52 equals 1. In the monohybrid cross, the probability of a dominant phenotype is 3/4, and the probability of a recessive phenotype is 1/4. Since we are certain that the F_2 will have either a dominant or recessive phenotype, adding the probabilities of both phenotypes $3/4 + 1/4 = 1$.

Rules for Combining Probabilities

Two rules of probability are useful in the analysis of genetics problems. The first is called the product rule, and is used when we wish to calculate the probability of two or more independent events occurring at the same time. The second is called the sum rule, and is used when two or more events are mutually exclusive, or are alternative events.

In the product rule, we are asking the probability of event A *and* event B occurring together. This rule can be summarized as follows: the probability that independent events will occur together is the product of their independent probabilities. As background, let us recall, that when a coin is tossed, the probability that it will be heads is 1/2, and the chance that it will be tails is 1/2. If we toss a coin four times, and it turns up heads each time, the probability that it will turn up tails on the fifth try is still 1/2. In other words, chance has no memory. The probability that two heterozygotes will have a child with cystic fibrosis is 1/4. This does not mean that if their first child has cystic fibrosis, they can be assured of having three unaffected children. It means that for each child, there is a 1 in 4 chance that it will have cystic fibrosis, no matter whether they have 1 child or 20 children. If they have four unaffected children, the chance that their fifth child will have cystic fibrosis is still 1/4.

If on the other hand, we want to ask what is the probability that we can toss a coin four times and get heads each time, the probability is $1/2 \times 1/2 \times 1/2 \times 1/2 = (1/2)^4 = 1/16$. Similarly, the probability that heterozygous parents will have four children affected with cystic fibrosis is $1/4 \times 1/4 \times 1/4 \times 1/4 = 1/256$.

In using the sum rule, we are asking how often one *or* the other of two mutually exclusive events can occur. For example, in rolling a die, what is the probability of a three or a five coming up? Since on a single throw, only one number can come up, it is impossible to get both numbers on a single throw. The probability of a three is 1/6, and the probability of a five is 1/6. If we want to know what is the probability of either a three or a five coming up on a single throw, we add the individual probabilities:

$$1/6 + 1/6 = 2/6 = 1/3$$

In considering the possible genotypic combinations in children of parents heterozygous for cystic fibrosis, what is the probability that a child will have either one or two copies of the dominant allele? The probability of being homozygous dominant (two copies) is 1/4, and the probability of being heterozygous (one copy) is 1/2. Therefore, the probability of being either heterozygous or homozygous dominant is $1/2 + 1/4 = 3/4$. This is in fact, the proportion of individuals with the dominant phenotype seen in the F_2 of a monohybrid cross.

In applying probability to the analysis of genetic problems, first determine whether you want to know the probabilities of event A *and* event B, or the probability of event A *or* event B. If you want A and B, use the product rule and multiply the probabilities of A and B. If you want the probability of A or B, use the sum rule, and add the probability of event A to the probability of event B. For example, the frequency of albinism is about 1/10,000, and the frequency of cystic fibrosis is about 1/2000. If we want to know the probability of having both albinism and cystic fibrosis, we multiply the probabilities:

$$\frac{1}{10{,}000} \times \frac{1}{2000} = \frac{1}{20{,}000{,}000}$$

If we want to know the probability of having either albinism or cystic fibrosis, we add the probabilities:

$$1/10{,}000 + 1/2000 = 6/10{,}000 = 1/1666$$

As you can see, the probabilities are very different and reflect whether or not we are asking that both events occur, or that one or the other event will occur.

APPENDIX B

ANSWERS TO QUESTIONS AND PROBLEMS

Chapter 2 Answers

1. (a) Chemical and physical cell barrier; controls flow of molecules
   (b) Generation of metabolic energy sources
   (c) Maintenance and allocation of genetic material
   (d) Protein synthesis

2. (a) Cytological structure identified as the gene repository
   (b) The complex of DNA, RNA, histones and nonhistone proteins that make up the chromosomes.
   (c) One of the two side-by-side constituents of a replicated chromosome, connected to its sister through the undivided centromere

3. (a) Chemical treatment of chromosomes resulting in unique banding patterns
   (b) Q banding with quinacrine and G banding with giemsa

4. 21, chromosome number; q, long arm; 1, region; 3, band

5. G_1 or a subset of G_1 known as G_0.

6. Prophase: chromosome condensation, spindle formation, centriole migration, nucleolar disintegration, nuclear membrane dissolution; metaphase: alignment of chromosomes on the equatorial plate, attachment of centromeres to spindles; anaphase: centromere division, daughter chromosomes migrate to opposite cell poles; telophase: cytoplasmic division, reformation of nuclear membrane and nucleoli, disintegration of the spindle apparatus

7. Both daughter cells have the normal diploid complement of all chromosomes except for 7. One cell has three copies of 7 and the other cell has 1 copy of 7.

8. Anaphase, telophase, and G_1 of interphase

9. Faithful chromosome replication during S phase; independent alignment of chromosomes on the equatorial plate during metaphase; centromere division and chromosome migration during anaphase

10. Meiosis II, the division responsible for separation of sister chromatids. Meiosis I, wherein homologs segregate, would still be necessary.

11. 2 chromosomes, 4 chromatids, and 2 centromeres should be present.

12. Chiasma (chiasmata)

13. Meiotic anaphase I: no centromere division, chromosomes consisting of 2 sister chromatids are the migrating species; meiotic anaphase II: centromere division, migrating species are the separating sister chromatids. Meiotic anaphase II more closely resem-

bles mitotic anaphase by the two criteria cited above.

14. Males: 4 morphologically indistinguishable haploid spermatids (sperm when mature). All 4 sperm are capable of fertilizing an ovum. Females: 3 relatively small, infertile, haploid polar bodies are produced. The relatively large and fertile secondary oocyte will not complete meiosis until after fertilization, and is therefore only arguably regarded as haploid. The size difference between the 4 products is the result of asymmetric cytokinesis.

15. Meiosis begins before birth of the parent and is completed shortly after fertilization. The time taken was therefore approximate. Shortest time: (July 1, 1994—Jan. 1, 1950) 0.95 = 42.3 years; longest time: (July 2, 1994—April 1, 1949) 1.05 = 45.4 years.

Chapter 3 Answers

1. *AB* – 50% and *Ab* – 50%

2. The genotypic ratio is 1 *AABb*:1 *AaBb*:1 *AAbb*:1 *Aabb* The phenotypic ratio is 1:1

3. Four, in equal proportions. If the genes are *AaBb*, the gametes are: *AB*, *Ab*, *aB* and *ab*.

4. 1/3

5. (a) Both are 3:1
   (b) 9:3:3:1
   (c) Swollen is dominant to pinched, yellow is dominant to green.
   (d) let P = swollen, and p = pinched; C = yellow and c = green. Then: $P_1 = PPcc \times ppCC$ or $PPCC \times ppcc$
   $F_1 = PpCc$

6. (a) Each affected parent is homozygous recessive *(aa)*. Since the unaffected parent has no family history of sickle cell anemia, each is probably homozygous dominant *(AA)*. Each member of the couple planning to marry is therefore heterozygous *(Aa)* for sickle cell anemia. There is a 25% chance that any children this couple has will be affected by sickle cell anemia.
   (b) If the man is affected, he is homozygous recessive *(aa)*, and the unaffected woman with no family history is probably homozygous dominant *(AA)*. None of their children will be affected by sickle cell anemia, but all will be heterozygous *(Aa)* for the trait.

7. Let S = smooth, and s = wrinkled and Y = yellow and y = green. The parents are: $SSYY \times ssyy$. The F_1 offspring are: *SsYy*.

8. Using the symbols from the problem above, the parents are: $SsYy \times SsYy$. The genotypes of the F_1 are: *SsYy*, *Ssyy*, *ssYy* and *ssyy*.

9. (a) All F_1 plants will be long-stemmed.
   (b) Let S = long-stemmed and s = short-stemmed. The long-stemmed P_1 genotype is *SS*, the short-stemmed P_1 genotype is *ss*. The long-stemmed F_1 genotype is: *Ss*.
   (c) Approximately 225 long-stemmed and 75 short-stemmed.
   (d) The expected genotypic ratio is: 1 *SS*:2 *Ss*:1 *ss*

10. The genotypic ratio in the offspring is: 1 AABB:2 AABb:1 AAbb:2 AaBB:4 AaBb:2 Aabb:1 aaBB:2 aaBb:1 aabb.

11. The P_1 generation is: $FF \times ff$. The F_1 generation is: *Ff*. The mode of inheritance is incomplete dominance.

12. The use of a P_1 generation with true-breeding parents is necessary to ensure that the organism defined as an F_1 is truly heterozygous for the trait in question.

13. 1/4 straight, round, 1/4 straight, wrinkled, 1/4 gnarled, round, 1/4 gnarled, wrinkled.

14. The 1:2:1 phenotypic ratio is suggestive of either incomplete dominance or codominance. The apparent contribution of both red and yellow to the orange phenotype suggests codominance as the mode of inheritance.

15. (a) All plants will bear round seeds.
    (b) The F_1 cross would be: $Ssss \times Ssss$.
    The male gametes produced would be: *S* (25%) and *s* (75%)
    The female gametes produced would be: *Sss* (75%) and *sss* (25%).
    From fertilizations in all combinations, 13/16 or 325 of the offspring will bear round seeds and 3/16 or 75 of the offspring will bear wrinkled seeds.

16. 9 of the 64 possible genotypic combinations will produce a phenotype of long, purple and wrinkled. Among 2048 progeny, 288 (9/64 of 2048) would be expected to show this phenotype.

17. All progeny were yellow, therefore the parental plant was homozygous for seed color. Examination of pod shape among the progeny reveals a 3:1 ratio of swollen to pinched. This suggests that the parent was heterozygous for pod shape. If we let P = swollen and p = wrinkled, and C = yellow, then the genotype of the self-crossed plant is: *CCPp*.

18. Since neither species produces progeny resembling a parent, simple dominance is ruled out. The species producing pink-flowered progeny from red and white (or very pale yellow) suggests incomplete dominance as a mode of inheritance. However, in the second species, the production of orange-colored progeny cannot be explained in this fashion. Orange would result from an equal production of red and yellow; instead in this case, codominance is suggested, with one parent producing bright red flowers and the other producing pale yellow flowers.

19. The hair color of the albino parent is white (lack of pigment). The albino parent's genotype for hair color is homozygous recessive for red hair *(rr)*. The genotype of the brown-haired parent is heterozygous *(Rr)*. The genotype of the brown-haired parent for skin color is heterozygous *(Aa)*. The genotype of the first child with respect to hair color is heterozygous *(Rr)*, and heterozygous for skin pigmentation *(Aa)*. The possible genotypes of the second child for hair color are *Rr* or *rr*. The phenotype of the second child for hair color is white (lack of pigmentation). The production of hair color depends on the synthesis and deposition of pigment in the hair. Even though the second child carries the genetic information for brown or red hair, this gene is not expressed because the gene for albinism prevents pigment formation. This is a form of gene interaction.

20. During meiotic prophase I, the replicated chromosomes synapse or pair with their homologues. These paired chromosomes align themselves on the equatorial plate during metaphase I. During anaphase I it is the homologues (each containing two chromatids) that separate from each other. There is no preordained orientation for this process—it is equally likely that a maternal or a paternal homologue will migrate to a given pole. This provides the basis for the law of random segregation. Independent assortment results from the fact that the migrational polarity of one set of homologues has absolutely no influence on the orientation of a second set of homologues. For example, if the maternal homologue of chromosome 1 migrated to a certain pole, it will have no bearing on whether the maternal or paternal homologue of chromosome 2 migrates to that same pole.

21. No conclusion can be reached, since no information is available on how handedness is inherited.

Chapter 4 Answers

1. (a) Female
(b) Male
(c) Consanguineous mating
(d) Heterozygous male
(e) Identical twins, both female.

2. Autosomal dominant with incomplete penetrance

3. Autosomal dominant

4. An autosomal dominant mode of inheritance is suggested by the pedigree. If the trait is highly penetrant, then the individual in question should show the trait.

5. The pedigree is consistent with an autosomal dominant mode of inheritance, but with incomplete penetrance. Specifically, the propositus must contain a dominant allele but does not exhibit the trait.

6. (a) The pedigree is consistent with an autosomal recessive mode of inheritance.
(b) Heterozygous

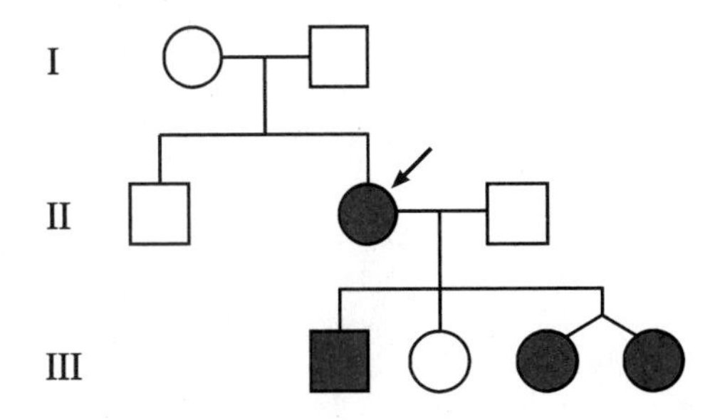

7. (a) This pedigree is consistent with autosomal recessive inheritance.
(b) If inheritance is autosomal recessive, the individual in question is heterozygous.

8. (a) I
(b) III
(c) II

9. Penetrance is a population parameter measuring the extent to which a phenotype that should be expressed is expressed. This is an all-or-none measure for any individual. Expressivity reflects the degree to which a phenotype is expressed in an individual.

10.

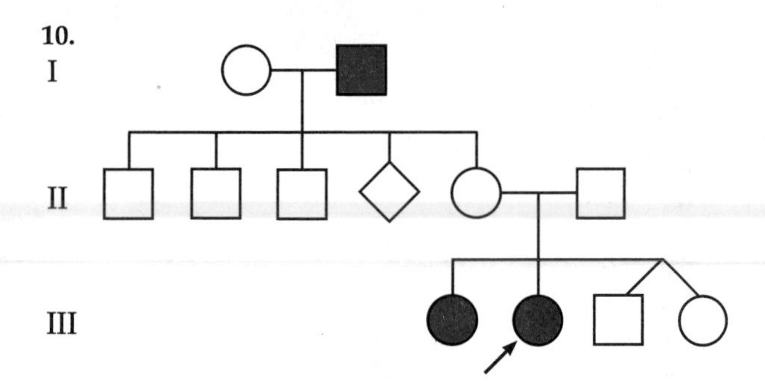

11. (a) NF is caused by an autosomal dominant allele. Due to the relative rarity of the disease, it is most likely that Sam's dad is heterozygous. This would give Sam a 50% chance of contracting the disease, since the mother is homozygous recessive. However, there is a chance that Sam's dad is homozygous. Although unlikely, this would ensure that Sam would contract the disease.
    (b) Either at least one of Sam's paternal grandparents must suffer from the disease, or Sam's father represents a new mutation. Because the mutation rate for NF is high, we cannot say anything for sure about the grandparents.

12.

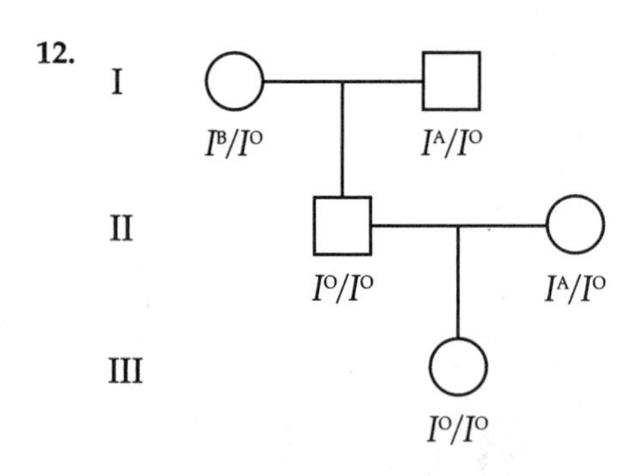

13. (a) *7—A, B, C, D, AB, AC, BC*
    (b) *—a/a, b/b, c/c, d/d, a/d, b/d, c/d, a/b, a/c, b/c*

14. Due to the rarity of the disease, we assume the maternal grandfather is heterozygous for the gene responsible for Huntington's disease. Then his daughter has a 1/2 chance of possessing the deleterious allele. In turn, should the woman have the HD allele, her child would have a 1/2 chance of inheriting it. Therefore, at present, the child has a $1/2 \times 1/2 = 1/4$ chance of having inherited the HD allele.

15. A polygenic mode of inheritance means that a trait is determined by the information contained in several genes while pleiotropy involves the converse concept that a single gene influences several traits.

16. Poor penetrance.

Chapter 5 Answers

1. Autosomal. Two copies of the gene in all individuals

2. (a) Maternal grandfather
   (b) No
   (c) No

3. (c) Sex-linked dominant

4. (a) Autosomal dominant

5. (d) Sex-linked recessive

6. (b) Autosomal recessive

7. (a) Sterile male
   (b) Sterile female
   (c) Fertile male

8. Presently unknown, but perhaps an XY chromosome constitution is necessary from conception to allow normal male development.

9. Different sex hormones are present in different concentrations in the two sexes, so different genes are turned on and off in males and females. These genes are the ones that control or code for the sex-limited and sex-influenced traits.

10. Female

11. (a) Black and white spotted daughters and white sons
    (b) Black and white spotted daughters and black sons
    (c) Black sons, white sons, black daughters, black and white spotted daughters.

12. Random inactivation in females, so the genes from both Xs are active in the body as a whole

13. The parents of the Duke of Saxe-Coburg-Gotha

14. (a) Man with type B blood
    (b) Undetermined
    (c) Man with type AB blood

Chapter 6 Answers

1. Condition 2 is most likely lethal. This condition involves a chromosomal aberration, trisomy. This has the potential for interfering with the action of all genes on the trisomic chromosome. Condition 1 involves an autosomal dominant lesion to a single gene, which is more likely to be tolerated by the organism.

2. (a) Trisomy: 3 copies of a single chromosome; triploidy: 3 copies of the entire chromosomal complement
   (b) Aneuploidy: a change in chromosome number involving less than an entire chromosome set; euploidy: one condition of having integral multiples of the haploid chromosome set (n, $2n$, $3n$ etc.).
   (c) Euploidy: see (b); polyploidy: a case of euploidy where $n > 2$.

3. (a) The cells were haploid
   (b) The cells were meiotic cells or gametes

4. Two or three possibilities should be considered. The child could be monosomic for the relevant chromosome. The child has the paternal copy carrying the allele for albinism (father is heterozygous) and a nondisjunction event resulted in failure to receive a chromosomal copy from the homozygous mother. The second possibility is that the maternal chromosome carries a small deletion, allowing the albinism to be expressed. The third possibility is that the child represents a new mutation, inheriting the albino allele and having the other by mutation. Since monosomy is lethal, either the second or third possibility seems likely.

5. Humans exhibit aneuploidy at levels 10 times that found in other mammals. A change in chromosome number is usually associated with a greatly reduced fitness and lowered reproductive success.

6. A mitotic, and not meiotic error—most likely the failure of a cell to undergo cytokinesis. However, inhibition of centromere division could also produce this condition. The error would have to occur in cells that are precursors to liver cells.

7. The sperm is produced from a tetraploid spermatogonial cell and is therefore diploid. Fertilization results in a triploid embryo.

8. The embryo will be tetraploid. Inhibition of centromere division results in nondisjunction of an entire chromosome set. After cytoplasmic division, some cytoplasm is lost in an inviable product lacking genetic material and the embryo develops from the tetraploid product.

9. Failure of chromosomes or chromatids to segregate during either meiosis or mitosis. The result is that both homologues (or sister chromatids) migrate to the same pole during anaphase.

10. The nondisjunction event must have taken place during meiosis II. Nondisjunction during meiosis I would result in retention of both a maternal and a paternal (grandparents to the trisomic individual) homologue not identical. However, nondisjunction during meiosis II results in retention of identical sister chromatids—either the maternal or paternal homologue, not both.

11. (a) Edwards syndrome, trisomy 18;
    (b) Turner syndrome, monosomy X;
    (c) Patau syndrome, trisomy 13.

12. One likely reason is the relative length of time taken to complete meiosis. Primary oocytes begin meiosis I before birth and do not complete meiosis until fertilization. Thus, years of radiant and corpuscular insults to these germ cells must be endured. In contrast, meiosis in males takes 48 days. Spermatogonial cells that suffer a mutation may never undergo meiosis.

13. Turner syndrome (45, X) is monosomy for the X chromosome. A paternal nondisjunction event could contribute a gamete lacking a sex chromosome to result in Turner's syndrome. The complementary gamete would contain both X and Y chromosomes. This gamete would contribute to Klinefelter's syndrome (47,XXY).

14. (a) Loss of part or all of a chromosome
    (b) Generation of an extra copy (copies) of part of a chromosome
    (c) Reversal of polarity of part of a chromosome relative to the remainder of the chromosome
    (d) Removal of part of a chromosome and reattachment of this segment to a different chromosome

Chapter 7 Answers

1. c

2. Proteins are found in the nucleus. Proteins are complex molecules composed of 20 different amino acids, nucleic acids are composed of only four different nucleotides. Cells contain hundreds or thousands of

different proteins, only two main types of nucleic acids.

3. c

4. Protease destroyed any small amounts of protein contaminants in the transforming extract. Similarly, treatment with RNAse destroyed any RNA present in the mixture.

5. Chargaff's Rule: A = T and C = G
If A = 27%, then T must equal 27%
If G = 23%, then C must equal 23%

Base composition:
A = 27%
T = 27%
C = 23%
G = 23%
100%

6. b

7. e

8. b

9. Two polynucleotide chains running in opposite directions, within each chain, bases are interior and sugar-phosphate groups are exterior. Bases in opposite chains are bonded to each other by hydrogen bonds. In forming such bonds, A always pairs with T, and C with G.

10. d

11.

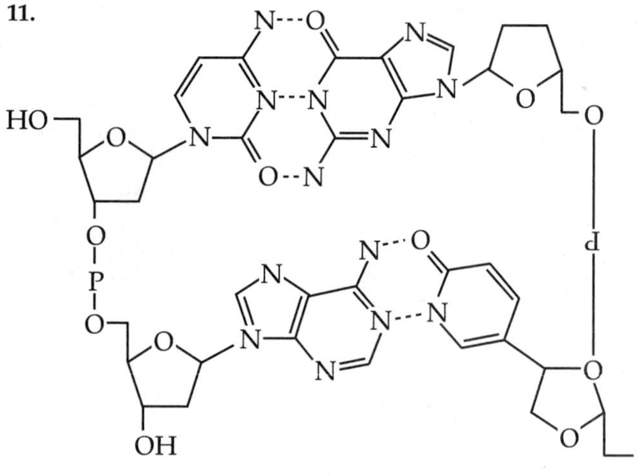

HO
O
N
N---O
N---N
O--N
N
N
N
O
O
O
P
O
O
N
N
N
N---O
N---N
N
O
O
O
P
OH

12. c

Chapter 8 Answers

1. d

2. b

3. The recognition sequence is AAGCTT. The enzyme that cuts this sequence is HindIII.

4. Because the human genome is very large, the vector chosen should be able to accept large inserts of human DNA. For this reason, a viral vector would be chosen. A plasmid vector would require several million plasmids to contain the library, while a viral vector could accommodate the library in less than a million viruses.

5. Because RFLPs behave as genes, some act as two allele systems, and others act as multiple allele systems. Those that act as multiple alleles have a better chance of being informative in a pedigree analysis. The second factor that is important is the distribution of alleles in the population. Rare alleles are useful in certain families where they are present, but in general, alleles with wide distributions are more useful. For example, in a two allele system, if each allele has a frequency of 50%, it is more likely to be informative than a distribution of 95% and 5%.

6. The separation of RFLPs from the gene causing the genetic disorder occurs by a recombination event (see Chapter 4). As with other genes, the greater the distance between the RFLP and the locus for the disorder, the greater the chance that a recombination event will occur and separate the two markers. Therefore, in using RFLPs for carrier detection, it is important to use RFLPs that are located very close to the gene in question.

7. In selecting target cells for gene therapy, it is important to consider that a prolonged or permanent effect is desired. Therefore the cells or tissue to be selected to receive the transferred gene should be long lived or mitotically active and self-renewing.

Chapter 9 Answers

1. Nucleotides in the DNA code for amino acids in proteins. A change in the nucleotide causes a change in the amino acid which changes the protein.

2. (a) Exon

(b) ribosome binding site
(c) Exon
(d) Intron
(e) 5′-flanking
(f) Cap and tail

3. pre-mRNA:
GCUAAAUGGCAaaauugccggaugacGCACAUUGA-CUCGGaaucgaGGUCAGAUGC
mRNA:
GCUAAAUGGCAGCACAUUGACUCGGGGUCA-GUAUGC

4. (a) DNA: TAC ACC ATA GCA CCA GGA TGT
mRNA: AUG UGG UAU CGU GGU CCU ACA
(b) Change the third base in the trp codon to U or C producing a termination codon
(c) Delete a G from the trp codon
(d) Change the first nucleotide in the gly codon to U
(e) Insert a fragment of DNA

5. mRNA: AUGCUUGUCACAGCCAAUGGGUAA
Polypeptide: met-leu-val-thr-ala-asn-gly

6. (a) DNA top strand
AAA TCC TCA ATA GAG CGA CTT
(b) DNA: bottom strand
TTT AGG AGT TAT CTC GCT GAA
(c) mRNA
UUU AGG AGU UAU CUC GCU GAA
(d) Polypeptide
phe arg ser tyr leu ala glu
The transcribed strand is the top strand (a).

7. Change in anticodon

8. Tripeptide composed of glycine

```
       H  O       H  O       H
       |  ||      |  ||      |   //O
$H_2N$ — C — C — N — C — C — N — C — C
       |      |   |       |   |   \
       H      H   H       H   H    OH
 N-terminus                  C-terminus
```

9. (a) Mutation in the rRNA or the proteins in the small subunit
(b) Mutation in the rRNA or the proteins in the large subunit
(c) Same as (a)

10. (a) UAC
(b) ACC
(c) UC(A/G) or AG(A/C/U/G)
(d) GA(U/C/A/G) or AA(U/C)

11. Transcription: DNA, TATA box, RNA polymerase, nucleotides, pre-mRNA, enhancer. Translation: ribosomes, tRNA, amino acyl synthetase, A site, anticodon, amino acids

12. (a) Less or no compound B
(b) No compound B

13. (a) No
(b) Yes

14. Biochemical processes control the phenotype. Each biochemical process is a series of individual interconnected reactions. Each reaction is due to a single enzyme. Genes control the production specificity and function of enzymes. If a gene is mutated the protein it codes for changes as does the protein's ability to carry out the biochemical reaction. (This idea can be extended to nonenzymatic proteins that function in the cell.)

Chapter 10 Answers

1. The mother's metabolism can compensate for the defect.

2. Accumulation of one or more precursors may be detrimental. Overuse of an alternative minor pathway may result in the accumulation of toxic intermediates. Deficiency of an important product may occur. Other reactions may be blocked.

3. Normal

4. No, because individuals who are G^D/G^D show 50% activity. The *g* allele reduces activity by 50% so heterozygotes appear normal. It is not until the level of activity falls below 50% that the mutant phenotype is observed.

5. (a) Daughters normal, ½ the sons have severe combined immunodeficiency disease
(b) Daughters normal, ½ the sons have severe combined immunodeficiency disease, and the other ½ have Lesch-Nyhan syndrome

6. (a) Buildup of substance A, no substance C
   (b) Buildup of substance B, no substance C

7. (a) Yes
   (b) The ratio would be 1:2:1. For this ratio, the substances would be as follows: substance A buildup, no C: substance B buildup, no C: normal

8. Alleles for enzyme 1: *A* (dominant, 50% activity); *a* (recessive, 0% activity). Alleles for enzyme 2: *B* (dominant, 50% activity); *b* (recessive, 0% activity).

	Enzyme 1	Enzyme 2	A	B	C
1*AABB*	100	100	N	N	N
2*AaBB*	50	100	N	N	N
4*AaBb*	50	50	N	N	N
2*AABb*	100	50	N	N	N
1*AAbb*	100	0	N	B	L
2*Aabb*	50	0	N	B	L
1*aaBB*	0	100	B	L	L
2*aaBb*	0	50	B	L	L
1*aabb*	0	50	B	L	L

N, normal; B, buildup; L, less.

9. Drugs act on proteins. Different people have different forms of proteins. Different proteins are inherited as different alleles of a gene.

10. (a) Most are in exons or at the border between exons and introns.
    (b) Any mutation that reduces or abolishes the production of beta globin will result in beta thalassemia. This can include mutations in exons, introns, or upstream regulatory regions.

11. (a) 1 and 6; 2 and 7; 3 and 8
    (b) yes
    (c) no

12. 5

13. and 14.

substrate →A →B →C →D →E

5 2/7 1/6 3/8 4

15. (a) ¼ normal, ¾ retarded
    (b) Of the children who could not produce E (and are retarded), ⅔ would build up A, ⅓ would build up D. Of all children from this mating, ¼ would be normal, ½ will be retarded and accumulate A, ¼ will be retarded and accumulate D.

Chapter 11 Answers

1. (a) A mutation caused by the addition or deletion of nucleotide residues from a coding portion of a gene that destroy the triplet reading frame (any additions or deletions that alter the frame except multiples of 3 bp).
   (b) A measure of the occurrence of mutations per individual per generation.
   (c) Mutations that arise in somatic (non-germline) cells and are not inherited.

2. (a) See pedigree below.
   (b) No. The mutation occurred in the germ line and was therefore not expressed in somatic tissue.
   (c) Approximately 50% of an affected individual's offspring will receive a parental chromatid bearing the dominant autosomal defect and will be affected.

3. Achondroplasia is a dominant disorder with high penetrance, thus an unaffected person does not carry the achondroplasia haplotype. Dominant germline mutations will occur in this individual, presumably at the average rate of 4×10^{-5} per generation.

4. The disorder is recessive, therefore the genetic defect responsible for the disease may be maintained in a heterozygous state in the gene pool of a population.

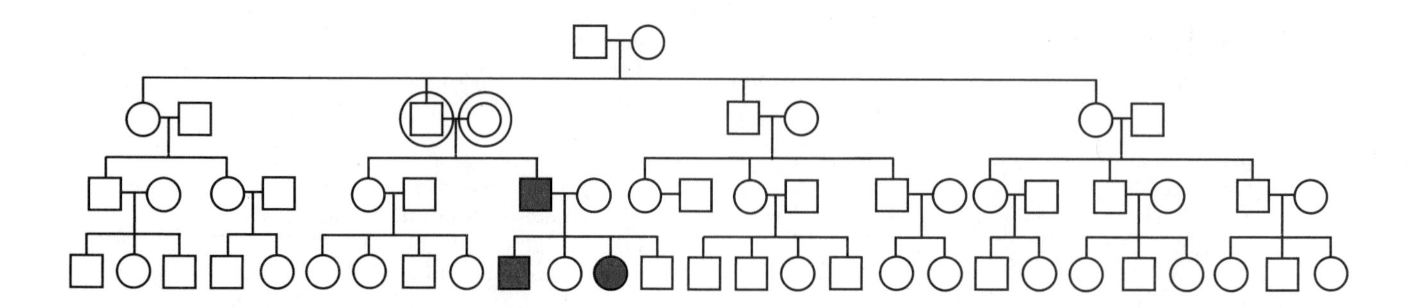

5.

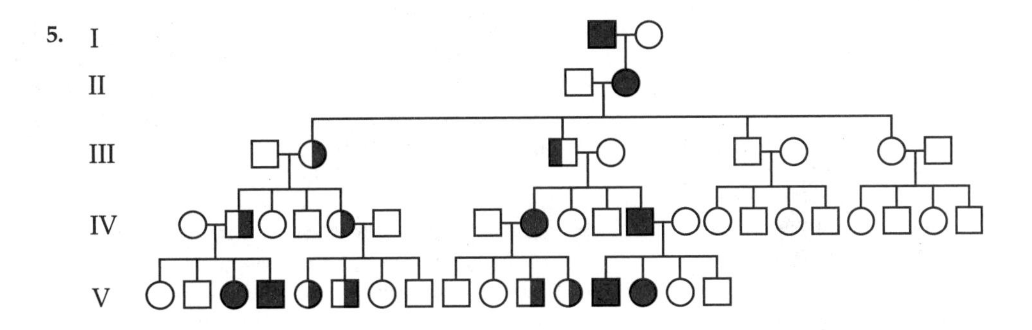

Chapter 12 Answers

1. $(3 \times 10^9 \text{ nt/sec})/(20 \text{ nt/sec}) = 1.5 \times 10^8 \text{ sec} = 4.7$ years. To shorten the time one could postulate more than one site of replication proceeding simultaneously.

2. The proofreading function removes nucleotides in the 3′ to 5′ direction.

3. Spontaneous mutations are those mutations arising without any intentional intervention. Induced mutations are those mutations arising from intervention involving the application of a mutagen—regardless of whether or not the mutagen has been shown to be responsible for the mutation.

4. Induced mutations require the application of a mutagen. Mutagens may create mutations by (1) binding directly to DNA, (2) breaking DNA, or (3) interfering with repair processes. Any 2 of the 3 answers are acceptable.

5. Inducing mutations will elevate the mutation rate, but won't direct mutations to gene X. Therefore, he'll have to do an equivalent amount of "wading" and the rationale is faulty.

6. X-rays produce intracellular ions that are highly reactive and can chemically interact with DNA to damage it. UV light contains frequencies absorbed by DNA bases. In this energized state the DNA bases can undergo self-destructive chemical reactions.

7. Yes. The atomic number has decreased by 2 due to the loss of the 2 protons contained in the α particle.

8. Roentgen: a unit that measures the air ionization potential of radiation; rad (radiation absorbed dose): a unit that measures the quantity of radiation absorbed by a biological system; rem: a unit that measures the potential of radiation to produce biological damage. From these definitions, rem units would be most useful for assessing the potential for genetic damage.

9. No preference—the sievert is the international standard unit proposed to replace the rem. As such, 0.1 sieverts produces the same amount of biological damage, regardless of the source of the radiation.

10. Cosmic rays, radioactive elements in or on the earth, and manufactured sources (medical testing, nuclear fallout, etc.). Any 2 of the 3 answers are acceptable.

11. Both should be equally concerned because the effects of radiation are cumulative.

12. Radiation preferentially kills mitotically active cells, i.e., dividing cells. Since cancer cells are those defined as growing without bound, radiation should preferentially kill these cells. However, preference is limited and many normal, mitotically active cells are killed.

13. (a) 5-Bromouracil
    (b) Nitrous acid

14. A frameshift mutation is the addition or loss of nucleotide pairs such that the normal translocational reading frame is dislodged. Intercalating agents interpose themselves between adjacent nucleotide pairs and this spatial distortion can lead to the misincorporation of "extra" nucleotides during replication or repair.

15. Compound A is the stronger mutagen. The Ames Test assess the mutagenicity of a compound by measuring the frequency at which it induces mutations in a His^- strain to generate a His^+ strain (a back mutation). The plates containing histidine are control plates to prove that equal quantities of cells are being applied to the plates. The experimental plates are those lacking histidine. In the absence of any mutations, no colonies should arise on these plates. Since equal quantities of the two chemicals were used, the number of colonies arising on the plates lacking histidine can be used to assess the relative mutagenicity of the two compounds. The data show that compound A induced more mutations and this compound is therefore the stronger mutagen.

Chapter 13 Answers

1. Carcinogenesis is often thought of as a multistep process—generally two or more distinct mutational events must occur. Genetic predisposition to acquire a particular form of tumor is often due to the genetic transmission of one or more of the mutant alleles through the germ line. Subsequently, individual somatic cells undergo secondary mutations to produce cancerous tissue.

2. (1) The existence of inherited predisposition toward certain forms of cancer implies the existence of Mendelian (chromosomal) mutations responsible for these diseases. (2) Mutagens known to act on DNA have been shown to be carcinogens as well. (3) Viral oncogenes have been identified as a causal agent in the transmission of some cancers (RSV, MMTV, HTLV-I); the active viral genes are often transduced cellular genes that have altered patterns of gene expression. (4) Gross chromosomal aberrations have been linked to certain forms of cancer (Wilm's tumor and chronic myelogenous leukemia).

3. Proto-oncogenes are thought to be involved in the regulation of cellular differentiation and growth. Deregulation of these processes results in uncontrolled proliferation of mutant (oncogenic) cells.

4. c-*myc* lies at the breakpoint of a translocation involving chromosome 8 and either chromosome 14, 22, or 2. The translocation places the *myc* gene in altered chromosomal milieu and thus disrupts its normal expression. Altered expression of c-*myc* is thought to be necessary for the production of Burkitt's lymphoma.

5. DNA from a malignant human cell line is transfected into a dish of partially transformed tissue culture cells (typically from the mouse). Transformed foci are identified and cultured. DNA is recovered from these transformed foci and the human genes responsible for the phenotype are identified by recombinant DNA techniques. The assay typically is used to identify dominant transforming genes.

6. Cigarette smoking, dietary fat vs. dietary fiber, asbestos, vinyl chloride, etc.

7. One of the parents has the mutant gene responsible for dominant familial retinoblastoma in his/her lineage. When present, there is a 90% chance that the gene is expressed in the form of retinoblastoma. Since both parents are phenotypically normal, there is a maximum 10% chance that one parent carries the defect. If carried, there is a 50% chance that the defective allele will be passed on to any particular child. If passed to a particular child, there is a 90% chance that the defect will be expressed. Therefore, the total probability of this couple having a child with retinoblastoma is $(0.1)(0.5)(0.9) = 0.045$ (4.5%).

8. Cancer is thought to be a multistep process; the probability of multiple steps occurring in the same somatic cell increases as a function of time. In addition, it is likely that the efficiency of mutational repair decreases as a function of age, thereby increasing the likelihood of the accumulation of multiple mutations.

9. Large-scale epidemiological studies were used to demonstrate a circumstantial association between individuals exposed to high levels of asbestos and the occurrence of certain forms of cancer, notably cancer of the lung and digestive tract. Laboratory experiments were then used to test the effect of asbestos exposure on animals under controlled conditions. The coupling of the two approaches allows researchers to define potentially hazardous environmental factors by statistical analysis of large populations and then directly test the carcinogenic potential of these factors in a controlled laboratory environment.

Chapter 14 Answers

1. Attenuation is the loss of viability of an infectious agent—usually a virus or bacterium. Active immunization requires exposure to the antigens of the infectious agent. This is most readily achieved by exposure to the agent itself. Since exposing people to viable infectious agents causes the disease one is trying to prevent, nonviable or attenuated agents are useful in this role.

2. *Phagocytes (macrophages and neutrophils):* recognize nonself organisms and cellular debris and engulf it. They then enzymatically degrade the materials, present antigens on the cell surface, and signal other elements of the immune system, alerting them to an infection. *T cells:* Helper T cells are involved in switching on the immune response and are crucial for signal transduction: killer T cells and natural killer cells recognize and destroy cells of the body harboring invading viruses and destroy cancer cells; suppressor T cells are involved in switching off the immune response. B cells produce antibodies but are more properly considered a part of humoral immunity.

3. Macrophages (phagocytes) can activate T cells by releasing interleukin 1.

4. The T cell receptor of T cells.

5. IgG molecules are monomers composed of two light chains and two heavy chains; therefore four polypeptide chains are required. IgM molecules have the same monomer structure as IgG molecules but are functional as pentamers; therefore $5 \times 4 = 20$ chains are required.

6. For any antibody class, the genes encoding the single heavy chain and the two light chains are complex. There are several hundred V or variable region gene segments, a few (5-10) J or joining region gene segments, and a single segment encoding the C or constant region. Recombination results in the juxtaposition of V, J, and C gene segments to form a functional gene. This process of recombination can therefore result in a large variety of functional antibody genes within the body's population of B cells.

7. (1) Migration inhibitory factor is released by T cells upon contact with a foreign antigen and, when received by macrophages, causes them to remain in the region of the infection. (2) Interleukin 1 is released by macrophages encountering an infection and activates T cells. (3) Interleukin 2 is released by activated T cells and stimulates more T cells to participate in the immune response.

8. $23 \times 47 \times 8 \times 14 \times 3 \times 6 = 2.2 \times 10^6$ haplotypes are possible. Since humans are diploid, there are $(2.2 \times 10^6)^2 = 4.84 \times 10^{12}$ possible genotypes.

9. Autosomal co-dominant.

10. The infant can only be type O and the genotype is therefore I^O/I^O. The man on trial can only be type AB and his genotype must therefore be I^A/I^B. The mother must be type A since she only possesses antibodies to type B blood. Since there is no question of maternity involved here, she contributed an I^O allele to the infant. Therefore the mother's genotype is I^A/I^O. The court can dismiss the suit because the father of the infant had to possess an I^O allele, and the man on trial doesn't have one.

11. No. Natural killer cells do not have to be activated by helper T cells to destroy virally infected cells.

12. 20% of all individuals carrying the HLA-B27 allele have ankylosing spondylitis. Also, males are affected 9:1 over females. Therefore, of 100 individuals who have B27, regardless of sex, twenty will be affected. Of these 100, half should be males, half should be females. If x is the rate in males, and y is the rate in females, $x = 9y$. In a group of 100 individuals with B27, $50x + 50y = 20$, or $450y + 50y = 20$; $y = 0.04$ and $x = 0.36$. This means that 36% of males with B27 will be affected, and 4% of females with B27 will be affected.

    (a) This boy has the B27 allele, therefore the probability is 0.36

    (b) This girl has the B27 allele and the probability is therefore 0.04

    (c) This boy also possesses the B27 allele and the probability is thus 0.36

    (d) This boy doesn't have the B27 allele. His probability of having ankylosing spondylitis is equal to the percentage of the population who lack B27 and have ankylosing spondylitis.

13. The hypothetical condition isn't feasible because the antibodies of the donated blood become too diluted in the recipient to initiate a significant reaction.

14. Lymphokines are the chemical messengers used by the immune system to coordinate a successful immune response. They primarily function to activate or deactivate components of the cellular immune system, although antibody producing B cells are also affected.

15. (a) Mother: Rh^-/Rh^-. Father: Almost certainly Rh^+/Rh^+ with a small chance that he contributed the Rh^+ allele 16 straight times but is genotypically Rh^+/Rh^-
    (b) Yes. An Rh^+ offspring presents the antigen to the mother at birth. Lacking the Rh antigen, the mother would produce antibodies to combat it. These antibodies could then enter the infant's bloodstream at birth and cause massive hemagglutination. Therefore all children except perhaps the eldest were at considerable risk.

Chapter 15 Answers

1. (a) Height in pea plants is determined by a single pair of genes with dominance and recessiveness. Height in humans is determined by polygenes.
    (b) For traits determined by polygenes, the offspring of matings between extremes in the population show a tendency to regress toward the mean expression of the trait in the population.
    (c) Differences in height may have reflected differences in the level of nutrition rather than differences in the genetic makeup of individuals. If this

was the case, many of the crosses ordered by Frederick William would have amounted to little more than random breeding in the population, which would not increase the genetic components for height in the offspring.

(d) Yes. Within some of the families it is likely that greater height was determined by genetic differences rather than by better nutrition. In these cases, brother-sister marriages would effectively homozygous the alleles for greater height in the offspring and subsequent generations, leading eventually to taller individuals.

2. (a) F_1 genotype = A′AB′B, phenotype = height of 6 ft.

(b) *A′AB′B* x *A′AB′B*

↓

Genotypes	Phenotypes
A′A′B′B′	7 ft
A′A′B′B	6 ft 6 in.
A′A′BB	6 ft
A′AB′B′	6 ft 6 in.
A′AB′B	6 ft
A′ABB	5 ft 6 in.
AAB′B′	6 ft
AAB′B	5 ft 6 in.
AABB	5 ft

3. In the case of polygenes, the expression of the trait depends on the interactions of many genes, each of which contributes a small effect to the expression of the trait. Thus, the differences between genotypes often are not clearly distinguishable. In the case of monogenic determination of a trait, the alleles of a single locus have major effects on the expression of the trait, and the differences between genotypes is usually easily discerned.

4. Genetic variance is the variation in phenotype exhibited by a population that is due to differences in the genotypes of the individuals of the population.

5. Environmental variance is any variation that occurs between individuals of the same genotype in a population.

6. Heritability is the proportion of variability in a population that is caused by genetic variance.

7. Relatives are used because the proportion of genes held in common by relatives is known.

8. Often closely related individuals share the same environment.

9. In this particular case, the individuals are genetically identical so any differences in the expression of a trait must be due to differences in their environments.

10. In this case, it is likely that the similarities in the expression of a trait between adopted and natural children are due to the sharing of a similar environment.

11. $H = 1$.

12. $H = 0$.

Chapter 16 Answers

1. Manic depression is a mood disorder, characterized by cycles of depression that alternate with periods of elation. Schizophrenia is a collection of disorders that affect the thought processes themselves, rather than mood. Both disorders that have strong genetic components, and both affect the ability of affected individuals to function in work and social settings.

2. As stated, the study shows that IQ is a strongly heritable trait. The relationship between IQ and intelligence is uncertain and questionable. Strictly speaking, the conclusions cannot be generalized, since conclusions about heritability are valid only for the group under study, and valid for this group only at the time of the study. In evaluating the conclusions of this study, additional information would be useful. This would include the family and economic conditions under which the twins were raised, whether the twins had contact with each other, and the IQs of the birth parents.

3. Epistasis is a form of nonreciprocal gene interaction, where one gene masks the expression of another. For example, if *A* masks the expression of *B* or *b*, the result is an increase in one phenotypic class (12:3:1 instead of 9:3:3:1), making one phenotype more common. If a recessive allele *a* masks the expression of *B* or *b*, the result is the appearance of a new phenotype in the offspring, in high frequency (9:3:3:4 instead of 9:3:3:1).

4. From the information provided, perfect pitch can be classified as a familial trait, but evidence for a genetic basis is lacking. To determine if the trait is genetic, further family studies and pedigree analysis is necessary, involving large families with several generations represented. See: Profitera, J. and Bidder, T. 1988. Perfect pitch. Am. J. Med. Genet. 29: 763-771.

5. The pedigree presented is consistent with autosomal dominant inheritance associated with imperfect penetrance. Perfect pitch and tune deafness could be alleles of the same gene, but further studies on both these conditions would be required to support this hypothesis.

6. Most single gene mutations that affect behavior act on the brain and/or the nervous system to alter a cellular function.

7. *Drosophila* has many advantages for the study of behavior. Mutagenesis and screening for behavior mutants allows the recovery of mutations that affect many forms of behavior. The ability to perform genetic crosses and recover large numbers of progeny over a short period of time also enhances the genetic analysis of behavior. This organism can serve as a model for human behavior, because cells of the nervous system in both *Drosophila* and humans use similar mechanisms to transmit impulses and store information.

Chapter 17 Answers

1.

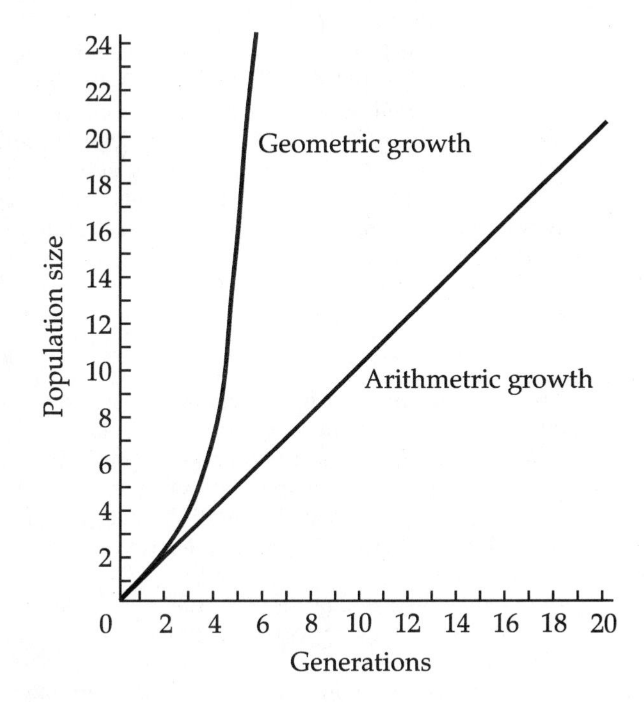

2. War, famine, and disease.

3. Population: Local groups of individuals occupying a given space at a given time. Gene pool: The set of genetic information carried by a population. Gene frequency: The frequency of occurrence of particular alleles in the gene pool of a population. Genotype frequency: The frequency of occurrence of particular genotypes among the individuals of a population.

4. (a) Evolution is the change over time of heritable characteristics of a population. Since characteristics that are heritable are determined by genes (by definition), evolution constitutes the change of gene frequencies in a population over time. Darwin and Wallace observed that those individuals in a population possessing favored characteristics (phenotypes) were responsible for a greater proportion of the surviving offspring in future generations than individuals with less favored characteristics. Since the favored characters are determined by genes, Darwin and Wallace were observing that natural selection tends to change the frequency of specific genes over the course of time.
   (b) No.

5. The simplest experiment consists of monitoring the frequencies of various alleles over a number of generations to determine if the gene frequencies are changing or are in equilibrium.

6. (a) Children: M = 0.5, N = 0.5. Adults: M = 0.5, N = 0.5.
   (b) Yes, allelic frequencies are unchanged.
   (c) No. The genotypic frequencies are changing within each generation.

7. The frequency of an allele has no relationship to its mode of expression. For example, a dominant allele may exist at a very low frequency in a population, and cannot ultimately overtake a recessive allele in frequency.

8. It will take one generation for the allelic frequencies to reach Hardy-Weinberg equilibrium, and two generations for the genotypic frequencies. After the first generation of random mating: $A = 0.225$, $a = 0.775$. $AA = 0.035$, $Aa = .38$, $aa = 0.585$. After the second generation of random mating: $A = 0.225$, $a = 0.775$. $AA = 0.050625$, $Aa = .34875$, $aa = 0.600625$.

9. In the case of multiple alleles, such as the ABO blood system, the greatest number of alleles that can be carried by an individual is two. A population, however, has the capacity to carry many different alleles for any given locus.

10. $A = 0.5$, $a = 0.5$

11. (a) $A = 0.95$, $a = 0.05$. Has to be X-linked.
    (b) Genotypes of males: $A/Y = 0.95$, $a/Y = 0.05$. Genotypes of females: $AA = 0.9025$, $Aa = 0.095$, $aa = 0.0025$.

12. O = 0.04 (Codominance)

13. U.S. Indians and Eskimos, and U.S. Blacks and Whites. The Indians and Eskimos are known to be derived from the same ancestral populations in Asia. Their allelic frequencies show some divergence. This is probably due to recent isolation of the two populations after their migrations to North America and Greenland. The similarity of the allelic frequencies in the U.S. Black and White populations may be due to genetic exchange between the populations after their arrival in North America.

14. All the populations are in agreement with the Hardy-Weinberg law.

15. $p = 0.5$, $q = 0.5$.

Chapter 18 Answers

1. All polymorphisms are mutations. When a mutation resulting in a particular allele is present in at least 1% of the population, it is termed a polymorphism. This percentage is chosen because it is clearly higher than can be accounted for by mutation alone. Because such genetic variations cannot be accounted for by mutation alone, other forces such as natural selection must be active. The study of such variation is important in understanding the process of evolution.

2. Mutation itself is a rare event, and as such, has little direct effect on gene frequency.

3. A founder effect is one example of genetic drift. Genetic drift is a chance event that alters gene frequencies. It acts through small populations, and founder effects involve a small number of individuals making a large contribution to the genes of the next generation.

4. No it is not an accurate description. Natural selection depends on fitness, the ability of a given genotype to survive and reproduce. It is the differential reproduction of some individuals that is the essence of natural selection.

5. Social customs and pressures have helped shape nonrandom mating patterns among humans, called assortative matings. Other forms of nonrandom matings include incest and consanguineous marriages. Such nonrandom matings lead to an increase in homozygouity and a decrease in genetic diversity. This produces a nonequilibrium situation.

6. The concept of race began in the late nineteenth century as a morphological "type" or model specimen. Individuals that varied from that type were placed into a separate race. In the 1930s and 1940s, a genetic concept of race began to replace the phenotypic concept. Currently, races are regarded as genetically distinct subgroups within a population.

7. Essay answer based on opinion.

8. Essay answer based on opinion.

9. Culture represents one form of adaptation, and is itself an evolutionary process. Culture tends to buffer populations from some forms of selection, and so alters the process of natural selection. Our present culture is affecting selection by permitting the survival and reproduction of genotypes that would otherwise not reproduce. Over the long run, this can lead to an increase in frequency of these genes.

Chapter 19 Answers

1. Prenatal screening: used to diagnose genetic defects and birth defects in the unborn.
   Newborn screening: used to diagnose genetic defects in neonatal and early infants stages of life.
   Carrier screening: used to identify phenotypically normal individuals who carry deleterious recessive traits.
   Occupational screening: used to identify those who are genetically susceptible to agents found in the workplace.
   Reproductive screening: used for identification of gamete donors carrying genetic or chromosomal defects.

2. The first recommendation would be for another α-fetoprotein test to determine whether the first test

may have been a false positive. If the second test is positive, amniocentesis is recommended. If the results from this procedure are ambiguous, fetoscopy may be considered, balancing the risks of this procedure against the benefits.

3. The essay answer depends on your point of view.

4. The essay answer depends on your point of view.

5. First, construct a pedigree to determine whether the trait is inherited as a sex-linked or an autosomal recessive trait, and whether the female could be or is a carrier. The second step is to educate the couple about the risks, if any, that their children may be affected.

6. Since NF is a dominant trait, and there is no evidence of this trait on either side of the family, the birth of the NF child is probably the result of a spontaneous mutation. However, because the phenotype of NF is so variable and some cases may escape detection, members of the immediate family should be examined by a specialist to rule out familial transmission. If it is confirmed that there is no family history, the recurrence risk is low.

GLOSSARY

5′ flanking region A nucleotide region adjacent to the 5′ end of a gene that contains regulatory sequences.

ABO blood groups Three alleles of a gene on human chromosome 9 that specify the presence and/or identity of certain molecules on the surface of red blood cells.

Acceptor stem The region of a tRNA molecule to which a specific amino acid is attached in an enzymatic reaction.

Achondroplasia An autosomal dominant disorder in which the long bones of the arms and legs fail to develop normally, producing a form of dwarfism.

Acquired immunodeficiency syndrome (AIDS) An infectious disease characterized by the loss of T4 helper lymphocytes, causing an inability to mount an immune response.

Acrocentric A chromosome with the centromere placed very close to one end.

Active site The portion of a protein that is required for enzymatic function.

Adaptation The process by which an organism adapts to the present environmental conditions. The degree to which an organism adapts can be under genetic control.

Adenine and guanine Purine nitrogenous bases found in nucleic acids.

Adenomas Tumors or growths found in glandular tissues.

Affect Pertaining to mood or feelings.

Affective disorders A group of mental diseases characterized by behavioral disorders, either alternation between manic and depressive behavior (bipolar disorder) or depression (unipolar disorder).

Agglutinate To clump together, as red blood cells in the presence of a specific antibody.

Albinism In humans, a complex of genetic disorders associated with the inability to synthesize the pigment melanin. The most common form is inherited as an autosomal recessive trait.

Alkaptonuria A relatively benign autosomal recessive genetic disorder associated with the excretion of high levels of homogentisic acid.

Allele One of the possible alternative forms of a gene, usually distinguished from other alleles by its phenotypic effects.

Allele frequency The percentage of all alleles of a given gene that are represented by a specific allele.

Alpha thalassemia Genetic disorder associated with an imbalance in the ratio of alpha and beta globin caused by reduced or absent synthesis of alpha globin.

Ames test A bioassay developed by Bruce Ames and his colleagues for identifying mutagenic compounds.

Amino group A chemical group (NH_4) found in amino acids and at one end of a polypeptide chain.

Amniocentesis A method of sampling the fluid surrounding the developing fetus by the insertion of a hollow needle and the withdrawal of suspended fetal cells and fluid. Used in the diagnosis of fetal genetic and developmental disorders. Usually performed in the 16th week of pregnancy.

Amnion A fluid filled sac within which human development occurs.

Anabolism The synthesis of complex molecules from building blocks or simpler components.

Anaphase A stage in mitosis during which the centromeres split and the daughter chromosomes begin to separate.

Aneuploidy A chromosome number that is not an exact multiple of the haploid set.

Ankylosing spondylitis An autoimmune disease that produces an arthritic condition of the spine; associated with the HLA allele B27.

Anthropological genetics The union of population genetics and anthropology to study the effects of culture on gene frequencies.

Antibodies A class of proteins produced by plasma cells that couple or bind specifically to the class of proteins that stimulate the immune response.

Anticodon loop The region of a tRNA molecule that contains the three base sequence (known as the anticodon) that pairs with a complementary sequence (known as a codon) in an mRNA molecule.

Antigens A foreign molecule or cell that stimulates the production of antibodies.

Antigen combining site The site at which the antibody will combine with or bind to the antigen; it is composed of the variable regions of the H chain and the L chain.

Antigen receptor Cell surface receptor on cells of the immune system that allow recognition of antigens.

Assortative mating Reproduction in which mate selection is not at random, but instead is based on physical, cultural or religious grounds.

Ataxia telangiectasia A heritable disorder associated with sensitivity to x-rays and the inability to repair DNA damage caused by ionizing radiation.

Autoradiography A method of identifying the location of radioactive molecules by using radioactive decay to produce a photographic image.

Autosomes Chromosomes other than the sex chromosomes.

B cells White blood cells that originate in bone marrow and mature in the bone marrow.

Background radiation Ionizing radiation in the environment that contributes to radiation exposure. Background radiation may be natural or man-made.

Bacteriophage A virus that infects bacterial cells. Genetically modified bacteriophage are used as cloning vectors.

Base analogue A purine or pyrimidine that differs in chemical structure from those normally found in nucleic acids, and which is incorporated into nucleic acids in place of the normal base.

Beta thalassemia Genetic disorder associated with an imbalance in the ratio of alpha and beta globin caused by reduced or absent synthesis of beta globin.

Biometrics The application of statistical methods to problems in biological sciences.

Bipolar disorder An emotional disorder characterized by mood swings that vary between manic activity and depression.

Bivalents The structure formed by a pair of synapsed, homologous chromosomes at prophase I of meiosis.

Bloom syndrome An autosomal recessive condition characterized by dwarfism, decreased immunity and sensitivity to sunlight. Chromosome fragility and translocations indicate defective DNA repair.

Cafe-au-lait spots Spots of abnormal skin pigmentation that are found on many individuals affected with neurofibromatosis.

Campodactyly A dominant human genetic trait that is expressed as immobile, bent fingers.

Cap A modified base (guanine nucleotide) that is attached to the 5′ end of eukaryotic mRNA molecules.

Carboxyl group A chemical group (COOH) found in amino acids and at one end of a polypeptide chain.

Catabolism The breakdown of complex molecules to simpler components.

Cell furrow A constriction of the cell membrane that forms at the point of cytoplasmic cleavage during cell division.

Cell cycle The sequence of events that takes place between successive mitotic divisions.

Cellular immunity Immune responses carried out by cells rather than by antibodies in the blood serum.

Central dogma The idea that DNA serves as a template for its own replication and for the transcription of RNA. In turn, the RNA is translated into the amino acid sequence of a protein.

Centromere A region of a chromosome to which fibers attach during cell division. Location of a centromere gives a chromosome its characteristic shape.

Chiasma (pl. Chiasmata) The crossing of nonsister chromatid strands seen in the first meiotic prophase. Chiasmata represent the structural evidence for crossing over.

Chorion A membrane outside and surrounding the amnion, from which projections known as villi extend to the uterine wall, forming the placenta.

Chorionic villus sampling (CVS) A method of sampling fetal chorionic cells by insertion of a catheter through the vagina into the uterus. Used in the diagnosis of biochemical and cytogenetic defects in the embryo. Usually performed in the 8th or 9th week of pregnancy.

Chromatid One of the subunits of a longitudinally divided chromosome, joined by a centromere to its sister chromatid.

Chromatin The complex of DNA, RNA, and proteins that make up a chromosome.

Chromosome theory of inheritance The theory that genes are carried on chromosomes and that the behavior of chromosomes during meiosis is the physical explanation for Mendel's observations on the segregation and independent assortment of genes.

Chromosomes The thread-like structures in the nucleus that carry genetic information.

Cline A gradient of genotypes and/or phenotypes along a geographical line.

Clinodactyly An autosomal dominant trait that produces a bent finger.

Clones Genetically identical organisms, cells or molecules all derived from a single ancestor. Cloning is the method used to produce such clones.

Coadaptation The accumulation of the fittest genes into a single genotype.

Cockayne syndrome A rare autosomal disorder characterized by sensitivity to light, mental retardation, and early death.

Codominant Full phenotypic expression of both members of a gene pair in the heterozygous condition.

Codon A triplet of bases in messenger RNA that encodes the information for the insertion of a specific amino acid in a protein.

Concordant trait If both members of a twin pair exhibit a trait, it is said to be concordant.

Consanguineous matings Matings between two individuals that share a common ancestor in the preceding two or three generations.

Conservative replication An early model of DNA replication that results in one daughter molecule consisting of old strands, and one consisting of newly synthesized strands.

Constant region (c) The region of an H or L chain closer to the C-terminus that is invariant within a given class of antibody.

Continuous variation A distribution of phenotypic characters that is distributed from one extreme to another in an overlapping or continuous fashion.

Correlation coefficient A measure of the way in which variables vary together.

Cri-du-chat syndrome A deletion of the short arm of chromosome 5 associated with an array of congenital malformations, the most characteristic of which is an infant cry that resembles a mewing cat.

Cristae The folded, inner membranes of mitochondria that contain the molecules engaged in energy formation.

Crossing over The process of exchanging parts between homologous chromosomes during meiosis; produces new combinations of genetic information.

C-terminus The end of a polypeptide or protein that has a free carboxyl group.

Cyclopia A birth defect found in mammals resulting in the production of a single eye, centrally placed in the forehead. In humans, some cases are genetically controlled.

Cystic fibrosis A fatal recessive genetic disorder common in the U.S. caucasion population associated with abnormal secretions of the exocrine glands.

Cytokinesis The process of cytoplasmic division that accompanies cell division.

Cytosine, thymine and uracil Pyrimidine nitrogenous bases found in nucleic acids.

Cytoskeleton A system of internal tubules and filaments that allows a cell to have a characteristic shape.

Deletion A chromosomal aberration in which a segment of a chromosome is deleted or missing.

Deoxyribonucleic acid (DNA) A molecule consisting of antiparallel strands of polynucleotides that is the primary carrier of genetic information.

Deoxyribose and ribose Pentose sugars found in nucleic acids. Deoxyribose is found in DNA, ribose in RNA.

Dermatoglyphics The study of the skin ridges on the fingers, palms, toes, and soles.

Dihybrid cross A mating between two individuals who are heterozygous at two loci (e.g., BbCc × BbCc).

Diploid The condition in which each chromosome is represented twice, as a member of a homologous pair.

Direct repair Mechanism of DNA repair involving reversal of damage. Light-induced breakage of improper bonds in thymine dimers is an example.

Discontinuous variation Phenotypes that fall into two or more distinct, nonoverlapping classes.

Disjunction The separation of chromosomes that occurs in anaphase of cell division.

Dispermy Fertilization of haploid egg by two haploid sperm forming a triploid embryo.

Dispersive replication A discarded model of DNA replication in which old and new segments of DNA are interspersed in the strands of DNA.

Dizygotic (DZ) twins Derived from two separate and nearly simultaneous fertilization events, each involving one egg and one sperm.

DNA ligase An enzyme that forms covalent bonds between the 5′ end of one polynucleotide chain, and the 3′ end of another chain.

DNA polymerase An enzyme that catalyzes the synthesis of DNA using a template DNA strand and nucleotides.

DNA replication The process of DNA synthesis.

DNA restriction fragment A segment of a longer DNA molecule produced by the action of a restriction endonuclease.

Dominant The trait expressed in the F_1 (or heterozygous) condition.

Dosage compensation A mechanism that regulates the expression of sex-linked gene products.

Dose rate The amount of ionizing radiation delivered to a specific tissue area or body per unit time.

Dosimetry The process of measuring radiation. For biological work, the usual units are the roentgen, rad and rem.

Doubling dose The dose of ionizing radiation that doubles the spontaneous mutation rate in the organism or species being studied.

Duplication A chromosomal aberration in which a segment of a chromosome is repeated and therefore present in more than one copy within the chromosomes.

Dyad The pair of sister chromatids that results from the separation of tetrads at the first meiotic division.

Endoplasmic reticulum (ER) A system of cytoplasmic membranes arranged into sheets and channels that functions in transport of gene products.

Environmental variance The phenotypic variance of a trait in a population that is attributed to exposure to differences in the environment by members of the population.

Enzymes Protein that catalyze a specific biochemical reaction.

Epidemiology The study of the factors that control the presence, absence or frequency of a disease.

Epigenesis The idea that an organism develops by the appearance and growth of new structures. Opposed to preformationism, which holds that development is the growth of structures already present in the egg.

Epistasis A form of gene interaction in which one gene affects the expression of another.

Equational division A cell division during which centromeres divide and sister chromatids are separated into different daughter cells.

Equilibrium density gradient centrifugation A method of separating macromolecules by means of centrifugal fields and solutions that contain regions of different densities.

Essential amino acids Amino acids that cannot be synthesized in the body and must be supplied in the diet.

Eukaryote An organism that is composed of one or more cells that contain membrane-bound nuclei and that undergo mitosis and meiosis.

Eugenics The improvement of the human species by selective breeding.

Evolution The appearance of new plant and animal specie from pre-existing species.

Excision repair Mechanism of DNA repair that cuts thymine dimers and some adjacent nucleotides out of a DNA strand, and repairs the gap using the other strand as a primer.

Exons DNA sequences that are transcribed and joined to other exons during mRNA processing, and are translated into the amino acid sequence of a protein.

Expressivity The range of phenotypes shown by a given genotype.

Familial hypercholesterolemia A dominant autosomal genetic condition associated with a defect in cellular receptors that function in cholesterol metabolism. Affected individuals are susceptible to heart disease and early death.

Familial adenomatus polyposis (FAP) A dominant condition associated with the development of growths known as polyps in the colon. These polyps often develop into malignant growths.

Fanconi anemia A rare autosomal disorder characterized by a reduction in circulating blood cells, and chromosome aberrations. Affected individuals are sensitive to X-rays and other ionizing radiation. DNA repair is defective.

Fetal alcohol syndrome A constellation of birth defects caused by maternal drinking during pregnancy.

Fitness A measure of the relative survival and reproductive success of a given individual or genotype.

Founder effects Gene frequencies established by chance in a population that is started by a small number of individuals (perhaps only a fertilized female).

Fragile X An X chromosome that carries a non-staining gap or break at band q27. Associated with mental retardation in hemizygous males.

Frameshift mutations Mutational events in which one or more bases is added to or removed from DNA, causing a shift in the codon reading frame.

Free radical An unstable and highly reactive molecule resulting from the interaction of ionizing radiation with water.

Fructosuria An autosomal recessive condition associated with the inability to metabolize the sugar fructose, which accumulates in the blood and urine.

Galactosemia A heritable trait associated with the inability to metabolize the sugar galactose. Left untreated, high levels of galactose-1-phosphate accumulate, causing cataracts and mental retardation.

Gamete A haploid reproductive cell, such as an egg or sperm.

Gene frequency The percentage of all alleles of a given gene that are represented by a specific allele.

Gene pool The set of genetic information carried by the members of a sexually reproducing organism.

Genes The fundamental units of heredity.

Genetic goitrous cretinism A hereditary disorder in which the failure to synthesize a needed hormone produces physical and mental abnormalities.

Genetic equilibrium The situation when the frequency of alleles for a given gene remain constant from generation to generation.

Genetic map The arrangement and distance between genes on a chromosome deduced from studies of genetic recombination.

Genetic drift The random fluctuation of gene frequencies from generation to generation that take place in small populations.

Genetic variance The phenotypic variance of a trait in a population that is attributed to genotypic differences.

Genetics The scientific study of heredity.

Genomic imprinting Phenomenon in which the expression of a gene depends on whether it is inherited from the mother or the father. Also known as genetic or parental imprinting.

Genotype The specific genetic constitution of an organism.

Geographic races Races or subspecies separated from one another by geographic barriers.

Gonial cells Cells in the ovary or testis that divide by mitosis and give rise to cells destined to undergo meiosis.

Gradualism The idea that evolution proceeds through small, cumulative steps over a very long period of time.

Graft versus host disease (GVHD) A situation that arises when a transplant such as bone marrow containing immunocompetent cells mounts an immune response against the host body that is unable to protect itself.

H chains The larger polypeptide in antibody molecules; specifies the class to which the immunoglobulin belongs.

Hairy pinnae A genetic trait causing the growth of hair along the rim of the ear, investigated as a possible holandric trait.

Haploid The condition in which each chromosome is represented once in an unpaired condition.

Haplotype A set of closely linked genes that tend to be inherited together, as the HLA complex.

Hardy-Weinberg law The statement that gene frequencies and genotype frequencies will remain constant from generation to generation when the population meets certain assumptions.

Hayflick limit The number of cell divisions that a cultured cell will undergo before dying.

Hemizygous A gene present in a single dose on the X chromosome that is expressed in males in both the recessive and dominant condition.

Hemoglobinopathies Disorders of hemoglobin synthesis and function.

Hemolytic disease of the newborn (HDN) A condition that results from Rh incompatibility and is characterized by jaundice, anemia, and an enlarged liver and spleen. Also known as erythroblastosis fetalis.

Hemophilia An X-linked recessive disorder characterized by the inability to properly form blood clots.

Hereditarianism The idea that all human traits are determined solely by the genotype, ignoring the contribution of the environment.

Heterogametic The production of gametes that contain different kinds of sex chromosomes. In humans, males produce gametes that contain X or Y chromosomes.

Heterozygous Carrying two different alleles for one or more genes.

Histones Small DNA binding proteins that function in the coiling of DNA to produce the structure of chromosomes.

HLA complex A cluster of genes located on chromosome 6 that are concerned with the acceptance or rejection of tissue in organ grafts and transplants.

Holandric The pattern of transmission from father to son expected for Y-linked genes.

Hominid A member of the family Hominidae which includes bipedal primates such as *Homo sapiens.*

Hominoid Members of the primate superfamily Hominoidea including the gibbons, great apes, and humans.

Homogametic The production of gametes that contain only one kind of sex chromosome. In humans, all gametes produced by females contain only an X chromosome.

Homologues Members of a chromosome pair.

Homozygous Having identical alleles for one or more genes.

Homunculus The miniature individual imagined by preformationists to be contained within the sperm or egg.

Human immunodeficiency virus (HIV) A human retrovirus that infects T4 lymphocytes, precipitating a loss in the ability to mount an immune response.

Humoral immunity Immune responses that are mediated by antibody (immunoglobulin) molecules.

Huntington disease A dominant genetic disorder characterized by involuntary movements of the limbs, mental deterioration, and death within 20 years of onset. Symptoms appear between 30 and 50 years of age.

Hydrogen bond A weak chemical bonding force that holds polynucleotide chains together in DNA.

Hypercalcemia A dominant trait that causes an elevated level of calcium in the blood.

Hypercholesterolemia An autosomal dominant genetic disorder associated with the deposition of cholesterol plaques in the arteries of heterozygotes and homozygous dominant individuals, which lead to heart disease and early death.

Hypophosphatemia A dominant X-linked condition that produces bone disorders such as rickets.

Ichthyosis A group of genetic diseases of the skin that produce scaling and dryness. The common form is inherited as a sex-linked recessive.

Immune response The activation of the immune system caused by the presence of a foreign substance or antigen.

Immune tolerance A condition of nonreactivity towards molecules or cells that might be expected to induce an immune response.

Immunological memory The capacity of the immune response to mount a rapid and vigorous response to a second contact with an antigen.

Inborn error of metabolism The concept advanced by Archibald Garrod that many genetic traits are the result of alterations in biochemical pathways.

Inbreeding Production of offspring by related parents.

Incest Sexual relations between parents and children or between brothers and sisters.

Incomplete dominance Failure of a dominant phenotype to be expressed in the heterozygous condition. Such heterozygotes have a phenotype that is intermediate between those of the homozygous forms.

Independent assortment The random distribution of members of homologous chromosome pairs during meiosis.

Initiation sites Chromosome regions where DNA replication begins.

Initiator codon A codon present in mRNA that signals the location for translation to begin. The codon AUG functions as an initiator codon.

Intelligence quotient (IQ) A score derived from standardized tests that is calculated by dividing the individual's mental age (determined by the test) by his or her chronological age.

Intercalating agent A molecule that inserts itself between the base pairs in a DNA molecule. Some intercalating agents disrupt the alignment and pairing of bases in DNA, resulting in mistakes during replication.

Interleukin 1 and 2 Lymphokines secreted by cells involved in the immune response.

Interphase The period of time in the cell cycle between mitotic divisions.

Introns Sequences present in some genes that are transcribed, but removed during processing, and therefore are not present in mRNA.

Inversion A chromosomal aberration in which the order of a chromosome segment is reversed.

Ionizing radiation Electromagnetic or corpuscular radiation that is capable of producing ions during the interaction with other matter, including biological molecules.

Karyotype The chromosome complement of a cell line or an individual photographed at metaphase and arranged in a standard sequence.

Klinefelter syndrome An aneuploidy of the sex chromosomes resulting in a male with an XXY chromosome constitution.

L chains The smaller polypeptide in antibody molecules.

LD_{50} The radiation dose that will kill half the members of a population within a specific time.

Lesch-Nyhan syndrome An X-linked recessive condition associated with a defect in purine metabolism that causes an overproduction of uric acid.

Leukemia A form of cancer associated with uncontrolled growth of leukocytes (white blood cells) or their precursors.

Library In recombinant DNA terminology, a collection of clones that contains all the genetic information in an individual. Also known as a gene bank.

Linkage A condition in which two or more genes do not show independent assortment rather they tend to be inherited together. Such genes are located on the same chromosome. By measuring the degree of recombination between such genes, the distance between them can be determined.

Locus The position occupied by a gene on a chromosome.

Lymphokines Glycoprotein molecules that are used as chemical signals to communicate between cells in the immune system.

Lyon hypothesis The proposal that dosage compensation in mammalian females is accomplished by the random inactivation of one of the two X chromosomes.

Macrophages Large, white blood cells that are phagocytic and involved in mounting an immune response.

Marfan syndrome An autosomal dominant genetic disorder that affects the skeletal system, the cardiovascular system, and the eye.

Meiosis The process of cell division during which one cycle of chromosome replication is followed by two successive cell divisions to produce four haploid cells.

Messenger RNA (mRNA) A single stranded complementary copy of the base sequence in a DNA molecule that constitutes a gene.

Metabolism The sum of all biochemical reactions by which living organisms generate and use energy.

Metacentric A chromosome with a centrally placed centromere.

Metaphase A stage in mitosis during which the chromosomes move and arrange themselves at the equator of the cell.

Metaphase plate The cluster of chromosomes aligned at the equator of the cell during mitosis.

Migration inhibition factor A lymphokine that inhibits the migration of macrophages.

Missense mutation A mutation that causes the substitution of one amino acid for another.

Mitochondria (sing. mitochondrion) A membrane bound organelle present in the cytoplasm of all cells of higher organisms. They are the sites of energy production within cells.

Mitosis Form of cell division that produces two cells, each with the same complement of chromosomes as the parent.

Molecular biology A branch of biology that seeks molecular explanations for biological phenomena.

Molecule A structure composed of two or more atoms held together by chemical bonds.

Monohybrid cross A mating between two individuals who are each heterozygous at a given locus (e.g., Bb × Bb)

Monomorphic Showing only one form.

Monosome A single ribosome, composed of two subunits, each of which contains ribosomal RNA and proteins.

Monosomy A condition in which one chromosome of a pair is missing having one less than the diploid number ($2n - 1$).

Monozygotic (MZ) twins Twins derived from a single fertilization event involving one egg and one sperm; such twins are genetically identical.

Mood A sustained emotion that influences perception of the world.

Mood disorder A group of behavior disorders associated with manic and/or depressive syndromes.

Mosaic A individual composed of two or more cell types of different genetic or chromosomal constitution. In this case, both cell lines originate from the same zygote.

Mullerian inhibiting hormone (MIH) A hormone produced by the developing testis that causes the breakdown of the Mullerian ducts in the embryo.

Multifactorial trait Trait that results from the interaction of one or environmental factors and two or more genes.

Muscular dystrophy A group of genetic diseases associated with progressive degeneration of muscles. One of these, Duchenne's muscular dystrophy, is inherited as a sex-linked recessive trait.

Mutation rate The number of events producing mutated alleles per locus per generation.

Natural killer cells White blood cells that originate in bone marrow and are able to kill invading microorganisms without activation by cells of the immune system.

Natural selection The differential reproduction shown by some members of a population that is the result of differences in fitness.

Neurofibromatosis A genetic disorder inherited in a dominant fashion that is associated with tumors of the nervous system.

Neutrophils White blood cells that ingest bacteria and small particles by phagocytosis.

Nitrogenous base A purine or pyrimidine that is a component of nucleotides.

Nondisjunction The failure of homologous chromosomes to properly separate during meiosis or mitosis.

Nonhistone proteins The array of proteins other than histones that are complexed with DNA in chromosomes.

Nonsense mutation A mutation that changes an amino acid-specifying codon to one of the three termination codons.

N-terminus The end of a polypeptide or protein that has a free amino group.

Nuclein A mixture of nucleic acids and proteins isolated from nuclei by Miescher.

Nucleolus (pl. nucleoli) A nuclear region that functions in the synthesis and assembly of ribosomes.

Nucleoside A purine or pyrimidine base attached to a ribose or deoxyribose sugar.

Nucleosomes A bead-like structure composed of histones wrapped by DNA.

Nucleotide A nucleoside attached to a phosphate group.

Nucleotide substitution Mutations that involve substitutions, insertions or deletions of one or more nucleotides in a DNA molecule.

Nucleus The membrane bounded organelle present in most cells that contains the chromosomes.

Oncogene A gene that induces or continues uncontrolled cell proliferation.

One gene-one polypeptide hypothesis A refinement of the one gene-one enzyme hypothesis made necessary by the discovery that some proteins are composed of subunits encoded by different genes.

One gene-one enzyme hypothesis The idea that individual genes control the synthesis and therefore the activity of a single enzyme. This idea provides the link between the gene and the phenotype.

Oogonia Mitotically active cells that produce primary oogonia.

Ootid The haploid cell produced by meiosis that will become the functional gamete.

Organelles A cytoplasmic structure having a specialized function.

Palindrome A word, phrase, or sentence that reads the same in both directions. Applied to a sequence of base pairs in DNA that reads the same in the 5′ to 3′ direction on complementary strands of DNA. Many recognition sites for restriction enzymes are palindromic sequences.

Pangenesis A discarded theory of development that postulated the existence of pangenes, small particles from all parts of the body that concentrated in the gametes, passing traits from generation to generation, blending the traits of the parents in the offspring.

Pedigree chart A diagram listing the members and ancestral relationships in a family, used in the study of human heredity.

Penetrance The proportion of individuals with a given genotype that show an expected and characteristic phenotype.

Pentose sugar A five carbon sugar molecule found in nucleic acids.

Pentosuria A relatively benign genetic disorder of sugar metabolism characterized by the accumulation of xylulose in the blood and urine.

Peptide bond A chemical link between the carboxyl group of one amino acid and the amino group of another amino acid.

Phagocyte A cell with the ability to surround and engulf viruses and microorganisms.

Pharmacogenetics A branch of genetics and pharmacology that is concerned with the inheritance of differences in response to drugs.

Phenotype The genetically controlled, observable properties of an organism.

Phenylketonuria (PKU) An autosomal recessive disorder of amino acid metabolism that results in mental retardation if untreated.

Philadelphia chromosome An abnormal chromosome produced by an exchange of portions of the long arms of chromosome 9 and 22.

Photoreactivation enzyme An enzyme that removes thymine dimers induced by ultraviolet light from DNA molecules.

Plasmids Extrachromosomal DNA molecules found naturally in bacterial cells. Modified plasmids are used as cloning vectors or vehicles.

Pleiotropy The appearance of several apparently unrelated phenotypic effects caused by a single gene.

Polar body A cell produced in the first or second division in female meiosis that contains little cytoplasm and will not function as a gamete.

Polygenic A phenotype that is dependent upon the interaction of a number of genes.

Polymorphism The occurrence of two or more genotypes in a population in frequencies that cannot be accounted for by mutation.

Polypeptide A polymer made of amino acids joined together by peptide bonds.

Polyploidy A chromosome number that is a multiple of the normal diploid chromosome set.

Polyps Growths attached to the substrate by small stalks. Commonly found in the nose, rectum, and uterus.

Polyribosome or polysome The complex formed with an mRNA molecule and several ribosomes.

Population A local group of organisms belonging to a single species, sharing a common gene pool; also called a deme.

Population bottleneck Fluctuations in size that occur when a large population is drastically reduced in size and then expands again; often results in an altered gene pool as a result of genetic drift.

Porphyria A genetic disorder inherited as a dominant trait that leads to intermittent attacks of pain and dementia, with symptoms first appearing in adulthood.

Postreplication repair A mechanism of DNA repair that utilizes the template strand to repair defects in DNA replication.

Prader-Willi syndrome A deletion of a small segment of the long arm of chromosome 15 that produces a syndrome characterized by uncontrolled eating and obesity.

Preformationism The idea that an organism develops by the growth of structures already present in the egg; in other words, the egg (or sperm) already contains a completely formed organism (the homunculus) that merely grows larger in the course of development.

Pre-mRNA The original transcript from a DNA strand, converted to mRNA molecules by removal of certain sequences, and addition of others.

Primary oocytes Cells in the ovary that undergo meiosis.

Primary spermatocytes Cells in the testis that undergo meiosis.

Primary structure The amino acid sequence in a polypeptide chain.

Prion An infectious protein that is the cause of several disorders, including Creutzfeld-Jacob syndrome.

Product The specific chemical compound that is the result of enzyme action. In biochemical pathways, a compound can serve as the product of one reaction and the substrate for the next reaction.

Progeria A genetic trait in humans associated with premature aging and early death.

Prokaryote An organism whose cells lack membrane-bound nuclei with true chromosomes. Cell division is usually by binary fission.

Promotor A region of a DNA molecule to which RNA polymerase binds and initiates transcription.

Promutagen A nonmutagenic compound that is a metabolic precursor to a mutagen.

Prophase A stage in mitosis during which the chromosomes become visible and split longitudinally except at the centromere.

Propositus An individual affected with a genetic disorder that led to the construction of a pedigree.

Proto-oncogene A cellular gene that functions in the regulation of cell growth.

Pseudoautosomal The pattern of inheritance for a gene located on both the X and Y chromosomes; in the absence of other information, such genes appear to be inherited in an autosomal fashion.

Pseudogenes A gene that closely resembles a gene at another locus, but is non-functional because of changes in its base sequences that prevent transcription or translation.

Pseudohermaphroditism An autosomal genetic condition that causes XY individuals to develop the phenotypic sex of females, but change to a male phenotype at puberty.

Punctuated equilibrium The idea that evolution can occur rapidly as the result of genetic changes in small, isolated populations.

Purines A class of double-ringed organic bases found in nucleic acids.

Pyrimidines A class of single-ringed organic bases found in nucleic acids.

Quaternary structure Structure formed by the interaction of two or more polypeptide chains in a protein.

R group A term used to indicate the position of an unspecified group in a chemical structure.

Race A genotypically distinct subgroup of a species.

Rad The *r*adiation *a*bsorbed *d*ose. A measurement of the radiation absorbed, as opposed to the amount produced. An amount equal to 100 ergs of energy absorbed per gram of irradiated tissue.

Radiation The process by which energy travels through space or a medium such as air.

Recessive The trait unexpressed in the F_1 but which is reexpressed in some members of the F_2 generation.

Reciprocal translocation A chromosomal aberration resulting in a positional change of a chromosome segment. This changes the arrangement of genes, but not the number of genes.

Recombinant DNA technology Techniques for joining DNA from two or more different organisms to produce hybrid or recombined DNA molecules.

Recombination The process of exchanging chromosome parts between homologous chromosomes during meiosis that produces new combinations of genetic information.

Reductional division A cell division that results in daughter cells that contain half the number of centromeres in the parental cell.

Regression to the mean In a polygenic system, the tendency of offspring of parents with extreme differences in phenotype to exhibit a phenotype that is the average of the two parental phenotypes.

Rem The *r*oentgen *e*quivalent *m*an. The amount of ionizing radiation that has the same biological effect as one rad of x-rays.

Restriction fragment length polymorphism (RFLP) Variations in the length of DNA fragments generated by a restriction endonuclease. Inherited in a codominant fashion, RFLPs are used as markers for specific chromosomes or genes.

Restriction enzymes Enzymes that recognize a specific base sequence in a DNA molecule and cleave or nick the DNA at that site.

Retinoblastoma A malignant tumor of the eye that arises in the retinal cells, usually occurring in children, with a frequency of 1 in 20,000. Associated with a deletion on the long arm of chromosome 13.

Retrovirus Viruses that use RNA as a genetic material. During their life cycle, the RNA is transcribed into DNA. The name retrovirus symbolizes this backward order of transcription.

Ribonucleic acid (RNA) A nucleic acid molecule that contains the pyrimidine uracil and the sugar ribose. The several forms of RNA function in gene expression.

Ribosomal RNA (rRNA) One of the components of the cellular organelles known as ribosomes.

Ribosomes Cytoplasmic particles composed of two subunits that are the site of gene product synthesis.

RNA polymerase An enzyme that catalyzes the formation of an RNA polynucleotide chain using a template DNA strand and ribonucleotides.

Robertsonian translocation Breakage in the short arms of acrocentric chromosomes followed by fusion of the long parts into a single chromosome.

Roentgen The amount of ionizing radiation that produces 2.083×10^9 ion pairs in 1 cubic centimeter of air.

Sandhoff disease A genetic disorder inherited as an autosomal recessive associated with a defect in production of hexosaminodase B, causing symptoms similar to those of Tay-Sachs disease.

Sarcoma A cancer of connective tissue. One type of sarcoma in chickens is associated with retrovirus known as the Rous sarcoma virus.

Secondary oocyte The haploid cell produced by meiosis that will become a functional gamete.

Secondary structure The pleated or helical structure in a protein molecule that is brought about by the formation of bonds between amino acids.

Segregation The separation of members of a gene pair from each other during gamete formation.

Selection The forces that bring about changes in the frequencies of alleles and genotypes in populations through differential reproduction.

Semiconservative replication A model of DNA replication that results in each daughter molecule containing one old strand and one newly synthesized strand. DNA replicates in this fashion.

Sense mutation A mutation that changes a termination codon into one that codes for an amino acid. Such mutations produce elongated proteins.

Severe combined immunodeficiency syndrome A disease characterized by the complete lack of ability to mount an immune response; inherited as an X-linked recessive and in another form as an autosomal recessive trait.

Sex chromosomes The chromosomes involved in sex determination. In humans, the X and Y chromosomes are the sex chromosomes.

Sex-influenced genes Loci that produce a phenotype that is conditioned by the sex of the individual.

Sex-linkage The pattern of inheritance that results from genes located on the X chromosome.

Sex-limited genes Loci that produce a phenotype that is produced in only one sex.

Sex ratio The relative proportion of males and females belonging to a specific age group in a population.

Sexual dimorphism The presence of morphological traits that distinguish males from females.

Sickle cell anemia A fatal recessive genetic disorder common in the U.S. black population associated with an abnormal type of hemoglobin, a blood transport protein.

Sickle cell trait The symptoms shown by those heterozygous for sickle cell anemia.

Sister chromatids Two chromatids joined by a common centromere.

Somatic cells All cells in the body other than those destined to become gametes.

Somatic cell hybridization A method of mapping human genes that uses hybrid cells produced by fusing together cells from two different organisms.

Species A group of actually or potentially reproducing organisms that are reproductively isolated from other populations.

Sperm Male haploid gametes produced by morphological transformation of spermatids.

Spermatids The four haploid cells produced by meiotic division of a primary spermatocyte.

Spermatogonia Mitotically active cells in the gonad that give rise to primary spermatocytes.

Stop codons Codons present in mRNA that signal the end of a growing polypeptide chain. UAA, UGA and UAG function as stop codons.

Submetacentric A chromosome with a centromere placed closer to one end than the other.

Substrate The specific chemical compound that is acted upon by an enzyme.

Synapsis The pairing of homologous chromosomes during prophase I of meiosis.

T cell immunodeficiency A disease characterized by a complete lack of T cell immunity but normal B cell immunity; may be inherited as an autosomal recessive trait.

T cells White blood cells that originate in bone marrow and undergo maturation in the thymus gland.

Tay-Sachs disease A fatal condition controlled by an autosomal recessive gene that causes deposition of unmetabolized intermediates in the nervous system.

Telocentric A chromosome with the centromere located at one end.

Telophase The last stage of mitosis during which division of the cytoplasm occurs and the chromosomes of the daughter cells condense and the nucleus reforms.

Template The single stranded DNA that serves to specify the nucleotide-sequence of a newly synthesized polynucleotide strand.

Teratogen Any physical or chemical agent that brings about an increase in congenital malformations.

Terminalization Movement of chiasmata toward the ends of chromosomes that occurs during diplotene stage of prophase I in meiosis.

Terminator codons Codons present in mRNA that signal the end of a growing polypeptide chain. UAA, UGA, and UAG function as terminator codons.

Tertiary structure The three-dimensional structure of a protein molecule brought about by folding on itself.

Testicular feminization An X-linked genetic trait that causes XY individuals to develop into phenotypic females.

Testis determining factor A gene located near the end of the short arm of the Y chromosome that plays a major role in causing the undifferentiated gonad to develop into a testis.

Testosterone A steroid hormone produced by the testis; the male sex hormone.

Tetrad The four homologous chromatids that are synapsed in the first meiotic prophase.

Tetraploidy A chromosome number that is four times the haploid number having four copies of all autosomes and four sex chromosomes.

Thalassemias A group of heritable hemoglobin disorders associated with an imbalance in the production of alpha and beta globins.

Thymine dimer A molecular lesion in which chemical bonds form between a pair of adjacent thymine bases in a DNA molecule.

Transcription Transfer of genetic information from the base sequence of DNA to the base sequence of RNA brought about by RNA synthesis.

Transfer RNA (tRNA) A small RNA molecule that contains a binding site for a specific type of amino acid and a three base segment known as an anticodon that recognizes a specific base sequence in messenger RNA.

Transformation The process of transferring genetic information between cells by means of DNA molecules.

Transgenic organism An organism whose genome has been modified by the introduction of external DNA sequences into the germ line.

Transitions Mutational events in which a purine is replaced by another purine or a pyrimidine is replaced by another pyrimidine.

Translation The process of converting the base sequence

in an RNA molecule into the linear sequence of amino acids in a protein.

Translocation A chromosomal aberration in which a chromosome segment is transferred to another, nonhomologous chromosome.

Transversions Mutational events in which a purine base in DNA is replaced by a pyrimidine, or a pyrimidine is replaced by a purine.

Trinucleotide repeat mutation A form of mutation associated with the expansion in copy number of a nucleotide triplet in or near a gene.

Triploidy A chromosome number that is three times the haploid number having three copies of all autosomes and three sex chromosomes.

Trisomy A condition in which one chromosome is present in three copies while all others are diploid having one more than the diploid number (2n + 1).

Trisomy 13 The presence of an extra copy of chromosome 13 that produces a distinct set of congenital abnormalities. Also called Patau syndrome.

Trisomy 18 The presence of an extra copy of chromosome 18 that results in a clinically distinct set of invariably lethal abnormalities. Also called Edwards syndrome.

Trisomy 21 An aneuploidy involving the presence of an extra copy of chromosome 21. Also called Down syndrome.

Tumor suppressor gene A gene that normally functions to suppress cell division.

Turner syndrome A monosomy of the X chromosome (45,X) that results in female sterility.

Unipolar disorder An emotional disorder characterized by prolonged periods of deep depression.

Variable region (V) A region at the N-terminus of an H or L chain that contains different amino acid sequences, even from one immunoglobulin molecule to another.

Vector A self-replicating DNA molecule used to transfer foreign DNA segments between host cells.

Werner's syndrome A Genetic trait in humans that causes aging to accelerate in adolescence, leading to death by about age 50.

Wilms tumor A malignant tumor of the kidney associated with a deletion on the short arm of chromosome 11.

Wobble hypothesis The idea that in the interaction between mRNA and tRNA, the first two bases of the codon are more important than the third base, so that some mismatching is allowed.

Xeroderma pigmentosum A class of genetic disorders characterized by sensitivity to sunlight, skin pigmentation, and skin cancer. Individuals with this disorder are unable to properly repair damage to DNA caused by ultraviolet light.

X-linkage The pattern of inheritance that results from genes located on the X chromosome

X-linked agammaglobulinemia (XLA) A rare sex-linked recessive trait characterized by the total absence of immunoglobulins and B cells.

XYY karyotype An aneuploidy of the sex chromosomes resulting in a male with an XYY chromosome constitution. Such males are disproportionately represented in penal institutions.

Y-linked Genes located on the Y chromosome.

Zygote The diploid cell resulting from the union of a male haploid gamete and a female haploid gamete.

INDEX

D

E

F

G

H

I

N

O

Y

Z

Figure and table credits (continued from copyright page)

Courtesy of Dr. Ira Rosenthal, Dept. of Pediatrics, Univ. of Illinois at Chicago. Figure 6.2: Reproduced by permission of Pediatrics Vol 74, p. 296, copyright 1984. Falix et al. 1984. Pediatrics 74: 296-299. Figure 6.5: Courtesy of Dr. Ira Rosenthal, Dept. of Pediatrics, Univ. of Illinois at Chicago. Figure 6.6: Courtesy of Dr. Ira Rosenthal, Dept. of Pediatrics, Univ. of Illinois at Chicago. Figure 6.7: Courtesy of Dr. Ira Rosenthal, Dept. of Pediatrics, Univ. of Illinois at Chicago. Figures 6.8 and 6.9: Reproduced, with permission, from The Annual Review of Genetics, Vol. 18, © 1984, by Annual Reviews, Inc. Figure 6.10: Courtesy of Dr. Ira Rosenthal, Dept. of Pediatrics, Univ. of Illinois at Chicago. Figure 6.11: From: Weiss, et al., 1982. Monozygotic twins discordant for Ulrich-Turner Syndrome. A. J. Med. Genet. *13*: 389-399. Figure 6.12: Martin M. Rotker, Taurus Photos. Figure 6.21: © Nancy Durrell McKena from *The Pregnancy and Birth Book* by Miriam Stoppard. Norling Kindersley, UK. 1985/Photo Researchers, Inc. Figure in Concepts & Controversies box on page 164: From Genest, et al. 1983. Ann Génétique *26*: 86-90. CHAPTER 7 Chapter Opener: © Lawrence Berkeley Laboratory, Photo Researchers. Figure 7.7: From: Franklin, R. and Gosling, R.G. 1953. Molecular configuration in sodium thymonucleate. Nature *171:* 740-741. Figure 7.11A Reprinted by permission. American Scientist. Nucleosomes: The structural quantum in chromosomes, by Olin, D. and Olin, A. *66:* 704-711. Figure 7.12: From: Harrison, C., et al. 1982. High resolution scanning electron microscopy of human metaphase chromosomes, J. Cell. Sci. *56:* 409-422. The Company of Biologists Limited. Figure 7.14: From: Taylor, J.H., Woods, P.S., and Hughes, W.L. 1957. The organization and duplicaton of chromosomes as revealed by autoradiographic studies using tritium-labled thymidine. Proc. Nat. Acad. Sci. *43*: 121-128. CHAPTER 8 Chapter Opener: Visuals Unlimited, Inc. Figure 8.3: Courtesy of Granada BioSciences, Inc. Figure 8.16: David Parker/Science Photo Library/Photo Researchers, Inc. Figure 8.20: Courtesy of Agricultural Research Service USDA. CHAPTER 9 Chapter Opener: Photograph courtesy of S.L. McKnight and O.L. Miller, Jr. Figure 9.4: Photo Researchers, Inc. Figure 9.13: Photograph courtesy of S.L. McKnight and O.O. Miller, Jr. CHAPTER 10 Chapter Opener: © CNRI/Phototake. Figure 10.15: Courtesy of B. Carragher, D. Bluemke, R. Josephs. From the Electron Microscopy and Image Processing Laboratory, University of Chicago. CHAPTER 11 Chapter Opener: Visuals Unlimited/© Ken Greer. Figure 11.1: From Haldane, J.B.S. and Poole, R. A new pedigree of recurrent bullous eruptions of the feet. J. Hered. *33*: 17-18. Copyright 1942, Am. Genet. Assn. Figure 11.7: From: Arlett, C.F. 1986. J. Inherited Metabolic Disease *9*:Suppl. 1, 69-84. By permission of MTP Press and SSIEM. CHAPTER 12 Chapter Opener: Visuals Unlimited. Figure 12.2: National Academy of Sciences. Tables 12.6 and 12.7: National Academy of Sciences. Figure 12.5: Courtesy of Dr. Robert Conard, Brookhaven Natl. Laboratory, Upton, NY. Figure 12.6: Wide World Photos. Figure 12.7: Wide World Photos. Figure 12.17: Courtesy of Dr. Lawrence Solomon, Dept. of Dermatology, Univ. of Illinois at Chicago. CHAPTER 13 Chapter Opener: Courtesy of Minnesota Timberwolves. Table 13.1: Courtesy of the American Cancer Society, Inc. Figure 13.9: From Kardos, G., et al. 1983. Leukemia in children with Down syndrome. Oncology *40*: 280-283. S. Karger, AG, Basel. Table 13.11: Courtesy of the American Cancer Society. Figure 13.13: Reprinted with permission. Wynder, E. and D. P. Rose, Diet and breast cancer. Hosp. Practice *19*: 73-88. Figure by Albert Miller. CHAPTER 14 Chapter Opener: Science Photo Library/Photo Researchers Inc. Figure 14.2: Andrejs Liepins/Science Photo Library/Photo Researchers Inc. Figure 14.10: From Singal, D. et al. 1969. Serotyping for homotransplantation. Transplantation *7*: 246-258. Figure 14.11: Wide World Photos. Figure 14.12: © Dr. Cecil Fox/Peter Arnold, Inc. CHAPTER 15 Chapter Opener: "Frederick the Great: Returning from Maneuvers" by Edward Francis Cunningham. The Granger Collection. Figure 15.8: From *Down's Anomaly*, by George Smith and J.M. Berg. Churchill, Livingstone, Edinburgh. Table 15.2: From British Med. J. *17*:247-250. Table 15.3 Reprinted from A twin study of human obesity by Albert Stunkard, Terryl T. Foch, and Zdenek Hrubec, J.A.M.A. *256*: 51-54. Copyright 1986, American Medical Association. Figure 15.10 From Davenport, C.B. The skin color of the races of mankind. Natural History *26*: 44-49. CHAPTER 16 Chapter Opener: Photo courtesy of Bettman Archives. Figure 16.6: Photo courtesy of Bettmen Archives. CHAPTER 17 Chapter Opener: Visuals Unlimited. Table 17.3: From Mourant, A.E. et al., 1976. *The Distribution of the Human Blood Groups and Other Polymorphisms.* Oxford Univ. Press. Figure 17.3: From Mourant, A.E. et al., 1976. *The Distribution of the Human Blood Groups and Other Polymorphisms.* Oxford Univ. Press. Table 17.5: From Mourant, A.E. et al., 1976. *The Distribution of the Human Blood Groups and Other Polymorphisms.* Oxford Univ. Press. Figure 17.6: From Friedlander, J. 1971. The population structure of South-Central Bougainville. Am. J. Phys. Anthropol. *35*: 13-26. Table 17.8 From Friedlander, J. 1971. The population structure of South-Central Bougainville. Am. J. Phys. Anthropol. *35*: 13-26. CHAPTER 18 Chapter Opener: Courtesy of Gamma Liaison. Figure 18.1: From Hirschfeld, L. and Hirschfeld, H. 1919. Serological differences between the blood of different races. Lancet *ii* 675-679. Figure 18.2: From Mourant, A.E. et al., 1976. *The Distribution of the Human Blood Groups and Other Polymorphisms.* Oxford Univ. Press.

Table 18.2: From Jenkins, T. et al., 1985. Ann. Human Biol. *12*: 363-371. Figure 18.5: Redrawn from *Genetics, Evolution, and Man* by W.F. Bodmer and L.L. Cavelli-Sforza © 1976 W.H. Freeman and Co. Figure 18.6. From Szeinberg, A. 1973. Investigation of genetic polymorphic traits in Jews. Israel J. Med. Sci. *9*: 1171-1179. Table 18.3: From Reed, S. 1980. *Counseling in Medical Genetics,* 3rd Edition. Alan R. Liss. Figure 18.12: Victor Englebert/Photo Researchers, Inc. Figure 18.13: Visuals Unlimited. Figure 18.14: Kolomburu Mission, N.W. Australia, Fr. Peter Weigand, O.S.B. Figure 18.15 From *Human Evolution* by J.B. Birdsell Copyright 1975, 1981 by Harper & Row, Publishers, Inc. Reprinted by permission of the publisher. Figure 18.16: From Reed, T.E. 1969. Caucasian genes in American Negroes. Science *165:* 762-768. 22 Aug. 1969. Copyright 1969 by A.A.A.S. CHAPTER 19 Chapter Opener: Courtesy of Dr. Michael Conneally, Indiana University. Figure 19.1a: Courtesy of Dr. McLone, Children's Memorial Hospital, Chicago, IL. Figure 19.1b: Custom Medical Stock Photo. Figure 19.2: Spencer Grant/Photo Researchers, Inc. Figure 19.3: Courtesy of Jason C. Birnholz, M.D. and Rush-Presbyterian-St. Luke's Medical Center, Chicago, Illinois. Table 19.6: From Selva, J. et al., 1986. Genetic screening for artifical insemination by donor (AID). Clin. Genet. *29*: 389-396. Table 19.7: Reprinted with permission from: Soc. Sci. Med. *20*: 601-607. Dagenais, D. et al. A cost benefit analysis of the Quebec Network of Genetic Medicine. Copyright 1985, Pergamon Journals, Ltd. CHAPTER 20 Chapter Opener: The Library of Congress/Photo Researchers Inc. Figure 20.1: Historical Picture Service, Chicago. Figure 20.3: Photo courtesy of American Philosophical Association. Figure 20.4: From Fehilly, C.B. et al. 1984. Interspecific chimerism between sheep and goat. Nature *307*: 634-636.

Color Insert I

"Tutankhamen on papyrus boat," 18th Dynasty, Egypt Museum, Cairo. Borromeo/Art Resource. (pg. 1).

Abraham van Wessel, *The Riddle of Nijmegan.* At Nijmegan, The Netherlands (pg. 2-3).

Francisco de Goya, *Mother with deformed child,* Musee du-Louvre, Paris, France. © Photo R.M.N. (pg. 4, left)

Juan Carreno de Miranda, *Eugenia Martinez Vellago.* All rights reserved © Museo del Prado, Madrid, Spain. (pg. 4, top right).

Edvard Munch. *Heritage II,* Oslo Kommunes Kunstsamlinger, Munch-Museet. (pg. 4, bottom right).

Petrus Gonsalvus, *Haarmenschen,* Kunsthistorisches Museum, Vienna (pg. 5, top left).

Henri Toulouse-Lautrec, *At the Moulin Rouge.* Oil on canvas, 48-3/8 × 55-1/4". Painted in 1892, Paris. Art Institute of Chicago, Helen Birch Bartlett Memorial Collection (pg. 5, bottom left).

Artist: A follower of Sanchez Coello, *Isabella Clara Eugenia with her dwarf.* All rights reserved © Museo del Prado, Madrid, Spain. (pg. 5, top right).

William Hogarth, *Marriage a la Mode: The Marriage Contract.* All rights reserved ©The National Gallery of London (pg. 6-7).

Artist unknown: *Portrait of Count Boruwalski.* Property of Prince Adam Karol Czartoryski de Borbon Orleans, deposited at the Czartoryski Foundation, National Museum in Cracow, Poland (pg. 7).

Antoine Dufour, *Vie des femmes celebres: Joan of Arc.* Nantes, Mus. Dobree. Giraudon/Art Resource (pg. 8, left).

Ivan Albright, Le Lorraine, *Among Those Left.* The Carnegie Museum of Art; Gift of the Artist, 49.24 (pg. 8, right).

Color Insert II

Pg. 1, top left: Chemical Design/Science Photo Library/Photo Researchers, Inc.

Pg. 1, top right: Courtesy of Oncor, Inc., Gaithersburg, MD and Dr. Huntington F. Willard, Stanford University, Stanford, CA.

Pg. 1, bottom: Lawrence Berkeley Laboratory

Pg. 2, top: ©Alexander Tsiarus/Science Source/Photo Researchers

Pg. 2, middle: Visuals Unlimited/©Elmer Koneman

Pg. 2, bottom: ©Andrew Leonard/The Stock Market

Pg. 3, top left: ©Ted Horowitz/The Stock Market

Pg. 3, top right: Prof. Stanley N. Cohen/Science Photo Library/Photo Researchers, Inc.

Pg. 3, bottom left: © Pfizer Inc./Phototake

Pg. 3, bottom right: VU/©K.G. Murti

Pg. 4, top left: © Roberto Poljak, P.F./Photo Researchers, Inc.

Pg. 4, middle left: Science VU/Visuals Unlimited

Pg. 4, middle right: CNRI/Science Photo Library/Photo Researchers, Inc.

Pg. 4, bottom right: Courtesy of Oncor, Inc.

Pg. 5, top left: Courtesy of Dr. Carl Withop

Pg. 5, top right: Kolomburu Mission, N.W. Australia, Fr. Peter Weigand, O.S.B.

Pg. 5, middle left: R. McNerling/Taurus Photos

Pg. 5, bottom right: Science VU/Visuals Unlimited

Pg. 6, top: Courtesy of Dr. Gerald Fishman, Dept. of Ophthalmology, Univ. of Illinois at Chicago.

Pg. 6, bottom left and right: Yvonne Alsip

Pg. 7, top: Courtesy of Dr. Raul Cano, California Polytechnical Institute

Pg. 8, top left and right: © 1984 Martin M. Rotker/Taurus Photos

Pg. 8, middle right: Courtesy of Dr. Michael Conneally, Indiana University.

Pg. 8, bottom: ©Ted Horowitz/The Stock Market

Organelle - A cytoplasmic structure having a specialised function

Vector - A self replicating DNA molecule used to transfer foreign DNA segments between host cells.

W9-CBU-738

Extraordinary USES*

*for ordinary things

Extraordinary USES*

*for ordinary things

FEATURING **Vinegar**, Baking Soda, **Salt**, Toothpaste, **String**, Plastic Cups, **Mayonnaise**, Nail Polish, **Tape**, and More Than 190 Other Common Household Items

2,317 WAYS TO SAVE MONEY **AND TIME**

Reader's Digest

The Reader's Digest Association, Inc.
Pleasantville, New York | Montreal

Project Staff

EDITOR Don Earnest

DESIGNERS Richard Kershner and Michele Laseau

CONTRIBUTING COPY EDITOR Jeanette Gingold

CONTRIBUTING INDEXER Nan Badgett

COVER ILLUSTRATION © Chuck Rekow

HUMOR ILLUSTRATION © Chuck Rekow

HOW-TO ILLUSTRATION © Bryon Thompson

Reader's Digest Home & Health Publishing

EDITOR IN CHIEF AND PUBLISHING DIRECTOR Neil Wertheimer

MANAGING EDITOR Suzanne G. Beason

ART DIRECTOR Michele Laseau

PRODUCTION TECHNOLOGY MANAGER Douglas A. Croll

MANUFACTURING MANAGER John L. Cassidy

MARKETING DIRECTOR Dawn Nelson

VICE PRESIDENT AND GENERAL MANAGER Keira Krausz

Reader's Digest Association, Inc.

PRESIDENT, NORTH AMERICA GLOBAL EDITOR-IN-CHIEF Eric W. Schrier

Text prepared especially for Reader's Digest by

NAILHAUS PUBLICATIONS, INC.

PUBLISHING DIRECTOR David Schiff

WRITERS Marilyn Bader, Serena Harding, Beth Kalet, Kathryn Kasturas, Michael Kaufman, Steven Schwartz, Anita Seline, Angelique B. Sharps, Delilah Smittle, and Amy Ziffer

© 2005 The Reader's Digest Association, Inc.

© 2005 The Reader's Digest Association (Canada) Ltd.

© 2005 Reader's Digest Association Far East Ltd.

Philippine Copyright 2005 Reader's Digest Association Far East Ltd.

All rights reserved. Unauthorized reproduction, in any manner, is prohibited.

Reader's Digest and the Pegasus logo are registered trademarks of The Reader's Digest Association, Inc.

ISBN: 0-88850-794-1

Library of Congress Cataloging-in-Publication Data
Extraordinary uses for ordinary things / Reader's Digest.-- 1st ed.
p. cm.
Includes index.
1. Home economics. I. Reader's Digest Association.
TX145.E95 2004
640--dc22
2004020058

Address any comments about *Extraordinary Uses for Ordinary Things* to:

The Reader's Digest Association (Canada) Ltd.
1100 René-Lévesque Blvd. West
Montreal, Quebec H3B 5H5

To order copies of *Extraordinary Uses for Ordinary Things*, call 1-800-465-0780.

Visit our website at www.rd.ca

Printed in the United States of America

06 07 08 / 5 4 3 2

Note to Readers

The information in this book has been carefully researched, and all efforts have been made to ensure accuracy and safety. Neither Nailhaus Publications, Inc. nor Reader's Digest Association, Inc. assumes any responsibility for any injuries suffered or damages or losses incurred as a result of following the instructions in this book. Before taking any action based on information in this book, study the information carefully and make sure that you understand it fully. Observe all warnings and Take Care notices. Test any new or unusual repair or cleaning method before applying it broadly, or on a highly visible area or valuable item. The mention of any brand or product in this book does not imply an endorsement. All prices and product names mentioned are subject to change and should be considered general examples rather than specific recommendations.

{Yes, Extraordinary}

Welcome to *Extraordinary Uses for Ordinary Things!* Inside you'll find thousands of ingenious, money-saving tips just like these. Jump in, raid your pantry, and start saving money today!

CLEAN YOUR DISHWASHER WITH KOOL-AID

Don't buy special powder to get rid of dishwasher iron deposits. Just dump in a packet of unsweetened Kool-Aid. It's a much cheaper way to make the inside of your dishwasher sparkle.

MAKE FLUFFY PANCAKES WITH CLUB SODA

Substitute club soda for the liquid called for in your favorite pancake or waffle recipe. You'll be amazed at how light and fluffy the breakfast treats will be.

REMOVE LIPSTICK STAINS WITH HAIR SPRAY

Got lipstick on your shirt? Apply hair spray and let it sit for a few minutes. When you wipe the spray off, the stain will come off with it.

HAVE A FACIAL WITH CAT LITTER

Make a deep-cleaning mud mask for your face with a couple of handfuls of cat litter. The clay in the litter detoxifies your skin by absorbing dirt and oil from the pores.

MELT SIDEWALK ICE WITH BAKING SODA

For an effective way to melt ice on steps and walkways, sprinkle them with generous amounts of baking soda mixed with sand. It won't stain or damage concrete surfaces.

GIVE CUT FLOWERS A LONGER LIFE WITH SODA POP

Pour about a quarter cup of soda pop into the water in that vase of flowers and the sugar in the drink will make the blossoms last longer.

TABLE OF CONTENTS

Most Useful Items
FOR JUST ABOUT ANYTHING

YOUR COMPLETE A-Z Guide

Denotes a SUPER ITEM: A household item with an amazingly large number of uses.

DISCOVER WHAT'S HIDING IN Your Cupboard

Once upon a time, in the days before computers, cable television, drive-through coffee bars, and carpet-sweeping robots, washing windows was a simple affair. Our parents poured a little vinegar or ammonia in a pail of water, grabbed a cloth, and in no time had a clear view of the outside world through gleaming glass. Then they would use the same combination to banish grime and grit from countertops, walls, shelves, fixtures, floors, and a good bit else of the house.

Some things, like window washing, shouldn't ever get more complicated. But somehow, they did. Today, store shelves are laden with a dazzling array of cleaning products, each with a unique use, a special formula, and a multimillion-dollar advertising campaign. Window cleaners alone take up shelves and shelves of space. The bottles are filled with colorful liquids and have labels touting their orange power, berry bouquet, or lemon or apple herbal scent. Ironically, many boast the added power of vinegar or ammonia as their "secret" ingredient.

This is the way of the world today. Every problem, every mess, every hobby, every daily task seems to require special tools, unique products, and extensive know-how. Why use a knife to chop garlic when there are 48 varieties of garlic presses available? Why use a rag for cleaning when you have specialized sponges, wipes, Swiffers, magnetically charged dusters, and HEPA-filter vacuums?

Which brings us to the point of this book: Why *not* just use a solution of vinegar or ammonia like our grandparents did to clean the windows? It works just as well as those fancy products—if not better. And it costs only about a quarter as much, sometimes less.

204 Everyday Items with Over 2,300 Uses

Making do with what you've already got. It's an honorable, smart, money-saving approach to life. And in fact, it can be downright fun. Sure you can buy a fancy lint brush to remove cat hairs from pants, but it's pretty amazing how a penny's worth of tape does the job even better. Yes, you can use strong kitchen chemicals to clean the inside of a vase that held its flower water a bit too long. But isn't it more entertaining—and easier—to use a couple of Alka-Seltzer tablets instead to fizz away the mess?

Welcome to *Extraordinary Uses for Ordinary Things*. On the following pages, we'll show you more than 2,300 ingenious ways to use 204 ordinary household products to restore, replace, repair, or revive practically everything in and around your home or to pamper yourself or entertain your kids. You'll save time and money—and you'll save shelf space because you won't need all those different kinds of specialized commercial preparations. You'll even save on gasoline, because you won't need to speed off to the mall every time you run out of a staple such as air freshener, shampoo, oven cleaner, or wrapping paper.

The household items featured in this book are not costly commercial concoctions. Rather, they are everyday items that you're likely to find in your home—in your kitchen, medicine cabinet, desk, garage, and even your wastebasket. And you'll be amazed by how much you can actually accomplish using just a few of the most versatile of these items, such as baking soda, duct tape, pantyhose, salt, vinegar, and WD-40. In fact, there's a popular maxim among handymen that whittles the list to a pair of basic necessities: "To get through life," the saying goes, "you only need two tools: WD-40 and duct tape. If it doesn't move, but should, reach for the WD-40. If it moves, but shouldn't, grab the duct tape."

Less Toxic and More Earth-Friendly Items

In addition to saving you time and money, there are other, less tangible advantages to using these everyday household products. For one thing, many of the items are safer to use and considerably more environmentally friendly than their off-the-shelf counterparts. Consider, for example, using vinegar and baking soda to clear a clogged bathroom or kitchen drain (page 64). It's usually just as effective as a commercial drain cleaner. The only difference is that the baking-soda-and-vinegar combination is far less caustic on your plumbing. Plus, you don't have to worry about getting it on your skin or in your eyes.

The hints in this book will also help you reduce household waste by giving you hundreds of delightful and surprising suggestions for reusing many of the items that you would otherwise toss in the trash or recycling bin. To name a few, these include lemon rinds and banana peels, used tea bags and coffee grinds, orphaned socks and worn-out pantyhose, plastic bags, empty bottles and jugs, cans, and newspapers.

At the end of the day, you'll experience the distinct pleasure that can only come from learning creative, new ways to use those familiar objects around your house that you always thought you knew so well. Even if you'll never use Alka-Seltzer tablets to lure fish onto your line, or need to plug a hole in your car radiator with black pepper, isn't it great to know you can?

Folk Wisdom for the 21st Century

As we noted earlier, much of the advice you'll find in *Extraordinary Uses for Ordinary Things* is not really new—it's just new to us. After all, "Waste not, want not" isn't merely a quaint adage from a bygone era; it actually defined a way of life for generations. In the days before mass manufacturing and mass marketing transformed us into a throwaway society, most folks knew perfectly well that salt and baking soda (or bicarbonate of soda, as it was commonly referred to in those times) had dozens upon dozens of uses.

Now, as landfills swell, and we realize that the earth's resources aren't really endless, there are signs of a shift back to thrift, so to speak. From recycling programs to energy-efficient appliances to hybrid cars, we're constantly looking for new ways to apply the old, commonsense values of our forebears. Even the International Space Station is an example of thrifty technologies at work today. When the station is completed, nearly every waste product and used item onboard the craft will be recycled for another purpose.

How We Put This Book Together

Of course, our number one priority was to provide you with the most reliable information available. To meet that goal, we conducted countless interviews with experts on everything from acne cures to yard care and scrutinized stacks of research materials. We also performed numerous hands-on tests in our own kitchens, living rooms, bathrooms, and other areas around our homes.

The result, we believe, is the most comprehensive and dependable guide to alternative uses for household products you can find. Like a great stew, we've combined some time-honored, traditional tips (such as using apple cider vinegar to kill weeds in your garden) with new tips given to us by various reliable sources (like recycling used fabric softener sheets to clean PC and TV screens), and a few tips that we came up with on our own (such as using bathtub appliqués to steady a legless PC case).

As in any comprehensive compilation such as this, the practical wisdom contained in this book is as much art as it is science. That is to say, although we employed trial-and-error methods wherever possible to provide you with specific amounts and clear directions for using household products and objects to obtain the desired results, we can't guarantee that these solutions will work in all situations. In other words, your mileage may vary.

Moreover, while we're confident that every one of the 204 products used for all the tips included here is generally safe and effective when used as directed, please pay close attention to our "Take Care" warnings about using, storing, and especially combining certain products, particularly *bleach* and *ammonia*. Under no conditions should you ever mix these two chemicals, or use them in poorly ventilated areas.

What's on the Following Pages

The main part of this book is arranged like an encyclopedia, with the 204 product categories organized in A-Z fashion (running from Adhesive Tape to Zucchini), to provide instant access to information as well as entertaining reading. But before that, in the first part of this book, you'll find a guide to the items that are most useful for certain areas, such as the garden or for cooking.

Scattered throughout, you'll also find hundreds of fascinating asides and anecdotes. Some highlight specific warnings and safety precautions, or offer advice about buying or using certain items. But many are just plain fun—providing quirky historical information about the invention or origins of products. Haven't you always wondered who invented the Band-Aid or how Scotch tape got its name? We've also included dozens of engaging and enlightening activities and simple science experiments you can do with your children or grandchildren (and not a single one requires a visit to the local toy store).

Whether you delight in discovering new ways to use commonplace household items, or if you simply hate to throw things away, we're sure you'll find the ideas in this book entertaining and enlightening. So, pull up a comfortable chair, settle back, and get ready to be dazzled by the incredible number of everyday problems that you'll soon be able to solve with ease. We're confident this is one book you'll return to over and over again for helpful hints, trustworthy advice, and even some good, old-fashioned inspiration.

—The Editors

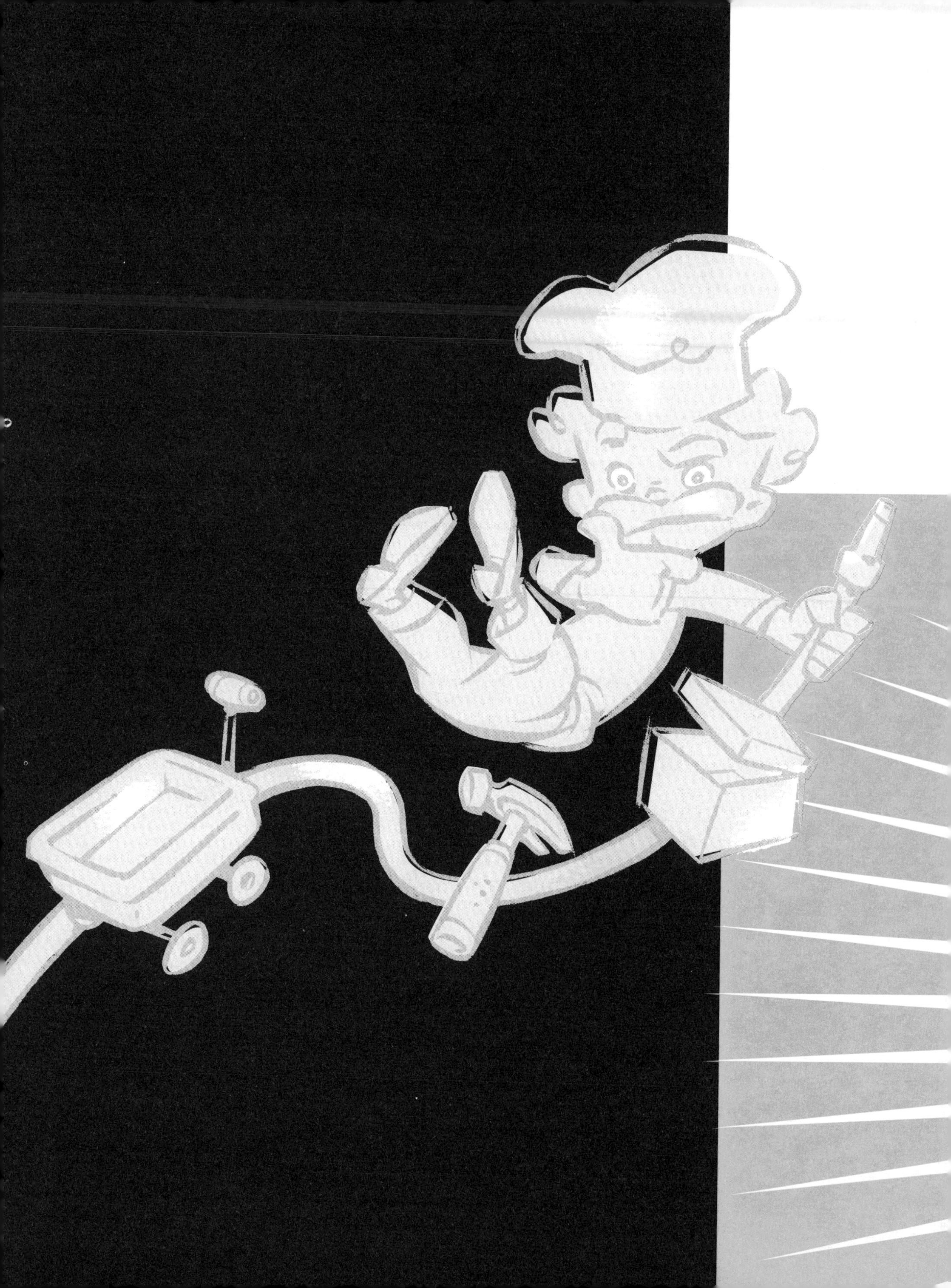

MOST Useful Items* FOR JUST ABOUT ANYTHING

* If you have a special interest, such as cooking or health and beauty, you'll soon discover that certain household items are especially useful. There are, for example, close to a dozen uses for plastic bottles in the garden. On these pages, you'll find these helpful items listed for most everyday areas of interest.

TOP 10 Most Useful Items for around the house

pad a package with a disposable diaper ...

p. 152

Carpet scraps { PAGE 115 }

Make an exercise mat, car mat, or knee pad. • Muffle appliance noise. • Protect floor under plants. • Cushion kitchen shelves. • Give your car traction. • Protect tools.

Compact discs {PAGE 138}

Use as holiday ornaments, driveway reflectors, or a circle template. • Catch candle drips. • Make an artistic bowl, decorative sun catcher, or clock.

Duct tape {PAGE 153}

Remove lint. • Repair toilet seats, screens, vacuum hoses, and frames. • Cover a book, pocket folder, or present. • Make a bandage or bumper sticker. • Catch flies. • Replace a grommet. • Reinforce a book binding. • Hem pants. • Hang Christmas lights. • Make Halloween costumes.

Fabric softener sheets {PAGE 167}

Freshen air and deodorize cars, dogs, gym bags, suitcases, and sneakers. • Pick up pet hairs. • Repel mosquitoes. • Stop static cling. • Make sheets smell good. • End tangled sewing thread.

Nail polish
{PAGE 217}

Mark hard-to-see items. • Mark thermostat and shower settings and levels in measuring cups and buckets. • Label sports gear and poison containers. • Seal envelopes and labels. • Stop shoe scuffs and keep laces, ribbons, and fabric from unraveling. • Make needle-threading easier. • Keep buckles and jewelry shiny. • Stop a stocking run. • Temporarily repair glasses. • Fix nicks in floors and glass. • Repair lacquered items. • Plug a hole in a cooler. • Fill washtub nicks.

Pantyhose
{PAGE 236}

Find and pick up small objects. • Buff shoes. • Keep hairbrush clean. • Remove nail polish. • Keep spray bottles clog-free. • Organize suitcases. Hang-dry sweaters. • Secure trash bags. • Dust under fridge. • Prevent soil erosion in houseplants.

Paper bags
{PAGE 242}

Pack on trips for souvenirs. • Dust off mops. • Carry laundry. • Cover textbooks. • Create a table decoration. • Use as gift bags and wrapping paper. • Reshape knits after washing. • Use as a pressing cloth. • Bag newspapers for recycling.

Plastic bags
{PAGE 261}

Keep mattresses dry. • Bulk curtains and stuff crafts. • Drain bath toys. • Clean pockets in the laundry. • Make bibs and a high-chair drop cloth. • Line a litter box. • Dispose of a Christmas tree.

Rubber bands
{PAGE 276}

Reshape your broom. • Childproof cabinets. • Keep thread from tangling. • Make a holder for car visor. • Use to grip paper. • Extend a button. • Use as a bookmark. • Cushion a remote control. • Secure bed slats and tighten furniture casters.

Sandwich and freezer bags
{PAGE 292}

Protect pictures and padlocks. • Dispense fabric softener. • Display baby teeth. • Carry baby wipes. • Mold soap. • Starch craft items. • Feed birds. • Make a funnel.

TOP

12 Most Useful Items for the cook

the secret to perfect poached eggs is vinegar ...

p. 354

Aluminum foil
{PAGE 39}
Bake a perfect piecrust. • Soften brown sugar. • Decorate a cake and create special-shaped pie pans. • Keep rolls and bread warm. • Make an extra-large salad bowl. • Make a toasted cheese sandwich with an iron.

Apples
{PAGE 53}
Keep a roast chicken moist and cakes fresh. • Ripen green tomatoes. • Fluff up hardened brown sugar. • Absorb excess salt in soups.

Baking soda
{PAGE 62}
Clean fruits and vegetables. • Remove fish smells. • Reduce the acidity of coffee and tomato-based sauces. • Reduce the gas-producing properties of beans. • Make fluffy omelets. • Replace yeast.

Coffee filters
{PAGE 133}
Cover food in microwave. • Filter cork crumbs from wine or food remnants from cooking oil. • Hold a taco, ice cream bar, or ice pop.

Ice cube trays
{PAGE 187}
Freeze eggs, pesto, chopped vegetables and herbs, chicken soup—even leftover wine—for future use.

Lemons
{PAGE 198}
Prevent potatoes from turning brown or rice from sticking. • Keep guacamole green. • Make soggy lettuce crisp. • Freshen the fridge and cutting boards.

Paper towels
{PAGE 248}
Microwave bacon, clean corn, and strain broth. • Keep vegetables crisp and vegetable bin clean. • Prevent soggy bread and rusty pots.

Plastic bags
{PAGE 264}
Cover a cookbook. • Bag hands to answer phone. • Crush graham crackers. • Use as mixing bowl or salad spinner. • Ripen fruit.

Rubber bands
{PAGE 276}
Keep spoons from sliding into bowls. • Secure casserole lids for travel. • Anchor a cutting board. • Get a better grip on twist-off lids and glasses.

Salt
{PAGE 283}
Prevent grease from splattering. • Speed cooking. • Shell hard-boiled eggs or pecans easier. • Test eggs for freshness and poach eggs perfectly. • Wash spinach better. • Keep salad crisp. • Revive wrinkled apples and stop cut fruit from browning. • Use to whip cream, beat eggs, and keep milk fresh. • Prevent mold on cheese.

Sandwich and freezer bags
{PAGE 293}
Store grater with cheese. • Make a pastry bag. • Dispose of cooking oil. • Color cookie dough. • Keep ice cream from forming crystals. • Soften marshmallows, melt chocolate, and save soda. • Grease pans.

Toothpicks
{PAGE 336}
Mark steaks for doneness. • Retrieve garlic cloves from marinade. • Prevent pots from boiling over. • Microwave potatoes faster. • Limit salad dressing. • Fry sausages better.

TOP

11 Most Useful Items for health and beauty

ease a backache with meat tenderizer ...

p. 210

Aspirin {PAGE 55} Dry up pimples. • Treat calluses. • Control dandruff. • Cut inflammation from bites and stings. • Restore hair color after swimming in chlorinated pool.

Baby oil {PAGE 58} Remove a bandage painlessly. • Treat cradle cap. • Make bath oil. • Make hot oil treatment for cuticles and calluses.

Baking soda {PAGE 69} Soothe minor burns, sunburn, poison ivy rash, bee stings, diaper rash, and other skin irritations. • Combat cradle cap. • Control dandruff. • Use as gargle or mouthwash. • Scrub teeth and clean dentures. • Alleviate itching in casts and athlete's foot. • Soothe tired, stinky feet. • Remove built-up hair gel, spray, or conditioner. • Use as an antiperspirant.

Butter {PAGE 101} Make pills easier to swallow. • Soothe aching feet. • Remove sap from skin. • Remove makeup. • Smooth legs after shaving. • Use as shaving cream. • Moisturize dry hair.

Chest rub
{PAGE 122}

Make calluses disappear. • Sooth aching feet. • Stop insect-bite itch. • Treat toenail fungus. • Repel biting insects.

Lemons
{PAGE 200}

Disinfect cuts and scrapes. • Soothe poison ivy rash. • Relieve rough hands and sore feet. • Remove warts. • Lighten age spots. • Create blond highlights. • Clean and whiten nails. • Cleanse and exfoliate your face. • Treat dandruff. • Soften dry elbows.

Mayonnaise
{PAGE 204}

Relieve sunburn pain. • Remove dead skin. • Condition hair. • Make a facial. • Strengthen fingernails.

Mustard
{PAGE 216}

Soothe aching back pain. • Relax stiff muscles. • Relieve congestion. • Make a facial mask.

Petroleum jelly
{PAGE 254}

Heal windburn. • Help diaper rash. • Protect baby's eyes from shampoo. • Moisturize lips. • Remove makeup. • Moisturize your face. • Create makeup. • Strengthen perfume. • Soften hands. • Do a professional manicure. • Smooth eyebrows.

Tea
{PAGE 327}

Relieve tired eyes. • Soothe bleeding gums. • Cool sunburn. • Relieve baby's pain from injection. • Reduce razor burn. • Condition dry hair and get the gray out. • Tan your skin. • Drain a boil. • Soothe nipples sore from nursing. • Soothe mouth pain.

Vinegar
{PAGE 355}

Control dandruff and condition hair. • Protect blond hair from chlorine. • Apply as antiperspirant. • Soak aching muscles. • Freshen breath. • Ease sunburn and itching. • Banish bruises. • Soothe sore throat. • Clear congestion. • Heal cold sores and athlete's foot. • Pamper skin. • Erase age or sun spots. • Soften cuticles. • Treat jellyfish or bee stings.

TOP

10 Most Useful Items for cleaning

clean fireplace doors with ashes …

p. 54

Ammonia
{PAGE 49}

Clean carpets, upholstery, ovens, fireplace doors, windows, porcelain fixtures, crystal, jewelry, and white shoes. • Remove tarnish and stains. • Fight mildew. • Strip floor wax.

Baking soda
{PAGE 62}

Clean baby bottles, thermoses, cutting boards, appliances, sponges and towels, coffeemakers, teapots, cookware, and fixtures. • Clear clogged drains. • Deodorize garbage pails. • Boost dishwashing liquid or make your own. • Remove stains. • Shine jewelry, stainless steel, chrome, and marble. • Wash wallpaper and remove crayon. •Remove must.

Borax
{PAGE 89}

Clear a clogged drain. • Remove stains. • Clean windows and mirrors. • Remove mildew from fabrics. • Sanitize your garbage disposal. • Eliminate urine odor.

Fabric softener sheets
{PAGE 167}

Lift burned-on food. • Freshen drawers. • Remove soap scum. • Repel dust from TV screen. • Freshen hampers and wastebaskets. • Buff chrome. • Keep dust off blinds. • Renew stuffed toys.

Lemons {PAGE 196}

Get rid of tough stains on marble. • Polish metals. • Clean the microwave. • Deodorize cutting boards, fridge, and garbage disposal.

Rubbing alcohol {PAGE 278}

Clean fixtures, venetian blinds, windows, and phones. • Remove hair spray from mirrors. • Prevent ring around the collar. • Remove ink stains.

Salt {PAGE 280}

Clean vases, discolored glass, flowerpots, artificial flowers, percolators, refrigerators, woks, and wicker. • Give brooms long life. • Ease fireplace or flour cleanup. Make metal polish. • Remove wine and grease from carpet, water marks from wood, and lipstick from glasses. • Restore a sponge. • Freshen the garbage disposal. • Remove baked-on food. • Soak up oven spills. • Remove stains from pans and clean cast iron.

Toothpaste {PAGE 334}

Clean piano keys and sinks. • Polish metal and jewelry. • Deodorize baby bottles. • Remove ink or lipstick from fabric, crayon from walls, and water marks from furniture.

Vinegar {PAGE 343}

Clean blinds, bricks, tile, paneling, carpets, piano keys, computers, appliances, and cutting boards. • Clean china, crystal, glassware, coffeemakers, and cookware. • Banish kitchen grease. • Deodorize drains and closets. • Polish metal. • Erase ballpoint pen marks. • Remove water rings and wax from furniture. • Revitalize leather. • Clean fixtures and purge bugs.

WD-40 {PAGE 377}

Remove carpet stains and floor scuffs. • Remove tea and tomato stains. • Clean toilet bowls. • Condition leather furniture. • Clean a chalkboard. • Remove marker and crayon from walls.

TOP

11 Most Useful Items for the

keep aphids off rosebushes with banana peels ...

p. 85

Aluminum foil
{PAGE 43}

Create a sun box for window plants or an incubator for seedlings. • Mix with mulch to deter insects. • Hang strips to scare crows and other birds. • Wrap tree trunks to prevent sunscald or to keep nibbling mice and rabbits away. • Keep cuttings from getting tangled.

Bottles
{PAGE 93}

Make a bird feeder, a gutter scoop, a watering can, or an individual drip irrigator for a plant. • Secure netting over flowerbeds. • Isolate weeds when spraying. • Cover seed-packet markers or make plant tags from cut strips. • Use as trash can on mower. • Use to space seeds. • Trap bugs.

Coffee cans
{PAGE 131}

Make a sprinkler to spread seeds and fertilizer. • Measure rain to ensure your garden is getting enough water • Make a bird feeder.

Milk cartons
{PAGE 212}

Make a bird feeder. • Use as a seed starter. • Make a collar to protect vegetables. • Collect kitchen scraps for compost.

Newspaper
{PAGE 224}

Protect and ripen end-of-season tomatoes. • Use as mulch or add to compost to remove odor. • Block weeds in flower and vegetable beds. • Get rid of earwigs.

Pantyhose
{PAGE 240}

Tie up tomatoes and beans. • Fill with hair clippings to repel deer. • Make a hammock for growing melons. • Store onions and off-season bulbs. • Prevent soil lost in houseplants. • Fill with soap scraps for cleaning hands at garden spigot.

Plastic bags
{PAGE 265}

Protect plants from frost and shoes from mud. • Speed budding of poinsettias and Christmas cactus. • Keep bugs off fruit on trees. • Store outdoor equipment manuals. • Bring a favorite cracked vase back into use. • Make disposable work aprons.

Plastic containers
{PAGE 267}

Make traps for slugs and wasps. • Stop ants from crawling up picnic table legs. • Use to start seedlings.

Salt
{PAGE 287}

Kill snails and slugs. • Inhibit the growth of weeds in walkway cracks. • Extend the life of cut flowers. • Clean flowerpots.

Tea
{PAGE 330}

Spur growth of rosebushes. • Water acid-loving plants. • Nourish houseplants. • Prepare a planter for potting. • Speed the decomposition of compost.

WD-40
{PAGE 379}

Keep animals out of flowerbeds and squirrels off bird feeders. • Keep tool handles from splintering. • Stop snow from sticking on shovel or snow thrower. Prevent wasps from building nests and repel pigeons. • Kill thistle plants.

TOP

11 Most Useful Items for outdoors

lubricate skate wheels with hair conditioner ...

p. 181

Aluminum foil

{PAGE 44}

Improve outdoor lighting. • Keep bees away from beverages. • Make an impromptu picnic platter or improvise a frying pan. • Make a drip pan for your barbecue and clean the grill. • Warm your toes when camping and keep your sleeping bag and matches dry. • Make a fishing lure.

Baking soda

{PAGE 74}

Keep weeds out of concrete cracks. • Clean plastic resin lawn furniture. • Feed your flowering plants. • Maintain proper pool alkalinity. • Scour the barbecue grill.

Bottles

{PAGE 93}

Use a plastic jug to make a scoop or bailer for your boat. • Make a plastic jug into an anchor for your boat and for landlubber jobs. • Make a bird feeder. • Fill a plastic jug with sand or cat litter for winter traction. • Keep your cooler cold.

Bubble pack

{PAGE 98}

Keep soft drinks cold. • Sleep on air while camping. • Cushion bleachers and benches.

Buckets {PAGE 99}
Boil lobster over the campfire. • Use as a footlocker. • Build a camp washing machine or make a camp shower.

Cat litter {PAGE 118}
Give your car traction on ice. • Prevent barbecue grease fires. • Keep tents and sleeping bags must-free. • Remove grease spots from driveway.

Cooking spray {PAGE 140}
Prevent grass from sticking to your mower. • Spray on fishing line for quicker casting. • Prevent snow from sticking to your shovel or your snow-thrower chute.

Duct tape {PAGE 157}
Seal out ticks. • Create a clothesline. • Stash a secret car key. • Patch a canoe or a pool. • Repair outdoor cushions and replace lawn furniture webbing. • Make bike streamers. • Tighten hockey shin guards and revive your hockey stick. • Preserve skateboarders' shoes. • Repair your ski gloves or a tent. • Waterproof your footwear.

Sandwich and freezer bags {PAGE 296}
Inflate to make valuables float when boating. • Make a hand cleaner for the beach. • Apply bug spray with ease.

Vinegar {PAGE 363}
Keep water fresh. • Clean outdoor furniture and decks. • Repel insects. • Trap flying insects. • Get rid of ants. • Clean off bird droppings.

WD-40 {PAGE 381}
Repel pigeons and wasps. • Waterproof shoes. • Remove wax from skis and snowboards. • Remove barnacles from boat and protect it from corrosion. • Untangle your fishing line and lure fish. • Clean and protect golf clubs. • Remove burrs from a horse's mane and protect hooves in winter. • Keep flies off cows.

TOP

13 Most Useful Items for storage

stash your valuables at the gym in a tennis ball …

p. 331

Baby wipe containers {PAGE 61}
Organize sewing supplies, recipe cards, coupons, craft supplies, old floppy disks, small tools, photos, receipts, bills, and more. • Store plastic shopping bags. • Store towels and rags.

Bottles {PAGE 91}
Store sugar. • Store small workshop items. • Use as a boot tree. • Make a bag or string dispenser.

Cans {PAGE 104}
Compartmentalize your tool pouch with juice cans. • Make a desk organizer. • Create pigeonholes to store silverware, nails, office supplies, and other odds and ends.

Candy tins {PAGE 107}
Make an emergency sewing kit. • Store broken jewelry. • Make a birthday keepsake. • Prevent jewelry chain tangles. • Organize a sewing box. • Store car fuses. • Keep earrings together. • Store workshop items.

Cardboard boxes {PAGE 109}
Make magazine holders from detergent boxes. • Make a home-office in-box. • Store hoes, rakes, and other long-handled garden tools. • Protect glassware or lightbulbs. • Store posters and artwork. • Store

Christmas ornaments. • Organize dowels, moldings, furring strips, 2x2s, and metal rods.

Cardboard tubes
{PAGE 112}
Store knitting needles and fabric scraps. • Keep Christmas lights tidy. • Preserve kids' artwork, important documents, and posters. • Keep linens crease-free, pants wrinkle-free, and electrical cords tangle-free. • Protect fluorescent lights. • Store string.

Clothespins
{PAGE 125}
Keep snacks fresh. • Organize workshop, kitchen, bathroom, and closets. • Keep gloves in shape.

Coffee cans
{PAGE 131}
Make a kids' bank. • Hold kitchen scraps. • Carry toilet paper when camping. • Store screws, nuts, and nails. • Organize and store belts. • Collect pocket stuff in the laundry.

Egg cartons
{PAGE 161}
Store and sort coins. • Organize buttons, safety pins, threads, bobbins, and fasteners. • Store golf balls or Christmas ornaments.

Film canisters
{PAGE 169}
Make a stamp dispenser or a sewing kit. • Organize pills. • Store fishing flies. • Carry change for tolls. • Stash jewelry at the gym. • Carry dressings, cooking spices, and condiments. • Carry nail polish remover.

Pantyhose
{PAGE 236}
Store wrapping paper. • Bundle blankets. • Store onions or flower bulbs.

Plastic bags
{PAGE 261}
Store wipes. • Collect used clothes. • Protect clothes. • Store skirts. • Keep purses in shape.

Sandwich and freezer bags
{PAGE 292}
Store breakables. • Save sweaters. • Create a sachet. • Add cedar to a closet. • Make a pencil bag. • De-clutter the bath.

TOP

15 Most Useful Items for kids

make finger paints from yogurt ...

p. 386

Aluminum pie pans {PAGE 47}
Use as mold for ice ornaments. • Minimize glitter mess and make trays for craft supplies.

Baking soda {PAGE 66}
Make watercolor paints or invisible ink. • Produce gas to blow up a balloon. • Clean crayon marks from walls and baby spit-up from clothing. • Combat cradle cap and diaper rash. • Wash chemicals out of new baby clothes.

Bathtub appliqués {PAGE 81}
Stick to bottom of kiddie pool. • Affix to sippy cups and high-chair seats.

Cardboard boxes {PAGE 110}
Make a medieval castle, a puppet theater, or a sundial. • Make a garage for toy vehicles. • Store tennis rackets, baseball bats, fishing poles, and other sporting goods. • Use as an impromptu sled. • Play beverage-box ski-ball.

Cardboard tubes {PAGE 112}
Make a kazoo or a megaphone. • Preserve your kids' artwork. • Build a toy log cabin. • Make no-gunpowder "English" firecrackers.

Compact discs {PAGE 138} Make a fun picture frame. • Make wall art for a teen's room. • Create spinning tops.

Corks {PAGE 141} Use burnt cork as Halloween face paint. • Create a craft stamp. • Make a cool bead curtain for a kid's room.

Duct tape {PAGE 155} Make Halloween costumes, a toy sword, or hand puppets. • Create bicycle streamers.

Jars {PAGE 190} Make a savings bank. • Dry kids' mittens. • Bring along baby treats and store baby-food portions. • Collect insects. • Create a miniature biosphere.

Margarine tubs {PAGE 205} Make a baby footprint paperweight. • Divvy up ice cream so kids can help themselves. • Bring fast food for baby. • Make a coin bank. • Give kids some lunch-box variety.

Paper bags {PAGE 242} Cover textbooks. • Make a kite. • Create a life-size body poster.

Paper plates {PAGE 248} Make index cards and Frisbee flash cards. • Make crafts, mobiles, and seasonal decorations.

Pillowcases {PAGE 258} Prepare travel pillows for kids. • Make wall hangings for kids' rooms. • Clean stuffed animals.

Sandwich and freezer bags {PAGE 292} Display baby teeth. • Make baby wipes. • Dye pasta for crafts. • Make kids' kitchen gloves. • Make a pencil bag. • Keep spare kids' clothes in car for mishaps. • Cure car sickness. • Play football while making pudding.

Tape {PAGE 326} Secure a baby's bib. • Create childproofing in a pinch. • Make a multicolor pen. • Make an unpoppable balloon.

TOP

14 Most Useful Items for quick repairs

repair your garden hose with a toothpick ...

p. 338

Adhesive tape {PAGE 36} Remove broken window glass safely. • Hang caulk tubes for storage. • Get a better grip on tools.

Aluminum foil {PAGE 45} Make flexible funnel for hard-to-reach places. • Reflect light for photography. • Reattach vinyl floor. • Make an artist's palette. • Prevent paint from skinning over. • Line roller pans and keep paint off doorknobs.

Baking soda {PAGE 73} Clean car-battery terminals and remove tar from car. • Use as a walkway de-icer. • Tighten cane chair seats. • Give deck a weathered look. • Clean air conditioner filters • Keep humidifier odor-free.

Basters {PAGE 78} Cure musty air conditioner. • Transfer paints and solvents. • Fix leaky refrigerator.

Bottles {PAGE 78} Make a neater paint bucket. • Store paints. • Make a workshop organizer. • Use as a level. • Make an anchor for weighting tarps and patio umbrellas.

Bubble pack {PAGE 78} Prevent toilet-tank condensation. • Insulate windows. • Cushion work surface and protect tools.

Buckets
{PAGE 99}

Hold paint and supplies when painting on a ladder and use lids to contain paint drips. • Make stilts for painting ceiling. • Organize extension cords. • Soak your saw to clean. • Use as a Christmas tree stand.

Cardboard boxes
{PAGE 111}

Make a temporary roof repair. • Protect fingers while hammering small nails. • Make an oil drip pan. • Identify fluid leaking from your car. • Make a bed tray. • Make an in-box. • Organize workshop. • Keep upholstery tacks straight.

Clothespins
{PAGE 125}

Clamp thin objects. • Make a clipboard. • Grip a nail to protect fingers. • Float paintbrushes in solvent.

Duct tape
{PAGE 155}

Temporarily fix a car taillight or water hose. • Repair siding. • Make a short-term roof shingle. • Create a clothesline. • Stash a secret car key. • Patch a canoe. • Repair a garbage can.

Garden hose
{PAGE 176}

Protect handsaw. • Make a rounded sanding block. • Make a paint-can grip.

Pantyhose
{PAGE 241}

Test a sanding job. • Apply stain in tight corners. • Patch holes in screens. • Strain paint.

Plastic bags
{PAGE 267}

Protect ceiling fan when painting ceiling. • Store paintbrushes. • Contain paint overspray.

Vinegar
{PAGE 369}

Wash concrete off skin. • Remove paint fumes. • Degrease grates, fans, and air-conditioner grilles. • Disinfect filters. • Help paint adhere to concrete. • Remove rust from tools. • Peel off wallpaper. • Slow plaster hardening. • Revive hardened paintbrushes.

YOUR COMPLETE

A-Z Guide*

*On the following pages you'll find 204 common household items that altogether have more than 2,300 unexpected uses—uses that are not just surprising and clever but can save you time, money, and effort. Some of the items, such as aluminum foil and vinegar, are labeled Super Items because they have so many extraordinary uses.

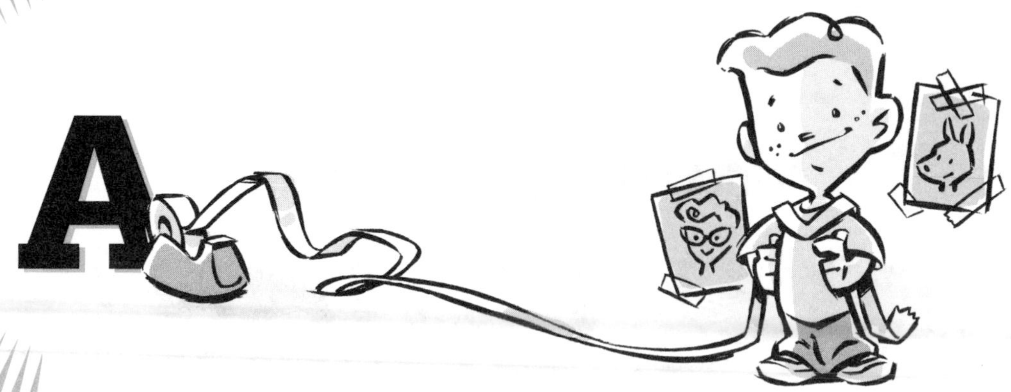

* Adhesive Tape

Remove a splinter Is a splinter too tiny or too deep to remove with tweezers? Avoid the agony of digging it out with a needle. Instead, cover the splinter with adhesive tape. After about three days, pull the tape off and the splinter should come out with it.

Stop ants in their tracks Is an army of ants marching toward the cookie jar on your countertop or some sweet prize in your pantry? Create a "moat" around the object by surrounding it with adhesive tape placed sticky side up.

Make a lint-lifter To lift lint and pet hair off clothing and upholstery, you don't need a special lint remover. Just wrap your hand with adhesive tape, sticky side out.

Reduce your hat size Got a hat that's a bit too big for your head? Wrap adhesive tape around the sweatband—it might take two or three layers depending on the size discrepancy. As a bonus, the adhesive tape will absorb brow sweat on hot days.

Clean a comb To remove the gunk that builds up between the teeth of your comb, press a strip of adhesive tape along the comb's length, and lift it off. Then dip the comb in a solution of alcohol and water, or ammonia and water, to sanitize it. Let dry.

Cover casters Prevent your furniture from leaving marks on your wood or vinyl floor by wrapping the furniture's caster wheels with adhesive tape.

Hang glue and caulk tubes Got an ungainly heap of glue and caulk tubes on your workbench? Cut a strip of adhesive or duct tape several inches long and fold it over the bottom of each tube, leaving a flap at the end. Punch a hole in the flap with a paper hole punch and hang the tube on a nail or hook. You'll free up counter space, and you'll be able to find the right tube fast.

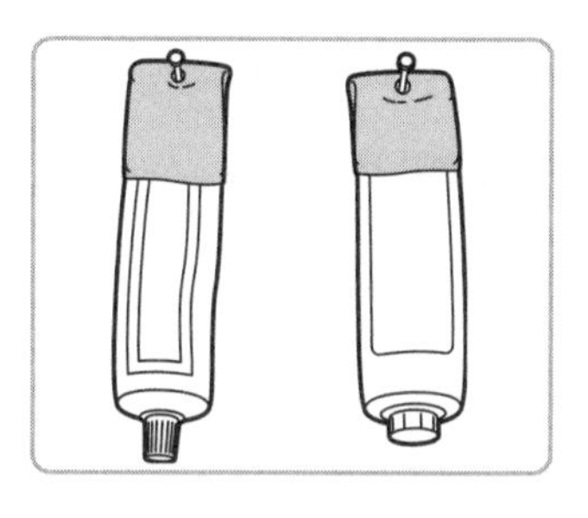

Safely remove broken window glass Removing a window sash to fix a broken pane of glass can be dangerous; there's always the possibility that a sharp shard will fall out and cut you. To prevent this, crisscross both sides of the broken pane with adhesive tape before removing the sash. And don't forget to wear heavy leather gloves when you pull the glass shards out of the frame.

Get a grip on tools Adhesive tape has just the right texture for wrapping tool handles. It gives you a positive, comfortable grip, and it's highly absorbent so that tools won't become slippery if your hand sweats. When you wrap tool handles, overlap each wrap by about half a tape width and use as many layers as needed to get the best grip. Here are some useful applications:

- Screwdriver handles are sometimes too narrow and slippery to grip well when you drive or remove stubborn screws. Wrap layers of adhesive tape around the handle until the tool feels comfortable in your hand—this is especially useful if you have arthritis in your fingers.
- Take a tip from carpenters who wrap wooden hammer handles that can get slippery with sweat. Wrap the whole gripping area of the tool. A few wraps just under the head will also protect the handle from damage caused by misdirected blows.
- Plumbers also keep adhesive tape in their tool kits: When they want to cut a pipe in a spot that's too tight for their hacksaw frame, they make a mini-hacksaw by removing the blade and wrapping one end of the blade to form a handle.

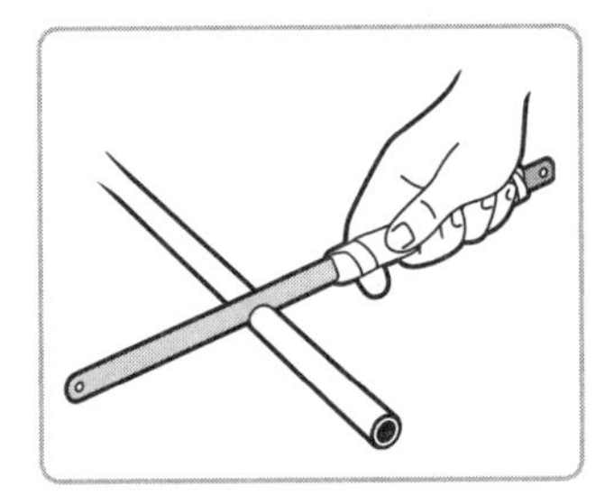

DID *You* KNOW?

In the 1920s, Josephine Dickson, an accident-prone New Jersey housewife, inspired the invention of the Band-Aid bandage. Her husband, Earle, who tended her various burns and wounds, hit upon the idea of sticking small squares of sterile gauze onto adhesive tape, covering it with a layer of crinoline, then rolling it back up so that Josephine could cut off and apply the ready-made bandages herself.

Earle's employer, Johnson & Johnson, soon began producing the first Band-Aids. By the time Earle died in 1961—by then a member of the company's board of directors—Band-Aid sales exceeded $30 million a year.

Alka-Seltzer

Clean your coffeemaker Fill your percolator or the water chamber of your drip coffeemaker with water and plop in four Alka-Seltzer tablets. When the Alka-Seltzer has dissolved, put the coffeemaker through a brew cycle to clean the tubes. Rinse the chamber out two or three times, then run a brew cycle with plain water before making coffee.

Clean a vase That stuck-on residue at the bottom of narrow-neck vases may seem impossible to scrub out, but you can easily bubble it away. Fill the vase halfway with water and drop in two Alka-Seltzer tablets. Wait until the fizzing stops, then rinse the vase clean. The same trick works for cleaning glass thermoses.

Clean glass cookware Say so long to scouring those stubborn stains off your ovenproof glass cookware. Just fill the container with water, add up to six Alka-Seltzer tablets, and let it soak for an hour. The stains should easily scrub away.

Clean your toilet The citric acid in Alka-Seltzer combined with its fizzing action is an effective toilet bowl cleaner. Simply drop a couple of tablets into the bowl and find something else to do for 20 minutes or so. When you return, a few swipes with a toilet brush will leave your bowl gleaming.

Clean jewelry Drop your dull-looking jewelry in a glass of fizzing Alka-Seltzer for a couple of minutes. It will sparkle and shine when you pull it out.

Unclog a drain Drain clogged again? Get almost instant relief: Drop a couple of Alka-Seltzer tablets down the opening, then pour in a cup of vinegar. Wait a few minutes and then run the hot water at full force to clear the clog. This is also a good way to eliminate kitchen drain odors.

Soothe insect bites Mosquito or other insect bite driving you nuts? To ease the itch, drop two Alka-Seltzer tablets in half a glass of water. Dip a cotton ball in the

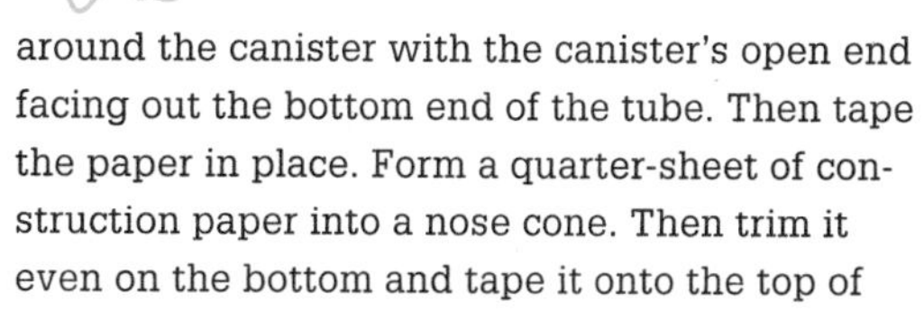

SCIENCE FAIR

Turn plop, plop, fizz, fizz into whoosh, whoosh, gee whiz with this **Alka-Seltzer rocket.** The rocket gets its thrust from the gas created when you drop a couple of Alka-Seltzer tablets into some water inside a film canister.

The type of film canister is key: Use a **Fuji 35mm plastic canister** that has a lid that fits inside the canister. Canisters with lids that fit around an outside lip won't work. You'll also need a couple of pieces of **construction paper, transparent tape,** and **scissors.**

To form the body of the rocket, wrap a piece of construction paper around the canister with the canister's open end facing out the bottom end of the tube. Then tape the paper in place. Form a quarter-sheet of construction paper into a nose cone. Then trim it even on the bottom and tape it onto the top of the rocket body.

To launch the rocket, fill the canister about halfway with refrigerated water—cold water is vital to a successful liftoff. Plop in two Alka-Seltzer tablets. Quickly pop on the lid, set the rocket on the ground, and stand back. The gas will quickly build up pressure in the canister, causing the canister lid to pop off and the rocket to launch several feet into the air.

glass and apply it to the bite. *Caution:* Don't do this if you are allergic to aspirin, which is a key ingredient in Alka-Seltzer.

Attract fish All avid anglers know fish are attracted to bubbles. If you are using a hollow plastic tube jig on your line, just break off a piece of Alka-Seltzer and slip it into the tube. The jig will produce an enticing stream of bubbles as it sinks.

Aluminum Foil

✱ ALUMINUM FOIL IN THE KITCHEN

Bake a perfect piecrust Keep the edges of your homemade pies from burning by covering them with strips of aluminum foil. The foil prevents the edges from getting overdone while the rest of your pie gets perfectly browned.

Create special-shaped cake pans Make a teddy bear birthday cake, a Valentine's Day heart cake, a Christmas tree cake, or whatever shaped cake the occasion may call for. Just form a double thickness of heavy-duty aluminum foil into the desired shape inside a large cake pan.

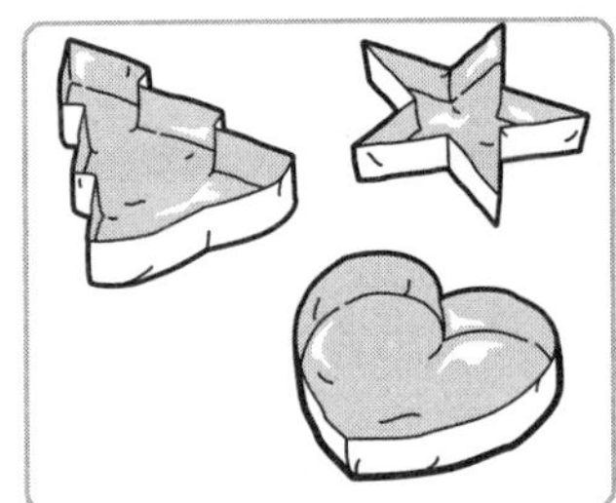

Soften up brown sugar To restore your hardened brown sugar to its former powdery glory, chip off a piece, wrap it in aluminum foil, and bake it in the oven at 300°F (150° C) for five minutes.

Decorate a cake No pastry bag handy? No problem. Form a piece of heavy-duty aluminum foil into a tube and fill it with free-flowing frosting. Bonus: There's no pastry bag to clean—simply toss out the foil when you're done.

Make an extra-large salad bowl You've invited half the neighborhood over for dinner, but don't have a bowl big enough to toss that much salad. Don't panic. Just line the kitchen sink with aluminum foil and toss away!

Keep rolls and breads warm Want to lock in the oven-fresh warmth of your homemade rolls or breads for a dinner party or picnic? Before you load up your basket, wrap your freshly baked goods in a napkin and place a layer of aluminum foil underneath. The foil will reflect the heat and keep your bread warm for quite some time.

Catch ice-cream cone drips Keep youngsters from making a mess of their clothes or your house by wrapping the bottom of an ice-cream cone (or a wedge of watermelon) with a piece of aluminum foil before handing it to them.

Toast your own cheese sandwich Next time you pack for a trip, include a couple of cheese sandwiches wrapped in aluminum foil. That way if you check into a hotel after the kitchen has closed, you won't have to resort to the cold, overpriced snacks in the mini-bar. Instead, use the hotel-room iron to press both sides of the wrapped sandwich and you'll have a tasty hot snack.

Polish your silver Is your silverware looking a bit dull these days? Try an ion exchange, a molecular reaction in which aluminum acts as a catalyst. All you have to do is line a pan with a sheet of aluminum foil, fill it with cold water, and add two teaspoons of salt. Drop your tarnished silverware into the solution, let it sit for two to three minutes, then rinse off and dry.

Keep silverware untarnished Store freshly cleaned silverware on top of a sheet of aluminum foil to deter tarnishing. For long-term storage of silverware, first tightly cover each piece in cellophane wrap—be sure to squeeze out as much air as possible—then wrap in foil and seal the ends.

Preserve steel-wool pads It's maddening. You use a steel-wool pad once, put it in a dish by the sink, and the next day you find a rusty mess fit only for the trash. To prevent rust and get your money's worth from a pad, wrap it in foil and toss it into the freezer. You can also lengthen the life of your steel-wool soap pads by crumpling up a sheet of foil and placing it under the steel wool in its dish or container. (Don't forget to periodically drain off the water that collects at the bottom.)

Scrub your pots Don't have a scrub pad? Crumple up a handful of aluminum foil and use it to scrub your pots.

Foil-Eating Acidic Foods

Think twice before ripping off a sheet of aluminum foil to wrap up your leftover meat loaf—particularly one that's dripping with tomato sauce. Highly acidic or salty foods such as lemons, grapefruits, ketchup, and pickles accelerate the oxidation of aluminum and can actually "eat" through foil with prolonged exposure. This can also leach aluminum into the food, which can affect its flavor and may pose a health risk. If you want to use foil for that meat loaf, however, cover it first with a layer or two of plastic wrap or wax paper to prevent the sauce from coming into contact with the foil.

Keep the oven clean Are you baking a bubbly lasagna or casserole? Keep messy drips off the bottom of the oven by laying a sheet or two of aluminum foil over the rack below. *Do not* line the bottom of the oven with foil; it could cause a fire.

Improve radiator efficiency Here's a simple way to get more heat out of your old cast-iron radiators without spending one cent more on your gas or oil bill: Make a heat reflector to put behind them. Tape heavy-duty aluminum foil to cardboard with the shiny side of the foil facing out. The radiant heat waves will bounce off the foil into the room instead of being absorbed by the wall behind the radiator. If your radiators have covers, it also helps to attach a piece of foil under the cover's top.

Keep pets off furniture Can't keep Snoopy off your brand-new sofa? Place a piece of aluminum foil on the seat cushions, and after one try at settling down on the noisy surface, your pet will no longer consider it a comfy place to snooze.

Protect a child's mattress As any parent of a potty-trained youngster knows, accidents happen. When they happen in bed, however, you can spare the mattress—even if you don't have a plastic protector available. First, lay several sheets of aluminum foil across the width of the mattress. Then, cover them with a good-sized beach towel. Finally, attach the mattress pad and bottom sheet.

Hide worn spots in mirrors Sometimes a worn spot adds to the charm of an old mirror; sometimes it's a distraction. You can easily disguise small flaws on a mirror's reflective surface by putting a piece of aluminum foil, shiny side facing out, on the back of the glass. To hold the foil in place, attach it to the backing behind the mirror or to the frame with masking tape. Don't tape it to the mirror itself.

Kids' Stuff Mixing **finger paints** is a great way for kids to learn first-hand how colors combine while also expressing their creativity. Unfortunately, their learning experience can be your "Excedrin moment." To **contain the mess,** cut down the sides of a **wide cardboard box** so that they are about three inches high. Line the inside of the box with **aluminum foil** and let the kids pour in the paint. With any luck, the paint should stay within the confines of the box, keeping splatters off walls and the floor.

Sharpen your scissors What can you do with those clean pieces of leftover foil you have hanging around? Use them to sharpen up your dull scissors! Smooth them out if necessary, and then fold the strips into several layers and start cutting. Seven or eight passes should do the trick. Pretty simple, huh? (See page 43 to find out how you can use the resulting scraps of foil for mulching or keeping birds off your fruit trees.)

Clean jewelry To clean your jewelry, simply line a small bowl with aluminum foil. Fill the bowl with hot water and mix in one tablespoon of bleach-free powdered

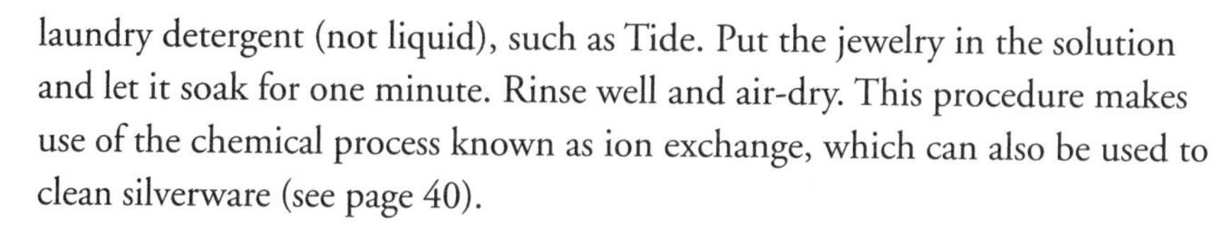

laundry detergent (not liquid), such as Tide. Put the jewelry in the solution and let it soak for one minute. Rinse well and air-dry. This procedure makes use of the chemical process known as ion exchange, which can also be used to clean silverware (see page 40).

Move furniture with ease To slide big pieces of furniture over a smooth floor, place small pieces of aluminum foil under the legs. Put the dull side of the foil down—the dull side is actually more slippery than the shiny side.

Fix loose batteries Is your flashlight, Walkman, or your kid's toy working intermittently? Check the battery compartment. Those springs that hold the batteries in place can lose their tension after a while, letting the batteries loosen. Fold a small piece of aluminum foil until you have a pad that's thick enough to take up the slack. Place the pad between the battery and the spring.

Don't dye your glasses You want to catch up on your reading during the time it takes to color your hair. But you can't read without your specs, and if you put them on, hair dye can stain them. Solution: Wrap the temples of your glasses with aluminum foil.

Clean out your fireplace Looking for an easy way to clean the ashes out of your fireplace? Place a double layer of heavy-duty aluminum foil across the bottom of the fireplace or under the wood grate. The next day—or once you're sure all the ashes have cooled—simply fold it up and throw it away or, even better, use the ashes as described on pages 54-55.

✱ ALUMINUM FOIL IN THE LAUNDRY ROOM

Speed your ironing When you iron clothing, a lot of the iron's heat is sucked up by the board itself—requiring you to make several passes to remove wrinkles. To speed things up, put a piece of aluminum foil under your ironing board cover. The foil will reflect the heat back through the clothing, smoothing wrinkles quicker.

Have you ever wondered why aluminum foil has one side that's shinier than the other? The answer has to do with how it's manufactured. According to Alcoa, the maker of Reynolds Wrap, the different shades of silver result during the final rolling process, when two layers of foil pass through the rolling mill simultaneously. The sides that contact the mill's heavy, polished rollers come out shiny, while the inside layers retain a dull, or matte, finish. Of course, the shiny side is better for reflecting light and heat, but when it comes to wrapping foods or lining grills, both sides are equally good.

Attach a patch An iron-on patch is an easy way to fix small holes in clothing—but only if it doesn't get stuck onto your ironing board. To avoid this, put a piece of aluminum foil under the hole. It won't stick to the patch, and you can just slip it out when you're finished.

Clean your iron Is starch building up on your clothes iron and causing it to stick? To get rid of it, run your hot iron over a piece of aluminum foil.

✱ ALUMINUM FOIL IN THE GARDEN

Put some bite in your mulch To keep hungry insects and slugs away from your cucumbers and other vegetables, mix strips of aluminum foil in with your garden mulch. As a bonus benefit, the foil will reflect light back up onto your plants.

Protect tree trunks Mice, rabbits, and other animals often feed on the bark of young trees during winter. A cheap and effective deterrent is to wrap the tree trunks with a double layer of heavy-duty aluminum foil in late fall. Be sure to remove the foil in spring.

Tip Prevent Sunscald on Trees

Wrapping young tree trunks with a couple of layers of aluminum foil during the winter can help prevent sunscald, a condition widely known as southwest disease, since it damages the southwest side of some young thin-barked trees—especially fruit trees, ashes, lindens, maples, oaks, and willows. The problem occurs on warm winter days when the sun's rays reactivate some dormant cells underneath the tree's bark. The subsequent drop in nighttime temperatures kills the cells and can injure the tree. In most regions, you can remove the aluminum wrapping in early spring.

Scare crows and other birds Are the birds eating the fruit on your trees? To foil them, dangle strips of aluminum foil from the branches using monofilament fishing line. Even better, hang some foil-wrapped seashells, which will add a bit of noise to further startle your fine-feathered thieves.

Create a sun box for plants A sunny window is a great place for keeping plants that love a lot of light. However, since the light always comes from the same direction, plants tend to bend toward it. To bathe your plants in light from all sides, make a sun box: Remove the top and one side from a cardboard box and line the other three sides and bottom with aluminum foil, shiny side out, taping or gluing it in place. Place plants in the box and set it near a window.

Build a seed incubator To give plants grown from seeds a healthy head start, line a shoe box with aluminum foil, shiny side up, allowing about two inches of foil to extend out over the sides. Poke several drainage holes in the bottom—penetrating the foil—then fill the box slightly more than halfway with potting soil, and plant the seeds. The foil inside the box will absorb heat to keep the seeds warm as they germinate, while the foil outside the box will reflect light onto the young sprouts. Place the box near a sunny window, keep the soil moist, and watch 'em grow!

Grow untangled cuttings Help plant cuttings grow strong and uncluttered by starting them in a container covered with a sheet of aluminum foil. Simply poke a few holes in the foil and insert the cuttings through the holes. There's even an added bonus: The foil slows water evaporation, so you'll need to add water less frequently.

✱ ALUMINUM FOIL IN THE GREAT OUTDOORS

Keep bees away from beverages You're about to relax in your backyard with a well-deserved glass of lemonade or soda pop. Suddenly bees start buzzing around your drink—which they view as sweet nectar. Keep them away by tightly covering the top of your glass with aluminum foil. Poke a straw through it, and then enjoy your drink in peace.

Make a barbecue drip pan To keep meat drippings off your barbecue coals, fashion a disposable drip pan out of a couple of layers of heavy-duty aluminum foil. Shape it freehand, or use an inverted baking pan as a mold (remember to remove the pan once your creation is finished). Also, don't forget to make your drip pan slightly larger than the meat on the grill.

Clean your barbecue grill After the last steak is brought in, and while the coals are still red-hot, lay a sheet of aluminum foil over the grill to burn off any remaining foodstuffs. The next time you use your barbecue, crumple up the foil and use it to easily scrub off the burned food before you start cooking.

Improve outdoor lighting Brighten up the electrical lighting in your backyard or campsite by making a foil reflector to put behind the light. Attach the reflector to the fixture with a few strips of electrical tape or duct tape—*do not* apply tape directly to the bulb.

DID *You* KNOW?

"Hand me the tinfoil, will ya?" To this day, it's not uncommon for folks to ask for tinfoil when they want to wrap leftovers. Household foil *was* made only of tin until 1947, when aluminum foil was introduced into the home, eventually replacing tinfoil in the kitchen drawer.

Make an impromptu picnic platter When you need a convenient disposable platter for picnics or church suppers, just cover a piece of cardboard with heavy-duty aluminum foil.

Improvise a frying pan Don't feel like lugging a frying pan along on a camping trip? Form your own by centering a forked stick over two layers of heavy-duty aluminum foil. Wrap the edges of the foil tightly around the forked branches but leave some slack in the foil between the forks. Invert the stick and depress the center to hold food for frying.

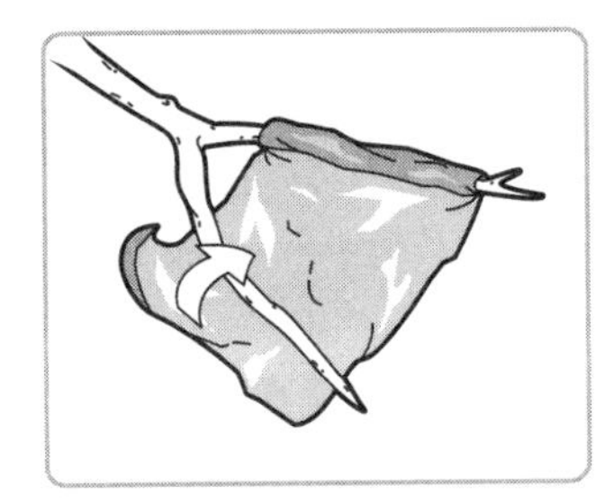

Warm your toes when camping Keep your tootsies toasty at night while cold-weather camping. Wrap some stones in aluminum foil and heat them by the campfire while you are toasting marshmallows. At bedtime, wrap the stones in towels and put them in the bottom of your sleeping bag.

Keep your sleeping bag dry Place a piece of heavy-duty aluminum foil under your sleeping bag to insulate against moisture.

Keep matches dry It's a tried-and-true soldier's trick worth remembering: Wrap your kitchen matches in aluminum foil to keep them from getting damp or wet on camping trips.

Lure a fish None of your fancy fishing lures working? You can make one in a jiffy that just might do the trick: Wrap some aluminum foil around a fishhook. Fringe the foil so that it covers the hook and wiggles invitingly when you reel in the line.

* ALUMINUM FOIL FOR THE DO-IT-YOURSELFER

Make a funnel Can't find a funnel? Double up a length of heavy-duty aluminum foil and roll it into the shape of a cone. This impromptu funnel has an advantage over a permanent funnel—you can bend the aluminum foil to reach awkward holes, like the oil filler hole tucked against the engine of your lawn tractor.

Re-attach a vinyl floor tile Don't become unglued just because a vinyl floor tile does. Simply reposition the tile on the floor, lay a piece of aluminum foil over it, and run a hot clothes iron over it a few times until you can feel the glue melting underneath. Put a pile of books or bricks on top of the tile to weight it down while the glue resets. This technique also works well to smooth out bulges and straighten curled seams in sheet vinyl flooring.

Make an artist's palette Tear off a length of heavy-duty aluminum foil, crimp up the edges, and you've got a ready-to-use palette for mixing paints. If you want to get a little fancier, cut a piece of cardboard into the shape of a palette, complete with thumb hole, and cover it with foil. Or if you already have a wooden

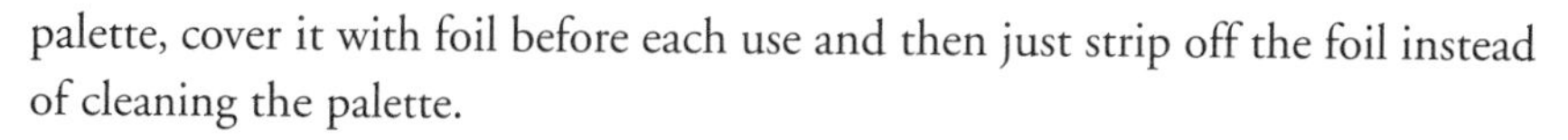

palette, cover it with foil before each use and then just strip off the foil instead of cleaning the palette.

Prevent paint from skinning over When you open a half-used can of paint, you'll typically find a skin of dried paint on the surface. Not only is this annoying to remove, but dried bits can wind up in the paint. You can prevent this by using a two-pronged attack when you close a used paint can: First, put a piece of aluminum foil under the can and trace around it. Cut out the circle and drop the aluminum foil disk onto the paint surface. Then take a deep breath, blow into the can, and quickly put the top in place. The carbon dioxide in your breath replaces some of the oxygen in the can, and helps keep the paint from drying.

Line roller pans Cleaning out paint roller pans is a pain, which is why a lot of folks buy disposable plastic pans or liners. But lining a metal roller pan with aluminum foil works just as well—and can be a lot cheaper.

Keep paint off doorknobs When you're painting a door, aluminum foil is great for wrapping doorknobs to keep paint off them. Overlap the foil onto the door when you wrap the knob, then run a sharp utility knife around the base of the knob to trim the foil. That way you can paint right up to the edge of the knob. In addition to wrapping knobs on the doors that you'll paint, wrap all the doorknobs that are along the route to where you will clean your hands and brushes.

Keep a paintbrush wet Going to continue painting tomorrow morning? Don't bother to clean the brush—just squeeze out the excess paint and wrap the brush tightly in aluminum foil (or plastic wrap). Use a rubber band to hold the foil tightly at the base of the handle. For extended wet-brush storage, think paintbrush Popsicle, and toss the wrapped brush in the freezer. But don't forget to defrost the brush for an hour or so before you paint.

Reflect light for photography Professional photographers use reflectors to throw extra light on dark areas of their subject and to even out the overall lighting. To make a reflector, lightly coat a piece of mat board or heavy cardboard with rubber cement and cover it with aluminum foil, shiny side out. You can make one single reflector, as large as you want, but it's better to make three panels and join them together with duct tape so that they stand up by themselves and fold up for handy storage and carrying.

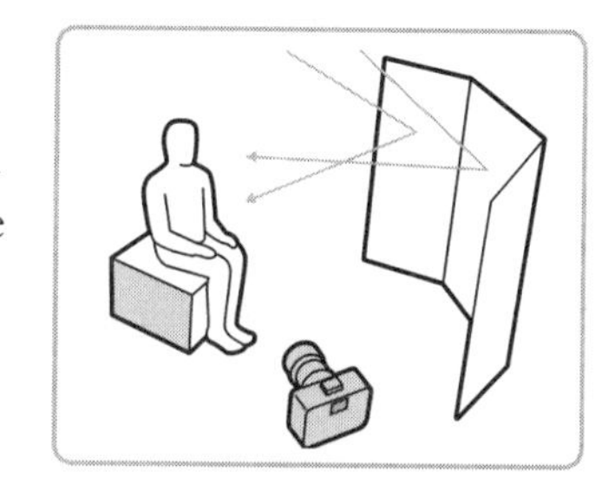

Shine your chrome For sparkling chrome on your appliances, strollers, golf club shafts, and older car bumpers, crumple up a handful of aluminum foil with the shiny side out and apply some elbow grease. If you rub real hard, the foil will even remove rust spots. *Note:* Most "chrome" on new cars is actually plastic—don't rub it with aluminum foil.

Aluminum Pie Pans

Make an instant colander Your pot of linguine is almost done when you realize you forgot to replace your broken colander. No need to panic. Just grab a clean aluminum pie pan and a small nail, and start poking holes. When you're done, bend the pan to fit comfortably over a deep bowl. Rinse your new colander clean, place it over the bowl, and *carefully* pour out your pasta.

Rein in splatters when frying Why risk burning yourself or anyone else with oil splatters from a hot frying pan? A safer way to fry is to poke a few holes in the bottom of an aluminum pie pan and place it upside down over the food in your frying pan. Use a pair of tongs or a fork to lift the pie pan and don't forget to wear a cooking glove.

Create a centerpiece Here's how to make a quick centerpiece for your table: Secure a pillar candle or a few votive candles to an aluminum pie pan by melting some wax from the bottom of the candles onto the pan. Add a thin layer of water or sand, and put in several rose petals or seashells.

Contain the mess from kids' projects Glitter is notorious for turning up in the corners and crevices of your home long after your youngster's masterpiece has been mailed off to Grandma. But you can minimize some messes by using an aluminum pie pan to encase projects involving glitter, beads, spray paint, feathers ...well, you get the picture.

Kids' Stuff Looking for a way to keep the kids busy indoors on a cold, wintry day? How about making an ice ornament that you can hang on a tree outside your house as a **homemade winter decoration?** All you'll need is an **aluminum pie pan,** some **water,** a piece of heavy **string** or a shoelace, and a mix of decorative—preferably biodegradable—materials, such as **dried flowers, dried leaves,** pinecones, seeds, shells, and twigs.

Let the children arrange the materials in the pie pan to their liking. Then fold the string or shoelace in half and place it in the pan. The fold should hang over the edges of the pan, while the two ends meet in the center. Slowly fill the pan with water, stopping just shy of the rim. You may have to place an object on the string to keep it from floating to the top.

If the temperatures outside your home are indeed below freezing, you can simply put the pan on your doorstep to freeze. Otherwise just pop it into your **freezer.** Once the water has frozen solid, slide off the pan and let your children choose the optimal outdoor location to display their artwork in ice.

Make trays for craft supplies Bring some order to your children's—or your own—inventory of crayons, beads, buttons, sequins, pipe cleaners, and such by sorting them in aluminum pie pans. To secure materials when storing the pans, cover each pan with a layer of plastic wrap.

Keep bugs out of pet dishes Use an aluminum pie pan filled with about a half-inch of water to create a metal moat around your pet's food dish. It should keep those marauding ants and roaches at bay.

Train your dog If Rover has a tendency to leap up on the sofa or kitchen counter, leave a few aluminum pie pans along the counter edges or the sofa back when you're not home. The resulting noise will give him a good scare when he jumps and hits them.

Keep squirrels and birds off your fruit trees Are furry and feathered fiends stealing the fruit off your trees? There's nothing better to scare off those pesky intruders than a few dangling aluminum pie pans. String them up in pairs (to make some noise), and you won't have to worry about finding any half-eaten apples or peaches come harvest time.

Make a mini-dustpan If you need a spare dustpan for your workplace or bathroom, an aluminum pie pan can fit the bill quite nicely. Simply cut one in half, and you're ready to go.

Use as a drip catcher under paint can Next time you have something that needs painting, place an aluminum pie pan under the paint can as a ready-made drip catcher. You'll save a lot of time cleaning up, and you can just toss the pan in the trash when you're done. Even better, rinse it off and recycle it for future paint jobs.

Store sanding disks and more Since they're highly resistant to corrosion, aluminum pie pans are especially well suited for storing sanding disks, hacksaw blades, and other hardware accessories in your workshop. Cut a pan in half and attach it (with staples or duct tape around the edges) open side up to a pegboard. Now get organized!

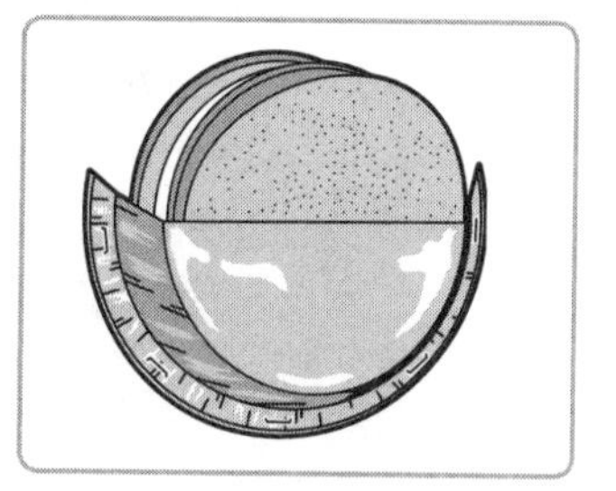

Use as an impromptu ashtray No ashtray on hand when you host a smoker in your home? No sweat. An aluminum pie pan—or even a piece of heavy-duty aluminum foil folded into a square with the sides turned up—should suffice.

Protect fingers during cookouts There's nothing like a cookout in the great outdoors. Whether you're planning a day trip or a longer excursion, be sure to pack a few aluminum pie pans. Put a small hole in the middle of each pan, then push them up the sticks used for roasting hot dogs or marshmallows. The pans deflect the heat of the fire, protecting your hands and your children's hands.

Ammonia

* AMMONIA IN THE KITCHEN

Clean your oven Here's a practically effortless way to clean an electric oven: First, turn the oven on, let it warm to 150°F (65°C), and then turn it off. Place a small bowl containing 1/2 cup ammonia on the top shelf and a large pan of boiling water on the bottom shelf. Close the oven door, and let it sit overnight. The next morning, remove the dish and pan, and let the oven air out awhile. Then wipe it clean using the ammonia and a few drops of dishwashing liquid diluted in a quart of warm water—even old burned-on grease should wipe right off. *Warning: Do not use this cleaning method with a gas oven unless the pilot lights are out and the main gas lines are shut off.*

TAKE CARE Never mix ammonia with bleach or any product containing chlorine. The combination produces toxic fumes that can be deadly. Work in a well-ventilated space and avoid inhaling the vapors. Wear rubber gloves and avoid getting ammonia on your skin or in your eyes. Always store ammonia out of the reach of children.

Clean oven racks Get the cooked-on grime off your oven racks by laying them out on an old towel in a large washtub. You can also use your bathtub, though you might need to clean it afterward. Fill the tub with warm water and add 1/2 cup ammonia. Let the racks soak for at least 15 minutes, then remove, rinse off, and wipe clean.

Make crystal sparkle Has the twinkle gone out of your good crystal? Bring back its lost luster by mixing several drops of ammonia in 2 cups water and applying

During the Middle Ages, ammonia was made in northern Europe by heating the scrapings of deer antlers, and was known as spirits of hartshorn. Before the start of World War I, it was chiefly produced by the dry distillation of nitrogenous vegetable and animal products.

Today most ammonia is made synthetically using the Haber process, in which hydrogen and nitrogen gases are combined under extreme pressures and medium temperatures. The technique was developed by Fritz Haber and Carl Bosch in 1909, and was first used on a large-scale basis by the Germans during World War I, primarily for the production of munitions.

with a soft cloth or brush. Rinse it off with clean water, then dry with a soft, dry cloth.

Repel moths Pesky kitchen moths seem to come out of nowhere! Send them back to wherever they came from by washing your drawers, pantry shelves, or cupboards with 1/2 cup ammonia diluted in 1 quart (1 liter) water. Leave drawers and cabinets open to thoroughly air-dry.

* AMMONIA AROUND THE HOUSE

Eliminate paint odors Your freshly painted home interior sure looks great, but that paint smell is driving you up the wall! There's no need to prolong your suffering, though. Absorb the odor by placing small dishes of ammonia in each room that's been painted. If the smell persists after several days, replenish the dishes. Vinegar or onion slices will also work.

Clean fireplace doors Think you'll need a blowtorch to remove that blackened-on soot from your glass fireplace doors? Before you get out the goggles, try mixing 1 tablespoon ammonia, 2 tablespoons vinegar, and 1 quart (1 liter) warm water in a spray bottle. Spray on some of the solution; let it sit for several seconds, then wipe off with an absorbent cloth. Repeat if necessary—it's worth the extra effort.

Clean gold and silver jewelry Brighten up your gold and silver trinkets by soaking them for 10 minutes in a solution of 1/2 cup clear ammonia mixed in 1 cup warm water. Gently wipe clean with a soft cloth and let dry. Note: Do not do this with jewelry containing pearls, because it could dull or damage their delicate surface.

Remove tarnish from brass or silver How can you put that sunny shine back in your tarnished silver or lacquered brass? Gently scrub it with a soft brush dipped

DID *You* KNOW?

Ammonia is a colorless, pungent gas. It is easily soluble in water, however, and the liquid ammonia products sold today contain the gas dissolved in water. Ammonia is one of the oldest cleaning compounds currently in use. It actually dates back to ancient Egypt. In fact, the word *ammonia* is derived from the Egyptian deity Ammon, whose temple in what is now Libya is credited with producing the earliest form of ammonia, sal ammoniac, by burning camel dung.

in a bit of ammonia. Wipe off any remaining liquid with a soft cloth—or preferably chamois.

Remove grease and soap scum To get rid of those ugly grease and soap-scum buildups in your porcelain enamel sink or tub, scrub it with a solution of 1 tablespoon ammonia in 1 gallon (3.7 liters) hot water. Rinse thoroughly when done.

Restore white shoes Brighten up your dingy white shoes or tennis sneakers by rubbing them with a cloth dipped in half-strength ammonia—that is, a solution made of half ammonia and half water.

Tip Testing Ammonia

Not sure if it's safe to put ammonia solution, or any other stain remover, on a particular fabric or material? Always test a drop or two on an inconspicuous part of the garment or object first. After applying, rub the area with a white terry-cloth towel to test colorfastness. If any color rubs off on the towel or if there is any noticeable change in the material's appearance, try another approach.

Remove stains from clothing Ammonia is great for cleaning clothes. Here are some ways you can use it to remove a variety of stains. Be sure to dilute ammonia with at least 50 percent water before applying it to silk, wool, or spandex.

- Rub out perspiration, blood, and urine stains on clothing by dabbing the area with a half-strength solution of ammonia and water before laundering.
- Remove most non-oily stains by making a mixture of equal parts ammonia, water, and dishwashing liquid. Put it in an empty spray bottle, shake well, and apply directly to the stain. Let it set for two or three minutes, and then rinse out.
- To erase pencil marks from clothing, use a few drops of undiluted ammonia and then rinse. If that doesn't work, put a little laundry detergent on the stain and rinse again.
- You can even remove washed-in paint stains from clothes by saturating them several times with a half-ammonia, half-turpentine solution, and then tossing them into the wash.

Clean carpets and upholstery Lift out stains from carpeting and upholstery by sponging them with 1 cup clear ammonia in 1/2 gallon (2 liters) warm water. Let dry thoroughly, and repeat if needed.

Brighten up windows Dirty, grimy windows can make any house look dingy. But it's easy to wipe away the dirt, fingerprints, soot, and dust covering your windows. Just wipe them down with a soft cloth dampened with a solution of 1 cup clear

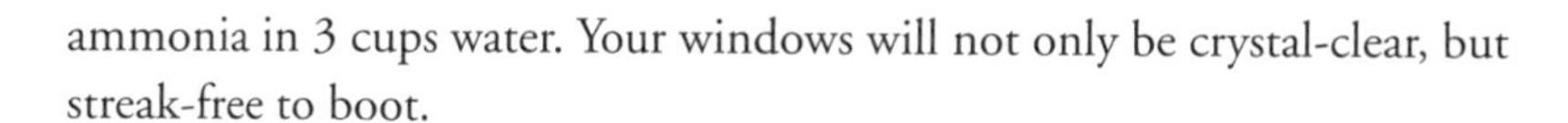

ammonia in 3 cups water. Your windows will not only be crystal-clear, but streak-free to boot.

Strip wax from resilient flooring Wax buildup on resilient flooring causes it to yellow in time. Remove old wax layers and freshen up your floor by washing it with a mixture of 1 cup ammonia in 1/2 gallon (2 liters) water. Let the solution sit for three to five minutes, then scrub with a nylon or plastic scouring pad to remove the old wax. Wipe away leftover residue with a clean cloth or sponge, then give the floor a thorough rinsing.

Clean bathroom tiles Make bathroom tiles sparkle again—and kill mildew on them—by sponging them with 1/4 cup ammonia in 1 gallon (3.7 liters) water.

✱ AMMONIA IN THE GARDEN

Use as plant food Give the alkaline-loving flowering plants and vegetables in your garden—such as clematis, lilac, hydrangea, and cucumbers—an occasional special treat with a shower of 1/4 cup ammonia diluted in 1 gallon (3.7 liters) water. They'll especially appreciate the boost in nitrogen.

Stop mosquito bites from itching If you forget to put on your insect repellent and mosquitoes make a meal of you, stop the itching instantly by applying a drop or two of ammonia directly to the bites. Don't use ammonia on a bite you've already scratched open, though; the itch will be replaced by a nasty sting.

Keep stray animals out of your trash Few things can be quite as startling as a raccoon leaping out of your garbage pail just as you're about to make your nightly trash deposit. Keep away those masked scavengers and other strays by spraying the outside and lids of your garbage bins with half-strength ammonia or by spraying the bags inside.

Remove stains from concrete Tired of those annoying discolorations on your concrete work? To get rid of them, scrub with 1 cup ammonia diluted in 1 gallon (3.7 liters) water. Hose it down well when you're done.

Fight mildew Ammonia and bleach are equally effective weapons in the battle against mold and mildew. However, each has its own distinct applications, and under no conditions should the two ever be combined.

Reach for the ammonia for the following chores, but be sure you use it in a well-ventilated area, and don't forget to wear rubber gloves:

- Clean the mildew off unfinished wooden patio furniture and picnic tables with a mixture of 1 cup ammonia, 1/2 cup vinegar, 1/4 cup baking soda, and 1 gallon (3.7 liters) water. Rinse off thoroughly and use an old terrycloth towel to absorb excess moisture.
- To remove mildew from painted outdoor surfaces, use the same combination of ingredients.

- To remove mildew from wicker furniture, wash it down with a solution of 2 tablespoons ammonia in 1 gallon (3.7 liters) water. Use an old toothbrush to get into hard-to-reach twists and turns. Rinse well and let air-dry.

* Apples

Roast a juicy chicken If your roasted chicken tends to emerge from the oven as dry as a snow boot on a summer's day, don't fret. The next time you roast a chicken, stuff an apple inside the bird before placing it in the roasting pan. When it's done cooking, toss the fruit in the trash, and get ready to sit down to a delicious—and juicy—main course.

Keep cakes fresh Want a simple and effective way to extend the shelf life of your homemade or store-bought cakes? Store them with a half an apple. It helps the cake maintain its moisture considerably longer than merely popping it in the fridge.

Ripen green tomatoes How's that? You just became the proud owner of a bunch of green tomatoes? No sweat. You can quickly ripen them up by placing them—along with an already-ripe apple—in a paper bag for a couple of days. For best results, maintain a ratio of about five or six tomatoes per apple.

Fluff up hardened brown sugar Brown sugar has the irritating habit of hardening up when exposed to humidity. Fortunately, it doesn't take much to make this a temporary condition. Simply place an apple wedge in a self-sealing plastic bag with the chunk of hardened brown sugar. Tightly seal the bag and put it in a dry place for a day or two. Your sugar will once again be soft enough to use.

Absorb salt in soups and stews Salting to taste is one thing, but it is possible to overdo it. When you find yourself getting heavy-handed with the saltshaker, simply

That old saying "One bad apple spoils the bunch" just might be true. Apples are among a diverse group of fruits—others include apricots, avocados, bananas, blueberries, cantaloupe, and peaches—that produce ethylene gas, a natural ripening agent. So the increased level of ethylene produced by a single rotten apple in a bag can significantly accelerate the aging process of the other apples around it.

Ethylene-producing fruits can help speed the ripening of something (like a green tomato, see hint on this page). But they can also have unwanted effects. Placing a bowl of ripe apples or bananas too close to freshly cut flowers, for instance, can cause them to wilt. And if your refrigerated potatoes seem to be sprouting buds too soon, they may be too close to the apples. Keep them at least one shelf apart.

drop a few apple (or potato) wedges in your pot. After cooking for another 10 minutes or so, remove the wedges—along with the excess salt.

Use as decorative candleholders Add a cozy, country feel to your table setting by creating a natural candleholder. Use an apple corer to carve a hole three-quarters of the way down into a pair of large apples, insert a tall decorative candle into each hole, surround the apples with a few leaves, branches, or flowers, and voilà! You have a lovely centerpiece.

* Ashes

Clean fireplace doors You normally wouldn't think of using dirty wood ashes to clean glass fireplace doors, but it works. Mix some ashes with a bit of water, and apply them with a damp cloth, sponge, or paper towel, or simply dip a wet sponge into the ashes. Rub the mixture over the doors' surfaces. Rinse with a wet paper towel or sponge, then dry with a clean cloth. The results will amaze you, but remember—wood ash was a key ingredient in old-fashioned lye soap.

Tip Selecting Firewood

For a hot-burning and long-lasting fire, you can't do much better than well-seasoned sugar maple. Green or wet wood burns poorly and builds up heaps of creosote (the leading cause of chimney fires) in your chimney; pine is another major producer of creosote. Never burn scraps of pressure-treated wood; it contains chemicals that can be extremely harmful when burned.

Don't be a fanatic about cleaning ashes from your fireplace. Leave a 1- to 2-inch (2.5 to 5 centimeters) layer of ash under the andiron to reflect heat back up to the burning wood and protect your fireplace floor against hot embers. Just be sure not to let the ashes clog up the space under the grate and block the airflow a good fire needs.

Reduce sun glare Pro ball players often wear that black stuff under their eyes to cut down glare from the sun or bright stadium lights. If you're troubled by sun glare while driving or hiking, you may want to try it too. Just put a drop or two of baby oil on your finger, dip it in some wood ashes, and apply under your eyes.

Use as plant food Wood ashes have a high alkaline content and trace amounts of calcium and potassium, which encourage blooms. If your soil tends to be acidic, sprinkle the ashes in spring around alkaline-loving plants such as clematis, hydrangea, lilac, and roses (but avoid acid-lovers like rhododendrons, blueberries, and azaleas). Avoid using ashes from easy-to-ignite, pre-formed logs, which may contain chemicals harmful to plants. And be sparing when adding ashes to your compost pile; they can counteract the benefits of manure and other high-nitrogen materials.

Repel insects Scatter a border of ashes around your garden to deter cutworms, slugs, and snails—it sticks to their bodies and draws moisture out of them. Also sprinkle small amounts of ashes over garden plants to manage infestations of soft-bodied insects. Wear eye protection and gloves; getting ashes in your eyes can be quite painful.

Clean pewter Restore the shine to your pewter by cleaning it with cigarette ashes. Dip a dampened piece of cheesecloth into the ashes and rub it well over the item. It will turn darker at first, but the shine will come out after a good rinsing.

Remove water spots and heat marks from wood furniture Use cigar and or cigarette ashes to remove those white rings left on your wooden furniture by wet glasses or hot cups. Mix the ashes with a few drops of water to make a paste, and rub lightly over the mark to remove it. Then shine it with your favorite furniture polish.

* Aspirin

Revive dead car batteries If you get behind the wheel only to discover that your car's battery has given up the ghost—and there's no one around to give you a jump—you may be able to get your vehicle started by dropping two aspirin tablets into the battery itself. The aspirin's acetylsalicylic acid will combine with the battery's sulfuric acid to produce one last charge. Just be sure to drive to your nearest service station.

The bark of the willow tree is rich in salicin, a natural painkiller and fever reducer. In the third century B.C. Hippocrates used it to relieve headaches and pain, and many traditional healers, including Native Americans, used salicin-containing herbs to treat cold and flu symptoms. But it wasn't until 1899 that Felix Hoffmann, a chemist at the German company Bayer, developed a modified derivative, acetylsalicylic acid, better known as aspirin.

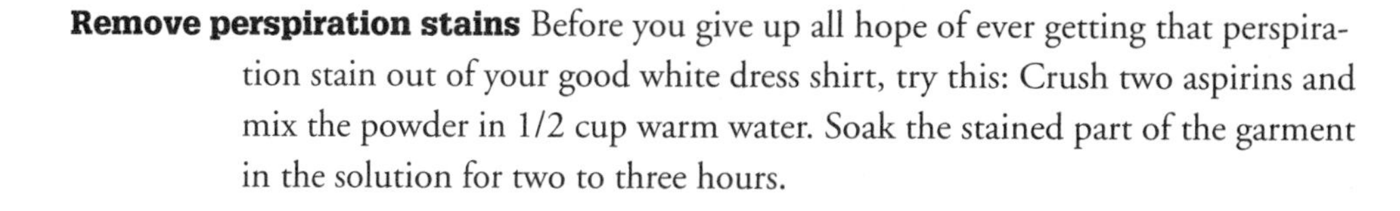

Remove perspiration stains Before you give up all hope of ever getting that perspiration stain out of your good white dress shirt, try this: Crush two aspirins and mix the powder in 1/2 cup warm water. Soak the stained part of the garment in the solution for two to three hours.

TAKE CARE About 10 percent of people with severe asthma are also allergic to aspirin—and, in fact, to all products containing salicylic acid, aspirin's key ingredient, including some cold medications, fruits, and food seasonings and additives. That percentage skyrockets to 30 to 40 percent for older asthmatics who also suffer from sinusitis or nasal polyps. Acute sensitivity to aspirin is also seen in a small percentage of the general population without asthma—particularly people with ulcers and other bleeding conditions.

Always consult your doctor before using any medication, and do not apply aspirin externally if you are allergic to taking it in internally.

Restore hair color Swimming in a chlorinated pool can have a noticeable, and often unpleasing, effect on your hair coloring if you have light-colored hair. But you can usually return your hair to its former shade by dissolving six to eight aspirins in a glass of warm water. Rub the solution thoroughly into your hair, and let it set for 10-15 minutes.

Dry up pimples Even those of us who are well past adolescence can get the occasional pimple. Put the kibosh on those annoying blemishes by crushing one aspirin and moistening it with a bit of water. Apply the paste to the pimple, and let it sit for a couple of minutes before washing off with soap and water. It will reduce the redness and soothe the sting. If the pimple persists, repeat the procedure as needed until it's gone.

Treat hard calluses Soften hard calluses on your feet by grinding five or six aspirins into a powder. Make a paste by adding 1/2 teaspoon each of lemon juice and water. Apply the mixture to the affected areas, then wrap your foot in a warm towel and cover it with a plastic bag. After staying off your feet for at least ten minutes, remove the bag and towel, and file down the softened callus with a pumice stone.

Control dandruff Is your dandruff problem getting you down? Keep it in check by crushing two aspirins to a fine powder and adding it to the normal amount of shampoo you use each time you wash your hair. Leave the mixture on your hair for 1-2 minutes, then rinse well and wash again with plain shampoo.

Apply to insect bites and stings Control the inflammation caused by mosquito bites or bee stings by wetting your skin and rubbing an aspirin over the spot. Of course, if you are allergic to bee stings—and have difficulty breathing, develop abdominal pains, or feel nauseated following a bee sting—get medical attention at once.

Help cut flowers last longer It's a tried-and-true way to keep roses and other cut flowers fresh longer: Put a crushed aspirin in the water before adding your flowers. Other household items that you can put in the water to extend the life of your flower arrangements include: a multivitamin, a teaspoon of sugar, a pinch of salt and baking soda, and even a copper penny. Also, don't forget to change the vase water every few days.

Use as garden aid Aspirin is not only a first-aid essential for you, but for your garden as well. Some gardeners grind it up for use as a rooting agent, or mix it with water to treat fungus conditions in the soil. But be careful when using aspirin around plants; too much of it can cause burns or other damage to your greenery. When treating soil, the typical dosage should be a half or a full aspirin tablet in 1 quart (1 liter) water.

Remove egg stains from clothes Did you drop some raw egg on your clothing while cooking or eating? First, scrape off as much of the egg as you can, and then try to sponge out the rest with lukewarm water. Don't use hot water—it will set the egg. If that doesn't completely remove the stain, mix water and cream of tartar into a paste and add a crushed aspirin. Spread the paste on the stain and leave it for 30 minutes. Rinse well in warm water and the egg will be gone.

* Baby Oil

Remove a bandage You can eliminate—or at least, significantly lessen—the "ouch" factor, and subsequent tears, when removing a youngster's bandage by first rubbing some baby oil into the adhesive parts on top and around the edges. If you see the bandage working loose, let the child finish the job to help him overcome his fear. Adults who have sensitive or fragile skin may also want to try this.

Make your own bath oil Do you have a favorite perfume or cologne? You can literally bathe in it by making your own scented bath oil. Simply add a few drops of your scent of choice to 1/4 cup baby oil in a small plastic bottle. Shake well, and add it to your bath.

Buff up your golf clubs Don't waste your money on fancy cleaning kits for your chrome-plated carbon steel golf club heads. Just keep a small bottle filled with baby oil in your golf bag along with a chamois cloth or towel. Dab a few drops of oil on the cloth and polish the head of your club after each round of golf.

Slip off a stuck ring Is that ring jammed on your finger again? First lubricate the ring area with a generous amount of baby oil. Then swivel the ring around to spread the oil under it. You should be able to slide the ring off with ease.

Clean your bathtub or shower Remove dirt and built-up soap scum around your bathtub or shower stall by wiping surfaces with 1 teaspoon baby oil on a moist cloth. Use another cloth to wipe away any leftover oil. Finally, spray the area with a disinfectant cleaner to kill any remaining germs. This technique is also great for cleaning soap film and watermarks off glass shower doors.

Shine stainless steel sinks and chrome trim Pamper your dull-looking stainless steel sinks by rubbing them down with a few drops of baby oil on a soft, clean cloth. Rub dry with a towel, and repeat if necessary. This is also a terrific way to remove stains on the chrome trim of your kitchen appliances and bathroom fixtures.

Polish leather bags and shoes Just a few drops of baby oil applied with a soft cloth can add new life to an old leather bag or pair of patent-leather shoes. Don't forget to wipe away any oil remaining on the leather when you're done.

Get scratches off dashboard plastic You can disguise scratches on the plastic lens covering the odometer and other indicators on your car's dashboard by rubbing over them with a bit of baby oil.

Remove latex paint from skin Did you get almost as much paint on your face and hands as you did on the bathroom you just painted? You can quickly get latex paint off your skin by first rubbing it with some baby oil, followed by a good washing with soap and hot water.

Treat cradle cap Cradle cap may be unsightly, but it is a common, usually harmless, phase in many babies' development. To combat it, gently rub in a little baby oil, and lightly comb it through your baby's hair. If your child gets upset, comb it a bit at a time, but do not leave the oil on for more than 24 hours. Then, thoroughly wash the hair to remove all of the oil. Repeat the process in persistent cases. *Note:* If you notice a lot of yellow crusting, or if the cradle cap has spread behind the ears or on the neck, contact your pediatrician instead.

* Baby Powder

Give sand the brush-off How many times have you had a family member return from a day at the beach only to discover that a good portion of the beach has been brought back into your living room? Minimize the mess by sprinkling some baby powder over sweaty, sand-covered kids (and adults) before they enter the house. In addition to soaking up excess moisture, the powder makes sand incredibly easy to brush off.

Cool sheets in summer Are those sticky, hot bed sheets giving you the summertime blues when you should be deep in dreamland? Cool things down by sprinkling a bit of baby powder between your sheets before hopping into the sack on warm summer nights.

When shopping for baby powder, you're invariably faced with three choices: ordinary, cornstarch, or medicated.

Ordinary baby powder is primarily talcum powder, which is not good for infants to breathe. Using talc on baby girls is particularly discouraged, since studies suggest that it could cause ovarian cancer later in life.

Pediatricians often recommend using a cornstarch-based powder—if one is needed at all—when changing diapers. Cornstarch powder is coarser than talcum powder but does not have the health risks. But it can promote fungal infection and should not be applied in skin folds or to broken skin.

Medicated baby powder has zinc oxide added to either talcum powder or cornstarch. It is generally used to soothe diaper rash and to prevent chafing.

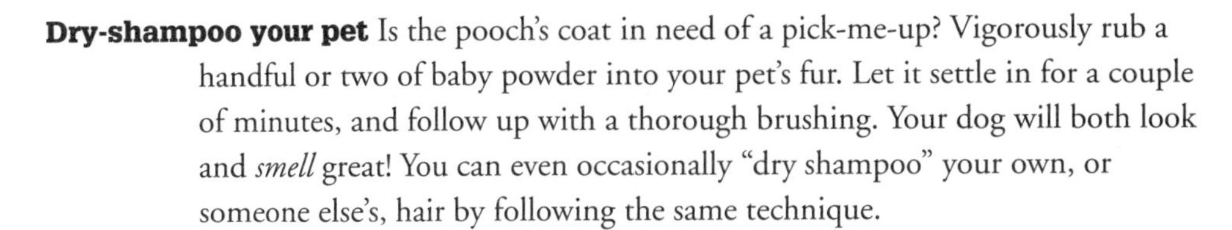

Dry-shampoo your pet Is the pooch's coat in need of a pick-me-up? Vigorously rub a handful or two of baby powder into your pet's fur. Let it settle in for a couple of minutes, and follow up with a thorough brushing. Your dog will both look and *smell* great! You can even occasionally "dry shampoo" your own, or someone else's, hair by following the same technique.

Absorb grease stains on clothing Frying foods can be dangerous business—especially for your clothes. If you get a grease splatter on your clothing, try dabbing the stain with some baby powder on a powder puff. Make sure you rub it in well, and then brush off any excess powder. Repeat until the mark is gone.

Clean your playing cards Here's a simple way to keep your playing cards from sticking together and getting grimy: Loosely place the cards in a plastic bag along with a bit of baby powder. Seal the bag and give it a few good shakes. When you remove your cards, they should feel fresh and smooth to the touch.

Slip on your rubber gloves Don't try jamming and squeezing your fingers into your rubber gloves when the powder layer inside the gloves wears out. Instead, give your fingers a light dusting with baby powder. Your rubber gloves should slide on good as new.

Remove mold from books If some of your books have been stored in a less than ideal environment and have gotten a bit moldy or mildewed, try this: First, let them thoroughly air-dry. Then, sprinkle some baby powder between the pages and stand the books upright for several hours. Afterward, gently brush out the remaining powder from each book. They may not be as good as new, but they should be in a lot better shape than they were.

Dust off your flower bulbs Many savvy gardeners use medicated baby powder to dust flower bulbs before planting them. Simply place 5-6 bulbs and about 3 tablespoons baby powder in a sealed plastic bag and give it a few gentle shakes. The medicated-powder coating helps both reduce the chance of rot and keep away moles, voles, grubs, and other bulb-munching pests.

* Baby Wipes

Use for quick, on-the-move cleanups Baby wipes can be used for more than just cleaning babies' bottoms. They're great for wiping your hands after pumping gas, mopping up small spills in the car, and cooling your sweaty brow after a run. In fact, they make ideal travel companions. So, next time you set off on the road, pack a small stack of wipes in a tightly closed self-sealing sandwich bag and put it in the glove compartment of your car or in your purse or knapsack.

Shine your shoes Most moms know that a baby wipe does a pretty good job of brightening Junior's white leather shoes. But did you ever think of using one to put the shine back in *your* leather pumps—especially with that 10 a.m. meeting fast approaching?

Recycle as dust cloths Believe it or not, some brands of baby wipes—Huggies, for instance—can be laundered and reused as dust cloths and cleaning rags for when you straighten up. It probably goes without saying, but only "mildly" soiled wipes should be considered candidates for laundering.

Buff up your bathroom Do you have company coming over and not much time to tidy up the house? Don't break out in a sweat. Try this double-handed trick: Take a baby wipe in one hand and start polishing your bathroom surfaces. Keep a dry washcloth in your other hand to shine things up as you make your rounds.

Remove stains from carpet, clothing, and upholstery Use a baby wipe to blot up coffee spills from your rug or carpet; it absorbs both the liquid *and* the stain. Wipes can also be effectively deployed when attacking various spills and drips on your clothing and upholstered furniture.

Clean your PC keyboard Periodically shaking out your PC's keyboard is a good way to get rid of the dust and debris that gathers underneath and in between the keys. But that's just half the job. Use a baby wipe to remove the dirt, dried spills, and unspecified gunk that builds up on the keys themselves. Make sure to turn off the computer or unplug the keyboard before you wipe the keys.

Soothe your skin Did you get a bit too much sun at the beach? You can temporarily cool a sunburn by gently patting the area with a baby wipe. Baby wipes can also be used to treat cuts and scrapes. Although most wipes don't have any antiseptic properties, there's nothing wrong with using one for an initial cleansing before applying the proper topical treatment.

Remove makeup It's one of the fashion industry's worst-kept secrets: Many models consider a baby wipe to be their best friend when it comes time to remove that stubborn makeup from their faces, particularly black eyeliner. Try it and see for yourself.

* Baby Wipes Containers

Organize your stuff Don't toss those empty wipes containers. These sturdy plastic boxes are incredibly useful for storing all sorts of items. And the rectangular ones are stackable to boot! Give the containers a good washing and let them dry thoroughly, then fill them with everything from sewing supplies, recipe cards, coupons, and craft and office supplies to old floppy disks, small tools, photos, receipts, and bills. Label the contents with a marker on masking tape, and you're set!

Make a first-aid kit Every home needs a first-aid kit. But you don't have to buy a ready-made one. Gather up your own choice of essentials (such as bandages, sterile

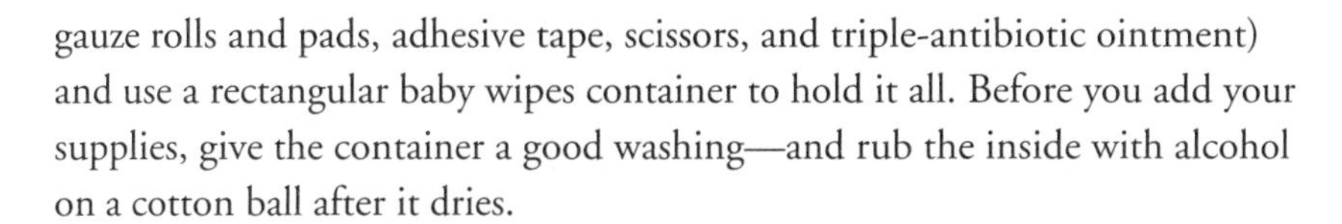

gauze rolls and pads, adhesive tape, scissors, and triple-antibiotic ointment) and use a rectangular baby wipes container to hold it all. Before you add your supplies, give the container a good washing—and rub the inside with alcohol on a cotton ball after it dries.

Use as a decorative yarn or twine dispenser A clean cylindrical wipes container makes a perfect dispenser for a roll of yarn or twine. Simply remove the container's cover, insert the roll, and thread it through the slot in the lid, then reattach the cover. Paint or paper over the container to give it a more decorative look.

Tip **Removing Labels**

Use a blow-dryer on a high setting to heat up the labels on baby wipes containers to make them easier to pull off. You can get rid of any leftover sticky stuff by applying a little WD-40 oil or orange citrus cleaner.

Store your plastic shopping bags Do you save plastic shopping bags for lining the small wastebaskets (or perhaps for pooper-scooper duty)? If so, bring order to the puffed-up chaos they create by storing the bags in cleaned, rectangular wipes containers. Each container can hold 40 to 50 bags—once you squeeze the air out of them. You can also use an empty 250-count tissue box—the kind with a perforated cutout dispenser—in a similar manner.

Make a piggy bank Well, maybe not a "piggy" bank, per se, but a bank nonetheless, and one that gives you a convenient place to dump your pocket change. Take a clean rectangular container and use a knife to cut a slot—be sure to make it wide enough to easily accommodate a quarter—on the lid. If you're making the bank for a child, you can either decorate it or let her put her own personal "stamp" on it.

Hold workshop towels or rags A used baby wipes container can be a welcome addition in the workshop for storing rags and paper towels—and to keep a steady supply on hand as needed. You can easily keep a full roll of detached paper towels or six or seven good-sized rags in each container.

Baking Soda

* BAKING SODA IN THE KITCHEN

Clean your produce You can't be too careful when it comes to food handling and preparation. Wash fruits and vegetables in a pot of cold water with 2-3 tablespoons baking soda; the baking soda will remove some of the impurities tap water leaves behind. Or put a small amount of baking soda on a wet sponge or vegetable brush and scrub your produce. Give everything a thorough rinsing before serving.

Tenderize meat Got a tough cut of meat on your hands? Soften it up by giving it rub-down in baking soda. Let it sit (in the refrigerator, of course) for three to five hours, then rinse it off well before cooking.

Soak out fish smells Get rid of that fishy smell from your store-bought flounder filets and fish steaks by soaking the raw fish for about an hour (inside your refrigerator) in 1 quart (1 liter) water with 2 tablespoons baking soda. Rinse the fish well and pat dry before cooking.

Reduce acids in recipes If you or someone in your family is sensitive to the high-acid content of tomato-based sauces or coffee, you can lower the overall acidity by sprinkling in a pinch of baking soda while cooking (or, in the case of coffee, before brewing). A bit of baking soda can also counteract the taste of vinegar if you happen to pour in a bit too much. Be careful not to overdo it with the soda, though—if you add too much, the vinegar-baking soda combination will start foaming.

Bake better beans Do you love baked beans but not their aftereffects? Adding a pinch of baking soda to baked beans as they're cooking will significantly reduce their gas-producing properties.

Fluff up your omelets Want to know the secret to making fluffier omelets? For every three eggs used, add 1/2 teaspoon baking soda. Shhhh! Don't let it get around.

Tip **Out of Baking Powder?**

> If you are out of baking powder, you can usually substitute 2 parts baking soda mixed with 1 part each cream of tartar and cornstarch. To make the equivalent of 1 teaspoon baking powder, for instance, mix 1/2 teaspoon baking soda with 1/4 teaspoon cream of tartar and 1/4 teaspoon cornstarch. The cornstarch slows the reaction between the acidic cream of tartar and the alkaline baking soda so that, like commercial baking powder, it maintains its leavening power longer.

Use as yeast substitute Need a stand-in for yeast when making dough? If you have some powdered vitamin C (or citric acid) and baking soda on hand, you can use a mixture of the two instead. Just mix in equal parts to equal the quantity of yeast required. What's more, the dough you add it to won't have to rise before baking.

Rid hands of food odors Chopping garlic or cleaning a fish can leave their "essence" on your fingers long after the chore is done. Get those nasty food smells off your hands by simply wetting them and vigorously rubbing with about 2 teaspoons baking soda instead of soap. The smell should wash off with the soda.

Clean baby bottles and accessories Here's some great advice for new parents: Keep all your baby bottles, nipples, caps, and brushes "baby fresh" by soaking them overnight in a container filled with hot water and half a box of baking soda. Be sure to give everything a good rinsing afterward, and to dry thoroughly before using. Baby bottles can also be boiled in a full pot of water and 3 tablespoons baking soda for three minutes.

Clean a cutting board Keep your wooden or plastic cutting board clean by occasionally scrubbing it with a paste made from 1 tablespoon each baking soda, salt, and water. Rinse thoroughly with hot water.

Clear a clogged drain Most kitchen drains can be unclogged by pouring in 1 cup baking soda followed by 1 cup hot vinegar (simply heat it up in the microwave for 1 minute). Give it several minutes to work, then add 1 quart (1 liter) boiling water. Repeat if necessary. If you know your drain is clogged with grease, use 1/2 cup each of baking soda and salt followed by 1 cup boiling water. Let the mixture work overnight; then rinse with hot tap water in the morning.

Boost potency of dishwashing liquid Looking for a more powerful dishwashing liquid? Try adding 2 tablespoons baking soda to the usual amount of liquid you use, and watch it cut through grease like a hot knife!

Make your own dishwashing detergent The dishwasher is fully loaded when you discover that you're out of your usual powdered dishwashing detergent. What do you do? Make your own: Combine 2 tablespoons baking soda with 2 tablespoons borax. You may be so pleased with the results you'll switch for good.

Deodorize your dishwasher Eliminate odors inside your automatic dishwasher by sprinkling 1/2 cup baking soda on the bottom of the dishwasher between loads. Or pour in half a box of baking soda and run the empty machine through its rinse cycle.

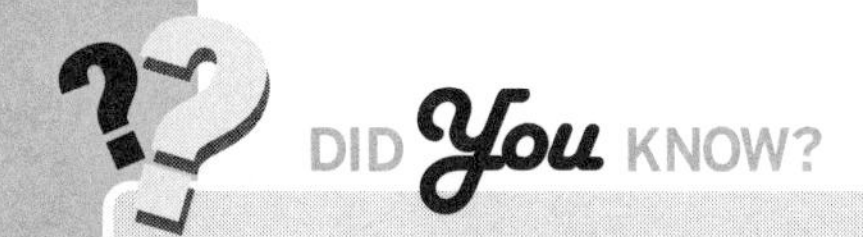

Baking soda is the main ingredient in many commercial fire extinguishers. And you can use it straight out of the box to extinguish small fires throughout your home. For quick access, keep baking soda in buckets placed strategically around the house.

Keep baking soda near your stove and barbecue so you can toss on a few handfuls to quell a flare-up. In the case of a grease fire, first turn off the heat, if possible, and try to cover the fire with a pan lid. Be careful not to let the hot grease splatter you.

Keep a box or two in your garage and inside your car to quickly extinguish any mechanical or car-interior fires.

Baking soda will also snuff out electrical fires and flames on clothing, wood, upholstery, and carpeting.

Clean your refrigerator To get rid of smells and dried-up spills inside your refrigerator, remove the contents, then sprinkle some baking soda on a damp sponge and scrub the sides, shelves, and compartments. Rinse with a clean, wet sponge. Don't forget to place a fresh box of soda inside when you're done.

Clean your microwave To clean those splatters off the inside of your microwave, put a solution of 2 tablespoons baking soda in 1 cup water in a microwave-safe container and cook on High for 2-3 minutes. Remove the container, then wipe down the microwave's moist interior with a damp paper towel.

Remove coffee and tea stains from china Don't let those annoying coffee and/or tea stains on your good china spoil another special occasion. Remove them by dipping a moist cloth in baking soda to form a stiff paste and gently rubbing your cups and saucers. Rinse clean and dry, then set your table with pride.

Clean a thermos To remove residue on the inside of a thermos, mix 1/4 cup baking soda in 1 quart (1 liter) water. Fill the thermos with the solution—if necessary, give it a going-over with a bottle brush to loosen things up—and let it soak overnight. Rinse clean before using.

Freshen a sponge or towel When a kitchen sponge or dish towel gets that distinctly sour smell, soak it overnight in 2 tablespoons baking soda and a couple of drops of antibacterial dish soap dissolved in 1 pint (450 milliliters) warm water. The following morning, squeeze out the remaining solution and rinse with cold water. It should smell as good as new.

Remove stains and scratches on countertops Is your kitchen countertop covered with stains or small knife cuts? Use a paste of 2 parts baking soda to 1 part water to "rub out" most of them. For stubborn stains, add a drop of chlorine bleach to the paste. Immediately wash the area with hot, soapy water to prevent the bleach from causing fading.

Shine up stainless steel and chrome trim To put the shine back in your stainless steel sink, sprinkle it with baking soda, then give it a rubdown—moving in the direction of the grain—with a moist cloth. To polish dull chrome trim on your appliances, pour a little baking soda onto a damp sponge and rub over the chrome. Let it dry for an hour or so, then wipe down with warm water and dry with a clean cloth.

Get rid of grease stains on stovetops Say good-bye to cooked-on grease stains on your stovetop or backsplash. First wet them with a little water and cover them with a bit of baking soda. Then rub them off with a damp sponge or towel.

Clean an automatic coffeemaker Properly caring for your automatic coffeemaker means never having to worry about bitter or weak coffee. Every two weeks or so, brew a pot of 1 quart (1 liter) water mixed with 1/4 cup baking soda, followed by a pot of clean water. Also, sweeten your coffeemaker's plastic basket by using an old toothbrush to give it an occasional scrubbing with a paste of 2 tablespoons baking soda and 1 tablespoon water. Rinse thoroughly with cold water when done.

Care for your coffeepots and teapots Remove mineral deposits in metal coffeepots and teapots by filling them with a solution of 1 cup vinegar and 4 tablespoons baking soda. Bring the mixture to a boil, then let simmer for five minutes. Or try boiling 5 cups water with 2 tablespoons soda and the juice of half a lemon. Rinse with cold water when done. To get off annoying exterior stains, wash your pots with a plastic scouring pad in a solution of 1/4 cup baking soda in 1 quart (1 liter) warm water. Follow up with a cold-water rinse.

Remove stains from nonstick cookware It may be called nonstick cookware, but a few of those stains seem to be stuck on pretty well. Blast them away by boiling 1 cup water mixed with 2 tablespoons baking soda and 1/2 cup vinegar for 10 minutes. Then wash in hot, soapy water. Rinse well and let dry, then season with a bit of salad oil.

Clean cast-iron cookware Although it's more prone to stains and rust than the nonstick variety, many folks swear by their iron cookware. You can remove even the toughest burned-on food remnants in your iron pots by boiling 1 quart (1 liter) water with 2 tablespoons baking soda for five minutes. Pour off most of the liquid, then lightly scrub it with a plastic scrub pad. Rinse well, dry, and season with a few drops of peanut oil.

Clean burned or scorched pots and pans It usually takes heavy-duty scrubbing to get scorched-on food off the bottom of a pot or pan. But you can make life much easier for yourself by simply boiling a few cups of water (enough to get the pan about 1/4 full) and adding 5 tablespoons baking soda. Turn off the heat, and let the soda settle in for a few hours or overnight. When you're ready, that burned-on gunk will practically slip right off.

Deodorize your garbage pail Does something smell "off" in your kitchen? Most likely, it's emanating from your trash can. But some smells linger even after you dispose of the offending garbage bag. So, be sure to give your kitchen garbage pail an occasional cleaning with a wet paper towel dipped in baking soda (you may want to wear rubber gloves for this). Rinse it out with a damp sponge, and let it dry before inserting a new bag. You can also ward off stinky surprises by sprinkling a little baking soda into the bottom of your pail before inserting the bag.

BAKING SODA AROUND THE HOUSE

Remove crayon marks from walls Has Junior redecorated your walls or wallpaper with some original artworks in crayon? Don't lose your cool. Just grab a damp rag, dip it in some baking soda, and lightly scrub the marks. They should come off with a minimal amount of effort.

Wash wallpaper Is your wallpaper looking a bit dingy? Brighten it up by wiping it with a rag or sponge moistened in a solution of 2 tablespoons baking soda in 1 quart (1 liter) water. To remove grease stains from wallpaper, make a paste of 1 tablespoon baking soda and 1 teaspoon water. Rub it on the stain, let it set for 5-10 minutes, then rub off with a damp sponge.

Clean baby spit-ups Infants do tend to spit up—and usually not at opportune moments. Never leave home without a small bottle of baking soda in your diaper bag. If your tyke spits up on his or her (or your) shirt after feeding, simply brush off any solid matter, moisten a washcloth, dip it in a bit of baking soda, and dab the spot. The odor (and the potential stain) will soon be gone.

Deodorize rugs and carpets How's this for a simple way to freshen up your carpets or rugs? Lightly sprinkle them with baking soda, let it settle in for 15 minutes or so, then vacuum up. Nothing to it!

Remove wine and grease stains from carpet What's that? Someone just dropped a slab of butter or a glass of cabernet on your beautiful white carpeting! Before you scream, get a paper towel, and blot up as much of the stain as possible. Then sprinkle a liberal amount of baking soda over the spot. Give the soda at least an hour to absorb the stain, then vacuum up the remaining powder. Now ... exhale!

Kids' Stuff Make **watercolor paints** for your kids using ingredients in your kitchen. In a small bowl, combine 3 tablespoons each of **baking soda, cornstarch,** and **vinegar** with 1 1/2 teaspoons **light corn syrup.** Wait for the fizzing to subside, then separate the mixture into several small containers or jar lids. Add eight drops of **food coloring** to each batch and mix well. Put a different color in each batch or combine colors to make new shades. Kids can either use the paint right away, or wait for them to harden, in which case, they'll need to use a wet brush before painting.

Freshen up musty drawers and closets Put baking soda sachets to work on persistent musty odors in dresser drawers, cabinet hutches, or closets. Just fill the toe of a clean sock or stocking with 3-4 tablespoons soda, put a knot about an inch above the bulge, and either hang it up or place it away in an unobtrusive corner. Use a few sachets in large spaces like closets and attic storage areas. Replace them every other month if needed. This treatment can also be used to rid closets of mothball smells.

Remove musty odor from books If those books you just took out of storage emerge with a musty smell, place each one in a brown paper bag with 2 tablespoons baking soda. No need to shake the bag, just tie it up and let it sit in a dry environment for about one week. When you open the bag, shake any remaining powder off the books, and the smell should be gone.

B

Polish silver and gold jewelry To remove built-up tarnish from your silver, make a thick paste with 1/4 cup baking soda and 2 tablespoons water. Apply with a damp sponge and gently rub, rinse, and buff dry. To polish gold jewelry, cover with a light coating of baking soda, pour a bit of vinegar over it, and rinse clean. *Note:* Do not use this technique with jewelry containing pearls or gemstones, as it could damage their finish and loosen the glue.

Get yellow stains off piano keys That old upright may still play great, but those yellowed keys definitely hit a sour note. Remove age stains on your ivories by mixing a solution of 1/4 cup baking soda in 1 quart (1 liter) warm water. Apply to each key with a dampened cloth (you can place a thin piece of cardboard between the keys to avoid seepage). Wipe again with a cloth dampened with plain water, and then buff dry with a clean cloth. (You can also clean piano keys with lemon juice and salt.)

Remove stains from fireplace bricks You may need to use a bit of elbow grease, but you can clean the smoke stains off your fireplace bricks by washing them with a solution of 1/2 cup baking soda in 1 quart (1 liter) warm water.

Remove white marks on wood surfaces Get those white marks—caused by hot cups or sweating glasses—off your coffee table or other wooden furniture by making a paste of 1 tablespoon baking soda and 1 teaspoon water. Gently rub the spot in a circular motion until it disappears. Remember not to use too much water.

Remove cigarette odors from furniture To eliminate that lingering smell of cigarette or cigar smoke on your upholstered furniture, simply lightly sprinkle your chairs or sofas with some baking soda. Let it sit for a few hours, then vacuum it off.

Shine up marble-topped furniture Revitalize the marble top on your coffee table or counter by washing it with a soft cloth dipped in a solution of 3 tablespoons baking soda and 1 quart (1 liter) warm water. Let it stand for 15 minutes to a half hour, then rinse with plain water and wipe dry.

Clean bathtubs and sinks Get the gunk off old enameled bathtubs and sinks by applying a paste of 2 parts baking soda and 1 part hydrogen peroxide. Let the paste set for about half an hour. Then give it a good scrubbing and rinse well; the paste will also sweeten your drain as it washes down.

Remove mineral deposits from showerheads Say so long to hard-water deposits on your showerhead. Cover the head with a thick sandwich-size bag filled with 1/4 cup baking soda and 1 cup vinegar. Loosely fasten the bag—you need to let some of the gas escape—with adhesive tape or a large bag tie. Let the solution work its magic for about an hour.

Then remove the bag and turn on your shower to wash off any remaining debris. Not only will the deposits disappear, but your showerhead will be back to its old shining self!

Absorb bathroom odors Keep your bathroom smelling fresh and clean by placing a decorative dish filled with 1/2 cup baking soda either on top of the toilet tank or on the floor behind the bowl. You can also make your own bathroom deodorizers by setting out dishes containing equal parts baking soda and your favorite scented bath salts.

Tidy up your toilet bowl You don't need all those chemicals to get your toilet bowl clean. Just pour half a box of baking soda into your toilet tank once a month. Let it stand overnight, then give it a few flushes in the morning. This actually cleans both the tank and the bowl. You can also pour several tablespoons of baking soda directly into your toilet bowl and scrub it on any stains. Wait a few minutes, then flush away the stains.

* BAKING SODA IN THE MEDICINE CABINET

Treat minor burns The next time you grab the wrong end of a frying pan or forget to use a pot holder, quickly pour some baking soda into a container of ice water, soak a cloth or gauze pad in it, and apply it to the burn. Keep applying the solution until the burn no longer feels hot. This treatment will also prevent many burns from blistering.

Cool off sunburn and other skin irritations For quick relief of sunburn pain, soak gauze pads or large cotton balls in a solution of 4 tablespoons baking soda mixed in 1 cup water and apply it to the affected areas. For a bad sunburn on your legs or torso—or to relieve the itching of chicken pox—take a lukewarm bath with a half to a full box of baking soda added to the running water. To ease the sting of razor burns, dab your skin with a cotton ball soaked in a solution of 1 tablespoon baking soda in 1 cup water.

Soothe poison ivy rashes Did you have an unplanned encounter with poison ivy when gardening or camping recently? To take away the itch, make a thick paste from 3 teaspoons baking soda and 1 teaspoon water and apply it to the affected areas. You can also use baking soda to treat oozing blisters caused by the rash. Mix 2 teaspoons baking soda in 1 quart (1 liter) water and use it to saturate a few sterile gauze pads. Cover the blisters with the wet pads for 10 minutes, four times a day. *Note:* Do not apply on or near your eyes.

Make a salve for bee stings Take the pain out of that bee sting—fast. Make a paste of 1 teaspoon baking soda mixed with several drops of cool water, and let it dry on the afflicted area. *Warning:* Many people have severe allergic reactions to bee stings. If you have difficulty breathing or notice a dramatic swelling, get medical attention at once. (You can also treat bee stings with meat tenderizer. See page 210.)

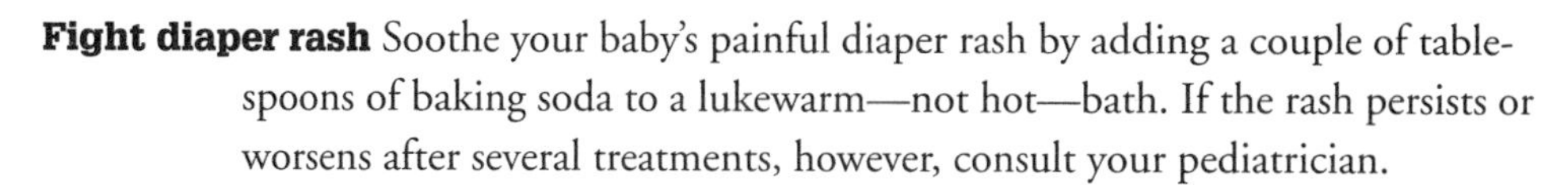

Fight diaper rash Soothe your baby's painful diaper rash by adding a couple of tablespoons of baking soda to a lukewarm—not hot—bath. If the rash persists or worsens after several treatments, however, consult your pediatrician.

Combat cradle cap Cradle cap is a commonplace, and typically harmless, condition in many infants. An old but often effective way to treat it is to make a paste of about 3 teaspoons baking soda and 1 teaspoon water. Apply it to your baby's scalp about an hour before bedtime and rinse it off the following morning. Do not use with shampoo. You may need to apply it several consecutive nights before the cradle cap recedes. (You can also treat cradle cap with baby oil. See page 59.)

Tip Baking Soda Shelf Life

How can you tell if the baking soda you've had stashed away in the back of your pantry is still good? Just pour out a small amount—a little less than a teaspoon—and add a few drops of vinegar or fresh lemon juice. If it doesn't fizz, it's time to replace it. By the way, a sealed box of baking soda has an average shelf life of 18 months, while an opened box lasts 6 months.

Control your dandruff Got a bit of a "flaky" problem? To get dandruff under control, wet your hair and then rub a handful of baking soda vigorously into your scalp. Rinse thoroughly and dry. Do this every time you normally wash your hair, but only use baking soda, no shampoo. Your hair may get dried out at first. But after a few weeks your scalp will start producing natural oils, leaving your hair softer and free of flakes.

Clean combs and brushes Freshen up your combs and hairbrushes by soaking them in a solution of 3 cups warm water and 2 teaspoons baking soda. Swirl them around in the water to loosen up all the debris caught between the teeth, then let them soak for about half an hour. Rinse well and dry before using.

Use as gargle or mouthwash Did the main course you ordered include a few too many onions or a bit too much garlic? Try gargling with 1 teaspoon baking soda in a half glass of water. The baking soda will neutralize the odors on contact. When used as a mouthwash, baking soda will also relieve canker-sore pain.

Scrub teeth and clean dentures If you run out of your regular toothpaste, or if you're looking for an all-natural alternative to commercial toothpaste, just dip your wet toothbrush in some baking soda and brush and rinse as usual. You can also use baking soda to clean retainers, mouthpieces, and dentures. Use a solution of 1 tablespoon baking soda dissolved in 1 cup warm water. Let the object soak for a half hour and rinse well before using.

Clean and sweeten toothbrushes Keep your family's toothbrushes squeaky clean by immersing them in a solution of 1/4 cup baking soda and 1/4 cup water. Let the brushes soak overnight about once every week or two. Be sure to give them a good rinsing before using.

Remove built-up gel, hair spray, or conditioner from hair When it comes to personal grooming, too much of a good thing can spell bad news for your hair. But a thorough cleansing with baking soda at least once a week will wash all of the gunk out of your hair. Simply add 1 tablespoon soda to your hair while shampooing. In addition to removing all the chemicals you put in your hair, it will wash away water impurities, and may actually lighten your hair.

Use as antiperspirant Looking for an effective, all-natural deodorant? Try applying a small amount—about a teaspoon's worth—of baking soda with a powder puff under each arm. You won't smell like a flower or some exotic spice. But then, you won't smell like anything from the opposite extreme, either.

Relieve itching inside a cast Wearing a plaster cast on your arm or leg is a misery any time of year, but wearing one in the summertime can be torture. The sweating and itchiness you feel underneath your "shell" can drive you nearly insane. Find temporary relief by using a hair dryer—on the coolest setting—to blow a bit of baking soda down the edges of the cast. *Note:* Have someone help you, to avoid getting the powder in your eyes.

Alleviate athlete's foot You can deploy wet or dry baking soda to combat a case of athlete's foot. First, try dusting your feet (along with your socks and shoes) with dry baking soda to dry out the infection. If that doesn't work, try making a paste of 1 teaspoon baking soda and 1/2 teaspoon water and rubbing it

SCIENCE FAIR

Use the gas produced by mixing **baking soda** and **vinegar** to blow up a balloon. First, pour 1/2 cup vinegar into the bottom of a narrow-neck **bottle** (such as an empty water bottle) or **jar.** Then insert a **funnel** into the mouth of an average-sized **balloon,** and fill it with 5 tablespoons baking soda. Carefully stretch the mouth of the balloon over the opening of the bottle, then gently lift it up so that the baking soda empties into the vinegar at the bottom of the bottle. The fizzing and foaming you see is actually a chemical reaction between the two ingredients. This reaction results in the release of **carbon dioxide gas**—which will soon **inflate the balloon!**

between your toes. Let it dry, and wash off after 15 minutes. Dry your feet thoroughly before putting on your shoes.

Soothe tired, stinky feet When your dogs start barking, treat them to a soothing bath of 4 tablespoons baking soda in 1 quart (1 liter) warm water. Besides relaxing your aching tootsies, the baking soda will remove the sweat and lint that gathers between your toes. Regular footbaths can also be an effective treatment for persistent foot odor.

Deodorize shoes and sneakers A smelly shoe or sneaker is no match for the power of baking soda. Liberally sprinkle soda in the offending loafer or lace-up and let it sit overnight. Dump out the powder in the morning. (Be careful when using baking soda with leather shoes, however; repeated applications can dry them out.) You can also make your own reusable "odor eaters" by filling the toes of old socks with 2 tablespoons baking soda and tying them up in a knot. Stuff the socks into each shoe at night before retiring. Remove the socks in the morning and breathe easier.

* BAKING SODA IN THE LAUNDRY

Boost strength of liquid detergent and bleach It may sound like a cliché, but adding 1/2 cup baking soda to your usual amount of liquid laundry detergent really will give you "whiter whites" and brighter colors. The baking soda also softens the water, so you can actually use less detergent. Adding 1/2 cup baking soda in top-loading machines (1/4 cup for front-loaders) also increases the potency of bleach, so you need only half the usual amount of bleach.

Remove mothball smell from clothes If your clothes come out of storage reeking of mothballs, take heed: Adding 1/2 cup baking soda during your washer's rinse cycle will get rid of the smell.

Wash new baby clothes Get all of the chemicals out of your newborn's clothing—without using any harsh detergents. Wash your baby's new clothes with some mild soap and 1/2 cup baking soda.

Rub out perspiration and other stains Pretreating clothes with a paste made from 4 tablespoons baking soda and 1/4 cup warm water can help vanquish a variety of stains. For example, rub it into shirts to remove perspiration stains; for really bad stains, let the paste dry for about two hours before washing. Rub out tar stains by applying the paste and washing in plain baking soda. For collar stains, rub in the paste and add a bit of vinegar as you're putting the shirt in the wash.

Wash mildewed shower curtains Just because your plastic shower curtain or liner gets dirty or mildewed doesn't mean you have to throw it away. Try cleaning it in your washing machine with two bath towels on the gentle setting. Add 1/2 cup baking soda to your detergent during the wash cycle and 1/2 cup vinegar during the rinse cycle. Let it drip-dry; don't put it in the dryer.

✱ BAKING SODA **FOR THE DO-IT-YOURSELFER**

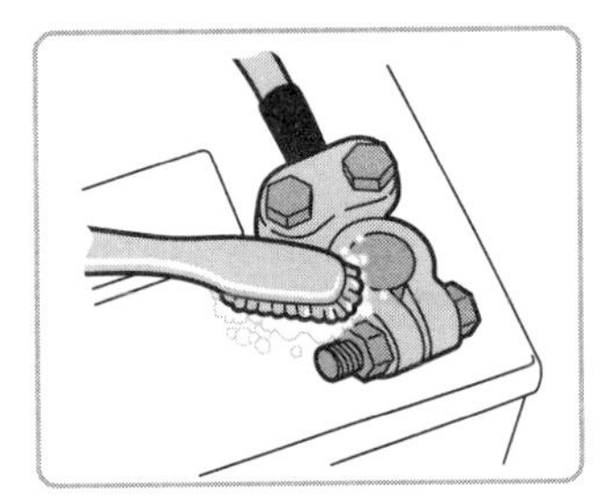

Clean battery terminals Eliminate the corrosive buildup on your car's battery terminals. Scrub them clean using an old toothbrush and a mixture of 3 tablespoons baking soda and 1 tablespoon warm water. Wipe them off with a wet towel and dry with another towel. Once the terminals have completely dried, apply a bit of petroleum jelly around each terminal to deter future corrosive buildup.

Use as deicer in winter Salt and commercial ice-melt formulations can stain—or actually eat away—the concrete around your house. For an equally effective, but completely innocuous, way to melt the ice on your steps and walkways during those cold winter months, try sprinkling them with generous amounts of baking soda. Add some sand for improved traction.

Tighten cane chair seats The bottoms of cane chairs can start to sag with age, but you can tighten them up again easily enough. Just soak two cloths in a solution of 1/2 cup baking soda in 1 quart (1 liter) hot water. Saturate the top surface of the caning with one cloth, while pushing the second up against the bottom of the caning to saturate the underside. Use a clean, dry cloth to soak up the excess moisture, then put the chair in the sun to dry.

> ***Kids' Stuff*** Spies use it and so can you. Send a message or draw a picture with **invisible ink.** Here's how you do it: Mix 1 tablespoon each of **baking soda** and **water.** Dip a **toothpick** or **paintbrush** in the mixture and write your message or draw a picture or design on a piece of **plain white paper.** Let the paper and the "ink" dry completely. To reveal your message or see your picture, mix 6 drops **food coloring** with 1 tablespoon water. Dip a clean **paintbrush** in the solution, and lightly paint over the paper. Use different food-coloring combinations for a cool effect.

Remove tar from your car It may look pretty bad, but it's not that hard to get road tar off your car without damaging the paint. Make a soft paste of 3 parts baking soda to 1 part water and apply to the tar spots with a damp cloth. Let it dry for five minutes, then rinse clean.

B

Give your deck the weathered look You can instantly give your wooden deck a weathered look by washing it in a solution of 2 cups baking soda in 1 gallon (3.7 liters) water. Use a stiff straw brush to work the solution into the wood, then rinse with cool water.

Clean air-conditioner filters Clean washable air-conditioner filters each month they're in use. First vacuum off as much dust and dirt as possible, then wash in a solution of 1 tablespoon baking soda in 1 quart (1 liter) water. Let the filters dry thoroughly before replacing.

Keep your humidifier odor-free Eliminate musty smells from a humidifier by adding 2 tablespoons baking soda to the water each time you change it. *Note:* Check your owner's manual or consult the unit's manufacturer before trying this.

* BAKING SODA IN THE GREAT OUTDOORS

Keep weeds out of cement cracks Looking for a safe way to keep weeds and grasses from growing in the cracks of your paved patios, driveways, and walkways? Sprinkle handfuls of baking soda onto the concrete and simply sweep it into the cracks. The added sodium will make it much less hospitable to dandelions and their friends.

Clean resin lawn furniture Most commercial cleaners are too abrasive to be used on resin lawn furniture. But you won't have to worry about scratching or dulling the surface if you clean your resin furniture with a wet sponge dipped in baking soda. Wipe using circular motions, then rinse well.

Use as plant food Give your flowering, alkaline-loving plants, such as clematis, delphiniums, and dianthus, an occasional shower in a mild solution of 1 tablespoon baking soda in 2 quarts (2 liters) water. They'll show their appreciation with fuller, healthier blooms.

Maintain proper pool alkalinity Add 1 1/2 pounds (680 grams) baking soda for every 10,000 gallons (38,000 liters) of water in your swimming pool to raise the total alkalinity by 10 ppm (parts per million). Most pools require alkalinity in the 80-150 ppm range. Maintaining the proper pool alkalinity level is vital for minimizing changes in pH if acidic or basic pool chemicals or contaminants are introduced to the water.

Scour barbecue grills Keep your barbecue grill in top condition by making a soft paste of 1/4 cup baking soda and 1/4 cup water. Apply the paste with a wire brush and let dry for 15 minutes. Then wipe it down with a dry cloth and place the grill over the hot coals for at least 15 minutes to burn off any residue before placing any food on top.

Make deodorizing dog shampoo The next time Rover rolls around in your compost heap, pull out the baking soda to freshen him up. Just rub a few handfuls of the powder into his coat and give it a thorough brushing. In addition to removing the smell, it will leave his coat shiny and clean.

Wash insides of pets' ears If your pet is constantly scratching at his ears, it could indicate the presence of an irritation or ear mites. Ease the itch (and wipe out any mites) by using a cotton ball dipped in a solution of 1 teaspoon baking soda in 1 cup warm water to gently wash the inside of his ears.

Keep bugs away from pets' dishes Placing a border of baking soda around your pet's food bowls will keep away six-legged intruders. And it won't harm your pet if he happens to lap up a little (though most pets aren't likely to savor soda's bitter taste).

Deodorize the litter box Don't waste money on expensive deodorized cat litter. Just put a thin layer of baking soda under the bargain-brand litter to absorb the odor. Or mix baking soda with the litter as you're changing it.

✱ Balloons

Protect a bandaged finger Bandaging an injury on your finger is easy; keeping the bandage dry as you go about your day can be a different story. But here's the secret to skipping those wet-bandage changes: Just slip a small balloon over your finger when doing dishes, bathing, or even simply washing your hands.

Keep track of your child Those inexpensive floating helium-filled balloons sold in most shopping malls can be more than just a treat for a youngster; they could be invaluable in locating a child who wanders off into a crowd. Even if you keep close tabs on your kids, you can buy a little peace of mind by simply tying (though not too tightly) a balloon to your child's wrist on those weekend shopping trips.

Make a party invitation How's this for an imaginative invitation? Inflate a balloon (for sanitary purposes, use an electric pump, if possible). Pinch off the end, but don't tie a knot in it. Write your invitation details on the balloon with a bright permanent marker; make sure the ink is dry before you deflate it. Place the balloon in an envelope, and mail one out to each guest. When your guests receive it, they'll have to blow it up to see what it says.

Transport cut flowers Don't bother with awkward, water-filled plastic bags and such when traveling with freshly cut flowers. Simply fill up a balloon with about 1/2 cup water and slip it over the cut ends of your flowers. Wrap a rubber band several times around the mouth of the balloon to keep it from slipping off.

Use as a hat mold To keep the shape in your freshly washed knit cap or cloth hat, fit it over an inflated balloon while it dries. Use a piece of masking tape to keep the balloon from tilting over or falling onto the ground.

Mark your campsite Bring along several helium-filled balloons on your next camping trip to attach to your tent or a post. They'll make it easier for the members of your party to locate your campsite when hiking or foraging in the woods.

Make an ice pack Looking for a flexible ice pack you can use for everything from icing a sore back to keeping food cold in your cooler? Fill a large, durable balloon with as much water as you need and put it in your freezer. You can even mold it to a certain extent into specific shapes—for example, put it under something flat like a box of pizza if you want a flat ice pack for your back. Use smaller latex balloons for making smaller ice packs for lunch boxes, etc.

Freeze for cooler punch To keep your party punch bowl cold and well filled, pour juice in several balloons (use a funnel) and place them in your freezer. When it's party time, peel the latex off the ice, and periodically drop a couple into the punch bowl.

Repel unwanted garden visitors Put those old deflated shiny metallic balloons—the ones lying around your house from past birthday parties—to work in your garden. Cut them into vertical strips and hang them from poles around your vegetables and on fruit trees to scare off invading birds, rabbits, and squirrels.

Protect your rifle A dirty rifle can jam up and just be downright dangerous to use. But you can keep dust and debris from accumulating in your rifle barrel by putting a sturdy latex balloon over the barrel's front end.

SCIENCE FAIR

You experience a discharge of **static electricity** when you touch a doorknob after shuffling across a carpet. But you rarely see this phenomenon, with the exception of lightning, which is static electricity on a grand scale. Here's an experiment that offers a dazzling display of static electricity in action:

Empty the contents of a package of nonflavored **gelatin** powder onto a piece of **paper.** Blow up a **balloon,** rub it on a **woolen sweater,** and then hold it about an inch over the powder. The **gelatin particles will arch up** toward the balloon. The slightly negatively charged electrons—the built-up static electricity on the balloon—are attracting the positively charged protons in the gelatin powder.

Bananas

Make a face mask Who needs Botox when you have bananas? That's right: You can use a banana as an all-natural face mask that moisturizes your skin and leaves it looking and feeling softer. Mash up a medium-sized ripe banana into a smooth paste, then gently apply it to your face and neck. Let it set for 10-20 minutes, then rinse it off with cold water. Another popular mask recipe calls for 1/4 cup plain yogurt, 2 tablespoons honey, and 1 medium banana.

Eat a frozen "banana-sicle" As a summer treat for friends and family, peel and cut four ripe bananas in half (across the middle). Stick a wooden ice-cream stick into the flat end of each piece. Place them all on a piece of wax paper, and then put it in the freezer. A few hours later, serve them up as simply yummy frozen banana-sicles. If you want to go all-out, quickly dip your frozen bananas in 6 ounces (170 grams) melted butterscotch or chocolate morsels (chopped nuts or shredded coconut are optional), then refreeze.

Tenderize a roast Banana leaves are commonly used in many Asian countries to wrap meat as it's cooking to make it more tender. Some folks in these areas say the banana itself also has this ability. So the next time you fear the roast you're cooking will turn tough on you, try softening it up by adding a ripe, peeled banana to the pan.

Polish silverware and leather shoes It may sound a bit like a lark, but using a banana peel is actually a great way to put the shine back into your silverware and leather shoes. First, remove any of the leftover stringy material from the inside of the peel, then just start rubbing the inside of the peel on your shoes or silver. When you're done, buff up the object with a paper towel or soft cloth. You might even want to use this technique to restore your leather furniture. Test it on a small section first before you take on the whole chair.

Brighten up houseplants Are the leaves on your houseplants looking dingy or dusty? Don't bother misting them with water—that just spreads the dirt around. Rather, wipe down each leaf with the inside of a banana peel. It'll remove all the gunk on the surface and replace it with a lustrous shine.

Deter aphids Are aphids attacking your rosebushes or other plants? Bury dried or cut-up banana peels an inch or two deep around the base of the aphid-prone plants, and soon the little suckers will pack up and leave. Don't use whole peels or the bananas themselves, though; they tend to be viewed as tasty treats by raccoons, squirrels, gophers, rabbits, and other animals, who will just dig them up.

Use as fertilizer or mulch Banana peels, like the fruit itself, are rich in potassium—an important nutrient for both you and your garden. Dry out banana peels on screens during the winter months. In early spring, grind them up in a food processor or blender and use it as a mulch to give new plants and seedlings a healthy start. Many cultivars of roses and other plants, like staghorn ferns, also benefit from the nutrients found in banana peels; simply cut up some peels and use them as plant food around your established plants.

Add to compost pile With their high content of potassium and phosphorus, whole bananas and peels are welcome additions to any compost pile—particularly in so-called compost tea recipes. The fruit breaks down especially fast in hot temperatures. But don't forget to remove any glued-on tags from the peels, and be sure to bury bananas deep within your pile—otherwise they may simply turn out to be a meal for a four-legged visitor.

Attract butterflies and birds Bring more butterflies and various bird species to your backyard by putting out overripe bananas (as well as other fruits such as mangos, oranges, and papayas) on a raised platform. Punch a few holes in the bananas to make the fruit more accessible to the butterflies. Some enthusiasts swear by adding a drop of Gatorade to further mush things up. The fruit is also likely to attract more bees and wasps as well, so make sure that the platform is well above head level and not centrally located. Moreover, you'll probably want to clear it off before sunset, to discourage visits from raccoons and other nocturnal creatures.

* Basters

Pour perfect batter To make picture-perfect pancakes, cookies, and muffins, simply fill your baster with batter so that you can pour just the right amount onto a griddle or cookie sheet or into a muffin pan.

Remove excess water from coffeemaker The perfect cup of coffee is determined by using the proper balance of water and ground coffee in your automatic coffeemaker. If you pour in too much water, however, you typically have to add more coffee or suffer through a weak pot. But there's another, often overlooked option: Simply use your kitchen baster to remove the excess water to bring it in at just the right level.

Water hard-to-reach plants Do you get drips all over yourself, the floor, or furniture when trying to water hanging plants or other difficult-to-reach houseplants? Instead, fill a baster with water and squeeze it directly into the pot. You can also use a baster to water a Christmas tree and to add small, precise amounts of water to cups containing seedlings or germinating seeds.

Refresh water in flower arrangements It's a fact: Cut flowers last longer with periodic water changes. But pouring out the old water and adding the new is not a particularly easy or pleasant task. Unless, that is, you use a baster to suck out the old water and then to squirt in fresh water.

TAKE CARE Never use your kitchen baster for tasks such as cleaning out a fish tank or spreading or transferring chemicals.

Basters are staples at discount stores, and it's worth a visit to pick up a few to keep around the house specifically for noncooking chores. Label them with a piece of masking tape to make sure you always use the same baster for the same task.

Place water in pet's bowl Are you getting tired of chasing the bunny, hamster, or other caged pet around the house whenever you change its water? Use a baster to fill the water dish. You can usually fit the baster between the slats without having to open the cage.

Clean your aquarium A baster makes it incredibly easy to change the water in your fish tank or to freshen it up a bit. Simply use the utensil to suck up the gunk that collects in the corners and in the gravel at the bottom of your tank.

Blow away roaches and ants If you've had it with sharing your living quarters with roaches or ants, give them the heave-ho by sprinkling boric acid along any cracks or crevices where you've spotted the intruders. Use a baster to blow small amounts of the powder into hard-to-reach corners and any deep voids you come across. *Note:* Keep in mind that boric acid can be toxic if ingested by young children or pets.

Transfer paints and solvents The toughest part of any touchup paint job is invariably pouring the paint from a large can into a small cup or container. To avoid the inevitable spills, and just to make life easier in general, use a baster to take the paint out of the can. In fact, it's a good idea to make a baster a permanent addition to your workshop for transferring any solvents, varnishes, and other liquid chemicals.

Cure a musty-smelling air conditioner If you detect a musty odor blowing out of the vents of your room air-conditioner, chances are it's caused by a clogged drain hole. First, unscrew the front of the unit and locate the drain hole. It's usually located under the barrier between the evaporator and compressor, or underneath the evaporator. Use a bent wire hanger to clear away any obstacles in the hole or use a baster to flush it clean. You may also need to use the baster to remove any water that may be pooling up at the bottom of the unit to gain access to the drain.

B

Fix a leaky refrigerator Is water leaking inside your refrigerator? The most likely cause is a blocked drain tube. This plastic tube runs from a drain hole in the back of the freezer compartment along the back of your fridge and drains into an evaporation pan underneath. Try forcing hot water through the drain hole in the freezer with a baster. If you can't access the drain hole, try disconnecting the tube on the back to blow water through it. After clearing the tube, pour a teaspoon of ammonia or bleach into the drain hole to prevent a recurrence of algae spores, the probable culprit.

* Bath Oil

Remove glue from labels or bandages Get rid of those sticky leftover adhesive marks from bandages, price tags, and labels. Rub them away with a bit of bath oil applied to a cotton ball. It works great on glass, metal, and most plastics.

Use as a hot-oil treatment Heat 1/2 cup bath oil mixed with 1/2 cup water on High in your microwave for 30 seconds. Place the solution in a deep bowl and soak your fingers or toes in it for 10-15 minutes to soften cuticles or calluses. After drying, use a pumice stone to smooth over calluses or a file to push down cuticles. Follow up by rubbing in hand cream until fully absorbed.

Pry apart stuck drinking glasses When moisture seeps in between stacked glasses, separating them can get mighty tough—not to mention dangerous. But you can break the "ties that bond" by applying a few drops of bath oil along the sides of the glasses. Give the oil a few minutes to work its way down, then simply slide your glasses apart.

Loosen chewing gum from hair and carpeting If your child comes home with chewing gum in his or her hair—or tracks a wad onto your rug or carpet—

It appears we *Homo sapiens* have had a penchant for perfumed body oils since the beginning of history. The first use of such oils is believed to have occurred in the Neolithic period (7000-4000 B.C.), when Stone Age people began combining olive and sesame seed oils with fragrant plants. The ancient Egyptians also used scented oils, primarily in religious rituals. And the use of body oils, such as myrrh and frankincense, for both religious and secular uses is documented in the Bible. Indeed, fragrant oils have been an integral part of most cultures—including those of Native Americans and many Asian peoples.

hold off on reaching for the scissors. Instead, rub a liberal amount of bath oil into the gum. It should loosen it up enough to comb out. On a carpet, test the oil on an inconspicuous area before applying to the spot.

Remove scuff marks You can get those annoying scuff marks off your patent-leather shoes or handbags. Apply a bit of bath oil to a clean, soft cloth or towel. Gently rub in the oil, then polish with another dry towel.

Soften a new baseball glove Apply several drops of bath oil in the midsection of the glove and a few more drops under each finger. Lightly spread the oil around with a soft cloth. Place a baseball in the pocket of the glove and fold the glove over the ball, keeping it in place with one or two belts or an Ace bandage. Let it sit for a couple of days, then release the constraints and remove any excess oil with a clean cloth. The glove should be noticeably more pliable.

Clean grease or oil from skin It doesn't take much tinkering around the inside of a car or mower engine to get your hands coated in grease or oil. But before you reach for any heavy-duty grease removers, try this: Rub a few squirts of bath oil onto your hands, then wash them in warm, soapy water. It works, and it's a lot easier on the dermis than harsh chemicals.

Revitalize vinyl upholstery Give your car's dreary-looking vinyl upholstery a makeover by using a small amount of bath oil on a soft cloth to wipe down the seats, dashboard, armrests, and other surfaces. Polish with a clean cloth to remove any excess oil. As an added bonus, a scented bath oil will make the interior smell better, too.

Slide together pipe joints Can't find the all-purpose lubricating oil or the WD-40 when you're trying to join pipes together? No problem. A few drops of bath oil should provide sufficient lubrication to fit pipe joints together with ease.

* Bathtub Appliqués

Place on the bottom of PC case Has your desktop computer case lost its "legs" (those four small rubber feet that invariably fall off over time from moving your PC around)? To steady your case, and to minimize vibrations, cut small squares from a bathtub appliqué, and apply them to the corners of your case where the feet used to be.

Apply to dance slippers, shoes, and pajamas Avoid nasty falls caused by slippery plastic dance slippers—and even new shoes. Cut small pieces of bathtub appliqués and apply them to the sole of each slipper or shoe. You can also sew cut pieces of an appliqué on the soles of your children's "feet" pajamas to prevent slips (and tears).

Stick to bottom of kids' wading pool A few bathtub appliqués applied to the floor of a kiddie pool will make it a lot less slippery for little feet and help prevent falls—especially when the water play turns rowdy. Also put a couple of appliqués along the edges of the pool to give kids easy places to grip onto.

Affix to sippy cups and high-chair seats Cut pieces of a bathtub appliqué and put them on toddlers' sippy cups to minimize spills. Also attach appliqués to high-chair seats to keep Junior from sliding down—or out.

* Beans (Dried)

Use for playing pieces We know you had your heart set on being the racing car in the next game of Monopoly, but if the car has taken a trip to parts unknown, would you settle for a bean? Beans work fine as replacement pieces for everything from checkers to Chutes and Ladders to bingo.

Treat sore muscles Is your bad back or tennis elbow acting up again? A hot beanbag may be just the cure you need. Place a couple of handfuls of dried beans in a cloth shoe bag, an old sock, or a folded towel (tie the ends tightly) and microwave it on High for 30 seconds to 1 minute. Let it cool for a minute or two, then apply it to your aching muscles.

Make a beanbag Pour 3/4 to 1 1/2 cups dried beans in an old sock, shaking them down to the toe section. Tie a loose knot and tighten it up as you work it down against the beans. Then cut off the remaining material about 1 inch (2.5 centimeters) above the knot. You now have a beanbag for tossing around or juggling. Or use it as a squeeze bag for exercising your hand muscles.

Practice your percussion Make a homemade percussion shaker or maraca for yourself or your youngster. Add 1/2 cup dried beans to a small plastic jar, or a soda or juice can—even an empty coconut shell. Cover any openings with adhesive or duct tape. You can use this noisemaker at sporting events or as a dog-training tool (give it a couple of shakes when the pooch misbehaves).

Decorate a jack-o'-lantern Embellish the fright potential of your Halloween jack-o'-lantern by gluing on various dried beans for the eyes and teeth.

Recycle a stuffed animal Make your own beanie creation by removing the stuffing from one of your child's old, unused stuffed animals. Replace the fluff with dried beans, and sew it closed. It's bound to rekindle your youngster's interest.

Beer

Use as setting lotion Put some life back into flat hair with some flat beer. Before you get into the shower, mix 3 tablespoons beer in 1/2 cup warm water. After you shampoo your hair, rub in the solution, let it set for a couple of minutes, then rinse it off. You may be so pleased by what you see, you'll want to keep a six-pack in the bathroom.

Soften up tough meat Who needs powdered meat tenderizer when you have some in a can? You guessed it: Beer makes a great tenderizer for tough, inexpensive cuts of meat. Pour a can over the meat, and let it soak in for about an hour before cooking. Even better, marinate it overnight in the fridge or put the beer in your slow cooker with the meat.

Polish gold jewelry Get the shine back in your solid gold (i.e., minus any gemstones) rings and other jewelry by pouring a bit of beer (*not* dark ale!) onto a soft cloth and rubbing it gently over the piece. Use a clean second cloth or towel to dry.

Clean wood furniture Have you got some beer that's old or went flat? Use it to clean wooden furniture. Just wipe it on with a soft cloth, and then off with another dry cloth.

Make a trap for slugs and snails Like some people, some garden pests find beer irresistible—especially slugs and snails. If you're having problems with these slimy invaders, bury a container, such as a clean, empty juice container cut lengthwise in half, in the area where you've seen the pests, pour in about half a can of warm, leftover beer, and leave it overnight. You're likely to find a horde of them, drunk and drowned, the next morning.

Remove coffee or tea stains from rugs Getting that coffee or tea stain out your rug may seem impossible, but you can literally lift it out by pouring a bit of beer right on top. Rub the beer lightly into the material, and the stain should disappear. You may have to repeat the process a couple of times to remove all traces of the stain.

Some popular brands of beer proclaim their New England or Rocky Mountain pedigrees, or boast being "Milwaukee's finest." But, in fact, Pennsylvania has been home to more breweries throughout its history than any other state. One of its earliest breweries was opened in 1680 by none other than William Penn, the state's founder. And the Keystone State is still the home of the U.S.'s oldest active brewery, D. G. Yuengling & Son of Pottsville, Pennsylvania, founded in 1829.

Berry Baskets

B

Keep peels out of drain Don't clog up your kitchen drain with peelings from potatoes or carrots. Use a berry basket as a sink strainer to catch those vegetable shavings as they fall.

Store soap pads and sponges Are you tired of throwing away prematurely rusted steel wool soap pads or smelly sponges? Place a berry basket near the corner of your kitchen sink and line the bottom with a layer of heavy-duty aluminum foil. Fashion a spout on a corner of the foil closest to the sink that can act as a drain to keep water from pooling up at the bottom of the basket. Now sit back and enjoy the added longevity of your soap pads and sponges.

Use as a colander Need a small colander to wash individual servings of fruits and vegetables or to drain off that child's portion of hot macaroni shells? Get your hands on an empty berry basket. It makes a dandy colander for these chores.

Kids' Stuff **Berry baskets** can be particularly useful for all sorts of children's crafts. For example, you can cut apart the panels, and carve out geometric shapes for kids to use as **stencils.** You can also turn one into an **Easter basket** by adding some **cellophane grass** and a (preferably pink) **pipe cleaner** for a handle. Or use one as a multiple **bubble maker;** simply dip it in some **water** mixed with **dishwashing liquid** and wave it through the air to create swarms of bubbles. Lastly, let kids decorate the baskets with **ribbons** or **construction paper** and use them to store their own little trinkets and toys.

Hold recycled paper towels Don't toss out those lightly used paper towels in your kitchen. You can reuse them to wipe down countertops or to soak up serious spills. Keep a berry basket in a convenient location in your kitchen to have your recycled towels at the ready when needed.

Use as dishwasher basket If the smaller items you place in your dishwasher (such as baby bottle caps, jar lids, and food-processor accessories) won't stay put, try putting them in a berry basket. Place the items inside one basket, then cover over with a second basket. Fasten them together with a thick rubber band and place on your dishwasher's upper rack.

Organize your meds A clean berry basket could be just what the doctor ordered for organizing your vitamins and medicine bottles. If you're taking several medications, a berry basket offers a convenient way to place them all—or prepackaged individual doses—in one, easy-to-remember location. You can also use baskets to organize medications in your cupboard or medicine cabinet according to their expiration dates or uses.

Arrange flowers Droopy or lopsided flower arrangements just don't cut it. That's why the pros use something known as a frog to keep cut flowers in place. To make your own, insert an inverted berry basket into a vase (cut the basket to fit, if necessary). It will keep your stalks standing tall.

Protect seedlings Help young plants thrive in your garden by placing inverted berry baskets over them. The baskets will let water, sunlight, and air in, but keep raccoons and squirrels out. Make sure the basket is buried below ground level and tightly secured (placing a few good-sized stones around it may suffice).

Make a bulb cage Squirrels and other rodents view freshly planted flower bulbs as nothing more than tasty morsels and easy pickings. But you can put a damper on their meal by planting bulbs in berry baskets. Be sure to place the basket at the correct depth, then insert the bulb and cover with soil.

Build a hanging orchid planter Orchids are said to be addictive: Once you start collecting them, you can't stop. If you've got the bug, you can at least save yourself a bit of money by making your own hanging baskets for your orchids. Fill up a berry basket with sphagnum moss mixed with a bit of potting soil and suspend it with a length of monofilament fishing line.

Fashion a string dispenser or screwdriver holder If you don't want to bother untangling knots every time you need a piece of string, twine, or yarn, build your own string dispenser with two berry baskets. Place the ball inside one berry basket. Feed the cord through the top of a second, inverted basket, then tie the two baskets together with twist ties. You can also mount an inverted berry basket on your workshop's pegboard and use it to hold and organize your screwdrivers; they'll fit neatly between the slats.

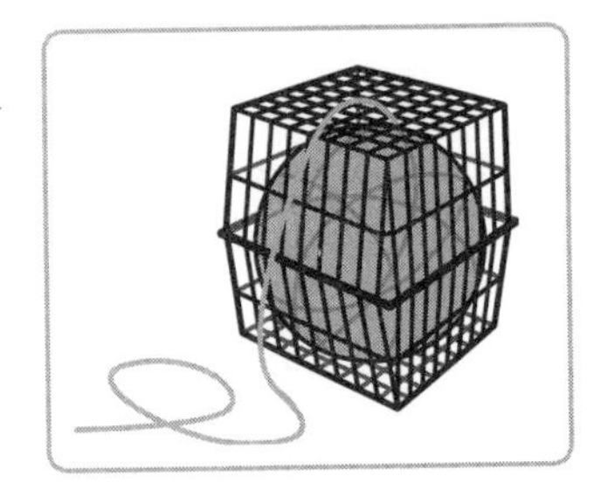

* Binder Clips

Strengthen your grip Does a weak grip or arthritis make it hard for you to open jars and do other tasks with your hands? Use a large binder clip to add some zip to your grip. Squeeze the folded-back wings of the clip, hold for a count of five, and relax. Do this a dozen or so times with each hand a few times a day. It will strengthen your grip and release tension too.

Mount a picture Here's a neat way to mount and hang a picture so that it has a clean frameless look. Sandwich the picture between a sheet of glass or clear plastic and piece of hardboard or stiff cardboard. Then use tiny binder clips along the edges to clamp the pieces together. Use two or three clips on each side. After the clips are in place, remove the clip handles at front. Tie picture wire to the rear handles for hanging the picture.

Keep your place A medium-sized binder clip makes an ideal bookmark. If you don't want to leave impression marks on the pages, tape a soft material like felt, or even just some adhesive tape, to the inside jaws of the clip before using.

Make a money clip To keep paper money in a neat bundle in your pocket or purse, stack the bills, fold them in half, and put a small binder clip over the fold.

Keep ID handy You're at the airport and you know you'll be asked to show your ID a few times. Instead of fishing in your wallet or trying to figure out which pocket you stuck your driver's license in, use a binder clip to firmly and conveniently attach your ID and other documents to your belt. You can also use a small binder clip to secure your office ID to your belt or a breast pocket.

* Bleach

Clean off mold and mildew Bleach and ammonia are both useful for removing mold and mildew both inside and outside your home. However, the two should *never* be used together. Bleach is especially suited for the following chores:

- Wash mildew out of washable fabrics. Wet the mildewed area and rub in some powdered detergent. Then wash the garment in the hottest water setting permitted by the clothing manufacturer using 1/2 cup chlorine bleach. If the garment can't be washed in hot water and bleach, soak it in a solution of 1/4 cup oxygen bleach (labeled "all fabric" or "perborate") in 1 gallon (3.7 liters) warm water for 30 minutes before washing.
- Remove mold and mildew from the grout between your bathroom tiles. Mix equal parts of chlorine bleach and water in a spray bottle, and spray it over grout. Let it sit for 15 minutes, then scrub with a stiff brush and rinse off. You can also do this just to make your grout look whiter.
- Get mold and mildew off your shower curtains. Wash them—along with a couple of bath towels (to prevent the plastic curtains from crinkling)—in warm water with 1/2 cup chlorine bleach and 1/4 cup laundry detergent. Let the washer run for a couple of minutes before loading. Put the shower

curtain and towels in the dryer on the lowest temperature setting for 10 minutes, then immediately hang-dry.

- Rid your rubber shower mat of mildew. Soak in a solution of 1/8 cup (3.7 liters) chlorine bleach in 1 gallon water for 3-4 hours. Rinse well.
- Get mildew and other stains off unpainted cement, patio stones, or stucco. Mix a solution of 1 cup chlorine bleach in 2 gallons (7.5 liters) water. Scrub vigorously with a stiff or wire brush and rinse. If any stains remain, scrub again using 1/2 cup washing soda (this is sodium carbonate, not baking soda) dissolved in 2 gallons (7.5 liters) warm water.
- Remove mildew from painted surfaces and siding. Make a solution of 1/4 cup chlorine bleach in 2 cups water and apply with a brush to mildewed areas. Let the solution set for 15 minutes, then rinse. Repeat as necessary.

Sterilize secondhand items Remember Mom saying, "Put that down. You don't know where it's been"? She had a point—especially when it comes to toys or kitchen utensils picked up at thrift shops and yard sales. Just to be on the safe side, take your used, waterproof items and soak them for 5-10 minutes in a solution containing 3/4 cup bleach, a few drops of antibacterial dishwashing liquid, and 1 gallon warm water. Rinse well, then air-dry, preferably in sunlight.

Clean butcher block cutting boards and countertops Don't even think about using furniture polish or any other household cleaner to clean a butcher block cutting board or countertop. Rather, scrub the surface with a brush dipped in a solution of 1 teaspoon bleach diluted in 2 quarts (2 liters) water. Scrub in small circles, and be careful not to saturate the wood. Wipe with a slightly damp paper towel, then immediately buff dry with a clean cloth.

TAKE CARE Never mix bleach with ammonia, lye, rust removers, oven or toilet-bowl cleaners, or vinegar. Any combination can produce toxic chlorine gas fumes, which can be deadly. Some people are even sensitive to the fumes of undiluted bleach itself. Always make sure you have adequate ventilation in your work area before you start pouring.

Brighten up glass dishware Put the sparkle back in your glasses and dishes by adding a teaspoon of bleach to your soapy dishwater as you're washing your glassware. Be sure to rinse well, and dry with a soft towel.

Shine white porcelain Want to get your white porcelain sink, candleholder, or pottery looking as good as new? In a well-ventilated area on a work surface protected by heavy plastic, place several paper towels over the item (or across the bottom of the sink) and carefully saturate them with undiluted bleach. Let soak for 15 minutes to a half hour, then rinse and wipe dry with a clean towel. *Note:*

Do not try this with antiques; you can diminish their value or cause damage. And never use bleach on colored porcelain, because the color will fade.

Make a household disinfectant spray Looking for a good, all-purpose disinfectant to use around the house? Mix 1 tablespoon bleach in 1 gallon (3.7 liters) hot water. Then fill a clean, empty spray bottle and use it on a paper towel to clean countertops, tablecloths, lawn furniture—basically, wherever it's needed. Just be sure not to use it in the presence of ammonia or other household cleaners.

TAKE CARE Some folks skip the bleach when cleaning their toilets, fearing that lingering ammonia from urine—especially in households with young children—could result in toxic fumes. Unless you are sure there is no such problem, you may want to stick with ammonia for this job.

Disinfect trash cans Even the best housekeepers must confront a gunked-up kitchen garbage pail every now and then. On such occasions, take the pail outside, and flush out any loose debris with a garden hose. Then add 1/2 to 1 cup bleach and several drops of dishwashing liquid to 1 gallon (3.7 liters) warm water. Use a toilet brush or long-handled scrub brush to splash and scour the solution on the bottom and sides of the container. Empty, then rinse with the hose, empty it again, and let air-dry.

Increase cut flowers' longevity Freshly cut flowers will stay fresh longer if you add 1/4 teaspoon bleach per quart (1 liter) of vase water. Another popular recipe calls for 3 drops bleach and 1 teaspoon sugar in 1 quart (1 liter) water. This will also keep the water from getting cloudy and inhibit the growth of bacteria.

Clean plastic lawn furniture Is your plastic-mesh lawn furniture looking dingy? Before you place it curbside, try washing it with some mild detergent mixed with 1/2 cup bleach in 1 gallon (3.7 liters) water. Rinse it clean, then air-dry.

Kill weeds in walkways Do weeds seem to thrive in the cracks and crevices of your walkways? Try pouring a bit of undiluted bleach over them. After a day or two, you can simply pull them out, and the bleach will keep them from coming back. Just be careful not to get bleach on the grass or plantings bordering the walkway.

Get rid of moss and algae To remove slippery and unsightly moss and algae on your brick, concrete, or stone walkways, scrub them with a solution of 3/4 cup bleach in 1 gallon (3.7 liters) water. Be careful not to get bleach on your grass or ornamental plants.

Sanitize garden tools You cut that diseased stalk off your rosebush with your branch clipper. Unless you want to spread the disease the next time you use the tool, sterilize it by washing it with 1/2 cup bleach in 1 quart (1 liter) water. Let the tool air-dry in the sun, then rub on a few drops of oil to prevent rust.

* Blow-Dryer

Get wax off wood furniture It may have been a romantic evening, but that hardened candle wax on your wooden table or bureau is not the sort of lingering memory you had in mind. Melt it with a blow-dryer on its slowest, hottest setting. Remove the softened wax with a paper towel, then wipe the area with a cloth dipped in equal parts vinegar and water. Repeat if necessary. You can also remove wax from silver candlestick holders with a blow-dryer: Use the blow-dryer to soften the wax, then just peel it off.

Clean off radiators Are those dusty cast-iron radiators around your house becoming something of an eyesore? To clean them, hang a large, damp cloth behind each radiator. Then use your blow-dryer on its highest, coolest setting to blow dust and hidden dirt onto the cloth.

Remove bumper stickers Want to remove those cutesy stickers your kids used to decorate your car bumper to "surprise" you? Use a blow-dryer on its hottest setting to soften the adhesive. Move the dryer slowly back and forth for several minutes, then use your fingernail or credit card to lift up a corner and slowly peel off.

Dust off silk flowers and artificial houseplants They may require less care than their living counterparts, but silk flowers and artificial houseplants are apt to collect dust and dirt. Use your blow-dryer on its highest, coolest setting for a quick, efficient way to clean them off. Since this will blow the dust onto the furniture surfaces and floor around the plant, do this just before you vacuum those areas.

* Borax

Clear a clogged drain Before you reach for a caustic drain cleaner to unclog that kitchen or bathroom drain, try this much gentler approach: Use a funnel to insert 1/2 cup borax into the drain, then slowly pour in 2 cups boiling water. Let the mixture set for 15 minutes, then flush with hot water. Repeat for stubborn clogs.

Rub out heavy sink stains Get rid of those stubborn stains—even rust—in your stainless steel or porcelain sink. Make a paste of 1 cup borax and 1/4 cup lemon juice. Put some of the paste on a cloth or sponge and rub it into the stain, then rinse with running warm water. The stain should wash away with the paste.

Clean windows and mirrors Want to get windows and mirrors spotless *and* streakless? Wash them with a clean sponge dipped in 2 tablespoons borax dissolved in 3 cups water.

Remove mildew from fabric To remove mildew from upholstery and other fabrics, soak a sponge in a solution of 1/2 cup borax dissolved in 2 cups hot water, and rub it into the affected areas. Let it soak in for several hours until the stain disappears, then rinse well. To remove mildew from clothing, soak it in a solution of 2 cups borax in 2 quarts (2 liters) water.

B

Kids' Stuff Help your children brew up some **slime**—that gooey, stretchy stuff kids love to play with. First, mix 1 cup water, 1 cup **white glue,** and 10 drops **food coloring** in a medium bowl. Then, in a second, larger bowl, stir 4 teaspoons **borax** into 1 1/3 cups **water** until the powder is fully dissolved. Slowly pour the contents of the first bowl into the second. Use a wooden **mixing spoon** to roll (don't mix) the glue-based solution around in the borax solution four or five times. Lift out the globs of glue mixture, then knead it for 2-3 minutes. Store your homemade slime in an **airtight container** or a **self-sealing plastic storage bag.**

Get out rug stains Remove stubborn stains from rugs and carpets. Thoroughly dampen the area, then rub in some borax. Let the area dry, then vacuum or blot it with a solution of equal parts vinegar and soapy water and let dry. Repeat if necessary. Don't forget to first test the procedure on an inconspicuous corner of the rug or on a carpet scrap before applying it to the stain.

Sanitize your garbage disposal A garbage disposal is a great convenience but can also be a great breeding ground for mold and bacteria. To maintain a more sanitary disposal, every couple of weeks pour 3 tablespoons borax down the drain and let it sit for 1 hour. Then turn on the disposal and flush it with hot water from the tap.

Clean your toilet Want a way to disinfect your toilet bowl *and* leave it glistening without having to worry about dangerous or unpleasant fumes? Use a stiff brush to scrub it using a solution of 1/2 cup borax in 1 gallon (3.7 liters) water.

Eliminate urine odor on mattresses Toilet training can be a rough experience for all the parties involved. If your child has an "accident" in bed, here's how to get rid of any lingering smell: Dampen the area, then rub in some borax. Let it dry, then vacuum up the powder.

Make your own dried flowers Give your homemade dried flowers the look of a professional job. Mix 1 cup borax with 2 cups cornmeal. Place a 3/4-inch (2-centimeter) coating of the mixture in the bottom of an airtight container, like a large flat plastic food storage container. Cut the stems off the flowers you want to dry, then lay them on top of the powder, and lightly sprinkle more of the mixture on top of the flowers (be careful not to bend or crush the petals or other flower parts). Cover the container, and leave it alone for

7-10 days. Then remove the flowers and brush off any excess powder with a soft brush.

Keep away weeds and ants Get the jump on those weeds that grow in the cracks of the concrete outside your house by sprinkling borax into all the crevices where you've seen weeds grow in the past. It will kill them off before they have a chance to take root. When applied around the foundation of your home, it will also keep ants and other six-legged intruders from entering your house. But be very careful when applying borax—it is toxic to plants (see Take Care warning).

TAKE CARE Borax, like its close relative, boric acid, has relatively low toxicity levels, and is considered safe for general household use, but the powder can be harmful if ingested in sufficient quantities by young children or pets. Store it safely out of their reach.

Borax is toxic to plants, however. In the yard, be very careful when applying borax onto or near soil. It doesn't take much to leach into the ground to kill off nearby plants and prevent future growth.

Control creeping Charlie Is your garden being overrun by that invasive perennial weed known as creeping Charlie (*Glechoma hederacea*, also known as ground ivy, creeping Jenny and gill-over-the-ground)? You may be able to conquer Charlie with borax. First, dissolve 8-10 ounces (230-280 grams) borax in 4 ounces (120 milliliters) warm water. Then pour the solution into 2 1/2 gallons (9.5 liters) warm water—this is enough to cover 1,000 square feet (93 square meters). Apply this treatment only one time in each of two years. If you still have creeping Charlie problems, consider switching to a standard herbicide. (See Take Care warning about using borax in the garden.)

Bottles

✱ BOTTLES AROUND THE HOUSE

Make a foot warmer Walking around on harsh winter days can leave you with cold and tired tootsies. But you don't need to shell out your hard-earned money on a heating pad or a hot-water bottle to ease your discomfort. Just fill up a 1- or 2-liter soda bottle with hot water, then sit down and roll it back and forth under your feet.

Use as a boot tree Want to keep your boot tops from getting wrinkled or folded over when you put them in storage? Insert a clean empty 1-liter soda bottle into each boot. For added tautness, put a couple of old socks on the bottles or wrap them in towels.

Recycle as a chew toy If Lassie has been chewing on your slippers instead of fetching them, maybe she's in need of some chew toys. A no-cost way to amuse your dog is to let her chew on an empty plastic 1-liter soda bottle. Maybe it's the crunchy sound they make, but dogs love them! Just be sure to remove the label and bottle cap (as well as the loose plastic ring under it). And replace it before it gets too chewed up—broken pieces of plastic are choke hazards.

Make a bag or string dispenser An empty 2-liter soda bottle makes the perfect container for storing and dispensing plastic grocery bags. Just cut off the bottom and top ends of the bottle, and mount it with screws upside down inside a kitchen cabinet or closet. Put washers under the screw heads to keep them from pulling through the plastic. Fill it with your recycled bags (squeeze the air out of them first) and pull them out as needed. You can make a twine dispenser the same way, using a 1-liter bottle and letting the cord come out the bottom.

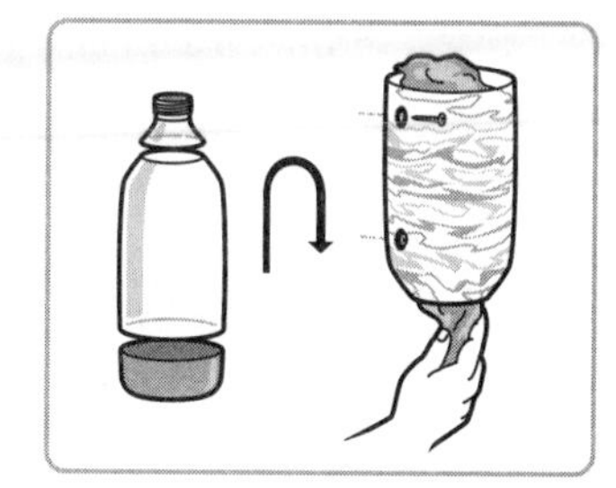

Place in toilet tank Unless your house was built relatively recently, chances are you have an older toilet that uses a lot of water each flush. To save a bit of money on your water bills, fill an empty 1-liter soda bottle with water (remove any labels first) and put it in the toilet tank to cut the amount of water in each flush.

Cut out a toy carryall If you're fed up with Lego or erector-set pieces underfoot, make a simple carryall to store them in by cutting a large hole in the side of a clean gallon jug with a handle. Cut the hole opposite the handle so you or your youngster can easily carry the container back to the playroom after putting the pieces away. For an easy way to store craft materials, crayons, or small toys, just cut the containers in half and use the bottom part to stash your stuff.

Store your sugar The next time you bring home a 5-pound (2.2-kilogram) bag of sugar from the supermarket, try pouring it into a clean, dry 1-gallon (3.7-liter) jug with a handle. The sugar is less likely to harden, and the handle makes it much easier to pour it out.

Tip Safe Rotary Cutter

Cutting plastic containers can be a tricky, dangerous business—especially when you reach for your sharpest kitchen knife. But you can greatly minimize the risk by visiting your local fabric or crafts store and picking up a rolling cutter knife (this is not the same device used to slice pizza, by the way). The device shown in the picture in the hint "Make a scoop or boat bailer," facing page, usually sells for between $6 and $10. Be careful, though. These knives use blades that are razor sharp, but they make life much easier when it's time to cut into a hard plastic container.

Fashion a funnel To make a handy, durable funnel, cut a cleaned milk jug, bleach, or liquid detergent container with a handle in half across its midsection. Use the top portion (with the spout and handle) as a funnel for easy pouring of paints, rice, coins, and so on.

✱ BOTTLES AWAY FROM HOME

Make a scoop or boat bailer Cut a clean plastic half-gallon (2-liter) jug with a handle diagonally from the bottom so that you have the top three-quarters of the jug intact. You now have a handy scoop that can be used for everything from removing leaves and other debris from your gutters, to cleaning out the litter box and poop-scooping up after your dog. Use it to scoop dog food from the bag, spread sand or ice-melt on walkways in winter, or bail water out of your boat (you might want to keep the cap on for this last application).

Keep the cooler cold Don't let your cooler lose its cool while you're on the road. Fill a few clean plastic jugs with water or juice and keep them in the freezer for use when transporting food in your cooler. This is not only good for keeping food cold; you can actually drink the water or juice as it melts. It's also not a bad idea to keep a few frozen jugs in your freezer if you have extra space; a full freezer actually uses less energy and can save money on your electric bill. When filling a jug, leave a little room at the top for the water to expand as it freezes.

Use for emergency road kit in winter Don't get stuck in your car the next time a surprise winter storm hits. Keep a couple of clean gallon (3.7-liter) jugs with handles filled with sand or kitty litter in the trunk of your car. Then you'll be prepared to sprinkle the material on the road surface to add traction under your wheels when you need to get moving on a slippery road. The handle makes it easier to pour them.

✱ BOTTLES IN THE GARDEN

Feed the birds Why spend money on a plastic bird feeder when you probably have one in your recycling bin? Take a clean 1/2-gallon (2-liter) juice or milk jug and carve a large hole on its side to remove the handle. (You might even drill a small hole under the large one to insert a sturdy twig or dowel for a perch.) Then poke a hole in the middle of the cap and suspend it from a tree with a piece of strong string or monofilament fishing line. Fill it up to the opening with birdseed, and enjoy the show.

Make a watering can No watering can? It's easy to make one from a clean 1-gallon (3.7-liter) juice, milk, or bleach jug with a handle. Drill about a dozen tiny (1/16-inch or 1.5-millimeter is good) holes just below the spout of the jug on the side opposite the handle. Or carefully punch the holes with an ice pick. Fill it with water, screw the cap on, and start sprinkling.

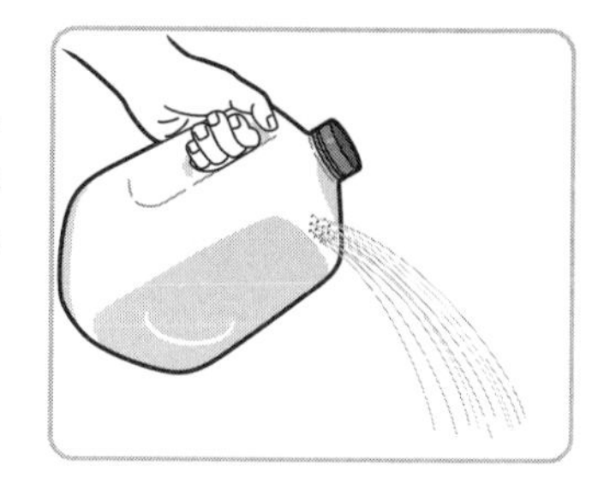

Create a drip irrigator for plants During dry spells, a good way to get water to the roots of your plants is to place several drip irrigators around your garden. You can make them from clean 1-gallon (3.7-liter) juice or detergent jugs. Cut a large hole in the bottom of a jug, then drill 2-5 tiny (about 1/16-inch or 1.5-millimeter) holes in or around the cap. Bury the capped jugs upside down about three-quarters submerged beneath the soil near the plants you need to water, and fill with water through the hole on top. Refill as often as needed.

Mark your plants Want an easy way to make ID badges for all the vegetables, herbs, and flowers in your garden? Cut vertical strips from a couple of clear 1-gallon (3.7-liter) water jugs. Make the strips the same width as your seed packets but double their length. Fold each strip over an empty packet to protect it from the elements, and staple it to a strong stick or chopstick.

Secure garden netting If you find yourself having to constantly re-stake the loose netting or plastic lining over your garden bed, place water-filled large plastic jugs around the corners to keep the material in place.

Use as an attachable trash can or harvest basket Here's a great tip for weekend gardeners and pros alike: Cut a large hole opposite the handle of a 1/2- or 1-gallon (2-or 3.7-liter) container, and loop the handle through a belt or rope on your waist. Use it to collect the debris—rocks, weeds, broken stems—you encounter as you mow the lawn or stroll through your garden. Use the same design to make an attachable basket for harvesting berries, cherries, and other small fruits or vegetables.

Space seeds in garden Want an easy way to perfectly space seeds in your garden? Use an empty soda bottle as your guide. Find the distance that the seed company recommends between seeds and then cut off the tapered top of the bottle so its diameter equals that distance. When you start planting, firmly press your bottle, cut edge down, into the soil and place a seed in the center of the circle it makes. Then line up the bottle so that its edge touches the curve of the first impression, and press down again. Plant a seed in the center, and repeat until you've filled your rows.

Build a bug trap Do yellow jackets, wasps, or moths swarm around you every time you set foot in the yard? Use an empty 2-liter soda bottle to make an environment-friendly trap for them. First, dissolve 1/2 cup sugar in 1/2 cup water in the bottle. Then add 1 cup apple cider vinegar and a banana peel (squish it up to fit it through). Screw on the cap and give the mixture a good shake before filling the bottle halfway with cold water. Cut or drill a 3/4-inch (2-centimeter) hole near the top of the bottle, and hang it from a tree branch where the bugs seem especially active. When the trap is full, toss it into the garbage and replace it with a new one.

Isolate weeds when spraying herbicides When using herbicides to kill weeds in your garden, you have to be careful not to also spray and kill surrounding plants. To isolate the weed you want to kill, cut a 2-liter soda bottle in half and place the top half over the weed you want to spray. Then direct your pump's spraying wand through the regular opening in the top of the bottle and blast away. After the spray settles down, pick up the bottle and move on to your next target. Always wear goggles and gloves when spraying chemicals in the garden.

Set up a backyard sprayer When temperatures soar outdoors, keep your kids cool with a homemade backyard sprayer. Just cut three 1-inch (2.5-centimeter) vertical slits in one side of a clean 2-liter soda bottle. Or make the slits at different angles so the water will squirt in different directions. Attach the nozzle of the hose to the bottle top with duct tape (make sure it's fastened on tight). Turn on the tap, and let the fun begin!

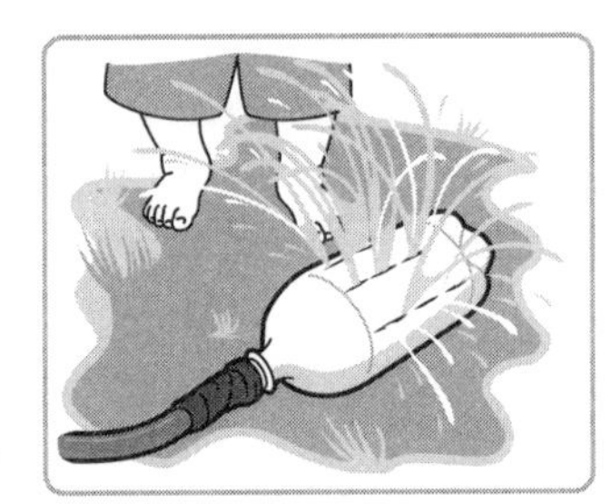

* BOTTLES FOR THE DO-IT-YOURSELFER

Build a paint bucket Tired of splattering paint all over as you work? Make a neater paint dispenser by cutting a large hole opposite the handle of a clean 1-gallon (3.7 liter) jug. Pour in the paint so that it's about an inch below the edge of the hole, and use the edge to remove any excess paint from your brush before you lift your brush. You can also cut jugs in half and use the bottom halves as disposable paint buckets when several people work on the same job.

Store your paints Why keep leftover house paints in rusted or dented cans when you can keep them clean and fresh in plastic jugs? Use a funnel to pour the paint into a clean, dry milk or water jug, and add a few marbles (they help mix the paint when you shake the container before your next paint job). Label each container with a piece of masking tape, noting the paint manufacturer, color name, and the date.

Use as workshop organizers Are you always searching for the right nail to use for a particular chore, or for a clothespin, picture hook, or small fastener? Bring some organization to your workshop with a few 1- or 1/2-gallon (3.7- or 2-liter) jugs. Cut out a section near the top of each jug on the side opposite the handle. Then use the containers to store and sort all the small items that seem to "slip through the cracks" of your workbench. The handle makes it easy to carry a jug to your worksite.

Use as a level substitute How can you make sure that shelf you're about to put up is straight if you don't have a level on hand? Easy. Just fill a 1-liter soda bottle about three-quarters full with water. Replace the cap, then lay the bottle on its side. When the water is level, so is the shelf

Make a weight for anchoring or lifting Fill a clean, dry gallon (3.7-liter) jug with a handle with sand and cap it. You now have an anchor that is great for holding down a paint tarp, securing a shaky patio umbrella, or steadying a table for repair. The handle makes it easy to move or attach a rope. Or use a pair of sand-filled bottles as exercise weights, varying the amount of sand to meet your lifting capacity.

* Bottle Openers

Remove chestnut shells An easy way to remove the shells from chestnuts is to use the pointed end of a bottle opener to pierce the tops and bottoms of the shells and then boil the chestnuts for 10 minutes.

Cut packing tape on cartons Can't wait to open that long-awaited package on your doorstep? If you don't have a penknife handy, just run the sharp end of a bottle opener along the tape. It should do the job quite nicely.

Deploy as a shrimp de-veiner If you don't have a small paring knife on hand when you're getting ready to de-vein a batch of shrimp, don't worry. Just use the sharp end of a bottle opener. It just happens to be the perfect shape to make this messy chore a breeze.

DID *You* KNOW?

The old-fashioned bottle opener with one flat end and one pointed end is often referred to as a "church key." Although no one is exactly sure how or when this association came into being, it originated years ago in the brewery industry and was used to describe a flat opener with a hooked cutout used to lever off bottle caps—it is widely believed that the term derived from the early openers' resemblance to the heavy, ornate keys used to unlock big, old doors, such as those found on churches. Ironically, the term is now applied only to openers with both flat and pointed ends.

Scrape barbecue grill Looking for an easy way to clean off the burned remnants of last weekend's meal from your barbecue grill? If you have a bottle opener and a metal file, you're in luck. Simply file a notch about 1/8-inch (3 millimeters) wide into the flat end of the opener and you're ready to go.

Loosen plaster or remove grout It may not be the carpenter's best friend, but the sharp end of a bottle opener can be handy for removing loose plaster from a wall before patching it. It's great for running along cracks, and you can use to undercut a hole—that is, make it wider at the bottom than at the surface—so that the new plaster will "key" into the old. The sharp end of the opener is equally useful for removing old grout between your bathroom tiles before regrouting.

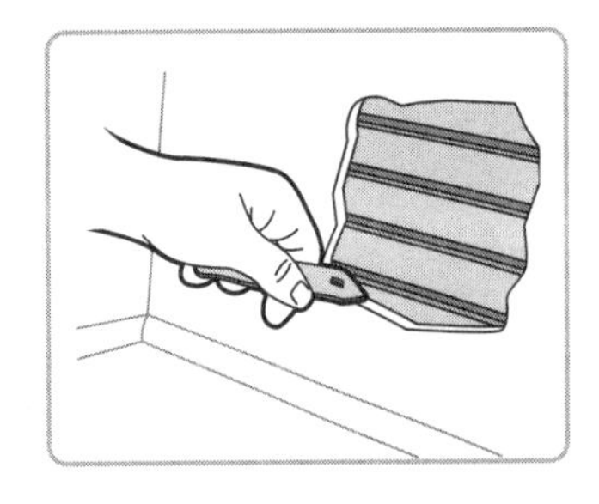

* Bread

Remove scorched taste from rice Did you leave the rice cooking too long and let it get burned? To get rid of the scorched taste, place a slice of white bread on top of the rice while it's still hot. Replace the pot lid and wait several minutes. When you remove the bread, the burned taste should be gone.

Soften up hard marshmallows You reach for your bag of marshmallows only to discover that they've gone stale. Put a couple of slices of fresh bread in the bag and seal it shut (you may want to transfer the marshmallows to a self-sealing plastic bag). Leave it alone for a couple of days. When you reopen the bag, your marshmallows should taste as good as new.

Absorb vegetable odors Love cabbage or broccoli, but hate the smell while it's cooking? Try putting a piece of white bread on top of the pot when cooking up a batch of "smelly" vegetables. It will absorb most of the odor.

Soak up grease and stop flare-ups To paraphrase a famous bear: Only you can prevent grease fires. One of the best ways to prevent a grease flare-up when broiling meat is to place a couple of slices of white bread in your drip pan to absorb the grease. It will also cut down on the amount of smoke produced.

Clean walls and wallpaper Most kids have a hard time understanding how easily the dirt on their hands can be transferred to walls. But you can remove most dirty or greasy fingerprints from painted walls by rubbing the area with a slice of white bread. Bread does a good job cleaning nonwashable wallpaper as well. Just cut off the crusts first to minimize the chance of scratching the paper.

Pick up glass fragments Picking up the large pieces of a broken glass or dish is usually easy enough, but getting up those tiny slivers can be a real pain (figuratively if not literally). The easiest way to make sure you don't miss any is to press a slice of bread over the area. Just be careful not to prick yourself when you toss the bread into the garbage.

Dust oil paintings You wouldn't want to try this with an original Renoir, or with any museum-quality painting for that matter, but you can clean off everyday dust and grime that collects on an oil painting by gently rubbing the surface with a piece of white bread.

* Bubble Pack

Prevent toilet-tank condensation If your toilet tank sweats in warm, humid weather, bubble pack could be just the right antiperspirant. Lining the inside of the tank with bubble pack will keep the outside of the tank from getting cold and causing condensation when it comes in contact with warm, moist air. To line the tank, shut off the supply valve under the tank and flush to drain the tank. Then wipe the inside walls clean and dry. Use silicone sealant to glue appropriate-sized pieces of bubble pack to the major flat surfaces.

Protect patio plants Keep your outdoor container plants warm and protected from winter frost damage. Wrap each container with bubble pack and use duct tape or string to hold the wrap in place. Make sure the wrap extends a couple of inches above the lip of the container. The added insulation will keep the soil warm all winter long.

DID *You* KNOW?

Bubble Wrap wallpaper? Yep, that's what inventors Alfred Fielding and Marc Chavannes had in mind when they began developing the product in Saddle Brook, New Jersey, in the late 1950s. Perhaps they had the padded-cell market in mind. In any case, they soon realized their invention had far greater potential as packaging material. In 1960 they raised $85,000 and founded the Sealed Air Corporation. Today Sealed Air is a Fortune 500 company with $3.5 billion in annual revenues. The company produces Bubble Wrap cushioning in a multitude of sizes, colors, and properties, along with other protective packaging materials such as Jiffy padded mailers.

Keep cola cold Wrap soft-drink cans with bubble pack to keep beverages refreshingly cold on hot summer days. Do the same for packages of frozen or chilled picnic foods. Wrap ice cream just before you leave for the picnic to help keep it firm en route.

Protect produce in the fridge Line your refrigerator's crisper drawer with bubble pack to prevent bruises to fruit and other produce. Cleanup will be easier, too—when the lining gets dirty, just throw it out and replace it with fresh bubble pack.

Add insulation Cut window-size pieces of wide bubble pack and duct-tape them to inside windows for added warmth and savings on fuel bills in winter. Lower the blinds to make it less noticeable.

Make a bedtime buffer Keep cold air from creeping into your bed on a chilly night by placing a large sheet of bubble pack between your bedspread or quilt and your top sheet. You'll be surprised at how effective it is in keeping warm air in and cold air out.

Cushion your work surface When repairing delicate glass or china, cover the work surface with bubble pack to help prevent breakage.

Protect tools Reduce wear and tear on your good-quality tools and extend their lives. Line your toolbox with bubble pack. Use duct tape to hold it in place.

Sleep on air while camping Get a better night's sleep on your next camping trip: Carry a 6-foot (2-meter) roll of wide bubble pack to use as a mat under your sleeping bag. No sleeping bag? Just fold a 12-foot-long (3.6-meter-long) piece of wide bubble pack in half, bubble side out, and duct-tape the edges. Then slip in and enjoy a restful night in your makeshift padded slumber bag.

Cushion bleachers and benches Take some bubble pack out to the ballgame with you to soften those hard stadium seats or benches. Or stretch a length along a picnic bench for more comfy dining.

* Buckets

Make a lobster pot If you don't have a large kettle, boil lobsters in an old metal bucket. Make sure to use pot holders and tongs when cooking and removing the lobster. Let the bucket cool before handling it again.

Kids' Stuff Add the beat of bucket **tom-toms** to create an exciting, fun atmosphere at your next family campfire. Cut **plastic buckets** to different lengths to create a distinct tone for each drum or use a mix of various-sized **plastic** and **galvanized buckets.** For more musical accompaniment, make a **broom-handle string bass** using a bucket as the sound box.

Create a food locker A tightly sealed 5-gallon (19-liter) bucket is an ideal waterproof (and animal-proof) food locker to bring with you on canoe trips.

Build a camp washing machine Here's a great way to wash clothes while camping. Make a hole in the lid of a 5-gallon (19-liter) plastic bucket and insert a new toilet plunger. Put in clothes and laundry detergent. Snap on the lid and move the plunger up and down as an agitator. You can safely clean even delicate garments.

Camp shower A bucket perforated with holes on the bottom makes an excellent campsite shower. Hang it securely from a sturdy branch, fill it using another bucket or jug, and then take a quick shower as the water comes out. Want to shower in warm water? Paint the outside of another bucket matte black. Fill it with water and leave it out in the sun all day.

Paint high Avoid messy paint spills when painting on a scaffold or ladder. Put your paint can and brush in a large bucket and use paint-can hooks to hang the bucket *and* the brush. If the bucket is large enough, you'll even have room for your paint scraper, putty knife, rags, or other painting tools you may need. A 5-gallon (19-liter) plastic bucket is ideal.

Paint low Use the lids from 5-gallon (19-liter) plastic buckets as trays for 1-gallon (3.78-liter) cans of paint. The lids act as platforms for the paint cans and are also large enough to hold a paintbrush.

Make stilts Make working on a ceiling less of a stretch. Use two sturdy buckets (minus handles) and a pair of old shoes to make your own mini-stilts. Drive screws through the shoe soles and into wood blocks inside the buckets. Or punch holes in the bucket bottoms and tie or strap down the shoes.

Keep extension cords tangle-free A 5-gallon (19-liter) bucket can help you keep a long extension cord free of tangles. Just cut or drill a hole near the bottom of the pail, making sure it is large enough for the cord's pronged end to pass through. Then coil the rest of the cord into the bucket. The cord will come right out when pulled and is easy to coil back in. Plug the ends of the cord together when it's not in use. You can use the center space to carry tools to a worksite.

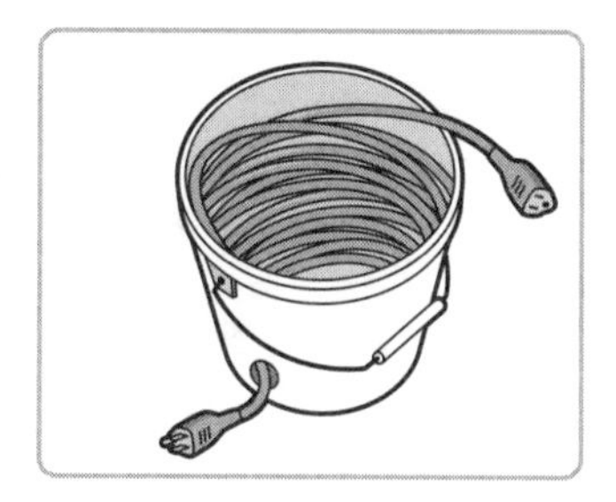

Soak your saw The best way to clean saw blades is to soak them in acetone or turpentine in a shallow pan, with a lid on the pan to contain the fumes. You can make your own shallow pan by cutting the bottom two inches or so off a plastic 5-gallon (19-liter) bucket with a utility knife. The bucket's lid can serve as the cover. Remember to wear rubber gloves and use a stick to lever out the sharp blades.

Garden in a bucket Use a 5-gallon (19-liter) plastic bucket as a minigarden or planter. Use another as a composter for scraps and cuttings. Bucket gardens are just the right size for apartment balconies.

Tip Where to Find Five-Gallon Buckets

Five-gallon (19-liter) plastic buckets are versatile, virtually indestructible, and offer a myriad of handy uses. And you can usually get them for free. Ask nicely and your local fast-food restaurant or supermarket deli section may be happy to give you the buckets shortening or coleslaw came in. Or check with neighborhood plasterers, who use 5-gallon buckets of drywall compound. Also keep an eye open for neighbors doing home improvements. Don't forget to get the lids, too. Wash a bucket with water and household bleach, then let it dry in the sun for a day or two. Put some scented kitty litter, charcoal, or a couple of drops of vanilla inside to remove any lingering odors.

Make a Christmas tree stand Fill a bucket partway with sand or gravel and insert the base of the tree in it. Then fill it the rest of the way and pour water on the sand or gravel to help keep the tree from drying out.

* Butter

Keep mold off cheese Why waste good cheese by letting the cut edges get hard or moldy? Give semi-hard cheeses a light coat of butter to keep them fresh and free of mold. Each time you use the cheese, coat the cut edge with butter before you rewrap it and put it back in the fridge.

Make cat feel at home Is the family feline freaked out by your move to a new home? Moving is often traumatic for pets as well as family members. Here's a good way to help an adult cat adjust to the new house or apartment: Spread a little butter on the top of one of its front paws. Cats love the taste of butter so much they'll keep coming back for more.

Get rid of fishy smell Your fishing trip was a big success, but now your hands reek of fish. What to do? Just rub some butter on your hands, wash with warm water and soap, and your hands will smell clean and fresh again.

Swallow pills with ease If you have difficulty getting pills to go down, try rolling them in a small amount of butter or margarine first. The pills will slide down your throat more easily.

B

Soothe aching feet To soothe tired feet, massage them with butter, wrap in a damp, hot towel, and sit for 10 minutes. Your feet will feel revitalized … and they'll smell like popcorn too.

Remove sap from skin You've just gotten home from a pleasant walk in the woods, but your hand is still covered with sticky tree sap that feels like it will never come off. Don't worry. Just rub butter on your hand and the gunky black sap will wash right off with soap and water.

Keep leftover onion fresh The recipe calls for half an onion and you want to keep the remaining half fresh as long as possible. Rub butter on the cut surface and wrap the leftover onion in aluminum foil before putting it in the fridge. The butter will keep it fresh longer.

Zap ink stain on doll's face Uh-oh, one of the kids used a pen to draw a new smile on that favorite doll's face. Try eliminating the kiddy graffiti by rubbing butter on it and leaving the doll face-up in the sun for a few days. Wash it off with soap and water.

Cut sticky foods with ease Rub butter on your knife or scissor blades before cutting sticky foods like dates, figs, or marshmallows. The butter will act as a lubricant and keep the food from sticking to the blades.

Emergency shave cream If you run out of shaving cream, try slathering some butter onto your wet skin for a smooth, close shave.

Prevent pots from boiling over You take your eye off the pasta for two seconds, and the next thing you know, the pot is boiling over onto the stovetop. Keep the boiling water in the pot next time by adding a tablespoon or two of butter.

Butter is the semi-solid material that results from churning cream—a process that is depicted on a Sumerian tablet from 2500 B.C. A butter-filled churn was found in a 2,000-year-old Egyptian grave, and butter was plentiful in King Tut's day, when it was made from the milk of water buffaloes and camels. The Bible also contains many references to butter—as the product of cow's milk. Later, the Vikings are believed to have introduced butter to Normandy, a region world-renowned for its butter.

In the U.S.A., butter was the only food ever defined by an Act of Congress prior to the enactment of the Food, Drug, and Cosmetic Act of 1938. Butter made in North America must contain at least 80 percent milk fat. The remaining 20 percent is composed of water and milk solids. It may be salted or unsalted (sweet). The salt adds flavor and also acts as a preservative. It takes 21 pounds of fresh cow's milk to make a pound of butter (10 kilograms to make 500 grams).

Kids' Stuff Making **butter** is fun and easy, especially when there are several kids around to take turns churning. All you need is a **jar,** a **marble,** and 1 to 2 cups of heavy **whipping cream** or double cream (preferably without carrageenan or other stabilizers added). Use the freshest cream possible and leave it out of the refrigerator until it reaches a temperature of about 60°F (15°C). Pour the cream into the jar, add the marble, close the lid, and let the kids take turns shaking (churning), about one shake per second. It may take anywhere from 5 to 30 minutes, but the kids will see the cream go through various stages from sloshy to coarse whipped cream. When the whipped cream suddenly seizes and collapses, fine-grained bits of butter will be visible in the liquid buttermilk. Before long a glob of yellowish **butter** will appear. Drain off the **buttermilk** and enjoy the delightful taste of fresh-made butter.

Treat dry hair Is your hair dry and brittle? Try buttering it up for a luxuriant shine. Massage a small chunk of butter into your dry hair, cover it with a shower cap for 30 minutes, then shampoo and rinse thoroughly.

* Buttons

Decorate a dollhouse Use buttons as sconces, plates, and wall hangings in a child's dollhouse. The more variety, the better.

Beanbag filler Use small buttons the next time you make beanbags and save the dried beans for the soup.

Make a necklace String attractive buttons on two strands of heavy-duty thread or dental floss. Make an attractive design by alternating large and small buttons of various colors.

Decorate a Christmas tree Give your Christmas tree an old-fashioned look. Make a garland by knotting large buttons on a sturdy length of string or dental floss.

Use as game pieces or poker chips Don't let lost pieces stop you from playing games like backgammon, bingo, or Parcheesi. Substitute buttons for the lost pieces and keep playing to your heart's content. For an impromptu game of poker, use buttons as chips, with each color representing a different value.

Keep tape unstuck You're trying to wrap a present and you can't find the end of the tape roll. Instead of scratching in frustration trying to find that elusive end every time you use the tape, stick a button on the end of the tape. As you use the tape, keep moving the button.

C

Cans

Keep tables together When you're having a large dinner party, lock card tables together by setting adjacent pairs of legs into empty cans. You won't have to clean up any spills caused by the tables moving this way and that.

Make light reflectors It's simple to make reflectors for your campsite or backyard lights. Just remove the bottom of a large empty can with a can opener and take off any label. Then use tin snips to cut the can in half lengthwise. You've just made two reflectors.

Quick floor patch Nail can lids to a wooden floor to plug knotholes and keep rodents out. If you can get access to the hole from the basement, nail the lid in place from underneath so the patch won't be obvious.

Tuna can egg poacher An empty 6-ounce (170-gram) tuna can is the perfect size to use as an egg poacher. Remove the bottom of the can as well as the top and remove any paper label. Then place the metal ring in a skillet of simmering water, and crack an egg into it.

Make a miniature golf course Arrange cans with both ends removed so the ball must go through them, go up a ramp into them, or ricochet off a board through them.

Tin cans are often described as "hermetically sealed," but do you know the origin of the term? The word *hermetic* comes from Hermes Trismegistus, a legendary alchemist who is reputed to have lived sometime in the first three centuries A.D. and to have invented a magic seal that keeps a vessel airtight.

The hermetically sealed can was invented in 1810 by British merchant Peter Durand. His cans were so thick they had to be hammered open! Two years later, Englishman Thomas Kensett set up America's first cannery on the New York waterfront to can oysters, meats, fruits, and vegetables.

Feed the birds A bird doesn't care if the feeder is plain or fancy as long as it is filled with suet. For a feeder that's about as basic as you can get, wedge a small can filled with suet between tree branches or posts.

Create decorative snowman Wrap an old soda can with white paper and tape with transparent tape. For a head use a styrene foam ball and tape it to the top of the can. Cover the body with cotton batting or cotton bandaging material and tape or glue it in place. Make a cone-shaped paper hat. Make eyes and a nose with buttons. To add arms, punch holes in the sides of the can and insert twigs. Use dots from a black marker pen to make buttons down the snowman's front. Make a scarf from a scrap of wooly fabric.

Make planters more portable Don't strain your back moving a planter loaded with heavy soil. Reduce the amount of soil and lighten the load by first filling one-third to one-half of the bottom of the planter with empty, upside-down aluminum cans. Finish filling with soil and add your plants. In addition to making the planter lighter, the rustproof aluminum cans also help it to drain well.

Protect young plants Remove both ends of an aluminum can and any paper label. Then push it into the earth to serve as a collar to protect young garden plants from cutworms. Use a soup can or a coffee can, depending on the size you need.

Make a tool tote Tired of fumbling around in your tool pouch to find the tool you need? Use empty frozen juice cans to transform the deep, wide pockets of a nail pouch into a convenient tote for wrenches, pliers, and screwdrivers. Make sure to remove the bottom of the can as well as the top. Glue or tape the cylinders together to keep them from shifting around, and slip them into the pouches to create dividers.

Make a pedestal Fill several wide identical-sized cans with rocks or sand and glue them together, one atop another. Screw a piece of wood into the bottom of the topmost can before attaching it, upside down, to the others. Paint your pedestal and place a potted plant, lamp, or statue on top. See the Tip on the following page for suggestions on the type of glue to use for this project.

Organize your desk If your office desk is a mess, a few empty cans can be the start of a nifty solution. Just attach several tin cans of assorted sizes together in a group to make an office-supplies holder for your desk. Start by cleaning and drying the cans and removing any labels. Then spray paint them (or wrap them in felt). When the paint is dry, glue them together using a hot glue gun. Your desk organizer is now ready to hold pens, pencils, paper clips, scissors, and such.

Tip Glue for Cans

When gluing cans and other metal pieces together, use a glue that adheres well to metal, such as polyvinyl chloride (PVC), liquid solder, or epoxy. If the joint won't be subject to stress, you can use a hot glue gun. Make sure to wash and dry the cans and to remove any labels first. Also let any paint dry thoroughly before gluing.

Make pigeonholes Assemble half a dozen or more empty cans and paint them with bright enamel. After they dry, glue the cans together and place them on their sides on a shelf. Then store silverware, nails, office supplies, or other odds and ends in them.

* Candles

Unstick a drawer If you have a desk or chest drawer that sticks, remove it and rub a candle on the runners. The drawer will open more smoothly when you slip it back in place.

Make a pincushion A wide candle makes an ideal pincushion. The wax will help pins and needles glide more easily through fabric too.

Weatherproof your labels After you address a package with a felt-tip pen, weatherproof the label by rubbing a white candle over the writing. Neither rain, nor sleet, nor snow will smear the label now.

Quiet a squeaky door If a squeaky door is driving you batty, take it off its hinges and rub a candle over the hinge surfaces that touch each other. The offending door will squeak no more.

Beeswax—the substance secreted by honeybees to make honeycombs—didn't make its debut in candles until the Middle Ages. Until then, candles were made of the rendered animal fat called tallow that produced a smoky flame and gave off acrid odors. Beeswax candles, by contrast, burned pure and clean. But they were not widely used at the time, being far too expensive for ordinary serfs and peasants.

The growth of whaling in the late 1700s brought a major change to candlemaking as spermaceti—a waxlike substance derived from sperm-whale oil—became available in bulk. The 19th century witnessed the advent of mass-produced candles and low-cost paraffin wax. Made from oil and coal shale, paraffin burned cleanly with no unpleasant odor.

Mend shoelace ends When the plastic or metal tips come off the ends of shoelaces, don't wait for the laces to fray. Do something to prevent the annoyance that comes from having to force a scraggly shoelace end through a teeny eyelet: Just dip the end into melted candle wax and the lace will hold until you can buy a new one.

Make a secret drawing Have a child make an "invisible" drawing with a white candle. Then let him or her cover it with a wash of watercolor paint to reveal the picture. The image will show up because the wax laid down by the candle will keep the paper in the areas it covers from absorbing the paint. If you have a few kids around, they can all make secret drawings and messages to swap and reveal.

Use puff-proof candle to ignite fires Don't let a draft blow out the flame when you're trying to light your fireplace or spark up the barbecue grill. Start your fire with one of those trick puff-proof birthday candles, designed to be a practical joke aid that prevents birthday celebrants from blowing out the candles on their cakes. Once your fire is up and roaring, smother the candle flame and save the trick candle for future use.

* Candy Tins

Emergency sewing kit A small candy tin is just the right size to hold a handy selection of needles, thread, and buttons in your purse or briefcase for on-the-spot repairs.

Store broken jewelry Don't lose all the little pieces of that broken jewelry you plan to have repaired someday. Keep the pieces together and safe in a small candy tin.

Prevent jewelry-chain tangles Keep necklaces and chain bracelets separate and tangle-free in their own individual tins.

Keep earrings together You are late for the party but you can only find one earring from the pair that matched your dress so nicely. To prevent pairs of small earrings from going their separate ways, store them together in a little candy tin and you'll be right on time for the next party.

Make a birthday keepsake Decorate the outside of a small candy tin, line it with felt or silk, and insert a penny or, if you can find one, a silver dollar from the birth year of your friend or loved one.

Organize your sewing gear Use a small candy tin to store snaps, sequins, buttons, and beads in your sewing box. Label the lids or glue on a sample for easy identification of the contents.

Store workshop accessories Candy tins are great for storing brads, glazing points, setscrews, lock washers, and other small items that might otherwise clutter up your workshop.

Store car fuses You'll always know where to find your spare car fuses if you store them in a little candy tin in the glove compartment of your car.

Cardboard Boxes

✱ CARDBOARD BOXES AROUND THE HOUSE

Make a bed tray Have breakfast in bed on a tray made from a cardboard box. Just remove the top flaps and cut arches from the two long sides to fit over your lap. Decorate the bottom of the box—which is now the top of your tray—with adhesive shelf paper and you're ready for those bacon and eggs.

Shield doors and furniture Use cardboard shields to protect doors and furniture from stains when you polish doorknobs and furniture pulls. Cut out the appropriate-sized shield and slide it over the items you are going to polish. This works best when you make shields that slip over the neck of knobs or knoblike pulls. But you can also make shields for hinges and U-shaped pulls.

Create gift-wrap suspense Take a cue from the Russians and their nesting *matryoshka* dolls. Next time you are giving a small but sure-to-be appreciated gift to a friend, place the gift-wrapped little box inside a series of increasingly bigger gaily wrapped boxes.

DID *You* KNOW?

The Chinese invented cardboard in the early 1500s, thus anticipating the demand for containers for Chinese take-out food by several hundred years.

In 1871 New Yorker Albert Jones patented the idea of gluing a piece of corrugated paper between two pieces of flat cardboard to create a material rigid enough to use for shipping. But it wasn't until 1890 that another American, Robert Gair, invented the corrugated cardboard box. His boxes were pre-cut flat pieces manufactured in bulk that folded into boxes, just like the cardboard boxes that surround us today.

Make dustcovers Keep dust and dirt out of a small appliance, power tool, or keyboard. Cut the flaps off a cardboard box that fits over the item, decorate it or cover it with self-adhesive decorative paper, and use it as a dustcover.

Tip A Good Box Source

Even if you don't drink alcohol, the proprietors of your local liquor store will often be happy to provide you with empty wine and liquor cartons. Don't forget to ask for the handy sections to be left intact.

Make an office in-box Making an in-box (or out-box) for your office desk is easy. Simply cut the top and one large panel off a cereal box; then slice the narrow sides at an angle. Wrap with self-adhesive decorative paper.

Make place mats Cut several 12 x 18-inch (30 x 45-centimeter) pieces of cardboard and cover them with colorful adhesive shelf paper or other decoration.

Play liquor box "ski ball" Transform your rec room or backyard into a carnival midway. Just leave the dividers in place in an empty wine or liquor carton. Place the carton at an angle and erect a small ramp in front (a rubber mat over a pile of books will do). Assign numbered values to each section of the carton, grab a few tennis or golf balls and you're ready to roll.

* CARDBOARD BOXES FOR STORING THINGS

Protect glassware or lightbulbs A good way to safely store fine crystal glassware is to put it in an empty wine or liquor carton with partitions. You can also use it for storing lightbulbs, but be sure to sort the bulbs by wattage so that it's easy to find the right one when you need a replacement.

Make a magazine holder Store your magazines in holders made from empty detergent boxes. Remove the top, then cut the box at an angle, from the top of one side to the bottom third of the other. Cover the holders with self-adhesive decorative paper.

Poster and artwork holder A clean liquor carton with its dividers intact is a great place to store rolled-up posters, drawings on paper, and canvases. Just insert the items upright between the partitions.

Store Christmas ornaments When you take down your Christmas tree, wrap each ornament in newspaper or tissue paper and store it in an empty liquor box with partitions. Each of the carton's segments can hold several of the wrapped holiday tree ornaments.

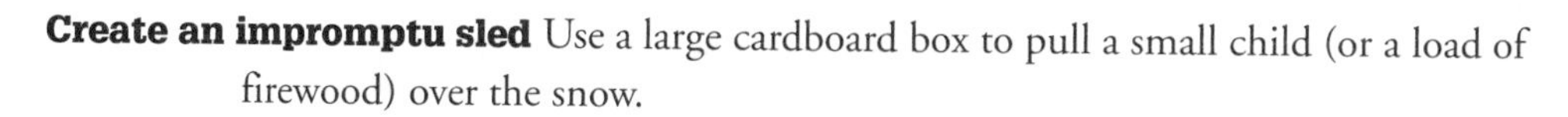

Create an impromptu sled Use a large cardboard box to pull a small child (or a load of firewood) over the snow.

Garage for toy vehicles Turn an empty large appliance box on its side and let the kids use it as a "garage" for their wheeled vehicles. They can also use a smaller box as a garage for miniature cars, trucks, and buses.

Make a puppet theater Stand a large cardboard box on end. Cut a big hole in the back for puppeteers to crouch in and a smaller one high up in the front for the stage. Decorate with markers or glue on pieces of fabric for curtains.

Organize kids' sporting goods Keep a decorated empty wine or liquor carton with partitions, and with the top cut off, in your child's room and use it for easy storage of tennis rackets, baseball bats, fishing poles, and such.

Make a play castle Turn a large appliance cardboard box into a medieval castle. Cut off the top flaps and make battlements by cutting notches along the top. To make a notch, use a utility knife to make a cut on either side of the section you want to remove, then fold the cut section forward and cut along the fold. To make a drawbridge, cut a large fold-down opening on one side that is attached at the bottom. Connect the top of the drawbridge to the sidewalls with ropes on either side, punching holes for the rope and knotting the rope on the other side. Use duct tape to reinforce the holes. Also cut

SCIENCE FAIR

Making a simple cardboard **sundial** is a great way for kids to observe how the sun's path changes every day. Just take a 10 x 10-inch (25 x 25-centimeter) piece of **cardboard** and poke a stick through the middle. If necessary, screw or nail a **small board** to the bottom of the stick to hold it upright. Place the sundial in a sunny spot. At **each hour,** have the kids **mark** where the **stick's shadow** falls on the cardboard. Check again the next day, and sure enough, the sundial seems pretty accurate. Check a week later, though, and the shadows won't align to the marks at the right times. When the curious kids start searching the Internet for a reason, give them a hint: The earth is tilted on its axis.

out narrow window slits in the walls. Let the kids draw stones and bricks on the walls.

✱ CARDBOARD BOXES FOR THE DO-IT-YOURSELFER

Repair a roof For temporary repair on your roof, put a piece of cardboard into a plastic bag and slide it under the shingles.

Organize your workshop A sectioned wine or liquor carton is a great place to store dowels, moldings, furring strips, weather stripping, and metal rods.

Store tall garden tools Turn three empty liquor cartons into a sectioned storage bin for your long-handled garden tools. Put a topless box on the floor with the dividers left in. Then cut the tops and bottoms off two similar boxes and stack them so the dividers match up. Use duct tape to attach the boxes to each other. Use the bin to store hoes, rakes, and other long-handled garden tools.

Protect work surfaces Keep work surfaces from being damaged. Flatten a large box or cut a large flat piece from a box and use it to protect your countertop, workbench, table, or desk from ink, paint, glue, or nicks from knives and scissors. Just replace it when it becomes messed up.

Protect your fingers Ouch! You just hammered your finger instead of the tiny nail you were trying to drive. To keep this from happening again, stick the little nail through a small piece of thin cardboard before you do your hammering. Hold the cardboard by an edge, position the nail, and pound it home. When you're done, use your bruise-free fingers to tear away the cardboard.

Keep upholstery tacks straight Reupholstering a chair or sofa? Here's a neat way to get a row of upholstery tacks perfectly straight and evenly spaced. Mark the spacing along the edge of a lightweight cardboard strip and press the tacks into it. After driving all of the tacks most of the way in, tug on the strip to pull the edge free before driving in the rest of the way.

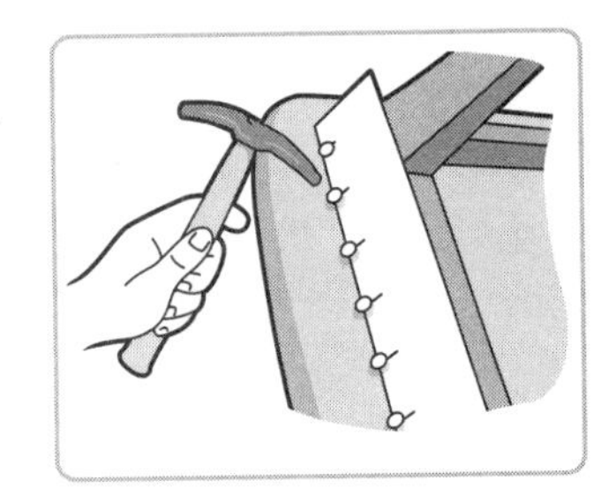

Make a drip pan Prevent an oil leak from soiling your garage floor or driveway. Make a drip pan by placing a few sheets of corrugated cardboard in a cookie sheet and placing the pan under your car's drip. For better absorption, sprinkle some cat litter, sawdust, or oatmeal into the pan on top of the cardboard. Replace with fresh cardboard as needed.

Help your mechanic *Something* is dripping from your car's engine, but you don't know what. Instead of blubbering helplessly to your mechanic about it, place a large piece of cardboard under the engine overnight and bring it with you when you take the car in for service. The color and location of the leaked fluid will help the mechanic identify the problem.

* Cardboard Tubes

C

Extend vacuum cleaner reach Can't reach that cobweb on the ceiling with your regular vacuum cleaner attachment? Try using a long, empty wrapping paper tube to extend the reach. You can even crush the end of the paper tube to create a crevice tool. Use duct tape to make the connection airtight.

Make a sheath Flatten a paper towel tube, duct tape one end shut, and you have a perfect sheath for a picnic/camp knife. Use toilet paper rolls for smaller cutlery.

Keep electrical cords tangle-free Keep computer and appliance cords tangle-free. Fanfold the cord and pass it through a toilet paper tube before plugging in. You can also use the tubes to store extension cords when they're not in use. Paper towel tubes will also work. Just cut them in half before using them to hold the cords.

Make a fly and pest strip Get rid of pesky flies and mosquitoes with a homemade pest strip. Just cover an empty paper towel or toilet paper roll with transparent tape, sticky side out, and hang where needed.

Use as kindling and logs Turn toilet paper and paper towel tubes into kindling and logs for your fireplace. For fire starter, use scissors to cut the cardboard into 1/8-inch (3-millimeter) strips. Keep the strips in a bin near the fireplace so they'll be handy to use next time you make a fire. To make logs, tape over one end of the tube and pack shredded newspaper inside. Then tape the other end. The tighter you pack the newspaper, the longer your log will burn.

Make boot trees To keep the tops of long, flexible boots from flopping over and developing ugly creases in the closet, insert cardboard mailing tubes into them to help them hold their shape.

DID *You* KNOW?

It took nearly 500 years for toilet paper to make the transition from sheets to rolls. Toilet paper was first produced in China in 1391 for the exclusive use of the emperor—in sheets that measured a whopping 2 x 3 feet (60 x 90 centimeters) each. Toilet paper in rolls was first made in the U.S.A. in 1890 by Scott Paper Company. Scott began making paper towels in 1907, thanks to a failed attempt to develop a new crepe toilet tissue. This paper was so thick it couldn't be cut and rolled into toilet paper, so Scott made larger rolls, perforated into 13 x 18-inch (33 x 45-centimeter), 13 x 18-inch (33 x 45-centimeter) sheets and sold them as Sani-Towels.

Make a plant guard It's easy to accidentally scar the trunk of a young tree when you are whacking weeds around it. To avoid doing this, cut a cardboard mailing tube in half lengthwise and tie the two halves around the trunk while you work around the tree. Then slip it off and use it on another tree.

Protect important documents Before storing diplomas, marriage certificates, and other important documents in your cedar chest, roll them tightly and insert them in paper towel tubes. This prevents creases and keeps the documents clean and dry.

Start seedlings Don't go to the garden supply store to buy biodegradable starting pots for seedlings. Just use the cardboard tubes from paper towels and toilet paper. Use scissors to cut each toilet paper tube into two pots, or each paper towel tube into four. Fill a tray with the cut cylinders packed against each other so they won't tip when you water the seedlings. This will also prevent them from drying out too quickly. Now fill each pot with seed-starting mix, gently pack it down, and sow your seeds. When you plant the seedlings, make sure to break down the side of the roll and make sure all the cardboard is completely buried.

Store knitting needles To keep your knitting needles from bending and breaking, try this: Use a long cardboard tube from kitchen foil or plastic wrap. Cover one end with cellophane tape. Pinch the other end closed and secure it tightly with tape. Slide the needles in through the tape on the taped end. The tape will hold them in place for secure, organized storage.

Store fabric scraps Roll up leftover fabric scraps tightly and insert them inside a cardboard tube from your bathroom or kitchen. For easy identification, tape or staple a sample of the fabric to the outside of the tube.

Store string Nothing is more useless and frustrating than tangled string. To keep your string ready to use, cut a notch into each end of a toilet paper tube. Secure one end of the string in one notch, wrap the string tightly around the tube, and then secure the other end in the other notch.

Keep linens crease-free Wrap tablecloths and napkins around cardboard tubes after laundering to avoid the creases they would get if they were folded. Use long tubes for tablecloths and paper towel or toilet paper tubes for napkins. To guard against stains, cover the tubes with plastic wrap first.

Keep pants crease-free You go to your closet for that good pair of pants you haven't worn in a while, only to find an ugly crease at the fold site from the hanger rack. It won't happen again if you cut a paper towel tube lengthwise, fold it in half horizontally, and place it over the rack before you hang up your pants. Before hanging pants, tape the sides of the cardboard together at the bottom to keep it from slipping.

Keep Christmas lights tidy Spending more time untangling your Christmas lights than it takes to put them up? Make yuletide prep easier by wrapping your lights around a cardboard tube. Secure them with masking tape. Put small strands of lights or garlands *inside* cardboard tubes, and seal the ends of the tubes with masking tape.

Tip Carpet Tubes

Carpet stores discard long, thick cardboard tubes that store workers will probably be happy to provide to you free for the asking. Because the tubes can be up to 12 feet (3.6 meters) long, you might want to ask for them to cut one to the size you want before you cart it away.

Protect fluorescent lights Keep fluorescent light tubes from breaking before you use them. They will fit neatly into long cardboard tubes sealed with tape at one end.

Make a kazoo Got a bunch of bored kids driving you crazy on a rainy day? Cut three small holes in the middle of a paper towel tube. Then cover one end of the tube with wax paper secured with a strong rubber band. Now hum into the other end, while using your fingers to plug one, two, or all three holes to vary the pitch. Make one for each kid. They may still drive you crazy, but they'll have a ball doing it!

Instant megaphone Don't shout yourself hoarse when you're calling outside for a child or pet to come home *right now*. Give your vocal cords a rest by using a wide cardboard tube as a megaphone to amplify your voice.

Make a hamster toy Place a couple of paper towel or toilet paper tubes in the hamster (or gerbil) cage. The little critters will love running and walking through them, and they like chewing on the cardboard too. When the tubes start looking ragged, just replace them with fresh ones.

Preserve kids' artwork You want to save some of your kids' precious artwork for posterity (or you don't want it to clutter up the house). Simply roll up the artwork and place it inside a paper towel tube. Label the outside with the child's name and date. The tubes are easy to store, and you can safely preserve the work of your budding young artists. Use this method to hold and store your documents, such as certificates and licenses, too.

Build a toy log cabin Notch the ends of several long tubes with a craft knife and then help the kids build log cabins, fences, or huts with them. Use different-sized tubes for added versatility. For added realism, have the kids paint or color the tubes before construction begins.

Make English crackers Keep the spirit of holiday firecrackers but cut out the dangers associated with burning explosives. Use toilet paper tubes to make English crackers, which "explode" into tiny gifts. For each cracker, tie a string about 8 inches (20 centimeters) long around a small gift such as candy, a balloon, or a figurine. After tying, the string should have about 6 inches (15 centimeters) to spare. Place the gift into the tube so the string dangles out one end. Cover the tube with bright-colored crepe paper or tissue and twist the ends. When you pull the string, out pops the gift.

* Car Wax

Fix skips on CDs Don't throw out that scratched compact disc. Try fixing it first with a small dab of car wax. Spread a cloth on a flat surface and place the CD on it damaged side up. Then, holding the disc with one hand, use the other to wipe the polish into the affected area with a soft cloth. Wait for it to dry and buff using short, brisk strokes along the scratch, not across it. A cloth sold to wipe eyeglasses or camera lenses will work well. When you can no longer see the scratch, wash the disc with water and let it dry before playing.

Keep bathroom mirrors fog-free Prevent your bathroom mirror from steaming up after your next hot shower. Apply a small amount of car paste wax to the mirror, let it dry, and buff with a soft cloth. Next time you step out of the shower, you'll be able to see your face in the mirror immediately. Rub the wax on bathroom fixtures to prevent water spots too.

Eliminate bathroom mildew To chase grime and mildew from your shower, follow these two simple steps: First clean the soap and water residue off the tiles or shower wall. Then rub on a layer of car paste wax and buff with a clean, dry cloth. You'll only need to reapply the wax about once a year. Don't wax the bathtub—it will become dangerously slippery.

Eradicate furniture stains Someone forgot to use a coaster and now there's an ugly white ring on the dining room table. When your regular furniture polish doesn't work, try using a dab of car wax. Trace the ring with your finger to apply the wax. Let it dry and buff with a soft cloth.

Keep snow from sticking When it's time to clear the driveway after a big snowstorm, you don't want snow sticking to your shovel. Apply two thick coats of car paste wax to the work surface of the shovel before you begin shoveling. The snow won't stick and there will be less wear and tear on your cardiovascular system. If you use a snow thrower, wax the inside of the chute.

* Carpet Scraps

Muffle clunky appliance noise Does your washer or dryer shake, rattle, and roll when you're doing a load? Put a piece of scrap carpet underneath it, and that may be all you need to calm things down.

Catch a falling sock You may never be able to stop socks and other articles of clothing from falling to the floor en route from washer to dryer. But you can make retrieval a lot easier by placing a narrow piece of carpet on the floor between the two appliances. When something falls, just pull out the strip and the article comes with it.

Keep Fido's home dry Don't let the raindrops keep falling on your dog's head. Weatherproof the doghouse: Make a rain flap by nailing a carpet remnant over the entrance to your doggy's domicile. In colder areas, you can also use small pieces of carpet to line interior walls and the floor to add insulation.

Make a scratching post If your cat is clawing up the living room sofa, this might do the trick. Make a scratching post by stapling carpet scraps to a post or board and place it near kitty's favorite target. If you want it to be freestanding, nail a board to the bottom of the post to serve as a base.

Keep garden paths weed-free Place a series of carpet scraps upside down and cover them with bark mulch or straw for a weed-free garden path. Use smaller scraps as mulch around your vegetable garden.

Exercise in comfort Make an instant exercise mat. Cut a length of old carpet around 3 feet (1 meter) wide and as long as your height. When you're not using it for yoga or sit-ups, roll it up and store it under your bed.

Make your own car mats Why buy expensive floor mats for your car when you can make your own? Cut carpet remnants to fit the floorboards of your car and drive off in comfort.

Protect your knees To protect your knees when you're washing the floor, weeding, or doing other work on all fours, make your own kneepads. Cut two pieces of carpeting 10 inches (25 centimeters) square and then cut two parallel slits or holes in each. Run old neckties or scarves through the slits and use them to tie the pads to your knees.

Keep floors dry Don't let the floor get soaked when you water your indoor plants. Place 12-inch (30-centimeter) round carpet scraps under houseplants to absorb any overwatering excess.

Prevent scratched floors Stop screeching chairs from scratching or making black marks on wood or vinyl floors. Glue small circles of carpet remnants to the bottom of chair and table legs.

Make a buffer Use epoxy resin to glue an old piece of carpet to a block of wood to make a buffer. Make several and use one to buff shoes, another to wipe blackboards, and one to clean window screens.

Cushion kitchen shelves Reduce the noisy clattering when putting away pots and pans: Cushion kitchen shelves and cabinets with pieces of carpet.

Add traction Keep good-sized carpet scraps in the trunk of your car to add traction when you're stuck in snow or ice. Keep one piece with your spare tire: When you have a flat, you won't have to kneel or lie on the dirty ground when you have to look under the carriage.

Protect workshop tools Does your workshop have a floor made of concrete or another hard material? If so, put down a few carpet remnants in the area closest to your workbench. Now when tools or containers accidentally fall to the floor, they will be far less likely to break.

* Castor Oil

Soften cuticles If you were ever forced to swallow castor oil as a child, this may be a pleasant surprise: The high vitamin-E content of that awful-tasting thick oil can work wonders on brittle nails and ragged cuticles. And you don't have to swallow the stuff. Just massage a small amount on your cuticles and nails each day and within three months you will have supple cuticles and healthy nails

Soothe tired eyes Before going to bed, rub odorless castor oil all around your eyes. Rub some on your eyelashes, too, to keep them shiny. Be careful not to get the oil in your eyes.

Lubricate kitchen scissors Use castor oil instead of toxic petroleum oil to lubricate kitchen scissors and other utensils that touch food.

Repel moles If moles are destroying your garden and yard, try using castor oil to get rid of them. Mix 1/2 cup castor oil and 2 gallons (7.5 liters) water and drench the molehill with it. It won't kill them, but it will get them out looking for another neighborhood to dig up.

Enjoy a massage Castor oil is just the right consistency to use as a soothing massage oil. For a real treat, warm the oil on the stovetop or on half-power in the microwave. Ahhh!

Castor oil is more than an old-fashioned medicine cabinet staple. It has hundreds of industrial uses. Large amounts are used in paints, varnishes, lipstick, hair tonic, and shampoo. Castor oil is also converted into plastics, soap, waxes, hydraulic fluids, and ink. And it is made into lubricants for jet engines and racing cars because it does not become stiff with cold or unduly thin with heat. North American manufacturers now use about 40 percent of the entire world's crop of castor oil.

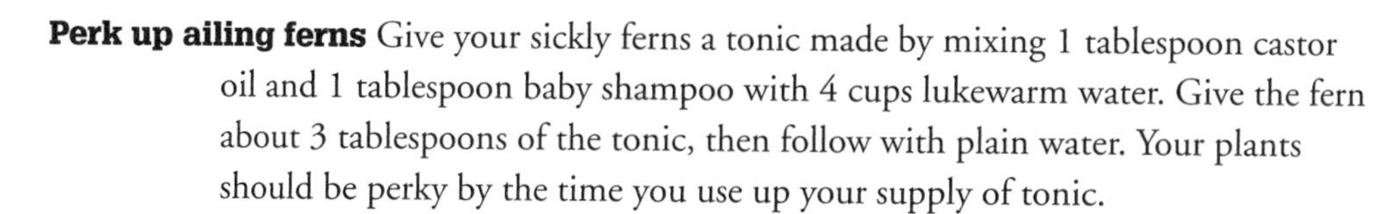

Perk up ailing ferns Give your sickly ferns a tonic made by mixing 1 tablespoon castor oil and 1 tablespoon baby shampoo with 4 cups lukewarm water. Give the fern about 3 tablespoons of the tonic, then follow with plain water. Your plants should be perky by the time you use up your supply of tonic.

Condition your hair For healthy, shiny hair, mix 2 teaspoons castor oil with 1 teaspoon glycerin and one egg white. Massage it into your wet hair, wait several minutes, and wash out.

* Cat Litter

Make a mud mask Make a deep-cleansing mud mask. Mix two handfuls of fresh cat litter with enough warm water to make a thick paste. Smear the paste over your face, let it set for 20 minutes, and rinse clean with water. The clay from cat litter detoxifies your skin by absorbing dirt and oil from the pores. When your friends compliment you on your complexion and ask how you did it, just tell them it's your little secret.

Sneaker deodorizer If your athletic shoes reek, fill a couple of old socks with scented cat litter, tie them shut, and place them in the sneakers overnight. Repeat if necessary until the sneakers are stink-free.

Add traction on ice Keep a bag of cat litter in the trunk of your car. Use it to add traction when you're stuck in ice or snow.

Prevent grease fires Don't let a grease fire spoil your next barbecue. Pour a layer of cat litter into the bottom of your grill for worry-free outdoor cooking.

Stop musty odors Get rid of that musty smell when you open the closet door. Just place a shallow box filled with cat litter in each musty closet or room. Cat litter works great as a deodorant.

Ed Lowe might not have gotten the idea for cat litter if a neighbor hadn't asked him for some sand for her cat box one day in 1947. Ed, who worked for his father's company selling industrial absorbents, suggested clay instead because it was more absorbent and would not leave tracks around the house. When she returned for more, he knew he had a winner. Soon he was crisscrossing the country, selling bags of his new Kitty Litter from the back of his Chevy Coupe. By 1990 Edward Lowe Industries, Inc., was the nation's largest producer of cat box filler with retail sales of more than $210 million annually.

Preserve flowers The fragrance and beauty of freshly cut flowers is such a fleeting thing. You can't save the smell, but you can preserve their beauty by drying your flowers on a bed of cat litter in an airtight container for 7-10 days.

Remove foul stench Just because your garbage cans hold garbage doesn't mean they have to smell disgusting. Sprinkle some cat litter into the bottom of garbage cans to keep them smelling fresh. Change the litter after a week or so or when it becomes damp. If you have a baby in the house, use cat litter the same way to freshen diaper pails.

Keep tents must-free Keep tents and sleeping bags fresh smelling and free of must when not in use. Pour cat litter into an old sock, tie the end, and store inside the bag or tent.

Repel moles Moles may hate the smell of soiled cat litter even more than you do. Pour some down their tunnels to send them scurrying to find new homes.

Make grease spots disappear Get rid of ugly grease and oil spots in your driveway or on your garage floor. Simply cover them with cat litter. If the spots are fresh, the litter will soak up most of the oil right away. To remove old stains, pour some paint thinner on the stain before tossing on the cat litter. Wait 12 hours and then sweep clean.

Freshen old books You can rejuvenate old books that smell musty by sealing them overnight in a can with clean cat litter.

* Chalk

Repel ants Keep ants at bay by drawing a line around home entry points. The ants will be repelled by the calcium carbonate in the chalk, which is actually made up of ground-up and compressed shells of marine animals. Scatter powdered chalk around garden plants to repel ants and slugs.

Polish metal and marble To make metal shine like new, put some chalk dust on a damp cloth and wipe. (You can make chalk dust by using a mortar to pulverize pieces of chalk.) Buff with a soft cloth for an even shinier finish. Wipe clean marble with a damp soft cloth dipped in powdered chalk. Rinse with clear water and dry thoroughly.

Keep silver from tarnishing You love serving company with your fine silver, but polishing it before each use is another story. Put one or two pieces of chalk in the drawer with your good silver. It will absorb moisture and slow tarnishing. Put some in your jewelry box to delay tarnishing there too.

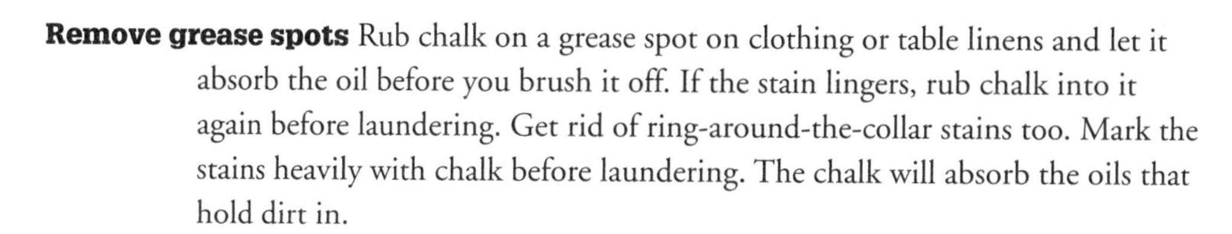

Remove grease spots Rub chalk on a grease spot on clothing or table linens and let it absorb the oil before you brush it off. If the stain lingers, rub chalk into it again before laundering. Get rid of ring-around-the-collar stains too. Mark the stains heavily with chalk before laundering. The chalk will absorb the oils that hold dirt in.

Stop screwdriver slips Does your screwdriver slip when you try to tighten a screw? It won't slip nearly as much if you rub some chalk on the tip of the blade.

Reduce closet dampness Tie a dozen pieces of chalk together and hang them up in your damp closet. The chalk will absorb moisture and help prevent mildew. Replace with a fresh bunch every few months.

Hide ceiling marks Temporarily cover up water or scuff marks on the ceiling until you have time to paint or make a permanent repair. Rub a stick of white chalk over the mark until it lightens or disappears.

Keep tools rust-free You can eliminate moisture and prevent rust from invading your toolbox by simply putting a few pieces of chalk in the box. Your tools will be rust-free and so will the toolbox.

The first "street painting" took place in 16th-century Italy, when artists began using chalk to make drawings on pavement. The artists often made paintings of the Virgin Mary (Madonna in Italian) and thus they became known as *madonnari.* The *madonnari* of old were itinerant artists known for a life of freedom and travel. But they always managed to attend the many regional holidays and festivals that took place in each Italian province. Today *madonnari* and their quaint street paintings continue to be a colorful part of the celebrations that take place every day in modern Italy.

* Charcoal Briquettes

Make a dehumidifier A humid closet, attic, or basement can wreak havoc on your health as well as your clothes. Get rid of all that humidity with several homemade dehumidifiers. To make one, just put some charcoal briquettes in a coffee can, punch a few holes in the lid, and place in the humid areas. Replace the charcoal every few months.

Keep root water fresh Put a piece of charcoal in the water when you're rooting plant cuttings. The charcoal will keep the water fresh.

Banish bathroom moisture and odors Hide a few pieces of charcoal in the nooks and crannies of your bathroom to soak up moisture and cut down on unpleasant odors. Replace them every couple of months.

Keep books mold-free Professional librarians use charcoal to get rid of musty odors on old books. You can do the same. If your bookcase has glass doors, it may provide a damp environment that can cause must and mold. A piece of charcoal or two placed inside will help keep the books dry and mold-free.

Henry Ford: father of ... *charcoal briquettes?* Yes, the first mass-produced charcoal briquettes were manufactured by the Ford Motor Company. They were made from waste wood from a Ford-owned sawmill in Kingsford, Michigan, built to provide wood for the bodies of Ford's popular "Woody" station wagons. The charcoal was manufactured into briquettes and sold as Ford Charcoal Briquettes. Henry Ford II closed the sawmill in 1951 and sold the plant to a group of local businessmen, who formed the Kingsford Chemical Company. The company continues to make the charcoal briquettes that are now a household name, despite relocating to Louisville, Kentucky, in 1961.

 Cheesecloth

Remove turkey stuffing with ease To keep turkey dressing from sticking to the bird's insides, pack the dressing in cheesecloth before you stuff it into the turkey's cavity. When the turkey is ready to serve, pull out the cheesecloth and the stuffing will slide out with it.

Make a homemade butterfly net Just sew cheesecloth into a bag and glue or staple it to a hoop formed from a wire coat hanger—and send the kids a-hunting. Or make a smaller cheesecloth net for when you take the kids fishing and let them use it as a bait-net to catch minnows. For an inexpensive Halloween costume, wrap a child in cheesecloth from head to toe and send your mini-mummy out to collect candy.

Convert a colander into a strainer If you can't find a strainer when you need one, a colander lined with cheesecloth will serve in a pinch.

Cut vacuuming time Here is a neat, time-saving way to vacuum the contents of a drawer filled with small objects without having to remove the contents. Simply cover the nozzle of your vacuum cleaner with cheesecloth, secured with a strong rubber band, and the vacuum will pick up only the dust.

Reduce waste drying herbs When drying fresh herbs, wrap them in cheesecloth to prevent seeds and smaller crumbled pieces from falling through.

Picnic food tent Keep bugs and dirt away from your picnic food serving plates. Wrap a piece of cheesecloth around an old wire umbrella form and place it over the plates. Use a hacksaw to remove the umbrella handle and tack the cheesecloth to the umbrella ribs with a needle and thread.

Make instant festive curtains Brighten any room with inexpensive, colorful, and festive cheesecloth curtains. Dye inexpensive cheesecloth (available in bulk from fabric vendors) in bright colors and cut it to the lengths and widths you need. Attach clip-on café-curtain hooks and your new curtains are ready to hang.

Chest Rub

Repel ticks and other bugs Going for a walk in the woods? Smear some chest rub on your legs and pants before you leave the house. It will keep ticks from biting and may spare you from getting Lyme disease. Pesky biting insects like gnats and mosquitoes will look elsewhere for victims if you apply chest rub to your skin before venturing outdoors. They hate the smell.

DID You KNOW?

Lunsford Richardson, the pharmacist who created Vick's VapoRub in 1905, also originated America's first "junk mail." Richardson was working in his brother-in-law's drugstore when he blended menthol and other ingredients into an ointment to clear sinuses and ease congestion. He called it Richardson's Croup and Pneumonia Cure Salve, but soon realized he needed something catchier to sell successfully. He changed the name to Vick's after his brother-in-law, Joshua Vick, and convinced the U.S. Post Office to institute a new policy allowing him to send advertisements addressed only to "Boxholder." Sales of Vick's first surpassed a million dollars during the Spanish flu epidemic of 1918.

Make calluses disappear Coat calluses with chest rub and then cover them with an adhesive bandage overnight. Repeat the procedure as needed. Most calluses will disappear after several days.

Soothe aching feet Are your feet aching after that long walk in the woods? Try applying a thick coat of chest rub and cover with a pair of socks before going to bed at night. When you wake up, your feet will be moisturized and rejuvenated.

Stop insect-bite itch fast Apply a generous coat of chest rub for immediate relief from itchy insect bites. The eucalyptus and menthol in the ointment are what do the trick.

Treat toenail fungus If you have a toenail fungus (onychomycosis), try applying a thick coat of chest rub to the affected nail several times a day. Many users and even some medical pros swear that it works (just check the Internet). But if you don't see results after a few weeks, consult a dermatologist or podiatrist.

* Chewing Gum

Retrieve valuables Oops, you just lost an earring or other small valuable down the drain. Try retrieving it with a just-chewed piece of gum stuck to the bottom of a fishing weight. Dangle it from a string tied to the weight, let it take hold, and reel it in.

Humans have been chewing gum a long time. Ancient Greeks chewed *mastiche*, a chewing gum made from the resin of the mastic tree. Ancient Mayans chewed *chicle*, the sap from the sapodilla tree. North American Indians chewed the sap from spruce trees, and passed the habit on to the Pilgrims. And other early American settlers made chewing gum from spruce sap and beeswax. John B. Curtis produced the first commercial chewing gum in 1848 and called it State of Maine Pure Spruce Gum. North Americans now spend more than $2 billion per year on chewing gum.

Lure a crab You'll be eating plenty of crab cakes if you try this trick: Briefly chew a stick of gum so that it is soft but still hasn't lost its flavor, then attach it to a crab line. Lower the line and wait for the crabs to go for the gum.

Fill cracks Fill a crack in a clay flowerpot or a dog bowl with piece of well-chewed gum.

Use as makeshift window putty Worried that a loose pane of glass may tumble and break before you get around to fixing it? Hold it in place temporarily with a wad or two of fresh-chewed gum.

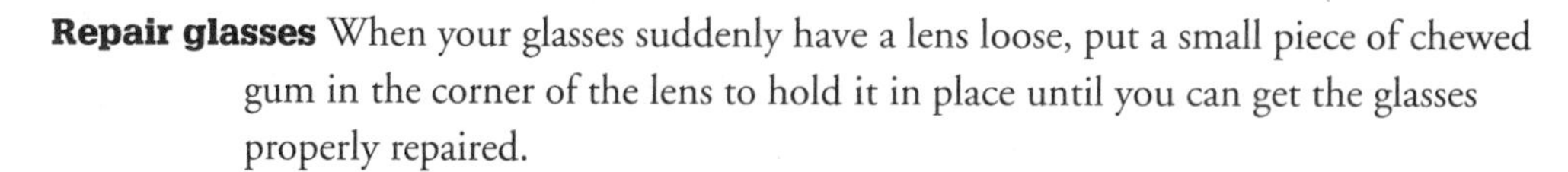

Repair glasses When your glasses suddenly have a lens loose, put a small piece of chewed gum in the corner of the lens to hold it in place until you can get the glasses properly repaired.

Treat flatulence and heartburn Settle stomach gases and relieve heartburn by chewing a stick of spearmint gum. The oils in the spearmint act as an antiflatulent. Chewing stimulates the production of saliva, which neutralizes stomach acid and corrects the flow of digestive juices. Spearmint also acts as a digestive aid.

* Chicken Wire

Repel deer Are the deer tearing up your garden again? Here's a simple method to keep them away: Stake chicken wire *flat* around the perimeter of your garden. Deer don't like to walk on it, and it is not an eyesore like a chicken-wire fence.

Crown catnip plants If you are growing catnip for your cat, put a crown of chicken wire over the plant, close to the ground. As the catnip grows through the wire and gets eaten, the roots will remain intact, growing new catnip. Make sure the edges of the wire are tucked in securely. Catnip is a hardy plant, even in frigid temperatures, so if the roots remain, you will see it year after year.

Protect bulbs from rodents Keep pesky burrowing rodents from damaging your flower bulbs. Line the bottom of a prepared bed with chicken wire, plant the bulbs, and cover with soil.

Flower holder Keep cut flowers aligned in a vase. Squish some chicken wire together and place it in the bottom of the vase before inserting the flowers.

Make a childproof corral Your garage or shed is full of dangerous tools and toxic substances. Keep kids away from these hazardous items by enclosing them in a childproof corral. Make it by first attaching standard-width chicken wire to the walls in a corner. Then staple 1 x 2s to the cut ends of the wire and install screw eyes in the wood to accommodate two padlocks.

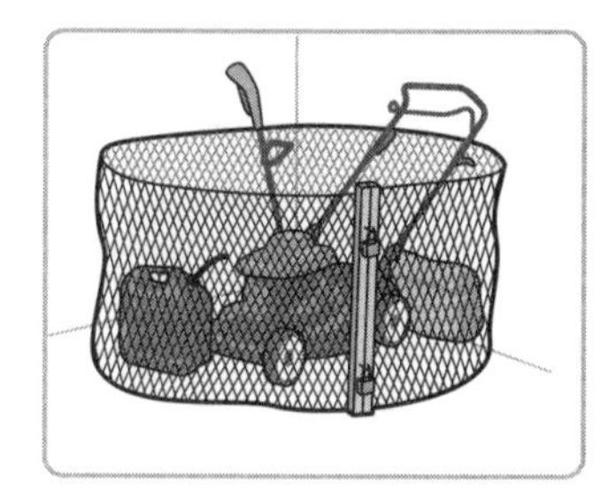

Firmer fence posts Before setting a fence post in concrete, wrap the base with chicken wire. This will make the anchoring firmer and the post more secure.

Secure insulation After you place fiberglass batting between roof rafters or floor joists, staple chicken wire across the joists to secure it and, in the case of the rafters, to keep it from sagging.

* Clipboards

Makeshift pants hanger Can't find a hanger for a pair of pants? Use a clipboard instead. Just suspend the clipboard from a hook inside the closet or on the bedroom door. Hang the trousers overnight by clipping the cuffs to the board.

Keep recipes at eye level When you are following a recipe clipped from a magazine or newspaper, it's hard to read and keep clean when the clipping is lying on the counter. Solve the problem by attaching a clipboard to a wall cabinet at eye level. Just snap the recipe of the day onto the clipboard and you are ready to create your kitchen magic.

Hold place mats Hang a clipboard inside a kitchen cabinet or pantry door and use the clamp as a convenient, space-saving way to store your place mats.

Keep sheet music in place Flimsy pages of sheet music are susceptible to drafts and sometimes seem to spend more time on the floor than on the music stand. To eliminate this problem, attach the music sheets to a clipboard before placing it in the stand. The pages will remain upright and in place.

Aid road-trip navigation Before starting out on a long motor trip, fold the map to the area you will be traveling in. Attach it to a clipboard and keep it nearby to check your progress at rest stops.

Organize your sandpaper Most of the time, sandpaper is still good after the first or second time you use it. The trick is to find that used sandpaper again. Hang a clipboard on a hook on your workshop pegboard. Just clip still-usable sandpaper to the board when you are done and the sandpaper will be handy next time you need it.

* Clothespins

Fasten Christmas lights Keep your outdoor Christmas lights in place and ready to withstand the elements. As you affix your lights to gutters, trees, bushes (or even your spouse!) fasten them securely with clip-on clothespins.

Make a clothespin clipboard Organize your workshop, kitchen, or bathroom with a homemade rack made with straight clothespins. Space several clothespins evenly apart on a piece of wood, and screw them on with screws coming through from the back of the board (pre-drill the holes so you don't split the clothespin). Now your rack is ready to hang.

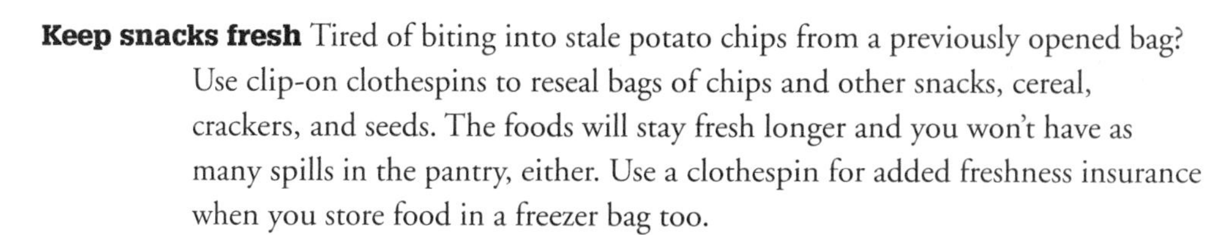

Keep snacks fresh Tired of biting into stale potato chips from a previously opened bag? Use clip-on clothespins to reseal bags of chips and other snacks, cereal, crackers, and seeds. The foods will stay fresh longer and you won't have as many spills in the pantry, either. Use a clothespin for added freshness insurance when you store food in a freezer bag too.

Organize your closet Okay, you found one shoe. Now, where the heck is the other one? From now on use clip-on clothespins to hold together pairs of shoes, boots, or sneakers, and put an end to those unscheduled hunting expeditions in your closets. It's a good idea for gloves, too.

Keep gloves in shape After washing wool gloves, insert a straight wooden clothespin into each finger. The clothespins will keep the gloves in their proper shape.

Prevent vacuum cord "snapback" Whoops! You're vacuuming the living room floor when suddenly the machine stops. You've accidentally pulled out the plug and the cord is automatically retracting and snapping back into the machine. To avoid similar annoyance in the future, simply clip a clothespin to the cord at the length you want.

Make an instant bib Make bibs for your child by using a clip-on clothespin to hold a dish towel around the child's neck. Use bigger towels to make lobster bibs for adults. It's much faster than tying on a bib.

Make clothespin puppets Traditional straight clothespins without the metal springs are ideal for making little puppets. Using the knob as a head, have kids paste on bits of yarn for hair, and scraps of cloth or colored paper for clothes to give each one its own personality.

Hold leaf bag open Ever try filling a large leaf bag all by your lonesome, only to see half the leaves fall to the ground because the bag won't stay open? Next time enlist a couple of clip-on clothespins as helpers. After you shake open the bag and

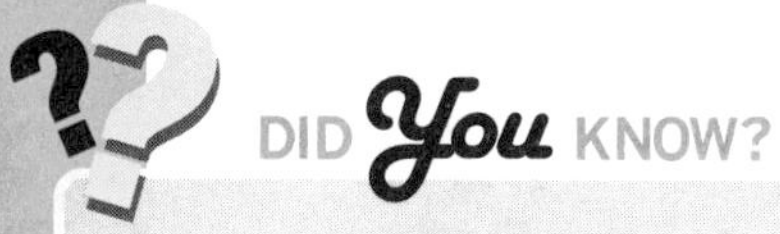

Between 1852 and 1857 the U.S. Patent Office granted patents for 146 different clothespins. Today wood clothespin makers have practically disappeared. "The small family-owned, secondary wood processing companies are dying off," says Richard Penley, who closed his family's clothespin plant in Maine in 2002. The company founded by Penley's grandfather and two brothers in 1923 survives, importing and distributing clothespins made overseas. Only two wood clothespin manufacturers are left in the United States: Diamond-Forster in Maine and National Clothespin Company in Vermont.

spread it wide, use the clothespins to clip one side of the bag to a chain-link fence or other convenient site. The bag will stay open for easy filling.

Mark a bulb spot What to do when a flower that blooms in the spring … doesn't? Just push a straight clothespin into the soil at the spot where it didn't grow. In the fall you will know exactly where to plant new bulbs to avoid gaps.

Grip a nail Hammer the nail and not your fingers. Just remember to use a clip-on clothespin to hold nails when hammering in hard-to-reach places.

Clamp thin objects Use clip-on clothespins as clamps when you're gluing two thin objects together. Let the clothespin hold them in place until the glue sets.

Keep paintbrush afloat Keep your paintbrush from sinking into the solvent residue when you soak it. Clamp the brush to the container with a clothespin.

Club Soda

Make pancakes and waffles fluffier If you like your pancakes and waffles on the fluffy side, substitute club soda for the liquid called for in the recipes. You'll be amazed at how light and fluffy your breakfast treats turn out.

Bubbling water has been associated with good health since the time of the ancient Romans, who enjoyed drinking mineral water almost as much as they liked bathing in it. The first club soda was sold in North America at the end of the 1700s. That's when pharmacists figured out how to infuse plain water with carbon dioxide, which they believed was responsible for giving natural bubbling water health-inducing qualities. Club soda and seltzer are essentially the same. However, seltzer is a natural effervescent water (named for a region in Germany where it is plentiful) whereas club soda is manufactured.

Give your plants a mineral bath Don't throw out that leftover club soda. Use it to water your indoor and outdoor plants. The minerals in the soda water help green plants grow. For maximum benefit, try to water your plants with club soda about once a week.

Remove fabric stains Clean grease stains from double-knit fabrics. Pour club soda on the stain and scrub gently. Scrub more vigorously to remove stains on carpets or less delicate articles of clothing.

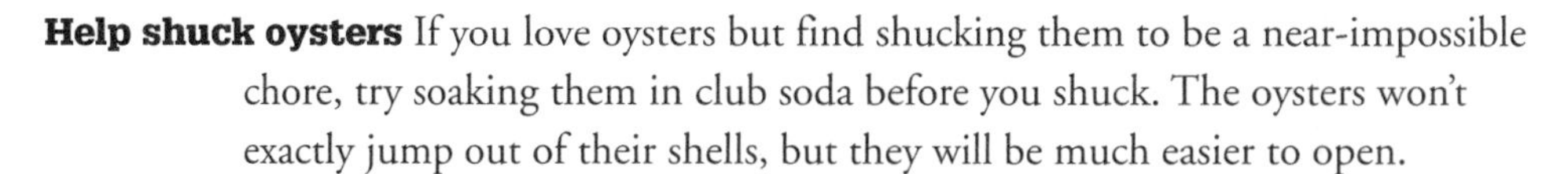

Help shuck oysters If you love oysters but find shucking them to be a near-impossible chore, try soaking them in club soda before you shuck. The oysters won't exactly jump out of their shells, but they will be much easier to open.

Clean precious gems Soak your diamonds, rubies, sapphires, and emeralds in club soda to give them a bright sheen. Simply place them in a glass full of club soda and let them soak overnight.

Clean your car windshield Keep a spray bottle filled with club soda in the trunk of your car. Use it to help remove bird droppings and greasy stains from the windshield. The fizzy water speeds the cleaning process.

Restore hair color If your blond hair turns green when you swim in a pool with too much chlorine, don't panic. Rinse your hair with club soda and it will change back to its original color.

Tame your tummy Cold club soda with a dash of bitters will work wonders on an upset stomach caused by indigestion or a hangover.

Clean countertops and fixtures Pour club soda directly on stainless steel countertops, ranges, and sinks. Wipe with a soft cloth, rinse with warm water, and wipe dry. To clean porcelain fixtures, simply pour club soda over them and wipe with a soft cloth. There's no need for soap or rinsing, and the soda will not mar the finish. Give the inside of your refrigerator a good cleaning with a weak solution of club soda and a little bit of salt.

Remove rust To loosen rusty nuts and bolts, pour some club soda over them. The carbonation helps to bubble the rust away.

Eliminate urine stains Did someone have an accident? After blotting up as much urine as possible, pour club soda over the stained area and immediately blot again. The club soda will get rid of the stain and help reduce the foul smell.

Ease cast-iron cleanup Food tastes delicious when it's cooked in cast iron, but cleaning those heavy pots and pans with the sticky mess inside is no fun at all. You can make the cleanup a lot easier by pouring some club soda in the pan while it's still warm. The bubbly soda will keep the mess from sticking.

* Coat Hangers

Stop caulk-tube ooze To prevent caulk from oozing from the tube once the job is done, cut a 3-inch (7.5-centimeter) piece of coat hanger wire; shape one end into a hook and insert the other, straight end into the tube. Now you can easily pull out the stopper as needed.

Secure a soldering iron Keeping a hot soldering iron from rolling away and burning something on your workbench is a real problem. To solve this, just twist a wire coat hanger into a holder for the iron to rest in. To make the holder, simply bend an ordinary coat hanger in half to form a large **V**. Then bend each half in half so that the entire piece is shaped like a **W**.

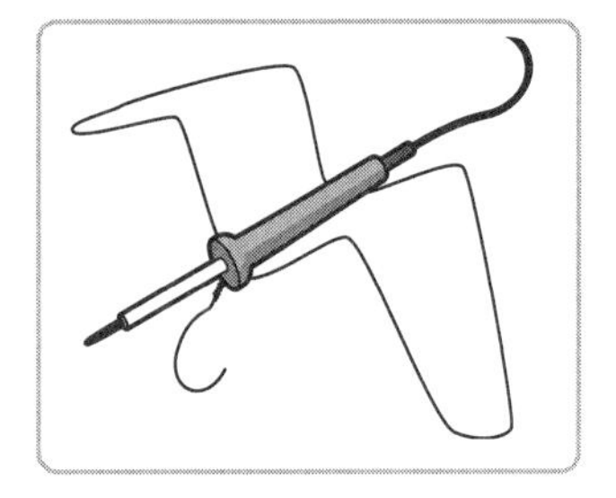

Extend your reach Can't reach that utensil that has fallen behind the refrigerator or stove? Try straightening a wire coat hanger (except for the hook at the end), and use it to fish for the object.

Make a giant bubble wand Kids will love to make giant bubbles with a homemade bubble wand fashioned from a wire coat hanger. Shape the hanger into a hoop with a handle and dip it into a bucket filled with 1 part liquid dishwashing detergent in 2 parts water. Add a few drops of food coloring to make the bubbles more visible.

Create arts and crafts Make mobiles for the kids' room using wire coat hangers; paint them in bright colors. Or use hangers to make wings and other accessories for costumes.

Unclog toilets and vacuum cleaners If your toilet is clogged by a foreign object, fish out the culprit with a straightened wire coat hanger. Use a straightened hanger to unclog a jammed vacuum cleaner hose.

Make a mini-greenhouse To convert a window box into a mini-greenhouse, bend three or four lengths of coat hanger wire into **U**'s and place the ends into the soil. Punch small holes in a dry-cleaning bag and wrap it around the box before putting it back in the window.

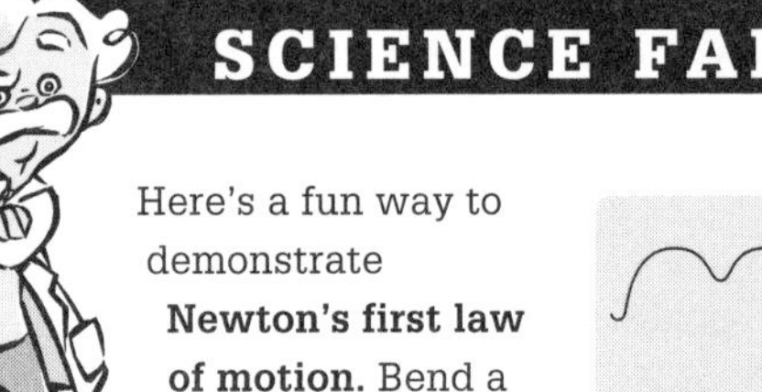

SCIENCE FAIR

Here's a fun way to demonstrate **Newton's first law of motion.** Bend a wire **coat hanger** into a large loopy **M**, as shown. Holding the wire in the middle, attach a same-sized **modeling-clay ball** to each of the hooks. Then place the low center point of the **M** on top of your head. If you **turn your head** to the left or right, the **inertia** of the balls will be enough to **keep them in place,** demonstrating Newton's law that "objects at rest tend to stay at rest." With practice, you can actually turn all the way around and the balls will remain still.

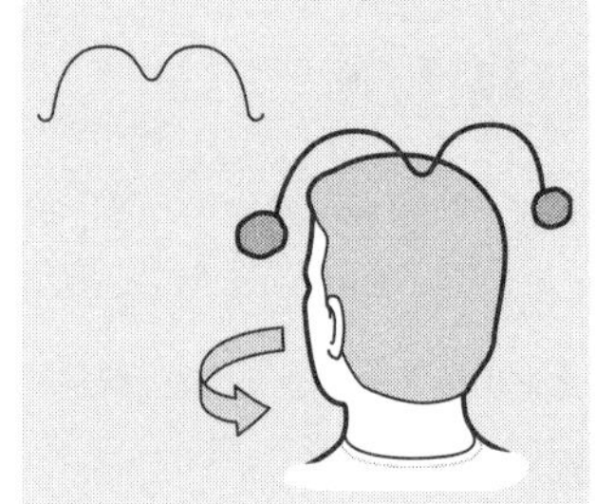

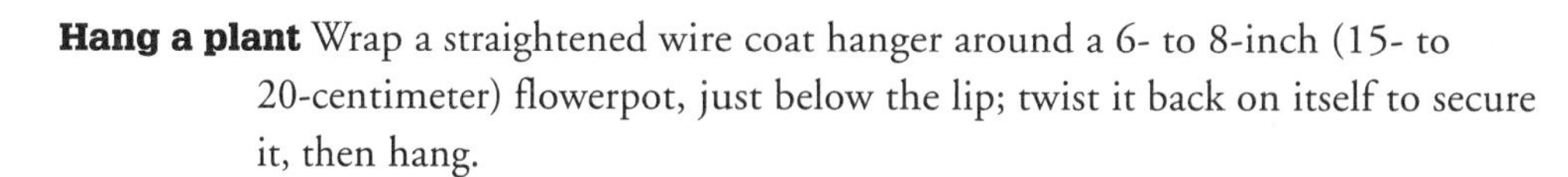

Hang a plant Wrap a straightened wire coat hanger around a 6- to 8-inch (15- to 20-centimeter) flowerpot, just below the lip; twist it back on itself to secure it, then hang.

Make plant markers Need some waterproof markers for your outdoor plants? Cut up little signs from a milk jug or similar rigid but easy-to-cut plastic. Write the name of the plant with an indelible marker. Cut short stakes from wire hangers. Make two small slits in each marker and pass the wire stakes through the slits. Neither rain nor sprinkler will obscure your signs.

Make a paint can holder When you are up on a ladder painting your house, one hand is holding on while the other is painting. How do you hold the paint can? Grab a pair of wire snips and cut the hook plus 1 inch (2.5 centimeters) of wire from a wire hanger. Use a pair of pliers to twist the 1-inch section firmly around the handle of your paint can. Now you have a handy hanger.

Light a hard-to-reach pilot light The pilot light has gone out way inside your stove or furnace. You'd rather not risk a burn by lighting a match and sticking your hand all the way in there. Instead, open up a wire hanger and tape the match to one end. Strike the match and use the hanger to reach the pilot.

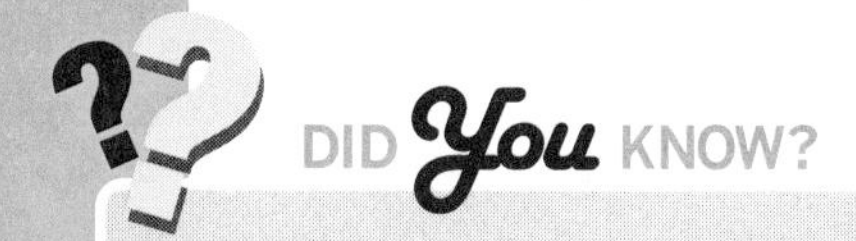

All Albert J. Parkhouse wanted to do when he arrived at work was to hang up his coat and get busy doing his job. It was 1903, and Albert worked at Timberlake Wire and Novelty Company in Jackson, Michigan. But when he went to hang his clothes on the hooks the company provided for workers, all were in use. Frustrated, Albert picked up a piece of wire, bent it into two large oblong hoops opposite each other, and twisted both ends at the center into a hook. He hung his coat on it and went to work. The company thought so much of the idea they patented it and made a fortune. Alas, Albert never got a penny for inventing the wire coat hanger.

* Coffee Beans

Freshen your breath What to do when you're all out of breath mints? Just suck on a coffee bean for a while and your mouth will smell clean and fresh again.

Remove foul odor from hands If your hands smell of garlic, fish or other strong foods you've been handling, a few coffee beans may be all you need to get rid of the odor. Put the beans in your hands and rub them together. The oil released from the coffee beans will absorb the foul smell. When the odor is gone, wash your hands in warm, soapy water.

Fill a beanbag They don't call them beanbags for nothing. Coffee beans are ideal as beanbag filler, but with the price of coffee nowadays it's a good idea to wait for a sale and then buy the cheapest beans available.

* Coffee Cans

Bake perfectly round bread Use small coffee cans to bake perfectly cylindrical loaves of bread. Use your favorite recipe but put the dough in a well-greased coffee can instead of a loaf pan. For yeast breads use two cans and fill each only half full. Grease the inside of the lids and place them on the cans. For yeast breads, you will know when it is time to bake when the rising dough pushes the lids off. Place the cans—without the lids—upright in the oven to bake.

Separate hamburgers Before you put those hamburger patties in the freezer, stack them with a coffee-can lid between each and put them in a plastic bag. Now, when the patties are frozen you'll be able to easily peel off as many as you need.

Hold kitchen scraps Line a coffee can with a small plastic bag and keep it near the sink to hold kitchen scraps and peelings. Instead of walking back and forth to the garbage can, you'll make one trip to dump all the scraps at the same time.

Make a bank To make a bank for the kids or a collection can for a favorite charity, use a utility knife to cut a 1/8-inch (3-millimeter) slit in the center of the plastic lid of a coffee can. Tape decorative paper or adhesive plastic to the sides of the kids' bank; for a collection can, use the sides of the can to highlight the charity you are helping.

Create a toy holder Make a decorative container for kids' miniature books and small toys. Wash and dry a coffee can and file off any sharp edges. Sponge on two coats of white acrylic paint, letting it dry between coats. Cut out a design from an old sheet or pillowcase to wrap around the can. Mix 4 tablespoons white glue with enough water to the consistency of paint. Paint on the glue mixture and gently press the fabric onto the can. Trim the bottom and tuck top edges

Ground coffee loses its flavor immediately unless it is specially packaged or brewed. Freshly roasted and ground coffee is often sealed in combination plastic-and-paper bags, but the coffee can is by far the most common container in North America. Vacuum-sealed cans keep coffee fresh for up to three years. The U.S. is the world's largest consumer of coffee, importing up to 2.5 million pounds (1.1 million kilograms) each year. More than half the U.S. population consumes coffee. The typical coffee drinker has 3.4 cups of coffee per day. That translates into 350 million cups of coffee guzzled daily.

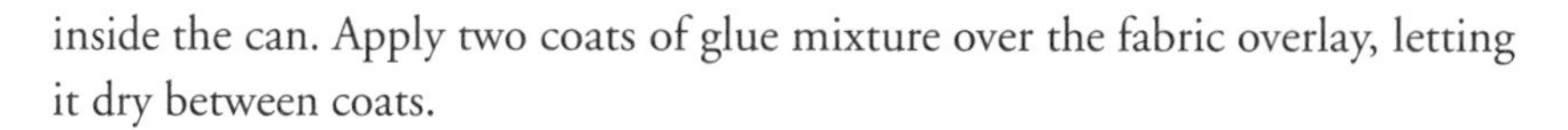

inside the can. Apply two coats of glue mixture over the fabric overlay, letting it dry between coats.

Store belts If you have more belts than places to hang them up, just roll them up and store them in a cleaned-out coffee can with a clear lid. Coffee cans are just the right size to keep belts from creasing, and clear lids will let you find each belt easily.

Keep the laundry room neat Have an empty coffee can nearby as you're going through the kids' pockets before putting up a load of wash. Use it to deposit gum and candy wrappers, paper scraps, and other assorted items that kids like to stuff into their pockets. Keep another can handy for coins and bills.

Make a dehumidifier If your basement is too damp, try this easy-to-make dehumidifier. Fill an empty coffee can with salt and leave it in a corner where it will be undisturbed. Replace the salt at monthly intervals or as needed.

Keep carpets dry Place plastic coffee-can lids under houseplants as saucers. They will protect carpets or wood floors and catch any excess water.

Keep toilet paper dry when camping Bring a few empty coffee cans with you on your next camping trip. Use them to keep toilet paper dry in rainy weather or when you're carrying supplies in a canoe or boat.

Gauge rainfall or sprinkler coverage Find out if your garden is getting enough water from the rain. Next time it starts to rain, place empty coffee cans in several places around the garden. When the rain stops, measure the depth of the water in the cans. If they measure at least an inch, there's no need for additional watering. This is also a good way to test if your sprinkler is getting sufficient water to the areas it is supposed to cover.

Make a coffee-can bird feeder To fashion a coffee can into a sturdy bird feeder, begin with a full can and open the top only halfway. (Pour the coffee into an airtight container.) Then open the bottom of the can halfway the same way. Carefully bend the cut ends down inside the can so the edges are not exposed to cut you. Punch a hole in the side of the can at both ends, where it will be the "top" of the feeder, and put some wire through each end to make a hanger.

Make a spot lawn seeder When it's time to reseed bare spots on your lawn, don't use a regular spreader. It wastes seed by throwing it everywhere. For precision seeding, fashion a spot seeder from an empty coffee can and a pair of plastic lids. Drill small holes in the bottom of the can, just big enough to let grass seeds pass through. Put one lid over the bottom of the can, fill the can with seeds, and cap it with the other lid. When you're ready to spread the seeds, take off the bottom lid. When you're finished, replace it to seal in any unused seed for safe storage.

Eliminate workshop clutter You want small items like screws, nuts, and nails to be handy, but you don't want them to take up workbench space. Here's a way to get the small stuff up out of the way. Drill a hole near the top of empty coffee cans so you can hang them on nails in your workshop wall. Label the cans with masking tape so you will know what's inside.

Soak a paintbrush An empty coffee can is perfect for briefly soaking a paintbrush in thinner before continuing a job the next day. Cut an **X** into the lid and insert the brush handles so the bristles clear the bottom of the can by about 1/2 inch (12 millimeters). If the can has no lid, attach a stick to the brush handle with a rubber band to keep the bristles off the bottom of the can.

Catch paint drips Turn the plastic lids from old coffee cans into drip catchers under paint cans and under furniture legs when you're painting. Protect cupboard shelves by putting them under jars of cooking oil and syrup too.

* Coffee Filters

Cover food in the microwave Coffee filters are microwave-safe. Use them to cover bowls or dishes to prevent splatter when cooking or baking in your microwave oven.

Filter cork crumbs from wine Don't let cork droppings ruin your enjoyment of a good glass of wine. If your attempt at opening the bottle results in floating cork crumbs, just decant the wine through a coffee filter.

The coffee filter was invented in 1908 by a housewife from Dresden, Germany. Melitta Bentz was looking for a way to brew a perfect cup of coffee without the bitterness often caused by overbrewing. She decided to try making a filtered coffee, pouring boiling water over ground coffee, and filtering out the grinds. Melitta experimented with different materials, until she found that the blotter paper that her son used for school worked best. She cut a round piece of blotting paper, put it in a metal cup, and the first Melitta coffee filter was born. Shortly thereafter, Melitta and her husband, Hugo, launched the company that still bears her name.

Line a sieve If you save your cooking oil for reuse after deep-fat frying, line your sieve with a basket-style coffee filter to remove smaller food remnants and impurities.

Hold a taco Serve tacos, hot dogs, popcorn, and other messy foods in cone or basket-style coffee filters. The filter is a perfect sleeve and will help keep fingers clean and cleanup a snap.

Catch ice-cream drips Next time the kids scream for ice-cream bars or ice pops, serve it to them with a drip catcher made from basket-style coffee filters. Just poke the stick through the center of two filters and the drips will fall into the paper, not on the child or your carpet.

Make an instant funnel Cut the end off a cone-style coffee filter to make an instant funnel. Keep a few in your car and use them to avoid spillage when you add a quart of oil or two.

Clean your specs Next time you clean your glasses, try using a coffee filter instead of a tissue. Good-quality coffee filters are made from 100 percent virgin paper, so you can use them to clean your glasses without leaving lint. You can also use them safely to polish mirrors and TV and computer-monitor screens.

Keep skillets rust-free Prolong the life of your good cast-iron cookware. Put a coffee filter in the skillet when it's not in use. The filter will absorb moisture and prevent rusting.

Prevent soil leakage When you're repotting a plant, line the pot with a coffee filter to keep the soil from leaking out through the drain hole.

Make an air freshener Fill a coffee filter with baking soda, twist-tie it shut, and you have just made an air freshener. Make several and tuck them into shoes, closets, the fridge, or wherever else they may be needed.

* Coffee Grounds

Don't raise any dust Before you clean the ashes out of your fireplace, sprinkle them with wet coffee grounds. They'll be easier to remove, and the ash and dust won't pollute the atmosphere of the room.

Deodorize a freezer Get rid of the smell of spoiled food after a freezer failure. Fill a couple of bowls with used or fresh coffee grounds and place them in the freezer overnight. For a flavored-coffee scent, add a couple of drops of vanilla to the grounds.

Fertilize plants Don't throw out those old coffee grounds. They're chock-full o' nutrients that your acidic-loving plants crave. Save them to fertilize rosebushes, azaleas, rhododendrons, evergreens, and camellias. It's better to use grounds from a drip coffeemaker than the boiled grounds from a percolator. The drip grounds are richer in nitrogen.

Keep worms alive A cup of used coffee grounds will keep your bait worms alive and wiggling all day long. Just mix the grounds into the soil in your bait box before you dump in the worms. They like coffee almost as much as we do, and the nutrients in the grounds will help them live longer.

Keep cats out of the garden Kitty won't think of your garden as a latrine anymore if you spread a pungent mixture of orange peels and used coffee grounds around your plants. The mix acts as great fertilizer too.

Boost carrot harvest To increase your carrot harvest, mix the seeds with fresh-ground coffee before sowing. Not only does the extra bulk make the tiny seeds easier to sow, but the coffee aroma may repel root maggots and other pests. As an added bonus, the grounds will help add nutrients to the soil as they decompose around the plants. You might also like to add a few radish seeds to the mix before sowing. The radishes will be up in a few days to mark the rows, and when you cultivate the radishes, you will be thinning the carrot seedlings and cultivating the soil at the same time.

Coffee grows on trees that reach a height of up to 20 feet (6 meters), but growers keep them pruned to about 6 feet (2 meters) to simplify picking and encourage heavy berry production. The first visible sign of a coffee tree's maturity is the appearance of small white blossoms, which fill the air with a heady aroma reminiscent of jasmine and orange. The mature tree bears cherry-size oval berries, each containing two coffee beans with their flat sides together. A mature coffee tree will produce one pound (450 grams) of coffee per growing season. It takes 2,000 hand-picked Arabica coffee berries (4,000 beans) to make a pound of roasted coffee.

* Coins

Test tire tread Let old Abe Lincoln's head tell you if it's time to replace the tires on your car. Insert an American penny into the tread. If you can't cover the top of Honest Abe's head inside the tread, it's time to head for the tire store. Check tires regularly and you will avoid the danger and inconvenience of a flat tire on a busy road.

Give carpet a lift When you move a chair, sofa, table, or bed, you will notice the deep indentations in your carpet made by the legs. To fluff it up again, simply hold a coin on its edge and scrape it against the flattened pile. If it still doesn't pop back up, hold a steam iron a couple of inches (5 centimeters) above the affected spot. When the area is damp, try fluffing again with the coin.

Keep cut flowers fresh Your posies and other cut flowers will stay fresh longer if you add a copper penny and a cube of sugar to the vase water.

Instant measure If you need to measure something but you don't have a ruler, just reach into your pocket and pull out a quarter. It measures exactly 1 inch (2.54 centimeters) in diameter. Just line up quarters to measure the length of a small object.

Make a noisemaker Drop a few coins into an empty aluminum soda can, seal the top with duct tape, and head for the stadium to root for your favorite team. Take your noisemaker with you when you walk the dog and use it as a training aid. When the pooch is naughty, just shake the noisemaker.

Decorate a barrette Use shiny pennies to decorate a barrette for a little girl. Gather enough pennies to complete the project (5 pennies for each large barrette; fewer for smaller barrettes). Arrange pennies as you like on the barrette and use hot-melt glue to attach them. Allow 24 hours to dry.

Hang doors perfectly Next time you hang an entry door, nickel-and-dime it to ensure proper clearance between the outside of the door and the inside of the frame. When the door is closed, the gap at the top should be the thickness of a nickel, and the gap at the sides should be that of a dime. If you do it right, you will keep the door from binding and it won't let in drafts.

SCIENCE FAIR

Scientists use **optical illusions** to show how the brain can be tricked. This simple experiment uses **two coins,** but you'll think you are seeing three. Hold two coins on top of each other between your **thumb and index finger.** Quickly slide the coins back and forth and you will **see a third coin!**

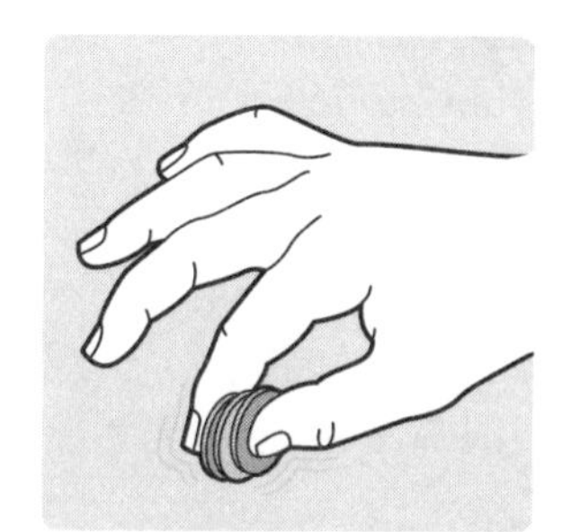

How it works: Scientists say that everything we see is actually light reflected from objects. Our eyes use the light to create images on our retinas, the light-sensitive linings in our eyeballs. Because images don't disappear instantly, when something moves quickly you may see both an object and an after-image of it at the same time.

Make a paperweight If you have ever traveled abroad, you have probably come home with a few odd-looking coins from foreign lands. Instead of leaving them lying around in a desk drawer, use them to make an interesting paperweight. Just put the coins into a small glass jar with a closable lid and cover the lid with decorative cloth or paper.

Coins were first produced around 700 B.C. by Lydians, a people who lived in what is now Turkey. From there, they spread to ancient Greece and Rome. However, worldwide use of coins (and paper money) took centuries to occur. Even in early America, bartering remained the popular way to exchange goods and services. On April 2, 1792, after the ratification of the U.S. Constitution, Congress passed the Mint Act. This established the coinage system in the United States and the dollar as the official U.S. currency. The first U.S. coins were produced by the Philadelphia Mint in 1793.

Colanders

Prevent grease splatters Sick of cleaning grease splatters on the stovetop after cooking your famous burgers? Prevent them by inverting a large metal colander over the frying pan. The holes will let heat escape but the colander will trap the splatters. Be careful! The metal colander will be hot—use an oven mitt to remove it.

Heat a pasta bowl Does your pasta get too cold too fast? To keep it warm longer, heat the bowl first. Place a colander in the bowl, pour the pasta and water into the colander, and let the hot water stand in the bowl for a few seconds to heat it. Pour out the water, add the pasta and sauce, and you're ready to serve.

Keep berries and grapes fresh Do your berries and grapes get moldy before you've had a chance to enjoy their sweet taste? To keep them fresher longer, store them in a colander, not a closed plastic container, in the refrigerator. The cold air will circulate through the holes in the colander, keeping them fresh for days.

Corral bathtub toys Don't let the bathtub look like another messy toy box. After each bath, corral your child's small playthings in a large colander and store it in the tub. The water will drain from the toys, keeping them ready for next time, and the bathtub will stay tidy.

Use as sand toy Forget spending money on expensive sand toys for your budding archeologist. A simple inexpensive plastic colander is perfect for digging at the beach or in the sandbox.

Cold Cream

Erase temporary tattoos Kids love temporary tattoos, but getting them off can be a painful, scrubbing chore. To make removal easier, loosen the tattoo by rubbing cold cream on it and then gently rub it off with a facecloth. Voilà!

Remove bumper stickers Does your bumper sticker still say "Honk for Gore"? Erase the past by rubbing cold cream on the sticker and letting it soak in. Once it does, you should be able to peel it right off.

Make face paint Need a safe, easy recipe for Halloween face paint? Mix 1 teaspoon cornstarch, 1/2 teaspoon water, 1/2 teaspoon cold cream, and 2 drops food coloring (depending on the costume) together. Use a small paintbrush to paint designs on your face or your child's. Remove it with soap and water.

Yes, cold cream really is cold. That's because it's made with a lot of water that evaporates and cools your warm face. Cold cream is the granddaddy of all facial ointments. It was invented by the Greek physician Galen in A.D. 157. It's not known exactly why Galen created his mixture, but according to Michael Boylan, a professor at Marymount College in Arlington, Virginia, and an expert on Galen, ancient medicine was based on treating with opposites. And Galen might have been seeking a cure for a hot and dry skin condition like eczema or psoriasis. But the women of ancient Greece soon discovered the soothing white cream was great for taking off makeup, which remains the primary use for cold cream to this day.

Compact Discs

Use as holiday ornaments Decorate your Christmas tree in style! Hang CDs shiny side out to create a flickering array of lights—or paint and decorate the label side to create inexpensive personalized ornaments. For variety, cut the CDs into stars and other shapes with sharp scissors. Drill a 1/4-inch (6-millimeter) hole through the CD and thread ribbon through to hang.

Make wall art in teen's room Old CDs make inexpensive and quirky wall art for your teenager's room. Attach the CDs with thumbtacks and use them to create a border at the ceiling or halfway up the wall. Or let your teen use them to frame his or her favorite posters.

Catch candle drips You should always use a candleholder specifically designed to catch melting wax. However, if one is not available, a CD is great in a pinch. Make sure it's a short candle that can stand on its own with a flat bottom. It should also be slightly larger than the CD hole. Place the candleholder on a stable, heat-resistant surface and keep a watchful eye on it.

Make artistic bowls Looking for a funky, decorative bowl? Place a CD in the oven on low heat over a metal bowl until the CD is soft. Wearing protective gloves, gently bend the CD into the shape desired. Seal the hole by gluing the bottom edge to another surface such as a flat dish using expoxy or PVC glue. Don't use the bowl for food.

Use as sidewalk/driveway reflectors Forget those ugly orange reflectors. Instead, drill small holes in a CD and screw it onto your mailbox post or onto a wood stake and push it into the ground. Install several of them to light a nighttime path to your front door.

Kids' Stuff Use an **old CD** to make a **picture frame** for someone you love. You need a CD, a **picture** of you that is larger than the CD hole, a **large bead, ribbon,** and **glue.** Glue the picture in the middle of the CD on the shiny side. If you wish, decorate the CD with **markers** or **stickers.** Use **hot-melt glue** to attach the bead at the top of the CD, let dry, and thread ribbon through the bead.

Use as template for perfect circle Need to draw a perfect circle? Forget tracing around cups or using cumbersome compasses. Every CD provides two circle sizes—trace around the inner hole or the outer circumference.

Tip **CD Repair**

Before throwing away or recycling a scratched-up CD, try to repair it. First, clean it thoroughly with a lint-free cloth or mild soap and a little bit of water. Hold the CD by the edge to keep from getting fingerprints on it. Polish it from the middle to the edge, not in a circular motion. If your CD still skips, try fixing it with a little non-gel toothpaste. Dab some toothpaste on the end of your finger and rub it lightly onto the entire CD. Use a damp paper towel to remove the toothpaste and dry it with a fresh paper towel. The fine abrasive in the toothpaste might smooth out the scratch. You might also want to use car wax on a scratch (see page 115).

Make a decorative sun catcher Sun catchers are attractive to watch, and all you need to make one is a couple of CDs. Glue two CDs together, shiny side out, wrap yarn or colored string through the hole, and hang them in a window. The prism will make a beautiful light show.

Create a spinning top Turn an old CD into a fun toy for the kids (and adults too!). With a knife, make two slits across from each other in the CD hole. Force a penny halfway through the hole, and then spin the CD on its edge.

Make a CD clock Old CDs can be functional! Turn a disc into a funky clock face for clockwork sold by arts-and-crafts stores. Paint and design one side of the CD and let it dry. Write or use stickers to create the numbers around its edge. Assemble the clockwork onto the CD.

Cover with felt and use as coasters CDs can help to prevent those unsightly stains from cups left on the table. Simply cut a round piece of felt to fit over the CD and glue it onto the label side of the CD so that the shiny side will face up when you use the coaster.

* Cooking Spray

Prevent rice and pasta from sticking Most cooks know that a little cooking oil in the boiling water will keep rice or pasta from sticking together when you drain it. If you run out of cooking oil, however, a spritz of cooking oil spray will do the job just as well.

Grating cheese Put less elbow grease into grating cheese by using a nonstick cooking spray on your cheese grater for smoother grating. The spray also makes for easier and faster cleanup.

Prevent tomato sauce stains Sick of those hard-to-clean tomato sauce stains on your plastic containers? To prevent them, apply a light coating of nonstick cooking spray on the inside of the container before you pour in the tomato sauce.

Keep car wheels clean. You know that fine black stuff that collects on the wheels of your car and is so hard to clean off? That's brake dust—it's produced every time you apply your brakes and the pads wear against the brake disks or cylinders. The

Ever wondered where PAM, the name of the popular cooking spray, comes from? It stands for "Product of Arthur Meyerhoff." The first patent for a nonstick cooking spray was issued in 1957 to Arthur and his partner, Leon Rubin, who began marketing PAM All Natural Cooking Spray in 1959. After appearing on local Chicago TV cooking shows in the early '60s, the product developed a loyal following, and it quickly became a household world. By the way, PAM is pretty durable stuff—it has a shelf life of two years.

next time you invest the elbow grease to get your wheels shiny, give them a light coating of cooking spray. The brake dust will wipe right off.

De-bug your car When those bugs smash into your car at 55 miles (88 kilometers), per hour, they really stick. Give your grille a spritz of nonstick cooking spray so you can just wipe away the insect debris.

Lubricate your bicycle chain Is your bike chain a bit creaky and you don't have any lubricating oil handy? Give it a shot of nonstick cooking spray instead. Don't use too much—the chain shouldn't look wet. Wipe off the excess with a clean rag.

Cure door squeak Heard that door squeak just one time too many? Hit the hinge with some nonstick cooking spray. Have paper towels handy to wipe up the drips.

Remove paint and grease Forget smelly solvents to remove paint and grease from your hands. Instead, use cooking spray to do the job. Work it in well and rinse. Wash again with soap and water.

Dry nail polish Need your nail polish to dry in a hurry? Spray it with a coat of cooking spray and let dry. The spray is also a great moisturizer for your hands.

Quick casting Pack a can of cooking spray when you go fishing. Spray it on your fishing line and the line will cast easier and farther.

Prevent grass from sticking Mowing the lawn should be easy, but cleaning stuck grass from the mower is tedious. Prevent grass from sticking on mower blades and the underside of the housing by spraying them with cooking oil before you begin mowing.

Prevent snow sticks Shoveling snow is hard enough, but it can be more aggravating when the snow sticks to the shovel. Spray the shovel with nonstick cooking spray before shoveling—the snow slides right off! If you use a snow thrower, spray inside the discharge chute to prevent it from clogging.

* Corks

Create a fishing bobber It's an idea that's as old as Tom Sawyer, but worth remembering: A cork makes a great substitute fishing bobber. Drive a staple into the top of the cork, then pull the staple out just a bit so you can slide your fishing line through it.

Make an impromptu pincushion Need a painless place to store pins while you sew? Save corks from wine bottles—they make great pincushions!

C

Prevent pottery scratches Your beautiful pottery can make ugly scratches on furniture. To save your tabletops, cut thin slices of cork and glue them to the bottom of your ceramic objects.

Replace soda bottle caps Lost the cap to your soda bottle and need a replacement? Cork it! Most wine corks fit most soda bottles perfectly.

Make a pour spout Don't have one of those fancy metal pour spouts to control the flow from your oil or vinegar bottle? You don't need one. Make your own spout by cutting out a wedge of the cork along its length. Use a utility or craft knife. Stick the cork in the bottle and pour away. When you're through, cover the hole with a tab of masking tape.

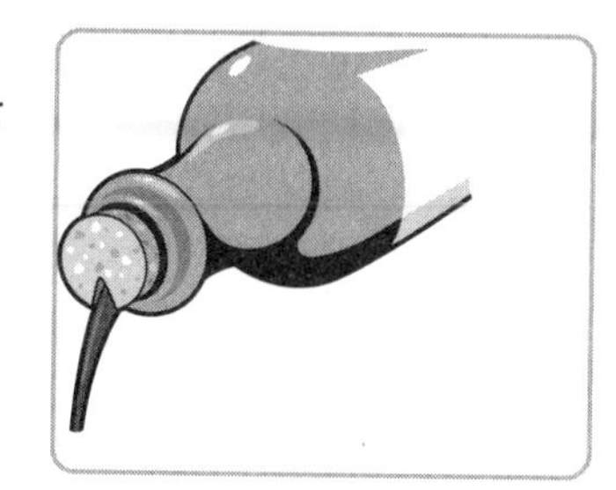

Use as Halloween face paint Kids love to dress up as a hobo for Halloween. To create that scruffy look, char the end of a piece of cork by holding it over a candle. Let it cool a little, then rub it on the kid's face.

Block sun glare In the olden days of football and baseball, players would burn cork and rub it under their eyes to reduce glare from the sun and stadium lights. These days, ballplayers use commercial products to do the same, but you can still use cork to get the job done.

Prevent chair scratches The sound of a chair scraping across your beautiful floor can make your skin crawl. Solve the problem by cutting cork into thin slices and attaching them to the bottom of the chair legs with a spot of wood glue.

Create craft stamps You can use cork to create a personalized stamp. Carve the end of a cork into any shape or design you want. Use it with ink from a stamp pad to decorate note cards. Or let the kids dip carved corks in paint to create artwork.

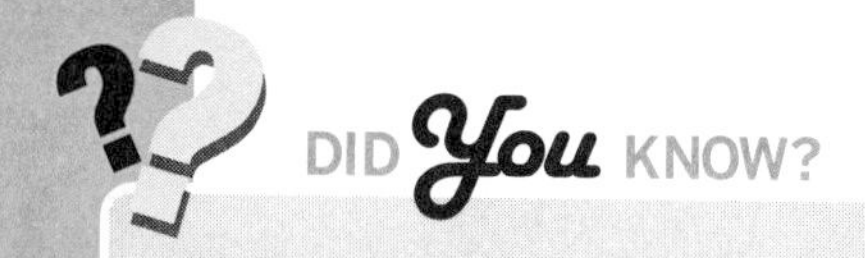

Cork, the bark of the cork oak, has been used to seal wine bottles and other vessels for more than 400 years. The bark has a unique honeycomb cell structure—each cell is sealed, filled with air, and not connected to any other cell. This makes it waterproof and a poor conductor of heat and vibration. Plus cork contains suberin, a natural waxy substance that makes it impermeable to liquids and gases and prevents the cork from rotting. No wonder it's still the material of choice for sealing up your favorite cabernet or bubbly.

Create a cool bead curtain. Want a creative, stylish beaded curtain for a child's or teen's room? Drill a hole through corks and string them onto a cord along with beads and other decorations. Make as many strings as you need and tie them onto a curtain rod.

Fasten earrings Earring backs always get lost, and you can't always find a perfect-sized stand-in when you need it. Instead, use a snippet of cork as a temporary substitute. Slice a small piece about the size of the backing and push it on. An eraser cut off the end of a pencil will also work.

TAKE CARE Should you use a corkscrew to open a bottle of wine? Yes, but don't use it on a bottle of champagne! Pushing a corkscrew down into a bottle of champagne against the pressure of the carbonation can actually make the bottle explode! If possible, wait a day before opening and let the carbonation settle a bit. Wrap the cork in a towel and twist the bottle, not the cork, slowly.

To open wine, use a traditional corkscrew that twists into the stopper. Peel the top of the plastic to expose the cork. Insert the corkscrew in the center. Twist it straight down and pull the cork straight out with even pressure.

Picture-perfect frames If you're always straightening picture frames on the wall, cut some small flat pieces of cork—all the same thickness—and glue them to the back of the frame. The cork will grip the wall and stop the sliding. It will also prevent the frame from marring the wall.

Mass-produce sowing holes Here's a neat trick for quickly getting your seeds sown in straight rows of evenly spaced holes. Mark out the spacing you need on a board. Drill drywall screws through the holes, using screws that will protrude about 3/4 inch (2 centimeters) through the board. Now twist wine corks onto the screws. Just press the board, corks down, into your garden bed, and voilà—instant seed holes.

* Cornstarch

Dry shampoo Fido needs a bath, but you just don't have time. Rub cornstarch into his coat and brush it out. The dry bath will fluff up his coat until it's tub time.

Untangle knots Knots in string or shoelaces can be stubborn to undo, but the solution is easy. Sprinkle the knot with a little cornstarch. It will then be easy to work the segments apart.

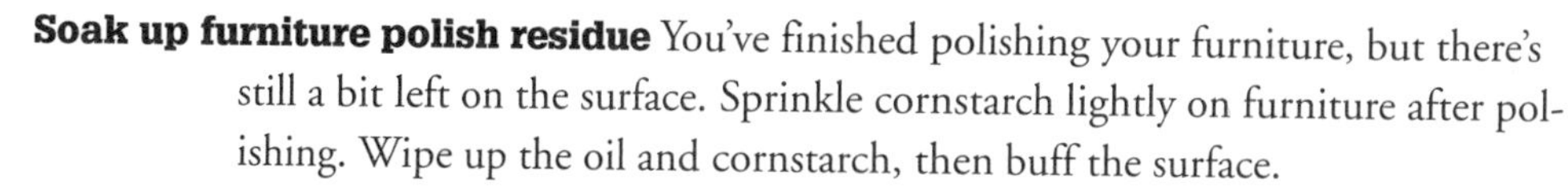

Soak up furniture polish residue You've finished polishing your furniture, but there's still a bit left on the surface. Sprinkle cornstarch lightly on furniture after polishing. Wipe up the oil and cornstarch, then buff the surface.

Remove ink stains from carpet Oh no, ink on the carpet! In this case a little spilt milk might save you from crying. Mix the milk with cornstarch to make a paste. Apply the paste to the ink stain. Allow the concoction to dry on the carpet for a few hours, then brush off the dried residue and vacuum it up.

Give carpets a fresh scent Before vacuuming a room, sprinkle a little cornstarch on your carpeting. Wait about half an hour and then vacuum normally.

Make your own paste The next time the kids want to go wild with construction paper and paste, save money by making the paste yourself. Mix 3 teaspoons cornstarch for every 4 teaspoons cold water. Stir until you reach a paste consistency. This is especially great for applying with fingers or a wooden tongue depressor or Popsicle stick. If you add food coloring, the paste can be used for painting objects.

Make finger paints This simple recipe will keep the kids happy for hours. Mix together 1/4 cup cornstarch and 2 cups cold water. Bring to a boil and continue boiling until the mixture becomes thick. Pour your product into several small containers and add food coloring to each container. You've created a collection of homemade finger paints.

Clean stuffed animals To clean a stuffed animal toy, rub a little cornstarch onto the toy, wait about 5 minutes, and then brush it clean. Or place the stuffed animal (or a few small ones) into a bag. Sprinkle cornstarch into the bag, close it tightly, and shake. Now brush the pretend pets clean.

Separate marshmallows Ever buy a bag of marshmallows only to find them stuck together? Here's how to get them apart: Add at least 1 teaspoon cornstarch to

Cornstarch has been made into biodegradable packing "peanuts" sold in bulk. If you receive an item shipped in this material, you can toss the peanuts on the lawn. They'll dissolve with water, leaving no toxic waste. To test if the peanuts are made from cornstarch, wet one in the sink to see if it dissolves.

the bag and shake. The cornstarch will absorb the extra moisture and force most of the marshmallows apart. Repackage the remaining marshmallows in a container and freeze them to avoid sticking in future.

Lift a scorch mark from clothing You moved the iron a little too slowly and now you have a scorch mark on your favorite shirt. Wet the scorched area and cover it with cornstarch. Let the cornstarch dry, then brush it away along with the scorch mark.

Remove grease spatters from walls Even the most careful cook cannot avoid an occasional spatter. A busy kitchen takes some wear and tear but here's a handy remedy for that unsightly grease spot. Sprinkle cornstarch onto a soft cloth. Rub the grease spot gently until it disappears.

Get rid of bloodstains The quicker you act, the better. Whether it's on clothing or table linens, you can remove or reduce a bloodstain with this method. Make a paste of cornstarch mixed with cold water. Cover the spot with the cornstarch paste and rub it gently into the fabric. Now put the cloth in a sunny location to dry. Once dry, brush off the remaining residue. If the stain is not completely gone, repeat the process.

Polish silver Is the sparkle gone from your good silverware? Make a simple paste by mixing cornstarch with water. Use a damp cloth to apply this to your silverware. Let it dry, then rub it off with cheesecloth or another soft cloth to reveal that old shine.

Make windows sparkle Create your own streak-free window cleaning solution by mixing 2 tablespoons cornstarch with 1/2 cup ammonia and 1/2 cup white vinegar in a bucket containing 3-4 quarts (3-4 liters) warm water. Don't be put off by the milky concoction you create. Mix well and put the solution in a trigger spray bottle. Spray on the windows, then wipe with a warm-water rinse. Now rub with a dry paper towel or lint-free cloth. Voilà!

Say good riddance to roaches There's no delicate way to manage this problem. Make a mixture that is 50 percent plaster of Paris and 50 percent cornstarch. Spread this in the crevices where roaches appear. It's a killer recipe.

* Correction Fluid

Cover scratches on appliances Daub small nicks on household appliances with correction fluid. Once it dries, cover your repair with clear nail polish for staying power. This works well on white china, too, but only for display. Now that correction fluid comes in a rainbow of colors, its uses go beyond white. You may easily find a match for your beige or yellow household stove or refrigerator.

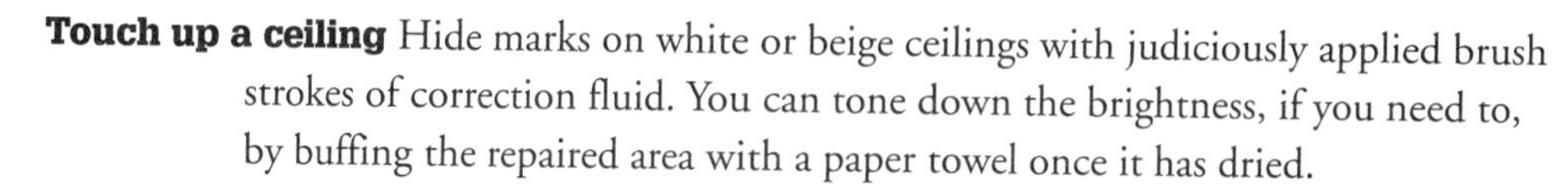

Touch up a ceiling Hide marks on white or beige ceilings with judiciously applied brush strokes of correction fluid. You can tone down the brightness, if you need to, by buffing the repaired area with a paper towel once it has dried.

Erase scuffs Need a quick fix for scuffed white shoes? Correction fluid will camouflage the offensive marks. On leather, buff gently once the fluid dries. No need to buff on patent leather.

Paint the town Decorate your windows for any occasion. Paint snowflakes, flowers, or Welcome Home signs using correction fluid. Later you can remove your art with nail polish remover, an ammonia solution, vinegar and water, or a commercial window cleaner. Or you can scrape it off with a single-edged razor blade in a holder made for removing paint from glass.

DID *You* KNOW?

Correction fluid was invented in 1951 by Bette Nesmith Graham, mother of Michael Nesmith of the Monkees musical group. Graham was working as an executive secretary in Texas. She used water-based paint and began supplying little bottles of it to other secretaries, calling it Mistake Out. Five years later, she improved the formula and changed the name to Liquid Paper. Despite its proven use, Graham was turned down when she tried to sell it to IBM, so she marketed it on her own. In the 1960s her invention began to generate a tidy profit; by 1979, when she sold the product to the Gillette Corp, she received $47.5 million plus a royalty on every bottle sold until 2000. Today, with the ease of correcting documents on a computer, correction fluid is no longer the office essential it once was.

* Cotton Balls

Scent the room Saturate a cotton ball with your favorite cologne and drop it into your vacuum cleaner bag. Now, as you vacuum, the scent will be expressed and gently permeate the room.

Deodorize the refrigerator Sometimes the refrigerator just doesn't smell fresh. Dampen a cotton ball with vanilla extract and place it on a shelf. You'll find it acts as a deodorizer, offering its own pleasant scent.

Fight mildew There are always hard-to-reach spots in the bathroom, usually around the fixtures, where mildew may breed in the grout between tiles. Forget about becoming a contortionist to return the sparkle to those areas. Soak a few cotton balls in bleach and place them in those difficult spots. Leave them to work their magic for a few hours. When you remove them, you'll find your job has been done. Finish by rinsing with a warm-water wash.

Protect little fingers Pad the ends of drawer runners with a cotton ball. This will prevent the drawer from closing completely and keep children from catching their fingers as the drawer slides shut.

Rescue your rubber gloves If your long, manicured nails sometimes puncture the fingertips of your rubber dishwashing gloves, here's a solution you'll appreciate. Push a cotton ball into the fingers of your gloves. The soft barrier should prolong the gloves' life.

* Crayons

Use as a floor filler Crayons make great fill material for small gouges or holes in resilient flooring. Get out your crayon box and select a color that most closely matches the floor. Melt the crayon in the microwave over wax paper on medium power, a minute at a time until you have a pliant glob of color. Now, with a plastic knife or putty knife, fill the hole. Smooth it over with a rolling pin, a book, or some other flat object. You'll find the crayon cools down quickly. Now wax the floor, to provide a clear protective coating over your new fill.

Fill furniture scratches Do your pets sometimes treat your furniture like … well, a scratching post? Don't despair. Use a crayon to cover scratches on wooden furniture. Choose the color most like the wood finish. Soften the crayon with a hair dryer or in the microwave on the defrost setting. Color over the scratches, then buff your repair job with a clean rag to restore the luster.

Carpet cover-up Even the most careful among us manage to stain the carpet. If you've tried to remove a stain and nothing works, here's a remedy you might be able

Jazberry Jam and Mango Tango. Those aren't ice-cream flavors, they are recently introduced Crayola crayon colors. When Edwin Binney and C. Harold Smith introduced the first crayons safe for children to use in 1903, a box of eight sold for a nickel and included colors with more pedestrian names: black, brown, blue, red, violet, orange, yellow, and green. Since then, the manufacturer, Binney and Smith, has introduced more than 400 colors, retiring many along the way. Currently there are 120 colors available. Inch Worm and Wild Blue Yonder are other recent introductions.

to live with. Find a crayon that matches or will blend with your carpet. Soften the crayon a bit with a hair dryer or in the microwave on the defrost setting. Now color over the spot. Cover your repair with wax paper and gently iron the color in. Keep the iron on a low setting. Repeat as often as necessary.

Colorful decoration Here's a fun project to do with the kids. Make a multicolored sun catcher by shaving crayons onto a 4- or 5-inch (10- or 12-centimeter) sheet of wax paper. Use a potato peeler or grater for this task. Place another sheet of wax paper over the top and press with a hot iron until the shavings melt together. Poke a hole near the top through the layers of wax and crayon while still warm. Once your ornament cools, peel away the papers and thread a ribbon through it to hang in a window.

* Cream of Tartar

Tub scrubber Let this simple solution of cream of tartar and hydrogen peroxide do the hard work of removing a bathtub stain for you. Fill a small, shallow cup or dish with cream of tartar and add hydrogen peroxide drop by drop until you have a thick paste. Apply to the stain and let it dry. When you remove the dried paste, you'll find that the stain is gone too.

Brighten cookware Discolored aluminum pots will sparkle again if you clean them with a mixture of 2 tablespoons cream of tartar dissolved into 1 quart (1 liter) water. Bring the mixture to a boil inside the pot and boil for 10 minutes.

Make play clay for kids Here's a recipe for fun dough that's like the famous commercial stuff: Add together 2 tablespoons cream of tartar, 1 cup salt, 4 cups plain flour (without rising agents), and 1-2 tablespoons cooking oil. Stir well with a wooden spoon to mix together, then slowly stir while adding 4 cups water. Cook the mixture in a saucepan over a medium flame, stirring occasionally until it thickens. It's ready when it forms a ball that is not sticky. Work in food coloring, if you want. Let it cool, then let the kids get creative. It dries out quicker than the commercial variety, so store it in an airtight container in the fridge.

* Curtain Rings

Get hooked On a camping trip or a hike, when you don't want to carry a backpack, it's easy to lash a few items to your belt loop with the help of a curtain ring. Mountain climbers rely on expensive carabiners, which they use to hold items and to control ropes. But you don't need to carry along anything so heavy. Attach your sneakers to your sleeping bag with a metal curtain ring; your gloves and canteen can dangle from a metal shower curtain ring or a brass key ring.

Keep curiosity at bay It's a natural stage of development, but not always one you want to encourage. Curious toddlers can't help poking around in your kitchen cupboards. If you've got a toddler visiting, lock up your accessible cupboards by clicking shower curtain rings over the latches. Then when baby leaves, it's easy to remove the rings.

Hold your hammer Sometimes you need three hands when you're doing household repair jobs. Attach a sturdy metal shower curtain ring to your belt and slip your hammer through it. Now you can climb a ladder or otherwise work with both hands and just grab the hammer when needed.

Store nuts and washers Keep nuts and washers on metal shower curtain rings hung from a hook in your workshop. The ring's pear shape and latching action ensure secure storage. Put nuts and washers of similar size on their own rings so that you can find the right size quickly.

Keep track of kids' mittens "Where are my mittens, Ma?" "Where did you leave them?" "I dunno." Something as simple as a curtain ring can help you do away with this dialogue: Drive a nail in the mudroom wall. Hand Junior a curtain ring and tell him to use it to clip his mittens together and hang them on the nail.

Dental Floss

Remove a stuck ring Here's a simple way to slip off a ring that's stuck on your finger. Wrap the length of your finger from the ring to the nail tightly with dental floss. Now you can slide the ring off over the floss "carpet."

Lift cookies off baking tray Ever fought with a freshly baked cookie that wouldn't come off the pan? Crumbled cookies may taste just as good as those in one piece, but they sure don't look as nice on the serving plate. Use dental floss to easily remove cookies from the baking tray. Hold a length of dental floss taut and slide it neatly between the cookie bottom and the pan.

Slice cake and cheese Use dental floss to cut cakes, especially delicate and sticky ones that tend to adhere to a knife. Just hold a length of the floss taut over the cake and then slice away, moving it slightly side to side as you cut through the cake. You can also use dental floss to cut small blocks of cheese cleanly.

Repair outdoor gear Because dental floss is strong and resilient but slender, it's the ideal replacement for thread when you are repairing an umbrella, tent, or backpack. These items take a beating and sometimes get pinhole nicks. Sew up the small holes with floss. To fix larger gouges, sew back and forth over the holes until you have covered the space with a floss patch.

Extra-strong string for hanging things Considering how thin it is, dental floss is strong stuff. Use it instead of string or wire to securely hang pictures, sun catchers, or wind chimes. Use it with a needle to thread together papers you want to attach or those you want to display, in clothesline fashion.

Secure a button permanently Did that button fall off again? This time, sew it back on with dental floss—it's much stronger than thread, which makes it perfect for reinstalling buttons on coats, jackets, and heavy shirts.

Separate photos Sometimes photos get stuck to each other and it seems the only way to separate them is to ruin them. Try working a length of dental floss between the pictures to gently pry them apart.

* Denture Tablets

Re-ignite your diamond's sparkle Has your diamond ring lost its sparkle? Drop a denture tablet into a glass containing a cup of water. Follow that with your ring or diamond earrings. Let it sit for a few minutes. Remove your jewelry and rinse to reveal the old sparkle and shine.

Vanish mineral deposits on glass Fresh flowers often leave a ring on your glass vases that seems impossible to remove no matter how hard you scrub. Here's the answer. Fill the vase with water and drop in a denture tablet. When the fizzing has stopped, all of the mineral deposits will be gone. Use the same method to clean thermos bottles, cruets, glasses, and coffee decanters.

Clean a coffeemaker Hard water leaves mineral deposits in the tank of your electric drip coffeemaker that not only slows the perking but also affects the taste of your brew. Denture tablets will fizz away these deposits and give the tank a bacterial clean-out too. The tablets were designed to clean and disinfect dentures, and they'll do the same job on your coffeemaker. Drop two denture tablets in the tank and fill it with water. Run the coffeemaker. Discard that potful of water and follow up with one or two rinse cycles with clean water.

Clean your toilet Looking for a way to make the toilet sparkle again? Porcelain fixtures respond to the cleaning agent in denture tablets. Here's a solution that does the job in the twinkling of an eye. Drop a denture tablet in the bowl. Wait about 20 minutes and flush. That's it!

Clean enamel cookware Stains on enamel cookware are a natural for the denture tablet cleaning solution. Fill the pot or pan with warm water and drop in a tablet or two, depending on its size. Wait a bit—once the fizzing has stopped, your cookware will be clean.

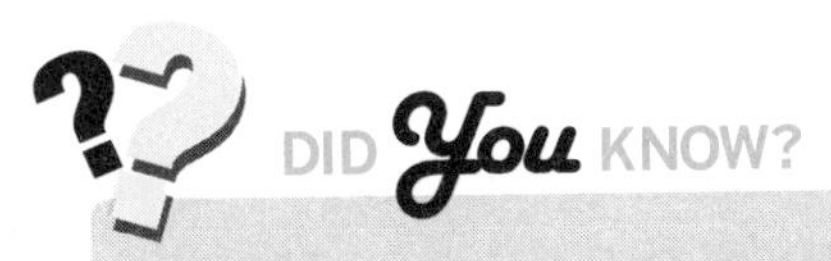

Bleaching agents are a common component of denture cleaner tablets, providing the chemical action that helps the tablets to remove plaque and to whiten and bleach away stains. This is what makes them surprisingly useful for cleaning toilets, coffeemakers, jewelry, and enamel cookware, among other things.

Unclog a drain Slow drain got you down? Reach for the denture tablets. Drop a couple of tablets into the drain and run water until the problem clears. For a more stubborn clog, drop 3 tablets down the sink, follow that with 1 cup white vinegar, and wait a few minutes. Now run hot water in the drain until the clog is gone.

* Disposable Diapers

Make a heating pad Soothe your aching neck. Or, for that matter, your aching back or shoulder. Use a disposable diaper's high level of absorbency to your advantage by creating a soft, pliant heating pad. Moisten a disposable diaper and place it in the microwave on medium-high setting for about 2 minutes. Check that it's not too hot for comfort and then apply to your achy part.

Keep a plant watered longer Before potting a plant, place a clean disposable diaper in the bottom of the flowerpot—absorbent side up. It will absorb water that would otherwise drain out the bottom and will keep the plant from drying out too fast. You'll also cut back on how often you have to water the plant.

Pad a package You want to mail your friend that lovely piece of china you know she'll love. But you don't have any protective wrapping on hand. If you have disposable diapers, wrap the item in the diapers or insert them as padding before sealing the box. Diapers cost more than regular protective packaging wrap, but at least you will have gotten the package out today, and you can be assured your gift will arrive in one piece.

It took a mother to invent disposable diapers. Looking for an alternative to messy cloth diapers, Marion Donovan first created a plastic covering for diapers. She made her prototype from a shower curtain and later parachute fabric. Manufacturers weren't interested, but when she created her own company and debuted the product in 1949 at Saks Fifth Avenue in New York City, it was an instant success. Donovan soon added disposable absorbent material to create the first disposable diaper and, in 1951, sold her company for $1 million.

Duct Tape

DUCT TAPE AROUND THE HOUSE

Temporarily hem your pants You've found a terrific pair of jeans, but the length isn't right. You expect a little shrinkage anyway, so why spend time hemming? Besides, thick denim jeans are difficult to sew through. Fake the hem with duct tape. The new hem will last through a few washes too.

Remove lint on clothing You're all set to go out for the night and suddenly you notice pet hairs on your outfit. Quick, grab the duct tape and in no time, you'll be ready to go. Wrap your hand with a length of duct tape, sticky side out. Then roll the sticky tape against your clothing in a rocking motion until every last hair has been picked up. Don't wipe, since that may affect the nap.

Make a bandage in a pinch You've gotten a bad scrape. Here's how to protect it until you get a proper bandage. Fold tissue paper or paper towel to cover the wound and cover this with duct tape. It may not be attractive, but it works in a jam.

Reseal bags of chips Tired of stale potato chips? To keep a half-finished bag fresh, fold up the top and seal it tight with a piece of duct tape.

Pocket folder protector Old pocket folders may lose their resiliency but are otherwise useful. Cover your old folder with duct tape; reinforce between sections and it's as good as new.

Bumper sticker Got something you want to say? Make your own bumper sticker. Cut a length of duct tape, affix it to your bumper and with a sharp marker, pen your message.

Keep a secret car key You'll never get locked out of your car again if you affix an extra key to the undercarriage with duct tape.

Catch pesky flies You've just checked into a rustic cabin on the lake and you're ready to start your vacation. Everything would be perfect if only the flying insects were not part of the deal. Grab your roll of duct tape and roll off a few foot-long strips. Hang them from the rafters as flypaper. Soon you'll be rid of the bugs and you can roll up the tape to toss it in the trash.

Replace a shower curtain grommet How many times have you yanked the shower curtain aside only to rip through one of the delicate eyelets? Grab the duct tape to make a simple repair. Once the curtain is dry, cut a rectangular piece and fold it from front to back over the torn hole. Slit the tape with a mat knife, razor blade, or scissors, and push the shower curtain ring back in place.

Repair a vacuum hose Has your vacuum hose cracked and developed a leak? It doesn't spell the end of your vacuum. Repair the broken hose with duct tape. Your vacuum will last until the motor gives out.

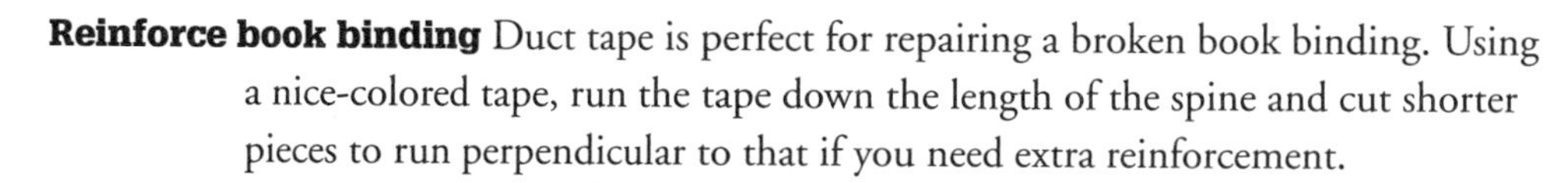

Reinforce book binding Duct tape is perfect for repairing a broken book binding. Using a nice-colored tape, run the tape down the length of the spine and cut shorter pieces to run perpendicular to that if you need extra reinforcement.

Cover a book Use duct tape in an interesting color to create a durable book cover for a school textbook or a paperback that you carry to the beach. Make a pattern for the cover on a sheet of newspaper; fit the pattern to your book, then cover the pattern, one row at a time, with duct tape, overlapping the rows. The resulting removable cover will be waterproof and sturdy.

Repair a photo frame Many people enjoy displaying family photos in easel-type frames on mantels and side tables throughout the house. But sometimes the foldout leg that holds a frame upright pulls away from the back of the frame and your photo won't stand up properly. Don't despair! Just use duct tape to reattach the broken leg to the frame back.

Hang Christmas lights Festive holiday lights are fun in season, but a real chore when it's time for them to come down. Use duct tape to hang your lights and the removal job will be much easier. Tear duct tape into thin strips. At intervals, wrap strips around the wire and then tape the strand to the gutter or wherever you hang your lights.

Wrap holiday presents Here's a novel way to wrap a special gift. Don't bother with the paper. Go straight for the tape. Press duct tape directly on the gift box. Make designs or cover in stripes and then add decorative touches by cutting shapes, letters, and motifs from tape to attach to the "wrapped" surface.

Duct tape really did start as duck tape. During World War II, the U.S. military needed a flexible, durable, waterproof tape. They called on Permacell, a division of Johnson & Johnson, which used its own medical tape as a base and added a strong polycoat adhesive and a polyethylene coating laminated to a cloth backing. The resulting strong flexible tape—colored army green and easy to rip into useful strips—was used for everything from sealing ammunition cases to repairing jeep windshields. GI's nicknamed it duck tape because it was waterproof, like a duck's back.

After the war, the tape—in a new silvery color—found use joining heating and air conditioning ductwork and became known as duct tape. It's rarely used to join ducts anymore, but it lives on—now in a rainbow of colors—as the handyman's best friend.

Make Halloween costumes Want to be the Tin Man for Halloween? How about a robot? These are just two ideas that work naturally with the classic silver duct tape. Make a basic costume from brown paper grocery bags, with openings in the back so the child can easily put on and take off the costume. Cover this pattern with rows of duct tape. For the legs, cover over an old pair of pants, again giving your little robot or Tin Man an easy way to remove the outfit for bathroom breaks. Duct tape comes in an array of colors, so let your imagination lead your creativity.

Make a toy sword Got a couple of would-be swashbucklers around the house? Make toy swords for the junior Errol Flynns by sketching a kid-size sword on a piece of cardboard. Use two pieces if you haven't got one thick enough. Be sure to make a handle the child's hand can fit around comfortably once it's been increased in thickness by several layers of duct tape. Wrap the entire blade shape in silver duct tape. Wrap the handle in black tape.

Make play rings and bracelets Make rings by tearing duct tape into strips about 1/2-inch (1.2-centimeter) wide, then folding the strips in half lengthwise—sticky sides together. Continue to put more strips over the first one until the ring is thick enough to stand on its own. You can adjust the size with a scissors and tape the ends closed. To make a stone for the ring, cover a small item such as a pebble and attach it to the ring. Make a bracelet by winding duct tape around a stiff paper pattern.

Make hand puppets Duct tape is great for puppet making. Use a small paper lunch bag as the base for the body of your puppet. Cover the bag with overlapping rows of duct tape. Make armholes through which your fingers will poke out. Create a head from a tape-covered ball of wadded paper and affix buttons or beads for eyes and mouth.

Make bicycle streamers Add snazzy streamers to your kids' handlebars. Make them using duct tape in various colors. Cut the tape into strips about 1/2-inch (1.2-centimeter) wide by 10 inches (25 centimeters) long. Fold each strip in half, sticky sides together. Once you have about half a dozen for each side, stick them into the end of the handlebar and secure them with wraps of duct tape. Be sure your child will still have a good grip on the handlebar.

✱ DUCT TAPE FOR THE DO-IT-YOURSELFER

Repair a taillight Someone just backed into your car and smashed the taillight! Here's a quick repair that will last until you have time to get to the repair shop.

Depending on where the cracks lie, use yellow or red duct tape to hold the remaining parts together. In some states this repair will even pass inspection.

Short-term auto hose fix Until you can get to your mechanic, duct tape makes a strong and dependable temporary fix for broken water hoses on your automobile. But don't wait too long. Duct tape can only withstand temperatures up to 200°F (93°C). Also, don't use it to repair a leak in your car's gas line—the gasoline dissolves the adhesive.

Make a temporary roof shingle If you've lost a wooden roof shingle, make a temporary replacement by wrapping duct tape in strips across a piece of 1/4-inch (6-millimeter) plywood you've cut to size. Wedge the makeshift shingle in place to fill the space. It will close the gap and repel water until you can repair the roof.

Fix a hole in your siding Stormy weather damaged your vinyl siding? A broken tree limb tossed by the storm, hailstones, or even an errant baseball can rip your siding. Patch tears in vinyl siding with duct tape. Choose tape in a color that matches your siding and apply it when the surface is dry. Smooth your repair by hand or with a rolling pin. The patch should last at least a season or two.

Replace lawn chair webbing Summertime is here, and you go to the shed to fetch your lawn furniture, only to discover the webbing on your favorite backyard chair has worn through. Don't throw it out. Colorful duct tape makes a great, sturdy replacement webbing. Cut strips twice as long as you need. Double the tape, putting sticky sides together, so that you have backing facing out on both sides. Then screw it in place with the screws on the chair.

Tape a broken window Before removing broken window glass, crisscross the broken pane with duct tape to hold it all together. This will ensure a shard doesn't fall out and cut you.

DID *You* KNOW?

As most Canadians know, the star of *The Red Green Show* on the Canadian Broadcasting Corporation—and the Public Broadcasting System in the United States—has been known to use duct tape for everything from fixing a spare tire to re-webbing a lawn chair. Red's real-life persona, Steve Smith, admits he doesn't use "the handyman's secret weapon" as much as his screen character. "I live in a pretty nice neighborhood, where duct tape is discouraged as a renovation tool," he says. Nevertheless, when he had to prevent his front door from locking, he put a small strip of duct tape across the bolt. He points out that this was the first time he'd used duct tape "to *stop* something from working."

Repair outdoor cushions Don't let a little rip in the cushions for your outdoor furniture bother you. Repair the tear with a closely matched duct tape and it will hold up for several seasons.

Repair a trash can Plastic trash cans often split or crack along the sides. But don't toss out the can with the trash. Repair the tear with duct tape. It's strong enough to withstand the abuses a trash can takes, and easy to manipulate on the curved or ridged surface of your can. Put tape over the crack both outside and inside the can.

Quick fix for a toilet seat You're giving a party and someone taps you on the shoulder to tell you the toilet seat has broken. You don't have to make a mad dash to the home center. Grab the duct tape and carefully wrap the break for a neat repair. Your guests will thank you.

Mend a screen Have the bugs found the tear in your window or door screen? Thwart their entrance until you make a permanent fix by covering the hole with duct tape.

* DUCT TAPE FOR SPORTS AND OUTDOOR GEAR

Tighten shin guards Hockey players need a little extra protection. Use duct tape to attach shin guards firmly in place. Put on all your equipment, including socks. Now split the duct tape to the width appropriate for your size—children might need narrower strips than adults—and start wrapping around your shin guard to keep it tight to your leg.

Add life to a hockey stick Street hockey sticks take a beating. If yours is showing its age, breathe a little more life into it by wrapping the bottom of the stick with duct tape. Replace the tape as often as needed.

Extend the life of skateboard shoes Kids who perform fantastic feats on their skateboards find their shoes wear out very quickly because a lot of the jumps involve sliding the toe or side of the foot along the board. They wear holes in new shoes fast. Protect their feet and prolong the life of their shoes by putting a layer or two of duct tape on the area that scrapes along the board.

Repair your ski gloves Ski glove seams tearing open? Duct tape is the perfect solution to ripped ski gloves because it's waterproof, incredibly adhesive, strong, and can easily be torn into strips of any width. Make your repair lengthwise or around the fingers and set out on the slopes again.

Repair a tent You open your tent at the campsite and oops—a little tear. No problem as long as you've brought your duct tape along. Cover the hole with a patch; for double protection mirror the patch inside the tent. You'll keep insects and weather where they belong.

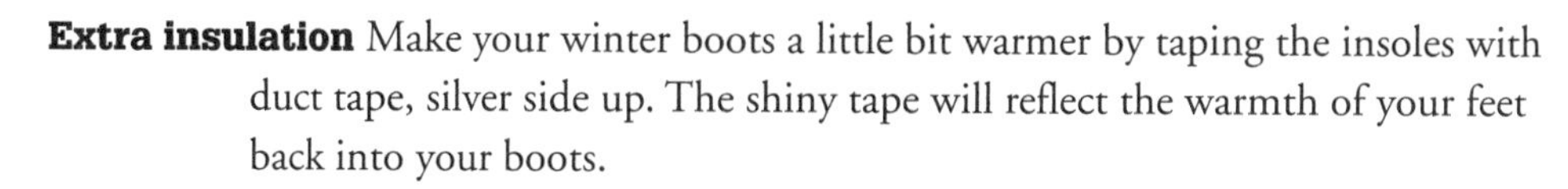

Extra insulation Make your winter boots a little bit warmer by taping the insoles with duct tape, silver side up. The shiny tape will reflect the warmth of your feet back into your boots.

Stay afloat You're out for a paddle, when you discover a small hole in your canoe. Thank goodness you thought to pack duct tape in your supply kit. Pull the canoe out of the water, dry the area around the hole, and apply a duct tape patch to the outside of the canoe. You're ready to finish your trip.

Waterproof footwear Need a waterproof pair of shoes for fishing, gardening, or pushing off the canoe into the lake? Cover an old pair of sneakers with duct tape, overlapping the edges of each row. As you round corners, cut little **V**'s in the edges of the tape so that you can lap the tape smoothly around the corner.

Pool patch Duct tape will repair a hole in your swimming pool liner well enough to stand up to water for at least a season. Be sure to cover the area thoroughly.

Protect yourself from ticks When you're out on a hike, on your way to your favorite fishing hole, or just weeding in the yard, protect your ankles from those pesky ticks. Wrap duct tape around your pant cuffs to seal out the bugs. This is a handy way to keep your pant leg out of your bicycle chain too!

Create a clothesline Whether you're out in the wilderness on a camping trip or in your own backyard, when you need a clothesline and you're without rope, think: duct tape. Twist a long piece of duct tape into a rope and bind it between trees for a clothesline. It makes a dandy jump rope as well or a basic rope sturdy enough to lash two items together. You can even use your creation to drag a child's wagon.

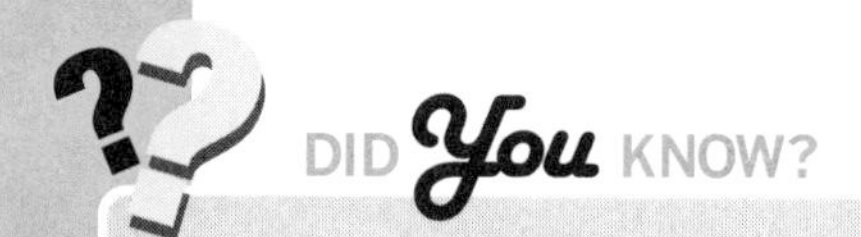

The folks at 3M's product information lines handle a lot of calls about duct tape. Three of the most commonly asked questions are:

1. Can duct tape be used for removing warts?
2. Can it be used to secure the duct from the household dryer to the outdoors?
3. Is it waterproof?

The official answers:

1. Duct tape is not recommended for removing warts, because it hasn't been scientifically tested.
2. The company does not recommend using duct tape for the dryer duct, because the temperatures may exceed 200°F (93°C), the maximum temperature duct tape can withstand.
3. The backing of the duct tape is waterproof, but the adhesive is not. Duct tape will hold up to water for a while, but eventually the adhesive will give out.

Protect your gas grill hose For some reason, mice and squirrels love to chew on rubber, and one of their favorite snacks is often the rubber hose that connects the propane tank to your gas grill. Protect the hose by wrapping it in duct tape.

Make an emergency sneaker lace You're enjoying a game of driveway hoops when you bust a sneaker lace. Ask for a brief time-out while you grab the duct tape from the garage. Cut off a piece of tape that's as long as you need and rip off twice the width you need. Fold the tape in half along its length, sticky side in. Thread your new lace onto your sneaker, tie it up, and you are ready for your next jump shot.

Repair your ski pants Oh no, you ripped your ski pants and the wind is whipping into the nylon outer layer. No need to pay inflated lodge shop prices for a new pair if you have a roll of duct tape in the car. Just slip a piece of tape inside the rip, sticky side out, and carefully press both sides of the rip together. The repair will be barely detectable.

* Dustpans

Decorate your door for fall Gather dried fall foliage, such as Indian corn, bittersweet branches with orange berries, and other decorative greens. Tie them together as a bouquet with a rubber band or tape. Spread them out in a fan shape and cover the binding with a ribbon. Now set this against a copper dustpan. Use super glue or a glue gun to attach your bouquet to the pan. Hang this homage to fall on your front door.

Enlist the littlest shoveler Youngsters enjoy mimicking their elders. While you shovel snow, let the little one help by your side using a dustpan as a shovel.

Use as a sand toy Pack a clean dustpan with your beach toys. It's a great sand scoop and will really help the castle builders in their task.

Speed toy cleanup Picking up all those little toys gets tiresome. Scoop them up with a dustpan and deposit them in the toy bin. It's a real time-saver, not to mention a back-saver.

E

Earrings

Use as a bulletin board tack Lend a little personal style to your bulletin board. Use mateless pierced earrings to tack up pictures, notes, souvenirs, and clippings.

Create a brooch Got a batch of mateless pierced earrings collecting dust in a box? Use wire cutters to snip off the stems and get creative: Arrange the earrings on a swatch of cardboard or foam core, and secure them with hot-melt glue. Add a pin backing and, voilà! a new brooch. Or use the same method to jazz up a plain picture frame.

Make a magnet Give your fridge some glitz. Use wire cutters to cut the stem off an orphan earring and glue it to a magnet. What a great way to emphasize how pleased you are with that perfect report card when you stick it on the refrigerator.

Clip your scarf Did you lose one of your very favorite earrings? Oh well, at least you can still work the survivor into an ensemble by using it to secure a scarf. Just tie the scarf as desired, then clip or pierce it with the earring.

Make an instant button. Oh, darn! You're dressed to go out and you discover a button missing. No need to re-invent your whole outfit. Just dip into your collection of clip-on earrings. Clip the earring on the button side of the clothing to create

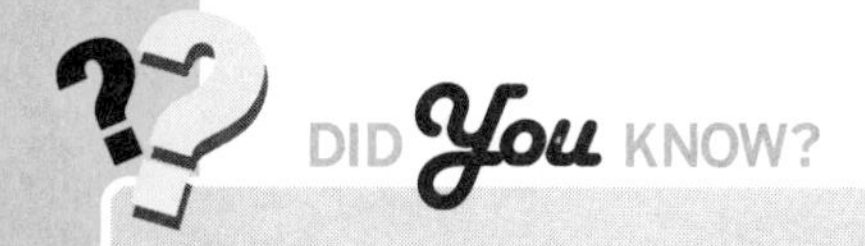

DID You KNOW?

People have been wearing—and probably losing—earrings for nearly 5,000 years. According to historians who have studied jewelry, the tiny baubles were likely introduced in western Asia in about 3000 B.C. The oldest earrings that have been discovered date to 2500 B.C. and were found in Iraq. The popularity of earrings over time has grown or receded, depending on hairstyles and clothing trends. The clip-on earring was introduced in the 1930s, and by the 1950s, fashionable women simply did not pierce their ears. But twenty years later, pierced ears were back.

a new "button," then button as usual with the buttonhole. If you have time, move the top button on that favorite blouse to replace the lost one and then use the earring at the top of the blouse.

Decorate your Christmas tree Scatter clip-on earrings around the boughs of your Christmas tree as an eye-catching accent to your larger tree decorations. Or use them as the main adornment on a small tree or wreath.

* Eggs

Make a facial Who has time or money to spend at the local day spa, paying someone to tell you how awful your skin looks? For a little pampering, head to the refrigerator and grab an egg. If you have dry skin that needs moisturizing, separate the egg and beat the yolk. Oily skin takes the egg white, to which a bit of lemon or honey can be added. For normal skin, use the entire egg. Apply the beaten egg, relax and wait 30 minutes, then rinse. You'll love your new fresh face.

Use as glue Out of regular white glue? Egg whites can act as a glue substitute when gluing paper or light cardboard together.

Add to compost Eggshells are a great addition to your compost because they are rich in calcium—a nutrient that helps plants. Crushing them before you put them in your compost heap will help them break down faster.

Water your plants After boiling eggs, don't pour the water down the drain. Instead, let it cool; then water plants with the nutrient-filled water.

Start seeds Plant seeds in eggshells. Place the eggshell halves in the carton, fill each with soil, and press seeds inside. The seeds will draw extra nutrients from the eggshells. Once the seedlings are about 3 inches (7.5 centimeters) tall, they are ready to be transplanted into your garden. Remove them from the shell before you put them in the ground. Then crush the eggshells and put them in your compost or plant them in your garden.

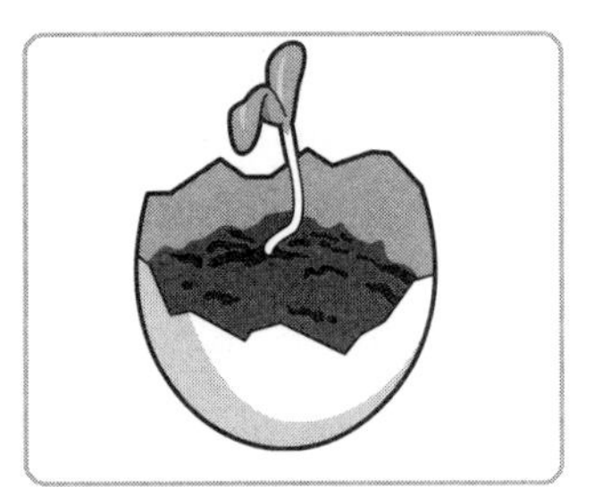

* Egg Cartons

Use for storing and organizing With a dozen handy compartments, egg cartons are a natural for storing and organizing small items. Here are some ideas to get you going. You're sure to come up with more of your own.

- Instead of emptying the coins in your pocket into a jar for later sorting, cut off a four-section piece of an egg carton and leave it on your dresser. Sort your quarters, dimes, nickels, and pennies as you pull them out of your

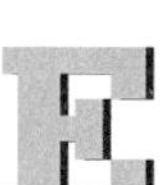

pockets. (Dump pennies in a larger container, such as a jar, or put them in a piggy bank.)

- Organize buttons, safety pins, threads, bobbins, and fasteners on your sewing table.
- Organize washers, tacks, small nuts and bolts, and screws on your workbench. Or use to keep disassembled parts in sequence.
- Keep small Christmas ornaments from being crushed in handy, stackable egg cartons.

Start a fire Fill a cardboard egg carton with briquettes (and a bit of leftover candle wax if it's handy), place in your barbecue grill, and light. Egg cartons can also be filled with tinder, such as small bits of wood and paper, and used as a fire starter in a fireplace or a woodstove.

Start seedlings An egg carton can become the perfect nursery for your seeds. Use a cardboard egg carton, not a polystyrene one. Fill each cell in the carton with soil and plant a few seeds in each one. Once the seeds have sprouted, divide the carton into individual cells and plant, cardboard cells and all.

Make ice Making a bunch of ice for a picnic or party? Use the bottom halves of clean polystyrene egg cartons as auxiliary ice trays.

Reinforce a trash bag Yuck! You pull the plastic trash bag out of the kitchen trash container and gunk drips out. Next time, put an opened empty egg carton at the bottom of the trash bag to prevent tears and punctures.

Create shippable homemade goodies Here's a great way to brighten the day of a soldier, student, or any faraway friend or loved one. Cover an egg carton with bright wrapping paper. Line the individual cells with candy wrappers or shredded coconut. Nestle homemade treats inside each. Include the carton in your next care package or birthday gift, and rest assured the treats will arrive intact.

Golf ball caddy An egg carton in your golf bag is a great way to keep golf balls clean and ready for teeing off.

* Emery Boards

Sand deep crevices If you are refinishing an elaborate piece of wood such as turned table legs or chair spindles, you can use emery boards to gently smooth those hard-to-reach crevices before applying stain or finish. These filelike nail sanders are easy to handle and provide a choice of two sanding grits.

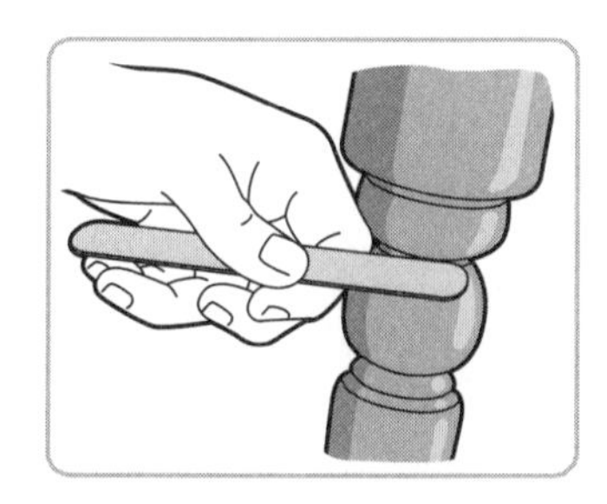

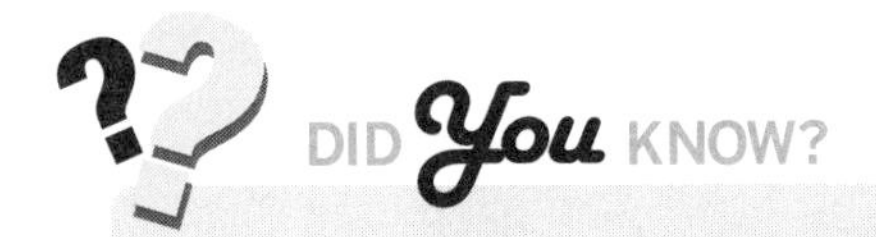

Ever wonder about emery boards, those little sticks that we tuck into our drawers and can never seem to find when we need to file down a torn nail? Emery is a natural mixture of corundum and magnetite. Diamond is the only mineral harder than corundum. Sapphires and rubies are also varieties of corundum. Makes it easy to understand why a manicurist's magazine would urge women to "treat their nails like jewels, not tools." Emery boards have changed a lot since 1910, when they were introduced. They now come with bright designs, give off scents, or are shaped as hearts and stars.

Remove dirt from an eraser Do you have a fussy student who doesn't like dirt on the end of the pencil? Take an emery board and rub lightly over the eraser until the dirt is filed off.

Prep seeds for planting Use an emery board to remove the hard coating on seeds before you plant them. This will speed sprouting and help them absorb moisture.

Save your suede Did somebody step on your blue suede shoes? Or worse, spill some wine on them? Don't check into Heartbreak Hotel. Rub the stain lightly with an emery board, and then hold the shoe over steam from a teakettle or pan to remove the stain. This works for suede clothing too.

* Envelopes

Shred old receipts faster The best way to get rid of receipts that may have your credit card number or other personal information is to shred them. But feeding tiny receipts into a shredder is tedious. Instead, place all the old receipts into a few old envelopes and shred the envelopes.

Make a small funnel You save money by buying your spices in bulk and you want to transfer them to smaller, handier bottles for use in the kitchen, but you don't have a small funnel to do the job. Make a couple of disposable funnels from an envelope. Seal the envelope, cut it in half diagonally, and snip off one corner on each half. Now you have two funnels for pouring spices into your smaller jars.

Sort and store sandpaper You know how sheets of sandpaper love to curl themselves up into useless tubes? Prevent that problem and keep your sandpaper sheets organized by storing them in standard letter-size cardboard mailing envelopes. Use one envelope for each grit and write the grit on the envelopes.

Make bookmarks Recycle envelopes by making them into handy bookmarks of different sizes. Cut off the gummed flap and one end of the envelope. Then slip the remainder over the corner of the page where you stopped reading for a quick placeholder that doesn't damage your book. Give a batch to the kids to decorate for their own set or to give as a homemade gift.

Make file folders Don't let papers get disorganized just because you ran out of file folders. Cut the short ends off a light cardboard mailing envelope. Turn it inside out so you have a blank cardboard on the outside. Cut a 3/4-inch (2-centimeter) wide strip lengthwise off the top of one side. The other edge becomes the place where you label your file.

* Epsom Salt

Get rid of raccoons Are the masked night marauders poking around your trash can, creating a mess and raising a din? A few tablespoons of Epsom salt spread around your garbage cans will deter the raccoons, who don't like the taste of the stuff. Don't forget to reapply after it rains.

Deter slugs Are you tired of visiting your yard at night only to find the place crawling with slimy slugs? Sprinkle Epsom salt where they glide and say good-bye to the slugs.

Fertilize tomatoes and other plants Want those Big Boys to be big? Add Epsom salt as a foolproof fertilizer. Every week, for every foot of height of your tomato plant, add one tablespoon. Your tomatoes will be the envy of the neighborhood. Epsom salt is also a good fertilizer for houseplants, roses and other flowers, and trees.

Make your grass greener How green is your valley? Not green enough, you say? Epsom salt, which adds needed magnesium and iron to your soil, may be the answer. Add 2 tablespoons to 1 gallon (3.7 liters) of water. Spread on your lawn and then water it with plain water to make sure it soaks into the grass.

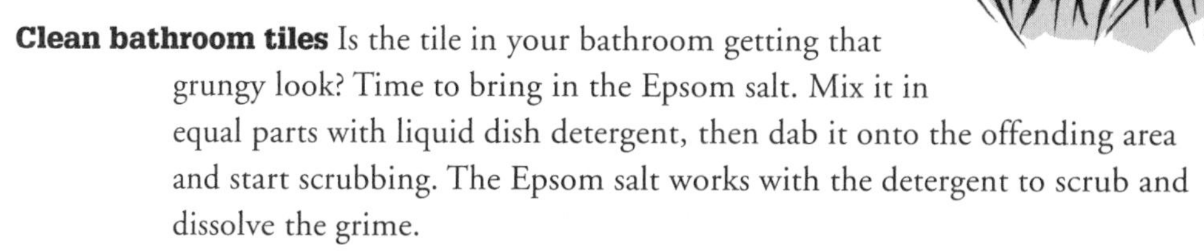

Clean bathroom tiles Is the tile in your bathroom getting that grungy look? Time to bring in the Epsom salt. Mix it in equal parts with liquid dish detergent, then dab it onto the offending area and start scrubbing. The Epsom salt works with the detergent to scrub and dissolve the grime.

Regenerate a car battery Is your car battery starting to sound as if it won't turn over? Worried that you'll be stuck the next time you try to start your car? Give your battery a little more life with this potion. Dissolve about an ounce of Epsom salt in warm water and add it to each battery cell.

Get rid of blackheads Here's a surefire way to dislodge blackheads: Mix 1 teaspoon Epsom salt and 3 drops iodine in 1/2 cup boiling water. When the mixture cools enough to stick your finger in it, apply it to the blackhead with a cotton ball. Repeat this three or four times, reheating the solution if necessary. Gently remove the blackhead and then dab the area with an alcohol-based astringent.

Frost your windows for Christmas If you are dreaming of a white Christmas, but the weather won't cooperate, at least you can make your windows look frosty. Mix Epsom salt with stale beer until the salt stops dissolving. Apply the mixture to your windows with a sponge—for a realistic look, sweep the sponge in an arc at the bottom corners. When the mixture dries, the windows will look frosted.

Kids' Stuff Here are two fun winter-inspired projects using Epsom salt for the holiday season:

Make snowflakes by folding a piece of **blue paper** several times and snipping shapes into the resulting square of paper. Unfold your snowflake. Brush one side with a thick mixture of **water** and **Epsom salt**. After it dries, turn it over and brush the other side. When it's finished, you'll have a **frosty-looking snowflake** you can hang in your window.

To make a **snowy scene**, use crayons to draw a picture on **construction paper**. Mix equal parts of **Epsom salt** and **boiling water**. Let it cool; then use a wide artist's paintbrush to paint the picture. When it dries, "snow" crystals will appear.

* Fabric Softener

End clinging dust on your TV Are you frustrated to see dust fly back onto your television screen, or other plastic surfaces, right after cleaning them? To eliminate the static cling that attracts dust, simply dampen your dust cloth with fabric softener straight from the bottle and dust as usual.

Remove old wallpaper Removing old wallpaper is a snap with fabric softener. Just stir 1 capful liquid softener into 1 quart (1 liter) water and sponge the solution onto the wallpaper. Let it soak in for 20 minutes, then scrape the paper from the wall. If the wallpaper has a water-resistant coating, score it with a wire-bristle brush before treating with the fabric softener solution.

Abolish carpet shock To eliminate static shock when you walk across your carpet, spray the carpet with a fabric softener solution. Dilute 1 cup softener with 2 1/2 quarts (2.5 liters) water; fill a spray bottle and lightly spritz the carpet. Take care not to saturate it and damage the carpet backing. Spray in the evening and let the carpet dry overnight before walking on it. The effect should last for several weeks.

Remove hair-spray residue Dried-on overspray from hair spray can be tough to remove from walls and vanities, but even a buildup of residue is no match for a solu-

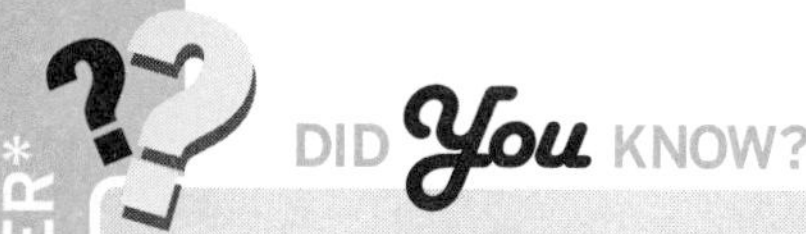

How does fabric softener reduce cling as well as soften clothes? The secret is in the electrical charges. Positively charged chemical lubricants in the fabric softener are attracted to your load of negatively charged clothes, softening the fabric. The softened fabrics create less friction, and less static, as they rub against each other in the dryer, and because fabric softener attracts moisture, the slightly damp surface of the fabrics makes them electrical conductors. As a result, the electrical charges travel through them instead of staying on the surface to cause static cling and sparks as you pull the clothing from the dryer.

tion of 1 part liquid fabric softener to 2 parts water. Stir to blend, pour into a spray bottle, spritz the surface, and polish it with a dry cloth.

Clean now, not later Clean glass tables, shower doors, and other hard surfaces, and repel dust with liquid fabric softener. Mix 1 part softener into 4 parts water and store in a squirt bottle, such as an empty dishwashing liquid bottle. Apply a little solution to a clean cloth, wipe the surface, and then polish with a dry cloth.

Float away baked-on grime Forget scrubbing. Instead, soak burned-on foods from casseroles with liquid fabric softener. Fill the casserole with water, add a squirt of liquid fabric softener, and soak for an hour, or until residue wipes easily away.

Keep paintbrushes pliable After using a paintbrush, clean the bristles thoroughly and rinse them in a coffee can full of water with a drop of liquid fabric softener mixed in. After rinsing, wipe the bristles dry and store the brush as usual.

Untangle and condition hair Liquid fabric softener diluted in water and applied after shampooing can untangle and condition fine, flyaway hair, as well as curly, coarse hair. Experiment with the amount of conditioner to match it to the texture of your hair, using a weaker solution for fine hair and a stronger solution for coarse, curly hair. Comb through your hair and rinse.

Remove hard-water stains Hard-water stains on windows can be difficult to remove. To speed up the process, dab full-strength liquid fabric softener onto the stains and let it soak for 10 minutes. Then wipe the softener and stain off the glass with a damp cloth and rinse.

Make your own fabric softener sheets Fabric softener sheets are convenient to use, but they're no bargain when compared to the price of liquid softeners. You can make your own dryer sheets and save money. Just moisten an old washcloth with 1 teaspoon liquid softener and toss it into the dryer with your next load.

* Fabric Softener Sheets

Pick up pet hair Pet hair can get a pretty tenacious grip on furniture and clothing. But a used fabric softener sheet will suck that fur right off the fabric with a couple of swipes. Just toss the fuzzy wipe into the trash.

End car odors Has that new-car smell gradually turned into that old-car stench? Tuck a new dryer fabric softener sheet under each car seat to counteract musty odors and cigarette smells.

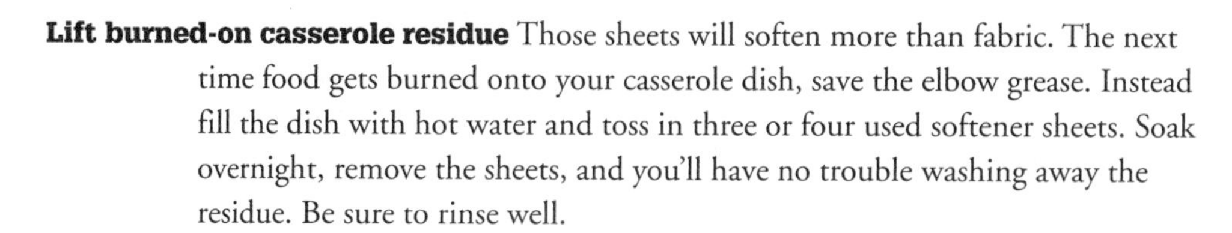

Lift burned-on casserole residue Those sheets will soften more than fabric. The next time food gets burned onto your casserole dish, save the elbow grease. Instead fill the dish with hot water and toss in three or four used softener sheets. Soak overnight, remove the sheets, and you'll have no trouble washing away the residue. Be sure to rinse well.

Freshen drawers There's no need to buy scented drawer-liner paper; give your dresser drawers a fresh-air fragrance by tucking a new dryer fabric softener sheet under existing drawer liners, or tape one to the back of each drawer.

Wipe soap scum from shower door Tired of scrubbing scummy shower doors? It's easy to wipe the soap scum away with a used dryer fabric softener sheet.

TAKE CARE People with allergies or chemical sensitivities may develop rashes or skin irritations when they come into contact with laundry treated with some commercial fabric softeners or fabric softener sheets. If you are sensitive to softeners, you can still soften your laundry by substituting 1/4 cup white vinegar or the same amount of your favorite hair conditioner to your washer's last rinse cycle for softer, fresher-smelling washables.

Repel dust from electrical appliances Because television and PC screens are electrically charged, they actually attract dust, making dusting them a never-ending chore, but not if you dust them with used dryer softener sheets. These sheets are designed to reduce static cling, so they remove the dust, and keep it from resettling for several days or more.

Do away with doggy odor If your best friend comes in from the rain and smells like a ... well ... wet dog, wipe him down with a used dryer softener sheet, and he'll smell as fresh as a daisy.

Freshen laundry hampers and wastebaskets There's still plenty of life left in used dryer fabric softener sheets. Toss one into the bottom of a laundry hamper or wastebasket to counteract odors.

Tame locker-room and sneaker smells Deodorizing sneakers and gym bags calls for strong stuff. Tuck a new dryer fabric softener sheet into each sneaker and leave overnight to neutralize odors (just remember to pull them out before wearing the sneaks). Drop a dryer sheet into the bottom of a gym bag and leave it there until your nose lets you know it's time to renew it.

Prevent musty odors in suitcases Place a single, unused dryer fabric softener sheet into an empty suitcase or other piece of luggage before storing. The bag will smell great the next time you use it.

Buff chrome to a brilliant shine After chrome is cleaned, it can still look streaky and dull, but whether it's your toaster or your hubcaps, you can easily buff up the shine with a used dryer softener sheet.

Use as a safe mosquito repellent For a safe mosquito repellent, look no farther than your laundry room. Save used dryer fabric softener sheets and pin or tie one to your clothing when you go outdoors to help repel mosquitoes.

Use an inconspicuous air freshener Don't spend hard-earned money on those plug-in air fresheners. Just tuck a few sheets of dryer fabric softener into closets, behind curtains, and under chairs.

Do away with static cling You'll never be embarrassed by static cling again if you keep a used fabric softener sheet in your purse or dresser drawer. When faced with static, dampen the sheet and rub it over your pantyhose to put an end to clinging skirts.

Keep dust off blinds Cleaning venetian blinds is a tedious chore, so make the results last by wiping them down with a used dryer fabric softener sheet to repel dust. Wipe them with another sheet whenever the effect wears off.

Renew grubby stuffed toys Wash fake-fur stuffed animals in the washing machine set on gentle cycle, then put the stuffed animals into the clothes dryer along with a pair of old tennis shoes and a fabric softener sheet, and they will come out fluffy and with silky-soft fur.

Substitute a dryer sheet for a tack cloth Sticky tack cloths are designed to pick up all traces of sawdust on a woodworking project before you paint or varnish it, but they are expensive and not always easy to find at the hardware store. If you find yourself in the middle of a project without a tack cloth, substitute an unused dryer fabric softener sheet; it will attract sawdust and hold it like a magnet.

Consolidate sheets and make them smell pretty To improve sheet storage, store the sheet set in one of the matching pillowcases, and tuck a new dryer fabric softener sheet into the packet for a fresh fragrance.

Abolish tangled sewing thread To put an end to tangled thread, keep an unused dryer fabric softener sheet in your sewing kit. After threading the needle, insert it into the sheet and pull all of the thread through to give it a nonstick coating.

* Film Canisters

Rattle toy for the cat Cats are amused by small objects that rattle and shake, and they really don't care what they look like. To provide endless entertainment for your cat, drop a few dried beans, a spoonful of dry rice, or other small objects that can't harm a cat, into an empty film canister, snap on the lid, and watch the fun begin.

Handy stamp dispenser To keep a roll of stamps from being damaged, make a stamp dispenser from an empty film canister. Hold the canister steady by taping it to a counter with duct tape, and use a utility knife to carefully cut a slit into the side of the canister. Drop the roll of stamps in, feed it out through the slit, snap the cap on, and it's ready to use.

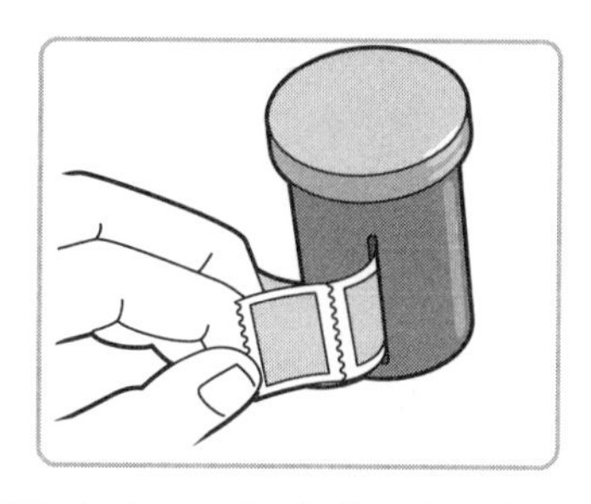

Use as hair rollers You can collect all the hair rollers you'll ever need if you save your empty plastic film canisters. To use, pop the top off, roll damp hair around the canister, and hold it in place by fastening a hair clip over the open end of the canister and your hair.

Emergency sewing kit You'll never be at a loss if you pop a button or your hem unravels if you fill an empty film canister with buttons, pins, and a pre-threaded needle. Make several; tuck one into each travel bag, purse, or gym bag, and hit the road.

On-the-road pill dispensers Use empty film canisters as travel-size pill bottles for your purse or overnight bag. If you take more than one medication, use a separate canister for each. Write the medication and dosage on a peel-and-stick label and attach to each canister. For at-a-glance identification, color the labels with different-colored highlighter pens.

Tip The Vanishing Film Canister

Plastic film canisters have myriad uses, from emergency ashtrays to spice bottles. But with the rise of digital cameras, these small wonders are rapidly going the way of the rotary dial phone or the phonograph needle. A good source for free film canisters has always been the neighborhood one-hour photo shop. But these days you may find that even they have a canister shortage. If so, check the yellow pages for a professional film developer, because most high-quality, professional photographers still use film—and film canisters.

Store fishing flies You can save a lot of money and grief by storing fishing flies and hooks in film canisters. They don't take up much room in a fishing vest, and if you do drop one in a stream, the airtight lid will keep it floating long enough for you to … well … fish it out.

Carry spices for camp cooking Just because you are roughing it, doesn't mean that you have to eat bland food. You can store a multitude of seasonings in individual film canisters to take along when you go camping, and you'll still have plenty of room for the food itself in your backpack or car trunk. It's a good idea for your RV or vacation cabin too.

Carry small change for laundry and tolls Film canisters are just the right size to hold quarters and smaller change. Tuck a canister of change into your laundry bag or your car's glove compartment, and you'll never have to hunt for change when you're at a self-service laundry or a tollbooth.

Bring your own diet aids If you are on a special diet, you can easily and discreetly transport your favorite salad dressings, artificial sweetener, or other condiments to restaurants in plastic film canisters. Clean, empty canisters hold single-sized servings, have snap-on, leakproof lids, and are small enough to tuck into a purse.

Keep jewelry close at hand An empty film canister doesn't take up much room in your gym bag, and it'll come in handy for keeping your rings and earrings from being misplaced while you work out.

Emergency nail polish remover Create a small, spillproof carry case for nail polish remover by tucking a small piece of sponge into a plastic film canister. Saturate the sponge with polish remover and snap on the lid. For an emergency repair, simply insert a finger and rub the nail against the fluid-soaked sponge to remove the polish.

Flour

Repel ants with flour Sprinkle a line of flour along the backs of pantry shelves and wherever you see ants entering the house. Repelled by the flour, ants won't cross over the line.

Freshen playing cards After a few games, cards can accumulate a patina of snack residue and hand oil, but you can restore them with some all-purpose flour in a paper bag. Drop the cards into the bag with enough flour to cover, shake vigorously, and remove the cards. The flour will absorb the oils, and it can be easily knocked off the cards by giving them a vigorous shuffle.

DID *You* KNOW?

Ever wondered why the word *flour* is pronounced exactly like the word *flower?* Well, you may be surprised to learn that *flour* is actually derived from the French word for flower, which is *fleur*. The French use the word to describe the most desirable, or floury (flowery) and protein-rich, part of a grain after processing removes the hull. And, because much of our food terminology comes from the French, we still bake and make sauces with the flower of grains, such as wheat, which we call flour.

F

Safe paste for children's crafts Look no farther than your kitchen canister for an inexpensive, nontoxic paste that is ideal for children's paper craft projects, such as papier-mâché and scrap-booking. To make the paste, add 3 cups cold water to a saucepan and blend in 1 cup all-purpose flour. Stirring constantly, bring the mixture to a boil. Reduce heat and simmer, stirring until smooth and thick. Cool and pour into a plastic squeeze bottle to use. This simple paste will keep for weeks in the refrigerator, and cleans up easily with soap and water.

Make modeling clay Keep the kids busy on a rainy day with modeling clay—they can even help you make the stuff. Knead together 3 cups all-purpose flour, 1/4 cup salt, 1 cup water, 1 tablespoon vegetable oil, and 1 or 2 drops food coloring. If the mixture is sticky, add more flour; if it's too stiff, add more water. When the "clay" is a workable consistency, store it until needed in a self-sealing plastic bag.

Polish brass and copper No need to go out and buy cleaner for your brass and silver. You can whip up your own at much less cost. Just combine equal parts of flour, salt, and vinegar, and mix into a paste. Spread the paste onto the metal, let it dry, and buff it off with a clean, dry cloth.

Bring back luster to a dull sink To buff your stainless steel sink back to a warm glow, sprinkle flour over it and rub lightly with a soft, dry cloth. Then rinse the sink to restore its shine.

* Flowerpots

Container for baking bread Want to give the staff of life an interesting shape? Take a new, clean medium-sized clay flowerpot, soak it in water for about 20 minutes, and then lightly grease the inside with butter. Place your bread dough,

For thousands of years people have been plopping plants into pots to transport a native plant to a new land or to bring an exotic plant home. In 1495 B.C., Egyptian queen Hatshepsut sent workers to Somalia to bring back incense trees in pots. And in 1787 Captain Bligh reportedly had more than 1,000 breadfruit plants in clay pots aboard the H.M.S. *Bounty*. The plants were destined for the West Indies, where they were to be grown as food for the slaves.

prepared as usual, in the pot and bake. The clay pot will give your bread a crusty outside and keep the inside moist.

Create a firewood container Who needs an expensive metal or brass rack to hold firewood by the fireplace? Spare yourself the expense and put an extra-large empty ceramic or clay flowerpot beside the hearth. It's a perfect—and cheap—place to keep kindling and small logs ready for when the weather outside gets frightful.

Unfurl yarn knot-free That sweater you're knitting will take forever if you're constantly stopping to pull out tangles in the yarn. To prevent this, place your ball of yarn under an upturned flowerpot and thread the end through the drain hole. Set it next to where you are sitting for more pleasurable purling.

Create an aquarium fish cave Some fish love to lurk in shadowy corners of their home aquariums, keeping themselves safe from imagined predators. Place a mini flowerpot on its side on the aquarium floor to create a cave for spelunking fish.

Kill fire ants If fire ants plague your yard or patio and you're tired of getting stung by the tiny attackers, a flowerpot can help you quench the problem. Place the flowerpot upside down over the anthill. Pour boiling water through the drain hole and you'll be burning down their house.

Help container plant roots The plants that you want to put in that beautiful new deep container you ordered for your patio have a shallow root system, and you don't want to go to the bother—and expense—of filling that huge container completely with potting soil. What do you do? When planting shallow-rooted plants in a deep container, one easy solution is to find another smaller flowerpot that will fit upside down in the base of the deeper pot and occupy a lot of that space. After you insert it, fill around it with soil before putting in your plants.

Keep soil in your flowerpot Soil from your houseplant won't slip-slide away if you place broken clay flowerpot shards in the bottom of the pot before re-planting. When watering your plants, you'll find that the water drains out, but not the soil.

* Foam Food Trays

Make knee pads for gardening If you find gardening is a pain in the knees when you tend your little patch of green, tape foam food trays to your knees. Or attach them to your legs using the top halves of old tube socks. The trays give you extra padding while you pull out weeds and fertilize your plants.

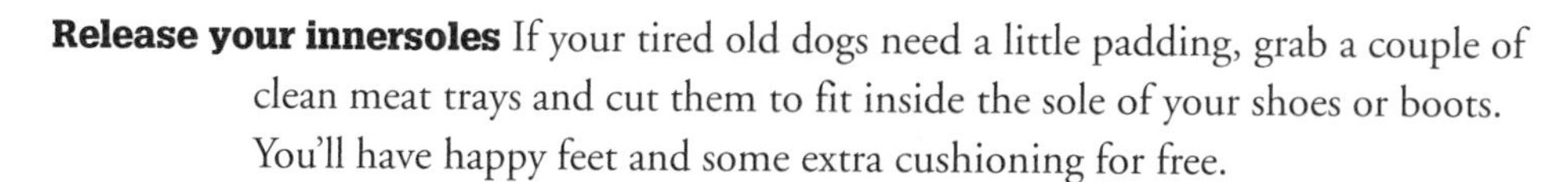

Release your innersoles If your tired old dogs need a little padding, grab a couple of clean meat trays and cut them to fit inside the sole of your shoes or boots. You'll have happy feet and some extra cushioning for free.

Produce a disposable serving dish If you need a quick disposable serving platter while you're on a cookout or camping trip, you can make one from a foam food tray. Wash it with soap and water, cover it entirely with foil, and load it up with food. Use these serving dishes to bring goodies to the church potluck, local bake sale, or sick neighbor. No worries about losing your own platters.

Provide an art palette Create a paint palette for your budding Picasso. A thoroughly cleaned and dried food tray is the perfect place for kids to squirt their tempera or oil paints. Are they experimenting with watercolors? Use two trays: Put watercolor paint in one and water in the other. At the end of the art session, you can just throw them away.

Protect pictures in the mail Why buy expensive padded envelopes to send photographs to loved ones? Cut foam trays slightly smaller than your mailing envelope. Insert your photographs between the trays, place in the envelope, and mail. The photos will arrive without creases or bends.

* Freezer

Eliminate unpopped popcorn Don't you just hate the kernels of popcorn that are left at the bottom of the bowl? Eliminate the popcorn duds by keeping your unpopped supply in the freezer.

Remove wax from candlesticks Grandma's heirloom silver candlesticks will get a new life if you place them in the freezer and then pick off the accumulated wax drippings. But don't do this if your candlesticks are made from more than one type of metal. The metals can expand and contract at different rates and damage the candlesticks.

Extend candle life Place candles in the freezer for at least two hours before burning. They will last longer.

Unstick photos Picture this: Water spills on a batch of photographs, causing them to stick together. If you pull them apart, your pictures will be ruined. Don't be so hasty. Stick them in the freezer for about 20 minutes. Then use a butter knife to gingerly separate the photos. If they don't come free, place them back in the freezer. This works for envelopes and stamps too.

Clean a pot Your favorite pot has been left on the stove too long, and now you've got a burned-on mess to clean up. Place the pot in the freezer for a couple of hours. When the burned food becomes frozen, it will be easier to remove.

Tip Freezer Tactics

Here are some ways to get the most out of your freezer or your refrigerator's freezer compartment:

- To prevent spoilage, keep your freezer at 0°F (-18°C). To check the temperature, stick a freezer thermometer (sold at hardware stores) between two frozen food containers.
- A full freezer runs the compressor less often and stays colder longer. Good to remember the next time there's a blackout.
- The shelves on a freezer door are a little warmer than the freezer interior, making them ideal for storing items such as bread and coffee.
- When defrosting your freezer, place a large towel or sheet on the bottom. Water drips onto it, making cleanup much easier.
- The next time you defrost your freezer, apply a thin coat of petroleum jelly to the walls to keep frost from sticking.

Remove odors Got a musty-smelling book or a plastic container with a fish odor? Place them in the freezer overnight. By morning they'll be fresh again. This works with almost any other small item that has a bad smell you want to get rid of.

* Funnels

Make a string dispenser Don't get yourself tied up in knots over tangled string. Nail a large funnel to the wall, with the stem pointing down. Place a ball of string in the funnel and thread the end through the funnel's stem. You have an instant knot-free string dispenser.

Separate eggs Want an egg-ceptional egg separator? Try a funnel. Simply crack the egg into the funnel. The white will slide out the spout into another container, while the yolk stays put. Of course, you have to be careful not to break the yolk when you're cracking the egg.

Make a kids' telephone Just because you choke every time you open your phone bill doesn't mean the kids have to, too. Use two small plastic funnels to make them a durable string telephone. For each funnel, tie a button to one end of a length of kite string and thread it through the large end of the funnel. Tie another button at the bottom of the spout to keep the string in place and let the kids start yakking.

Garden Hose

Snake decoy to scare birds If flocks of annoying, messy birds are invading your yard, try replicating their natural predator to keep them away. Cut a short length of hose, lay it in your grass—poised like a snake—and the birds will steer clear.

Stabilize a tree A short length of old garden hose is a good way to tie a young tree to its stake. You'll find that the hose is flexible enough to bend when the tree does, but at the same time, it's strong enough to keep the tree tied to its stake until it can stand on its own. Also, the hose will not damage the bark of the young tree as it grows.

Capture earwigs Pesky garden earwigs will find their final resting place in that leaky old hose. Cut the hose into 12-inch (30-centimeter) lengths, making sure the inside is completely dry. Place the hose segments where you have seen earwigs crawling around and leave them overnight. By the morning the hoses should be filled with the earwigs and ready for disposal. One method is to dunk the hoses in a bucket of kerosene.

Tip Buying a Hose

It's just a garden-variety hose, right? Actually, there are a few important points to keep in mind when you buy this important outdoor tool:

- To determine how long a hose you need, measure the distance from the faucet to the farthest point in your yard. Add several feet to allow for watering around corners; this will help you avoid annoying kinks that cut water pressure.
- Vinyl and rubber hoses are generally more sturdy and weather resistant than ones made of cheaper forms of plastic. If a hose flattens when you step on it, it is not up to gardening duties.
- Buy a hose with a lifetime warranty; only good-quality hoses have one.

Unclog a downspout When leaves and debris clog up your rainspout and gutters, turn to your garden hose to get things flowing again. Push the hose up the spout and poke through the blockage. You don't even have to turn the hose on, because the water in the gutters will flush out the dam.

Cover swing set chains No parent wants to see his or her child hurt on the backyard swing set. Put a length of old hose over each chain to protect little hands from getting pinched or twisted. If you have access to one end of the chains, just slip the chain through the hose. Otherwise, slit the hose down the middle, and slip it over the swing set chains. Close the slit hose with a few wraps of duct tape.

Make a play phone Transform your old garden hose into a fun new telephone for the kids. Cut any length of hose you desire. Stick a funnel at each end and attach it with glue or tape. Now the kids can talk for as long as they want, with no roaming charges.

Protect your handsaw and ice skate blades Keep your handsaw blade sharp and safe by protecting it with a length of garden hose. Just cut a piece of hose to the length you need, slit it along its length, and slip it over the teeth. This is a good way to protect the blades of your ice skates on the way to the rink and your cooking knives when you pack them for a camping trip.

Make a paint can grip You don't want that heavy paint can to slip and spill. Plus those thin wire handles can really cut into your hand. Get a better grip by cutting a short length of hose. Slit it down the middle and encase the paint can handle.

Make a sander for curves If you've got a tight concave surface to sand—a piece of cove molding, for example—grab a 10-inch (25-centimeter) length of garden hose. Split open the hose lengthwise and insert one edge of the sand-paper. Wrap it around the hose, cut it to fit, and insert the other end in the slit. Firmly close the slit with a bit of duct tape. Get stroking!

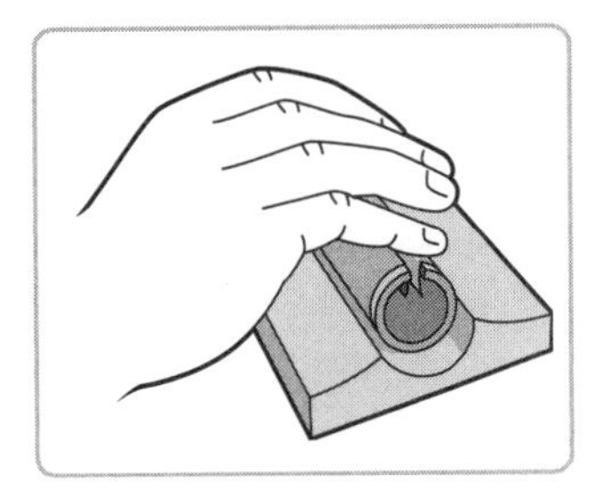

* Gloves

Grip a stubborn jar lid It's a jarring experience when you can't open a jar of peanut butter or olives. If the lid just won't come loose, don some rubber gloves. You'll get a better grip to unscrew the top.

Make an ice pack If you need an ice pack in a hurry, fill a kitchen rubber glove with ice. Close the wrist with a rubber band to contain water from the melting ice. When you're done, turn the glove inside out to dry.

Paper-sorting finger Don't fancy licking your finger when you riffle through a stack of papers or dollar bills? Cut off the index finger piece from an old rubber glove and you have an ideal sheath for your finger the next time you have to quickly sort through some papers.

Make strong rubber bands If you need some extra-strong rubber bands, cut up old rubber gloves. Make horizontal cuts in the finger sections for small rubber bands and in the body of the glove for large ones.

Latex surgical gloves for extra insulation You've got a good pair of gloves or mittens, but your hands still get cold while shoveling the snow or doing other outdoor activities. Try slipping on a pair of latex surgical gloves underneath your usual mittens or gloves. The rubber is a super insulator, so your hands will stay toasty, and dry too.

Clean your knickknacks Need to dust that collection of glass animals or other delicate items? Put on some fabric gloves—the softer the better—to clean your bric-a-brac thoroughly.

Dust a chandelier If your chandelier has become a haven for spiderwebs and dust, try this surefire dusting tip. Soak some old fabric gloves in window cleaner. Slip them on and wipe off the lighting fixture. You'll beam at the gleaming results.

Remove cat hair Here's a quick and easy way to remove cat hair from upholstery: Put on a rubber glove and wet it. When you rub it against fabric, the cat hair will stick to the glove. If you are worried about getting the upholstery slightly damp, test it in an inconspicuous area first.

* Glycerin

Make your own soap Homemade soap is a great gift and snap to make if you have glycerin and a microwave. Here's how: Cut the glycerin material, usually sold in blocks, into 2-inch (5-centimeter) cubes. Using a microwave set at half-power, zap several cubes in a glass container for 30 seconds at a time—checking and stirring as needed—until the glycerin melts. Add drops of color dye or scents at this point,

Glycerin is a clear, colorless thick paste that is a by-product of the soap-making process, in which lye is combined with animal or vegetable fat. Commercial soap makers remove the glycerin when they make soap so that they can use it in more profitable lotions and creams. Glycerin works well in lotions because it is essentially a moisturizing material that dissolves in alcohol or water. Glycerin is also used to make nitroglycerin and candy and to preserve fruit and laboratory specimens. Look for glycerin in the hand lotion aisle of your local drugstore or at a craft store that carries soap-making products.

if you wish. Pour the melted glycerin into soap or candy molds. If you don't have any molds, fill the bottom 3/4 inch (2 centimeters) of a polystyrene cup. Let harden for 30 minutes.

Clean a freezer spill Spilled sticky foods that are frozen to the bottom of your freezer don't have a chance against glycerin. Unstick the spill and wipe it clean with a rag dabbed with glycerin, a natural solvent.

Remove tar stains Do you think it's impossible to remove a tar or mustard stain? It's not, if you use glycerin. Rub glycerin into the spot and leave it for about an hour. Then, with paper towels, gently remove the spot using a blot-and-lift motion. You may need to do this several times.

Make new liquid soap Wondering what to do with those little leftover slivers of soap? Add a bit of glycerin and crush them together with some warm water. Pour the mixture into a pump bottle. You'll have liquid soap on the cheap.

* Golf Gear

Make a golf-tee tie rack If your ties are scattered about the closet or your room, try using golf tees to get them organized. Sand and paint a length of pine board. Drill 1/8-inch (3-millimeter) holes every 2 inches (5 centimeters). Dip the tip of each tee in yellow carpenter's glue and tap it into a hole. Hang the tie rack on the closet wall or inside the door. A perfect gift for the golfer in your life.

Aerate your lawn Kill two birds with one stone by wearing your golf shoes to aerate your lawn the next time you mow. The grip that a golf shoe gives you is also a good idea if you have to push the mower up a hill.

Fill stripped screw holes You're replacing a rusty door hinge when you discover that a screw won't grip because its hole has gotten too big. The fix is easy. Dip the tip of a golf tee in yellow carpenter's glue and tap the tee into the hole. Cut the tee flush with the door frame surface with a utility knife. When the glue dries, you can drill a new pilot hole for the screw in the same spot.

* Hair Conditioner

Take off makeup Put your face first. Why buy expensive makeup removers when a perfectly good substitute sits in your shower stall? Hair conditioner quickly and easily removes makeup for much less money than name-brand makeup removers.

Unstick a ring Grandma's antique ring just got stuck on your middle finger. Now what? Grab a bottle of hair conditioner and slick down the finger. The ring should slide right off.

Protect your shoes in foul weather Here's a way to keep salt and chemicals off your shoes during the winter: Lather your shoes or boots with hair conditioner to protect them from winter's harsh elements. It's a good leather conditioner too.

Lubricate a zipper You're racing out the door, throwing on your jacket, and dang! Your zipper's stuck, so you yank and pull until it finally zips up. A dab of hair conditioner rubbed along the zipper teeth can help you avoid this bother next time.

Smooth shave-irritated legs After you shave your legs, they may feel rough and irritated. Rub on hair conditioner; it acts like a lotion and can soothe the hurt away.

DID *You* KNOW?

Hair conditioner has been around for about 50 years. While researching ways to help World War II burn victims, Swiss chemists developed a compound that improved the health of hair. In the 1950s other scientists developing fabric softeners found that the same material could soften hair.

Despite our efforts to keep hair healthy with hair conditioner, we still lose on average between 50 and 100 strands a day. For most of us, thankfully, there are still many more strands left: People with blond hair have an average of 140,000 strands of hair, brown-haired people, 100,000, and redheads, 90,000.

Smooth-sliding shower curtain Tired of yanking on the shower curtain? Instead of closing smoothly, does it stutter along the curtain rod, letting the shower spray water onto the floor? Rub the rod with hair conditioner, and the curtain will glide across it.

Prevent rust on tools Every good do-it-yourselfer knows how important it is to take care of the tools in your toolbox. One way to condition them and keep rust from invading is to rub them down with hair conditioner.

Clean and shine your houseplants Do your houseplants need a good dusting? Feel like your peace lily could use a makeover? Put a bit of hair conditioner on a soft cloth and rub the plant leaves to remove dust and shine the leaves.

Oil skate wheels Do your child's skateboard wheels whine? Or are the kids complaining about their in-line and roller skates sticking? Try this trick: Rub hair conditioner on the axles of the wheels, and they'll be down the block with their rehabilitated equipment in no time.

Shine stainless steel Forget expensive stainless steel polishers. Apply hair conditioner to your faucets, golf clubs, chrome fixtures, or anything else that needs a shine. Rub it off with a soft cloth, and you'll be impressed with the gleam.

Clean silk garments Do you dare to ignore that "dry clean only" label in your silk shirt? Here's a low-cost alternative to sending it out. Fill the sink with water (warm water for whites and cold water for colors). Add a tablespoon of hair conditioner. Immerse the shirt in the water and let it sit for a few minutes. Then pull it out, rinse, and hang it up to dry. The conditioner keeps the shirt feeling silky smooth.

* Hair Spray

Exterminate houseflies An annoying, buzzing housefly has been bobbing and weaving around your house for two days. Make it bite the dust with a squirt of hair spray. Take aim and fire. Watch the fly drop. But make sure the hair spray is water-soluble so that, if any spray hits the walls, you'll be able to wipe it clean. Works on wasps and bees too.

Reduce runs in pantyhose Often those bothersome runs in your pantyhose or stockings start at the toes. Head off a running disaster by spraying hair spray on the toes of a new pair of pantyhose. The spray strengthens the threads and makes them last longer.

Remove lipstick from fabric Has someone been kissing your shirts? Apply hair spray to the lipstick stain and let it sit for a few minutes. Wipe off the hair spray and the stain should come off with it. Then wash your shirts as usual.

Preserve a Christmas wreath When you buy a wreath at your local Christmas tree lot, it's fresh, green, and lush. By the time a week has gone by, it's starting to shed needles and look a little dry. To make the wreath last longer, grab your can of hair spray and spritz it all over as soon as you get the fresh wreath home. The hair spray traps the moisture in the needles.

Protect children's artwork Picture this: Your preschooler has just returned home with a priceless work of art demanding that it find a place on the refrigerator door. Before you stick it up, preserve the creation with hair spray, to help it last longer. This works especially well on chalk pictures, keeping them from being smudged so easily.

Preserve your shoes' shine After you've lovingly polished your shoes to give them the just-from-the-store look, lightly spray them with hair spray. The shoe polish won't rub off so easily with this coat of protection.

Keep recipe cards splatter-free Don't let the spaghetti sauce on the stove splatter on your favorite recipe card. A good coating of hair spray will prevent the card from being ruined by kitchen eruptions. With the protection, they wipe off easily.

Here are some great moments in hair-spray history:

- A Norwegian inventor developed the technology that became the aerosol can in the early 1900s. What would hair spray be without aerosol cans?
- L'Oréal introduced its hair spray, called Elnett, in 1960. The next year Alberto VO5 introduced its version.
- In 1964 hair spray surpassed lipstick as women's most popular cosmetic aid. Must have been all those beehive hairdos.
- Hair spray makes possible the bumper sticker that reads "The Higher the Hair, the Closer to God."
- In 1984 the hair spray on Michael Jackson's hair ignited while he was rehearsing a commercial for Pepsi.

Keep drapes dirt-free Did you just buy new drapes or have your old ones cleaned? Want to keep that like-new look for a while? The trick is to apply several coats of hair spray, letting each coat dry thoroughly before the next one.

Remove ink marks on garments Your toddler just went wild with a ballpoint pen on your white upholstery and your new shirt. Squirt the stain with hair spray and the pen marks should come right off.

Extend the life of cut flowers A bouquet of cut flowers is such a beautiful thing, you want to do whatever you can to postpone wilting. Just as it preserves your hairstyle, a spritz of hair spray can preserve your cut flowers. Stand a foot away from the bouquet and give them a quick spray, just on the undersides of the leaves and petals.

Hydrogen Peroxide

Remove stains of unknown origin Can't tell what that stain is? Still want to remove it? Try this sure-fire remover: Mix a teaspoon of 3% hydrogen peroxide with a little cream of tartar or a dab of non-gel toothpaste. Rub the paste on the stain with a soft cloth. Rinse. The stain, whatever it was, should be gone.

TAKE CARE Hydrogen peroxide is considered corrosive—even in the relatively weak 3% solution sold as a household antiseptic. Don't put it in your eyes or around your nose. Don't swallow it or try to set it on fire either.

Remove wine stains Hydrogen peroxide works well to remove wine stains so don't worry if you spill while you quaff.

Remove grass stains If grass stains are ruining your kids' clothes, hydrogen peroxide may bring relief. Mix a few drops of ammonia with just 1 teaspoon 3% hydrogen peroxide. Rub on the stain. As soon as it disappears, rinse and launder.

Remove mildew The sight and smell of mildew is a bathroom's enemy. Bring out the tough ammunition: a bottle of 3% hydrogen peroxide. Don't water it down, just attack directly by pouring the peroxide on the offending area. Wipe it clean. Mildew surrender.

Remove bloodstains This works only on fresh bloodstains: Apply 3% hydrogen peroxide directly to the stain, rinse with fresh water, and launder as usual.

Sanitize your cutting board Hydrogen peroxide is a surefire bacteria-killer—just the ally you need to fight the proliferation of bacteria on your cutting board, especially after you cut chicken or other meat. To kill the germs on your cutting board, use a paper towel to wipe the board down with vinegar, then use another paper towel to wipe it with hydrogen peroxide. Ordinary 3% peroxide is fine.

Hydrogen peroxide, (H_2O_2) was discovered in 1818. The most common household use for it is as an antiseptic and bleaching agent. (It's the key ingredient in most teeth-whitening kits and all-fabric oxygen bleaches, for example). Textile manufacturers use higher concentrations of hydrogen peroxide to bleach fabric.

During World War II, hydrogen peroxide solutions fueled torpedoes and rockets.

* Ice-Cream Scoops

Scoop meatballs and cookie dough If you want uniform-size meatballs every time, use an ice-cream scoop to measure out the perfect orbs. This method works well for cookies too. Dip the scoop in the dough, and plop the ball on the cookie sheet. You'll end up with cookies all the same size—no tiffs over which one is the largest.

Make butter balls At your next large family gathering, scoop out large globes of butter or margarine to serve to your guests. A smaller scoop, or melon baller, can create individual-size balls of butter.

Create sand castles On your next trip to the beach, throw an ice-cream scoop into your bag. Your kids will have a fun tool for making their sand castles down by the shore. The scoop allows them to make interesting rounded shapes with the sand.

Spade, dipper, spatula, or spoon—the styles of ice-cream scoop you can buy are almost as varied as the flavors of ice cream you'll put in them. One Web site lists 168 choices of the device! Here's some more dish on ice-cream scoops that you might not know:

- A scoop introduced during the Depression, called the slicer, helped the ice-cream parlor owner scoop out the same amount every time and not give away any extra.
- Many ways have been developed to help the ice cream plop out of a scoop. Some scoops split apart; others have a wire scraper to nudge the stuff out. Still others have antifreeze in the handle or a button on the back to make it pop out.
- Some ice-cream scoops, also called molds, can imprint symbols on the ice cream, for fraternal organizations and others.
- Most scoops come in two standard sizes—#10 and #20, indicating the number of level scoops you'll get from a quart of ice cream. But since most people make rounded scoops, it's more practical to think of a #10 as giving you about seven rounded scoops and a #20, about 12.

Plant seeds If you're out in the garden faced with a plot of earth that needs seeding, turn to your kitchen drawer for help. An ice-cream scoop will make equal-sized planting holes for the seeds for your future harvest.

Repot a houseplant Does dirt scatter everywhere when you are repotting your houseplants? An ice-cream scoop is the perfect way to add soil to the new pot without making a mess.

Pre-scoop ice cream If you're tiring of constantly being bugged by your kids for a scoop of ice cream, try this tip. Scoop several scoops of ice cream onto a wax-paper-lined cookie sheet, spaced apart. Place the sheet with the scoops back in your freezer to re-harden. Remove the scoops from the wax paper and pile them up in a self-sealing plastic bag. The next time the kids want a scoop of strawberry ice cream, they can help themselves.

* Ice Cubes

Water hanging plants and Christmas trees If you're constantly reaching for the step stool to water hard-to-reach hanging plants, ice cubes can help. Just toss several cubes into the pots. The ice melts and waters the plants and does it without causing a sudden downpour from the drain hole. This is also a good way to water your Christmas tree, whose base may be hard to reach with a watering can.

Remove dents in carpeting If you've recently rearranged the furniture in your living room, you know that heavy pieces can leave ugly indents in your carpet. Use ice cubes to remove them. Put an ice cube, for example, on the spot where the chair leg stood. Let it melt, then brush up the dent. Rug rehab completed.

Smooth caulk seams You're caulking around the bathtub, but the sticky caulk compound keeps adhering to your finger as you try to smooth it. If you don't do something about it, the finished job will look pretty awful. Solve the problem by running an ice cube along the caulk line. This forms the caulk into a nice even bead and the caulk will never stick to the ice cube.

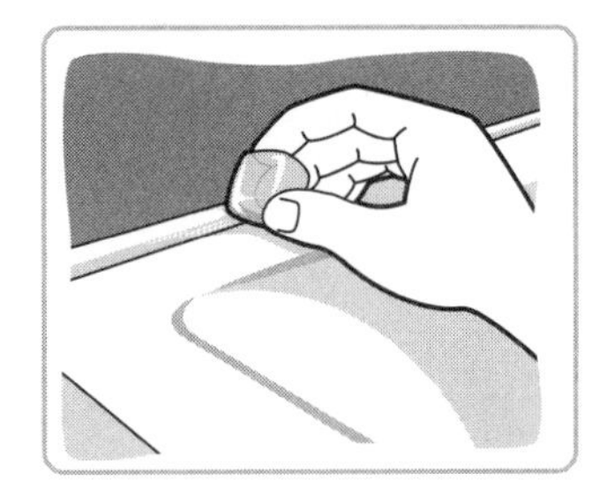

Help iron out wrinkles So your ready-to-wear shirt is full of wrinkles and there's no time to wash it again. Turn on the iron and wrap an ice cube in a soft cloth. Rub over the wrinkle just before you iron and the shirt will smooth out.

Mask the taste of medicine No matter what flavor your local pharmacist offers in children's medicine, kids can still turn up their noses at the taste. Have them suck on an ice cube before taking the medicine. This numbs the taste buds and allows the medicine to go down, without the spoonful of sugar.

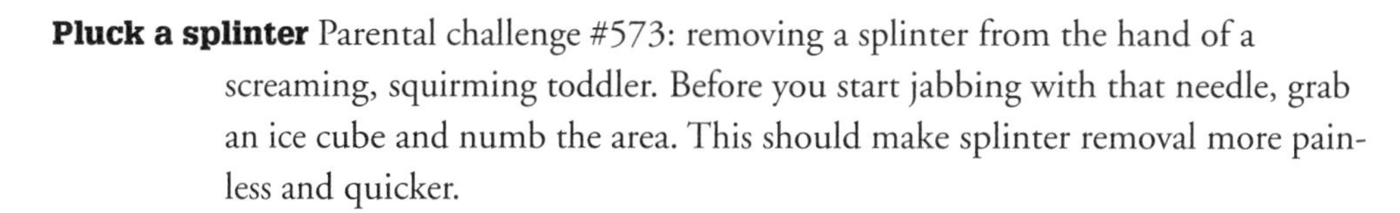

Pluck a splinter Parental challenge #573: removing a splinter from the hand of a screaming, squirming toddler. Before you start jabbing with that needle, grab an ice cube and numb the area. This should make splinter removal more painless and quicker.

Prevent a blister from a burn Have you burned yourself? An ice cube applied to the burn will stop it from blistering.

Cool water for your pets Imagine what it's like to wear a fur coat in the middle of summer. Your rabbits, hamsters, and gerbils will love your thoughtfulness if you place a few cubes in their water dish to cool down. This is also a good tip for your cat, who's spent the hot morning lounging on your bed, or your dog, who's just had a long romp in the park.

Unstick a sluggish disposal If your garbage disposal is not working at its optimum because of grease buildup (not something stuck inside), ice cubes may help. Throw some down the disposal and grind them up. The grease will cling to the ice, making the disposal residue-free.

Make creamy salad dressing Do you want to make your homemade salad dressing as smooth and even as the bottled variety? Try this: Put all the dressing ingredients in a jar with a lid, then add a single ice cube. Close the lid and shake vigorously. Spoon out the ice cube and serve. Your guests will be impressed by how creamy your salad dressing is.

Stop sauces from curdling Imagine this: Your snooty neighbors are over for a Sunday brunch featuring eggs Benedict. But when you mixed butter and egg yolks with lemon juice to make hollandaise sauce for the dish, it curdled. What do you do? Place an ice cube in the saucepan, stir, and watch the sauce turn back into a silky masterpiece.

Here are some cold, hard facts about ice cubes:

- To make clear ice cubes, use distilled water and boil it first. It's the air in the water that causes ice cubes to turn cloudy.
- A British Columbian company sells fake ice cubes that glow and blink in your drink.
- Those aren't ice cubes in that inviting drink in the print advertisement, because they'd never last under hot studio lights. They're plastic or glass.
- The word ice cube has been commercially co-opted over the years. To name two examples: a vintage candy (the chocolate pat wrapped in silver paper) and a well-known rapper/actor.

De-fat soup and stews Want to get as much fat as possible out of your homemade soup or stew as quickly as possible? Fill a metal ladle with ice cubes and skim the bottom of the ladle over the top of the liquid in the soup pot. Fat will collect on the ladle.

Reheat rice Does your leftover rice dry out when you reheat it in the microwave? Try this: Put an ice cube on top of the rice when you put it in the microwave. The ice cube will melt as the rice reheats, giving the rice much-needed moisture.

Remove gum from clothing You're just about to walk out the door when Junior points to the gum stuck to his pants. Keep your cool and grab an ice cube. Rub the ice on the gum to harden it, then scrape it off with a spoon.

* Ice Cube Trays

Divide a drawer If your junk drawer is an unsightly mess, insert a plastic ice cube tray for easy, low-cost organization. One "cube" can hold paper clips, the next, rubber bands, another, stamps. It's another small way to bring order to your life.

Organize your workbench If you're looking through your toolbox for that perfect-sized fastener that you know you have somewhere, here's the answer to your problem. An ice cube tray can help you organize and store small parts you may need at one time or another, such as screws, nails, bolts, and other diminutive hardware.

Keep parts in sequence You're disassembling your latest swap-meet acquisition that has lots of small parts and worry that you'll never be able to get them back together again in the correct sequence. Use an old plastic ice cube tray to help keep the small parts in the right order until you get around to reassembling it. If you really want to be organized, mark the sequence by putting a number on a piece of masking tape in each compartment. The bottom half of an egg carton will also work.

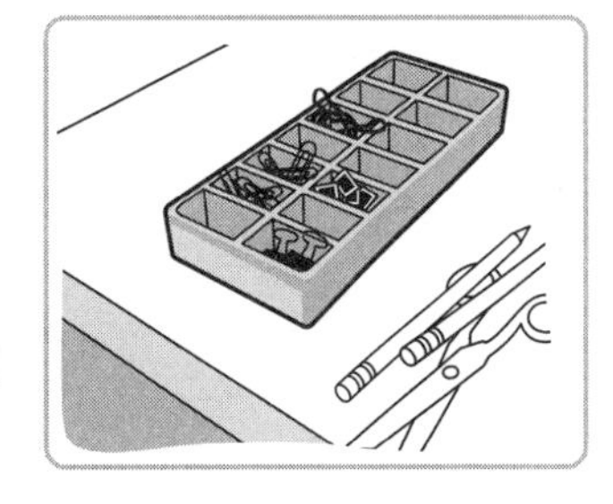

A painter's palette Your child, a budding Mary Cassatt or Picasso, requires a palette to mix colors. A plastic ice cube tray provides the perfect sturdy container for holding and mixing small amounts of paints and watercolors.

Freeze extra eggs Are you overstocked on bargain-priced eggs? Freeze them for future baking projects. Medium eggs are just the right size to freeze in plastic ice cube trays with one egg in each cell, with no spillover. After they freeze, pop them out into a self-sealing plastic bag. Defrost as many as you need when the time comes.

Freeze foods in handy cubes An ice cube tray is a great way to freeze small amounts of many different kinds of food for later use. The idea is to freeze the food in the tray's cells, pop out the frozen cubes, and put them in a labeled self-sealing plastic bag for future use. Some ideas:

- Your garden is brimming with basil, but your family can't eat pesto as quickly as you're making it. Make a big batch of pesto (without the cheese) and freeze it in ice cube trays. Later, when you're ready to enjoy summer's bounty in the middle of winter, defrost as many cubes as you need, add cheese, and mix with pasta.
- There's only so much sweet potato your growing baby will eat at one sitting. Freeze the rest of it in trays for a future high-chair meal.
- The recipe calls for 1/2 cup chopped celery, but you have an entire head of celery and no plans to use it soon. Chop it all, place in an ice cube tray, add a little water, and freeze. The next time you need chopped celery, it's at your fingertips. This works well for onions, carrots, or any other vegetable you'd use for stew and such.
- Are you always throwing out leftover parsley? Just chop it up, put it in an ice cube tray with a little water, and freeze for future use. Works with other fresh herbs.
- There's a bit of chicken soup left in the bottom of the pot. It's too little for another meal, but you hate to throw it out. Freeze the leftovers, and the next time you make soup or another dish that needs some seasoning, grab a cube or two.
- If you are cooking a homemade broth, make an extra-large batch and freeze the excess in ice cube trays. You'll have broth cubes to add instant flavor to future no-time-to-cook dishes. You can do the same with a leftover half-can of broth.
- Here's what to do with that half-drunk bottle of red or white wine: Freeze the wine into cubes that can be used later in pasta sauce, casseroles, or stews.

Kids' Stuff This is a great summertime project. Collect a bunch of small objects around your house: **buttons, beads, tiny toys**. Then get an **ice cube tray** and place one or more of the items in each tray cube. Fill the tray with water. Then cut a length of **yarn** (long enough to make a comfortable **necklace, bracelet, or anklet**). Lay the yarn in the ice cube tray, making sure it hits every cube and is submerged. **Freeze**. When frozen, pop out and tie on the jewelry. The kids will cool off while they see how long their creation takes to melt.

* Ice Scrapers

Remove splattered paint If you just painted your bathroom and have gotten paint splatters all over your acrylic bathtub, use an ice scraper to remove them without scratching the tub surface. Use ice scrapers to remove paint specks from any other nonmetallic surfaces.

Smooth wood filler Do you have small gouges in your wood floors? Want to use wood filler to make them smooth again? An ice scraper can help you do the job right. Once you've packed wood filler into a hole, the ice scraper is the perfect tool to smooth and level it.

Remove wax from skis Every experienced skier knows that old wax buildup on skis can slow you down. An ice scraper can swiftly and neatly take off that old wax and prepare your plows for the next coat.

Scrape out your freezer Your windshield isn't the only place ice and frost build up. If the frost is building up in your freezer and you want to delay the defrosting chore for a while, head out to the car and borrow the scraper.

Clean up bread dough No matter how much flour you put on your work surface, some of that sticky bread dough always seems to stick to it. A clean ice scraper is just the tool for skimming the sticky stuff off the work surface. In a pinch, a plastic scraper can also substitute for a spatula for nonstick pans.

J

* Jars

Waterproof camping storage When you're boating or camping, keeping things like matches and paper money dry can be a challenge. Store items that you don't want to get wet in clear jars with screw tops that can't pop off. Even if you're backpacking, plastic peanut butter jars are light enough not to weigh you down, plus they provide more protection for crushable items than a resealable plastic bag.

Create workshop storage Don't let workshop hardware get mixed up. Keep all your nails, screws, nuts, and bolts organized by screwing jar lids to the underside of a wooden or melamine shelf. (Make sure the screw won't poke through the top of the shelf.) Then put each type of hardware in its own jar, and screw each jar onto its lid. You'll keep everything off the counters, and by using clear jars, you can find what you need at a glance. Works great for storing seeds in the potting shed too!

Stamp out cookies Just about any clean, empty wide-mouthed jar is just the right size for cutting cookies out of any rolled dough.

Use to dry gloves or mittens You took a break from shoveling snow to come in for soup and a sandwich, and want to get back to work. To help your gloves or mittens dry out during lunch, pull each one over the bottom of an empty jar, then stand the jar upside down on a radiator or hot-air vent. Warm air will fill the jar and radiate out to dry damp clothing in a jiffy.

Make a piggy bank You can encourage thriftiness in your child by making a piggy bank out of any jar with a metal lid. Take the lid off the jar, place it on a flat work surface like a cutting board, and tap a screwdriver with a hammer to carefully punch a slot hole in the center. Then use the hammer or a rasp to smooth the rough edges on the underside of the slot to protect fingers from scratches. Personalizing the mini-bank with paints or collage makes a fun rainy-day project.

Collect insects Help the kids observe nature by gently collecting fireflies and other interesting bugs in clear jars. Punch a few small airholes in the lids for ventilation. Don't make the holes too large, or your bugs will escape! Don't forget to let the critters go after you've admired them.

Make baby-food portions Take advantage of the fact that baby-food jars are already the perfect size for baby's portions. Clean them thoroughly before reuse, and fill them with anything from pureed carrots to vanilla pudding. Attach a spoon with a rubber band, and you've got a perfect take-along meal when you travel with your little one.

Bring along baby's treats Dry cereal can be a nutritious snack for your baby. No need to bring the whole box when you leave the house; pack individual servings in clean, dry baby-food jars. If they get spilled, the mess is minimal.

SCIENCE FAIR

Turn a large **wide-mouthed jar** into a **miniature biosphere**. Clean the jar and lid, then place a handful of **pebbles** and **charcoal chips** in the bottom. Add several trowelfuls of slightly damp, sterilized **potting soil**. Select a few **plants** that like similar conditions (such as ferns and mosses, which both like moderate light and moisture). Add a few colorful **stones, seashells,** or a piece of **driftwood**. Add **water** to make the terrarium humid. Tighten the lid and place the jar in dim light for two days. Then display in bright light but not direct sunlight. You shouldn't need to add water—it cycles from the plants to the soil and back again.

It's important to use sterilized soil to avoid introducing unwanted organisms. The charcoal chips filter the water as it recycles.

* Jar Lids

Make safety reflectors Is your driveway difficult to maneuver after dark? With some scrap wood and jar lids you can make inexpensive reflectors to guide drivers. Spray the lids with reflective paint, screw them to the sides of stakes cut from the scrap wood, and drive the stakes into the ground. Voilà! No more dinged fenders or flattened flowers!

Save half-eaten fruit Got half a peach, apple, or orange you'd like to save for later? Wrap a jar lid in plastic wrap or wax paper and then set the fruit cut side down on it in the refrigerator. A bit of lemon juice on the cut surface of the fruit will help prevent discoloration. And why throw out the contents of a partially consumed glass of milk or juice? Just cover it with a lid and refrigerate to keep it fresh.

Cut biscuits Yum, homemade biscuits! Lids with deep rims or canning jar bands (the part with the cut-out center) make impromptu biscuit cutters. Use different-sized lids for Papa, Mama, and Baby biscuits. Dip the bottom edge of the lid in flour to keep it from sticking when you press it into the dough. Avoid lids whose rims are rolled inward; the dough can get stuck inside and be hard to extract.

Create a spoon rest Place a jar lid on the stove or the countertop next to the stove while cooking. After stirring a pot, rest the spoon on the lid, and there'll be less to clean up later.

Drip catcher under honey jar Honey is delicious, but it can be a sticky mess. At the table, place the honey jar on a plastic lid to stop drips from getting on the tabletop. Store it that way, too, and your cabinet shelf will stay cleaner.

Make coasters to protect furniture Wet drinking glasses and hot coffee mugs can really do a number on furniture finishes. The simple solution is to keep plenty of coasters on hand. Glue rounds of felt or cork to both sides of a jar lid (especially flat canning jar lids, which shouldn't be reused for canning, anyhow), and keep a stack wherever cups and glasses accumulate in your house. Your furniture will thank you!

Kids' Stuff What's on your fridge door? Probably a bunch of magnets and children's artwork. Combine the two and you've got something both useful and beautiful. For a stimulating **craft project**, set out a bunch of **fun materials**—paints, glue, fabrics, family photos, googly eyes, glitter, pompoms, or even just paper and markers—and let your child decorate several **jar lids**. Glue some strong **magnets** from the hardware store on the backs (**hot-melt glue** works well), and when they're done, have an unveiling—with refreshments, of course!

Saucers for potted plants Lids with a rim are perfect for catching excess water under small potted plants, and unlike your ceramic saucers, if they get encrusted with minerals, you won't mind throwing them out.

Organize your desk Corral those paper clips and other small office items that clutter up your desk, by putting them in jar lids with deep rims. Works great to hold loose change or earrings on your dresser or bureau too. A quick coat of matte spray paint and an acrylic sealant will make them more attractive and water-resistant.

* Ketchup

Get rid of chlorine green If chlorine from swimming pools is turning your blond tresses green or just giving your hair an unwanted scent, eliminate the problem with a ketchup shampoo. To avoid a mess, do it in the shower. Massage ketchup generously into your hair and leave it for fifteen minutes, then wash it out, using baby shampoo. The odor and color should be gone.

Make copper pots gleam When copper pots and pans—or decorative molds—get dull and tarnished, brighten them with ketchup. It's cheaper than commercial tarnish removers and safe to apply without gloves. Coat the copper surface with a thin layer of the condiment. Let it sit for five to thirty minutes. Acids in the ketchup will react with the tarnish and remove it. Rinse the pan and dry immediately.

Keep silver jewelry sparkling Let ketchup do the work of shining tarnished silver. If your ring, bracelet, or earring has a smooth surface, dunk it in a small bowl of ketchup for a few minutes. If it has a tooled or detailed surface, use an old toothbrush to work ketchup into the crevices. To avoid damaging the silver, don't leave the ketchup on any longer than necessary. Rinse your jewelry clean, dry it, and it's ready to wear.

Ketchup originated in the Far East as a salty fish sauce. The word *ketchup* (also spelled *catsup*) probably comes from Chinese or Malay. Brought to the West, it was transformed by the 1700s into a huge variety of sauces with vegetable and animal main ingredients. To this day, you can still find banana ketchup, mushroom ketchup, and other variants. Tomato ketchup is a relative newcomer, first sold in 1837, but it is now found in more than 90 percent of North American homes.

Keys

Weigh down drapery Need to keep your draperies hanging properly? Just slip a few old keys in the hems. If you are worried about them falling out, tack them in place with a few stitches going through the holes in the keys. You can also keep blind cords from tangling by using keys as weights on their bottoms.

Make fishing sinkers Old unused keys make great weights for your fishing line. Since they already have a hole in them, attaching them to the line is a cinch. Whenever you come across an unidentified key, toss it into your tackle box.

Create an instant plumb bob You are getting ready to hang wallpaper and you need to draw a perfectly vertical line on the wall to get you started. Take a length of cord or string and tie a key or two to one end. You've got a plumb bob that will give you a true vertical. You can do the same with a pair of scissors too.

Kool-Aid

Clean your dishwasher Is the inside of your dishwasher rusty brown? The cause is a high iron content in your water. Dump a packet of unsweetened lemonade Kool-Aid into the soap drawer and run the washer through a hot-water cycle. When you open the door, the inside will be as white as the day you bought the machine.

Clean rust from concrete Nasty rust stains on your concrete? Mix unsweetened lemonade Kool-Aid with hot water. Scrub and the rust stain should come right out.

Color wall paints Mix any flavor of unsweetened Kool-Aid into water-based latex paint to alter its color. Or mix unsweetened Kool-Aid with water to create your own watercolors, but don't give them to the kids—Kool-Aid stains can be tough to remove.

Make play makeup lip gloss Make some tasty lip gloss for little girls playing dress-up. Let the girls pick their favorite presweetened Kool-Aid flavor. Blend a package of the drink mix with 3 tablespoons vegetable shortening, then microwave for one minute. Transfer to a 35mm film canister and refrigerate overnight.

Ladders

Make display shelving Convert a short wooden stepladder to shelving for displaying plants and collectibles. It's as easy as one, two, three:

1. Remove the folding metal spreader that holds the front and rear legs of the ladder together. Then position the ladder's rear legs upright against the wall and attach two 1 x 2 cleats to fix the distance between the front and rear legs. Position the cleats so that their tops are level with the top of a rung.

2. Each shelf will be supported at front by an existing rung. To support the back of each shelf, attach a cleat between the rear legs, positioning it at the same level as a rung.

3. Cut plywood or boards to fit as shelves and screw them to the rungs and cleats. Now screw the centermost rear cleat to the wall and you're done.

Construct a rustic indoor trellis Give your vines and trailing plants something to climb on. Using wall anchors, attach vinyl-covered hooks to your wall and hang an attractive straight ladder (or a segment of one) from the hooks, positioning the ladder's legs on the floor a couple of inches from the wall. It's easy to train potted plants to grow up and around this rustic support. It looks nice on a porch too.

Display quilts and more Don't let your fancy stitching languish in the closet! For that homespun feel, a ladder is a great way to display lacework, crochet, quilts, and throws. To prevent rough surfaces from damaging delicate fabrics, smooth wooden ladder rungs with sandpaper or metal rungs with steel wool if necessary.

Create a garden focal point Got some old wooden straight ladders around that you no longer trust? Show your whimsical side by using them to create a decorative garden archway. Cut two sections of old ladder to the desired height and position them opposite one another along a path. Screw the legs of each one to two

L

strong posts sunk deeply into the soil. Cut a third ladder section to fit across the top of the two others and tie it to them using supple grapevine, young willow twigs, or heavy jute twine. Festoon your archway with fun and fanciful stuff, such as old tools, or let climbing plants clamber up and over it. It also works well as the entryway to an enclosed area.

Plop it down and plant it When a ladder is truly on its last legs, it can still be of service lying down. On the ground, a straight ladder or the front part of a stepladder makes a shallow planter with ready-made sections that look sweet filled with annuals, herbs, or salad greens. After a couple of years of contact with soil, a wooden ladder will decompose, so don't expect to use it again.

Make a temporary table The big family picnic is a summertime staple, but where to put all the food? You can cook up a makeshift table in no time by placing a straight ladder across two sawhorses. Top it with plywood and cover it with a tablecloth. The ladder will provide strength to support your buffet, as well as any guests who might lean on it.

Make a pot rack Accessorize your country kitchen with a pot rack made from a sawed-off section of a wooden straight ladder with thin, round rungs. Sand the cut ends smooth; then tie two pieces of sturdy rope to the rungs at either end. To suspend your pot rack, screw four large metal eye hooks into the ceiling, going into the joists; then tie the other ends of the ropes to them. Hang some S-hooks from the rungs to hold your kitchenware. Leave the rack unfinished if you want a rustic look. Or paint or stain it if you want a more finished look.

Lemons

✱LEMONS AROUND THE HOUSE

Eliminate fireplace odor There's nothing cozier on a cold winter night than a warm fire burning in the fireplace—unless the fire happens to smell horrible. Next time you have a fire that sends a stench into the room, try throwing a few lemon peels into the flames. Or simply burn some lemon peels along with your firewood as a preventive measure.

Get rid of tough stains on marble You probably think of marble as stone, but it is really petrified calcium (also known as old seashells). That explains why it is so porous and easily stained and damaged. Those stains can be hard to remove. If washing won't remove a stubborn stain, try this: Cut a lemon in half, dip the exposed flesh into some table salt, and rub it vigorously on the stain. But do this only as a last resort; acid can damage marble. Rinse well.

Make a room scent/humidifier Freshen and moisturize the air in your home on dry winter days. Make your own room scent that also doubles as a humidifier. If you have a wood-burning stove, place an enameled cast-iron pot or bowl on top, fill with water, and add lemon (and/or orange) peels, cinnamon sticks, cloves, and apple skins. No wood-burning stove? Use your stovetop instead and just simmer the water periodically.

Neutralize cat-box odor You don't have to use an aerosol spray to neutralize foul-smelling cat-box odors or freshen the air in your bathroom. Just cut a couple of lemons in half. Then place them, cut side up, in a dish in the room, and the air will soon smell lemon-fresh.

With all due respect to Trini Lopez and his rendition of "Lemon Tree," a lemon tree actually isn't very pretty—and its flower isn't sweet either. The tree's straggly branches bear little resemblance to an orange tree's dense foliage, and its purplish flowers lack the pleasant fragrance of orange blossoms. Yes, the fruit of the "poor lemon" is sour—thanks to its high citric acid content—but it is hardly "impossible to eat." Sailors have been sucking on vitamin-C-rich lemons for hundreds of years to prevent scurvy. To this day, the British navy requires ships to carry enough lemons so that every sailor can have one ounce of juice daily.

Deodorize a humidifier When your humidifier starts to smell funky, deodorize it with ease: Just pour 3 or 4 teaspoons lemon juice into the water. It will not only remove the off odor but will replace it with a lemon-fresh fragrance. Repeat every couple of weeks to keep the odor from returning.

Clean tarnished brass Say good-bye to tarnish on brass, copper, or stainless steel. Make a paste of lemon juice and salt (or substitute baking soda or cream of tartar for the salt) and coat the affected area. Let it stay on for 5 minutes. Then wash in warm water, rinse, and polish dry. Use the same mixture to clean metal kitchen sinks too. Apply the paste, scrub gently, and rinse.

Polish chrome Get rid of mineral deposits and polish chrome faucets and other tarnished chrome. Simply rub lemon rind over the chrome and watch it shine! Rinse well and dry with a soft cloth.

L

Prevent potatoes from turning brown Potatoes and cauliflower tend to turn brown when boiling, especially when you're having company for dinner. You can make sure the white vegetables stay white by squeezing a teaspoon of fresh lemon juice into the cooking water.

Freshen the fridge Remove refrigerator odors with ease. Dab lemon juice on a cotton ball or sponge and leave it in the fridge for several hours. Make sure to toss out any malodorous items that might be causing the bad smell.

Brighten dull aluminum Make those dull pots and pans sparkle, inside and out. Just rub the cut side of half a lemon all over them and buff with a soft cloth.

Kids' Stuff Kids love to send and receive secret messages, and what better way to do it than by writing them in **invisible ink**? All they need is **lemon juice** (fresh-squeezed or bottled) to use as ink, a **cotton swab** to write with, and a sheet of **white paper** to write on. When the ink is dry and they are ready to read the invisible message, have them hold the paper up to bright sunlight or a lightbulb. The heat will cause the writing to darken to a pale brown and the message can be read! Make sure they don't overdo the heating and ignite the paper.

Keep rice from sticking To keep your rice from sticking together in a gloppy mass, add a spoonful of lemon juice to the boiling water when cooking. When the rice is done, let it cool for a few minutes, then fluff with a fork before serving.

Refresh cutting boards No wonder your kitchen cutting board smells! After all, you use it to chop onions, crush garlic, cut raw and cooked meat and chicken, and prepare fish. To get rid of the smell and help sanitize the cutting board, rub it all over with the cut side of half a lemon or wash it in undiluted juice straight from the bottle.

Keep guacamole green You've been making guacamole all day long for the big party, and you don't want it to turn brown on top before the guests arrive. The solution: Sprinkle a liberal amount of fresh lemon juice over it and it will stay fresh and green. The flavor of the lemon juice is a natural complement to the avocados in the guacamole. Make the fruit salad hours in advance too. Just squeeze some lemon juice onto the apple slices, and they'll stay snowy white.

Make soggy lettuce crisp Don't toss that soggy lettuce into the garbage. With the help of a little lemon juice you can toss it in a salad instead. Add the juice of half a lemon to a bowl of cold water. Then put the soggy lettuce in it and refrigerate for 1 hour. Make sure to dry the leaves completely before putting them into salads or sandwiches.

Keep insects out of the kitchen You don't need insecticides or ant traps to ant-proof your kitchen. Just give it the lemon treatment. First squirt some lemon juice on door thresholds and windowsills. Then squeeze lemon juice into any holes or cracks where the ants are getting in. Finally, scatter small slices of lemon peel around the outdoor entrance. The ants will get the message that they aren't welcome. Lemons are also effective against roaches and fleas: Simply mix the juice of 4 lemons (along with the rinds) with 1/2 gallon (2 liters) water and wash your floors with it; then watch the fleas and roaches flee. They hate the smell.

Clean your microwave Is the inside of your microwave caked with bits of hardened food? You can give it a good cleaning without scratching the surface with harsh cleansers or using a lot of elbow grease. Just mix 3 tablespoons lemon juice into 1 1/2 cups water in a microwave-safe bowl. Microwave on High for 5-10 minutes, allowing the steam to condense on the inside walls and ceiling of the oven. Then just wipe away the softened food with a dishrag.

Deodorize your garbage disposal If your garbage disposal is beginning to make your sink smell yucky, here's an easy way to deodorize it: Save leftover lemon and orange peels and toss them down the drain. To keep it smelling fresh, repeat once every month.

*LEMONS IN THE LAUNDRY

Bleach delicate fabrics Ordinary household chlorine bleach can cause the iron in water to precipitate out into fabrics, leaving additional stains. For a mild, stain-free bleach, soak your delicates in a mixture of lemon juice and baking soda for at least half an hour before washing.

Remove unsightly underarm stains Avoid expensive dry-cleaning bills. You can remove unsightly underarm stains from shirts and blouses simply by scrubbing them with a mixture of equal parts lemon juice (or white vinegar) and water.

Tip Before You Squeeze

To get the most juice out of fresh lemons, bring them to room temperature and roll them under your palm against the kitchen counter before squeezing. This will break down the connective tissue and juice-cell walls, allowing the lemon to release more liquid when you squeeze it.

Boost laundry detergent To remove rust and mineral discolorations from cotton T-shirts and briefs, pour 1 cup lemon juice into the washer during the wash cycle. The natural bleaching action of the juice will zap the stains and leave the clothes smelling fresh.

L

Rid clothes of mildew You unpack the clothes you've stored for the season and discover that some of the garments are stained with mildew. To get rid of mildew on clothes, make a paste of lemon juice and salt and rub it on the affected area, then dry the clothes in sunlight. Repeat the process until the stain is gone. This works well for rust stains on clothes too.

Whiten clothes Diluted or straight, lemon juice is a safe and effective fabric whitener when added to your wash water. Your clothes will also come out smelling lemon-fresh.

*LEMONS FOR HEALTH AND BEAUTY

Lighten age spots Before buying expensive medicated creams to lighten unsightly liver spots and freckles, try this: Apply lemon juice directly to the area, let sit for 15 minutes, and then rinse your skin clean. Lemon juice is a safe and effective skin-lightening agent.

Create blond highlights For blond highlights worthy of the finest beauty salon, add 1/4 cup lemon juice to 3/4 cup water and rinse your hair with the mixture. Then sit in the sun until your hair dries. Lemon juice is a natural bleach. Don't forget to put on plenty of sunscreen before you sit out in the sun. To maximize the effect, repeat once daily for up to a week.

Clean and whiten nails Pamper your fingernails without the help of a manicurist. Add the juice of 1/2 lemon to 1 cup warm water and soak your fingertips in the mixture for 5 minutes. After pushing back the cuticles, rub some lemon peel back and forth against the nail.

SCIENCE FAIR

Turn a **lemon into a battery**! It won't start your car, but you will be able to feel the current with your tongue. Roll the lemon on a flat surface to "activate" the juices. Then cut two small slices in the lemon about 1/2 inch (1.25 centimeters) apart. Place a **penny** into one slot and a **dime** into the other. Now touch your **tongue** to the penny and the dime at the same time. You'll feel a slight **electric tingle**. Here's how it works: The acid in the lemon reacts differently with each of the two metals. One coin contains positive electric charges, while the other contains negative charges. The charges create current. Your tongue conducts the charges, causing a small amount of electricity to flow.

Cleanse your face Clean and exfoliate your face by washing it with lemon juice. You can also dab lemon juice on blackheads to draw them out during the day. Your skin should improve after several days of treatment.

Freshen your breath Make an impromptu mouthwash using lemon juice straight from the bottle. Rinse with the juice and then swallow it for longer-lasting fresh breath. The citric acid in the juice alters the pH level in your mouth, killing the bacteria that cause bad breath. Rinse after a few minutes, because long-term exposure to the acid in the lemon can harm tooth enamel.

Treat flaky dandruff If itchy, scaly dandruff has you scratching your head, relief may be no farther away than your refrigerator. Just massage 2 tablespoons lemon juice into your scalp and rinse with water. Then stir 1 teaspoon lemon juice into 1 cup water and rinse your hair with it. Repeat this daily until your dandruff disappears. No more itchy scalp, and your hair will smell lemon-fresh.

Soften dry, scaly elbows It's bad enough that your elbows are dry and itchy, but they look terrible too. Your elbows will look and feel better after a few treatments with this regimen: Mix baking soda and lemon juice to make an abrasive paste. Then rub the paste into your elbows for a soothing, smoothing, and exfoliating treatment.

Remove berry stains Sure it was fun to pick your own berries, but now your fingers are stained with berry juice that won't come off with soap and water. Try washing your hands with undiluted lemon juice. Wait a few minutes and wash with warm, soapy water. Repeat if necessary until the stain is completely gone.

Disinfect cuts and scrapes Stop bleeding and disinfect minor cuts and scrapes. Pour a few drops of lemon juice directly on the cut or apply the juice with a cotton ball and hold firmly in place for one minute.

Soothe poison ivy rash You won't need an ocean of calamine lotion the next time poison ivy comes a-creeping around. Just apply lemon juice full-strength directly to the affected area to soothe itching and alleviate the rash.

Relieve rough hands and sore feet You don't have to take extreme measures to soothe your extremities. If you have rough hands or sore feet, rinse them in a mixture of equal parts lemon juice and water, then massage with olive oil and dab dry with a soft cloth.

Remove warts You've tried countless remedies to get rid of your warts, and nothing seems to work. Next time, try this: Apply a dab of lemon juice directly to the wart, using a cotton swab. Repeat for several days until the acids in the lemon juice dissolve the wart completely.

Lighter Fluid

Wipe away rust Rust marks on stainless steel will come off in a jiffy. Just pour a little lighter fluid onto a clean rag and rub the rust spot away. Use another rag to wipe away any remaining fluid.

Get gum out of hair It happens to the best of us, not to mention the kids. Gum in the hair is a pain in the neck to remove. Here is an easy solution that really works: Apply a few drops of lighter fluid directly to the sticky area, wait a few seconds, and comb or wipe away the gum. The solvents in the fluid break down the gum, making it easy to remove from many surfaces besides hair.

Remove labels with ease Lighter fluid will remove labels and adhesives from almost any surface. Use it to quickly and easily remove the strapping tape from new appliances or to take stickers off book covers.

Take out crayon marks Did the kids leave their mark with crayons on your walls during that last visit? No problem. Dab some lighter fluid on a clean rag and wipe till the marks vanish.

TAKE CARE Lighter fluid is inexpensive, easy to find (look for a small plastic bottle next to the larger bottles of barbecue starter) and has many surprising uses. But it is highly flammable and can be hazardous to your health if inhaled or ingested. Always use it in a well-ventilated area. Do not smoke around it or use it near an open flame.

Remove heel marks from floors You don't have to scrub to remove those black heel marks on the kitchen floor. Just pour a little lighter fluid on a paper towel and the marks will wipe right off.

Rid cooking-oil stain from clothes When cooking-oil stains won't wash out of clothes, try pouring a little lighter fluid directly onto the stain before washing it the next time. The stain will come out in the wash.

Lip Balm

Prevent windburn You love to ski, but you hate wearing a ski mask. Next time you go snow skiing, try rubbing lip balm, such as ChapStick, on your face before you hit the slopes. The lip balm will protect your skin from windburn.

Remove a stuck ring No need to pull and tug on your poor beleaguered finger to try to remove that stuck ring. Simply coat the finger with lip balm and gently wriggle the ring loose.

Groom wild eyebrows Use lip balm as a styling wax to groom unruly mustaches, eyebrows, or other wild hairs.

Tip Lip Balm and Lipstick

During the dry winter months you may be tempted to apply a layer of lip balm before you put on your lipstick. Beauty experts say this is not a good idea because the lip balm could interfere with the adherence of the lipstick. Instead of using lip balm during the day, the experts recommend that you switch to a moisturizing lipstick. Save the lip balm for moisturizing your lips before you go to bed.

Zap bleeding from shaving cuts Ouch! You just cut yourself shaving and you've no time to spare. Just dab a bit of lip balm directly onto the nick and the bleeding from most shaving cuts will quickly stop.

Lubricate a zipper Rub a small amount of lip balm up and down the teeth of a sticky or stuck zipper. Then zip and unzip it a few times. The lip balm will act as a lubricant to make the zipper work smoothly.

Simplify carpentry Rub some lip balm over nails and screws being drilled or pounded into wood. The lip balm will help them slide in a little easier.

Keep a lightbulb from sticking Outdoor lightbulbs, which are exposed to the elements, often get stuck in place and become hard to remove. Before screwing a lightbulb into an outdoor socket, coat the threads on the bulb with lip balm. This will prevent sticking and make removal easier.

Lubricate tracks for sliding things Apply lip balm to the tracks of drawers and windows, or to the ridges on a medicine cabinet, for easier opening and shutting.

M

* Magazines

No-cost gift wrap Cut out pages with colorful magazine advertisements and use them to make lovely gift wrap for small gifts.

Keep wet boots in shape Roll up a couple of old magazines and use them as boot trees inside a pair of damp boots. The magazines will help the boots maintain their shape as they dry.

Use in kids' craft projects Save up your old magazines for use in rainy-day craft projects with the kids. Let them go through the magazines to find pictures and words to use in collages. Suggest themes for the collages if you like.

Line drawers Pages from large magazines with heavy coated paper make wonderful liners for small dresser and desk drawers. Look for advertisements with especially colorful designs or pictures. Clip the page, place inside the drawer, and press around the edges to define where to trim with scissors.

* Magnets

Clean up a nail spill Keep a strong magnet on your workbench. Next time you spill a jar of small items like nails, screws, tacks, or washers, save time and energy and let the magnet help pick them up for you.

DID *You* KNOW?

Ancient Chinese and Greeks discovered that certain rare stones, called lodestones, seem to magically attract bits of iron and always pointed in the same direction when allowed to swing freely.

Manmade magnets come in many shapes and sizes, but every magnet has a north pole and a south pole. If you break a magnet into pieces, each piece, no matter how small, will have a north and south pole. The magnetic field, which every magnet creates, has long been used to harness energy, although scientists still don't know for sure what it is!

Prevent a frozen car lock Here's a great way to use refrigerator magnets during the bitter cold of winter. Place them over the outside door locks of your car overnight and they will keep the locks from freezing.

Keep desk drawer neat Are your paper clips all over the place? Place a magnet in your office desk drawer to keep the paper clips together.

Store a broom in a handy place Why run to the hall closet every time you need to sweep the kitchen? Instead, just use a screw to attach a magnet about halfway down the broom handle. Then store the broom attached to the side of your refrigerator between the fridge and the wall, where it will remain hidden until you are ready to use it.

* Margarine Tubs

Corral those odds and ends Loose thumbtacks in every room? Odd bolts and nails in a broken cup? Stray superball under the couch? These are just some of the items waiting to be organized into your extra plastic margarine tubs. Get your board game going faster and easier by storing the loose pieces in a tub until the next time. You've sorted out all the sky pieces for a puzzle, so keep them separate and safe in their own tub. With or without their lids, a few clean margarine tubs can do wonders for a junk drawer in need of organization.

Make a baby footprint paperweight Make an enduring impression of your baby's foot, using quick-drying modeling clay—which comes in lots of great colors. Put enough clay in a margarine tub to hold a good impression. Put a thin layer of petroleum jelly on baby's foot and press it firmly into the clay. Let the clay dry as directed, then flex the tub away from the edges until the clay comes free. Years from now you'll be able to show Johnny that his size 13s were once smaller than the palm of your hand! You can also preserve your pet's paw print the same way.

Use as a paint container Want to touch up the little spots here and there in the living room, but don't want to lug around a gallon of paint? Pour a little paint into a margarine tub to carry as you make your inspection. Hold it in a nest of paper towels to catch any possible drips. The tubs with lids are also perfect for storing that little bit of leftover paint for future touch-ups.

Make individual ice-cream portions Small margarine tubs are just the right size for a quick ice-cream snack. And when it comes home from the store, a gallon of ice cream is the perfect consistency to portion out into the tubs. No more time-consuming getting out the bowls, finding the scoop, and waiting for the ice cream to soften up enough to dish out. When Johnny and Janey want their ice cream *now*, they can get it themselves and everyone has an equal portion—they just go to the freezer and pull out a tub.

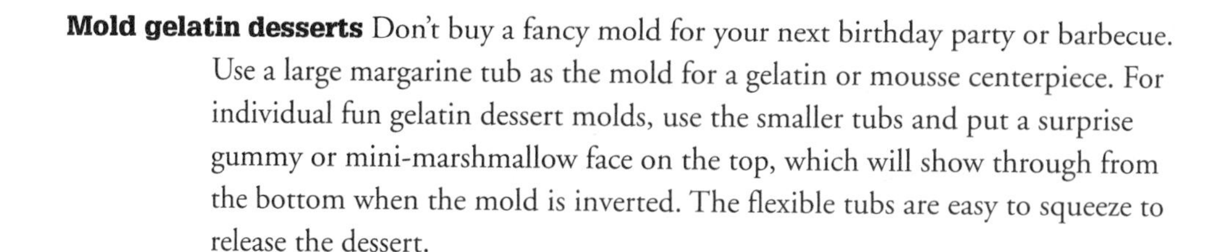

Mold gelatin desserts Don't buy a fancy mold for your next birthday party or barbecue. Use a large margarine tub as the mold for a gelatin or mousse centerpiece. For individual fun gelatin dessert molds, use the smaller tubs and put a surprise gummy or mini-marshmallow face on the top, which will show through from the bottom when the mold is inverted. The flexible tubs are easy to squeeze to release the dessert.

Make frugal freezer storage Reuse your clean, sturdy margarine and other plastic containers for freezing measured portions of soups and stocks, and to break up leftovers into single servings. A 2-pound (1 kilogram) container, for example, stores the perfect amount of sauce for 1 pound (.5 kilogram) of pasta. *Hint:* Before freezing, let the food cool just enough to reduce condensation.

Give kids some lunch box variety As a break from the usual sandwich, put some fruit salad, rice mix, or other interesting fare in one or two recycled margarine tubs for your child's lunch. The tubs are easy to open and will keep the food from getting crushed.

Bring fast food for baby Need to bring your home cooking for Junior on the road? Use a disposable margarine tub for a container that won't break in your baby bag. It's also a handy food bowl, and you won't have to wrap it up and bring it home for cleaning.

Make a piggy bank Use a tall tub as a homemade bank for your little one. Cut out a piece of paper that will fit wrapped around the side, tape it in place and encourage him or her to decorate it with flair. Cut a slit in the top, and start saving!

Travel light with your pet Lightweight, disposable margarine tubs make the perfect pet food containers and double as food and water bowls. And those valuable dog cookies won't get crushed if you put them in a plastic tub. If your pet is vacationing at a friend's house, make things a little easier for the caregiver by putting one serving in each container, to be used and discarded as needed.

Create thrifty seed starters Starting your seeds indoors is supposed to save you money, so don't spend your savings on lots of big seed trays. Take a margarine tub, poke a few holes in the bottom, add moistened seed-starting mix, and sow your seeds following packet instructions. Use permanent marker on the side of the tub to help you remember what you've sown, and use the tub's lid as a drip saucer. Small tubs are space savers as well, especially if you want to start only one or two of each type of plant.

* Marshmallows

Separate toes when applying polish Get the comfort of a salon treatment when giving yourself a home pedicure. Just place marshmallows between your toes to separate them before you apply the nail polish.

Keep brown sugar soft Ever notice how brown sugar seems to harden overnight once you've opened the bag? Next time you open a bag of brown sugar, add a few marshmallows to the bag before closing it. The marshmallows will add enough moisture to keep the sugar soft for weeks.

Stop ice-cream drips Here's an easy way to keep a leaky ice-cream cone from staining your clothes. Just place a large marshmallow in the bottom of the cone before you add the ice cream.

Keep wax off birthday cakes If one of your birthday wishes is to keep candle wax off the frosting on the cake, try this trick: Push each candle into a marshmallow and set the marshmallow atop the frosting. The wax will melt onto the marshmallow, which you can discard. Meanwhile, the marshmallows will add a festive look to the cake.

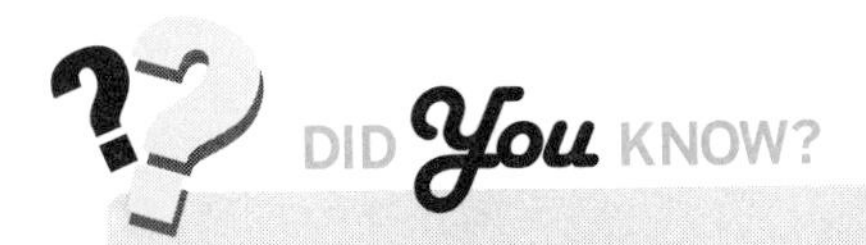

Ancient Egyptians made the first marshmallow candy—a honey-based concoction flavored and thickened with the sap of the root of the marshmallow plant (*Althaea officinalis*). Marshmallow grows in salt marshes and on banks near large bodies of water. Its sap was used to make marshmallow candy and medicine until the mid-1800s. Today's commercial marshmallows are a mixture of corn syrup or sugar, gelatin, gum arabic, and flavoring.

Impromptu cupcake frosting You're already mixing the batter for the cupcakes when you realize you're out of frosting. No problem—if you happen to have some marshmallows on hand. Just pop a marshmallow on top of each cupcake about a minute or so before they come out of the oven. It will make a delicious, instant, gooey frosting.

* Masking Tape

Label foods and school supplies You don't need to buy labels or a fancy machine that makes them. Use inexpensive masking tape instead to mark food containers and freezer bags before putting them in the refrigerator or freezer, and don't forget to write the date! You can also use masking tape to conveniently mark kids' schoolbooks and supplies.

Fix a broken umbrella rib If a strong wind breaks a rib on your umbrella, it's easy to fix it. Use a piece of masking tape and a length of wire cut from a coat hanger to make a splint.

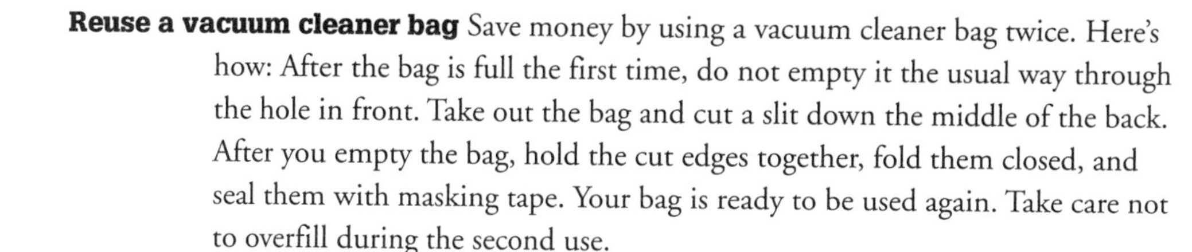

Reuse a vacuum cleaner bag Save money by using a vacuum cleaner bag twice. Here's how: After the bag is full the first time, do not empty it the usual way through the hole in front. Take out the bag and cut a slit down the middle of the back. After you empty the bag, hold the cut edges together, fold them closed, and seal them with masking tape. Your bag is ready to be used again. Take care not to overfill during the second use.

Hang party streamers Use masking tape instead of transparent tape to put up streamers and balloons for your next party. The masking tape won't leave a residue on the wall like transparent tape does. Always remember to remove the masking tape within a day or two. If you wait too long, it could take paint off the wall when it comes off.

Keep paint can neat To prevent paint from filling the groove at the top of a paint can, simply cover the rim of the can with masking tape.

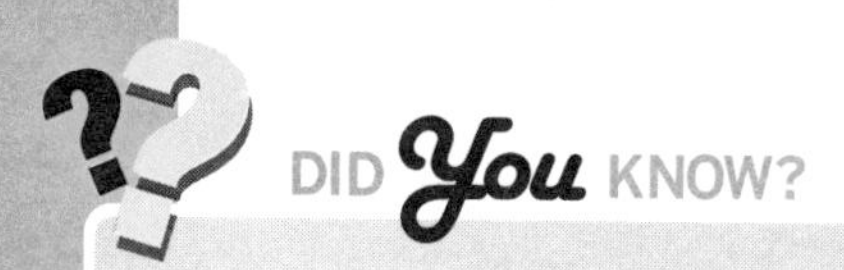

In 1923, when Richard Drew joined 3M as an engineer, the company made only sandpaper. Drew was testing one of 3M's sandpapers at a local auto body shop when he noticed that workers found it difficult to make clean dividing lines on two-color paint jobs. This inspired Drew to look for a solution, despite direct orders from 3M's president to devote his attention solely to sandpaper. Luckily for 3M, he didn't listen. In 1925 Drew invented the first masking tape, a 2-inch-wide tan paper with a pressure-sensitive adhesive backing. Five years later he invented Scotch cellophane tape.

Make a road for toy cars Make a highway for those tiny toy cars your little ones love to play with. Just tape two strips of masking tape to a floor or tabletop. Add a little handmade cardboard stop sign or two and they're off to the races. Carefully guiding a toy car along the taped roadway is more than just fun for small children, it also helps them gain motor control of their fingers for skills they will need later, such as writing.

* Mayonnaise

Condition your hair Hold the mayo … and massage it into your hair and scalp just as you would any fine conditioner! Cover your head with a shower cap, wait several minutes, and shampoo. The mayonnaise will moisturize your hair and give it a lustrous sheen.

Give yourself a facial Why waste money on expensive creams when you can treat yourself to a soothing facial with whole-egg mayonnaise from your own refrigerator? Gently spread the mayonnaise over your face and leave it on for about 20 minutes. Then wipe it off and rinse with cool water. Your face will feel clean and smooth.

Strengthen your fingernails To add some oomph to your fingernails, just plunge them into a bowl of mayonnaise every so often. Keep them bathed in the mayo for about 5 minutes and then wash with warm water.

Relieve sunburn pain Did someone forget to put on sunscreen? To treat dry, sunburned skin, slather mayonnaise liberally over the affected area. The mayonnaise will relieve the pain and moisturize the skin.

Remove dead skin Soften and remove dead skin from elbows and feet. Rub mayonnaise over the dry, rough tissue, leave it on for 10 minutes, and wipe it away with a damp cloth.

Safe way to kill head lice Many dermatologists now recommend using mayonnaise to kill and remove head lice from kids instead of toxic prescription drugs and over-the-counter preparations. What's more, lice are becoming more resistant to such chemical treatments. To treat head lice with mayonnaise, massage a liberal amount of mayonnaise into the hair and scalp before bedtime. Cover with a shower cap to maximize the effect. Shampoo in the morning and then use a fine-tooth comb to remove any remaining lice and nits. To completely eradicate the infestation, repeat the treatment in 7-10 days.

Make plant leaves shiny Professional florists use this trick to keep houseplant leaves shiny and clean. You can do the same thing at home. Just rub a little mayonnaise on the leaves with a paper towel, and they will stay bright and shiny for weeks and even months at a time.

Remove crayon marks Did the kids leave crayon marks on your wood furniture? Here's a simple way to remove them that requires hardly any elbow grease: Simply rub some mayonnaise on the crayon marks and let it soak in for several minutes. Then wipe the surface clean with a damp cloth.

Clean piano keys If the keys to your piano are starting to yellow, just tickle the ivories with a little mayonnaise applied with a soft cloth. Wait a few minutes, wipe with a damp cloth, and buff. The piano keys will look like new.

Remove bumper stickers Time to get rid of that Nixon for President bumper sticker on your car? Instead of attacking it with a razor and risk scratching the bumper, rub some mayonnaise over the entire sticker. Let it sit for several minutes and wipe it off. The mayonnaise will dissolve the glue.

Get tar off your car To get road tar or pine sap off your car with ease, slather some mayonnaise over the affected area, let it sit for several minutes, and wipe it away with a clean, soft rag.

M

* Meat Tenderizer

Ease backache To relieve your aching back, mix sufficient water with meat tenderizer to make a paste and rub it on your back where it hurts. The enzymes in the tenderizer will help soothe those aching back muscles.

Relieve wasp-sting pain Make a paste of meat tenderizer and water and apply it directly to the sting from a bee or wasp. Be careful not to push any remaining part of the stinger deeper into your skin. The enzymes in the meat tenderizer will break down the proteins in the insect venom.

Remove protein-based stains Try a little tenderness to remove protein-based stains like milk, chocolate, and blood from clothes. For fresh wet stains, sprinkle on enough meat tenderizer to cover the area and let it sit for an hour. Then brush off the dried tenderizer and launder as usual. For stains that are already set, mix water and meat tenderizer to make a paste and rub it into the stain. Wait an hour before laundering as usual.

DID *You* KNOW?

When Lloyd Rigler and Lawrence Deutsch dined at Adolph Rempp's Los Angeles restaurant one night in 1949, they had no clue that their fortunes were about to change forever. But the two partners were so impressed by the tender, flavorful meat that they soon bought the rights to the product now known as Adolph's Meat Tenderizer. Rigler toured the country—visiting 63 cities in 60 days—demonstrating the product, winning over skeptical food critics by sending them home with tenderized meat to cook themselves. Rave reviews led to windfall sales and profits, allowing the partners to sell the business in 1974 and turn their attention to philanthropy and the arts for the rest of their lives.

"Tenderize" tough perspiration stains Tenderize away hard-to-remove perspiration stains. Before you wash that sweat-stained sweatshirt (or any other perspiration-stained garment) dampen the stain and then sprinkle some meat tenderizer on it. Then just wash as usual.

* Milk

Make frozen fish taste fresh If you want fish from your freezer to taste like it was fresh caught, try this trick: Place the frozen fish in a bath of milk until it thaws. The milk will make it taste fresher.

Boost corn on the cob flavor Here's a simple way to make corn on the cob taste sweeter and fresher. Just add 1/4 cup powdered milk to the pot of boiling water before you toss in the corn.

Repair cracked china Before you throw out that cracked plate from your grandmother's old china set, try mending it with milk. Place the plate in a pan, cover it with milk (fresh or reconstituted powdered milk), and bring to a boil. As soon as it starts to boil, lower the heat and simmer for about 45 minutes. The protein in the milk will miraculously meld most fine cracks.

Polish silverware Tarnished silverware will look like new with a little help from some sour milk. If you don't have any sour milk on hand, you can make some by adding vinegar to fresh milk. Then simply soak the silver in the milk for half an hour to loosen the tarnish, wash in warm, soapy water, and buff with a soft cloth.

Soothe sunburn and bug bites If your skin feels like it's burning up from too much sun exposure or if itchy bug bites are driving you crazy, try using a little milk paste for soothing relief. Mix one part powdered milk with two parts water and add a pinch or two of salt. Dab it on the burn or bite. The enzymes in the milk powder will help neutralize the insect-bite venom and help relieve sunburn pain.

Impromptu makeup remover When you run out of makeup remover and you can't get to the store, use powdered milk instead. Just mix 3 tablespoons powdered milk with 1/3 cup warm water in a jar and shake well. Add more water or powder as necessary to achieve the consistency of heavy cream. Now you are ready to apply your makeshift makeup remover with a facecloth. When you're done, wipe it off and rinse with water.

Give yourself a facial Here's another way to give yourself a fancy spa facial at home. Make a mask by mixing 1/4 cup powdered milk with enough water to form a thick paste. Thoroughly coat your face with the mixture, let dry completely, then rinse with warm water. Your face will feel fresh and rejuvenated.

Soften skin Treat yourself to a luxurious foamy milk bath. Toss 1/2 cup or so of powdered milk into the tub as it fills. Milk acts as a natural skin softener.

Clean and soften dirty hands You come back from the garden with stained and gritty hands. Regular soap just won't do, but this will: Make a paste of oatmeal and milk and rub it vigorously on your hands. The stains will be gone and the oatmeal-and-milk mixture will soften and soothe your skin.

Clean patent leather Make your patent-leather purses or shoes look like new again. Just dab on a little milk, let it dry, and buff with a soft cloth.

Remove ink stains from clothes To remove ink stains from colored clothes, an overnight milk bath will often do the trick. Just soak the affected garment in milk overnight and launder as usual the next day.

* Milk Cartons

Make ice blocks for parties Keep drinks cold at your next barbecue or party with ice blocks made from empty milk cartons. Just rinse out the old cartons, fill them with water, and put them in the freezer. Peel away the container when you're ready to put the blocks in the cooler or punch bowl.

Make a lacy candle Here's an easy way to make a delicate, lacy candle. Coat the inside of a milk carton with cooking spray, put a taper candle in the middle, anchoring it with a base of melted wax, then fill it with ice cubes. Pour in hot wax; when the wax cools, peel off the carton. The melting ice will form beautiful, lacy voids in the wax.

Instant kids' bowling alley Make an indoor bowling alley for the kids with pins made from empty milk and juice cartons. Just rinse the cartons (use whatever sizes you like) and let them dry. Then take two same-sized cartons and slide one upside down into the other, squeezing it a little to make it fit. Once you've made ten, set your pins up at the end of the hall and let the kids use a tennis ball to roll for strikes and spares.

Feed birds in winter To make an attractive wintertime treat for feathered visitors, combine melted suet and birdseed into an empty milk carton. Suet is beef fat; you can get it from a butcher. To render it, chop or grind the fat and heat it over a

John Van Wormer, an Ohio toy factory owner, didn't cry over spilt milk when he dropped a bottle of the stuff on the floor one morning in 1915. Instead, he was inspired to patent a paper-based milk carton he named Pure-Pak. It took him 10 years to perfect a machine that coated the paper with wax and sealed the carton with animal glues. Those early waxy containers were slow to catch on with skeptical consumers and bore little resemblance to today's milk cartons, but now some 30,000 million Pure-Pak cartons are sold annually.

low flame until it melts. Then strain it through cheesecloth into the carton. Insert a loop of string into the mixture while it is still melted. After it hardens, tear away the carton and tie your new mass of bird food to a branch. Do this only in cold weather. Once the temperature gets above about 70°F (20°C), the suet will turn rancid and melt.

Make seed starters Milk cartons are the perfect size to use for seed starters. Simply cut off the top half of a carton, punch holes in the bottom, fill with potting mix, and sow the seeds according to instructions on the packet.

Make vegetable garden collars Use empty milk cartons to discourage grubs and cutworms from attacking your young tomato and pepper plants. Just cut off the tops and bottoms of the containers, and when the ground is soft, push them into the ground around the plants when you set them out.

Collect food scraps for compost Keep an empty milk carton handy near the kitchen sink and use it to collect food scraps for your compost heap.

Disposable paint holder If you have a small paint project and you don't want to save the leftover paint (or lug a heavy can), an empty milk carton can help. Just cut off the top of the carton and pour in the amount of paint you need. When the job is finished, throw the carton into the trash, leftover paint and all.

* Mothballs

Rinse woolens for storage Of course it is a good idea to store woolens with mothballs to ward off moths. To give your favorite sweaters even more protection, dissolve a few mothballs in the final rinse when you wash them before storage.

SCIENCE FAIR

Make **mothballs dance** and give children a basic science lesson too. Just fill a **glass jar** about 2/3 full with **water**. Add about 1/4 to 1/3 cup **vinegar** and 2 teaspoons **baking soda**. Stir gently, then toss in a few **mothballs** and watch them bounce up and down. The vinegar and baking soda create carbon dioxide bubbles, which cling to the irregular surfaces of the mothballs. When enough bubbles accumulate to lift the weight of a mothball, it rises to the surface of the water. There, some of the bubbles escape into the air, and the mothball sinks to the bottom of the jar to start the cycle again. The effect will last longer if the container is sealed.

Kill bugs on potted plants To exterminate bugs on a potted plant, put the plant in a clear plastic bag, such as a cleaning bag, add a few mothballs, and seal for a week. When you take the plant out of the bag, your plant will be bug-free. It will also keep moths away for a while.

Repel mice from garage or shed Don't let mice spend their winter vacation in your garage. Place a few mothballs around the garage, and the mice will seek other quarters. To keep mice out of your potting shed, put the mothballs around the base of wrapped or covered plants.

Keep dogs and cats away from garden Don't throw out old mothballs. Scatter them around your gardens and flowerbeds to keep cats, dogs, and rodents away. Animals hate the smell!

Keep bats at bay Bats won't invade your belfry (or attic) if you scatter a few mothballs around. Add some mothballs to the boxes you store in the attic and silverfish will stay away too.

 Mouse Pads

Pad under table legs When you get a new mouse pad for your computer, don't throw out the old one. Use it to make pads for table legs and chairs to prevent them from scratching wood and other hard-surface floors. Just cut the foam and cloth pad into small pieces and superglue each piece to the bottom of a leg.

Make knee pads for the garden Old computer mouse pads are just the right size to cushion your knees when you're working in the garden. Kneel on them loose as is or attach them directly to your pant legs with duct tape.

Pad under houseplants Keep potted plant containers from scratching or damaging your hard floors. Just set the pot atop an old mouse pad and your floor will remain scratch-free. Use four pads for large pots.

Hot pad for the table Protect your table from hot casseroles, coffeepots, and serving dishes. Use old PC mouse pads as hot pads. The cloth-topped foam mouse pad is the perfect size to hold most hot containers you bring to the table.

 Mouthwash

Clean computer monitor screen Out of glass cleaner? A strong, alcohol-based mouthwash will work as well as, or better than, glass cleaner on your computer monitor or TV screen. Apply with a damp, soft cloth and buff dry. Remember to use only on glass screens, not liquid crystal displays! The alcohol can damage the material used in LCDs.

Cleanse your face An antiseptic mouthwash makes a wonderful astringent for cleansing your face. Check the ingredients to make sure it does not contain sugar, then use as follows. Wash your face with warm, soapy water and rinse. Dab a cotton ball with mouthwash and gently wipe your face as you would with any astringent. You should feel a pleasant, tingling sensation. Rinse with warm water followed by a splash of cold water. Your face will look and feel clean and refreshed.

Treat athlete's foot A sugarless antiseptic mouthwash may be all you need to treat mild cases of athlete's foot or toenail fungus. Use a cotton ball soaked in mouthwash to apply to the affected area several times a day. Be prepared: It will sting a bit! Athlete's foot should respond after a few days. Toenail fungus may take up to several months. If you do not see a response by then, make an appointment with a dermatologist or podiatrist.

Add to wash water Smelly gym socks are often full of bacteria and fungi that may not all come out in the wash—unless you add a cup of alcohol-based, sugarless mouthwash during the regular wash cycle.

Tip Homemade Mouthwash

Freshen your breath with your own alcohol-free mouthwash. Place 1 ounce (30 grams) whole cloves and/or 3 ounces (85 grams) fresh rosemary in a pint-size (half-liter) jar and pour in 2 cups boiling water. Cover the jar tightly and let it steep overnight before straining. Need a mouthwash immediately? Dissolve 1/2 teaspoon baking soda in 1/2 cup warm water.

Cure underarm odor Regular deodorants mask unpleasant underarm odors with a heavy perfume smell but do little to attack the cause of the problem. To get rid of the bacteria that cause perspiration odor, dampen a cotton ball with a sugarless, alcohol-based mouthwash and swab your armpits. If you've just shaved your armpits, it's best to wait for another day to try this.

Disinfect a cut When you need to clean out a small cut or wound, use an alcohol-based mouthwash to disinfect your skin. Remember that before it became a mouthwash, it was successfully used as an antiseptic to prevent surgical infections.

Get rid of dandruff To treat a bad case of dandruff, wash your hair with your regular shampoo; then rinse with an alcohol-based mouthwash. You can follow with your regular conditioner.

Clean your toilet All out of your regular toilet bowl cleaner? Try pouring 1/4 cup alcohol-based mouthwash into the bowl. Let it stand in the water for 1/2 hour, then swish with a toilet brush before flushing. The mouthwash will disinfect germs as it leaves your toilet bowl sparkling and clean.

* Mustard

Soothe an aching back Take a bath in yellow mustard to relieve an aching back or arthritis pain. Simply pour a regular 6- to 8-ounce (175- to 240-milliliter) bottle of mustard into the hot water as the tub fills. Mix well and soak yourself for 15 minutes. If you don't have time for a bath, you can rub some mustard directly on the affected areas. Use only mild yellow mustard and make sure to apply it to a small test area first. Undiluted mustard may irritate your skin.

Relax stiff muscles Next time you take a bath in Epsom salt, throw in a few tablespoons yellow mustard too. The mustard will enhance the soothing effects of the Epsom salt and also help to relax stiff, sore muscles.

Relieve congestion Relieve congestion with a mustard plaster just like Grandma used to make. Rub your chest with prepared mustard, soak a washcloth in hot water, wring it out, and place it over the mustard.

Make a facial mask Pat your face with mild yellow mustard for a bracing facial that will soothe and stimulate your skin. Try it on a small test area first to make sure it will not be irritating.

Remove skunk smell from car You didn't see the skunk in the road until it was too late, and now your car exudes that foul aroma. Use mustard powder to get rid of those awful skunk odors. Pour 1 cup dry mustard into a bucket of warm water, mix well, and splash it on the tires, wheels, and underbody of the car. Your passengers will thank you.

Remove odor from bottles You've got some nice bottles you'd like to keep, but after washing them, they still smell like whatever came in them. Mustard is a sure way to kill the smell. After washing, just squirt a little mustard into the bottle, fill with warm water, and shake it up. Rinse well, and the smell will be gone.

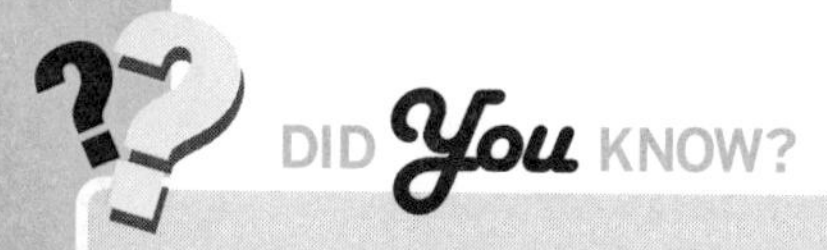

Ancient Romans brought mustard back from Egypt and used the seeds to flavor unfermented grape juice, called *must*. This is believed to be how the mustard plant got its name. The Romans also made a paste from the ground seeds for medicinal purposes and may have used it as a condiment. But the mustard we use today was first prepared in Dijon, France, in the 13th century. Dijon-style mustard is made from darker seeds than yellow mustard.

Nail Polish

NAIL POLISH AROUND THE HOUSE

Make buttons glow in the dark It happens all the time. The lights are dimmed, you grab the remote control to increase the TV volume, and darn, you hit the wrong button and change the channel instead. To put an end to video flubs, dab glow-in-the-dark nail polish onto frequently used remote buttons. You can also use phosphorescent polish to mark keys and keyholes and other hard-to-spot items.

Mark your thermostat setting When you wake up with a chill and don't have your glasses, it's easy to return to your comfort zone if you've marked your dial-type thermostat. Simply set it to your preferred temperature and then make a thin mark with colored nail polish from the dial into the outside ring.

Tip Using Nail Polish

- To keep nail polishes fresh and easy to use, store them in the refrigerator. Keep them together in a little square plastic container.
- Shaking a polish bottle to mix the color can cause bubbles. Roll the bottle between your palms instead.
- Wipe the inside threads of your nail polish bottle and cap with a cotton swab dipped in polish remover before closing them. It'll open more easily.

Mark temperature settings on shower knobs Don't waste precious shower time fiddling with the water temperature. With the shower on, select your ideal settings, then turn off the flow to the shower and make a small mark with bright nail polish onto the stationary lip of both the hot and cold knob indicating the handle position that's best. Once it's set, no sweat!

Make cup measurements legible Find your measuring cup markings faster, especially if you like to measure "on the fly" while cooking. Use a very visible color of nail polish to trace over the basic measurement levels. This also works great for those dimly lit, late-night bottle feedings, when you need to see how well

Junior has tanked up. And you won't have to squint to find the correct dosage on little plastic medicine cups if you first mark them with a thin line of dark polish.

N

Mark levels inside a bucket When you're mixing in a big bucket, you don't typically have the opportunity to lift the bucket to check the quantity. Besides, the bucket you use for mixing might not have the measurements clearly marked at all. Make sure you know you're using the right amounts by marking pint, quart, and gallon (or half, full, and other liter) levels with lines of nail polish. Use a color that stands out against the bucket's color.

Label your sports gear You share a lot of interests with your golf partner, including the same brand of golf balls. Make it clear who got on the green first, by putting a dot of bright nail polish on your ball supply. This also works well with batting gloves and other items that don't have enough room to fit your name.

Label poison containers If everyone in your home has easy access to your cupboard, prevent someone from grabbing dangerous items in haste. Use dark red or other easily visible nail polish to label the poisons. Draw an unmistakable X on the label as well as the lid or spout.

Seal an envelope Do you have a mild distrust of those self-sealing envelopes? Brush a little nail polish along the underside of the flap, seal it, and it won't even open over a teakettle! Add some flair to a special card by brushing your initial (or any design) in nail polish over the sealed flap tip, as a modern type of sealing wax that doesn't need to be melted first.

Smudgeproof important drug labels Preserve the important information on your prescription medicine and other important medicine labels with a coat of clear polish, and they won't be smudged as you grab them after getting your glass of water.

Waterproof address labels When you're sending a parcel on a rainy day, a little clear polish brushed over the address information will make sure your package goes to the right place.

Prevent rust rings from metal containers If your guests are going to peek into your medicine cabinet, you don't want them to see rust rings on your shelves. Brush nail lacquer around the bottom of shaving cream cans and other metal containers to avoid those unsightly stains.

Make a gleaming paperweight To create paperweights that look like gemstones, or interesting rocks for the base of your potted cactus, try this: Find some palm-size, smooth clean rocks. Put about 1/2 inch (1.25 centimeters) water into a pie pan, and put 1 drop clear nail polish onto the water. The polish will spread out over the water surface. Holding a rock with your fingertips, slowly roll it in the water to coat it with the polish. Set the rock on newspaper to dry.

Prevent rusty toilet seat screws If you're installing a new toilet seat, keep those screws from quickly rusting. Paint them with a coat or two of clear nail lacquer; it will also help prevent seat wobble by keeping the screws in place.

Paint shaker holes to restrict salt If your favorite saltshaker dispenses a little too generously, paint a few of the holes shut with nail polish. It is a good idea for those watching their salt.

Tarnish-proof costume jewelry Inexpensive costume jewelry can add sparkle and color to an everyday outfit, but not if it tarnishes and the tarnish rubs off the jewelry and onto your skin. To keep your fake jewelry and your skin sparkling clean, brush clear nail polish onto the back of each piece and allow it to dry before wearing.

Protect your belt buckle's shine Cover new or just-shined belt buckles with a coat of clear polish. You'll prevent oxidation and guarantee a gleaming first impression.

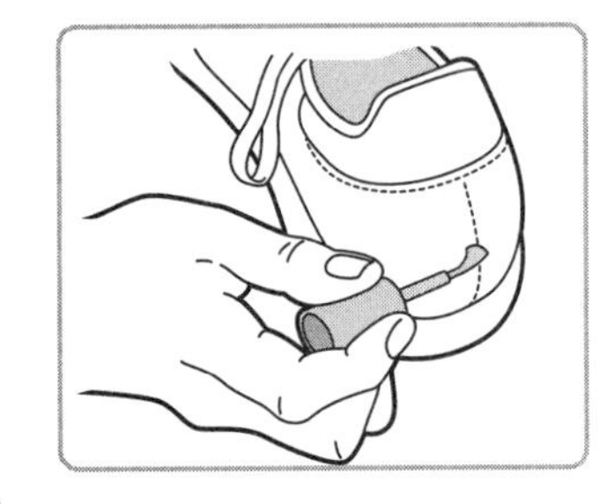

Seal out scuffs on shoes On leather shoes, it's the back and toes that really take the brunt of the wear and tear that leaves scratches on the surface. Next time you buy a new pair of shoes—especially ones for a kid or an active adult—give these areas the extra measure of protection they need. Paint a little clear nail polish on the outside of the back seam and over the toes. Rub the polish in a little to feather out the shine of the polish. After it dries, you'll be a step ahead of those perennial shoe problems "driver's heel" and "jump rope toe."

Keep laces from unraveling Neaten the appearance of frayed shoelaces, and extend their life. Dip the ends in clear nail polish and twist the raveled ends together. Repair laces in the evening so that the polish will dry overnight.

Nail polish is certainly not a recent concept. As early as 3000 B.C., ancient Chinese nobility are believed to have colored their long nails with polishes, made from gum arabic, beeswax, gelatin, and pigments. The nobility wore shades of gold, silver, red, or black, while lesser classes were restricted to pastel shades. Colored nails were also popular with ancient Egyptians, who often dyed their nails with henna or stained them with berries. Polish wasn't just for women. In Egypt and Rome, military commanders painted their nails red before going into battle.

Get rid of a wart Warts are unsightly, embarrassing, and infectious. In order to get rid of warts and prevent spreading the virus to others, cover them with nail polish. The wart should be gone or greatly diminished in one week.

*NAIL POLISH IN THE SEWING ROOM

Protect pearl buttons Delicate pearl buttons will keep their brand-new sparkle with a protective coat of clear nail polish. It will keep costume pearl buttons from peeling as well.

Prevent loss of buttons Keep that brand-new shirt in good shape by putting a drop of clear nail polish on the thread in the buttons. It prevents the thread from fraying, so taking this precaution in advance could save you some embarrassment later. Put a dab on just-repaired buttons as well.

DID *You* KNOW?

Unless you work in a lab, you probably don't know that clear nail polish is the respected workhorse used in mounting microscopic slides. Officially referred to as NPM (nail polish mountant) it is the preferred and inexpensive substance used around a cover glass to seal it onto a slide, protecting the specimen from air and moisture.

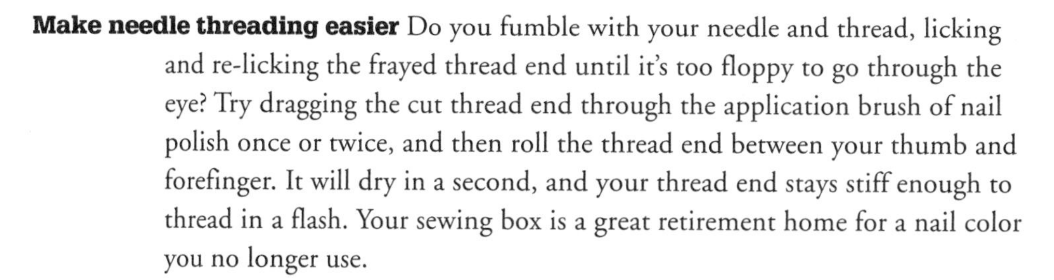

Make needle threading easier Do you fumble with your needle and thread, licking and re-licking the frayed thread end until it's too floppy to go through the eye? Try dragging the cut thread end through the application brush of nail polish once or twice, and then roll the thread end between your thumb and forefinger. It will dry in a second, and your thread end stays stiff enough to thread in a flash. Your sewing box is a great retirement home for a nail color you no longer use.

Prevent frayed fabric from unraveling Do you have wisps peeking out from the bottom of your skirt? Is the nylon lining of your jacket fraying at the cuffs? You can tame those fraying strays by brushing them into place with some clear nail polish.

Keep ribbons from fraying The gift is perfect, so make sure the wrapping is just as nice. Brush the cut ends of ribbon with a little clear nail polish to stop them from unraveling. This is also the perfect solution for your little girl's hair ribbons on special occasions; at least one part of her will stay together all day!

Stop a run in your hose It's a helpless feeling, realizing that a small run in your stocking is about to turn into a big embarrassment. Happily you can stop runs permanently and prolong the life of fragile stockings with a dab of clear nail polish. Simply apply polish to each end of a run (no need to remove hose), and let it dry. This invisible fix stops runs, and lasts through many hand launderings.

Mend a fingernail You just split a nail, but don't have a nail repair kit handy? Grab an unused tea bag instead. Cut the bag open, dump the tea, cut a piece of the bag into the shape of your nail, and cover it with clear nail polish. Press it onto your nail, then apply colored nail polish. You'll be good to go until the break grows out.

Temporarily repair eyeglasses So you sat on your glasses and one lens has a small crack, but you can't get to the optometrist right away? Seal the crack on both sides with a thin coat of clear nail polish. That will hold it together until you can see your way to the doctor's office.

Stop a windshield crack from spreading If you've developed a small crack in your windshield, stop it cold with some clear polish. Working in the shade, brush the crack on both sides of the glass with polish to fill it well. Move the car into the sun so the windshield can dry. You will eventually need to repair your windshield, but this will give you time to shop around for the best estimate.

Fill small nicks on floors and glass Have the children been playing hockey on your hardwood floors? Fill those little nicks by dabbing them with some clear nail polish. It will dry shiny, so sand the spot gently with some 600-grit sandpaper. A thick coat of clear nail polish also helps to soften the sharp edge of a nicked mirror or glass pane.

Reset loose jewelry stones If your jewelry has popped a stone or two, you don't have to put it in the "play dress-up" box yet. The stone can be reset using a little drop of clear nail polish as the "glue." It dries quickly, and the repair will be invisible.

Repair lacquered items Did you chip a favorite lacquered vase or other lacquered item? Try mixing colors of nail polish to match the piece. Paint over the chipped area to make it less noticeable. *Caution:* You may lower the value of an antique by doing this, so you probably only want to try this with inexpensive items.

Plug a hole in your cooler A small hole inside your cooler doesn't make it trash-worthy yet. Seal the hole with two coats of nail polish to hold in ice and other melted substances.

Fill washtub nicks It's a mystery how they got there, but your washing machine tub has one or two nicks near the holes, and now you're concerned about snags in your clothes or even rust spots. Seal those nicks with some nail polish, feathering the edges so there is no lip.

Keep chipped car paint from rusting If your car suffers small dings and chips, you can keep them from rusting or enlarging by dabbing clear nail polish onto the damaged areas.

Smooth wooden hangers If you've noticed a few splinters or nicks in your wooden hangers, no need to toss them out. Brush some nail polish over the rough edges to smooth the surface again and keep your coat linings safe.

Tighten loose screws You're not rough with your drawers and cabinets, but you find yourself tightening certain pull screws once too often. Keep them in place by brushing a little clear polish on the screw threads, insert the screws, and let dry before using again. This is also a great solution if you've been keeping a Phillips screwdriver in the kitchen for loose pot handles. You can also use clear nail polish to keep nuts on machine screws or bolts from coming loose, and if you need to take the nuts off, a twist with a wrench will break the seal.

Mend holes in window screens You notice a small hole has been poked in your window or door screen. If the hole is no more than about 1/4 inch (6 millimeters) in diameter, you can block the bugs and keep the hole from getting bigger by dabbing on a bit of clear nail polish.

Fix torn window shades Got a little tear in your window shade? Don't worry. You can usually seal it with a dab of clear nail polish.

* Nail Polish Remover

Remove stains from china Your bone china has assorted stains from years of use. Spruce up your set by rubbing soiled areas with nail polish remover. Clean spots with a cotton swab and then wash dishes as usual.

TAKE CARE Frequent use of nail polish remover containing acetone—check the label—can cause dry skin and brittle nails. All nail polish removers are flammable and potentially hazardous if inhaled for a long time; use them in a well-ventilated area away from flames. And work carefully; they can damage synthetic fabrics, wood finishes, and plastics.

Eliminate ink stains If the ink stains on your skin won't come off with soap and water, they are probably not water-soluble. Try using nail polish remover instead. Take a cotton ball and wipe the affected areas with the solution. Once the ink stains are gone, wash skin with soap and water. Nail polish remover can also eliminate ink stains on the drum of your clothes dryer.

Rub paint off windows Spare your nails the next time you want to remove paint on a window. Working in a well-ventilated area, dab on nail polish remover in small sections. Let the solution remain on the painted areas for a few minutes before rubbing it off with a cloth. Once finished, take a damp cloth and go over the areas again.

Remove stickers from glass Scraping price stickers from glass objects can be messy, and it often leaves behind a gummy adhesive that attracts dirt and is sticky to the touch. Remove the stickers and clean up the residual glue by wiping the area with acetone-based nail polish remover. The same method can be used for removing stickers and sticky residue from metal surfaces.

Dissolve melted plastic Ever get too close to a hot metal toaster with a plastic bag of bread or bagels? The resulting mess can be a real cleaning challenge. But don't let a little melted plastic ruin a perfectly good appliance. Eliminate the sticky mess with nail polisher remover. First unplug the toaster and wait for it to cool. Then pour a little nail polish remover on a soft cloth and gently rub over the damaged areas. Once the melted plastic is removed, wipe with a damp cloth and dry with a paper towel. Your toaster is now ready for the next round of bagels. The same solution works for melted plastic on curling irons.

Unhinge superglue Superglue will stick tenaciously to just about anything, including your skin. And trying to peel it off your fingers can actually cause skin damage. Instead, soak a cotton ball with acetone-based nail polish remover and hold it on the skin until the glue dissolves.

Clean vinyl shoes Patent-leather shoes may not reflect up, but they *do* show off scuff marks, as will white or other light-colored vinyl shoes. To remove the marks,

SCIENCE FAIR

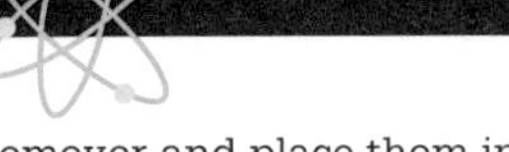

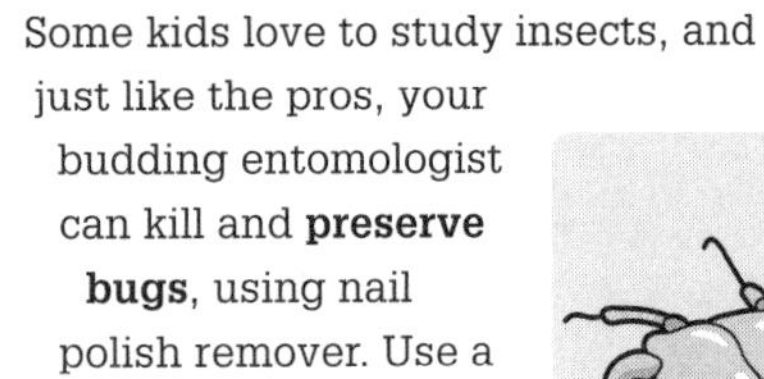

Some kids love to study insects, and just like the pros, your budding entomologist can kill and **preserve bugs**, using nail polish remover. Use a **nail polish remover** that contains the solvent acetone. (Check the label or sniff it for a banana-like odor.) To preserve the insects, soak some **cotton balls** with the polish remover and place them in a **glass or plastic jar** along with several **tissues** and the selected insects. A wide-mouthed peanut butter jar works well. The tissues prevent the insects from damaging their wings. Seal the jar tightly with a lid, and the specimens will quickly dehydrate. Use a **straight pin** stuck through the insect's body to mount it on a corkboard or corrugated **cardboard**.

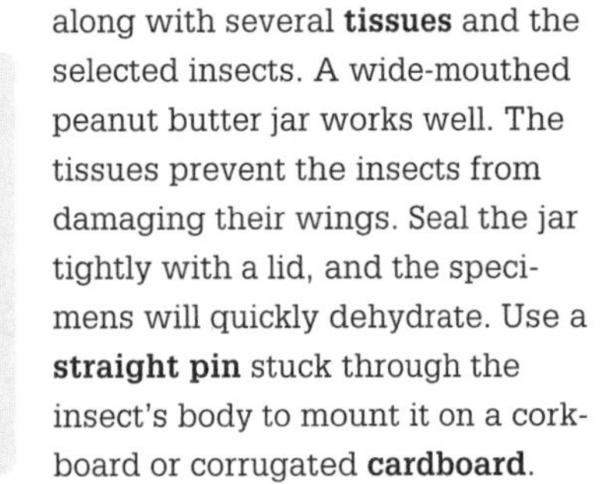

rub them lightly but briskly with a soft cloth or paper towel dipped in nail polish remover. Afterward, remove any residue with a damp cloth.

Keep watches clean Tired of looking at your watch and seeing unsightly scratches when you check the time? Get rid of them with nail polish remover. If the face of your watch is made from unbreakable plastic, rub the remover over the scratches until they diminish or disappear.

Clean computer keyboards You can keep computer keyboards clean with nail polish remover and an old toothbrush. Simply moisten the brush with remover and lightly rub the keys.

Dilute correction fluid To take the goop out of correction fluid or old nail polish, dilute it with nail polish remover. Pour just a few drops into the bottle and then shake. Add a little more polish remover to the solution, if needed, to attain the desired consistency.

Prep brass for re-lacquering Old or damaged lacquer coatings on brass can be safely removed with nail polish remover. Take a soft cloth and pour a small amount of remover on it. Rub the brass object until the old lacquer has been lifted. Your brass item is now ready to be polished or professionally re-lacquered.

* Newspaper

Encase your glassware for moving Are you relocating or packing up items for long-term storage? Use several sheets of soaking-wet newspapers to wrap up your glass dishes, bowls, drinking glasses, and other fragile items, and then let them thoroughly dry before packing. The newspaper will harden and form a protective cast around the glass that will dramatically improve its chances of surviving the move without breaking.

Store sweaters and blankets Don't treat moths to a fine meal of your homemade or store-bought woolen sweaters and blankets. When putting them into storage,

DID *You* KNOW?

America's first newspaper was called *Publick Occurrences, Both Foreign and Domestick.* It was a folded three-panel sheet of paper printed in Boston on September 25, 1690, by Richard Pierce and Benjamin Harris. Unfortunately, British colonial authorities closed it after its first and only issue and quickly issued a decree banning "unlicensed" publications. Ironically, the only known copy isn't found in the United States. It's in the Public Records Office in London.

wrap your woolens in a few sheets of newspaper (be sure to tape up the corners). It will keep away the moths, and keep out dust and dirt.

Clean and polish your windows If you're like most folks, you probably use a lot of absorbent paper towels for drying off your just-washed windows. Did you know that crumpled-up newspaper dries and polishes windows even better than paper towels? And it's a lot cheaper too.

Deodorize luggage and containers Do you have a plastic container or wooden box with a persistent, unpleasant odor? Stuff in a few sheets of crumpled newspaper and seal it closed for three or four days. You can also use this technique to deodorize trunks and suitcases (using more newspaper, of course).

Dry wet shoes If your shoes get soaked after walking through the rain or slogging through the snow, stuff them with dry, balled-up newspaper to prevent any long-term damage. Place the shoes on their sides at room temperature so the moisture can be thoroughly absorbed. For severe sogginess, you may need to replace the stuffing a few times.

Make an impromptu ironing board If you always take along a travel iron—just in case you end up in a motel that doesn't provide irons and ironing boards—it's a cinch to make your own on-the-road ironing board. Simply fill a pillowcase with a short stack of newspapers, keeping it as level as possible. Then place it on a countertop or the floor and get pressing.

Create an emergency splint If someone you're with takes a nasty fall—and you suspect there may be a bone injury to an arm or leg—it's important to immobilize the limb to prevent pain and additional damage. Fashion a makeshift splint by folding up several sheets of newspaper until stiff and attach it beneath the limb using a few pieces of adhesive tape. You may need to overlap a couple of folded sheets to make a splint long enough for a leg injury.

Remove oven residue They may call it a self-cleaning oven, but when it's done cleaning, you always have to contend with mopping off that ashlike residue. Don't waste a roll of paper towels on the flaky stuff; clean it up with a few sheets of moistened, crumpled newspaper.

Pick up broken glass shards Okay, so everyone breaks a big glass bowl at least once in a lifetime. It's no big thing. A safe way to get up the small shards of glass that remain after you remove the large pieces is to blot the area with wet newspapers. The tiny fragments will stick to the paper, which makes for easy disposal. Just carefully drop the newspaper in your garbage can.

Unscrew a broken lightbulb To remove a broken lightbulb, wad up several sheets of newspaper, press the paper over the bulb, and turn it counterclockwise. (Make sure you're wearing protective gloves and that the power is off.) The bulb should loosen up enough to remove from the socket. Wrap it in the paper and toss it into the garbage.

N

Slow-ripen tomatoes in late fall Is there an early frost predicted and you still have a bunch of tomatoes on the vine? Relax. Pick your tomatoes and wrap each one in a couple of sheets of newspaper. Store them in airtight containers inside a dark cabinet or closet at room temperature. Check each one every three to four days; they will all eventually ripen to perfection.

Line your trash compactor Putting a few layers of newspapers at the bottom of your trash compactor will not only soak up the nasty odors caused by rotting foods, it will also protect the unit against damage caused by any sharp objects that manage to squeeze through.

Use as mulch Newspaper makes terrific mulch for veggies and flowers. It's excellent at retaining moisture and does an equally fine job at fighting off and suffocating weeds. Just lay down several sheets of newspaper. Then cover the paper with about 3 inches (7.5 centimeters) of wood mulch so it doesn't blow away. *Warning:* Avoid using glossy stock and colored newsprint for mulching (or composting); color inks may contain lead or harmful dyes that can leach into the ground. To check your newsprint, contact your local paper and ask about the inks they use; many papers now use only safe vegetable-based inks.

Add to compost Adding moderate amounts of wet, shredded newsprint—printed in black ink only—to your compost heap is a good and relatively safe way to reduce odor and to give earthworms a tasty treat.

Adios, earwigs If your garden is under siege by earwigs—those creepy-looking insects with the sharp pincers on their hindquarters—get rid of them by making your own environmentally friendly traps. Tightly roll up a wet newspaper, and put a rubber band around it to keep it from unraveling. Place it in the area you've seen the insects, and leave it overnight. By morning, it will be

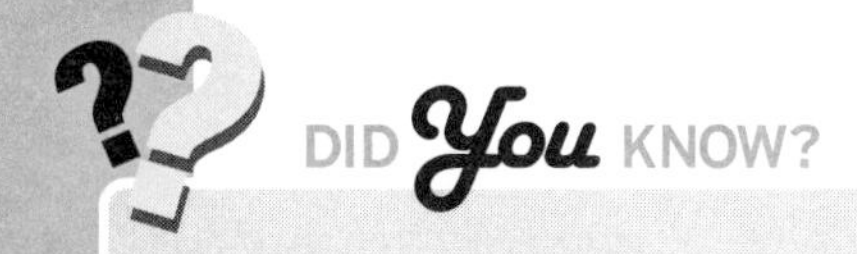

DID *You* KNOW?

What we call newsprint—the type of paper used by newspapers around the world—was invented around 1838 by a Nova Scotia teenager, Charles Fenerty. After hearing frequent complaints from local paper mills about maintaining adequate supplies of rags to make rag paper, Fenerty hit upon the idea of making paper from spruce pulp. Unfortunately, Fenerty didn't go public with his discovery until 1844. By then, a consortium of European investors had already patented a process for creating paper solely from wood fiber.

standing room only for the bugs. Place the newspaper in a plastic grocery bag, tie a knot at the top of the bag, and toss it into the trash. Repeat until your traps are free of earwigs.

Winterize outdoor faucets If you live in an older home without frost-free outdoor spigots, it's a good idea to insulate the outdoor faucets. To prevent damage from ice and cold temperatures, make sure you shut off the valve to each faucet, and drain off any excess water from the spigots. Then insulate each faucet by wrapping it with a few sheets of newspaper covered with a plastic bag (keep the bag in place by wrapping it with duct tape or a few rubber bands).

Protect windows when painting Don't bother buying thick masking or carpenter tape when painting around the windows of your home. Simply wet several long strips of newspaper and place them on the glass alongside the wood you're painting. The newspaper will easily adhere to the surface and keep the paint off the glass or frames, and it is much easier to remove than tape.

Roll your own fireplace logs Bad winter on the way? Bolster your supply of fireplace logs by making a few of your own out of old newspapers. Just lay out a bunch of sheets end to end, roll them up as tightly as you can, tie up the ends with twine or wire, and wet them in a solution of slightly soapy water. Although it will take a while, let them dry thoroughly, standing on end, before using. *Note:* Do not use newspaper logs in a woodstove unless the manufacturer specifies that it is okay.

Put traction under your wheels Unless your vehicle has four-wheel drive, it's always a good idea to keep a small stack of newspapers in the trunk of your car during the winter months to prevent getting stranded on a patch of ice or slush. Placing a dozen or two sheets of newspaper under each rear wheel will often provide just the traction you need to get your car back on the road.

* Oatmeal

Treat itchy poison ivy or chicken pox Take the itch out of a case of chicken pox or a poison ivy rash with a relaxing, warm oatmeal bath. Simply grind 1 cup oatmeal in your blender until it is a fine powder, then pour it into a piece of cheesecloth, the foot section of a clean nylon stocking, or the leg of an old pantyhose. Knot the material, and tie it around the faucet of your bathtub so the bag is suspended under the running water. Fill the tub with lukewarm water and soak in it for 30 minutes. You may find additional relief by applying the oatmeal pouch directly to the rash or pox.

Add luxury to a regular bath You don't have to have itchy skin to make a luxurious bath mix with oatmeal. And it beats buying expensive bath oils. All you need is 1 cup oatmeal and your favorite scented oil, such as rose or lavender. Grind the oatmeal in a blender, put it in a cheesecloth bag, add a few drops of the scented oil, and suspend the bag under the running water as you fill your bathtub. You'll not only find it sweetly soothing, you can also use the oatmeal bag as a washcloth to exfoliate your skin.

Make a facial mask If you're looking for a quick pick-me-up that will leave you feeling *and* looking better, give yourself an oatmeal facial. Combine 1/2 cup hot—

Thirty minutes. Five minutes. One minute! Oatmeal cooking times depend on how the oats were made into oatmeal. After the inedible hull is removed, the oat is called a groat. If the groats are just cut into about four pieces, the oatmeal takes up to 30 minutes to cook. If the groats are steamed and rolled but not cut, it takes about five minutes. If they are steamed, rolled, *and* cut, the cooking time drops to a minute or so. Steaming, rolling, and cooking breaks down the fiber, so if you want a lot of fiber, use 30-minute oatmeal and cook it until it is chewy, not mushy.

not boiling—water and 1/3 cup oatmeal. After the water and oatmeal have settled for two or three minutes, mix in 2 tablespoons plain yogurt, 2 tablespoons honey, and 1 small egg white. Apply a thin layer of the mixture to your face, and let it sit for 10-15 minutes. Then rinse with warm water. (Be sure to place a metal or plastic strainer in your sink to avoid clogging the drain with the granules.)

Make a dry shampoo Do you sometimes need to skip washing your hair in order to get to work on time? Keep a batch of dry shampoo on hand in an airtight container specifically for those occasions when your alarm clock "malfunctions." Put 1 cup oatmeal in the blender and grind it into a fine powder. Add 1 cup baking soda, and mix well. Rub a bit of the mixture into your hair. Give it a minute or two to soak up the oils, then brush or shake it out of your hair (preferably over a towel or bag to avoid getting it all over). This dry shampoo mixture is also ideal for cleaning the hair of bedridden people who are unable to get into a shower or bathtub. Plus, it's equally effective for deodorizing that big ol' bath-hating mutt of yours.

* Olive Oil

Remove paint from hair Did you get almost as much paint in your hair as you did on the walls in your last paint job? You can easily remove that undesirable tint by moistening a cotton ball with some olive oil and gently rubbing it into your hair. The same approach is also effective for removing mascara—just be sure to wipe your eyes with a tissue when done.

Make your own furniture polish Restore the lost luster of your wooden furniture by whipping up some serious homemade furniture polish that's just as good as any of the commercial stuff. Combine 2 parts olive oil and 1 part lemon juice or white vinegar in a clean recycled spray bottle, shake it up, and spritz on. Leave on the mixture for a minute or two, then wipe off with a clean terrycloth or paper towel. In a hurry? Get fast results by applying olive oil straight from the bottle onto a paper towel. Wipe off the excess with another paper towel or an absorbent cloth.

Use as hair conditioner Is your hair as dry and brittle as sagebrush in the desert? Put the moisture back into it by heating 1/2 cup olive oil (don't boil it), and then liberally applying it to your hair. Cover your hair with a plastic grocery bag, then wrap it in a towel. Let it set for 45 minutes, then shampoo and thoroughly rinse.

Clear up acne Okay, the notion of applying oil to your face to treat acne does sound a bit wacky. Still, many folks swear this works: Make a paste by mixing 4 tablespoons salt with 3 tablespoons olive oil. Pour the mixture onto your hands and fingers and work it around your face. Leave it on for a minute or two, then rinse it off with warm, soapy water. Apply daily for one week, then cut back to two or three times weekly. You should see a noticeable improvement in your condition. (The principle is that the salt cleanses the pores by exfoliation, while the olive oil restores the skin's natural moisture.)

O

Substitute for shaving cream If you run out of shaving cream, don't waste your time trying to make do with soap—it could be rough on your skin. Olive oil, on the other hand, is a dandy substitute for shaving cream. It not only makes it easier for the blade to glide over your face or legs, but it will moisturize your skin as well. In fact, after trying this, you may swear off shaving cream altogether.

Tip **Buying Olive Oil**

Expensive extra virgin olive oil is made from olives crushed soon after harvest and processed without excessive heat. It's great for culinary uses where the taste of the oil is important. But for everyday cooking and non-food applications, lower grades of olive oil—light, extra light, or just plain olive oil—work fine and save you money.

Clean your greasy hands To remove car grease or paint from your hands, pour 1 teaspoon olive oil and 1 teaspoon salt or sugar into your palms. Vigorously rub the mixture into your hands and between your fingers for several minutes; then wash it off with soap and water. Not only will your hands be cleaner, they'll be softer as well.

Recondition an old baseball mitt If your beloved, aging baseball glove is showing signs of wear and tear—cracking and hardening of the leather—you can give it a second lease on life with an occasional olive oil rubdown. Just work the oil into the dry areas of your mitt with a soft cloth, let it set for 30 minutes, then wipe off any excess. Your game may not improve, but at least it won't be your glove's fault. Some folks prefer to use bath oil to recondition their mitts (see page 81).

* Onions

Remove rust from knives Forget about using steel wool or harsh chemicals—how's this for an easy way to get the rust off your kitchen or utility knives? Plunge your rusty knife into a large onion three or four times (if it's very rusty, it may require a few extra stabs). The only tears you shed will be ones of joy over your rust-free blade.

Eliminate new paint smell Your bedroom's new shade of paint looks great, but the smell is keeping you up all night. What to do? Place several freshly cut slices of onion in a dish with a bit of water. It will absorb the smell within a few hours.

Correct pet "mistakes" If Rover or Kitty is still not respecting your property—whether it be by chewing, tearing, or soiling—you may be able to get the message across by leaving several onion slices where the damage has been done. Neither cats nor dogs are particularly fond of *"eau de onion,"* and they'll avoid returning to the scene of their crimes.

Soothe a bee sting If you have a nasty encounter with a bee at a barbecue, grab one of the onion slices intended for your burger and place it over the area where you got stung. It will ease the soreness. (If you are severely allergic to bee or other insect stings, seek medical attention at once.)

Use as smelling salts If you happen to be with someone at a party or in a restaurant who feels faint—and you don't normally carry smelling salts in your pocket—reach for a freshly cut onion. The strong odor is likely to bring him around.

Use as a natural pesticide Whip up an effective insect and animal repellent for the flowers and vegetables in your garden. In a blender, puree 4 onions, 2 cloves garlic, 2 tablespoons cayenne pepper, and 1 quart water (1 liter). Set the mixture aside. Now dilute 2 tablespoons soap flakes in 2 gallons (7.5 liters) water. Pour in the contents of your blender, shake or stir well, and you have a potent, environment-friendly solution to spray on your plants.

How can you keep your eyes from tearing when cutting onions? Suggestions range from wearing protective goggles while chopping, to placing a fan behind you to blow away the onion's tear-producing vapors, to rubbing your hands with vinegar before you start slicing. The National Onion Association, however, advises chilling onions in the freezer for 30 minutes prior to slicing them. The association also suggests cutting off the top portion and peeling off the outer layers. The idea is to leave the root end intact, because it has the highest concentrations of the sulfur compounds that cause your eyes to tear.

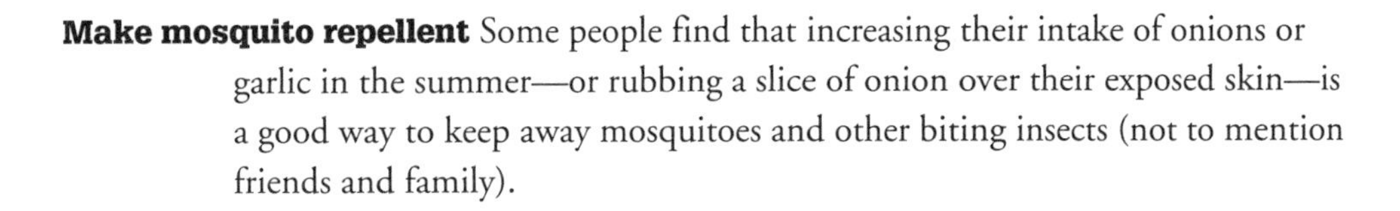

Make mosquito repellent Some people find that increasing their intake of onions or garlic in the summer—or rubbing a slice of onion over their exposed skin—is a good way to keep away mosquitoes and other biting insects (not to mention friends and family).

* Oranges

Use for kindling Dried orange and lemon peels are a far superior choice for use as kindling than newspaper. Not only do they smell better and produce less creosote than newspaper, but the flammable oils found inside the peels enable them to burn much longer than paper.

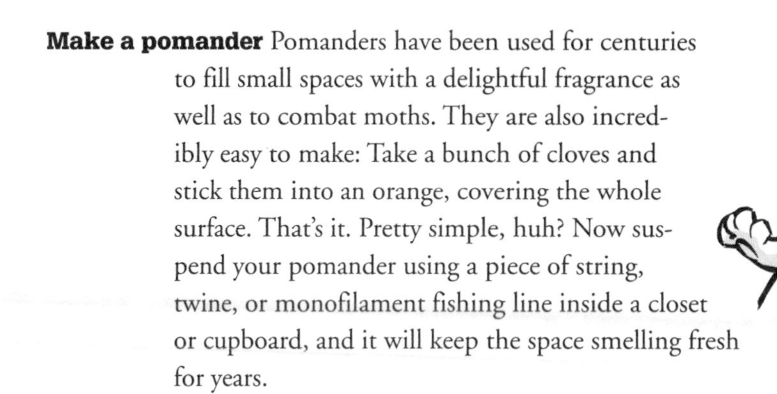

Make a pomander Pomanders have been used for centuries to fill small spaces with a delightful fragrance as well as to combat moths. They are also incredibly easy to make: Take a bunch of cloves and stick them into an orange, covering the whole surface. That's it. Pretty simple, huh? Now suspend your pomander using a piece of string, twine, or monofilament fishing line inside a closet or cupboard, and it will keep the space smelling fresh for years.

Simmer for stovetop potpourri Fill your abode with a refreshing citrus scent by simmering several orange and/or lemon peels in 1-2 cups of water in an aluminum pot for a few hours. Add water as needed during the simmering. This process freshens up the pot as well as the air in your home.

Keep kitties off your lawn Are the neighbor's cats still mistaking your lawn for their litter box? Gently point them elsewhere by making a mixture of orange peels and coffee grounds and distributing it around the cats' "old haunts." If they don't take the hint, lay down a second batch and try moistening it with a bit of water.

Apply as mosquito repellent If you're not crazy about the idea of rubbing onions all over yourself to keep away mosquitoes (see previous page), you may be happy to know that you can often get similar results by rubbing fresh orange or lemon peels over your exposed skin. It's said that mosquitoes and gnats are totally repulsed by either scent.

Show ants the door Get rid of the ants in your garden, on your patio, and along the foundation of your home. In a blender, make a smooth puree of a few orange peels in 1 cup warm water. Slowly pour the solution over and into anthills to send the little pests packing.

* Oven Cleaner

Put the style back in your curling iron Is your curling iron buried under a layer of caked-on styling gel or conditioner? Before the next time you use it, spray on a light coating of oven cleaner. Let it sit for one hour, then wipe it off with a damp rag, and dry with a clean cloth. *Warning:* Do not use iron until it is thoroughly dry.

Wipe away bathtub ring Got a stubborn stain or ring around your white porcelain tub that refuses to come clean? Call out the big guns by spraying it with oven cleaner. Let it sit for a few hours, then give it a thorough rinsing. *Warning:* Do not apply oven cleaner to colored porcelain tubs; it could cause fading. And be careful not to get the oven cleaner on your shower curtain; it can ruin both plastic and fabric.

Clean grimy tile grout lines Ready for an all-out attack on grout grunge? First, make sure you have plenty of ventilation—it's a good idea to use your exhaust fan to suck air out of a small bathroom. Put on your rubber gloves and spray oven cleaner into the grout lines. Wipe the cleaner off with a sponge within five seconds. Rinse thoroughly with water to reveal sparkling grout lines.

TAKE CARE Most oven cleaners contain highly caustic lye, which can burn the skin and damage the eyes. Always wear long rubber gloves and protective eyewear when using oven cleaner. The mist from oven cleaner spray can irritate nasal membranes. Ingestion can cause corrosive burns to the mouth, throat, and stomach that require immediate medical attention. Store oven cleaner well out of children's reach.

Clean ovenproof glass cookware You've tried everything to scrub those baked-on stains off your Pyrex or CorningWare cookware. Now try this: Put on rubber gloves and cover the cookware with oven cleaner. Then place the cookware in a heavy-duty garbage bag, close it tightly with twist ties, and leave overnight. Open the bag outdoors, keeping your face away from the dangerous fumes. Use rubber gloves to remove and wash the cookware.

Clean a cast-iron pot If you need to clean and re-season that encrusted secondhand cast-iron skillet you found at a yard sale, start by giving it a good spraying with oven cleaner and placing it in a sealed plastic bag overnight. (This keeps the cleaner working by preventing it from drying.) The next day, remove the pot and scrub it with a stiff wire brush. Then, wash it thoroughly with soap and water, rinse well, and immediately dry it with a couple of clean, dry cloths. *Note:* This technique eliminates built-up gunk and grease, but not rust. For that, you'll need to use vinegar. Don't leave it on too long, though. Prolonged exposure to vinegar can damage your cast-iron utensil.

Remove stains from concrete Get those unsightly grease, oil, and transmission fluid stains off your concrete driveway or garage floor. Spray them with oven cleaner. Let it settle for 5-10 minutes, then scrub with a stiff brush and rinse it off with your garden hose at its highest pressure. Severe stains may require a second application.

Strip paint or varnish For an easy way to remove paint or varnish from wooden or metal furniture, try using a can of oven cleaner; it costs less than commercial paint strippers and is easier to apply (that is, if you spray rather than brush it on). After applying, scrub off the old paint with a wire brush. Neutralize the stripped surface by coating it with vinegar, and then wash it off with clean water. Allow the wood or metal to thoroughly dry before repainting. Don't use oven cleaner to strip antiques or expensive furnishings; it can darken the wood or discolor the metal.

Oven Mitts

O

Use as beverage cozy or egg warmer Keep that mug of java or tea from getting cold when you're called away by placing an oven mitt over it. The glove's insulation will keep it warm until you get back. You can also use an oven mitt to keep boiled eggs warm for up to half an hour. Conversely, an oven mitt will help keep a cold drink colder longer.

Use for dusting and polishing Although oven mitts are typically confined to kitchen duty, they're actually great for dusting and polishing around your house. Use one side of the mitt to apply wax or polish to your furniture, and the other side to buff it up. It's a great way to use old mitts or all those extra ones you've collected.

When pruning thorny plants Although oven mitts may be a bit too awkward to use for weeding or planting seedlings in your garden, they can come in awfully handy when it comes time to prune trees, hedges, and bushes—particularly those thorny devils such as holly, firethorn, and rosebushes.

Remove hot engine parts Keeping an oven mitt in your car's glove compartment or trunk can make life a lot easier when you need to handle hot radiator caps and the like during an on-the-road emergency.

Change a hot lightbulb Did the lightbulb on your reading lamp just blow out? Don't scorch your fingers when replacing it. Once you've removed the lampshade, put on an oven mitt, remove the dead bulb from the socket, and toss it into the garbage. That way, you won't still be blowing on your fingertips when screwing in the new bulb.

* Paintbrushes

Use for delicate dusting A feather duster or dust rag is fine for cleaning shelves and such, but neither one is much good when you need to get into the tiny cracks and crevices of chandeliers, wicker furniture or baskets, and all sorts of knick-knacks. That's when a small natural-bristle paintbrush can be indispensable. The soft bristles are perfect for cleaning out areas that are otherwise impossible to reach. It's also excellent for dusting delicate items such as porcelain or carved-wood figurines.

Brush off beach chairs Keep a clean, dry paintbrush in your car specifically for those return trips from the beach. Use it to remove sand from beach chairs, towels, toys, the kids, and even yourself before you open the car door or trunk. You'll wind up with a lot less to vacuum the next time you clean your vehicle.

Brush on the sauce A small synthetic-bristle paintbrush can be invaluable in the kitchen. You can use it to brush on pie glaze, marinades, and sauces while baking or roasting. You can also use it to paint on the barbecue sauce when grilling burgers and steaks in the great outdoors. To top it all off, a paintbrush is easier to clean than most conventional pastry brushes.

Apply stain remover to clothes Let's face it: Pouring detergent or stain remover onto a soiled garment is often a hit-or-miss proposition—and when you miss, it usually involves grabbing the paper towels to soak up a spill. Make life easier for yourself. Use a small paintbrush to apply liquid stain remover to dirty shirt collars and such. It's neater and a lot more accurate.

Clean your window screens Are your window screens screaming out for a good cleaning? Use a large, clean paintbrush to give them a good dusting. Then shake off the brush, dip it into a small dish of kerosene, and "paint" both sides of your screens. Dry off the mesh with a clean cloth.

Cover up seeds when sowing Sow your seeds with a little TLC. When planting seeds in rows, use a large paintbrush to gently brush them over with soil. This lets you distribute the exact amount of soil needed and prevents overpacking.

P

Buying and Cleaning Paintbrushes

Natural-bristle brushes work best with alkyd/oil-based paint. But use a synthetic-bristle brush with latex paint because the water in the paint can ruin natural bristles. Before cleaning any brush, wipe excess paint onto newspaper. Clean a brush used with alkyd paint in mineral spirits until you get out all the paint, then shake out. To clean out latex paint, wash the brush thoroughly with soapy water, rinse clean, and shake out. Many latex paints contain acrylics that won't wash out completely with soap and water. In this case, finish up with mineral spirits.

Pantyhose

* PANTYHOSE AROUND THE HOUSE

Find lost small objects Have you ever spent hours on your hands and knees searching through a carpet for a lost gemstone, contact lens, or some other tiny, precious item? If not, count yourself among the lucky few. Should you ever be faced with this situation, try this: Cut a leg off an old pair of pantyhose, make sure the toe section is intact, and pull it up over the nozzle of your vacuum cleaner hose. (If you want additional security, you can even cut off the other leg and slip that over as well.) Secure the stocking in place with a tightly wound rubber band. Turn on the vacuum, *carefully* move the nozzle over the carpet, and you'll soon find your lost valuable attached to the pantyhose filter.

Vacuum your fish tank If you have a wet-dry shop vacuum, you can change the water in your fish tank without disturbing the gravel and tank accessories. (You'll still have to relocate the fish, of course.) Just pull the foot of an old nylon stocking over the end of the vacuum's nozzle, secure it with a rubber band, and you are ready to suck out the water.

Buff your shoes Bring out the shine in your freshly polished shoes by buffing them with a medium-length strip of pantyhose. It works so well, you may retire that chamois cloth for good.

Keep your hairbrush clean If you dread the prospect of cleaning out your hairbrush, here's a way to make the job much easier. Cut a 2-inch (5-centimeter) strip from the leg section of a pair of pantyhose, and stretch it over and around the bristles of your new (or newly cleaned) hairbrush. If necessary, use a bobby pin or a comb to push the hose down over the bristles. The next time your brush needs cleaning, simply lift up and remove the pantyhose layer—along with all the dead hair, lint, etc. on top—and replace it with a fresh strip.

Wrap up wrapping paper Keep your used rolls of wrapping paper from tearing and unraveling by storing them in tubes made by cutting the leg sections off old pairs of pantyhose. (Don't forget to leave the foot section intact.) Or, if you have a bunch of used rolls, you can simply put one in each leg of a pair of pantyhose and hang them over a hanger in your closet.

Remove nail polish Can't find the cotton balls? Moisten strips of recycled pantyhose with nail polish remover to take off your old nail polish. Cut the material into 3-inch (7.5-centimeter) squares, and store a stack of them in an old bandage container or makeup case.

Keep spray bottles clog-free If you recycle your spray bottles to use with homemade cleaners or furniture polishes, you can prevent any potential clogs by covering the open end of the tube—the part that goes inside the bottle—with a small, square-cut piece of pantyhose held in place with a small rubber band. This works especially well for filtering garden sprays that are mixed from concentrates.

Substitute for stuffing Is your kid's teddy bear or doll losing its stuffing? Get out a needle and thread and prepare the patient for an emergency "stuffing transplant." Replace the lost filler with narrow strips of clean, worn-out pantyhose (ball them up, if possible). Stitch the hole up well, and a complete recovery is guaranteed. This works well with throw pillows and seat cushions too.

Organize your suitcase As any seasoned traveler knows, you can squeeze more of your belongings into any piece of luggage by rolling up your clothes. To keep your bulkier rolls from unwrapping, cover them in flexible nylon tubes. Simply cut the legs off a pair of old pantyhose, snip off the foot sections, and stretch the stockings over your rolled-up garments. Happy travels!

Take a citrus bath Make your own scented bath oil by drying and grinding up orange and/or lemon peels and then pouring them into the foot section of a recycled pantyhose. Put a knot about 1 inch (2.5 centimeters) above the peels, and leave another 6 inches (15 centimeters) or so of hose above that before cutting off

Nylon, the world's first synthetic fiber, was invented at E. I. DuPont de Nemours, Inc., and unveiled on October 28, 1938. Instead of calling a press conference, company vice president Charles Stine chose to make the landmark announcement to 3,000 women's club members at the New York World's Fair, introducing it with live models wearing nylon stockings. Stine's instincts were right on the money: By the end of 1940, DuPont had sold 64 million pairs of stockings. Nylon actually made its big-screen debut a year earlier, when it was used to create the tornado that lifted Dorothy out of Kansas in *The Wizard of Oz*.

the remainder. Tie the stocking to the bathtub faucet with the peels suspended below the running water. In addition to giving your bath a fresh citrus fragrance, you can use the stocking to exfoliate your skin.

Hold mothballs or potpourri Looking for an easy way to store mothballs in your closet or to make sachets of potpourri to keep in your dresser drawers? Pour either ingredient into the toe section of your recycled nylons. Knot off the contents, then cut off the remaining hose. If you plan to hang up the mothballs, leave several inches of material before cutting.

Make a ponytail scrunchy Why buy a scrunchy for your ponytail when you can easily make one for nothing? Just cut a horizontal strip about 3-inches (7.5 centimeters) wide across a pantyhose leg, wrap it a few times around your ponytail, and you're done.

Use to hang-dry sweaters Avoid getting clothespin marks on your newly washed sweaters by putting an old pair of pantyhose through the neck of the sweater and running the legs out through the arms. Then hang the sweater to dry on your clothesline by clipping the clothespins onto the pantyhose instead of the wool.

Bundle blankets for storage For an effortless and foolproof way to keep blankets and quilts securely bundled before they go into temporary storage, wrap them up in large "rubber bands" made from the waistbands from your used pantyhose. You can reuse the bands year after year if needed.

Tie up boxes, newspapers, magazines If you run out of twine (or need something stronger—say, for a large stack of glossy magazines), tie up your bundles of boxes, newspapers, and other types of recyclable paper goods using an old pair of pantyhose. Cut off the legs and waistband, and you'll be able to get everything curbside without any snags.

* PANTYHOSE IN THE KITCHEN

Store onions in cutoff bundles Get the maximum shelf life out of your onions by hanging them in nylon holders that provide the good air circulation needed to keep them fresh. Place the onions one at a time into the leg of a clean pair of pantyhose. Work the first one down to the foot section. Tie a knot above it and add the next one—repeat until done. Cut off the remaining hose, and then hang the stocking in a cool, dry area of your kitchen. You can easily remove your onions when needed by snipping off each knot, starting from the bottom.

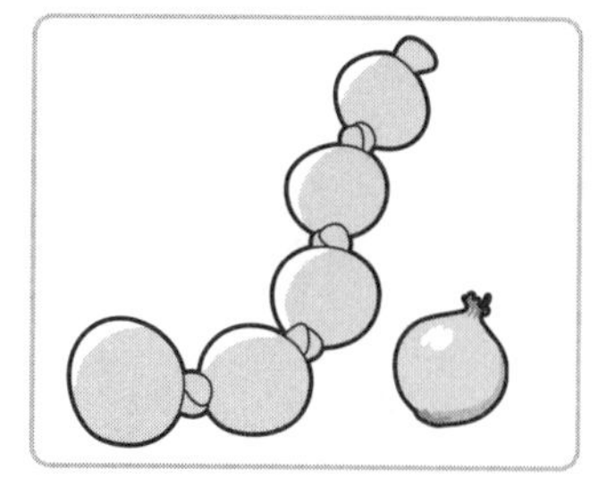

Make a pot or dish scrubber Clean those stains off your nonstick cookware by making a do-it-yourself scrub pad. Crumple up a pair of clean old pantyhose, moisten it with a bit of warm water, add a couple of drops of liquid dishwashing detergent, and you're good to go. You can also make terrific scrubbers for dishes—as well as walls and other nonporous surfaces—by cutting off the foot or toe section, fitting it over a sponge, and knotting off the end.

Make a flour duster Looking for a simple way to dust baking pans and surfaces with exactly the right amount of flour? Just cut the foot section off a clean old pantyhose leg, fill it with flour, tie a knot in it, and keep it in your flour jar. Give your new flour dispenser a few gentle shakes whenever you need to dust flour onto a baking pan or prepare a surface for rolling out dough for breads or pastries.

Keep a rolling pin from sticking Getting pie dough to the perfect consistency is an art form in itself. Although you can always add water to dough that's too dry, it often results in a gluey consistency that winds up sticking to your rolling pin. Avoid the hassle of scraping clean your rolling pin by covering it with a piece of pantyhose. It will hold enough flour to keep even the wettest pie dough from sticking to the pin.

Secure trash bags How many times have you opened your kitchen trash can only to discover that the liner has slipped down (and that someone in your house has covered it over with fresh garbage anyway)? You can prevent such "accidents" by firmly securing the garbage bag or liner to your trash can with the elastic waistband from a recycled pair of pantyhose; tie a knot in the band to keep it tight. You can also use this method to keep garbage bags from slipping off the edge of your outdoor garbage bins.

Dust under the fridge Having trouble catching those dust bunnies residing underneath and alongside your refrigerator? Round them up by balling up a pair of old pantyhose and attaching it with a rubber band to a coat hanger or yardstick. The dust and dirt will cling to the nylon, which can easily be washed off before being called back for dusting duty.

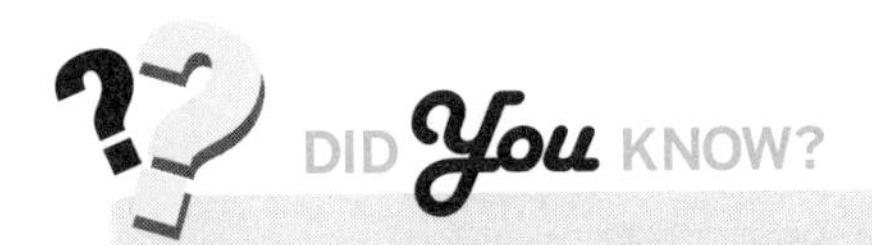

You have probably heard that you can temporarily replace a broken fan belt with pantyhose in an emergency. Well, don't believe it—it won't work! Jim Kerr, automotive technician instructor at the Saskatchewan Institute of Applied Arts and Sciences and "Tech Talk" columnist for CanadianDriver.com, says, "Pulleys in most vehicles require flat belts, not the rounded shape pantyhose would present. Even on a V-belt pulley, they fly off as soon as the engine starts. We know; we've tried it." A much better idea is to replace the belts before they get in bad condition.

P

Stake delicate plants Give your young plants and trees the support they need. Use strips of pantyhose to attach them to your garden stakes. The nylon's flexibility will stretch as your seedlings or saplings fill out and mature—unlike string or twine, which can actually damage plant stalks if you tie it too tightly.

Store flower bulbs in winter Pantyhose legs make terrific sacks for storing your flower bulbs over winter, since they let air freely circulate around the bulbs to prevent mold and rot. Simply cut a leg off a pair of pantyhose and place your bulbs inside, knot off the end, and place ID tags on each sack using a strip of masking tape. Hang them up in a cool, dry space, and they'll be ready for planting in the spring

Prevent soil erosion in houseplants When moving a houseplant to a larger or better accommodation, put a piece of pantyhose at the bottom of the new pot. It will act as a liner that lets the excess water flow out without draining the soil along with it.

Support melons Keep small melons such as cantaloupe and muskmelons off the ground—and free of pests and disease—by making protective sleeves for them from your old pantyhose. Cut the legs off the pantyhose. As your young melons start to develop, slide each one into the foot section, and tie the leg to a stake to suspend the melon above the ground. The nylon holders will stretch as the melons mature, while keeping them from touching the damp soil, where they would be susceptible to rot or invasion by hungry insects and other garden pests.

Keep deer out of your garden If you've been catching Bambi and her friends nibbling on your crops, put up a "No Trespassing" sign they will easily understand. Simply fill the foot sections of some old pantyhose with human hair clippings collected from hairbrushes or your local barbershop—or, even better, use Rover's fur after a good brushing. Tie up the ends, and hang up the nylon satchels where the deer tend to snack. They won't be back for seconds. The hair or fur will lose its scent after a while, so replace every four or five days as needed.

Clean up after gardening Here are two recycling tips in one: Save up your leftover slivers of soap, and place them in the foot section of an old nylon stocking. Knot it off, and hang it next to your outdoor faucet. Use the soap-filled stocking to quickly wash off your hands after gardening and other outdoor work without worrying about getting dirt on door handles or bathroom fixtures inside your house.

Cover a kids' bug jar What child doesn't like to catch fireflies—and hopefully release them—on a warm summer night? When making a bug jar for your youngster, don't bother using a hammer and nail to punch holes in the jar's metal lid (in fact, save the lids for other projects). It's much easier to just cut a 5- or 6-inch (15-centimeter) square from an old pair of pantyhose and affix it to the jar with a rubber band. The nylon cover lets plenty of air enter the jar, and makes it easier to get the bugs in and out.

* PANTYHOSE FOR THE DO-IT-YOURSELFER

Apply stain to wood crevices Getting wood stain or varnish into the tight corners and crevices of that unfinished bookcase or table that you just bought can be a maddening task. Your brush just won't fit into them and give them an even coating. But there's really nothing to it once you know the secret. Just cut a strip from an old pair of pantyhose, fold it over a few times, and use a rubber band to affix it to the tip of a wooden Popsicle stick. Dip your homemade applicator into the stain or varnish, and you'll have no trouble getting it into those hard-to-reach spots.

Test a sanded surface for snags Think you did a pretty good job sanding down that woodworking project? Put it to the pantyhose test. Wrap a long piece of pantyhose around the palm of your hand and rub it over the wood. If the pantyhose snags onto any spots, sand them until you're able to freely move the nylon over the surface without any catches.

Patch a hole in a screen Don't invite the bugs in for a bite; use a small square of pantyhose to temporarily patch that hole in your window screen. You can secure the

According to the Toy Industry Association, legendary doll maker Madame Alexander came up with the concept for pantyhose in the early 1950s, when she started sewing tiny pairs of silk stockings onto her dolls' underpants to keep them from slipping down. But Allen Gant, Sr., of Burlington, North Carolina, invented pantyhose as we know them, and they were first produced in 1959 by Glen Raven Mills, his family's textile business. Another pantyhose pioneer is Hollywood actress Julie Newmar, best known as the original Catwoman on the old *Batman* TV series in the late 1960s, who holds a patent for "ultra-sheer, ultra-snug" pantyhose.

patch by simply applying some rubber cement around the hole before pressing the patch in place. When you're ready to fix the hole with a piece of screening, peel off the nylon and the glue. If you want the patch to last a bit longer, sew it onto the screen with thread.

P

Clean your pool Want a more effective way to skim the debris off the surface of your pool water? Cut a leg off a pair of pantyhose and fit it over your pool's skimmer basket. It will catch a lot of tiny dirt particles and hairs that would otherwise make their way into—and possibly clog—your pool's filter unit.

Make a paint strainer Strain your paint like the pros: Use a pantyhose filter to remove the lumps of paint from an old can of paint. First, cut a leg off a pair of old pantyhose, clip the foot off the leg, and make a cut along the leg's length so that you have a flat piece of nylon. Then cut the leg into 12- to 14-inch (30- to 32-centimeter) sections to make the filters. Stretch the nylon over a clean bucket or other receptacle and hold it in place with a rubber band or perhaps even the waistband from that pair of pantyhose. Now slowly pour the paint into the bucket.

Paper Bags

✱ PAPER BAGS AROUND THE HOUSE

Dust off your mops Dust mops make it a breeze to get up the dust balls and pet hair around your home, but how do you get the stuff off your mop? Place a large paper bag over the mop head; use a piece of string or a rubber band to keep it from slipping off. Now give it several good shakes (a few gentle bumps wouldn't hurt either). Lay the mop on its side for a few minutes to let the dust in the bag settle. Then carefully remove the bag for easy disposal of your dusty dirt.

Clean artificial flowers Authentic silk flowers are actually pretty rare these days; most are now made of nylon or some other man-made material. But regardless of whether they're silk or something else, you can easily freshen them up by placing them in a paper bag with 1/4 cup salt. Give the bag a few gentle shakes, and your flowers will emerge as clean as the day you purchased them.

Carry your laundry If your laundry basket is already overflowing, or (gasp!) the plastic handle suddenly gives out, you can always use a sturdy shopping bag to pick up the slack. A bag with handles will probably make the job easier, but any large bag will do in a pinch. Just be sure your laundry is completely dry before using the bag on the return trip. Otherwise, your freshly cleaned clothes could wind up under your feet.

Cover your kids' textbooks Helping your children make book covers for their textbooks isn't only fun, it's also a subtle way to teach kids to respect public property. And few materials rival a paper bag when it comes to making a rugged book cover. First, cut the bag along its seams to make it a flat, wide rectangle, then place the book in the center. Fold in the top and bottom edges so the bag is only slightly wider than the book's height. Next, fold over the sides to form sleeves over the book covers. Cut off the excess, leaving a couple of inches on either side to slide over the front and back covers. Put a piece of masking tape on the top and bottom of each sleeve (over the paper, not the book) to keep it on tight, and you're done. Lastly, let your child put his or her personal design on each cover.

Create a table decoration Use a small designer shopping bag with handles to make an attractive centerpiece for your dining room table or living room mantel. Fill a small cup with some water and place it in the middle of the bag. Place a few fresh-cut flowers in the glass, and presto! All done.

Make your own wrapping paper Need to wrap a present in a hurry? You don't have to rush out to buy wrapping paper. Just cut a large paper bag along the seams until it's a flat rectangle. Position it so that any printing is facing up at you, put your gift on top and fold, cut, and tape the paper around your gift. If you wish, personalize your homemade wrapping paper by decorating it with markers, paint, or stickers.

Are paper shopping bags better for the environment than their plastic counterparts? Not really, according to the U.S. Environmental Protection Agency. Paper bags generate 70 percent more air pollutants and 50 times more water pollutants than plastic bags. What's more, it takes four times as much energy to make a paper bag as it does to manufacture a plastic bag, and 91 percent more energy to recycle paper than plastic. On the other hand, paper bags come from a renewable resource (trees), while most plastic bags are made from nonrenewable resources (polyethylene, a combination of crude oil and natural gas). So what's the answer? Bring your own cloth shopping bags with you to the market!

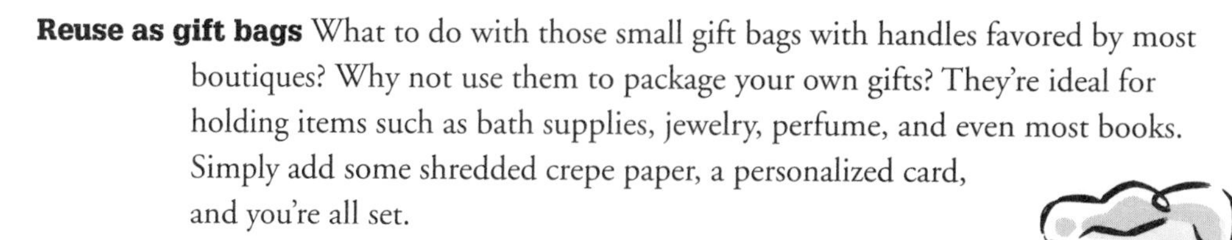

Reuse as gift bags What to do with those small gift bags with handles favored by most boutiques? Why not use them to package your own gifts? They're ideal for holding items such as bath supplies, jewelry, perfume, and even most books. Simply add some shredded crepe paper, a personalized card, and you're all set.

Recycle as towel or tissue dispensers Add a simple but elegant touch to your guest bathrooms by using small "boutique bags" as paper towel or tissue dispensers. You can even embellish them with your own personal touches, such as ribbons or stickers that match your decor.

Reshape knits after washing Put the shape back into your wool sweater or mittens by tracing the contours of the item on a paper bag before you wash it. Then use your outline to stretch the item back to its original shape after washing it.

Store linen sets Have you ever emptied the contents of your linen closet looking for the flat sheet to match the fitted one you just pulled out? You can easily spare yourself some grief by using medium-sized paper bags to store your complete linen sets. Not only will your shelves be better organized, but you can also keep your linens smelling fresh by placing a used fabric softener sheet in each bag.

Use as a pressing cloth If your ironing board's cover appears to have seen its last steam iron, don't sweat it. You can easily make a temporary pressing cloth by splitting open one or two paper bags. Dampen the bags and lay them over your ironing board to get those last few shirts or skirts pressed for the workweek.

Pack your bags Getting ready to leave on a family vacation? Don't forget to pack a few large shopping bags—the kind with handles—in your luggage. They're guaranteed to come in handy to bring home the souvenirs you pick up, or perhaps your soiled laundry or beach towels.

Bag your recycled newspapers Double up on your recycling efforts by using large paper bags to hold your newspapers for collection. It not only spares you the time and effort needed to tie up your bundles with string, but it also makes it easier to sort out your magazines, newsprint, and glossy pages.

* PAPER BAGS IN THE KITCHEN

Make cleanups easier Cut open one or two paper bags and spread them out over your countertop when peeling vegetables, husking corn, shelling peas, or any other messy task. When you're done, simply fold up the paper, and toss it into the trash for a fast and easy cleanup.

Keep bread fresh If you live in a high-humidity area, your bread will stay fresher when stored inside a paper bag rather than a plastic one. The paper's ability to "breathe" will keep the bread's crust crisp while allowing the center of the loaf to stay soft and moist.

Use to ripen fruit Many fruits—including avocados, bananas, pears, peaches, and tomatoes—will ripen better when placed in a paper bag. To hasten the ripening process of any fruit, place an already ripe apple or banana peel in the same bag and store it at room temperature. To ripen green bananas, wrap them in a damp dishtowel before placing them in the bag. Once your fruits have adequately ripened, you can halt the process by putting them in the refrigerator.

Store mushrooms Remove your store-bought mushrooms from their mesh packaging and place them in a paper bag inside your refrigerator to keep them fresh for up to five days.

* PAPER BAGS IN THE GARDEN

Store geraniums in winter Although they're considered to be annuals, geraniums are easy to overwinter. First, remove the plants from their pots or carefully dig them up from your garden bed, shake off as much soil as possible, and place each plant in its own paper bag. Cover each bag with a second paper bag turned upside down and store them in a cool, dry place. When spring arrives, cut off all but 1 inch (2.5 centimeters) of stem and repot. Place them in a sunny spot, water regularly, and watch your plants "spring" back to life.

Feed your plants Bonemeal is an excellent source of nutrients for all the plants in your garden. You can easily make your own by first drying your leftover chicken bones in a microwave oven (depending on the quantity, cook them for 1-4 minutes on High). Then place the dried bones in a sturdy paper bag and grind them up using a mallet, hammer, or rolling pin. When done, distribute the powder around your plants and watch them thrive.

Add to compost Brown paper bags are a great addition to any garden compost heap. Not only do they contain less ink and pigment than newsprint, but they will also attract more earthworms to your pile (in fact, the only thing the worms like better than paper bags is cardboard). It's best to shred and wet the bags before adding to your pile. Also, be sure to mix them in well to prevent them from blowing away after they dry.

Dry your herbs To dry fresh herbs, first wash each plant under cold water and dry thoroughly with paper towels. Make sure the plants are completely dry before you proceed to reduce the risk of mold. Take five or six plants, remove the lower leaves, and place them upside down inside a large paper bag. Gather the end of the bag around the stems and tie it up. Punch a few holes in the bag for ventilation, then store it in a warm, dry area for at least two weeks. Once the plants have dried, inspect them carefully for any signs of mold. If you find any, toss out the whole bunch. You can grind them up, once you've removed the stems, with a rolling pin or a full soda bottle, or keep them whole to retain the flavor longer. Store your dried herbs in airtight containers and away from sunlight.

P

Kids' Stuff Make a **life-size body poster** of your child. Start by cutting up 4-6 **paper bags** so they lie completely flat (any print should be facing down). Arrange them into one big square on the floor and **tape** the undersides together. Then have your child lie down in the middle, and use a **crayon** to trace the outline of his or her entire body. Give him or her **crayons or watercolor paints** to fill in the face, clothing, and other details. When your kid is finished, hang it up in his or her room as a terrific **wall decoration**.

* PAPER BAGS FOR THE DO-IT-YOURSELFER

Move snow off your windshield If you're tired of having to constantly scrape ice and snow off your car's windshield during the winter months, keep some paper bags on hand. When there's snow in the forecast, go out to your car and turn on the wipers. Then, shut off the engine with the wipers positioned near the middle of your windshield. Now, split open a couple of paper bags and use your car's wipers to hold them in place. After the last snowflake falls, pull off the paper to instantly clear your windshield. *Note:* To prevent damaging your car's wipers, *do not* turn on the ignition until you've removed the snow and paper from the windshield.

Make a fire starter Looking for an easy way to get a fire going in your fireplace? Simply fill a paper bag with some balled-up newspaper and perhaps some bits of candle wax. Stick the bag under your logs, light it, then sit back and enjoy your roaring fire.

Spray-paint small items You don't have to make a mess every time you need to spray-paint a small item. Just place the object to be painted inside a large shopping bag and spray away; the bag will contain the excess spray. Once the item has dried, simply remove it and toss away the bag.

Build a bag kite Make a simple bag kite for your children to play with by folding over the top of a paper bag to keep it open. Glue on pieces of party streamers under the fold. Reinforce the kite by gluing in some strips of balsa wood or a few thin twigs along the length of the bag. Poke a couple of holes above the opened end, and attach two

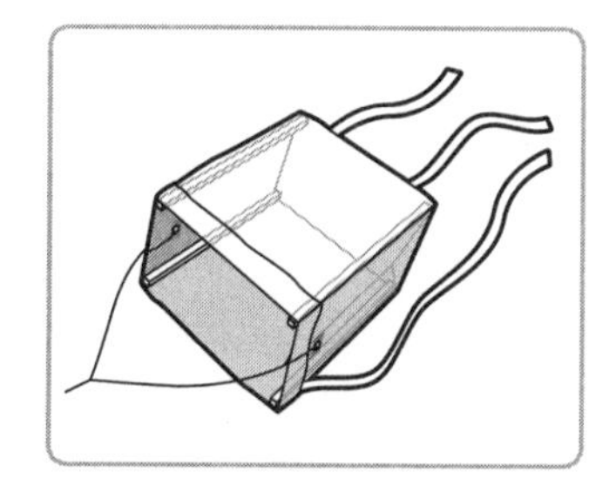

pieces of string or yarn (put a piece of masking or transparent tape over the holes to prevent them from tearing) and tie the ends onto a roll of kite string. It should take off when the kids start running.

* Paper Clips

Open shrink-wrapped CDs Opening shrink-wrap, especially on CDs, can be a test of skill and patience! Save your fingernails and teeth from destruction. Twist out the end of a paper clip and slice the wrap. To prevent scratches, slip the clip under the folded section of wrap and lift up.

Use as hooks for hanging Paper clips make great impromptu hooks. Making a hanging ceramic plaque? Insert a large, sturdy paper clip on the back before the clay hardens.

Use as zipper pull Don't throw away your jacket or pocketbook just because the zipper pull broke. Untwist a small paper clip enough to slip it through the hole. Twist it closed and zip! For a more decorative look, thread beads over the paper clip or glue on sequins before closing.

Hold the end of transparent tape Got a roll of transparent tape without a dispenser? Don't drive yourself nuts trying to locate and lift the end of the tape. Stick a paper clip under the end the next time you use the roll.

Make a bookmark Paper clips make great bookmarks because they don't fall out. A piece of ribbon or colorful string attached to the clip will make it even easier to use and find.

Pit cherries Need a seedless cherry for a recipe? Don't like to pit the cherry while you're eating it? Clip it to pit it! Over a bowl or sink, unfold a clean paper clip at the

SCIENCE FAIR

Want to amaze your friends? Challenge them to make a **paper clip float on water**. Give them a cup of water and a paper clip. When they fail, you show them how to do it. Tear off a piece of **paper towel**—larger than the clip—and place it on top of the water. Put the **paper clip** on top of the paper towel and wait a few seconds. The towel will sink, leaving the clip floating. It's magic! Actually, it's the surface tension of the water that allows the clip to float. As the paper towel sinks, it lowers the paper clip onto the water without breaking the surface tension.

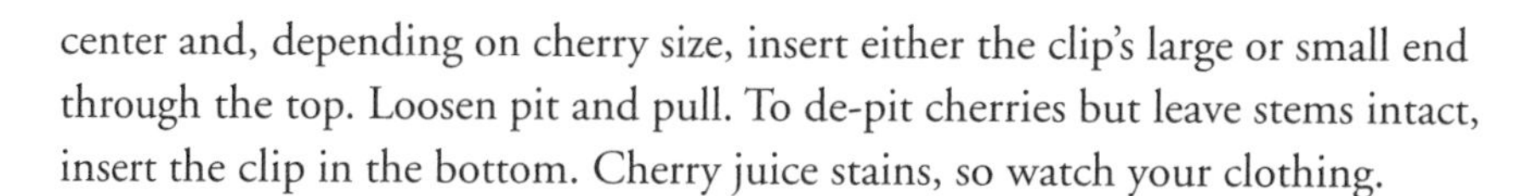

center and, depending on cherry size, insert either the clip's large or small end through the top. Loosen pit and pull. To de-pit cherries but leave stems intact, insert the clip in the bottom. Cherry juice stains, so watch your clothing.

Extend a ceiling fan chain Put away the step stool and put an end to your ballet routine while trying to reach a broken or too-short ceiling fan chain. To extend the chain, just fasten a chain of paper clips to its end.

* Paper Plates

Make index cards It's inevitable—at the eleventh hour your child will say, "I need index cards for school tomorrow." If you don't have any, use paper plates and a ruler. Measure out a 3 x 5 or 4 x 6 (A7 or A6) card on the plate and cut. Use the first card as a template for the rest.

Protect stored dishes Prevent stored dishes from clattering and breaking, especially when you are moving, by inserting a paper plate between each dish when packing.

Paint can drip catcher Painters scrape the paintbrush on the side of the can to remove excess paint. To prevent drips from falling on the floor, place a paper plate under the can.

Make Frisbee flash cards Drilling your kids with flash cards can be a drag, but here's a way to make it fun. Write the numbers, letters, words or shapes you are teaching on paper plates and let the kids toss them like Frisbees across the room when they get a correct answer.

Make a snowman decoration When the cold wind blows and cabin fever peaks, paper plates can provide an inexpensive, creative outlet for kids. They can use them to make masks, mobiles, and seasonal decorations. To create a cute winter snowman, use two paper plates. Cut the rim off one plate to make it smaller. Staple the smaller plate to the larger plate, creating a head and body. Make boots and hat out of black construction paper and mittens out of red paper and glue on. Decorate the face with googly eyes, buttons, pipe cleaners, or draw on features with crayon or marker.

* Paper Towels

Mess-free bacon zapping Here's a sure-fire way to cook bacon in your microwave oven. Layer two paper towels on the bottom of your microwave. Lay slices of bacon side by side, on the paper towels. Cover with two more paper towels. Run your microwave on High at 1-minute intervals, checking for crispness. It should take 3-4 minutes to cook. There's no pan to clean, and the towels absorb the grease. Toss them for easy cleanup.

Clean silk from fresh corn If you hate picking the silk off a freshly husked ear of corn, a paper towel can help. Dampen one and run it across the ear. The towel picks up the silk, and the corn is ready for the boiling pot or the grill.

Strain grease from broth That pot of chicken broth has been bubbling for hours, and you don't want to skim off the fat. Instead, use a paper towel to absorb the fat. Place another pot in the sink. Put a colander (or a sieve) in the new pot and put a paper towel in the colander. Now pour the broth through the towel into the waiting pot. You'll find that the fat stays in the towel, while the cleaner broth streams through. Of course, be sure to wear cooking mitts or use potholders to avoid burning your hands with the boiling-hot liquid.

Keep produce fresh longer Don't you hate it when you open the vegetable bin in the refrigerator and find last week's moldy carrots mixed with the now-yellow lettuce? Make your produce last long enough so you can eat it. Line your vegetable bins with paper towels. They absorb the moisture that causes your fruits and vegetables to rot. Makes cleaning up the bin easier too.

Keep frozen bread from getting soggy If you like to buy bread in bulk from the discount store, this tip will help you freeze and thaw your bread better. Place a paper towel in each bag of bread to be frozen. When you're ready to eat that frozen loaf, the paper towel absorbs the moisture as the bread thaws.

Clean a can opener Have you ever noticed that strange gunk that collects on the cutting wheel of your can opener? You don't want that in your food. Clean your can opener by "opening" a paper towel. Close the wheel on the edge of a paper towel, close the handles, and turn the crank. The paper towel will clean off the gunk as the wheel cuts through it.

SCIENCE FAIR

Learn how all other colors are actually mixes of the **primary colors** of red, blue, and yellow. Cut a **paper towel** into strips. With a **marker**, draw a rectangle or large circle on one end of each strip. Try interesting shades of **orange, green, purple, or brown**. Black is good too. Place the other end of the strip into a **glass jar** filled with **water**, leaving the colored end dry and draped over the side of the jar. As the water from the jar slowly (about 20 minutes) moves down the towel and into the color blot, you'll see the **colors separate**. This also demonstrates **capillary attraction**— the force that allows the porous paper to soak up the water and carry it over the side of the jar.

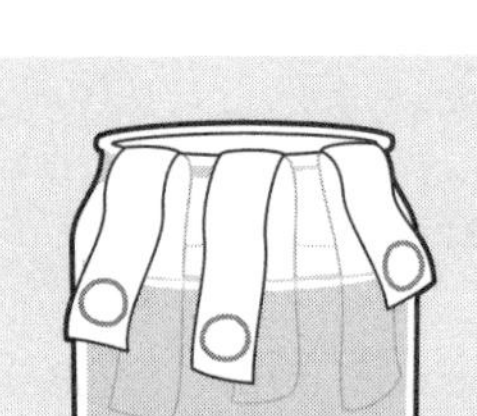

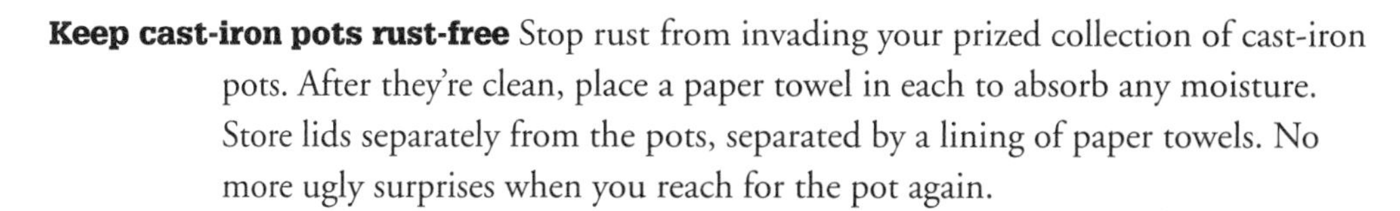

Keep cast-iron pots rust-free Stop rust from invading your prized collection of cast-iron pots. After they're clean, place a paper towel in each to absorb any moisture. Store lids separately from the pots, separated by a lining of paper towels. No more ugly surprises when you reach for the pot again.

Make a place mat for kids Your darling grandchildren are coming for an extended visit, and though they are adorable, they're a disaster at mealtime. Paper towels can help you weather the storm. Use a paper towel as a place mat. It will catch spills and crumbs during the meal and makes cleaning up easy.

Test viability of old seeds You've just found a packet of watermelon seeds dated from two springs back. Should you bother to plant them or has their shelf life expired and they're best planted in the garbage can? To find out for sure, dampen two paper towels and lay down a few seeds. Cover with two more dampened paper towels. Over the next two weeks, keep the towels damp and keep checking on the seeds. If most of the seeds sprout, then plant the rest of the batch in the garden.

Clean a sewing machine Your sewing machine is good as new after its recent tune-up, but you're worried about getting grease from the machine onto the fabric for that new vest you're sewing. Thread the sewing machine and stitch several lines up a paper towel first. That should take care of any residual grease so you'll be ready to resume your sewing projects.

Make a beautiful kids' butterfly Use colored markers to draw a bold design on a paper towel. Then lightly spray water on the towel. It should be damp so that the colors start to run, but do not soak. When the towel is dry, fold it in half, open it up, and then gather it together using the fold line as your guide. Loop a pipe cleaner around the center to make the body of the butterfly and twist it closed. To make antennae, fold another pipe cleaner into a **V** shape and slip it under the first pipe cleaner at the top of the butterfly.

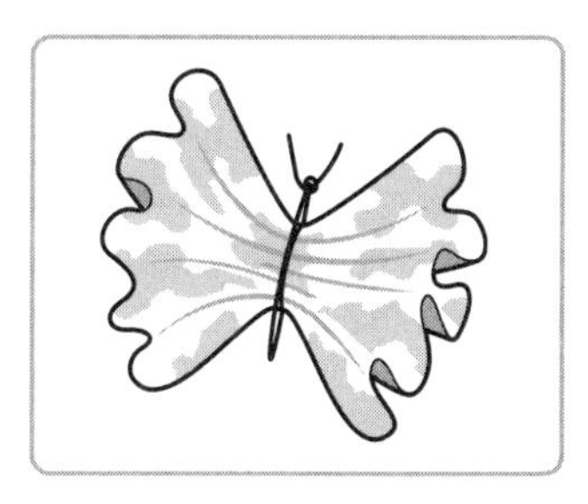

* Peanut Butter

Get chewing gum out of hair There it is. The piece of chewing gum you gave to your child not ten minutes ago is now a wadded mess in his hair. Apply some peanut butter and rub the gum until it comes out. Your child's hair may smell like peanut butter until you shampoo it, but it's better than cutting the gum out.

Remove price-tag adhesive You've removed the price tag from that new vase you've just purchased but you're left with that pesky gummy glue on the glass. Remove it easily by rubbing peanut butter on it.

Bait a mouse trap You know the mice are out there, scurrying around your kitchen at night. It's time to get tough. Lay traps, but bait them with peanut butter. They can't resist, and it's nearly impossible for them to swipe without tripping the trap. You'll be rid of the critters in no time.

Eliminate stinky fish smell If you're trying to eat more fish for health reasons, but hate the smell that hangs in your house after you've cooked it, try this trick. Put a dollop of peanut butter in the pan with the frying fish. The peanut butter absorbs the odor instead of your furnishings.

Plug an ice-cream cone Ice-cream cones are fun to eat but a bit messy too. Here's a delectable solution: Plug up the bottom of an ice-cream cone with a bit of peanut butter. Now, when munching through that scoop of double chocolate fudge, you'll be protected from leaks. And there's a pea-nutty surprise at the end of the treat.

Pencils

Ease a new key into a lock You just had a new house key made, but you can't seem to fit it into your front door lock. Rub a pencil over the teeth of the key. The graphite powder should help the key open the door.

Use as hair accessory Take a pencil to school to help with your … hair. A pencil can help give lift to curly hair if you don't have a pick. Two pencils crossed in an **X** also can stabilize and decorate a hair bun, plus provide you a new writing tool if you lose yours during the day.

Decorate a picture frame Dress up the frame for this year's class picture with pencils. Glue two sharpened pencils lengthwise to the frame. Sharpen down two other pencils to fit the width of the frame.

SCIENCE FAIR

Are your eyes deceiving you? Cut a small **piece of paper**, about 2 inches (5 centimeters) square. Turn the square so it's a diamond. On one side, draw an animal or a person. On the other side, draw a setting for the animal, or a hat and hair for the person. Examples are a cheetah and grasslands or a boy with a hat. Next **tape** the bottom point of the diamond onto the point of a **pencil**. Then, holding the pencil so the picture is upright, **twirl** the pencil rapidly between your hands. You should see **both images** from the two sides of the paper **at the same time**.

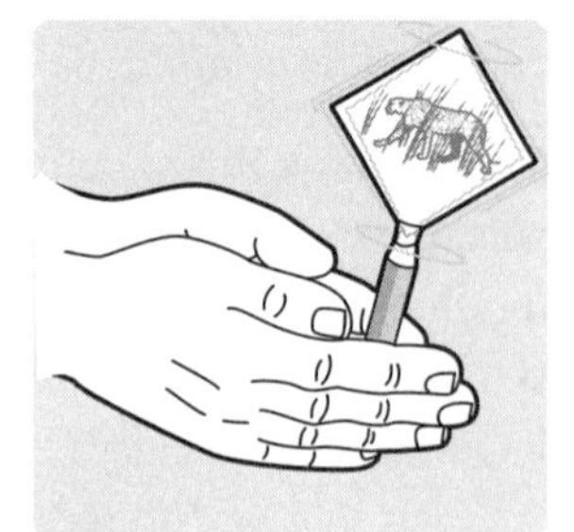

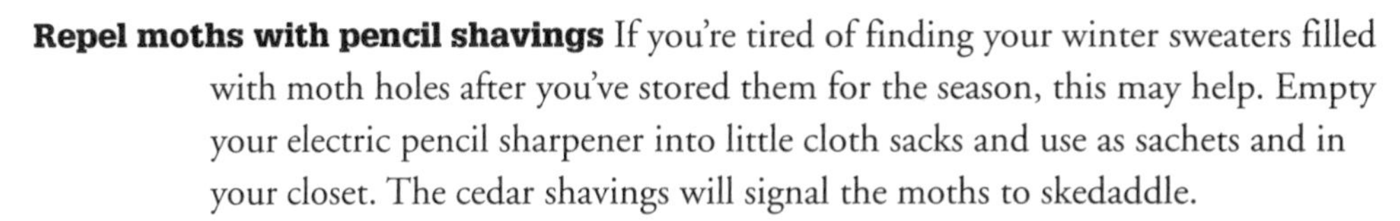

Repel moths with pencil shavings If you're tired of finding your winter sweaters filled with moth holes after you've stored them for the season, this may help. Empty your electric pencil sharpener into little cloth sacks and use as sachets and in your closet. The cedar shavings will signal the moths to skedaddle.

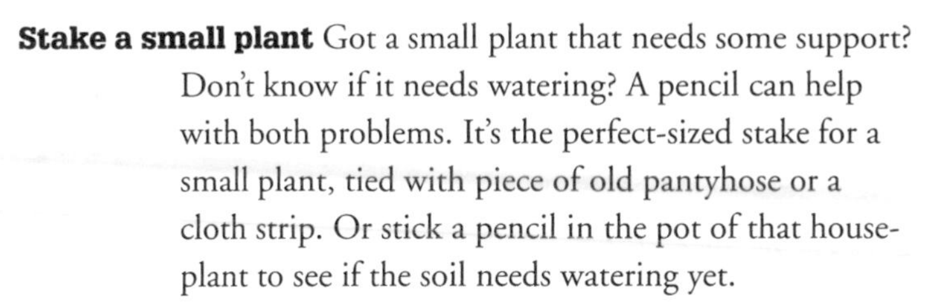

Stake a small plant Got a small plant that needs some support? Don't know if it needs watering? A pencil can help with both problems. It's the perfect-sized stake for a small plant, tied with piece of old pantyhose or a cloth strip. Or stick a pencil in the pot of that houseplant to see if the soil needs watering yet.

Lubricate a sticky zipper Your zipper is refusing to budge, no matter how hard you tug and pull. Pick up a pencil and end your frustration. Run the pencil lead along the teeth of the zipper to unstick it. In no time you'll be zipping out the door with your jacket safely zipped up.

* Pencil Erasers

Shine your coins If you've just inherited a rather grimy coin collection from your uncle, but you'd like to see it with more luster, try an eraser to shine up the coins. But don't do this to rare and valuable coins—you can erase their value along with their surface patina.

Store pins or drill bits A box is not the most handy place to keep sewing pins. What if the box spills? What if you have a hard time grabbing just one when you're laying out pattern pieces? Here's the solution: Stick pins in an eraser. They won't fall out, and it's easy to grab the ones you need. This is also a good tip if you're storing several small drill bits.

Clean off crayon marks Your toddler has gone wild with the crayons, but he drew on the walls and not on paper. You've tried everything to get it off, but not this: an eraser. Try "erasing" the crayon marks to get the wall back to a clean slate.

Remove scuff marks on vinyl floors Those new shoes of your husband's have left black streak marks all over the kitchen floor. An eraser will take them off in no time. Give him the eraser and have him do it.

Clean your piano keys Whether it's a baby grand piano that fills the corner of the living room, a more conventional upright, or just a fold-away electronic keyboard, cleaning the keys can be a nightmare project of dust and finger marks. And when you clean it, it's hard to reach some spots to remove dirt. The sides of the black keys are especially difficult to clean. Find an

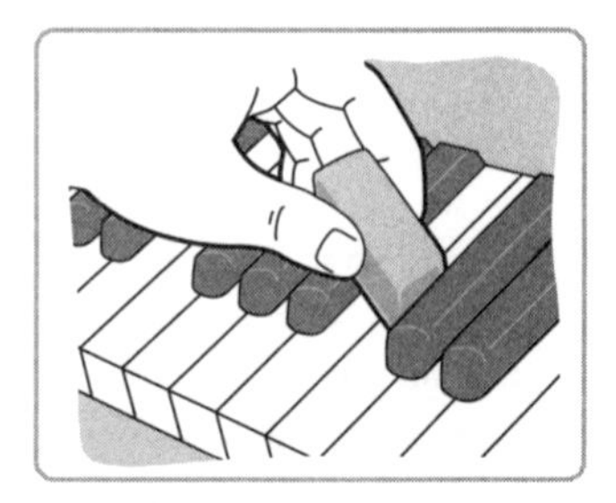

eraser that fits between the ivories and the black keys and you'll have 86'ed the dirt on the 88s. This works well whether you have a piano with real ivory keys or the more common plastic ones.

Cushion picture frames Don't you hate it when that heavy mirror or picture frame gets slightly crooked? Tired of worrying about the black marks and scrapes the frame is making on the wall? Glue erasers to the bottom corners of the frame. The pictures will now hang straighter and not leave their mark.

Remove residue from stick-on labels That gray gummy substance on the new picture frame you just purchased is a sight to behold and not coming off with plain soap and water. Rub the residue with an eraser and watch the stuff just peel away.

* Pepper

Stop a car radiator leak A heat wave has hit your town, and your aging, leaking car radiator isn't too happy about it. If it's overheating because of a small leak, pepper can help. Before you bring your car to a mechanic for a more thorough repair, pour a handful of pepper into your radiator. It will temporarily plug the leaks until you can get some help.

Use as a decongestant Is your nose stopped up? Are your ears plugged? Do you have a cold? Forget the over-the-counter medications. Nothing gets things flowing again faster than some cayenne pepper. Sprinkle it on your food and grab some tissues.

Keep colors bright That new cherry-red shirt you just purchased is fantastic, but just think how faded the color will look after the shirt has been washed a few times. Add a teaspoon of pepper to the wash load. Pepper keeps bright colors bright and prevents them from running too.

Get bugs off plants There's nothing more frustrating than a swarm of bugs nibbling at your fledgling garden. Just when things are starting to pop up, the bugs are there, chowing down. Mix black pepper with flour. Sprinkle around your plants. Bugs take a hike.

The black pepper on your table actually starts out as a red berry on a bush. When the pea-size berry is placed in boiling water for ten minutes, it shrinks and turns black, becoming the familiar pepper-corn that we fill our pepper grinders with. Of course, some of us skip that step and buy pepper already ground. Either way, black pepper is an ancient spice and the most common one in the world.

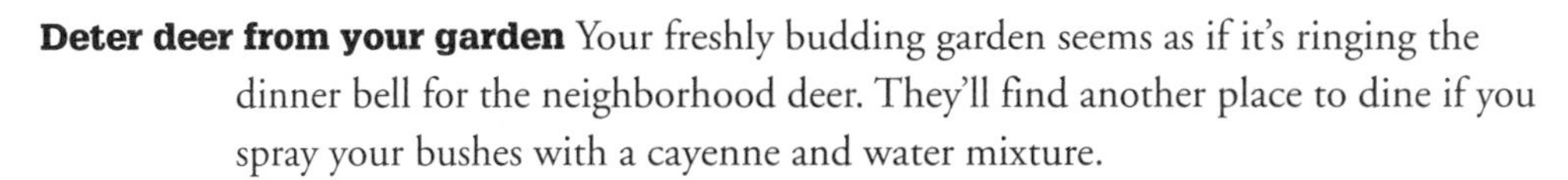

Deter deer from your garden Your freshly budding garden seems as if it's ringing the dinner bell for the neighborhood deer. They'll find another place to dine if you spray your bushes with a cayenne and water mixture.

Keep ants out of the kitchen Two or three of your annual summer visitors have invaded your kitchen. Those ants are looking for sugar. Give them some pepper instead. Cayenne pepper sprinkled in spots where the ants are looking, such as along the backs of your countertops or on your baseboards, will tell them that no sugar is ahead.

Kill an ant colony If you find the ants' home colony a little too close to yours and it is causing them to relocate to your kitchen, cayenne pepper can help get rid of it. Pour the pepper down the ant hole and say so long to ants.

super item 31 uses!

Petroleum Jelly

*PETROLEUM JELLY FOR PERSONAL GROOMING

Moisturize your lips and more If you don't want to pay a lot for expensive lip balm, makeup remover, or even facial moisturizer, then your answer is a tube of petroleum jelly. It can soothe lips, take off foundation, eye shadow, mascara, and more. It will even act as a moisturizer on your face.

Make emergency makeup Oh no! You've run out of your favorite shade of eye shadow. What do you do now? It's easy—make your own. Add a bit of food coloring to petroleum jelly and apply as usual. This is a quick way to make stopgap blush, lipstick, or eye shadow.

Lengthen the life of perfume You've picked out a great scent to wear on your night out, but it's got to last. Worry not. Dab a bit of petroleum jelly on your pulse points. Then spray on the perfume. Now you can dance the night away and not worry about your perfume turning in early.

Remove a stuck ring Is your wedding ring stuck? Trying to get it off can take a lot of tug and pull. Apply some petroleum jelly and it will glide right off.

Soften chapped hands If you're constantly applying hand lotion to your tired, chapped hands, but then taking it off again so you can get more work done, try this tip. Apply a liberal amount of petroleum jelly to your hands just before you go to bed. By morning, they'll be soft and smooth.

No more messy manicures During home manicures, it's hard to keep the nail polish from running over on your cuticles. Petroleum jelly can help your manicures look more professional. Dab some along the base of your nails and the sides. If polish seeps off the nail during the manicure, all you do is wipe off the petroleum jelly and the sloppy nail polish is gone.

Smooth wild eyebrow hairs If you have runaway eyebrows—the ones where the hairs won't lie flat but curl up instead, control the wildness with some petroleum jelly. Rub a dab into your brows. They'll calm down and behave.

Stop hair dye runs There's nothing more embarrassing than a home hair color job gone awry. Imagine finishing applying that new auburn shade to your tresses when you notice that you've dyed your hairline and part of your forehead too. Next time, run a bit of petroleum jelly across your hairline. If dye seeps off your hair, the petroleum jelly will catch it.

Heal windburned skin You've just had a glorious hike through the countryside in autumn. And as much as you enjoyed the changing colors of the season, the hike has left you with an unpleasant souvenir: windburn. Grab a jar of petroleum jelly and apply it liberally to your face or wherever you've been chapped. The jelly helps relieve the pain.

Help prevent diaper rash It's so heartbreaking to hear a baby experiencing the pain of diaper rash. Help is just a few moments away. Petroleum jelly sets up a protective coat on the skin so the rash can heal. No more pain.

No more shampoo tears Thinking of buying special no-tears shampoo for your child? Forget about it. If you have some petroleum jelly, you have the solution. Rub a fair amount into your baby's eyebrows. It acts as a protective shield against shampoo running down into his eyes.

*PETROLEUM JELLY AROUND THE HOUSE

Smoother closing shower curtains Stop the water from squirting out onto the bathroom floor. Get that shower curtain into place quickly. Lubricate the curtain rod with petroleum jelly and you'll whip that curtain across the shower in no time.

Take out lipstick stains You set the table at that lovely dinner party with your favorite cloth napkins, but your girlfriends left their mark all over them. Now dotted with lipstick stains, those napkins may be headed for the trash. But try this first. Before you wash them, blot petroleum jelly on the stain. Launder as usual and hopefully you will kiss the stains good-bye.

Eject wax from candlesticks The long red tapers you used at last night's candlelit dinner were a beautiful sight until you saw the candle wax drippings left in the candleholders. Next time apply petroleum jelly to the insides of the holders before you put the candles in. The wax will pop out for easy cleaning.

Remove chewing gum from wood Did you discover bubble gum stuck under the dining room table or behind the headboard of Junior's bed? Trouble yourself about it no further. Squeeze some petroleum jelly on the offending wad, rub it in until the gum starts to disintegrate, then remove.

Make vacuum parts fit together smoothly It's nice that your vacuum cleaner comes with so many accessories and extensions. But it's frustrating when the parts get stuck together and you have to yank them apart. Apply a bit of petroleum jelly to the rims of the tubes and the parts will easily slide together and apart.

Shine patent-leather shoes You've got a great pair of patent-leather shoes and a dynamite bag to match. The luster stays longer if you polish the items with petroleum jelly.

Restore leather jackets You don't need fancy leather moisturizer to take care of your favorite leather jacket. Petroleum jelly does the job just as well. Apply, rub it in, wipe off the excess, and you're ready to go.

Keep ants away from pet food bowls Poor Fido's food bowl has been invaded by ants. Since she prefers her food without them, help her out with this idea. Ring her food bowl with petroleum jelly. The ants will no longer be tempted by the kibble if they have to cross that mountain of petroleum jelly.

Grease a baseball mitt Got a new baseball mitt, but it's as stiff as a dugout bench? Soften it up with petroleum jelly. Apply liberal amounts. Work it into the glove, then tie it up with a baseball inside. Do this in the winter, and by the spring you'll be ready to take the field.

Keep a bottle lid from sticking If you're having a hard time unscrewing that bottle of glue or nail polish, remember this tip for when you finally do get it open. Rub a little petroleum jelly along the rim of the bottle. Next time, the top won't stick.

Soothe sore pet paws Sometimes your cat's or dog's paw pads can get cracked and dry. Give a little tender loving care to your best friend. Squirt a little petroleum jelly on the pads to stop pain. They'll love you for it.

Although there are many generic versions of petroleum jelly, the only real major brand is Vaseline. Because of this, the online American Package Museum has included a jar of Vaseline petroleum jelly in its collection of classic packaging designs—a collection that also includes Alka-Seltzer and Bayer Aspirin. Why did the site include Vaseline? "Vaseline is a well-established brand with a 145-year-old history," says Ian House, who created the site. "It also seems to enjoy popularity as a cultural icon for humorous reasons. Nobody knows quite what to do with it ... but nobody can seem to live without it!"

Mask doorknobs when painting You're about to undertake painting the family room. But do you really want to fiddle with removing all the metal fixtures, including doorknobs? Petroleum jelly rubbed on the metal will prevent paint from sticking. When you're done painting, just wipe off the jelly and the unwanted paint is gone.

Stop battery terminal corrosion It's no coincidence that your car battery always dies on the coldest winter day. Low temperatures increase electrical resistance and thicken engine oil, making the battery work harder. Corrosion on the battery terminals also increases resistance and might just be the last straw that makes the battery give up. Before winter starts, disconnect the terminals and clean them with a wire brush. Reconnect, then smear with petroleum jelly. The jelly will prevent corrosion and help keep the battery cranking all winter long.

Protect stored chrome If you're getting ready to store the kids' bikes for the winter, or stow that stroller until your next baby comes along, stop a moment before you stash. Take some petroleum jelly and apply it to the chrome parts of the equipment. When it's time to take the items out of storage, they'll be rust-free. The same method works for machinery stored in your garage

Keep an outdoor lightbulb from sticking Have you ever unscrewed a lightbulb and found yourself holding the glass while the metal base remains in the socket? It won't happen again if you remember to apply petroleum jelly to the base of the bulb before screwing it into the fixture. This is an especially good idea for lightbulbs used outdoors.

Seal a plumber's plunger Before you reach for that plunger to unclog the bathroom toilet, find some petroleum jelly. Apply it along the rim of the plunger and it will help create a tighter seal. Whoosh, clog's gone.

Lubricate cabinets and windows Can't stand to hear your medicine cabinet door creak along its runners? Or how about that window that you have to force open every time you want a breeze in the house? With a small paintbrush, apply petroleum jelly to the window sash channel and cabinet door runners. Let the sliding begin.

Stop squeaking door hinges It's so annoying when a squeaky door makes an ill-timed noise when you're trying to keep quiet. Put petroleum jelly on the hinge pins of the door. No more squeaks.

Remove watermarks on wood Your most recent party left lots of watermark rings on your wood furniture. To make them disappear, apply petroleum jelly and let it sit overnight. In the morning, wipe the watermark away with the jelly.

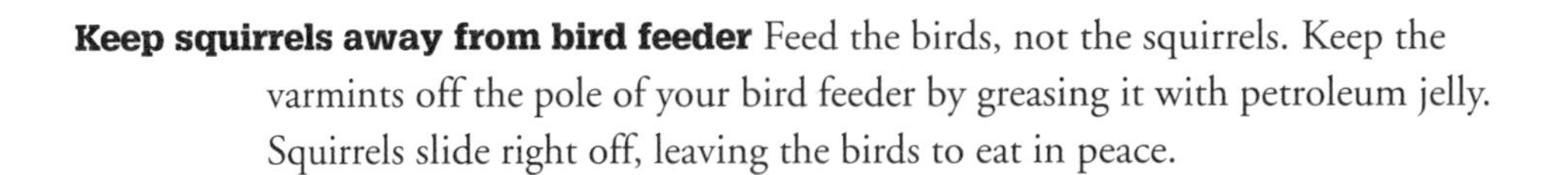

Keep squirrels away from bird feeder Feed the birds, not the squirrels. Keep the varmints off the pole of your bird feeder by greasing it with petroleum jelly. Squirrels slide right off, leaving the birds to eat in peace.

Pillowcases

Dust ceiling fan blades Have you ever seen dust bunnies careening off your ceiling fan when you turn it on for the first time in weeks? Grab an old pillowcase and place it over one of the ceiling fan blades. Slowly pull off the pillowcase. The blades get dusted and the dust bunnies stay in the pillowcase, instead of parachuting to the floor.

Clear out cobwebs There's a cobweb way up high in the corner of your dining room. Before you take a broom to it, cover the broom with an old pillowcase. Now you can wipe away the cobweb without scratching the wall paint. It's also easier to remove the cobweb from the pillow than to pull it out of the broom bristles.

Cover a baby's changing table Have you priced those expensive changing-table covers lately? Forget about it! Pick up a few of the cheapest white pillowcases you can find and use those to cover the changing table pad. When one is soiled, just slip it off and replace with a clean one.

Make a set of linen napkins Who needs formal linen napkins that need to be pressed every time you use them? Pillowcases are available in a wide array of colors and designs. Pick a color or design you like, and start cutting. If you're really ambitious, sew a 1/2-inch (1.25-centimeter) hem on each edge. You'll have a new set of colorful napkins for a fraction of the cost of regular cloth napkins.

Prepare travel pillows Road trips can be a lot of fun, but a little dirty too. Your youngsters may want to bring their own pillows along, but they'll stain them with candy, food, and markers. Take their favorite pillows and layer several pillowcases on each. When the outside one gets dirty, remove it for a fresh start.

Use for wrapping paper Trying to wrap a basketball or an odd-shaped piece of art? Is your wrapping paper not doing the trick? Place the gift in a pillowcase and tie closed with a ribbon.

Store your sweaters Stored in plastic, winter sweaters can get musty. But stored just in a closet, they're prey to moths. The solution can be found among your linens. Put the sweaters in a pillowcase for seasonal storage. They will stay free from dust but the pillowcase fabric will allow them to breathe.

Protect clothing hanging in a closet You've just laundered a favorite dress shirt or skirt and you know you won't be wearing it again for a while. To protect the garment, cut a hole in the top of an old pillowcase and slip it over the hanger and clothing.

Stash your leather accessories You reach up to pull a leather purse or suede shoes down from a shelf. Of course, the item is dusty and now you have to clean it. Save yourself the time and hassle next time by storing infrequently used items in a pillowcase. They'll be clean and ready to use when the occasion arises.

Keep matching sheets together Solve this host nightmare. Your recently arrived overnight guests want to go to bed, but it's not made. You run to the linen closet, but you can't find a matched set of sheets. Next time, file away your linens. Place newly laundered and folded sheets in their matching pillowcase before putting them in the closet.

Machine-wash your delicates Sweaters and pantyhose can get pulled out of shape when they twist around in the washer. To protect these garments during washing, toss them into a pillowcase and close with string or rubber band. Set the machine on the delicate setting, add the soap, and worry not about knots.

Machine-wash stuffed animals Your child's Beanie Babies collection is cute but mighty dusty. Time for a bath. Place them in a pillowcase and put them in the washer. The pillowcase will ensure they get a gentle but thorough wash. If any parts fall off the stuffed animals, they'll be caught in the pillowcase so you can reattach them after their washing machine bath.

Kids' Stuff Kids love to personalize their bedrooms. You can help kids as young as 4 or 5 do just that by making a **pillowcase into a wall hanging**. Let the youngster choose a pillowcase color, and then slit a hole about 1 inch long in each side seam. Use **fabric paints** to create a design or scene or let him **rubber-stamp** a picture on the pillowcase. Stick a **dowel** through the seam openings. Cut a length of **yarn** about 30 inches (75 centimeters) long. Tie one end of the yarn to each end of the dowel. Hang on the wall and let the little one collect the compliments.

Use as a traveling laundry bag When you travel, you always want to keep your dirty laundry separate from your clean clothes. Stick a pillowcase in your suitcase and toss in the dirty laundry as it accumulates. When you get home, just empty the pillowcase into the washer and throw in the pillowcase as well.

Wash a lot of lettuce in washing machine Expecting a large crowd for an outdoor salad luncheon? Do you have 20 heads of lettuce to wash? Here's your solution: Place one pillowcase inside another. Pull apart the lettuce heads and fill the inside case with lettuce leaves. Close both pillowcases with string or rubber band, and throw the whole package in the washing machine with another large item, such as a towel, to balance it. Now run the spin cycle a few times. Your leaves come out rinsed and dried. It's better than a salad spinner.

Pipe Cleaners

Decorate a ponytail Need a fresh look for your hair? Tired of using plain old ribbons? Once you have your hair in a ponytail, twist a pipe cleaner around the hair band. Twist a couple together for an even brighter effect.

Use as an emergency shoelace Your shoelace broke and you're about to go out on the court to play that grudge match in basketball. A pipe cleaner is a good stopgap tie-up. Just thread in your shoe as you would a shoelace and twist it up at the top.

Safety pin holder Safety pins come in so many sizes, it's hard to keep them together and organized. Thread safety pins onto a pipe cleaner, through the bottom loop of each pin, for easy access.

Clean gas burners Have you noticed that your stovetop burners are not firing on all their jets? Do you see an interrupted circle of blue when you turn a burner on? Poke a pipe cleaner through the little vents. This cleans the burner and allows it to work more efficiently. This works for pressure cooker safety valves, or pressure cookers' too.

Make napkin rings Colorful pipe cleaners are an easy and fast resource for making napkin rings. Just twist around the napkin and place on the table. If you want to get adventurous, use two, one for the napkin ring, attach the other pipe cleaner to it, shaped into a heart, shamrock, flower, curlicue, or something else.

Use as twist tie You're ready to close up that smelly bag of kitchen garbage when you discover you are out of twist ties. Grab a pipe cleaner instead.

DID *You* KNOW?

According to reliable sources, pipe cleaners were invented in the late 19th century in Rochester, New York, by J. Harry Stedman, who also came up with the idea of issuing transfers between local streetcars. Although pipe smoking has declined dramatically since then, pipe cleaners have flourished, having been co-opted by the arts-and-crafts community. Technically, the pipe cleaners used in classrooms, summer camps, Scout troops, and so on are called chenille stems. These stems come in various colors and widths and are fuzzier than the real pipe cleaners smokers use.

Use as a travel toy If you're worried about having a bored, wiggly child on your hands during your next long car or plane ride, throw a bunch of pipe cleaners into your bag. Whip them out when the "Are we there yet?" questions start coming your way. Colorful pipe cleaners can be bent and shaped into fun figures, animals, flowers, or whatever. They even make cool temporary bracelets and necklaces.

Use as a mini-scrubber Pipe cleaners are great for cleaning in tight spaces. Use one to remove dirt from the wheel of a can opener or to clean the bobbin area of a sewing machine.

Decorate a gift To give a special touch to a birthday or holiday present, shape a pipe cleaner into a bow or a heart. Poke one end through a hole in the card, and affix it to the package with a dab of glue.

super item 44 uses!

Plastic Bags

*PLASTIC BAGS AROUND THE HOUSE

Line a cracked flower vase Grandmother's beautiful flower vase is a sight to behold when it's filled with posies. The problem is the vase leaks from a large crack that runs its length. Line the vase with a plastic bag before you fill it with water and add the bouquet, giving fresh life to a treasured heirloom.

Bulk up curtain valances You've picked out snazzy new curtain balloon valances for your bedroom. The problem is the manufacturer has only sent you enough stuffing to make the valances look a bit better than limp. Recycle some plastic bags by stuffing them in the valances for a resilient pouf.

Stuff crafts or pillows There are a number of ways to stuff a craft project: with beans, rice, fabric filler, plastic beads, pantyhose, and so on. But have you ever tried stuffing a craft item or throw pillow with plastic bags? There are plenty on hand, so you don't have to worry about running out, and you're recycling.

Make party decorations Here is an easy way to create streamers for a party using plastic bags. Cut each bag into strips starting from the open end and stopping short of the bottom. Then attach the bag bottom to the ceiling with tape.

Drain bath toys Don't let Rubber Ducky and all of the rest of your child's bath toys get moldy and create a potential hazard in the tub. Instead, after the bath is done, gather them up in a plastic bag that has been punctured a few times. Hang the bag by its handles on one of the faucets to let the water drain out. Toys are collected in one place, ready for the next time.

Keep kids' mattresses dry There's no need to buy an expensive mattress guard if bed-wetting is a problem. Instead, line the mattress with plastic garbage bags. Big bags are also useful to protect toilet-training toddlers' car seats or car upholstery for kids coming home from the swimming pool.

Make a laundry pocket pickin's bag You may think that the laundry's all done, until you open the dryer to find a tissue paper left in someone's pocket has shredded and now is plastered all over the dryer drum. Hang a plastic bag near where you sort laundry. Before you start the wash, go through the pockets and dump any contents in the bag for later sorting.

Treat chapped hands If your hands are cracked and scaly, try this solution. Rub a thick layer of petroleum jelly on your hands. Place them in a plastic bag. The jelly and your body's warmth will help make your hands supple in about 15 minutes.

*PLASTIC BAGS FOR STORING STUFF

Store extra baby wipes Shopping at the warehouse grocer, you picked up a jumbo box of baby wipes at a great price. You've got enough wipes to last for several months, as long as they don't dry out before you can use them. To protect your good investment, keep the opened carton of wipes in a plastic bag sealed with a twist tie.

Collect clothes for thrift shop If you're constantly setting aside clothes to give to charity, but then find them back in your closet or drawers, try this solution: Hang a large garbage bag in your closet. That way, the next time you find something you want to give, you just toss it in the bag. Once it's full, you can take it to the local donation center. Don't forget to hang a new bag in the closet.

Cover clothes for storage You'd like to protect that seersucker suit for next season. Grab a large, unused garbage bag. Slit a hole in the top and push the hanger through for an instant dustcover.

Store your skirts If you find you have an overstuffed closet but plenty of room to spare in your dresser, conduct a clothes transfer. Roll up your skirts and place them

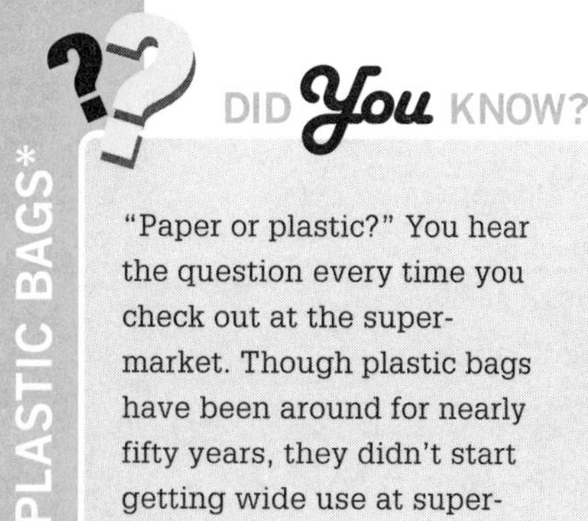

"Paper or plastic?" You hear the question every time you check out at the supermarket. Though plastic bags have been around for nearly fifty years, they didn't start getting wide use at supermarkets until 1977. Within two decades, plastic had replaced paper bags as the most common grocery bag. Today four out of five bags used at the grocer's are plastic. Thankfully, supermarkets have implemented recycling programs for the proliferating bags.

each in a plastic bag. That will help them stay wrinkle-free until you're ready to wear one.

Keep purses in shape Ever notice that if you've changed purses and leave an empty one in your closet, it deflates and loses its shape? Fill your purse with plastic bags to retain its original shape.

Tip Storing Plastic Bags

All those shopping bags are spilling out of the utility drawer in your kitchen. Here are some better ways to store them:

- Stuff them inside an empty tissue box for easy retrieval.
- Poke a bunch down a cardboard tube, such as a paper towel or mailing tube or even a section of a carpet tube.
- Fill a clean, empty gallon (4-liter) plastic jug. Cut a 4-inch (10-centimeter) hole in the bottom. Stuff with bags and hang by its handle on a hook. Pull the bags out of the spout.
- Make a bag "sock." Fold a kitchen towel lengthwise with the wrong side facing out. Stitch the long edges together. Sew 1/2-inch (1.25-centimeter) casings around the top and bottom openings and thread elastic through them, securing the ends. Turn the sock right side out, sew a loop of ribbon or string on the back to hang it up, stuff bags into the top opening, and pull them out from the bottom one.

*PLASTIC BAGS KEEPING THINGS CLEAN

Protect hand when cleaning toilet When cleaning your toilets with a long-handled brush or a shorter tool, first wrap your hand in a used plastic bag. You'll be able to do the appropriate scrubbing without your hand getting dirty in the process.

Prevent steel wool from rusting A few days ago you got a new steel wool pad to clean a dirty pot. Now that steel pad is sitting useless in its own pool of rust. Next time, when you're not using the pad, toss it into a plastic bag where it won't rust and you'll be able to use it again.

Make bibs for kids The grandkids just popped in, and they're hungry. But you don't have any bibs to protect their clothes while they eat. Make some by tying a plastic bag loosely around the kids' necks so their clothes stay free of stains. You can make quick aprons this way too.

Create a high-chair drop cloth Baby stores are quite happy to sell you an expensive drop cloth to place under your child's high chair. Why spend the money on a sheet of plastic when you have all those large garbage bags that can do the job? Split the seams of a bag and place it under the high chair to catch all the drips and dribbles. When it gets filthy, take it outside and shake, or just toss it.

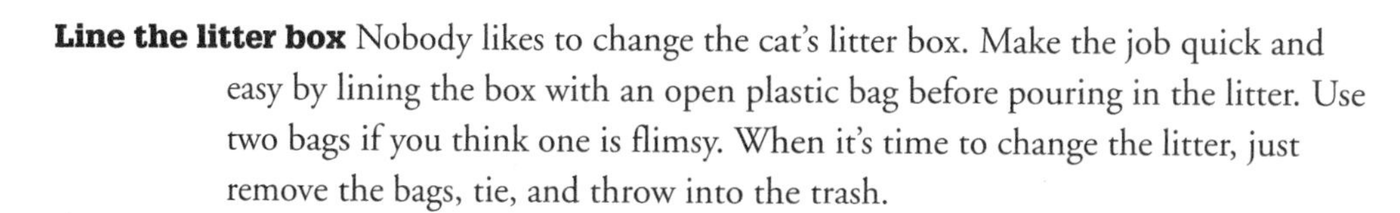

Line the litter box Nobody likes to change the cat's litter box. Make the job quick and easy by lining the box with an open plastic bag before pouring in the litter. Use two bags if you think one is flimsy. When it's time to change the litter, just remove the bags, tie, and throw into the trash.

Needle-free Christmas tree removal O Christmas tree, how lovely are thy branches! Until those needles start dropping. When it's time to take down your tree, place a large garbage bag over the top and pull down. If it doesn't fit in one bag, use another from the bottom and pull up. You can quickly remove the tree without needles trailing behind you.

Keep polish off your hand You want to polish up your scruffy white sandals. The problem is, you're going to get more polish on your hands than your shoes. Before you polish, wrap your hand in a plastic bag before inserting it into the sandal. Then when polish runs off the sandal straps, your hand is protected. Leave the bag in the sandal until the polish is dry.

*PLASTIC BAGS IN THE KITCHEN

Cover a cookbook You're trying a new recipe from a borrowed cookbook that you don't want to get splattered during your creation. Cover the book with a clear plastic bag. You'll be able to read the directions, while the book stays clean.

Bag the phone Picture this: You're in the middle of making your famous snickerdoodle cookies. You're up to your elbows in dough. The phone rings. Now what? Wrap your hands in a plastic bag and answer the phone. You won't miss a call or have to clean the phone when you're done.

Scrape dishes Your extended family of 25 has just finished their Sunday dinner. Time to clean the dishes. Here's an easy way to get rid of the table scraps: Line a bowl with a plastic bag and scrape scraps into it. Once it's full, just gather up the

SCIENCE FAIR

It is said that a plastic bag can carry about 20 pounds (9 kilograms) of groceries before you need to double-bag. Find out **how much a bag can hold** without its handles breaking. For this experiment, you will need a **kitchen scale**, a **plastic bag**, and a **bunch of rocks**. Place the bag on the scale. Fill it with rocks until the scale reads 10 pounds (4.5 kilograms). Lift the bag. Does it hold? Add more rocks in 2-pound (.9-kilogram) increments, testing the bag's strength after each addition. When the handles start to tear, you'll know the bag's actual strength.

handles and toss. Place the bowl in a prominent place in your kitchen so everyone can scrape their own dishes when bringing them to the sink.

Crush graham crackers Don't spend hard-earned grocery dollars on a box of pre-crushed graham crackers or a ready-to-fill graham cracker crust. It's much cheaper and a real snap to crush graham crackers yourself. Just crumble several graham crackers into a plastic bag. Lay the bag on the kitchen counter and go over it several times with a rolling pin. In no time, you'll have as many graham cracker crumbs as you need, plus the remainder of a box of crackers to snack on as well.

Replace a mixing bowl If you're cooking for a crowd and are short on mixing bowls, try using a plastic bag instead. Place all the dry ingredients to be mixed in the bag, gather it up and gently shake. If the ingredients are wet, use your hands to mix.

Spin dry salad greens The kids will enjoy helping you with this one. Wash lettuce and shake out as much water as you can in the sink. Then place the greens in a plastic grocery bag that has been lined with a paper towel. Grab the handles and spin the bag in large circles in the air. After several whirls, you'll have dry lettuce.

Ripen fruit Some of the fruit from that bushel of peaches you just bought at the local farm stand are hard as rocks. Place the fruit with a few already ripe pieces or some ripe bananas in a plastic bag. The ripe fruit will help soften the others through the release of their natural gas. But don't leave them for more than a day or two or you'll have purple, moldy peaches.

*PLASTIC BAGS IN THE YARD

Protect plants from frost When frost threatens your small plants, grab a bunch of plastic bags to protect them. Here's how: Cut a hole in the bottom of each bag. Slip one over each plant and anchor it inside using small rocks. Then pull the bags over the plants, roll them closed, and secure them with clothespins or paper clips. You can open the bags up again if the weather turns warm.

Start poinsettia buds for Xmas You want that Christmas poinsettia to look gorgeous by the time the holidays arrive. You can speed up Mother Nature by placing the poinsettia in a large, dark garbage bag for several weeks to wake up the plant's buds.

Protect fruit on the tree Are there some apples in your orchard you want to protect or some plums that need a little more time on the tree? Slip the fruit into clear plastic bags while still on the trees. You'll keep out critters while the fruit continues to ripen.

Protect your shoes from mud It rained hard last night, and you need to get out in the garden to do your regular weeding. But you're worried about getting mud all over your shoes. Cover them in plastic bags. The mud gets on the bag, not on the shoes, and your feet stay dry so you can stay out longer in the garden.

Clean a grill easily That neighborhood barbecue was a blast, but your grill is a sorry mess now. Take the racks off and place them in a garbage bag. Spray oven cleaner on the grill and close up the bag. The next day, open the bag, making sure to keep your face away from the fumes. All that burned-on gunk should wipe right off.

Cover garage-sale signs If you've gone to the trouble of advertising your upcoming garage sale with yard signs but worry that rain may hurt your publicity campaign before even the early birds show up, protect the signs by covering them with pieces cut from clear plastic bags. Passersby can still see the lettering, which will be protected from smearing by the rain.

Store outdoor equipment manuals Your weed-whacker spindle just gave out and you have to replace it. But how? Stash all your outdoor equipment's warranties and owner's manuals in a plastic bag and hang it in your garage. You'll know exactly where to look for help.

Protect your car mirrors A big snowstorm is due tonight, and you've got a doctor's appointment in the morning. Get a step ahead by covering your car's side mirrors with plastic bags before the storm starts. When you're cleaning off the car the next morning, just remove the bag. No ice to scrape off.

Make a jump rope "I'm bored!" cries your child as you're trying to finish your yard work. Here's a simple solution: Make a jump rope by twisting up several plastic bags and tying them together end to end. Talk about cheap fun.

*PLASTIC BAGS ON THE GO

Pack your shoes Your next cruise requires shoes for all types of occasions, but you worry that packing them in the suitcase will get everything else dirty. Wrap each pair in its own plastic bag. It will keep the dirt off the clothes, and you can rest assured you've packed complete pairs.

DID You KNOW?

Worried about the growing number of plastic bags filling landfills, countries across the globe have started putting restrictions on the seemingly indispensable item. Bangladesh has banned plastic bags, blaming them for clogging drainage pipes and causing a flood. Some Australian towns also have banned plastic bags, and the country is pushing stores to halve their use of bags (estimated at 7 billion annually) in a few years. If you want to use a plastic bag in Ireland, you'll be charged about 19 cents a bag. In Taiwan, it's 34 cents a bag.

Protect your hands when pumping gas You've stopped at the gas station for a fill-up while on your way to meet friends for lunch. The last thing you want is to greet them with hands that smell of gasoline. Grab one of those plastic bags you keep in your car and cover your hands with it while you pump.

Stash your wet umbrella When you're out in the rain and running to your next appointment, who wants to deal with a soggy umbrella dripping all over your clothes and car? One of those plastic bags that newspapers are delivered in is the perfect size to cover your umbrella the next time it rains. Just fold the umbrella up and slip it into the bag.

Make an instant poncho Leave a large garbage bag in your car. The next time it rains unexpectedly, cut some arm slits and one for your head. Slip on your impromptu poncho and keep dry.

Scoot in the snow Your neighborhood just got 6 inches (15 centimeters) of snow and the kids are hoping to take advantage of it right now. Grab some garbage bags, tie one around one each of their waists, and let them fanny-slide down the hills.

*PLASTIC BAGS FOR THE DO-IT-YOURSELFER

Cover ceiling fans You're painting the sun porch ceiling, and you don't want to remove the ceiling fans for the process. Cover the blades with plastic bags to protect them from paint splatters. Use masking tape to keep the bags shut.

Store paintbrushes You're halfway through painting the living room, and it's time to break for lunch. No need to clean the paintbrush. Just stick it in a plastic bag and it will remain wet and ready to use when you return. Going to finish next weekend, you say? Stick the bag-covered brush in the freezer. Defrost next Saturday and you are ready to go.

Contain paint overspray If you've got a few small items to spray-paint, use a plastic bag to control the overspray. Just place one item at a time in the bag, spray-paint, and remove to a spread-out newspaper to dry. When you're done, toss the bag for a easy cleanup.

* Plastic Containers

Trap plant-eating slugs Sock it to those slugs eating your newly planted vegetable plants. Dig a hole the size of a plastic container near the plant. Place the container, flush with the ground, in the hole. Fill the container with beer or salted water and place cut potatoes around the rim to attract the slugs. Slugs crawl in, but they don't crawl out.

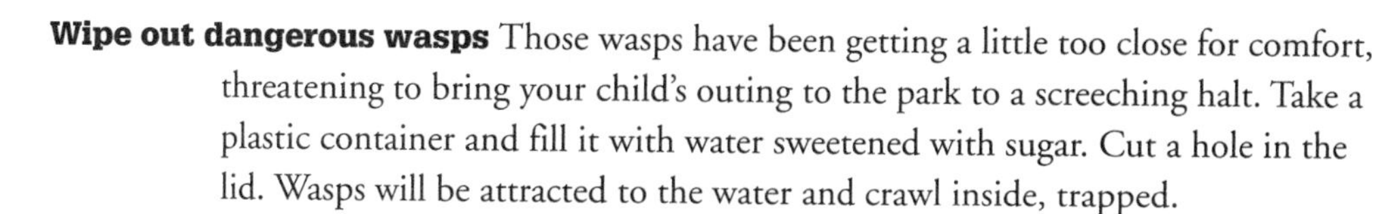

Wipe out dangerous wasps Those wasps have been getting a little too close for comfort, threatening to bring your child's outing to the park to a screeching halt. Take a plastic container and fill it with water sweetened with sugar. Cut a hole in the lid. Wasps will be attracted to the water and crawl inside, trapped.

Keep ants away from picnic table You watch helplessly as the ants march up the picnic table leg, onto the tabletop, and into the picnic meal. Here's a foolproof way to stop them in their tracks: Place a plastic container on the bottom of each picnic table leg. Fill with water. The ants won't be able to crawl past.

Use as a portable dog dish The next time you go out for a hike with your dog, pack a portion of its food in a plastic container. Of course, you can pack another container with a snack for yourself. An empty container also makes a great water bowl on the go.

Organize your sewing area You're sitting down at your sewing area to start on your Christmas craft projects. But instead of sewing, you're hunting for that extra bobbin or the right-color thread. Plastic containers can help you bring order to your sewing area. Fill several with thread spools, others with implements such as seam rippers and measuring tapes. Yet another can be filled with pins.

* Plastic Lids

Stop a sink or tub If your drain stopper has disappeared, but you need to stop the water in the sink or bathtub, here's a stopgap solution: Place a plastic lid over the drain. The vacuum created keeps the water from slip-sliding away.

Keep the fridge clean Drippy bottles and containers with leaks can create a big mess on your refrigerator shelves. Create coasters from plastic lids to keep things clean. Place the lids under food containers to stop any potential leaks. If they get

The tremendous success of Tupperware plastic bowls is due to one man's inventiveness and one woman's understanding of American society. The man was Earl Tupper of New Hampshire, who in 1942 saw that the durable, flexible new plastic called polyethylene could be molded into plastic bowls with tight-fitting lids. But Tupperware sales were meager until 1948, when Tupper met Brownie Wise, a divorced mother from Detroit. Wise saw that while there was a stigma against women going to work, it was perfectly acceptable for them to have a "party" where Tupperware would be sold. Tupper wisely put Wise in charge of Tupperware sales, turning the product into what the *Guinness Book of World Records* called "one of the enduring symbols of our era."

dirty, throw them in the dishwasher, while your fridge shelves stay free of a sticky mess.

Use as kids' coasters Entertaining a crowd of kids and want to make sure your tabletops survive? (Or at least give them a fighting chance!) Give kids plastic lids to use as coasters. Write their names on the coasters so they won't get their drinks mixed up.

Use as coasters for plants Plastic lids are the perfect water catcher for small houseplants. One under each plant will help keep watermarks off your furniture.

Scrape nonstick pans We all know that stuff often sticks to so-called nonstick pans. And, of course, using steel wool to get it off is a no-no. Try scraping off the gunk with a plastic lid.

Separate frozen hamburgers The neighborhood block party is next week, but you have the hamburger meat now. Season the meat as desired and shape it into patties. Place each patty on a plastic lid. Then stack them up, place in a plastic bag, and freeze. When the grill is fired up, you'll have no trouble separating your pre-formed hamburgers.

Prevent paintbrush drips Worried about getting messy paint drips all over yourself while you're touching up a repair job on the ceiling? Try this trick. Cut a slot in the middle of a plastic lid. The kind of plastic lid that comes on coffee cans is the perfect size for most paintbrushes. Insert the handle of your paintbrush through the lid so that the lid is on the narrow part of the handle just above your hand. The lid will catch any paint drips. Even with this shield, always be careful not to put too much paint on your brush when you are painting overhead.

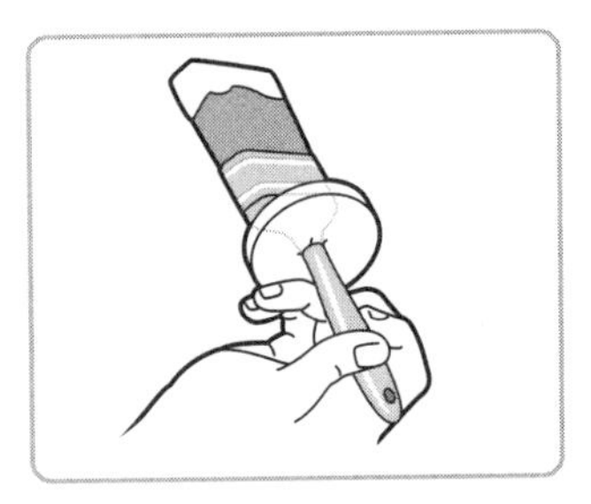

Close a bag Out of twist ties? Need to get that smelly garbage bag out of the house? Grab a plastic lid, cut a slit in it, gather the top of the bag, and thread it through. The bag's now completely sealed and ready for disposal.

* Plastic Tablecloths

Make a shower curtain A colorful tablecloth can make a great-looking shower curtain to match your bathroom decor. Punch holes about 6 inches (15 centimeters) apart and 1/2 inch (1.25 centimeters) from one edge of a hemmed tablecloth. Insert shower curtain rings or loop strings through the holes and loosely tie to the curtain rod.

Make a high-chair drop cloth Bombs away! It's par for the course for a baby to get more food on the floor than in his or her mouth. Catch the debris and protect your floor by spreading a plastic tablecloth under the high chair.

P

Collect leaves Save all that bending at leaf-raking time. Don't pick the leaves up to put them in a wheelbarrow to transport them to the curb or leaf pile. Just rake the leaves onto an old plastic tablecloth, gather up the four corners, and drag the tablecloth to the curb or pile.

* Plastic Wrap

Treat a hangnail Get rid of a hangnail overnight while you sleep. Before going to bed, apply hand cream to the affected area, wrap the fingertip with plastic wrap, and secure in place with transparent tape. The plastic wrap will confine the moisture and soften the cuticle.

Treat psoriasis Here is a method often recommended by dermatologists to treat individual psoriasis lesions. After you apply a topical steroid cream, cover the area with a small piece of plastic wrap and use adhesive tape to affix the wrap to your skin. The wrap will enhance the effect of the steroid, seal in moisture, and inhibit proliferation of the rash.

Enhance liniment effect For pain in your knee or other sore spots, rub in some liniment and wrap the area with plastic wrap. The wrap will increase the heating effect of the liniment. Make sure to test on a small area first to make sure your skin does not burn.

Keep ice cream smooth Ever notice how ice crystals form on ice cream in the freezer once the container has been opened? The ice cream will stay smooth and free of those annoying, yucky crystals if you rewrap the container completely in plastic wrap before you return it to the freezer.

Keep fridge top forever clean Make the next time you clean the top of your refrigerator the last time. After you've gotten it all clean and shiny, cover the top with over-

DID *You* KNOW?

Saran (polyvinylidene chloride) was accidentally discovered in 1933 by Ralph Wiley, a Dow Chemical Company lab worker. One day Ralph came across a vial coated with a smelly, clear green film he couldn't scrub off. He called it eonite, after an indestructible material featured in the "Little Orphan Annie" comic strip. Dow researchers analyzed the substance and dubbed the greasy film Saran. Soon the armed forces were using it to spray fighter planes, and carmakers used it to protect upholstery. Dow later got rid of the green color and unpleasant odor and transformed it into a solid material, which was approved for food packaging after World War II and as a contact food wrap in 1956.

lapping sheets of plastic wrap. Next time it's due for a cleaning, all you need do is remove the old sheets, toss them in the trash, and replace with new layers of wrap.

Protect computer keyboard Hooray! You are off on vacation and won't be tapping that computer keyboard for a couple of weeks. Cover your computer keyboard with plastic wrap during long periods of inactivity to keep out dust and grime.

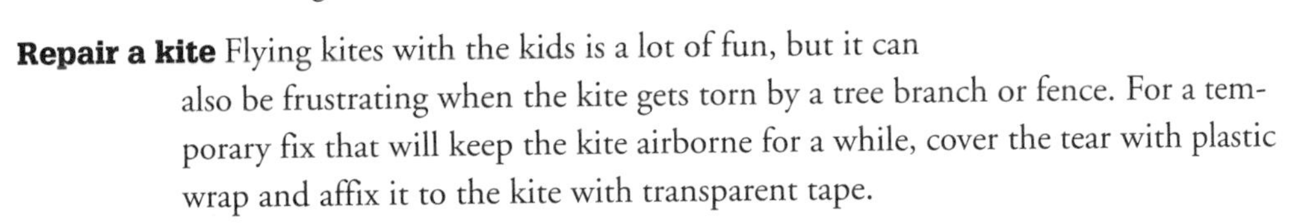

Repair a kite Flying kites with the kids is a lot of fun, but it can also be frustrating when the kite gets torn by a tree branch or fence. For a temporary fix that will keep the kite airborne for a while, cover the tear with plastic wrap and affix it to the kite with transparent tape.

Keep stored paint fresh Your leftover paint will stay fresh longer if you stretch a sheet of premium plastic wrap over the top of the can before tightly replacing the lid.

TAKE CARE When microwaving foods covered with plastic wrap, always turn back a corner of the wrap or cut a slit in it to let steam escape. Never use plastic wrap when microwaving foods with a high sugar content; they can become extremely hot and melt the wrap.

* Plungers

Remove car dents Before forking over big bucks to have an auto body repair shop pull a dent out of your car, try this: Wet a plumber's plunger, push it over the dent, and then pull it out sharply.

Catch chips when drilling ceiling Before you use a star drill to make an overhead hole, remove the handle from a plunger and place the cup over the shank of the drill. The cup will catch falling chunks of plaster, cement, or brick.

Use as an outdoor candleholder Looking for a place to put one of those bug-deterring citronella candles? Plant a plunger handle in the ground and put the candle in the rubber cup.

* Popsicle Sticks

Emergency splint for a finger You've got a kid with an apparent broken finger and you're on the way to the emergency room. Use a Popsicle stick as a temporary splint for the finger. Tape it on with adhesive tape to help stabilize the finger until it can be set.

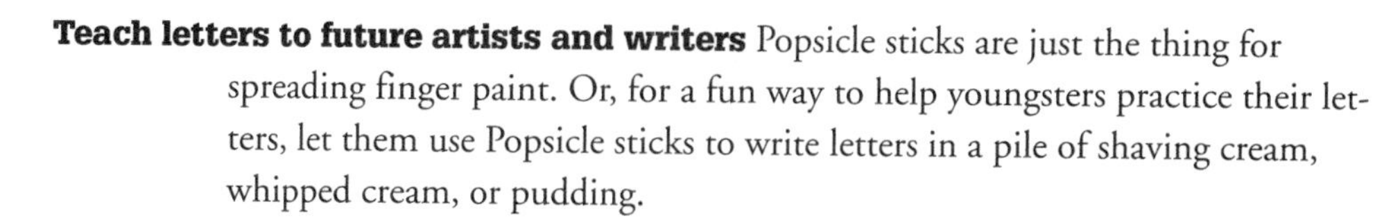

Teach letters to future artists and writers Popsicle sticks are just the thing for spreading finger paint. Or, for a fun way to help youngsters practice their letters, let them use Popsicle sticks to write letters in a pile of shaving cream, whipped cream, or pudding.

Skewer kids' food It's more fun to eat food if you get to play with it first, as the parents of many picky eaters know. Popsicle sticks are good to have in your bag of tricks at mealtime. Skewer bites of hot dog, pineapple, melon, and more. Or give kids a stick and have them spread their own peanut butter and jelly.

Label your plantings Is that parsley, sage, rosemary, or thyme popping up from your garden? Remember what you planted by using Popsicle sticks as plant labels. Just write the type of seeds you planted on the stick with indelible marker.

Keep track of paint colors You're out in the garage searching through the cans of left-over paint. Did you paint the living room with a paint called Whipped Cream or Sand? Don't get mixed up again. After you've painted a room, dip a Popsicle stick in the can. Let it dry. Write the name of the paint and the room where it was used on the stick. Now you'll know what color to use when it's time to paint again. These guides can also help a home decorator pick out fabrics and decorative items.

Kids' Stuff Here's a great little gift to encourage young readers. All you need is a **Popsicle stick**, **paint**, **craft foam**, **glue**, and a **marker**. Paint the Popsicle stick in a bright color, such as red. When it's dry, write on one side a reading slogan such as "I love to read" or "I'll hold the page." Then cut a shape out of the craft foam, such as a heart, flower, a cat, or a dog. Glue the shape to the top of the Popsicle stick, and you have a **homemade bookmark**.

* Pots and Pans

Catch draining engine oil No need to run out and buy an oil-collecting pan. When it is time to change the oil in your car engine, just stick an old 5-quart (4.75-liter) or larger pot beneath the drain plug.

Create an instant birdbath You can quickly provide feathered visitors to your backyard with a place to refresh themselves. Just set an old pan atop a flowerpot and keep it filled with water.

Use as a large scoop Leave those 50-pound (20-kilogram) sacks of fertilizer and grass seed in the garden shed and use a pot to carry what you need to the place you're tending. A small pot with a handle also makes a terrific boat bailer or dog-food scoop.

Make an extra grill You've got a big barbecue planned, and your grill is not big enough to handle all those burgers and dogs. Improvise an auxiliary grill by building a fire in an old, large pot. Cook on a cake rack placed over the pot. After you are finished, put the pot's cover on to choke out the fire and save the charcoal for another cookout.

* Potatoes

Make a decorative stamp Forget those expensive rubber stamps that go for up to $10 or more apiece. A potato can provide the right medium for making your own stamp for decorating holiday cards and envelopes. Cut a potato in half widthwise. Carve a design on one half. Then start stamping as you would with a wooden version.

Remove stains on hands Your family's favorite carrot soup is simmering on the stove, and you've got the orange hands to show for it. Otherwise hard-to-remove stains on hands from peeling carrots or handling pumpkin come right off if you rub your hands with a potato.

Extract salt for soup Hm, did you go a bit overboard when salting the soup? No problem. Just cut a few potatoes into large chunks. Toss them into the soup pot still on the stove. When they start to soften, in about 10 minutes, remove them and the excess salt they have absorbed. Save them for another use, such as potato salad.

Remove a broken lightbulb You're changing a lightbulb in the nightstand lamp, and it breaks off in your hand. So now the glass is off, but the stem's still inside. Unplug the lamp. Cut a potato widthwise and place it over the broken bulb. Twist, and the rest of the lightbulb should come out easily.

Remove tarnish on silverware High tea is being served at your house later today, and you're out of silver polish. Grab a bunch of potatoes and boil them up. Remove them from the water and save them for another use. Place your silverware in the remaining water and let it sit for an hour. Then remove the silverware and wash. The tarnish should be gone.

Keep ski goggles clear You can't keep a good lookout for trees and other skiers through snow goggles that fog up during your downhill descent. Rub raw potato over the goggles before you get on the ski lift, and the ride down should be crystal clear.

End puffy morning eyes We all hate waking up in the morning and looking at our mug in the mirror. What are those puffy spots on your face? Oh yeah, those are your eyes. A little morning TLC is what you need. Apply slices of raw, cold potatoes to your peepers to make the puffiness go away.

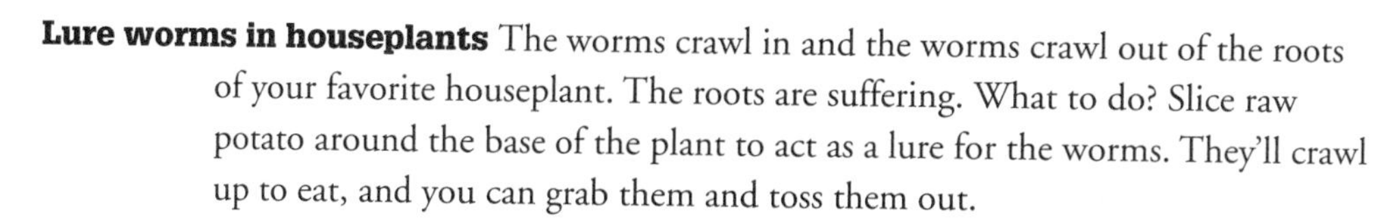

Lure worms in houseplants The worms crawl in and the worms crawl out of the roots of your favorite houseplant. The roots are suffering. What to do? Slice raw potato around the base of the plant to act as a lure for the worms. They'll crawl up to eat, and you can grab them and toss them out.

Feed new geraniums A raw potato can give a fledgling geranium all the nutrients it could desire. Carve a small hole in a potato. Slip a geranium stem into the hole. Plant the whole thing, potato and all.

Hold a floral arrangement in place If you have a small arrangement of flowers that you'd like to stabilize but have none of that green floral foam on hand to stick the flower stems in, try a large baking potato. Cut it in half lengthwise and place it cut side down. Poke holes where you want the flowers and then insert the stems.

Restore old, beat-up shoes Try as you might, your old shoes are just too scuffed to take a shine anymore. They don't have holes, and they are so nice and comfy that you hate to throw them away. Before you give them the brush-off, cut a potato in half and rub those old shoes with the raw potato. After that, polish them; they should come out nice and shiny.

Make a hot or cold compress Potatoes retain heat and cold well. The next time you need a hot compress, boil a potato, wrap it in a towel, and apply to the area. Refrigerate the boiled potato if you need a cold compress.

SCIENCE FAIR

Here's a way to demonstrate the **power of air pressure**. Grasp a **plastic straw** in the middle and try to **plunge it into a potato**. It crumples and bends, unable to penetrate the potato. Now grasp another straw in the middle, but this time, put your **finger** over the top. The straw will plunge right into the tuber. When the air is trapped inside the straw, it presses against the straw's sides, stiffening the straw enough to plunge into the potato. In fact, the deeper the straw plunges, the less space there is for the air and the stiffer the straw gets.

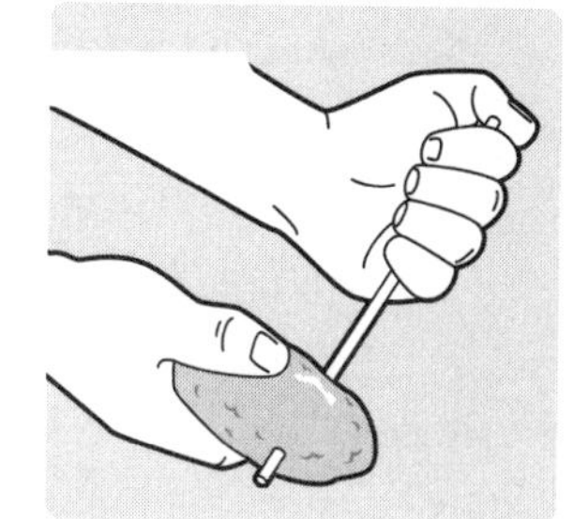

* Return Address Labels

Tag your bags Address labels aren't only for sticking on envelopes. They can be an effective, inexpensive way of making sure your lost items stand a chance of finding their way home. Place an address label—covered with a small piece of transparent or clear packing tape to prevent wear—inside your laptop PC bag, designer eyewear case, gym bag, knapsack, and all pieces of luggage—whether they are tagged or not.

Label your stuff Few folks bother to take out insurance on their collection of personal electronics equipment, but replacing a high-end PDA, camera, camcorder, or MP3 player can run into some serious money. Still, a tape-covered address label conspicuously placed on your gear just may facilitate its safe return. Of course, there are no guarantees in these matters, but at least the policy is cheap enough.

Hang on to your umbrella A well-made umbrella can last for years, but that won't help if it's left behind on a bus or train. Minimize the risk of your loss becoming someone else's gain by sticking an address label on the umbrella handle and wrapping it once around with clear packing tape. This protects the label from the elements—and makes it considerably more difficult to remove.

Secure school supplies It's one of the universal truths of parenthood: Kids' pencil cases, folders, markers, and other school supplies are forever disappearing. You may be able to lessen the losses, however, by affixing address labels with a piece of transparent tape to the contents of your child's desk and backpack.

Identify items for repair Do you suffer separation anxiety when you bring your beloved stereo equipment or another precious item into the repair shop? You may feel better if you place an address label on the base or some other unobtrusive, undamaged area. *Note:* This practice is not recommended for all personal treasures—you probably wouldn't want to label paper documents, paintings, photos, and such.

R

Rubber Bands

Stop sliding spoons Plop! The spoon slipped into the mixing bowl again, and now you have to fish it out of the messy batter. This time, after you rinse off the spoon, wrap a rubber band around the top of the handle to catch the spoon and avoid the mess.

Secure your casserole lids Don't spill it! That's what you say when you hand somebody your lovingly prepared casserole dish to carry in the car on the way to that potluck dinner. You won't have to worry if you secure the top to the base with a couple of wide rubber bands.

Anchor your cutting board Do you find yourself chasing your cutting board around the counter when you're chopping up veggies? Give the board some traction by putting a rubber band around each end.

Get a grip on twist-off tops Ouch! The tops on most beer bottles these days are supposed to be twist-off, but for some reason they still have those sharp little crimps from the bottle-opener days. And those little crimps can really dig into your hand. Wrap the top in a rubber band to save the pain. The same trick works great for smooth, tough-to-grip soda bottle tops too.

Get a grip on drinking glasses Does arthritis make it tough for you to grasp a drinking glass securely, especially when it is wet with condensation? Wrap a couple of rubber bands around the glass to make it easier to grip. Works great for kids, too, whose small hands sometimes have a hard time holding a glass.

Reshape your broom No need to toss out that broom because the bristles have become splayed with use. Wrap a rubber band around the broom a few inches from the bottom. Leave it for a day or so to get the bristles back in line.

The first rubber band was patented in 1845 by Stephen Perry, who owned a manufacturing company in London.

A key ingredient in making rubber bands is sulfur. When it is added to the rubber and heated—a process known as vulcanization—it makes the rubber strong and stretchy and prevents it from rotting. The process of making rubber bands is surprisingly similar to making a loaf of bread. First the dry ingredients are mixed with natural rubber. The resulting friction and chemical reaction heats and partially vulcanizes the rubber. The rubber is cooled, then rolled out like bread dough. It's extruded into a long tube, and the tube is heated to finish the vulcanization. Then it's rinsed, cooled, and sliced into bands.

Childproof kitchen and bath cabinets The grandkids are coming! Time to get out the rubber bands and temporarily childproof the bathroom and kitchen cabinets you don't want them to get into. Just wrap the bands tightly around pairs of handles.

Keep thread from tangling Tired of tangled thread in your sewing box? Just wrap a rubber band around the spools to keep the thread from unraveling.

Make a holder for your car visor Snap a couple of rubber bands around the sun visors of your car. Now you have a handy spot to slip toll receipts, directions, maybe even your favorite CD.

Thumb through papers with ease Stop licking your finger. Just wrap a rubber band around your index finger a few times the next time you need to shuffle papers. Not too tight, though! You don't want to cut off circulation to your fingertip.

Extend a button Having trouble breathing? Maybe that top shirt button is a tad too tight. Stick a small rubber band through the buttonhole, then loop the ends over the button. Put on your tie and breathe easy.

Use as a bookmark Paper bookmarks work fine, until they slip out of the book. Instead, wrap a rubber band from top to bottom around the part of the book you've already read. You won't lose your place, even if you drop the book.

Cushion your remote control To protect your fine furniture from scratches and nicks, wrap a wide rubber band around both ends of the television remote control. You'll be protecting the remote too—it will be less likely to slide off a table and be damaged.

Secure bed slats Do the slats under your mattress sometimes slip out? Wrap rubber bands around their ends to make them stay in place.

Tighten furniture casters Furniture leg casters can become loose with wear. To tighten up a caster, wrap a rubber band around the stem and reinsert.

Gauge your liquids Hm, just how much finish is left in that can up on the shelf anyway? Snap a band around the liquid containers in your workshop to indicate how much is left and you'll always know at a glance.

Wipe your paintbrush Every time you dip your paintbrush, you wipe the excess against the side of the can. Before you know it, paint is dripping off the side of the can and the little groove around the rim is so full of paint that it splatters everywhere when you go to hammer the lid back on. Avoiding all this mess is easy. Just wrap a rubber band around the can from top to

bottom, going across the middle of the can opening. Now, when you fill your brush, you can just tap it against the rubber band and the excess paint will fall back into the can.

R

* Rubber Jar Rings

Keep rug from slipping If you have a throw rug that tends to skate across the floor, keep it in place by sewing a rubber jar ring or two on the underside in each corner.

Play indoor quoits What else can you do to keep antsy pre-schoolers happily occupied on a rainy day? Turn a stool or small table upside down and let them try to toss rubber jar rings over the legs.

Protect tabletops Protect your tabletops from scratches and watermarks by placing a rubber jar ring under vases and lamps.

* Rubbing Alcohol

Clean bathroom fixtures Just reach into the medicine cabinet the next time you need to clean chrome bathroom fixtures. Pour some rubbing alcohol straight from the bottle onto a soft, absorbent cloth and the fixtures. No need to rinse—the alcohol just evaporates. It does a great job of making chrome sparkle, plus it will kill any germs in its path.

Remove hair spray from mirrors When you are spritzing your head with hair spray, some of it inevitably winds up on the mirror. A quick wipe with rubbing alcohol will whisk away that sticky residue and leave your mirror sparkling clean.

Clean venetian blinds Rubbing alcohol does a terrific job of cleaning the slats of venetian blinds. To make quick work of the job, wrap a flat tool—a spatula or maybe a 6-inch (15-centimeter) drywall knife—in cloth and secure with a rubber band. Dip in alcohol and go to work.

Keep windows sparkling and frost-free Do your windows frost up in the wintertime? Wash them with a solution of 1/2 cup rubbing alcohol to 1 quart (1 liter) water to prevent the frost. Polish the windows with newspaper after you wash them to make them shine.

Dissolve windshield frost Wouldn't you rather be inside savoring your morning coffee a little longer instead of scrape, scrape, scraping frost off your car windows? Fill a spray bottle with rubbing alcohol and spritz the car glass. You'll be able to wipe the frost right off. Ah, good to the last drop!

Prevent ring around the collar To prevent your neck from staining your shirt collar, wipe your neck with rubbing alcohol each morning before you dress. Feels good too.

TAKE CARE Don't confuse denatured alcohol with rubbing alcohol. Denatured alcohol is ethanol (drinking alcohol) to which poisonous and foul-tasting chemicals have been added to render it unfit for drinking. Often, the chemicals used in denatured alcohol are not ones you should put on your skin. Rubbing alcohol is made of chemicals that are safe for skin contact—most often it's 70 percent isopropyl alcohol and 30 percent water.

Clean your phone Is your phone getting a bit grubby? Wipe it down with rubbing alcohol. It'll remove the grime and disinfect the phone at the same time.

Remove ink stains Did you get ink on your favorite shirt or dress? Try soaking the spot in rubbing alcohol for a few minutes before putting the garment in the wash.

Erase permanent markers. Did Junior decide to decorate your countertop with a permanent marker? Don't worry, most countertops are made of a nonpermeable material such as plastic laminate or marble. Rubbing alcohol will dissolve the marker back to a liquid state so you can wipe it right off.

Remove dog ticks Ticks hate the taste of rubbing alcohol as much as they love the taste of your dog. Before you pull a tick off Fido, dab the critter with rubbing alcohol to make it loosen its grip. Then grab the tick as close to the dog's skin as you can and pull it straight out. Dab again with alcohol to disinfect the wound. This works on people too.

Get rid of fruit flies The next time you see fruit flies hovering in the kitchen, get out a fine-misting spray bottle and fill it with rubbing alcohol. Spraying the little flies knocks them out and makes them fall to the floor, where you can sweep them up. The alcohol is less effective than insecticide, but it's a lot safer than spraying poison around your kitchen.

Make a shapeable ice pack The problem with ice packs is they won't conform to the shape of the injured body part. Make a slushy, conformable pack by mixing 1 part rubbing alcohol with 3 parts water in a self-closing plastic bag. The next time that sore knee acts up, wrap the bag of slush in a cloth and apply it to the area. Ahhh!

Stretch tight-fitting new shoes This doesn't always work, but it sure is worth a try: If your new leather shoes are pinching your feet, try swabbing the tight spot with a cotton ball soaked in rubbing alcohol. Walk around in the shoes for a few minutes to see if they stretch enough to be comfortable. If not, the next step is to take them back to the shoe store.

S

Salt

SALT AROUND THE HOUSE

Clear flower residue in a vase Once your beautiful bouquet is gone, the souvenir it leaves behind is not the kind of reminder you want: deposits of minerals on the vase interior. Reach inside the vase, rub the offending ring of deposits with salt, then wash with soapy water. If your hand won't fit inside, fill the vase with a strong solution of salt and water, shake it or brush gently with a bottle brush, then wash. This should clear away the residue.

Clean artificial flowers You can quickly freshen up artificial flowers—whether they are authentic silk ones or the more common nylon variety—by placing them in a paper bag with 1/4 cup salt. Give the bag a few gentle shakes, and your flowers will emerge as clean as the day you bought them.

Hold artificial flowers in place Salt is a great medium for keeping artificial flowers in the arrangement you want. Fill a vase or other container with salt, add a little cold water, and arrange your artificial flowers. The salt will solidify, and the flowers will stay put.

Keep wicker looking new Wicker furniture can yellow with age and exposure to the sun and elements. To keep your wicker natural-looking, scrub it with a stiff brush dipped in warm salt water. Let the piece dry in the sun. Repeat this process every year or every other year.

Give brooms a long life A new straw broom will last longer if you soak its bristles in a bucket of hot, salty water. After about 20 minutes, remove the broom and let it dry.

Ease fireplace cleanup When you're ready to turn in for the night but the fire is still glowing in the hearth, douse the flames with salt. The fire will burn out more quickly, so you'll wind up with less soot than if you let it smolder. Cleanup is easier, too, because the salt helps the ashes and residue gather into easy sweepings.

Make your own brass and copper polish When exposure to the elements dulls brass or copper items, there's no need to buy expensive cleaning products. To shine your candlesticks or remove green tarnish from copper pots, make a paste by mixing equal parts salt, flour, and vinegar. Use a soft cloth to rub this over the item, then rinse with warm, soapy water and buff back to its original shine.

Remove wine from carpet Argggh! Red wine spilled on a white carpet is the worst. But there's hope. First, while the red wine is still wet, pour some white wine on it to dilute the color. Then clean the spot with a sponge and cold water. Sprinkle the area with salt and wait about 10 minutes. Now vacuum up the whole mess.

Clean grease stains from rugs Did that football-watching couch potato knock his greasy nachos onto your nice white carpet? Before you kill him, mix up 1 part salt to 4 parts rubbing alcohol and rub it hard on the grease stain, being careful to rub in the direction of the rug's natural nap. Or better yet, have him do it. Then you can kill him.

Remove watermarks from wood Watermarks left from glasses or bottles on a wood table really stand out. Make them disappear by mixing 1 teaspoon salt with a few drops of water to form a paste. Gently rub the paste onto the ring with a soft cloth or sponge and work it over the spot until it's gone. Restore the luster of your wood with furniture polish.

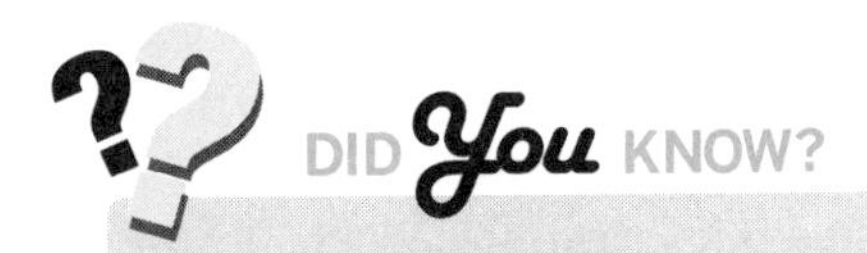

Salt may be the key to life on Mars. Thanks to Mars missions, scientists have confirmed two things about the Red Planet: There's plenty of ice and plenty of salt. Of course, life as we know it requires water. And while temperatures on Mars are either too high or too low for fresh water, the presence of salt makes it possible that there is life-sustaining salt water below the planet's surface.

Restore a sponge Hand sponges and mop sponges usually get grungy beyond use long before they are really worn out. To restore sponges to a pristine state, soak them overnight in a solution of about 1/4 cup salt per quart (liter) of water.

Relieve stings, bites, and poison ivy Salt works well to lessen the pain of bee stings, bug bites, and poison ivy:

- Stung by a bee? Immediately wet the sting and cover with salt. It will lessen the pain and reduce the swelling. Of course, if you are allergic to bee stings, you should get immediate medical attention.
- For relief from the itching of mosquito and chigger bites, soak the area in salt water, then apply a coating of lard or vegetable oil.

- When poison ivy erupts, relieve the itching by soaking in hot salt water. If the case is very unfortunate, you might want to immerse yourself in a tub full of salt water.

Keep windows and windshields frost-free As you probably know, salt greatly decreases the temperature at which ice freezes. You can use this fact to keep the windows in your home frost-free by wiping them with a sponge dipped in salt water, then letting them dry. In the winter, keep a small cloth bag of salt in your car. When the windshield and other windows are wet, rub them with the bag. The next time you go out to your car, the windows won't be covered with ice or snow.

Deodorize your sneakers Sneakers and other canvas shoes can get pretty smelly, especially if you wear them without socks in the summertime. Knock down the odor and soak up the moisture by occasionally sprinkling a little salt in your canvas shoes.

Make a scented air freshener Buying fragranced air fresheners can get expensive. Here is a wonderful way to make your room smell like a rose any time of the year: Layer rose petals and salt in a pretty jar with a tight-fitting lid. Remove the lid to freshen the room.

Give goldfish a parasite-killing bath The next time you take your goldfish out of its tank to change the water, put Goldie in an invigorating saltwater bath for 15 minutes while you clean the tank. Make the bath by mixing 1 teaspoon plain (noniodized) salt into 1 quart (1 liter) freshwater. (Just like the tank water, you should let tap water sit overnight first to let the chlorine evaporate.) The salt water kills parasites on the fish's scales and helps the fish absorb electrolytes. Don't add salt to the fish's tank, though. Goldfish are freshwater fish and can't spend a lot of time in salt water.

Clean your fish tank To remove mineral deposits from hard water in your fish tank, rub the inside of the tank with salt, then rinse the tank well before reinstalling the fish. Use only plain, not iodized, salt.

The United States is the world leader in salt production. In 2002 the U.S. produced 43.9 million metric tons of salt, according to the Salt Institute. China was second, with 35 million metric tons. Among other nations producing significant amounts of salt were Germany, with 15.7 million metric tons; India, with 14.8 million metric tons; and Canada, with 13 million metric tons.

Repel fleas in pet habitats If Fido enjoys his doghouse, chances are fleas do too. Keep fleas from infesting your pet's home by washing down the interior walls and floor every few weeks with a solution of salt water.

End the ant parade If ants are beating a path to your home, intercept them by sprinkling salt across the door frame or directly on their paths. Ants will be discouraged from crossing this barrier.

*SALT IN THE KITCHEN

Freshen your garbage disposal Is an unpleasant odor wafting from your garbage disposal? Freshen it up with salt. Just dump in 1/2 cup salt, run the cold water, and start the disposal. The salt will dislodge stuck waste and neutralize odors.

Remove baked-on food Yes, you can remove food that has been baked onto cooking pans or serving plates. In fact, it's easy. Baked-on food can be "lifted" with a pre-treatment of salt. Before washing, sprinkle the stuck-on food with salt. Dampen the area, let it sit until the salt lifts the baked-on food, then wash it away with soapy water.

Soak stains off enamel pans You can run out of elbow grease trying to scrub burned-on stains off enamel pans. Skip the sweat. Soak the pan overnight in salt water. Then boil salt water in the pan the next day. The stains should lift right off.

Keep oven spills from hardening The next time food bubbles over in your oven, don't give it a chance to bake on and cool. Toss some salt on the stuff while it is still liquid. When the oven cools, you'll be able to wipe up the spill with a cloth. The same technique works for spills on the stovetop. The salt will remove odors too, and if you'd like to add a pleasant scent, mix a little cinnamon in with the salt.

Scrub off burned milk Burned milk is one of the toughest stains to remove, but salt makes it a lot easier. Wet the burned pan and sprinkle it with salt. Wait about 10 minutes, then scrub the pan. The salt absorbs that burned-milk odor too.

Clean greasy iron pans Grease can be tough to remove from iron pans, because it is not water-soluble. Shortcut the problem by sprinkling salt in the pan before you wash it. The pan will absorb most of the grease. Wipe the pan out and then wash as usual.

Clean discolored glass Did your dishwasher fail to remove those stubborn stains from your glassware? Hand-scrubbing failed too? Try this: Mix a handful of salt in a quart of vinegar and soak the glassware overnight. The stains should wipe off in the morning.

Clean your cast-iron wok No matter how thoroughly you dry them, cast-iron woks tend to rust when you wash them in water. Instead, when you're done cooking, but while your wok is still hot, pour in about 1/4 cup salt and scrub it with a stiff wire brush. Wipe it clean, then apply a light coating of sesame or vegetable oil before stowing it. Don't clean a wok with a nonstick coating this way, because it will scratch the coating.

Remove lipstick marks from glassware Lipstick smudges on glassware can be hard to remove, even in the dishwasher. That's because the emollients designed to help lipstick stay on your lips do a good job sticking to glassware too. Before washing your stemware, rocks glasses, or water tumblers, rub the edges with salt to erase lipstick stains.

Brighten up your cutting boards After you wash cutting boards and breadboards with soap and water, rub them with a damp cloth dipped in salt. The boards will be lighter and brighter in color.

Clean the refrigerator We all have to do it sometime, and today it's your turn. You've removed all the food and the racks from the fridge. Now mix up a handful of salt in 1 gallon (3.7 liters) or so of warm water and use it with a sponge to clean the inside of the refrigerator. The mixture isn't abrasive, so it won't scratch surfaces. And you won't be introducing chemical fumes or odors.

Speed cleanup of messy dough Here's a way to make short work of cleanup after you've rolled out dough or kneaded breads. Sprinkle your floury countertop with salt. Now you can neatly wipe away everything with a sponge. No more sticky lumps.

Erase tea and coffee stains Tea and coffee leave stains on cups and in pots. You can easily scrub away these unattractive rings by sprinkling salt onto a sponge and rubbing in little circles across the ring. If the stain persists, mix white vinegar with salt in equal proportions and rub with the sponge.

Shine your teapot spout Teapots with seriously stained spouts can be cleaned with salt. Stuff the spout with salt and let it sit overnight or at least several hours. Then run boiling water through the pot, washing away the salt and revealing the old sparkle. If the stain persists, treat the rim with a cotton swab dipped in salt.

Salt was surely the first food seasoning. Prehistoric people got all the salt they needed from the meat that made up a large portion of their diet. When humans began turning to agriculture as a more reliable food source, they discovered that salt—most likely from the sea—gave vegetables that salty taste they craved. As the millennia passed, salt gradually made life more comfortable and certain as people learned to use it to preserve food, cure hides, and heal wounds.

Clean your coffee percolator If your percolated coffee tastes a bit bitter these days, try this: Fill the percolator with water and add 4 tablespoons salt. Then percolate as usual. Rinse the percolator and all of its parts well and the next pot you make should have that delicious flavor we all love.

Revive overcooked coffee You made a pot of coffee and then got distracted for an hour. Meanwhile, the coffee continued to cook in the pot and now it's bitter. Before you throw out the brew, try adding a pinch of salt to a cup.

Kids' Stuff Here's a **craft dough** easily fashioned into detailed **ornaments, miniature foods,** and **dolls.** In a bowl, slowly stir 1 cup **salt** into 1 cup boiling water. After the salt dissolves, stir in 2 cups **white all-purpose flour.** Turn the dough out onto a work surface and knead until smooth. If the dough sticks, add flour by the tablespoon until it is pliant. It should be easy to shape into balls, tubes, wreaths, and other shapes. Air-dry your creations or bake them in a 200°F (95°C) oven for up to two hours; time depends on thickness. Or microwave on High 1-2 minutes. Apply **paint;** protect and shine with **clear nail polish** or varnish.

Prevent grease splatters How many times have you been burned by splattering grease while cooking bacon when all you wanted was a hearty breakfast? Next time, add a few dashes of salt to the pan before beginning to fry foods that can splatter. You'll cook without pain and you won't have to clean grease off your cooktop.

Speed up cooking time In a hurry? Add a pinch or two of salt to the water you are boiling food in. This makes the water boil at a higher temperature so the food you are cooking will require less time on the stovetop. Keep in mind: Salt does not make the water boil faster.

Shell hard-boiled eggs with ease Ever wonder whether there's a secret to peeling hard-boiled eggs without breaking the shell into a million tiny pieces? There is, and now it's out of the box! Add a teaspoon of salt to your water before placing the eggs in it to boil.

Make perfect poached eggs You *know* it's possible to keep the whites intact when you poach eggs—you've had them in a restaurant. But no matter how careful you are, the whites always diffuse into the water when you poach eggs at home. Here's the secret the restaurant chefs know: Sprinkle about 1/2 teaspoon salt into the water just before you put in your eggs. This helps to "set" the whites in a neat package. A dash of vinegar also helps, and improves the taste of the eggs too.

Test an egg's freshness In doubt about whether your eggs are fresh? Add 2 teaspoons salt to 1 cup water and gently place the egg in the cup. A fresh egg will sink. An old one floats.

Shell pecans easier Pecans can be tough nuts to crack. And once you do crack them, it can be tough to dig out the meat. Soak the nuts in salt water for several hours before shelling, and the meat will come cleanly away from the shells.

Wash spinach more easily Fresh spinach leaves are lovely to look at, but their curving, bumpy surface makes it difficult to wash away all the dirt that collects in the crevices. Try this trick: Wash spinach leaves in salted water. Dirt is driven out along with salt in the rinse water, and you can cut the rinses down to just one.

Keep salad crisp Do you need to prepare leafy salad in advance of a dinner party? Lightly salt the salad immediately after you prepare it, and it will remain crisp for several hours.

Revive wrinkled apples Do your apples need a face-lift? Soak them in mildly salted water to make the skin smooth again.

Stop cut fruit from browning You're working ahead, making fruit salad for a party and you want to make sure your fresh-cut fruit looks appetizing when you serve the dish. To ensure that cut apples and pears retain their color, soak them briefly in a bowl of lightly salted water.

Use to whip cream and beat eggs The next time you whip cream or beat eggs, add a pinch of salt first. The cream will whip up lighter. The eggs will beat faster and higher, and they'll firm up better when you cook them.

Keep your milk fresh Add a pinch of salt to a carton of milk to make it stay fresh longer. Works for cream too.

Prevent mold on cheese Cheese is much too expensive to throw away because it has become moldy. Prevent the mold by wrapping the cheese in a napkin soaked in salt water before storing it in the refrigerator.

Extinguish grease fires Store your box of salt next to the stove. Then, should a grease fire erupt, toss the salt on it to extinguish the flames. *Never* pour water on a grease fire—it will cause the grease to splatter and spread the fire. Salt is also the solution when the barbecue flames from meat drippings get too high. Sprinkling salt on the coals will quell the flames without causing a lot of smoke and cooling the coals as water does.

The concentration of salt in your body is nearly one-third of the concentration found in seawater. This is why blood, sweat, and tears are so salty. Many scientists believe that humans, as well as all animals, need salt because all life evolved from the oceans. When the first land dwellers crawled out of the sea, they carried the need for salt—and a bit of the supply—with them and passed it on to their descendants.

Pick up spilled eggs If you've ever dropped an uncooked egg, you know what a mess it is to clean up. Cover the spill with salt. It will draw the egg together and you can easily wipe it up with a sponge or paper towel.

✱SALT IN THE LAUNDRY

Clean your iron's metal soleplate It seems to happen on a regular basis. No matter how careful you are while ironing, something melts onto the iron, forming a rough surface that is difficult to remove. Salt crystals are the answer. Turn your iron onto high. Sprinkle table salt onto a section of newspaper on your ironing board. Run the hot iron over the salt, and you'll iron away the bumps.

Make a quick pre-treatment You're out to a restaurant dining with friends and notice that a little salad dressing has spotted your slacks. You know it can't be checked with water, but here's an idea that will stop the stain from ruining your clothing. Drown the spot in salt to absorb the grease. When you get home, wash as usual.

Remove perspiration stains Salt's the secret to getting rid of those stubborn yellow perspiration stains on shirts. Dissolve 4 tablespoons salt in 1 quart (1 liter) hot water. Just sponge the garment with the solution until the stain disappears.

One of the several words salt has added to our language is *salary*. It comes from the Latin word *salarium*, which means "salt money" and refers to that part of a Roman soldier's pay that was made in salt or used to buy salt—a life-preserving commodity. This is also the origin of the phrase "worth one's salt."

Set the color in new towels The first two or three times you wash new colored towels, add 1 cup salt to the wash. The salt will set the colors so your towels will remain bright much longer.

✱SALT IN THE GARDEN

Stop weeds in their tracks Those weeds that pop up in the cracks of your walkways can be tough to eradicate. But salt can do the job. Bring a solution of about 1 cup salt in 2 cups water to a boil. Pour directly on the weeds to kill them. Another equally effective method is to spread salt directly onto the weeds or unwanted grass that come up between patio bricks or blocks. Sprinkle with water or just wait until rain does the job for you.

Rid your garden of snails and slugs These little critters are not good for your plants. But there's a simple solution. Take a container of salt into the garden and douse the offenders. They won't survive long.

Clean flowerpots without water Need to clean out a flowerpot so that you can reuse it? Instead of making a muddy mess by washing the pot in water, just sprinkle in a little salt and scrub off the dry dirt with a stiff brush. This method is especially handy if your potting bench is not near a water source.

*SALT IN THE BATH

A pre-shampoo dandruff treatment The abrasiveness of ordinary table salt works great for scrubbing out dandruff before you shampoo. Grab a saltshaker and shake some salt onto your dry scalp. Then work it through your hair, giving your scalp a massage. You'll find you've worked out the dry, flaky skin and are ready for a shampoo.

Condition your skin You've heard of bath salts, of course. Usually this conjures images of scented crystals that bubble up in your tub and may contain coloring and other stuff that leave a dreaded bathtub ring. Now strip that picture to its core, and you've got salt. Dissolve 1 cup table salt in your tub and soak as usual. Your skin will be noticeably softer. Buy sea salt for a real treat. It comes in larger chunks and can be found in health food stores or the gourmet section of a grocery store.

Give yourself a salt rubdown. Try this trick to remove dead skin particles and boost your circulation. Either while still in the tub, or just after stepping out of the tub—while your skin is still damp—give yourself a massage with dry salt. Ordinary salt works well; the larger sea salt crystals also do the job.

Freshen your breath the old-fashioned way Store-bought mouthwash can contain food coloring, alcohol, and sweeteners. Not to mention the cost! Use the recipe Grandma used and your breath will be just as sweet. Mix 1 teaspoon salt and 1 teaspoon baking soda into 1/2 cup water. Rinse and gargle.

Open hair-clogged drains It's tough to keep hair and shampoo residues from collecting in the bathtub drain and clogging it. Dissolve the mess with 1 cup salt, 1 cup baking soda, and 1/2 cup white vinegar. Pour the mixture down the drain. After 10 minutes, follow up with a 1/2 gallon (2 liters) boiling water. Run your hot-water tap until the drain flows freely.

Remove spots on tub enamel Yellow spots on your enamel bathtub or sinks can be lessened by mixing up a solution of salt and turpentine in equal parts. Using rubber gloves, rub away the discoloration and then rinse thoroughly. Don't forget to ventilate the bathroom while performing this cleaning task.

Saltshakers

Cut back on sugar You can cut back on sugar but still keep your sweet tooth happy if you fill a saltshaker with sugar. For sugar-restricted diets, use your sugar shaker as an alternative to dipping into the sugar bowl, and sprinkle lightly over food.

Use as a cinnamon/sugar dispenser Cinnamon toast is a great comfort food, and everyone likes it made a certain way. Mix sugar and cinnamon to your taste in a saltshaker. Once you've found the proportions you like, you can make it easily and consistently every time. Your cinnamon/sugar shaker is also perfect for sprinkling a little flavor on cereal.

Use for flour-dusting Baking is sometimes a messy job, so make at least one part of it tidier by putting flour into a large saltshaker. It's perfect for dusting your cake pans or muffin cups. Keep it neat and keep it handy in the cupboard, especially if you have an aggressively helpful junior chef!

Tip Colored Salt

Want to bring a little unexpected fun to your dinner table? Try colored salt! Put a few tablespoons of salt into a plastic sandwich bag and add a few drops of food coloring. Work it gently with your fingers to mix, and let dry in the open bag for about a day. Just cut a hole in the corner of the bag to pour the festive salt into your shaker. As a bonus, your colored table or kosher salt is wonderful homemade glitter!

Use to apply dry fertilizer If you use dry fertilizer, try putting it in a saltshaker to use when fertilizing seedlings. It gives you lots of application control so you can prevent fertilizer burn on your tender babies.

Sand

Protect and store garden tools Your gardening tools are meant to last longer than your perennials, so keep them clean and protected from the elements. Fill a 5-gallon (19-liter) bucket with builder's sand (available at masonry supply and home centers) and pour in about 1 quart (1 liter) of clean motor oil. Plunge shovels and other tools into the sand a few times to clean and lubricate them. To prevent rust, you can leave the tool blades in the bucket of sand for storage. A coffee can filled with sand and a little motor oil will give the same protection to your pruners and hand trowels.

Clean a narrow-neck vase You've held on to your bouquet as long as possible, but it's finally time to toss it along with the water it was sitting in. Now the vase needs cleaning, but the opening is too narrow for your hand. Put a little sand and warm, sudsy water in the vase, and swish gently. The sand will do the work of cleaning the residue inside for you!

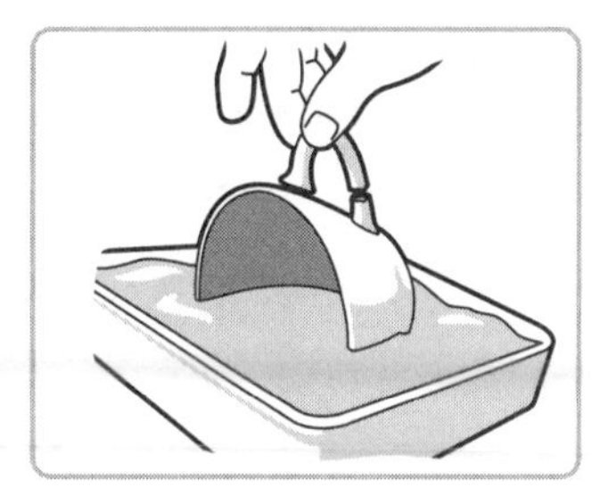

Hold items while gluing Repairing small items, such as broken china, with glue would be easy if you had three hands—one for each piece along with one to apply the glue. Since you only have two hands, try this: Stick the biggest part of the item in a small container of sand to hold it steady. Position the large piece so that when you set the broken piece in place, the piece will balance. Apply glue to both edges and stick on the broken piece. Leave the mended piece there, and the sand will hold it steady until the glue dries.

Carry in trunk for traction A bag of sand in the trunk of your car is good insurance in icy weather against getting stuck or spinning out from a parking spot. Throw in a clean margarine tub as well, to use as a scoop. For those with rear-wheel-drive vehicles, a bag or two of sand will also give you some extra traction.

DID *You* KNOW?

Sand sculpture festivals have been springing up in coastal towns around the world and have become huge summer tourist events. Typically, they are built around a theme. For example, one festival in Zeebrugge, Belgium, had a huge, sprawling sculpture representing more than 100 Hollywood icons and characters. Like most festivals, it was produced by a multinational team of sculptors, who worked for weeks on "the pile"—tons of trucked-in special sculpting sand that contains 8-10 percent clay for stability.

* Sandpaper

Sharpen sewing needles Think twice before throwing out a used piece of fine-grit sandpaper; the unused edges or corners are perfect for tucking into your sewing box. Poking your sewing needles through sandpaper a few times, or twisting them inside a folded piece of sandpaper, will make them sharper than ever.

Sharpen your scissors Are your scissor cuts less than crisp? Try cutting through a sheet of fine-grit sandpaper to finish off the edge and keep your cuts clean.

Remove fuzzy pills on sweaters If you're fighting a losing battle with the fuzz balls on your sweaters, a little sandpaper will handle them. Use any grit, and rub lightly in one direction.

Remove scorches on wool Take some medium-grit sandpaper to any small scorch spots on your woolen clothing. The mark left by a careless spark will be less noticeable with some light sanding around the edges.

Hold pleats while ironing If you're a pleat perfectionist, keep some fine- or medium-grit sandpaper handy with your iron. Put the sandpaper under the pleat to hold it in place while you iron a nice sharp fold.

Kids' Stuff You or your little Leonardo can make a beautiful **one-of-a-kind T-shirt**. Have the youngster use **crayons** to draw a bold design on the rough side of a sheet of **sandpaper.** Lay the T-shirt on your **ironing board** and slip a sheet of **aluminum foil** inside, between the front and back of the shirt. Place the sandpaper onto the T-shirt, design side down. Using an **iron** on the warm setting, press the back of the sandpaper in one spot for about ten seconds and then move on to the next spot until the entire design has been pressed. Let the shirt cool to set the design, launder on a cool setting, then hang to dry.

Roughen slippery leather soles New shoes with slippery soles can send you flying, so take a little sandpaper and a little time to sand across the width of the soles and roughen up the slick surface. It's thriftier and easier than taking your new shoes to a repair shop to have new rubber soles put on.

Remove ink stains and scuff marks from suede A little fine-grit sandpaper and a gentle touch is great for removing or at least minimizing an ink stain or small scuff mark on suede clothing or shoes. Afterward, bring up the nap with a toothbrush or nailbrush. You might avoid an expensive trip to the dry cleaner!

Use to deter slugs Slugs are truly the unwelcome guests that will never leave, but you can stop them from getting into your potted plants in the first place. Put those used sanding disks to work under the bases of your pots, making sure the sandpaper is wider than the pot base.

Remove stubborn grout stains Sometimes your bathroom abrasive cleaner is just not abrasive enough. Get tough on grout stains with fine-grit sandpaper. Fold the sandpaper and use the folded edge to sand in the grout seam. Be careful not to sand the tile and scratch the finish.

Open a stuck jar Having a tough time opening a jar? Grab a piece of sandpaper and place it grit side down on the lid. The sandpaper should improve your grip enough to do the job.

Make an emery board If you don't have an emery board handy the next time you need to smooth your nails, just raid the sandpaper stash in the garage workshop. Look for a piece marked 120 grit or 150 grit on the back.

Sandwich and Freezer Bags

S

Protect your pictures You just picked up a batch of beautiful photos of your newest grandchild. Before you pass them around your bridge party, encase each in a small, clear sandwich bag. Then you can hear the oohs and aahs without smudges on your pictures.

Freeze a washcloth for a cold pack It's hard to predict when someone in your household will next suffer a burn, teething pain, or another bump or scrape. Be ready. Freeze a wet washcloth in a sandwich or freezer bag. Pull it out of the freezer the next time someone needs some cold care.

Protect your padlocks When the weather is cold enough to freeze your padlocks on the outdoor shed or garage, remember that a sandwich bag can help. Slip one over the lock and you'll avoid frozen tumblers.

Make a fabric-softener dispenser Who can ever remember to add the fabric softener to the wash at the right time? You won't have to again. Punch some pinholes in a sealable plastic bag and, holding it over the washer basin, fill it with fabric softener. Seal the bag and toss into the laundry. The softener dispenses slowly through the pinholes during the wash and you won't have to remember that extra step.

Display baby teeth Your daughter has lost her first tooth and wants to show it off. You don't want to lose that precious memento of this important rite of passage. Place it in a sealable plastic bag. She can easily display it, and you won't worry about the tooth getting lost.

Make baby wipes for pennies You could buy the outrageously expensive baby wipes at the store or purchase some in bulk and hope they don't dry out before you use them up. Or you can just take the thrifty parent's way out: Make your own baby wipes by placing soft paper towels in a sealable bag with a mixture of 1 tablespoon gentle antibacterial soap, 1 teaspoon baby oil, and 1/3 cup water. Use enough of the mixture just to get the wipes damp, not drenched.

DID *You* KNOW?

Of course, we haven't always carried our ham-and-cheese sandwiches to work in plastic bags. Sandwich-size plastic bags were first introduced in 1957. Seven years later, Mobil Corp. introduced sandwich bags with tuck-in flaps, also known as Baggies. Just in case you are too young to remember, before plastic bags, we wrapped our sandwiches in paper or wax paper, like they still do in most delis.

Mold soap scraps into a new bar The thrifty among us hate to throw out a sliver of soap. Yet they're impossible to use when they get small. Instead, start collecting them all in a sealable plastic bag. When you have several, place the bag in a pan of warm, not boiling, water. Watch the soap pieces melt. When the mixture cools, you have a new bar of soap.

Starch craft items You've just completed that handmade Christmas stocking for your grandchild. But the last fabric ornaments to attach need to be starched. Throw them in a sealable plastic bag that contains a bit of starch. Shake until covered, remove, and let dry. Save the starch in the bag for your next craft project.

Feed the birds Be kind to the birds in your yard during the lean winter months. Mix some birdseed with peanut butter in a sealable plastic bag. Seal the bag and mix the ingredients by kneading the outside of the bag. Then place the glob in a small net bag or spread on a pinecone. Attach to a tree and await the grateful flock.

✱ SANDWICH AND FREEZER BAGS IN THE KITCHEN

Store grated cheese Pasta or pizza is always better with a dash of freshly grated Parmesan cheese. But who wants to bother with getting the grater out every time you want that taste? Instead, take a wedge of Parmesan cheese, grate the whole thing at once, and then double bag it in two self-closing bags to protect the freshness. Or stick the grater in the bag with the cheese wedge and pull it out for a short grate when the pesto gets to the table. That way you won't have to clean the grater after each use.

Make a pastry bag Pastry bags can be cumbersome, expensive, and hard to clean. Stop scrounging around the kitchen drawer for the pastry bag tip. Place the food to be piped, be it deviled-egg mix or decorating frosting, into a sealable bag. Squish out the air and close the top. Snip off a corner of the bag to the size you want—start conservatively—and you are ready to begin squeezing.

Dispose of cooking oil Unless you want the plumber for a best friend, don't clog your kitchen drain with used cooking oil. Instead, wait for it to cool, then dump it in a sealable plastic bag. Toss the bag into the trash.

Color cookie dough without stained hands Experienced bakers know what a mess your hands can be after coloring cookie dough. Here's a clean idea: Place your prepared dough in a bag, add the drops of food coloring, and squish around until the color is uniform. You can use the dough now or stick it in the freezer ready to roll out when the next occasion arises.

Stop ice crystals on ice cream It's truly annoying to open up that container of mint chocolate chip ice cream from the freezer to find unappetizing crystals forming on the frozen dessert. Place your half-full ice-cream container in a sealable bag and no crystals will form.

Store extra ice cubes It's a common experience. You open the freezer to grab some ice cubes from the ice cube maker and they're all stuck together, sometimes clogging the ice cube dispenser on the front of the fridge. When your tray fills up, toss the cubes in a sealable freezer bag. They won't stick together and you'll have easy access to the ice.

Soften hard marshmallows You're about to pull out that bag of marshmallows from your kitchen cabinet to make s'mores around the dying grill when you notice that the once-fluffy puffs have turned hard as rocks. Warm some water in a pan. Place the marshmallows in a sealable plastic bag, seal, and place in the pan. The warmth will soften them up in no time.

Melt chocolate without a mess Melting chocolate in a microwave or double boiler leaves you with a messy bowl or pot to wash. Here's a mess-free method: Warm some water in a pan (do not boil). Place the chocolate you want to melt in a sealable freezer bag. Seal and place the bag in the pan. In a few moments, you have melted chocolate, ready to bake or decorate with. You can even leave the bag sealed and snip off a bottom corner of the bag to pipe the chocolate onto a cake. When you are done, just toss the bag.

Keep soda from going flat You have to run out for a few errands and you don't want to take that soda with you. Leave the opened bottle or can at home zipped up in a large self-closing bag. That should help keep the fizz in until you get back.

Grease your pans If you're never quite sure how to handle shortening and butter when greasing a cake pan or cookie sheet, here's a tip: Place a sandwich bag over your hand, scoop up a small amount of shortening or butter from the tub, and start greasing. You can leave the bag in the canister of shortening for next time.

Kids' Stuff **Dyed dry pasta** in different shapes and sizes is great for getting kids' creative juices flowing. They can use it to make string jewelry on yarn, or to decorate a picture frame or pencil canister, for example. To dye the pasta, put a handful of **pasta** in a **sealable plastic bag**. Add several drops of **food coloring**. Next squirt in a few drops of **rubbing alcohol**. Seal the bag. Shake it up so the coloring dyes the pasta. Spread the pasta out on foil and let dry.

Use as kids' gloves There's nothing more welcome than helping hands in the kitchen. But when they're little hands that tend to get dirty and leave prints all over the place, then something must be done. Before they start "helping" you make those chocolate chip cookies, place small sandwich bags over their hands. These instant gloves are disposable for easy cleanup.

Make a funnel That handiest of kitchen tools, the funnel, can be replicated easily with a small sandwich bag. Fill the bag with the contents you need funneled. Snip off the end and transfer into the needed container. Then just toss the bag when the funneling is done.

* SANDWICH AND FREEZER BAGS FOR STORING THINGS

Protect your fragile breakables There's a precious family heirloom, a statue, a vase, or a trinket that needs some extra padding when storing. Here's what to do: Place it gently in a self-closing bag, close the bag most of the way, blow it up with air, then seal it. The air forms a protective cushion around the memento.

Save your sweaters You're about to put away that pile of winter sweaters for the season. Don't just throw them in a box without protection. Place each sweater in a sealable plastic bag and seal. They'll be clean and moth-free when the cold weather rolls around again. Save the bags for next spring when the sweaters need to be stored again.

Create a sachet If your drawers are starting to smell musty, a sealable bag can be your dresser's best friend. Fill the bag with potpourri—for example, flower petals along with a few crushed fragrant leaves and a couple of drops of aromatic oil. Punch a bunch of small holes in the bag. Then place in the drawer. Your drawers will smell fresh again soon.

Add cedar to your closet Cedar closets smell great, and, more important, they repel moths. If you aren't lucky enough to have a cedar closet, you can easily create the next best thing. Fill a sealable bag with cedar chips—the kind you buy at a pet store for the hamster cage. Zip it closed, then punch several small holes in it. Hang the bag in your closet (a pants hanger is handy for this) and let the cedar smell do its work.

Make a pencil bag Do the kids have trouble keeping track of their school pencils, pens, and rulers? Puncture three holes along the bottom edge of a sealable freezer bag so it will fit in a three-ring binder. Now the young scholars can zip their supplies in and out of the bag.

The self-sealing plastic bag became a part of our lives in 1969, when Dow Chemical introduced the Ziploc bag. A wide variety of sealable bags has been developed since then, including snack, sandwich, quart (liter), gallon (3.7 liter), and 2-gallon (7.4 liter) sizes, and double-strength freezer bags. There are even ones with flat bottoms to make them easy to pour into. The vegetable bag, with holes in it to help keep veggies fresh, was short-lived, however. You can even get sealable bags in small pouches of 10 and 8 bags, so you can take a bagful of bags with you.

✱ SANDWICH AND FREEZER BAGS IN THE BATH

De-clutter the bathroom Here's a quick cleanup solution: Guests are coming over and the bathroom is strewn with Hubby's razor, shave cream, and more. Quickly gather up all the supplies in one clear sealable bag. That way, he will know where his shaving supplies are and you don't have to deal with them. Now, if we could just do something about the whiskers in the sink!

Make a bath pillow Ready for a nice hot bath? Want to luxuriate in the warm water with bubbles and champagne? Well, here's the perfect, and cheap, thing to make your bath experience complete: Blow up a gallon-size (3.7-liter) sealable plastic bag and you'll have a comfortable pillow during your soak.

Clean your dentures No more dentures in a cup by your bedside. Toss your teeth in with their cleaner in a sealable plastic bag. They'll be clean and ready to go in the morning.

Organize your makeup Many of us have scads of makeup. Pats of ill-advised eye shadow and samples of powder and blush from department stores fill our makeup cases. Problem is, there are only a few cosmetics we really use every single day. Stash those favorites in a sealable plastic bag so you don't have to hunt around for them every morning.

✱ SANDWICH AND FREEZER BAGS OUT AND ABOUT

Stash dirty clothes Chocolate ice cream is careering down your child's white Sunday-best shirt. If you can keep the stain from drying, it will be a lot easier to get out. Change your child's shirt and spray the stained shirt with stain remover if you have a small bottle handy or just soak in water if you don't. Then seal the shirt in a sealable plastic bag, and it will be ready for the wash when you get home.

Hold spare clothes Toilet training a child? Need to be ready for meal mishaps? Put a change of clothes for your son or daughter in a sealable plastic bag, and keep it in the trunk of your car. You won't have to think twice the next time you have an "accident."

Carry detergent for washing If you're planning a trip to a friend's beach house and think you'll be doing a few loads of laundry while you're there, premeasure some detergent in a bag that you can pour out when the time comes. Beats lugging a big box of detergent down to the shore.

Carry wet washcloth for cooling off Going for a long trip on a hot and sticky day? Use a sealable bag to take along a wet washcloth that has been soaked in water and lemon juice so that everyone can get a refreshing wipe-off. This is a good trick for fast on-the-road face and hand cleanups anytime.

Keep your valuables dry and afloat Whoops! You tipped the canoe and got dunked. No biggie, until that sinking feeling hits—your car keys and cell phone are at the bottom of the lake. Avoid this disaster by putting your valuables in a sealable bag. Blow air into it before you seal the bag so it will float. A sealable bag is perfect for keeping valuables dry at the water park or beach too.

Kids' Stuff This is a great and **yummy activity** if you're outdoors. Pour a small box of instant **pudding mix** in a **sealable plastic bag** and add the amount of **milk** called for on the box. Seal it up and then seal that bag in **another sealable bag**. Now you're ready to **play football**. Toss the bag of pudding around with your friends until it mixes and the pudding forms. Open the first bag and remove the second bag. Pour the pudding into flat-bottomed **ice-cream cones** and chow down.

Create a beach hand cleaner You're sitting on the beach and it's time for lunch. But before you reach into your cooler, you want to get the grit off your hands. Baby powder in a sealable plastic bag is the key. Place your hands in the bag, then remove them and rub them together. The sand is gone.

Apply bug spray to your face It's difficult to cleanly apply bug spray to your face without squirting yourself in the eyes or getting it on your hands. Instead, throw some cotton balls into a plastic bag, squirt in the bug spray, seal, and shake. Now use the cotton balls to apply the bug spray.

Cure car sickness The last thing you need in your car is a child throwing up. Make your child feel better and head off the mess and stench. Place a few cotton balls in a sealable plastic bag. Squirt in 2 drops lavender oil. If motion sickness strikes, the child can open the bag and take a few whiffs of the oil.

Use as a portable water dish Your furry best friend has happily hiked alongside you during your trek in the great outdoors. You take a break, and he gives you one of those longing looks as you draw on your canteen. No problem. You pull a sealable plastic bag full of water from your pack and hold it open while Buddy laps his fill.

* Screening

Store pierced earrings Keep your pierced and hook earrings organized and ready at-a-glance with a spare piece of window screen. Cut a square of screen with metal shears or utility scissors and cover the edges of the square with duct or cloth tape. Then push the earrings through the holes in the screen. If you like, you can hang the screen square on the wall by attaching string or floral wire to the top corners.

Get rid of paint lumps You want to do a touch-up paint job, but your used can of paint has some lumps in it. Instead of going through the bother of straining the paint into another container, try this: Cut a circle of screening sized to fit inside the can (use the lid as a guide). Place the screen circle on top of the paint and push it gently down to the bottom with your stir stick. The lumps will now be trapped at the bottom of the can. Stir up the paint and get to work.

Protect newly planted seeds Who knows what is walking around your garden at night, so protect newly planted seeds by covering them with a sheet of screen material. It also might deter the neighborhood felines from using your nice, fluffy soil as a cat box. When the seedlings emerge, you can bend the screening to make cages.

* Shampoo

Revitalize leather shoes and purses You don't need expensive mink oil to bring life back to your leather shoes and purses. A little shampoo and a clean rag will do the job. Rub shampoo into worn areas in circles to clean and bring back the color of your accessories. It will protect your shoes from salt stains as well.

Lubricate a zipper If your zipper gets stuck, don't yank on it until it breaks. Put a drop of shampoo on a cotton swab and dab it onto the zipper. The shampoo will help the zipper to slide free, and any residue will come out in the next wash.

Resize a shrunken sweater Oh no, you've shrunk your favorite sweater! Don't panic, you can bring it back to full size again with baby shampoo and warm water. Fill a basin with warm water, squirt in some baby shampoo, and swish once with your hand. Lay the sweater on top of the water and let it sink on its own and soak for 15 minutes. Gently take your sweater out without wringing it and put it in a container, then fill the sink again with clean water. Lay the sweater on top and let it sink again to rinse. Take the sweater out, place it on a

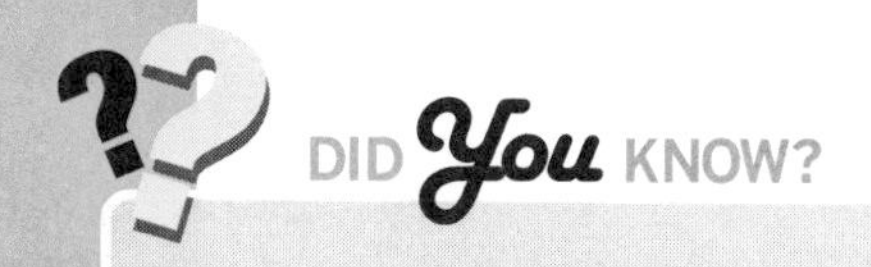

One of the longest-running advertising campaigns in history, "the Breck Girl," was the brainchild of Edward Breck, a member of the family that started Breck Shampoo Co. The ads, featuring wholesome, beautiful girls with gorgeous hair, began in 1936, during the Great Depression, although they didn't go national until 1947. Only two artists were used during the 40-year campaign. The best known was Ralph William Williams, who took over the job in 1957. Among the models for Williams's Breck girls were Cybill Shepherd, Kim Basinger, and Brooke Shields—all unknowns at the time. The campaign ceased soon after Williams's death in 1976.

towel, and roll the towel to take out most of the moisture. Lay the sweater on a dry towel on a flat surface and gently start to reshape it. Come back to the sweater while it's drying to reshape a little more each time. Your patience will be rewarded!

Wash houseplant leaves Houseplants get dusty too, but unlike furniture they need to breathe. Make a soapy solution with a few drops of shampoo in a pot of water, dunk in a cloth and wring it out, and wipe those dusty leaves clean.

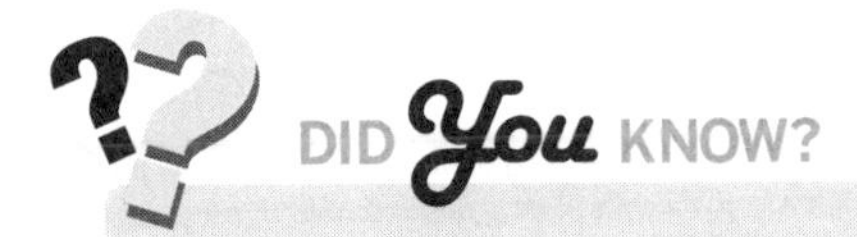

In the early 1900s, Martha Matilda Harper invented the reclining chair used when shampooing hair at beauty salons—unfortunately, she never patented it. But Harper was still a success. She emigrated from Canada to the United States as a young girl, bringing her own recipe for a hair "tonic" (shampoo). Eventually she went from making her tonic in a shed to opening her own shop, where she offered the Harper Method. She enticed wealthy women to leave their homes for a health-conscious salon experience where they would be shampooed and pampered by professionals. She was her own best advertisement, with hair that reached down past her feet.

Clean your car The grease-cutting power of shampoo works on the family grease monkey's baby as well. Use about 1/4 cup shampoo to a bucket of water and sponge up the car as usual. Use a dab of shampoo directly on a rag or sponge for hard-to-remove tar spots.

Remove sticky gunk from pet fur Did Rex or Fluffy step on tar or roll in what you hope is gum? Rub a tiny amount of shampoo on the spot and gently draw out the sticky stuff toward the end of the fur. Rinse with a wet cloth.

Lubricate stubborn nuts and bolts Got a nut and bolt that won't come apart? If your spot lubricant isn't handy or you've run out, try a drop of shampoo. Let it seep into the threads and the bolt will be much more cooperative.

Remove bandages painlessly Now you don't have to say "Ready?" when removing a bandage. Rub just a drop of shampoo on and around the bandage to let it seep through the air holes. It will come off with no muss and definitely no fuss.

Revitalize your feet Give your feet a pick-me-up while you sleep. Rub a little shampoo all over your feet and put on a light pair of cotton socks. When you wake up, your feet will feel smooth and silky.

Remove your eye makeup You can't beat no-tears baby shampoo for a thrifty eye makeup remover. Put a drop on a damp cotton pad to gently remove the makeup, then rinse clear. No frills, no tears!

S

Give yourself a bubble bath Shampoo makes a nice and sudsy bubble bath. It's especially relaxing if you love the scent of your favorite shampoo, and the tub will rinse cleaner.

Substitute for shaving cream You're on the road and discover you forgot to bring your shaving cream. Don't use soap to lather up. With its softening agents, shampoo is a much better alternative.

Clean grimy hands In place of soap, some straight shampoo works wonders for cleaning stubborn or sticky grime from your hands. It even works well to remove water-based paint.

Remove hair spray from walls If you've been using hair spray to kill flies, or you've just noticed hair spray buildup on your bathroom walls, reach for the shampoo. Put some on a wet sponge to clean, and wipe off suds with a clean, wet sponge. Shampoo is tailor-made to handle hair product buildup.

Clean the tub and faucets Need to do a quick tub cleanup before guests arrive? Grab the handiest item—your shampoo! It does a great job on soap scum because it rinses clean. You can use it to buff a shine into your chrome faucets as well.

Use to wash delicates Shampoo makes a great cleanser for your delicates. It suds up well with just a drop, and you get two cleaning products for the price of one!

Clean brushes and combs Skin oils can build up on your combs and brushes faster than you realize. And if you're tucking them into your purse or pocket, they're accumulating dust and dirt as well. Give them a fresh start in a shampoo bath. First comb any loose hair out of the brush, then rub a little shampoo around the bristles or along the teeth of the comb. Put a small squirt of shampoo in a tall glass of water, let the comb and brush sit for a few minutes, swish, and rinse clean.

Johnson & Johnson introduced the world's first shampoo made specifically for infants in 1955—containing its now-famous No More Tears formula. The company has promoted its baby shampoo to be "as gentle to the eyes as pure water." But, in fact, like most baby shampoos, Johnson's contains many of the same ingredients found in adult formulations, including citric acid, PEG-80 sorbitan laurate, and sodium trideceth sulfate. The lack of baby tears has less to do with the shampoo's purity than it does with maintaining a relatively neutral pH.

Shaving Cream

Use to clean hands The next time your hands get dirty on a camping trip, save that hard-lugged water for cooking and drinking. Squirt a little shaving cream in your hands and rub as you would liquid soap. Then wipe your hands off with a towel.

Prevent bathroom mirror fog-up Before you shower, wipe some shaving cream onto your bathroom mirror. It will keep it from fogging up so you don't have to wait to get to work with your toiletries or shaving after you get out of the shower.

From the late 1920s through the early 1960s, one of the best things about a long, tedious car ride were the signs every hundred yards or so advertising Burma-Shave, a brushless shaving cream. Here are some of the more memorable ones:

Shaving Brushes ...
You'll Soon See 'Em ...
On The Shelf ...
In Some Museum ...
Burma-Shave.

Are Your Whiskers ...
When You Wake ...
Tougher Than ...
A Two-Bit Steak? ...
Try ... Burma-Shave.

Within This Vale ...
Of Toil ... And Sin ...
Your Head Grows Bald ...
But Not Your Chin—Use ...
Burma-Shave.

Golfers!
If Fewer Strokes ...
Are What You Crave ...
You're Out Of The Rough ...
With ... Burma-Shave.

Remove stains from carpeting Junior is very sorry for spilling a little juice on the carpet, so make it "all better" with some shaving cream on the spot. Blot the stain, pat it with a wet sponge, squirt some shaving cream on it, and then wipe clean with a damp sponge. Use the same technique on your clothes for small stains; shaving cream can remove that spot of breakfast you discovered you're wearing during your once-over in the bathroom.

Silence a squeaky door hinge A squeaky door hinge can ruin a peaceful nap-time. With its ability to seep into nooks and crannies, a little shaving cream on the hinge will let you check on the baby undetected.

Sheets

Make a beanbag bull's-eye Are the kids rained out of their ball game? Here's one way to ease their disappointment and let them give their pitching arms a workout anyway. Draw a large bull's-eye on a sheet. Tape the sheet to a wall and let the kids pitch beanbags at it.

Use as a tablecloth It is your turn to host the whole clan for Thanksgiving. You're using every table in the house, but you don't have enough tablecloths. A patterned sheet makes an attractive festive table covering.

Repel deer from your yard Circle the garden with a cord about 3 feet (1 meter) above the ground, then tie strips of white sheets to it every 2 feet (60 centimeters); a tail-height flash of white is a danger sign to a deer.

Scoop up all those fall leaves No reason to strain your back by constantly lifting piles of leaves into a wheelbarrow or bag. Just rake the leaves onto a sheet laid on the ground. Then gather the four corners and drag the leaves to the curb or leaf pile.

Wrap up the old Christmas tree After removing holiday decorations, wrap an old sheet around the tree so that you can carry or pull it out of the house without leaving a trail of pine needles.

* Shoe Bags

Organize your utility closet A hanging shoe bag is a great organizer in the utility closet. Use its pockets to store sponges, scrub brushes, and other cleaning utensils—and even some bottles of cleaning products. It's also good for separating your clean, lemon-oil, and lint-free rags so you'll always have the right one for the job.

Organize your office area Free up some valuable drawer space in your office with an over-the-door shoe holder. Its pockets can store lots of supplies that you need to keep handy, like scissors, staples, and markers. You can use the pockets to organize bills and other "to do" items as well.

Organize your bathroom A shoe bag can keep lots of everyday bathroom items handy and neat. Brushes, shampoo, hand towels, hair spray—almost everything can be stored at your fingertips instead of cluttering the shower or counter.

Organize your child's room A shoe bag hung over their bedroom door is a great way to help your kids organize their small toys. Whether your child likes dolls, dinosaurs, or different-colored blocks, a shoe bag puts the toys on display and kids can keep them sorted themselves.

Organize car-trip toys and games Cut a shoe bag to fit the back of your car seat, and let your children make their own choices for back seat entertainment.

Organize your bedroom Instead of lifting a hanger to get a belt, or rummaging through your drawer for a scarf, try organizing your clothing accessories with an over-the-door shoe bag. The pockets can be used in the bedroom for keeping socks, gloves, and much more than shoes handy.

Shoe Boxes

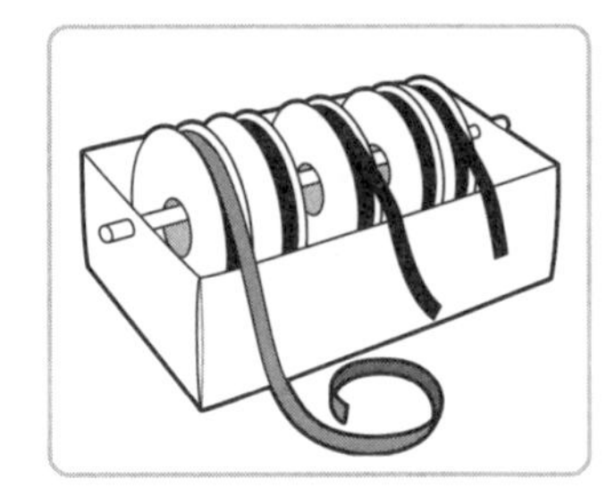

Make a gift ribbon dispenser You will thank yourself each time you look for ribbon to wrap a present, if you use a shoe box to make this handy ribbon dispenser. Take a used broom handle or piece of a bamboo garden stake—anything you can use as a small dowel—and cut it a little longer than the length of the shoe box. Cut two holes for the dowel, one in each short end of the box, at a height where a spool of ribbon slipped onto the dowel would spin freely. Slip your ribbon spools onto the dowel as you poke it from one end of the shoe box through to the other. Once the dowel is in place, you can duct tape it at either short end to keep it from slipping out. You could also cut holes along one long side of the shoe box for each spool of ribbon, and pull a little bit of each ribbon through the hole. Now you're ready to wrap!

Use for play bricks Kids can get creative using a collection of shoe boxes as building bricks. Tape the lids on for them. You can even let the little ones color the "bricks" with poster paint.

Get your stuff organized There are lots of ways shoe boxes can help you get organized besides collecting old photos and receipts. Label the boxes and use them to store keepsakes, canceled checks, bills to be paid, and other items you want to keep track of. For a neater appearance, cover the boxes with contact paper or any other decorative self-adhesive paper.

Pack yummy gifts Shoe boxes are the perfect size for loaves of homemade bread, but of course, you can also pack cookies in them.

Use as a whelping box Puppies or kittens on the way! To reduce the risk of the mother rolling onto a newborn and smothering it, place one or several puppies or kittens in a towel-lined shoe box while the others are being born.

* Shortening

Clean ink stains Next time a leaky pen leaves your hands full of ink, reach for a can of shortening. To remove ink stains from your hands and also from vinyl surfaces, rub on a dollop of shortening and wipe the stains away with a rag or paper towel.

Remove sticky adhesives Don't wear down your fingernails trying to scratch off resistant sticky labels and price tags. Instead use shortening to remove them (and their dried glue and gum residue) from glass, metals, and most plastics. Simply coat the area with shortening, wait 10 minutes, and scrub clean with a gentle scrub-sponge.

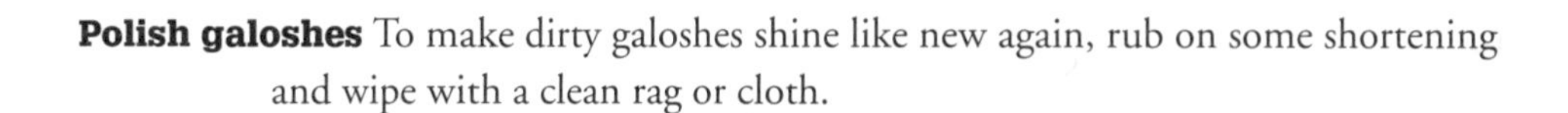

Polish galoshes To make dirty galoshes shine like new again, rub on some shortening and wipe with a clean rag or cloth.

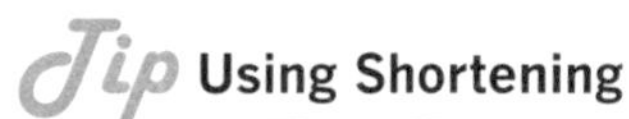

Tip Using Shortening

- Keep shortening away from sunlight to keep it from turning rancid.
- Never leave shortening unattended while frying.
- The most efficient temperature for frying with shortening is 325°F-350°F (165°C-180°C). Do not overheat shortening or it will burn. If shortening starts to smoke, turn off the heat and let it cool.
- If shortening catches fire, cover the pan with its lid, turn off the heat, and let it cool. Never put water on burning or hot shortening: It may splatter and burn you.

Soothe and prevent diaper rash Next time the baby is fussing from a painful case of diaper rash, rub some shortening on his bottom for fast relief. It will soothe and moisturize his sensitive skin.

Remove tar from fabric Tar stains on clothing are icky and tough to remove, but you can make the job easier with a little help from some shortening. After scraping off as much of the tar as you can, put a small glob of shortening over the remaining spot. Wait 3 hours, and then launder as usual.

Keep snow from sticking to shovel Before you dig out the car or shovel the driveway after a snowstorm, coat the blade of your snow shovel with shortening or liquid vegetable oil. It will not only keep snow from sticking but also make shoveling less tiring and more efficient.

Remove makeup All out of your regular makeup remover? Don't fret: Just use a dab of shortening instead. Your face won't know the difference.

Moisturize dry skin Why pay for fancy creams and lotions to moisturize your skin when ordinary shortening can do the trick at a fraction of the cost? Some hospitals even use shortening to keep skin soft and moist, and you can too. Next time your hands are feeling dry and scaly, just rub in a little shortening. It's natural and fragrance-free.

Repel squirrels Keep pesky squirrels from getting at a bird feeder. Just grease the pole with a liberal amount of shortening and the rodents won't be able to get a claw hold to climb up.

* Shower Curtains

Line cabinet shelves Don't discard your old vinyl shower curtains or tablecloths. Turn them into easy-to-clean shelf liners instead. Simply cut to shelf size and set in place, using some rubber cement to hold them if you prefer. When it's time for cleaning, just wipe with a damp sponge.

Make a protective apron For those extra messy jobs around the house, wear a homemade apron made from an old shower curtain. Make a cobbler's style apron, with a vest as well as a skirt. Use pinking shears to cut the vinyl to size. Poke two holes at the top of the vest for cords or ribbons to tie around your neck, and make two more holes in the sides for tying it around your waist.

Durable painting drop cloth Save an old shower curtain liner and use it as a drop cloth the next time you paint a room. The material is heavier and more durable than that used in commercially sold plastic drop cloths!

Protect floor under high chair Even the best-behaved and cutest babies leave a mess on the floor when they eat. Protect your floor or carpet and make cleanup a breeze. Cut a 36- to 48-inch (about 1-meter) square from an old shower curtain and place it under the baby's high chair. You can use the leftover scraps to make bibs too.

Cover picnic tables and benches Don't let a yucky table or sticky benches spoil your next picnic. Use an old shower curtain as a makeshift tablecloth (or as a tablecloth liner). Bring an extra shower curtain and fold it over a sticky or dirty picnic bench before you sit down to eat.

In the early 1920s, Waldo Semon, a rubber scientist, was none too thrilled when he first discovered polyvinyl chloride, commonly known as vinyl. He was trying to develop a new adhesive, and this stuff just didn't stick at all! But Waldo was quick to recognize the material's potential and began experimenting with it, even making golf balls and shoe heels out of it. Soon vinyl products like raincoats and shower curtains reached the consumer. Today vinyl is the second-largest selling plastic in the world, and the vinyl industry employs more than 100,000 people in the United States alone.

Protect table when cutting fabric Next time you're cutting a pattern on your dining room table, put a shower curtain or plastic tablecloth under it before you cut. The scissors will glide more easily across the surface and you'll protect the tabletop from an accidental nick.

Block weeds in mulched beds Those old shower curtains will also come in handy next time you do any landscaping with gravel or bark chips. Just place the shower curtain under the mulching material to prevent annoying weeds from poking through.

Skateboards

Use as a laundry cart If your home has a laundry chute, keep a basket atop a skateboard directly below the chute. When you're ready to do the laundry, simply roll the load over to the washer.

Make a shelf Is your kid an avid skateboarder? When he or she is ready for a new skateboard, turn the old one into a shelf for his or her room. Support it on a couple of metal shelf brackets. You can remove the wheels or leave them on.

Use as a painter's scooter. Crawling along the floor to paint a baseboard can get real old real fast. Borrow your kid's skateboard and save your knees. Sit cross-legged on the skateboard and roll along with your paintbrush and can.

Soap

Loosen stuck zippers Zipper stuck? Rub it loose with a bar of soap along the zipper's teeth. The soap's lubrication will get it moving.

Unstick furniture drawers If your cabinet or dresser drawers are sticking, rub the bottom of the drawer and the supports they rest on with a bar of soap.

Lubricate screws and saw blades A little lube with soap makes metal move through wood much more easily. Twist a screw into a bar of soap before driving it and rub some on your handsaw blade.

Remove a broken lightbulb If a bulb breaks while still screwed in, don't chance nicks and cuts trying to remove it. First, turn off the power. Insert the corner of a large, dry bar of soap into the socket. Give it a few turns and that base will unscrew.

Say farewell to fleas Fed up with those doggone fleas? Put a few drops of dish soap and some water on a plate. Place the plate on the floor next to a lamp. Fleas love light—they will jump on the plate and drown.

DID *You* KNOW?

Contrary to popular belief, hanging a perfumed bar of soap won't necessarily keep deer off your property. Stephen Vantassel, who runs Wildlife Damage Control in Springfield, Massachusetts, says that how long or how well soap works depends on a number of factors, including the type of plant you are protecting and the location of the soap. However, studies have shown that soap, especially tallow-based soap, will stop deer from making lunch out of your shrubbery. Local gardening centers can advise you about commercial spray repellents.

Deodorize your car Want your car to smell nice, but tired of those tree-shaped pine deodorizers? Place a little piece of your favorite-smelling soap in a mesh bag and hang it from your rearview mirror.

Mark a hem Forget store-bought marking chalk. A thin sliver of soap, like the ones left when a bar is just about finished, works just as well when you are marking a hem, and the markings wash right out.

Make a pin holder Here's an easy-to-make alternative to a pincushion. Wrap a bar of soap in fabric and tie the fabric in place with a ribbon. Stick in your pins. As a bonus, the soap lubricates the pins, making them easier to insert.

Prevent cast-iron marks Nip cookout cleanup blues in the bud. Rub the bottom of your cast-iron pot with a bar of soap before cooking with it over a sooty open flame. Look, Ma! No black marks!

Tip Homemade Soap

Handcrafted soap makes a great gift and is easy to make. You need a solid bar of glycerin (from a drugstore); soap molds (from a crafts store); a clean, dry can; a double boiler; food coloring; and essential oil. Place the glycerin in the can and put the can in a double boiler, which has water in the top as well as the bottom, and heat until the glycerin melts. For color, mix in food coloring. Spray a mold with nonstick cooking spray and fill it halfway with melted glycerin. Add a few drops of essential oil and fill the rest with glycerin. Let it harden.

Keep stored clothes fresh Pack a bar of your favorite scented soap when you store clothes or luggage. It will keep your clothes smelling fresh till next season and prevent musty odor in your luggage.

Save those soap slivers When your soap slivers get too tiny to handle, don't throw them away. Just make a small slit in a sponge and put the slivers inside. The soap will last for several more washings. Or make a washcloth that's easy for little hands to hold by putting the soap slivers in a sock.

* Socks

Protect stored breakables Want to protect Grandma's precious vase or your bobble head collection? Wrap it up! Slip the item into a sock to help protect it from breaking or chipping.

Cover kids' shoes when packing Does Junior insist on bringing along his favorite old sneakers? Before you toss them into the suitcase, cover each one with an adult-size sock to protect the rest of the clothing.

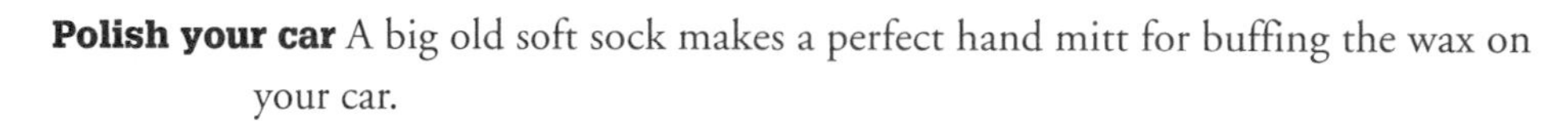

Polish your car A big old soft sock makes a perfect hand mitt for buffing the wax on your car.

Keep hands clean changing tire If you ever get a flat tire on your way to a fancy party or job interview, you'll thank yourself for having the foresight to throw a pair of socks in the trunk. Slip the socks on your hands while handling the tire, and they'll be clean when you arrive.

Protect floor surfaces The next time you need to move a heavy table or sofa across a smooth floor, put socks over the legs and just slide the piece.

Store your work goggles Shop goggles won't fit in an eyeglass case, so just slip them into a sock to protect them from getting scratched. You can even nail or screw the sock to the wall or bench so you will always know where the goggles are.

Wash bag for dainty lingerie Protect your precious delicates in the washing machine. Slip them into a sock and tie the ends.

Use as cleaning mitts Save those old or solo socks to use as cleaning mitts. Slip them on, and they are great for cleaning in tight corners and crevices.

Clean shutter and blind slats Forget wasting money on those expensive gadgets and gizmos for cleaning venetian blind slats. Just slip a sock over your hand and gently rub the dust off. You can use some dusting spray on the sock, if you like.

Clean rough plaster walls Use nylon or Ban-Lon socks, instead of a sponge or cloth, to clean rough plaster walls. No small pieces of material will be left behind.

Wash small stuffed animals Does your child's favorite stuffed fuzzy need a bath? Slip small stuffed animals into a sock and tie the end to prevent buttons, eyes, and other decorative items from coming loose.

Protect a wall from ladder marks To prevent marks on the wall when you're leaning your ladder on it, slip socks over the ladder top ends. For safety reasons, however, make sure someone is holding the ladder.

* Soda Pop

Clean car battery terminals Yes, it's true, the acidic properties of soda pop will help to eliminate corrosion from your car battery. Nearly all carbonated soft drinks contain carbonic acid, which helps to remove stains and dissolve rust deposits. Pour some soda pop over the battery terminals and let it sit. Remove the sticky residue with a wet sponge.

Loosen rusted-on nuts and bolts Stop struggling with rusted-on nuts and bolts. Soda pop can help to loosen any rusted-on nuts and bolts. Soak a rag in the soda pop and wrap it around the bolt for several minutes.

Remove rust spots from chrome Are you babying an older car—you know, one of those babies that has real chrome on the outside? If the chrome is developing small rust spots, you can remove them by rubbing the area with a crumpled piece of aluminum foil dipped in cola.

Make cut flowers last longer Don't throw away those last drops of soda pop. Pour about 1/4 cup into the water in a vase full of cut flowers. The sugar in the soda will make the blossoms last longer. *Note:* If you have a clear vase and want the water to remain clear, use a clear soda pop, such as Sprite or 7-Up.

Clean your toilet Eliminate dirt and odor with a simple can of soda. Pour into the toilet, let sit for an hour, then scrub and flush.

DID *You* KNOW?

Ginger has long been a traditional remedy for nausea, and recent scientific research has shown that ginger ale does work better than a placebo. Ginger ale was first marketed about 100 years ago by a Toronto pharmacist named John McLaughlin. McLaughlin kept trying new formulas until he patented what is now known throughout the world as Canada Dry Ginger Ale. Soda pops—carbonated beverages—originally were served only to customers at drugstore soda fountains. McLaughlin was one of the pioneers of the technology of mass bottling, which allowed customers to take soda pop home in bottles.

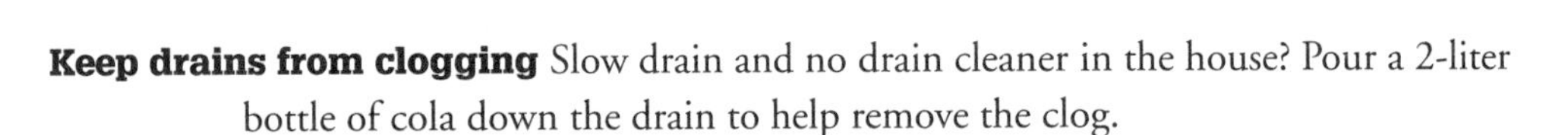

Keep drains from clogging Slow drain and no drain cleaner in the house? Pour a 2-liter bottle of cola down the drain to help remove the clog.

Get gum out of hair It's inevitable—kids get gum in their hair. Put the gummy hair section in a bowl with some cola. Let soak for a few minutes and rinse.

Make a roast ham moist Want to make your ham juicier? Pour a can of cola over your traditional ham recipe and follow regular baking instructions. Yum!

Clean your coins Who wants dirty money? If coin collecting is your hobby, use cola to clean your stash. Place the coins in a small dish and soak in cola for a shimmering shine. Of course, you don't want to do this with very rare and valuable coins.

Remove oil stains from concrete Here's how to remove oil stains from concrete driveways and garage floors: Gather up a small bag of cat litter, a few cans of cola, a stiff bristle broom, bucket, laundry detergent, bleach, eye protection, and rubber gloves. Cover the stain with a thin layer of cat litter and brush it in. Sweep up the litter and pour cola to cover the area. Work the cola in with a bristle broom, and leave the cola for about twenty minutes. Mix 1/4 cup laundry detergent with 1/4 cup bleach in 1 gallon (3.7 liters) warm water and use it to mop up the mess.

Spices

Make a hair tonic You can spice up your hair care regimen with a homemade tonic that will enhance your natural color and impart shine. For dark hair, use 1 tablespoon crumbled sage or 1 sprig chopped fresh rosemary or a mixture of 1 teaspoon allspice, 1 teaspoon ground cinnamon, and 1/2 teaspoon ground cloves. For blond hair, use 1 tablespoon chamomile. Pour 1 cup boiling water over the herb or spice mix, let it steep for 30 minutes, strain it through a coffee filter, and let it cool. Pour it repeatedly over your hair (use a dishpan to catch the runoff) as a final rinse after shampooing.

Treat minor cuts If you nick your finger while chopping vegetables for dinner, you may not even need to leave the kitchen for first aid. Alum, the old-fashioned pickling salt at the back of your spice cupboard, is an astringent. In a pinch, sprinkle some on a minor cut to stanch the flow of blood.

Keep feet smelling sweet If you use sage only to stuff turkeys, then you've been missing out. Sage is great for preventing foot odor because it kills the odor-causing bacteria that grow on your feet in the warm, moist environment inside your shoes. Just crumble a leaf or two into your shoes before you put them on. At the end of the day, just shake the remains into the trash.

Deodorize bottles for reuse You'd like to reuse those wonderful wide-mouthed pickle jars, but simply washing them with soap and water doesn't get rid of the pickle smell. What to do? Add 1 teaspoon dry mustard to 1 quart (1 liter) water, fill the jar, and let it soak overnight. It'll smell fresh by morning. This solution banishes the odor of tomatoes, garlic, and other foods with strong scents.

Keep your thermos fresh You just uncapped the thermos bottle you haven't used for six months, and the inside smells musty. To keep it from happening the next time, place a whole clove inside the thermos before capping it. A teaspoon of salt works too. Be sure to empty and rinse the thermos before using it.

Scent your home What could be more welcoming than the smell of something good cooking? Instead of using commercial air fresheners, simply toss a handful of

DID *You* KNOW?

What's the difference between a spice and an herb? The basic guideline is this: If it's made from a plant's leaf, it's an herb; if it's made from the bark, fruit, seed, stem, or root, it's a spice. Parsley and basil are typical herbs because we eat their leaves. Cinnamon (the bark of a tree) and pepper (the fruit of a vine) are considered spices. Then what to call salt, perhaps the most essential food enhancer of all but not a plant product at all? *Seasoning* is the word that describes anything used to flavor foods, regardless of its origin.

whole cloves or a cinnamon stick in a pot of water and keep it simmering on the stove for half an hour. Or place a teaspoon or two of the ground spices on a cookie sheet and place it in a 200°F (93°C) oven with the door ajar for 30 minutes. Either way, your house will naturally smell spicy good.

Keep woolens whole Woolen clothing can last a lifetime—if you keep moths away. If you don't have a cedar-lined chest or closet, preserve your cold-weather clothing using clove sachets. Purchase some small drawstring muslin bags at a tea shop or health food store, and fill each one with a handful of whole cloves. To prevent any transfer of oils or color to clothes and to contain any spills, put the sachet in a small plastic bag, but don't seal it. Attach it to a hanger in your closet or tuck one in your sweater chest for woolens without holes.

Keep ants at bay Flour, sugar, and paprika can all fall prey to ants. Keep these cooking essentials safe by slipping a bay leaf inside your storage containers. If you're concerned about the flour or sugar picking up a bay leaf flavor, tape the leaf to the inside of the canister lid. This trick works inside cabinets, too, where sachets of sage, bay, stick cinnamon, or whole cloves will smell pleasant while discouraging ants.

Stamp out silverfish These pesky critters frequent places with lots of moisture, such as kitchens, baths, and laundry rooms. Hang an aromatic sachet containing apple pie spices, sage, or bay leaves on a hook in your bathroom vanity and behind the washer, or keep a few in decorative baskets along baseboards.

Control insects in the garden You don't have to use harsh pesticides to control small-insect infestation outdoors. If ants are swarming on your garden path, add 1 tablespoon ground black pepper (or another strong-smelling ground spice, such as ground cloves or dry mustard) to 1 cup sifted white flour and sprinkle the mixture on and around the pests. They'll vanish within the hour. Sweep the dry mix into the garden or yard instead of trying to hose it off; water will just make it gooey.

Deter plant-eating animals Everyone knows that hot peppers make your mouth burn. So if rodents are attacking your ornamental plants, the solution may be to make them too "hot" for the critters. In fact, hot peppers are the basis for many commercial rodent repellents. Chop up the hottest pepper you can find (habañero is best) and combine it with 1 tablespoon ground cayenne pepper and 1/2 gallon (2 liters) water. Boil the mix for 15-20 minutes, then let it cool. Strain it through cheesecloth, add 1 tablespoon dishwashing liquid, and pour it into a spray bottle. Spray vulnerable plants liberally every five days or so. The spray works best for rabbits, chipmunks, and woodchucks, but may also deter deer, especially if used in combination with commercial products.

Shield your vegetable garden For centuries, gardeners have used companion planting to repel insect pests. Aromatic plants such as basil, tansy, marigolds, and sage are all reputed to send a signal to bugs to go elsewhere, so try planting some near your prized vegetables. Mint, thyme, dill, and sage are old-time favorites near cabbage family plants (cabbage, broccoli, cauliflower, and brussels sprouts) for their supposed ability to fend off cabbage moths. Best of all, you can eat the savory herbs!

Tip Toothache and Oil of Cloves

If you have a toothache, get to a dentist as soon as possible. Meanwhile, oil of cloves may provide temporary relief. Place a drop directly into the aching tooth or apply it with a cotton swab. But don't put it directly on the gums. An active ingredient in the spice oil, eugenol, is a natural pain reliever.

* Sponges

Make flowerpots hold water longer If your potted houseplants dry out too quickly after watering, when you repot them, try this simple trick for keeping the soil moist longer. Tuck a damp sponge in the bottom of the pot before filling it with soil. It'll act as a water reservoir. And it will also help prevent a gusher if you accidentally overwater.

An unwelcome mat for garden Anyone who's ever cleaned a floor with ammonia knows that the smell of this strong, everyday household cleaner is overpowering. Throw browsing animals "off the scent" of ripening vegetables in your garden by soaking old sponges in your floor-cleaning solution and distributing them wherever you expect the next garden raid.

Keep your veggies fresh Moisture that collects at the bottom of your refrigerator bins hastens the demise of healthful vegetables. Extend their life by lining bins with dry sponges. When you notice that they're wet, wring them out and let them dry before putting them back in the fridge. Every now and then, between uses, let them soak in some warm water with a splash of bleach to discourage the growth of mold.

Sop up umbrella overrun It's raining, and the family has been tramping in and out with umbrellas all day. Your umbrella stand has only a shallow receptacle to catch drips. Suddenly there's a waterfall coming out of it! Protect your flooring from umbrella stand overflow with a strategically placed sponge in its base. If you forget to squeeze it out, it'll dry on its own as soon as the weather clears.

Stretch the life of soap A shower is so refreshing in the morning—until you reach for the soap and are treated to the slimy sensation of a bar that's been left to marinate in its own suds. You'll enjoy bathing more and your soap will last longer if you park a sponge on the soap dish. It'll absorb moisture so soap can dry out.

> ***Kids' Stuff*** Making seeds grow seems magical to young children. For an easy and renewable **play garden** with a minimum of mess, all you need is an **old soap dish,** a **sponge,** and **seeds** of a plant such as lobelia, flax, or chia (yes, the same "chia" as in Chia Pet; look for the seed at health food stores). Cut the sponge to fit the dish, add water until it's moist but not sopping, and sprinkle the seed liberally over the top. Prop an **inverted glass bowl** over it until the seeds begin to grow. A bright window and a daily watering will keep it going for weeks.

Protect fragile items If you're shipping or storing small, fragile valuables that won't be harmed by a little contact with water, sponges are a clever way to cushion them. Dampen a sponge, wrap it around the delicate item, and use a rubber band to secure it. As it dries, the sponge will conform to the contours of your crystal ashtray or porcelain figurine. To unpack it, just dip the item in water again. You'll even get your sponge back!

Lift lint from fabric To quickly remove lint and pet fur from clothes and upholstery, give the fabric a quick wipe with a dampened and wrung-out sponge. Just run your fingers over the sponge and the unwanted fuzz will come off in a ball for easy disposal.

* Spray Bottles

Mist your houseplants Keep your houseplants healthy and happy by using an empty trigger-type spray bottle as a plant mister. Clean the bottle by filling it with equal parts water and vinegar—don't use liquid soap, as you may not be able to get it all out—let the solution sit for an hour, and rinse it out thoroughly with cold water. Repeat if necessary. Then, fill the bottle with lukewarm warm water, and use it to frequently give your plants a soothing, misty shower.

Help with the laundry An empty spray bottle can always be put to good use around your laundry room. Use clean, recycled bottles to spray water on your clothes as you're ironing. Or fill a spray bottle with stain remover solution so that you can apply it to your garments without having to blot up drips.

Cool off in summer Whether you're jogging around the park, taking a breather between volleyball matches, or just sitting out in the sun, a recycled spray bottle filled

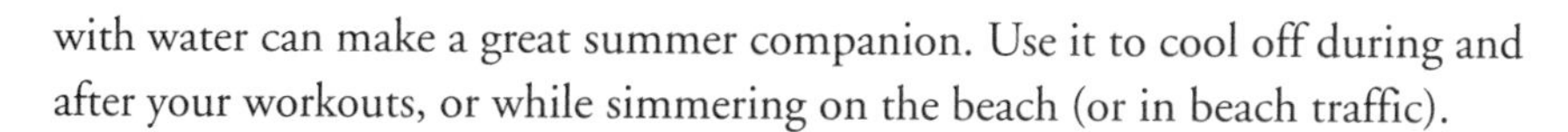

with water can make a great summer companion. Use it to cool off during and after your workouts, or while simmering on the beach (or in beach traffic).

Keep car windows clean Be sure to include a recycled spray bottle filled with windshield cleaner in the trunk of your car as part of your roadside emergency kit. Use it to clean off your car's headlights, mirrors, and of course, windows whenever needed. During winter months, mix in 1/2 teaspoon antifreeze, and you can spray it on to melt the ice on your windshield or mirrors.

Spray away garden pests Keep a few recycled spray bottles on hand to use around the yard. Here are two immediate uses:

- Fill one with undiluted white vinegar to get rid of the weeds and grass poking out of the cracks in your concrete, as well as ants and other insects—but be careful not to spray it on your plants; the high acidity could kill them.
- For an effective homemade insecticide recipe that works on most soft-bodied pests, but won't harm your plants, mix several cloves crushed garlic, 1/4 cup canola oil, 3 tablespoons hot pepper sauce, and 1/2 teaspoon mild liquid soap in 1 gallon (3.7 liters) water. Pour some into your spray bottle, and shake well before using.

* Squirt Bottles

Stop cooking oil drips Tired of cleaning up the oil spills around your kitchen? Fill a cleaned, recycled squirt bottle with olive oil or another favorite cooking oil. It's a lot easier to handle than a jar or bottle, and you can pour precisely the right amount of oil over your salads or into your frying pan without having to worry about drips or spills.

Substitute for a baster If you can't find your kitchen baster, or one that's in working condition, a cleaned squeeze bottle makes a dandy substitute. Simply squeeze out some air first, and use it to suck up the fat from your roasts and soups. You can even effectively use it to distribute marinades and drippings over meat.

Put the squeeze on condiments Recycled squirt bottles are great for storing condiments and other foodstuffs that are typically sold in jars—such as mayonnaise, salad dressing, jelly and jams, or honey. In addition to having fewer sticky or messy jars in your refrigerator, you'll also be lightening the load in your dishwasher by eliminating the need for knives or spoons. Make sure you give the bottles a thorough cleaning before using.

Clean out crevices A clean, empty squeeze bottle may be just the cleaning tool you need to get the dust out of the corners of your picture frames and other tight spaces. Use it to give a good blast of air to blow out the dirt you can't otherwise reach.

Let the children play Fill up a few clean squeeze bottles with water, then give them to your kids to squirt each other with in the yard on those hot summer days. It will keep them cool while they burn off some energy.

* Steel Wool

Turn nasty sneakers nice If your sneakers are looking so bad that the only thing you'd do in them is, well, *sneak* around, some steel wool may keep them from the trash can. Moisten a steel wool soap pad and gently scrub away at stains and stuck-on goo. Wipe them clean with a damp sponge or send them through the washer, and you may be able to enjoy many more months of wear.

Crayons begone Your toddler just created a work of crayon art on paper. Unfortunately, it's on the *wall*paper. Use a bit of steel wool soap pad to just skim the surface, making strokes in one direction instead of scrubbing in a circle, and your wall will be a fresh "canvas" in no time.

"Shoo" heel marks away Those black marks that rubber soles leave behind just don't come off with a mop, no matter how long you try. To rid a vinyl floor of unsightly smudges, gently rub the surface with a moistened steel wool soap pad. When the heel mark is gone, wipe the floor clean with a damp sponge.

Tip No Steel Wool on Stainless Steel

An oft-repeated advice is to clean stainless steel with steel wool. Yet stainless steel manufacturers caution against using any abrasive on stainless steel. Steel wool may make stainless steel look better, but it scratches the surface and ultimately hastens rusting. The safest way to care for stainless steel is to wash with a sponge and mild soap and water.

Sharpen your scissors Sometimes you just want a small piece of a steel wool soap pad for a minor job. Cutting it in half with a pair of scissors will help keep the scissors sharp while giving you the pint-size pad you need for your project.

Rebuff rodents Mice, squirrels, and bats are experts at finding every conceivable entry into a house. When you discover one of their entry points, stuff it full of steel wool. Steel wool is much more effective than foam or newspaper because even dedicated gnawers are unlikely to try to chew through such a sharp blockade.

Keep garden tools in good shape Nothing will extend the life of your gardening tools like a good cleaning at the end of each growing season. Grab a wad of fine steel wool from your woodshop (000, or "three aught," would be a good choice), saturate it with the same ordinary household oil you use on squeaky door hinges, and rub rust off your shears, loppers, shovels, and anything else with metal parts. Wipe them clean with a dry rag, sharpen any blades, and reapply a bit of oil before storing them for the winter.

Straws

Keep jewelry chains unknotted You're dressing for dinner out, and you reach into your jewelry box for your best gold chain only to find that it's tangled and kinked. Next time, run it through a straw cut to the proper length and close the clasp before putting it away. It'll always be ready to wear.

Give flowers needed height Your flower arrangement would be just perfect, except a few of the flowers aren't tall enough. You can improve on nature by sticking each of the too-short stems into plastic straws, trimming the straw to get the desired height, and inserting them into the vase.

Get slow ketchup flowing Anticipation is great, but ketchup that comes out of the bottle while your burger and fries are still hot is even better. If your ketchup is recalcitrant, insert a straw all the way into the bottle and stir it around a little to get the flow started.

Improvise some foamy fun To make enough cheap and easy toys for even a large group of children, cut the ends of some plastic straws at a sharp angle and set out a shallow pan of liquid dish soap diluted with a little bit of water. Dip a straw in the soap and blow through the other end. Little kids love the piles of bubbles that result.

Make a pull-toy protector Pull toys are perennial favorites of young children, but you can spend all day untying the knots that a toddler will inevitably put in the pull string. By running the string through a plastic straw (or a series of them), you can keep it untangled.

DID You KNOW?

The quest for a perfectly cold mint julep led to the invention of the drinking straw. Mint juleps are served chilled, and their flavor diminishes as they warm up. Holding a glass heats the contents, so the custom was to drink mint juleps through natural straws made from a section of hollow grass stem, usually rye. But the rye imparted an undesirable "grassy" flavor. In 1888 Marvin Stone, a Washington, D.C., manufacturer of paper cigarette holders, fashioned a paper tube through which to sip his favorite libation. When other mint julep aficionados began clamoring for paper straws, he realized he had a hot new product on his hands.

Have seasonings, will travel Maybe you're on a low-sodium diet and need potassium salt that most restaurants don't keep on the table, or perhaps you want salt and pepper to season your brown-bag lunch just before you eat it. Straws provide an easy way to take along small amounts of dry seasonings. Fold one end over and tape it shut, fill it, and fold and tape the other end. If moisture is a concern, use a plastic straw.

Fix loose veneer The veneer from a favorite piece of furniture has lost its grip near the edge of the piece. A bit of yellow carpenter's glue is the obvious solution for re-adhering the veneer, but how do you get the glue under there? Veneer can be very brittle, and you don't want to break off a piece by lifting it up. The solution: Cut a length of plastic drinking straw and press it to flatten it somewhat. Fold it in half and fill one half with glue, slowly dripping the glue in from the top. Slip the filled half under the veneer and gently blow in the glue. Wipe off any excess, cover the area with wax paper and a wood block, and clamp overnight.

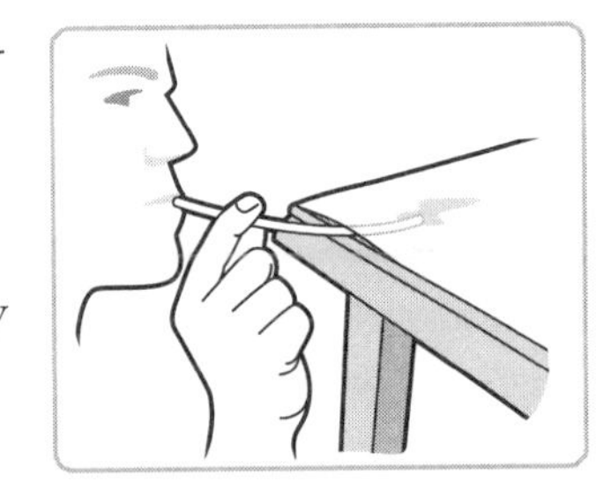

* String

Polish silverware more easily Remember Dad or Grandpa taking a photograph of the dining room table beautifully set for holiday meals? The rich shine of polished silverware was part of what made it so beautiful. To get your silverware looking that good, run a length of string through some silver polish and use it to get at those hard-to-reach spots between fork tines.

Make wicks for watering plants Keeping plants watered while you're on a short trip is easier than you think. Fill a large container with water and place it next to your potted plants. Cut pieces of string so they're long enough to hang down to the bottom of the container at one end and be buried a few inches in the soil of the pots at the other. Soak the strings until they're thoroughly wet and put them in position. As the soil begins to dry, capillary action will draw water from the reservoir to the pots through the strings.

Stop the sound of a dripping faucet If a leaky faucet going "plop … plop … plop" is keeping you awake at night, there's a way to silence it until the plumber arrives. Tie a piece of string to the fixture with one end right where the water is oozing out and the other end hanging down to the bottom of the sink basin. Water droplets will travel down the string silently instead of driving you to distraction.

Use as a straight-line guide Trimming a long hedge straight is a near-impossible feat unless you use a visual guide. Drive two stakes into the ground, one at each end of the hedge. Measure the height you want for the trimmed hedge, then run the string between the two stakes, tying it to each one at that exact height. As you clip away, cut down to the string line but no farther; the top of your hedge will be straight as an arrow.

Make a quick package opener Next time you're preparing a box for mailing, take a second to make it easier for the recipient to open. Place a piece of string along the center and side seams before you tape, allowing a tiny bit to hang free at one end. That way, the recipient just needs to pull the strings to sever the tape without resorting to sharp blades that might damage delicate contents. Do the same for packing boxes when you move.

Outline garden features When you're making a new garden, lay common bright white string on the ground to outline paths and beds. From an upstairs window or other high vantage point, you'll be able to tell at a glance if borders are straight and whether the layout is pleasing.

Measure irregular objects A cloth tape measure is the ideal tool for measuring odd-shaped objects, but you may not have one if you don't sew. Wrap a plain piece of string around the item instead, then hold it up to a ruler to get the measurement you need.

Plant perfectly straight rows It's harder than it looks to make straight garden rows freehand. Use string two ways to keep plants in line:

- For planting heavy seeds such as beans, put sticks in the ground at each end of a row and run a piece of string between to guide you as you plant.
- To plant dozens of lightweight seeds in a snap, cut string to the length of a row, wet it thoroughly, then sprinkle the seed directly on it. The moisture will make seeds stick long enough to lay the string in a prepared furrow. Just cover the string with soil and you're done!

Stop slamming doors Is a slamming door getting on your nerves? Here are two ideas for using string to control the way a door closes:

- A piece of light twine tied to both sides of a knob and running around the door edge provides just enough friction to slow it down and prevent a loud slam when it shuts.
- Use thicker rope the same way to temporarily prop open a door that automatically locks when it closes or to make sure pets don't get trapped in one room of the house.

Styrofoam

Keep nail polish nice When applying nail polish, a foam pellet or a small chunk cut from a block of foam packaging placed between each finger or toe will help spread them apart and keep the polish unblemished until it can dry.

Make your own shipping pellets You'd like to use foam to ship some fragile things, but all you've got is sheets or blocks of foam, not pellets. No problem. Just break up what you have into pieces small enough to fit in a blender and pulse it on and off to shred the foam into perfect packing material.

Hold treats for freezing and serving To prepare a quantity of snow cones or ice-cream cones in advance, cut a foam block to size so it will fit flat in your freezer. Cut holes just large enough and close enough to hold cones so they won't touch, fall over, or poke through the bottom. Fill the cones and slip them into the waiting holes. Then pop the whole thing into the freezer ready for serving at a moment's notice.

Ask anyone what kind of material those coffee cups, packing material, picnic coolers, and other white foam products are made from, and they'll reply Styrofoam. But strictly speaking, Styrofoam, a trademark of the Dow Chemical Company, refers only to durable extruded polystyrene, like that used in the blue insulation boards familiar to anyone who's been around a construction site. Those other more commonplace white foam items that you can easily tear or crumble are made of cheaper expanded polystyrene and are better referred to as foam plastic or styrene foam.

Make a buoyant tray for the pool Styrofoam is nearly unsinkable. Use the scraps from a construction project to make a drink holder or tray that will float in your pool:

- To make a soda-can holder, cut two pieces to the size you want the finished holder to be, then cut holes the same size as a soda can in one piece. Glue the piece with holes on top of the other piece, using a glue gun with hot-melt glue.
- To make a tray with a rim, just glue small strips of foam that are at least 1 inch (2.5 centimeters) high around the edge of a larger tray-size section of the material.

Make a kickboard A sharp kitchen knife is all you need to cut a scrap of Styrofoam insulation into a kickboard for your swimming pool.

S

Help shrubs withstand winter Sometimes shrubs need a little help to survive winter's ravages. Leftover sheets of extruded tongue-and-groove Styrofoam insulation are perfect for the job. They're rigid, waterproof, and block wind and road salt. Here are two ways to use the material:

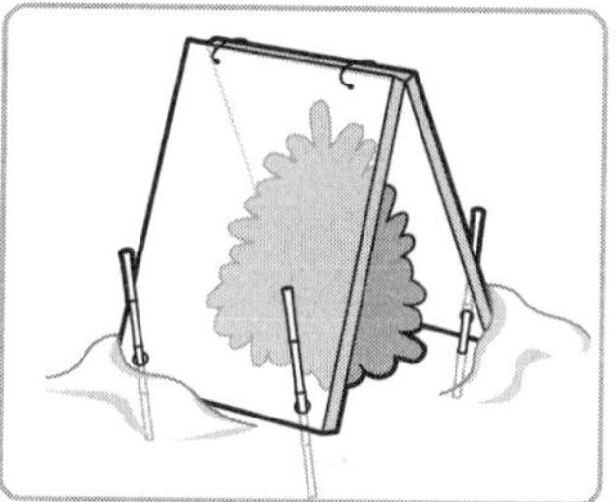

- To give moderate protection, cut two Styrofoam sheets and lash them together to form a pup tent over the plant. To hold the pieces in place, drive bamboo garden stakes through the bottom of each piece into the ground.
- For something more substantial, fit pieces together to box in the plants on four sides. Put a stake inside each corner and join the pieces with duct or packing tape.

Plants in containers that overwinter outdoors are more likely to survive with Styrofoam protection too.

Tip Recycling Foam Pellets

Even with lots of creative reuse, sometimes foam packing pellets just come in faster than they go out again. If you've got more than you can handle, remember that packing and shipping businesses often accept clean pellets for reuse. Just call to confirm before you bring them in.

* Sugar

Keep cut flowers fresh Make your own preservative to keep cut flowers fresh longer. Dissolve 3 tablespoons sugar and 2 tablespoons white vinegar per quart (liter) of warm water. When you fill the vase, make sure the cut stems are covered by 3-4 inches (7-10 centimeters) of the prepared water. The sugar nourishes the plants, while the vinegar inhibits bacterial growth. You'll be surprised how long the arrangement stays fresh!

Nix nematode worms in garden If your outdoor plants look unhealthy, with ugly knots at the roots, chances are they've been victims of an attack of the nematodes! The nematode worm, nemesis of many an otherwise healthy garden, is a microscopic parasite that pierces the roots of plants and causes knots. You can prevent nematode attacks by using sugar to create an inhospitable environment for the tiny worms. Apply 5 pounds (2 kilograms) sugar for every 250 square feet (25 square meters) of garden. Microorganisms feeding on the sugar will increase the organic matter in the soil, thereby eliminating those nasty little nematodes.

Kids' Stuff Here's an easy way to make **old-fashioned rock candy** with the kids, with no strings, paper clips, sticks, or thermometers needed: Make syrup by stirring 2 1/2 cups **sugar** into 1 cup **hot water.** Pour the syrup into several **open dishes** and set aside. Add a grain of sugar to act as a seed crystal in each container. Within days or weeks you should be able to collect glittering crystals of rock candy. Use a spoon to scoop it out, then rinse and dry the candy before you eat it.

Clean greasy, grimy hands Your work is done for the day, but your hands are still covered with grease, grime, or paint. To clean filthy hands easily and thoroughly, pour equal amounts of olive oil and sugar into the cupped palm of one hand, and then gently rub your hands together for several minutes. Rinse thoroughly and dry. The grit of the sugar acts as an abrasive to help the oil remove grease, paint, and grime. Your hands will look and feel clean, soft, and moisturized.

Make a nontoxic fly trap Keep your kitchen free of flies with a homemade fly trap that uses no toxic chemicals. In a small saucepan, simmer 2 cups milk, 1/4 pound (113 grams) raw sugar, and 2 ounces (56 grams) ground pepper for about 10 minutes, stirring occasionally. Pour into shallow dishes or bowls and set them around the kitchen, patio, or anywhere the flies are a problem. The pesty bugs will flock to the bowls and drown!

Exterminate roaches If you hate smelly, noxious pesticides as much as you loathe cockroaches, don't call an exterminator. Instead, when you have a roach infestation, scatter a mixture of equal parts sugar and baking powder over the infested area. The sugar will attract the roaches, and the baking powder will kill them. Replace it frequently with a fresh mixture to prevent future infestations.

Soothe a burned tongue That slice of piping-hot pizza sure looked great, but ouch! You burned your tongue when you bit into it. To relieve a tongue burned by hot pizza, coffee, tea, or soup, reach for the sugar bowl and sprinkle a pinch or two of sugar over the affected area. The pain will begin to subside immediately.

Keep desserts fresh You used sugar to sweeten the cake batter; now use it to keep the finished cake fresh and moist. Store the cake in an airtight container with a couple of sugar cubes, and it will stay fresh for days longer. Store a few lumps of sugar with cheese the same way to prevent the cheese from molding.

T

* Talcum Powder

Keep ants away For an effective organic ant repellent, scatter talcum powder liberally around house foundations and known points of entry, such as doors and windows. Other effective organic repellents include cream of tartar, borax, powdered sulfur, and oil of cloves. You can also try planting mint around the house foundations.

Fix a squeaky floor Don't let squeaky floorboards drive you crazy. For a quick fix, sprinkle talcum powder or powdered graphite between the boards that squeak. If that doesn't do the trick, squirt in some liquid wax.

Remove bloodstains from fabric To remove fresh bloodstains from clothing or furniture, make a paste of water and talcum powder and apply it to the spot. When it dries, brush away the stain. Substitute cornstarch or cornmeal if you are out of talcum powder.

Get rid of greasy carpet stain A greasy stain can spoil the look of the most luxurious carpeting. You can remove greasy stains from a carpet with a combination of talcum powder and patience. Just cover the affected area with talcum powder and wait at least 6 hours for the talcum to absorb the grease. Then vacuum the stain away. Baking soda, cornmeal, or cornstarch may be substituted for the talcum powder.

TAKE CARE Health care experts warn that scented talcum powder may cause skin allergies and worsen body odors. They recommend only unscented powder, used on dry skin. Women are advised not to use talcum powder in the vaginal or anal areas, where excessive powdering has been linked to an increased risk of ovarian cancer.

Degrease polyester stains Your favorite polyester shirt or blouse may come back in style someday, but you'll have to get rid of that ugly grease stain before you wear it. To get rid of grease stains on polyester, sprinkle some talcum powder directly onto the spot and rub it in with your fingers. Wait 24 hours, and gently brush. Repeat as necessary until the stain is completely gone.

Loosen tangles and knots Don't break a fingernail trying to untie that knot in your shoelace. Sprinkle some talcum powder on shoelaces (or any knotted cords) and the knots will pull apart more easily. Use talcum powder to help untangle chain necklaces too.

Tape

✱TAPE IN THE KITCHEN AND DINING ROOM

Safely pick up glass shards Why risk cutting yourself picking up bits of broken glass from the kitchen floor? Just hold a long piece of transparent tape tightly at each end and use it to blot up all the shards.

Create a no-fly zone Make your own fly and pest strips that are free of polluting toxic chemicals and poisons. Cover empty paper towel or toilet paper rolls with transparent tape, sticky side out, and hang them in the kitchen or wherever else you need them.

Mark start of plastic-wrap roll If you've ever had trouble finding the beginning of a roll of plastic food wrap, you'll appreciate this time-saving trick: Put a piece of transparent tape on your finger, sticky side out, and dab your finger on the roll until you find the edge. Use a short piece of tape to lift the edge and pull gently.

Prevent salt and pepper spills Many salt and pepper shakers, especially ceramic ones, have to be filled through a hole in the bottom. Before you refill one of these shakers, tape over the holes on top. That way you won't have any wasteful spills when you turn the shaker upside down to fill it. Also, remember to tape the tops when moving to a new home, or even when you're just transporting the shakers to and from a picnic.

Keep hands free at grocery Next time you go food shopping, bring some tape with you and use it to affix your shopping list to the handle of your grocery cart. This will free both your hands and you won't keep misplacing or dropping your list.

Make candles fit snugly Don't let loose candles spoil the romantic mood or cause a fire at your next candlelight dinner. If the candles don't fit snugly into the holder, wrap layers of tape around their bottom edges until they fit just right.

✱TAPE AROUND THE HOUSE

Keep spare batteries handy You won't be behind the times for long if you remember to tape extra batteries to the back of your wall clock! When the clock stops and it's time to replace the batteries, they'll be readily at hand.

Code your keys Are you always groping around to find the right key when you get home in the dark? Just wrap some tape around the top of your house key, and you'll be able to feel for the right key when it is too dark to see. Or if you have several similar-looking keys that you can't tell apart, color-code them with different-colored tape.

T

Prevent jewelry tangles To keep fine chains from tangling when you travel, cover both sides of each chain with transparent tape. You can also use tape to keep a pair of earrings from separating.

Mark a phone number for quick reference Use transparent tape to highlight numbers in the phone book that you often look up. The tape will make the page easier to find and you will also be able to find the number easily without having to search the whole page.

Contain grease stains on paper You may never be able to get rid of grease spots on books or important papers, but you can keep them from spreading with a little help from some transparent tape. Affix tape over both sides of the spot to keep the grease from seeping through to other pages or papers.

Keep papers from blowing in the wind If you have to make a speech or accept an award at an outdoor event, bring a roll of transparent tape with you. When it's your turn to talk, place some tape on the lectern, sticky side out, to prevent your papers from blowing away.

Find your favorite photo negative Before framing a favorite photograph, tape the negative to the back of the picture. If you ever want to make copies of the photo, you won't have to go searching through piles of old negatives to find the right one.

Deter cat from scratching Stop naughty cats and kittens from scratching your fine furniture! Sprinkle ground red pepper on a strip of tape and attach it to the areas you don't want them to scratch. They hate the smell, and they'll quickly get the message.

Keep flowers upright in vase To keep cut flowers from sagging in their vase, crisscross several pieces of transparent tape across the mouth of the vase, leaving spaces where you can insert the flowers. The flowers will look perky and fresh for a few extra days.

Scotch tape got its name from an insult hurled at Richard Drew, the 3M company engineer who invented it. In 1925, five years before he invented the world's first and best-known transparent cellophane tape, Drew invented masking tape. He was field-testing his first masking-tape samples to find the right amount of adhesive when a frustrated body-shop painter exclaimed, "Take this tape back to those Scotch bosses of yours and tell them to put more adhesive on it!"

Make sewing easier Let transparent tape simplify your sewing: Use it to hold a zipper in place when you're making a garment. (You can sew through the tape and remove it when you're done.) Keep badges, patches, or name tags in place when sewing them onto shirts, uniforms, or caps. Tape hooks, eyes, and snaps to garments when sewing so they won't slip. Just pull the tape off when you're done. Tape a pattern to the material; when you cut the pattern, you'll have a reinforced edge.

End loose ends on thread spools Put an end to time-wasting frustrating searches for loose ends of thread. Just tape the ends to the top or bottom of the spool when you're done sewing and they'll be at your fingertips and ready to use next time you sew.

Remove lipstick from silk Why pay an expensive dry-cleaning bill to remove a lipstick spot from a silk scarf or dress when you can do it yourself for free? Just place a piece of transparent tape (or masking tape) over the spot and yank it off. If you can still see some of the lipstick color, sprinkle on some talcum powder or chalk and dab until the powder and the remaining lipstick disappear.

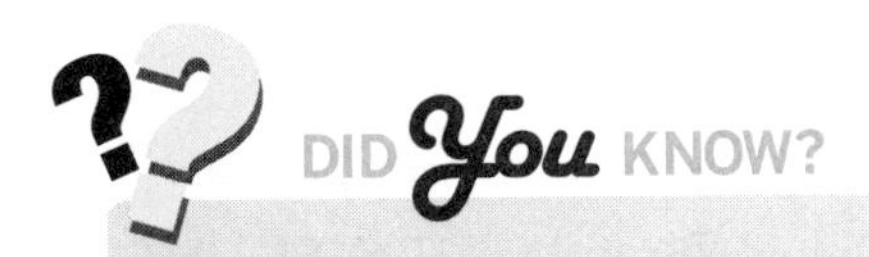

DID *You* KNOW?

Transparent tape is made from an acetate film derived from wood pulp or cotton fibers. It is formed into paper-thin sheets and wound onto giant rolls before being coated with adhesive. Twenty-nine raw materials go into the adhesive used in transparent tape. Even after the film and adhesive have been produced, 10 separate steps remain to be done before a roll of tape is manufactured.

Clean your nail file Clean a nail file easily and effectively. Simply place a piece of transparent tape over it, press, and pull off. The tape will pick up all the dirt embedded in the surface of the file.

*TAPE FOR THE DO-IT-YOURSELFER

Keep picture nails from damaging wall Before driving a nail into the wall, put a piece of tape on the wall at the site. This will prevent the paint from peeling off if you have to remove the nail, and it will prevent wallboards from cracking too.

Keep screws handy When doing household repairs, place loose screws, nuts, and bolts directly on a piece of transparent tape so they won't roll around and get lost.

Stick some double-sided tape on your workbench and use it to hold screws and such in place while you're working on a project.

Mend a broken plant stem Use transparent tape to add support to a broken plant stem. Just wrap the stem in tape at the damaged area and leave the tape on until it mends. The taped plant will keep growing as long as moisture can continue to travel up the stem.

Make a seed strip Make your own seed strip to create perfectly straight rows in your garden with almost no effort. Sprinkle some seeds on a piece of wax paper and use your fingers to arrange and align them. After removing the excess, take a strip of transparent tape and place it over the seeds. Then just bury the tape in the garden and you will soon have perfect rows.

Catch cricket invaders If noisy crickets have invaded your basement or garage, try trapping them with packaging tape. Take a strip of the tape and place on the floor, sticky side up. Later, just release your catch into the wild or feed them to a cricket-eating pet.

*TAPE FOR SAFETY'S SAKE

Make safety markers for a car emergency You'll be a lot safer when your car breaks down at night if you have safety markers on hand to warn oncoming drivers. You can make your own safety markers easily at home. Just put strips of brightly colored reflector tape on some old coffee cans. Keep them in the trunk of your car for use in an emergency.

Mark dark stairways Stop stumbling on those poorly lit cellar or outdoor stairways or worrying about guests tripping and falling. Simply apply reflector tape along the edges of the steps and you and your guests will be able to see exactly where you're stepping.

Make pets visible at night Don't let your beloved family pet get hit by a car during the night. Put reflector tape on Rover's collar so drivers will be able to see him immediately in the dark.

*TAPE FOR THE KIDS

Secure baby's bib Stop bits of food from getting under a child's bib by taping the edges of the bib to her clothes.

Makeshift childproofing If you bring a baby or small child with you when visiting a home that isn't childproofed, bring a roll of transparent tape along too. Use it to cover electrical outlets as a temporary safety measure. Although it will not confer a lot of protection, it could give you the extra time you need to remove a child from a potentially hazardous situation.

Kids' Stuff Kids will be delighted and amazed when you do this **easy party trick** at a birthday gathering. Secretly place a piece of **transparent tape** over an area of a **blown-up balloon**. When you are ready, get the children's attention and hold up the balloon in one hand and a **pin** in the other. For added effect, tell them to cover their ears. Pierce the balloon with the pin at the taped spot and remove it. The **balloon will not pop**! Then pop the balloon in another area. Guaranteed to leave them laughing and scratching their heads.

Make multicolored designs Tape a few different-colored markers or pencils together and give them to the kids to draw multicolored designs. Be careful not to use too many, so the children can maintain control of the drawings.

Tea

TEA FOR HEALTH AND BEAUTY

Cool sunburned skin What can you do when you forget to use sunscreen and have to pay the price with a painful burn? A few wet tea bags applied to the affected skin will take out the sting. This works well for other types of minor burns (i.e., from a teapot or steam iron) too. If the sunburn is too widespread to treat this way, put some tea bags in your bathwater and soak your whole body in the tub.

Relieve your tired eyes Revitalize tired, achy, or puffy eyes. Soak two tea bags in warm water and place them over your closed eyes for 20 minutes. The tannins in the tea act to reduce puffiness and soothe tired eyes.

Reduce razor burn Ouch! Why didn't you remember to replace that razor blade *before* you started to shave? To soothe razor burn and relieve painful nicks and cuts,

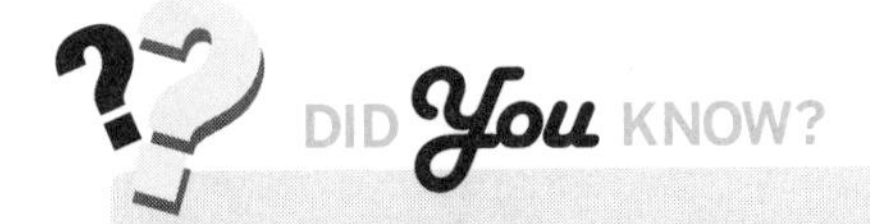

DID *You* KNOW?

Legend has it that tea originated some 5,000 years ago with the Chinese emperor Shen Nung. A wise ruler and creative scientist, the emperor insisted that all drinking water be boiled as a health precaution. One summer day, during a rest stop in a distant region, servants began to boil water for the royal entourage to drink when some dried leaves from a nearby bush fell into the pot. As the water boiled, it turned brown. The emperor's scientific curiosity was aroused, and he insisted on tasting the liquid. It was just his cup of tea.

apply a wet tea bag to the affected area. And don't forget to replace the blade before your next shave.

Get the gray out Turn gray hair dark again without an expensive trip to the salon or the use of chemical hair dyes. Make your own natural dye using brewed tea and herbs: Steep 3 tea bags in 1 cup boiling water. Add 1 tablespoon each of rosemary and sage (either fresh or dried) and let it stand overnight before straining. To use, shampoo as usual, and then pour or spray the mixture on your hair, making sure to saturate it thoroughly. Take care not to stain clothes. Blot with a towel and do not rinse. It may take several treatments to achieve desired results.

Condition dry hair To give a natural shine to dry hair, use a quart (liter) of warm, unsweetened tea (freshly brewed or instant) as a final rinse after your regular shampoo.

Tan your skin with tea Give pale skin a healthy tan appearance without exposure to dangerous ultraviolet rays. Brew 2 cups strong black tea, let it cool, and pour into a plastic spray bottle. Make sure your skin is clean and dry. Then spray the tea directly onto your skin and let it air-dry. Repeat as desired for a healthy-looking glowing tan. This will also work to give a man's face a more natural look after shaving off a beard.

Drain a boil Drain a boil with a boiled tea bag! Cover a boil with a wet tea bag overnight and the boil should drain without pain by the time you wake up next morning.

Soothe nipples sore from nursing When breast-feeding the baby leaves your nipples sore, treat them to an ice-cold bag of tea. Just brew a cup of tea, remove the bag, and place it in a cup of ice for about a minute. Then place the wet tea bag on the sore nipple and cover it with a nursing pad under your bra for several minutes while you enjoy a cup of tea. The tannic acid in the wet tea leaves will soothe and help heal the sore nipple.

Soothe those bleeding gums The child may be all smiles later when the tooth fairy arrives, but right now those bleeding gums are no fun whatsoever. To stop the bleeding and soothe the pain from a lost or recently pulled tooth, wet a tea bag with cool water and press it directly onto the site.

Relieve baby's pain from injection Is the baby *still* crying from that recent inoculation shot? Try wetting a tea bag and placing it over the site of the injection. Hold it gently in place until the crying stops. The tannic acid in the tea will soothe the soreness. You might try it on yourself the next time an injection leaves your arm sore.

Dry poison ivy rash Dry a weepy poison ivy rash with strongly brewed tea. Simply dip a cotton ball into the tea, dab it on the affected area, and let it air-dry. Repeat as needed.

Stop foot odor Put an end to smelly feet by giving them a daily tea bath. Just soak your tootsies in strongly brewed tea for 20 minutes a day and say good-bye to offensive odors.

Make soothing mouthwash To ease toothache or other mouth pain, rinse your mouth with a cup of hot peppermint tea mixed with a pinch or two of salt. Peppermint is an antiseptic and contains menthol, which alleviates pain on contact with skin surfaces. To make peppermint tea, boil 1 tablespoon fresh peppermint leaves in 1 cup water and steep for several minutes.

*TEA AROUND THE HOUSE

Tenderize tough meat Even the toughest cuts of meat will melt in your mouth after you marinate them in regular black tea. Here's how: Place 4 tablespoons black tea leaves in a pot of warm (not boiling) water and steep for 5 minutes. Strain to remove the leaves and stir in 1/2 cup brown sugar until it dissolves. Set aside. Season up to 3 pounds (1.5 kilograms) meat with salt, pepper, onion, and garlic powder, and place it in a Dutch oven. Pour the liquid over the seasoned meat and cook in a preheated 325°F (165°C) oven until the meat is fork tender, about 90 minutes.

Clean wood furniture and floors Freshly brewed tea is great for cleaning wood furniture and floors. Just boil a couple of tea bags in a quart (liter) of water and let it cool. Dip a soft cloth in the tea, wring out the excess, and use it to wipe away dirt and grime. Buff dry with a clean, soft cloth.

Create "antique" fashions Soak white lace or garments in a tea bath to create an antique beige, ecru, or ivory look. Use 3 tea bags for every 2 cups of boiling water and steep for 20 minutes. Let it cool for a few minutes before soaking the material for 10 minutes or more. The longer you let it soak, the darker the shade you will get.

Tip Dyeing with Herbal Teas

Using regular tea to dye fabrics has been around for a long time and was first used to hide stains on linens. But you can also use herbal teas to dye fabric different colors and subtle hues. Try using hibiscus to achieve red tones and darker herbal teas like licorice for soft brown tints. Always experiment using fabric scraps until you obtain the desired results.

Shine your mirrors To make mirrors sparkle and shine, brew a pot of strong tea, let it cool, and then use it to clean the mirrors. Dampen a soft cloth in the tea and wipe it all over the surface of the mirrors. Then buff with a soft, dry cloth for a sparkly, streak-free shine.

T

Control dust from fireplace ash Keep dust from rising from the ashes when you clean out your fireplace. Before you begin cleaning, sprinkle wet tea leaves over the area. The tea will keep the ashes from spreading all over as you lift them out.

Perfume a sachet Next time you make a sachet, try perfuming it with the fragrant aroma of your favorite herbal tea. Just open a few used herbal tea bags and spread the wet tea on some old newspaper to dry. Then use the dry tea as stuffing for the sachet.

* TEA IN THE GARDEN

Give roses a boost Sprinkle new or used tea leaves (loose or in tea bags) around your rosebushes and cover with mulch to give them a midsummer boost. When you water the plants, the nutrients from the tea will be released into the soil, spurring growth. Roses love the tannic acid that occurs naturally in tea.

Feed your ferns Schedule an occasional teatime for your ferns and other acid-loving houseplants. Substitute brewed tea when watering the plants. Or work wet tea leaves into the soil around the plants to give them a lush, luxuriant look.

Prepare planter for potting For healthier potted plants, place a few used tea bags on top of the drainage layer at the bottom of the planter before potting. The tea bags will retain water and leach nutrients to the soil.

Enhance your compost pile To speed up the decomposition process and enrich your compost, pour a few cups of strongly brewed tea into the heap. The liquid tea will hasten decomposition and draw acid-producing bacteria, creating desirable acid-rich compost.

* Tennis Balls

Fluff down-filled clothes and comforters Down-filled items like jackets, vests, comforters, and pillows get flat and soggy when you wash them. You can fluff them up again by tossing a couple of tennis balls into the dryer when you put them in.

Sand curves in furniture Wrap a tennis ball in sandpaper and use it to sand curves when you're refinishing furniture. The tennis ball is just the right size and shape to fit comfortably in your hand.

Cover your trailer hitch To protect chrome trailer hitches from scratches and rust, cut a tennis ball and slip it over your hitching ball. The tennis ball will keep moisture out and rust away.

Massage your back Give yourself a relaxing and therapeutic back massage. Simply fill a long tube sock with a few tennis balls, tie the end, and stretch your homemade massager around your back just as you would a towel after a shower or bath.

Keep swimming pool oil-free Float a couple of tennis balls in your swimming pool to absorb body oils from swimmers. Replace the balls every couple of weeks during high-use periods.

Make bike kickstand for soft soil To prevent a bicycle kickstand from sinking into soft grass, sand, or mud, cut a slit in a tennis ball and put it on the end of the kickstand.

Store your valuables at gym Here's a neat way to hide and store valuables when you're working out at the gym. Make a 2-inch (5-centimeter) slit along one seam of a tennis ball and insert the valuables inside. Keep the ball in your gym bag among other sporting gear. Just remember not to use the doctored ball next time you're out on the tennis court!

Park right every time Make parking your car in the garage easier. Hang a tennis ball on a string from the garage ceiling so it will hit the windshield at the spot where you should stop the car. You'll always know exactly where to park.

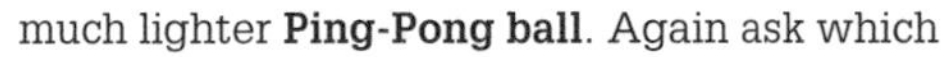

SCIENCE FAIR

Teach kids about **gravity**. Stand on a chair holding **two tennis balls**, one in each hand, and extend your arms so they're at the same distance from the floor. Ask the kids to observe as you release both balls at once. Did they hit the floor at the same time? Now repeat using a **tennis ball** and a much lighter **Ping-Pong ball**. Again ask which will land first. Most will guess the heavier tennis ball, but they'll **land at the same time** because gravity exerts the same force on all objects regardless of their weight. Of course, if you try this with a ball and a feather, the kids will also learn that less dense objects fall slower due to air resistance.

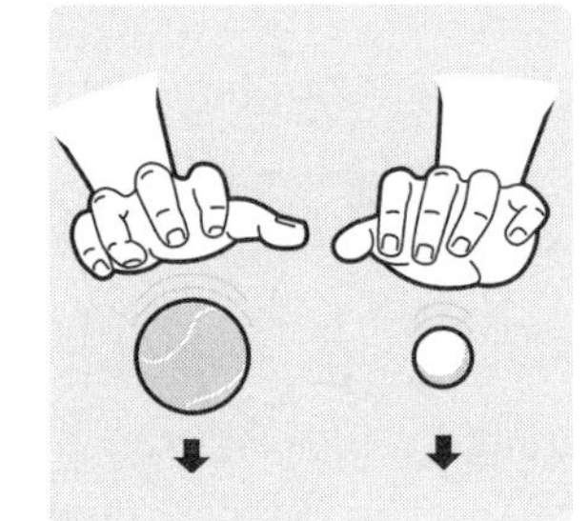

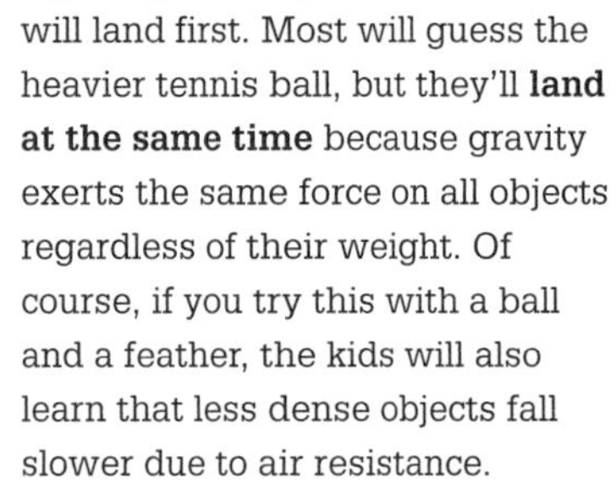

Massage your sore feet For a simple but amazingly enjoyable and therapeutic foot massage, take your shoes off, place a tennis ball on the floor, and roll it around under your feet.

Get a better grip on bottle caps If your hands are weakened by arthritis or other ailments, you probably have a difficult time removing twist-off bottle caps. An old tennis ball may help. Simply cut a ball in half and use one of the halves to enhance your grip.

Tires

Protect your vegetables Plant your tomatoes, potatoes, eggplants, peppers, or other vegetables inside tires laid on the ground. The tires will protect the plants from harsh winds, and the dark rubber will absorb heat from the sun and warm the surrounding soil.

Make a wading pool for kids To make an impromptu wading pool for toddlers, drape a shower curtain over the center of a large truck tire and fill it with water.

Make a classic tire swing A swing made from an old tire is a timeless source of pleasure for children of all ages. To make one in your backyard, drill a few drainage holes in the bottom of the tire. Drill two holes for bolts in the top, bolt two chains to the tire, and suspend it by the chains from a healthy branch of a hardwood tree. Use 3/16-inch (18-millimeter) playground chain. Put some wood chips or other soft material under and around the swing to cushion falls.

Tip Tire Checkup

Spending five minutes a month to check your tires can protect against avoidable breakdowns and crashes, improve vehicle handling, increase gas mileage, and extend the life of your tires. Here are some guidelines:

- Check tire air pressure at least once a month and before going on a long trip. Don't forget the spare.
- Inspect for uneven wear on tire treads, cracks, foreign objects, or other signs of wear or trauma. Remove bits of glass and other objects wedged in the tread.
- Make sure your tire valves have caps.
- Do not overload your vehicle.

Store plumbing snakes An old bicycle tire is just the right size to store metal snakes used to clean plumbing lines or "fish wires" used to run electrical cables inside walls. Just lay the snake or fish wire inside the tire, where it will expand to the shape of the tire and become encased within it. Then you can hang the tire conveniently on a hook in your workshop, garage, basement, or shed.

Tomato Juice

Deodorize plastic containers To remove foul odors from a plastic container, pour a little tomato juice onto a sponge and wipe it around the inside of the container. Then wash the container and lid in warm, soapy water, dry well, and store them separated in the freezer for a couple of days. The container will be stench-free and ready to use again.

Remove skunk stink from dog Is there a dog alive that hasn't been sprayed by a skunk at least once? If your dog gets skunked, douse the affected area thoroughly with undiluted tomato juice. Make sure to sponge some of the juice over the pet's face, too, avoiding its eyes. Wait a few minutes for the acids from the tomatoes to neutralize the skunk smell and then give the dog a shampoo or scrub with soap and water. Repeat as necessary over several days until the smell is completely gone.

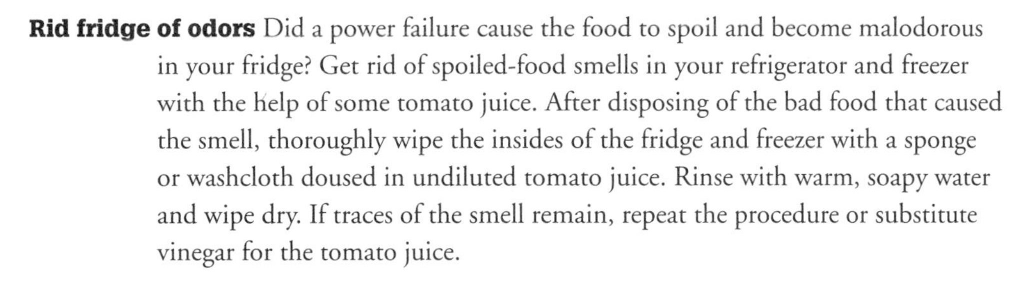

Rid fridge of odors Did a power failure cause the food to spoil and become malodorous in your fridge? Get rid of spoiled-food smells in your refrigerator and freezer with the help of some tomato juice. After disposing of the bad food that caused the smell, thoroughly wipe the insides of the fridge and freezer with a sponge or washcloth doused in undiluted tomato juice. Rinse with warm, soapy water and wipe dry. If traces of the smell remain, repeat the procedure or substitute vinegar for the tomato juice.

Restore blond hair color If you're a blonde who has ever gone swimming in a chlorine-treated pool, you know it can sometimes give your hair an unappealing green tint. To restore the blond color to your hair, saturate it with undiluted tomato juice, cover it with a shower cap, and wait 10-15 minutes. Then rinse thoroughly, shampoo, and soon you'll be ready to have more fun.

Relieve a sore throat For temporary relief of sore throat symptoms, gargle with a mixture of 1/2 cup tomato juice and 1/2 cup hot water, plus about 10 drops hot pepper sauce.

* Toothbrushes

Use as all-purpose cleaners Don't throw out your old toothbrushes. Instead, use them to clean a host of diverse items and small or hard-to-reach areas and crevices. Use a toothbrush to clean artificial flowers and plants, costume jewelry, combs, shower tracks, crevices between tiles, and around faucets. Also clean computer keyboards, can-opener blades, and around stove burners. And don't forget the seams on shoes where the leather meets the sole.

Brush your cheese grater Give the teeth of a cheese grater a good brushing with an old toothbrush before you wash the grater or put it in the dishwasher. This will make it easier to wash and will prevent clogs in your dishwasher drain by getting rid of bits of cheese or any other item you may have grated.

Remove tough stains Removing a stain can be a pain, especially one that has soaked deep down into soft fibers. To remove those deep stains, try using a soft-bristled nylon toothbrush, dabbing it gently to work in the stain-removing agent (bleach or vinegar, for example) until the stain is gone.

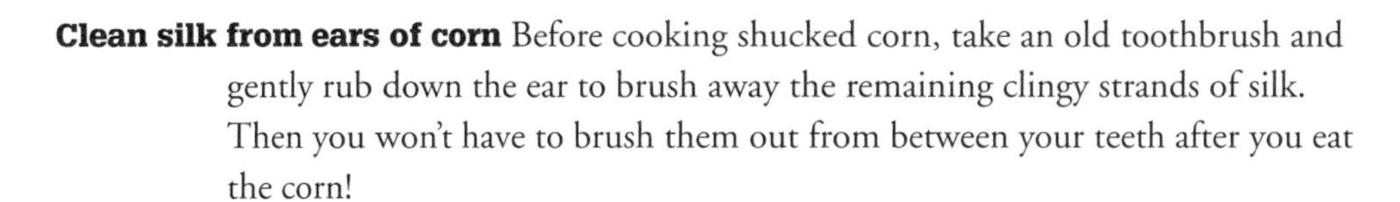

Clean silk from ears of corn Before cooking shucked corn, take an old toothbrush and gently rub down the ear to brush away the remaining clingy strands of silk. Then you won't have to brush them out from between your teeth after you eat the corn!

Clean and oil your waffle iron A clean, soft toothbrush is just the right utensil to clean crumbs and burned batter from the nooks and crannies of a waffle iron. Use it to spread oil evenly on the waffle iron surface before the next use too.

Apply hair dye Dyeing your hair at home? Use an old toothbrush as an applicator. It's the perfect size.

Clean gunk from appliances Dip an old toothbrush in soapy water and use it to clean between appliance knobs and buttons, and raised-letter nameplates.

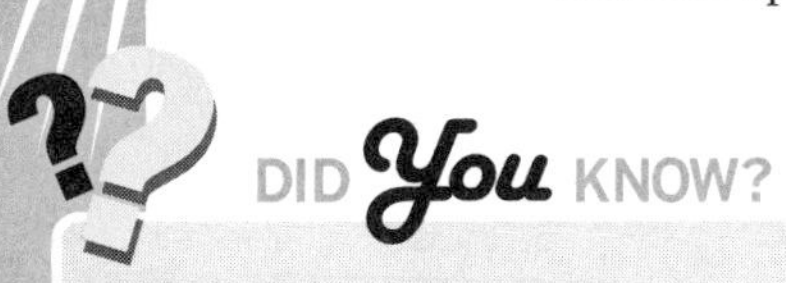

DID *You* KNOW?

The ancient Chinese were apparently the first to use toothbrushes, which they made with bristles from the necks of cold-climate pigs. The first toothbrushes in America were manufactured in the late 19th century, but toothbrushing did not become a two- or three-times-a-day habit for many common folks until after World War II, when returning GIs brought home their army-enforced habits. By then DuPont had invented the nylon bristle, which, unlike the natural bristles used earlier, dried completely between brushings and was resistant to the growth of bacteria. Nylon bristles are still used in most toothbrushes made today.

* Toothpaste

Remove scuffs from shoes A little toothpaste does an amazing job of removing scuffs from leather shoes. Just squirt a dab on the scuffed area and rub with a soft cloth. Wipe clean with a damp cloth. The leather will look like new.

Clean your piano keys Has tickling the ivories left them a bit dingy? Clean them up with toothpaste and a toothbrush, then wipe them down with a damp cloth. Makes sense, since ivory is essentially elephant teeth. However, toothpaste will work just as well on modern pianos that usually have keys covered with plastic rather than real ivory.

Spiff up your sneakers Want to clean and whiten the rubber part of your sneakers? Get out the non-gel toothpaste and an old toothbrush. After scrubbing, clean off the toothpaste with a damp cloth.

Clean your clothes iron The mild abrasive in non-gel toothpaste is just the ticket for scrubbing the gunk off the bottom plate of your clothes iron. Apply the toothpaste to the cool iron, scrub with a rag, then rinse clean.

Polish a diamond ring Put a little toothpaste on an old toothbrush and use it to make your diamond ring sparkle instead of your teeth. Clean off the residue with a damp cloth.

Deodorize baby bottles Baby bottles inevitably pick up a sour-milk smell. Toothpaste will remove the odor in a jiffy. Just put some on your bottle brush and scrub away. Be sure to rinse thoroughly.

Prevent fogged goggles Whether you are doing woodworking or going skiing or scuba diving, nothing is more frustrating (and sometimes dangerous) than fogged goggles. Prevent the problem by coating the goggles with toothpaste and then wiping them off.

Prevent bathroom mirrors from fogging Ouch! You cut yourself shaving and it's no wonder—you can't see your face clearly in that fogged-up bathroom mirror. Next time, coat the mirror with non-gel toothpaste and wipe it off before you get in the shower. When you get out, the mirror won't be fogged.

Shine bathroom and kitchen chrome They make commercial cleaners with a very fine abrasive designed to shine up chrome, but if you don't have any handy, the fine abrasive in non-gel toothpaste works just as well. Just smear on the toothpaste and polish with a soft, dry cloth.

Clean the bathroom sink Non-gel toothpaste works as well as anything else to clean the bathroom sink. The tube's sitting right there, so just squirt some in, scrub with a sponge, and rinse it out. Bonus: The toothpaste will kill any odors emanating from the drain trap.

Remove crayon from walls Did crayon-toting kids get creative on your wall? Roll up your sleeves and grab a tube of non-gel toothpaste and a rag or—better yet—a scrub brush. Squirt the toothpaste on the wall and start scrubbing. The fine

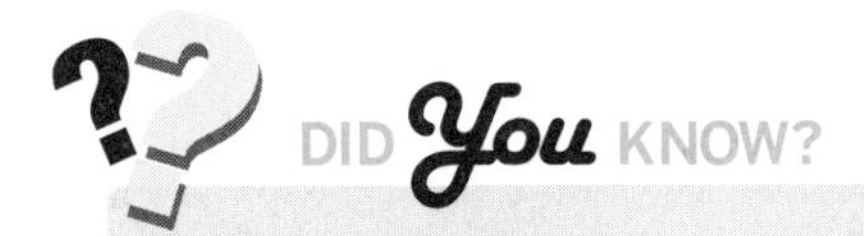

Ancient Egyptians used a mixture of ox-hoof ashes, burned eggshells, myrrh, pumice, and water to clean their teeth. And for most of history, tooth-cleaning concoctions were used mostly by the wealthy. That began to change in 1850, when Dr. Washington Sheffield of New London, Connecticut, developed a formula we would recognize as toothpaste. He called it Dr. Sheffield's Creme Dentifrice. It was his son, Dr. Lucius Tracy Sheffield, who observed collapsible metal tubes of paint and thought, Why not toothpaste? To this day, Sheffield Laboratories, the company Dr. Washington Sheffield founded in 1850, continues to make toothpaste and put it in tubes.

abrasive in the toothpaste will rub away the crayon every time. Rinse the wall with water.

Remove ink or lipstick stains from fabric Oh no, a pen opened up in the pocket of your favorite shirt! This may or may not work, depending on the fabric and the ink, but it is certainly worth a try before consigning the shirt to the scrap bin. Put non-gel toothpaste on the stain and rub the fabric vigorously together. Rinse with water. Did some of the ink come out? Great! Repeat the process a few more times until you get rid of all the ink. The same process works for lipstick.

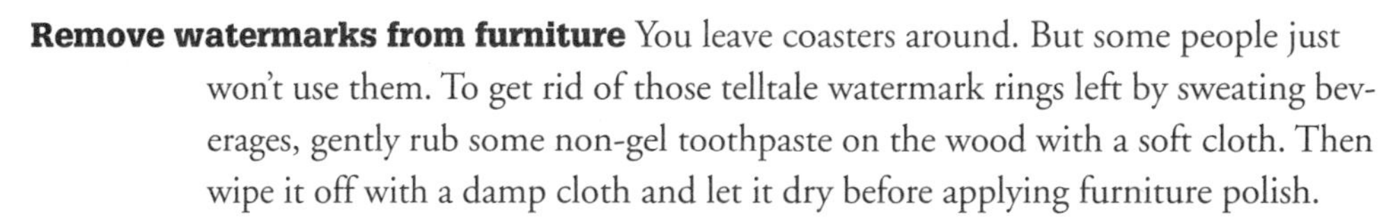

Remove watermarks from furniture You leave coasters around. But some people just won't use them. To get rid of those telltale watermark rings left by sweating beverages, gently rub some non-gel toothpaste on the wood with a soft cloth. Then wipe it off with a damp cloth and let it dry before applying furniture polish.

Remove beach tar Getting that black beach tar on your feet can put a small crimp in your vacation, but it is easy enough to remove. Just rub it with some non-gel toothpaste and rinse.

Clear up pimples Your teenager is bemoaning a prominent pimple, and the day before the dance too! Tonight, have her or him dab a bit of non-gel, nonwhitening toothpaste on the offending spot, and it should be dried up by morning. The toothpaste dehydrates the pimple and absorbs the oil. This remedy works best on pimples that have come to a head. Caution: This remedy may be irritating to sensitive skin.

TAKE CARE All toothpaste, including gels, contains abrasives. The amount varies, but too much can damage your tooth enamel. People with sensitive teeth in particular should use a low-abrasive toothpaste. Ask your dentist which is the best toothpaste for you.

Clean smells from hands The ingredients in toothpaste that deodorize your mouth will work on your hands as well. If you've gotten into something stinky, wash your hands with toothpaste, and they'll smell great.

* Toothpicks

Mark rare, medium, and well done Your guests want their steaks done differently at the family cookout, but how do you keep track of who gets what? Easy. Just use different-colored toothpicks to mark them as rare, medium, and well done and get ready for the accolades.

Stick through garlic clove for marinade If you marinate foods with garlic cloves, stick a toothpick through the clove so you can remove it easily when you are ready to serve the food.

Keep pots from boiling over Oh darn! It seems like all you have to do is turn around for one minute and the pot is boiling over, making a mess on the stovetop. Next time, just stick a toothpick, laid flat, between the lid and pot. The little space will allow enough steam to escape to prevent the pot from boiling over. This also works with a casserole dish that's cooking in the oven.

Microwave potatoes faster The next time you microwave a potato, stick four toothpick "legs" in one side. The suspended potato will cook much faster because the microwaves will reach the bottom as well as the top and sides.

- Buddhist monks used toothpicks as far back as the 700s, and researchers have even found toothpick grooves in the teeth of prehistoric humans.
- Toothpicks were first used in the United States at the Union Oyster House, the oldest restaurant in Boston, which opened in 1826.
- In 1872 Silas Noble and J. P. Cooley patented the first toothpick-manufacturing machine.
- One cord of white birch wood, (also known as the toothpick tree) can make 7.5 million toothpicks.

Control your use of salad dressing Restrict your intake of carbs and calories from salad dressing. Instead of removing the foil seal when you open the bottle, take a toothpick and punch several holes in the foil. This will help prevent overuse of the dressing and make it last longer.

Keep sausages from rolling around When cooking sausages, insert toothpicks between pairs to make turning them over easy and keep them from rolling around in the pan. They'll cook more evenly and only need to be turned over once.

Mark start of tape roll Instead of wasting time trying to find the beginning of a tape roll, just wrap it around a toothpick whenever you are done using the tape, and the start of the roll will always be easy to find. No more frustration, and you can use the time you just saved to attack something else on your to-do list.

Use to light candles When a candle has burned down and the wick is hard to reach, don't burn your fingers trying to use a small match to light it. Light a wooden toothpick instead and use it to light the wick.

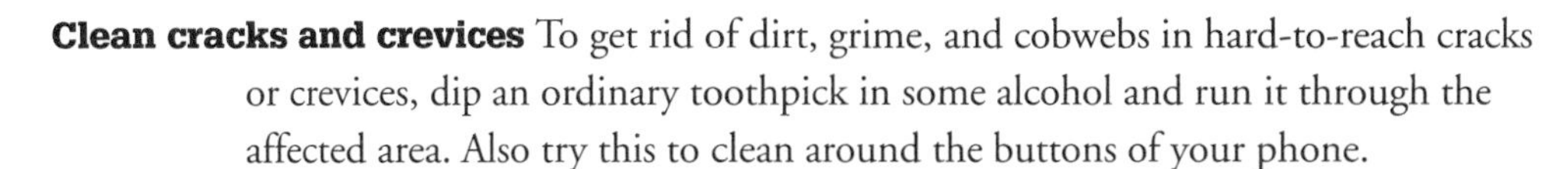

Clean cracks and crevices To get rid of dirt, grime, and cobwebs in hard-to-reach cracks or crevices, dip an ordinary toothpick in some alcohol and run it through the affected area. Also try this to clean around the buttons of your phone.

Apply glue to sequins If you're working on a project that calls for gluing on sequins or buttons, squirt a little glue on a piece of paper and dip in a toothpick to apply small dabs of glue. You won't make a mess, and you won't waste glue.

Make sewing easier Make sewing projects easier and complete them faster. Just use a round toothpick to push fabrics, lace, or gatherings under the pressure foot as you sew.

Touch up furniture crevices The secret to a good paint touch-up job is to use as little paint as possible, because even if you do have the right paint, the stuff in the can may not exactly match the sun-faded or dirty paint on the furniture. The solution: Dip the end of a toothpick in the paint and use it to touch up just the crevice. Unlike a brush, the toothpick won't apply more paint than you need, and you won't have a brush to clean.

Repair small holes in wood Did you drive a finish nail or brad into the wrong spot in your pine project? Don't panic. Dip the tip of a toothpick into white or yellow glue. Stick the toothpick in the hole and break it off. Sand the toothpick flush to the surface and you will never notice the repair.

Repair a loose hinge screw You took the door off and removed the hinges before you painted it. Now as you reattach the hinges, a screw just keeps turning without tightening—the hole is stripped. The fix is easy: Put some glue on the end of a toothpick and stick it in the hole. Break it off. Add one or two more toothpicks with glue until the hole is tightly filled, breaking each one off as you go. Re-drill the hole and you're ready to screw the hinge in place.

Repair a bent plant stem If the stem of your favorite plant has folded over, it by no means dooms the plant. Straighten the stem and support it by placing a toothpick against the stem and wrapping the toothpick on with tape. Water the plant. Keep your eye on the plant—depending on how fast it grows, the stem will regain its strength and you'll need to remove the splint so you don't strangle the stem.

TAKE CARE Overuse of toothpicks can damage tooth enamel and lacerate gums. If you have bonding or veneers, be extra careful to avoid breakage. Toothpicks can also cause wear to tooth roots, especially in the elderly whose gums have pulled away, exposing the roots.

Repair a leaky garden hose If your garden hose springs a leak, don't go out and buy another one; just find the hole and insert a toothpick in it. Cut off the excess part of the toothpick. The water will make the wood swell, plugging up the leak every time.

Foil those cutworms Cutworms kill seedlings by encircling the stem and severing it. To protect your seedlings, stick a toothpick in the soil about 1/4 inch (60 millimeters) from each stem. This prevents a cutworm from encircling it.

* Twist Ties

Organize electrical cords Does the top of your computer desk look like wild vines have overtaken it? Is there a thicket of wires behind your entertainment center? Tame the jungle of electrical wires by rolling each one up neatly and securing the extra length with a twist tie.

Make a trellis All you need are some twist ties and some of those plastic rings from soda six-packs to make a terrific trellis for climbing annual vines such as peas or morning glories. Just use the twist ties to join together as many of the six-pack rings as you want. Attach the trellis between two stakes, also using twist ties. You can even add sections to the trellis as the plant grows so that it looks like the plant is climbing on its own. At the end of the season, just roll the trellis up for storage and you can use it again next year.

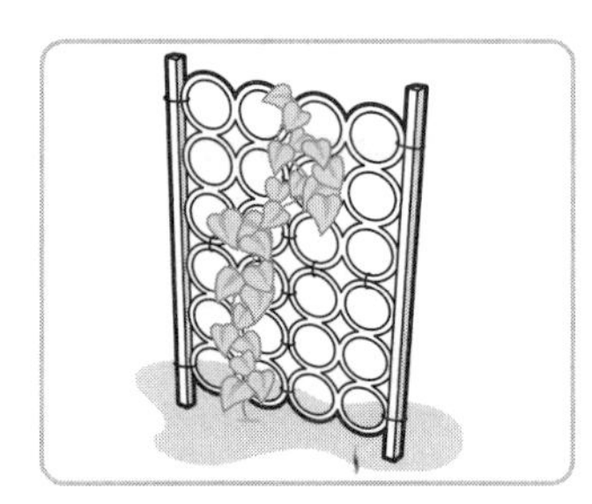

Tie up plant stems Twist ties are handy for securing a drooping plant stem to a stake or holding vines to a trellis. Don't twist the ties too tight, because you might injure the stem and restrict its growth.

Temporarily repair eyeglasses Whoops! Your specs slip off because that tiny screw that holds the earpiece fell out. Secure the earpiece temporarily with a twist tie. Trim the edges off the tie so that you just have the center wire. After you insert and tie it off, snip the excess with scissors.

Use for emergency shoelace Don't have a replacement shoelace handy? Try some twist ties. Use one tie across each opposing pair of eyelets.

Code your keys Got several similar-looking keys on your chain? Quickly identify them with twist ties of different colors secured through the holes in the keys.

Hang Christmas tree ornaments Some of those Christmas tree ornaments have been in the family for generations. As extra insurance against breakage, secure them to the tree with twist ties.

Make an emergency cuff link Oh, brother, you are in trouble! You packed a nice shirt with French cuffs to wear to the wedding, but you forgot your cuff links. Well, it's not so bad. Secure the cuffs with twist ties. Pull the ties through so the twist is discreetly hidden inside the cuff.

Bind loose-leaf paper Hold sheets of loose-leaf paper together by inserting twist ties in the holes.

U

* Umbrellas

Use as a drying rack An old umbrella makes a handy drying rack. Just strip off the fabric and hang the frame upside down from your shower bar. Attach wet clothing with clothespins. Plus, your new drying rack easily folds up for storage.

Clean a chandelier The next time you climb up there to clean the chandelier or ceiling fan, bring an old umbrella with you. Open the umbrella and hook its handle on the fixture so that it hangs upside down to catch any drips or dust.

Cover your picnic food To keep flies from feasting on your picnic before you do, open an old umbrella and cut off the handle. Place the umbrella over the dishes. It will shield your repast from the sun too.

Signal in a crowd The next time you and your sweetie go to a crowded event, carry a couple of identical bright-colored umbrellas. If you get separated, you can hold the umbrellas over your head and open them up to find each other in a flash.

Block plant overspray Houseplants love to be misted with water, but your walls don't love to get soaked with overspray. Stick an open umbrella between the plants and the wall and give your plants a shower.

Make plant stakes The wind caught your umbrella, turned it inside out, and ripped the fabric. Before you toss it into the trash, remove the umbrella's ribs. They make excellent supports for top-heavy garden plants, such as peonies.

Make an instant trellis Remove the fabric from an old umbrella and insert the handle into the ground to support climbing vines such as morning glories. The umbrella's shape, covered with flowers, will look terrific in the garden.

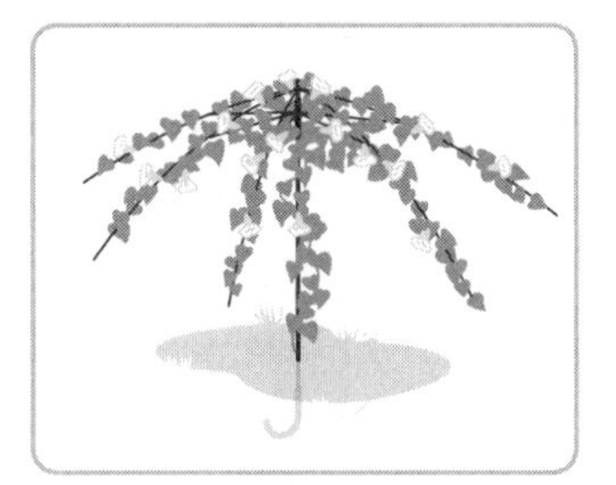

Shield your seedlings You thought you waited long enough before planting your seedlings outside, but now they're predicting a killer frost tonight. Sacrifice an old umbrella to save the seedlings. Open the umbrella, then cut off the handle. Place the umbrella over the seedlings to keep the frost them.

V

* Vanilla Extract

Freshen up the fridge Having trouble getting rid of that bad odor in your refrigerator, even after scrubbing it out? Wipe down the inside of the fridge with vanilla extract. To prolong the fresh vanilla scent, soak a cotton ball or a piece of sponge with vanilla extract and leave it in the refrigerator.

Deodorize your microwave Is the odor of fish, or some other strong smell, lingering in your microwave? Pour a little vanilla extract in a bowl and microwave on High for one minute. Now, that's better.

Neutralize the smell of fresh paint If you would rather not have the unpleasant smell of fresh paint in your house, mix 1 tablespoon vanilla extract into the paint can when you open it. The house will smell delicious!

Sweeten the smell of your home It's an old Realtor's trick. Put a drop or two of vanilla extract on a lightbulb, turn on the light, and your house will be filled with the appealing scent of baked goods in the oven.

Use as perfume Try it! Just put a dab of vanilla extract on each wrist. You'll smell delicious, and many people find the scent of vanilla to be very relaxing.

Repel bugs Everybody likes the smell of vanilla. Everybody but bugs, that is. Dilute 1 tablespoon vanilla extract with 1 cup water and wipe the mixture on your exposed skin to discourage mosquitoes, blackflies, and ticks.

Relieve minor burns Yee-oow! You accidentally grabbed a hot pot or got splattered with grease in the kitchen. Grab the vanilla extract for quick pain relief. The evaporation of the alcohol in the vanilla extract cools the burn.

* Vegetable Oil

Help remove a splinter That stubborn splinter just won't come out. Take a break from poking at your finger for a few minutes and soak it in vegetable oil. The oil will soften up your skin, perhaps just enough to ease that splinter out with your tweezers.

Remove labels and stickers Used jars—both plastic and glass—are always handy to have around. But removing the old labels always leaves a stubborn sticky residue. Soak the label with vegetable oil and the label will slide right off. Works great for sticky price tags too.

Separate stuck glasses When stacked drinking glasses get stuck together, it seems like nothing you can do will separate them. But the solution is simple: Just pour a little vegetable oil around the rim of the bottom glass and the glasses will pull apart with ease.

Oiling Cutting Boards

To restore and preserve dried-out wooden kitchen items such as cutting boards and salad bowls and tongs, use salad bowl oil—a food-safe oil that won't get rancid and is designed to protect wood that will come into contact with food. Don't use regular vegetable oil. Vegetable oil will soak into dried-out wood and make it look much better, but it never really dries and can get rancid after soaking into the wood.

Smooth your feet Rub your feet with vegetable oil before you go to bed and put on a pair of socks. When you awaken, your tootsies will be silky-smooth and soft.

Prevent clippings from sticking to your mower The next time you turn over your lawn mower to remove the stuck-on grass clippings, rub some vegetable oil under the housing and on the blade. It will take a lot longer for the clippings to build up next time.

Control mosquitoes in the birdbath It's so satisfying to watch birds enjoying the garden bath you provide. But unfortunately, that still water is a perfect breeding ground for mosquitoes. Floating a few tablespoons of vegetable oil on the surface of the water will help keep mosquitoes from using the water, and it won't bother the birds. But it's still important to change the water twice a week so any larvae don't have time to hatch.

Season cast-iron cookware After washing and thoroughly drying a cast-iron skillet or wok, use a paper towel to wipe it down with vegetable oil. Just leave a very thin layer of oil. It will prevent the pan from rusting and season it for the next time you use it.

 Vegetable Peelers

Slice slivers of cheese or chocolate When you need cheese slivers that are thinner than you can cut with a knife, or you want to decorate a cake with fine curlicues of chocolate, reach for the vegetable peeler.

Sharpen your pencils No pencil sharpener handy? A vegetable peeler will do a fine job of bringing your pencil to a point.

Soften hard butter fast You're ready to add the butter to your cake mix when you discover that the only sticks you have are as hard as a rock. When you need to soften cold, hard butter in a hurry, shave off what you need with a vegetable peeler. You'll have soft butter in moments.

Renew scented soaps Ornamental scented soaps are a great addition to the powder room because they make the room smell great as well as adding a decorative touch. But after a while, the surface of exposed ornamental soaps dries out, causing the scent to fade. To renew the scent, use a vegetable peeler to skim off a thin layer, revealing a new moist and fragrant surface.

Vinegar

* VINEGAR AROUND THE HOUSE

Clear dirt off PCs and peripherals Your computer, printer, fax machine, and other home office gear will work better if you keep them clean and dust-free. Before you start cleaning, make sure that all your equipment is shut off. Now mix equal parts white vinegar and water in a bucket. Dampen a clean cloth in the solution—never use a spray bottle; you don't want to get liquid on the circuits inside—then squeeze it out as hard as you can, and start wiping. Keep a few cotton swabs on hand for getting to the buildups in tight spaces (like around the keys of your PC keyboard).

Clean your computer mouse If you have a mouse with a removable tracking ball, use a 50/50 vinegar-water solution to clean it. First, remove the ball from underneath the mouse by twisting off the cover over it. Use a cloth, dampened with the solution and wrung out, to wipe the ball clean and to remove fingerprints and dirt from the mouse itself. Then use a moistened cotton swab to clean out the gunk and debris from inside the ball chamber (let it dry a couple of hours before reinserting the ball).

Tip Buying Vinegar

Vinegar comes in a surprising number of varieties—including herbal organic blends, Champagne, rice, and wine—not to mention bottle sizes. For household chores, however, plain distilled white vinegar is the best and least expensive choice, and you can buy it by the gallon (3.7 liters) to save even more money. Apple cider vinegar runs a close second in practicality and is also widely used in cooking and home remedies. All other types of vinegar are strictly for ingestion and can be pretty pricey as well.

Clean your window blinds You can make the job of cleaning mini-blinds or venetians considerably less torturous by giving them "the white glove treatment." Just put on a white cotton glove—the kind sold for gardening is perfect—and moisten the fingers in a solution made of equal parts white vinegar and hot tap water. Now simply slide your fingers across both sides of each slat and prepare to be amazed. Use a container of clean water to periodically wash off the glove.

Unclog and deodorize drains The combination of vinegar and baking soda is one of the most effective ways to unclog and deodorize drains. It's also far gentler on your pipes (and your wallet) than commercial drain cleaners.

- To clear clogs in sink and tub drains, use a funnel to pour in 1/2 cup baking soda followed by 1 cup vinegar. When the foaming subsides, flush with hot tap water. Wait five minutes, and then flush again with cold water. Besides clearing blockages, this technique also washes away odor-causing bacteria.
- To speed up a slow drain, pour in 1/2 cup salt followed by 2 cups boiling vinegar, then flush with hot and cold tap water.

Get rid of smoke odor If you've recently burned a steak—or if your chain-smoking aunt recently paid you a surprise visit—remove the lingering smoky odor by placing a shallow bowl about three-quarters full of white or cider vinegar in the room where the scent is strongest. Use several bowls if the smell permeates your entire home. The odor should be gone in less than a day. You can also quickly dispense of the smell of fresh cigarette smoke inside a room by moistening a cloth with vinegar and waving it around a bit.

Wipe away mildew When you want to remove mildew stains, reach for white vinegar first. It can be safely used without additional ventilation and can be applied to almost any surface—bathroom fixtures and tile, clothing, furniture, painted surfaces, plastic curtains, and more. To eliminate heavy mildew accumulations, use it full strength. For light stains, dilute it with an equal amount of water. You can also prevent mildew from forming on the bottoms of rugs and carpeting by misting the backs with full-strength white vinegar from a spray bottle.

Clean chrome and stainless steel To clean chrome and stainless steel fixtures around your home, apply a light misting of undiluted white vinegar from a recycled spray bottle. Buff with a soft cloth to bring out the brightness.

Shine your silver Make your silverware—as well as your pure silver bracelets, rings, and other jewelry—shine like new by soaking them in a mixture of 1/2 cup white vinegar and 2 tablespoons baking soda for two to three hours. Rinse them under cold water and dry thoroughly with a soft cloth.

Polish brass and copper items Put the shimmer back in your brass, bronze, and copper objects by making a paste of equal parts white vinegar and salt, or vinegar and baking soda (wait for the fizzing to stop before using). Use a clean, soft cloth or paper towel to rub the paste into the item until the tarnish is gone. Then rinse with cool water and polish with a soft towel until dry.

TAKE CARE

- Do not apply vinegar to jewelry containing pearls or gemstones because it can damage their finish or, in the case of pearls, actually disintegrate them.
- Do not attempt to remove tarnish from antiques, because it could diminish their value.

Erase ballpoint-pen marks Has the budding young artist in your home just decorated a painted wall in your home with a ballpoint original? Don't lose your cool. Rather, dab some full-strength white vinegar on the "masterpiece" using a cloth or a sponge. Repeat until the marks are gone. Then go out and buy your child a nice big sketch pad.

Unglue stickers, decals, and price tags To remove a sticker or decal affixed to painted furniture or a painted wall, simply saturate the corners and sides of the sticker with full-strength white vinegar and carefully scrape it off (using an expired credit card or a plastic phone card). Remove any sticky remains by pouring on a bit more vinegar. Let it sit for a minute or two, and then wipe with a clean cloth. This approach is equally effective for removing price tags and other stickers from glass, plastic, and other glossy surfaces.

Burnish your scissors When your scissor blades get sticky or grimy, don't use water to wash them off; you're far more likely to rust the fastener that holds the blades together—or the blades themselves—than get them clean. Instead, wipe down the blades with a cloth dipped in full-strength white vinegar, and then dry it off with a rag or dish towel.

Get the salt off your shoes As if a winter's worth of ice, slush, and snow wasn't rough enough on your shoes and boots, the worst thing, by far, is all the rock salt that's used to melt it. In addition to leaving unsightly white stains, salt can actually cause your footwear to crack and even disintegrate if it's left on indefinitely. To remove it and prevent long-term damage, wipe fresh stains with a cloth dipped in undiluted white vinegar.

Clean your piano keys Here's an easy and efficient way to get those grimy fingerprints and stains off your piano keys. Dip a soft cloth into a solution of 1/2 cup white vinegar mixed in 2 cups water, squeeze it out until there are no drips, then gently wipe off each key. Use a second cloth to dry off the keys as you move along, then leave the keyboard uncovered for 24 hours.

Deodorize lunch boxes, footlockers, and car trunks Does your old footlocker smell like, well, an old footlocker? Or perhaps your child's lunch box has taken on the bouquet of week-old tuna? What about that musty old car trunk? Quit holding your breath every time you open it. Instead, soak a slice of white bread in white vinegar and leave it in the malodorous space overnight. The smell should be gone by morning.

Freshen a musty closet Got a closet that doesn't smell as fresh as you'd like? First, remove the contents, then wash down the walls, ceiling, and floor with a cloth dampened in a solution of 1 cup each of vinegar and ammonia and 1/4 cup baking soda in 1 gallon (3.7 liters) water. Keep the closet door open and let the interior dry before replacing your clothes and other stuff. If the smell persists, place a small pan of cat litter inside. Replenish every few days until the odor is gone.

Brighten up brickwork How's this for an effortless way to clean your brick floors without breaking out the polish? Just go over them with a damp mop dipped in 1 cup white vinegar mixed with 1 gallon (3.7 liters) warm water. Your floors will look so good you'll never think about cleaning them with anything else. You can also use this same solution to brighten up the bricks around your fireplace.

Revitalize wood paneling Does the wood paneling in your den look dull and dreary? Liven it up with this simple homemade remedy: Mix 1 pint warm water, 4 tablespoons white or apple cider vinegar, and 2 tablespoons olive oil in a container, give it a couple of shakes, and apply with a clean cloth. Let the mixture soak into the wood for several minutes, then polish with a dry cloth.

Restore your rugs If your rugs or carpets are looking worn and dingy from too much foot traffic or an excess of kids' building blocks, toy trucks, and such, bring them back to life by brushing them with a clean push broom dipped in a solution of 1 cup white vinegar in 1 gallon (3.7 liters) water. Your faded threads will perk up, and you don't even need to rinse off the solution.

Remove carpet stains You can lift out many stains from your carpet with vinegar:

- Rub light carpet stains with a mixture of 2 tablespoons salt dissolved in 1/2 cup white vinegar. Let the solution dry, then vacuum.
- For larger or darker stains, add 2 tablespoons borax to the mixture and use in the same way.
- For tough, ground-in dirt and other stains, make a paste of 1 tablespoon vinegar with 1 tablespoon cornstarch, and rub it into the stain using a dry cloth. Let it set for two days, then vacuum.
- To make spray-on spot and stain remover, fill a spray bottle with 5 parts water and 1 part vinegar. Fill a second spray bottle with 1 part nonsudsy ammonia and 5 parts water. Saturate a stain with the vinegar solution. Let it settle for a few minutes, then blot thoroughly with a clean, dry cloth. Then spray and blot using the ammonia solution. Repeat until the stain is gone.

Tip: Vinegar and Floor Cleaning

Damp-mopping with a mild vinegar solution is widely recommended as a way to clean wood and no-wax vinyl or laminate flooring. But, if possible, check with your flooring manufacturer first. Even when diluted, vinegar's acidity can ruin some floor finishes, and too much water will damage most wooden floors. If you want to try vinegar on your floors, use 1/2 cup white vinegar mixed in 1 gallon (3.7 liters) warm water. Start with a trial application in an inconspicuous area. Before applying the solution, squeeze out the mop thoroughly (or just use a spray bottle to moisten the mop head).

* VINEGAR IN THE GARAGE

Remove bumper stickers If those tattered old bumper stickers on your car make you feel more nauseated than nostalgic, it's time to break out the vinegar. Saturate the top and sides of the sticker with undiluted distilled vinegar and wait 10-15 minutes for the vinegar to soak through. Then use an expired credit card (or one of those promotional plastic cards that come in the mail) to scrape it off. Use more full-strength vinegar to get rid of any remaining gluey residue. Use the same technique to detach those cute decals your youngster used to decorate the back windshield.

Clean windshield wiper blades When your windshield actually gets blurrier after you turn on your wipers during a rainstorm, it usually means that your wiper blades are dirty. To make them as good as new, dampen a cloth or rag with some full-strength white vinegar and run it down the full length of each blade once or twice.

Keep car windows frost-free If you park your car outdoors during the cold winter months, a smart and simple way to keep frost from forming on your windows is by wiping (or, better yet, spraying) the outsides of the windows with a solution of 3 parts white vinegar to 1 part water. Each coating may last up to several weeks—although, unfortunately, it won't do much in the way of warding off a heavy snowfall.

Care for your car's carpets A good vacuuming will get up the sand and other loose debris from your car's carpeting, but it won't do diddly for stains or ground-in dirt. For that, mix up a solution of equal parts water and white vinegar and sponge it into the carpet. Give the mixture a couple of minutes to settle in; then blot it up with a cloth or paper towel. This technique will also eliminate salt residues left on car carpets during the winter months.

V

Remove candle wax Candles are great for creating a romantic mood, but the mood can quickly sour if you wind up getting melted candle wax on your fine wood furniture. To remove it, first soften the wax using a blow-dryer on its hottest setting and blot up as much as you can with paper towels. Then remove what's left by rubbing with a cloth soaked in a solution made of equal parts white vinegar and water. Wipe clean with a soft, absorbent cloth.

Give grease stains the slip Eliminate grease stains from your kitchen table or counter by wiping them down with a cloth dampened in a solution of equal parts white vinegar and water. In addition to removing the grease, the vinegar will neutralize any odors on the surface (once its own aroma evaporates, that is).

Conceal scratches in wood furniture Got a scratch on a wooden tabletop that grabs your attention every time you look at it? To make it much less noticeable, mix some distilled or cider vinegar and iodine in a small jar and paint over the scratch with a small artist's brush. Use more iodine for darker woods; more vinegar for lighter shades.

TAKE CARE Don't use vinegar—or alcohol or lemon juice—on marble tabletops, countertops, or floors. Vinegar's acidity can dull or even pit the protective coating—and possibly damage the stone itself. Also, avoid using vinegar on travertine and limestone; the acid eats through the calcium in the stonework.

Get rid of water rings on furniture To remove white rings left by wet glasses on wood furniture, mix equal parts vinegar and olive oil and apply it with a soft cloth while moving with the wood grain. Use another clean, soft cloth to shine it up. To get white water rings off leather furniture, dab them with a sponge soaked in full-strength white vinegar.

Wipe off wax or polish buildup When furniture polish or wax builds up on wood furniture or leather tabletops, get rid of it with diluted white vinegar. To get built-up polish off a piece of wood furniture, dip a cloth in equal parts vinegar and water and squeeze it out well. Then, moving with the grain, clean away the polish. Wipe dry with a soft towel or cloth. Most leather tabletops will come clean simply by wiping them down with a soft cloth dipped in 1/4 cup vinegar and 1/2 cup water. Use a clean towel to dry off any remaining liquid.

Revitalize leather furniture Has your leather sofa or easy chair lost its luster? To restore it to its former glory, mix equal parts white vinegar and boiled linseed oil in a recycled spray bottle, shake it up well, and spray it on. Spread it evenly over your furniture using a soft cloth, give it a couple of minutes to settle in, then rub it off with a clean cloth.

Refresh your refrigerator Did you know that vinegar might be an even more effective safe cleanser for your refrigerator than baking soda? Use equal parts white vinegar and water to wash both the interior and exterior of your fridge, including the door gasket and the fronts of the vegetable and fruit bins. To prevent mildew growth, wash the inside walls and bin interiors with some full-strength vinegar on a cloth. Also use undiluted vinegar to wipe off accumulated dust and grime on top of your refrigerator. Of course, you'll still want to put that box of baking soda inside your refrigerator to keep it smelling clean when you're done.

Steam-clean your microwave To clean your microwave, place a glass bowl filled with a solution of 1/4 cup vinegar in 1 cup water inside, and zap the mixture for five minutes on the highest setting. Once the bowl cools, dip a cloth or sponge into the liquid and use it to wipe away stains and splatters on the interior.

Taken literally, vinegar is nothing more than wine that's gone bad; the word derives from the French *vin* (wine) and *aigre* (sour). But, in fact, anything used to make alcohol can be turned into vinegar, including apples, honey, malted barley, molasses, rice, sugarcane, and even coconuts. Vinegar's acidic, solvent properties were well known even in ancient times. According to one popular legend, Cleopatra is said to have wagered she could dispose of a fortune in the course of a single meal. She won the bet by dissolving a handful of pearls in a cup of vinegar ... and then consuming it.

Disinfect cutting boards To disinfect and clean your wood cutting boards or butcher block countertop, wipe them with full-strength white vinegar after each use. The acetic acid in the vinegar is a good disinfectant, effective against such harmful bugs as *E. coli*, *Salmonella,* and *Staphylococcus.* Never use water and dishwashing detergent, because it can weaken surface wood fibers. When your wooden cutting surface needs deodorizing as well as disinfecting, spread some baking soda over it and then spray on undiluted white vinegar. Let it foam and bubble for five to ten minutes, then rinse with a cloth dipped in clean cold water.

Deodorize your garbage disposal Here's an incredibly easy way to keep your garbage disposal unit sanitized and smelling clean: Mix equal parts water and vinegar in a bowl, pour the solution into an ice cube tray, and freeze it. Then simply drop a couple of "vinegar cubes" down your disposal every week or so, followed by a cold-water rinse.

Wash out your dishwasher To keep your dishwasher operating at peak performance and remove built-up soap film, pour 1 cup undiluted white vinegar into the bottom of the unit—or in a bowl on the top rack. Then run the machine through a full cycle without any dishes or detergent. Do this once a month, especially if you have hard water. *Note:* If there's no mention of vinegar in your dishwasher owner's manual, check with the manufacturer first.

Clean china, crystal, and glassware Put the sparkle back in your glassware by adding vinegar to your rinse water or dishwater.

- To keep your everyday glassware gleaming, add 1/4 cup vinegar to your dishwasher's rinse cycle.
- To rid drinking glasses of cloudiness or spots caused by hard water, heat up a pot of equal parts white vinegar and water (use full-strength vinegar if your glasses are very cloudy), and let them soak in it for 15-30 minutes. Give them a good scrubbing with a bottle brush, then rinse clean.
- Add 2 tablespoons vinegar to your dishwater when cleaning your good crystal glasses. Then rinse them in a solution of 3 parts warm water to 1 part vinegar and allow them to air-dry. You can also wash delicate crystal and fine china by adding 1 cup vinegar to a basin of warm water. Gently dunk the glasses in the solution and let dry.
- To get coffee stains and other discolorations off china dishes and teacups, try scrubbing them with equal parts vinegar and salt, followed by rinsing them under warm water.

Clean a coffeemaker If your coffee consistently comes out weak or bitter, odds are, your coffeemaker needs cleaning. Fill the decanter with 2 cups white vinegar and 1 cup water. Place a filter in the machine, and pour the solution into the coffeemaker's water chamber. Turn on the coffeemaker and let it run through a full brew cycle. Remove the filter and replace it with a fresh one. Then run clean water through the machine for two full cycles, replacing the filter again for the second brew. If you have soft water, clean your coffeemaker after 80 brew cycles—after 40 cycles if you have hard water.

Clean a teakettle To eliminate lime and mineral deposits in a teakettle, bring 3 cups full-strength white vinegar to a full boil for five minutes and leave the vinegar in the kettle overnight. Rinse out with cold water the next day.

Cut the grease Every professional cook knows that distilled vinegar is one of the best grease cutters around. It even works on seriously greasy surfaces such as the fry vats used in many food outlets. But you don't need to have a deep fryer to find plenty of ways to put vinegar to good use:

- When you're finished frying, clean up grease splatters from your stovetop, walls, range hood, and surrounding countertop by washing them with a sponge dipped in undiluted white vinegar. Use another sponge soaked in cold tap water to rinse, then wipe dry with a soft cloth.
- Pour 3-4 tablespoons white vinegar into your favorite brand (especially bargain brands) of liquid dishwashing detergent and give it a few shakes. The added vinegar will not only increase the detergent's grease-fighting capabilities, but also provide you with more dishwashing liquid for the money, because you'll need less soap to clean your dishes.
- Boiling 2 cups vinegar in your frying pan for 10 minutes will help keep food from sticking to it for several months at a time.
- Remove burned-on grease and food stains from your stainless steel cookware by mixing 1 cup distilled vinegar in enough water to cover the stains (if they're near the top of a large pot, you may need to increase the vinegar). Let it boil for five minutes. The stains should come off with some mild scrubbing when you wash the utensil.
- Get that blackened, cooked-on grease off your broiler pan by softening it up with a solution of 1 cup apple cider vinegar and 2 tablespoons sugar. Apply the mixture while the pan is still hot, and let it sit for an hour or so. Then watch in amazement as the grime slides off with a light scrubbing.
- Got a hot plate that looks more like a grease pan? Whip it back into shape by washing it with a sponge dipped in full-strength white vinegar.
- Fight grease buildups in your oven by wiping down the inside with a rag or sponge soaked in full-strength white vinegar once a week. The same treatment gets grease off the grates on gas stoves.

Homemade Wine Vinegar

Contrary to popular belief, old wine rarely turns into vinegar; usually a half-empty bottle just spoils due to oxidation. To create vinegar, you need the presence of *Acetobacter,* a specific type of bacteria. You can make your own wine vinegar, though, by mixing one part leftover red, white, or rosé wine with 2 parts cider vinegar. Pour the mixture into a clean, recycled wine bottle and store it in a dark cabinet. It just might taste as good, if not better, on your salads than some of those fancy wine vinegars sold at upscale food shops.

Brush-clean can opener blades Does that dirty wheel blade of your electric can opener look like it's seen at least one can too many? To clean and sanitize it, dip an old toothbrush in white vinegar, and then position the bristles of the brush around the side and edge of the wheel. Turn on the appliance, and let the blade scrub itself clean.

Remove stains from pots, pans, and ovenware Nothing will do a better job than vinegar when it comes to removing stubborn stains on your cookware. Here's how to put the power of vinegar to use:

- Give those dark stains on your aluminum cookware (caused by cooking acidic foods) the heave-ho by mixing in 1 teaspoon white vinegar for every cup of water needed to cover the stains. Let it boil for a couple of minutes, then rinse with cold water.
- To remove stains from your stainless steel pots and pans, soak them in 2 cups white vinegar for 30 minutes, then rinse them with hot, soapy water followed by a cold-water rinse.
- To get cooked-on food stains off your glass ovenware, fill them with 1 part vinegar and 4 parts water, heat the mixture to a slow boil, and let it boil at a low level for five minutes. The stains should come off with some mild scrubbing once the mixture cools.
- They call it nonstick, but no cookware is stainproof. For mineral stains on your nonstick cookware, rub the utensil with a cloth dipped in undiluted distilled vinegar. To loosen up stubborn stains, mix 2 tablespoons baking soda, 1/2 cup vinegar, and 1 cup water and let it boil for 10 minutes.

Clear the air in your kitchen If the smell of yesterday's cooked cabbage or fish stew is hanging around your kitchen longer than you'd like, mix a pot of 1/2 cup white vinegar in 1 cup water. Let it boil until the liquid is almost gone. You'll be breathing easier in no time.

Refresh your ice trays If your plastic ice trays are covered with hard-water stains—or if it's been a while since you've cleaned them—a few cups of white vinegar can help you in either case. To remove the spots or disinfect your trays, let them soak in undiluted vinegar for four to five hours, then rinse well under cold water and let dry.

Make all-purpose cleaners For fast cleanups around the kitchen, keep two recycled spray bottles filled with these vinegar-based solutions:

- For glass, stainless steel, and plastic laminate surfaces, fill your spray bottle with 2 parts water, 1 part distilled white vinegar, and a couple of drops of dishwashing liquid.
- For cleaning walls and other painted surfaces, mix up 1/2 cup white vinegar, 1 cup ammonia, and 1/4 cup baking soda in 1 gallon (3.7 liters) water and

pour some into a spray bottle. Spritz it on spots and stains whenever needed and wipe off with a clean towel.

Make an all-purpose scrub for pots and pans How would you like an effective scouring mix that costs a few pennies, and can be safely used on all of your metal cookware—including expensive copper pots and pans? Want even better news? You probably already have this "miracle mix" in your kitchen. Simply combine equal parts salt and flour and add just enough vinegar to make a paste. Work the paste around the cooking surface and the outside of the utensil, then rinse off with warm water and dry thoroughly with a soft dish towel.

Sanitize jars, containers, and vases Do you cringe at the thought of cleaning out a mayonnaise, peanut butter, or mustard jar to reuse it? Or worse, getting the residue out of a slimy vase, decanter, or container? There is an easy way to handle these jobs. Fill the item with equal parts vinegar and warm, soapy water and let it stand for 10-15 minutes. If you're cleaning a bottle or jar, close it up and give it a few good shakes; otherwise use a bottle brush to scrape off the remains before thoroughly rinsing.

Clean a dirty thermos To get a thermos bottle clean, fill it with warm water and 1/4 cup white vinegar. If you see any residue, add some uncooked rice, which will act as an abrasive to scrape it off. Close and shake well. Then rinse and let it air-dry.

Purge bugs from your pantry Do you have moths or other insects in your cupboard or pantry? Fill a small bowl with 1 1/2 cups apple cider vinegar and add a couple of drops of liquid dish detergent. Leave it in there for a week; it will attract the bugs, which will fall into the bowl and drown. Then empty the shelves, and give the interior a thorough washing with dishwashing detergent or 2 cups baking soda in 1 quart (1 liter) water. Discard all wheat products (breads, pasta, flour, and such), and clean off canned goods before putting them back.

Trap fruit flies Did you bring home fruit flies from the market? You can make traps for them that can be used anywhere around your house by filling an old jar about halfway with apple cider. Punch a few holes in the lid, screw it back on, and you're good to go.

* VINEGAR FOR THE COOK

Tenderize and purify meats and seafood Soaking a lean or inexpensive cut of red meat in a couple of cups of vinegar breaks down tough fibers to make it more tender—and in addition, kills off any potentially harmful bacteria. You can also use vinegar to tenderize seafood steaks. Let the meat or fish soak in

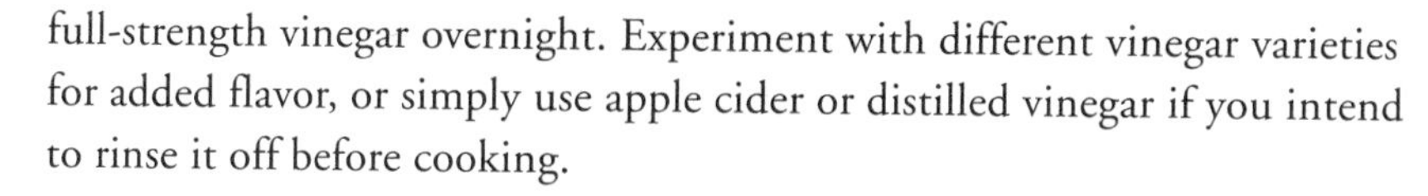

full-strength vinegar overnight. Experiment with different vinegar varieties for added flavor, or simply use apple cider or distilled vinegar if you intend to rinse it off before cooking.

Keep corned beef from shrinking Ever notice how the corned beef that comes out of the pot is always smaller than the one that went in? Stop your meat from shrinking by adding a couple of tablespoons of apple cider vinegar to the water when boiling your beef.

Make better boiled or poached eggs Vinegar does marvelous things for eggs. Here are the two most useful "egg-samples":

- When you are making hard-boiled eggs, adding 2 tablespoons distilled vinegar for every quart (liter) of water will keep the eggs from cracking and make them easier to shell.
- When you are poaching eggs, adding a couple of tablespoons of vinegar to the water will keep your eggs in tight shape by preventing the egg whites from spreading.

Wash store-bought produce You can't be too careful these days when it comes to handling the foods you eat. Before serving your fruits and vegetables, a great way to eliminate the hidden dirt, pesticides, and even insects, is to rinse them in 4 tablespoons apple cider vinegar dissolved in 1 gallon (3.7 liters) cold water.

Authentic balsamic vinegar comes solely from Modena, Italy, and is made from Trebbiano grapes, a particularly sweet white variety grown in the surrounding hills. Italian law mandates that the vinegar be aged in wooden barrels made of chestnut, juniper, mulberry, or oak. There are only two grades of true balsamic vinegar—which typically sells for $100-$200 for a 100-milliliter bottle: *tradizionale vecchio*, vinegar that is at least 12 years old, and *tradizionale extra vecchio*, vinegar that's aged for at least 25 years (some balsamic vinegars are known to have been aged for more than 100 years).

Remove odors from your hands It's often difficult to get strong onion, garlic, or fish odors off your hands after preparing a meal. But you'll find these scents are a lot easier to wash off if you rub some distilled vinegar on your hands before and after you slice your vegetables or clean your fish.

Get rid of berry stains You can use undiluted white vinegar on your hands to remove stains from berries and other fruits.

Control your dandruff To give your dandruff the brush-off, follow up each shampoo with a rinse of 2 cups apple cider vinegar mixed with 2 cups cold water. You can also fight dandruff by applying 3 tablespoons vinegar onto your hair and massaging into your scalp before you shampoo. Wait a few minutes, then rinse it out and wash as usual.

Condition your hair Want to put the life back into your limp or damaged hair? You can whip up a terrific hair conditioner by combining 1 teaspoon apple cider vinegar with 2 tablespoons olive oil and 3 egg whites. Rub the mixture into your hair, then keep it covered for 30 minutes using plastic wrap or a shower cap. When time's up, shampoo and rinse as usual.

Protect blond hair from chlorine Keep your golden locks from turning green in a chlorinated pool by rubbing 1/4 cup cider vinegar into your hair and letting it set for 15 minutes before diving in.

Apply as antiperspirant Why not put the deodorizing power of vinegar to use where it matters most? That's right, you don't need a roll-on or spray to keep your underarms smelling fresh. Instead, splash a little white vinegar under each arm in the morning, and let it dry. In addition to combating perspiration odor, this method also does away with those deodorant stains on your garments.

Soak away aching muscles Got a sore back, a strained tendon in your shoulder or calf, or maybe you're just feeling generally rundown? Adding 2 cups apple cider vinegar to your bathwater is a great way to soothe away your aches and pains, or to simply to take the edge off a stressful day. Adding a few drops of peppermint oil to your bath can lend an able assist as well.

Freshen your breath After you consume a fair portion of garlic or onions, a quick and easy way to sweeten your breath is to rinse your mouth with a solution made by dissolving 2 tablespoons apple cider vinegar and 1 teaspoon salt in a glass of warm water.

Ease sunburn and itching You can cool a bad sunburn by gently dabbing the area with a cotton ball or soft cloth saturated with white or cider vinegar. (This treatment is especially effective if it's applied before the burn starts to sting.) The same technique works to instantly stop the itch of mosquito and other insect bites, as well as the rashes caused by exposure to poison ivy or poison oak.

Banish bruises If you or someone you care about has a nasty fall, you can speed healing and prevent black-and-blue marks by soaking a piece of cotton gauze in white or apple cider vinegar and leaving it on the injured area for one hour.

Soothe a sore throat Here are three ways that you can make a sore throat feel better:

- If your throat is left raw by a bad cough, or even a speaking or singing engagement, you'll find fast relief by gargling with 1 tablespoon apple cider

vinegar and 1 teaspoon salt dissolved in a glass of warm water; use several times a day if needed.

- For sore throats associated with a cold or flu, combine 1/4 cup cider vinegar and 1/4 cup honey and take 1 tablespoon every four hours.
- To soothe both a cough and a sore throat, mix 1/2 cup vinegar, 1/2 cup water, 4 teaspoons honey, and 1 teaspoon hot sauce. Swallow 1 tablespoon four or five times daily, including one before bedtime. *Warning:* Children under one year old should never be given honey.

Breathe easier Adding 1/4 cup white vinegar to the water in your hot-steam vaporizer can help ease congestion caused by a chest cold or sinus infection. It can also be good for your vaporizer: The vinegar will clear away any mineral deposits in the water tubes resulting from the use of hard water. *Note:* Check with the manufacturer before adding vinegar to a cool-mist vaporizer.

Treat an active cold sore The only thing worse than a bad cold is a bad cold sore. Fortunately, you can usually dry up a cold sore in short order by dabbing it with a cotton ball saturated in white vinegar three times a day. The vinegar will quickly soothe the pain and swelling.

Make a poultice for corns and calluses Here's an old-fashioned, time-proven method to treat corns and calluses: Saturate a piece of white or stale bread with 1/4 cup white vinegar. Let the bread soak in the vinegar for 30 minutes, then break off a piece big enough to completely cover the corn. Keep the poultice in place with gauze or adhesive tape, and leave it on overnight. The next morning, the hard, callused skin will be dissolved, and the corn should be easy to remove. Older, thicker calluses may require several treatments.

Get the jump on athlete's foot A bad case of athlete's foot can drive you hopping mad. But you can often quell the infection, and quickly ease the itching, by rinsing your feet three or four times a day for a few days with undiluted apple cider vinegar. As an added precaution, soak your socks or stockings in a mixture of 1 part vinegar and 4 parts water for 30 minutes before laundering them.

Pamper your skin Using vinegar as a skin toner dates back to the time of Helen of Troy. And it's just as effective today. After you wash your face, mix 1 tablespoon apple cider vinegar with 2 cups water as a finishing rinse to cleanse and tighten your skin. You can also make your own facial treatment by mixing 1/4 cup cider vinegar with 1/4 cup water. Gently apply the solution to your face and let it dry.

Say good-bye to age or sun spots Before you take any drastic measures to remove or cover up those brown spots on your skin caused by overexposure to the sun or hormonal changes, give vinegar a try. Simply pour some full-strength apple cider vinegar onto a cotton ball and apply it to the spots for 10 minutes at least twice a day. The spots should fade or disappear within a few weeks.

Soften your cuticles You can soften the cuticles on your fingers and toes before manicuring them by soaking your digits in a bowl of undiluted white vinegar for five minutes.

Make nail polish last longer Your nail polish will have a longer life expectancy if you first dampen your nails with some vinegar on a cotton ball and let it dry before applying your favorite polish.

Clean your eyeglasses When it's more difficult to see with your glasses on than it is with them off, it's a clear indication that they're in need of a good cleaning. Applying a few drops of white vinegar to your glass lenses and wiping them with a soft cloth will easily remove dirt, sweat, and fingerprints, and leave them spotless. Don't use vinegar on plastic lenses, however.

Treat a jellyfish or bee sting A jellyfish can pack a nasty sting. If you have an encounter with one, pouring some undiluted vinegar on the sting will take away the pain in no time, and let you scrape out the stinger with a plastic credit card. The same treatment can also be used to treat bee stings. But using vinegar on stings inflicted by the jellyfish's cousin the Portuguese man-of-war is now discouraged because vinegar may actually increase the amount of toxin released under the skin. *Warning:* If you have difficulty breathing or the sting area becomes inflamed and swollen, get medical attention at once; you could be having an allergic reaction.

* VINEGAR IN THE BATHROOM

Wash mildew from shower curtains Clean those ugly mildew stains off your plastic shower curtain by putting it and a couple of soiled towels in your washing machine. Add 1/2 cup laundry detergent and 1/2 cup baking soda to the load,

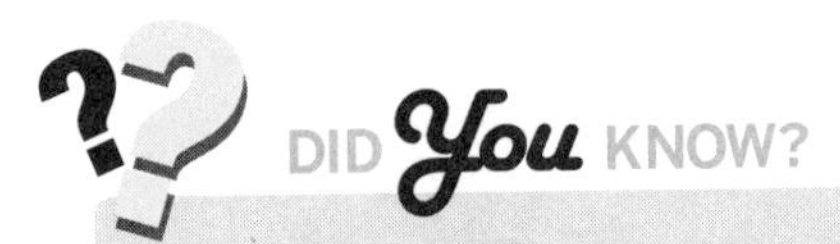

The world's only museum dedicated to vinegar, the International Vinegar Museum, is located in Roslyn, South Dakota. Housed in a building that was the former town hall, the museum is operated by Dr. Lawrence J. Diggs, an international vinegar consultant also known as the Vinegar Man (visit him online at www.vinegarman.com). The museum showcases vinegars from around the world, has displays on the various methods used to make vinegar, and even lets visitors sample different types of vinegars. It's also among the world's least expensive museums: Admission for adults is $2; for those under 18, $1; and "instant scholarships for those too poor to pay."

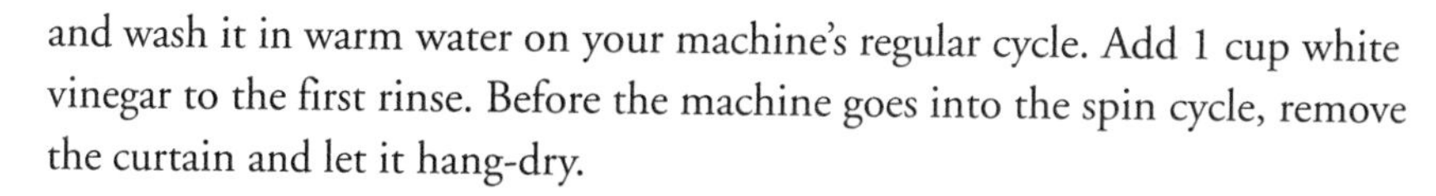

and wash it in warm water on your machine's regular cycle. Add 1 cup white vinegar to the first rinse. Before the machine goes into the spin cycle, remove the curtain and let it hang-dry.

Shine ceramic tiles If soap scum or water spots have dulled the ceramic tiles around your sink or bath, bring back the brightness by scrubbing them with 1/2 cup white vinegar, 1/2 cup ammonia, and 1/4 cup borax mixed in 1 gallon (3.7 liters) warm water. Rinse well with cool water and let air-dry.

Whiten your grout Has the grout between the tiles of your shower or bathtub enclosure become stained or discolored? Restore it to its original shade of white by using a toothbrush dipped in undiluted white vinegar to scrub away the dinginess.

Clean sinks and bathtubs Put the shine back in your porcelain sinks and bathtubs by giving them a good scrubbing with full-strength white vinegar, followed by a rinse of clean cold water. To remove hard-water stains from your tub, pour in 3 cups white vinegar under running hot tap water. Let the tub fill up over the stains and allow it to soak for four hours. When the water drains out, you should be able to easily scrub off the stains.

Shine up your shower doors To leave your glass shower doors sparkling clean—and to remove all of those annoying water spots—wipe them down with a cloth dipped in a solution of 1/2 cup white vinegar, 1 cup ammonia, and 1/4 cup baking soda mixed in 1 gallon (3.7 liters) warm water.

Disinfect shower door tracks Use vinegar to remove accumulated dirt and grime from the tracks of your shower doors. Fill the tracks with about 2 cups full-strength white vinegar and let it sit for three to five hours. (If the tracks are really dirty, heat the vinegar in a glass container for 30 seconds in your microwave first.) Then pour some hot water over the track to flush away the gunk. You may need to use a small scrub brush, or even a recycled toothbrush, to get up tough stains.

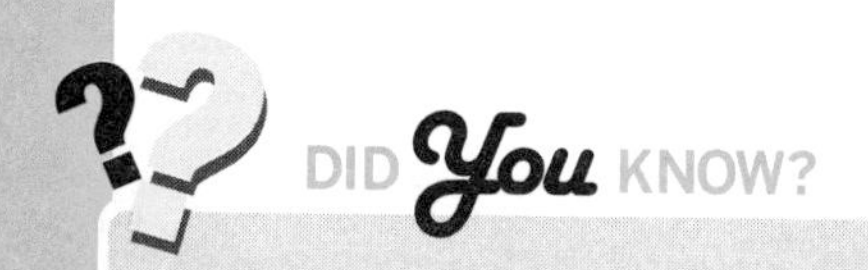

Some researchers believe vinegar will ultimately be adopted as a simple and inexpensive way to diagnose cervical cancer in women—especially those living in impoverished nations. In tests conducted over a two-year period, midwives in Zimbabwe used a vinegar solution to detect more than 75 percent of potential cancers in 10,000 women (the solution turns tissue containing precancerous cells white). Although the test is not as accurate as a Pap smear, doctors believe it will soon be an important screening tool in developing countries, where only 5 percent of women are currently tested for this often fatal disease.

Remove mineral deposits from showerheads Wash away blockages and mineral deposits from removable showerheads by placing them in 1 quart (1 liter) boiling water with 1/2 cup distilled vinegar for 10 minutes (use hot, not boiling, liquid for plastic showerheads). When you remove it from the solution, the obstructions should be gone. If you have a nonremovable showerhead, fill a small plastic bag half full with vinegar and tape it over the fixture. Let it sit for about 1 hour, then remove the bag and wipe off any remaining vinegar from the showerhead.

Wipe down bathroom fixtures Don't stop at the shower when you're cleaning with vinegar! Pour a bit of undiluted white vinegar onto a soft cloth and use it to wipe your chrome faucets, towel racks, bathroom mirrors, doorknobs, and such. It'll leave them gleaming.

TAKE CARE Combining vinegar with bleach—or any other product containing chlorine, such as chlorinated lime (sold as bleaching powder)—may produce chlorine gas. In low concentrations, this toxic, acrid-smelling gas can cause damage to your eyes, skin, or respiratory system. High concentrations are often fatal.

Fight mold and mildew To remove and inhibit bathroom mold and mildew, pour a solution of 3 tablespoons white vinegar, 1 teaspoon borax, and 2 cups hot water into a clean, recycled spray bottle and give it a few good shakes. Then spray the mixture on painted surfaces, tiles, windows, or wherever you see mold or mildew spots. Use a soft scrub brush to work the solution into the stains or just let it soak in.

Disinfect toilet bowls Want an easy way to keep your toilet looking and smelling clean? Pour 2 cups white vinegar into the bowl and let the solution soak overnight before flushing. Including this vinegar soak in your weekly cleaning regimen will also help keep away those ugly water rings that typically appear just above the water level.

Clean your toothbrush holder Get the grime, bacteria, and caked-on toothpaste drippings out of the grooves of your bathroom toothbrush holder by cleaning the openings with cotton swabs moistened with white vinegar.

Wash out your rinse cup If several people in your home use the same rinse cup after brushing their teeth, give it a weekly cleaning by filling it with equal parts water and white vinegar, or just full-strength vinegar, and let it sit overnight. Rinse thoroughly with cold water before using.

* VINEGAR IN THE LAUNDRY

Soften fabrics, kill bacteria, eliminate static, and more There are so many benefits to be reaped by adding 1 cup white vinegar to your washer's rinse cycle that it's

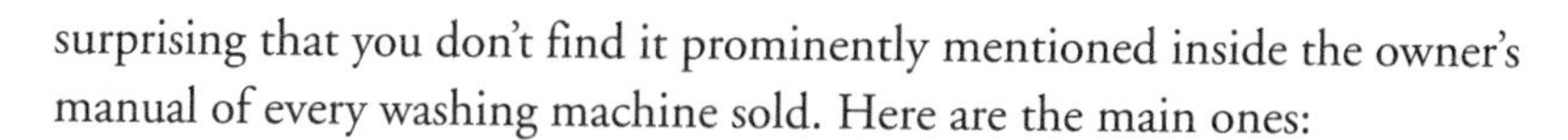

surprising that you don't find it prominently mentioned inside the owner's manual of every washing machine sold. Here are the main ones:

- A single cup of vinegar will kill off any bacteria that may be present in your wash load, especially if it includes cloth diapers and the like.
- A cup of vinegar will keep your clothes coming out of the wash soft and smelling fresh—so you can kiss your fabric-softening liquids and sheets good-bye (unless, of course, you happen to like your clothes smelling of heavy perfumes).
- A cup of vinegar will brighten small loads of white clothes.
- Added to the last rinse, a cup of vinegar will keep your clothes lint- and static-free.
- Adding a cupful of vinegar to the last rinse will set the color of your newly dyed fabrics.

Clean your washing machine An easy way to periodically clean out soap scum and disinfect your clothes washer is to pour in 2 cups vinegar, then run the machine through a full cycle without any clothes or detergent. If your washer is particularly dirty, fill it with very hot water, add 2 gallons (7.5 liters) vinegar, and let the agitator run for 8-10 minutes. Turn off the washer and let the solution stand overnight. In the morning, empty the basin and run your washer through a complete cycle.

Stop reds from running Unless you have a fondness for pink-tinted clothing, take one simple precaution to prevent red—or other brightly dyed—washable clothes from ruining your wash loads. Soak your new garments in a few cups of undiluted white vinegar for 10-15 minutes before their first washing. You'll never have to worry about running colors again!

Brighten your loads Why waste money on that costly all-color bleach when you can get the same results using vinegar? Just add 1/2 cup white vinegar to your machine's wash cycle to brighten up the colors in each load.

Make new clothes ready to wear Get the chemicals, dust, odor, and whatever else out of your brand-new or secondhand clothes by pouring 1 cup white vinegar into the wash cycle the first time you wash them.

Whiten your dingy crew socks If it's getting increasingly difficult to identify the white socks in your sock drawer, here's a simple way to make them so bright you can't miss them. Start by adding 1 cup vinegar to 1 1/2 quarts (1.5 liters) tap water in a large pot. Bring the solution to a boil, then pour it into a bucket and drop in your dingy socks. Let them soak overnight. The next day, wash them as you normally would.

Get the yellow out of clothing To restore yellowed clothing, let the garments soak overnight in a solution of 12 parts warm water to 1 part vinegar. Wash them the following morning.

Soften up your blankets Add 2 cups white vinegar to your washer's rinse water (or a washtub filled with water) to remove soap residue from both cotton and wool blankets before drying. This will also leave them feeling fresh and soft as new.

Spray away wrinkles In a perfect world, laundry would emerge from the dryer freshly pressed. Until that day, you can often get the wrinkles out of clothes after drying by misting them with a solution of 1 part vinegar to 3 parts water. Once you're sure you didn't miss a spot, hang it up and let it air-dry. You may find this approach works better for some clothes than ironing; it's certainly a lot gentler on the material.

Flush your iron's interior To eliminate mineral deposits and prevent corrosion on your steam iron, give it an occasional cleaning by filling the reservoir with undiluted white vinegar. Place the iron in an upright position, switch on the steam setting, and let the vinegar steam through it for 5-10 minutes. Then refill the chamber with clean water and repeat. Finally, give the water chamber a good rinsing with cold, clean water.

Clean your iron's soleplate To remove scorch marks from the soleplate of your iron, scrub it with a paste made by heating up equal parts vinegar and salt in a small pan. Use a rag dipped in clean water to wipe away the remaining residue.

Sharpen your creases You'll find the creases in your freshly ironed clothes coming out a lot neater if you lightly spray them with equal parts water and vinegar before ironing them. For truly sharp creases in slacks and dress shirts, first dampen the garment using a cloth moistened in a solution of 1 part white vinegar and 2 parts water. Then place a brown paper bag over the crease and start ironing.

TAKE CARE Keep cider vinegar out of the laundry. Using it to pretreat clothes or adding it to wash or rinse water may actually create stains rather than remove them. Use only distilled white vinegar for laundering.

Make old hemlines disappear Want to make those needle marks from an old hemline disappear for good? Just moisten the area with a cloth dipped in equal parts vinegar and water, then place it under the garment before you start ironing.

Erase scorch marks Did your iron get too hot under the collar—or perhaps on a sleeve or pant leg? You can often eliminate slight scorch marks by rubbing the spot with a cloth dampened with white vinegar, then blotting it with a clean towel. Repeat if necessary.

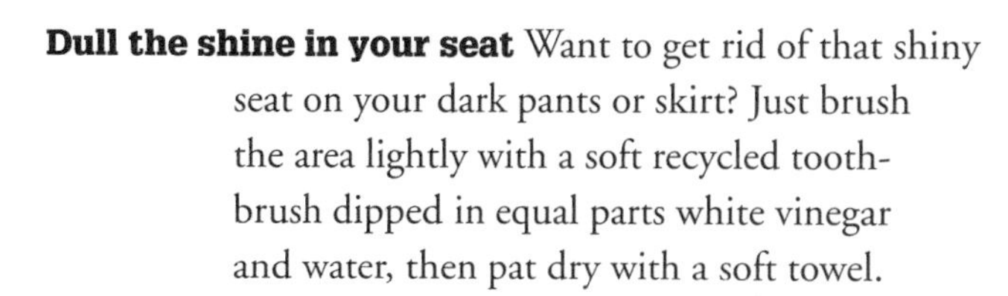

Dull the shine in your seat Want to get rid of that shiny seat on your dark pants or skirt? Just brush the area lightly with a soft recycled toothbrush dipped in equal parts white vinegar and water, then pat dry with a soft towel.

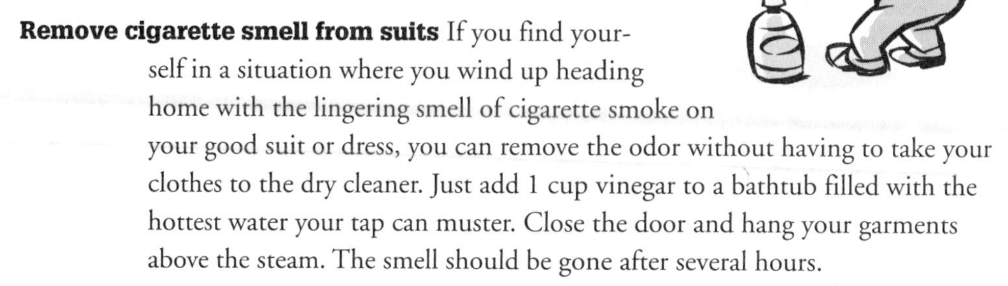

Remove cigarette smell from suits If you find yourself in a situation where you wind up heading home with the lingering smell of cigarette smoke on your good suit or dress, you can remove the odor without having to take your clothes to the dry cleaner. Just add 1 cup vinegar to a bathtub filled with the hottest water your tap can muster. Close the door and hang your garments above the steam. The smell should be gone after several hours.

Reshape your woolens Shrunken woolen sweaters and other items can usually be stretched back to their former size or shape after boiling them in a solution of 1 part vinegar to 2 parts water for 25 minutes. Let the garment air-dry after you've finished stretching it.

✱ VINEGAR FOR REMOVING STAINS

Brush off suede stains To eliminate a fresh grease spot on a suede jacket or skirt, gently brush it with a soft toothbrush dipped in white vinegar. Let the spot air-dry, then brush with a suede brush. Repeat if necessary. You can also generally tone up suede items by lightly wiping them with a sponge dipped in vinegar.

Pat away water-soluble stains You can lift out many water-soluble stains—including beer, orange and other fruit juices, black coffee or tea, and vomit—from your cotton-blend clothing by patting the spot with a cloth or towel moistened with undiluted white vinegar just before placing it in the wash. For large stains, you may want to soak the garment overnight in a solution of 3 parts vinegar to 1 part cold water before washing.

Unset old stains Older, set-in stains will often come out in the wash after being pretreated with a solution of 3 tablespoons white vinegar and 2 tablespoons liquid detergent in 1 quart (1 liter) warm water. Rub the solution into the stain, then blot it dry before washing.

Sponge out serious stains Cola, hair dye, ketchup, and wine stains on washable cotton blends should be treated as soon as possible (that is, within 24 hours). Sponge the area with undiluted vinegar and launder immediately afterward. For severe stains, add 1-2 cups vinegar to the wash cycle as well.

Get the rust out To remove a rust stain from your cotton work clothes, moisten the spot with some full-strength vinegar and then rub in a bit of salt. If it's warm outdoors, let it dry in the sunlight (otherwise a sunny window will do), then toss it in the wash.

Clear away crayon stains Somehow or other, kids often manage to get crayon marks on their clothing. You can easily get these stains off by rubbing them with a recycled toothbrush soaked in undiluted vinegar before washing them.

Remove rings from collars and cuffs Are you tired of seeing those old sweat rings around your shirt collars? What about the annoying discoloration along the edges of your cuffs? Give them the boot by scrubbing the material with a paste made from 2 parts white vinegar to 3 parts baking soda. Let the paste set for half an hour before washing. This approach also works to remove light mildew stains from clothing.

Pretreat perspiration stains Want to see those sweat marks disappear from shirts and other garments? Just pour a bit of vinegar directly onto the stain, and rub it into the fabric before placing the item in the wash. You can also remove deodorant stains from your washable shirts and blouses by gently rubbing the spot with undiluted vinegar before laundering.

Make pen ink disappear Did someone in your house come home with a leaky pen in his pocket? Treat the stain by first wetting it with some white vinegar, then rub in a paste of 2 parts vinegar to 3 parts cornstarch. Let the paste thoroughly dry before washing the item.

Kids' Stuff Making **tie-dyed clothing** is tons of fun for kids of all ages. Start with a few **white T-shirts**, then use as many colors as allowed by your local supermarket's selection of unsweetened Kool-Aid powder mixes. Dissolve each package of **Kool-Aid** into 1 ounce **vinegar** in its own bowl or container. Use **rubber bands** to twist your shirts into unusual shapes, then dip them into the bowls (snap on a pair of rubber gloves beforehand). After drying, set your colors by placing a pillowcase or thin dish towel over each shirt and ironing it with a **medium-hot iron**. Wait at least 24 hours, then wash each shirt separately.

Soak out bloodstains Whether you nick yourself while shaving, or receive an unexpected scratch, it's important to treat the stains on your clothing as soon as possible; bloodstains are relatively easy to remove before they set but can be nearly impossible to wash out after 24 hours. If you can get to the stain before it sets, treat it by pouring full-strength white vinegar on the spot. Let it soak in for 5-10 minutes, then blot well with a cloth or towel. Repeat if necessary, then wash immediately.

* VINEGAR IN THE GREAT OUTDOORS

Use as insect repellent Planning a camping trip? Here's an old army trick to keep away the ticks and mosquitoes: Approximately three days before you leave, start taking 1 tablespoon apple cider vinegar three times a day. Continue using the vinegar throughout your trek, and you just might return home without a bite. Another time-honored approach to keep gnats and mosquitoes at bay is to moisten a cloth or cotton ball with white vinegar and rub it over your exposed skin.

Maintain fresh water when hiking Keep your water supply fresh and clean tasting when hiking or camping by adding a few drops of apple cider vinegar to your canteen or water bottle. It's also a good idea to use a half-vinegar, half-water rinse to clean out your water container at the end of each trip to kill bacteria and remove residue.

Clean outdoor furniture and decks If you live in a hot, humid climate, you're probably no stranger to seeing mildew on your wooden decks and patio furniture. But before you reach for the bleach, try these milder vinegar-based solutions:

- Keep some full-strength white vinegar in a recycled spray bottle and use it wherever you see any mildew growth. The stain will wipe right off most surfaces, and the vinegar will keep it from coming back for a while.
- Remove mildew from wood decks and wood patio furniture by sponging them off with a solution of 1 cup ammonia, 1/2 cup white vinegar, and 1/4 cup baking soda mixed in 1 gallon (3.7 liters) water. Keep an old toothbrush on hand to work the solution into corners and other tight spaces.
- To deodorize and inhibit mildew growth on outdoor plastic mesh furniture and patio umbrellas, mix 2 cups white vinegar and 2 tablespoons liquid dish soap in a bucket of hot water. Use a soft brush to work it into the grooves of the plastic as well as for scrubbing seat pads and umbrella fabric. Rinse with cold water; then dry in the sun.

Make a trap to lure flying insects Who wants to play host to a bunch of gnats, flies mosquitoes, or other six-legged pests when you're trying to have a cookout in your yard? Keep the flying gate-crashers at bay by giving them their own VIP section. Place a bowl filled with apple cider vinegar near some food, but away from you and guests. By the evening's end, most of your uninvited guests will be floating inside the bowl.

Give ants the boot Serve the ants on your premises with an eviction notice. Pour equal parts water and white vinegar into a spray bottle. Then spray it on anthills and around areas where you see the insects. Ants hate the smell of vinegar. It won't take long for them to move on to better-smelling quarters. Also keep the spray bottle handy for outdoor trips or to keep ants away from picnic or children's play areas. If you have lots of anthills around your property, try pouring full-strength vinegar over them to hasten the bugs' departure.

Clean off bird droppings Have the birds been using your patio or driveway for target practice again? Make those messy droppings disappear in no time by spraying them with full-strength apple cider vinegar. Or pour the vinegar onto a rag and wipe them off.

✱ VINEGAR IN THE GARDEN

Test soil acidity or alkalinity To do a quick test for excess alkalinity in the soil in your yard, place a handful of earth in a container and then pour in 1/2 cup white vinegar. If the soil fizzes or bubbles, it's definitely alkaline. Similarly, to see if your soil has a high acidity, mix the earth with 1/2 cup water and 1/2 cup baking soda. This time, fizzing would indicate acid in the soil. To find the exact pH level of your soil, have it tested or pick up a simple, do-it-yourself kit or meter.

Clean a hummingbird feeder Hummingbirds are innately discriminating creatures, so don't expect to see them flocking around a dirty, sticky, or crusted-over sugar-water feeder. Regularly clean your feeders by thoroughly washing them in equal parts apple cider vinegar and hot water. Rinse well with cold water after washing, and air-dry them outdoors in full sunlight before refilling them with food.

Speed germination of flower seeds You can get woody seeds, such as moonflower, passionflower, morning glory, and gourds, off to a healthier start by scarifying them—that is, lightly rubbing them between a couple of sheets of fine sandpaper—and soaking them overnight in a solution of 1/2 cup apple cider vinegar and 1 pint (half liter) warm water. Next morning, remove the seeds from the solution, rinse them off, and plant them. You can also use the solution (minus the sandpaper treatment) to start many herb and vegetable seeds.

Keep cut flowers fresh Everyone likes to keep cut flowers around as long as possible, and there are several good methods. One way is to mix 2 tablespoons apple cider vinegar and 2 tablespoons sugar with the vase water before adding the flowers. Be sure to change the water (with more vinegar and sugar, of course) every few days to enhance your flowers' longevity.

Tip A Myth About Vinegar

It's a rural legend that you can substantially lower your soil's pH (which is the same as raising its acidity), by simply pouring a vinegar-water solution around your yard. In fact, it takes a lot of hard work to lower the pH of high-alkaline soil. You can, however, use vinegar around the garden to help existing plants (see the tips below for treating plant diseases and encouraging blooms on azaleas and gardenias). But even that takes diligence—and repeated applications. Also, vinegar loses most of its potency after a rainfall. So you'll need to reapply any treatments after those surprise downpours.

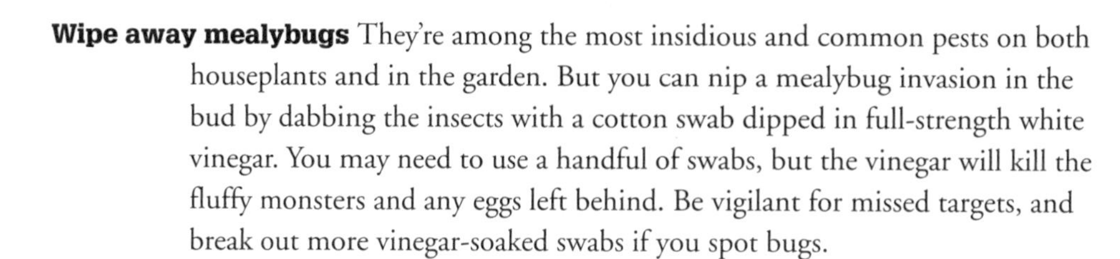

Wipe away mealybugs They're among the most insidious and common pests on both houseplants and in the garden. But you can nip a mealybug invasion in the bud by dabbing the insects with a cotton swab dipped in full-strength white vinegar. You may need to use a handful of swabs, but the vinegar will kill the fluffy monsters and any eggs left behind. Be vigilant for missed targets, and break out more vinegar-soaked swabs if you spot bugs.

Eliminate insects around the garden If the bugs are feasting on the fruits and vegetables in your garden, give them the boot with this simple, nonpoisonous trap. Fill a 2-liter soda bottle with 1 cup apple cider vinegar and 1 cup sugar. Next, slice up a banana peel into small pieces, put them in the bottle, add 1 cup cold water, and shake it up. Tie a piece of string around the neck of the bottle and hang it from a low tree branch, or place it on the ground, to trap and kill the six-legged freeloaders. Replace used traps with new ones as needed.

Encourage blooms on azaleas and gardenias A little bit of acid goes a long way toward bringing out the blooms on your azalea and gardenia bushes—especially if you have hard water. Both bushes do best in acidic soils (with pH levels between 4 and 5.5). To keep them healthy and to produce more flowers, water them every week or so with 3 tablespoons white vinegar mixed in 1 gallon (3.7 liters) water. Don't apply the solution while the bush is in bloom, however; it may shorten the life of the flowers or harm the plant.

Stop yellow leaves on plants The sudden appearance of yellow leaves on plants accustomed to acidic soils—such as azaleas, hydrangeas, and gardenias—could signal a drop in the plant's iron intake or a shift in the ground's pH above a comfortable 5.0 level. Either problem can be resolved by watering the soil around the afflicted plants once a week for three weeks with 1 cup of a solution made by mixing 2 tablespoons apple cider vinegar in 1 quart (1 liter) water.

Treat rust and other plant diseases You can use vinegar to treat a host of plant diseases, including rust, black spot, and powdery mildew. Mix 2 tablespoons apple cider vinegar in 2 quarts (2 liters) water, and pour some into a recycled spray bottle. Spray the solution on your affected plants in the morning or early

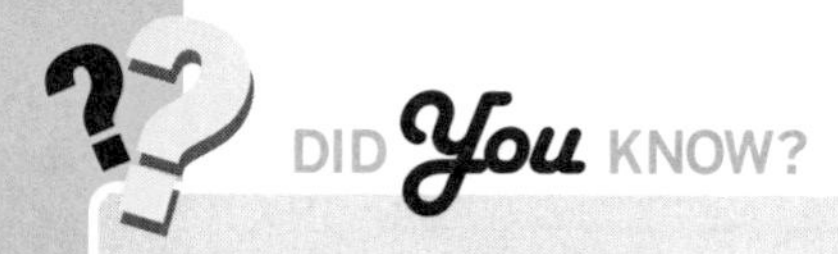

Looking for a nontoxic alternative to commercial weed killers? Vinegar is the way to go. In field and greenhouse studies, researchers at the Agricultural Research Service in Beltsville, Maryland, proved vinegar effective at killing five common weeds—including the all-too-common Canada thistle—within their first two weeks aboveground. The vinegar was hand-sprayed in concentrations varying between 5 and 10 percent. But that's old news to seasoned gardeners who've been using undiluted apple cider vinegar for ages to kill everything from poison ivy to crabgrass (and, regrettably, the occasional ornamental plant that grew too close to the target).

evening (when temperatures are relatively cool and there's no direct light on the plant) until the condition is cured.

Clean your lawn mower blades Grass, especially when it's damp, has a tendency to accumulate on your lawn mower blades after you cut the lawn—sometimes with grubs or other insects hiding inside. Before you park your mower back in the garage or toolshed, wipe down the blades with a cloth dampened with undiluted white vinegar. It will clean off leftover grass on the blades, as well as any pests that may have been planning to hang out awhile.

Keep out four-legged creatures Some animals—including cats, deer, dogs, rabbits, and raccoons—can't stand the scent of vinegar even after it has dried. You can keep these unauthorized visitors out of your garden by soaking several recycled rags in white vinegar, and placing them on stakes around your veggies. Resoak the rags about every 7-10 days.

SCIENCE FAIR

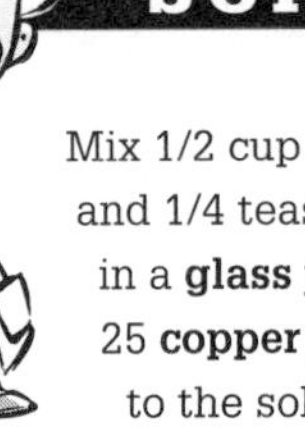

Mix 1/2 cup **vinegar** and 1/4 teaspoon **salt** in a **glass jar**. Add 25 **copper pennies** to the solution, and let them sit for five minutes. While you're waiting, take a large **iron nail** and clean it off with some baking soda applied to a damp sponge. Rinse off the nail, and place it into the solution. After 15 minutes, the **nail will be coated with copper**, while the pennies will shine like new. This is a result of the acetic acid in the vinegar combining with the copper on the pennies to form copper acetate, which then accumulates on the nail.

Exterminate dandelions and unwanted grass Are dandelions sprouting up in the cracks of your driveway or along the fringes of your patio? Make them disappear for good by spraying them with full-strength white or apple cider vinegar. Early in the season, give each plant a single spritz of vinegar in its midsection, or in the middle of the flower before the plants go to seed. Aim another shot near the stem at ground level so the vinegar can soak down to the roots. Keep an eye on the weather, though; if it rains the next day, you'll need to give the weeds another spraying.

✱ VINEGAR PET CARE

Keep the kitties away If you want to keep Snowball and Fluffy out of the kids' playroom, or discourage them from using your favorite easy chair as a scratching

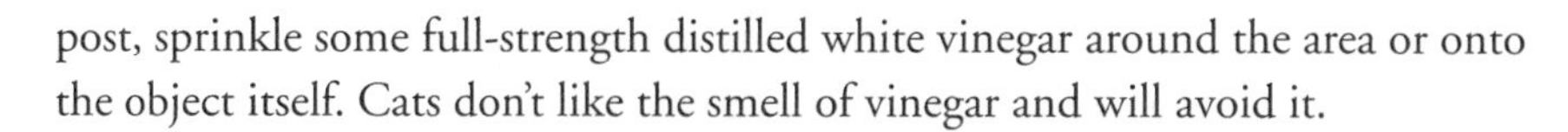

post, sprinkle some full-strength distilled white vinegar around the area or onto the object itself. Cats don't like the smell of vinegar and will avoid it.

Unmark your pet's spots When housebreaking a puppy or kitten, it'll often wet previously soiled spots. After cleaning up the mess, it's essential to remove the scent from your floor, carpeting, or sofa. And nothing does that better than vinegar:

- On a floor, blot up as much of the stain as possible. Then mop with equal parts white vinegar and warm water. (On a wood or vinyl floor, test a few drops of vinegar in an inconspicuous area to make sure it won't harm the finish.) Dry with a cloth or paper towel.
- For carpets, rugs, and upholstery, thoroughly blot the area with a towel or some rags. Then pour a bit of undiluted vinegar over the spot. Blot it up with a towel, then reapply the vinegar—let it air-dry. Once the vinegar dries, the spot should be completely deodorized

Add to pet's drinking water Adding a teaspoon of apple cider vinegar to your dog or cat's drinking water provides needed nutrients to its diet, gives it a shinier, healthier-looking coat, and acts as a natural deterrent to fleas and ticks.

Directly protect against fleas and ticks To give your dog effective flea and tick protection, fill a spray bottle with equal parts water and vinegar and apply it directly to the dog's coat and rub it in well. You may have more trouble doing this with cats, because they really hate the smell of the stuff.

Clean your pet's ears If you've noticed that Rover has been scratching around his ears a lot more than usual lately, a bit of vinegar could bring him some big relief. Swabbing your pet's ears with a cotton ball or soft cloth dabbed in solution of 2 parts vinegar and 1 part water will keep them clean and help deter ear mites and bacteria. It also soothes minor itches from mosquito bites and such. *Warning*: Do not apply vinegar to open lacerations. If you see a cut in your pet's ears, seek veterinary treatment.

Remove skunk odor If Fido has an unpleasant encounter with an ornery skunk, here are some ways to help him get rid of the smell:

- Bathe your pet in a mixture of 1/2 cup white vinegar, 1/4 cup baking soda, and 1 teaspoon liquid soap in 1 quart (1 liter) 3% hydrogen peroxide. Work the solution deep into his coat, give it a few minutes to soak in, then rinse him thoroughly with clean water.
- Bathe your pet in equal parts water and vinegar (preferably outdoors in a large washtub). Then repeat the procedure using 1 part vinegar to 2 parts water, followed by a good rinsing.

- If you happen to have an unscheduled meeting with skunk, use undiluted vinegar to get the smell out your own clothes. Let the affected clothing soak in the vinegar overnight.

* VINEGAR FOR THE DO-IT-YOURSELFER

Wash concrete off your skin Even though you wear rubber gloves when working with concrete, some of the stuff inevitably splashes on your skin. Prolonged contact with wet concrete can cause your skin to crack, and may even lead to eczema. Use undiluted white vinegar to wash dried concrete or mortar off your skin, then wash with warm, soapy water.

Remove paint fumes Place a couple of shallow dishes filled with undiluted white vinegar around a freshly painted room to quickly get rid of the strong paint smell.

Degrease grates, fans, and air-conditioner grilles Even in the cleanest of homes, air-conditioner grilles, heating grates, and fan blades eventually develop a layer of dust and grease. To clean them, wipe them with full-strength white vinegar. Use an old toothbrush to work the vinegar into the tight spaces on air-conditioner grilles and exhaust fans.

Disinfect air-conditioner and humidifier filters An air-conditioner or humidifier filter can quickly become inundated with dust, soot, pet dander, and even potentially harmful bacteria. Every 10 days or so, clean your filter in equal parts white vinegar and warm water. Let the filter soak in the solution for an hour, then simply squeeze it dry before using. If your filters are particularly dirty, let them soak overnight.

Keep the paint on your cement floors Painted cement floors have a tendency to peel after a while. But you can keep the paint stuck to the cement longer by giving the floor an initial coat of white vinegar before you paint it. Wait until the

You just came across an old, unopened bottle of vinegar, and you wonder: Is it still any good? The answer is an unqualified yes. In fact, vinegar has a practically limitless shelf life. Its acid content makes it self-preserving and even negates the need for refrigeration (although many people mistakenly believe in refrigerating their open bottles). You won't see any changes in white vinegar over time, but some other types may change slightly in color or develop a hazy appearance or a bit of sediment. However, these are strictly cosmetic changes; the vinegar itself will be virtually unchanged.

vinegar has dried, then begin painting. This same technique will also help keep paint affixed to galvanized metal._

Get rid of rust If you want to clean up those rusted old tools you recently unearthed in your basement or picked up at a tag sale, soak them in full-strength white vinegar for several days. The same treatment is equally effective at removing the rust from corroded nuts and bolts. And you can pour vinegar on rusted hinges and screws to loosen them up for removal.

Peel off wallpaper Removing old wallpaper can be messy, but you can make it peel off easily by soaking it with a vinegar solution. Spray equal parts white vinegar and water on the wallpaper until it is saturated and wait a few minutes. Then zip the stuff off the wall with a wallpaper scraper. If it is stubborn, try carefully scoring the wallpaper with the scraper before you spritz.

Slow hardening of plaster Want to keep your plaster pliable a bit longer to get it all smoothed out? Just add a couple of tablespoons of white vinegar to your plaster mix. It will slow down the hardening process to give you the extra time you need.

Revive your paintbrushes To remove dried-on paint from a synthetic-bristle paintbrush, soak it in full-strength white vinegar until the paint dissolves and the bristles are soft and pliable, then wash in hot, soapy water. Does a paintbrush seem beyond hope? Before you toss it, try boiling it in 1-2 cups vinegar for 10 minutes, followed by a thorough washing in soapy water.

Vodka

Make your own vanilla extract Here's an unusual homemade treat you can use to spice up a gift basket, and it takes only minutes to make. Get one real dried vanilla bean (available at specialty food stores) and slice it open from top to bottom. Place it in a glass jar and cover it with 3/4 cup vodka. Seal the jar, and let it rest

Essential to James Bond's martini—and so intrinsic to Russian culture that its name derives from the Russian word for water *(voda)*—vodka was first made in the 1400s as an antiseptic and painkiller before it was drunk as a beverage. But what exactly is it? Classically, vodka starts as a soupy mixture of ground wheat or rye that's fermented (sugars in the grain are converted into alcohol by yeast), then distilled (heated until the alcohol evaporates and then condenses). Flavorings such as citrus were originally added to mask the taste of impurities, but are used today for enhancement and brand identification.

in a kitchen cabinet for 4-6 months, shaking it occasionally. Filter your homemade vanilla extract through an unbleached coffee filter or cheesecloth into a decorative bottle and watch the face of your favorite cook light up with pleasure!

Clean glass and jewelry In a pinch, a few drops of vodka will clean any kind of glass or jewelry with crystalline gemstones. So although people might look at you askance, you could dip a napkin into your vodka on the rocks to wipe away the grime on your eyeglasses or dunk your diamond ring for a few minutes to get it sparkling again. But don't try this with contact lenses! Also avoid getting alcohol on any gemstone that's not a crystal. Only diamonds, emeralds, and the like will benefit from a vodka bath.

Use as a hygienic soak Vodka is an alcohol, and like any alcohol, it kills germs. If you don't have ordinary rubbing alcohol on hand, use vodka instead. You can use it to soak razor blades you plan to reuse, as well as to clean hairbrushes, toothbrushes, and pet brushes, or on anything else that might spread germs from person to person or animal to animal.

Keep cut flowers fresh The secret to keeping cut flowers looking good as long as possible is to minimize the growth of bacteria in the water and to provide nourishment to replace what the flower would have gotten had it not been cut. Add a few drops of vodka (or any clear spirit) to the vase water for antibacterial action along with 1 teaspoon sugar. Change the water every other day, refreshing the vodka and sugar each time.

Kill weeds in the yard For a quick and easy weed killer, mix 1 ounce (30 milliliters) vodka, a few drops liquid dish soap, and 2 cups water in a spray bottle. Spray it on the weed leaves until the mixture runs off. Apply it at midday on a sunny day to weeds growing in direct sunlight, because the alcohol breaks down the waxy cuticle covering on leaves, leaving them susceptible to dehydration in sunlight. It won't work in shade.

W

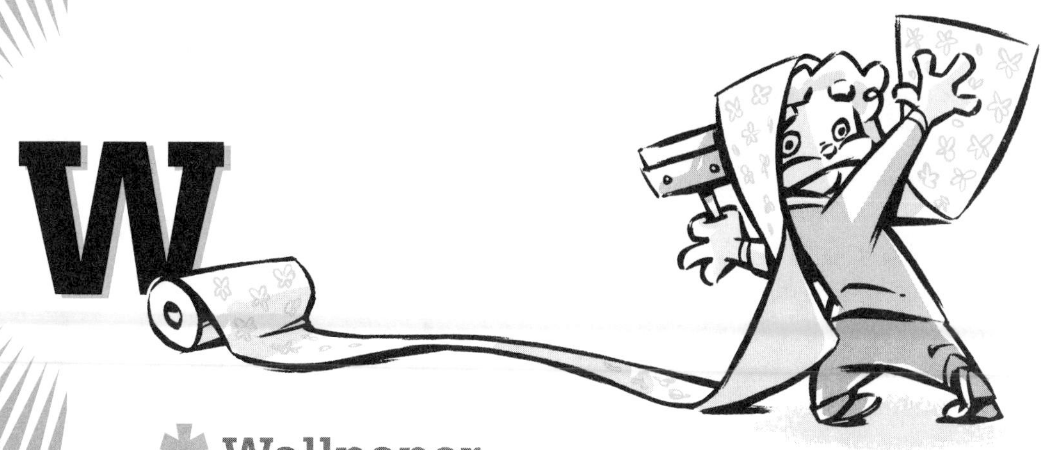

* Wallpaper

Line your drawers Wallpaper remnants can be a great substitute for shelf liner paper when used to line dresser drawers or closet shelves—especially designs with raised patterns or fabrics, which may add a bit of friction to prevent things from moving around. Cut the wallpaper into strips sized in both length and width to accommodate the space.

Restore a folding screen If you have an old folding screen that's become torn or stained over the years, give it a new, younger look by covering it with leftover wallpaper. Use masking tape to hold the strips at top and bottom if you don't want to glue it on top of the original material.

Protect schoolbooks If your child goes through book covers on textbooks on a semiregular basis, get your hands on some old rolls of wallpaper. Book covers made of wallpaper are typically more rugged than even the traditional brown paper bag sleeves; they can hold their own against pens and pencils, and are much better at handling the elements, especially rain and snow.

Make a jigsaw puzzle What to do with your leftover wallpaper? Why not use a piece to make a jigsaw puzzle? Simply cut off a medium-sized rectangular piece and glue it onto a piece of thin cardboard. Once it's dried, cut it up into a bunch of curvy and angular shapes. It'll give you, or the kids, something to do on a rainy or snowbound day.

* Wax Paper

Fail-safe cake decorating You made a special birthday cake, and now comes the moment of truth: Can you pipe out the lettering in frosting on the first try? Not many of us can, so try this trick to make it easier. Cut a piece of wax paper the same size as your cake, using the cake pan as a guide. Then pipe the name and the message onto the paper and freeze it. After just half an hour it should be easy to handle. Loosen the frosting and slide it off onto the cake using a spatula. Everyone will think you're a cake-decorating professional!

Funnel spices into jars Filling narrow-mouthed spice jars can make a big mess on your kitchen counter. Roll a piece of wax paper into a funnel shape and pour spices into your jars without spilling a single mustard seed. In a pinch, you can even funnel liquids by using a couple of layers of wax paper offset so the seams in the layers don't line up.

Speed kitchen cleanup Wax paper can help keep all kinds of kitchen surfaces clean.

- Line vegetable and meat bins with a layer of wax paper. When it needs replacement, just wad it up and throw it in the trash or, if it's not stained with meat juices, the compost pile.
- If your kitchen cabinets don't extend to the ceiling, a layer of wax paper on top will catch dust and grease particles. Every month or two, just fold it up, discard it, and put a fresh layer down.
- If you're worried about meat juices getting into the pores of your cutting board, cover it with three layers of wax paper before slicing raw meat and throw the paper out immediately. It beats scrubbing the cutting board with bleach!

Tame the waffle-eating waffle iron Having trouble extricating waffles from your waffle iron? Nonstick surfaces don't last forever. You can't fix the problem permanently, but if you just want to get it to work today, put a layer of wax paper in between the plates of your waffle iron for a few minutes while it heats up. The wax will be transferred to the plates, temporarily helping waffles pop out again.

Uncork bottles with ease If you keep a bottle of cooking wine in your kitchen, you probably uncork it and recork it many times before using it up. Instead of struggling with the cork each time, wrap some wax paper around the cork before reinserting it. It'll be easier to remove the next time, and the paper helps keep little bits of cork from getting into the wine.

Wax paper is an example of how important packaging can be to a product's success. Before 1927, wax paper was sold in pre-cut sheets in envelopes, but housewives and deli owners alike were frustrated by the tendency of the sheets to stick together on warm summer days. Then an enterprising inventor, Nicholas Marcalus, put wax paper on a roll in a box with a built-in cutter, and the product as we know it today came to be. In fact, Reynolds wax paper is called Cut-Rite after this packaging innovation, which was awarded a patent and is used for a multitude of products today.

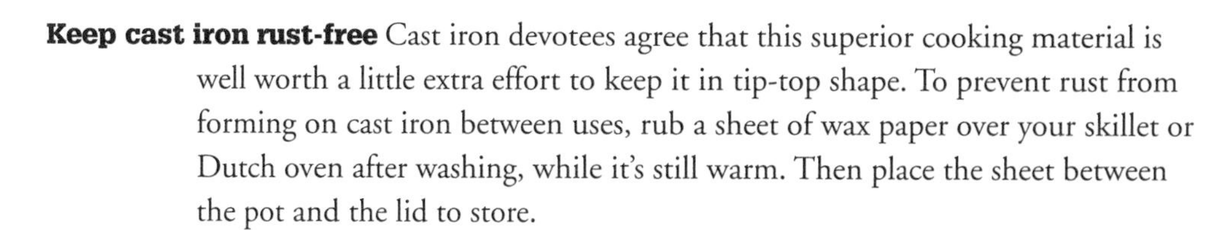

Keep cast iron rust-free Cast iron devotees agree that this superior cooking material is well worth a little extra effort to keep it in tip-top shape. To prevent rust from forming on cast iron between uses, rub a sheet of wax paper over your skillet or Dutch oven after washing, while it's still warm. Then place the sheet between the pot and the lid to store.

Keep candles from staining table linens Candles in colors that coordinate with your dining room linens make a lovely finishing touch to table settings—and it's helpful to store them all together—but if you store the candles with table linens, the candle color can rub off on the linens. To avoid this, wrap colorful candles in plain wax paper before storage. Avoid paper with holiday patterns, which can also stain linens.

Store delicate fabrics Treasured lace doilies and other linens handed down in your family can decay quickly if not stored with care. A sheet of wax paper between each fabric piece will help block extraneous light and prevent transfer of dyes without trapping moisture.

Stop water spotting Company's coming, and you want every room of the house to look its best. To keep bathroom fixtures temporarily spotless, rub them with a sheet of wax paper after cleaning them. The wax that transfers will deflect water droplets like magic—at least until the next cleaning.

Give your car antenna a smooth ride If you have a newer vehicle, your car antenna probably retracts each time you turn off the ignition, carrying grime with it that can eventually bring your antenna (and your reception) to a grinding halt. Every now and then, rub the antenna with a piece of wax paper to coat the shaft and help it repel dirt.

Make a snow slide go faster Everyone knows, the more slippery the slide, the more fun it is! Keep tots swooshing on their tushes by balling up a large piece of wax paper and rubbing it all over the slide surface.

Kids' Stuff What kid wouldn't like **homemade "stained glass"** art that takes only minutes to make? First, make **crayon shavings** using a **vegetable peeler**. Keep each color separate. Put a paper towel or bag on your counter. Place a sheet of **wax paper** on top, sprinkle it with crayon shavings, and cover with another layer of wax paper and paper towel. Press it for a few minutes using a **warm iron** and remove the paper towel layers. Cut your new see-through art into **sun-catching medallions** or **colorful bookmarks** using craft scissors to create a decorative edge.

Protect surfaces from glue Woodworkers know that there's enough glue in a wood joint if some squeezes out when they clamp the joint. They also know that excess glue will be a real pain to remove if it drips on the workbench or, worse, bonds the clamping blocks to the project. To prevent this, cover the bench with strips of wax paper and put pieces of wax paper between the clamping blocks and the project. The glue won't adhere to, or soak through, the wax.

Make educational place mats One way to make learning fun is with personalized place mats featuring math facts or other lessons your child is trying to memorize. Take several flash cards and sandwich them between layers of wax paper cut to place-mat size. Sandwich that between two layers of paper towels and press it all with a warm iron to "laminate" the flash cards in place. Remove the paper towels before use.

WD-40

WD-40 AROUND THE HOUSE

Treat your shoes Spray WD-40 on new leather shoes before you start wearing them regularly. It will help prevent blisters by softening the leather and making the shoes more comfortable. Keep the shoes waterproof and shiny by spraying them periodically with WD-40 and buffing gently with a soft cloth. To give the old "soft shoo" to squeaky shoes, spray some WD-40 at the spot where the sole and heel join and the squeaks will cease.

Separate stuck glassware What can you do when you reach for a drinking glass and get two locked together, one stuck tightly inside the other? You don't want to risk breaking one or both by trying to pull them apart. Stuck glasses will separate with ease if you squirt some WD-40 on them, wait a few seconds for it to work its way between the glasses, and then gently pull the glasses apart. Remember to wash the glasses thoroughly before you use them.

Free stuck Lego blocks When Junior's construction project hits a snag because some of the plastic blocks are stuck together, let WD-40 help get them unstuck. Spray a little on the blocks where they are locked together, then wiggle them gently and pull them apart. The lubricant in WD-40 will penetrate into the fine seam where the blocks are joined.

Tone down polyurethane shine A new coat of polyurethane can sometimes make a wood floor look a little *too* shiny. To tone down the shine and cut the glare, spray some WD-40 onto a soft cloth and wipe up the floor with it.

Remove strong glue You didn't wear protective gloves when using that super-strong glue and now some of it is super-stuck to your fingers! Don't panic. Just reach for the WD-40, spray some directly on the sticky fingers, and rub your hands together until your fingers are no longer sticky. Use WD-40 to remove the glue from other unwanted surfaces as well.

TAKE CARE

- Do not spray WD-40 near an open flame or other heat source, or near electrical currents or battery terminals. Always disconnect appliances before spraying.
- Do not place a WD-40 can in direct sunlight or on hot surfaces. Never store it in temperatures above 120°F (50°C) or puncture the pressurized can.
- Use WD-40 in well-ventilated areas. Never swallow or inhale it (if swallowed, call a physician immediately).

Get off that stuck ring When pulling and tugging can't get that ring off your finger, reach for the WD-40. A short burst of WD-40 will get the ring to slide right off. Remember to wash your hands after spraying them with WD-40.

Free stuck fingers Use WD-40 to free Junior's finger when he gets it stuck in a bottle. Just spray it on the finger, let it seep in, and pull the finger out. Be sure to wash Junior's hand and the bottle afterward.

Loosen zippers Stubborn zippers on jackets, pants, backpacks, and sleeping bags will become compliant again after you spray them with WD-40. Just spray it on and pull the zipper up and down a few times to distribute the lubricant evenly over all the teeth. If you want to avoid getting the WD-40 on the fabric, spray it on a plastic lid; then pick it up and apply it with an artist's brush.

Exterminate roaches and repel insects Don't let cockroaches, insects, or spiders get the upper hand in your home.

- Keep a can of WD-40 handy, and when you see a roach, spray a small amount directly on it for an instant kill.
- To keep insects and spiders out of your home, spray WD-40 on windowsills and frames, screens, and door frames. Be careful not to inhale the fumes when you spray and do not do this at all if you have babies or small children at home.

Keep puppies from chewing Your new puppy is adorable, but will he *ever* stop chewing up the house? To keep puppies from chewing on telephone and television-cable lines, spray WD-40 on the lines. The pups hate the smell.

Clean and lubricate guitar strings To clean, lubricate, and prevent corrosion on guitar strings, apply a small amount of WD-40 after each playing. Spray the WD-40 on a rag and wipe the rag over the strings rather than spraying directly on the strings—you don't want WD-40 to build up on the guitar neck or body.

Keep wooden tool handles splinter-free No tools can last forever, but you can prolong the life of your wood-handled tools by preventing splintering. To keep wooden handles from splintering, rub a generous amount of WD-40 into the wood. It will shield the wood from moisture and other corrosive elements and keep it smooth and splinter-free for the life of the tool.

Unstick wobbly shopping-cart wheels Attention supermarket shoppers: Keep a can of WD-40 handy whenever you go food shopping. Then when you get stuck with a sticky, wobbly-wheeled shopping cart, you can spray the wheels to reduce friction and wobbling. Less wobbling means faster shopping.

Remove chewing gum from hair It's one of an adult's worst nightmares: chewing gum tangled in a child's hair. You don't have to panic or run for the scissors. Simply spray the gummed-up hair with WD-40, and the gum will comb out with ease. Make sure you are in a well-ventilated area when you spray and take care to avoid contact with the child's eyes.

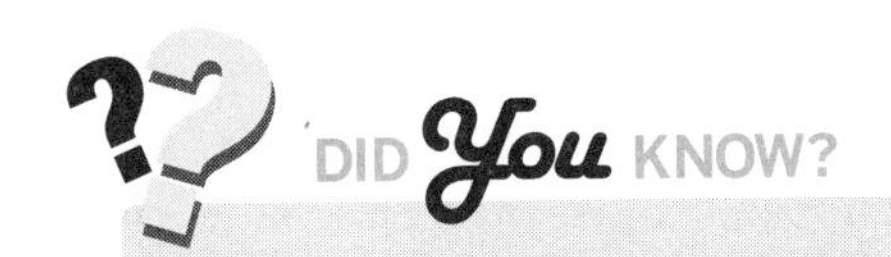

- In 1953 Norm Larsen founded the Rocket Chemical Company in San Diego and, with two employees, set out to develop a rust-preventing solvent and degreaser for the aerospace industry. On the fortieth try, they succeeded in creating a "water displacement" compound. The name WD-40 stands for "Water Displacement—40th Try."
- In 1958, a few years after WD-40's first industrial use, the company put it in aerosol cans and sold it for home use—inspired by employees who snuck cans out of the plant to use at home.
- In 1962, when U.S. astronaut John Glenn circled the Earth in *Friendship VII*, the space capsule was coated with WD-40 and so was the Atlas missile used to boost it into space.
- In 1969 Rocket Chemical renamed itself the WD-40 Company after the product.

Break in a new baseball glove Use WD-40 instead of neat's-foot oil to break in a new baseball glove. Spray the glove with WD-40, put a baseball in the palm, and fold it sideways. Take a rubber band or belt and tie it around the folded glove. The WD-40 will help soften the leather and help it form around the baseball. Keep the glove tied up overnight, and then wear it for a while so it will begin to fit the shape of your hand.

✱ WD-40 FOR CLEANING THINGS

Remove tough scuff marks Those tough black scuff marks on your kitchen floor won't be so tough anymore if you spray them with WD-40. Use WD-40 to help

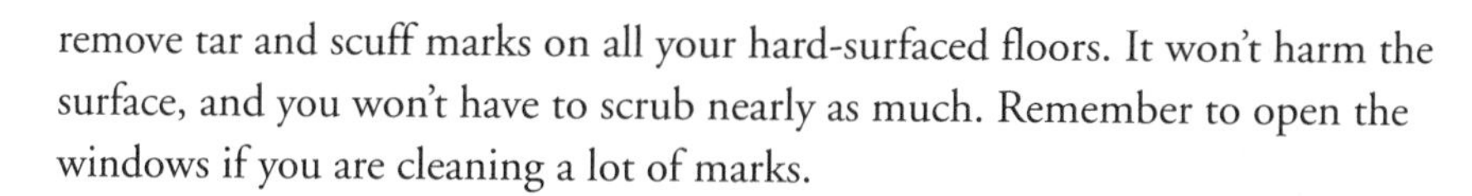

remove tar and scuff marks on all your hard-surfaced floors. It won't harm the surface, and you won't have to scrub nearly as much. Remember to open the windows if you are cleaning a lot of marks.

Clean dried glue Clean dried glue from virtually any hard surface with ease: Simply spray WD-40 onto the spot, wait at least 30 seconds, and wipe clean with a damp cloth.

Degrease your hands When you're done working on the car and your hands are greasy and blackened with grime, use WD-40 to help get them clean. Spray a small amount of WD-40 into your hands and rub them together for a few seconds, then wipe with a paper towel and wash with soap and water. The grease and grime will wash right off.

Remove decals You don't need a chisel or even a razor blade to remove old decals, bumper stickers, or cellophane tape. Just spray them with WD-40, wait about 30 seconds, and wipe them away.

Remove stickers from glass What were the manufacturers thinking when they put that sticker on the glass? Don't they know how hard it is to get off? When soap and water doesn't work and you don't want to ruin a fingernail or risk scratching delicate glass with a blade, try a little WD-40. Spray it on the sticker and glass, wait a few minutes, and then use a no-scratch spatula or acrylic scraper to scrape the sticker off. The solvents in WD-40 cause the adhesive to lose its stickiness.

Wipe away tea stains To remove tea stains from countertops, spray a little WD-40 on a sponge or damp cloth and wipe the stain away.

Clean carpet stains Don't let ink or other stains ruin your fine carpet. Spray the stain with WD-40, wait a minute or two, and then use your regular carpet cleaner or gently cleanse with a sponge and warm, soapy water. Continue until the stain is completely gone.

Get tomato stains off clothes That homegrown tomato looked so inviting you couldn't resist. Now your shirt or blouse has a big, hard-to-remove tomato stain! To

A million cans of WD-40 are produced each week in the United States alone. The secret recipe, which has had the same basic ingredients for more than 50 years, is known only by a handful of people within the WD-40 Company. A lone "brew master" mixes the product at corporate headquarters in San Diego. Although the company refuses to reveal the ingredients, it will say, "WD-40 does *not* contain silicone, kerosene, water, wax, graphite, chlorofluorocarbons (CFCs), or any known cancer-causing agents." That's good enough for most Americans: WD-40 can be found in 4 out of 5 American homes.

remove stains from fresh tomatoes or tomato sauce, spray some WD-40 directly on the spot, wait a couple of minutes, and wash as usual.

Clean toilet bowls You don't need a bald genie or a specialized product to clean ugly gunk and lime stains from your toilet bowl. Use WD-40 instead: Spray it into the bowl for a couple of seconds and swish with a nylon toilet brush. The solvents in the WD-40 will help dissolve the gunk and lime.

Clean your fridge When soap and water can't get rid of old bits of food stuck in and around your refrigerator, it's time to reach for the WD-40. After clearing all foodstuffs from the areas to be treated, spray a small amount of WD-40 on each resistant spot. Then wipe them away with a rag or sponge. Make sure you wash off all the WD-40 before returning food to the fridge.

Condition leather furniture Keep your favorite leather recliner and other leather furniture in tip-top shape by softening and preserving it with WD-40. Just spray it on and buff with a soft cloth. The combination of ingredients in WD-40 will clean, penetrate, lubricate, and protect the leather.

Pretreat blood and other stains Oh no! Your kid fell down and cut himself while playing, and there's blood all over his brand-new shirt. After you tend to the wound, give some first aid to the shirt too. Pretreat the bloodstains with WD-40. Spray some directly on the stains, wait a couple of minutes, and then launder as usual. The WD-40 will help lift the stain so that it will come out easily in the wash. Try to get to the stain while it is still fresh, because once it sets, it will be harder to get rid of. Use WD-40 to pretreat other stubborn stains on clothing, such as lipstick, dirt, grease, and ink stains.

Clean chalkboards When it comes to cleaning and restoring a chalkboard, WD-40 is the teacher's pet. Just spray it on and wipe with a clean cloth. The chalkboard will look as clean and fresh as it did on the first day of school.

Remove marker and crayon marks Did the kids use your wall as if it was a big coloring book? Not to worry! Simply spray some WD-40 onto the marks and wipe with a clean rag. WD-40 will not damage the paint or most wallpaper (test fabric or other fancy wall coverings first). It will also remove marker and crayon marks from furniture and appliances.

* WD-40 IN THE YARD

Rejuvenate the barbecue grill To make a worn old barbecue grill look like new again, spray it liberally with WD-40, wait a few seconds, and scrub with a wire brush. Remember to use WD-40 only on a grill that is not in use and has cooled off.

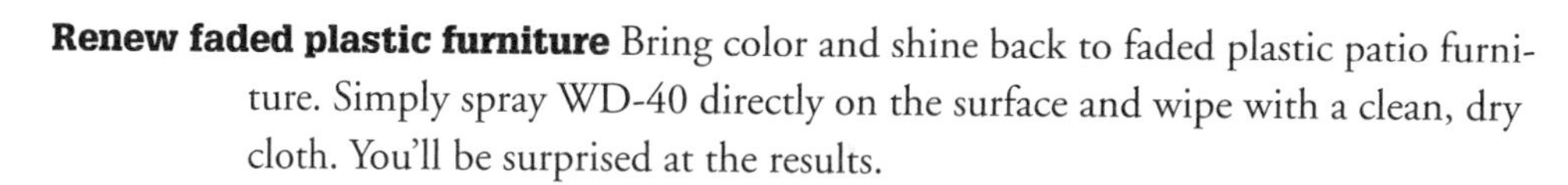

Renew faded plastic furniture Bring color and shine back to faded plastic patio furniture. Simply spray WD-40 directly on the surface and wipe with a clean, dry cloth. You'll be surprised at the results.

W

Prevent snow buildup on windows Does the weather forecast predict a big winter snowstorm? You can't stop the snow from falling, but you can prevent it from building up on your house's windows. Just spray WD-40 over the outside of your windows before the snow starts and the snow won't stick.

Keep shovel or chute snow-free Here is a simple tip to make shoveling snow quicker and less strenuous by keeping the snow from sticking to your shovel and weighing it down. Spray a thin layer of WD-40 on the shovel blade, and the snow will slide right off. If you have a snow thrower, spray WD-40 on the inside of the chute so snow won't stick and clog the chute.

Protect a bird feeder To keep squirrels from taking over a bird feeder, spray a generous amount of WD-40 on top of the feeder. The pesky squirrels will slide right off.

Remove cat's paw marks Your cat may seem like a member of the family most of the time, but that isn't what you are thinking about when you have to clean a slew of paw marks off patio furniture or the hood of your car. To remove the paw marks, spray some WD-40 on them and wipe with a clean rag.

Keep animals from flowerbeds Animals just love to play in your garden, digging up your favorite plants you worked so hard to grow. What animals *don't* love is the smell of WD-40. To keep the animals out and your flowers looking beautiful all season, spray WD-40 evenly over the flowerbeds one or more times over the course of the season.

Tip **The Little Red Straw**

"I lost the red straw!" has been a common cry among countless users of WD-40 over the years. In response, the company introduced a notched cap, designed to hold the straw in place across the top of the can when not in use. Because the straw exceeds the width of the can by a substantial margin, the notch may be of little use to those with limited storage space. To save space, store the straw by bending it inside the lip of the can or simply tape it to the side as it was when first purchased. A snug rubber band will also work well.

Repel pigeons Are the pigeons using your balcony more than you are? If pigeons and their feathers and droppings are keeping you from enjoying the view from your balcony, spray the entire area, including railings and furniture, with WD-40. The pigeons can't stand the smell and they'll fly the coop.

Keep wasps from building nests Don't let yellow jackets and other wasps ruin your spring and summer fun. Their favorite place to build nests is under eaves. So next spring mist some WD-40 under all the eaves of your house. It will block the wasps from building their nests there.

Remove doggie-doo Uh-oh, now you've stepped in it! Few things in life are more unpleasant than cleaning doggie-doo from the bottom of a sneaker, but the task will be a lot easier if you have a can of WD-40 handy. Spray some on the affected sole and use an old toothbrush to clean the crevices. Rinse with cold water and the sneakers will be ready to hit the pavement again. Now, don't forget to watch where you step!

Kill thistle plants Don't let pesky prickly weeds like bull and Russian thistle ruin your yard or garden. Just spray some WD-40 on them and they'll wither and die.

* WD-40 IN THE GREAT OUTDOORS

Winterproof boots and shoes Waterproof your winter boots and shoes by giving them a coat of WD-40. It'll act as a barrier so water can't penetrate the material. Also use WD-40 to remove ugly salt stains from boots and shoes during the winter months. Just spray WD-40 onto the stains and wipe with a clean rag. Your boots and shoes will look almost as good as new.

Remove old wax from skis and snowboards To remove old wax and dirt from skis and snowboards, spray the base sparingly with WD-40 before scraping with an acrylic scraper. Use a brass brush to further clean the base and remove any oxidized base material.

Protect your boat from corrosion To protect your boat's outer finish from salt water and corrosion, spray WD-40 on the stern immediately after each use. The short time it takes will save you from having to replace parts, and it will keep your boat looking like it did on the day you bought it for a long time to come.

Remove barnacles on boats Removing barnacles from the bottom of a boat is a difficult and odious task but you can make it easier and less unpleasant with the help of some WD-40. Spray the area generously with WD-40, wait a few seconds, and then use a putty knife to scrape off the barnacles. Spray any remnants with WD-40 and scrape again. If necessary, use sandpaper to get rid of all of the remnants and corrosive glue still left by the barnacles.

Spray on fishing lures Salmon fishermen in the Pacific Northwest spray their lures with WD-40 because it attracts fish and disguises the human odor that can scare them off and keep them from biting. You can increase the catch on your next fishing trip by bringing a can of WD-40 along with you and spraying it on *your* lures or live bait before you cast. But first check local regulations to make sure the use of chemical-laced lures and bait is legal in your state.

Don't Overdo the WD-40

When you need to apply tiny amounts of WD-40 to a specific area, such as the electrical contacts on an electric guitar, an aerosol spray is overkill. Instead, store some WD-40 in a clean nail-varnish bottle (with cap brush) and brush on as needed.

Untangle fishing lines To loosen a tangled fishing line, spray it with WD-40 and use a pin to undo any small knots. Also use WD-40 to extend the life of curled (but not too old) fishing lines. Just take out the first 10 to 20 feet of line and spray it with WD-40 the night before each trip.

Clean and protect golf clubs Whether you're a duffer or a pro, you can protect and clean your clubs by spraying them with WD-40 after each use. Also use WD-40 to help loosen stuck-on spikes.

Remove burrs To remove burrs from a horse's mane or tail without tearing its hair out (or having to cut any of its hair off!) just spray on some WD-40. You'll be able to slide the burrs right out. This will work for dogs and cats, too.

Protect horses' hooves Winter horseback riding can be fun if you are warmly dressed but it can be downright painful to your horse if ice forms on the horseshoes. To keep ice from forming on horseshoes during cold winter rides, spray the bottom of the horse's hooves with WD-40 before you set out.

Keep flies off cows If flies are tormenting your cows, just spray some WD-40 on the cows. Flies hate the smell and they'll stay clear. Take care not to spray any WD-40 in the cows' eyes.

✱ WD-40 FOR YOUR HEALTH

Relieve arthritis symptoms For occasional joint pain or arthritis symptoms in the knees or other areas of the body, advocates swear by spraying WD-40 on the affected

WD-40 is one of the few products with its own fan club. The official WD-40 Fan Club has more than 63,000-members and is growing. Over the years members have contributed thousands of unique and sometimes strange uses for the product. But according to Gary Ridge, president and CEO of the WD-40 Company, the strangest use of all occurred in China. "In Hong Kong some time ago there was a python caught in the suspension of a public bus," he said. "They used WD-40 to get that little slippery guy out of there!"

area and massaging it in, saying it provides temporary relief and makes movement easier. For severe, persistent pain, consult a health care professional.

Clean your hearing aid To give your hearing aid a good cleaning, use a cotton swab dipped in WD-40. Do not use WD-40 to try to loosen up the volume control (it will loosen it too much).

Relieve bee-sting pain For fast relief of pain from a bee, wasp, or hornet sting, reach for the WD-40 can and spray it directly on the bite site. It will take the "ouch" right out.

Remove stuck prostheses If you wear a prosthetic device, you know how difficult it can be to remove at times, especially when no one is around to help. Next time you get stuck with a stuck prosthesis, spray some WD-40 at the junction where it attaches. The chemical solvents and lubricants in WD-40 will help make it easier to remove.

* WD-40 IN THE GARAGE

Keep dead bugs off car grille It's bad enough that your car grille and hood have to get splattered with bugs every time you drive down the interstate, but do they have to be so darn tough to scrape off? The answer is no. Just spray some WD-40 on the grille and hood before going for a drive and most of the critters will slide right off. The few bugs that are left will be easy to wipe off later without damaging your car's finish.

Clean and restore license plate To help restore a license plate that is beginning to rust, spray it with WD-40 and wipe with a clean rag. This will remove light surface rust and will also help prevent more rust from forming. It's an easy way to clean up lightly rusted plates and it won't leave a greasy feel.

Remove stuck spark plugs To save time replacing spark plugs, do it the NASCAR way. NASCAR mechanics spray WD-40 on stuck plugs so they can remove them quickly and easily. Perhaps that's one reason why WD-40 has been designated as NASCAR's "official multi-purpose problem-solver."

Coat a truck bed For easy removal of a truck-bed liner, spray the truck bed with WD-40 before you install the liner. When it comes time to remove it, the liner will slide right out.

Remove "paint rub" from another car You return to your parked car to find that while you were gone, another vehicle got a bit too close for comfort. Luckily there's no dent, but now your car has a blotch of "paint rub" from the other car on it. To remove paint-rub stains on your car and restore its original finish, spray the affected area with WD-40, wait a few seconds, and wipe with a clean rag.

Revive spark plugs Can't get your car to start on a rainy or humid day? To get your engine purring, just spray some WD-40 on the spark-plug wires before you try starting it up again. WD-40 displaces water and keeps moisture away from the plugs.

Clean oil spots from driveway Did a leaky oil pan leave a big ugly spot in the middle of your concrete driveway? To get rid of an unsightly oil spot, just spray it with a generous amount of WD-40 and then hose it down with water.

W

* Weather Stripping

Keep appliances in place Affixing small pieces of weather stripping to the bottom of telephones, electric can openers, PC speakers, and similar items will help keep them from sliding off counters or desktops.

Add traction to boots Some rubber boots may be great at keeping out moisture, but don't prevent you from slipping on ice-, snow-, or slush-covered surfaces. But you can usually improve the traction of your waterproof footwear by gluing a few strips of flat weather stripping onto the toe, middle, and heel sections.

Get a grip on tools Wrapping the handles of tools such as hammers, axes, and wrenches with flat weather stripping will not only give you a better and more comfortable grip on them, but it might even prevent wooden handles from getting damaged. Spiral the weather stripping around the handle, overlapping it half a width.

Fix leaky car windows Use small slivers of household weather stripping to patch up the dented weather stripping around car windows to prevent wind and water from getting inside your car. You can also use it to firm up sagging rubber gaskets around your car's trunk or doors.

* Window Cleaner

Reduce swelling from bee stings Spritzing some window cleaner on a bee sting is a quick way to reduce the swelling and pain. But first be sure to remove any stinger. Flick it sideways to get it out—don't tweeze it—then spray. Use only spray-on window cleaner that contains ammonia and never use a concentrated product. It is the small amount of ammonia that does the work, and beekeepers have known for years that a very dilute solution of ammonia helps relieve stings.

Get off that stuck ring That ring felt a little tight going on, and now … oops!… it's stuck. Spray a little window cleaner on your finger for lubrication and ease the ring off.

Clean your jewelry Use window cleaner to spruce up jewelry that is all metal or has crystalline gemstones, such as diamonds or rubies. Spray on the cleaner, then use an old toothbrush for cleaning. But don't do this if the piece has opaque stones such as opal or turquoise or organic gems such as coral or pearl. The ammonia and detergents in the cleaner can discolor these porous lovelies.

Remove stubborn laundry stains If laundering with detergent isn't enough to get tough stains such as blood, grass, or tomato sauce out of a fabric, try a clear ammonia-based spray-on window cleaner instead. (It's the ammonia in the window cleaner that does the trick, and you want uncolored cleaner to avoid staining the fabric.) Spray the stain with the window cleaner and let it sit for up to 15 minutes. Blot with a clean rag, rinse with cool water, and launder again. A few tips:

- Do a test on a seam or other inconspicuous part of the garment to see if the color runs.
- Use cool water and don't put the garment in the dryer until the stain is completely gone.
- Don't use this on silk, wool, or their blends.
- If the fabric color seems changed after using window cleaner on it, moisten the fabric with white vinegar and rinse it with water. Acidic vinegar will neutralize alkaline ammonia.

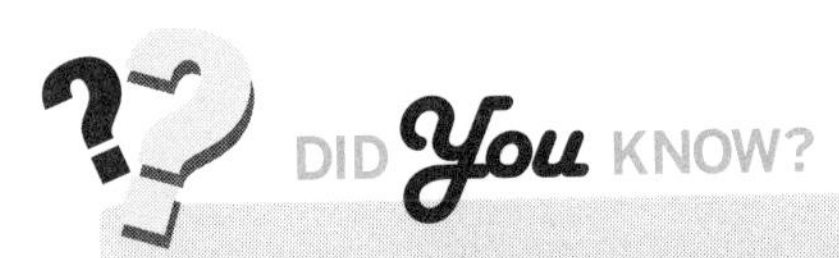

Will spraying window cleaner on a pimple really help make it go away? Although the formula varies by brand, window cleaners generally contain ammonia, detergents, solvents, and alcohol. This combination will clean, disinfect, and dry out skin. So as long as you keep it out of your eyes and have no allergies to the ingredients, it probably will help suppress pimples.

Y

* Yogurt

Make moss "paint" for the garden Wouldn't it be nice to simply paint some moss between the cracks of your stone walkway, on the sides of flowerpots, or anywhere else you want it to grow? Well, you can. Just dump a cup of plain active-culture yogurt into your blender along with a handful of common lawn moss and about a cup of water. Blend for about 30 seconds. Use a paintbrush to spread the mixture wherever you want moss to grow—as long as the spot is cool and shady. Mist the moss occasionally until it gets established.

Make a facial mask You don't have to go to a spa to give your face a quick assist:

- To cleanse your skin and tighten the pores, slather some plain yogurt on your face and let it sit for about 20 minutes.
- For a revitalizing facial mask, mix 1 teaspoon plain yogurt with the juice from 1/4 slice of orange, some of the orange pulp, and 1 teaspoon aloe. Leave the mixture on your face for at least five minutes before rinsing it off.

Relieve sunburn For quick, temporary relief of mild sunburn, apply cold plain yogurt. The yogurt adds much needed moisture and, at the same time, its coldness soothes. Rinse with cool water.

Make play finger paint Ready for some messy rainy-day fun? Mix food coloring with yogurt to make finger paints and let the little ones go wild. You can even turn it into a lesson about primary and secondary colors. For example, have the kids put a few drops of yellow food coloring and a few drops of blue in the yogurt to make green finger paint. Or mix red and blue to produce purple.

Cure dog or cat flatulence If Bowser has been a bit odoriferous lately, the problem may be a lack of the friendly digestive bacteria that prevent gas and diarrhea. The active culture in plain yogurt can help restore the helpful bacteria. Add 2 teaspoons yogurt to the food for cats or small dogs weighing up to 14 pounds (6 kilograms). Add 1 tablespoon for medium-sized dogs weighing15-34 pounds (7-15 kilograms). Add 2 tablespoons for large dogs weighing 35-84 pounds (16-38 kilograms). Add 3 tablespoons for dogs larger than that.

Z

* Zippers

Secure your valuables Nothing ruins a vacation like reaching into your pocket and discovering it has been picked. To keep your wallet, passport, and other valuables safe, sew a zipper into the inside pocket of your jacket to keep items safely zipped inside.

Make a sock puppet Create a happy sock puppet that will keep kids amused for hours. Just sew on buttons for the nose and eyes and some yarn for hair, and use a small smiling upturned zipper for the mouth.

Create convertible pants Here's a great idea for hikers and bikers who like to travel light. Cut the legs off a pair of jeans or other comfortable pants above the knee. Then reattach the legs with zippers. Zip off the legs when it gets warm, zip them back on for cool mornings and evenings. Besides lightening your load, you won't need to search for a place to change.

* Zucchini

Use as a rolling pin A large zucchini works great for rolling out dough for biscuits or piecrust. The zucchini has just the right shape and weight, and the dough won't stick to its smooth skin.

Use as exercise weights Zucchini come in a large range of sizes, so you'll have no trouble finding a couple that are just the right weight for your arm exercises.

Use as Mr. Zucchini Head Got some bored kids on your hands? Dust off that old Mr. Potato Head set and hand the youngsters a couple of large zucchini. The new vegetable is sure to renew their interest in the old toy.

Juggle 'em! Okay, okay, we're impressed. You can juggle three balls. Now let's see how you do with three zucchini!

Index

Note: **Boldface** entries and page numbers refer to main A-Z headings in text.

* Indicates a Super Item.

D

E

F

G

H

I

J

K

L

M

Q

R

These simple experiments for kids demonstrate basic science using everyday items.